TABLE 6-3 Summary of Basic Boolean Identities

<table>
<tr><td colspan="3" align="center">Fundamental laws</td></tr>
<tr><td>OR</td><td>AND</td><td>NOT</td></tr>
<tr><td>$A + 0 = A$</td><td>$A0 = 0$</td><td>$A + \overline{A} = 1$</td></tr>
<tr><td>$A + 1 = 1$</td><td>$A1 = A$</td><td>$A\overline{A} = 0$</td></tr>
<tr><td>$A + A = A$</td><td>$AA = A$</td><td>$\overline{\overline{A}} = A$</td></tr>
<tr><td>$A + \overline{A} = 1$</td><td>$A\overline{A} = 0$</td><td></td></tr>
</table>

Associative laws
$$(A + B) + C = A + (B + C) \qquad (AB)C = A(BC)$$

Commutative laws
$$A + B = B + A \qquad AB = BA$$

Distributive law
$$A(B + C) = AB + AC$$

De Morgan's laws
$$\overline{AB \cdots} = \overline{A} + \overline{B} + \cdots$$
$$\overline{A + B + \cdots} = \overline{A}\overline{B} \cdots$$

Auxiliary identities
$$A + AB = A \qquad A + \overline{A}B = A + B$$
$$(A + B)(A + C) = A + BC$$

D0215701

TABLE 8-2 FLIP-FLOP Truth Tables

SR			J-K			D		T		Direct inputs			
S_n	R_n	Q_{n+1}	J_n	K_n	Q_{n+1}	D_n	Q_{n+1}	T_n	Q_{n+1}	Ck	Cr	Pr	Q
0	0	Q_n	0	0	Q_n	1	1	1	$\overline{Q_n}$	0	1	0	1
1	0	1	1	0	1	0	0	0	Q_n	0	0	1	0
0	1	0	0	1	0					1	1	1	*
1	1	?	1	1	Q_n								
Fig. 8-9			Fig. 8-11			Fig. 8-14		Fig. 8-15					

*Refer to truth table SR, J-K, D, or T for Q_{n+1} as a function of the inputs.

MICROELECTRONICS

McGRAW-HILL SERIES IN ELECTRICAL ENGINEERING

Consulting Editor
Stephen W. Director, *Carnegie-Mellon University*

CIRCUITS AND SYSTEMS
COMMUNICATIONS AND SIGNAL PROCESSING
CONTROL THEORY
ELECTRONICS AND ELECTRONIC CIRCUITS
POWER AND ENERGY
ELECTROMAGNETICS
COMPUTER ENGINEERING
INTRODUCTORY
RADAR AND ANTENNAS
VLSI

Previous Consulting Editors

Ronald N. Bracewell, Colin Cherry, James F. Gibbons, Willis W. Harman,
Hubert Heffner, Edward W. Herold, John G. Linvill, Simon Ramo,
Ronald A. Rohrer, Anthony E. Siegman, Charles Susskind,
Frederick E. Terman, John G. Truxal, Ernst Weber, and John R. Whinnery

ELECTRONICS AND ELECTRONIC CIRCUITS

Consulting Editor
Stephen W. Director, *Carnegie-Mellon University*

MICROELECTRONICS

SECOND EDITION

JACOB MILLMAN, PH.D.

Charles Batchelor Professor, Emeritus
Columbia University

ARVIN GRABEL, SC.D.

Professor of Electrical and Computer Engineering
Northeastern University

McGRAW-HILL BOOK COMPANY

New York   St. Louis   San Francisco   Auckland   Bogotá   Hamburg
London   Madrid   Mexico   Milan   Montreal   New Delhi
Panama   Paris   São Paulo   Singapore   Sydney   Tokyo   Toronto

MICROELECTRONICS

Copyright © 1987 by McGraw-Hill, Inc. All rights reserved.
Formerly published under the title of *Microelectronics: Digital
and Analog Circuits and Systems,* copyright © 1979 by McGraw-Hill, Inc.
All rights reserved.
Printed in the United States of America. Except as permitted under the
United States Copyright Act of 1976, no part of this publication may be
reproduced or distributed in any form or by any means, or stored in a data
base or retrieval system, without the prior written permission of the
publisher.

2 3 4 5 6 7 8 9 0   DOCDOC   8 9 2 1 0 9 8 7

ISBN 0-07-042330-X

This book was set in Times Roman by General Graphic Services.
The editors were Sanjeev Rao and David A. Damstra; the designer was Elliot Epstein;
the cover designer was Rafael Hernandez; the production supervisor was Diane Renda.
The drawings were done by J & R Services, Inc.
R. R. Donnelly & Sons Company was printer and binder.

Library of Congress Cataloging-in-Publication Data

Millman, Jacob. (date)
  Microelectronics.

  (McGraw-Hill series in electrical engineering.
Electronics and electronic circuits)
  Includes bibliographies and index.
  1. Microelectronics.   2. Integrated circuits.
3. Electronic circuit design.   4. Digital electronics.
I. Grabel, Arvin.   II. Title.   III. Series.
TK7874.M527   1987   621.381′7   86-18546
ISBN 0-07-042330-X
ISBN 0-07-042331-8 (solutions manual)

ABOUT THE AUTHORS

Jacob Millman is Professor Emeritus of Columbia University, where he held the Charles Batchelor Chair in Electronics. He obtained his B.S. (1932) and Ph.D. (1935) in physics from the Massachusetts Institute of Technology. His first graduate year was spent in Munich, Germany (1932–1933). He taught electrical engineering at the City College of New York (1936–1941 and 1946–1951) and at Columbia (1952–1975). During World War II he worked on the development of radar systems at the Radiation Laboratory of MIT (1942–1945).

Dr. Millman is the author or coauthor of eight textbooks; *Electronics* (1941 and revised in 1952); *Pulse and Digital Circuits* (1956); *Vacuum-Tube and Semiconductor Electronics* (1958); *Pulse, Digital, and Switching Waveforms* (1965); *Electronic Devices and Circuits* (1967); *Integrated Electronics* (1972); *Electronic Fundamentals and Applications* (1975); and *Microelectronics* (1979). These books have been translated into ten languages.

Professor Millman is a Fellow and Life Member of the IEEE and a Fellow of the American Physical Society. He was given the Great Teachers Award of Columbia University (1967), the Education Medal of the IEEE (1970), and was elected to the IEEE Centenial Hall of Fame as one of the All-Time Top Electrical Engineering Educators (1984).

He has lectured extensively outside the United States: Italy, Spain, Israel, Greece, Brazil, Uruguay, Germany, Holland, and France.

Arvin Grabel has served on the faculty at Northeastern University since 1964 and is currently Professor of Electrical and Computer Engineering. He obtained all three degrees from New York University. As Instructor of Electrical Engineering he taught at the New York University Graduate Center at Bell Laboratories. Professor Grabel has been Visiting Professor at the University of California at Santa Barbara and The Cooper Union for the Advancement of Art and Science. ''Basic Electrical Engineering,'' of which he is a coauthor, is now in its fifth edition and has been translated into six languages.

CONTENTS

Part Two
DIGITAL CIRCUITS AND SYSTEMS

6   Basic Logic (Digital) Circuits   207

7   Combinatorial Digital Circuits   264

PREFACE

The primary objective of this book—as it was for the previous edition—is to serve as the text in modern electronics courses for electrical engineering, computer engineering, and computer science students. Our approach is to stress the fundamental concepts that underlie the physical operation, analysis, and design of integrated circuits and systems. By coupling this approach with a diversity of applications we hope to convey both the substance and flavor of the subject. The breadth and depth of treatment also makes this volume a valuable adjunct to the continuing education of practicing engineers, scientists, and professionals in fields akin to electrical and computer engineering.

This book is an extensive rewritten version of the first edition, and a substantial amount of new material has been added to reflect changes in technology and curricula. The text, divided into five major sections, is organized to provide maximum pedagogical flexibility without loss of continuity. Thus, the individual faculty member can adapt the material to a number of different courses which suit the needs and interests of both students and professors.

Part 1 (Chapters 1 to 5) focuses on the behavior of the major *semiconductor devices* used in integrated circuits (ICs). The five chapters outline the properties of semiconductors and explain the physical operation and circuit characteristics of junction diodes, bipolar transistors (BJTs), and field-effect transistors (FETs). The last chapter describes IC fabrication techniques. Its intent is to provide the student with an overview of the processes employed and the constraints imposed by fabrication on circuit design.

These first five chapters are intended for students who have had no previous course in electronics. They supply the fundamental material required for an understanding of the rest of the book. The mathematics and physics background obtained in the first year or two of a typical engineering program is the only prerequisite for Part 1. Most students also have had a course in circuit analysis prior to studying electronics. While this preparation is valuable, it is not essential because the elementary circuit analysis used in these chapters is explained in Appendix C.

Part 2 (Chapters 6 to 9) treats *digital circuits and systems,* and Part 3 (Chapters 10-14) deals with *amplifier circuits and systems.* Part 3 (analog) may be taken up before Part 2 (digital) if desired. The material contained in Part 1 provides the necessary background for either section. Our reasons for introducing digital material first are twofold:

1. In many universities, computer engineering and science majors are required to take only one electronics course. Clearly, for these students such a course should concentrate on digital electronics. Judicious selection of topics in Parts 1 and 2 can create a satisfactory one-term course, and this opportunity gives the instructor the freedom and the incentive to consider the subject matter most appropriate to his or her objectives.

2. Only elementary circuit theory of the level described in Appendix C is required. Consequently, the student need not have any electrical engineering prerequisites to handle these topics.

The first chapter in Part 2 treats the basic logic-gate building blocks used in digital systems. We concentrate on the operation and performance of the four major IC technologies. The two field-effect transistor (FET) families are NMOS and CMOS, whereas transistor-transistor logic (TTL) and emitter-coupled logic (ECL) are the standard bipolar junction transistor (BJT) logic families investigated. The standard small-scale integration (SSI), medium-scale integration (MSI), and large-scale integration (LSI) circuits and systems derived from these logic gates are developed in the next two chapters (combinatorial and sequential circuits and systems). The last chapter deals with very large scale integration (VLSI) systems in which static and dynamic random-access memory (RAM) cells are discussed. Technologies used only in VLSI systems, such as integrated-injection logic (I^2L), CMOS domino logic, and charge-coupled devices (CCD), are also introduced.

The development in Part 3 (Chapters 10 to 14) on *amplifier circuits and systems* parallels that in Part 2. The initial pair of chapters focuses on the properties of basic BJT and FET amplifier stages. Methods by which IC transistors are biased are presented; the use of small-signal models to evaluate the performance of single-stage and cascaded amplifiers is discussed. The operational amplifier (Op-Amp) as a basic building block is also described. These amplifiers are used to form feedback-amplifier systems. The fundamental feedback concepts and techniques developed are used in the analysis and design of the four basic feedback-amplifier topologies. The internal architecture and performance of modern Op-Amp chips are examined in Chapter 14 and serve to bring together many of the concepts previously encountered in Part 3.

Part 4 (Chapters 15 and 16) examines the circuit and systems aspect of *data acquisition and signal processing*. Many of these circuits are used in both digital and analog systems and employ both logic gates and Op-Amps. Circuits for the generation of sinusoidal, ramp, and pulse waveforms and the conversion of analog signals to digital signals (or vice versa) are discussed. Integrators, active filters including switched-capacitor filters, multipliers, and logarithmic amplifiers are among the signal-conditioning circuits described.

The last part (Chapter 17) exposes the student to electronic *power circuits and systems*. The conversion of ac to dc is treated and leads to a discussion of monolithic voltage regulators. High-voltage and high-power amplifier devices and circuits are also examined.

This text contains enough material for 2 or 3 one-semester electronics devices-circuits-systems courses. With the ever-increasing component density on an IC chip, the difference between an electronic device, circuit, and system has become quite blurred, and in this book no attempt is made to distinguish between them. An entire monolithic package, such as an Op-Amp, is often referred to as a *device*. Of course, a single transistor is clearly a *device* and a large-scale microelectronic chip merits the designation *system* or at least *subsystem*.

A brief historical survey of electronics is in the Prologue (following this preface). It is hoped that both the instructor and student will read this fascinating history before beginning the study of the text.

Most electronic engineers design a new product, subsystem, or system by interconnecting standard IC chips so that the overall assembly achieves the desired external objectives. Clearly these engineers must know what IC chips are commercially available, what function they perform, and what their limitations are. Chip designers must be aware of what functions need to be performed and what limitations most affect the performance of the systems in which the chips are used.

From this perspective, the goal of this book is to take the reader step by step from a qualitative knowledge of semiconductor properties to an understanding of the operation of solid-state devices and finally to an appreciation for how these are combined to form ICs with distinct and useful input-output characteristics. A very broad variety of IC chips are studied in this book. We describe not only what is fabricated, but also attempt to convey a deep understanding of the digital and/or analog functions performed by the chip. After each circuit or system is studied, reference is made to a specific commercially available chip which realizes the desired function. Practical limitations of real rather than ideal devices and circuits are explained. To appreciate nonideal behavior, manufacturer's specifications of representative devices and integrated circuits are given in Appendix B. The depth of discussion, the broad choice of topics, and the practical emphasis combine to prepare the student to do useful engineering immediately upon graduation.

The attention given to pedagogy is reflected in the explanation of device-circuit-system behavior and in the context in which the specific topics are discussed. We have been diligent in our efforts to ensure that new concepts are introduced by the use of familiar analytic techniques and that the development of new methods of analysis relies only on concepts previously encountered. Also, considerable care was exercised in the selection of the many illustrative examples and numerical calculations incorporated in the body of the text.

Many of the methods of analysis discussed lead to the "pencil-and-paper" calculations which an engineer often performs. Such computations are invaluable as they help develop insight into the behavior of the circuit or system being designed. When used in conjunction with computer simulations, they provide the engineer with powerful design tools. Students should be encouraged to use circuit simulations such as SPICE and MICROCAP II, both of which are available for use with personal computers.

The review questions at the end of each chapter are a significant adjunct to the approximately 800 problems given in Appendix D. Many of these problems are new, and the majority of the problems used in the earlier edition have been modified. Used together, the questions and problems test the students' grasp of fundamental concepts and provide experience in the design and analysis of electronic circuits. Realistic parameter values are used in virtually all numerical problems.

The review questions test the students' qualitative knowledge of the text material. These can also be used very effectively as part of a quiz or an exam.

A solutions manual is available to an instructor who has adopted the text. As an added pedagogical aid, transparencies of over 100 involved figures in the book are also available to the instructor.

We have had the benefit of valuable advice and suggestions from the many professors and engineers working in industry who used the first edition as either instructor, student, or practicing engineer. All these individuals have influenced this edition and to them, we express our thanks and appreciation. We are especially grateful to Professor Arthur Dickerson whose comments and insight were invaluable in the preparation of this book.

We are indebted to David Damstra, who from manuscript through production contributed much to this book; to the comments of and reviews provided by Sanjeev Rao; and to Mary Rosenberg, whose proofing of final pages was a great help.

Jacob Millman
Arvin Grabel

P.S. I am among the two generations of electrical engineers who have studied electronics from Jacob Millman's books. I had the pleasure of once again being Professor Millman's ''student'' when we worked closely together in the planning and organization of this book and in the detailed preparation of the first six chapters. I have attempted to convey the guiding spirit of this truly remarkable teacher and writer in the remainder of this text. The last eleven chapters were my sole responsibility and therefore a reflection of the quality of the student and not the mentor.

I am indebted to Jacob Millman for the opportunity of collaborating with him. His influence and style have contributed immeasurably to me as teacher and author.

Arvin Grabel

Prologue
A BRIEF HISTORY OF ELECTRONICS

Electronics—to most of us this brings to mind a variety of things from "chips" and computers to television and transistors. Yet, while we agree on specific items that constitute electronics, its definition is elusive. In the next several paragraphs and the remainder of this prologue, we define electronics as used in this book not in the dictionary sense, but in a manner which attempts to convey the flavor and nuance of the discipline. We have chosen history as the vehicle to accomplish this because it is the efforts of individuals who contribute and have contributed to the field that truly define the discipline.

Electronics, in the strictest sense, is the science and technology of the motion of charges in a gas, vacuum, or semiconductor. Note that charge motion confined to a metal is not considered electronics. This was an historical division used early in the twentieth century to separate the already flourishing field of electrical engineering from the new and emerging field of electronic engineering. At that time electrical engineering dealt with devices that depended solely on the motion of electrons in metals, such as motors, generators, light bulbs, and wire communication systems (telephone and telegraph). As we approach the end of the twentieth century, however, the historical division between electrical and electronic engineering no longer serves its original function.

Today practicing electrical engineers perform diverse functions (design, development, production, research, and even teaching) with varied applications. They deal with systems by which we can communicate with one another worldwide, by which vast quantities of data are manipulated, and by which highly complex manufacturing processes are automated and with the elements used to realize them. The province of electrical engineering also includes the devices, circuits, and systems used for the generation, distribution, and conversion of electric energy. The group mentioned in the first of the two previous sentences possesses the common property of processing information; the group mentioned in the second can be regarded as processing energy. This distinction between information processing and energy processing serves to separate electronics from the rest of electrical engineering. Consequently, we view the nature of the discipline of electronics comprising the four C's—communication, computation, control, and components.

The prologue to this text provides a brief, coordinated historical review of modern electronics. The principal focus is on the development and applications of electronic devices and the growth of the industries resulting from the exploitation of these devices in practical circuits and systems.

This history is divided into two major periods of time, referred to as the *vacuum-tube era* and the *transistor era*. The former encompasses developments in the first half of the twentieth century, and the latter era begins with the invention of the transistor in 1948. The concluding section contains a brief speculation on the future direction of electronics. These descriptions provide a framework for and foreshadow the particular technical topics treated in the text.

BACKGROUND The origins of electrical engineering are based on the achievements of such scientific giants as Ampere, Coulomb, Faraday, Gauss, Henry, Kirchhoff, Maxwell, and Ohm. The first practical uses of their work, in the context of modern electronics, was in the development of commmunications systems. In 1837, Samuel Morse, Professor of Fine Arts at New York University, demonstrated the telegraph system.[1] The significance of electric telegraphy was the introduction of an effective method of coding information into electrical signals. The dots and dashes of the Morse code represented the first use of digital (binary) signals.

Nearly 40 years later (1876), Bell invented the telephone and introduced a method for coding information (speech) as a continuous electrical signal and then decoding these signals at a receiver. Edison's invention of the phonograph in 1877 demonstrated that electrical signals could be stored and subsequently recovered. The phonograph record can be considered the first electrical read-only memory (ROM).

The introduction of radio communication is based on the major contribution of James Clerk Maxwell, who in 1865 codified earlier researches into a consistent theory of electromagnetism, now referred to as *Maxwell's equations*. The major leap forward was provided by Maxwell's prediction of the existence of electromagnetic waves which could be propagated in space. Here is a case of theory preceding experiment, as it was not until 23 years later that Hertz produced such waves in laboratory experiments. Marconi was the first to exploit the use of *Hertzian waves*, as they were then called. In 1896 Marconi succeeding in transmitting these waves and detecting them 2 miles (mi) away. Wireless telegraphy had its humble origin in these experiments.

---

[1]While Morse is credited with the invention, the first widespread system in operation was in Great Britain and based on the efforts of William Thomson, later to become Lord Kelvin, and Sir Charles Wheatstone.

THE VACUUM-TUBE ERA   The vacuum-tube era spans the first half of the twentieth century; modern electronics took shape technologically in this period.

The origin of the term "electronics" can be attributed to H. A. Lorentz, who in 1895 postulated the existence of discrete charges he called *electrons* (reintroducing the word for "amber" used by the ancient Greeks). Two years later J. J. Thompson experimentally verified the existence of electrons. In that same year Braun built the first electron tube, a primitive cathode-ray tube (CRT).

The Discovery of Vacuum Tubes   In 1904 Fleming invented a two-element device, the diode, which he called the *valve*. It consisted of a heated wire, the filament, which emitted electrons (the Edison effect) and was separated by a short distance from a metallic plate. The entire structure was encapsulated in a vacuum. A positive plate-to-cathode (filament) voltage produced a current, whereas a negative applied voltage reduced the current to zero. This unilateral property of the valve made it useful as a detector of wireless (radio) signals.

Two years later, Pickard used a silicon crystal with a "cat's whisker" (a pointed wire pressed into the silicon) as a detector. This was the first semiconductor diode; however, it was unreliable and was soon abandoned. Thus semiconductor electronics appeared to have died a premature death in 1906.

The invention of the *audion* (triode) by DeForest in 1906 was the seminal achievement in the earliest days of electronics. Indeed, one can strongly argue that electronics as we know it today would not exist without the invention of the triode. DeForest's audion consisted of a third electrode (the grid) inserted between the plate and the cathode of the Fleming valve. The grid voltage controlled the charge flow between plate and cathode. A small change in grid voltage resulted in a larger plate-voltage change, making the audion the first amplifier.

The triode was the first device to exhibit the circuit property we now refer to as a *controlled* or *dependent source*. Because it retained the unilateral property of the valve, the triode also provided the properties of a controlled switch. Today, virtually all electronic circuits exploit device characteristics which display either controlled-source or controlled-switch behavior.

Initial Circuit Applications   By 1911, technological improvements—a better vacuum and an oxide-coated cathode—made the audion a reliable device, thus ushering in the age of practical electronics.[1] The first applications of vacuum tubes were to telephone and radio communication, and simultaneously the Institute of Radio Engineers (IRE) was founded in 1912 in the United States. It is a tribute to the imagination and foresight of the early engineers who immediately realized the significance of radio and formed their own professional society. The American Institute of Electrical Engineers (AIEE), which focused on the conventional

---

[1]By coincidence, Professor Millman was born the same year.

interests of electrical engineers, was founded in 1884. In 1963 both societies merged into a single organization, the Institute of Electrical and Electronic Engineers (IEEE), a move which reflected a half century of development in the profession.

Through use of the simple diodes and triodes available, the ingenuity of the early engineers resulted in the invention of many new circuits. Notable among them were cascaded amplifiers, regenerative amplifiers (Armstrong,[1] 1912), oscillators (DeForest, 1912), heterodyning (Armstrong, 1917), and multivibrators (Eccles-Jordan, 1918). The oscillator was the first instance by which electronics signals were generated by solely electronic means. The increased gain of both regenerative (positive-feedback) and cascaded amplifiers in conjunction with the frequency translation provided by heterodyning improved signal processing and enhanced the detection of weak signals. The early multivibrators were the forerunners of modern flip-flops and clock generators (timing circuits).

Electronics Industries The amplifier had almost immediate commercial application in long-distance telephony. The advances in tube technology made by telephone companies helped to provide the impetus for a major new industry—commercial radio broadcasting. Station KDKA in Pittsburgh, Pennsylvania was created by Westinghouse Electric Corporation in 1920. Four short years later, there were 500 stations in the United States, and by 1926, network broadcasting was a reality. Simultaneously, radio broadcasting was introduced throughout the industrial world.

Electronic industries[2] fit one or more of the groups: components, communications, control, and computation.

Components Initially, component companies came into existence to produce the various types of electron devices as well as passive circuit elements (resistors, capacitors, inductors, transformers, etc.). Engineers and scientists in these organizations made significant advances in developing new and better devices. These included the indirectly heated cathode, the tetrode and pentode tubes into which a fourth and then a fifth electrode were introduced into the triode, and gas tubes such as the thyratron. With new and improved devices, new circuits were invented which provided single-knob tuning, automatic gain control (AGC), and multiband receiver operation.

Communications Radio signals are most conveniently transmitted at frequencies above 500 kHz. As the frequency of the signals representing the information are most often considerably below 500 kHz, these signals must be encoded into and shifted to the higher-transmitting frequencies by a process referred to as *modulation*. The first broadcast radio systems used amplitude modulation (AM). To improve

---

[1]Armstrong was an undergraduate at Columbia University at this time.

[2]Many companies' activities fit into more than one category, often with subsidiaries or divisions being identified with one group.

fidelity and reduce the effect of atmospheric interference, Armstrong[1] (1930) conceived of and developed frequency modulation (FM).

Black-and-white television began in 1930 and was based on Zworykin's iconoscope and kinescope (the first television camera tube and picture tube, respectively). By 1940, television in the United States was in modest use; its widespread adoption was delayed by World War II. The development of color television began about 1950 and during the 1960s became the dominant television system.

The techniques used in radio broadcasting were adapted to fit other applications. Telephone systems were transformed into one of the major forms of electronic communication. In turn, circuits developed for electronic telephony were widely used in radio-receiver systems. Radar and Loran (developed during World War II) utilized radio communication as aids in both air and sea navigation.

Each of the aforementioned innovations required that new circuits be invented. Significant achievements included the negative-feedback amplifier invented by Black (1927), the FM limiter, and the FM discriminator. Another major circuit development was the sawtooth generator, which provided the linear time base for the early oscilloscopes and for deflection systems in television. Many of the newer communications systems utilized discrete signals (pulses) rather than continuous signals. Consequently, a variety of pulse circuits were developed for the timing and synchronization needed in television, radar, and so on and for pulse generation and modulation. Furthermore, new communications systems were operating at higher frequencies and based on new microwave devices such as the klystron, the magnetron, and traveling-wave tubes.

Computers

Although the transistor and the integrated circuit gave impetus to the extraordinary growth of the computer industry, their origins are found in the vacuum-tube era. There has been a great deal of interest in computing machines for over 300 years. In 1633 Schickard described (in correspondence with his friend Kepler, the astronomer) a mechanical computer to perform addition, subtraction, multiplication, and division. He designed a wheel with 10 spokes on it, one spoke of which was longer than the others, and this wheel was placed mechanically next to another similar wheel. After the first wheel made 10 angular increments, which corresponded to the 10 digits, the large spoke engaged the next wheel, and it would turn by one increment. In other words, he invented the *carry* in arithmetic. About the same time Pascal (1642) and Leibnitz (1671) had similar ideas. But the first really serious effort to build a mechanical calculator was made about 200 years later (1833) by Babbage, a mathematics professor in England. The "analytic engine," as Babbage's computer was called, contained all the elements of a modern digital computer. It used punched cards—invented 30 years earlier by Jacquard, a French tapestry maker—for

---

[1]Dates indicate initial disclosure and not necessarily public awareness or patent issue.

input and output, contained both memory and an arithmetic unit, and was a stored-program machine. However, the technology simply was not available to convert his ideas into a practical machine.[1]

The first working calculator was electromechanical, not electronic, and was built by IBM engineers under the direction of Professor Aiken of Harvard University in 1930. It was called the "IBM automatic sequence controlled calculator, Mark I." It was 17 meters (m) long and 3 m high and was very clumsy. Yet it was used to make calculations for over 15 years. The first electronic calculator was completed in 1946 by Eckert and Mauchly at the Moore School of Electrical Engineering at the University of Pennsylvania. It was called the ENIAC, an acronym for electronic numerical integrator and computer. It was used for computation of ballistic tables for the armed forces, and it was not a general-purpose calculator. It contained 18,000 vacuum tubes, occupied 40 racks of equipment, and filled a room that was about 10×13 m. Von Newmann, a consultant for this project, suggested that the computer use binary numbers and boolean logic and contain a stored program for basic operations.

In 1946 IBM introduced the first small commercial electronic computer, the type 603. Two years later, the first general-purpose digital computer, the IBM 604, was brought out and over 4000 machines were sold in 12 years. Thus 1948 can be considered as the beginning of the computer industry. (Coincidentally, the transistor was invented in that same year.)

At this time, a number of institutions, including Harvard, Princeton, Pennsylvania Universities, the Massachusetts Institute of Technology (MIT), the Courant Institute at New York University, and the Institute for Advanced Studies, were engaged in computer research. Funded by several government agencies, these engineers and scientists developed both hardware and software concepts that were subsequently used in commercial general-purpose computers.

The IBM 650, considered the workhorse of the industry, was introduced in 1954. This and other vacuum-tube machines marketed by other companies are known as *first-generation digital computers*.

Analog computers were also developed during the latter part of the vacuum-tube era. Such machines, used to solve large systems of differential equations, are based on the construction of electronic circuits whose behavior is governed by an analogous set of equations as those to be solved. The differential analyzer, developed by Bush at MIT, was the first electromechanical analog computer. Electronic versions became a reality with the invention of the operational amplifier.[2]

---

[1] Babbage's efforts were not wholly in vain. His attempt to construct the computer resulted in many advances in machine-tool operations which had a significant impact on manufacturing in Victorian England.

[2] The term "operational amplifier" was coined by J. R. Ragazzini, a colleague of Millman's at Columbia University and later one of Grabel's professors at New York University.

Controls   The origin of the electronic control industries was in "industrial electronics," which may be defined as "the use of electronic devices in the control of machines (other than in communication and computation)." Thyratrons, gaseous diodes, mercury arc rectifiers, and high-voltage high-power tubes were the devices used. These devices were used in circuits which provided high-voltage and high-power alternating-current-to-direct-current (ac-to-dc) conversion (rectifiers), dc-to-ac conversion (inverters), and high-voltage transmitting circuits. Applications included motor-speed control, voltage regulation, induction and dielectric heating, and a variety of industrial process controls. Also, the first use of computers (analog) in control systems appeared at this time.

Analyses and Theory   In addition to industrial growth, significant analytical and theoretical progress was made. The following is a brief indication of the range of achievements.

Circuit analysis and synthesis techniques were developed, notably by groups at Bell Laboratories and MIT. Bode and Nyquist developed feedback amplifier theory and transformed Black's circuit concept into one enjoying widespread use and import.

Shannon in the United States and Kotelnikov in the Soviet Union independently developed information theory which was to have great impact on data transmission. One particular application was to the pulse-code-modulation (PCM) technique proposed by Reeves.

The use of boolean algebra in the analysis and design of switching circuits was another of Shannon's contributions (1937). In Great Britain, the concept of a universal computing machine was proposed by Turing, and Wilkes developed microprogramming.

Sampled-data systems, introduced by Ragazzini and Zadeh, were applied in control applications, paving the way for control systems based on digital-computer processing.

Studies of materials, particularly the application of quantum mechanics to solids, led to new devices and later helped to lead to the invention of the transistor. Transducers, devices by which light, sound, pressure, temperature, and other variables are converted to and from electrical signals, were introduced to make use of the advantages afforded by electronics.

New forms of instrumentation (oscilloscopes, vacuum-tube voltmeters, etc.) evolved to both use electronics in measurement and for testing electronic equipment.

The 1950s was a decade of transition. It marked the end of the development of sophisticated vacuum-tube systems and the beginning of the transistor age. Today, the entire field is dominated by semiconductor devices except for high-voltage high-power applications. Indeed, vacuum tubes are omitted from virtually all electrical engineering curricula.

THE TRANSISTOR ERA The age of semiconductor electronics began with the invention of the transistor in 1948. However, this era originated in earlier work performed between 1920 and 1945. During that time the study of the electromagnetic properties of semiconductors and metals was principally the province of physicists. Notable contributions were made by Block, Davydov, Lark-Horovitz, Mott, Schottky, Slater, Sommerfeld, Van Vleck, Wigner, Wilson, and others at universities throughout the world.[1] There were even attempts at manufacturing solid electronic devices. Lillienthal and Heil in the 1930s each received a patent for a solid-state amplifying device, the precursors of the junction and MOS field-effect transistors (FETs). However, these devices performed poorly, no apparent need for them existed, and in all probability neither inventor could explain the theory underlying the devices.

Major impetus for the development of solid-state devices was not forthcoming until 1945.[2] Vacuum tubes had major limitations: power was consumed even when they were not in use and filaments burned out, requiring tube replacement. M. J. Kelly, then director of research and later president of Bell Laboratories, had the foresight to recognize that reliable, expanded telephone communication required electronic, rather than electromechanical, switching and better amplifiers. He formed a solid-state research group consisting of theoretical and experimental physicists, an electrical engineer, and a physical chemist who worked with the metallurgists in the laboratory. The following quote is from the authorization for work in this group: ''The research carried out in this case has as its purpose the obtaining of new knowledge that can be used in the development of completely new and improved components and apparatus elements of communication systems.'' One of the most important goals was to try to develop a solid-state amplifier which would eliminate the shortcomings of the vacuum tube.

Discovery of the Bipolar Junction Transistor An experiment was performed in December 1947 in which two closely spaced gold-wire probes were pressed into the surface of a germanium crystal. It was observed that the output voltage at the ''collector'' probe with respect to the germanium ''base'' was greater than the input voltage to the ''emitter'' probe. Brattain and Bardeen recognized that this was the effect for which they were looking, and the solid-state amplifier in the form of the point-contact transistor[3] was born.[4] The performance of the first transistors was very poor. They had low gain and bandwidth and were noisy, and their characteristics varied widely from device to device.

---

[1] Slater and Sommerfeld were two of Millman's graduate professors.

[2] Semiconductor diodes, however, were widely used in microwave communications during World War II.

[3] J. R. Pierce, later to direct the first communications satellite projects, coined the term ''transistor'' as a contraction of transfer resistor.

[4] The invention was announced at a press conference on June 30, 1948 and relegated to the back pages of the few newspapers that carried the item.

Shockley, the group leader, recognized that the difficulties were with the point contact. He proposed the junction transistor and almost immediately developed the theory of its operation. The new devices depended on charge carriers of both polarities; thus they were bipolar devices. The two carriers were the well-known electrons and other "strange particles." These strange particles could be explained only quantum mechanically and behaved as if they were positive charges. They were called "holes" because they represented sites in the crystal where electrons were missing. Shockley's theory predicted that large current densities could be achieved for small applied potentials. The possibility of obtaining important practical devices *without heated filaments* was immediately apparent.

The electrical properties of transistors depended on a carefully controlled specific impurity content (in the order of 1 impurity atom per 100 million germanium atoms). Consequently, reliable devices could not be fabricated without exceptionally pure crystals to which the desired impurities could be added. Teal at Bell Labs (1950) grew single crystals of germanium with an inherent impurity content of less than 1 part per billion (ppb). From this development, the first grown-junction transistors were made and were followed 1 year later by alloy-junction transistors. Thus, in 1951, 3 years after the discovery of amplification in a solid, transistors were produced commercially.

American Telephone and Telegraph (AT&T)[1] made a monumental decision—not to keep these discoveries secret. Its technical staff members held symposia to share their knowledge with professors (who passed it on to their students) and with engineers and scientists from other companies. Patent licenses were offered to any company interested in manufacturing transistors. The tube companies, such as RCA, Raytheon, General Electric, Westinghouse, and Western Electric (the manufacturing arm of AT&T), were the first to fabricate transistors. Other existing and newly formed companies who recognized the potential of these devices were soon fabricating them.

One of these companies, Texas Instruments, in its newly formed solid-state laboratory headed by Teal, announced the production of silicon transistors in 1954. Silicon afforded operation to 200 degrees Celsius (°C), whereas variations in characteristics limit germanium devices to 75°C. Today, an overwhelming majority of semiconductor devices are fabricated in silicon.

Bardeen, Brattain, and Shockley were awarded the Nobel Prize in Physics in 1956 for their invention of the transistor and contributions to the understanding of semiconductors. This was the first Nobel award given for an engineering device in nearly 50 years.

Invention of the Integrated Circuit Shortly after joining Texas Instruments in 1958, Kilby conceived of the monolithic-circuit concept, that is, the idea of using germanium or silicon to build an entire circuit. Resistors were to be formed with the bulk

---

[1]Bell Labs is the research arm of AT&T.

semiconductor or by diffusing one semiconductor into another. Using a metallic layer and the semiconductor for the plates and an oxide layer for the dielectric, Kilby formed capacitors. (He also thought of the junction capacitor.) To demonstrate the feasibility of the concept, he built both an oscillator and a multivibrator from germanium, making circuit interconnections from thermally bonded gold wire. However, in the patent disclosure he indicated that component connections could be achieved by deposition of a conducting layer. Kilby announced his *solid circuit* [later called the *integrated circuit* (IC)] at an IRE convention in 1959.

About this same time Noyce[1] also had the monolithic-circuit idea for making "multiple devices on a single piece of silicon in order to make interconnections between devices as part of the manufacturing process and thus reduce size, weight, etc., as well as the cost per active element." He indicated how resistors and capacitors could be fabricated, how *pn* junctions could be used to isolate devices from one another,[2] and how interconnections between circuit components could be achieved by evaporating metal into windows in an oxide layer.

The real keys to IC manufacture were the planar transistor and batch processing. The planar process used transistors in which the base and emitter regions were diffused into the collector. The first diffused transistors were developed by Hoerni at Fairchild (1958). A major step in successful production was the passivation of the junctions by a surface oxide layer. The fabrication techniques used were production lithography and the diffusion process developed earlier by Noyce and Moore. Batch processing permitted many IC "chips," as ICs were colloquially known, to be made from a single silicon wafer. By 1961, both Fairchild and Texas Instruments were producing ICs commercially, followed soon afterward by other companies.

The Microelectronic Industries   Today, in addition to individual circuits, subsystems and even entire systems containing thousands of components can be fabricated on a single silicon chip. The term "microelectronics" refers to the design and fabrication of these high-component-density ICs. Moore[3] noted in 1964 that the number of components on a chip had doubled every year since 1959, when the planar transistor was introduced. He predicted correctly that this trend would continue. A large IC chip is only about 3×5 millimeters (mm) in area and 0.3 mm thick (about three times the thickness of a human hair). By 1984 such chips could contain upward of 400,000 components, corresponding to 30,000 components/mm^2 (about 15 components/mil^2). These numbers are difficult to fathom, particularly since ICs are produced in an industrial plant and

---

[1]Noyce was the director of research and development at Fairchild Semiconductor at the time. Subsequently, he was one of the founders and board chairman of Intel.

[2]Lehovec, research director at the Sprague Electric Company, independently conceived of this idea and was awarded a patent in 1959.

[3]Moore was then director of research at Fairchild and later was a founder and president of Intel.

not under laboratory conditions. The following approximate dates give some indication of the increasing component count per chip:

1951—*discrete transistors*
1960—small-scale integration (SSI), fewer than 100 components
1966—medium-scale integration (MSI), 100 to 1000 components
1969—large-scale integration (LSI), 1000 to 10,000 components
1975—very-large-scale integration (VLSI),[1] more than 10,000 components

One can divide the electronic industries into chip manufacturers and chip users. The IC manufacturers are the major segment of the component industries, whereas chip users are most often the companies producing communication, control, and computer equipment. Since the invention of the IC, many innovations have contributed to the growth of microelectronics. Several of these are described in the remainder of this section.

The Field-Effect Transistor   Much of the work leading to the invention of the bipolar transistor involved studies of the effect an applied electric field had on the conductivity of semiconductors. Shockley proposed the junction field-effect transistor (JFET) in 1951, but early attempts at fabrication failed because a stable surface could not be obtained. This difficulty was overcome with the introduction of the planar process and silicon dioxide (SiO_2) passivation. In 1958, the first JFET was produced by Teszner in France.

The techniques used to make reliable JFETs led to an even more important device, the metal-oxide-semiconductor field-effect transistor (MOSFET). The structure consists of a metallic electrode (the gate) placed on the SiO_2 between two electrodes in the semiconductor (source and drain). The current in the "channel" between source and drain can be controlled by applying an appropriate voltage between the gate and the semiconductor. Atalla and Kahng (1960) at Bell Laboratories reported the first such device. Two years later, Hofstein and Heiman at RCA were awarded a patent for their development of MOSFETs suitable for IC fabrication. Subsequent improvements in processing and device design and the growth of the computer industry have made MOS devices the most widely used transistors.

Digital Integrated Circuits   The growth of the computer industry spurred new IC development; in turn, new IC concepts resulted in new computer architectures. Two of the major advances were in new circuit configurations and semiconductor memories.

Speed, power consumption, and component density are important considerations in digital ICs. An early bipolar logic family was transistor-coupled transistor logic, invented by Buie (1961) of Pacific Semiconductor,[2] from which

---

[1] By 1984, most VLSI chips had 100,000 or more components.

[2] Pacific Semiconductor is now part of TRW.

the standard *transistor-transistor logic* (TTL) evolved. A major feature of TTL is its use of transistors with multiple emitters to improve component density. A high-speed bipolar product line, known as *emitter-coupler logic* (ECL), was introduced by Motorola in 1962. Very high density bipolar chips were achieved by using transistors with multiple collectors (1972). Developed simultaneously by Hart and Slob at Phillips (Netherlands) and by Berger and Wiedman at IBM (Germany), this new technology is called *integrated-injection logic* (I^2L).

The use of MOSFETs was immediately attractive because very high component densities are obtainable. Originally, reliable fabrication employed PMOS devices—that is, MOSFETs whose operation depended on hole flow. Improved fabrication methods led to the use of *n*-channel metal-oxide-semiconductor (NMOS) devices. Conduction in these transistors is by electrons and result in higher speed performance. Presently, NMOS technology is predominant.

The complementary metal oxide semiconductor (CMOS), a circuit configuration employing both *p*-channel metal-oxide-semiconductor (PMOS) and NMOS devices, was first applied in digital watches because of its extremely low power consumption. Recent advances including the use of polysilicon gates and reductions in device size have made CMOS circuits a major digital technology of the 1980s. It is expected that CMOS technology will prevail over NMOS by 1990.

It is in semiconductor memories, however, that MOSFETs are a major force. *Random-access memories* (RAMs), capable of both storing and retrieving data (write and read, respectively), were first developed by using bipolar transistors and marketed in 1970. These early RAMs stored approximately 1000 bits of information. With the use of MOS technology, 16,000-bit RAMs were available in 1973, 64,000-bit chips in 1978, and 288,000-bit RAMS were reported in 1982. 1,000,000-bit chips became available in 1986.

Read-only memories (ROM), used for look-up tables in computers (e.g., to obtain the values of sin *x*), were first introduced in 1967. Subsequent developments included programmable ROMs (PROMs) and erasable PROMs (EPROM) in which data stored could be removed (erased) and new data stored.

More than half of the MOS ICs produced in 1970 went into manufacture of calculators. In an effort to standardize chip design while maintaining proprietary circuits demanded by customers, several IC manufacturers proposed partitioning calculator architecture into its circuit functions. This concept led to the *microprocessor* first developed by M. E. Hoff at Intel (1969). Four-bit microprocessors were introduced by Intel (1971), followed 1 year later by an 8-bit device. Soon, other manufacturers were also producing microprocessors, and 16-bit units were available by the late 1970s. The development of the microprocessor led to the "computer on a chip." Cochran and Boone of Texas Instruments were awarded a patent, filed in 1971, for such a single-chip microcomputer, although the Intel 8048 was the first commercially available product.

Another development arising from MOS technology is the *charge-coupled device* (CCD). The CCD, invented by Boyle and Smith at Bell Laboratories

(1970), consists of an MOS device in which a long chain of closely spaced gates is formed between drain and source. Charges introduced into the channel under the gates can be transferred from one gate electrode to the next when appropriate gate voltages are applied. Such devices have been used for memories and registers with a 64,000-bit RAM fabricated in 1977. Recently, CCDs have found application in camera manufacture, image processing, and communications.

Analog Circuits   The first major development in analog ICs came in 1964 when Widlar, then with Fairchild Semiconductor, developed the first *operational amplifier* (the μA709). Since then the operational amplifier has become the "workhorse" in analog signal processing. Other circuits and subsystems have been subsequently developed and include analog multipliers, digital-to-analog (D/A) and analog-to-digital (A/D) converters, and active filters. Most of these circuits employ bipolar transistors, but MOS devices have also been used since the late 1970s.

Fabrication Techniques   The increase in component density owes much to those who improved fabrication processes. These advances include epitaxial growth (1960), electron-beam mask production (1969), and ion implantation (1971). Minimum linewidths on IC chips, 25 micrometers (μm) in 1961, are presently 2 μm, and it is predicted that 1 μm will be used in production by 1990. Since area decreases and density increases as the square of the linear dimension, circuits with component densities over 600 times that of the earliest ICs are expected by the end of this decade. Another area of contributions to reliable IC design and production was the development of computer-aided design (CAD) and automated testing. The programs SPICE and SUPREM, developed at the University of California at Berkeley and Stanford University, respectively, are two widely used CAD tools.

From the few companies manufacturing ICs in the early 1960s, the industry has experienced tremendous growth. As an illustration, in Silicon Valley,[1] 24 new microelectronics companies were formed between 1967 and 1969 alone. By 1984, more than 100 U.S. firms were involved in IC manufacture.

Communications and Controls Industries   These industries adopted solid-state electronics, slowly at first, but now almost all equipment, except that involving high voltage or high power, is transistorized. Both discrete transistors and ICs are utilized. Discrete transistors are used primarily in medium voltage or power applications which include traditional "industrial" and consumer electronics (audio output stages, automobile ignition systems, power switches for tape drives, power supplies, etc.). The IC is employed in most other applications.

The communications industry has changed drastically because of microelectronics. In 1970 data transmission constituted a very small fraction of the

[1]Silicon Valley is the region at the southern end of San Francisco Bay in Santa Clara County, California.

total volume of all communications. Since 1980, however, digital transmission has equaled or surpassed analog transmission. The widespread adoption of PCM transmission can be directly attributed to microelectronics. The telephone systems now employ digital ICs for switching and memory. Active filters for both voice and detection of touch-tone frequency pairs are realized with analog ICs. Obviously, communications satellites became feasible and economically viable because of microelectronics.

The introduction of digital communication has led to many circuit innovations. Some of these are clever modifications by which traditional circuits have been adapted to new technologies and uses. Others are new; among them are switched-capacitor filters and digital filters. A whole new area of electronics, called *digital signal processing*, has evolved because ICs have made the "marriage" of communications and computation possible.

Similarly, the control industry has been drastically affected by the introduction of semiconductor electronics. In some traditional areas, such as motor-speed control and power rectifiers and inverters, the silicon controlled rectifier (SCR), a four-layer three-junction bipolar device, has replaced the thyratron. Early in the transistor era, small, dedicated computers were used in numerical control of machine tools. Automation of industrial processes was made possible by large electronic computers.

The introduction of microprocessors, microcomputers, and other digital ICs has led to "smart" instruments and continuously increasing variety of digital control systems. With microelectronics, computers have become integral components of control systems.

The Computer Industry   The most dramatic outgrowth of the microelectronics industry has been the virtual creation of an entirely new industry—the modern computer industry. While the origins of the electronic computer were vacuum-tube-based, the impact of semiconductor technology was felt immediately.

The first transistorized, special-purpose computer was developed by Cray[1] (1956). The IBM 7090/7094 (1959) was the first general-purpose second-generation computer, that is, a transistorized machine. Hybrid ICs (many discrete transistors on a single substrate) characterized the third-generation computer (IBM 360 series in 1964). Simultaneously, other manufacturers, including Burroughs, Control Data, and Univac, also introduced medium and large computers containing ICs. Semiconductor memories were subsequently used in third-generation machines (IBM 370 series, 1970).

In 1965 another revolution in the computer industry began when the Digital Equipment Corporation introduced its PDP8 minicomputer, the first machine to sell for under $20,000. Since then, the minicomputer has become a major segment of the industry, one involving many companies worldwide.

---

[1]Cray is a founder of Control Data Corporation and later founded Cray Computers, developers of the fastest "supercomputers" currently available.

Currently, in the 1980s, the fourth generation of machines is being developed and introduced. These computers employ VLSI chips for both processing and memory. Today, electronic computers are available in a variety of sizes ranging from the simplest of microprocessors to supercomputers capable of executing tens of millions of instructions per second.

To achieve higher speeds, increased computational capability, and more flexible processing, many innovations were made. In addition to faster high-density chips, these include parallel processing, pipelining, and new concepts in compilers and assemblers. In addition, time-sharing and distributed computation have had an important effect on computer usage.

The impact of microelectronics was dramatically expressed by Noyce in 1977: "Today's microcomputer at a cost of perhaps $300 has more computing capacity than the first large electronic computer, ENIAC. It is 20 times faster, has a larger memory, is thousands of times more reliable, consumes the power of a light bulb rather than that of a locomotive, occupies 1/30,000 the volume and costs 1/10,000 as much. It is available by mail order or from your local hobby shop."

THE FUTURE

Throughout most of your (the readers') lives, it has been possible to have live television communication from anywhere on Earth or, in fact, from millions of miles in space. What is amazing is not that these are nearly everyday events, but that they can be achieved at all. It is an awesome accomplishment when we consider that someone at the Johnson Space Center can throw a switch and instruct a space vehicle 1 *billion* kilometers (km) away to turn on its TV camera, focus it, and send pictures back to Earth. (Even at the speed of light it takes nearly 2 hours to transmit the instruction and receive the signals.) None of this is possible without the advances in electronics, culminated by the IC, that were described in the two previous sections. However, even as this is history, the achievement is indicative of the future direction of electronics.

The ability to transmit television pictures from a spacecraft requires that communications, computation, and control equipment act in unison as a single entity. It is apparent that the areas of electronics are merging and the "intelligent" electronic systems that result are at the core of the information age.[1]

The marriage of extensive communications and inexpensive computers has already begun to penetrate nearly every aspect of society. In addition to traditional industrial applications, the ability and relative ease by which information can be stored, retrieved, manipulated, and transmitted has affected us in our homes, our places of work, and our means for getting from one to the other. Office automation (word processors, electronic mail, etc.) is transforming how and where we work. Energy management, appliance control, security

---

[1]The era of the 1980s extending into the twenty-first century has been called the "information age," partly because more than 50 percent of the U.S. workforce can be classified as "information workers."

systems, cable television, and personal computers are some microelectronic applications in the home. The computer-controlled Bay Area Rapid Transit (BART) system in metropolitan San Francisco and electronic ignition, emission control, and safety systems in automobiles are examples of the impact of electronics on transportation. So widespread will the impact of microelectronics be that Noyce has used the analogy that by the end of this century electronics will be like the electric motor of today—largely unnoticed.

We believe that the electronics industries will continue to be the four C's—components, communications, computation, and control. It will be exceedingly difficult to recognize them as separate entities because they will be merged to an even greater degree. Similarly, the distinction between device, circuit, and system will become increasingly blurred. For the next decade, silicon-based technology will dominate electronics. However, the results of research on new materials, particularly gallium arsenide (GaAs), will, in all probability, begin to play a significant role.[1]

The future impact of microelectronics is readily apparent from the following statistics and projections for the U.S. electronics market (in *billions* of dollars):

	1985	1990
Electronic sales	215	400
Integrated circuit sales	11	35

These projections indicate that the creativity and ingenuity of the engineers and scientists of the past will be the springboard for the engineering talent of the future.

GENERAL REFERENCES

1 Fiftieth Anniversary Issue: *Electronics*, vol. 53, no. 9, April 1980.

2 Fiftieth Anniversary Issue: *Proceedings of the IRE*, vol. 50, no. 5, May 1962.

3 Special Issue: Historical Notes on Important Tubes and Semiconductor Devices, *IEEE Trans. Electron Devices*, vol. ED-23, no. 7, July 1976.

4 Weiner, C.: How the Transistor Emerged, *IEEE Spectrum*, pp. 24–33, January 1973.

5 Forester, T.: "The Microelectronics Revolution", MIT Press, Cambridge, Mass., 1981.

6 Special Issue: Microelectronics, *Scientific American*, September 1977.

7 Wolff, M. F.: The Genesis of the Integrated Circuit, *IEEE Spectrum*, pp. 45–53, August 1976.

8 Mayo, J. S.: Technical Requirements of the Information Age, *Bell Labs Record*, vol. 60, no. 55, 1982.

9 Kidder, T.: *The Soul of a New Machine*, Little, Brown and Company, Boston, 1981.

[1]There is even speculation that organic materials, such as DNA, may find use in electronics by the end of this century.

Part One
SEMICONDUCTOR DEVICES

Semiconductor devices are the central components used to process the electrical signals that arise in communication, computation, and control systems. It is the electrical behavior of these devices that provides the controlled sources and controlled switches needed in signal-processing circuits. In the five chapters of this section, the physical operation and characteristics of the major semiconductor devices are developed. Elementary circuit applications are introduced to illustrate how the device characteristics are exploited in switches and amplifiers. Chapter 1 deals with the concepts which govern the electrical properties of semiconductors. Chapters 2 to 4 treat junction diodes and bipolar and field-effect transistors. An overview of integrated-circuit (IC) fabrication is presented in Chap. 5.

Chapter 1
SEMICONDUCTORS

The controlled flow of charged particles is fundamental to the operation of all electronic devices. Consequently, the materials used in these devices must be capable of providing a source of mobile charges and the processes which govern the flow of charges must be amenable to control. The physical properties of semiconductors as they relate to electronic devices are described in this chapter. In particular, the electrical characteristics of materials which allow us to distinguish semiconductors from insulators and conductors and the use of doping a semiconductor with impurities to control its electrical behavior are discussed.

Two principal charge-transport processes are also investigated: (1) *drift*, which is the motion of charges produced by an electric field, and (2) *diffusion*, which is motion resulting from a nonuniform charge distribution.

1-1 FORCES, FIELDS, AND ENERGY   In this section we introduce the basic quantities used to describe the effects of charged particles. For most students, it is a brief review of material previously treated in physics courses.

Charged Particles   The *electron* is the principal negatively charged particle whose *charge*, or quantity of electricity, has been determined as 1.60×10^{-19} coulombs (C).[1] The number of electrons per coulomb is the reciprocal of the electronic charge or approximately 6×10^{18}. Since a current of 1 ampere (A) is 1 coulomb per second (C/s), a current of 1 picoampere (pA, or 10^{-12} A) represents the motion

---

[1] International System of Units (SI) units are used almost exclusively in this text. Abbreviations for units are used with numeric values and after symbolic equations; otherwise they are spelled out. Units, technical terms, and their abbreviations are listed in App. A-1. The values of many important physical constants are given in App. A-2, and conversion factors and prefixes are listed in App. A-3.

of approximately 6 million electrons. Yet a current of 1 pA is so small that considerable difficulty is experienced in attempting its measurement.

In dealing with atoms, it is often convenient to consider the positive nucleus and inner electronic bands as an equivalent positive charge (the ionic core) whose magnitude is an integral multiple of the charge on an electron. The number of *valence electrons*, that is, those in the outermost electronic band, provides a negative charge to make the atom neutral. Under certain conditions, one or more valence electrons may be removed from the atom, leaving a positive ion. Similarly, one or more electrons can be added to the valence band, creating a negative ion. For example, the sodium and chlorine ions in common salt are singly ionized particles, with each having a magnitude of charge equal to that of the electron. The sodium ion is positive and the chlorine ion negative, resulting from the removal and addition, respectively, of one valence electron. Doubly ionized particles have ionic charges equal to twice that of the electron.

In a silicon crystal, each silicon ion shares a pair of electrons with each of four ionic neighbors. This configuration is called a *covalent bond*. Circumstances may cause an electron to be missing from this structure, leaving a "hole" in the bond.[1] These vacancies may move from ion to ion in the crystal and produce an effect equivalent to that of the motion of positive charges. The magnitude of charge associated with the hole is that of an electron.

Field Intensity

In the vicinity of a charged particle an electric field is said to exist; that is, a charged particle exerts a force on other charged particles as given by Coulomb's law. For the one-dimensional case[2] in which charge q_1 is at x_0, the force exerted on charge q_2 at an arbitrary distance x in newtons (N) is

$$F_x = \frac{q_1 q_2}{4\pi\epsilon(x - x_0)^2} \quad \text{N} \tag{1-1}$$

where ϵ is the permittivity of the medium in which the charges reside. From Newton's third law, an equal but opposite force acts on q_1.

The motion of q_2 is obtained by applying Newton's second law and is

$$F_x = \frac{q_1 q_2}{4\pi\epsilon(x - x_0)^2} = \frac{d}{dt}(m_2 v_x) \quad \text{N} \tag{1-2}$$

in which m_2 is the mass of q_2 and v_x is its velocity in the x direction. For a nonrelativistic system (m_2 is constant), Eq. (1-2) reduces to

$$F_x = m_2 \frac{dv_x}{dt} = m_2 a_x \quad \text{N} \tag{1-3}$$

where $a_x = dv_x/dt$ is the acceleration.

---

[1] This brief introduction to the hole as a charge carrier is discussed further in Sec. 1-3.

[2] The electric field and the forces it exerts are three-dimensional in general. However, in many electronic structures a uniform cross section permits one-dimensional representation.

A convenient method for describing the effect of charged particles is the use of the *electric field intensity* $\mathscr{E}$, defined as the force exerted on a unit positive charge. Thus the force on a charge q in an electric field is, in one dimension,

$$F_x = q\mathscr{E}_x \quad \text{N} \tag{1-4}$$

Potential

By definition, the potential V [in volts (V)] of point B with respect to point A is the work done against the field in moving a unit positive charge from A to B. In one dimension, with A at x_0 and B at an arbitrary distance x, it follows that[1]

$$V \equiv -\int_{x_0}^{x} \mathscr{E}_x \, dx \quad \text{V} \tag{1-5}$$

where $\mathscr{E}_x$ represents the x component of the field. Differentiation of Eq. (1-5) gives

$$\mathscr{E} = -\frac{dV}{dx} \quad \text{V/m} \tag{1-6}$$

The minus sign indicates that the electric field is directed from the region of higher potential to the region of lower potential.

By definition, the potential energy U equals the potential multiplied by the charge q under consideration, or [in joules (J)]

$$U \equiv qV \quad \text{J} \tag{1-7}$$

If an electron is being considered, q is replaced by $-q$ (where q is the magnitude of the electronic charge).

Because the energy associated with a single electron is so small, it is convenient to introduce the *electron volt* (eV) unit of energy (work), defined as

$$1 \text{ eV} = 1.60 \times 10^{-19} \text{ J}$$

Of course, any type of energy, whether it is electrical, mechanical, thermal, or similar, may be expressed in electron volts.

Equation (1-7) indicates that if an electron falls through a potential of 1 V, its kinetic energy increases and its potential energy decreases by 1.60×10^{-19} J or 1 eV. Although each electron possesses a small amount of energy, an enormous number of electrons is required for even a small current. Consequently, electron devices can handle reasonable amounts of power.

The law of conservation of energy states that the total energy W, which equals the sum of the potential energy U and the kinetic energy $\frac{1}{2}mv^2$, remains constant. Thus at any point

$$W = U + \tfrac{1}{2}mv^2 = \text{const} \tag{1-8}$$

---

[1] The symbol $\equiv$ is used to designate "equal to by definition."

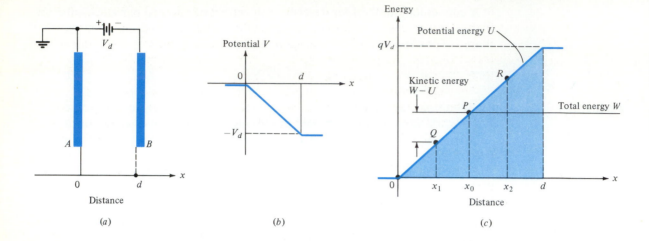

FIGURE 1-1
(a) Parallel-plate system showing an electron leaving A with an initial velocity v_0 and moving in a retarding field. (b) The potential and (c) the potential-energy barrier.

As an illustration of this law, consider two parallel plates A and B separated by a distance d, as shown in Fig. 1-1a, and with B at a negative potential V_d with respect to A. An electron leaves the surface of A toward B with an initial velocity v_0 in the x direction. What speed will the electron have if it reaches B?

From the definition Eq. (1-5), it is clear that only differences of potential have meaning; hence, let us arbitrarily ground A, that is, consider it to be at zero potential. Then the potential at B is $V = -V_d$, and the potential energy is $U = -qV_d$. If we equate the total energy at A to that at B, we obtain

$$W = \tfrac{1}{2}mv_0^2 = \tfrac{1}{2}mv^2 + qV_d \quad \text{J} \tag{1-9}$$

This equation indicates that v must be less than v_0, which is obviously correct since the electron is moving in a repelling field. The final velocity attained by the electron in this conservative system is independent of the form of the variation of the field distribution between the plates and depends only on the magnitude of the potential difference V_d. Note that, if the electron is to reach electrode B, its initial velocity must be large enough so that $\tfrac{1}{2}mv_0^2 > qV_d$. Otherwise, Eq. (1-9) leads to the impossible result that v is imaginary. We now wish to elaborate on these considerations.

The Concept of a Potential-Energy Barrier For the configuration of Fig. 1-1a in which the electrodes are large in comparison with the separation d, we can draw (Fig. 1-1b) a linear plot of potential V versus distance x (in the interelectrode space). The corresponding potential energy U versus x is indicated in Fig. 1-1c and is obtained from the curve by multiplying each ordinate by the charge on the electron (a negative number). The total energy W of the electron remains constant and is represented as a horizontal line. The kinetic energy at any distance x_1 equals the difference between the total energy W and the potential energy U at this point. This difference is greatest at 0, indicating that the kinetic energy is a maximum when the electron leaves the electrode A. At point P this difference is zero, which means that no kinetic energy exists, so that the particle is at rest at this point. This distance x_0 is the maximum that the electron can travel from A. At point P (where $x = x_0$) it comes momentarily to rest and then reverses its motion and returns to A.

Consider a point such as x_2 which is at a greater distance than x_0 from electrode A. Here the total energy W is less than the potential energy U, so that the difference, which represents the kinetic energy, is negative. This is an impossible physical condition, however, since negative kinetic energy ($\frac{1}{2}mv^2 < 0$) implies an imaginary velocity. We must conclude that the particle can never advance a distance greater than x_0 from electrode A. At point P this difference is zero, which means that no kinetic energy exists, so that the particle is at rest at this point. This distance x_0 is the maximum that the electron can travel. The foregoing analysis leads to the very important conclusion that the shaded portion of Fig. 1-1c can never be penetrated by the electron. Thus, at point P, the particle acts as if it had collided with a solid wall, hill, or barrier and the direction of its flight had been altered. Potential-energy barriers of this sort play an important role in the analysis of semiconductor devices.

It must be emphasized that the words "collides with a potential hill" constitute a convenient descriptive phrase and that an actual encounter between two material bodies is not implied.

1-2   CONDUCTION IN METALS

In a metal the outer, or valence, electrons of an atom are as much associated with one ion as with another, so that the electron attachment to any individual atom is almost zero. Depending on the metal, at least one (and sometimes two or three) electron (or electrons) per atom is (are) free to move throughout the interior of the metal under the action of applied fields.

Figure 1-2 is a two-dimensional schematic diagram of the charge distribution within a metal. The shaded regions represent the net positive charge of the

FIGURE 1-2
Schematic arrangement of the atoms in one plane in a metal, drawn for monovalent atoms. The black dots represent the electron gas, each atom having contributed one electron to this gas.

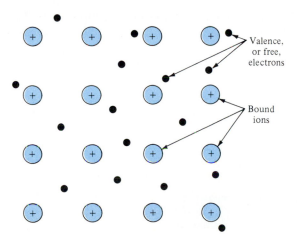

Valence, or free, electrons

Bound ions

nucleus and the tightly bound inner electrons. The black dots represent the outer, or valence, electrons in the atom. It is these electrons that cannot be said to belong to any particular atom; instead, they have completely lost their individuality and can wander freely about from atom to atom in the metal. Thus a metal is visualized as a region containing a periodic three-dimensional array of heavy, tightly bound ions permeated with a swarm of electrons that may move about quite freely. This scheme is known as the *electron-gas* description of a metal.

According to the electron-gas theory of a metal, the electrons are in continuous motion, the direction of flight being changed at each collision with the heavy, almost stationary ions. The average distance between collisions is called the *mean free path*. Since the motion is random, the net number of electrons crossing a unit area in the metal during a given time is, on the average, zero. Consequently, the average current is zero.

Let us now see how the situation is changed if a constant electric field $\mathscr{E}$ is applied to the metal. As a result of this electrostatic force, the electrons would be accelerated and the velocity would increase indefinitely with time, were it not for the collisions with the ions. At each inelastic collision with an ion, however, an electron loses energy and changes direction. The probability that an electron moves in a particular direction after a collision is equal to the probability that it travels in the opposite direction. Hence the velocity of an electron increases linearly with time between collisions and is, on an average, reduced to zero at each collision. A steady-state condition is reached when an average value of *drift velocity* v_d is attained; its direction is opposite to that of the electric field. The speed at time t between collisions is at, where the acceleration a equals q/m. As a result, the drift velocity is proportional to $\mathscr{E}$ and is given by

$$v_d = \mu\mathscr{E} \qquad \text{m/s} \tag{1-10}$$

where the proportionality constant μ is called the *mobility* of the electron.[1]

According to the foregoing theory, a steady-state drift velocity has been superimposed on the random thermal motion of the electrons. Such a directed flow of electrons constitutes a current which we can now calculate.

Current Density In Fig. 1-3, N electrons are distributed uniformly throughout a conductor of length L and cross-sectional area A. An electron, under the influence of an

---

[1] A subscript is usually added to μ where more than one type of charge carrier is present. The dimensions of mobility are given in square meters per volt-second [$m^2/(V \cdot s)$].

FIGURE 1-3
Conductor used to calculate current density.

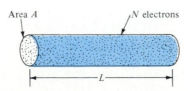

Area A N electrons

L

electric field $\mathscr{E}$ travels L meters in T seconds, thus making the drift velocity v_d equal to L/T. The current I is, by definition, the total charge passing through any area per unit time and is the charge per carrier multiplied by the number of carriers per second crossing the area. Hence [in amperes (A)]

$$I = \frac{qN}{T} \cdot \frac{L}{L} = \frac{qNv_d}{L} \quad A \tag{1-11}$$

The current density, denoted by the symbol J, is the current per unit area in the conducting medium; that is, assuming a uniform current density

$$J \equiv \frac{I}{A} \quad A/m^2 \tag{1-12}$$

substitution of Eq. (1-11) into Eq. (1-12) yields

$$J = \frac{qNv_d}{LA} \quad A/m^2 \tag{1-13}$$

From Fig. 1-3 it is evident that LA is the volume containing the N electrons. The volume concentration of electrons or simply the electron concentration n is then

$$n = \frac{N}{LA} \quad m^{-3} \tag{1-14}$$

and Eq. (1-13) reduces to

$$J = qnv_d = \rho_v v_d \quad A/m^2 \tag{1-15}$$

where $\rho_v = qn$ is the volume charge density in coulombs per cubic meter (C/m³).

This derivation is independent of the form of the conduction medium. Consequently, Fig. 1-3 does not necessarily represent a metal conductor. It may represent equally well a portion of a gaseous-discharge tube or a volume element of a semiconductor. Furthermore, neither ρ_v nor v_d must be constant but may vary with time.

Conductivity   We have already seen that v_d is proportional to $\mathscr{E}$. Combination of Eqs. (1-10) and (1-15) results in

$$J = qnv_d = qn\mu\mathscr{E} = \sigma\mathscr{E} \quad A/m^2 \tag{1-16}$$

where

$$\sigma = qn\mu \quad (\Omega \cdot m)^{-1} \tag{1-17}$$

is the conductivity of the material. By recalling that $\mathscr{E}L = V$ is the applied voltage across the conductor, we can obtain the current I from Eq. (1-16) and recognize it as Ohm's law. Thus

$$I = JA = \sigma\mathscr{E}A \cdot \frac{L}{L} = \frac{\sigma A}{L} V = \frac{V}{R} \quad A \tag{1-18}$$

The resistance R of the conductor is given [in ohms (Ω)] by

$$R = \frac{L}{\sigma A} = \rho \frac{L}{A} \quad \Omega \tag{1-19}$$

where the resistivity ρ is the reciprocal of the conductivity.

As already mentioned, the energy acquired by the electrons from the applied field is, as a result of collisions, given to the lattice ions. Consequently, power is dissipated within the metal and the power density [Joule heat in watts per cubic meter (W/m^3)] is given by $J = \sigma\mathcal{E}^2$. (This relation is analogous to $P = VI = V^2/R$.)

Example 1-1

A conducting line on an IC chip is 2.8 millimeters (mm) long and has a rectangular cross section 1×4 micrometers (μm). A current of 5 mA produces a voltage drop of 100 mV across the line. Determine the electron concentration given that the electron mobility is 500 $cm^2/(V \cdot s)$.

Solution

The electron concentration can be obtained from σ in Eq. (1-17). The conductivity is determined by solving Eq. (1-18) for σ as

$$\sigma = \frac{IL}{VA} = \frac{5 \times 10^{-3} \times 2.8 \times 10^{-3}}{0.1 \times (10^{-6} \times 4 \times 10^{-6})} = 3.50 \times 10^7 \; (\Omega \cdot m)^{-1}$$

Then, from Eq. (1-17), we obtain

$$n = \frac{\sigma}{q\mu} = \frac{3.5 \times 10^7}{1.60 \times 10^{-19} \times 500 \times 10^{-4}}$$
$$= 4.38 \times 10^{27} \; m^{-3} = 4.38 \times 10^{21} \; cm^{-3}$$

As seen in Eq. (1-17), conductivity is proportional to the concentration of charge carriers. The free-electron concentration found in Example 1-1 is a typical value for conductors. Few carriers are available in insulators, and electron concentrations are in the order of $10^7 \; m^{-3}$. Materials whose carrier concentrations lie between those of conductors and insulators are called *semiconductors*, the properties of which are discussed in the next two sections.

1-3 THE INTRINSIC SEMICONDUCTOR Silicon, germanium, and gallium arsenide are the three most widely used semiconductors. Because of the predominance of silicon devices, we confine our discussion to it.

The crystal structure of silicon consists of a regular repetition in three dimensions of a unit cell having the form of a tetrahedron with an atom at each vertex. A two-dimensional symbolic representation of this structure is illustrated in Fig. 1-4. Silicon has a total of 14 electrons in its atomic structure, four of which are valence electrons, so that the atom is tetravalent. The inert ionic core of the silicon atom has a charge of $+4$ measured in units of electronic charge. The binding forces between neighboring atoms result from the fact that

FIGURE 1-4
Two-dimensional representation of silicon crystal.

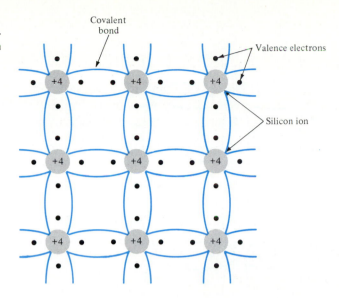

each valence electron of a silicon atom is shared by one of its four nearest neighbors. This covalent bond is represented in Fig. 1-4 by the two lines which join each ion to each of its neighbors. The valence electrons serve to bind one atom to the next and this results in these electrons being tightly bound to the nucleus. Hence, in spite of the availability of four valence electrons, few of these are free to contribute to conduction.

The Hole

At a very low temperature (say, 0 K) the ideal structure shown in Fig. 1-4 is approached, and the crystal behaves as an insulator, since no free carriers of electricity are available. At room temperature, however, some of the covalent bonds will be broken because of the thermal energy supplied to the crystal, and conduction is made possible. This situation is illustrated in Fig. 1-5. Here an electron, which usually forms part of a covalent bond, is pictured as being dislodged and is thus free to wander in a random fashion throughout the crystal. The energy E_G required to break such a covalent bond is about 1.1 eV for silicon at room temperature. The absence of the electron in the covalent bond is represented by the small circle in Fig. 1-5, and such an incomplete covalent bond is called a *hole*. The importance of the hole is that it may serve as a carrier of electricity comparable in effectiveness with the free electron.

The mechanism by which a hole contributes to the conductivity is qualitatively described as follows. When a bond is incomplete so that a hole exists, it is relatively easy for a valence electron in a neighboring atom to leave its covalent bond to fill this hole. An electron moving from a bond to fill a hole leaves a hole in its initial position. Hence the hole effectively moves in the direction opposite to that of the electron. This hole, in its new position, may now be filled by an electron from another covalent bond, and the hole will

FIGURE 1-5
Silicon crystal with a
broken covalent bond.

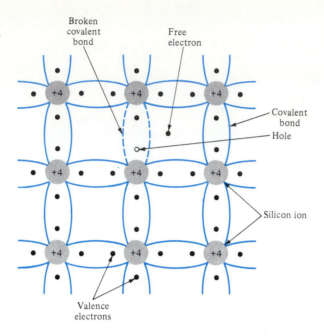

correspondingly move one more step in the direction opposite to the motion
of the electron. Here we have a mechanism for the conduction of electricity
which does not involve free electrons. This phenomenon is illustrated schematically in Fig. 1-6, where a circle with a dot in it represents a completed
bond and an empty circle designates a hole. Figure 1-6a shows a row of 10
ions, with a broken bond, or hole, at ion 6. Now imagine that an electron from
ion 7 moves into the hole at ion 6, so that the configuration seen in Fig. 1-6b
results. If we compare this figure with Fig. 1-6a, it looks as if the hole in Fig.
1-6a has moved toward the right in Fig. 1-6b (from ion 6 to ion 7). This discussion
indicates that the motion of the hole in one direction actually means the transport of a negative charge an equal distance in the opposite direction. As far as
the flow of electric current is concerned, the hole behaves like a positive charge
equal in magnitude to the electronic charge. We can consider that the holes
are physical entities whose movement constitutes a flow of current. The heuristic argument that a hole behaves as a free positive charge carrier may be
justified by quantum mechanics.

Conduction in Intrinsic Semiconductors   The crystal structure displayed in Figs. 1-4 and
1-5 assumed a pure sample of silicon; that is, the sample contains no foreign
atoms. Such pure crystals are called *intrinsic semiconductors*. As shown in

FIGURE 1-6
The mechanism by
which a hole contrib-
utes to the conductiv-
ity.

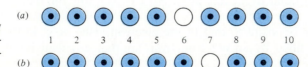

Fig. 1-5, breaking a covalent bond results in both a free electron and a hole. Consequently, the hole concentration p and electron concentration n must be equal and

$$p = n = n_i \qquad (1\text{-}20)$$

where the intrinsic concentration is denoted by n_i. Thermal agitation generates new electron-hole pairs, whereas other electron-hole pairs disappear as a result of recombination. The value of n_i is temperature-dependent; this variation is discussed in Sec. 1-5.

Both holes and electrons contribute to the conduction process. Because the mechanisms by which holes and electrons move about in the crystal differ, the mobilities of these carriers are different. The subscripts p and n are used to distinguish hole and electron values. These carriers move in opposite directions in an electric field, but as they are of opposite sign, the current of each is in the same direction. The current density J that results from an electric field $\mathscr{E}$ is obtained from Eq. (1-16), modified to include both carriers, and is

$$J = q(n\mu_n + p\mu_p)\mathscr{E} = \sigma\mathscr{E} \qquad \text{A/m}^2 \qquad (1\text{-}21)$$

The conductivity is

$$\sigma = q(n\mu_n + p\mu_p) \qquad (\Omega\cdot\text{m})^{-1} \qquad (1\text{-}22)$$

For intrinsic semiconductors, $p = n = n_i$ and Eq. (1-22) reduces to

$$\sigma_i = qn_i\,(\mu_n + \mu_p) \qquad (\Omega\cdot\text{m})^{-1} \qquad (1\text{-}23)$$

Values of important properties of silicon are given in Table 1-1. Note that silicon has on the order of 10^{22} atoms/cm^3, while at room temperature (300 K),

TABLE 1-1   Properties of Intrinsic Silicon

Property	Value
Atomic number	14
Atomic weight	28.1
Density (g/cm^3)	2.33
Relative permittivity (dielectric constant)	11.9
Atoms/cm^3	5.0×10^{22}
Energy gap E_{G0} at 0 K (eV)	1.21
Energy gap E_G at 300 K (eV)	1.12
Resistivity at 300 K ($\Omega\cdot$cm)	2.30×10^5
Electron mobility μ_n at 300 K [cm^2/(V·s)]	1500
Hole mobility μ_p at 300 K [cm^2/(V·s)]	475
Intrinsic concentration at 300 K (cm^{-3})	1.45×10^{10}
Electron diffusion constant D_n at 300 K (cm^2/s)	34
Hole diffusion constant D_p at 300 K (cm^2/s)	13

From S. M. Sze (ed.), ''VSLI Technology,'' McGraw-Hill Book Company, New York, 1983.

$n_i \simeq 10^{10}$ cm^{-3}. Hence only 1 atom in about 10^{12} contributes a free electron (and also a hole) to the crystal because of broken covalent bonds.

Example 1-2

An intrinsic silicon bar is 3 mm long and has a rectangular cross section 50 × 100 μm. At 300 K, determine the electric field intensity in the bar and the voltage across the bar when a steady current of 1 μA is measured.

Solution

The field intensity can be obtained from the current density and conductivity as

$$\mathcal{E} = \frac{J}{\sigma} = \frac{I}{A} \times \frac{1}{\sigma} = \frac{I}{A} \cdot \rho \qquad \text{V/m}$$

Using the value of ρ given in Table 1-1, we obtain

$$\mathcal{E} = \frac{10^{-6}}{50 \times 10^{-6} \times 100 \times 10^{-6}} \times 2.30 \times 10^{5} \times 10^{-2}$$

where the factor 10^{-2} converts the resistivity from $\Omega \cdot$cm to $\Omega \cdot$m:

$$\mathcal{E} = 4.60 \times 10^{5} \text{ V/m} = 4.60 \times 10^{3} \text{ V/cm}$$

The voltage across the bar is

$$V_{\text{bar}} = \mathcal{E}L = 4.60 \times 10^{5} \times 3 \times 10^{-3} = 1380 \text{ V}$$

The result obtained in Example 1-2 indicates that an extremely large voltage is needed to produce a small current (1 μA). This, however, is not surprising since the intrinsic carrier concentration is much closer to that of an insulator than it is to a conductor. Thus intrinsic semiconductors are not suitable for electron devices. In Sec. 1-4 we investigate one method by which carrier concentration can be increased.

1-4 EXTRINSIC SEMICONDUCTORS A common expedient used to increase the number of carriers is to introduce a small carefully controlled impurity content into an intrinsic semiconductor. The addition of impurities, most often trivalent or pentavalent atoms, forms an *extrinsic*, or *doped*, semiconductor. Each type of impurity establishes a semiconductor which has a predominance of one kind of carrier. The usual level of doping is in the range of 1 impurity atom for 10^6 to 10^8 silicon atoms. Thus most physical and chemical properties are essentially those of silicon and only the electrical properties change markedly.

n-Type Semiconductors Figure 1-7 depicts the crystal structure obtained when silicon is doped with a pentavalent impurity. Four of the five valence electrons occupy covalent bonds, and the fifth will be nominally unbound and will be available as a carrier of current. The energy required to detach this fifth electron from the atom is of the order of only 0.05 eV for silicon and is considerably less than the energy required to break a covalent bond. Suitable pentavalent impurities are anti-

FIGURE 1-7
Crystal lattice with a silicon atom displaced by a pentavalent impurity atom.

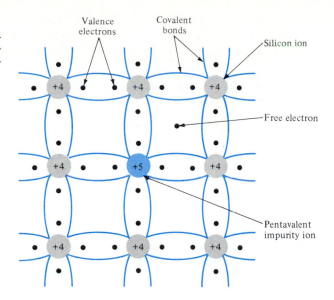

Valence electrons · Covalent bonds · Silicon ion · Free electron · Pentavalent impurity ion

mony, phosphorus, and arsenic. Such impurities donate excess electron carriers and are referred to as *donor*, or *n-type*, impurities.

If intrinsic semiconductor material is doped with *n*-type impurities, not only does the number of electrons increase, but the number of holes decreases below that which would be available in the intrinsic semiconductor. The number of holes decreases because the larger number of electrons present causes the rate of recombination of electrons with holes to increase. Consequently, the dominant carriers are the negative electrons and doping with donors results in an *n-type semiconductor*.

p-Type Semiconductors Boron, gallium, and indium are trivalent atoms which, when added to intrinsic semiconductors, provide electrons to fill only three covalent bonds. The vacancy that exists in the fourth bond constitutes a hole as illustrated in Fig. 1-8. This type of impurity makes positive carriers available because it creates holes which can accept electrons. Thus trivalent impurities are called *acceptors* and form *p-type semiconductors* in which holes are the predominant carrier.

The Mass-Action Law We noted previously that the addition of *n*-type impurities causes the number of holes to decrease. Similarly, doping with *p*-type impurities decreases the concentration of free electrons below that in the intrinsic semiconductor. A theoretical analysis (Sec. 1-7) leads to the result that, under thermal equilibrium, the product of the free negative and positive concentrations is a constant independent of the amount of donor and acceptor impurity doping. This relationship is called the *mass-action law* and is given by

$$np = n_i^2 \qquad (1\text{-}24)$$

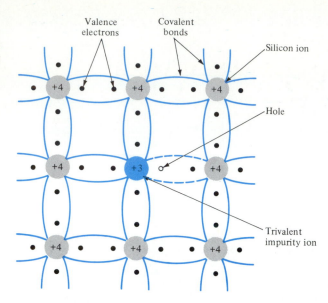

The intrinsic concentration n_i is a function of temperature (Sec. 1-5).

We have the important result that the doping of an intrinsic semiconductor not only increases the conductivity, but also serves to produce a conductor in which the charge carriers are either predominantly holes or predominantly electrons. In an *n*-type semiconductor, the electrons are called the *majority carriers*, and the holes are called the *minority carriers*. In a *p*-type material, the holes are the majority carriers and the electrons are the minority carriers.

Carrier Concentrations   We have previously indicated that only a small amount of energy is needed to ionize the impurity atoms. The temperatures at which electronic devices normally operate (>200 K) provide sufficient thermal energy to ionize virtually all impurities. This fact, in conjunction with the mass-action law, permits us to determine the charge densities in a semiconductor.

Let N_D be the concentration of donor atoms and N_A the concentration of acceptor atoms. Since these impurities are practically all ionized, they produce positive-ion and negative-ion densities of N_D and N_A, respectively. To maintain the electric neutrality of the crystal, the total positive charge density must equal the concentration of negative charges as expressed in Eq. (1-25):

$$N_D + p = N_A + n \tag{1-25}$$

Let us now consider an *n*-type material having $N_A = 0$. Since the number of electrons is much greater than the number of holes in an *n*-type semiconductor ($n \gg p$), Eq. (1-25) reduces to

$$n \simeq N_D \tag{1-26}$$

In an n-type material the free-electron concentration is approximately equal to the density of donor atoms.

The concentration p of holes in the n-type semiconductor is obtained from Eq. (1-24). Thus

$$p = \frac{n_i^2}{N_D} \tag{1-27}$$

Similarly, in a p-type semiconductor, with $N_D = 0$, we have

$$p \simeq N_A \tag{1-28}$$

and

$$n = \frac{n_i^2}{N_A} \tag{1-29}$$

Example 1-3

An n-type silicon sample is 3 mm long and has a rectangular cross section $50 \times 100 \ \mu$m. The donor concentration at 300 K is 5×10^{14} cm^{-3} and corresponds to 1 impurity atom for 10^8 silicon atoms. A steady current of 1 μA exists in the bar. Determine the electron and hole concentrations, the conductivity, and the voltage across the bar. (Note that this is an n-type sample that has the same dimensions and current as does the intrinsic silicon in Example 1-2.)

Solution[1]

From Eqs. (1-26) and (1-27), and using the values of n_i and μ_n in Table 1-1, we obtain

$$n = N_D = 5 \times 10^{14} \text{ cm}^{-3}$$

and

$$p = \frac{(1.45 \times 10^{10})^2}{5 \times 10^{14}} = 4.2 \times 10^5 \text{ cm}^{-3}$$

As $n \gg p$, only the electron concentration need be considered in Eq. (1-22), so that the conductivity is

$$\sigma = qn\mu_n = 1.60 \times 10^{-19} \times 5 \times 10^{14} \times 1.5 \times 10^3 = 0.12 \ (\Omega \cdot \text{cm})^{-1}$$

The voltage across the bar, obtained from $\mathscr{E} \cdot L$, where, by means of Eq. (1-21), $\mathscr{E} = J/\sigma$, is

$$V_{\text{bar}} = \frac{J}{\sigma} L = \frac{IL}{A\sigma} = \frac{10^{-6} (0.3)}{(5 \times 10^{-3}) (10^{-2}) \times 0.12} = 0.05 \text{ V}$$

The efficacy of using extrinsic semiconductors in electronic devices is readily apparent when the results of Examples 1-2 and 1-3 are compared. To produce the small current of 1 μA, 1380 V must be applied to the intrinsic sample,

---

[1]All calculations in this example are made by using centimeters instead of meters in all equations.

whereas only 50 mV is required for the n-type sample. This reduction of voltage by a factor of 28,000 exactly equals the decrease in resistivity (from 2.30×10^5 to $1/\sigma = 8.33$ Ω·cm). Yet the dramatic increase in the number of free electrons (1.45×10^{10} to 5×10^{14} cm^{-3}) occurs when only 1 silicon atom in 100 million is replaced by an impurity atom!

It is possible to add donors to a p-type crystal or, conversely, to add acceptors to n-type material. If equal concentrations of donors and acceptors permeate the semiconductor, the semiconductor remains intrinsic. The hole of the acceptor combines with the conduction electron of the donor to give no additional free carriers. Thus, from Eq. (1-25) with $N_D = N_A$, we observe that $p = n$, and from Eq. (1-24), $n^2 = n_i^2$, or $n = n_i =$ intrinsic concentration.

An extension of the preceding argument indicates that if the concentration of donor atoms added to a p-type semiconductor exceeds the acceptor concentration ($N_D > N_A$), the specimen is changed from a p-type to an n-type semiconductor. Conversely, the addition of a sufficient number of acceptors to an n-type sample results in a p-type semiconductor. This is precisely what is done in fabricating IC transistors. To determine the carrier concentrations under these circumstances, N_D is replaced by $N_D - N_A$ in Eqs. (1-26) and (1-27) when p-type material is changed to an n-type semiconductor. Similarly, when an n-type semiconductor is converted to p type, N_A in Eqs. (1-28) and (1-29) is replaced by $N_A - N_D$.

Generation and Recombination of Charges   In an intrinsic semiconductor the number of holes is equal to the number of free electrons. Thermal agitation, however, continues to generate g new hole-electron pairs per unit volume per second, while other hole-electron pairs disappear as a result of recombination; in other words, free electrons fall into empty covalent bonds, resulting in the loss of a pair of mobile carriers. On an average, a hole (an electron) will exist for τ_p (τ_n) seconds before recombination. This time is called the *mean lifetime* of the hole (electron). These parameters are very important in semiconductor devices because they indicate the time required for electron and hole concentrations which have been caused to change to return to their equilibrium concentrations.

1-5   VARIATIONS IN THE PROPERTIES OF SILICON   The conductivity of a semiconductor, given in Eq. (1-22), depends on both hole and electron concentration and mobility. Because semiconductor devices are subject to a wide range of operating temperatures, the variations of these parameters with temperature are important.

Intrinsic Concentration   With increasing temperature, the density of hole-electron pairs increases in an intrinsic semiconductor. Theoretically it is found that the intrinsic concentration n_i varies with T as

$$n_i^2 = A_0 T^3 \epsilon^{-E_{G0}/kT} \tag{1-30}$$

where E_{G0} is the energy gap (the energy required to break a covalent bond) at 0 K in electron volts, k is the Boltzmann constant in electron volts per degree kelvin (eV/K), and A_0 is a constant independent of T.

The increase in n_i^2 with temperature also has an effect on the charge densities in extrinsic semiconductors. For example, consider an n-type sample with a donor concentration N_D which is subjected to a temperature increase from 300 to 400 K. The electron density n at 400 K does not change appreciably from its value at 300 K because the ionized donor impurities contribute nearly all of the carriers. However, the mass-action law indicates that the hole concentration p increases. Similarly, for p-type semiconductors, n increases with modest rises in temperature, and $p \simeq N_A$ remains constant.

Mobility

The variation with temperature (100 to 400 K) of electron and hole mobilities is proportional to T^{-m}. For silicon, $m = 2.5$ for electrons and 2.7 for holes. Mobility μ decreases with temperature because more carriers are present and these carriers are more energetic at higher temperatures. Each of these facts results in an increased number of collisions and μ decreases.

Mobilities are also functions of the electric field intensity and doping levels. In n-type silicon, μ is constant at a given temperature only if $\mathscr{E} < 10^3$ V/cm. For $\mathscr{E} > 10^4$ V/cm, μ_n is inversely proportional to $\mathscr{E}$ and drift velocities approach 10^7 cm/s (the saturation velocity). Between 10^3 and 10^4 V/cm, μ_n varies approximately as $\mathscr{E}^{-1/2}$.

Conductivity

The conductivity of an intrinsic semiconductor increases with increasing temperature because the increase in hole-electron pairs is greater than the decrease in their mobilities. For extrinsic semiconductors, in the temperature range 100 to 600 K, the number of majority carriers is nearly constant but diminished mobility causes the conductivity to decrease with temperature.

1-6 DIFFUSION

In addition to a conduction current, the transport of charges in a semiconductor may be accounted for by a mechanism called *diffusion* (not ordinarily encountered in metals). The essential features of diffusion are now discussed.

It is possible to have a nonuniform concentration of particles in a semiconductor. As indicated in Fig. 1-9, the concentration p of holes varies with distance

FIGURE 1-9
Representation of a nonuniform hole density and the resultant diffusion current density.

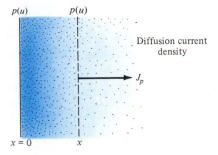

x in the semiconductor, and there exists a concentration gradient dp/dx in the density of carriers. The existence of a gradient implies that if an imaginary surface (indicated by dashed line) is drawn in the semiconductor, the density of holes immediately on one side of the surface is larger than the density on the other side. The holes are in a random motion as a result of their thermal energy. Accordingly, holes will continue to move back and forth across this surface. We may then expect that, in a given time interval, more holes will cross the surface from the side of greater concentration to the side of smaller concentration than in the reverse direction. This net transport of holes across the surface constitutes a current in the positive x direction. It should be noted that this net transport of charge is not the result of mutual repulsion among charges of like sign, but is simply the result of a statistical phenomenon. This diffusion is exactly analogous to that which occurs in a contained neutral gas if a concentration gradient exists.[1] The diffusion hole-current density J_p is proportional to the concentration gradient and is given by

$$J_p = -qD_p \frac{dp}{dx} \qquad \text{A/m}^2 \tag{1-31}$$

where D_p (in square meters per second) is called the *diffusion constant* for holes. Since p in Fig. 1-9 decreases with increasing x, dp/dx is negative and the minus sign in Eq. (1-31) is needed, so that J_p will be positive in the positive x direction. A similar equation exists for diffusion electron current density [p is replaced by n, and the minus sign is replaced by a plus sign in Eq. (1-31)].

The Einstein Relationship Since both diffusion and mobility are statistical thermodynamic phenomena, D and μ are not independent. The relationship between them is given by the Einstein equation

$$\frac{D_p}{\mu_p} = \frac{D_n}{\mu_n} = V_T \tag{1-32}$$

where V_T is the "volt equivalent of temperature," defined by

$$V_T \equiv \frac{\bar{k}T}{q} = \frac{T}{11,600} \qquad \text{V} \tag{1-33}$$

where $\bar{k}$ is the Boltzmann constant in joules per Kelvin. Note the distinction between $\bar{k}$ and k; the latter is the Boltzmann constant in electron volts per Kelvin. (Values of $\bar{k}$ and k are given in App. A-1. From Sec. 1-3 it follows that $\bar{k} = 1.60 \times 10^{-19}k$.) At room temperature (300 K), $V_T = 0.0259$ V, and $\mu = 38.6D$. (Measured values of μ and computed values of D for silicon are given in Table 1-1.)

---

[1] It is also the process by which the scent of flowers can permeate a room.

Total Current   It is possible for both a potential gradient and a concentration gradient to exist simultaneously within a semiconductor. In such a situation the total hole current is the sum of the drift current [Eq. (1-16), with n replaced by p] and the diffusion current [Eq. (1-31)], or

$$J_p = q\mu_p p \mathscr{E} - qD_p \frac{dp}{dx} \qquad \text{A/m}^2 \qquad (1\text{-}34)$$

Similarly, the net electron current is

$$J_n = q\mu_n n \mathscr{E} + qD_n \frac{dn}{dx} \qquad \text{A/m}^2 \qquad (1\text{-}35)$$

1-7 GRADED SEMICONDUCTORS   The semiconductor sample shown in Fig. 1-10a has a hole concentration that is a function of x; that is, the doping is graded (nonuniform). The electron density must also vary with x, a consequence of the mass-action law. Let us assume thermal equilibrium and that no carriers are injected into the sample from any *external* source (zero excitation). Under these conditions there can be no *steady* charge motion, only the random motion due to thermal agitation. Hence the total hole current must be zero; also, the total electron current is zero. Since p is not constant, we expect a nonzero hole diffusion current. For the total hole current to vanish, a hole drift current must exist which is equal and opposite to the diffusion current. However, a conduction current requires an electric field, and we conclude that, as a result of the nonuniform doping, an electric field is generated within the semiconductor. We shall now find this field and the corresponding potential variation throughout the bar.

Setting $J_p = 0$ in Eq. (1-34) and using the Einstein relationship $D_p = \mu_p V_T$ [Eq. (1-32)], we obtain

$$\mathscr{E} = \frac{V_T}{p} \frac{dp}{dx} \qquad \text{V/m} \qquad (1\text{-}36)$$

FIGURE 1-10
(a) A graded semiconductor; $p(x)$ is not constant. (b) A pn junction in which p and n are uniformly doped with impurity concentrations N_A and N_D, respectively.

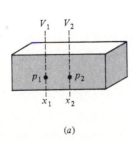

V_1 V_2

p_1 p_2

x_1 x_2

(a)

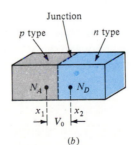

Junction

p type n type

N_A N_D

x_1 x_2

V_0

(b)

If the doping concentration $p(x)$ is known, this equation allows the built-in field $\mathcal{E}(x)$ to be calculated. From $\mathcal{E} = -dV/dx$, we can calculate the potential as

$$dV = -V_T \frac{dp}{p} \tag{1-37}$$

Integration of Eq. (1-36) from x_1, where the concentration is p_1 and the potential is V_1, to x_2 where $p = p_2$ and $V = V_2$, yields

$$V_{21} \equiv V_2 - V_1 = V_T \ln \frac{p_1}{p_2} \qquad \text{V} \tag{1-38}$$

Note that the potential difference between two points depends only on the concentration at these two points and is independent of their separation $x_2 - x_1$. Equation (1-38) may be expressed in the form

$$p_1 = p_2 \epsilon^{+V_{21}/V_T} \tag{1-39}$$

This is the Boltzmann relationship of kinetic gas theory.

Mass-Action Law Starting with $J_n = 0$ and proceeding as above, we obtain the Boltzmann equation for electrons:

$$n_1 = n_2 \epsilon^{-V_{21}/V_T} \tag{1-40}$$

Multiplication of Eq. (1-39) by Eq. (1-40) gives

$$n_1 p_1 = n_2 p_2 \tag{1-41}$$

This equation states that the product np is a constant independent of x, and hence the amount of doping, under thermal equilibrium. For an intrinsic semiconductor, $n = p = n_i$, and $np = n_i^2$, which is the law of mass action introduced in Eq. (1-24).

An Open-Circuited Step-Graded Junction Consider the special case indicated in Fig. 1-10b. The left half of the bar is p type with a constant concentration N_A, whereas the right half is n type with a uniform density N_D. The dashed plane is a metallurgical (pn) junction separating the two sections with different concentration. This type of doping, where the density changes abruptly from p to n type, is called *step grading*, and the junction is located at the plane where the concentration is zero. As described earlier, theory indicates that there is a built-in potential between these two sections called the *contact difference of potential* V_0. Equation (1-38) allows us to calculate V_0 as

$$V_0 = V_{21} = V_T \ln \frac{p_{p0}}{p_{n0}} \qquad \text{V} \tag{1-42}$$

because $p_1 = p_{p0} = $ thermal-equilibrium hole concentration in the p side and $p_2 = p_{n0} = $ thermal equilibrium hole concentration in the n side. From Eq.

(1-28), $p_{p0} = N_A$, and from Eq. (1-27) $p_{n0} = n_i^2/N_D$, so that

$$V_0 = V_T \ln \frac{N_A N_D}{n_i^2} \quad \text{V} \tag{1-43}$$

The same expression for V_0 is obtained from an analysis corresponding to that given above and based on equating the total electron current I_n to zero (Prob. 1-18). The *pn* junction is studied in detail in Chap. 2.

REFERENCES

1 Shockley, W.: "Electrons and Holes in Semiconductors," D. Van Nostrand, Princeton, N.J. (reprinted), 1963.

2 Yang, E. S.: "Fundamentals of Semiconductor Devices," McGraw-Hill Book Company, New York, 1978.

3 Sze, S. M.: "Physics of Semiconductor Devices," 2d ed., John Wiley & Sons, New York, 1979.

4 Adler, R. B., A. C. Smith, and R. L. Longini: "Introduction to Semiconductor Physics," vol. 1, SEEC, John Wiley & Sons, New York, 1965.

REVIEW QUESTIONS

1-1 Define electric field intensity.

1-2 Define potential energy.

1-3 Define an electron volt.

1-4 What is meant by the electron-gas description of a metal?

1-5 Define mobility.

1-6 Define conductivity.

1-7 Why does an intrinsic semiconductor behave as an insulator at 0 K?

1-8 What is a hole? How does it contribute to conduction?

1-9 (*a*) What is the intrinsic concentration of holes?
(*b*) What is the relationship between the density in Rev. 1-9*a* and the electron concentration?

1-10 What is the distinction between intrinsic and extrinsic semiconductors?

1-11 Show a two-dimensional representation of a silicon crystal containing a donor impurity atom.

1-12 Repeat Rev. 1-11 for an acceptor impurity atom.

1-13 What type of semiconductor results when silicon is doped with (*a*) donor and (*b*) acceptor impurities?

1-14 State the mass-action law.

1-15 A semiconductor has donor and acceptor concentrations of N_D and N_A, respectively. What relationships must be used to determine the electron n and hole p concentrations?

1-16 Describe recombination.

1-17 Define mean lifetime of a carrier.

1-18 Does the resistance of an extrinsic semiconductor increase or decrease with temperature? Explain briefly.

1-19 Repeat Rev. 1-18 for an intrinsic semiconductor.

1-20 Define the volt equivalent of temperature.

1-21 What condition(s) must exist for diffusion to occur?

1-22 Define the diffusion constant for (*a*) holes, and (*b*) electrons.

1-23 Are diffusion and drift related? If so, how?

1-24 Write an equation for the net electron current in a semiconductor and give the physical significance of each term.

1-25 Define a graded semiconductor.

1-26 Why must an electric field exist in a graded semiconductor?

1-27 On what parameters does the contact difference of potential in an open-circuited step-graded *pn* junction depend?

Chapter 2
THE *pn* JUNCTION DIODE

The *pn* junction is the basic building block on which the operation of all semiconductor devices depends. On the basis of the semiconductor properties described in Chap. 1, the behavior of the *pn* junction is developed. Attention is focused on the volt-ampere characteristics and the circuit models useful in representing junction operation. Because the *pn* junction is itself a two-element device (the diode), we also investigate its use as a circuit element.

2-1 THE OPEN-CIRCUITED JUNCTION

A *pn* junction is formed (Fig. 2-1) when a single crystal of semiconductor is doped with acceptors on one side and donors on the other. In Fig. 2-1, donor ions are represented by plus signs and the electrons they "donate" are indicated by the small, filled circles. Holes are depicted by unfilled, small circles and acceptor ions by minus signs. It is assumed that the junction in Fig. 2-1 has reached equilibrium conditions and that the semiconductor has uniform cross section.

FIGURE 2-1
A schematic representation of a *pn* junction.

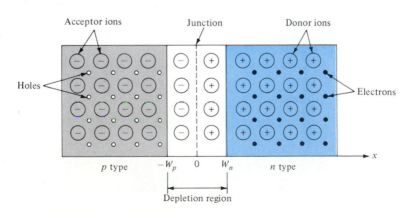

The Depletion Region Initially, a concentration gradient exists across the junction, causing holes to diffuse to the right and electrons to the left. We see that the holes which neutralized the acceptor ions near the junction in the *p*-type silicon have disappeared as a result of combination with electrons which have diffused across the junction. Similarly, electrons in the *n*-type silicon have combined with holes which have crossed the junction from the *p* material. The unneutralized ions in the neighborhood of the junction, referred to as *uncovered charges*, result in a charge density ρ_v as shown in Fig. 2-2a. Since this region is depleted of mobile charges it is called the *depletion region*, the *space-charge region*, or the *transition region*. The width of this region is in the order of a few tenths of a micrometer (about the wavelength of visible light). Carriers exist only outside the depletion region; to the left, they are predominantly holes ($p \simeq N_A$), and to the right, they are electrons ($n \simeq N_D$).

FIGURE 2-2

(*a*) Charge density; (*b*) electric field intensity; (*c*) electrostatic potential; (*d*) potential barrier for electrons in the depletion region of a *pn* junction.

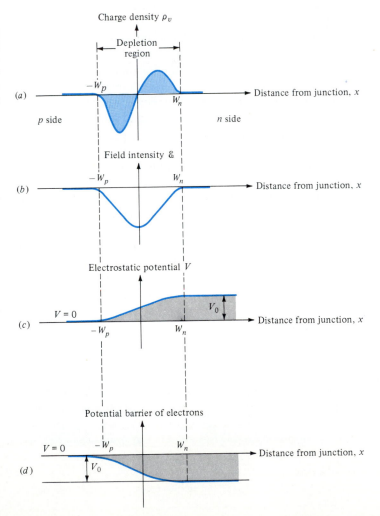

In Sec. 1-7 we demonstrated that an electric field and contact potential difference result from a nonuniform charge concentration. The charge distribution, which is zero at the junction, constitutes an electric dipole; that is, it is positive on one side and negative on the other side. The particular shape of ρ_v versus x depends on how the junction is graded. (A step-graded junction is examined in detail in Sec. 2-13.)

The electric field intensity and potential variations are obtained from the charge distribution and Poisson's equation:

$$\frac{d^2V}{dx^2} = -\frac{\rho_v}{\epsilon} \tag{2-1}$$

where ϵ is the permittivity (dielectric constant) of the medium. Usually ϵ is expressed as $\epsilon = \epsilon_r\epsilon_0$, where ϵ_r is the relative permittivity and ϵ_0 is the dielectric constant of free space. Recalling that $\mathcal{E} = -(dV/dx)$, we see that integration of Eq. (2-1) gives

$$\mathcal{E}(x) = \int_{-W_p}^{x} \frac{\rho_v(x')}{\epsilon} \, dx' \tag{2-2}$$

As displayed in Fig. 2-2b, $\mathcal{E}$ is negative because the field is directed from right (plus) to left (minus). Note that $\mathcal{E}(-W_p) = \mathcal{E}(W_n) = 0$; that is, it is assumed that no field exists outside the depletion region.

The electrostatic-potential variation in the depletion region is shown in Fig. 2-2c and is the negative integral of $\mathcal{E}(x)$ of Fig. 2-2b. This variation constitutes a potential-energy barrier (Sec. 1-1) against the further diffusion of holes across the barrier. The form of the potential-energy barrier against the flow of electrons from the n side across the junction is shown in Fig. 2-2d. It is similar to that shown in Fig. 2-2c, except that it is inverted, since the charge on an electron is negative. Note the existence, across the depletion layer, of the contact potential V_0, discussed in Sec. 1-7.

Under open-circuited conditions the net hole current must be zero. If this statement were not true, the hole density at one end of the semiconductor would continue to increase indefinitely with time, a situation which is obviously physically impossible. Since the concentration of holes in the p side is much greater than that in the n side, a very large hole diffusion current tends to cross the junction from the p to the n material. Hence an electric field must build up across the junction in such a direction that a hole drift current will tend to cross the junction from the n to the p side to counterbalance the diffusion current. This equilibrium condition of zero resultant hole current allows us to calculate the height of the potential barrier V_0 [Eq. (1-43)] in terms of the donor and acceptor concentrations. For doping densities usually encountered, the numerical value for V_0 is of the order of magnitude of several tenths of a volt.

The net electron current must also be zero; consequently, electron diffusion from n type to p type must be counteracted by electron drift from p type to n type.

FIGURE 2-3

(*a*) The forward-biased and (*b*) the reverse-biased *pn* junction.

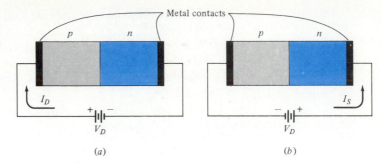

(*a*)

(*b*)

2-2 THE BIASED *pn* JUNCTION The essential electrical characteristic of a *pn* junction is that carrier flow is permitted in one direction and is virtually eliminated in the other. We now consider how this *unilateral*, or *rectifier*, action comes about by investigating the behavior of the junction to an externally applied voltage.

The Forward-Biased *pn* Junction In Fig. 2-3*a* the voltage V_D is applied to the junction with the positive terminal of the battery connected to the *p* side and the negative terminal to the *n* side. In accordance with Fig. 2-2, we assume that no voltage drop appears across the body of the semiconductor outside the depletion region or across the metal contacts. Consequently, the applied voltage reduces the potential barrier by an amount qV_D, thus disturbing the equilibrium established between the diffusion and drift of carriers across the junction. The result of decreasing the junction potential is to permit holes to diffuse from the *p* side to the *n* side of the junction. Similarly, electrons can now diffuse from the *n* to the *p* side. Holes traveling to the right and electrons to the left constitute a current in the same direction. Hence the resultant current crossing the junction is the sum of the hole and electron currents. Once they cross the junction, the electrons (holes) become minority carriers in the *p* (*n*) region and constitute an injected minority current. This diffusion current can be large as the number of carriers available is substantial. The applied voltage of the polarity indicated in Fig. 2-3*a* that produces this current is called *forward bias* and the junction is said to be *forward-biased*.

The Reverse-Biased *pn* Junction The polarity of the applied voltage in Fig. 2-3*b* (opposite to that shown in Fig. 2-3*a*) *reverse-biases* the junction. The effect of this voltage is to increase the potential barrier by qV_D and, as a result, decreases the flow of majority carriers (holes in *p* type and electrons in *n* type). However, minority carriers (electrons in *p* type and holes in *n* type), since they "fall down" the potential hill, are unaffected by the increase in barrier potential. The initial equilibrium conditions are, nevertheless, perturbed and a small current exists from *n* to *p* across the junction (opposite that forward bias). This current, designated by I_S and called the *reverse saturation current*, is very small as few minority carriers are available. From the preceding discussion we expect I_S to be independent of the magnitude of the reverse bias.

The conduction mechanism under reverse bias can be described alternatively as follows: the polarity of V_D is such that it causes both holes in the *p* type and electrons in the *n* type to move away from the junction. Consequently, the negative charge-density region extends further to the left of the junction (see Fig. 2-2a) and the region of positive charge density spreads to the right. This process cannot continue indefinitely because a steady hole flow to the left requires that holes be supplied across the junction from the *n*-type silicon. As there are only a few such carriers, the resultant current is virtually zero. The small saturation current that does exist results from thermally generated electron-hole pairs. Holes so formed in the *n*-type silicon "wander" over to the junction and are "pulled" across by the electric field. An analogous statement also applies to electrons thermally generated in the *p*-type material.

Ohmic Contacts In discussing forward and reverse bias we assumed that the external bias voltage V_D appeared directly across the junction. Thus V_D had the effect of raising or lowering the electrostatic potential across the junction. To justify this assumption, we must specify how electric contact is made to the semiconductor from the external bias circuit. In Fig. 2-3 we indicate metal contacts with which the homogeneous *p*- and *n*-type materials are provided. Thus we have introduced two metal-semiconductor junctions, one at each end of the diode, and expect a contact potential across these additional junctions. However, we shall assume that the metal-semiconductor contacts shown in Fig. 2-3 have been fabricated so that they are nonrectifying. In other words, the contact potential across these junctions is constant, independent of the direction and magnitude of the current. A contact of this type is referred to as an *ohmic contact*. Inasmuch as the voltage across the metal-semiconductor ohmic contacts remains constant and the voltage drop across the bulk of the crystal is neglected, approximately the entire applied voltage will indeed appear as a change in the height of the potential barrier at the *pn* junction.

The Short-Circuited and Open-Circuited *pn* Junction If the voltage V_D in Fig. 2-3 were set equal to zero, the *pn* junction would be short-circuited. Under these conditions, no current exists ($I = 0$) and the electrostatic potential V_0 remains unchanged and equal to the value under open-circuit conditions. If there were a current ($I \neq 0$), the metal would become heated. As no external source of energy is available, the energy required to heat the metal wire would have to be supplied by the *pn* bar. The bulk semiconductor, therefore, would have to cool off. Clearly, under thermal equilibrium, the simultaneous heating of the metal and cooling of the bar is impossible, and we conclude that $I = 0$. Because the sum of the voltages around the closed loop must be zero, the junction potential V_0 must be exactly compensated for by the metal semiconductor contact potentials at the ohmic contacts.

Large Forward Voltages Consider the situation when V_D in Fig. 2-3a is increased until it approaches V_0. When V_D equals V_0, the barrier disappears and the current tends to be arbitrarily large. As a practical matter, the barrier can never be reduced to zero because the bulk resistance of the crystal, as well as the resistance of the ohmic contacts, limits the current. Under this condition we can no longer assume that all of V_D appears across the junction. We conclude that, as V_D becomes comparable with V_0, the current in a real *pn* junction is governed by the ohmic-contact and semiconductor-bulk resistance.

2-3   THE VOLT-AMPERE CHARACTERISTIC   The approximately unilateral nature of the *pn* junction was developed in Sec. 2-2. We now wish to describe quantitatively the *volt-ampere characteristic* which relates the voltage applied to the junction to the current it produces. One significant feature of the volt-ampere characteristic is that action occurring in the vicinity of the junction is referred to accessible external terminal quantities. The *pn* junction and its ohmic contacts, that is, its terminals, form a two-element device called a *junction diode*.

Theoretical analysis of a *pn* junction[1] results in the relationship given in Eq. (2-3) and displayed in Fig. 2-4:

$$I_D = I_S(\epsilon^{V_D/\eta V_T} - 1) \qquad \text{A} \qquad (2\text{-}3)$$

The positive sense of I_D is from *p* type to *n* type (within the semiconductor), and V_D is positive for a forward-biased junction. One of the factors on which the parameter η, in Eq. (2-3), depends is the kind of semiconductor used; for silicon, η is approximately 2 at normal currents. The volt-equivalent of temperature V_T is given in Eq. (1-33) and restated in Eq. (2-4) for convenience:

$$V_T = \frac{T}{11,600} \qquad \text{V} \qquad (2\text{-}4)$$

At room temperature ($T = 293$ K), $V_T = 25$ mV.

The reverse saturation current I_S depends on the hole and electron concentrations and the junction area. Thus I_S serves as a "scale factor" for junction

---

[1]See Refs. 1 to 4 at the end of the chapter.

FIGURE 2-4
(*a*) Junction-diode characteristic and (*b*) the volt-ampere characteristic redrawn to show the order of magnitude of the currents and breakdown voltage.

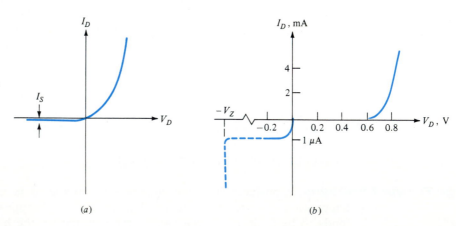

currents; for specified carrier densities, an increase in the area results in an increase in the current capacity of the junction.

Examination of Eq. (2-3) indicates that for forward bias, and for V_D several times V_T which makes $\epsilon^{V_D/\eta V_T} \gg 1$, I_D varies exponentially with applied voltage. For this case, Eq. (2-3) can be approximated by

$$I_D = I_S \, \epsilon^{V_D/\eta V_T} \quad \text{A} \qquad (2\text{-}5)$$

This result is expected as a decrease in the potential barrier permits carriers to diffuse more readily across the junction. Similarly, I_D is negative with magnitude I_S for reverse bias when V_D has a magnitude which is several times V_T. Both the negative sign, indicating current from n to p, and the constant current value for reverse bias are consistent with the discussion in Sec. 2-2.

Often, because forward and reverse currents differ by several orders of magnitude, two different current scales are used to display the junction characteristic as in Fig. 2-4b. The dashed portion of the reverse-biased characteristic (note the broken scale) indicates that at a voltage $-V_Z$ the junction exhibits an abrupt departure from Eq. (2-3). At this voltage a large reverse current may exist and the junction is in its *breakdown* region. This phenomenon is discussed in Sec. 2-11.

The forward characteristic of the 1N4153, a fast-switching silicon diode, is depicted in Fig. 2-5. A noteworthy feature of this characteristic is the existence of a *cut-in*, *offset*, or *turn-on* voltage V_γ below which the current is small (less than 1 percent of rated current). From Fig. 2-5, V_γ is approximately 0.6 V; beyond V_γ, the current increases rapidly. The significance of the offset voltage is that the diode characteristic can be approximated as having negligible current for applied voltages less than V_γ.

The parameter η can be determined from the exponential nature of the volt-ampere characteristic. From Eq. (2-5), we have

$$\log I_D = \log I_S + \frac{0.43 V_D}{\eta V_T} \qquad (2\text{-}6)$$

FIGURE 2-5

The forward volt-ampere characteristic of an IN4153 silicon diode at 25°C.

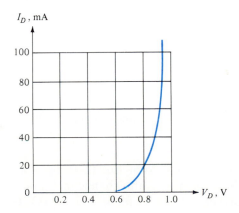

FIGURE 2-6

Logarithmic characteristic of a 1N4153 silicon diode at 25°C.

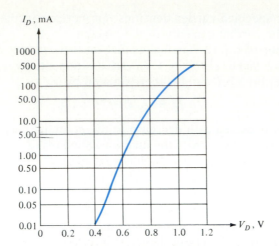

A plot of log I_D versus V_D results in a straight line of slope $0.434/\eta V_T$ from which η is obtained. In practice, this linear relation is observed at low current levels. For the logarithmic characteristic of the IN4153 at $T = 25$°C, shown in Fig. 2-6, the linear range occurs for $I_D < 25$ mA. From the slope, η is approximately 2. At higher current levels, the slope decreases because the total applied voltage does not appear across the junction but also includes the ohmic voltage drop in the contacts and bulk semiconductor. Furthermore, it is found that η approaches unity and $I_D = I_S \epsilon^{V_D/V_T}$ for high current levels.

Example 2-1

Determine the change in diode voltage corresponding to a 10:1 change in I_D for a diode operating at 300 K.

Solution

From Eq. (2-5), we obtain

$$I_{D1} = I_S \epsilon^{V_{D1}/\eta V_T} \quad \text{and} \quad I_{D2} = I_S \epsilon^{V_{D2}/\eta V_T}$$

Thus

$$\frac{I_{D2}}{I_{D1}} = \epsilon^{(V_{D2} - V_{D1})/\eta V_T}$$

from which

$$\log \frac{I_{D2}}{I_{D1}} = \frac{0.434(V_{D2} - V_{D1})}{\eta V_T} \quad \text{and} \quad V_{D2} - V_{D1} = 2.303 \eta V_T \log \frac{I_{D2}}{I_{D1}}$$

At $T = 300$ K, $V_T = 26$ mV from Eq. (2-4) and for $I_{D2}/I_{D1} = 10$, $V_{D2} - V_{D1} = 60\eta$ mV. Thus, for $\eta = 2$, the change in V_D needed to produce a 10:1 change in diode current is 120 mV; for $\eta = 1$, 60 mV is required.

2.4 TEMPERATURE DEPENDENCE OF THE V-I CHARACTERISTIC

The diode characteristic in Eq. (2-3) has two terms, V_T and I_S, which are heavily dependent on temperature. Equation (2-4) expresses the functional relationship between V_T and temperature. Theoretical analysis of a silicon junction indicates that I_S changes by 8 percent/°C. Practical diodes only approximate this result because a component of the reverse saturation current is due to surface leakage. Experimental data shows that the change in I_S is 7 percent/°C, and, since $(1.07)^{10} \simeq 2$, we conclude that *the reverse saturation current doubles for every 10°C rise in temperature.*

By knowing I_S at temperature T_1, we can compute I_S at any temperature T from

$$I_S(T) = I_S(T_1) \times 2^{(T - T_1)/10} \qquad \text{A} \qquad (2\text{-}7)$$

With a fixed applied voltage, an increase in temperature causes an increase in I_S. As T increases, however, the voltage can be reduced to maintain the original value of current. For temperatures in the vicinity of room temperature, it is found that

$$\frac{dV_D}{dT} \simeq -2.2 \text{ mV/°C} \qquad (2\text{-}8)$$

to maintain a constant value of I. Of note is that dV_D/dT decreases with increasing temperature.

2-5 GERMANIUM DIODES

Junction diodes fabricated from germanium are commercially available and employed in circuit applications. The basis for their operation is the same as that described for silicon diodes and the volt-ampere characteristics as given in Eq. (2-3). Two differences exist: (1) $\eta = 1$, and (2) the value of I_S for a germanium diode is three to four orders of magnitude greater than for a silicon device of the same dimensions and doping densities. Another distinguishing feature of the volt-ampere characteristic for germanium is that the cut-in voltage $V_\gamma \simeq 0.2$ V.

2-6 THE DIODE AS A CIRCUIT ELEMENT

In this section we begin to examine the circuit properties of diodes. As a point of departure, the characteristics of an ideal diode are described.

The Ideal Diode An ideal diode is a two-element device which has the circuit symbol and volt-ampere characteristic shown in Fig. 2-7.[1] Inspection of the curve indicates that current in the device is in one direction only so that the ideal diode is a unilateral circuit element. This unilateral behavior is important in switching as it provides

---

[1] The use of lowercase variables indicates that v_D and i_D are functions of time. Uppercase variables are used to indicate constant values. This convention is used throughout the text.

FIGURE 2-7
(a) Circuit symbol and
(b) volt-ampere char-
acteristic of an ideal
diode.

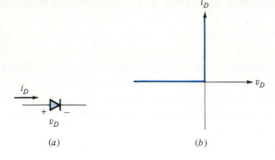

an ON-OFF characteristic. Note that when v_D is zero, i_D can have any positive value and when i_D is zero, v_D can have any negative value, a condition corresponding to a switch. This property of the diode is also widely used in rectification and waveshaping as only appropriate signal polarities may be transmitted and processed.

The *pn* junction described in the previous sections of this chapter only approximates the ideal. Comparison of Figs. 2-4a and 2-7b indicates that "real" diodes have a small, but nonzero, reverse current and that a voltage drop exists for forward bias. Also, the nonlinear characteristic of a *pn* junction often requires graphical methods of analysis of the type described in the remainder of this section.

The Load-Line Concept   The circuit shown in Fig. 2-8a contains a practical *pn* diode; its volt-ampere characteristic is depicted in Fig. 2-8b. While the symbol for a physical diode is the same as that for an ideal diode, in this text we distinguish them as shown in Figs. 2-7a and 2-8a.

For the circuit in Fig. 2-8a, Kirchhoff's voltage law (KVL)[1] requires that

---

[1] A summary of the circuit theory used in the analysis of the electronic circuits in this book is given in App. C.

FIGURE 2-8
(a) Diode circuit and (b)
diode characteristic and
load line for circuit.

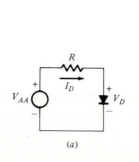

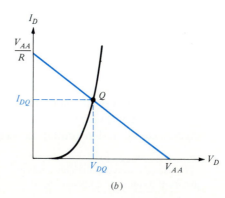

$$-V_{AA} + I_D R + V_D = 0$$

or, solving for I_D

$$I_D = -\frac{1}{R} V_D + \frac{V_{AA}}{R} \qquad \text{A} \qquad (2\text{-}9)$$

Equation (2-9) defines a straight line, called the *load line*, which is also indicated in Fig. 2-8*b*. Note that the slope and intercepts of the load line depend only on R and V_{AA}. Both Eq. (2-9) and the diode characteristic must be satisfied simultaneously; the point Q at their intersection, called the *quiescent* or *operating point*, is the only such point. The values of current in and voltage across the diode are denoted by I_{DQ} and V_{DQ}, respectively. The subscript Q is used to indicate quiescence values that exist in the circuit.

The concept of the load line has wider applicability than that indicated for the diode. Consider any device in series with a resistor R and supply V_{AA} as illustrated in Fig. 2-9. Applying KVL and solving for i_x, we obtain

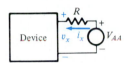

FIGURE 2-9
A device supplied by a source through a series resistor.

$$i_x = -\frac{1}{R} v_x + \frac{V_{AA}}{R} \qquad \text{A} \qquad (2\text{-}10)$$

This is the equation of the load line; its intersection with the volt-ampere characteristic of the box labeled Device determines the operating values of the voltages and current in the circuit. Note that the load line passes through the points $i_x = 0$, $v_x = V_{AA}$ and $i_x = V_{AA}/R$, $v_x = 0$ *independently of the device characteristics* and has a slope equal to $-1/R$.

It is instructive to investigate the changes that occur in the operating point for the circuit of Fig. 2-8*a* as V_{AA} and R vary. These changes are displayed in Fig. 2-10*a* for the case where R is constant and V_{AA} varies; Fig. 2-10*b* indicates the results when R varies with V_{AA} held constant. From Fig. 2-10*a* we observe that as V_{AA} increases (decreases), I_{DQ} also increases (decreases). Note that for

FIGURE 2-10
Changes in operating point when (*a*) V_{AA} varies and (*b*) R varies.

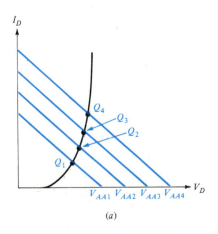

(*a*)

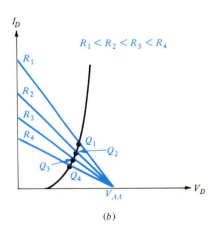

(*b*)

small changes in V_{AA}, the portion of the diode characteristic between adjacent Q points is approximately linear. However, for large changes in V_{AA}, as from V_{AA1} to V_{AA4}, the portion of the diode curve is not linear. It is evident in Fig. 2-10b that an increase in R causes I_{DQ} to decrease.

2-7 LARGE-SIGNAL DIODE MODELS
It is convenient to represent the diode by a combination of ideal, linear circuit elements called an *equivalent circuit* or *circuit model*. As the diode is used with other circuit elements or devices, the model allows us to evaluate the currents and voltages in the network using standard circuit analysis methods.

The ideal diode (Fig. 2-7) is a *binary* device in the sense that it exists in only one of two possible states; that is, the diode is either ON or OFF at a given time. Consider a real diode such as that whose characteristic is shown in Fig. 2-11. If the applied voltage across this diode exceeds the cut-in voltage V_γ, with the anode A (the p side) more positive than the cathode K (the n side), the diode is forward-biased and is in the ON state. The OFF state exists when the applied voltage is less than V_γ and, in effect, reverse-biases the diode.

As shown in Fig. 2-11a, the two line segments approximate the forward characteristic of the diode. This piecewise representation is modeled by a voltage source V_γ in series with a resistance R_f (usually 5 to 50 Ω for silicon diodes) as depicted in Fig. 2-11b. This piecewise linear characteristic has value because for $v_D < V_\gamma$, the forward current is sufficiently small that it can be neglected. Furthermore, the diode voltage drop is generally small in comparison with the applied voltages in the circuit so that the difference between the straight line and the actual characteristic introduces negligible error. In effect, the ON state can be regarded as an ideal diode in series with a battery V_γ and a resistor R_f. It is important to note, however, that the only accessible terminals for measurement are A and K.

For the OFF state, the diode characteristic is approximated by the straight line passing through the origin depicted in Fig. 2-12a, the slope of which is $1/R_r$. This representation gives rise to the equivalent circuit in Fig. 2-12b. As

FIGURE 2-11
(a) Piecewise linear diode forward characteristic; (b) diode model for forward bias.

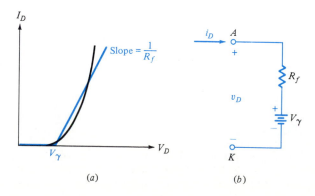

(a) (b)

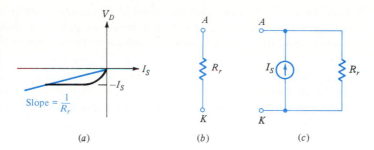

FIGURE 2-12
(*a*) Piecewise linear reverse characteristic of diode; (*b*) diode model based on piecewise linear representation; (*c*) model to include surface leakage.

R_r is generally several hundreds of kilohms or more, we can often assume that it is infinite and consider the reverse-biased diode as an open circuit. Where higher degrees of accuracy are required, the model shown in Fig. 2-12*c* is useful. The current source I_S is used to indicate the constant reverse saturation current. The resistance R_r in Fig. 2-12*c* may also take into account the increase in reverse current with increasing reverse voltage caused by surface leakage.

Analysis of Diode Circuits Using the Large-Signal Model   A general method of analysis of a circuit containing several diodes, resistors, and sources consists in assuming (guessing) the state of each diode.[1] For the ON state, the diode is replaced by the circuit in Fig. 2-11*b* and the model in Fig. 2-12*b* is used for all diodes in the OFF state. Once the diodes have been replaced by their equivalent circuits, the entire network is linear and all voltages and currents can be calculated by using Kirchhoff's laws. The assumption that a diode is ON is verified by observing whether the current in it is in the forward direction. If the current is from anode to cathode, the initial guess is justified. If the current is in the reverse direction (from cathode to anode), however, the assumption that the diode is ON is incorrect and the analysis must resume with the diode assumed to be OFF.

In an analogous fashion, we test the assumption that the diode is OFF by finding the voltage across it. If this voltage is either in the reverse direction or less than V_γ in the forward direction, the diode is OFF. If this voltage is greater than V_γ in the forward direction, our assumption is incorrect and this diode must be assumed ON when the analysis resumes.

The method of analysis outlined in the preceding two paragraphs will be employed in the following example and throughout the text.

Example 2-2   Determine the output voltage v_o in the circuit in Fig. 2-13*a* for the following values of input voltages: (*a*) $v_1 = v_2 = 5$ V; (*b*) $v_1 = 5$ V, $v_2 = 0$; and (*c*) $v_1 = v_2 = 0$. A silicon diode is used and has $R_f = 30\ \Omega$, $V_\gamma = 0.6$ V, $I_S = 0$, and $R_r \to \infty$.

---

[1] You should not be taken aback by this statement; with experience, these are "educated" guesses. If your guess is incorrect, the analysis described above will so indicate.

Solution

It is important to note that the implied reference node (ground) is not shown in Fig. 2-13*a*. All indicated voltages are measured with respect to the reference node with voltage drops considered positive. The circuit in Fig. 2-13*a* is redrawn in Fig. 2-13*b* for which connections to the reference node are included.

(*a*) With $v_1 = v_2 = 5$ V, we assume that *D*1 and *D*2 are OFF. Replacement of the diodes by the model of Fig. 2-12*b*, with $R_r \to \infty$, results in the circuit shown in Fig. 2-14*a*. Inspection of this circuit makes it evident that no current exists. Consequently, there is a zero voltage drop across any resistance and, from KVL, $V_{D1} = V_{D2} = 0 (< V_\gamma)$, verifying our assumption. Thus Fig. 2-12*b* is a valid representation for this situation and we see that $v_o = 5$ V.

FIGURE 2-13
(*a*) Circuit for Example 2-2; (*b*) alternative schematic for circuit in part *a*.

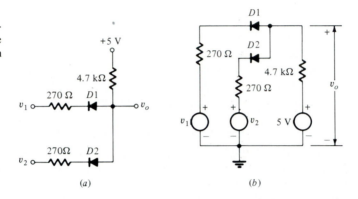

(*a*)

(*b*)

FIGURE 2-14
Equivalent circuits for Example 2-2; (*a*) both diodes OFF; (*b*) diode *D*1 OFF and diode *D*2 ON; (*c*) both diodes ON.

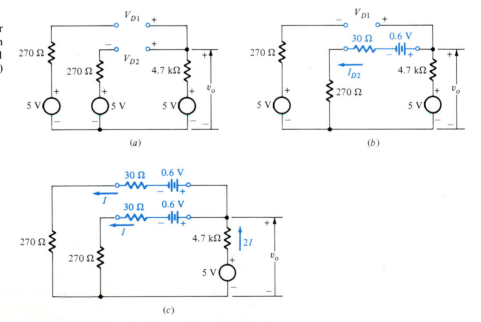

(*a*)

(*b*)

(*c*)

(b) We assume that D1 is OFF and D2 is ON with $v_1 = 5$ V and $v_2 = 0$. Using the models in Figs. 2-11b and 2-12b, we obtain the circuit in Fig. 2-14b. The KVL expression for the inner loop is

$$-5 + 4700I_{D2} + 0.6 + 30I_{D2} + 270I_{D2} = 0$$

Solving for I_{D2}, we obtain

$$I_{D2} = \frac{5 - 0.6}{4700 + 30 + 270} = 0.88 \text{ mA}$$

As I_{D2} is positive (in the forward direction), our assumption that D2 is ON is justified. No current exists in the outer loop, so that

$$v_o = 5 - 4700I_{D2} = 5 - 4700 \times 0.88 \times 10^{-3} = 0.864 \text{ V}$$

Alternatively, v_o may be calculated from

$$v_o = 0.6 + 300I_{D2} = 0.6 + 300 \times 0.88 \times 10^{-3} = 0.864 \text{ V}$$

No current exists in the outer loop, so that

$$V_{D1} = v_o - 5 = 0.864 - 5 = -4.136 \text{ V}$$

The negative value of V_{D1} confirms our guess that D1 is OFF. Thus the circuit in Fig. 2-14b represents the conditions in the circuit and the computed value of $v_o = 0.864$ V is the output voltage. If we had assumed D2 to be OFF instead of ON, I_{D2} would be zero. With no current in either diode, $v_o = 5$ V (as in part a above), thus making $V_{D2} = 5$ V. As this is greater than $V_\gamma = 0.6$ V, our guess is erroneous. Similarly, if D1 is considered ON with D2 OFF, the KVL expression for the outer loop becomes

$$-5 + 4700I_{D1} + 0.6 + 30I_{D1} + 270I_{D1} + 5 = 0$$

Evaluation yields a negative value of I_{D1}, and our initial assumption is wrong.

(c) The equivalent circuit of Fig. 2-14c applies when $v_1 = v_2 = 0$, assuming that both D1 and D2 are ON. Because of symmetry, the same current I exists in each diode; KCL requires a current of $2I$ be supplied to these branches. For the inner loop, the KVL expression is

$$-5 + 4700 \times 2I + 0.6 + (30 + 270)I = 0$$

from which $I = 0.454$ mA.

The positive value indicates our assumption to be correct. Then

$$v_o = 0.6 + (30 + 270)I = 0.6 + 300 \times 4.54 \times 10^{-4} = 0.736 \text{ V}$$

We observe in Example 2-2 that disparate values of v_o exist depending on the state of the diodes. When both inputs are "high" (i.e., 5 V), the output is also high. The output is "low" when either (or both) input is also low. Circuits with this type of behavior are referred to as AND gates and are discussed in detail in Chap. 6.

FIGURE 2-15
(*a*) Half-wave rectifier
and (*b*) current and voltage waveforms.

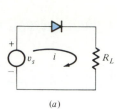

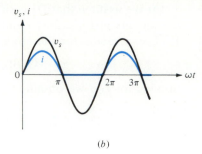

(*a*)   (*b*)

2-8 ELEMENTARY DIODE APPLICATIONS   Several classes of circuits exploit the ON-OFF behavior of diodes to markedly alter electrical waveforms. We now introduce the basis of such circuits; specific applications are treated in subsequent chapters.

Rectifiers   Consider the circuit in Fig. 2-15*a* in which the series combination of an ideal diode and load resistance is excited by a sinusoidal voltage. During the first half-cycle of the input the diode is ON and a current v_s/R_L exists. The diode is OFF during the negative half-cycle of v_s so that the current is zero (see Fig. 2-15). As current exists for only one-half of the cycle, the circuit in Fig. 2-15*a* is called a *half-wave rectifier*. It is significant that the average value (the dc component) of current over one period is not zero whereas the average voltage over one cycle is zero. This factor is the basis of the rectifier circuits used to convert the generally available alternating current to the direct current required by most electronic systems.

    The circuit in Fig. 2-16*a* uses the capacitor C as a simple filter to convert the waveform in Fig. 2-15*b* to the nearly constant (dc) level displayed in Fig. 2-16*b*. The following is a qualitative description of the effect of the capacitor on the circuit response. In the steady state at $t = t_1$, the voltage across the capacitor is V_1 and at that time the input voltage equals V_1, turning on the diode. Subsequent to t_1, the capacitor voltage, that is, the output voltage v_o,

FIGURE 2-16
(*a*) Rectifier with capacitor filter; (*b*) output
voltage of circuit in
part *a*.

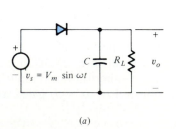

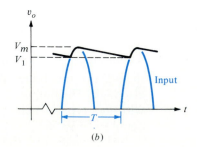

(*a*)   (*b*)

FIGURE 2-17
(a) Rectifier equivalent
circuit; (b) current waveform showing ignition
and extinction angles.

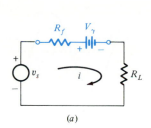

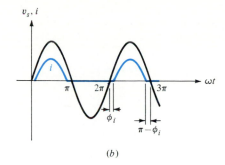

(a)   (b)

follows the input voltage until t_2 for which v_o is V_m. After t_2, the input voltage
decreases at a faster rate than the discharge of the capacitor, turning off the
diode. This restricts the discharge of C to be through R_L. If the time constant
$R_L C$ is much greater than the period T of the input waveform, the discharge is
slow. Thus only a small decrease in v_o occurs between t_2 and t_3, at which time
the process is repeated. In practical rectifier-filter circuits (Sec. 17-4) the variation (ripple) in the output waveform is considerably smaller than that shown
in Fig. 2-16b and v_o is virtually constant.

Use of a real diode in Fig. 2-15a gives rise to the equivalent circuit in Fig.
2-17a, which is valid for forward bias. The current i is obtained from KVL as

$$i = \frac{v_s - V_\gamma}{R_L} = \frac{V_m \sin \omega t - V_\gamma}{R_L} \qquad \text{A} \tag{2-11}$$

and is greater than zero only when $v_s > V_\gamma$. Thus the current waveform, depicted
in Fig. 2-17b, does not start when $\omega t = 0$ but has a *cut-in* or *ignition angle* ϕ_i
given by

$$\phi_i = \sin^{-1} \frac{V_\gamma}{V_m} \tag{2-12}$$

Similarly, an *extinction angle* exists at the end of the positive half-cycle; its
value is $\pi - \phi_i$.

This circuit may be used as a battery charger (for calculators) by replacing
R_L in Fig. 2-15a by V_{BB}, the battery voltage, in series with a current-limiting
resistance R_s. In this case, the ignition angle is given in Eq. (2-12) with V_γ
replaced by $V_\gamma + V_{BB}$.

Clipping and Clamping Circuits   Clipping circuits are used to select for transmission that part
of a waveform which lies above or below some reference level. In this sense,
the rectifier in Fig. 2-15a is a clipping circuit as only input voltage levels above
V_γ are transmitted to the output. Most often, the diodes employed are biased
by a reference voltage which determines the portion of the signal to be transmitted.

The circuit in Fig. 2-18a is a simple clipping circuit. Assuming that the diode
is ideal, we see that v_o equals V_R when the diode D is ON and v_o is v_i when the

FIGURE 2-18
(*a*) Diode clipping circuit and (*b*) its output waveform.

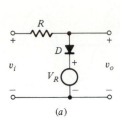

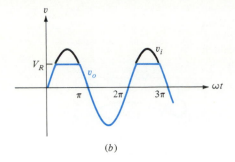

(*a*)   (*b*)

diode is OFF. The transition from OFF to ON occurs at the time the input voltage reaches the reference voltage V_R as displayed by the waveforms in Fig. 2-18*b*.

The equivalent circuits in Fig. 2-19 characterize the clipper of Fig. 2-18*a* when a practical diode is employed. In Fig. 2-19*b* it is evident that v_o is v_i when D is OFF. From KVL, the current i is obtained as

$$i = \frac{v_i - V_R - V_\gamma}{R_f + R} \quad \text{A} \tag{2-13}$$

from which

$$v_o = iR_f + V_\gamma + V_R = \frac{R_f}{R_f + R} v_i + \frac{R}{R_f + R}(V_R + V_\gamma) \quad \text{V} \tag{2-14}$$

As forward bias requires $i > 0$, we observe in Eq. (2-13) that the transition from OFF to ON occurs when v_i equals $V_\gamma + V_R$. This transition, or breakpoint, indicates the abrupt change in slope in the plot of v_o versus v_i, called the *transfer characteristic*, shown in Fig. 2-20. The slope in Fig. 2-20 is unity when D is OFF and shows that $v_o = v_i$. For the ON state, the slope is given by the coefficient of v_i in Eq. (2-14). The waveforms in Fig. 2-20 illustrate how the transfer characteristic is used to determine the output voltage from the input signal.

Note that the transfer characteristic in Fig. 2-20 is derived from the piecewise linear approximation to the diode characteristic which assumes an abrupt tran-

FIGURE 2-19
Models for circuit in Fig. 2-18*a* for (*a*) forward bias and (*b*) reverse bias.

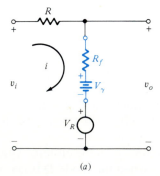

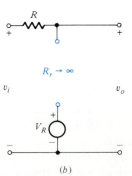

(*a*)   (*b*)

FIGURE 2-20
Transfer characteristic of clipping circuit showing input waveform and resultant output waveform.

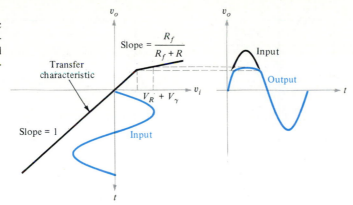

sition from reverse to forward bias. Actually, this transition is not abrupt but is gradual. Consequently, the breakpoint is not sharp but occurs over a small voltage range, usually 0.1 or 0.2 V. The small break region causes the output waveform to differ somewhat from that displayed in Fig. 2-20. Fortunately, in many applications v_i is significantly larger than the few tenths of a volt break region, so that the piecewise linear representation is valid.

Example 2-3

For the circuit in Fig. 2-21a, for which the input voltage is the sawtooth waveform displayed in Fig. 2-21b, sketch (a) the transfer characteristic v_o versus v_s and (b) the output waveform v_o. The diode parameters are $R_f = 10\ \Omega$, $V_\gamma = 0.6$ V, and $I_S = 0$.

Solution

The equivalent circuits for forward and reverse bias are given in Figs. 2-22a and 2-22b. For reverse bias, the output relation is $v_o = v_s$ as no current exists. In Fig. 2-22a, the output is expressable as

$$v_o = V_R - V_\gamma - iR_f = 6 - 0.6 - 10i = 5.4 - 10i \quad \text{V}$$

The current, obtained by solving the KVL equation for the loop, is

$$i = \frac{V_R - V_\gamma - v_s}{R_f + R} = \frac{6 - 0.6 - v_s}{10 + 1000} = \frac{5.4 - v_s}{1010} \quad \text{A}$$

FIGURE 2-21
(a) Circuit for Example 2-3; (b) input waveform for Example 2-3.

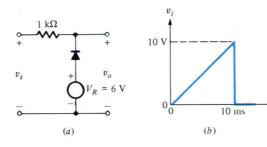

FIGURE 2-22
(*a*) Forward-bias and (*b*) reverse-bias models for Example 2-3; (*c*) transfer characteristic of circuit in Fig. 2-21*a*; (*d*) output waveform for the input in Fig. 2-21*b*.

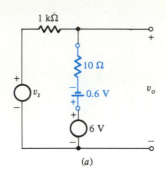

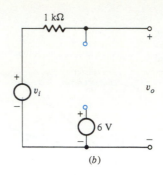

(*a*)   (*b*)

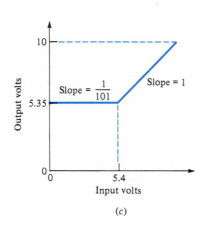

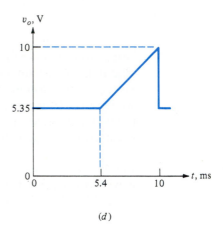

(*c*)   (*d*)

The breakpoint is derived from the current relation and is the value of v_s at which i is zero. Thus

$$v_s = V_R - V_\gamma = 6 - 0.6 = 5.4 \text{ V}$$

The diode is ON for $v_s < 5.4$ V and OFF for $v_s > 5.4$ V. Substitution for i in the equation for v_o yields

$$v_o = \frac{R_f}{R_f + R} v_s + \frac{R(V_R - V_\gamma)}{R_f + R}$$

$$= \frac{10 v_s}{10 + 1000} + \frac{1000(6 - 0.6)}{10 + 1000} = \frac{v_s}{101} + 5.35 \quad \text{V}$$

The transfer characteristic is depicted in Fig. 2-22*c*; the output waveform, derived from the input waveform and transfer characteristic, is illustrated in Fig. 2-22*d*.

Three features of the output waveform can be discerned. First, the circuit in Fig. 2-21*a* is that shown in Fig. 2-18*a* with the diode connection reversed. As a result, the circuit in this example transmits input voltages above a given

FIGURE 2-23
(*a*) Two-level clipping circuit; (*b*) circuit illustrating use of clamping (catching) diodes.

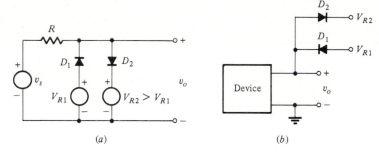

(*a*)   (*b*)

level (approximately $V_R - V_\gamma$), whereas the portion of the input signal above $V_R + V_\gamma$ is clipped in Fig. 2-18*a*. The second feature shows that for $R \gg R_f$, as is the case here, the output voltage is essentially constant at $V_R - V_\gamma = 5.4$ V when the diode is ON. The actual difference is about 1 percent, which corresponds to the ratio R_f/R. As a result, it is often convenient to simply model the diode by its cut-in voltage and neglect the effect of R_f. The third observation is that the output voltage and transfer characteristic have the same shape. Consequently, we can use this technique in the laboratory to obtain the transfer characteristic by applying a sawtooth input and displaying the output waveform on an oscilloscope.[1]

Combination of the properties of the clipping circuits in Figs. 2-18*a* and 2-21*a* results in the two-level clipper illustrated in Fig. 2-23*a*. If v_s and R represent the Thévenin equivalent at the output terminals, Fig. 2-23*b* is identical to Fig. 2-23*a*. For these circuits, the output voltage is restricted to be between V_{R1} and V_{R2} (approximately). The analysis of this circuit is part of Prob. 2-26. Because $D1$ and $D2$ in Fig. 2-23*b* prevent the output from exceeding V_{R2} or dropping below V_{R1}, they are often called *clamping* or *catching* diodes.

2-9 SMALL-SIGNAL DIODE MODELS   The circuits described in the previous section utilized the ON-OFF behavior of diodes. In these applications the applied signal (usually time varying) is large in comparison to the bias level (the constant reference voltage) and the models in Figs. 2-11 and 2-12 are used to describe the diode. We now consider the situation where the signal amplitude is small compared to the bias. It is convenient to use *small-signal* or *incremental* equivalent circuits to represent the diode in order to enable us to relate the response component due to the applied (excitation) signal $v_s(t)$. The circuit in Fig. 2-24*a* is useful in developing the small-signal models.

In the circuit of Fig. 2-24*a*, $V_m < V_{AA}$, so that the diode remains forward-biased at all times. The instantaneous value of the voltage $v(t)$ applied to the diode-resistance combination is

$$v(t) = V_{AA} + v_s(t) = V_{AA} + V_m \sin \omega t \qquad \text{V} \qquad (2\text{-}15)$$

---

[1]This method yields only approximate results for most circuits because of capacitive effects and slight offset voltages that exist.

FIGURE 2-24
(*a*) Diode circuit with constant and sinusoidal excitation; (*b*) variation in load line and input and output waveforms for current in part *a*.

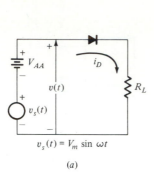

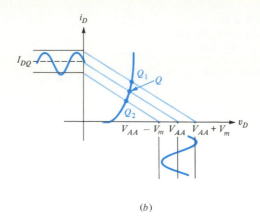

(*a*)

(*b*)

At each instant of time, we can draw a load line (Fig. 2-24*b*) in the manner described in Sec. 2-5. Maximum and minimum values of $v(t)$ are $V_{AA} + V_m$ and $V_{AA} - V_{am}$, respectively, and, for $\omega t = n\pi$ where n is an integer, $v = V_{AA}$. As shown in Fig. 2-24*b*, the current i_D is composed of a sinusoidal component superimposed on the quiescent level I_{DQ} and is expressed as

$$i_D = I_{DQ} + i_d(t) = I_{DQ} + I_d \sin \omega t \qquad \text{A} \qquad (2\text{-}16)$$

In Eq. (2-16), i_D is the instantaneous value of diode current and I_{DQ} the dc component of i_D and i_d is the time-varying component of i_D whose peak value is given by I_d.* The form of the current expressed in Eq. (2-16) results from the fact that the diode characteristic between Q_1 and Q_2 can be approximated by a straight line whose slope is that of the diode volt-ampere relation evaluated at Q. In this region, therefore, the diode behaves linearly and, in effect, superposition applies. That is, the quiescent (dc) value I_{DQ} is established by the constant bias supply V_{AA} and the sinusoidal component $i_d(t)$ is produced by the excitation $v_s(t)$.

The time-varying components of the voltages and currents in the circuit in Fig. 2-24*a* can be determined analytically (instead of graphically as in Fig. 2-24*b*) by applying Kirchhoff's laws to the small-signal equivalent circuit shown in Fig. 2-25. Here the diode is replaced by its *incremental resistance* $r_d \equiv 1/g_d$, where the *incremental conductance* g_d is given by

$$g_d \equiv \left. \frac{di_D}{dv_D} \right|_Q \qquad \text{℧} \qquad (2\text{-}17)$$

---

*The use of upper- and lowercase variables and subscripts in Eq. (2-16) illustrates the standard manner by which current and voltage components in electronic circuits are described. Repeated subscripts, such as V_{AA}, are used to indicate supply voltages. We use this notation throughout this text.

FIGURE 2-25
Small-signal equivalent
circuit of Fig. 2-24a.

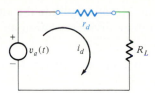

Note that g_d is simply the slope of the diode characteristic evaluated at the operating point Q, and consequently the value of r_d is a function of the quiescent current. To make use of the circuit in Fig. 2-25, we must *first* establish the quiescent values of diode current and voltage.

For a junction diode, Eq. (2-17) becomes, using Eq. (2-3)

$$g_d = \frac{I_S \, \epsilon^{V_{DQ}/\eta V_T}}{\eta V_T} = \frac{I_{DQ} + I_S}{\eta V_T} \quad \text{℧} \tag{2-18}$$

Most often, $I_{DQ} \gg I_S$, so that Eq. (2-18) reduces to

$$r_d = \frac{1}{g_d} \simeq \frac{\eta V_T}{I_{DQ}} \quad \Omega \tag{2-19}$$

and we observe that the incremental resistance varies inversely with current. At $T = 20°C$, $V_T = 25 \, \text{mV}$; therefore, $r_d = 25\eta/I_{DQ}$, where I_{DQ} is in milliamperes and r_d in ohms. For a silicon diode ($\eta = 2$) and $I_{DQ} = 5 \, \text{mA}$, $r_d = 10 \, \Omega$.

Example 2-4

The circuit in Fig. 2-24a is used at 20°C with $V_{AA} = 9 \, \text{V}$, $V_m = 0.2 \, \text{V}$, and $R_L = 2 \, \text{k}\Omega$. In the large-signal model of the diode, $V_\gamma = 0.6 \, \text{V}$, $R_f = 10 \, \Omega$, and $\eta = 2$. Determine (*a*) the alternating component of the voltage across R_L and (*b*) the total voltage across R_L.

Solution

(*a*) First, the bias levels must be determined from the dc model in Fig. 2-26a. From the KVL expression for the loop, I_{DQ} is determined as

$$I_{DQ} = \frac{9 - 0.6}{2000 + 10} = 4.18 \, \text{mA}$$

The incremental resistance, from Eq. (2-19), is

$$r_d = \frac{2 \times 25}{4.18} = 12.0 \, \Omega$$

FIGURE 2-26
(*a*) Dc and (*b*) incremental models for Example 2-4.

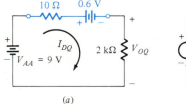

(a)

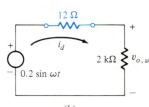

(b)

Using the small-signal model in Fig. 2-26*b*, we obtain the alternating component of the output voltage from the voltage-divider relation as

$$v_{o,\text{ac}} = \frac{2000}{2000 + 12} \times 0.2 \sin \omega t = 0.199 \sin \omega t \qquad \text{V}$$

(*b*) The total load voltage $v_o(t)$ is the sum of the quiescent and time-varying components. The quiescent voltage across R_L is

$$V_{OQ} = I_{DQ}R_L = 4.18 \times 10^{-3} \times 2 \times 10^3 = 8.36 \text{ V}$$

Thus

$$v_o(t) = 8.36 + 0.199 \sin \omega t \qquad \text{V}$$

The waveforms in Fig. 2-27 are what we would observe if the output were displayed on an oscilloscope. In Fig. 2-27*a*, with the selector knob set on dc, the oscilloscope displays v_o, the total instantaneous output voltage. We observe that the alternating component is almost indistinguishable. However, in Fig. 2-27*b* with the selector knob set to ac and the sensitivity increased, the alternating component is conveniently displayed for measurement. In effect, the small-signal model performs an analogous function to switching the selector knob from dc to ac. By removing the quiescent (dc) level, we can focus on the effect that the time-varying input has on the time-varying response.

FIGURE 2-27
Oscilloscope displays of output voltage in Example 2-4: (*a*) selector knob set to dc and (*b*) selector knob set to ac. The dashed line indicates 0 V.

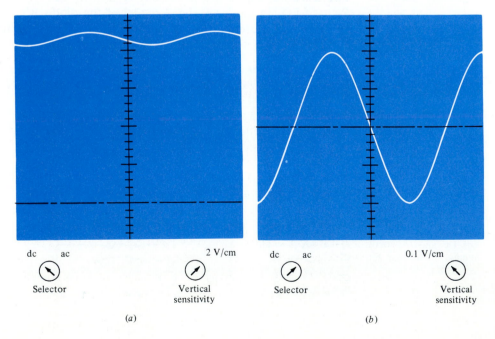

dc   ac   2 V/cm   dc   ac   0.1 V/cm

Selector   Vertical   Selector   Vertical
                      sensitivity   sensitivity

(*a*)   (*b*)

FIGURE 2-28
Small-signal diode models for (a) forward bias and (b) reverse bias.

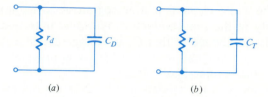

(a)   (b)

Diffusion Capacitance   To obtain the results in Example 2-4, we assumed that the frequency of the sinusoidal excitation was sufficiently low that charge storage in the diode had negligible effect. At higher excitation frequencies charge-storage effects are accounted for by means of the *diffusion capacitance* C_D in the small-signal diode model shown in Fig. 2-28a. The origin of C_D can be qualitatively described as follows. For a forward-biased junction, holes diffuse from the p side to the n side. Consequently, in the vicinity of the junction on the n side, we have a greater hole concentration than normally exists because of the diffusion. This "excess" hole density can be considered as charge storage in the neighborhood of the junction. The amount of excess charge is established by the degree of forward bias. As we move further from the junction, the excess hole concentration decreases because of recombination with the majority electrons. Similar statements apply to electrons which diffuse into the p region. Now, if a signal is applied which increases the forward bias by ΔV, the increased hole (electron) diffusion causes a change ΔQ in the charge stored near the junction. The ratio $\Delta Q/\Delta V$ in the limit defines the diffusion capacitance C_D. For a junction diode, where one side is much more heavily doped than the other (as is often the case), C_D (derived in Sec. 2-13) is found to be [in farads (F)]

$$C_D = \frac{dQ}{dV}\bigg|_Q = \frac{\tau I_{DQ}}{V_T} = \frac{\tau}{r_d} \qquad \text{F} \qquad (2\text{-}20)$$

The *mean carrier lifetime* τ in Eq. (2-20) is a measure of the recombination time for "excess" minority carriers. As $\tau = r_d C_D$, carrier lifetime can be regarded as a "diffusion time constant."

Transition Capacitance   The equivalent circuit in Fig. 2-28b is used to model the reverse-biased diode. The resistance r_r is the incremental resistance [as defined in Eq. (2-19)], with the subscript r signifying the diode is reverse-biased. The element C_T, called the *depletion, transition, barrier,* or *space-charge capacitance,* represents the change in charge stored in the depletion region with respect to a change in junction voltage. Earlier (Sec. 2-2) we indicated that increase in the level of reverse bias caused the width of the depletion region W to increase. An increase in W is accompanied by additional uncovered ions in the space-charge region. Because positive ions exist on one side of the junction and negative ions on the other, C_T is analogous to a parallel-plate capacitor for which

$$C_T = \frac{\epsilon A}{W} \qquad \text{F} \qquad (2\text{-}21)$$

where W is the width of the depletion region, A is the junction area, and ϵ is the permittivity of the semiconductor. We must note that W is a function of the reverse-biased voltage so that C_T is voltage-dependent. For a step-graded junction (discussed in detail in Sec. 2-13), W is inversely proportional to the square root of the reverse-bias voltage.

The capacitances in the models in Fig. 2-28 approximate to a high degree the charge-storage effects in a diode. However, both depletion and diffusion capacitances exist for both forward and reverse bias. Under forward-biased conditions, the value of depletion capacitance is so small compared to C_D that it is generally neglected. Similarly, a small amount of carrier diffusion exists in a reverse-biased diode, but this capacitance is negligible when compared to C_T.[1]

2-10 JUNCTION-DIODE SWITCHING TIMES

The transient response of a diode driven from an ON to an OFF state, or in the opposite direction, signifies that an interval of time elapses before the diode reaches its new steady state. Because it represents an important practical limitation, the reverse recovery, that is, switching from ON to OFF, is treated in the following paragraphs.

The sequence of events which accompanies the reverse-biasing of a conducting diode is depicted in Fig. 2-29. We consider that the step input voltage v_i in Fig. 2-29*b* is applied to the diode-resistor circuit in Fig. 2-29*a* and that for a long time prior to $t = 0$ the voltage $v_i = V_F$ forward-biases the diode. At $t = 0$, the applied voltage abruptly changes to $-V_R$ and remains at this level for $t > 0$. If we assume that R_L and V_F are much larger than R_f and V_γ, respectively, the circuit current $i_D \simeq V_F/R_L$. This value is indicated for $t \le 0$ in Fig. 2-29*c*. The forward bias causes a large number of carriers to diffuse across the junction so that the excess minority-carrier density is high. Under reverse-biased conditions the excess minority carriers in the vicinity of the junction is virtually zero. Consequently, sudden reversal of the voltage cannot be accompanied by a change in the state of the diode until the number of excess minority carriers is reduced to zero. That is, these carriers must be swept back across the junction to the side from which they originated. This charge motion produces a current in the reverse direction. The period of time during which the excess minority carriers decrease to zero, between $t = 0$ and $t = t_1$, is called the *storage time* t_s. During this time interval the diode conducts easily; the current, determined by the applied voltage and external load resistance, is $-V_R/R_L$. The voltage drop across the diode is decreased slightly because of the change in current in the ohmic resistance of the diode but does not reverse (Fig. 2-29*d*). At $t = t_1 = t_s$ the excess minority-carrier density becomes zero. Subsequent to this time, the diode voltage begins to reverse toward $-V_R$ and the magnitude of the current decreases toward I_S.

---

[1]Certain computer-aided circuit analysis programs include these capacitances in device models for both completeness and greater accuracy.

FIGURE 2-29
(*a*) Diode-resistor cir-
cuit; (*b*) input waveform
applied to circuit in part
a showing abrupt change
from forward to reverse
bias; (*c*) current and (*d*)
diode voltage wave-
forms displaying stor-
age and transition times.

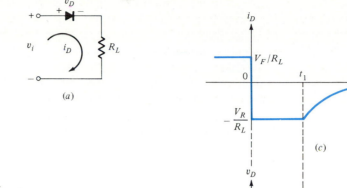

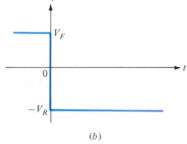

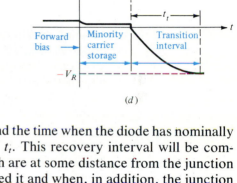

The time which elapses between t_1 and the time when the diode has nominally recovered is called the *transition time* t_t. This recovery interval will be completed when the minority carriers which are at some distance from the junction have diffused to the junction and crossed it and when, in addition, the junction transition capacitance across the reverse-biased junction has charged through R_L to the voltage $-V_R$.

Manufacturers normally specify the *reverse recovery time* t_{rr} of a diode in a typical operating condition in terms of the current waveform in Fig. 2-29*c*. The time t_{rr} is the interval from the current reversal at $t = 0$ until the diode has recovered to a specified extent in terms of either the diode current or the diode resistance. If the specified value of R_L is larger than several hundred ohms, ordinarily the manufacturers will specify the capacitance C_L shunting R_L in the measuring circuit which is used to determine t_{rr}. Commercial switching-type diodes are available with times t_{rr} in the range from less than 1 nanosecond (ns) up to as high as 1 microsecond (μs) in diodes intended for switching large currents. For the 1N4153, the reverse recovery time is a few nanoseconds under the test conditions given in App. B-2.

The *forward recovery time* t_{rf} is the time required for the diode voltage to change from 10 to 90 percent of its final value when the diode is switched from OFF to ON. Since $t_{rf} \ll t_{rr}$, it is usual practice to neglect t_{rf}.

2-11   ZENER DIODES   Avalanche multiplication and Zener breakdown are the two processes which produce the breakdown region in the reverse-biased characteristic in Fig. 2-4*b* and redrawn in Fig. 2-30*a*. Diodes which have adequate power-dis-

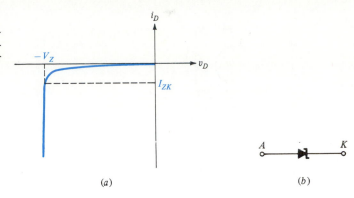

FIGURE 2-30
(*a*) Reverse-bias characteristic showing breakdown region; (*b*) Zener diode symbol.

sipation capabilities to operate in the breakdown region are commonly called *Zener diodes*, whose circuit symbol is indicated in Fig. 2-30*b*. (The term "Zener diode" is used independently of the breakdown mechanism.) These devices are employed as voltage regulators and in other applications in which a constant reference voltage is required.

Avalanche Multiplication Consider the following situation in a reverse-biased diode. A thermally generated carrier (part of the reverse saturation current) falls down the junction barrier and acquires energy from the applied potential. This carrier collides with a crystal ion and imparts sufficient energy to disrupt a covalent bond. In addition to the original carrier, a new electron-hole pair has now been generated. These carriers may also pick up sufficient energy from the applied field, collide with another crystal ion, and create still other electron-hole pairs. Thus each new carrier may, in turn, produce additional carriers through collision and the action of disrupting bonds. This cumulative process is referred to as *avalanche multiplication*. It results in large reverse currents, and the diode is said to be in the region of *avalanche breakdown*.

Zener Breakdown Even if the initially available carriers do not acquire sufficient energy to disrupt bonds, it is possible to initiate breakdown through a direct rupture of the bonds. Because of the existence of the electric field at the junction, a sufficiently strong force may be exerted on a bound electron by the field to tear it out of its covalent bond. The new hole-electron pair which is created increases the reverse current. Note that this process, called *Zener breakdown*, does not involve collisions of carriers with the crystal ions.

The field intensity $\mathcal{E}$ increases as the impurity concentration increases, for a fixed applied voltage. It is found that Zener breakdown occurs at a field of approximately 2×10^7 V/m. This value is reached at or below about 6 V for heavily doped junctions. For lightly doped diodes, the breakdown voltage is higher and avalanche multiplication is the predominant effect. Silicon diodes operated in avalanche breakdown are available with maintaining voltages from several volts to several hundred volts and with power ratings up to 50 W.

FIGURE 2-31
(a) Dc and (b) small-signal models of Zener diode.

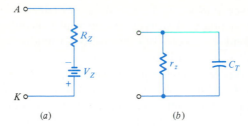

(a)   (b)

Zener Diode Models   The Zener diode characteristic can be approximated by a piecewise linear volt-ampere relationship in much the same fashion as were forward-biased diodes. The model in Fig. 2-31a results from such a representation. In many instances the characteristic shown in Fig. 2-30a at breakdown is virtually vertical so that the static resistance $R_Z = 0$.

The dynamic, or small-signal, model is shown in Fig. 2-31b. The *dynamic resistance* r_z is the reciprocal of the slope of the volt-ampere characteristic in the operating range. It relates the change in operating current ΔI_Z to the change in operating voltage ΔV_Z by $\Delta V_Z = r_z \Delta I_Z$. Ideally, $r_z = 0$, corresponding to a vertical characteristic in the breakdown region. For values of V_Z in the order of a few volts, r_z is in the order of a few ohms. However, for currents below I_{ZK} in Fig. 2-30a, on the "knee" of the curve, r_z may be in the range of hundreds of ohms. These values of r_z are also obtained for $V_Z > 10$ V and at low voltage levels, particularly for currents below about 1 mA.

The capacitance across a breakdown diode is the transition capacitance and hence varies inversely as some power of the voltage. Since C_T is proportional to the cross-sectional area of the diode, high-power avalanche diodes have very large capacitances. Values of C_T from 10 to 10,000 picofarads (pF) are common.

A simple Zener Regulator   The Zener diode in Fig. 2-32a is used to maintain a constant output voltage $V_o = V_Z$ independent of variations in load resistance R_L and the unregulated voltage $V_S > V_Z$. The analysis of the regulator utilizes the equivalent circuit in Fig. 2-32b, in which we assume $R_Z = 0$. With reference to Ohm's law and Kirchhoff's laws, the equation which describes this circuit is

$$I_Z = I_S - I_L = \frac{V_S - V_Z}{R_S} - \frac{V_Z}{R_L} \quad \text{A} \quad (2\text{-}22)$$

FIGURE 2-32
(a) Simple Zener regulator circuit and (b) its equivalent circuit.

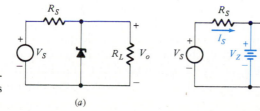

(a)   (b)

In Eq. (2-22) we see that $I_L = V_Z/R_L$ increases (decreases) with decreasing (increasing) load resistance. However, the current I_S is independent of R_L; thus I_Z changes with load variations but the output remains constant at V_Z. The range of I_Z is limited at both high- and low-current values. The high-current limitation results from the maximum power-dissipation capability of the Zener diode. The current I_{ZK} (Fig. 2-30*a*) represents the lowest value of diode current for which regulation occurs. Below I_{ZK}, regulation is poor and the output voltage deviates from V_Z. The permissible range of values for I_Z restricts the values of load for which regulation is achieved. For a given diode, these limits on I_Z also restrict the minimum and maximum values of V_S for proper circuit operation.

Some manufacturers specify the value of I_{ZK} as the minimum current at which the diode can be used. A sharp knee in the characteristic is exhibited by several commercially available diodes even in the microampere region. A common rule of thumb for cases where I_{ZK} is not given is to choose I_{ZK} to be a few percent (5 to 10) of the maximum rated current.

Temperature Characteristics   A matter of interest in connection with Zener diodes, as with semiconductor devices generally, is their temperature sensitivity. The temperature coefficient is given as the percentage change in reference voltage per Celsius degree change in diode temperature. This data is supplied by the manufacturer. The coefficient may be either positive or negative and will normally be in the range ± 0.1 percent/°C. If the reference voltage is above 6 V, where the physical mechanism involved is avalanche multiplication, the temperature coefficient is positive. Below 6 V, however, where true Zener breakdown is involved, the temperature coefficient is negative.

Temperature-compensated reference diodes provide reference voltages that are virtually constant over a wide temperature range. These devices consist of a reverse-biased Zener diode with a positive temperature coefficient, combined in a single package with a forward-biased diode whose temperature coefficient is negative. As an example, the Motorola 1N8241 silicon 6.2-V reference diode has a temperature coefficient of ± 0.005 percent/°C at 7.5 mA over the range -55 to $+100$°C. The dynamic resistance is only 10 Ω. The voltage stability with time of some of these reference diodes is comparable with that of conventional standard cells.

2-12   SCHOTTKY BARRIER DIODES   The junction formed by a metal and extrinsic semiconductor can be either rectifying or ohmic. Because of the differences in carrier concentrations in the two materials, a potential barrier exists. Ohmic contacts, used to make connections to semiconductor devices, exist when care is exerted to eliminate the effect of the barrier. Such is the case of a junction between aluminum and heavily doped silicon used in IC fabrication. When lightly doped silicon (or gallium arsenide) is used, however, the aluminum-silicon junction

FIGURE 2-33
Schottky barrier diode
(*a*) characteristic and (*b*)
circuit symbol.

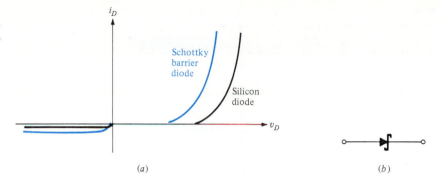

Schottky
barrier
diode

Silicon
diode

(*a*)

(*b*)

is rectifying and the devices so formed are called *Schottky barrier* or simply *Schottky diodes*.

Figure 2-33*a* displays a comparison of the volt-ampere characteristics of a Schottky barrier diode and a silicon junction diode. We note that the characteristics have similar shape, so that Eq. (2-3) also governs the behavior of the Schottky barrier diode. Two major differences between the two characteristics are also observed: (1) the cut-in voltage V_γ is lower, and (2) the reverse saturation current is higher in the Schottky diode. Both features result from the high electron concentration in the metal. With a greater number of carriers available, comparable currents are obtained at lower voltages. Similarly, this number of carriers makes the saturation current higher.

A principal use of Schottky barrier diodes in ICs is that it switches faster than does a junction diode. Because it is a majority-carrier device (recall there are no minority carriers in a metal), the storage time is negligible and the reverse recovery time includes only the transition time shown in Fig. 2-29*c*.

2-13 THE STEP-GRADED JUNCTION DIODE

In this section we present a more quantitative approach to several of the concepts described earlier in this chapter. The step-graded junction, first introduced in Sec. 1-7, is used to develop relationships for depletion (transition) capacitance C_T, variations in minority-carrier density, and diffusion capacitance C_D.

Depletion Capacitance

A step-graded junction is formed when there is an abrupt change from acceptor ions on one side to donor ions on the other side. Such a junction is formed between emitter and base of a planar transistor. It is unnecessary that the donor and acceptor concentrations be equal. In fact, it is often advantageous to have an asymmetrical junction. Figure 2-34*b* is a graph of the charge density as a function of distance from the junction in which the acceptor concentration N_A is assumed to be much greater than the donor concentration N_D. Since the net charge must be zero, it follows that

$$N_A W_p = N_D W_n \qquad \text{m}^{-2} \qquad (2\text{-}23)$$

FIGURE 2-34

(*a*) A reverse-biased step-graded *pn* junction; (*b*) the charge density, (*c*) the field intensity; (*d*) the potential variation with distance *x* from the junction.

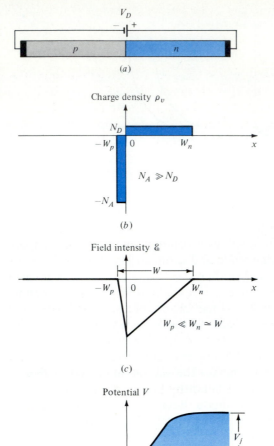

If $N_A \gg N_D$, then $W_p \ll W_n \simeq W$. The relationship between potential and charge density is given by Eq. (2-1):

$$\frac{d^2V}{dx^2} = \frac{-qN_D}{\epsilon} \qquad \text{V/m}^2 \qquad (2\text{-}24)$$

The electric lines of flux start on the positive donor ions and terminate on the negative acceptor ions. Hence there are no flux lines to the right of the boundary $x = W_n$ in Fig. 2-34, and $\mathscr{E} = -dV/dx = 0$ at $x = W_n \simeq W$. Integration of Eq. (2-24) subject to this boundary condition yields

$$\frac{dV}{dx} = \frac{-qN_D}{\epsilon}(x - W) = -\mathscr{E}(x) \qquad \text{V/m} \qquad (2\text{-}25)$$

Neglecting the small potential drop across W_p, we may arbitrarily choose $V = 0$ at $x = 0$. Integration of Eq. (2-25) subject to this condition gives

$$V = \frac{-qN_D}{2\epsilon}(x^2 - 2Wx) \qquad \text{V} \qquad (2\text{-}26)$$

The linear variation in field intensity and the quadratic dependence of potential on distance are plotted in Figs. 2-34c and 2-34d. These graphs should be compared with the corresponding curves in Fig. 2-2.

At $x = W$, $V = V_j =$ the junction, or barrier, potential. Thus

$$V_j = \frac{qN_D W^2}{2\epsilon} \qquad \text{V} \qquad (2\text{-}27)$$

In this section we have used the symbol V to represent the potential at any distance x from the junction. Let us now introduce V_D as the externally applied diode voltage. Since the barrier potential represents a reverse voltage, it is lowered by an applied forward voltage. Thus

$$V_j = V_0 - V_D \qquad \text{V}$$

where V_D is a negative number for an applied reverse bias and V_0 is the contact potential (Fig. 2-2d). This equation and Eq. (2-27) confirm our qualitative conclusion that the thickness of the depletion layer increases with applied voltage. We now see that W varies as $V_j^{1/2} = (V_0 - V_D)^{1/2}$.

If A is the area of the junction, the charge Q in the distance W is

$$Q = qN_D WA \qquad \text{C}$$

The incremental depletion capacitance C_T is

$$C_T = \frac{dQ}{dV_D} = qN_D A \frac{dW}{dV_j} \qquad \text{F} \qquad (2\text{-}28)$$

From Eq. (2-27), $dW/dV_j = \epsilon/qN_D W$, and hence

$$C_T = \frac{\epsilon A}{W} \qquad \text{F} \qquad (2\text{-}29)$$

It is interesting to recall that this formula is exactly the expression we would obtain for a parallel-plate capacitor of area A and plate separation W containing a dielectric of permittivity ϵ, a result already anticipated in Eq. (2-21).

Solving for W in Eq. (2-27) and substituting the result in Eq. (2-29), we obtain

$$C_T = A \left[\frac{q\epsilon N_D}{2(V_0 - V_D)} \right]^{1/2} \qquad \text{F} \qquad (2\text{-}30)$$

It is sometimes convenient to rewrite this equation as

$$C_T = A\,C_0 \left(1 - \frac{V_D}{V_0}\right)^{-1/2} \qquad \text{F} \qquad (2\text{-}31)$$

where C_0 is the junction capacitance per unit area at zero bias ($V_D = 0$).

Analytic Expressions for Minority-Carrier Concentrations If the voltage across a diode is applied in the forward direction, the potential barrier at the junction is lowered and holes from the p side enter the n region. Similarly, electrons from the n type move into the p side. Define p_n as the hole concentration in the n-type semiconductor. If the small value of the thermally generated hole concentration is designated by p_{no}, the *injected*, or *excess*, hole concentration p'_n is defined by $p'_n \equiv p_n - p_{no}$. As the holes diffuse into the n side, they encounter a plentiful supply of electrons and recombine with them. Hence $p_n(x)$ decreases with the distance x into the n material. It is found that the excess hole density falls off exponentially with x:

$$p'_n(x) = p'_n(0)\,\epsilon^{-x/L_p} = p_n(x) - p_{no} \qquad \text{m}^{-3} \qquad (2\text{-}32)$$

where $p'_n(0)$ is the value of the injected minority concentration at the junction $x = 0$. The parameter L_p is called the *diffusion length for holes* and is related to the diffusion constant D_p (Sec. 1-6) and the mean lifetime τ_p by

$$L_p = (D_p\tau_p)^{1/2} \qquad \text{m} \qquad (2\text{-}33)$$

We see that L_p represents the distance from the junction at which the injected concentration has fallen to $1/\epsilon$ of its value at $x = 0$. It can be demonstrated that L_p also equals the average distance that an injected hole travels before recombining with an electron. Hence L_p is the *mean free path for holes*.

The exponential behavior of the excess minority-carrier density as a function of distance on either side of the junction is shown in Fig. 2-35a. The shaded area under the curve in the n type (p type) is proportional to the injected hole (electron) charge. Note that n_p denotes the electron concentration in the p-type material at a distance x from the junction and $n_p(0)$ is the value of this density at $x = 0$.

FIGURE 2-35
Minority-carrier density as a function of the distance x from the junction for (a) forward and (b) reverse bias. The depletion region is assumed to be so small relative to the diffusion length that it is not shown. Note that the curves are not drawn to scale since $p_n(0) \gg p_{no}$.

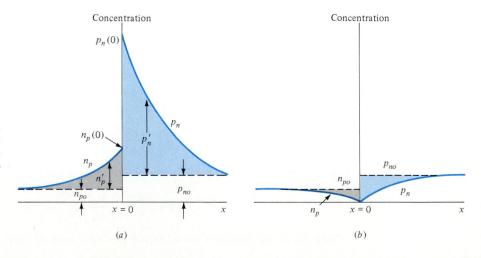

In Sec. 2-2 it is pointed out that a forward bias V lowers the barrier height and allows more carriers to cross the junction. Hence $p_n(0)$ must be a function of V. From the Boltzmann relationship [Eq. (1-39)], it seems reasonable that $p_n(0)$ should depend exponentially on V. It is found that

$$p_n(0) = p_{no}\epsilon^{V/V_T} \qquad \text{m}^{-3} \tag{2-34}$$

This relationship, which gives the hole concentration at the edge of the n region (at $x = 0$, just outside the transition region) in terms of the thermal equilibrium minority-carrier concentration p_{no} (far away from the junction) and the applied potential V, is called the *law of the junction*. A similar equation with p and n interchanged gives the electron concentration at the edge of the p region in terms of V.

When an external voltage reverse-biases the junction, the steady-state density of minority carriers is as shown in Fig. 2-35b. Far from the junction the minority carriers are equal to their thermal-equilibrium values p_{no} and n_{po}, as is also the situation in Fig. 2-35a. As the minority carriers approach the junction, they are rapidly swept across, and the density of minority carriers diminishes to zero at this junction. This result follows from the law of the junction [Eq. (2-34)] since the concentration $p_n(0)$ reduces to zero for a negative junction potential V.

The injected charge under reverse bias is given by the shaded area in Fig. 2-35b. This charge is negative since it represents less charge than is available under conditions of thermal equilibrium with no applied voltage.

Charge-Control Description of a Diode   From Eq. (1-34) it follows that the hole diffusion current $I_p(0)$ crossing the junction under forward bias is proportional to the slope at the origin of the p_n curve in Fig. 2-35a. The corresponding electron diffusion current $I_n(0)$ is proportional to the slope at the origin of the n_p curve in Fig. 2-35a. Theoretically, it can be shown that the minority-carrier drift current crossing the junction is negligible compared with the minority-carrier diffusion current. Hence $I_p(0)$ represents the total current of holes moving from left to right across the junction, and $I_n(0)$ is the total current of electrons traveling from right to left across the junction. Therefore, the total diode current I is the sum of these two currents, or

$$I = I_p(0) + I_n(0) \qquad \text{A} \tag{2-35}$$

The reverse saturation hole (electron) current is proportional to the slope at $x = 0$ of the $p_n(n_p)$ curves in Fig. 2-35b. The total reverse saturation current is the sum of these two currents and is negative.

For simplicity of discussion we assume that one side of the diode, say, the p material, is so heavily doped in comparison with the n side that the current I crossing the junction is due entirely to holes moving from the p to the n side, or $I = I_p(0)$. From Eq. (1-31)

$$I_p(x) = -AqD_p\frac{dp_n}{dx} = \frac{AqD_p p_n'(0)}{L_p}\epsilon^{-x/L_p} \qquad \text{A} \tag{2-36}$$

where Eq. (2-32) is used for $p_n(x)$. The hole current I is given by $I_p(x)$ in Eq. (2-36), with $x = 0$, or

$$I = \frac{AqD_p p'(0)}{L_p} \qquad \text{A} \qquad (2\text{-}37)$$

The excess minority charge Q exists only on the n side and is given by the shaded area in the n region in Fig. 2-35a multiplied by the diode cross section A and the electronic charge q. Hence, from Eq. (2-32), we obtain

$$Q = \int_0^\infty Aqp'(0)\, \epsilon^{-x/L_p}\, dx = AqL_p p'(0) \qquad \text{C} \qquad (2\text{-}38)$$

Elimination of $p'(0)$ from Eqs. (2-37) and (2-38) yields

$$I = \frac{Q}{\tau} \qquad \text{A} \qquad (2\text{-}39)$$

where $\tau \equiv L_p^2/D_p \equiv \tau_p$ = mean lifetime for holes [Eq. (2-33)].

Equation (2-38) is an important relationship, referred to as the *charge-control description of a diode*. It states that the diode current (which consists of holes crossing the junction from the p side to the n side) is proportional to the stored charge Q of excess minority carriers. The factor of proportionality is the reciprocal of the decay time constant (the mean lifetime τ) of the minority carriers. Thus, in the steady state, *the current I supplies minority carriers at the rate at which these carriers are disappearing because of the process of recombination*.

The charge-control characterization of a diode describes the device in terms of the current I and the stored charge Q, whereas the equivalent-circuit characterization uses the current I and the junction voltage V. One immediately apparent advantage of this charge-control description is that the exponential relationship between I and V is replaced by the linear dependence of I on Q. The charge Q also makes a simple parameter, the sign of which determines whether the diode is forward- or reverse-biased. The diode is forward-biased if Q is positive and reverse-biased if Q is negative.

Diffusion Capacitance   In Sec. 2-9 we introduced the diffusion capacitance C_D as a means for modeling minority-carrier storage in the vicinity of a forward-biased diode. We can now derive this element based on the charge-control description just concluded. From Eqs. (2-39) and (2-17)

$$C_D \equiv \frac{dQ}{dV} = \tau \frac{dI}{dV} = \tau g_d = \frac{\tau}{r_d} \qquad \text{F} \qquad (2\text{-}40)$$

where the diode incremental conductance is $g_d \equiv dI/dV$. Substitution of the expression for the diode incremental resistance $r_d = 1/g_d$ given in Eq. (2-18) into Eq. (2-40) yields

$$C_D = \frac{\tau I_D}{\eta V_T} \qquad \text{F} \qquad (2\text{-}41)$$

We see that *the diffusion capacitance is proportional to the current I_D.* In the derivation given above we have assumed that the diode current I_D is due to holes only. If this assumption is not satisfied, Eq. (2-40) gives the diffusion capacitance C_{Dp} due to holes only, and a similar expression can be obtained for the diffusion capacitance C_{Dn} due to electrons. The total diffusion capacitance can then be obtained as the sum of C_{Dp} and C_{Dn}.

REFERENCES

1 Gray, P. E., D. DeWitt, A. R. Boothroyd, and J. F. Gibbons: "Physical Electronics and Circuit Models of Transistors," vol. 2, SEEC, John Wiley & Sons, New York, 1964.

2 Millman, J., and C. C. Halkias: "Integrated Electronics," McGraw-Hill Book Company, New York, 1972.

3 Yang, E. S.: "Fundamentals of Semiconductor Devices," McGraw-Hill Book Company, New York, 1978.

4 Muller, R. S., and T. I. Kamins: "Device Electronics for Integrated Circuits," John Wiley & Sons, New York, 1977.

5 Ghausi, M. S.: "Principles and Design of Linear Active Circuits," McGraw-Hill Book Company, New York, 1965.

REVIEW QUESTIONS

2-1   Where is the magnitude of the electric field intensity maximum in a *pn* junction? Explain why.

2-2   (*a*)   What is the depletion region?
   (*b*)   Do holes and electrons constitute the space charge?
   (*c*)   Do donor and acceptor ions constitute the space charge?

2-3   (*a*)   What is the mechanism for the major portion of current in a forward-biased junction?
   (*b*)   What external voltage polarity is needed for forward bias?

2-4   (*a*)   For reverse bias, does the width of the depletion region increase or decrease?
   (*b*)   What happens to the junction potential?

2-5   (*a*)   What is an ohmic contact?
   (*b*)   Can you measure directly the contact potential of an ohmic contact?

2-6   Write the volt-ampere relation for a *pn* junction and give the significance of each term.

2-7   Why can the reverse-saturation current be considered a scale factor for diode current ratings?

2-8   What is meant by the cut-in voltage?

2-9   How can η be determined from the logarithmic characteristic?

2-10   (*a*)   How does the reverse saturation current of a *pn* diode vary with temperature?
   (*b*)   At constant current, how does the diode voltage vary with temperature?

2-11 What parameters in a germanium diode differ from those in a silicon diode?

2-12 (*a*) Sketch the volt-ampere characteristic of an ideal diode.
(*b*) Explain how this resembles a switch.

2-13 What is the significance of the load line?

2-14 Draw a large-signal diode model for forward bias.

2-15 Explain how a diode acts as a rectifier.

2-16 Explain the action of the capacitor filter in a rectifier circuit.

2-17 Describe the operation of a diode clipper.

2-18 What is meant by the transfer characteristic?

2-19 (*a*) Draw the small-signal model of a *pn* diode for both forward and reverse bias.
(*b*) Explain the physical significance of each element.

2-20 Under forward-biased conditions and for increasing diode current, do (*a*) the incremental resistance and (*b*) the diffusion capacitance increase or decrease?

2-21 In words, explain how the small-signal equivalent circuit is used to determine the response in a circuit containing a diode.

2-22 Why doesn't the constant bias source appear in the small-signal model of a circuit containing a diode?

2-23 Does the depletion capacitance increase or decrease with increasing magnitude of the reverse voltage?

2-24 Explain the physical significance of storage time and transition time.

2-25 Describe the physical mechanisms which produce (*a*) avalanche breakdown and (*b*) Zener breakdown.

2-26 (*a*) Sketch the volt-ampere characteristic of a Zener diode.
(*b*) Indicate on your sketch the knee of the curve.
(*c*) What is the significance of the knee.

2-27 Draw both the large- and small-signal Zener diode models.

2-28 How does a Schottky barrier diode differ from a silicon junction diode?

2-29 Sketch the minority-carrier density of a step-graded *pn* junction diode for forward and reverse bias.

2-30 What is meant by the charge-control description of a diode?

Chapter 3
BIPOLAR JUNCTION TRANSISTORS

The bipolar junction transistor is one of two principal semiconductor devices used for amplification and switching. The objectives of this chapter are to describe the physical principles which govern bipolar junction transistor operation and to characterize this device as a circuit element. The development of the bipolar junction transistor volt-ampere characteristics and small-signal and dc equivalent circuits is based on the corresponding concepts discussed in Chap. 2 for the junction diode. In particular, the Ebers-Moll equations which describe the active, saturation, cutoff, and inverted modes of bipolar transistor operation are presented. The operation of a simple common-emitter stage as both an amplifier and switch is investigated. Because of its importance in integrated circuits, the emitter-coupled (differential) pair is also introduced.

3-1 THE IDEAL CURRENT-CONTROLLED SOURCE

electronic amplifiers and switches exploit the properties of controlled sources to achieve their function. Transistors are extensively used in such circuits because they exhibit controlled-source characteristics. To focus attention on some important aspects of transistor operation, it is beneficial to examine the circuit properties of controlled sources and how these can be used as amplifying and switching devices.

An ideal current-controlled current source, depicted in Fig. 3-1a, is a circuit element consisting of three terminals, one of which is common to both input and output.[1] The input terminal pair (1–3) provides the control current i_1, and

---

[1]More precisely, controlled sources can have two input and two output terminals. However, most electronic devices have one terminal that is common to both input and output.

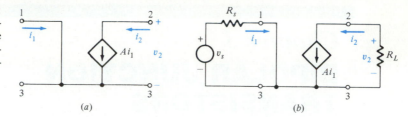

(*a*)   (*b*)

a current source of strength Ai_1 acts as the output terminal pair (2–3).[1] Parameter A relates the strength of the source to the control current and is often referred to as the *current gain*. Physically, A is related to the processes which occur within the device used to obtain the controlled source.

It is evident in Fig. 3-1*a* that the effect of i_1 is transmitted to the output by means of the source, whereas signals applied at the output terminals do not affect the control current. This *unilateral* behavior allows the portions of a circuit where the control signal is applied to be isolated from circuit elements connected to the output.

In the circuit shown in Fig. 3-1*b*, a signal source v_s in series with a resistance R_s is connected to the input and a load resistance R_L is placed across the output of the controlled source. The control current is $i_1 = v_s/R_s$, and the output voltage v_2 is

$$v_2 = -Ai_1R_L = -\frac{AR_L}{R_s}v_s \qquad (3\text{-}1)$$

From Eq. (3-1) we see that when $|AR_L/R_s| > 1$, then $|v_2| > |v_s|$ and voltage amplification is achieved. In addition, if $A > 1$ as is usually the case, current gain is realized as the output current is greater than the input current. It is also true that the power dissipated by R_L is greater than the power supplied by v_s. Thus the controlled source is capable of providing power gain. One consequence of amplification is that the power required for control is less than the amount of power controlled. This, coupled with the unilateral property, enables the controlled source to be operated as a controlled switch.[2]

The output volt-ampere characteristics are a convenient method for displaying the dependence of source strength on the control variable. These char-

---

[1]A controlled source is designated by the diamond symbol as is depicted in Fig. 3-1. For a current source, the arrow in the diamond indicates the direction of the current, whereas the + and − signs placed at the top and bottom of the diamond indicate the polarity of the voltage source. In either case, the strength of the source is shown algebraically adjacent to the diamond.

[2]We are aware that it is desirable to control substantial quantities of energy with a small expenditure of energy. Consider the mechanical energy necessary to turn the switch which activates a 120-V 10-A window-mounted air conditioner. Roughly, this requires that a 0.05-kilogram (kg) switch be moved 1 cm in 0.25 s and is about 5×10^{-3} J at a rate of 0.02 W. Yet the energy and power controlled by this effort are equivalent to lifting the unit from the floor to the window sill in 0.25 s.

FIGURE 3-2
Current-controlled current source volt-ampere characteristics with resultant waveforms produced by step excitation.

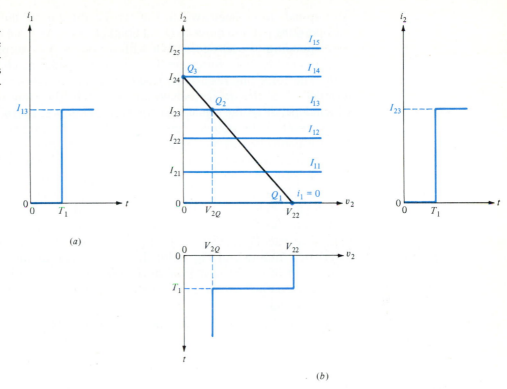

(a)

(b)

acteristics for the controlled source in Fig. 3-1a are shown in Fig. 3-2b as the family of curves of i_2 versus v_2 for different values of i_1. The horizontal characteristic indicates that i_2 is independent of v_2. (This is true for any ideal current source.) To demonstrate switching behavior, consider the circuit shown in Fig. 3-3a. The box labeled Device has the output characteristics displayed in Fig. 3-2b on which the load line, representing KVL for the output loop, is drawn. The input voltage waveform is given in Fig. 3-3b and the corresponding input current waveform, in Fig. 3-2a. Let us assume that the value $i_1 = V_s/R_s$ corresponds to the current I_{13} in Fig. 3-2a. When $v_s = 0$ ($0 \le t \le T_1$), $i_1 = 0$, the operating point is at Q_1 and results in $v_2 = V_{22}$ and $i_2 = 0$. This situation

FIGURE 3-3
(a) Circuit using current-controlled current source; (b) input waveform.

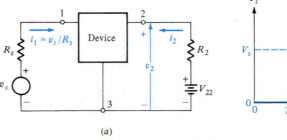

(a)

(b)

corresponds to an open switch. For $t > T_1$, the control current is I_{13}, causing the operating point to move to Q_2; at point Q_2, $v_2 = V_{2Q}$ and $i_2 = I_{23}$, a condition corresponding to a closed switch with a voltage V_{2Q} across the switch. The resultant output waveforms are illustrated alongside the characteristics in Fig. 3-2b. The foregoing discussion leads us to the conclusion that at the output terminals 2–3, the device behaves as a switch, the state (open or closed) of which depends on the signal applied to terminals 1–3. Furthermore, the voltage and current at the output, dependent only on the external elements V_{22} and R_L, are controlled by the input current i_1. Note that if it is desired that the voltage drop across the controlled switch be zero, the control current V_s/R_s must be selected to equal I_{14}. For this input current, the output current is V_{22}/R_L and $v_2 = V_{22} - i_2 R_L = 0$ and is indicated by point Q_3 in Fig. 3-2b.

Amplifier behavior can also be demonstrated by the use of the circuit in Fig. 3-3a. Consider that $v_s = V_{11} + V_m \sin \omega t$, where $V_m < V_{11}$. The voltage V_{11} is used to bias the device at an operating point Q shown on the output characteristics (redrawn in Fig. 3-4b). In addition, we assume that the sinusoidal component of v_s, the signal, produces the current i_1 as indicated in Fig. 3-4a. This variation in i_1 produces the current i_2 and voltage v_2 displayed. Under the conditions stated earlier in this section, the amplitude of the sinusoidal component of v_2 is greater than V_m, again demonstrating that the *signal* is amplified. It is important to note that the bias voltage V_{11} is essential to the amplification

FIGURE 3-4
Load line for circuit in Fig. 3-3a; waveforms of i_2 and v_2 produced by sinusoidal input signal.

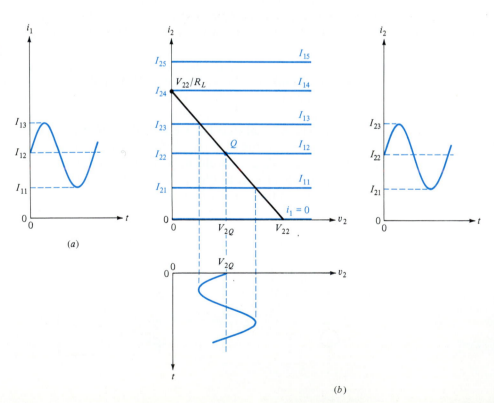

process. Without V_{11}, the output current becomes zero whenever the input sinusoid is negative. This causes the circuit to behave as a clipping or rectifying circuit, and the output consists of only a portion of a sinusoid. Thus, in contrast to the controlled-switch operation, amplifier performance requires that for the signal levels used, the operating point be restricted to the "middle" of the volt-ampere characteristics. If the input signal levels are much smaller than the bias levels, analysis of amplifier circuits lends itself to the use of small-signal models (as discussed in Sec. 3-10).

We must realize that practical devices only approximate ideal characteristics. For use in amplification, however, devices which exhibit constant current (or voltage) characteristics over a range of operating values are desirable.

3-2 THE JUNCTION TRANSISTOR   The *bipolar junction transistor* (BJT), also called the *junction transistor* or *bipolar transistor,* is a three-element device formed from two junctions which share a common semiconductor layer. The two types of bipolar transistors are pictorially represented in Fig. 3-5. In the *pnp* transistor in Fig. 3-5*a*, the common *n*-type region is sandwiched between two *p*-type layers. Analogously, a *p* region is common to two *n*-type layers in the *npn* transistor in Fig. 3-5*b*. The three elements of a BJT are referred to as the *emitter, base,* and *collector* and are indicated in the circuit symbols in Fig. 3-6. The arrow on the emitter lead specifies the direction of current when the emitter-base junction is forward-biased. In both cases, the assumed positive direction for the terminal currents I_E, I_B, and I_C is *into* the transistor. The voltage between each pair of terminals is also indicated in Fig. 3-6 by means of the double-subscript notation. For example, V_{CB} represents the voltage drop between collector and base.

The representation of the bipolar transistor in Fig. 3-5*b* depicts a symmetrical structure and allows us to choose either *n* region as the emitter. In a practical transistor, however, as in the *planar npn* IC transistor in Fig. 3-7, the emitter and collector regions differ markedly. Figure 3-7*a* represents a *cross section*

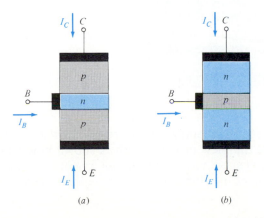

FIGURE 3-5
Positive current convention in (*a*) *pnp* and (*b*) *npn* transistors.

FIGURE 3-6
Circuit symbols for (*a*)
pnp and (*b*) *npn* transistors.

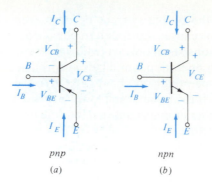

pnp

(*a*)

npn

(*b*)

FIGURE 3-7
Structure of an IC *npn*
transistor including *p*
substrate and isolation
island and aluminum
contacts.

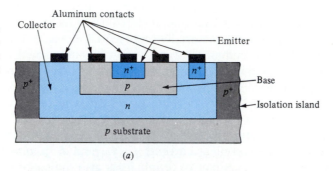

(*a*)

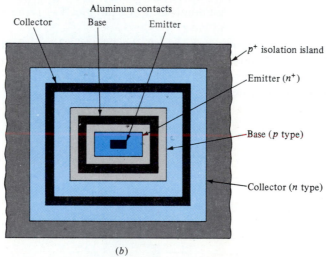

(*b*)

of the transistor with shading used to represent the several impurity concentrations. In the top view in Fig. 3-7*b* it is shown how aluminum contacts are made to the collector, base, and emitter regions. Note that because of the selective doping of the silicon block, a *p* region is sandwiched between two *n* sections to form an *npn* transistor. The term "planar transistor"[1] refers to the fact that the three output leads *C, B,* and *E* are connected to the aluminum contacts for the collector, base, and emitter, respectively, *with all three contacts lying in a single plane.*

As illustrated in Fig. 3-7, the emitter area is considerably smaller than the collector area. This difference is due mainly to the fact that in the most prevalent uses of the BJTs, the collector region must handle more power than the emitter. Hence more surface area is required for heat dissipation. A second difference is the doping densities of the emitter and collector regions. The emitter generally serves as the source of mobile charges. A high doping density is used (thus the n^+ designation) to enhance this ability to make many carriers available. That is, when the emitter-base junction is forward-biased, the emitter injects electrons into the base where they move toward the collector. If the collector-base diode is reverse-biased, the minority-carrier electrons in the base are swept into the collector region, where they become a major component of collector current. The collector, as its name implies, is normally not required to provide many carriers, and so its doping levels need not be as high as that for the emitter. (The small n^+ region in the collector helps to form a good ohmic contact.) For reasons discussed in Chap. 5, the base region is doped at a level between the emitter and collector concentrations.

The Physical Behavior of a Bipolar Transistor   The essential features of a BJT as an active circuit element can be appreciated by considering the situation depicted in Fig. 3-8. Here a *pnp* transistor is shown with voltage sources which serve to bias the emitter-base junction in the forward direction (V_{EB} positive) and the collector-base junction in the reverse direction (V_{CB} negative). In our discussion of the *pn* diode in Chap. 2, we showed that V_{EB} (V_{CB}) appears across the very small emitter (collector) space-charge region. The electric field is confined to the depletion region, and the field is zero in the rest of the semiconductor. Hence the potential is constant within each region (emitter, base, or collector) and no conduction (drift) currents exist. Consequently, *the current components in a BJT are all diffusion currents.*

---

[1]A more complete description of transistor construction and the fabrication techniques employed appears in Chap. 5. In this chapter the designation "integrated-circuit" (IC) transistor is explained.

FIGURE 3-8
A common-base circuit showing bias supplies V_{EE} and V_{CC}.

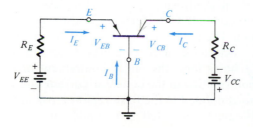

FIGURE 3-9

Transistor carrier components for a forward-biased emitter-base junction and reverse-biased collector-base junction.

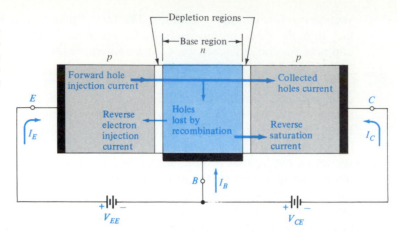

Let us for the moment assume an idealized transistor whose base is so lightly doped compared to the emitter region that we can neglect all currents due to electrons. We also assume that the width of the base region is small compared with the diffusion length so that recombination in this region can be neglected. For this ideal transistor, a forward voltage V_{EB} injects holes into the base, and all these holes travel through the base and into the collector region. This action results in the collector current being equal to the emitter current $|I_C| = |I_E|$ for all values of reverse collector voltage V_{CB}. This transistor exhibits precisely the characteristic of the current-controlled current source described in Sec. 3-1, with unity current gain ($A = 1$). The output characteristics of this ideal transistor are those indicated in Fig. 3-2b with $i_2 = -I_C$, $i_1 = I_E$, and $v_2 = -V_{CB}$.

Now let us consider the behavior of a practical (nonideal) transistor in the circuit shown in Fig. 3-8. We can no longer neglect recombination or the effects of the electron concentration in a real transistor. Figure 3-9 shows the various current components of a *pnp* transistor biased to correspond to the circuit shown in Fig. 3-8. For this situation we again assume that no electric field exists in the semiconductor regions outside the depletion regions so that the voltages V_{EB} and V_{CB} appear across the emitter and collector junctions, respectively. The forward-biased emitter-base junction injects many holes into the base (forward injection), where they become minority carriers. Electrons crossing the junction from base to emitter (reverse injection) is kept small in transistor design by doping the base less heavily than the emitter. In the narrow base region, holes diffuse toward the collector-base junction. A small number of injected holes recombine with electrons in the base and constitute a portion of the base current. The holes that reach the collector-base junction are swept into the collector because of the reverse bias. Under the bias conditions shown in Fig. 3-9, these holes form the major component of I_C. However, there is another (small) contribution to the collector current due to thermally generated carriers. Holes so generated in the base region cross into the collector section and electrons, thermally generated in the collector, cross the junction into the

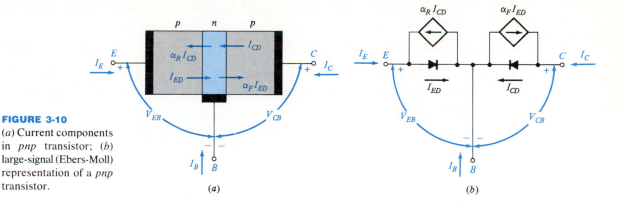

FIGURE 3-10
(*a*) Current components in *pnp* transistor; (*b*) large-signal (Ebers-Moll) representation of a *pnp* transistor.

base layer. These two thermal currents constitute the reverse saturation current of the collector-base junction indicated in Fig. 3-9. Consequently, as shown in Fig. 3-9, it is clear that the collector current consists of two components, one due to injected holes at the emitter-base diode and the other attributed to thermally generated carriers crossing the collector-base junction.

3-3 THE EBERS-MOLL REPRESENTATION OF THE BJT
The behavior of the bipolar transistor can be described in both conceptual and quantitative terms by observing in Fig. 3-5 that this device consists of two coupled *pn* junctions. The base region is common to both junctions and forms the coupling between them. Bipolar transistors are constructed with very narrow base regions (considerably smaller than one diffusion length). Consequently, significant electrical interaction exists between the junctions as explained in Sec. 3-2; this interaction is often referred to as *transistor action*. The current components which comprise the terminal currents I_E and I_C are shown for a *pnp* transistor in Fig. 3-10*a*. The voltages V_{EB} and V_{CB} are the voltage drops from emitter to base and collector to base, respectively. Assuming that there is no voltage drop across the bulk semiconductors forming the emitter, base, and collector regions, these voltages appear across the respective junctions. With both voltages measured with respect to the base, this connection is called the *common-base configuration*. The emitter current in Fig. 3-10*a* has two components. The current associated with the emitter-base diode is designated I_{ED} and that with the collector-base diode, I_{CD}. The component $\alpha_R I_{CD}$ is the portion of I_{CD} that is coupled through the base to the emitter. Similarly, $\alpha_F I_{ED}$ is the fraction of I_{ED} coupled into the collector.

On the basis of the considerations in the preceding paragraph we can construct the Ebers-Moll model in Fig. 3-10*b*. The two back-to-back diodes (whose cathodes are connected) represent the junctions of the bipolar transistor, whereas the two controlled sources indicate the coupling between junctions. The currents I_{ED} and I_{CD} are related to V_{EB} and V_{CB} by the diode volt-ampere relation given in Eq. (2-3).[1] Thus I_E and I_C can be expressed in terms of the two diode currents as

$$I_E = I_{ED} - \alpha_R I_{CD} = I_{ES}(\epsilon^{V_{EB}/V_T} - 1) - \alpha_R I_{CS}(\epsilon^{V_{CB}/V_T} - 1) \qquad (3\text{-}2)$$

$$I_C = -\alpha_F I_{ED} + I_{CD} = -\alpha_F I_{ES}(\epsilon^{V_{EB}/V_T} - 1) + I_{CS}(\epsilon^{V_{CB}/V_T} - 1) \qquad (3\text{-}3)$$

The relationships expressed in Eqs. (3-2) and (3-3) are known as the *Ebers-Moll equations*.

The quantities I_{ES} and I_{CS} in Eqs. (3-2) and (3-3) are the reverse saturation currents of the emitter-base and collector-base junctions, respectively. The parameters α_F and α_R are each less than unity, as not all the current from one diode is coupled to the other junction. The subscripts refer to forward (F) transmission from emitter to collector and reverse (R) transmission from collector to emitter. The four quantities I_{ES}, I_{CS}, α_F, and α_R are functions of doping densities and transistor geometry. These quantities are not independent but are found from theory to be related by

$$\alpha_F I_{ES} = \alpha_R I_{CS} \qquad (3\text{-}4)$$

This equation is often called the *reciprocity condition* for the BJT.

The base current is obtained by realizing that the sum of the terminal currents must be zero (KCL). Thus

$$I_B = -(I_E + I_C) \qquad (3\text{-}5)$$

It is instructive to indicate the typical values of the quantities in the Ebers-Moll equations. For IC transistors (Fig. 3-7) of the nominally small dimensions used

$$0.98 \le \alpha_F \le 0.998 \qquad \text{and} \qquad 0.40 \le \alpha_R \le 0.8$$

and I_{ES} and I_{CS} are in the order of 10^{-15} A. Both I_{ES} and I_{CS} depend on their respective junction areas. Consequently, for specified donor and acceptor doping levels, transistor currents can be scaled by changing device dimensions. This is typically done in IC design to realize transistors with differing current-handling capability. By this method I_{ES} and I_{CS} can be increased to about 10^{-13} and 10^{-12} A, respectively. The value of α_F remains essentially unchanged, and [from Eq. (3-4)] α_R can be reduced below 0.1. Scaling device dimensions are utilized in discrete transistors to achieve higher current and power levels than are possible on a chip.

The Ebers-Moll equations for an *npn* device are obtained from Eqs. (3-2) and (3-3) once we recognize that the forward current in each diode is from *p* to *n*, and that forward bias requires a positive voltage from *p* to *n*. Consequently, the directions of all current components and junction voltages for an *npn* transistor are reversed from those for a *pnp* device as depicted in Fig.

---

[1]Most IC transistors operate at currents which are typically at least nine orders of magnitude greater than saturation currents. Consequently, $\eta = 1$ as described in Sec. 2-3.

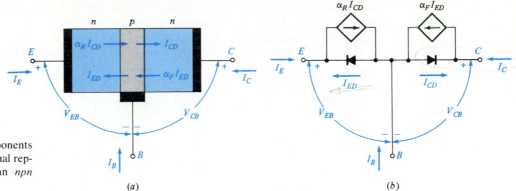

FIGURE 3-11
(*a*) Current components and (*b*) large-signal representation of an *npn* transistor.

3-11*a*. From this argument it follows that Eqs. (3-2) and (3-3) are valid for an *npn* transistor provided a minus sign is inserted before V_{EB}, V_{CB}, and each current component. The results are stated in Eqs. (3-6) and (3-7):

$$I_E = -I_{ES} (\epsilon^{-V_{EB}/V_T} - 1) + \alpha_R I_{CS} (\epsilon^{-V_{CB}/V_T} - 1) \qquad (3\text{-}6)$$

$$I_C = \alpha_F I_{ES} (\epsilon^{-V_{EB}/V_T} - 1) - I_{CS} (\epsilon^{-V_{CB}/V_T} - 1) \qquad (3\text{-}7)$$

On the basis of Eqs. (3-6) and (3-7), the model shown in Fig. 3-11*b* is obtained.

Large-Signal Current Gains Let us consider an *npn* transistor for the situation that the emitter-base diode is forward-biased ($V_{EB} < 0$) and that the collector and base terminals are short-circuited ($V_{CB} = 0$). From Eqs. (3-6) and (3-7) under these conditions, we obtain

$$-I_E = I_{ES} (\epsilon^{-V_{EB}/V_T} - 1) \qquad \text{and} \qquad I_C = \alpha_F I_{ES} (\epsilon^{-V_{EB}/V_T} - 1)$$

Hence $I_C = -\alpha_F I_E$ and α_F is defined as

$$\alpha_F \equiv -\frac{I_C}{I_E}\bigg|_{V_{CB}=0} \qquad (3\text{-}8)$$

The quantity α_F as given in Eq. (3-8) is the *common-base forward short-circuit current gain*.[1]

Similarly, when $V_{CB} < 0$, the *common-base reverse short-circuit current* gain α_R is determined as

$$\alpha_R \equiv -\frac{I_E}{I_C}\bigg|_{V_{EB}=0} \qquad (3\text{-}9)$$

---

[1]Sometimes α_F is designated by its corresponding *h* parameter (h_{FB}) (see App. C), where the *B* subscript indicates common base.

Note that in Eq. (3-9) it is the collector-base junction that is forward-biased and the emitter-base diode that is short-circuited, thus reversing the roles of collector and emitter from the forward conditions in Eq. (3-8).

The definitions in Eqs. (3-8) and (3-9) apply to both *npn* and *pnp* transistors. For an *npn* device, I_C is positive while I_E is negative; the reverse is true for *pnp* transistors. As a result, α_F and α_R are always positive.

When $V_{CB} = 0$ and $V_{EB} < 0$, the base current [Eq. (3-5)] is expressable as

$$I_B = -(1 - \alpha_F)I_E \tag{3-10}$$

Since typical values of α_F are almost unity (indicated previously in this section), we see that I_B is quite small compared with I_E and the magnitudes of I_C and I_E are virtually equal.

It is often convenient to express the collector and emitter currents in terms of the much smaller base current. Combining Eqs. (3-8) and (3-10), we obtain

$$I_C = \frac{\alpha_F}{1 - \alpha_F} I_B = \beta_F I_B \tag{3-11}$$

$$I_E = \frac{-I_B}{1 - \alpha_F} = -(\beta_F + 1)I_B \tag{3-12}$$

where

$$\beta_F = \frac{\alpha_F}{1 - \alpha_F} \tag{3-13}$$

The quantity β_F is the *common-emitter forward short-circuit current gain* (also designated by h_{FE}).

A similar analysis for the reverse condition yields the *common-emitter short-circuit reverse current gain* as

$$\beta_R = \frac{\alpha_R}{1 - \alpha_R} \tag{3-14}$$

Typical values of β_F for IC transistors lie between 50 and 250; β_R is usually between 1 and 5.

Modes of Transistor Operation   Each junction in the BJT can be forward- or reverse-biased independently. Thus four modes of operation exist and are defined in Table 3-1. A detailed analysis of BJT operation in these modes appears in subsequent sections of this chapter. Our purpose here is to examine briefly these modes and highlight some of their features.

In the *forward-active* region, the bipolar transistor behaves as a controlled source. This conclusion can be reached from Eqs. (3-6) and (3-7) for the conditions listed in Table 3-1. For junction bias voltages of at least several tenths of a volt and assuming that I_{CS} is negligibly small, as is almost always the case, $I_C = -\alpha_F I_E$. Thus control of the input current I_E specifies the output current I_C. This is the action of a current-controlled current source as changes in the

TABLE 3-1 Operating Modes in a Bipolar Transistor

| Mode | *Junction bias condition* | |
	Emitter-base	*Collector-base*
Forward-active	Forward	Reverse
Cutoff	Reverse	Reverse
Saturation	Forward	Forward
Reverse-active	Reverse	Forward

emitter-base bias level adjust the value of I_E and hence I_C. With controlled-source characteristics obtainable, the BJT can be used as an amplifier and the forward-active mode is prevalent in analog circuits.

Both junctions are reverse-biased in the *cutoff mode;* both I_E and I_C are in the order of the diode reverse saturation currents (Prob. 3-5). Here the situation is one of nearly zero current with "large" reverse junction voltages ($V_{CB} \gg V_T$), and behavior approximates an open switch. With both diodes forward-biased in *saturation,* the collector current may be appreciable, but only a small voltage exists across the collector junction. This condition is nearly that of a closed switch. Operation of the BJT between cutoff and saturation corresponds to the action of a switch. (Compare this with the discussion in Sec. 3-1.)

The *reverse-active* or *inverted mode* is similar to the forward-active mode with a significant difference. Although behavior in the reverse-active region is that of a controlled source ($I_E = -\alpha_R I_C$), the smaller value of the current gain α_R compared to α_F makes this mode unsuitable, in general, for amplification. However, the inverted mode has application in digital circuits (Chap. 6) and certain analog switching circuits.

Example 3-1

An *npn* transistor is operated with the collector-base junction reverse-biased by at least a few tenths of a volt and with the emitter open-circuited. Determine (*a*) the mode of operation, (*b*) the collector and base currents, and (*c*) the values of I_C and V_{EB} at room temperature for $I_{ES} = 10^{-15}$ A, $I_{CS} = 2 \times 10^{-15}$ A, and $\alpha_F = 0.99$.

Solution

(*a*) With the collector-base diode reverse-biased, we see in Table 3-1 that the mode of operation is either cut off or forward-active. Which condition exists is determined by the state of the emitter-base junction. Using Eq. (3-6), with $I_E = 0$ (open circuit), we obtain

$$I_E = 0 = -I_{ES}(\epsilon^{-V_{EB}/V_T} - 1) - \alpha_R I_{CS}$$

from which

$$\epsilon^{-V_{EB}/V_T} = 1 - \frac{\alpha_R I_{CS}}{I_{ES}} = 1 - \alpha_F \tag{1}$$

where the reciprocity condition, $\alpha_R I_{CS} = \alpha_F I_{ES}$ from Eq. (3-4) has been used. Inverting and taking the logarithm of both sides, we have

$$\frac{V_{EB}}{V_T} = \ln \frac{1}{1 - \alpha_F} = \ln (\beta_F + 1) \tag{2}$$

In (2) we observe that V_{EB} is positive, thus reverse-biasing the emitter junction. Consequently, the transistor is cut off.

(b) With $I_E = 0$, KCL requires $I_C = -I_B$. The collector current is obtained from Eq. (3-7), into which (1) is substituted:

$$I_C = -I_B = -\alpha_F \alpha_R I_{CS} + I_{CS} = (1 - \alpha_F \alpha_R) I_{CS} \tag{3}$$

(c) Substitution of values in (2) gives

$$\frac{V_{EB}}{25 \times 10^{-3}} = \ln \frac{1}{1 - 0.99} \quad \text{and} \quad V_{EB} = 115 \text{ mV}$$

The value of α_R, from the reciprocity condition, is

$$\alpha_R = \alpha_F \frac{I_{ES}}{I_{CS}} = 0.99 \frac{10^{-15}}{2 \times 10^{-15}} = 0.495$$

Then, using (3), we obtain

$$I_C = -I_B = (1 - 0.99 \times 0.495) \times 2 \times 10^{-15} = 1.02 \times 10^{-15} \text{ A}$$

With $I_E = 0$, the result indicates that the transistor, between the base and collector terminals, behaves as a diode and the current determined is the effective reverse saturation collector current for an open-circuited emitter. Although the value of I_C obtained is small, it increases markedly with temperature.

The current given by (3) in Example 3-1 is often referred to as the *reverse collector current*. As we see in the next section, this is an important quantity in BJTs and is usually designated by I_{CO}. Using a similar analysis, with the collector open-circuited and the emitter-base diode reverse-biased, the reverse emitter current I_{EO} is obtained. The two results are stated in Eq. (3-15):

$$I_{CO} = (1 - \alpha_F \alpha_R) I_{CS} \qquad I_{EO} = (1 - \alpha_F \alpha_R) I_{ES} \tag{3-15}$$

Minority-Carrier Concentrations   The excess minority-carrier concentration in the base region, due primarily to forward injection, is depicted in Fig. 3-12. The width W of the base region is defined as the distance between the base side of the emitter-base and collector-base depletion regions. Ideally, the excess minority-carrier density decreases linearly across the base region. In reality, the distribution is given by the dashed line, which takes recombination into account. The concentration is zero at the collector-base boundary as minority carriers reaching this point are swept into the collector.

FIGURE 3-12
Minority-carrier concentration in the base for cutoff, forward-active, and saturation regions.

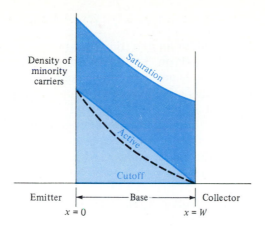

Density of minority carriers

Saturation

Active

Cutoff

Emitter | Base | Collector

$x = 0$ $x = W$

Also shown in Fig. 3-12 are the minority-carrier densities in cutoff and saturation. As expected, reverse-biasing the emitter-base junction effectively precludes forward injection, so that this density is virtually zero. In saturation, an excess of minority carriers exists above the level for the forward-active mode. This excess is attributed to carrier injection into the base by the forward-biased collector-base diode.

3-4 THE COMMON-BASE (*CB*) CHARACTERISTICS

We are now in position to graphically display the volt-ampere characteristics of the BJT based on the Ebers-Moll equations and the modes of operation described in the previous section. To illustrate these characteristics we use a 2N2907A *pnp* transistor. In the next section dealing with the common-emitter circuit, we use a 2N2222A *npn* transistor. These transistors are complementary; that is, their characteristics and rating are virtually identical except that the algebraic signs of the voltages and currents reflect the differences between *pnp* and *npn* devices.

For a *pnp* transistor, the major current components comprise holes. Since holes flow from emitter to collector and out of the base, then, referring to the polarity conventions in Fig. 3-5, we see that I_E is positive and both I_C and I_B are negative. The junction voltages V_{EB} and V_{CB} are positive for forward bias and negative for reverse bias. For *npn* transistors, all current and voltage polarities are the negative of those for *pnp* devices. Note that for both transistor types the signs of I_B and I_C are the same and both are opposite to that for I_E.

The Output Characteristics   It is convenient to recast the Ebers-Moll equations directly in terms of I_E and I_C as follows: for a *pnp* transistor, solve for $I_{CS}(\epsilon^{V_{CB}/V_T} - 1)$ from Eq. (3-3), substitute this value into Eq. (3-2), and identify I_{EO} from Eq. (3-15b). The result (Prob. 3-6) is

$$I_E = I_{EO} (\epsilon^{V_{EB}/V_T} - 1) - \alpha_R I_C \qquad (3-16a)$$

Proceeding in a similar fashion, we find

$$I_C = -\alpha_F I_E + I_{CO}(\epsilon^{V_{CB}/V_T} - 1) \tag{3-16b}$$

These equations are valid for an *npn* transistor provided a minus sign is added before I_C, I_E, V_{EB}, and V_{CB} (Prob. 3-6). From Eq. (3-16b) we see that I_C depends only on the input current I_E and the output voltage V_{CB}. The *output characteristics* which display this relationship are shown in Fig. 3-13 and constitute the family of curves of I_C versus V_{CB} for different values of I_E. To better illustrate behavior in the different modes of operation, only the portion of the characteristics in the vicinity of $V_{CB} = 0$ is shown. These characteristics can be measured by using the circuit shown in Fig. 3-8, where it is assumed that we can vary the amplitudes of each power supply and the values of the two resistances.

In the forward-active region (Table 3-1), I_E is positive, I_C is negative, and V_{CB} is negative. Note that it is customary, as in Fig. 3-13, to plot increasing values of $|I_C|$ in the positive y direction and increasing values of the magnitude of the reverse-bias voltage V_{CB} in the positive x direction. The collector current in the forward-active region is independent of V_{CB} and thus constant for a given value of I_E. This is evident from Eq. (3-16b), which, evaluated in the forward-active mode, yields

$$I_C = -\alpha_F I_E - I_{CO} \tag{3-17}$$

Equation (3-17) is valid for an *npn* transistor if $-I_{CO}$ is changed to $+I_{CO}$. If $I_E = 0$, then from Eq. (3-17), $I_C = -I_{CO}$ and the BJT is cut off. The characteristic for $I_E = 0$ is technically not coincident with the V_{CB} axis but appears so because I_{CO} is extremely small. Note that since $\alpha_F \approx 1$, $|I_C| \approx |I_E|$.

The curves indicate that increasing V_{CB} so that we forward-bias this junction ($V_{CB} \geq 0.6$ V), the collector current increases (I_C becomes less negative). With both diodes forward-biased, the transistor is saturated.

The output characteristics of the inverted BJT display I_E versus V_{EB} for different I_C values. Under these conditions I_C (acting as the emitter current) is

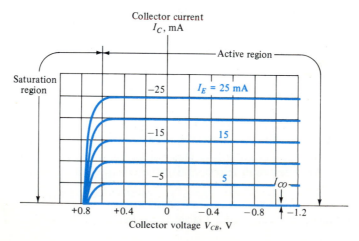

FIGURE 3-13

Common-base output characteristics for a 2N2907A *pnp* silicon transistor in the vicinity of zero collector voltage. Note that the positive and negative V_{CB} axes are reversed from normal.

FIGURE 3-14
(*a*) Common-base input characteristics (V_{EB} versus I_E) for the 2N2907A *pnp* transistor; (*b*) the same characteristics plotted as I_E versus V_{EB}. Note the similarity to a diode curve.

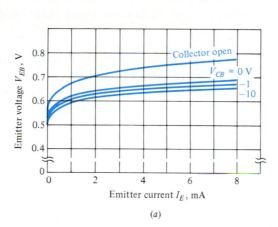

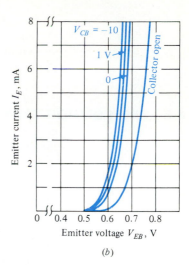

(*a*)

(*b*)

positive and I_E (acting as the collector current) is negative. On the basis of Eq. (3-16), a family of curves (not shown) similar to those in Fig. 3-13 is obtained.

The Input Characteristics The input volt-ampere characteristics are plots of I_E versus V_{EB} for various values of V_{CB}. As seen in Fig. 3-14, these curves represent the characteristics of the emitter-base diode at different collector-base voltages. An evident feature of these characteristics is the existence of a *cut-in, turn-on,* or *threshhold voltage* $V_\gamma = 0.5$ V below which I_E is extremely small. If, with the collector open-circuited, we plotted the reverse-bias characteristic ($V_{EB} < 0$), we would observe a saturation current equal to I_{EO}. A second feature of this curve is that the emitter-base diode characteristic is affected by changing V_{CB}. We now consider the phenomenon which accounts for the shape of the curves in Fig. 3-14.

The Early Effect or Base-Width Modulation In Sec. 2-13 we indicated that the width of the depletion region of a junction increases as the magnitude of the reverse-bias voltage increases. We need only consider effects due to the collector-base junction as the emitter-base diode is forward-biased. Consequently, the effective base width W in Fig. 3-12 decreases with increasing V_{CB}; this modulation of base width is known as the *Early effect.* We can attribute three consequences to base-width modulation: (1) the narrower base width means that there is less chance for recombination, causing α_F to increase as $|V_{CE}|$ increases; (2) the concentration gradient of minority carriers within the base increases (as diffusion current is proportional to the concentration gradient, I_E increases with the reverse-bias voltage at the collector-base diode); and (3) for extremely large voltages, W may be reduced to zero, causing voltage breakdown of the BJT. This phenomenon, referred to as *punch-through,* is discussed in Sec. 3-13.

At a constant value of V_{EB}, the Early effect predicts that I_E increases as we increase $|V_{CB}|$. This conclusion accounts for the shift in the input characteristics shown in Fig. 3-14. In Sec. 3-5 we observe other manifestations of base-width modulation.

3-5 THE COMMON-EMITTER (CE) CONFIGURATION

Most BJT circuits employ the common-emitter configuration shown in Fig. 3-15. This is due mainly to the fact that it is desirable to use the small base current as the control quantity rather than the emitter current. In the CE configuration, I_B, the input current, and V_{CE}, the output voltage, are the independent variables, whereas the input voltage V_{BE} and the output current I_C are the dependent variables.

We feel that the physical operation of the BJT is somewhat easier to understand if reference is made to *pnp* devices. Hence the preceding discussions of the CB configuration and the Ebers-Moll equations focused on the *pnp* transistor. However, *npn* devices are most prevalent in both ICs and discrete-component circuits which employ BJTs. Our discussion of the CE configuration thus concentrates on *npn* transistors, and, as previously stated, we utilize the 2N2222A, a discrete transistor which is a widely used industry standard.

The Output Characteristics[1]

The common-emitter output characteristics is the family of curves shown in Fig. 3-16 in which I_C versus V_{CE} is plotted for various values of I_B. A load line corresponding to $R_C = 500\ \Omega$ and a supply voltage $V_{CC} = 10$ V has been superimposed on these characteristics. Construction of the load line is based on KVL for the output loop and is identical to the method explained in Sec. 2-4. The output characteristics display three regions of operation, as did the common-base characteristics. The active region is discussed here; cutoff and saturation are considered in the next section.

In the active region, for an *npn* transistor, Eq. (3-17) must be modified to $I_C = -\alpha_F I_E + I_{CO}$. Combination of this equation with Eq. (3-5) yields

---

[1]Transistor output and input characteristics are no longer supplied by the transistor manufacturer since they are seldom used in either digital or analog design. However, these characteristics are necessary for an understanding of the transistor. The device characteristics shown in this chapter were obtained experimentally.

FIGURE 3-15
A common-emitter circuit employing an *npn* transistor.

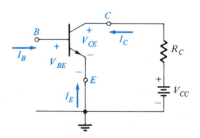

FIGURE 3-16
Common-emitter output characteristics of a 2N2222A *npn* silicon transistor. A load line corresponding to $V_{CC} =$ 10 V and $R_C = 500 \, \Omega$ is superimposed.

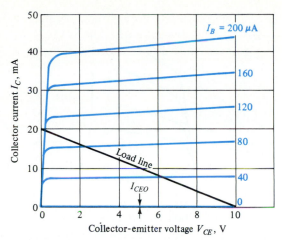

Identification of $\beta_F = \alpha_F/(1 - \alpha_F)$ from Eq. (3-13) allows Eq. (3-18) to be written as

$$I_C = \frac{\alpha_F I_B}{1 - \alpha_F} + \frac{I_{CO}}{1 - \alpha_F} \tag{3-18}$$

$$I_C = \beta_F I_B + (\beta_F + 1) I_{CO} \tag{3-19}$$

Typical of BJT operation in the active region is the fact that $I_B \gg I_{CO}$. Thus

$$I_C = \beta_F I_B \tag{3-20}$$

is an excellent approximation of the collector current and is used extensively. Equation (3-20) indicates controlled-source behavior in the active mode. By controlling the input current I_B, we can specify the output current I_C.

The *dc forward current gain* h_{FE} is a quantity specified by device manufacturers and is defined as

$$h_{FE} \equiv \frac{I_C}{I_B} \approx \beta_F \tag{3-21}$$

The subscripts F and E denote "forward transfer" and common emitter, respectively. Because I_{CO} is, in general, negligible compared with other currents in the active region, h_{FE} and β_F have essentially the same values.[1]

If α_F were truly constant, then, according to Eq. (3-18), I_C would be independent of V_{CE} and the curves of Fig. 3-16 would be horizontal. Assume that, because of the Early effect, α_F increases by only 0.1 percent, from 0.995 to 0.996, as $|V_{CE}|$ increases from a few volts to 10 V. Then the value of β_F increases

---

[1]The dc current gain is sometimes called β_{dc}. As $h_{FE} = \beta_{dc} \approx \beta_F$, these quantities are often used interchangeably in the literature.

FIGURE 3-17
Common-emitter out-
put characteristics for
an *npn* transistor with
V_{BE} as a parameter. The
curves are extended
back (dashed lines) to
the negative V_{CE} axis
and intersect at the Early
voltage.

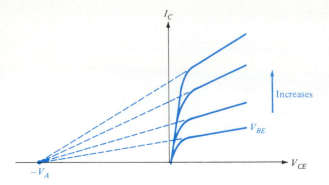

FIGURE 3-17
Common-emitter output characteristics for an *npn* transistor with V_{BE} as a parameter. The curves are extended back (dashed lines) to the negative V_{CE} axis and intersect at the Early voltage.

from $0.995/(1 - 0.995) = 200$ to $0.996/(1 - 0.996) = 250$, or about 25 percent. This numerical example illustrates that a very small change (0.1 percent) in α_F is reflected in a very large change (25 percent) in the value of β_F. It should also be clear that a slight change in α_F has a large effect on β_F and hence on the common-emitter curves. Therefore, the common-emitter characteristics are normally subject to a wide variation even among transistors of a given type. This variation in β_F is an important consideration in circuit design.

The influence of the Early effect on the *CE* output curves is demonstrated graphically in Fig. 3-17. In Fig. 3-17 we plot curves of I_C versus V_{CE} for various values of V_{BE} for a "typical" *npn* transistor. If we extend the linear portion of these curves back to the V_{CE} axis, as is indicated by the dashed lines, they would all meet at the common point $-V_A$. The voltage V_A is called the *Early voltage* for which typical values lie between 50 and 100 V. The Early voltage determines the slope of the I_C versus V_{CE} characteristic (in Fig. 3-17) for a given operating value of V_{BE}. The reciprocal of this slope has the dimensions of ohms, and, in succeeding sections dealing with BJT models, this effect will manifest itself as a resistance associated with the controlled source.

The common-emitter current gain $\beta_F \approx h_{FE}$ also varies with collector current as illustrated in Fig. 3-18 for a typical IC transistor and in Fig. 3-19 for the 2N2222A. Note that I_C is plotted on a logarithmic scale in both Figs. 3-18 and 3-19. We observe in Fig. 3-18 that β_F decreases from its midcurrent levels at

FIGURE 3-18
Normalized β_F variation with collector current I_C for an IC transistor. Note the logarithmic scale for I_C.

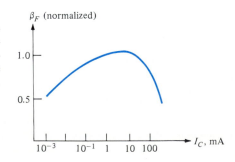

FIGURE 3-19

Plots of β_F (normalized to unity at $V_{CE} = 1$ V, $I_C = 30$ mA at $T = 25°C$) versus collector current at three different junction temperatures T_j for the 2N2222A transistor. (*Courtesy of Motorola, Inc.*)

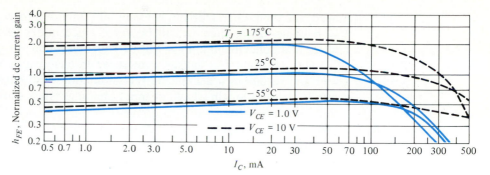

both high and low values of I_C.[1] The majority of bipolar ICs utilize transistors in their midcurrent range, where β_F is nearly constant. The curves in Fig. 3-19 also display the decrease in $h_{FE} = \beta_F$ at high and low current levels. Over a wide range of values, however, the normalized current gain is fairly constant. Note the increase in h_{FE} as V_{CE} is increased from 1.0 to 10 V because of the Early effect. Both sets of curves show a wide variation in β_F even for a transistor of a particular type. Device manufacturers generally specify minimum and maximum values for particular values of V_{CE} and temperature.

The Input Characteristics The input characteristics (shown in Fig. 3-20) are curves which display the relationship between I_B and V_{BE} for different values of V_{CE}. We observe that, with the collector shorted to the emitter and the emitter forward-

---

[1] This is due to parasitic surface effects at low currents and second-order effects associated with high carrier injection for large values of I_C. The details of this variation are beyond the scope of this book and are usually treated in texts on device physics.

FIGURE 3-20

(*a*) Common-emitter input characteristics (V_{BE} versus I_B) for the 2N2222A transistor; (*b*) the curves in part *a* plotted as I_B versus V_{BE}. Note the similarity with the diode characteristic.

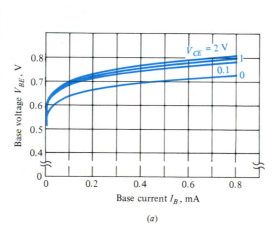

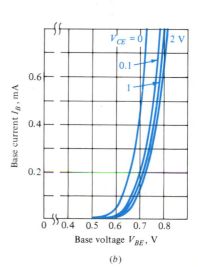

(*a*)

(*b*)

biased, the input characteristic is essentially that of a forward-biased diode. If V_{BE} becomes zero, then I_B will be zero, since under these conditions both emitter and collector junctions will be short-circuited. In general, increase in $|V_{CE}|$ with constant V_{BE} causes a decrease in base width W due to the Early effect and results in a decreasing recombination base current. These considerations account for the shape of input characteristics shown in Fig. 3-20.

The Reverse-Active Mode   The input and output characteristics of the inverted transistor have the same general shape as in Figs. 3-20 and 3-16. The reverse-active input characteristics display the behavior of the forward-biased collector-base junction. Recall that in the reverse-active mode, α_R and β_R have lower values than α_F and β_F, respectively. Consequently, for a given value of I_B, lower values of I_E exist in the reverse-active mode than in the forward-active region.

3-6   CUTOFF AND SATURATION MODES   We indicated qualitatively in Sec. 3-2 that operation of a BJT in cutoff and saturation approximates the state of an open and closed switch, respectively. In this section we investigate the behavior of the transistor in these modes in more quantitative terms.

Cutoff   Both junctions are reverse-biased in cutoff. In the *CB* configuration we showed that cutoff occurs when the input current $I_E = 0$ and, consequently, $I_C = -I_B = I_{CO}$. For the *CE* configuration, we now examine operation when the input current $I_B = 0$. It is important to note that, theoretically, both junctions are not reverse-biased if the base is open-circuited (Prob. 3-5). From Eq. (3-5), if $I_B = 0$, $I_E = -I_C$, and use of Eq. (3-18) gives

$$I_C = -I_E = \frac{I_{CO}}{1 - \alpha_F} \equiv I_{CEO} \qquad (3\text{-}22)$$

The rationale for the subscripts of the current I_{CEO} in Eq. (3-22) is that this is the current from C to E when B (the missing terminal) is *Open*. At $I_C \approx I_{CO}$, α_F is very nearly zero because of recombination in the emitter-base depletion region. Hence, from Eq. (3-22), we find that $I_C = -I_E = I_{CEO} = I_{CO}$, and, for practical purposes, the BJT is very nearly cut off. In Example 3-1,[1] the relationship given by Eq. (2) indicates that for α_F approaching zero, $V_{EB} \approx 0$. Thus cutoff ($I_E = 0$) *for a silicon transistor requires a reverse voltage V_{BE} that is virtually zero and $-I_B = I_C = I_{CO}$.*

The collector current in a physical transistor (a real, nonidealized, commercial device) when the emitter current is zero is designated by the symbol I_{CBO}. Two factors cooperate to make I_{CBO} larger than I_{CO}: (1) there exists a leakage current which flows, not through the junction, but around it and across the surfaces—this current is proportional to the voltage across the junction;

---

[1] In this example, we assumed that α_F was constant at all current levels.

and (2) I_{CBO} exceeds I_{CO} because new carriers may be generated by collision in the collector-junction transition region, leading to avalanche multiplication. But even before breakdown is approached, this multiplication component of current may attain considerable proportions (Fig. 3-40).

At 25°C, I_{CBO} for a silicon transistor whose power dissipation is in the range of some hundreds of milliwatts is of the order of nanoamperes. Typical, small-dimension IC BJTs have values of I_{CBO} in the order of tens and hundreds of picoamperes.

A germanium transistor has an I_{CBO} in the range of microamperes. The temperature sensitivity of I_{CBO} is the same as that of the reverse saturation current I_S of a pn diode (Sec. 2-4). Specifically, it is found that I_{CBO} approximately doubles for every 10°C increase in temperature for silicon. However, because of the lower absolute value of I_{CBO} in silicon, these transistors may be used up to a junction temperature of about 200°C, whereas germanium transistors are limited to about 100°C.

In addition to the change of reverse saturation current with temperature, there may also be a wide variability (by a factor of ≥ 100) of I_{CBO} among samples of a given discrete transistor type. Accordingly, the manufacturers' specification sheets (App. B-3) list the maximum value of I_{CBO}. A low-power silicon transistor is considered "leaky" if I_{CBO} exceeds 10 nA at 25°C.

Cutoff in the Inverted Transistor   Reverse-biasing the emitter-base diode and open-circuiting the collector corresponds to cutoff in the inverted transistor. Under these conditions, the *emitter cutoff current* is I_{EBO}. For specified values of V_{CE} and V_{BE} (reverse-biased), the *collector cutoff current* and *base cutoff current* are designated by I_{CEX} and I_{BL}, respectively. The maximum values of these currents are also listed in the specification sheets and are of the same order of magnitude as I_{CBO}.

The CE Saturation Region   In the saturation region the collector junction (as well as the emitter junction) is forward-biased by at least the cut-in voltage. Since the voltage V_{BE} (or V_{BC}) across a forward-biased junction has a magnitude of only a few tenths of a volt, $V_{CE} = V_{BE} - V_{BC}$ is also only a few tenths of a volt at saturation. Hence in Fig. 3-16 the saturation region is very close to the zero-voltage axis, where all the curves merge and fall rapidly toward the origin. A load line has been superimposed on the characteristics shown in Fig. 3-16 corresponding to a resistance $R_C = 500\ \Omega$ and a supply voltage of 10 V. We note that in the saturation region the collector current is approximately independent of base current for given values of V_{CC} and R_C. Hence we may consider that the onset of saturation takes place at the knee of the transistor curves in Fig. 3-16.

We are not able to read the collector-to-emitter voltage $V_{CE(\text{sat})}$ with any precision from the curves in Fig. 3-16. We refer instead to Fig. 3-21, in which the characteristics in the 0- to 0.5-V region in Fig. 3-16 are expanded. Again, the same load line corresponding to $R_C = 500\ \Omega$ and $V_{CC} = 10$ V is superimposed. We observe in saturation that both I_C and V_{CE} are nearly independent

FIGURE 3-21
The characteristics of the 2N2222A transistor in and near the saturation region. A load line corresponding to $V_{CC} = 10$ V and $R_C = 500\ \Omega$ is superimposed.

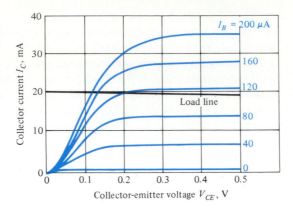

of I_B. A change in I_B from 120 to 160 μA (Fig. 3-21) indicates about a 50-mV change in $V_{CE(\text{sat})}$ and, even on an expanded scale, an imperceptible change in I_C. By contrast, in Fig. 3-16, a 40-μA change in I_B, from 40 to 80 μA, is accompanied by significant changes in both I_C and V_{CE}. This is the active region for which $I_C = \beta_F I_B$. In saturation, I_B no longer "controls" I_C so that α_F no longer relates the two. It is convenient to introduce the parameter

$$\beta_{\text{forced}} \equiv \left. \frac{I_C}{I_B} \right|_{\text{in saturation}} \tag{3-23}$$

to relate I_C and I_B in saturation. Note that $\beta_{\text{forced}} < \beta_F$; a value of $\beta_{\text{forced}} = \beta_F$ corresponds to the active region. Both Figs. 3-16 and 3-21 show that in saturation I_C is determined by the external elements V_{CC} and R_C and is approximately V_{CC}/R_C.

We also observe in Fig. 3-21 that the collector-emitter voltage $V_{CE(\text{sat})}$ does vary somewhat with I_B. In digital circuits which exploit the BJT as a switch, the value of $V_{CE(\text{sat})}$ is important. [One can consider $V_{CE(\text{sat})}$ as a measure of how closely a practical switch approximates the ideal.] The Ebers-Moll equations are used to determine analytically the value of $V_{CE(\text{sat})}$. The details of this analysis are left to the student (Prob. 3-7). The method is as follows. Starting with Eqs. (3-6) and (3-7), obtain the expression for I_B. Solve for I_{ED} and I_{CD} from the equations for I_C and I_B. Take the logarithm of I_{ED}/I_{CD} and identify $V_{CE} = V_{CB} - V_{EB}$ and $\beta_{\text{forced}} = I_C/I_B$. The result is

$$V_{CE} = V_{CE(\text{sat})} = V_T \ln \frac{1/\alpha_R + \beta_{\text{forced}}/\beta_R}{1 - \beta_{\text{forced}}/\beta_F} \tag{3-24}$$

Table 3-2 illustrates the variation of $V_{CE(\text{sat})}$ with β_{forced} at room temperature for an IC transistor having $\beta_F = 100$ and $\beta_R = 1$.

TABLE 3-2 Variation in $V_{CE(\text{sat})}$ with β_{forced}

β_{forced}	99.9	99	75	50	25	10	5	1	0.1	0.01
$V_{CE(\text{sat})}$ (mV)	286	231	143	116	86	65	50	28	19	18

It is evident in Table 3-2 that $V_{CE(sat)}$ decreases as β_{forced} decreases and we drive the BJT more heavily into saturation. The derivation of Eq. (3-24) neglects the bulk resistance of the semiconductor forming the collector region.[1] Even with as low a resistance as 5 Ω, a 10-mA current produces a 50-mV drop so that $V_{CE(sat)}$ is generally assumed to be about 0.2 V. We also observe that as β_{forced} approaches β_F, $V_{CE(sat)}$ is about 0.3 V.[2] A value of $V_{CE(sat)} = 0.3$ V is commonly used as the boundary between the active and saturated regions. Transistors operating at this boundary are referred to as being barely saturated or at the edge of saturation.

Most manufacturers of discrete transistors display the variation of $V_{CE(sat)}$ with I_C at a value of $\beta_{forced} = 10$. Such a curve is shown for the 2N2222A in Fig. 3-22. At high currents we note an increase in $V_{CE(sat)}$ due to bulk resistance effects.[3] Throughout the midcurrent range, the values of $V_{CE(sat)}$ are comparable with those for IC transistors. Also shown in Fig. 3-22 is the variation with I_C of the emitter-base voltage in saturation $V_{BE(sat)}$.

A quantity sometimes used to describe the BJT in saturation is $V_{CE(sat)}/I_C$, called the *common-emitter saturation resistance* and designated by R_{CES}, R_{CS}, or $R_{CE(sat)}$. To specify R_{CES} properly, we must indicate the operating point at which it is determined. Note that when R_{CES} is obtained from measured values, bulk resistance effects are included. The usefulness of R_{CES} stems from the fact (as apparent in Fig. 3-21) that to the left of the knee each curve, for a fixed I_B, can be approximated by a straight line.

Summary of BJT Voltages   Typical values of transistor operating voltages are listed in Table 3-3. We will use these values throughout the text.

---

[1] The emitter bulk resistance also has an effect. However, doping density and physical size make the collector resistance the predominant concern.

[2] The limit of $V_{CE(sat)}$ as $\beta_{forced} \to \beta_F$ is, from Eq. (3-24), infinite. However, $\beta_{forced} = \beta_F$ corresponds to the active region, and the assumptions in deriving Eq. (3-24) are no longer valid.

[3] Construction of discrete BJTs often permits the bulk resistances to be smaller than for IC devices.

FIGURE 3-22

Saturation voltages for the 2N2222A transistor versus collector current for $I_C/I_B = 10$. Note that I_C is plotted on a logarithmic scale. (*Courtesy of Motorola, Inc.*)

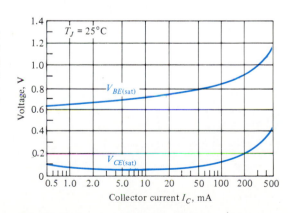

TABLE 3-3   Typical Junction Voltages at 25°C

Quantity	V_{CE} at edge of saturation	$V_{CE(\text{sat})}$	V_{BE}			
			Cut-in	Active	Saturation	Cutoff
Value (in V)	0.3	0.2	0.5	0.7	0.8	0

It is reasonable to expect that the temperature variation of the voltage across a forward-biased junction is the same as that for a diode, namely, -2.2 mV/°C. In saturation the transistor consists of two forward-biased diodes back to back in series opposing. Hence it is to be anticipated that the temperature-induced voltage change in one junction will be canceled by the change in the other junction. We do indeed find such to be the case for $V_{CE(\text{sat})}$ whose temperature coefficient is about one-tenth that of $V_{BE(\text{sat})}$.

The values of circuit currents and voltages obtained from ''pencil-and-paper'' calculations based on the data in Table 3-3 correlate well with experimental values. The reader should be aware, however, that these values are typical and not exact. A variety of reasons exist in the design, fabrication, and manufacture of circuits which require the designer to have more accurate results. In these situations, computer simulations, such as SPICE, are widely employed. Pencil-and-paper calculations are still used to indicate nominal values of circuit quantities.

3-7   DC MODELS   We can construct a dc model for each operating region of the BJT from the previous discussions of the Ebers-Moll equations. Our focus is on the *CE* configuration, but the models apply equally well to the *CB* circuit.

The model for the forward-active region is displayed in Fig. 3-23*a* and is based on Eq. (3-11). Because reverse saturation currents are generally negligibly small, their effect is usually omitted. The battery in the base-emitter circuit is V_{BE} and, from Table 3-3, is usually 0.7 V. The controlled current source $\beta_F I_B$ relates I_C and I_B in the active region. The resistance R_0, indicated by the dashed symbol, is the consequence of the Early effect. Generally, R_0 is sufficiently

FIGURE 3-23

Large-signal (dc) equivalent circuits for an *npn* transistor for (*a*) forward-active and (*b*) saturation-region operation.

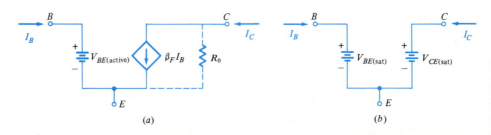

(*a*)   (*b*)

larger than the external resistances used so it is neglected in many pencil-and-paper calculations.[1]

The current I_{CBO} enters the collector terminal and leaves the base terminal in the cutoff region ($I_E = 0$). Typical values of the voltage drops across the external base and collector resistances produced by I_{CBO} are less than a few millivolts at room temperature. Consequently, it is often convenient to represent cutoff by open circuits between each pair of transistor terminals.

In saturation, the equivalent circuit in Fig. 3-23b is useful for determining currents and voltages in a circuit. The two batteries represent the saturation values of the terminal voltages $V_{BE(\text{sat})}$ and $V_{CE(\text{sat})}$.

The use of these models in the dc analysis of BJT circuits requires that we know the region of operation. As with the diode circuits in Sec. 2-7, the method involves that we assume a particular operating region and, by analysis, verify our assumption. Observation of the circuit configuration and the bias supplies (and some experience) helps us to "guess" correctly. The following four examples illustrate the methodology used in analysis. Each circuit in these examples is commonly incorporated as a part of the analog and digital circuits described later in this text.

Example 3-2 Determine the region of operation and the values of I_B, I_C, and V_{CE} for the circuit in Fig. 3-24a for R_B equal to (a) 300 kΩ and (b) 150 kΩ. The transistor used has $\beta_F = 100$. Neglect reverse saturation currents.

Solution Observation of the circuit in Fig. 3-24a makes it evident that, with the base returned to a positive potential and the emitter at ground, $V_{BE} > 0$. Thus we can safely state that the emitter-base junction is forward-biased. Consequently, the BJT is either in its forward-active mode or saturated. We assume forward-active operation and use the model in Fig. 3-23a to obtain the equivalent circuit in Fig. 3-24b. Note that in Fig. 3-24a the terminal $+V_{CC}$ signifies a connection

---

[1]It is not neglected in computer simulations.

FIGURE 3-24
(a) Schematic diagram of the *CE* configuration; (b) equivalent circuit of part a.

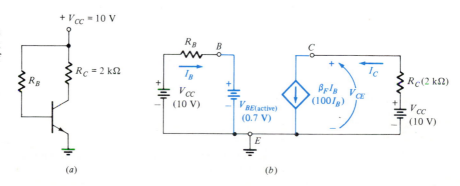

(a)   (b)

to the positive terminal of the supply voltage. It is implied that the negative terminal of the supply is grounded.

To verify our assumption, the value of V_{CE} is computed; if $V_{CE} > 0.3$ V, our assumption is correct. A value of $V_{CE} < 0.3$ V indicates (see Table 3-2) an incorrect guess, the BJT is in saturation, and we must recompute the currents and voltages using the BJT model given in Fig. 3-23b.

(a) For the circuit in Fig. 3-24b, the KVL expression for the emitter-base loop is

$$-V_{CC} + I_B R_B + V_{BE} = 0$$

Solving for I_B and substituting values[1] gives

$$I_B = \frac{V_{CC} - V_{BE}}{R_B} = \frac{10 - 0.7}{300} = 0.031 \text{ mA} = 31.0 \text{ } \mu\text{A}$$

The collector-loop relations are

$$I_C = \beta_F I_B \qquad \text{and} \qquad -V_{CC} + I_C R_C + V_{CE} = 0$$

from which

$$I_C = 100 \times 0.031 = 3.10 \text{ mA}$$

and

$$V_{CE} = V_{CC} - I_C R_C = 10 - 3.1 \times 2 = 3.80 \text{ V}$$

With $V_{CE} > 0.3$ V, our initial assumption is verified.

(b) For $R_B = 150$ kΩ and using the relations in part a, we obtain

$$I_B = \frac{10 - 0.7}{150} = 0.062 \text{ mA} = 62.0 \text{ } \mu\text{A} \qquad I_C = 100 \times 0.062 = 6.20 \text{ mA}$$

and

$$V_{CE} = 10 - 6.2 \times 2 = -2.40 \text{ V}$$

As $V_{CE} < 0.3$ V, our assumption is invalid and the BJT is in saturation. Indeed, with a positive collector supply voltage and the emitter at ground, it is physically impossible to obtain a negative value of V_{CE}!

In saturation, $V_{BE(\text{sat})} = 0.8$ V and $V_{CE(\text{sat})} = 0.2$ V. Use of these values gives

$$I_B = \frac{10 - 0.8}{150} = 0.0613 \text{ mA}$$

$$I_C = \frac{V_{CC} - V_{CE(\text{sat})}}{R_C} = \frac{10 - 0.2}{2} = 4.90 \text{ mA}$$

---

[1]It is convenient to use current in milliamperes and resistance in kilohms in calculations. We will do so unless otherwise noted.

FIGURE 3-25
(*a*) Circuit for Example 3-3; (*b*) circuit in part *a* with the transistor replaced by its dc model (Fig. 3-23*a*).

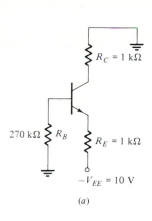

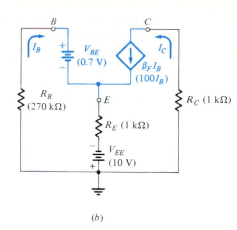

(*a*)   (*b*)

Example 3-3

For the circuit in Fig. 3-25*a*, determine the region of operation and the values of I_B, I_C, and V_{CE}. The BJT has $\beta_F = 100$.

Solution

This circuit differs from that in Fig. 3-24*a* in two respects: (1) an emitter resistance has been added; and (2) the base and collector resistances are returned to ground and the emitter is connected, through R_E, to a negative supply voltage. We will assume operation in the forward-active mode; the equivalent circuit is depicted in Fig. 3-25*b*. Then KVL for the base-emitter loop gives

$$I_B R_B + V_{BE} - I_E R_E - V_{EE} = 0$$

Since KCL requires $I_E = -(I_B + I_C)$ and $I_C = \beta_F I_B$, the KVL equation becomes

$$I_B [R_B + (1 + \beta_F)R_E] + V_{BE} - V_{EE} = 0$$

Solving for I_B and substituting numerical values yields

$$I_B = \frac{V_{EE} - V_{BE}}{R_B + (1 + \beta_F)R_E} = \frac{10 - 0.7}{270 + (1 + 100)(1)} = 0.0251 \text{ mA}$$

The KVL expression for the collector-emitter loop is

$$I_C R_C + V_{CE} - I_E R_E - V_{EE} = 0$$

From KCL at the emitter mode we replace $-I_E$ by $I_B + \beta_F I_B = (1 + \beta_F)I_B$ and with $I_C = \beta_F I_B$ obtain

$$\beta_F I_B R_C + V_{CE} + (\beta_F + 1)I_B R_E - V_{EE} = 0$$

Solving for V_{CE} and using the known values, we obtain

$$V_{CE} = V_{EE} - \beta_F I_B \left(R_C + \frac{\beta_F + 1}{\beta_F} R_E \right)$$

$$= 10 - 100 \times 0.0251 \left(1 + \frac{100 + 1}{100} \times 1 \right) = 4.96 \text{ V}$$

Obviously, $V_{CE} > 0.3$ V and verifies our assumption that the operation is in the forward-active region. Hence

$$I_C = \beta_F I_B = 100 \times 0.0251 = 2.51 \text{ mA}$$

Example 3-4

(a) Determine I_C and V_{CE} for the circuit in Fig. 3-26a. The transistor has $\beta_F = 150$. (b) What is the minimum value of R_C for which the transistor is just barely saturated?

Solution

(a) For convenience, Fig. 3-26a is redrawn as is shown in Fig. 3-26b. The base-bias network indicated in Fig. 3-26b can be replaced by its Thévenin equivalent and is so indicated in Fig. 3-26c, where

$$V_{BB} = \frac{R_2}{R_1 + R_2} V_{CC} = \frac{11 \times 12}{110 + 11} = 1.09 \text{ V}$$

$$R_B = R_1 \parallel R_2 = \frac{R_1 R_2}{R_1 + R_2} = \frac{110 \times 11}{110 + 11} = 10 \text{ k}\Omega$$

Note the similarity of this circuit with that in Fig. 3-24a (with the addition of R_E). We will again assume operation in the forward-active region for which the

FIGURE 3-26
(a) Circuit for Example
3-4; (b) circuit in part a
showing base-bias net-
work. The base-bias
network is replaced by
its Thévenin equivalent
in part c. The transistor
is represented in its forward-active model in
part d.

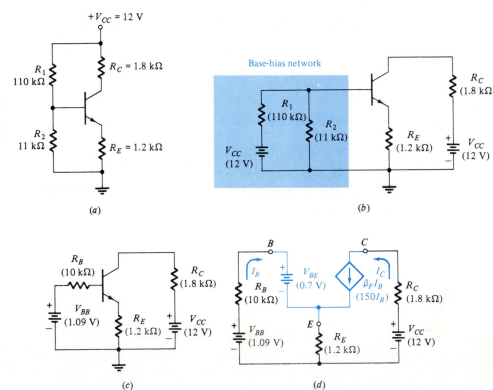

(a)

(b)

(c)

(d)

model is given in Fig. 3-26d. Proceeding as in Example 3-3, with V_{BB} replacing V_{EE} in the base loop and V_{CC} replacing V_{EE} in the collector loop, we obtain

$$I_B = \frac{1.09 - 0.7}{10 + (150 + 1) \times 1.2} = 2.04 \times 10^{-3} \text{ mA} = 2.04 \text{ } \mu\text{A}$$

$$V_{CE} = 12 - 150 \times 2.04 \times 10^{-3} \left[1.80 + \frac{150 + 1}{150} \times 1.20 \right] = 11.1 \text{ V}$$

The BJT is biased forward-active as $V_{CE} > 0.3$ V. Thus

$$I_C = 150 \times 2.04 \times 10^{-3} = 0.306 \text{ mA}$$

(b) At the edge of saturation, $V_{CE} = 0.3$ V and $\beta_{\text{forced}} = \beta_F = 150$. With no changes in the base-bias network, I_B remains unchanged from the value determined in part a. For $\beta_F = 150$, the value of I_C is that obtained in part a and equals 0.306 mA. Then, from the KVL expression for the collector-emitter loop in part a, R_C is determined as

$$R_C = \frac{V_{CC} - V_{CE}}{I_C} - \frac{\beta_F + 1}{\beta_F} R_E$$

$$= \frac{12 - 0.3}{0.306} - \frac{150 + 1}{150} \times 1.20 = 37.0 \text{ k}\Omega$$

This is the value of R_C at the edge of saturation. Any value of R_C larger than 37.0 kΩ will reduce I_C and, consequently, for a constant I_B, will drive the BJT further into saturation.

Example 3-5

Determine I_C and V_{CE} for the circuit in Fig. 3-27a. The transistor has $\beta_F = 125$ and $\beta_R = 2$.

Solution

Let us assuue forward-active operation. The base-emitter loop KVL equation is, with $I_E = -(\beta_F + 1)I_B$

$$I_B R_B + V_{BE} + (\beta_F + 1)I_B R_E = 0$$

Examination of this equation indicates that for $V_{BE} > 0$, I_B is negative. This is an impossible condition in an *npn* transistor. With $I_{CBO} \approx 0$, I_B must always be greater than or equal to zero. Thus, the emitter-base junction cannot be forward-biased; with reverse bias the BJT is either cut off or in its reverse-active mode. If we assume that the transistor is cut off, then $I_B = I_C = I_E = 0$. Consequently, the voltage drop from base to ground is $V_{EE} = 5$ V and the drop from collector to ground is 0 V. These values make V_{BC} positive (5 V), thus forward-biasing the collector-base diode. Therefore, the BJT can be only in the reverse-active mode.

Figure 3-27b gives the equivalent circuit for reverse-active operation. From this figure we see that KVL for the base-collector loop requires

$$-V_{EE} + I_B R_B + V_{BC} - I_C R_C = 0$$

FIGURE 3-27
(*a*) The circuit for Ex-
ample 3-5; the reverse-
active equivalent circuit
of the transistor is used
in part *b*.

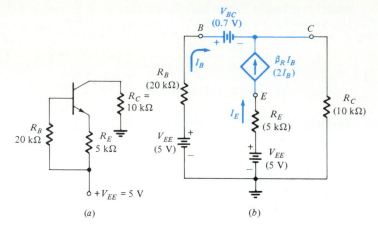

(*a*)   (*b*)

In the reverse-active mode $-I_C = (\beta_R + 1)I_B$; thus, solving for I_B results in

$$I_B = \frac{V_{EE} - V_{BC}}{R_B + (\beta_R + 1)R_C} = \frac{5 - 0.7}{20 + (2 + 1)10} = 0.086 \text{ mA}$$

The collector and emitter current become

$$I_C = -(2 + 1) \times 0.086 = -0.258 \text{ mA}$$
$$I_E = \beta_R I_B = 2 \times 0.086 = 0.172 \text{ mA}$$

The value of V_{CE} is determined from the KVL equation for the collector-emitter loop as

$$I_C R_C + V_{CE} - I_E R_E + V_{EE} = 0$$

Substitution of values yields

$$V_{CE} = -V_{EE} + I_E R_E - I_C R_C$$
$$= -5 + 0.172 \times 5 - (-0.258) 10 = -1.56 \text{ V}$$

Note that in the reverse-active mode V_{CE} is negative and $|V_{CE}| > 0.3$ V as the collector and emitter roles are reversed. A value of $|V_{CE}| < 0.3$ V, with V_{CE} negative, is indicative of saturation of the inverted transistor.

3-8 THE BJT AS A SWITCH

The circuit in Fig. 3-28*a* is a simple *CE* switch. The input voltage waveform v_s shown in the figure is used to control the state of the switch (between collector and emitter). For $t < T_1$, $v_s = -V_1$ and the emitter-base diode is reverse-biased. If we neglect the reverse-current components,[1] since the collector-base diode is also reverse-biased, the BJT is cut off and no current exists anywhere in the circuit. Consequently, $v_o = V_{CC}$, and with i_C

---

[1] From now on we shall neglect these currents in calculations unless we state otherwise.

FIGURE 3-28
(a) A simple BJT switch with input waveform; (b) the waveforms for v_o and i_C displaying the rise time, fall time, delay, and storage time during switching.

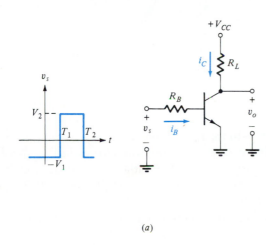

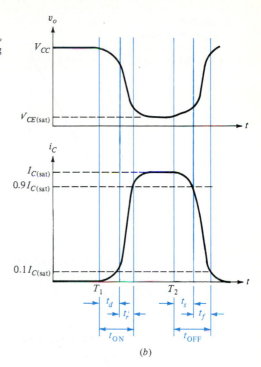

(a)

(b)

$= 0$, this is an open switch. Actually, $i_C \approx I_{CO}$ and $v_o = V_{CC} - I_{CO}R_L$. However, with I_{CO} of the order of a nanoampere and R_L of the order of kilohms, v_o differs from V_{CC} by only a few microvolts. Thus, for practical purposes, $v_o = V_{CC}$.

The input voltage becomes V_2 for $T_1 < t < T_2$. The value of V_2 is selected to ensure that the BJT is at least at the edge of saturation. From Table 3-2, $v_{CE} = v_o = V_{CE(\text{sat})} \leq 0.3$ V and $i_C = (V_{CC} - V_{CE(\text{sat})})/R_L$; these values approximate the closed switch. Note that the current in the closed switch is determined by the external elements V_{CC} and R_L. For $V_{CC} \gg 0.3$ V, $i_C = V_{CC}/R_L$.

At $t = T_2$, the input waveform switches back to $-V_1$, eventually causing the BJT to return to cutoff. Sketches of both v_o and i_C are depicted in Fig. 3-28. The causes of the switching transients are described later in this section.

The nature of the switching characteristics is readily discernible from the transfer characteristic, a graph of v_o versus v_s, for the circuit. The following example develops this characteristic.

Example 3-6

The circuit in Fig. 3-28a employs a 2N2222A transistor, $V_{CC} = 10$ V, $R_L = 500$ Ω, and $R_B = 47$ kΩ. (a) Plot the transfer characteristic of the circuit. (b) Sketch the output waveform for $t \leq 10$ ms for the input voltage shown in Fig. 3-29a.

Solution

(a) The load line for this circuit is superimposed on the BJT output characteristics in Fig. 3-16; the input volt-ampere curves are displayed in Fig. 3-20. From Fig. 3-20 we observe that no appreciable base current exists until the cut-in

FIGURE 3-29
(*a*) The input waveform and (*b*) transfer characteristic for Example 3-6.

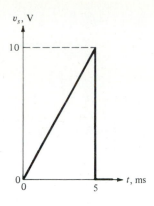

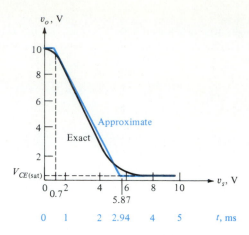

voltage is exceeded. In Fig. 3-16 for $I_B \approx 0$, we see that $v_o = v_{CE} = 10$ V. We indicated in Sec. 3-5 that increasing V_{CE} shifts the curves in Fig. 3-20b to the right. Thus we can assume, for $V_{CE} = 10$ V, that cut in occurs near $V_{BE} = 0.7$ V. This is expected as increasing v_s causes a transition in BJT operation from cutoff to the forward-active region.

Once the transistor is in the active region, $V_{BE} = 0.7$ V and, from the KVL expression for the base-emitter loop,

$$ i_B = \frac{v_s - V_{BE}}{R_B} = \frac{v_s - 0.7}{47} \text{ mA} $$

As i_B increases linearly with v_s, we see that along the load line v_{CE} decreases almost linearly until the transistor approaches saturation. The onset of saturation (shown in Fig. 3-16) occurs as I_B approaches 120 μA. Let us interpolate this value of I_B from Fig. 3-16 as 110 μA. The corresponding value of v_s is

$$ 0.110 = \frac{v_s - 0.7}{47} \qquad \text{and} \qquad v_s = 5.87 \text{ V} $$

Further increase in v_s, and hence i_B, has no effect on the output.

We obtain the points on the transfer characteristic by first determining i_B for various values of v_s. We then use these values to find the corresponding values of v_{CE} from the load line in Fig. 3-16 in the forward-active region. In saturation, $v_{CE} < 0.3$ V.

The exact transfer characteristic in Fig. 3-29b is rounded near both cutoff and saturation. We can attribute this to the fact that the BJT does not turn on abruptly in the vicinity of cut in but, as seen in Fig. 3-20b, there is a knee in the input characteristics. Similarly, the knee in the output characteristics of Fig. 3-16 near saturation show the curves for various values of I_B are bunched.

The approximate piecewise linear transfer characteristic assumes abrupt transitions from cutoff to the active region and from the active mode to saturation. As shown in Fig. 3-29b, this is a good approximation of the exact characteristic and, therefore, is often used.

(b) The response to an input waveform that varies linearly with time has the same shape as the transfer characteristic. Thus the curves in Fig. 3-29b indicate the "exact" and approximate output waveforms versus time. The time scale corresponds to the slope of the input waveform of 2 V/ms, and $v_o = V_{CE(sat)}$ at $t = 2.94$ ms.

The approximate transfer characteristic illustrates the behavior of the circuit. The two horizontal segments indicate the open (OFF) and closed (ON) states of the switch. Along each of these segments the output is unaffected by changes in the input as the BJT is cut off or saturated. The line segment that connects the horizontal portions of the characteristic illustrates the linear dependence of the output on the input. This is the controlled-source behavior needed for amplification and is provided by the BJT biased in the forward-active region.

BJT Switching Speed   Our description of the circuit in Fig. 3-28a at the beginning of this section focused on the ON and OFF states of the switch. We now consider the switching transients indicated in the waveforms in Fig. 3-28.

As seen in Fig. 3-28, the current does not immediately respond to the input signal. Instead, there is a delay, and the time that elapses during this delay, together with the time required for the current to rise to 10 percent of its maximum (saturation) value, is called the delay time t_d. The current waveform has a nonzero rise time t_r, which is the time required for the current to rise through the active region from 10 to 90 percent of $I_{C(sat)}$. The total turn-on time t_{ON} is the sum of the delay and rise time, $t_{ON} \equiv t_d + t_r$. When the input signal returns to its initial state at $t = T_2$, the current again fails to respond immediately. The interval which elapses between the transition of the input waveform and the time when i_C has dropped to 90 percent of I_{CS} is called the *storage time* t_s. The storage interval is followed by the fall time t_f, which is the time required for i_C to fall from 90 to 10 percent of $I_{C(sat)}$. The turn-off time t_{OFF} is defined as the sum of the storage and fall times, $t_{OFF} \equiv t_s + t_f$. We shall consider now the physical reasons for the existence of each of these times. The exact calculations of these times is complex; approximate methods for the time intervals associated with the active region are given in Sec. 11-5 of this text.

Three factors contribute to the delay time: (1) when the driving signal is applied to the transistor input, a nonzero time is required to charge up the emitter-junction transition capacitance so that the transistor may be brought from cutoff to the active region; (2) even when the transistor has been brought to the point where minority carriers have begun to cross the emitter junction into the base, a time interval is required before these carriers can cross the base region to the collector junction and be recorded as collector current; and (3) some time is required for the collector current to rise to 10 percent of its maximum.

The rise time and the fall time are due to the fact that if a base-current step is used to saturate the transistor or return it from saturation to cutoff, the transistor collector current must traverse the active region. The collector current increases or decreases along an exponential curve whose time constant τ_r can be shown to be given by $\tau_r = \beta_o(C_\mu R_L + 1/\omega_T)$, where C_μ is the collector transition capacitance and ω_T is the radian frequency at which the current gain is unity.

The failure of the transistor to respond to the trailing edge of the driving pulse for the time interval t_s results from the fact that a transistor in saturation has excess minority carriers stored in the base. The transistor cannot respond until this excess charge has been removed. The stored charge density in the base is indicated in Fig. 3-12 under various operating conditions. The effect of excess minority carriers in the base is similar to the turn-off transient of the *pn* diode discussed in Sec. 2-10.

Consider that the transistor is in its saturation region and that at $t = T_2$ an input step is used to turn the transistor off, as in Fig. 3-28. Since the turn-off process cannot begin until the abnormal carrier density (the heavily shaded area of Fig. 3-12) has been removed, a relatively long storage delay time t_s may elapse before the transistor responds to the turn-off signal at the input. In an extreme case this storage-time delay may be several times the rise or fall time through the active region. It is clear that when transistor switches are to be used in an application where speed is at a premium, it is advantageous to reduce the storage time. A method for preventing a transistor from saturating, and thus eliminating storage time, is the use of a Schottky diode in conjunction with the BJT. These compound devices are called *Schottky transistors* and are described in Sec. 5-3.

3-9 THE BJT AS AN AMPLIFIER The circuit in Fig. 3-30 is an elementary *CE* amplifier stage. For purposes of demonstrating amplifier operation, we employ a 2N2222A transistor with $V_{CC} = 10$ V and $R_L = 500 \ \Omega$. The output characteristics and load line, first displayed in Fig. 3-16, are redrawn for convenience in Fig. 3-31. In Fig. 3-30, $R_B = 232.5$ kΩ is selected to bias the device in the forward-active region at Q, corresponding to $I_B = 40 \ \mu$A, $I_{CQ} = 8$ mA, and $V_{CEQ} = $

FIGURE 3-30
Elementary common-emitter amplifier stage.

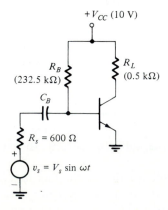

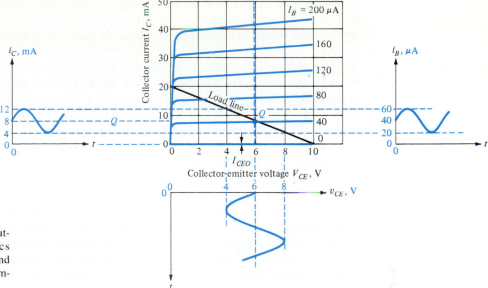

FIGURE 3-31
Common-emitter output characteristics showing load line and sinusoidal signal components.

6 V. The capacitor C_B (called a *blocking* capacitor) is used to isolate the dc bias from the signal source $v_s = V_{sm} \sin \omega t$ and its source resistance R_s. This capacitor acts as an open-circuit under quiescent conditions (no input signal) because the reactance of a capacitor is infinite at zero frequency (dc). We assume that at the angular (radian) frequency of the signal, the reactance of C_B is sufficiently small compared with R_s that the series combination of these elements is R_s. Consequently, the effect of the capacitor on the signal transmitted from the source v_s to the amplifier input is negligible. The amplitude V_{sm} is chosen to provide a signal component of base current $i_b = I_{bm} \sin \omega t$, where $I_{bm} = 20 \ \mu A$. The total instantaneous base current i_B is the superposition of the dc bias level and the signal current. Hence

$$i_B = I_{BQ} + i_b = 40 + 20 \sin \omega t \qquad \mu A$$

As is seen in Fig. 3-31, the effect of this signal causes both i_C and v_{CE} to vary (approximately) sinusoidally about their quiescent levels. These quantities are expressable as

$$i_C = I_{CQ} + i_c = I_{CQ} + I_{cm} \sin \omega t \qquad A \qquad (3\text{-}25)$$

$$v_{CE} = V_{CEQ} + v_{ce} = V_{CEQ} + V_{cem} \sin \omega t \qquad V \qquad (3\text{-}26)$$

The total instantaneous output voltage v_{CE} would be observed on an oscilloscope if the selector switch is on DC (Fig. 2-27a). With the selector switch on AC, only the sinusoidal output—$V_{cem} \sin \omega t$—is seen on the screen (Fig. 2-27b).

We note from Fig. 3-31 that the small change in i_B caused by the signal (I_{bm} = 20 μA) results in I_{cm} = 4 mA and V_{cem} = 2 V. The increased signal levels at the output are an indication of the amplification provided by the circuit.

Notation

At this point it is important to make a few remarks on transistor symbols. Specifically, instantaneous values of quantities which vary with time are represented by lowercase letters (i for current, v for voltage, and p for power). Maximum, average (dc), and effective, or root-mean-square (rms), values are represented by the uppercase letter of the proper symbol (I, V, or P). Average (dc) values and instantaneous total values are indicated by the uppercase subscript of the proper electrode symbol (B for base, C for collector, E for emitter). Varying components from some quiescent value are indicated by the lowercase subscript of the proper electrode symbol. A single subscript is used if the reference electrode is clearly understood. If there is any possibility of ambiguity, the conventional double-subscript notation should be used. For example, in Fig. 3-31, we show collector and base currents and voltages in the common-emitter transistor configuration, employing the notation just described. The collector and emitter current and voltage component variations from the corresponding quiescent values are

$$i_c = i_C - I_C = \Delta i_C \qquad v_C = v_C - V_C = \Delta v_C$$
$$i_b = i_B - I_B = \Delta i_B \qquad v_b = v_B - V_B = \Delta v_B \tag{3-27}$$

The magnitude of the supply voltage is indicated by repeating the electrode subscript. This notation is summarized in Table 3-4.

It is apparent from the previous paragraph that we are interested in changes in voltage and current about the operating point that occur because of the applied signal. In the next section we show that V_{sm} = 26.5 mV is required to make I_{bm} = 20 μA. Thus the voltage gain (or amplification) A_V is

$$|A_V| = \frac{V_{cm}}{V_{sm}} = \frac{2}{26.5 \times 10^{-3}} = 75.5$$

TABLE 3-4 Notation Summarized

Notation	Base (collector) voltage with respect to emitter	Base (collector) current toward electrode from external circuit
Instantaneous total value	$v_B(v_C)$	$i_B(i_C)$
Quiescent value	$V_B(V_C)$	$I_B(I_C)$
Varying component value	$v_b(v_c)$	$i_b(i_c)$
Effective value of varying component (phasor, if a sinusoid)	$V_b(V_c)$	$I_b(I_c)$
Supply voltage (magnitude)	$V_{BB}(V_{CC})$	

and the current gain A_I is given by

$$|A_I| = \frac{I_{cm}}{I_{bm}} = \frac{4 \times 10^{-3}}{20 \times 10^{-6}} = 200$$

We note that I_{cm}/I_{bm} is the change in collector current Δi_C to the change in base current Δi_B about the operating point Q.

Also evident in Fig. 3-31 is that v_{ce} and i_c are 180° out of phase. This phase reversal of signal voltage and current indicates that the BJT behaves as a source[1] controlled by i_b. As stated previously, the controlled source is the basis for amplification; it must be emphasized that *only the signal* is amplified.

The signal power delivered to the load resistance R_L is

$$P_L = (I_c)^2 R_L = \left(\frac{4 \times 10^{-3}}{\sqrt{2}}\right)^2 500 = 4 \text{ mW}$$

The total power supplied by both the bias and signal sources is

$$P_S = \frac{V_{sm}}{\sqrt{2}} \cdot \frac{I_{bm}}{\sqrt{2}} + V_{CC} (I_{CQ} + I_{BQ}) \qquad \text{W}$$

$$= \frac{2.65 \times 10^{-2}}{\sqrt{2}} \times \frac{2 \times 10^{-5}}{\sqrt{2}} + 10 (8 \times 10^{-3} + 4 \times 10^{-5}) = 80.4 \text{ mW}$$

It is evident that that the total power supplied to the circuit is considerably greater than the output signal power. However, the input signal power $V_{sm}I_{bm}/2$ is only 0.265 μW, whereas the output signal power is 4 mW or more than that provided by the signal source.

Let us consider how we might determine the output signal levels if the input signal amplitude were reduced, for example, to correspond to $I_b = 1\ \mu$A. It is obvious that we cannot resolve such small changes on the output characteristics shown in Fig. 3-31. In Example 2-4 and Sec. 2-9, we showed that this situation is best handled by the use of a small-signal model of the device. An additional factor is that the volt-ampere characteristics are dc characteristics and intrinsically eliminate any charge-storage effects that are present. These too can be incorporated in the small-signal model of the BJT which we treat in the next section.

3-10   THE BJT SMALL-SIGNAL MODEL   The small-signal equivalent circuit of the BJT is developed from the small-signal diode models of Sec. 2-9 and the Ebers-Moll representation in Sec. 3-3. The elements forming the equivalent circuits relate the *changes in voltages and currents about the operating point*. Each

---

[1]Standard circuit convention is to consider power dissipated as positive if a current in an element is directed from plus to minus (a voltage drop). In a source providing power, the current is directed from minus to plus (a voltage rise). It is in this sense that we view the phase reversal of v_{ce} relative to i_c as indicating power is dissipated in the load.

FIGURE 3-32

Small-signal hybrid-π equivalent circuit of a BJT. The resistance r_μ (shown dashed) is usually omitted as it generally has negligible effect.

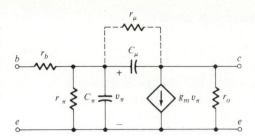

element in the model is a function of the quiescent voltages and currents established by the bias. As the input signal causes these changes, the equivalent circuit permits us to relate the output signal to the input signal.[1]

The *hybrid-π equivalent circuit* for the *CE*-connected BJT is displayed in Fig. 3-32. We can identify the elements in this model with the coupled-diode representation of the transistor. The forward-biased emitter-base junction is modeled by r_π and C_π, where C_π is essentially the diffusion capacitance and r_π is related to the incremental resistance[2] of the emitter-base diode. Typically, r_π has values ranging from a few hundred to several thousand ohms. The capacitance C_μ is the depletion-region capacitance of the reverse-biased collector-base junction. The incremental resistance r_μ of this diode is indicated by the dashed line in Fig. 3-32. This resistance accounts for the feedback (base-width modulation) between input and output due to the Early effect (Sec. 3-4). Because of its extremely high value (several megohms), r_μ is almost always neglected in computation. (We will do so in this text unless otherwise stated.) Coupling between the junctions is modeled by the controlled current source $g_m v_\pi$ and is proportional to the input current i_b. The output resistance r_o is also the result of the Early effect and equals the reciprocal of the slope of the dashed lines in Fig. 3-17. Typical values of r_o are tens of kilohms to hundreds of kilohms.

The resistance r_b is the *base-spreading resistance* and accounts for the voltage drop in the path between the base contact and the active base region (between *b* and *e*) under the emitter (see Fig. 3-7). This resistance decreases with increasing current levels and has typical values between 40 and 400 Ω. Because of the larger cross-sectional area of the collector region (Fig. 3-7), the *collector-spreading* resistance is in the order of magnitude of 1 Ω and is usually neglected (except in high-current discrete transistors or in computer simulations).

---

[1] This model is also referred to as an "ac" equivalent circuit to distinguish it from dc or large-signal models.

[2] Note that small-signal resistances are identified by lowercase letters. This notation allows us to distinguish incremental quantities from dc parameters and physical resistors.

FIGURE 3-33
The low-frequency hy-
brid-π model.

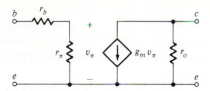

The Low-Frequency Model   The capacitances C_π and C_μ can be calculated as shown in Chap. 2 by the use of Eqs. (2-20) and (2-29). It is evident from these equations that both C_π and C_μ are dependent on operating-point values of BJT voltages and currents. At typical quiescent levels, for both IC and low-power discrete transistors, values of C_π are in the order of tens of picofarads to one or two hundred picofarads. Values of C_μ are generally a few picofarads (1 to 5 pF). At low signal frequencies, the reactances of both capacitors are extremely high. For example, at $\omega = 10^4$ rad/s, the magnitude of the reactance of $C_\pi = 50$ pF is 2 MΩ and that for $C_\mu = 2$ pF is 50 MΩ. At such frequencies the effects of C_π and C_μ are negligible and, consequently, can be approximated by open circuits. This leads to the low-frequency model shown in Fig. 3-33.

We observe in Fig. 3-33 that

$$v_\pi = r_\pi i_b$$

For $v_{ce} = 0$, no current exists in r_o and

$$i_c = g_m v_\pi = g_m r_\pi i_b \qquad \text{or} \qquad \frac{i_c}{i_b} = g_m r_\pi$$

It is convenient to introduce

$$\beta_o = \left. \frac{\Delta i_C}{\Delta i_B} \right|_{V_{CE} = \text{const} = V_{CEQ}} = \left. \frac{i_c}{i_b} \right|_{v_{ce} = 0} \tag{3-28}$$

The parameter β_o is the incremental (ac) *common-emitter forward short-circuit current gain*[1] and is evaluated at the operating point. The constant value of v_{CE} is indicative of no incremental change in this quantity, and thus only $v_{ce} = 0$. (The condition $v_{ce} = 0$ and $i_c \neq 0$ represents a short circuit between collector and emitter relative to the *signal*. It does not, however, indicate a physical short-circuit connection between these terminals.) The vertical dashed line through Q in Fig. 3-31 can be used to evaluate β_o. This is done in Example 3-7 later in this section. From Eq. (3-28) and the analysis of the model in Fig. 3-33

$$\beta_o = g_m r_\pi \tag{3-29}$$

---

[1]Manufacturers' data sheets identify this quantity by h_{fe}; however, we will use β_o as it is the more common symbol in the literature.

The parameter $g_m = i_c/v_\pi$, called the *transconductance*, reflects the incremental changes in i_C about the operating point produced by incremental changes in the emitter-base voltage. The voltage drop $i_b r_b$ is small so that changes in the base-emitter terminal voltage can be assumed to appear across the junction. Quantitatively, g_m is expressable as

$$g_m = \left.\frac{\Delta i_C}{\Delta v_{BE}}\right|_{v_{CE}=\text{const}=V_{CEQ}} = \left.\frac{\partial i_C}{\partial v_{BE}}\right|_{V_{CE}=0} \tag{3-30}$$

We believe it is important to reiterate the statements concerning the meaning of $v_{ce} = 0$ made earlier in this section. Maintaining v_{CE} constant infers that no incremental change in v_{CE} exists. Hence $v_{ce} = 0$, and, in Fig. 3-33, this is equivalent to short-circuiting the collector and emitter. Note that we are not physically shorting terminals c and e in the real transistor; this means only that the signal component of v_{CE} is zero.

From Eq. (3-17), $i_C = -\alpha_F i_E$ for either an *npn* or a *pnp* transistor and Eq. (3-30) becomes[1]

$$g_m = -\alpha_F \left.\frac{\partial i_E}{\partial v_{BE}}\right|_{v_{ce}=0} \tag{3-31}$$

We wish to relate g_m to the conductance of the emitter-base diode. The incremental conductance of the diode is given by Eq. (2-17) as

$$g_d = \frac{di_D}{dv_D}$$

where i_D and v_D are the forward current and voltage of the diode. For an *npn* transistor, v_{BE} forward-biases the emitter diode and $v_{BE} = v_D$; however, i_E is in the opposite direction of i_D (from n to p) so that $i_E = -i_D$. Therefore, $\partial i_E/\partial v_{BE} = -di_D/dv_D$ and

$$g_m = \alpha_F g_d \tag{3-32}$$

Equation (3-32) remains valid for a *pnp* transistor because forward-biasing the emitter junction makes $i_E = i_D$ and $v_{BE} = -v_D$.

The emitter-diode conductance g_d is expressed in Eq. (2-19) with $\eta = 1$. Hence $g_d = -I_{EQ}/V_T$ for an *npn* transistor and $g_d = +I_{EQ}/V_T$ for a *pnp* device. For the *npn* (*pnp*) transistor, I_{EQ} is negative (positive); thus g_d is positive in both instances and can be written as $g_d = |I_{EQ}|/V_T$. Using Eqs. (3-31) and (3-17) and neglecting I_{CO} compared with I_{CQ}, we obtain the following simple expression for the transconductance:

$$g_m = \frac{\alpha_F |I_{EQ}|}{V_T}$$
$$= \frac{|I_{CQ}|}{V_T} \tag{3-33}$$

---

[1] We assume that α_F is independent of v_{BE} in Eq. (3-31).

Equation (3-33) indicates that g_m is directly proportional to the quiescent collector current and inversely proportional to temperature. At room temperature, with I_{CQ} expressed in milliamperes, we obtain

$$g_m \approx \frac{|I_{CQ}|}{25} \quad \text{mA/mV} = \mho \qquad (3\text{-}34)$$

The use of Eqs. (3-29) and (3-34) allow us to determine r_π since β_o is specified by the manufacturer. Once we know r_π, we can calculate r_b from the input resistance. From Fig. (3-33) we see that

$$r_i = r_b + r_\pi \qquad (3\text{-}35)$$

Most device manufacturers specify r_i as h_{ie} at a given operation point.[1]

Example 3-7 In the circuit shown in Fig. 3-30, determine (*a*) the magnitude of V_s that produces an output signal of peak amplitude 2 V and (*b*) the magnitude of the output signal for $V_s = 2$ mV and (*c*) repeat part *b* for $V_s = 265$ mV. Assume low-frequency operation at room temperature.

Solution The initial step is to construct the low-frequency small-signal model of the circuit. This is accomplished by first replacing the transistor by the equivalent circuit in Fig. 3-33. We now add to this model only the circuit elements external to the BJT which influence incremental values of currents and voltages. The result is depicted in Fig. 3-34*a*. Note that the bias supply does not appear in

---

[1]At different quiescent conditions, r_i is obtained from laboratory measurement.

FIGURE 3-34
Low-frequency small-signal equivalent of circuit in Fig. 3-30 in Example 3-7.

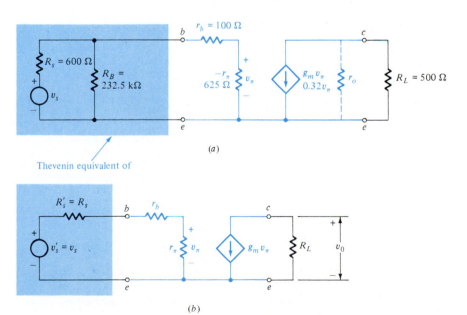

the model as it does not contribute to *changes* in voltage and current. Furthermore, as the incremental voltage across V_{CC} is zero, it acts as a short circuit.

The value of g_m is determined from Eq. (3-34) at $I_C = 8$ mA (the quiescent current given in Sec. 3-9) as

$$g_m = \tfrac{8}{25} = 0.32 \ \mho$$

The value of β_o is evaluated from the characteristics in Fig. 3-31 along the vertical dashed line and is

$$\beta_o = \frac{(12 - 4) \times 10^{-3}}{(60 - 20) \times 10^{-6}} = 200$$

(Note that β_o is the current gain A_i obtained in Sec. 3-9.) Use of Eq. (3-29) gives

$$r_\pi = \frac{\beta_o}{g_m} = \frac{200}{0.32} = 625 \ \Omega$$

The base-spreading resistance r_b, obtained from measurement, is 100 Ω.[1] From the slope of the collector characteristic at Q in Fig. 3-31, the value of r_o is found to be in excess of 5 kΩ. Thus the effect of r_o is negligible; since it appears in parallel with $R_L = 500 \ \Omega$, the parallel combination r_o and R_L is 500 Ω.

(*a*) Now that the model is completed, it is convenient to replace v_s, R_s, and R_B by a Thévenin equivalent as shown in Fig. 3-34*b*:

$$v'_s = \frac{R_B}{R_B + R_s} \, v_s \qquad R'_s = R_s \| R_B = \frac{R_s R_B}{R_s + R_B}$$

As $R_B = 232.5$ k$\Omega \gg R_s = 600 \ \Omega$, $v'_s = v_s$ and $R'_s = R_s$. In Fig. 3-34*b*

$$v_o = -g_m v_\pi R_L$$

and use of the voltage-divider relationship gives

$$v_\pi = \frac{r_\pi}{R_s + r_b + r_\pi} \, v_s$$

Combination of these equation yields

$$v_o = \frac{-g_m r_\pi R_L}{R_s + r_b + r_\pi} \, v_s = \frac{-\beta_o R_L}{R_s + r_b + r_\pi} \, v_s = A_V v_s$$

The ratio v_o/v_s is called the *voltage gain* A_V.[2] The minus sign in the expression

---

[1] The data sheet in App. B-3 lists the value of h_{ie} at an operating point significantly different from that in this example, so it cannot be used.

[2] Both *G* (for gain) and *A* (for amplification) are used extensively to identify this type of transfer function.

for A_V (and v_o) indicates the phase reversal discussed in Sec. 3-9. Evaluation gives

$$A_V = \frac{-200 \times 500}{600 + 100 + 625} = -75.5$$

and noting that the output voltage $V_{om} = V_{cm}$

$$V_{sm} = \frac{V_{om}}{|A_V|} = \frac{2}{75.5} = 26.5 \text{ mV}$$

This is the value indicated in Sec. 3-9.

(b) For an input signal $V_{sm} = 2$ mV, the output amplitude is

$$V_{om} = |A_V| \, V_{sm} = 75.5 \times 2 \times 10^{-3} = 151 \text{ mV}$$

(c) With $V_{sm} = 265$ mV, we obtain

$$V_{om} = 75.5 \times 0.265 = 20.0 \text{ V}$$

This answer is obviously *wrong* and corresponds to a physically impossible situation. With an input signal 10 times that in part *a* we would, if linear operation is assumed, expect $I_{bm} = 10 \times 20 \, \mu A = 200 \, \mu A$. A quick inspection of the load line in Fig. 3-31 indicates that a sinusoidal signal of amplitude 200 μA about $I_{BQ} = 40 \, \mu A$ would drive the BJT heavily into saturation during the positive half-cycle and into cutoff during the negative half-cycle. We use this part of the example to point out that the use of the small-signal model is restricted to BJT operation in the linear portion of the active region. In Fig. 3-29b we displayed the transfer characteristic of a similar circuit. There we noted that amplifier usage was restricted to the linear segment connecting cutoff and saturation. If we were to use the transfer characteristic with the given input, we would observe a highly distorted and clipped output.

The results in Example 3-7 show that amplification is achieved only for a limited range of input signals. Further restrictions on amplifier performance exist. In our analysis we assumed that the effects of C_π and C_μ were negligible at the signal frequencies. This is not the case when the frequency is increased. Consequently, the gain is affected and results in limitations on the high-frequency range over which the stage can be used. Low-frequency performance limits arise when we remove the assumption that C_B has negligible reactance. The frequency response of amplifier circuits is treated in Chap. 11.

3-11   THE BJT AS A DIODE   Fabrication efficiency and the facility with which device characteristics can be matched are often the motives for using the BJT as a diode in IC design. In several earlier sections of this chapter, we observed that by either short-circuiting two of the transistor terminals or by leaving one open-circuited, the BJT behaved as a reverse-biased diode. We now consider forward-biased behavior of one of the more widely employed diode connections of the BJT.

FIGURE 3-35
Diode-connected tran-
sistor.

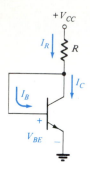

The base and collector of the transistor in Fig. 3-35 are connected; this configuration uses the emitter-base junction as the diode. The remainder of the circuit in Fig. 3-35 shows the collector (and base) returned to the supply voltage V_{CC} through the resistor R. This positive voltage forward-biases the emitter-base junction, and as $V_{BC} = 0$ is less than the cut-in voltage, the collector-base diode is reverse-biased. Consequently, the BJT is in the forward-active region. From KVL, we obtain

$$I_R = \frac{V_{CC} - V_{BE}}{R} \tag{3-36}$$

With $V_{BE} = 0.7$ V in the active mode, the current I_R is a constant and is dependent only on the supply voltage V_{CC} and resistance R.

We can use KCL to relate I_R to the transistor currents. Since $I_C = \beta_F I_B$, $I_R = I_C + I_B$ becomes

$$I_R = (\beta_F + 1)I_B = \left(1 + \frac{1}{\beta_F}\right)I_C \tag{3-37}$$

As I_R is constant and with $\beta_F \gg 1$, $I_C \approx I_R$ is also constant. This observation is the basis for the current source described in the following example.

Example 3-8

The circuit shown in Fig. 3-36*a* is a *current source* or *current mirror* and is extensively used to bias BJTs in analog ICs. The transistors $Q1$ and $Q2$ are identical; that is, they are fabricated to have matched characteristics. (*a*) Determine I_C in terms of the circuit parameters. (*b*) Evaluate I_C for $V_{CC} = 10$ V, $R = 10$ kΩ, and $\beta_F = 100$. (*c*) Repeat part *b* for $\beta_F = 200$.

Solution

(*a*) The current I_R is given by Eq. (3-36). The base-emitter voltage V_{BE} of each transistor, as a consequence of KVL, is the same. Because $Q1$ and $Q2$ are identical transistors and operate at the same value of V_{BE}, the base and collector

FIGURE 3-36
(*a*) Current mirror and
(*b*) its representation.

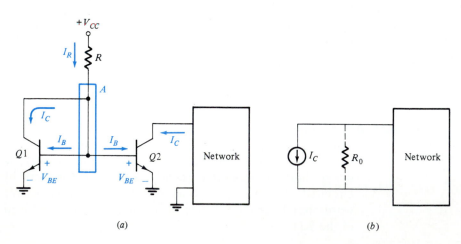

(*a*)   (*b*)

currents in each are equal. From KCL at the node where both bases and the collector of $Q1$ are connected

$$I_R = I_C + 2I_B = \frac{V_{CC} - V_{BE}}{R}$$

Substitution of $I_B = I_C/\beta_F$ and solving for I_C yields

$$I_C = \frac{\beta_F}{\beta_F + 2} \cdot \frac{V_{CC} - V_{BE}}{R}$$

(b) Substitution of values gives

$$I_C = \frac{100}{100 + 2} \cdot \frac{10 - 0.7}{10} = 0.912 \text{ mA}$$

(c) For $\beta_F = 200$

$$I_C = \frac{200}{200 + 2} \cdot \frac{10 - 0.7}{10} = 0.921 \text{ mA}$$

The results in Example 3-8 illustrate that even with a 100 percent change in β_F, the variation in I_C is about 1 percent. The collector current of $Q2$ is essentially constant and independent of the transistor parameters. The value of I_C is dependent only on V_{CC} and R. This is the behavior of a constant current source and permits us to model the circuit in Fig. 3-36a as is depicted in Fig. 3-36b. The resistance R_0 indicated by the dashed line is the output resistance of $Q2$ and is due primarily to the Early effect.

3-12 THE EMITTER-COUPLED PAIR The *emitter-coupled* or *differential pair* in Fig. 3-37 is among the most important transistor configurations employed in ICs. In Fig. 3-37, the current source I_{EE} is realized by the current mirror of Fig. 3-36a or

FIGURE 3-37
Emitter-coupled (differential) pair.

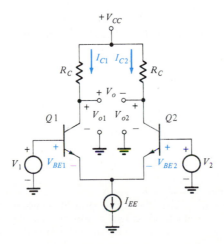

other similar circuits (Sec. 10-3). Furthermore, we assume $Q1$ and $Q2$ to be identical transistors[1] and both collector resistances are fabricated with equal values. Our objective in this section is to demonstrate that the differential pair can be utilized as either an amplifier or a switch. To accomplish this objective, we develop the dc transfer characteristic of the circuit.

The KVL expression for the loop containing the two emitter-base junctions is

$$-V_1 + V_{BE1} - V_{BE2} + V_2 = 0 \tag{3-38}$$

With the transistors biased in the forward-active mode, the reverse saturation current of the collector-base junction is negligible. The collector currents I_{C1} and I_{C2} are given by Eq. (3-7) as

$$I_{C1} = \alpha_F I_{ES} \, \epsilon^{V_{BE1}/V_T} \tag{3-39}$$

$$I_{C2} = \alpha_F I_{ES} \, \epsilon^{V_{BE2}/V_T} \tag{3-40}$$

In Eqs. (3-39) and (3-40) we assume that $\epsilon^{V_{BE}/V_T} \gg 1$ ($V_{BE} \geq V_T$) and the reverse saturation current components of I_{C1} and I_{C2} are negligible. We now form the ratio I_{C1}/I_{C2} as

$$\frac{I_{C1}}{I_{C2}} = \epsilon^{(V_{BE1} - V_{BE2})/V_T} = \epsilon^{V_d/V_T} \tag{3-41}$$

From Eq. (3-38) we see that $V_{BE1} - V_{BE2} = V_1 - V_2 = V_d$, where V_d is the difference (hence the name differential) between the two input voltages. The KCL constraint at the emitter node requires

$$-(I_{E1} + I_{E2}) = I_{EE} = \frac{I_{C1}}{\alpha_F} + \frac{I_{C2}}{\alpha_F} \tag{3-42}$$

Division of both sides of Eq. (3-42) by I_{C1}/α_F results in

$$\frac{\alpha_F I_{EE}}{I_{C1}} = \frac{I_{C2}}{I_{C1}} + 1 \tag{3-43}$$

Substitution of Eq. (3-41) in Eq. (3-43) and solving for I_{C1} yields

$$I_{C1} = \frac{\alpha_F I_{EE}}{1 + \epsilon^{-V_d/V_T}} \tag{3-44}$$

A similar analysis gives

$$I_{C2} = \frac{\alpha_F I_{EE}}{1 + \epsilon^{+V_d/V_T}} \tag{3-45}$$

---

[1] Several manufacturers fabricate as many as five virtually identical transistors within a single package (see App. B). Also available are several transistors, sometimes with both *npn* and *pnp* devices, in one package where one pair is connected as shown in Fig. 3-37. These are also useful in simulating IC designs.

FIGURE 3-38
Transfer characteristic
(I_C versus V_d) for the
emitter-coupled pair.

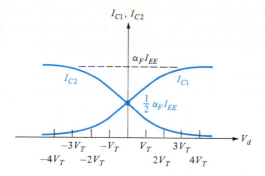

We observe in Eqs. (3-44) and (3-45) that increasing positive values of V_d of magnitude greater than $4V_T$ cause I_{C1} and I_{C2} to approach $\alpha_F I_{EE}$ and zero, respectively. Alternatively, a negative value of V_d with $|V_d| > 4V_T$ causes I_{C1} to approach zero and I_{C2} tends toward $\alpha_F I_{EE}$. On the basis of Eqs. (3-44) and (3-45) we can construct the transfer characteristic displayed in Fig. 3-38.

The voltages V_{o1} and V_{o2} are defined by

$$V_{o1} \equiv V_{CC} - I_{C1} R_C \qquad (3\text{-}46)$$

$$V_{o2} \equiv V_{CC} - I_{C2} R_C \qquad (3\text{-}47)$$

and result in the characteristics depicted in Fig. 3-39. Also shown in Fig. 3-39 is the differential (difference) output $V_o = V_{o1} - V_{o2}$.

We can interpret the transfer characteristics in Figs. 3-38 and 3-39 as follows. First, application of $V_d > 4V_T = 100$ mV makes $I_{C1} \approx \alpha_F I_{EE}$ and $I_{C2} \approx 0$. Simultaneously, $V_{o2} = V_{CC}$ and $V_{o1} = V_{CC} - \alpha_F I_{EE} R_C$ can be made "small" by proper choice of R_C.[1] Thus we can approximate the output of $Q1$ as a closed

---

[1] V_{o1} and V_{o2} are always selected to maintain $Q1$ and $Q2$ in the active region.

FIGURE 3-39
Voltage transfer char-
acteristic (V_o versus V_d)
for the emitter-coupled
pair.

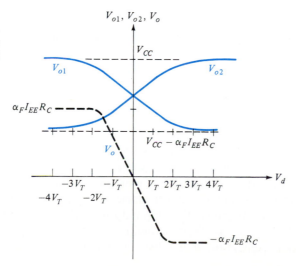

switch and that at $Q2$ as an open switch. The states of these switches are changed by applying $V_d < -4V_T$. The differential output also displays two distinct output levels, one positive and one negative, for a change in V_d of about $4V_T$.

A second very important observation is that in the range $-2V_T \leq V_d \leq 2V_T$, the quantities I_{C1}, I_{C2}, V_{o1}, V_{o2}, and V_o all respond in a nearly linear fashion to changes in V_d. For this range of inputs, the circuit behaves as a controlled source (i.e., an amplifier). Both the switching and amplifying properties of the emitter-coupled pair are employed extensively. The switching characteristic is exploited in digital circuits (Part 2), whereas the amplifying property plays a prominent role in analog circuits (Part 3).

3-13 TRANSISTOR RATINGS

The transistors used in the circuits described in this chapter were always assumed to be operating within allowable limits for current, voltage, and power dissipation. We now discuss the ratings (listed in the manufacturers' specification sheets) which must not be exceeded in BJT usage.

Maximum Collector Current

Even if power and voltage ratings are not exceeded, there is an absolute maximum value of current-handling capacity of the collector, associated with the junction area and the wire bonds that connect the transistor terminals to the external leads. This rating, which determines the maximum allowable saturation current, is 800 mA for the 2N2222A transistor.

Maximum Power Dissipation

Device destruction can occur if the collector-base junction is subjected to excessive power. The *maximum power dissipation* P_D is the rating used to indicate the limits on the power-handling capability of the collector. For the 2N2222A, $P_D = 0.5$ W at an ambient temperature of 25°C. At higher ambient temperatures, P_D must be derated by 12 mW/°C. Quantitatively, this means that

$$P_D(T) = 500 - 12(T - 25) \quad \text{mW}$$

where $P_D(T)$ is the maximum power dissipation at temperature T (in Celsius).

Maximum Output Voltage Rating

There is an upper limit to the maximum allowable collector-junction voltage since there is the possibility of voltage breakdown in the transistor at high voltages. Two types of breakdown are possible, *avalanche* breakdown, discussed in Sec. 2-11, and *punch-through,* discussed in this section.

The maximum reverse-biasing voltage which may be applied before breakdown between the collector and emitter terminals of the transistor, under the condition that the base lead be open-circuited, is represented by the term BV_{CEO}. Breakdown may occur because of avalanche multiplication of the current I_{CO} that crosses the collector junction. For 2N2222A the *CE* characteristics extending into the breakdown region are shown in Fig. 3-40, and $BV_{CEO} \simeq$ 50 V. The specification sheets list the minimum value of BV_{CEO} at 40 V. For

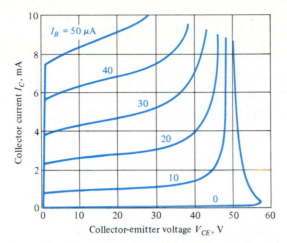

FIGURE 3-40
Common-emitter characteristics of the 2N2222A transistor extended into the breakdown region.

the CB configuration, the output breakdown voltage BV_{CBO} is usually about twice BV_{CEO}. If the base is returned to the emitter through a resistor R, the breakdown voltage, designated by BV_{CER}, will lie between BV_{CEO} and BV_{CBO}. In other words, the maximum allowable collector-emitter voltage depends not only on the transistor, but also on the circuit in which it is used.

Punch-Through The second mechanism by which the usefulness of a transistor may be terminated as the collector voltage is increased is called punch-through and results from the increased width of the collector-junction transition region with increased collector-junction voltage (the Early effect).

The transition region at a junction is the region of uncovered charges on both sides of the junction at the positions occupied by the impurity atoms. As the voltage applied across the junction increases, the transition region penetrates deeper into the base. Since the base is very thin, it is possible that, at moderate voltages, the transition region will have spread completely across the base to reach the emitter junction. This punch-through has the effect of lowering the barrier at the emitter-base junction, and, as a consequence, the emitter current can become excessive. Thus an upper limit on the magnitude of the collector-base voltage exists.

Punch-through differs from avalanche breakdown in that it takes place at a fixed voltage [given by V_j in Eq. (2-27), with $W = W_B$] between collector and base and is not dependent on circuit configuration. In a particular transistor, the voltage limit is determined by punch-through or avalanche breakdown, whichever occurs at the lower voltage.

Maximum Input Voltage Rating Consider the circuit configuration in Fig. 3-41, where V_{BB} represents a biasing voltage intended to keep the transistor cut off. Assume that the transistor is just at the point of cutoff, with $I_E = 0$, so that $I_B = I_{CBO}$. If we require that at cutoff $V_{BE} \approx 0$ V, the condition of cutoff requires that

$$V_{BE} = -V_{BB} + R_B I_{CBO} < 0 \qquad (3\text{-}48)$$

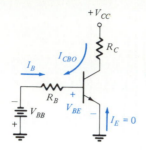

FIGURE 3-41
Common-emitter stage
biased to maintain tran-
sistor in the cutoff re-
gion.

As an extreme example, consider that R_B is, say, as large as 100 kΩ and that we want to allow for the contingency that I_{CBO} may become as large as 100 μA, as might occur with a large-geometry power transistor or with a medium-power device at elevated temperatures. Then V_{BB} must be at least 10 V. When I_{CBO} is small, the magnitude of the voltage across the base-emitter junction will be 10 V. Hence we must use a transistor whose maximum allowable reverse base-emitter junction voltage before breakdown exceeds 10 V. It is with this contingency in mind that a manufacturer supplies a rating for the reverse breakdown voltage between emitter and base, represented by the term BV_{EB0}. The subscript 0 indicates that BV_{EB0} is measured under the condition that the collector current is zero. Breakdown voltages BV_{EB0} may be as high as some tens of volts or as low as 0.5 V. If $BV_{EB0} = 6$ V (as it is for the 2N2222A transistor), V_{BB} must be chosen to have a maximum value of 6 V.

Small-geometry IC transistors are subject to the same maximum ratings. Typical values for these ratings are of the same order as those indicated for the 2N2222A.

REFERENCES

1 Hodges, D. E., and H. G. Jackson: "Analysis and Design of Digital Integrated Circuits," McGraw-Hill Book Company, New York, 1983.

2 Gray, P. R., and R. G. Meyer: "Analysis and Design of Analog Integrated Circuits," 2d ed., John Wiley and Sons, New York, 1984.

3 Yang, E. S.: "Fundamentals of Semiconductor Devices," McGraw-Hill Book Company, New York, 1978.

4 Grebene, A. B.: "Bipolar and MOS Analog Integrated Circuit Design," John Wiley and Sons, New York, 1984.

The following three papers are classics in the field:

5 Schockley, W.: The Theory of *p-n* Junctions in Semiconductors and *p-n* Junction Transistors, *Bell Systems Tech. J.*, vol. 28, pp. 435–489, July 1949.

6 Ebers, J. J., and J. L. Moll: Large-Signal Behavior of Junction Transistors, *Proc. IRE*, vol. 42, pp. 1761–1772, December 1954.

7 Early, J. M.: Effects of Space-Charge Layer Widening in Junction Transistors, *Proc. IRE*, vol. 40, pp. 1401–1406, November 1952.

REVIEW QUESTIONS

3-1   List three characteristics of a current-controlled current source.

3-2   Draw a load line on the output characteristics of an ideal current-controlled current source and indicate the point on the load line where operation approximates (*a*) an open switch, (*b*) a closed switch, and (*c*) a controlled source

3-3   Draw the circuit symbol for an *npn* transistor and indicate the reference directions for the three currents and the reference polarities for three voltages.

3-4   Repeat Rev. 3-3 for a *pnp* transistor.

3-5   Indicate the electron and hole current components for a *pnp* transistor biased in the forward-active region.

3-6   (*a*)   Write the Ebers-Moll equations for a *pnp* transistor.
         (*b*)   Draw a circuit model based on the Ebers-Moll equations.

3-7   State the reciprocity condition.

3-8   Repeat Rev. 3-6 for an *npn* transistor.

3-9   Define α_F and α_R and briefly describe the significance of each.

3-10   Define the common-emitter short-circuit current gain in words and by an equation.

3-11   What is the significance of the reverse short-circuit current gain?

3-12   Define the four modes of BJT operation and indicate the principal behavior in each mode.

3-13   For a *pnp* transistor in the active region, what is the sign (positive or negative) of I_E, I_C, I_B, V_{CB}, and V_{EB}?

3-14   Repeat Rev. 3-13 for an *npn* transistor.

3-15   (*a*)   Sketch the *CB* output characteristics for a transistor and indicate the active, cutoff, and saturation regions.
         (*b*)   Qualitatively explain the shape of these curves.

3-16   Sketch the *CB* input characteristics and explain their shapes.

3-17   Explain base-width modulation (the Early effect).

3-18   (*a*)   Draw a circuit of a transistor in the *CE* configuration.
         (*b*)   Sketch the output characteristics and explain their shapes.

3-19   (*a*)   What is the order of magnitude of the current I_{CBO}?
         (*b*)   How does I_{CBO} vary with temperature?
         (*c*)   Why does I_{CBO} differ from I_{CO}?

3-20   (*a*)   Define β_{forced}.
         (*b*)   How does β_{forced} differ from β_F?

3-21   (*a*)   List the typical values of $V_{CE(\text{sat})}$ and $V_{BE(\text{sat})}$.
         (*b*)   What are the values of V_{CE} at the edge of saturation, V_{BE} at cut in, and V_{BE} in the active region?

3-22   Draw the dc models of a transistor in each of the four regions of operation.

3-23   Explain how a BJT can be used as a switch.

3-24   (*a*)   List the factors which determine the switching speed of a BJT.
         (*b*)   Explain the significance of each term in Rev. 3-24*a*.

3-25 Explain how a BJT can be used as an amplifier.

3-26 Indicate whether each of the following is a dc, time-varying, or instantaneous quantity: v_{ce}, V_{CE}, v_{CE}, and V_{ce}.

3-27 (*a*) Draw the hybrid-π equivalent circuit.
(*b*) Explain the origin (or physical process) which gives rise to each term.

3-28 Draw the low-frequency equivalent circuit of a BJT.

3-29 (*a*) Define the transconductance g_m.
(*b*) Write an equation which relates g_m to the current gain β_o.

3-30 (*a*) Show by means of a circuit diagram how a BJT can be used as a diode.
(*b*) What is the region of operation of the BJT in Rev. 3-30*a*.

3-31 (*a*) Sketch a circuit diagram of a current source.
(*b*) Briefly explain why this is a current mirror.

3-32 Draw the circuit configuration of an emitter-coupled pair.

3-33 Explain briefly how the differential pair can be used as an amplifier and switch.

3-34 What limits the maximum current a transistor is capable of handling?

3-35 (*a*) Describe punch-through.
(*b*) What limitation on transistor operation is attributable to punch-through?

Chapter 4
FIELD-EFFECT TRANSISTORS

The *field-effect transistor,* or simply the FET, is a three-terminal semiconductor device used extensively in digital and analog circuits. There are two types of such devices, the MOSFET and JFET, acronyms for metal oxide semiconductor and junction field-effect transistors, respectively.

An important feature of the FET is that it is often simpler to fabricate and occupies less space on a chip than does a BJT. The resultant component density can be extremely high, often exceeding 100,000 MOSFETs per chip. A second desirable property is that MOS devices can be connected as resistors and capacitors. This makes possible the design of systems consisting exclusively of MOSFETs and no other components. Exploitation of these features makes the MOSFET the dominant device in very-large-scale-integration (VLSI) systems. High input resistance and low noise are two of the properties of JFETs used in signal-processing circuits.

In contrast to the BJT treated in Chap. 3, the FET is a majority-carrier device. Its operation depends on the use of an applied electric field to control device current. Thus the FET is a voltage-controlled current source. The physical principles which govern both types of FET are examined in this chapter and are used to develop the volt-ampere characteristics. The behavior of the FET as a switching and amplifying device is also investigated. The JFET is described first as its operation is developed directly from the *pn* junction and semiconductor properties. The characteristics of the MOSFET are more easily understood once the volt-ampere relationships of the JFET are presented.

4-1 THE IDEAL VOLTAGE-CONTROLLED CURRENT SOURCE

In the opening paragraphs of this chapter we alluded to the fact that the FET behaves as a voltage-controlled current source. Just as we introduced the current-controlled current source (Sec. 3-1) prior to treating the BJT, it is convenient to describe the properties of a voltage-controlled current source before we discuss the FET.

FIGURE 4-1
(*a*) Circuit representation and (*b*) volt-ampere characteristics of an ideal voltage-controlled current source.

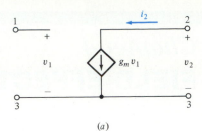

(*a*)

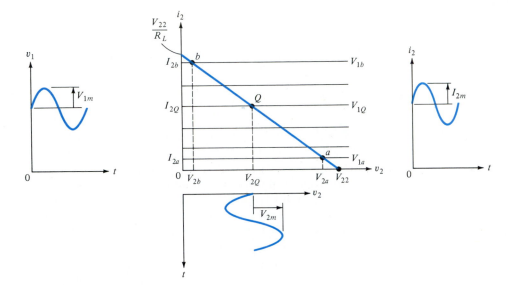

(*b*)

The ideal voltage-controlled current source, depicted in Fig. 4-1*a*, is a three-terminal element in which the control voltage v_1 is applied to terminals 1–3 and the current source $g_m v_1$ acts at terminals 2–3. The parameter g_m, called the *transconductance* or *mutual conductance,* relates the strength of the source to the control voltage (refer to Fig. 3-33). The output characteristics of the voltage-controlled current source are displayed in Fig. 4-1*b*, on which a load line corresponding to R_L and V_{22} is drawn. The load line represents the KVL equation for the output loop (containing terminals 2–3) in the circuit shown in Fig. 4-2.

At point *a* on the load line, corresponding to $v_1 = V_{1a}$, the voltage v_2 is "large" whereas the current i_2 is "small." These values approximate the behavior of an open switch. Similarly, circuit operation at *b* on the load line, for which $v_1 = V_{1b}$, indicates a "high" current through and a "low" voltage drop across the device. This situation approximates the action of a closed switch. Consequently, a signal voltage $v_i = v_1$ applied between terminals 1 and 3 controls the state of the switch at terminals 2–3. That is, if v_i is changed from V_{1a} to V_{1b}, the action observed between terminals 2 and 3 is that of closing the

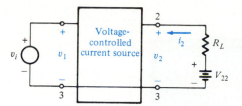

FIGURE 4-2
Circuit using a voltage-controlled current source as either a switch or amplifier. The load line is displayed in Fig. 4-1b.

initially open switch. Similarly, a change in v_i from V_{1b} to V_{1a} causes the switch to open.

In Fig. 4-1b, we also show the behavior of the circuit in Fig. 4-2 when $v_i = V_{1Q} + V_{1m} \sin \omega t$. This is indicated by the sinusoidal variation of v_1 about the quiescent level V_{1Q}. The corresponding output voltage v_2 is also a sinusoid of peak amplitude V_{2m} superimposed on the quiescent level V_{2Q}. Similarly, i_2 is the superposition of a dc component I_{2Q} and a sinusoid of peak amplitude I_{2m}. Most often, the amplitude of the sinusoidal component of the output voltage V_{2m} is greater than V_{1m}, and the amplitude of the sinusoidal input component and voltage gain (amplification) is achieved.

This brief introduction demonstrates that the voltage-controlled current source can be used as both a switch and amplifier. In the next several sections we discuss the physical operation and characteristics of FETs and show that these devices exhibit voltage-controlled current source properties.

4-2 THE JUNCTION FIELD-EFFECT TRANSISTOR

The basic structure of an *n*-channel JFET is illustrated in Fig. 4-3.[1] The *drain* and *source* terminals are made by the ohmic contacts at the ends of an *n*-type semiconductor bar. Majority-carrier electrons can be caused to flow along the length of the bar by means of a voltage applied between drain and source. The third terminal, called the *gate*, is formed by electrically connecting the two shallow p^+ regions. The *n*-type region be-

---

[1]The actual structure and fabrication of JFETs is described in Chap. 5.

FIGURE 4-3

(*a*) Structure and (*b*) symbol of an *n*-channel junction field-effect transistor (JFET).

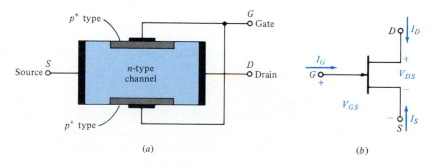

(*a*)

(*b*)

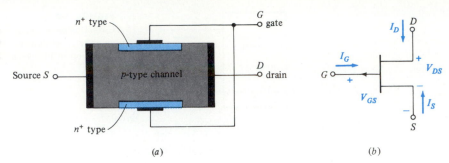

FIGURE 4-4

p-Channel junction field-effect transistor (*a*) structure and (*b*) circuit symbol.

tween the two p^+ gates is called the *channel* through which majority carriers move between source and drain.

The standard conventions for positive terminal currents and voltages are identified in Fig. 4-3*b*, in which the circuit symbol for an *n*-channel JFET is depicted. The corresponding structure and circuit symbol for the *p*-channel JFET are shown in Fig. 4-4.

The structures shown in Figs. 4-3 and 4-4 are convenient representations by which JFET operation can be described. The cross section of a *planar IC n-channel* JFET is shown in Fig. 4-5. The top view in Fig. 4-5 indicates how the aluminum contacts are made to the source, drain, and gate regions.

JFET Operation   Consider the pictorial representation of an *n*-channel device displayed in Fig. 4-3 and redrawn for convenient reference in Fig. 4-6*a*. The schematic diagram in Fig. 4-6*b* corresponds to Fig. 4-6*a* and illustrates the *common-source* configuration. While our discussion focuses on an *n*-channel device, it applies equally to the *p*-channel JFET if we recognize that voltage polarities and current directions in *p*-channel devices are opposite to those for *n*-channel JFETs.

We observe that the gate regions and the channel constitute a *pn* junction which, in JFET operation, is maintained in a reverse-biased state. Application of a negative gate-to-source voltage reverse-biases the junction, as does application of a positive drain-to-source voltage. It is necessary to recall that on the two sides of the reverse-biased *pn* junction (the depletion region) there are

FIGURE 4-5

Planar IC *n*-channel JFET structure.

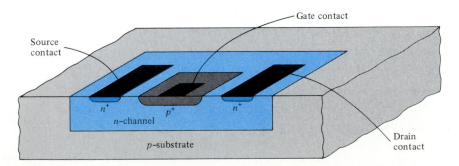

FIGURE 4-6

Biased *n*-channel JFET structure showing depletion region constricting the channel.

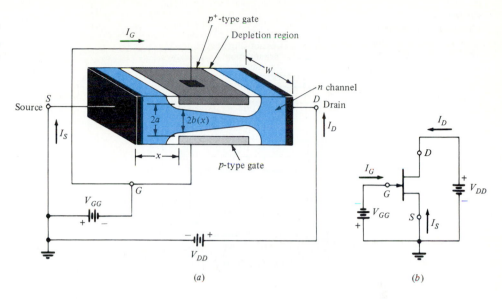

(a)   *(b)*

space-charge regions (Sec. 2-1). The current carriers have diffused across the junction, leaving only uncovered positive ions on the *n* side and negative ions on the *p* side. As the reverse bias across the junction increases, so does the thickness of the region of immobile uncovered charges. We can now indicate the rationale for the use of the p^+ gate regions. In Sec. 2-13, we showed that the depletion region extends further into the region of lower doping. Thus the use of p^+ results in a space-charge layer that lies almost entirely in the *n* channel. The conductivity of this region is nominally zero because of the unavailability of current carriers. Hence we see that the effective width of the channel in Fig. 4-6 will decrease with increasing reverse bias. At a gate-to-source voltage $V_{GS} = V_p$, called the "pinch-off" voltage, the channel width is reduced to zero because all the free charge has been removed from the channel. Accordingly, for a fixed drain-to-source voltage, the drain current will be a function of the reverse-biasing voltage across the gate junction. The term "field effect" is used to describe this device because the mechanism of current control is the effect of the extension, with increasing reverse bias, of the field associated with the depletion region.

4-3 THE JFET VOLT-AMPERE CHARACTERISTICS

The drain characteristics for a typical discrete *n*-channel FET shown in Fig. 4-7 give I_D against V_{DS}, with V_{GS} as a parameter. To see qualitatively why the characteristics have the form shown, consider first the case for which $V_{GS} = 0$. For $I_D = 0$, the channel between the gate junctions is entirely open. In response to a small applied voltage V_{DS}, the *n*-type bar acts as a simple semiconductor resistor, and the current I_D increases linearly with V_{DS}. With increasing current, the ohmic voltage drop

FIGURE 4-7
Output characteristics of a 2N4869 *n*-channel JFET. (*Courtesy of Siliconix, Inc.*)

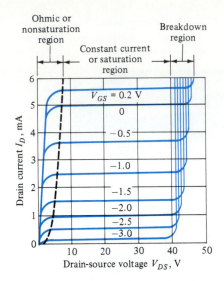

along the *n*-type channel region reverse-biases the gate junction, and the conducting portion of the channel begins to constrict. Because of the ohmic drop along the length of the channel itself, the constriction is not uniform, but is more pronounced at distances farther from the source, as indicated in Fig. 4-6. Eventually, a voltage V_{DS} is reached at which the channel is "pinched off." This is the voltage, not too sharply defined in Fig. 4-7, where the current I_D begins to level off and approach a constant value. It is, of course, in principle not possible for the channel to close completely and thereby reduce the current I_D to zero. Indeed, if such could be the case, the ohmic drop required to provide the necessary back bias would itself be lacking. Note that each characteristic curve has an ohmic or nonsaturation region for small values of V_{DS}, where I_D is proportional to V_{DS}. Each also has a constant-current or current saturation region for large values of V_{DS}, where I_D responds very slightly to V_{DS}.

With $V_{GS} = 0$, the voltage needed to reverse-bias the junction is provided by V_{DS}. If a negative V_{GS} is applied, the depletion region that results reduces the channel width even with $V_{DS} = 0$. Thus pinch-off occurs at a smaller value of V_{DS} and the maximum value of I_D is reduced as noted in Fig. 4.7. At $V_{GS} = V_p$, the pinch-off voltage, $I_D = 0$ as the channel is completely constricted for all values of $V_{DS} \geq 0$.

Note that a curve for $V_{GS} = +0.2$ V, which is in the direction of forward bias, is also given. Recall that the gate current is very small as this voltage is less than the cut-in voltage $V_\gamma = 0.5$ V for silicon. The gate current, for $V_{GS} \leq 0$, is virtually zero and is most often neglected.

Four regions of JFET operation, apparent in Fig. 4-7, are the ohmic, *saturation, breakdown,* and *cutoff* regions, each of which is now examined in more detail.

The Ohmic Region   In the ohmic, or voltage-variable-resistance, region of the JFET, V_{DS} is small and I_D can be appreciable. This situation corresponds to a closed switch. We now describe analytically, the volt-ampere relationship in the ohmic region.

Assume, first, that a small voltage V_{DS} is applied between drain and source. The resulting small drain current I_D will then have no appreciable effect on the channel profile. Under these conditions we may consider the effective channel cross section A to be constant throughout its length. Hence $A = 2bW$, where $2b$ is the channel width corresponding to zero drain current for a specified V_{GS} and W is the channel dimension perpendicular to the b direction, as indicated in Fig. 4-6.

Since no current flows in the depletion region, then, using Ohm's law [Eq. (1-21)], we obtain for the drain current

$$I_D = AqN_D\mu_n\mathscr{E}_x = 2bWqN_D\mu_n \frac{V_{DS}}{L} = 2bqN_D\mu_n \left(\frac{W}{L}\right) V_{DS} \qquad (4\text{-}1)$$

where L is the length of the channel. Equation (4-1) describes the volt-ampere characteristics of Fig. 4-7 for very small V_{DS}, and it suggests that under these conditions the FET behaves like an ohmic resistance whose value is determined by V_{GS}. The ratio V_{DS}/I_D at the origin is called the *on drain resistance* $r_{DS(ON)}$. With $V_{GS} = 0$ which makes $b = a$, from Eq. (4-1), we obtain

$$r_{DS(ON)} = \frac{1}{2aqN_D\mu_n} \cdot \left(\frac{L}{W}\right) \qquad (4\text{-}2)$$

The parameter $r_{DS(ON)}$ is important in switching applications as it is a measure of how much the FET deviates from an ideal switch, for which the ON resistance is zero. For commercially available n-channel FETs and MOSFETs, values of $r_{DS(ON)}$ ranging from a few ohms to several hundred ohms are listed in the manufacturers' specification sheets. Since the mobility for holes is less than that for electrons, $r_{DS(ON)}$ is much higher for p- than n-channel FETs. Increased mobility also signifies increased switching speed. These factors contribute to the prevalence of n-channel devices rather than p-channel FETs.

The notion that in the ohmic region the JFET is a voltage-variable resistance can be discerned from Eq. (4-1) and Fig. 4-6 as follows. The channel width b is a function of the reverse-biasing voltage V_{GS}. Increase in $|V_{GS}|$ results in decrease in b and I_D at a specified value of V_{DS}. Consequently, the slope of the I_D versus V_{DS} characteristic at the origin decreases as $|V_{GS}|$ increases.

Note that I_D depends on the W/L ratio. This quantity is important in FET design as it serves as a scale factor for device current. For given doping densities, adjustment of the W/L ratio permits FETs with different current-handling capabilities to be fabricated on the same chip. Furthermore, from Eq. (4-2), $r_{DS(ON)}$ can be controlled by selecting the W/L ratio.

The Saturation or Pinch-Off Region   We now consider the situation where V_{DS} is used to establish an electric field $\mathscr{E}_x$ along the x axis at a specified value of $|V_{GS}| < |V_p|$. If a substantial drain current I_D flows, the drain end of the gate is more reverse

FIGURE 4-8
After pinch off, as V_{DS} is increased, L' increases, but δ and I_D remain essentially constant (G_1 and G_2 are tied together).

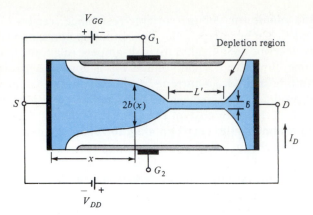

biased than the source end, and hence the boundaries of the depletion region are not parallel to the longitudinal axis of the channel, but converge as shown in Fig. 4-8. A qualitative explanation is given in the following paragraph of what takes place within the channel as the applied drain voltage is increased and pinch-off is approached.

As V_{DS} increases, $\mathscr{E}_x$ and I_D increase, whereas $b(x)$ decreases because the channel narrows, and hence the current density $J = I_D/2b(x)W$ increases. We now see that complete pinch-off ($b = 0$) cannot take place, because if it did, J would become infinite, which is a physically impossible condition. If J were to increase without limit, it follows that [from Eq. (4-1)] $\mathscr{E}_x$ would also increase, provided μ_n remains constant. It is found experimentally, however, that the mobility is a function of electric field intensity and remains constant only for $\mathscr{E}_x < 10^3$ V/cm in n-type silicon. For moderate fields, 10^3 to 10^4 V/cm, the mobility is approximately inversely proportional to the square root of the applied field. For still higher fields, such as are encountered at pinch-off, μ_n is inversely proportional to $\mathscr{E}_x$. In this region the drift velocity of the electrons ($v_x = \mu_n \mathscr{E}_x$) remains constant, and Ohm's law is no longer valid. From Eq. (4-1) we now see that both I_D and b remain constant, thus explaining the constant-current portion of the V-I characteristic illustrated in Fig. 4-7.

What happens if V_{DS} is increased beyond pinch-off, with V_{GS} held constant? As explained above, the minimum channel width $b_{min} = \delta$ has a nonzero constant value. This minimum width occurs at the drain end of the bar. As V_{DS} is increased, this increment in potential causes an increase in $\mathscr{E}_x$ in an adjacent channel section toward the source. Referring to Fig. 4-8, the velocity-limited region L' increases with V_{DS}, whereas δ remains at a fixed value.

The student must take care not to confuse the different meanings of pinch-off and saturation used in describing semiconductor devices. Along the constant-current portion of the characteristic, pinch-off refers to the fact that V_{DS} is used to constrict the channel almost entirely. The pinch-off voltage V_p refers to the voltage applied to the gate which totally blocks the channel independent of V_{DS}. Saturation in an FET refers to the limiting value of drift velocity. Thus the number of carriers that can be transported through the channel per unit time is limited or saturated and I_D remains constant. This is a very different meaning of saturation than that encountered in Chap. 3 for the BJT.

The constant-current characteristics shown in Fig. 4-7 indicate that the FET approximates the ideal voltage-controlled current source in Fig. 4-1, where v_1, i_2, and v_2 correspond, respectively, to V_{GS}, I_{DS}, and V_{DS}.

Breakdown

The maximum voltage that can be applied between any two terminals of the FET is the lowest voltage that will cause avalanche breakdown (Sec. 2-11) across the gate junction. From Fig. 4-7 it is seen that avalanche occurs at a lower value of $|V_{DS}|$ when the gate is reverse-biased than for $V_{GS} = 0$. This is caused by the fact that the reverse-bias gate voltage adds to the drain voltage and hence increases the effective voltage across the gate junction. The breakdown voltage between drain and source with the gate short-circuited to the source, designated by BV_{DSS}, is given in the manufacturers' specification sheets (App. B-5). Its voltage range is from several volts for IC devices to more than 50 V for power FETs.

Cutoff

With a physical FET device the same leakage drain current $I_{DS(OFF)}$ still exists even under the cutoff condition $|V_{GS}| > |V_p|$. The gate reverse current, also called the *gate cutoff current*, designated I_{GSS}, gives the gate-to-source current, with the drain short-circuited to the source for $|V_{GS}| > |V_p|$. A manufacturer specifies maximum values of $I_{DS(OFF)}$ and I_{GSS}. Each of these is in the range from about 1 pA in IC devices to tens of nanoamperes in large-geometry discrete FETs. At temperatures of about 150°C these quantities increase by a factor of about 1000.

We note that in cutoff, with $|V_{GS}| > |V_p|$, $I_D \approx 0$ and V_{DS} can be "large." This is the behavior of an open switch.

4-4 THE JFET TRANSFER CHARACTERISTIC The volt-ampere characteristics in Fig. 4-7 indicate that in the saturation region, the values of drain current I_D depend on the reverse-biasing voltage V_{GS}. The *transfer characteristic*, a plot of I_D versus V_{GS} at a constant value of V_{DS} is a convenient method of displaying this relationship. Figure 4-9 illustrates the transfer characteristic of a 2N4869 *n*-channel JFET at $V_{DS} = 10$ V. The drain current at $V_{GS} = 0$ is designated by the symbol I_{DSS} and, for the 2N4869, is 5 mA. For commercially fabricated JFETs, values of I_{DSS} range from tens of microamperes to hundreds of milliamperes. The lower values of I_{DSS} are typical of IC JFETs; the higher values are common in power devices.

The transfer characteristic can be expressed analytically as given in Eq. (4-3):

$$I_D = I_{DSS} \left(1 - \frac{V_{GS}}{V_p}\right)^2 \tag{4-3}$$

FIGURE 4-9
Transfer characteristic
(I_C versus V_{GS}) for a
2N4869 n-channel JFET
with $V_{DS} = 10$ V.

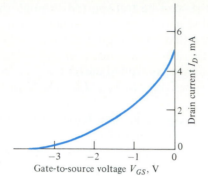

For an n-channel JFET, both V_{GS} and V_p are negative; both quantities are positive in p-channel devices. Thus Eq. (4-3) is valid for both types of JFETs.

The JFET transfer characteristic again demonstrates the controlled-source behavior of the device. If we regard V_{GS} as the input and I_D as the output, Eq. (4-3) and Fig. 4-9 describe a voltage-controlled current source. Operation as a controlled source is the basis of JFET amplifier circuits.

4-5   THE MESFET   The MESFET is a JFET fabricated in gallium arsenide (GaAs) which employs a metal-semiconductor gate region (a Schottky diode). The principles of operation and characteristics of the MESFET are similar to those for the silicon JFET described in Secs. 4-2 to 4-4. Electron mobility in GaAs is 5 to 10 times higher than it is in silicon and allows MESFET operation at frequencies higher than can be achieved with silicon devices. As hole mobility in GaAs is less than that encountered in silicon, n-channel MESFETs are almost always used.

Applications of MESFETs were initially in microwave circuits for high-frequency performance, generally in the range of 1 to 10 gigahertz (GHz). However, starting in 1984, high-speed logic circuits employing MESFETs have been produced commercially. These logic circuits are designed to be compatible with the high-speed bipolar logic family called *emitter-coupled logic* (ECL) (Sec. 6-14).

4-6   THE ENHANCEMENT MOSFET   In a junction field-effect transistor, the effective size of the channel is controlled by an electric field applied to the channel through a *pn* junction. A basically different field-effect device is obtained by using a metal gate electrode separated by an oxide layer from the semiconductor channel. This metal-oxide-semiconductor (MOS) arrangement allows the channel characteristics to be controlled by an electric field established by applying a voltage between the gate and body of the semiconductor and transmitted through the oxide layer. Such a device is called a MOSFET or MOS transistor. Its importance is underscored by the fact that more ICs are fabricated with MOS devices than with any other kind of semiconductor device.

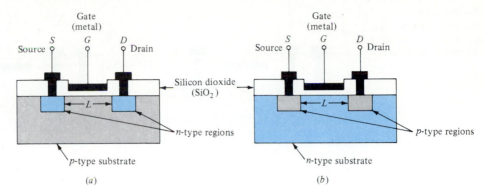

FIGURE 4-10
Enhancement (MOS-FET) structures: (a) n channel; (b) p channel.

There are two types of MOS transistors. The *depletion MOSFET* has a behavior similar to that of the JFET; at zero gate voltage and a fixed drain voltage, the current is a maximum and then decreases with applied gate potential (of the proper polarity) as in Fig. 4.7. The second kind of device, called the *enhancement MOSFET*, exhibits no current at zero gate voltage and the magnitude of the output current increases with an increase in the magnitude of the gate potential. Both types can exist in either the p- or n-channel variety. We consider the characteristics of an n-channel enhancement type in this and the succeeding section and the depletion MOS in Sec. 4-7.

The Enhancement MOS Structure   A simplified form of the structure of an n-channel enhancement MOSFET is shown in Fig. 4-10a and that for a p-channel device, in Fig. 4-10b. The devices depicted in Fig. 4-10 are commonly referred to as *NMOS* and *PMOS transistors*. As indicated in Fig. 4-10a, the two n-type regions embedded in the p-type substrate (the body) are the source and drain electrodes. The region between source and drain is the channel, which is covered by a thin silicon dioxide (SiO_2) layer. The gate is formed by the metal electrode placed over the oxide layer. At present, MOSFET fabrication technology utilizes a polysilicon[1] conducting layer for the gate rather than the metal gate displayed in Fig. 4-10. The physical principles which govern MOSFET operation, however, are the same for both types of gate.

The metal area of the gate, in conjunction with the insulating dielectric oxide layer and the semiconductor channel, form a parallel-plate capacitor. The insulating layer of silicon dioxide is the reason why this device is also called the *insulated-gate field-effect transistor* (IGFET). This layer results in an extremely high input resistance (10^{10} to 10^{15} Ω) for the MOSFET.

Physical Behavior of the Enhancement MOSFET   In Fig. 4-11a we show an NMOS transistor in which the source and substrate are grounded and the drain-to-source voltage V_{DS} is set to zero. The positive voltage applied to the gate establishes

---

[1]Polysilicon refers to doped silicon in which the individual parts of the crystalline structure are randomly oriented in space. The behavior of polysilicon is similar to that of a metal.

FIGURE 4-11
Biased NMOS enhancement transistor showing induced channel with (a) $V_{DS} = 0$ and (b) $V_{DS} > 0$.

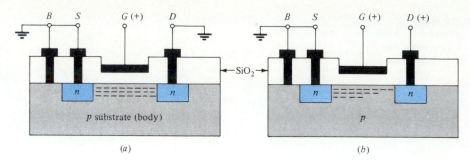

(a)　　　　　　　　　(b)

an electric field which is directed perpendicularly through the oxide. This field will end on "induced" negative charges near the semiconductor surface, as shown in Fig. 4-11a. Since the p-type substrate contains very few electrons, the positive surface charges are primarily electrons obtained from the n-type source and drain. These mobile negative charges, which are minority carriers in the p-type substrate, form an "inversion layer." Such an inversion layer is formed only if V_{GS} exceeds a threshold level V_T.[1] The induced charges beneath the oxide constitute an n channel. As the voltage on the gate increases beyond V_T, the number of induced negative charges in the semiconductor increases. Consequently, the conductivity of the channel increases. Application of a positive potential between drain and source produces a current in the induced channel between drain and source. Thus, the drain current is enhanced by the positive gate voltage and the device is called an *enhancement-type MOSFET*.

Let us now consider the situation where V_{DS} is increased from zero with V_{GS} maintained at a constant positive value greater than V_T (that is, $V_{GS} - V_T > 0$). For small values of V_{DS} ($V_{DS} < V_{GS} - V_T$), an increase in V_{DS} is accompanied by an increase in drain current I_D. The behavior of the MOSFET is that of a resistance, and this region is referred to as the *ohmic region*. As V_{DS} increases, the drop across the channel increases in magnitude, and hence the voltage across the gate oxide at the drain side of the channel $V_{GD} = V_{DS} - V_{GS}$ decreases. This reduced potential difference lowers the field across the drain end of the dielectric, which results in fewer inversion charges in this portion of the induced channel. The channel is being "pinched off," and I_D increases much more slowly with respect to increases in V_{DS} than in the ohmic region near the origin. Ideally, once pinch-off is achieved, a further increase in V_{DS} produces no change in I_D and current saturation exists. This *saturation region* is similar in nature to velocity saturation in a JFET. The value of I_D attained in saturation depends on the value of V_{GS}. Increases in $V_{GS} > V_T$ result in increasing saturation values of I_D.

---

[1] In this chapter the threshold voltage should not be confused with the volt-equivalent of temperature described in Sec. 1-6.

4-7 THE ENHANCEMENT MOSFET VOLT-AMPERE CHARACTERISTICS

Manufacturers of IC MOS transistors do not provide curves of the volt-ampere characteristics. Where needed or desirable, these curves are generated from the analytic expressions for MOSFET behavior in each region of operation.

Analytic Expressions for the Volt-Ampere Characteristics

An inversion channel exists between source and drain, with $V_{DS} = 0$, only if $V_{GS} > V_T$. For $V_{GS} < V_T$, there are no mobile carriers at the drain end of the channel and $I_D = 0$. Thus V_T is analogous to the pinch-off voltage in a JFET. The condition that $V_{GS} < V_T$ and $I_D = 0$ signifies that the MOSFET is *cut off* and corresponds to an open switch.

Ohmic Region

As described in the previous section, for $V_{GS} > V_T$, the channel conductivity is controlled by V_{DS} in the *ohmic* (also called the *nonsaturation* or *triode*) region. More precisely, the ohmic region is defined by $V_{GS} - V_T > V_{DS}$ (or $V_{GD} = V_{GS} - V_{DS} > V_T$). Theoretical analysis[1] of the ohmic region leads to the result that the drain characteristic is given by

$$I_D = k \left(\frac{W}{L} \right) [2 \, (V_{GS} - V_T) \, V_{DS} - V_{DS}^2] \tag{4-4}$$

where L is the channel length, W the channel width (perpendicular to L),[2] and k the process parameter in microamperes per square volt. The process parameter $k = \mu_n C_o / 2$, where μ_n is the electron mobility and C_o is the gate capacitance per unit area (and equals ϵ/T_{ox}, the ratio of the permittivity and thickness of the oxide layer). Of note is that V_T also depends on C_o as well as the doping densities of the *n*-type drain and source and *p* substrate.

Saturation Region

Ideally, I_D is constant and independent of V_{DS} in the saturation region for which $V_{GS} - V_T < V_{DS}$ (but greater than zero). The value of I_D depends only on the effective control voltage $V_{GS} - V_T$ as given in Eq. (4-5):

$$I_D = k \left(\frac{W}{L} \right) (V_{GS} - V_T)^2 \equiv I_{DS} \tag{4-5}$$

where the subscript S added to I_D denotes that the drain current in the saturation region is under consideration.

The dividing line between the ohmic and saturation regions is given by $V_{GS} - V_T = V_{DS}$. Substitution of this value of V_{DS} into Eq. (4-4) results in Eq. (4-5). The dashed curve in Fig. 4-12, which indicates the boundary between the ohmic and saturation regions, is given by

$$I_D = k \left(\frac{W}{L} \right) V_{DS}^2 \tag{4-6}$$

---

[1]See Refs. 1 to 5 at the end of this chapter.

[2]In Figs. 4-10 and 4-11, W is measured into the paper.

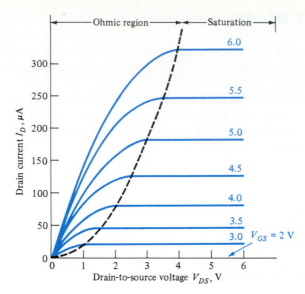

FIGURE 4-12
Enhancement NMOS
output characteristics.

Several observations concerning the expressions in Eqs. (4-4) and (4-5) are noteworthy. First, the *aspect ratio* W/L is an important design parameter as it serves as a scale factor for the drain current. Thus two (or more) MOSFETs having the same value of V_T but with different current capabilities can be fabricated on the same chip by using two (or more) different values of W/L. Second, the parameter k has typical values which lie in the range of 10 to 50 μA/V^2 in present-day commercial NMOS processes. Consequently, high values of I_D (several milliamperes) are obtainable only in devices with high W/L ratios, that is, devices which consume a large area.

The volt-ampere characteristics in Fig. 4-12 are obtained from Eqs. (4-4) and (4-5) for an *n*-channel enhancement MOSFET having $k = 20$ μA/V^2, $W/L = 1$, and $V_T = 2$ V. The dividing line between the ohmic and saturation regions indicated in Fig. 4-12 is obtained by plotting the parabola in Eq. (4-6). Note that if W/L is multiplied by a factor F, then I_D is also multiplied by F for the same pair of values V_{DS} and V_{GS}.

The MOSFET transfer characteristic is a plot of I_D versus V_{GS} at constant V_{DS} in the saturation region. The curve in Fig. 4-13 is the transfer characteristics for the MOSFET given in Fig. 4-12.

The volt-ampere characteristics in Fig. 4-12 are for an ideal MOSFET. In reality, I_D increases slightly with V_{DS} in the saturation region. The cause of this is "channel-length modulation," an effect analogous to base-width modulation in the BJT. As depicted in Fig. 4-14, if the actual characteristics are extended back into the second quadrant, they all meet at $V_{DS} = -1/\lambda$. Because of the similarity with the Early effect in BJTs, the quantity $1/\lambda$ is also referred

FIGURE 4-13
Transfer characteristics
of NMOS enhancement
transistor in Fig. 4-12.

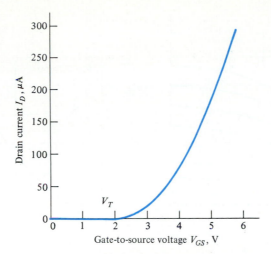

to as the *Early voltage*. Typical values of λ are in the range 0.01 to 0.03 V^{-1}.
To account for channel-length modulation, Eq. (4-5) is modified by the factor
$(1 + \lambda V_{DS})$ as given in Eq. (4-7):

$$I_D = k \frac{W}{L} (V_{GS} - V_T)^2 (1 + \lambda V_{DS}) \qquad (4\text{-}7)$$

The effect of the term $(1 + \lambda V_{DS})$ is usually negligible in digital circuits but
can be important in analog circuits.

FIGURE 4-14
Extension of output
characteristics of NMOS
transistor showing ef-
fect of channel-length
modulation.

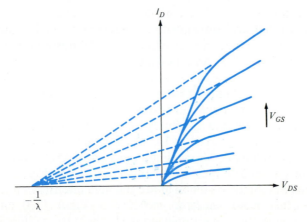

FIGURE 4-15
(a) The drain characteristics and (b) the transfer characteristic for $V_{DS} = -30$ V of the 3N133 enhancement PMOS. (Courtesy of Siliconix, Inc.)

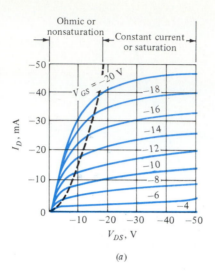

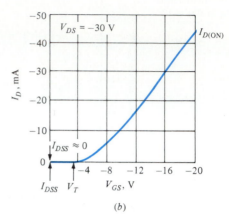

(a)

(b)

p-Channel Enhancement MOSFET Characteristics

The PMOS transistor, depicted in Fig. 4-10b, consists of two p-type regions in an n-type substrate. The physical principles of a p-channel enhancement MOSFET are the same as in NMOS devices. Equations (4-4) to (4-7) are applicable once we recognize that all voltage polarities and current directions in a PMOS device are opposite to the corresponding quantities in NMOS transistors. Also, in evaluating the process parameter k, μ_n is replaced by the hole mobility μ_p.

The volt-ampere characteristics of a discrete PMOS transistor is shown in Fig. 4-15a; its transfer characteristic is depicted in Fig. 4-15b. Note that the general shape of these curves is similar to the NMOS characteristics in Figs. 4-12 and 4-13. However, observe the differences in current and voltage levels for these devices. In discrete MOSFET fabrication, device dimensions can be made larger than are convenient in IC MOSFETs. Consequently, discrete MOS transistors which have higher current capabilities and which can operate with larger applied voltages are obtainable.

Comparison of PMOS and NMOS Transistors

Historically, p-channel enhancement transistors were used first in MOS systems because they were more easily produced with greater yields and reliability than n-channel devices. Improvements in fabrication methods have led to the dominance of NMOS transistors, and—with the exception of CMOS technology (Sec. 4-15)—have made PMOS devices almost obsolete. The reasons for this are described in the next paragraph.

The hole mobility in silicon at normal field intensities is about 500 cm²/(V·s). On the other hand, electron mobility is about 1300 cm²/(V·s). Thus, for devices having the same dimensions (1) the current in a PMOS transistor is less than half of that in an NMOS device and (2) the ON resistance of a *p*-channel MOSFET is nearly three times that for an *n*-channel MOSFET. Alternatively, to achieve the same values of current and ON resistance as in an NMOS transistor, the *W/L* ratio of a PMOS device must be increased to account for the lower hole mobility [see Eq. (4-5)]. This results in PMOS devices requiring nearly three times the area of an equivalent NMOS transistor. Thus NMOS circuits are smaller than PMOS circuits of the same complexity. The higher packing density of the *n*-channel MOS also makes it faster in switching applications due to the smaller junction areas. The operating speed is limited primarily by the internal *RC* time constants, and the capacitance is directly proportional to the junction cross sections. For all the reasons stated in this paragraph, NMOS devices are used almost exclusively.

4-8   THE DEPLETION MOSFET

A second type of MOS transistor can be made if, between the *n*-type regions for drain and source, a narrow *n* channel is embedded into the substrate. Let us consider the operation of such an *n*-channel structure as shown in Fig. 4-16. The minus signs in Fig. 4-16 are intended to indicate free electrons in the channel near the interface with the oxide layer. With $V_{DS} = 0$, a negative gate voltage induces positive charge into the channel. The recombination of induced positive charge with the existing negative charge in the channel causes a depletion of majority carriers. This action accounts for the designation "depletion MOSFET." If the gate voltage is made more negative, majority carriers can be virtually depleted, and, in effect, the channel is eliminated. Under these circumstances, the drain current is zero. The least negative value of V_{GS} for which the channel is depleted of majority carriers is called the *threshold voltage* V_T (analogous to the pinch-off voltage in a JFET).

With $V_{GS} = 0$, application of a positive V_{DS} produces an appreciable drain current denoted by I_{DSS}. As V_{GS} decreases toward the threshold, the drain current decreases. At fixed V_{GS}, increasing values of V_{DS} cause the drain current to saturate as the channel becomes pinched off. The reasons for this are similar to the causes of saturation in enhancement devices. Note in Fig. 4-16*b* that because of the voltage drop along the channel due to I_D, the region of the channel nearest the drain is depleted more than is the region in the vicinity of

FIGURE 4-16

Structure of an *n*-channel depletion-mode MOSFET with (*a*) $V_{GS} = 0$ and (*b*) $V_T < V_{GS} < 0$.

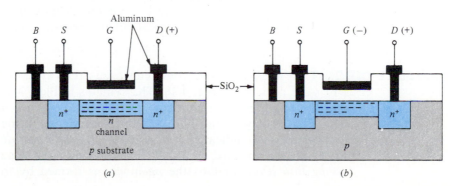

(*a*)   (*b*)

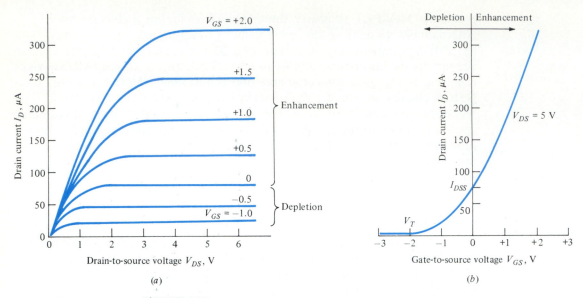

(a) *(b)*

FIGURE 4-17

(*a*) Output and (*b*) transfer characteristics of an NMOS depletion transistor.

the source. This phenomenon is analogous to pinch-off occurring in a JFET at the drain and of the channel (Fig. 4-6).

Depletion MOSFETs exhibit both ohmic and saturated regions. These regions are described analytically by Eqs. (4-4) and (4-5). Note, however, that in the depletion NMOS transistor, V_T is negative.

A MOSFET of the type just described may also be operated in an enhancement mode. It is only necessary to apply a positive gate voltage so that negative charges are induced into the *n*-type channel. The additional negative charge induced in the channel increases ("enhances") the number of majority carriers already present. Thus, for positive V_{GS}, the drain current I_D is greater than I_{DSS}. This is illustrated in Fig. 4-17*a*, in which the volt-ampere characteristics of an *n*-channel depletion MOSFET having $k = 20 \ \mu\text{A/V}^2$, $W/L = 1$, and $V_T = -2$ V are displayed. The transfer function for this device at $V_{DS} = 5$ V is depicted in Fig. 4-17*b*. Note that channel-length modulation effects [Eq. (4-7)] are not included in the characteristics shown in Fig. 4-17.

4-9 MOSFET CIRCUIT SYMBOLS Four commonly used circuit symbols for *n*-channel MOSFETs are depicted in Fig. 4-18. The symbols in Figs. 4-18*a* and 4-18*b* can be used for either enhancement or depletion devices. The circuit symbol in Fig. 4-18*c* is used only for the enhancement-mode device. If the body, or substrate, connection is not indicated, it is assumed that the substrate is either connected to the source terminal or that *B* is tied to the most negative potential. This connection reverse-biases the *pn* junctions formed by the drain and source

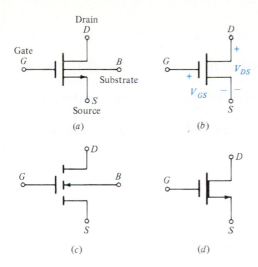

(a)

(b)

(c)

(d)

FIGURE 4-18
Circuit symbols for an NMOS transistor. Both enhancement- and depletion-mode transistors can be represented by parts a and b. The symbols in parts c and d represent only enhancement- and depletion-mode MOSFETS, respectively. For PMOS devices, the arrows are reversed. (In part b note that the arrow is often omitted.)

regions and the substrate. [The most negative potential is ground (0 V) in most instances.] Where both enhancement and depletion devices are employed in the same circuit, we distinguish depletion MOSFETs by the use of the symbol given in Fig. 4-18d. Most often we use Fig. 4-18b as the n-channel MOSFET circuit symbol with standard substrate connections implied.

The positive sense of all terminal currents is into the device. Thus, for an n-channel MOSFET, I_D is positive and I_S is negative. Since I_G is virtually zero, $I_D = I_S$. The voltage drop between drain and source is designated by V_{DS}; V_{GS} is used to indicate the voltage drop from gate to source. Both quantities are positive for n-channel enhancement MOSFETs. Depletion-mode operation requires negative values of V_{GS} and positive values of V_{DS}.

For p-channel MOSFETs, the circuit symbols in Fig. 4-18 are used with the direction of the arrow reversed. Terminal currents and voltages are the negatives of the corresponding quantities in n-channel MOSFETs. The source and substrate are short-circuited in the standard p-channel MOSFET. These, in turn, are connected to the most positive potential to ensure that junctions formed with the n-type body remain reverse-biased.

4-10   THE DC ANALYSIS OF FETS

The techniques described in this section apply equally to JFETs and MOSFETs. Furthermore, the methods of analysis are valid for both p- and n-channel devices.

The Bias Line

Let us consider the circuit in Fig. 4-19 in which the source resistance R_S is used to establish V_{GS} without requiring an additional power supply (Fig. 4-6). Because $I_G = 0$, there is no voltage drop across R_G and the KVL relation for the gate-source loop is

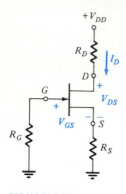

FIGURE 4-19
Self-biased JFET stage.

$$V_{GS} = -I_D R_S \quad \text{or} \quad I_D = \frac{-V_{GS}}{R_S} \tag{4-8}$$

Equation (4-8) defines a straight line called the *bias line* and is plotted on the JFET transfer characteristic shown in Fig. 4-20. The intersection of the transfer characteristic and the bias line determines the operating (quiescent) values of drain current I_{DQ} and gate-to-source voltage V_{GSQ}.

The drain-to-source voltage V_{DSQ} is evaluated from the KVL equation for the drain-source loop. This expression is

$$-V_{DD} + I_D R_D + V_{DS} + I_D R_S = 0 \tag{4-9}$$

Substitution of I_{DQ} into Eq. (4-9) gives the quiescent value of drain-to-source voltage V_{DSQ} that exists in the circuit. Note that Eq. (4-9) specifies the load line for the circuit. By constructing the load line on the output characteristics, we can also determine the value of V_{DSQ} from the intersection of the load line with the characteristic for V_{GSQ}.

Example 4-1

Consider the circuit in Fig. 4-21a, which employs an *n*-channel enhancement MOSFET whose transfer characteristic is shown in Fig. 4-13 and is repeated for convenience in Fig. 4-22. Determine I_{DQ}, V_{DSQ}, and the output quiescent voltage V_{OQ}.

Solution

First replace the gate-bias resistances R_1 and R_2 and the drain supply voltage V_{DD} by their Thévenin equivalents as shown in Fig. 4-21b. (Note that this is analogous to the analysis of the BJT circuit in Example 3-4.)

The bias-line equation is obtained from the KVL expression for the gate-source loop in Fig. 4-21b and is

$$V_{GS} = V_{GG} - I_D R_S \quad \text{or} \quad I_D = -\frac{1}{R_S} V_{GS} + \frac{V_{GG}}{R_S}$$

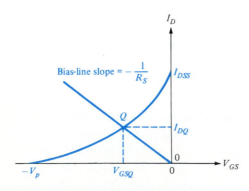

FIGURE 4-20
The bias line, determined by R_S, is drawn on the transfer characteristic. The intersection Q is the quiescent point, and the drain current and the gate-to-source voltage that exist in the circuit are indicated by I_{DQ} and V_{GSQ}, respectively.

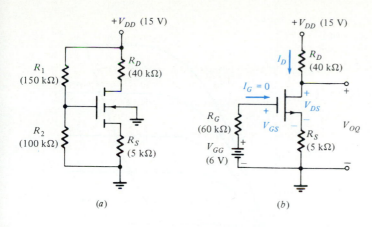

(a)   (b)

FIGURE 4-21
(a) Self biased enhancement MOSFET circuit; (b) the equivalent of part a in which gate-bias circuit (V_{DD}, R_1, and R_2) is replaced by its Thévenin equivalent V_{GG} and R_G.

It is convenient to express current in milliamperes and resistance in kilohms, and, unless otherwise stated, we shall use these quantities in numerical calculations. Substitution of values yields

$$I_D = -\frac{V_{GS}}{5} + \frac{6}{5}$$

The resultant bias line is depicted in Fig. 4-22, from which

$$I_{DQ} = 0.19 \text{ mA}$$

Use of Eq. (4-9) and substitution of values results in

$$V_{DSQ} = 15 - 0.19 \times 40 - 0.19 \times 5 = 6.45 \text{ V}$$

The voltage from drain to ground V_{OQ} is

$$V_{OQ} = V_{DSQ} + I_{DQ}R_S = 6.45 + 0.19 \times 5 = 7.40 \text{ V}$$

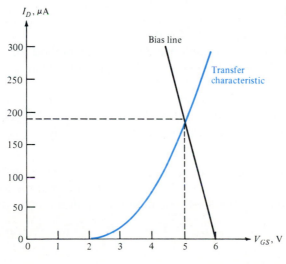

FIGURE 4-22
Metal oxide semiconductor field-effect transistor transfer characteristic and bias line for Example 4-1.

FIGURE 4-23

(a) An enhancement NMOS transistor connected as a resistance; (b) the nonlinear resistance characteristic of the circuit in part a.

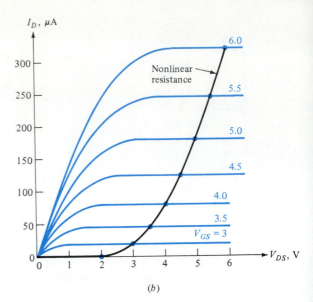

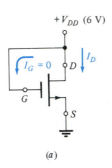

(a)

(b)

4-11 THE MOSFET AS A RESISTANCE

In the introduction to this chapter we indicated that one advantage of the MOSFET is that it can be used as a capacitor or a resistor as well as a three-terminal active element. We showed in Sec. 4-5 that the insulating oxide layer between gate and channel forms a parallel-plate capacitor. Now we consider the use of the enhancement MOSFET as a resistive element.

One such connection is shown in Fig. 4-23a for which the MOSFET has the output characteristics displayed in Fig. 4-23b. By connection of the gate to the drain, $V_{GS} = V_{DS}$. The resistance characteristic is indicated on Fig. 4-23b as the locus of points for which $V_{GS} = V_{DS}$. As is evident in Fig. 4-23b, this MOSFET connection results in a nonlinear resistance. Note that the MOSFET operates in the saturated region since $V_{GS} - V_T < V_{DS}$; furthermore, even with $I_D = 0$, $V_{DS} = V_{GS} = V_T$ for this connection.

In the following example we describe the method for constructing the load line for a MOSFET having a nonlinear drain resistance.

Example 4-2

The circuit in Fig. 4-24a employs a MOSFET $Q1$ having output characteristics displayed in Fig. 4-24b. The load MOSFET $Q2$ has the resistance characteristic given in Fig. 4-23b. Plot the transfer characteristic $v_o = V_{DS1}$ versus $v_i = V_{GS1}$ for the circuit.

Solution

We must first construct the load line for the circuit and then, from the load line, determine V_{DS1} as V_{GS1} is varied. Recalling that the load line is a graphical representation of the KVL equation for the drain loop, we obtain

$$V_{DS1} + V_{DS2} = V_{DD} \quad \text{or} \quad V_{DS1} = V_{DD} - V_{DS2} \quad (4\text{-}10)$$

FIGURE 4-24

(a) Metal oxide semiconductor field-effect transistor circuit with nonlinear MOSFET load resistance; (b) MOSFET output characteristics showing nonlinear load line for Example 4-2.

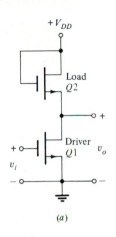

(a)

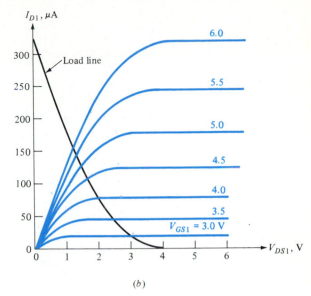

(b)

Since $I_G = 0$ for both $Q1$ and $Q2$, $I_{D1} = I_{D2}$.

The load characteristic in Fig. 4-23b is a plot of I_{D2} versus V_{DS2}. From this figure we note that when $I_{D2} = 320$ μA, $V_{GS2} = V_{DS2} = 6$ V. Consequently, $V_{DS1} = 6 - 6 = 0$ V. This determines one point on the load line shown in Fig. 4-24b ($I_{D1} = 320$ μA, $V_{DS1} = 0$ V). Similarly, when $I_{D2} = 80$ μA, $V_{GS2} = V_{DS2} = 4$ V, so that the point $I_{D1} = 80$ μA, $V_{DS1} = 6 - 4 = 2$ V is also on the load line in Fig. 4-24b. Thus, for each value of I_{D2} in Fig. 4-23b for which $V_{GS2} = V_{DS2}$, we obtain a value of V_{DS2}; this pair of values determines one point on the load line of Fig. 4-24b because of the constraints imposed by KVL and KCL.

Now that the load line is constructed (Fig. 4-24b), we can determine V_{DS1} (the output) as a function of V_{GS1} (the input). For values of $v_i = V_{GS1} \leq V_T = 2$ V, the current I_{D1} is zero and $V_{DS1} = 4$ V. Increasing v_i to 5 V results in $V_{DS1} = 1.5$ V and is determined from the intersection of the load line and the characteristic for $V_{GS1} = 5$ V. The transfer characteristic, displayed in Fig. 4-25, is obtained by determining the value of V_{DS1} at the intersection of the load line with the characteristic for each value of V_{GS1}.

Depletion devices also can be connected, as depicted in Fig. 4-26a, to obtain resistance characteristics. For this circuit, $V_{GS} = 0$ as the gate and source terminals are connected. Thus the resultant resistance characteristic is indicated by the heavy curve for $V_{GS} = 0$ on the MOSFET characteristics in Fig. 4-26b. The analysis of MOSFET circuits using depletion MOSFET load resistances is analogous to that given in Example 4-2. Several such circuits are included in the problems at the end of this chapter.

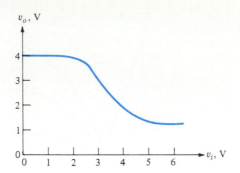

FIGURE 4-25
Voltage transfer characteristic (v_o versus v_i) for Example 4-2.

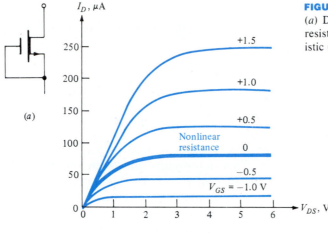

FIGURE 4-26
(a) Depletion-mode MOSFET connected as a resistance; (b) nonlinear resistance characteristic ($V_{GS} = 0$).

4-12 THE FET AS A SWITCH Metal oxide semiconductor field-effect transistors are used extensively in digital circuits which exploit the behavior of these devices as switches. We demonstrate the controlled-switch operation of the MOSFET using the circuit shown in Fig. 4-24a and Example 4-2. The input voltage waveform is the step displayed in Fig. 4-27a. For $t < T$, the input voltage is 1.5 V; thus, from the transfer characteristic in Fig. 4-25, we observe that $v_o = 4$ V. The current in the circuit I_{D1} is zero as determined from the load line in Fig. 4-24b. This is the characteristic of an open switch as the voltage across the switch is "appreciable" while the current is zero.

For $t > T$, the input voltage is 5.0 V, resulting in $v_o = 1.5$ V (Fig. 4-25) and $I_{D1} = 125$ μA (Fig. 4-24b). This condition approximates a closed switch as "appreciable" current exists with a low voltage across the switch terminals (D and S). The output waveform is displayed in Fig. 4-27b.

The two states of the switch can be discerned from the transfer characteristic given in Fig. 4.25. As long as $v_i \leq V_T = 2$ V, the output voltage is 4 V and,

FIGURE 4-27

(*a*) Input step-voltage waveform (v_i) applied to circuit in Fig. 4-24*a* and (*b*) resultant output voltage (v_o) waveform.

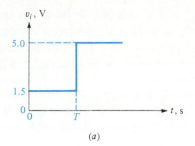

(*a*)

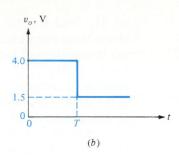

(*b*)

from Fig. 4-24*b*, the current is zero. Input voltages greater than about 5 V produce only small changes in both v_o and I_{D1}. Consequently, the output current is determined almost entirely by the load characteristic and drain supply V_{DD}. The small changes in output voltage with input voltage changes is reflected in the "flattening" of the transfer characteristic in Fig. 4-25 for $v_i \geq 5$ V.

Practical switches cannot change state instantaneously (Fig. 4-27*b*). We treat the transient response of the switch in Sec. 6-6.

Example 4-3

The circuit in Fig. 4-28*a* utilizes a depletion MOSFET whose characteristics are shown in Fig. 4-28*b*. (*a*) Sketch the output waveform for the input waveform given in Fig. 4-28*a*, assuming that $R_D = 36$ kΩ. (*b*) What are the new output levels if R_D is changed to 50 kΩ?

FIGURE 4-28

(*a*) Depletion-mode MOSFET circuit and input voltage waveform; (*b*) MOSFET output characteristics [the load lines correspond to $V_{DD} = 10$ V and $R_D = 36$ kΩ (solid) and $R_D = 50$ kΩ (dashed)] (*c*) the output voltage waveform.

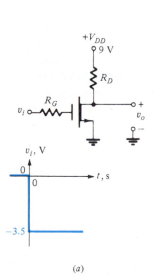

(*a*)

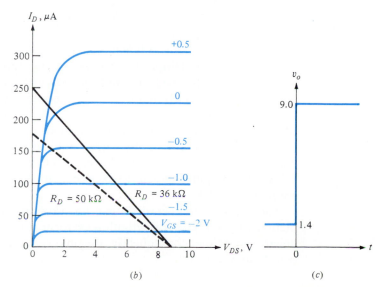

(*b*)

(*c*)

Solution

(a) The load line for $V_{DD} = 9$ V and $R_D = 36$ kΩ is drawn on the output characteristics in Fig. 4-28b. When $t < 0$, the input voltage is zero, and, from the intersection of the load line and the $V_{GS} = 0$ characteristic, $V_{DS} = v_o = 1.4$ V. Similarly, the output voltage is 9 V when $v_i = V_{GS} = -3.5$ V. The resultant waveform is displayed in Fig. 4-28c. (b) Changing R_D to 50 kΩ results in the dashed load line shown in Fig. 4-28b. For $t > 0$, the MOSFET is cut off as the magnitude of v_i is greater than $|V_T|$. Hence $v_o = 9$ V. With $v_i = 0$, as is the case for $t < 0$, the output voltage is 0.8 V, obtained at the intersection of the dashed line and $V_{GS} = 0$.

We note that increasing R_D, for a fixed V_{DD}, results in a decrease in the voltage across the "closed" switch. However, the current in the switch also decreases as I_D is approximately V_{DD}/R_D.

Example 4-4

The circuit in Fig. 4-29a employs a depletion MOSFET $Q2$ as the load for the enhancement MOSFET $Q1$. This configuration is in common usage in digital ICs. The MOSFET load $Q2$ has $k = 20$ μA/V^2, $W/L = \frac{1}{4}$, and $V_T = -2$ V, and its output characteristics are displayed in Fig. 4-29b. The enhancement MOSFET characteristics originally given in Fig. 4-12 and corresponding to a device having $k = 20$ μA/V^2, $W/L = 1$, and $V_T = 2$ V, are redrawn for convenience in Fig. 4-29c. Sketch the transfer function v_o versus v_i for this circuit.

Solution

The load characteristic is shown in Fig. 4-29b for $V_{GS2} = 0$ and results in the load line depicted in Fig. 4-29c. The construction of the load line follows the method used to obtain Fig. 4-24b and makes use of the facts that $I_{D1} = I_{D2}$ and $V_{DS1} + V_{DS2} = V_{DD} = 5$ V [Eq. (4-10)]. Since $I_{D2} = 20$ μA for V_{DS2} in the range of 5 to 2 V, $I_{D1} = 20$ μA as V_{DS1} changes from 0 to 3 V. As V_{DS2} decreases from 2 to 0 V, V_{GS1} increases from 3 to 5 V and $I_{D1} = I_{D2}$ decreases from 20 to 0 μA. These values of I_{D1} versus V_{DS1} are plotted as the load line in Fig. 4-29c. The transfer function, depicted in Fig. 4-30, is obtained by varying $v_i = V_{GS1}$ and obtaining the corresponding values of $v_o = V_{DS1}$ from the intersection of the load line and the volt-ampere characteristics of $Q1$. For $v_i \leq 2$ V, $Q1$ is cut off and $v_o = 5$ V (the intersection of the load line and the output characteristics occurs at $V_{DS1} = 5$ V and $I_{D1} = 0$). As v_i is increased, the operating point moves along the load line toward the I_D axis. For each value of v_i, we determine the corresponding value of v_o; for example, at $v_i = 2.5$ V, $v_o = 4.7$ V. Similarly, for $v_i = 3.0$ V, $v_o = 3.0$ V. When v_i exceeds 3.5 V, the output characteristics bunch together along the load line and v_o approaches 0.2 V.

The transfer characteristic can be obtained both analytically and experimentally by applying a sawtooth waveform at the input. This serves to vary v_i linearly with time; consequently, the time variation of v_o is the transfer function (see Example 2-3).

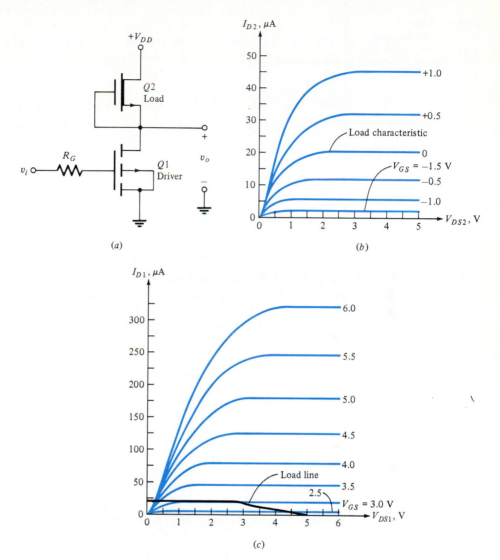

FIGURE 4-29
(a) Circuit for Example 4-4 using NMOS enhancement transistor with an NMOS depletion transistor connected as the load; (b) the depletion load resistance characteristic; (c) the load line corresponding to part b plotted on the output curves of Q1.

The transfer function in Fig. 4-30 shows that the circuit in Fig. 4-29a exhibits the properties of a controlled switch. This circuit displays a steeper slope in the transition region between the open and closed switch than does the transfer function of the enhancement-load circuit in Fig. 4-24. The practical consequences of this difference in performance of the two circuits is discussed in Sec. 6-5.

FIGURE 4-30
Voltage transfer characteristic of circuit in Fig. 4-29a and Example 4-4.

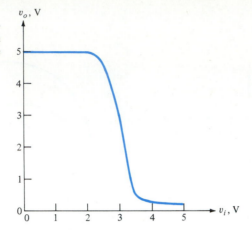

4-13 THE FET AS AN AMPLIFIER

Field-effect transistor amplifier circuits exploit the voltage-controlled current-source nature of these devices. As discussed earlier in this chapter, it is in the saturation region where I_D depends (approximately) only on V_{GS}, thus making available controlled-source behavior. We demonstrate the use of the FET as an amplifier by considering the common-source circuit in Fig. 4-31.

The signal to be amplified in Fig. 4-31 is v_s, whereas V_{GG} provides the necessary reverse-bias between gate and source of the JFET. The volt-ampere characteristics of the JFET are shown in Fig. 4-32 upon which a load line corresponding to $V_{DD} = 30$ V and $R_D = 6$ kΩ is constructed. The value of V_{GG} is selected as 1.5 V so that the transistor is biased at point Q and results in $V_{DSQ} = 19$ V and $I_{DQ} = 1.8$ mA.

The instantaneous gate-to-source voltage is $v_{GS} = v_s - V_{GG}$. Assuming that v_s is a sinusoid of peak voltage $V_m = 0.5$ V, the variation with time in v_{GS}, shown in Fig. 4-32, is a sinusoid superimposed on the quiescent level. The resultant waveforms for i_D and v_{DS} are displayed alongside the characteristics.

FIGURE 4-31
Common-source JFET amplifier stage.

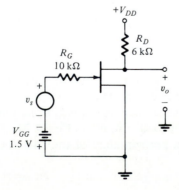

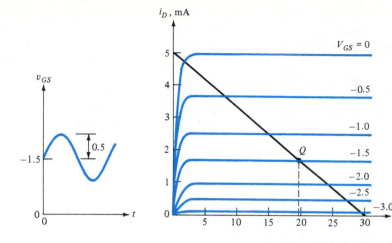

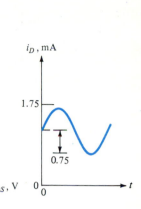

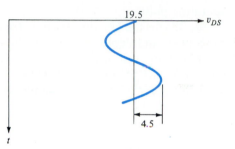

FIGURE 4-32
Output characteristics of JFET in Fig. 4-31. The load line corresponds to $V_{DD} = 30$ V and $R_D = 6$ kΩ. The Q point is established by $V_{GG} = -1.5$ V. The sinusoidal signals superposed on the quiescent levels are displayed for i_D, v_{DS}, and v_{GS}.

We note that both quantities can be considered as sinusoids superimposed on the respective dc values. Thus

$$v_{GS} = -V_{GG} + v_{gs} = -1.5 + 0.5 \sin \omega t \qquad (4\text{-}11)$$

$$i_D = I_{DQ} + i_d = 1.75 + 0.75 \sin \omega t \qquad \text{mA} \qquad (4\text{-}12)$$

$$v_o = v_{DS} = V_{DSQ} + v_{ds} = 19.5 - 4.5 \sin \omega t \qquad (4\text{-}13)$$

We observe in Eq. (4-12) and Fig. 4-32 that the output signal is greater than the input signal, thus demonstrating amplification. The negative sign in Eq. (4-13) indicates the phase reversal of the output signal voltage relative to the input signal voltage. This signifies that an increase in v_{GS} causes a decrease in v_{DS} and accounts for the power-delivering capability of the controlled source. This situation is analogous to that for the BJT amplifier described in Sec. 3-9.

The magnitude of the voltage gain $|A_V|$ is the ratio of the output signal amplitude V_{om} to the input signal amplitude V_{sm}. For the circuit in Fig. 4-31, this gain is $|A_V| = 4.5/0.5 = 9$. Observe that only the input signal is amplified. The higher signal power at the output (relative to the input signal power) is obtained only at the expense of the bias power $V_{DD}I_{DQ}$ that is supplied. Indeed, the bias power is significantly higher than the output power for this circuit.

It is noteworthy to recognize that we biased the JFET in the "middle" of the characteristics. If we selected the operating point either close to the ohmic region or near the pinch-off voltage, the output sinusoid would be clipped during either the positive or negative half-cycles of the input signal. Similarly, with the operating point at Q in Fig. 4-32, the maximum input signal that can be amplified without appreciable distortion is restricted to values of v_{GS} that correspond to the segment of the load line above pinch-off and below the ohmic region. (Compare this with the discussion in Sec. 4-1.)

The amplifying behavior of the FET can be related to the transfer characteristic of the circuit. In Fig. 4-30 we demonstrated that the nearly horizontal segments approximate an open or closed switch. The portion of the curve connecting these two segments indicates that a change in v_i produces a change in v_o. In particular, the change in v_o is larger than the change in v_i over most of this range, which indicates amplification. In fact, one can determine the gain of the circuit by determining the slope of the transfer characteristic at the operating point.

The discussion in this section treated a JFET circuit. Because both enhancement and depletion-type MOSFETs display similar volt-ampere characteristics, they, too, may be viewed as voltage-controlled current sources. Consequently, MOSFETs can also be used as amplifying devices and with the foregoing discussion equally applicable to them.

4-14 SMALL-SIGNAL FET MODELS The small-signal equivalent circuit, valid for both the JFET and MOSFET, is used to relate incremental changes in transistor currents and voltages about the quiescent point. In Sec. 4-13 we showed that i_D, v_{DS}, and v_{GS} each comprise the superposition of a dc and an ac component. The ac component represented the change about the operating point produced by the application of a sinusoidal signal. Thus, paralleling the notation developed for the BJT (Sec. 3-9), we find that

$$
\begin{aligned}
i_d &= i_D - I_{DQ} = \Delta i_D \\
v_{ds} &= v_{DS} - V_{DSQ} = \Delta v_{DS} \\
v_{gs} &= v_{GS} - V_{GSQ} = \Delta v_{GS}
\end{aligned}
\tag{4-14}
$$

The Low-Frequency Model The small-signal FET model is a circuit which is used to show the relationships that exist between i_D, v_{DS}, and v_{GS}. The low-frequency equivalent circuit of the FET is shown in Fig. 4-33. Capacitive elements, that is, energy storage effects, are not indicated in Fig. 4-33 as these elements influence performance only at high frequencies (see Sec. 3-10).

The elements in Fig. 4-33 are related to the physical processes which occur in the FET. The voltage-controlled current source $g_m v_{gs}$ indicates the dependence of i_d on v_{gs} when the FET is operated in the saturated region (pinch-off).

FIGURE 4-33
Low-frequency small-signal equivalent circuit of a field-effect transistor.

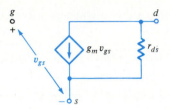

This parameter g_m is the slope of the transfer characteristic (Fig. 4-9) evaluated at quiescent conditions. The output resistance r_{ds} is the slope of the output characteristic evaluated at the operating point. Physically, this parameter is attributed to channel-length modulation (Sec. 4-7). The open circuits ($r \rightarrow \infty$) that appear between g and s and g and d reflect the fact that the junction formed by the gate and channel in a JFET is reverse-biased. As we have indicated previously (Sec. 2-9), the effect on circuit performance of the large value of the incremental resistance of a reverse-biased junction is almost always negligible. In the MOSFET the paths between gate and source and gate and drain are through the insulating oxide layer. Consequently, the extremely high resistance of this path has negligible effect on device and circuit performance.

The value of g_m can be determined analytically from the expressions for the drain current in Eqs. (4-3) and (4-5) for the JFET and the MOSFET, respectively. The transconductance g_m is defined as

$$g_m \equiv \left. \frac{\partial i_D}{\partial v_{GS}} \right|_{v_{DS}=V_{DSQ}} = \left. \frac{i_d}{v_{gs}} \right|_{v_{ds}=0} \tag{4-15}$$

Since i_D represents the total drain current and v_{GS} is the total gate-to-source voltage (see Fig. 4-32), Eq. (4-3) becomes

$$i_D = I_{DSS} \left(1 - \frac{v_{GS}}{V_p} \right)^2$$

and, using Eq. (4-15), we obtain

$$g_m = \frac{-2I_{DSS}}{V_p} \left(1 - \frac{V_{GSQ}}{V_p} \right) \tag{4-16}$$

Recall that, for n-channel JFETS, V_p and V_{GS} are both negative and I_{DSS} is positive; for p-channel devices, V_p and V_{GS} are positive and I_{DSS} is negative. Also, $|V_{GS}| < |V_p|$. Hence V_{GSQ}/V_p is positive and its value is less than unity, and I_{DSS}/V_p is negative. Consequently, g_m has a positive value for either an n- or p-channel JFET.

Use of Eq. (4-3), evaluated at the operating point, permits $1 - (V_{GSQ}/V_p)$ to be written as $\pm(I_{DQ}/I_{DSS})^{1/2}$ so that Eq. (4-16) becomes

$$g_m = \pm \frac{2}{V_p} \sqrt{I_{DQ}I_{DSS}} \tag{4-17}$$

Since we have demonstrated that g_m is always positive, this equation can be written in the alternative form

$$g_m = -\frac{2I_{DSS}}{V_p}\sqrt{\frac{I_{DQ}}{I_{DSS}}} = g_{mo}\sqrt{\frac{I_{DQ}}{I_{DSS}}} \tag{4-18}$$

The term $g_{mo} = 2I_{DSS}/V_p$ is the value of g_m when $V_{GSQ} = 0$ for which $I_{DQ} = I_{DSS}$.

Similarly, for an NMOS transistor, g_m is expressable as

$$g_m = 2\sqrt{k\left(\frac{W}{L}\right)I_{DQ}} \tag{4-19}$$

[The derivation of Eq. (4-19) is left to the student as an exercise in Prob. 4-39.]

Because r_{ds} reflects the effect of channel-length modulation, Eq. (4-7) is used to relate i_D and v_{DS}. For a MOSFET, the output conductance g_{ds} is expressed as

$$g_{ds} \equiv \frac{1}{r_{ds}} = \left.\frac{\partial i_D}{\partial v_{DS}}\right|_{v_{GS}=V_{GSQ}} = \left.\frac{i_d}{v_{ds}}\right|_{v_{gs}=0} \tag{4-20}$$

Application of Eq. (4-20) in conjunction with Eq. (4-7) yields

$$g_{ds} = \lambda k\left(\frac{W}{L}\right)(V_{GSQ} - V_T)^2 = \frac{\lambda I_{DQ}}{1 + \lambda V_{DSQ}} \tag{4-21}$$

Hence

$$r_{ds} = \frac{1 + \lambda V_{DSQ}}{\lambda I_{DQ}} \tag{4-22}$$

For IC FETs, Eq. (4-22) is usually evaluated at $V_{DSQ} = 0$ and reduces to

$$r_{ds} = \frac{1}{\lambda I_{DQ}} \tag{4-23}$$

Normal operation of IC FETs is at drain-to-source voltages in the order of a few volts. Consequently, the term λV_{DSQ} in Eq. (4-22) is much less than unity so that Eq. (4-23) is a good approximation of Eq. (4-22). For discrete FETs, particularly those which are used at moderate voltage and power levels, r_{ds} is evaluated by using Eq. (4-22). (See Figs. 4-7 and 4-32 for the 2N4869 n-channel JFET used in Example 4-5.)

Equation (4-22) is also valid for a JFET as the channel-length modulation term $(1 + \lambda v_{DS})$ can also be incorporated into Eq. (4-3).

Example 4-5

Determine the voltage gain of the JFET amplifier stage shown in Fig. 4-31. The JFET has $I_{DSS} = 5$ mA, $V_p = -3.6$ V, and $\lambda = 0.01$ V^{-1}.

FIGURE 4-34
Small-signal equivalent
of circuit in Fig. 4-31.

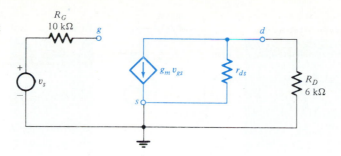

Solution

The equivalent circuit is shown in Fig. 4-34 in which the JFET is represented by the model given in Fig. 4-33. From the load line and characteristics illustrated in Fig. 4-32, $I_{DQ} = 1.8$ mA, $V_{GSQ} = -1.5$ V, and $V_{DSQ} = 19$ V. Use of Eq. (4-16) gives

$$g_m = \frac{2 \times 5}{3.6}\left(1 - \frac{1.5}{3.6}\right) = 1.62 \times 10^{-3} \, \mho = 1.62 \text{ mA/V}$$

The value of r_{ds}, from Eq. (4-22), is

$$r_{ds} = \frac{1 + 0.01 \times 19.5}{0.01 \times 1.80} = 66.4 \text{ k}\Omega$$

The output voltage is $v_o = -g_m R_L v_{gs}$, where R_L is $r_{ds} \parallel R_D$ and $v_{gs} = v_s$. The value of R_L is

$$R_L = \frac{6 \times 66.4}{6 + 66.4} = 5.50 \text{ k}\Omega$$

and

$$A_V = \left|\frac{v_o}{v_s}\right| = g_m R_L = 1.62 \times 5.50 = 8.91$$

This value of gain obtained analytically is in excellent agreement with the value found graphically in Sec. 4-13.

Example 4-6

The supply voltage V_{DD} in the circuit in Fig. 4-21a and Example 4-1 changes by $+0.3$ V. What change occurs in V_{DSQ} because of the variation in V_{DD}? Use the small-signal model and take $r_{ds} = 50$ kΩ.

Solution

The change in the supply voltage can be represented as shown in Fig. 4-35a. By obtaining the Thévenin equivalent of the gate bias circuit, the MOSFET stage is represented as depicted in Fig. 4-35b (see also Fig. 4-21b). It is difficult to resolve the small changes in the supply voltage in either the bias line in Fig. 4-22 or from the output characteristics for the transistor. Therefore, it is convenient to consider these changes as a small (incremental) dc signal applied in the circuit. The small-signal model is displayed in Fig. 4-36a; conversion of the

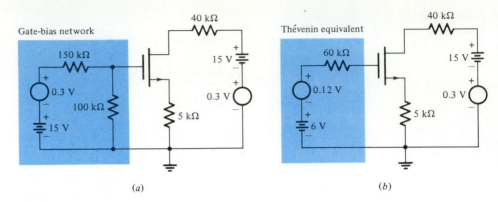

(a)   *(b)*

FIGURE 4-35
(a) Metal oxide semiconductor field-effect transistor circuit showing change in supply voltage as a signal source (0.3 V); *(b)* circuit in part *a* with the gate-bias network replaced by its Thévenin equivalent.

current source $g_m v_{gs}$ in parallel with r_{ds} to a voltage source $g_m r_{ds} v_{gs} = \mu v_{gs}$ in series with r_{ds} is shown in Fig. 4-36*b*. The symbol μ^1 is called the *amplification factor* and

$$\mu = g_m r_{ds} \qquad (4\text{-}24)$$

In Fig. 4-36*b* and through use of KVL for the gate and drain loops

$$v_{gs} = 0.12 - 5 i_d$$

and $\qquad -0.3 + 40\, i_d + 50 i_d - \mu v_{gs} + 5 i_d = 0$

The value of μ is obtained by first evaluating g_m. From Example 4-1, $I_{DQ} = 190\ \mu\text{A}$ and for $k = 20\ \mu\text{A/V}^2$ and $W/L = 1$, use of Eq. (4-19) gives

$$g_m = 2\ \sqrt{20 \times 10^{-6}\,(1)\ 190 \times 10^{-6}} = 1.23 \times 10^{-4}\ \text{℧} = 0.123\ \text{mA/V}$$

and $\mu = 0.123 \times 50 = 6.15$. Substitution of the expression for v_{gs} into the drain-loop KVL equation and solving for i_d yields

$$i_d = \frac{0.3 + 0.12 \times 6.15}{90 + 5\,(1 + 6.15)} = 8.25 \times 10^{-3}\ \text{mA} = 8.25\ \mu\text{A}$$

The voltage v_{ds} is then

$$\begin{aligned}
v_{ds} &= -40 i_d + 0.3 - 5 i_d \\
&= -40 \times 8.25 \times 10^{-3} + 0.3 - 5 \times 8.25 \times 10^{-3} \\
&= -0.0713\ \text{V} = -71.3\ \text{mV}
\end{aligned}$$

The total voltage $v_{DS} = V_{DSQ} + v_{ds} = 6.45 - 0.0713 = 6.39\ \text{V}$. This result indicates that a 2 percent change in V_{DD} results in a 1.1 percent change in v_{DS}.

---

[1] The use of μ for the amplification factor should not be confused with carrier mobility.

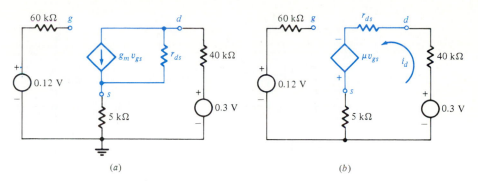

FIGURE 4-36

(*a*) Small-signal model for circuit in Fig. 4-35*b*; (*b*) circuit in part *a* with the current source $g_m v_{gs}$ in parallel with r_{ds} converted to its voltage-source equivalent.

The High-Frequency Model At high frequencies, the capacitive effects associated with the reverse-biased junction and the oxide layer must also be included in the FET small-signal equivalent circuit (Fig. 4-37). Because the junction exists between both gate and source and gate and drain, each of the capacitances C_{gs} and C_{gd} contain a component of the capacitance associated with the depletion region. In addition, these capacitances contain components attributed to the capacitances formed by the dielectric oxide layer, the metallic contact regions, and the semiconductor layers. For pencil-and-paper calculations, it is convenient to combine these effects as illustrated in Fig. 4-37. The models used in computer simulations depict these effects as separate capacitive elements.

4-15 CMOS DEVICES A two-transistor compound device consisting of one NMOS and one PMOS FET is used extensively in digital ICs. Furthermore, these compound transistors are becoming increasingly important in analog circuit applications. The combination of the NMOS and PMOS transistors on the same chip are called *complementary MOS* devices or simply *CMOS* devices. (The fabrication of such devices and the circuits utilizing them is referred to as *CMOS technology*.) In this section our aim is to provide a brief introduction to the CMOS configuration so that we may illustrate the circuit properties that are exploited. Specific circuit applications of CMOS technology are discussed in several sections of Parts 2, 3, and 4 of this text.

FIGURE 4-37

High-frequency FET small-signal model.

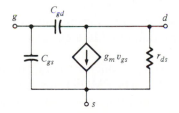

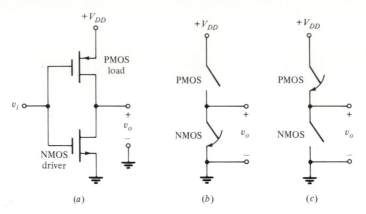

FIGURE 4-38
(*a*) Complementary metal oxide semiconductor switch containing NMOS driver and PMOS load. Ideal switch representation of CMOS circuit in part *a* when NMOS transistor shown in part *b* is ON and (*c*) when PMOS transistor conducts.

The CMOS circuit in Fig. 4-38*a*, employed in digital applications, consists of an NMOS transistor (the driver) to which a PMOS transistor is connected as the load. The gate terminals of the two transistors are connected to one another. Let us consider that the magnitude of the threshold voltage V_T of each transistor is the same and equals $V_{DD}/2$. Application of a positive value of $v_i > V_T$ simultaneously turns on the NMOS transistor and cuts off the PMOS device. (Recall that a positive gate voltage is required in *n*-channel transistors and negative gate voltages are needed in *p*-channel devices.) Because the drain and source terminals of both transistors are series-connected, no current exists in the NMOS device (as a result of the cutoff condition in the PMOS transistor). Consequently, the output voltage is virtually zero. This situation is idealized in Fig. 4-38*b* in which the closed switch represents the NMOS devise and the open switch, the PMOS device.

Similarly, when a negative (or zero) input voltage is applied, the PMOS load is turned on, but the NMOS driver is cut off. The switch conditions in Fig. 4-38*c* approximate this situation. Again, because one transistor is cut off, no current exists in the circuit. The output voltage is high; it is V_{DD} in the idealized case shown in Fig. 4-38*c* and is approximately V_{DD} in Fig. 4-38*a*.

The action described is that of opening and closing a switch by means of the input control voltage. However, because no current exists in either switch state, the power dissipated by the transistors is virtually zero. (Indeed, power is consumed in the CMOS device only during the switching interval.) The extremely small power consumption of CMOS circuits is the major motivation for their widespread use.

Complementary metal oxide semiconductor analog circuits often employ the configuration depicted in Fig. 4-39*a*. The PMOS transistor provides the resistive load for the NMOS transistor, which functions as the controlled source.

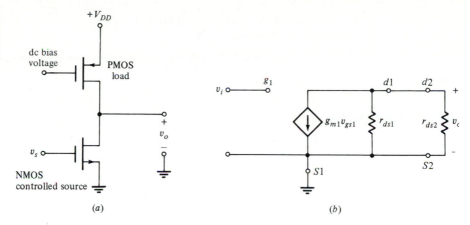

FIGURE 4-39

(*a*) Circuit configuration and (*b*) low-frequency equivalent of CMOS amplifier stage.

The small-signal (incremental) model of the circuit in Fig. 4-39*a* is shown in Fig. 4-39*b*. Note the absence of the source $g_{m2}v_{gs2}$ in the model of the PMOS device because $v_{gs2} = 0$ (the source and gate are each at constant potential). The load resistance r_{ds2} is generally in the order of tens of kilohms. If we were to use a resistor of value r_{ds2} as a load, such a resistance would require a very much larger area of the chip than does the PMOS transistor. This is one major advantage of CMOS technology in analog circuits.

A second advantage of this circuit is the dual function performed by the PMOS transistor: (1) it provides the dc resistance in the circuit (analogous to Fig. 4-26), and (2) it provides the ac (small-signal) load resistance. These resistance values can be markedly different as the bias requirements and signal-processing requirements differ. For example, consider that $r_{ds} = 20$ kΩ is needed to provide the desired voltage gain. If this resistance were required to handle a dc current of 0.5 mA as well, the quiescent voltage drop across it would be $0.5 \times 20 = 10$ V. Often, a voltage drop of this size would require a supply voltage that is larger than can be accommodated readily by the devices employed. However, the PMOS transistor permits the use of more practical supply voltage levels while still providing the 20-kΩ incremental resistance needed to achieve the voltage gain. Recall that r_{ds2} is the slope of the output characteristic at the quiescent point and, as operation is in the saturated region, where the curve is nearly horizontal, r_{ds2} can be appreciable (of the order of tens of kilohms).

REFERENCES

1 Hodges, D. E. and H. G. Jackson: "Analysis and Design of Digital Integrated Circuits," McGraw-Hill Book Company, New York, 1983.

2 Gray, P. R. and R. G. Meyer: "Analysis and Design of Analog Integrated Circuits," 2d ed., John Wiley and Sons, New York, 1984.

3 Yang, E. S.: "Fundamentals of Semiconductor Devices," McGraw-Hill Book Company, New York, 1978.

4 Grebene, A. B.: "Bipolar and MOS Analog Integrated Circuit Design," John Wiley and Sons, New York, 1984.

5 Muller, R. S. and T. I. Kamins: "Device Electronics for Integrated Circuits," John Wiley and Sons, New York, 1977.

REVIEW QUESTIONS

4-1 State three properties of an ideal voltage-controlled current source.

4-2 (*a*) Sketch the basic structure of an *n*-channel JFET.
(*b*) Draw the circuit symbol for the JFET.

4-3 Draw the family of drain characteristics of an *n*-channel JFET and qualitatively explain their shape.

4-4 Define the pinch-off voltage V_p.

4-5 How does the JFET behave for (*a*) small values of $|V_{DS}|$ and (*b*) large values of $|V_{DS}|$?

4-6 Sketch the cross section of an NMOS enhancement transistor.

4-7 Repeat Rev. 4-6 for a PMOS enhancement transistor.

4-8 (*a*) Sketch the output and transfer characteristics of an NMOS enhancement transistor.
(*b*) Qualitatively explain the shape of the characteristics in Rev. 4-8*a*.

4-9 Why are NMOS devices preferred over PMOS transistors?

4-10 Repeat Rev. 4-8 for (*a*) an NMOS depletion transistor and (*b*) a PMOS depletion transistor.

4-11 What is the significance of the threshold voltage V_T in (*a*) enhancement-mode and (*b*) depletion-mode MOSFETs?

4-12 What is the significance of the W/L ratio on the drain current in a MOSFET?

4-13 (*a*) Explain in words what is meant by channel-length modulation.
(*b*) What effect does channel-length modulation have on the drain current?

4-14 Explain the use of the bias line in determining the quiescent voltages and currents in an FET circuit.

4-15 Draw the circuit symbols used for MOSFETs with and without the substrate connection.

4-16 (*a*) Show the circuit diagram of an NMOS enhancement device connected as a resistance.
(*b*) Sketch the resistance characteristic.

4-17 Repeat Rev. 4-16 for an NMOS depletion transistor.

4-18 Explain how an FET can be used as a switch.

4-19 Explain how an FET can be used as an amplifier.

4-20 Draw the low-frequency small-signal model of an FET and explain the significance of each element.

4-21 (*a*) What elements must be added to the model in Rev. 4-20 so that the equivalent circuit is valid at high frequencies?

(*b*) What is the physical origin of these elements?

4-22 (*a*) Draw the small-signal equivalent circuit of a common-source stage.

(*b*) Why doesn't the supply voltage V_{DD} appear in this model?

4-23 What is meant by CMOS technology?

4-24 (*a*) Draw the schematic diagram of a CMOS circuit used in digital applications.

(*b*) Explain the advantages of this circuit.

(*c*) Are there any disadvantages? If so, list them.

4-25 Repeat Rev. 4-24 for a CMOS analog circuit.

Chapter 5
INTEGRATED-CIRCUIT FABRICATION

An integrated circuit (IC) consists of a single crystal chip of silicon, typically 0.25 mm thick and covering a surface area 1 to 10 mm by 1 to 10 mm containing both active and passive elements. The processes used to fabricate ICs are described qualitatively in this chapter. These processes include wafer preparation, epitaxial growth, impurity diffusion, ion implantation, oxide growth, photolithography, chemical etching, and metallization. Batch processing is employed and offers excellent repeatability for the production of large numbers of ICs at low cost.

Each step in the fabrication contributes to both the capabilities and limitations of the circuits produced. The object of this discussion is to present an overview of IC technology and the implications of this technology on circuit design. Specifically, both bipolar and MOS fabrication are treated.

5-1 MONOLITHIC INTEGRATED-CIRCUIT (MICROELECTRONIC) TECHNOLOGY

The term "monolithic" is derived from the Greek words *monos* (meaning "single") and *lithos* (meaning "stone"). Thus a monolithic integrated circuit is built into a single "stone" or single crystal of silicon. The word "integrated" refers to the fact that all the circuit components—transistors, diodes, resistors, capacitors, and their interconnections—are fabricated as a single entity. Note that inductors are not included; one of the consequences of semiconductor IC construction is that practical values of inductance cannot be realized.[1]

The variety of manufacturing processes by which ICs are fabricated take place through a single plane and hence can be conveniently called *planar technology*. The structure of a simple bipolar IC is shown in Fig. 5-1a and is the

---

[1]In Parts 3 and 4 we introduce some circuits whose performance characteristics are similar (or identical) to those containing inductors.

FIGURE 5-1
(a) Cross section of planar IC realization of the current-mirror in part b.

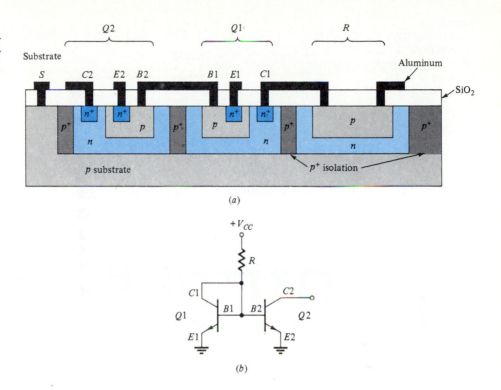

(a)

(b)

realization of the circuit in Fig. 5-1b. (This is the current source first introduced in Sec. 3-11.) The NMOS circuit in Fig. 5-2a, first described in Sec. 4-12, is constructed as depicted in Fig. 5-2b. As seen in Figs. 5-1a and 5-2b, these structures consist of several layers of material. The various layers are n- and p-doped silicon regions, silicon dioxide (SiO_2; also referred to as the *oxide layer*), and the metallic regions.

The silicon layers form the elements of the devices and the *substrate*, or *body*, on (or into) which the IC is constructed. In addition, silicon regions are also used to isolate one component from another. Three different processes used to form the silicon layers are epitaxy, diffusion, and ion implantation.

The oxide layer is used to both protect the surface of the chip from external contaminants and to allow for the selective formation of the n and p regions. The oxide is removed by chemical etching which exposes particular portions of the silicon surface where the n and p regions are to be formed. The regions to be etched are defined by photolithographic techniques.

The thin metallic layer is usually obtained by the chemical vapor deposition of aluminum over the chip surface. Photolithography is used for pattern definition, and etching removes the unwanted aluminum, leaving only the desired connections to and between components.

The representations in Figs. 5-1 and 5-2 are each part of a more complex IC. Many such circuits are fabricated simultaneously on a single wafer of

FIGURE 5-2

(*a*) *n*-Channel MOS circuit with depletion load and (*b*) versus implementation as an IC.

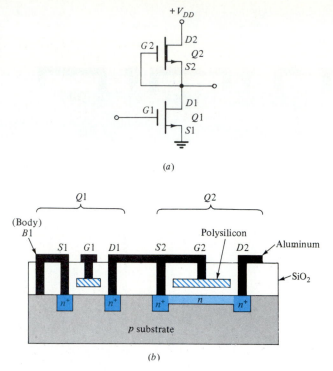

silicon. The silicon crystal (the wafer) forms the substrate on which all the circuit components are made.

In 1985, standard IC manufacture utilizes a 10-cm wafer such as that shown in Fig. 5-3*a*. (Wafers of 15-cm diameter are being introduced for commercial production and 20-cm technology should be available by 1990.) The wafer thickness, about 0.2 to 0.3 mm, provides the chip with sufficient mechanical rigidity to preclude bending. (This dimension exceeds that required to achieve the electrical characteristics of the components.) On completion of the fabrication processes, the wafer is sawed into 100 to 8000 rectangular chips having sides of 1 to 10 mm (for the largest chips). Each chip is a single IC as depicted in Fig. 5-3*b*, which contains as few as tens of components to as many as several hundred thousand components. The photomicrograph in Fig. 5-3*c* depicts a few of the components on the chip in Fig. 5-3*b*.

We are now in a position to appreciate some of the significant advantages of microelectronic technology. If we process twenty 10-cm wafers in a single batch, we can manufacture as many as 160,000 ICs simultaneously. If the average component count per IC were only 700, a batch would contain more than 100 million components. Some of the chips will contain faults due to imperfections in the manufacturing process, but if the *yield* (the percentage of fault-free chips per wafer) is only 10 percent, 16,000 good chips are mass-produced in a single batch!

(a)

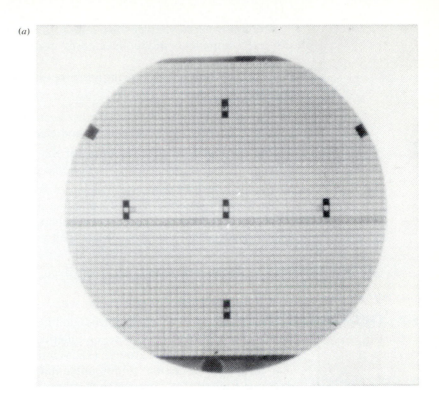

FIGURE 5-3

The 10-cm wafer in (a) contains many identical chips (b). The photomicrograph in (c) is a small segment of the chip in (b) magnified many times.

(b)

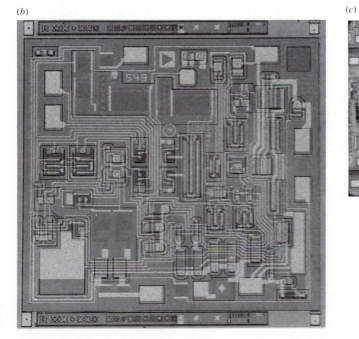

(c)

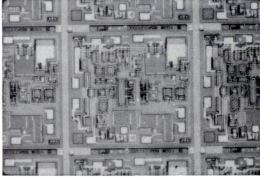

The following advantages are offered by IC technology as compared with discrete components interconnected by conventional techniques:

1. Low cost (due to large quantities processed).
2. Small size.
3. High reliability. All components are fabricated simultaneously, and with no soldered joints, both mechanical and electrical failure are reduced.
4. Improved performance. Because of the low cost, more complex circuitry may be used to obtain better functional characteristics.
5. Matched devices. Since all transistors are manufactured simultaneously by the same processes, the corresponding parameters of these devices as well as the temperature variation of their characteristics have essentially the same magnitudes (the parameters track well with temperature).

5-2 THE PLANAR PROCESSES   As briefly introduced in Sec. 5-1, planar technology for IC fabrication consists of six or seven independent processes: (1) crystal growth of the wafer, (2) epitaxial growth, (3) oxidation, (4) photolithography and chemical etching, (5) diffusion, (6) ion implantation, and (7) metallization. We now examine each of these processes in more detail.

Crystal Growth of the Substrate   A tiny seed crystal of silicon is attached to a rod and lowered into a crucible of molten silicon to which acceptor impurities have been added. As the rod is very slowly pulled out of the melt under carefully controlled conditions, a single p-type crystal ingot of the order of 4 in (10 cm) in diameter and 20 in (50 cm) long is grown. This technique is referred to as the *Czochralski* or *CZ* process. The ingot is subsequently sliced into round wafers approximately 0.2 mm thick to form the substrate on which all integrated components will be fabricated. One side of each wafer is lapped and polished to eliminate surface imperfections before proceeding with the next process.

Epitaxial Growth   The word "epitaxy" comes from the Greek words *epis,* meaning layered and *taxis,* meaning ordered. In IC fabrication, the epitaxial process is used to grow a layer of single-crystal silicon as an extension of an existing crystal wafer of the same material. Epitaxial growth is performed in a special furnace called a *reactor* into which the finished silicon wafers are inserted and heated to 900 to 1000°C. Current production technology uses the hydrogen reduction of the gases silane (SiH_4) or silicon tetrachloride ($SiCl_4$) as the source of the silicon to be grown. Silane has the advantages of requiring lower temperature and having a faster growth rate than does $SiCl_4$.

The chemical reaction for the hydrogen reduction of $SiCl_4$ is

$$SiCl_4 + 2H_2 \xrightarrow{1200°C} Si + 4HCl \tag{5-1a}$$

and that for silane is

$$SiH_4 \xrightarrow[1000°C]{H_2 \text{ atmosphere}} Si + 2H_2 \tag{5-1b}$$

An n-type epitaxial layer, typically 5 to 25 μm (1 $\mu m = 10^{-6} m = 10^{-3} mm$)[1] thick, is grown into a p-type substrate which has a resistivity of approximately 10 Ω·cm, corresponding to $N_A = 1.4 \times 10^{15}$ atoms/cm^3. The epitaxial process described above indicates that the resistivity of the n-type epitaxial layer can be chosen independently of that of the substrate. Values of 0.1 to 0.5 Ω·cm are chosen for the n-type layer.

Since it is required to produce epitaxial films of specific impurity concentrations, it is necessary to introduce impurities such as phosphine (PH_3) for n-type doping or diborane (B_2H_6) for p-type doping into the $SiCl_4$-hydrogen gas stream. One apparatus (which allows simple and precise impurity control) for the production of an epitaxial layer consists of a long cylindrical quartz tube encircled by a radio-frequency induction coil. The silicon wafers are placed on a rectangular graphite rod called a *boat*. The boat is inserted into the reactor and the graphite is heated to about 1200°C. A control console permits the introduction and removal of various gases required for the growth of appropriate epitaxial layers. Thus it is possible to form an almost abrupt step pn junction similar to the junction shown in Fig. 2-1.

Oxidation

Fundamental to the overwhelming use of silicon technology is the ability to grow an oxide layer on the silicon surface. The salient characteristics of SiO_2 as a passivating layer are

1. It is capable of being etched by hydrogen fluoride (HF) to which the underlying silicon is impervious.
2. The impurities used to dope the silicon do not penetrate the silicon dioxide. Thus, when used with the masking techniques (described under the section entitled "Photolithography"), selective doping of specific regions of the chip are accomplished.

Most often, thermal oxidation of silicon is achieved in the presence of water vapor. The chemical reaction is

$$Si + 2H_2O \longrightarrow SiO_2 + 2H_2 \tag{5-2}$$

---

[1] Since the wafer thickness is approximately 0.25 mm, the epitaxial depth is one-tenth or less of the substrate.

The thickness of the oxide layers is generally in the order of 0.02 to 2 μm; the specific value selected depends on the barrier required to prevent dopant penetration. Process temperature, impurity concentration, and processing time are several of the factors that influence the thickness of the SiO_2 layer.

Often silicon nitride (Si_3N_4) is used as a passivating layer because of its superior masking properties. A common use of Si_3N_4 is as a sandwich between two SiO_2 layers. The Si_3N_4 provides the necessary barrier to prevent dopant penetration of the thin underlying SiO_2 layer (essential in MOS devices). The outer silicon dioxide layer, obtained by chemical vapor deposition (CVD), completely covers the chip and protects it from scratches and mechanical damage.

Photolithography   The monolithic technique described in Sec. 5-1 requires the selective removal of the SiO_2 to form openings through which impurities may be diffused. The photoetching method used for this removal is illustrated in Fig. 5-4. During the photolithographic process the wafer is coated with a uniform film of a photosensitive emulsion. A large black-and-white layout of the desired pattern of openings is made and then reduced photographically. This negative, or stencil, of the required dimensions is placed as a mask over the photoresist, as shown in Fig. 5-4. By exposure of the emulsion to ultraviolet (UV) light through the mask, the photoresist becomes polymerized under the transparent regions of the stencil. The mask is now removed, and the wafer is "developed" by using a chemical (such as trichloroethylene) which dissolves the unexposed (unpolymerized) portions of the photoresist film and leaves the surface pattern as shown in Fig. 5-4b. The emulsion which was not removed in development is now *fixed*, or *cured*, so that it becomes resistant to the corrosive etches used next. The chip is immersed in an etching solution of hydrofluoric acid, which removes the oxide from the areas through which dopants are to be diffused. Those portions of the SiO_2 which are protected by the photoresist are unaffected by the acid (Fig. 5-4c). After diffusion of impurities, the resist mask is removed (stripped) with a chemical solvent (such as hot H_2SO_4) coupled with a mechanical abrasion process. A *negative photoresist* is used in the process described above. *Positive photoresists* are also employed in which the exposed portion of the polymer is washed away and thus retains the unexposed material. The remainder of the processing steps are identical and independent of the type of photoresist used.

FIGURE 5-4
Photolithographic technique: (*a*) masking and exposure to ultraviolet radiation. The photoresist after (*b*) development and (*c*) etching.

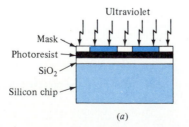
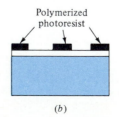
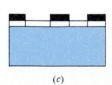

(*a*)   (*b*)   (*c*)

The making of a photographic mask involves complicated and expensive processes. After the circuit layout has been determined, a large-scale drawing is made showing the locations of the openings to be etched in the SiO_2 for a particular process step. Invariably, chip layout is obtained by computer-aided design. The drawing is made to a magnified scale of about 500:1 and results in dimensions more easily managed by a drafter. This technique allows a production process control of 1 μm and a resolution between adjacent lines of 2 μm.

The composite drawing of the circuit is partitioned into several levels, called *masking levels,* used in fabricating the chip. For example, the gate patterns for the MOS devices are on one level, the source and drain contact windows on another, etc. Computer-driven optical-pattern generators convert the patterns into digital information and transfer the geometric layout of the chip onto a photosensitive glass plate. The plate, on which the pattern is reduced by about $100\times$, can be used directly on the chip or in conjunction with a step-and-repeat camera for second-stage reduction ($5\times$ to $10\times$ reduction). The final two-dimensional images on the several glass plates are the masks used in each step of the IC fabrication process (ion implantation, oxidation, metallization, etc.).

The smallest features that can be formed by the photolithographic process described in the foregoing paragraphs is limited by a wavelength of light. Electron beams have much smaller wavelengths than optical radiation and are capable of defining much smaller areas. Hence electron-beam lithography is now used in the production of masks. A narrow electron beam scans a mask covered with an electron-sensitive resist. In this manner the pattern is written on the mask, with the scanning controlled by a computer. The advantages of this method of mask preparation are higher resolution, the elimination of two photographic reduction steps, and shorter production time. In commercial production, the higher cost of the equipment for electron-beam lithography is often offset by the advantages that accrue.

Diffusion

Historically, diffusion of impurities into silicon was the basic step in the planar process. Even before the introduction of integrated circuits, this method was used to fabricate discrete transistors. The introduction of controlled impurity concentrations is performed in a diffusion furnace at a temperature of about 1000°C over a 1- to 2-h period. The diffusion oven usually accommodates 20 wafers in a quartz carrier inside a quartz tube. The temperature must be carefully controlled so that it is uniform over the entire hot zone of the furnace. Impurity sources can be gases, liquids, or solids and are brought into contact with the exposed silicon surface in the furnace. Gaseous impurities used are generally the hydrides of boron, arsenic, and phosphorus. An inert gas (nitrogen) transports the impurity atoms to the surface of the wafers where they diffuse into the silicon.

For the sake of simplicity, the cross-sectional diagrams in this chapter are all shown with vertical diffusion edges (Fig. 5-5a). However, when a window is opened in the SiO_2 and impurities are introduced, they will diffuse laterally

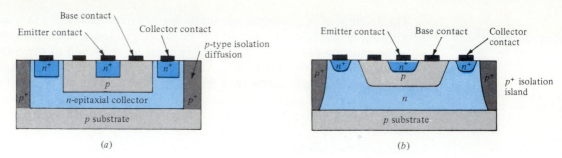

FIGURE 5-5

Cross sections of an IC transistor: (*a*) idealized and (*b*) actual.

the same distance that they do vertically. Hence the impurity will spread out under the passivating oxide surface layer, and the junction profiles should be drawn more realistically as shown in Fig. 5-5*b*.

In a bipolar transistor, two diffusions of impurities are often used. For an *npn* device, the first is the diffusion of the *p*-type base into the *n*-type epitaxially grown collector; the second is the *n*-type emitter region into the *p*-type base. A typical impurity profile for a double-diffused *npn* monolithic transistor is displayed in Fig. 5-6.

The background, or epitaxial-collector, concentration N_{BC} is represented by the dashed line in Fig. 5-6. The concentration N of boron is high (5×10^{18} atoms/cm^3) at the surface and falls off with distance into the silicon as indicated

FIGURE 5-6

A typical impurity profile in a monolithic double-diffused planar transistor. Note that $N(x)$ (in atoms per cubic centimeter) is plotted on a logarithmic scale.

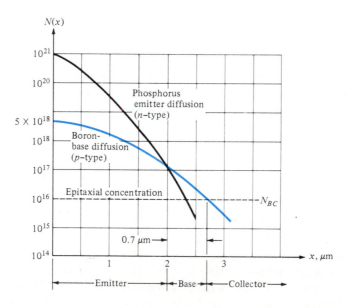

in Fig. 5-6. At that distance $x = x_j$, at which N equals the concentration N_{BC}, and the net impurity density is zero. For $x < x_j$, the net impurity concentration is positive, and for $x > x_j$, it is negative. Hence x_j represents the distance from the surface at which the collector junction is formed. For the transistor whose impurity profile is indicated in Fig. 5-6, $x_j = 2.7 \ \mu m$.

The emitter diffusion (phosphorus) starts from a much higher surface concentration (close to the solid solubility) of about 10^{21} atoms/cm^3 and is diffused to a depth of 2 μm, where the emitter junction is formed. This junction corresponds to the intersection of the base and emitter distributions of impurities. We now see that the base thickness for this monolithic transistor is 0.7 μm. The emitter-base junction is usually treated as a step-graded junction, whereas the base-collector junction is considered a linearly graded junction, because of the slower rate of change of concentration with respect to distance.

Ion Implantation   A second method of introducing impurities is ion implantation. In a vacuum, a beam of appropriate ions (boron for p type and phosphorus for n type) are accelerated by energies between 30 and 200 keV. The depth of penetration of these ions is determined by the accelerating energy and the beam current (the concentration of dopant ions). This process is frequently used where thin layers of doped silicon are required such as the emitter region of a BJT, the channel in a MOSFET, and the gate region of a JFET. For such narrow regions, ion implantation permits the doping concentrations to be more readily controlled than does diffusion.

The SiO$_2$ passivation layer forms an effective barrier against implanted ions, so that only the photolithographically defined regions are doped. A second advantage of ion implantation is that it is performed at low temperatures. Consequently, previously diffused (or implanted) regions have a lesser tendency for lateral spreading.

Another feature of the ion-implantation process is that both the accelerating potential and the beam current are electrically controlled outside of the apparatus in which the implants occur. By contrast, the temperature in the diffusion process must be controlled over the large area inside of the oven. All these advantages have made ion implantation a major process in IC fabrication.

Metallization   The metallization process is used to form the interconnections of the components on the chip. These are formed by the deposition of a thin layer of aluminum (the conductor most frequently used) over the entire surface of the chip. Deposition is achieved by high-vacuum evaporation inside a bell jar. The aluminum is heated until it vaporizes; the gaseous molecules formed uniformly radiate in all directions and completely cover the wafer surface. A mask is used to define the connection pattern between components and the unwanted aluminum is etched and removed.

In this section we described the planar technique involved in fabricating a monolithic IC. We have encountered the seven processes listed at the beginning of Sec. 5-2. The following four sections describe the sequences of these processes that are required to fabricate transistors, diodes, resistors, and capacitors.

5-3 BIPOLAR TRANSISTOR FABRICATION

Our purpose in this section is to describe the fabrication of planar BJTs for monolithic circuits using the processes treated in Sec. 5-2. To illustrate the fabrication sequence, we focus on the construction of the two *npn* transistors in the current source shown in Fig. 5-1*b*. In Sec. 5-8, we describe how the resistor is fabricated.

Transistor Fabrication Once the wafer (the *p*-type substrate) is prepared, an *n*-type epitaxial layer is grown as shown in Fig. 5-7*a*. This layer forms the collector regions of the transistors. Subsequently, the oxide layer is deposited to cover the surface.

The regions for each transistor must now be isolated from one another. To accomplish this, three windows (Fig. 5-7*b*) are formed in the SiO_2 by means of photolithography and etching. A p^+ region is diffused into the exposed epitaxial layer until it reaches the substrate. This process establishes an *isolation island* around each transistor as depicted in the top view of Fig. 5-7*c*. *Electrical isolation is achieved by connecting the substrate to the most negative voltage in the circuit.* Thus it is guaranteed that the *pn* junctions between the collectors and the substrate remain reverse-biased.

After completion of the isolation diffusion, the wafer surface is again covered by an SiO_2 layer. A second mask is used to form the windows into which the *p*-type bases are diffused. This is illustrated in Fig. 5-7*d*, and the base regions are defined in the top view of Fig. 5-7*e*.

A SiO_2 layer is grown to cover the wafer subsequent to the base diffusion. A third masking and etching process removes the SiO_2 in preparation for the shallow emitter diffusion (Fig. 5-7*f*). Note that an n^+ region is also diffused into the collector region of each transistor. The aluminum collector contacts will be made here, and the n^+ areas help to form good ohmic contacts (see Sec. 5-7). Another SiO_2 layer is grown over the wafer surface after the emitter diffusion.

Metallization is the last processing step. The oxide layer is etched by using a fourth mask to expose the wafer where contacts are required. Aluminum is evaporated to cover the entire surface; the excess is chemically removed (a sixth mask) and leaves the desired contacts and interconnection pattern. The cross-sectional view of Fig. 5-7*g* and the top view of Fig. 5-7*h* show the result of the metallization sequence. Figure 5-7*g* is identical with Fig. 5-1*a* for $Q1$ and $Q2$.

The dimensions indicated in Fig. 5-7 are typical of those used for ''small-geometry'' BJTs in commercial fabrication. By constructing both transistors simultaneously and in close physical proximity, they have virtually identical electrical characteristics. To fabricate transistors having different electrical properties, one usually alters the geometry of the device. In particular, to obtain

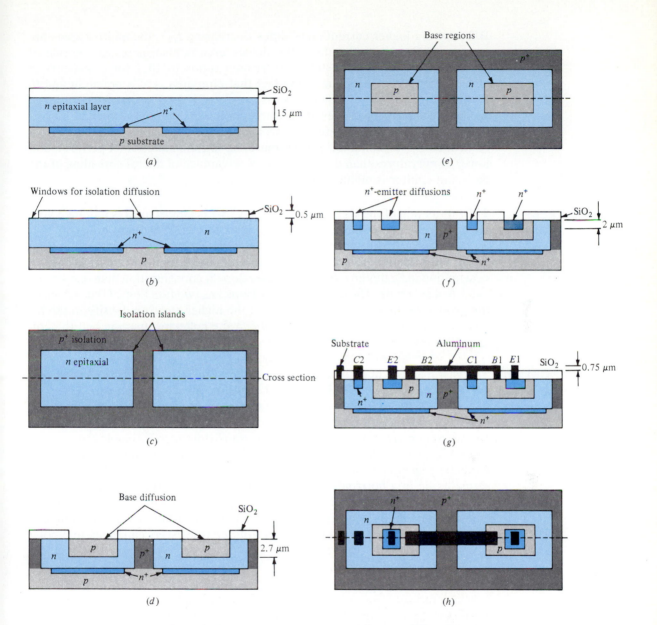

FIGURE 5-7

Steps in the fabrication of an *npn* transistor: (*a*) *n*-type epitaxial growth and oxidation (SiO$_2$); (*b*) masking and etching to expose *n*-type surface for isolation diffusion; (*c*) top view after *p*-type isolation diffusion; (*d*) base diffusion; (*e*) top view after base diffusion; (*f*) *n*$^+$ emitter diffusion (both the emitters of the transistors and collector-contact regions are diffused simultaneously); (*g*) metallization and passivation (SiO$_2$); (*h*) top view of chip showing contacts and interconnections between components. Note that the SiO$_2$ layer is assumed to be transparent so that base, emitter, and collector regions can be seen. The dimensions indicated are typical of modern commercial IC processes.

BJTs having higher current capabilities (increasing I_{ES}), the emitter areas are increased. Consequently, the entire device area is also increased. A rule of thumb often used is to restrict emitter-area ratios to 10:1 for transistors in close proximity. This is due to the fact that chip area is at a premium and to the limitations of the diffusion process.

Commercial IC processes often employ ion implantation of the emitter and base areas. These regions are shallow, and their depths can be more accurately controlled by implantation. Furthermore, as ion implantation is performed at lower temperatures than diffusion, the disadvantage of lateral spreading of the base and emitter is minimized.

Buried Layer

The fabrication of the BJTs indicated in Fig. 5-7 is almost always modified with an additional processing step, as identified in Fig. 5-8 (and shown in Fig. 5-7). The two n^+ regions, referred to as the *buried layer,* at the interface between the n and p layers is deposited prior to epitaxial growth. Recall that the n^+ designation indicates an n region of higher dopant concentration than one simply identified as n type. The use of the n^+ regions has two functions: (1) it enhances the growth of the epitaxial layer, and (2) the higher electron density in the n^+ area reduces the series resistance between the collector junction and the collector terminal (see Sec. 3-7).

pnp Fabrication

The vast majority of IC bipolar transistors are *npn*. However, in some circuits, *pnp* devices are required. For example, in the emitter-coupled pair described in Sec. 3-12, the collector resistors are generally realized by using a pair of *pnp* transistors in a current-source configuration. The *lateral pnp* and *vertical* or *substrate pnp* are the two kinds of such transistors usually employed.

In Fig. 5-8 we see that the base, collector, and isolation regions form a parasitic *pnp* transistor. The term "lateral" refers to the fact that the three elements lie in a horizontal plane in contrast to the vertical plane of the *npn* transistors. Similarly, a parasitic vertical *pnp* device is formed by the base and collector of the *npn* transistor and the *p*-type substrate. These observations lead to the construction of the two types of *pnp* transistors employed in bipolar ICs.

FIGURE 5-8
IC transistors showing buried layer.

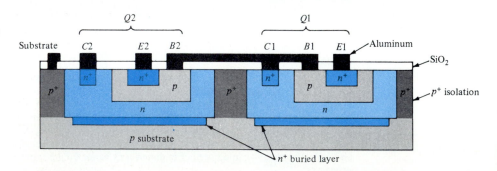

FIGURE 5-9
Cross section of a lateral *pnp* transistor.

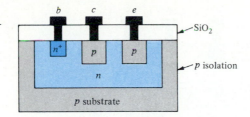

The *lateral pnp*, whose cross section is shown in Fig. 5-9, is formed by implanting the *p*-type emitter and collector regions at the same time that the bases of the *npn* devices are fabricated. Similarly, the n^+ base contact of this *pnp* transistor and the n^+ emitters of the *npn* BJTs are formed simultaneously. Thus we see that both the *npn* and *pnp* transistors are fabricated by using the same sequence of processes. All that is required, for the lateral *pnp* device, are additional windows in the masking steps. The lateral *pnp* transistor has a considerably lower value of β_F than does a *npn* device. This is due to the fact that the *p*-type emitter cannot inject minority carriers into the *n*-type base with the same efficiency as does the n^+ emitter into the *p*-type base of an *npn* BJT. Furthermore, the larger base area and the fact that some injected holes migrate to the substrate cause the number of holes reaching the collector to decrease. Consequently, lateral *pnp* transistors are generally used in circuits having low collector currents.

The *vertical* or *substrate pnp* transistor is used where higher current and power are required. The construction of this transistor is depicted in Fig. 5-10. As seen in Fig. 5-10, this device can also be fabricated simultaneously and by the same processes as are employed with the *npn* transistors. The two simultaneously performed steps are (1) the fabrication of the *p*-emitter regions of the *pnp* transistor and the bases of the *npn* devices and (2) the fabrication of the n^+ base region of the substrate *pnp* device and the emitters of the *npn* transistors.

We have already emphasized that the substrate must be connected to the most negative potential in the circuit. Hence a vertical *pnp* transistor can be

FIGURE 5-10
Vertical or substrate *pnp* transistor cross section.

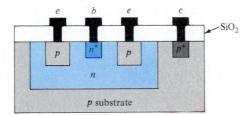

FIGURE 5-11
(*a*) Cross section and (*b*)
top view of multiple-
emitter transistor.

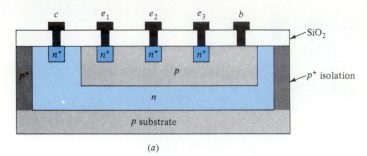

(*a*)

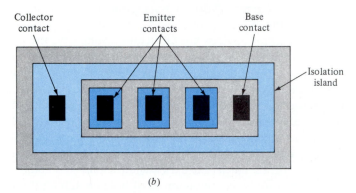

(*b*)

used only if its collector is at a fixed negative voltage. Such a configuration is called an *emitter follower,* and is discussed in Sec. 10-10.

Multiple-Emitter Transistors   Component density of an IC is enhanced by the efficient utilization of chip area. "Merged" transistors, that is, two or more devices which share one or more common regions, provide one method of space saving on a chip. The most widely used merged transistor is the *multiple-emitter transistor.* This arrangement, displayed in Fig. 5-11, is the basis of the *transistor-transistor logic* (TTL or T²L) family discussed in Sec. 6-11. In Fig. 5-11, each emitter stripe can be considered as the emitter of a separate transistor. Each of these devices shares a common base and collector. In effect, this arrangement simulates the two equivalent configurations in Fig. 5-12. Multiple-emitter transistors have been fabricated with more than 60 emitter stripes.

The Schottky Transistor   To obtain the fastest circuit operation, a transistor must be prevented from entering saturation (Sec. 3-8). This condition can be achieved, as indicated in Fig. 5-13*a,* by using a Schottky diode as a clamp between the base and the collector. If an attempt is made to saturate this transistor by increasing the base current, the collector voltage drops, *D* conducts, and the base-to-collector voltage is limited to about 0.4 V. Since the collector junction is forward-biased by less than the cut-in voltage (0.5 V), the transistor does *not* enter saturation (Sec. 3-6).

As indicated in Fig. 5-13*b,* the aluminum metallization for the base lead is allowed to make contact also with the *n*-type collector region (but without an

FIGURE 5-12
(a) Three transistors with common collector and base terminals and its (b) single multiple-emitter transistor equivalent.

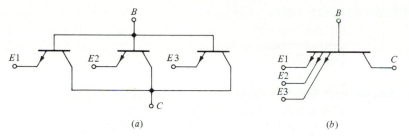

(a)

(b)

intervening n^+ section). This simple procedure results in formation of a metal-semiconductor diode between base and collector. The device in Fig. 5-13b is equivalent to the circuit shown in Fig. 5-13a. This is referred to as a *Schottky transistor* and is represented by the symbol shown in Fig. 5-13c. Note that since the metal-semiconductor junction is formed during the metallization process, the Schottky transistor requires the same number of fabrication processes as does an *npn* device.

The Schottky transistor is used in digital circuits to increase switching speed. There is a delay (storage time) in switching a transistor from ON (saturation) to OFF (cutoff) because the excess minority carriers in the base must first be removed. Connection of the Schottky diode between base and collector prevents the transistor from saturating and thus virtually eliminates storage time. (Recall that minority-carrier storage in a Schottky diode is essentially zero.)

Super-β Transistors   The typical value of β_F for an *npn* monolithic transistor is in the order of 150. By implantation (or diffusion) of the emitter region further into the *p*-type base region (see Fig. 5-7f), the value of β_F can be increased to 2000 to 5000. Devices fabricated in this manner are called *super-β transistors*. The increased values of β_F are the result of a narrower base layer than is usually obtained and are accompanied by a reduction in the emitter-base junction breakdown voltage. Consequently, these transistors are only used in circuits in which the emitter-base junction is subject to low voltages.

FIGURE 5-13
(a) Schottky diode connected between base and collector to form a Schottky transistor; (b) fabrication and (c) circuit symbol of Schottky transistor.

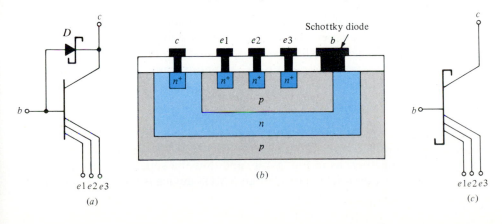

(a)

(b)

(c)

5-4 FABRICATION OF FETs The fabrication of MOS transistors is demonstrated by describing the process sequence used to construct the NMOS enhancement and depletion devices as shown in Fig. 5-2. In this section, JFET construction is also included.

NMOS Enhancement Fabrication The first step in the sequence is to coat the entire surface of a p-type wafer with silicon nitride (Si_3N_4). In Sec. 5-2 we noted that Si_3N_4 is more impervious to the dopants used than is SiO_2. The first mask and etching step is used to define an area large enough to include the source, gate, and drain. The Si_3N_4 is chemically removed from the surface outside of the transistor region. Next, a p^+ layer is ion-implanted near the surface of the exposed p substrate. The p^+ implant serves to isolate adjacent devices as explained in Sec. 5-3. This part of the processing sequence is completed by growing a thick (1-μm) SiO_2 layer, called the *field oxide,* over the p^+-implanted regions as illustrated in Fig. 5-14a. (The Si_3N_4 region is unaffected by oxidation.)

In the second phase of processing, the remaining Si_3N_4 (but not the SiO_2) is removed by selective etching and a thin SiO_2 layer (800 to 1000 A) is thermally grown over the transistor areas (Fig. 5-14b). This process provides the oxide layer that lies under the gates of the transistors.

Polycrystalline silicon, colloquially referred to as *polysilicon* or simply *poly,* is now deposited over the entire wafer. A second photolithographic process defines the gate region and the excess polysilicon is removed by etching. The resulting cross section and top view of the chip are depicted in Figs. 5-14c and 5-14d. The use of polysilicon gates reduces the threshold voltage V_T below the values obtainable with metal gates. As a result, lower supply voltages can be used with polysilicon gates. Consequently, most commercial MOS ICs are fabricated with polysilicon gates.

The n^+ source and drain regions are usually obtained by ion implantation. The field oxide and polysilicon gate prevent the penetration of dopants below these regions. The thin oxide layer, however, is penetrated by the dopant and allows the formation of drain and source. As a result of ion implantation, there is a *self-alignment* of both gate and source and gate and drain. Since there is no overlap of these electrodes, the capacitances between them, C_{gs} and C_{gd}, are drastically reduced.

After the implantation of source and drain, the entire wafer is covered with a protective insulating layer (usually SiO_2). A third mask is used to define the contacts to the device (including the body B) and etching exposes the contact surfaces. Aluminum is evaporated over the entire wafer and a fourth mask is used to pattern the circuit connections (Figs. 5-14e and 5-14f). Note that Fig. 5-14e corresponds to $Q1$ in Fig. 5-1b.

Self-Isolation The p^+ implant in Figs. 5-14e and 5-14f serves as a low resistance for the contact B to the substrate (body) of the MOSFET. Usually the source and body are tied together, as in Fig. 5-14e, and thus the source-substrate diode is cut off. The drain potential polarity for an NMOS device is positive with respect

FIGURE 5-14

Fabrication of enhancement MOSFETs: (*a*) p^+ implantation and thick oxide growth; (*b*) selective etching of Si₃N₄ and thin-oxide growth; (*c*) deposition of polysilicon gate; (*d*) top view showing gate aspect ratio (*W/L*) and implanted n^+ drain and source regions; (*e*) and (*f*) cross section and top view showing metallization and interconnection between substrate and source.

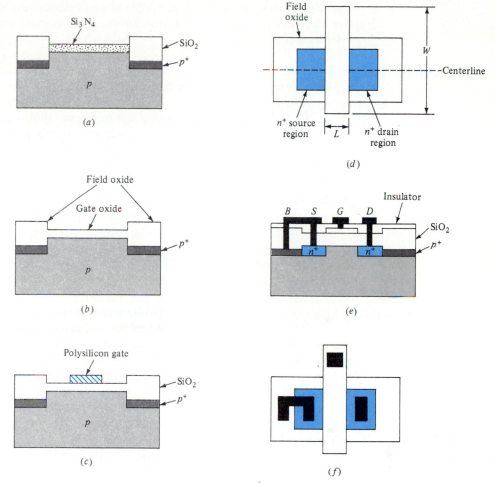

to the source and hence with respect to the *p* substrate. Consequently, the drain-to-substrate diode is also cut off (Fig. 5-14*e*). Clearly, *no isolation island is required for a MOS transistor* and the current is confined to the channel between *D* and *S*. For a BJT the isolation diffusion occupies an extremely large percentage of the transistor area. It is the lack of an isolation border that accounts for the fact that the packing density of MOSFETs is about 20 times that of bipolar transistor ICs.

NMOS Depletion Transistors The fabrication of depletion-mode MOSFETs is similar to that for enhancement devices. The one additional step required is the implantation of the *n* channel (see *Q*2 of Fig. 5-2*b*). This is performed prior to the deposition of the polysilicon gate layer and involves an additional mask and etching step.

  The four-mask NMOS process described is the simplest fabrication sequence that can be employed. To improve yield and obtain more effective control of the electrical properties of the transistors, most commercial processes utilize a minimum of seven masks.

Gate Length and Width   To obtain transistors of different current-handling capacity, usually only the device geometry can be adjusted. In Eq. (4-5), the drain current I_D is seen to vary with W/L, the ratio of gate width to length. Most high-density chips (VLSI) employ devices using the minimum dimensions that can be resolved (2 μm in 1986). For $W/L = 1$, both gate width and length can be 2 μm. To make $W/L = \frac{1}{4}$, as in Example 4-4, minimum gate width is used and L is increased by a factor of 4 (resulting in a 2 $\times$ 8 μm gate).

The devices fabricated in this manner are all low-current devices (50 to 300 μA as shown in Figs. 4-12 and 4-17). To increase current levels to about 1 mA, for example, MOSFETs can be constructed with $W/L = 4$ or $W = 8$ μm and $L = 2$ μm. Theoretically, W/L can be increased to provide any desired current level. However, increase in the gate area also results in increase in the device capacitance and adversely affects the speed of operation. Consequently, it is rare to fabricate MOSFETs having W/L ratios higher than 10.

JFET Fabrication   An n-channel JFET is constructed by means of the bipolar fabrication process. The epitaxial layer, which served as the collector of the BJT, now becomes the n channel of the JFET. As illustrated in Fig. 5-15, isolation islands are diffused into the n-epitaxial layer to separate individual devices. The p^+ gate region is ion-implanted (or diffused) into the n channel and a thin oxide layer is grown. The entire wafer is now covered with SiO_2. Masking and etching define the contact surfaces for the device terminals. Often n^+ regions are implanted under the drain and source contact regions to provide good ohmic contacts. The entire wafer is now covered by an aluminum layer, and the final mask defines the desired interconnection pattern. Removal of the excess aluminum by etching completes the process.

5-5   CMOS TECHNOLOGY   Complementary metal oxide semiconductor circuits require that both NMOS and PMOS enhancement transistors be fabricated on the same chip. To accomplish this, at least two additional steps are needed. The CMOS

FIGURE 5-15
Junction field-effect transistor fabrication and structure.

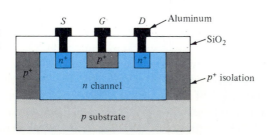

FIGURE 5-16
(a) Cross section of CMOS IC used to implement configuration in part b.

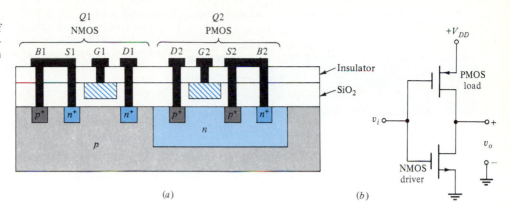

(a)

(b)

circuit shown in Fig. 4-38a, repeated for convenience in Fig. 5-16b, is fabricated as shown in Fig. 5-16a. As is depicted in Fig. 5-16a, the PMOS transistor is fabricated within an n-type well that is implanted or diffused into the p substrate. This n-type region acts as the body $B2$ or the substrate of the PMOS transistor, and at least one additional masking and etching process is required to obtain this n region. The second additional processing step needed is the ion implantation of the p-type source and drain regions of the PMOS device. The processes by which the oxide regions, the polysilicon gate areas, and metallization are accomplished are identical to those for the NMOS enhancement transistor.

As always, the circuit configuration determines the metallization mask. For example, the inverter shown in Fig. 5-16b requires that connections be formed between $D1$ and $D2$, as well as between $G1$ and $G2$.

Separate substrate connections $B1$ and $B2$ are provided. Observe that $B1$ is tied to $S1$ and is connected to the lowest voltage (0 V in Fig. 5-16b), whereas $B2$ is connected to $S2$ and is maintained at the most positive voltage V_{DD}. Since $B1$ is the p-type well and $B2$ is the n-type well, the pn diode formed by these regions is reverse-biased. Thus isolation between the NMOS and PMOS transistors is automatically achieved.

It is important to observe that the PMOS transistor occupies more chip area than does the NMOS device. This is a result of the fact that hole mobility is less than one-half of the electron mobility. The process factor k in Eq. (4-4) is directly proportional to mobility, and, for both transistors to carry the same current, W/L for the PMOS must be larger than that for the NMOS transistor.

5-6 MONOLITHIC DIODES   In the current source of Fig. 5-1b, the base of $Q1$ is short-circuited to the collector. Consequently, a junction diode exists between the emitter and base terminals. This technique is one of five possible connections. The three most popular diode configurations are shown in Fig. 5-17. They are obtained from a BJT structure by using the emitter-base diode, with the collector short-circuited to the base (Fig. 5-17a); the emitter-base diode, with the collector open (Fig. 5-17b); and the collector-base diode, with the emitter open-circuited

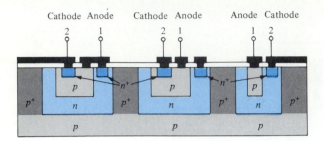

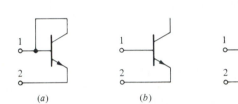

FIGURE 5-17

Cross section and connections for IC diodes: (*a*) emitter-base diode with collector short-circuited to base; (*b*) emitter-base diode with collector open; (*c*) collector-base diode (no emitter diffusion or implant).

(*a*)   (*b*)   (*c*)

(or not fabricated at all) (Fig. 5-17*c*). The choice of the diode type used depends on the application and circuit performance desired. Collector-base diodes have the higher collector-base voltage-breakdown rating of the collector junction (~12 V minimum), and they are suitable for common-cathode diode arrays diffused within a single isolation island, as shown in Fig. 5-18*a*. Common-anode arrays can also be made with the collector-base diffusion, as shown in Fig. 5-18*b*. A separate isolation is required for each diode, and the anodes are connected by metallization.

The emitter and base regions are very popular for the fabrication of diodes, provided the reverse-voltage requirement of the circuit does not exceed the lower base-emitter breakdown voltage (~7 V). Common-anode arrays can easily be made with the emitter and base diffusions by using the multiple-emitter

FIGURE 5-18

Diode pairs: (*a*) common-cathode, (*b*) common-anode.

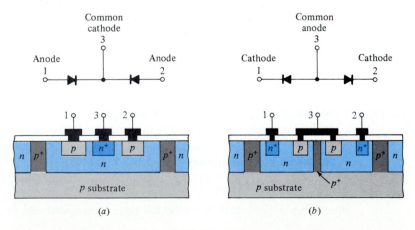

(*a*)   (*b*)

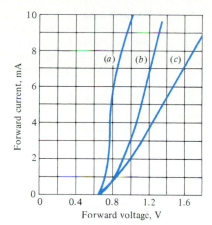

FIGURE 5-19
Typical diode volt-ampere characteristics for the three diode types in Fig. 5-17: (*a*) base-emitter junction with collector short-circuited to base; (*b*) base-emitter diode (collector open); (*c*) collector-base junction (emitter open). (*Courtesy of Fairchild Semiconductor.*)

transistor within a single isolation area. This is the same structure as shown in Fig. 5-11. The collector may be shorted to the base as in Fig. 5-17*a*, or the collector may be left floating (open) as in Fig. 5-17*b*.

Diode Characteristics   The forward volt-ampere characteristics of the three diode types discussed above are shown in Fig. 5-19. It will be observed that the diode-connected transistor (emitter-base diode with collector short-circuited to the base) provides the highest conduction for a given forward voltage. The reverse recovery time for this diode is also smaller, one-third to one-fourth that of the collector-base diode.

5-7 THE METAL-SEMICONDUCTOR CONTACT   Two types of metal-semiconductor junctions are possible, *ohmic* and *rectifying*. The former is the type of contact desired when a lead is to be attached to a semiconductor. On the other hand, the rectifying contact results in a metal-semiconductor diode called a *Schottky barrier* (see Sec. 2-12).

As mentioned in Sec. 5-1, aluminum acts as a *p*-type impurity when in contact with silicon. If aluminum is to be attached as a lead to *n*-type silicon and an ohmic contact is desired, the formation of a *pn* junction must be prevented. It is for this reason that n^+ diffusions are made in the *n* regions near the surface where the aluminum is deposited (Fig. 5-7*g*). On the other hand, if the n^+ diffusion is omitted and the aluminum is deposited directly on the *n*-type silicon, an equivalent *pn* structure is formed, resulting in an excellent metal-semiconductor diode. In Fig. 5-20*a* contact 1 is a Schottky barrier, whereas contact 2 is an ohmic (nonrectifying) contact, and a Schottky diode exists between these two terminals, as indicated in Fig. 5-20*b*.

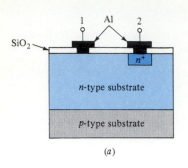

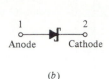

(a)

(b)

FIGURE 5-20
(a) An IC Schottky diode— the aluminum and the lightly doped n region form a rectifying junction, whereas the heavily doped n^+ region and metal form an ohmic contact; (b) Schottky diode symbol.

5-8 INTEGRATED-CIRCUIT RESISTORS

Resistors in monolithic integrated circuits are most often obtained by utilizing the bulk resistivity of one of the transistor regions. The most common technique is to use the diffused or implanted p-type base region of the bipolar transistor. Resistors are also fabricated from the epitaxial layer and the n^+-emitter region. In MOS technology, the polysilicon layer is sometimes used. The n-type substrate of the PMOS transistor in CMOS fabrication is also employed for realizing resistors. Thin-film deposition is a completely different technique by which IC resistors are fabricated. In this section we briefly describe these methods.

Sheet Resistance The semiconductor layers used to form resistors are very thin. Consequently, it is convenient to introduce a quantity called the *sheet resistance R_S*.

In Fig. 5-21, if the width W equals the length L, we have a square $L \times L$ of material with resistivity ρ, thickness t, and cross-sectional area $A = Lt$. The resistance of this conductor (in ohms per square, often indicated as $\Omega/\square$), is

$$R_S = \frac{\rho L}{Lt} = \frac{\rho}{t} \tag{5-2}$$

Note that R_S is independent of the size of the square. Typically, the sheet resistance of the base and emitter diffusions whose profiles are given in Fig. 5-6, is 200 and 5 $\Omega/\square$, respectively.

Diffused Resistors The construction of a base-diffused resistor is shown in Fig. 5-1a and is repeated in Fig. 5-22a. A top view of this resistor is shown in Fig. 5-22b. The

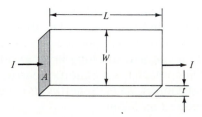

FIGURE 5-21
Pertaining to the sheet resistance (ohms per square).

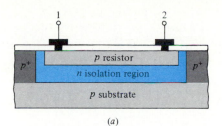

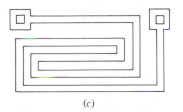

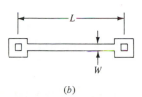

FIGURE 5-22
(a) Cross section and (b) top view of a diffused p-type resistance. (c) A method for increasing the length of the resistor.

resistance value may be computed from

$$R = \frac{\rho L}{tW} = R_s \frac{L}{W} \tag{5-3}$$

where L and W are the length and width of the diffused area, respectively, as shown in the top view. For example, a base-diffused strip 25 μm wide and 250 μm long contains 10 (25×25 μm) squares, and its value is $10 \times 200 = 2000$ Ω. Empirical corrections for the end contacts are usually included in calculations of R.

Observe that, in Fig. 5-22a, the n-type epitaxial layer (the collector region) serves to isolate the p-type resistor from other chip components.

The structure of the n^+-type emitter-diffused resistor is similar to that for the base-diffused resistor. An n^+ diffusion into the p-type base layer is performed simultaneously with the diffusion for the emitters of the BJTs on the chip. The base region acts to isolate this resistance from other components.

Since the sheet resistance of the base and emitter regions is fixed by the fabrication process, the only variables available to control resistance values are stripe length and stripe width. Stripe widths of less than 5 μm are rarely used because slight errors in mask construction or alignment or in photographic resolution can result in significant resistance-value variations. A method for increasing the length of the resistor, and hence its resistance value, is displayed in Fig. 5-22c. By folding the path back on itself, the length can be increased appreciably without utilizing excessive chip area.

The range of values obtainable with diffused resistors is limited by the size of the area required by the resistor. Practical range of resistance is 20 Ω to 30 kΩ for a base-diffused resistor and 10 Ω to 1 kΩ for an emitter-diffused resistor. The tolerance which results from profile variations and surface geometry errors is as high as ±20 percent of the nominal value with ratio tolerance of ±2 percent for minimum width geometry. For resistor widths in the order

FIGURE 5-23
The equivalent circuit
of a diffused resistor.

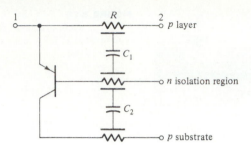

of 50 μm, matching tolerance is about 0.2 percent. For this reason, IC design should, where possible, utilize *resistance ratios rather than absolute values*. The values of diffused resistors increase with temperature. For base-diffused resistors, this variation is in the order of 2000 ppm/°C (parts per million per degree Celsius); variations of 600 ppm/°C are typical of emitter-diffused resistors.

The equivalent circuit of the diffused resistor R is shown in Fig. 5-23, where the parasitic capacitances of the base-isolation C_1 and isolation-substrate C_2 junctions are included. In addition, it can be seen that a parasitic *pnp* transistor exists, with the substrate as collector, the isolation *n*-type region as base, and the resistor *p*-type material as the emitter. The collector is reverse-biased because the *p*-type substrate is at the most negative potential. It is also necessary that the emitter be reverse-biased to keep the parasitic transistor at cutoff. This condition is maintained by placing all resistors in the same isolation region and *connecting the n-type isolation region surrounding the resistors to the most positive voltage present in the circuit*. Typical values of β_F for this parasitic transistor range from 0.5 to 5.

Ion-Implanted Resistors   As the base and emitter regions are often formed by ion implantation, this process is also used to form resistors having the same structure as shown in Fig. 5-22. Implanted *n*-type resistors can be fabricated by using an MOS process similar to that used to form the channel in an NMOS depletion transistor. Ion-implanted resistors are fabricated having resistance values comparable to those achieved with base diffusion. However, tolerances and temperature variations are well below those obtained for diffused resistors. Implanted resistance values can be controlled to three percent and temperature coefficients can be as low as 100 ppm/°C. Matching tolerances are also improved by about 25 percent compared with diffused resistors.

Epitaxial Resistors   The sheet resistance of the collector epitaxial region is about six times that of the base diffusion. Hence it is possible to fabricate higher value resistances by using the epitaxial layer. Such a resistor is defined by the isolation diffusion which surrounds the resistor (Fig. 5-22). These sidewall effects become important, and, to maintain accuracy of resistance values, the isolation diffusion must be carefully controlled. The temperature variation of epitaxial resistors is about 3000 ppm/°C, and absolute and matching tolerances are in the order of 30 and 5 percent, respectively.

Pinch Resistors Consider what happens to the resistance illustrated in Fig. 5-22 if an emitter diffusion is added as in Fig. 5-24. The *n*-type material does not contribute to the conduction because to do so, the current from 1 to 2 would have to flow through the *np* diode at contact 2 in the reverse direction. In other words, only the very small diode reverse saturation current passes through the *n*-emitter material. With the reduction in the conduction path cross section of the *p* material (*pinching*), the resistance must increase. Resistances in excess of 50 kΩ may result, although their exact values are not highly controllable (about ±50 percent with a ±10 percent matching tolerance). Pinch resistors are non-linear since the resistance depends on the impressed voltage, a situation analogous to the variation with voltage of the resistance of the channel in an FET.

The same voltage limitation of reverse base-emitter breakdown BV_{BE0} (~6 V) must apply for pinch resistors, since they are identical in construction to the base-emitter junction with the emitter lead absent. This is not a serious problem since such a resistor is usually used in low-voltage biasing applications across a forward-biased base-emitter junction.

High resistance values can be obtained in a small area to operate at higher voltages by use of an *epitaxial pinch resistor*. The structure is that of an *n*-type epitaxial resistor into which a *p*-base diffusion or implant is made. The *p*-type base constricts the conduction path in the epitaxial layer, thereby increasing the resistance. The junction between the *p* base and the epitaxial layer is essentially the collector-base junction of a transistor. This junction has a higher reverse breakdown voltage than does the emitter-base junction.

MOS Resistors Metal-oxide-semiconductor circuits generally utilize diffused or implanted resistors of the type previously described. Three other resistance structures can also be used. The first is a *polysilicon resistor* and is formed simultaneously with the gate regions of the MOS transistors. Polysilicon resistors have tolerances and temperature coefficients comparable to diffused resistors.

The *well resistor* makes use of the *n*-type diffusion, which forms the substrate of the PMOS transistor in CMOS technology. In effect, this is analogous to

FIGURE 5-24
Cross section of a pinch resistor.

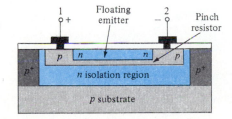

the epitaxial resistor in bipolar technology. Well resistors have high temperature coefficients and poor tolerances.

The third type of resistor is the MOS transistor itself. When biased in the ohmic region, the MOSFET behaves as a resistance (nonlinear). In addition, as described in Sec. 4-11 and Example 4-4, both enhancement and depletion MOSFETs are used as nonlinear resistors in the saturated region.

Thin-Film Resistors   A technique of vapor thin-film deposition can also be used to fabricate resistors for integrated circuits. The metal [usually nichrome (NiCr)] film is deposited (to a thickness of <1 μm) on the silicon dioxide layer, and masked etching is used to produce the desired geometry. The metal resistor is then covered by an insulating layer, and apertures for the ohmic contacts are opened through this insulating layer. Typical sheet-resistance values for nichrome thin-film resistors are 40 to 400 $\Omega/\square$, resulting in resistance values from about 20 Ω to 50 kΩ. The temperature coefficient and tolerance of nichrome thin-film resistors are comparable to the values obtained for implanted resistors.

Other materials, such as tantalum, are also used to fabricate thin-film resistors and result in sheet resistances as high as 2 k$\Omega/\square$ and temperature coefficients as low as 10 ppm/°C.

Diffused or implanted resistors cannot be adjusted after they are fabricated. However, the ohmic value of a thin-film resistor may be trimmed precisely by cutting away part of the resistor with a laser beam. *Laser trimming,* as this technique is called, is a costly process and is used only where precise component values are required. One such application is to the fabrication of the active filters (Sec. 16-7) needed in modems and telephone communication.

5-9   INTEGRATED-CIRCUIT CAPACITORS

Capacitors in ICs are fabricated by utilizing either the depletion-region capacitance of a reverse-biased *pn* junction, the MOS transistor, or thin-film deposition.

Junction Capacitors   A cross-sectional view of a junction capacitor is shown in Fig. 5-25a. The capacitor is formed by the reverse-biased junction J_2, which separates the

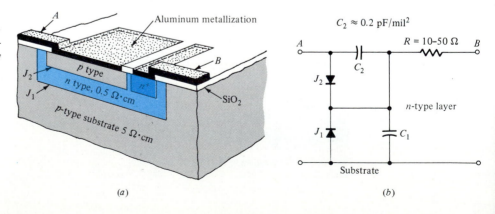

FIGURE 5-25
(a) Junction-type IC capacitor and (b) equivalent circuit. (*Courtesy of Motorola, Inc.*)

epitaxial n-type layer from the upper p-type diffusion area. An additional junction J_1 appears between the n-type epitaxial plane and the substrate, and a parasitic capacitance C_1 is associated with this reverse-biased junction. The equivalent circuit of the junction capacitor is shown in Fig. 5-25b, where the desired capacitance C_2 should be as large as possible relative to C_1. The value of C_2 depends on the junction area and impurity concentration. This junction is essentially linearly graded. The series resistance R (10 to 50 Ω) represents the resistance of the n-type layer.

It is clear that the substrate must be at the most negative voltage so as to minimize C_1 and isolate the capacitor from other elements by keeping junction J_1 reverse-biased. It should also be pointed out that the junction capacitor C_2 is polarized since the pn junction J_2 must always be reverse-biased.

MOS and Thin-Film Capacitors   An MOS nonpolarized capacitor is indicated in Fig. 5-26a. This structure is a parallel-plate capacitor with SiO_2 (whose thickness is 500 Å) as the dielectric. A surface thin film of metal (aluminum) is the top plate. The bottom plate consists of the heavily doped n^+ region that is formed during either the emitter diffusion (implantation) in bipolar processes or the implantation of the drain and source regions in MOS processes. The equivalent circuit of the MOS capacitor is shown in Fig. 5-26b, where C_1 denotes the parasitic capacitance of the collector-substrate junction and R is the small series resistance of the n^+ region. Note that the upper plate need not be metal but can be the polysilicon layer used to obtain the gate regions of the MOS transistors.

Some commercial MOS fabrication processes utilize two layers of polysilicon, thus providing an additional layer of interconnections between components. The two polysilicon layers are separated by a thin SiO_2 region and thus form a capacitor (Fig. 5-27). Capacitors realized in this manner are referred to as *poly-poly capacitors*.

Thin-film capacitors are constructed in a fashion similar to that for MOS capacitors. A thin conducting film (the upper plate) is evaporated onto the SiO_2 layer (the dielectric medium), and the heavily doped n^+ region below the oxide is the lower plate.

FIGURE 5-26

(a) Structure and (b) equivalent circuit of an MOS capacitor.

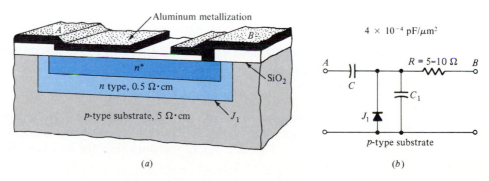

(a)

(b)

FIGURE 5-27
MOS capacitor formed from two polysilicon layers.

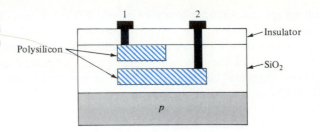

The capacitance of MOS or junction capacitors is quite small, generally about 4×10^{-4} pF/μm^2. Thus a 40-pF capacitor requires an area of 10^5 μm^2 or covers a rectangle 1×0.1 mm of chip surface. Most IC capacitors used are typically less than 100 pF. Values in excess of 500 pF have been obtained, but only at the expense of using a large portion of the chip area.

The use of tantalum films can increase the capacitance per unit area by a factor of 10. A controlled growth of tantalum pentoxide (Ta_2O_5) is used for the dielectric, and metallic tantalum is deposited for the top plate (since aluminum is soluble in Ta_2O_5). The increased capacitance is obtained at the expense of additional processing steps.

5-10 INTEGRATED-CIRCUIT PACKAGING The fabrication sequence is completed when all the processes required to realize all the components and their interconnections are performed. Each wafer is cut into chips (Fig. 5-3), so that the individual microelectronic systems are obtained. The chips are then placed into the package in which they will be used. Wire bonding connects the leads from the chip to the pins of the package. These are the terminals to which external components can be connected to the ICs. In general, external connections are determined by the system in which the circuits are used. Signal inputs and outputs, supply voltages, ground, and components that are not fabricated on the chip are typically applied to the external connections.

FIGURE 5-28
Dual-in-line (DIP) IC package.

A commonly used IC package is the *dual-in-line* (DIP) package shown in Fig. 5-28. The type of standard package can contain as few as 8 pins and as many as 40 pins. The particular number of pin connections is determined by the function the circuit must perform. In any given IC, however, not all pin connections are utilized.

5-11 CHARACTERISTICS OF INTEGRATED-CIRCUIT COMPONENTS From our discussion of IC technology, we can summarize the significant characteristics of ICs as follows:

1. Standard ICs (those stocked by manufacturers) are very inexpensive; for example, the National Semiconductor LM741 Op-Amp containing 21 transistors, 1 diode, and 12 resistors sells for less than 50 cents (in quantity).

However, specially designed chips (small-quantity production) are relatively expensive.

2. The small size of ICs allows complicated systems (consisting of several hundred chips) to be packaged into an instrument of practical volume.

3. Since all components are fabricated simultaneously under controlled conditions and because there are no internal soldered joints, a microelectronic device is very reliable.[1]

4. Because of the low cost, complex circuitry on a chip may be used to obtain improved performance characteristics. The addition of a transistor to an IC increases the cost by less than 1 cent!

5. Device parameters are matched and track well with temperature.

6. A restricted range of values exists for resistors and capacitors. Typically, $10 \ \Omega < R < 50 \ k\Omega$ and $C < 200 \ pF$.

7. Poor tolerances are obtained in fabricating resistors and capacitors of specific magnitudes. For example, ± 20 percent of absolute values is typical, except for ion-implanted components. Resistance-ratio tolerance can be specified to no worse than ± 2 percent because all resistors are made at the same time using the same techniques.

8. Components may have high temperature coefficients and also be voltage-sensitive.

9. High-frequency response is limited by parasitic capacitances.

10. No practical inductors or transformers can be integrated.

11. Because extra steps are required in the fabrication of thin-film resistors and capacitors, their cost increases and the yield decreases. Hence these thin-film devices should be used only if their special characteristics are required. In particular, where precise resistance values are needed, thin-film components are utilized because they can be laser-trimmed.

5-12 MICROELECTRONIC CIRCUIT LAYOUT It is also useful to describe some of the IC layout techniques often employed to use chip area efficiently.

Bipolar Circuits The following represent some reasonable layout rules used in bipolar circuit fabrication:

1. In the layout, allow an isolation border equal to twice the epitaxial thickness to take lateral diffusion into account.

2. Since the isolation diffusion occupies an appreciable area of the chip, the number of isolation islands should be minimized.

---

[1]Early in the U.S. space program, electronic circuit failure was most often attributed to failure of soldered joints. The use of ICs reduced this failure considerably.

3. Place all p-type resistors in the same isolation island and return that isolation region to *the most positive potential* in the circuit. For *n*-type resistors, isolation regions are connected to *the most negative potential* in the circuit.

4. For resistors, use the widest possible designs consistent with die-size limitations. Resistances which must have a close ratio must have the same width and be placed close to one another.

5. Place all transistors *whose collectors are tied together* into the same isolation island. For most circuits, each transistor will be in a separate island.

6. Connect the substrate to *the most negative potential* of the circuit.

7. Use minimum dimensions for emitter regions, base regions, and contacts consistent with device current requirements.

8. Determine component and metallization geometries from the performance requirements of the circuit. For example, the transistor in the output stage of an amplifier would have a larger area than the other transistors if the output stage were to supply the maximum current.

9. Keep all metallization runs as short and wide as possible, particularly at the emitter and collector connections of a saturating transistor.

10. Optimize the layout arrangement for the smallest possible chip size.

11. Use an alignment pattern of the artwork so as to simplify the registration of successive masks.

12. Minimize the number of crossovers.

MOS Circuits

Many of the layout rules also apply to MOS-circuit fabrication. Note, again, that isolation islands are not needed, thereby increasing component densities. One important factor in large-scale ICs is the utilization of minimum gate dimensions consistent with desired current levels. The use of polysilicon gates is an aid in using small devices. Because polysilicon is an effective barrier to dopants, the implantation of the drain and source regions is self-aligned and mask registration errors are minimized.

Crossovers

Very often the layout of a monolithic circuit requires two conducting paths to cross over each other. This crossover cannot be made directly because it will result in electric contact between two parts of the circuit. Since all resistors are protected by the SiO_2 layer, any resistor may be used as a crossover region. In other words, if aluminum metallization is run over a resistor, no electric contact will take place between the resistor and the aluminum.

Sometimes the layout is so complex that additional crossover regions may be required. A diffused structure, useful in bipolar circuits, which allows a crossover is obtained as follows. During the emitter fabrication, n^+ impurities are diffused along a line in the epitaxial region and contact windows are opened at each end of the line. This process forms a "diffused wire." Aluminum is deposited on the insulating SiO_2 (between the two end contacts) in a line perpendicular to the diffused section so as to form a connecting wire for some other part of the circuit. Thus the two wires (one of aluminum and the other of n^+ material) cross over each other without making electric contact. The diffused wire is called a "buried crossover."

In MOS fabrication, the equivalent of the buried crossover is accomplished by the use of a second polysilicon layer (see Fig. 5-27). Consequently, component interconnection can be made by using a polysilicon buried layer as well as with the aluminum metallization.

Computer-Aided Design   The components in an IC cannot be modified once the chip is fabricated. Consequently, much more analysis of a given design is required prior to fabrication than is generally needed for discrete-component circuits. Computer-aided design (CAD) and tools for circuit analysis (SPICE), device fabrication (SUPREM), and circuit layout are employed extensively. These tools are not used to design ICs but provide the information necessary to evaluate the effectiveness of a given design. No commercial ICs are fabricated without such analyses.

REFERENCES   1 Grebene, A. B.: "Bipolar and MOS Analog Integrated Circuit Design," John Wiley and Sons, New York, 1984.

2 Sze, S. M., ed.: "VLSI Technology," McGraw-Hill Book Company, New York, 1983.

3 Ghandi, S.K.: "VLSI Fabrication Principles," John Wiley and Sons, New York, 1983.

4 Colclasser, R. A., and S. Diehl-Nagle: "Materials and Devices," McGraw-Hill Book Company, New York, 1985.

5 Hodges, D. E., and H. G. Jackson: "Analysis and Design of Digital Integrated Circuits," McGraw-Hill Book Company, New York, 1983.

6 Gray, P. R., and R. G. Meyer: "Analysis and Design of Analog Integrated Circuits," 2d ed., John Wiley and Sons, New York, 1984.

7 Yang, E. S.: "Fundamentals of Semiconductor Devices," McGraw-Hill Book Company, New York, 1978.

8 Oldham, W. G.: The Fabrication of Microelectronic Circuits, *Scientific American*, vol. 287, no. 3, pp. 111–128, September 1977.

REVIEW QUESTIONS

5-1   What are five advantages of ICs?

5-2   List the steps involved in fabricating a monolithic IC.

5-3   Describe epitaxial growth.

5-4   Describe the photoetching process.

5-5   (*a*)   Describe the diffusion process.
         (*b*)   What is meant by an impurity profile?

5-6   (*a*)   How is the surface layer of SiO_2 formed?
         (*b*)   What are the reasons for forming the SiO_2 layers?

5-7   Explain how isolation between components is obtained in an IC.

5-8 How are the components interconnected in an IC?

5-9 Describe the ion-implantation process.

5-10 Sketch the cross section of an IC bipolar transistor.

5-11 Define buried layer. Why is it used?

5-12 Describe a lateral *pnp* transistor. Why is its current gain low?

5-13 Describe a vertical *pnp* transistor. Why is it of limited use?

5-14 Describe a super-β transistor.

5-15 Sketch the cross section of an *n*-channel JFET.

5-16 Sketch the cross section of an enhancement-type NMOS transistor.

5-17 Repeat Rev. 5-16 for a depletion-type NMOS transistor.

5-18 (*a*) What is meant by polysilicon?
(*b*) What is the effect of using a polysilicon gate?

5-19 Sketch the cross section of a CMOS composite transistor.

5-20 (*a*) How are IC diodes fabricated?
(*b*) Draw the circuit diagram of two types of emitter-base diodes.

5-21 Sketch the top view of a multiple-emitter transistor. Show the isolation, collector, base, and emitter regions.

5-22 How is an aluminum contact made with *n*-type silicon so that it is (*a*) ohmic; (*b*) rectifying?

5-23 Why is storage time eliminated in a metal-semiconductor diode?

5-24 What is a Schottky transistor? Why is storage time eliminated in such a transistor? Are there any extra fabrication steps required to produce such a transistor? Explain.

5-25 Sketch the cross section of an IC Schottky transistor.

5-26 (*a*) Define sheet resistance R_S.
(*b*) Sketch the cross section of an IC resistor.
(*c*) What are the order of magnitudes of the smallest and the largest values of an IC resistance?

5-27 (*a*) Sketch the equivalent circuit of a base-diffused resistor, showing all parasitic elements.
(*b*) What must be done (externally) to minimize the effect of the parasites?

5-28 Describe a thin-film resistor.

5-29 (*a*) Sketch the cross section of a junction capacitor.
(*b*) Draw the equivalent circuit, showing all parasitic elements.

5-30 Repeat Rev. 5-29 for an MOS capacitor.

5-31 What are the two basic distinctions between a junction and an MOS capacitor?

5-32 (*a*) To what voltage is the substrate connected? Why?
(*b*) Repeat (*a*) for the isolation islands containing the resistors.
(*c*) Can several transistors be placed in the same isolation island? Explain.

5-33 List six important characteristics of integrated components.

5-34 List six design rules for monolithic-circuit layout.

Part Two
DIGITAL CIRCUITS AND SYSTEMS

Digital circuits exploit the controlled-switch behavior of electronic devices to process electrical signals which represent numerical or coded data. These digital signals are generally binary in nature; that is, they are signals which have only two distinct levels. Binary signals are used extensively in communication, control, and instrumentation systems as well as in computers. In this section we treat the circuits and systems utilized to process digital signals.

Even in a large-scale digital system, only a few different operations must be performed, although these may be repeated many times. Logic, arithmetic, and memory circuits in conjunction with input and output devices are the five constituent elements of a digital system. In Chap. 6 we introduce logic circuits, also referred to as *logic gates*. Our focus is on the operation of the fundamental building blocks which comprise the four major fabrication technologies in current use. These are NMOS, CMOS, transistor-transistor logic (TTL), and emitter-coupled logic (ECL). Combinatorial and sequential circuits, that is, the interconnection of several or many logic gates, are treated in Chaps. 7 and 8, respectively.

The physical realizations of most of the circuits described in Chaps. 6 to 8 are often referred to as *small-scale integration* (SSI) or *medium-scale integration* (MSI). In Chap. 9, *large-scale integration* (LSI) and *very-large-scale integration* (VLSI) systems are discussed. Included are memory systems, logic arrays, and microprocessors.

Chapter 6
BASIC LOGIC (DIGITAL) CIRCUITS

Boolean algebra is a system for the mathematical analysis of logic and is named for the nineteenth-century English mathematician, George Boole. Logic gates refer to digital circuits used to implement boolean algebraic equations. The basic logic gates, NOT, AND, and OR, as well as their complements NAND and NOR, are discussed in this chapter. Our principal objective is to describe quantitatively the realization of these gates using integrated circuits (ICs). The two logic families that utilize FETs are NMOS and CMOS; ECL and TTL are the two most important bipolar logic families. All four logic families depend for their operation on the capability of both the FET and BJT to act as a binary device (i.e., a switch).

For convenience, we open this chapter with a brief discussion of binary numbers and their representation as electrical signals. This is followed by an introduction to boolean algebra. For many readers this may be a review of material treated elsewhere in the curriculum.

6-1 THE BINARY SYSTEM

A binary signal or device or circuit exists in one of two permissible states. For example, consider the circuit in Fig. 6-1. The voltage V_o is 5 V when the switch S is open and 0 V when S is closed. No other values of V_o are possible. As both the switch and V_o exist in either of two states, they operate in a binary manner. In Chaps. 3 and 4 we demonstrated that both the BJT and FET exhibit controlled-switch characteristics and thus function as binary devices. Two-state transistor switching circuits are fast, reliable, and inexpensive and can be manufactured in large quantities. Consequently, contemporary digital systems operate on a *binary* or *base 2* number system. Because boolean algebra is a two-state logic representation, the binary system is used for both logic and arithmetic operations. Consequently, the same circuits are used to implement both functions in a digital system.

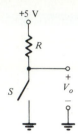

FIGURE 6-1
Binary circuit.

Various designations are used for the two quantized states. Numerically, the binary digits are 1 and 0; in logic systems, the two states are either true or false or yes and no. Electronically, the states are often represented as either ON and OFF, or HI and LO. The HI and LO symbols generally correspond to the voltage or current levels in a switching element, and the designations $V(1)$ and $V(0)$ are used to indicate the voltage levels corresponding to the binary digits 1 and 0, respectively. Because two states are available, each binary digit, or bit, is capable of conveying information. Binary *words* are groups of bits that have collective significance, with a group of 8 bits called a *byte*.

The representation of numbers in binary exactly parallels that used in the decimal (base 10) system. The decimal number 378 is, in reality, simply 300 + 70 + 8 or $3 \times 10^2 + 7 \times 10^1 + 8 \times 10^0$. Each place in the decimal number represents a power of 10; each digit is the number of times the power of 10 is counted. A binary number is composed of a sequence of the digits 1 and 0, each of which multiplies a power of 2. The number 101011 is $1 \times 2^5 + 0 \times 2^4 + 1 \times 2^3 + 0 \times 2^2 + 1 \times 2^1 + 1 \times 2^0$ and equals the decimal number 43. Three decimal digits can be used to represent 1000 different numbers from 0 to 999, with the highest number $10^3 - 1$. Similarly, a 6-bit number can be used to represent 2^6 different values from 0 to $2^6 - 1$. This relationship is generalized and is given in Eq. (6-1).

$$N = 2^n - 1 \tag{6-1}$$

where n is the number of bits and N is the largest decimal number represented by n bits.

A decimal number D can be converted to a binary number B by the following technique:

1. Form two rows of numbers as shown in Table 6-1.
2. Start at the extreme right, divide D by 2 and place the whole-number portion of the quotient D_1 in the first column of the D row.
3. Place the remainder R_1, if any, in the first column of the B row. (The value of R_1 will be 1 or 0 as D is either odd or even.)
4. Divide D_1 by 2 and place quotient D_2 in column 2 of the D row.
5. Place the remainder R_2, a 0 or 1, in column 2 of the B row.

TABLE 6-1 Array for Decimal-to-Binary Conversion

Column $k + 1$	Column k	Column 2	Column 1	
0	$D_k = D_{k-1}/2$	$D_2 = D_1/2$	$D_1 = D/2$	Decimal No. D (D row)
0	R_k	R_2	R_1	Binary No. B (B row)

6. Repeat steps 4 and 5 until the resultant quotient is zero. The digits in the B row, when read left to right, are the binary representation of the decimal number D. The leftmost bit, representing the place value of the highest power of 2, is the *most-significant bit* (MSB), whereas the rightmost bit is the *least-significant bit* (LSB).

Example 6-1

Convert the decimal number 73 into binary.

Solution

Set up an array similar to Table 6-1, as shown in Table 6-2. The binary representation of 73 is the 7-bit number 1001001. This result is readily checked as

$$1001001 = 1 \times 2^6 + 0 \times 2^5 + 0 \times 2^4 + 1 \times 2^3 + 0 \times 2^2$$

$$+ 0 \times 2^1 + 1 \times 2^0$$

$$= 64 + 8 + 1 = 73$$

TABLE 6-2 Array for Example 6-1

8	*7*	*6*	*5*	*4*	*3*	*2*	*1*	
0	$\frac{1}{2} = 0$	$\frac{2}{2} = 1$	$\frac{4}{2} = 2$	$\frac{9}{2} = 4$	$\frac{18}{2} = 9$	$\frac{36}{2} = 18$	$\frac{73}{2} = 36$	$D = 73$
0	1	0	0	1	0	0	1	B

The method outlined in Table 6-1 can be extended to convert a decimal number D to a base number B. The remainders $R_1, R_2, \ldots, R_n$, read from left to right, form the digits of the base B number. If, $B = 5$, for example, R can only have the value 0, 1, 2, 3, or 4. Often a subscript designating the base used to express a number is indicated. Thus N_{10} is a decimal number and N_2 is a binary number.

Just as the decimal point separates positive and negative powers of 10, the binary point separates place values of positive and negative powers of 2. The binary number $101.011 = 1 \times 2^2 + 0 \times 2^1 + 1 \times 2^0 + 0 \times 2^{-1} + 1 \times 2^{-2} + 1 \times 2^{-3}$ and has a decimal equivalent of 5.375.

Negative numbers are represented by adding a *sign bit* to the left of the MSB in a binary word. A zero (0) designates a positive number and a one (1), a negative number. Thus $0_\wedge 1001001$ has the decimal equivalent $+73$, and $1_\wedge 1001001$ denotes -73. The caret ($\wedge$) is sometimes used to indicate that the first bit is the sign bit.

A variety of binary-derived number representations and binary codes are used in digital systems. Several of these are introduced in Sec. 7-3 when we treat binary arithmetic.

6-2 BOOLEAN ALGEBRA

Boolean algebra is a two-state (binary) symbolic logic. A boolean variable A assumes one of two permissible values, 0 or 1. Thus A may be 1 ($A = 1$) or A may be 0 ($A = 0$). If A is not 1, A must be zero. A self-consistent

FIGURE 6-2
Waveforms for (a) pos-
itive logic and (b) neg-
ative logic.

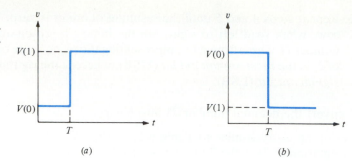

boolean algebra involving several variables requires only three basic logic functions called AND, OR, and NOT. A *logic gate* is a circuit used to realize a basic
logic function. Combinations of logic gates serve to implement complex boolean
equations. These combinatorial circuits are treated in Chap. 7.

Logic Systems

The implementation of gate depends on the manner by which a binary signal
is defined. In a *dc,* or *level-logic,* system a bit is implemented as one of two
voltage levels. If, as in Fig. 6-2a, the more positive voltage is the 1 level and
the other is the 0 level [$V(1) > V(0)$], the system is said to employ dc *positive
logic.* On the other hand, a dc *negative*-logic system, as in Fig. 6-2b, is one
which designates the more negative voltage state of the bit as the 1 level and
the more positive as the 0 level [$V(0) > V(1)$]. It should be emphasized that
the absolute values of the two voltages are of no significance in these definitions.
In particular, the 0 state need not represent a zero voltage level (although in
some systems it might).

In a *dynamic,* or *pulse-logic,* system a bit is recognized by the presence or
absence of a pulse. A 1 signifies the existence of a positive pulse in a dynamic
positive-logic system; a negative pulse denotes a 1 in a dynamic negative-logic
system. In either system a 0 at a particular input (or output) at a given instant
of time designates that no pulse is present at that particular moment.

The OR Gate

An OR gate has two or more inputs and a single output and operates in accordance with the following definition: The *output of an* OR *assumes the* 1 *state
if one or more inputs assume the* 1 *state.* The *n* inputs to a logic circuit will
be designated by $A, B, \ldots, N$ and the output, by Y. It is to be understood
that each of these variables may assume one of two possible values, either 0
or 1. A standard symbol for the OR circuit is given in Fig. 6-3a, together with
the boolean expression for this gate. The equation is to be read "Y equals A
or B or . . . or N." Instead of defining a logical operation in words, an alternative
method is to give a *truth table* which contains a tabulation of all possible input
values and their corresponding outputs. It should be clear that the 2-input truth
table in Fig. 6-3b is equivalent to the preceding definition of the OR operation.

Assume waveforms A and B have the binary levels as a function of time as
illustrated in Fig. 6-3c. Then the output waveform Y in Fig. 6-3c corresponds

to the truth table in Fig. 6-3b for a positive-logic OR gate. Note that with $V(0) = 0$, the OR operation is achieved in both level logic and dynamic logic.

If it is remembered that A, B, and C can take on only the value 0 or 1, the following equations from boolean algebra pertaining to the OR $(+)$ operation are easily verified:

$$A + B + C = (A + B) + C = A + (B + C) \tag{6-2}$$

$$A + B = B + A \tag{6-3}$$

$$A + A = A \tag{6-4}$$

$$A + 1 = 1 \tag{6-5}$$

$$A + 0 = A \tag{6-6}$$

These equations may be justified by referring to the definition of the OR operation or to a truth table.

We have already encountered a simple diode OR gate; the circuit in Fig. 2-13 obeys the truth table in Fig. 6-3b for negative logic.

The AND Gate

An AND gate has two or more inputs and a single output and operates in accordance with the following definition: The *output of an* AND *assumes the* 1 *state if and only if all the inputs assume the* 1 *state*. A symbol for the AND circuit is given in Fig. 6-4a, together with the boolean expression for this gate. The equation is to be read "Y equals A *and* B *and* . . . *and* N." [Sometimes

FIGURE 6-3
The OR gate: (*a*) circuit symbol, (*b*) truth table, (*c*) waveforms for positive logic.

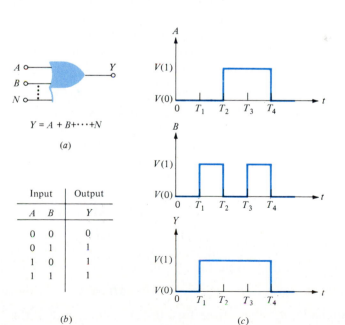

$$Y = A + B + \cdots + N$$

(*a*)

Input		Output
A	B	Y
0	0	0
0	1	1
1	0	1
1	1	1

(*b*)

(*c*)

FIGURE 6-4
The AND gate: (*a*) circuit symbol, (*b*) truth table, (*c*) waveforms for positive logic.

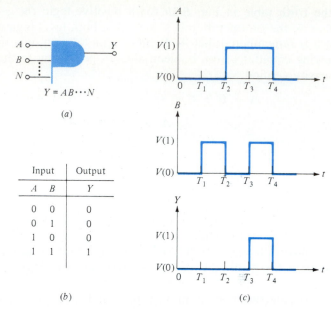

$$Y = AB \cdots N$$

(*a*)

Input		Output
A	*B*	*Y*
0	0	0
0	1	0
1	0	0
1	1	1

(*b*)

(*c*)

a dot (·) or a cross (X) is placed between symbols to indicate the AND operation.] It may be verified that the 2-input truth table in Fig. 6-4*b* is consistent with the preceding definition of the AND operation. The waveforms in Fig. 6-4*c* correspond to the truth table in Fig. 6-4*b* for positive logic. Historically, this circuit was called a *coincidence circuit* because all inputs must be present at the same time to produce an output pulse.

Since *A*, *B*, and *C* can have only the value 0 or 1, the following expressions involving the AND operation may be verified:

$$ABC = (AB)C = A(BC) \tag{6-7}$$

$$AB = BA \tag{6-8}$$

$$AA = A \tag{6-9}$$

$$A1 = A \tag{6-10}$$

$$A0 = 0 \tag{6-11}$$

$$A(B + C) = AB + AC \tag{6-12}$$

These equations may be proved by reference to the definition of the AND operation, to a truth table, or to the behavior of the AND circuits discussed above. Also, by using Eqs. (6-10), (6-12), and (6-5), it can be shown that

$$A + AB = A \tag{6-13}$$

Similarly, it follows from Eqs. (6-12), (6-9), and (6-5) that

$$A + BC = (A + B)(A + C) \tag{6-14}$$

We shall have occasion to refer to the last two equations later in the Chapter.

The circuit in Fig. 2-13, analyzed in Example 2-2, is a diode-resistor implementation of a positive-logic AND gate. Reversing the diodes is all that is required to realize a negative-logic AND gate.

Note that the same circuit (Fig. 2-13) can be used to implement both the positive-logic AND gate and the negative-logic OR gate. This results from the fact that $V(0)$ in positive logic and $V(1)$ in negative logic each signify the lower of the two voltage levels. Similarly, $V(1)$ and $V(0)$ are the higher voltages in positive and negative logic, respectively. Thus we conclude that a *negative* OR *gate is the same circuit as a positive* AND *gate*. This result is not restricted to diode logic; it is valid independently of the hardware used to implement the circuit.

The NOT (Inverter) Gate   The NOT circuit has a single input and a single output and performs the operation of *logic negation* in accordance with the following definition: *The output of a* NOT *circuit takes on the 1 state if and only if the input does* not *take on the 1 state*. The standard to indicate a *logic negation* is a small circle drawn at the point where a signal line joins a logic symbol. Negation at the input of a logic block is indicated in Fig. 6-5a and at the output in Fig. 6-5b. The symbol for a NOT gate and the boolean expression for negation are given in Fig. 6-5c. The equation is to be read "Y equals NOT A" or "Y is the complement of A." [Sometimes a prime (') is used instead of the bar ($-$) to indicate the NOT operation.] The truth table is given in Fig. 6-5d.

A circuit which accomplishes a logic negation is called a NOT *circuit*, or, since it inverts the sense of the output with respect to the input, it is also known as an *inverter*. In a truly binary system only two levels $V(0)$ and $V(1)$ are recognized, and the output, as well as the input, of an inverter must operate between these two voltages. When the input is at $V(0)$, the output must be at $V(1)$, and vice versa. Ideally, then, a NOT circuit inverts a signal while preserving its shape and the binary levels between which the signal operates.

The bipolar and field-effect transistor (FET) switches discussed in Secs. 3-8 and 4-12 are inverters. We observed in Figs. 3-29 and 4-25 that an input transition from a low voltage to a high voltage produced the opposite transition in the output signal. Furthermore, the ideal controlled sources, introduced in Secs. 3-1 and 4-1, are inverter circuits when acting as controlled switches.

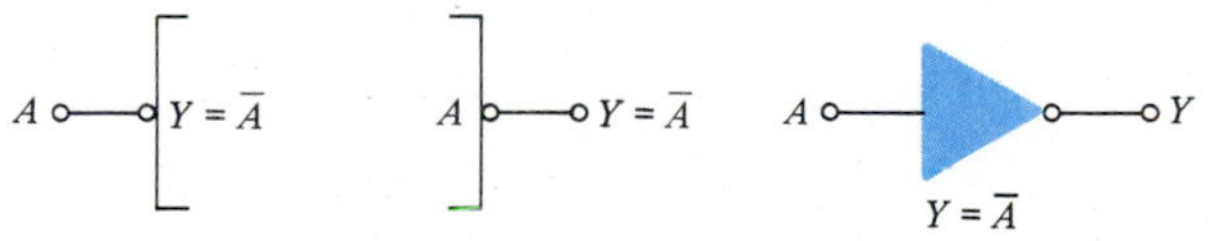

FIGURE 6-5

Logic negation at (a) the input and (b) the output of a logic block, (c) circuit symbol for an inverter (NOT gate), (d) the truth table.

Input	Output
A	Y
0	1
1	0

(a) $A \circ\!\!-\!\!\circ Y = \overline{A}$ (b) $A \circ\!\!-\!\!\circ Y = \overline{A}$ (c) $A \circ\!\!-\!\!\triangleright\!\!-\!\!\circ Y$ $Y = \overline{A}$ (d)

From the basic definitions of the NOT, AND, and OR connectives we can verify the following boolean identities:

$$\overline{\overline{A}} = A \tag{6-15}$$

$$\overline{A} + A = 1 \tag{6-16}$$

$$\overline{A}A = 0 \tag{6-17}$$

$$A + \overline{A}B = A + B \tag{6-18}$$

Example 6-2

Verify Eq. (6-18).

Solution

Since $B + 1 = 1$ and $A1 = A$, it follows that

$$A + \overline{A}B = A(B + 1) + \overline{A}B = AB + A + \overline{A}B = (A + \overline{A})B + A = B + A$$

where use is made of Eq. (6-16).

The Inhibit (Enable) Operation   A NOT circuit preceding one terminal (S) of an AND gate acts as an *inhibitor*. This modified AND circuit implements the logical statement. *If $A = 1, B = 1, \ldots, M = 1$, then $Y = 1$ provided that $S = 0$. However, if $S = 1$, then the coincidence of $A, B, \ldots, M$ is inhibited (disabled), and $Y = 0$*. Such a configuration is also called an *anticoincidence* circuit. The logical block symbol is drawn in Fig. 6-6a, together with its boolean equation. The equation is to be read "Y equals A *and* B *and* . . . *and* M *and not* S." The truth table for a 3-input AND gate with one inhibitor terminal (S) is given in Fig. 6-6b.

The terminal S is also called a *strobe* or an *enable input*. The enabling bit $S = 0$ allows the gate to perform its AND logic, whereas the inhibiting bit $S = 1$ causes the output to remain at $Y = 0$, independently of the values of the input bits.

It is possible to have a 2-input AND, one terminal of which is inhibiting. This circuit satisfies the logic: "The output is true (1) if input A is true (1) provided that B is not true (0) [or equivalently, provided that B is false (0)]." Another possible configuration is an AND with more than one inhibit terminal.

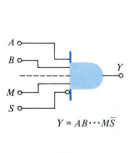

(a)

	Input			Output
	A	B	S	Y
1	0	0	0	0
2	0	1	0	0
3	1	0	0	0
4	1	1	0	1
5	0	0	1	0
6	0	1	1	0
7	1	0	1	0
8	1	1	1	0

(b)

FIGURE 6-6
(*a*) Circuit symbol and boolean expression for an AND gate with an enable terminal S. (*b*) The truth table for $Y = AB\overline{S}$.

$Y = AB\cdots M\overline{S}$

6-3 THE EXCLUSIVE-OR, NAND, and NOR GATES The three gates discussed in this section are simple combinations of AND, OR, and inverter gates. Because of their importance in processing digital signals and, in general, their direct circuit implementation using transistors, they are considered basic gates.

The Exclusive-OR Gate An exclusive-OR gate obeys the following definition: The *output of a 2-input exclusive-OR assumes the 1 state if one and only one input assumes the 1 state*. The standard symbol for an exclusive-OR is given in Fig. 6-7a and the truth table, in Fig. 6-7b. The circuit in Fig. 6-3 is referred to as an inclusive-OR if it is to be distinguished from the exclusive-OR.

The above definition is equivalent to the statement: "If $A = 1$ or $B = 1$ but not simultaneously, then $Y = 1$." In boolean notation, this is expressed

$$Y = (A + B)(\overline{AB}) \tag{6-19}$$

This logic function can be implemented by using the basic logic gates as shown in Fig. 6-8a.

A second logic statement equivalent to the definition of the exclusive-OR is the following: "If $A = 1$ and $B = 0$, or if $B = 1$ and $A = 0$, then $Y = 1$." The boolean expression is

$$Y = A\overline{B} + B\overline{A} \tag{6-20}$$

The block diagram which satisfies this logic is indicated in Fig. 6-8b.

An exclusive-OR is employed within the arithmetic section of a computer. Another application is as an *inequality comparator, matching circuit*, or *detector* because, as can be seen from the truth table, $Y = 1$ only if $A \neq B$. This

FIGURE 6-7
The exclusive-OR gate:
(*a*) circuit symbol, (*b*) truth table, (*c*) waveforms for positive logic.

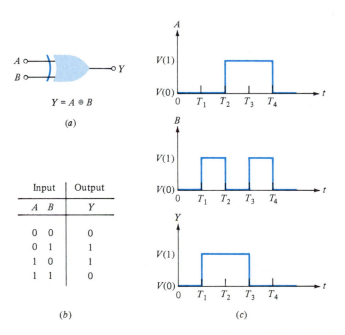

$Y = A \oplus B$

(a)

Input		Output
A	B	Y
0	0	0
0	1	1
1	0	1
1	1	0

(b)

(c)

FIGURE 6-8
Two implementations for the exclusive-OR gate.

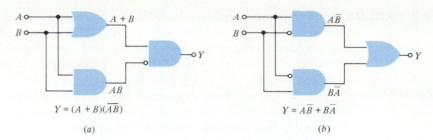

$$Y = (A + B)(\overline{AB})$$

(a)

$$Y = A\overline{B} + B\overline{A}$$

(b)

property is used to check for the inequality of two bits. If bit A is not identical with bit B, then an output is obtained. Equivalently, "If A and B are both 1 or if A and B are both 0, then no output is obtained, and $Y = 0$." This latter statement may be put into boolean form as

$$Y = \overline{AB + \overline{A}\,\overline{B}} \tag{6-21}$$

This equation leads to a third implementation for the exclusive-OR block, which is indicated by the logic diagram in Fig. 6-9a. An *equality detector* gives an output $Z = 1$ if A and B are both 1 or if A and B are both 0, and hence

$$Z = \overline{Y} = AB + \overline{A}\,\overline{B} \tag{6-22}$$

where use was made of Eq. (6-15). If the output Z is desired, the negation in Fig. 6-9a may be omitted or an additional inverter may be cascaded with the output of the exclusive-OR.

A fourth possibility for this gate is

$$Y = (A + B)(\overline{A} + \overline{B}) \tag{6-23}$$

which may be verified from the definition or from the truth table. This logic is depicted in Fig. 6-9b.

It should be noted that a 2-input exclusive-OR behaves as a *controlled inverter* or an inverter with a strobe input. Thus, if A is the input and $B = S$ is the strobe, then from the truth table in Fig. 6-7 it follows that $Y = \overline{A}$ if $S = 1$, whereas $Y = A$ if $S = 0$.

We have demonstrated that there often are several ways to implement a logical circuit. In practice one of these may be realized more advantageously than the others. Boolean algebra is sometimes employed for manipulating a

FIGURE 6-9
Two additional implementations of the exclusive-OR gate.

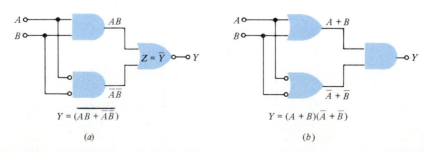

$$Y = \overline{(AB + \overline{A}\,\overline{B})}$$

(a)

$$Y = (A + B)(\overline{A} + \overline{B})$$

(b)

logic equation so as to transform it into a form which is better from the point of view of implementation in hardware. In the next section we shall verify through the use of boolean algebra that the four expressions given above for the exclusive-OR are equivalent.

De Morgan's Laws The following two binary equations are known as De Morgan's theorems:

$$\overline{ABC\cdots} = \overline{A} + \overline{B} + \overline{C} + \cdots \tag{6-24}$$

$$\overline{A + B + C + \cdots} = \overline{A}\overline{B}\overline{C}\cdots \tag{6-25}$$

To verify Eq. (6-24), note that if all inputs are 1, each side of the equation equals 0. On the other hand, if one (or more than one) input is 0, each side of Eq. (6-24) equals 1. Hence, for all possible inputs the right-hand side of the equation equals the left-hand side. Equation (6-25) is verified in a similar manner. De Morgan's laws complete the list of basic boolean identities. For future reference, all these relationships are summarized in Table 6-3.

   With the aid of boolean algebra we shall now demonstrate the equivalence of the four exclusive-OR circuits of the preceding section. Using Eq. (6-24), it is immediately clear that Eq. (6-19) is equivalent to Eq. (6-23). Now the latter equation can be expanded with the aid of Table 6-3 as follows:

$$(A + B)(\overline{A} + \overline{B}) = A\overline{A} + B\overline{A} + A\overline{B} + B\overline{B} = B\overline{A} + A\overline{B} \tag{6-26}$$

This result shows that the exclusive-OR in Eq. (6-21) is equivalent to that in Eq. (6-23).

TABLE 6-3 Summary of Basic Boolean Identities

Fundamental laws		
OR	AND	NOT
$A + 0 = A$	$A0 = 0$	$A + \overline{A} = 1$
$A + 1 = 1$	$A1 = A$	$A\overline{A} = 0$
$A + A = A$	$AA = A$	$\overline{\overline{A}} = A$
$A + \overline{A} = 1$	$A\overline{A} = 0$	

Associative laws
$(A + B) + C = A + (B + C) \qquad (AB)C = A(BC)$

Commutative laws
$A + B = B + A \qquad AB = BA$

Distributive law
$A(B + C) = AB + AC$

De Morgan's laws
$\overline{AB\cdots} = \overline{A} + \overline{B} + \cdots$
$\overline{A + B + \cdots} = \overline{A}\overline{B}\cdots$

Auxiliary identities
$A + AB = A \qquad A + \overline{A}B = A + B$
$(A + B)(A + C) = A + BC$

It follows from De Morgan's laws that *to find the complement of a boolean function change all* OR *to* AND *operations, all* AND *to* OR *operations, and negate each binary symbol.* If this procedure is applied to Eq. (6-21), the result is Eq. (6-23), if use is made of the identity $\overline{\overline{A}} = A$.

With the aid of De Morgan's law we can show that *an* AND *circuit for positive logic also operates as an* OR *gate for negative logic.* Let Y be the output and $A, B \ldots, N$ be the inputs to a positive AND so that

$$Y = AB \cdots N \tag{6-27}$$

Then, by Eq. (6-24),

$$\overline{Y} = \overline{A} + \overline{B} + \cdots + \overline{N} \tag{6-28}$$

If the output and all inputs of a circuit are complemented so that a 1 becomes a 0 and vice versa, then positive logic is changed to negative logic (refer to Fig. 6-2). Since Y and $\overline{Y}$ represent the *same* output terminal, A and $\overline{A}$ the *same* input terminal, etc., the circuit which performs the positive AND logic in Eq. (6-27) also operates as the negative OR gate in Eq. (6-28). Similar reasoning is used to verify that the same circuit is either a negative AND or a positive OR, depending on how the binary levels are defined. We verified this result for the diode circuit in Fig. 2-13 but the present proof is independent of how the circuit is implemented.

It should now be clear that it is really not necessary to use all three connectives OR, AND, and NOT. The OR and the NOT are sufficient because, from De Morgan law in Eq. (6-24), the AND can be obtained from the OR and the NOT, as displayed in Fig. 6-10a. Similarly, the AND and the NOT may be chosen as the basic logic circuits, and from the De Morgan law in Eq. (6-25), the OR may be constructed as shown in Fig. 6-10b. This figure makes clear once again that an OR (AND) circuit negated at input and output performs the AND (OR) logic.

The NAND Gate

In Fig. 6-8a the negation before the second AND could equally well be put at the output of the first AND without changing the logic. Such an AND-NOT sequence is also present in Fig. 6-10b and in many other logic operations. This negated AND is called a NOT-AND, or a NAND, gate. The logic symbol, boolean equation, truth table, and waveforms are displayed in Fig. 6-11.

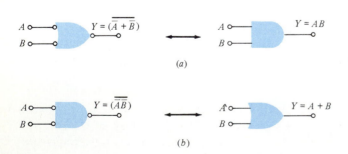

(a)

(b)

FIGURE 6-10

(a) Conversion of an OR gate into an AND gate by inverting all inputs. (b) An AND gate is converted to an OR gate if all inputs are inverted and the output is negated.

FIGURE 6-11

The NAND gate: (*a*) circuit symbol, (*b*) truth table, (*c*) waveforms for positive logic.

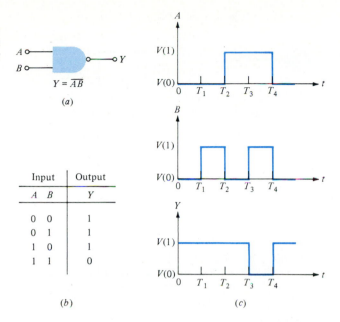

$$Y = \overline{AB}$$

(*a*)

Input		Output
A	*B*	*Y*
0	0	1
0	1	1
1	0	1
1	1	0

(*b*)

(*c*)

The NAND gate can be constructed by placing a transistor inverter (the switch in Fig. 3-28*a*) after a diode AND gate (Fig. 2-13). This gate realization is called *diode-transistor logic* (DTL) and was one of the early semiconductor logic families developed. It has been supplanted by other logic families, treated later in this chapter, which provide significantly improved performance.

The NOR Gate A negation following an OR gate is called NOT-OR, or simply a NOR gate. The logic symbol, boolean expression, truth table, and waveforms are given in Fig. 6-12. A DTL NOR gate is implemented by using a transistor inverter following a diode OR gate.

Because transistor switches are inherently inverters, NAND and NOR gates are extensively used to implement logic functions. De Morgan's laws afford a convenient methodology by which these gates are utilized in logic design. Indeed, entire logic systems can be implemented by using either only NAND gates or only NOR gates. This feature is illustrated in Fig. 6-13 utilizing NOR gates. In Fig. 6-13*a*, both input leads are tied together, and, consequently, the single input is negated (a NOT gate). The NOT gate is used to invert the output of the NOR gate in Fig. 6-13*b* to obtain an OR gate. Negation of both inputs in Fig. 6-13*c* converts the NOR gate into an AND gate (De Morgan's law). The corresponding analysis using NAND gates is left to the student (Prob. 6-18).

6-4 LOGIC GATE CHARACTERISTICS The design and fabrication of logic gates using real (commercially available) transistors give rise to circuits whose input and output waveforms can only approximate those illustrated in Sec. 6-2. Transitions from

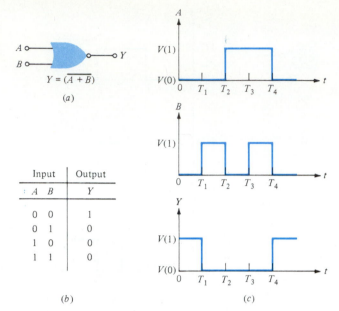

FIGURE 6-12
The NOR gate: (a) circuit
symbol, (b) truth table,
(c) waveforms for pos-
itive logic.

$$Y = \overline{(A + B)}$$

(a)

Input		Output
A	B	Y
0	0	1
0	1	0
1	0	0
1	1	0

(b)

(c)

$V(0)$ to $V(1)$ and vice versa cannot occur instantaneously. Also, manufacturing tolerances and temperature and other environmental changes can cause the voltage levels to vary. Since input(s) are derived from the output level(s) of other gates, each logic circuit serves as the load of previous stages. These loading effects can cause degradation of the logic levels. Each of these deviations from the ideal imposes limits on the performance of practical logic circuits. Our discussion centers on the inverter, as we have already demonstrated that the basic BJT and FET switches behave as inverters.

The Ideal Inverter The ideal inverter, fed from a single supply V_{DD} and deriving its output from a single input, has the transfer characteristic (v_o vs. v_i) displayed in Fig. 6-14a. Note that the transition between states occurs abruptly at an input voltage $v_i = V_{DD}/2$. Thus the output state is uniquely determined for all input voltages (except $V_{DD}/2$), and no uncertainty in the output state exists.

FIGURE 6-13
Use of NOR gates to im-
plement (a) an inverter,
(b) an OR gate, (c) an AND
gate.

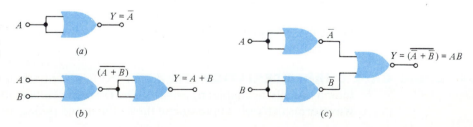

FIGURE 6-14
(a) Transfer characteristic and (b) voltage and current waveforms of an ideal inverter.

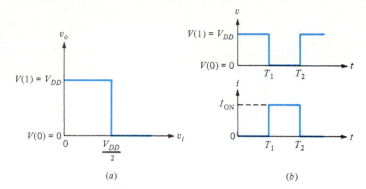

(a)   (b)

The output current and voltage waveforms corresponding to transitions from $V(1)$ to $V(0)$ and back to $V(1)$ that exist in an ideal controlled switch are shown in Fig. 6-14b. The static power dissipated in either state is zero because when $v_o \neq 0$, $i = 0$ (the ideal open switch) and when $i \neq 0$, $v_o = 0$ (the ideal closed switch). In addition, because the transition between states is instantaneous (Fig. 6-14b), the dynamic power dissipation, that is, the power consumed during switching, is also zero.

Two other characteristics of the ideal inverter are related to the interconnection of these gates: (1) the input circuit of the ideal inverter has no loading effect on the driving signal (the output of a previous gate), and (2) the inverter output is capable of driving an arbitrary number of similar gates with no degradation in output level.

The ideal inverter characteristics apply equally to multiple-input gates (AND, OR, NAND, NOR). In addition, such ideal gates can accept an arbitrary number of inputs without loading effects disturbing the driving stages. The preceding discussion leads us to observe that the following characteristics are important in evaluating the performance of a practical gate:

1. The range of voltages which correspond to the logic levels $V(0)$ and $V(1)$

2. The uncertainty region or range of input voltages for which the output state is undefined

3. The switching speed

4. Static and dynamic power dissipation

5. Input and output loading effects

The Practical Inverter Transfer Characteristic   The transfer characteristic of a real, commercially available inverter circuit in Fig. 6-15 indicates several deviations from the ideal characteristic in Fig. 6-14a. These differences are that the voltages $V(1)$ and $V(0)$ are not constant, that they can differ from the supply voltage and zero, respectively, and that the transition between states is not abrupt. The general shape of the transfer curve is similar to the transfer characteristics

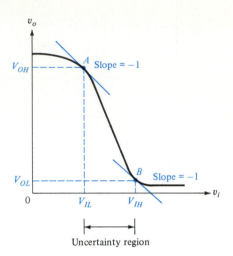

FIGURE 6-15
The voltage transfer characteristic of a practical inverter. The points at which the slope is -1 define the high and low logic levels.

of the BJT and FET switches displayed in Figs. 3-29 and 4-30, respectively.[1] For the BJT switch, the characteristic shows that the transistor is cut off for $v_i < V_\gamma$, the cut-in voltage, and $v_o = V_{CC}$, the supply voltage. Operation is in the forward-active region (controlled-source behavior) for $v_i > V_\gamma$ and v_o decreases with increasing v_i until the onset of saturation. A further increase in v_i drives the BJT more heavily into saturation and limits v_o to $V_{CE(\text{sat})} = 0.2$ V. Similarly, a MOSFET switch is cut off for $v_i < V_T$, the threshold voltage. Increasing v_i beyond V_T causes the MOSFET to operate in its saturated region and is accompanied by a decrease in v_o. As v_i continues to increase, the MOSFET is ultimately driven into the ohmic region and v_o displays only small changes in value.

The transfer characteristic is useful in defining the voltage ranges which correspond to $V(1)$ and $V(0)$. In Fig. 6-15, the magnitude of the slope of the transfer characteristic is unity at points A and B. To the left of A and to the right of B, the magnitude of the slope is less than unity, whereas between A and B, the magnitude of the slope is greater than unity. As the slope is a measure of the voltage gain between input and output, *we conclude that the transition between states requires a voltage gain greater than 1*. Note that the negative slope is indicative of logical negation.

The ordinate and abscissa values at A are designated by V_{OH} and V_{IL}, respectively. The quantity V_{IL} indicates the maximum value of v_i identifiable as logic 0 that defines the output state as logic 1. Thus V_{OH} is the minimum value that $V(1)$ can assume. Similarly, at B, V_{OL} is the maximum value of v_o corresponding to $V(0)$ and V_{IH} is the minimum input voltage (logic 1) needed to produce this output state.

---

[1] The reader may find it useful to review Secs. 3-8 and 4-12 before continuing with this section and the remainder of the chapter.

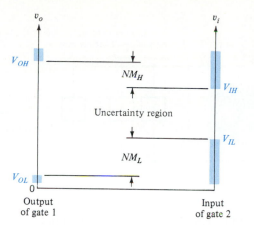

FIGURE 6-16
Input and output voltage levels used to define noise margins (NM) and the uncertainty region.

The range of input and output voltages corresponding to $V(0)$ and $V(1)$ are depicted in Fig. 6-16. Since the output of gate 1 serves as the input to gate 2, it is necessary that $V_{OH} > V_{IH}$ and $V_{OL} < V_{IL}$ for the gate to perform its logic function. If, for example, $V_{OH} < V_{IH}$ there would exist values of the driving signal corresponding to $V(1)$ which would not cause a transition to $V(0)$ in gate 2. Hence a logic error would be introduced into the circuit. A similar argument is used to verify that $V_{OL} < V_{IL}$.

The previous description of inverter characteristics also applies to NAND and NOR gates. For these gates, one must interpret v_i as the combination of inputs necessary to produce a transition between states. Noninverting logic circuits, such as AND and OR gates, have transfer characteristics that have positive slopes. Here, we identify points analogous to A and B for which the slope of the characteristic is $+1$. With this change, the conclusions reached in the preceding paragraphs apply to noninverting logic circuits.

Noise Margins
Noise in an electronic circuit is the presence of any unwanted signal. There are many sources of noise, including power-supply ripple and electromagnetic radiation (e.g., from fluorescent lights, or radio and television signals). As noise is always present, it is essential that logic gates do not respond to them and introduce logic errors. The term "noise margin," designated NM, is a measure of the immunity a logic circuit has to unwanted signals. The quantities NM_H and NM_L correspond to the noise margins for $V(1)$ and $V(0)$, respectively. From Fig. 6-16, we see that

$$NM_H = V_{OH} - V_{IH} \qquad NM_L = V_{IL} - V_{OL} \qquad (6\text{-}29)$$

The significance of the noise margin is that an unwanted signal of amplitude less than NM will not alter the logic state. Noise amplitudes exceeding NM result in input signals in the uncertainty region or cause unwanted transitions.

FIGURE 6-17
(*a*) Circuit diagram and
(*b*) equivalent circuit illustrating fan-out.

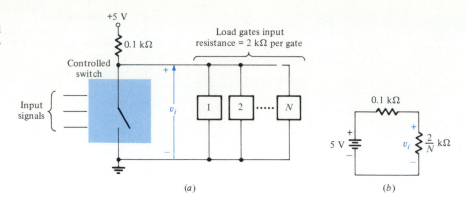

(*a*)   (*b*)

Fan-Out

A logic gate must be capable of providing the input to several similar circuits. *Fan-out* is the term used to indicate the number of circuits a gate can drive. The following example illustrates the loading effect on the output of a gate.

Example 6-3

A gate circuit in its logic 1 state is represented in Fig. 6-17*a*. The load consists of N identical gate inputs, each having an input resistance of 2 kΩ. To assure proper operation, the input voltage v_i to the load gates must be at least 3.5 V. Determine the number of gates that can be driven; that is, determine the fan-out.

Solution

The circuit in Fig. 6-17*b* is the equivalent circuit of that given in Fig. 6-17*a*. The parallel combination of the N identical resistances is $2/N$ kΩ and is the equivalent load on the gate. From Fig. 6-17*b*

$$v_i = \frac{2/N}{0.1 + 2/N} \times 5 \geq 3.5 \text{ V}$$

Solving for N gives $N = 8.57$. Since the number of stages is an integer, the fan-out is 8. Note that selection of $N = 9$ makes $v_i = 3.44$ V, a value below the specification.

Fan-In

The term "fan-in" is the number of inputs a logic circuit can accept. If the fan-in is exceeded, a logic gate will produce either an undefined or an incorrect output state. Furthermore, the input signals may be deteriorated because of excessive loading.

Power Dissipation Typical voltage and current waveforms for a practical gate circuit are displayed in Fig. 6-18. We observe that in either logic state, neither v nor i is zero. Consequently, the dc or static power dissipation is nonzero and the gate consumes energy in either state. Furthermore, even if we make $V(0) = 0$ and $I_{\text{OFF}} = 0$ (reducing static power dissipation to zero), dynamic power dissipation is nonzero because of the finite transition time between states. Note that during the switching intervals, $T_1 < t < T_2$ and $T_3 < t < T_4$, both v and i differ from zero.

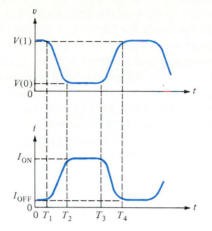

FIGURE 6-18
Practical current and voltage waveforms in a switch. Power dissipation exists in both logic states (static) and during the transition between states (dynamic).

Both static and dynamic dissipation contribute to the total power consumption of a gate. Sometimes static dissipation is the dominant factor. In other instances such as certain VLSI systems fabricated in CMOS technology, dynamic power dissipation is the major portion of total power consumption.

Speed of Operation   The speed at which a gate can be operated is influenced by the time for a signal to propagate from input to output and the transition time between states. In Fig. 6-19 we display typical input and output waveforms of an inverter. The *rise* and *fall* times t_r and t_f, respectively, are measures of the transition times between logic states. Both terms are defined by the time difference between

FIGURE 6-19

Input and output waveforms for one cycle showing rise time, fall time, cycle time, and propagation delay.

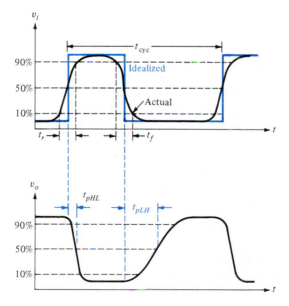

the 10 and 90 percent values of the voltage difference $V(1) - V(0)$ as indicated in Fig. 6-19. The rise and fall times are important quantities as the leading and trailing edges of the signal are often used to trigger other circuits. (This is particularly significant in timing and synchronization discussed in Chap. 8.)

The *propagation delay* t_p is the difference between the times for which the input and output voltages are at their 50 percent values. Note that because transitions from $V(0)$ to $V(1)$ and from $V(1)$ to $V(0)$ are not necessarily equal, the two delay times differ and are often distinguished by the additional subscripts *HL* and *LH*. Thus t_{pHL} and t_{pLH} denote propagation delay times from high to low and low to high, respectively.

The rectangular wave in Fig. 6-19 represents an ideal input signal that makes an instantaneous transition ($t_r = 0$) at the time the actual input signal is at its 50 percent value. This input allows the propagation delay to be computed (Sec. 6-5) more readily than from the actual input signal. As indicated in Fig. 6-19, the terms t_{pHL} and t_{pLH} are only measures of the output response to the rectangular input pulses.

The time for a logic circuit to make two successive transitions (so that it returns to its original state) is called the *cycle time* t_{cyc} as indicated in Fig. 6-19. Often, t_{cyc} is specified by its reciprocal, the clock frequency f_{CK}. Typically, digital systems operate at a cycle time in the order of 20 to 50 times the gate propagation delay.

The *delay-power product*[1] is frequently used as a measure of gate performance and is simply the product of the propagation delay and power dissipation of the gate. However, while two types of logic circuit realizations may have the same delay-power product, the circuit designer may often have to decide whether faster operation or lower power consumption is the more important requirement in the particular application under consideration. Manufacturers usually specify typical values of both t_p and P_{av}.

6-5   THE NMOS INVERTER   The NMOS logic family is one of the four major technologies used to realize digital circuits. The only component used in fabrication is the NMOS transistor which can be utilized as both a controlled switch and a resistance. The small chip area required for each transistor, coupled with the simplicity of the circuit configuration, results in NMOS circuits having the highest component density. This high "packing" density is exploited in VLSI systems where, at present, NMOS technology dominates. Although NMOS logic circuits are not commercially available as SSI or MSI packages as is true for the three other logic families we will treat, we choose to discuss this family first because its behavior most resembles that of the simple switch-resistance combination in Fig. 6-1.

---

[1]Sometimes this is incorrectly referred to as the *speed-power product*.

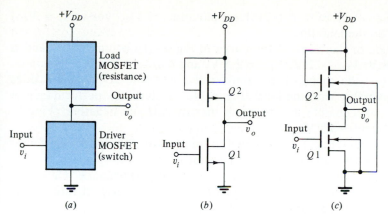

FIGURE 6-20
A MOSFET inverter: (*a*) basic structure, (*b*) circuit diagram with an enhancement load, (*c*) circuit in (*b*) in which substrate connections are indicated.

The Saturated Enhancement Load   The basic structure of an NMOS inverter is shown in Fig. 6-20*a*. The driver is an enhancement device, and several different forms of the NMOS load resistance are possible. One of these loads uses an enhancement transistor (depicted in Fig. 6-20*b*) whose substrate connections are indicated in Fig. 6-20*c*. This circuit was analyzed in Example 4-2 for the case where $V_{DD} = 6$ V, and both transistors are characterized by $k = 20$ μA/V^2, $V_T = 2$ V, and $W/L = 1$. The load resistance characteristic, load line, and transfer characteristic, given in Figs. 4-23*b*, 4-24, and 4-25, are repeated for convenience in Fig. 6-21. Recall that the load resistance characteristic (Fig. 6-21*a*) is I_{D2} versus V_{DS2} when $V_{GS2} = V_{DS2}$. The load line is constructed from the KVL and KCL constraints, viz., that $I_{D1} = I_{D2}$ and $V_{DS1} = V_{DD} - V_{DS2}$.

FIGURE 6-21
(*a*) Nonlinear resistance characteristic and load line for an enhancement (saturated) load. Both the driver and the load have the same aspect ratio ($W/L = 1$). (*b*) The transfer characteristic of the inverter. At points *A* and *B* the slope is -1.

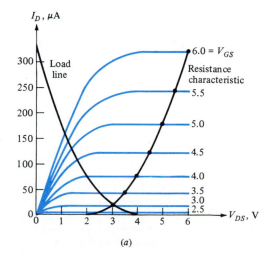

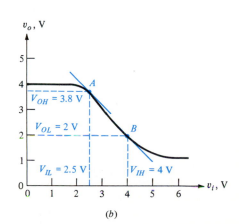

The transfer characteristic (Fig. 6-21b) is a plot of $v_o = V_{DS1}$ versus $v_i = V_{GS1}$ for each point on the load line.

The magnitude of the slope of the curve in Fig. 6-21b is unity[1] at A and B and $V_{OH} = 3.8$ V, $V_{OL} = 2$ V, $V_{IL} = 2.5$ V, and $V_{IH} = 4$ V are determined. It is apparent that the shape of the transfer characteristic in Fig. 6-21b does not compare favorably with the ideal (Fig. 6-14a). In fact, the performance of this circuit is unacceptable as the noise margin NM_H is negative ($V_{OH} - V_{IH} = 3.8 - 4.0 = -0.2$ V).

Only three parameters can be adjusted in this circuit: the processing factor k, V_T, and the aspect ratio W/L. It can be shown (Probs. 6-31 and 6-33) that variation of k and V_T has little or no effect on circuit performance. However, considerable improvement can be obtained by altering the aspect ratio of the load transistor. This is demonstrated in Example 6-4.

Example 6-4

The driver MOSFET has $W/L = 1$, but the load MOSFET in Fig. 6-20b is changed so that $W/L = \frac{1}{4}$. Both $Q1$ and $Q2$ have $k = 20$ μA/V^2 and $V_T = 2$ V. The supply voltage is $V_{DD} = 6$ V. (a) Sketch the transfer characteristic. (b) Determine the noise margins.

Solution

(a) The characteristics for $Q1$ are displayed in Fig. 6-21a. Those for $Q2$ are also given in Fig. 6-21a, except that the scale for I_D must be multiplied by the aspect ratio of $\frac{1}{4}$. The load characteristic is given by I_{D2} versus V_{DS2} for $V_{GS2} = V_{DS2}$. The values are given in the following table:

$V_{DS2} = V_{GS2}$ (V)	2.0	2.5	3.0	3.5	4.0	4.5	5.0	5.5	6.0
I_{D2} (μA)	0	1.25	5.0	11.3	20.0	31.3	45.0	61.3	80.0

This load resistance results in the load line in Fig. 6-22a and is based on $I_{D1} = I_{D2}$ and $V_{DS1} = V_{DD} - V_{DS2}$. The transfer characteristic in Fig. 6-22b ($v_o = V_{DS1}$ versus $v_i = V_{GS1}$) is obtained from the load line.

(b) From the transfer characteristic, the unity-slope (magnitude) points are approximated as indicated in Fig. 6-22b, and $V_{OH} = 3.9$ V, $V_{OL} = 0.9$ V, $V_{IH} = 3.6$ V, and $V_{IL} = 2.1$ V are identified. Use of Eq. (6-29) gives

$$NM_H = 3.9 - 3.6 = 0.3 \text{ V} \qquad \text{and} \qquad NM_L = 2.1 - 0.9 = 1.2 \text{ V}$$

The transfer characteristic in Fig. 6-22b more closely approximates the ideal than does that in Fig. 6-21b (with $Q2$ having $W/L = 1$). Comparison of Figs. 6-21b and 6-22b shows that if the aspect ratio is decreased, $V(1)$ and $V(0)$ are defined more sharply and the slope in the transition region is higher. The effect of reducing the aspect ratio is observed in comparing the load lines in Figs. 6-21a and 6-22a. In Fig. 6-22a, the load line intersects more of the output

---

[1] A 45° degree right triangle can be used to find these unity-slope points.

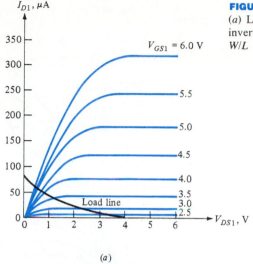

FIGURE 6-22
(a) Load line and (b) voltage transfer characteristic for inverter in Example 6-4. The enhancement load has $W/L = \frac{1}{4}$ and the aspect ratio of the driver is unity.

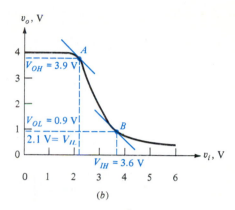

characteristics in the ohmic region than does that in Fig. 6-21a and causes the transition between states to be more abrupt.

Further decrease in the aspect ratio results in improved noise margins and more sharply delineated logic levels. However, this can be achieved only at the expense of additional chip area as the gate length of $Q2$ must be increased. Circuit designers, therefore, must make the trade-off between improved performance and reduced component densities. Current practice indicates that aspect ratios of $\frac{1}{4}$ or $\frac{1}{5}$ provide acceptable design compromises.

The Linear (Nonsaturated) Load   A second technique for improving the performance of an NMOS inverter is to use a linear (ohmic or nonsaturating) load resistance. From the discussion in Sec. 4-7 it follows that a MOSFET will operate in the ohmic region if

$$V_{GS} - V_{DS} > V_T \qquad (6\text{-}30)$$

In Fig. 6-23a, the gate of $Q2$ is connected to a separate supply $V_{GG} = 9$ V. Applying KVL around the loop $G_2 - S_2 - D_2 - \text{ground} - G_2$, we obtain

$$V_{GS2} - V_{DS2} + V_{DD} - V_{GG} = 0 \qquad (6\text{-}31)$$

Hence $V_{GS2} - V_{DS2} = V_{GG} - V_{DD} = 9 - 6 = 3$ V, and, for $V_T = 2$ V, Eq. (6-30) is satisfied so that $Q2$ is constrained to operate in its linear region.

Consider the inverter shown in Fig. 6-23, in which $Q1$ has the characteristics illustrated in Fig. 6-21a and $Q2$ has the characteristics displayed in Fig. 6-23b. From Eq. (6-31) we find

$$V_{DS2} = V_{GS2} - 3 \qquad (6\text{-}32)$$

FIGURE 6-23

(a) An NMOS inverter with linear load, (b) load resistance characteristic, (c) load line, (d) voltage transfer characteristic.

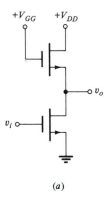

(a)

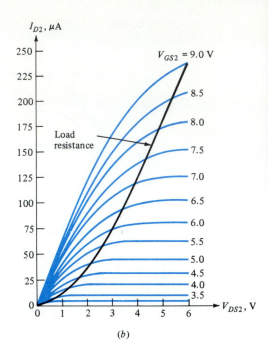

(b)

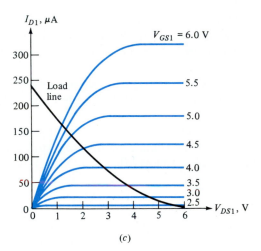

(c)

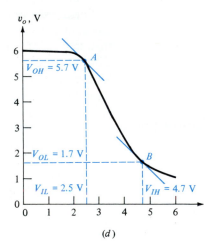

(d)

For each value of V_{GS2} in Fig. 6-23b, V_{DS2} is determined by Eq. (6-32). The current I_{D2} for each pair of values V_{GS2} and V_{DS2} is plotted against V_{DS2}, resulting in the load characteristic shown in Fig. 6-23b. Observe that because Q2 operates in its ohmic region, the load characteristic is nearly linear. The load line drawn using $V_{DS1} = 6 - V_{DS2}$ and $I_{D1} = I_{D2}$ is displayed in Fig. 6-23c. The resultant transfer characteristic ($v_o = V_{DS1}$ versus $v_i = V_{GS1}$) is given in Fig. 6-23d. This curve can be seen to resemble more nearly the ideal characteristic than does the simple enhancement load. Note that by selecting V_{GG} greater than V_{DD} by at least V_T, $V(1) \simeq V_{DD}$. The noise margins are $NM_H = 5.7 - 4.7 =$

1.0 V and $NM_L = 2.5 - 1.7 = 0.8$ V. The major disadvantage of using the linear load is that two different supply voltages are needed. Many systems, both analog and digital, are required to operate from a single supply for a variety of reasons (cost, size, availability, power dissipation, etc.).

The NMOS Depletion Load The use of a depletion-mode NMOS load transistor (Fig. 6-24) is a third effective technique for obtaining improved inverter performance. This configuration is first illustrated in Fig. 4-29a and used in Example 4-4. The depletion characteristics of $Q2$ are shown in Fig. 4-29b and are repeated in Fig. 6-24b for convenient reference. The enhancement MOSFET $Q1$ is the same one used for the other inverters discussed in this section (Fig. 6-23c or 6-24c). Proceeding as in Example 4-4, the load line in Fig. 6-24c and the transfer characteristic in Fig. 6-24d are obtained. As with the linear load, $V_o(1) \simeq V_{DD}$ but $V_o(0)$ is only a few tenths of a volt. Note that the transfer curve very nearly

FIGURE 6-24

An NMOS inverter with depletion load: (a) circuit configuration, (b) depletion volt-ampere characteristics, (c) load line, (d) transfer characteristic.

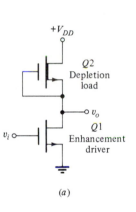

(a)

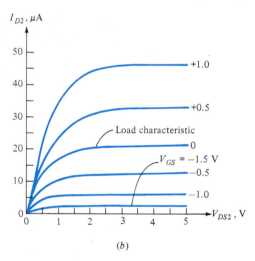

(b)

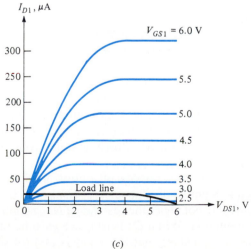

(c)

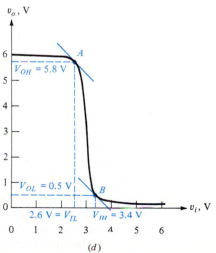

(d)

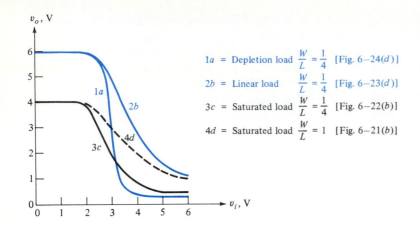

FIGURE 6-25
Comparison of voltage transfer characteristics for an NMOS inverter with two
different enhancement loads, a linear load,
and a depletion load.

$1a$ = Depletion load $\dfrac{W}{L} = \dfrac{1}{4}$ [Fig. 6–24(d)]

$2b$ = Linear load $\dfrac{W}{L} = \dfrac{1}{4}$ [Fig. 6–23(d)]

$3c$ = Saturated load $\dfrac{W}{L} = \dfrac{1}{4}$ [Fig. 6–22(b)]

$4d$ = Saturated load $\dfrac{W}{L} = 1$ [Fig. 6–21(b)]

approximates the ideal characteristic. The noise margins obtained are $NM_H =$ 2.4 V and $NM_L = 2.1$ V.

Although additional processing steps are required to fabricate both depletion and enhancement devices on the same chip, the marked improvement in performance obtained leads to the widespread utilization of this configuration in present-day NMOS logic circuits.

Summary

The three methods used to enhance inverter performance are decrease the aspect ratio, use a linear (nonsaturating) load, and use a depletion load. The transfer characteristics for each of the four cases presented in this section are displayed in Fig. 6-25. These curves amply demonstrate that the depletion load (curve $1a$) is the most effective technique.

6-6 PROPAGATION DELAY OF AN NMOS INVERTER

The propagation delay of an NMOS inverter is limited by the speed that the MOSFET and load capacitors in the circuit can be charged or discharged during a transition between states. Associated with the device are the gate-to-drain, drain-to-substrate, and source-to-substrate capacitances (where these are not tied together). In addition, the relatively thick field oxide introduces "sidewall" capacitances with each of the device elements. (See Fig. 5-2, and note that the metallization layer, field oxide, and either gate or drain or source form the three layers of a capacitor.) All the device capacitors are voltage-dependent, making computer simulation a necessity for obtaining accurate results. For pencil-and-paper calculations, however, all capacitive effects can be summed to form a single total capacitance, C_{tot} as shown in Fig. 6-26a.

The inverter, in its simplest form, consists of a controlled switch, a load resistance R_L, and the equivalent capacitance C_{tot} displayed in Fig. 6-26b. Because the switching element is not ideal, the resistance R_{ON} is included during the interval when $Q1$ is conducting. When $Q1$ is OFF, we assume that the open switch has infinite resistance. Consider that the input is at $V(0)$ so that $v_o =$

FIGURE 6-26
(a) An NMOS deple-tion-load inverter with capacitive load. (b) Equivalent circuit rep-resentation.

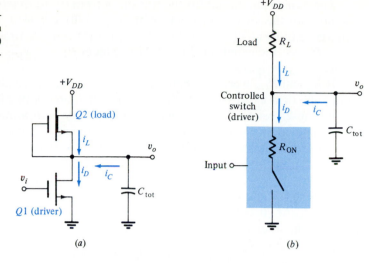

(a) (b)

$V(1) = V_{DD}$; thus C_{tot} is charged to V_{DD}. The switch is now closed at $t = 0$ because of an instantaneous input transition from $V(0)$ to $V(1)$. The output is required to make the opposite transition and C_{tot} must be discharged toward $V(0)$. The equivalent circuit for $t \geq 0$ is displayed in Fig. 6-26b with the switch closed (R_{ON} connected to ground). The capacitor will discharge to $V(0) = R_{\text{ON}} V_{DD}/(R_{\text{ON}} + R_L)$ with a time constant $\tau_{HL} = C_{\text{tot}} R_{\text{ON}} R_L/(R_{\text{ON}} + R_L)$.[1] The output waveform for $v_o(t)$ is shown in Fig. 6-27a. The propagation delay t_{pHL} is defined, as indicated in Fig. 6-27a, as the time for v_o to fall from $V(1)$ to the 50 percent voltage V' [halfway between $V(1)$ and $V(0)$]. Note that

$$V' = V(0) + \tfrac{1}{2} [V(1) - V(0)] = \tfrac{1}{2} [V(1) + V(0)] \qquad (6\text{-}33)$$

[1]To verify these results, the reader need only obtain a Thévenin equivalent of the portion of the circuit containing R_{ON}, R_L, and V_{DD}.

FIGURE 6-27
Output waveforms for Fig. 6-26b showing (a) transition from $V(1)$ to $V(0)$ and (b) the transi-tion from $V(0)$ to $V(1)$.

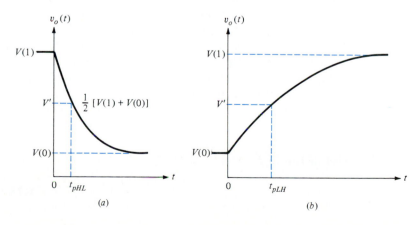

(a) (b)

The equivalent circuit for the transition from ON to OFF is that given in Fig. 6-26b with the switch open and with $v_o = V(0)$ at $t = 0$. The capacitor will charge exponentially from $V(0)$ toward $V(1)$ with a time constant $\tau_{LH} = C_{tot} R_L$ as indicated in Fig. 6-27b. Since $R_L \gg R_{ON}$, $\tau_{LH} \gg \tau_{HL}$ and, hence, $t_{pLH} \gg t_{pHL}$.

If R_L and R_{ON} were constant, t_{pLH} and t_{pHL} could be readily calculated from the analytical expressions for the exponential curves in Fig. 6-27. However, since these resistances are nonlinear (they vary with voltage), a simpler approximate method for calculating propagation delays is used. This calculation, based on the amount of charge transferred to (or from) C_{tot} follows.

For a transition from $V(1)$ to $V(0)$, the current available to discharge C_{tot} is $i_C = i_D - i_L$ (Fig. 6-26). This current is time-varying, and we shall designate the average value of i_C by I_{av}. In a time interval Δt the change of charge on C_{tot} is $|I_{av}|\Delta t$. If the change in output voltage during this time interval is Δv_o, the change in charge is also given by $C_{tot} \Delta v_o$. Hence

$$|I_{av}|\Delta t = C_{tot} \Delta v_o \tag{6-34}$$

For $\Delta t = t_{pHL}$ in Fig. 6-27a, $\Delta v_o = \frac{1}{2}[V(1) - V(0)]$, and from Eq. (6-34)

$$t_{pHL} = \frac{C_{tot}}{2 |I_{av}|} [V(1) - V(0)] \tag{6-35}$$

The magnitude of I_{av} is obtained by calculating $i_C = i_D - i_L$ measured at V_{OH} and also at $V' = \frac{1}{2}(V_{OH} + V_{OL})$ and averaging the two values. Thus

$$I_{AV} = \frac{1}{2}[(i_D - i_L)|_{V_{OH}} + (i_D - i_L)|_{V'}] \tag{6-36}$$

Example 6-5

Determine t_{pHL} for the inverter circuit shown in Fig. 6-24a. Assume that $C_{tot} = 0.2$ pF and that the input signal has $V(0) = 0.3$ V and $V(1) = 6$ V.

Solution

To calculate t_{pHL}, note from Fig. 6-24d that the output is at $V_{OH} = V_{DS1} = 5.8$ V and that the input has changed to $V(1) = V_{GS1} = 6$ V. For this pair of voltages, the current $i_D = 320$ μA is obtained from Fig. 6-24c. For the load transistor, when $V_{DS1} = 5.8$ V, $V_{DS2} = V_{DD} - V_{DS1} = 6 - 5.8 = 0.2$ V. The load MOSFET characteristics shown in Fig. 6-24b indicate that $i_L = i_{D2} = 5$ μA at $V_{DS2} = 0.2$ V and $V_{GS2} = 0$.

At the 50 percent voltage V' [Eq. (6-33)], $V_{DS1} = V'_o = \frac{1}{2}(5.8 + 0.5) = 3.15$ V, and V_{GS1} remains at 6 V. Corresponding to this pair of voltage values we obtain (Fig. 6-24c) $i_D = i_{D1} = 310$ μA. From the load characteristic shown in Fig. 6-24b for which $V_{GS2} = 0$ and $V_{DS2} = 2.85$ V, $i_L = I_{D2} = 20$ μA. Then, from Eq. (6-36), we obtain

$$I_{av} = \frac{(320 - 5) + (310 - 20)}{2} = 303 \ \mu A$$

Use of Eq. (6-35) yields

$$t_{pHL} = \frac{0.2 \times 10^{-12} \, (5.8 - 0.5)}{2 \times 303 \times 10^{-6}} = 1.75 \text{ ns}$$

Carrying out an analogous calculation (Prob. 6-40) for the propagation delay as the output rises when the input changes abruptly from $V(1)$ to $V(0)$, we obtain a value of $t_{pLH} = 26.5$ ns. Note that, as expected, $t_{pLH} \gg t_{pHL}$. The analysis illustrated in Example 6-5 is only approximate but for a pencil-and-paper calculation gives surprisingly good agreement (≈ 25 percent) with computer simulations.

6-7 NMOS LOGIC GATES

The NMOS inverters in Sec. 6-5 can be modified to form NAND and NOR gates by using multiple drivers feeding a single load. The circuit in Fig. 6-28a is a 2-input (fan-in of 2) NOR gate and consists of two identical NMOS enhancement drivers and one depletion load. For ideal inverters, the behavior of this circuit is analogous to that shown in Fig. 6-28b, in which the switches are open for inputs of $V(0)$ and closed for an input $V(1)$. Consequently, if either one (or both) of the inputs in Fig. 6-28b is at $V(1)$, a switch is closed and $v_o = V(0) = 0$. The output Y is $V(1) = V_{DD}$ only when both inputs A and B are at $V(0)$ (both switches open).

For the actual NOR circuit in Fig. 6-28a, one or more input signals greater than or equal to V_{IH} cause the output to become less than or equal to V_{OL}. Only when both inputs are below V_{IL} is $v_o \geq V_{OH}$. In Fig. 6-16 the ranges of values of the output and input voltages are defined by $V_o(1) \geq V_{OH}$, $V_o(0) \leq V_{OL}$, $V_i(1) \geq V_{IH}$, and $V_i(0) \leq V_{IL}$. Using this notation, NOR logic is displayed in the truth table of Fig. 6-28c.

FIGURE 6-28
(a) An NMOS NOR gate, (b) idealized representation of (a), (c) truth table.

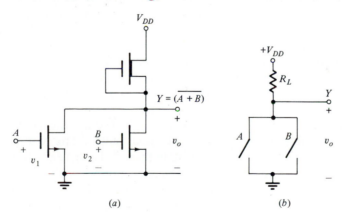

(a) (b)

A		B		Y	
v_1	State	v_2	State	v_o	State
$\leq V_{IL}$	0	$\leq V_{IL}$	0	$\geq V_{OH}$	1
$\leq V_{IL}$	0	$\geq V_{IH}$	1	$\leq V_{OL}$	0
$\geq V_{IH}$	1	$\leq V_{IL}$	0	$\leq V_{OL}$	0
$\geq V_{IH}$	1	$\geq V_{OL}$	1	$\leq V_{OL}$	0

(c)

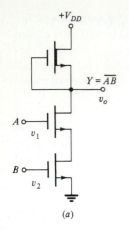

(a)

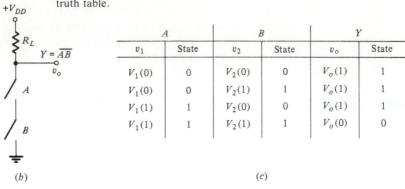

(b)

FIGURE 6-29

(a) A 2-input NMOS NAND gate. (b) An idealized representation of (a), and (c) the truth table.

	A		B		Y	
v_1	State	v_2	State	v_o	State
$V_1(0)$	0	$V_2(0)$	0	$V_o(1)$	1
$V_1(0)$	0	$V_2(1)$	1	$V_o(1)$	1
$V_1(1)$	1	$V_2(0)$	0	$V_o(1)$	1
$V_1(1)$	1	$V_2(1)$	1	$V_o(0)$	0

(c)

The fan-in can be increased from 2 by placing additional drivers in parallel. The maximum fan-in is limited by the current capability of the load MOSFET and input loading effects (such as C_{tot}).

The NAND gate in Fig. 6-29a is constructed by series-connecting the drivers and the load. Its idealized equivalent is shown in Fig. 6-29b, and the truth table is given in Fig. 6-29c. In Fig. 6-29b, a current path exists only if both switches are closed; that is, both A and B must be at logic 1 for the output to be $V(0) = 0$. All other input combinations result in zero current in R_L and, consequently, $v_o = V_{DD}$ and $Y = 1$. The NMOS NAND gate in Fig. 6-29a will produce $Y = 0$ corresponding to $v_o \leq V_{OL}$, if and only if both A and B are 1 (input voltages greater than or equal to V_{IH}).

Note that when one or both driver transistors are cut off, little power is consumed by either circuit. When both drivers are conducting, however, static power dissipation is not negligible. The NAND gate consumes power during only one of the four possible input conditions, whereas the NOR gate dissipates power during three of the four input states. The NOR gate, however, has an advantage over the NAND gate in that all the driver source terminals in the NOR gate are grounded. This connection allows each substrate to be connected directly to its source terminal and makes for easier fabrication.

An important feature in the fabrication of NMOS circuits is the fact that only one load transistor is required independent of the fan-in. This factor adds appreciably to the component densities that can be achieved. In addition, even though the aspect ratio of the depletion MOSFET requires greater area than is needed for each driver, this factor does not significantly increase the overall chip area of the circuit because only one load transistor is needed.

The MOSFET circuit configurations discussed in this section are a form of *direct-coupled transistor logic* (DCTL) that was introduced early in the development of bipolar logic circuits. Another example of DCTL using MOSFETs is the AND-OR-INVERT (AOI) gate discussed in Sec. 7-1.

FIGURE 6-30
(a) The circuit diagram of a CMOS inverter and (b) its equivalent switch representation. (c) The voltage transfer characteristic for (a) with $V_{DD} = 5$ V and threshold voltages of 2 V ($Q1$) and -2 V ($Q2$).

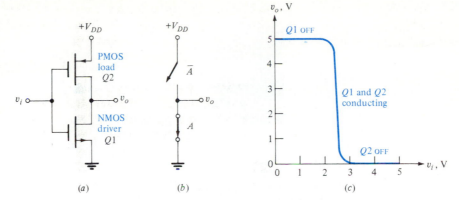

(a)

(b)

(c)

6-8   THE CMOS INVERTER

Complementary metal-oxide-semiconductor digital circuits are widely used because they have the distinct advantage of having virtually no static power dissipation in either the logic 1 or logic 0 state. The basic CMOS inverter, initially described in Sec. 4-15, has the circuit configuration shown in Fig. 6-30. The series-connected NMOS driver and PMOS load are both enhancement transistors. Their drains are connected, and the output signal is taken at this node. The input signal is applied simultaneously to both devices at the common gate terminal formed by connecting both gates. The input voltage v_i varies from $V(0) = 0$ V to $V(1) = V_{DD}$. When $v_i = 0$, then $V_{GS1} = 0$ and $Q1$ is OFF, whereas $V_{GS2} = -V_{DD}$ and the PMOS device $Q2$ is ON. However, since the two FETs are in series, the current in $Q2$ equals that in $Q1$ ($I_{D1} = -I_{D2} = 0$), even though the gate voltage has a magnitude that nominally causes conduction. In other words, $Q2$ operates at the origin of the PMOS output characteristic corresponding to the gate voltage $V_{GS2} = -V_{DD}$. Since $V_{DS2} = 0$, it follows that $v_o = V_{DD}$. Inverter action has been verified because $v_o = V(1)$ for $v_i = V(0)$.

Consider now that $v_i = V_{DD} = V_{GS1}$; then $Q1$ is ON but $Q2$ with $V_{GS2} = 0$ is OFF. Hence $I_{D1} = -I_{D2} = 0$ and $Q1$ operates at the origin of the NMOS drain characteristics independent of V_{GS1}. Since the voltage across $Q1$ is zero, $v_o = 0$. Again the NOT property is obtained; $v_o = V(0)$ for $v_i = V(1)$. In either logic state $Q1$ or $Q2$ is OFF and the quiescent power dissipation is theoretically zero. In reality, the *standby power* equals the product of the OFF leakage current and V_{DD} and equals a few nanowatts per gate.

From the discussion in the previous paragraphs, we can see that the switching circuit in Fig. 6-30b is analogous to CMOS operation. Because one switch must always be open, no current path exists between the supply voltage and ground. Consequently, the power is always zero. In the ideal situation, switching is instantaneous and no *dynamic power* is consumed. In the next paragraph we demonstrate, that the dynamic power is not zero in practice.

Consider the circuit in Fig. 6-30a for which $V_{DD} = 5$ V, $Q1$ has $V_T = 2$ V, and $Q2$ has $V_T = -2$ V. Let us assume that the processing factor k and the

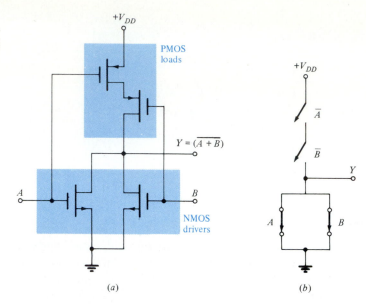

aspect ratio for both the PMOS and NMOS devices are equal. From our previous discussion we recognize that for $v_i \leq 2$ V, $Q1$ is cut off and with $V_{GS2} \leq -3$ V, $Q2$ is conducting. The output voltage for this condition is $v_o = V_{DD} = 5$ V. Similarly, for $v_i \geq 3$ V, $V_{GS2} \geq -2$ V, cutting off $Q2$ and turning on $Q1$ so that the output is $V(0) = 0$ V. However, for v_i increasing from zero, then in the range $2 < v_i < 3$ V, both devices are conducting with $I_{D1} = -I_{D2}$, and v_o decreases from 5 to 0 V. At $v_i = 2.5$ V, $v_o = V_{DD}/2 = 2.5$ V as shown in the transfer characteristic in Fig. 6-30c. Note that the curve in Fig. 6-30c is very nearly the characteristic of the ideal inverter in Fig. 6-14a. With both devices conducting during the transition between states ($2 < v_i < 3$ V), current exists in the circuit and dynamic power is consumed. This often becomes a significant factor in VLSI systems employing CMOS technology.

High-speed operation of a digital system, corresponding to short cycle times t_{cyc} and high clock frequencies f_{CK} (Fig. 6-19), is usually desirable. Dynamic power dissipation increases the more frequently a logic gate is required to switch states. Hence, the average power consumed by a CMOS gate is proportional to the clock frequency f_{CK}.

6-9 CMOS LOGIC GATES Complementary metal-oxide-semiconductor NAND and NOR gates can be constructed from the basic inverter in much the same fashion as in NMOS technology. However, the one notable difference is that *each NMOS driver requires its own PMOS load*.[1] Consequently CMOS ICs do not have component densities as great as NMOS circuits.

---

[1] In Sec. 9-3, a form of CMOS circuit called *domino logic* in which the number of PMOS load transistors is reduced is discussed.

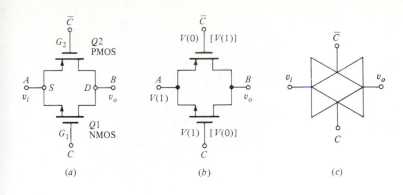

FIGURE 6-32
(*a*) A CMOS transmission gate. (*b*) The input voltage is $V(1)$ and the control voltage is considered first to be $V(1)$ and then $V(0)$. (*c*) The circuit symbol.

(*a*)   (*b*)   (*c*)

The NOR Gate

A 2-input CMOS NOR circuit is displayed in Fig. 6-31*a*, and its representation by ideal switches is given in Fig. 6-31*b*. The drivers are parallel-connected (as in NMOS), but the loads are series-connected. The need for this arrangement is discernible by analyzing the circuit in Fig. 6-31*b*. With either A or B at logic 1, the output is at ground. No path is permissible between the output node and the power supply V_{DD}. If such a path existed it would make $v_o = V_{DD}$, a violation of KVL. By placing the loads in series, if either or both inputs are at 1, one or both load switches remain open. Similarly, if both inputs are at $V(0)$, both load switches are closed providing the path from the output node to V_{DD}.

The NAND Gate

The CMOS NAND gate is fabricated by series-connecting the drivers and parallel-connecting the loads. This configuration is discussed in Prob. 6-48.

The Transmission Gate

The configuration of complementary MOSFETs in Fig. 6-32*a* acts as a (digital or analog) transmission gate controlled by the complementary gate voltages C and $\bar{C}$. Consider positive logic with the two logic levels $V(0)$ and $V(1)$. Assume that $C = 1$ so that $v_{G1} = V(1)$ and $v_{G2} = V(0)$, as indicated in Fig. 6-32*b*. (Ignore for now the values in the brackets.) If $A = V(1)$, then $v_{GS1} = V(1) - V(1) = 0$ and $Q1$ is OFF. However, $|v_{GS2}| = V(1) - V(0) > V_T$ and v_{GS2} is negative, causing the PMOS $Q2$ to conduct. Since there is no applied drain voltage, $Q2$ operates in the ohmic region where $v_{DS2} \approx 0$. In other words, $Q2$ behaves as a small resistance connecting the output to the input and $B = V(1) = A$. In a similar manner, it can be shown that, if $A = V(0)$, then $Q2$ is OFF whereas $Q1$ conducts and $B = V(0) = A$.

Now consider the case $C = 0$ so that $v_{G1} = V(0)$ and $v_{G2} = V(1)$, as indicated by the bracketed values in Fig. 6-32*b*. If the input is $V(1)$ as shown, then v_{GS1} is negative and the NMOS $Q1$ is OFF whereas $v_{GS2} = 0$ and $Q2$ is also OFF. Since both FETs are nonconducting, there is an open circuit between input and output and, hence, transmission through the gate is inhibited. If the input is $V(0)$, it is again found that both devices are OFF. *In summary*, if $C = 1$, the gate transmits the input to the output so that $B = A$, whereas if $C = 0$, no transmission is possible.

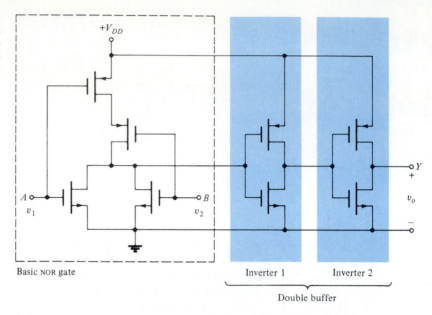

Basic NOR gate   Inverter 1   Inverter 2

Double buffer

The n well of the PMOS is tied to $V(1)$, the most positive voltage in the circuit and the p substrate of the NMOS is tied to $V(0)$, the most negative voltage. The symbol for the transmission gate is indicated in Fig. 6-32c. The control C is binary (it can have only one of two values), but the input v_i may be either digital, as discussed in the foregoing paragraphs, or an analog signal whose instantaneous value must lie between $V(0)$ and $V(1)$. For example, if $V(0) = -5$ V and $V(1) = +5$ V, a sinusoidal input signal (whose peak value does not exceed 5 V) appears at the output if $C = 1$ ($v_{G1} = +5$ V) but is not transmitted through the gate if $C = 0$ ($v_{G1} = -5$ V).

CMOS Logic Families   Complementary metal-oxide-semiconductor logic circuits are commercially available as both SSI and MSI subsystems. The most common CMOS logic families are designated as the 4000B, 74C, and 74HC series and all are fabricated by using polysilicon gates. The circuits in this series may be operated at supply voltages as low as 3 V to as high as 20 V, a feature which affords the designer considerable flexibility. When operated from a 5-V supply, each output is capable of driving one input of a TTL series 74LS gate. (This series is treated in Sec. 6-13.)

The basic structure of the NOR gate in these logic families is shown in Fig. 6-31a. The outputs of these gates are double-buffered; that is, the output of the basic gate becomes the input to the first of a pair of cascaded inverters, as illustrated for a NOR gate in Fig. 6-33. The cascaded inverters do not affect the logic function realized. However, the dimensions of the MOSFETs in inverter 2 are increased, so that its output is capable of driving many similar gates (a fan-out in excess of 50 is common). The double-buffered output of Fig. 6-33 can also drive a load which is not on the same chip as the NOR gate (called an "off-chip" load) and has increased wiring capacitance.

The newest family of CMOS circuits is the 74HC series which utilizes more advanced fabrication methods to reduce device dimensions. Consequently, the propagation delays are reduced in comparison to other CMOS families. This higher-speed CMOS family is also capable of driving the input of a TTL 74LS gate and has comparable propagation delay. However, as with all CMOS circuits, static gate dissipation is extremely small. As the transfer characteristic, particularly with double-buffering, is most nearly that of the ideal logic circuit, noise margins are large and CMOS gates display excellent noise immunity.

6-10 THE BJT INVERTER The BJT inverter, depicted in Fig. 6-34a, is simply the transistor switch described in Secs. 3-5, 3-6, and 3-8. The transfer characteristic of this circuit can be developed as follows:

1. For $v_i \leq 0.5 \text{ V} = V_\gamma$, the cut-in voltage (see Table 3-1), the BJT is cut off and, neglecting I_{CO}, the output voltage is V_{CC}.

2. The transistor is just barely conducting for v_i slightly greater than 0.5 V, and the small collector current produced causes v_o to decrease from V_{CC} by $I_C R_C$.

3. In the forward-active region, $V_{BE(ON)} = 0.7 \text{ V}$, $I_C = \beta_F I_B$, and

$$I_B = \frac{v_i - V_{BE(ON)}}{R_B} \tag{6-37}$$

The output voltage is given by

$$v_o = V_{CC} - I_C R_C = V_{CC} - \frac{\beta_F R_C}{R_B}[v_i - V_{BE(ON)}] \tag{6-38}$$

Consequently, from Eq. (6-38), v_o decreases linearly with v_i.

FIGURE 6-34
(a) The circuit and (b) the transfer characteristic for a BJT inverter.

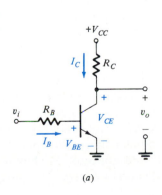

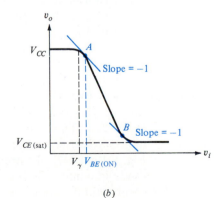

(a)

(b)

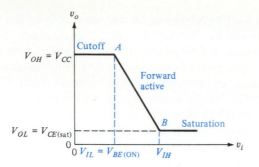

FIGURE 6-35
Straight-line representation of the transfer characteristic of a BJT inverter.

4. Continued increase in v_i causes the transistor to just barely saturate and $v_o = V_{CE} = 0.3$ V. With $v_o = 0.3$ V, the value of v_i is obtained from Eq. (6-38) and that for I_B from Eq. (6-37).

5. Further increase in v_i drives the transistor heavily into saturation and $v_o = V_{CE(sat)} = 0.2$ V.

The resultant transfer characteristic is displayed in Fig. 6-34b. Also indicated in this figure are the points at which the slope is -1 so that V_{OH}, V_{OL}, V_{IL}, and V_{IH} can be determined. We observe that $V_{OH} \simeq V_{CC}$ and V_{IL} is approximately $V_{BE(ON)} = 0.7$ V. Note that V_{OL} is nearly $V_{CE(sat)} = 0.2$ V and V_{IH} is the value of v_i corresponding to $v_o = 0.2$ V, obtained from Eq. (6-38). On the basis of these observations it is often convenient to represent the transfer characteristic as illustrated in Fig. 6-35. The three straight-line segments represent the state of the circuit when the BJT is in the cutoff, forward-active, and saturation regions. The values of V_{OH}, V_{IL}, V_{OL}, and V_{IH} are determined at the two breakpoints A and B. Point A represents the transition from cutoff to forward-active, whereas the transistor is at the edge of saturation at B. The approximate characteristic in Fig. 6-35 provides adequate accuracy for pencil-and-paper calculations, with more accurate results obtained from computer simulations.

The characteristic in Fig. 6-35, with values of $V_{OL} = 0.2$ V and $V_{IL} = 0.7$ V, gives somewhat optimistic noise margins. Conservative designs use $V_{OL} = 0.3$ V (V_{CE} at the edge of saturation) and $V_{IL} = 0.5$ V, the cut-in voltage. The value of V_{IH} can be controlled by adjusting either R_B or R_C. For a specified value of R_C, reduction of R_B will cause the BJT to saturate at a lower value of v_i [see Eq. (6-38)]. Increase in R_C for a fixed R_B produces the same effect.

The Schottky Transistor Inverter   One limitation of driving the BJT into saturation is that the propagation delay is increased because of minority-carrier storage time (Sec. 3-8). This effect is not present in a MOSFET because the FET is a majority-carrier device. Hence storage time is not taken into consideration in Fig. 6-19 or 6-27. To prevent the bipolar transistor from saturating, a Schottky diode is connected between base and collector as shown in Fig. 5-13a. This combination is referred to as a *Schottky transistor*, whose circuit symbol is given in Fig. 5-13c. The fabrication of this device and the explanation of why a Schottky transistor cannot be driven into saturation are described in Sec. 5-3.

The improvement in switching speed obtained with Schottky transistors is the basis for several of the most popular TTL circuits discussed in Secs. 6-11 to 6-13.

6-11   THE TTL NAND GATE

The most widely used SSI technology over the last two decades (1966–1985) has been the *transistor-transistor logic* (TTL) family. The NAND gate is the basic TTL building block, and its development evolved from an earlier bipolar IC logic family called *diode-transistor logic* (DTL). As an aid in the analysis of the TTL NAND gate, the following example describing the DTL NAND circuit is useful.

Example 6-6

The positive-logic DTL NAND gate in Fig. 6-36 is essentially a diode AND circuit (Fig. 2-13) in cascade with a BJT inverter. The binary inputs A, B, and C have logic levels corresponding to the $V(0)$ and $V(1)$ outputs of similar gates. The BJT parameters are $V_\gamma = 0.5$ V, $V_{BE(ON)} = 0.7$ V, $V_{BE(sat)} = 0.8$ V, and $V_{CE(sat)} = 0.2$ V. The diode cut-in voltage is 0.6 V and, when conducting, has a 0.7-V drop across it. Assume that Q is unloaded by the following stage. (*a*) Verify that the circuit functions as a NAND gate for $\beta_F > \beta_{F(min)}$. (*b*) Determine $\beta_{F(min)}$. (*c*) Will the circuit operate if $D2$ is not used?

Solution

(*a*) The output levels of the BJT inverter (see Fig. 6-35) are $V(0) = V_{CE(sat)} = 0.2$ V and $V(1) = V_{CC} = 5$ V. If at least one input is at $V(0)$, its diode conducts and $V_P = 0.2 + 0.7 = 0.9$ V. Since a voltage of $(2)(0.7) = 1.4$ V is required for $D1$ and $D2$ to be conducting, these diodes are cut off, and $V_{BE} = 0$. Since the cut-in voltage of Q is $V_\gamma = 0.5$ V, Q is OFF, the output rises to 5 V, and $Y = 1$. This confirms the first three rows of the NAND truth table in Fig. 6-11b.

If all inputs are at $V(1) = 5$ V, we shall assume that all input diodes are OFF, that D_1 and D_2 conduct, and that Q is in saturation. If these conditions

FIGURE 6-36

A positive-logic diode-transistor (DTL) NAND gate.

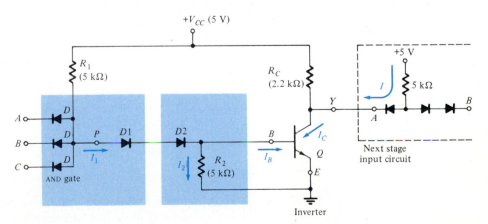

Inverter

are true, the voltage at P is the sum of two diode drops plus $V_{BE(sat)}$ or $V_P = 0.7 + 0.7 + 0.8 = 2.2$ V. The voltage across each input diode is $5 - 2.2 = 2.8$ V in the reverse direction, thus justifying the assumption that D is OFF. We now determine the base current of Q:

$$I_1 = \frac{V_{CC} - V_P}{R_1} = \frac{5 - 2.2}{5} = 0.560 \text{ mA}$$

$$I_2 = \frac{V_{BE(sat)}}{R_2} = \frac{0.8}{5} = 0.160 \text{ mA}$$

$$I_B = I_1 - I_2 = 0.560 - 0.160 = 0.400 \text{ mA}$$

Assuming that $\beta_F > \beta_{F(min)}$, this value of I_B saturates Q and makes $v_o = V(0) = V_{CE(sat)}$ and verifies the last line of the NAND truth table.

(b) The value of $\beta_{F(min)}$ is $I_{C(sat)}/I_B$:

$$I_{C(sat)} = \frac{V_{CC} - V_{CE(sat)}}{R_C} = \frac{5 - 0.2}{2.2} = 2.182 \text{ mA}$$

$$\beta_{F(min)} = \frac{I_{C(sat)}}{I_B} = \frac{2.182}{0.400} = 5.46$$

Thus, for $\beta_F \geq 5.46$, the assumption that Q is saturated is valid.

(c) If at least one input is at $V(0)$, then $V_P = 0.2 + 0.7 = 0.9$ V. Hence, if only one diode $D1$ is used between P and the base B, then $V_{BE} = 0.9 - 0.6 = 0.3$ V, where 0.6 V represents the diode cut-in voltage. Since the cut-in base voltage is $V_\gamma = 0.5$ V, theoretically Q is cut off. However, this is not a very conservative design because a small (>0.2 V) spike of noise will turn Q on. An even more conservative design uses three diodes in series, instead of the two indicated in Fig. 6-36.

In the foregoing discussion we have unrealistically assumed that the NAND gate is unloaded. If it drives N similar gates, we say that the *fan-out* is N. The output transistor now acts as a *sink* for the current in the input to the gates it drives. In other words, when Q is in saturation ($Y = 0$), the input current I in Fig. 6-36 of a following stage adds to the collector current of Q. Assume that all the input diodes to a following stage (which is now considered to be a *current source*) are high except the one driven by Q. Then the current in this diode is $I = (5 - 0.9)/5 = 0.820$ mA. This current is called a *standard load*. The total collector current of Q is now $I_C = 0.820N + 2.182$ mA, where 2.182 mA is the unloaded collector current found in part *a* of the preceding example. Since the base current is almost independent of loading, I_B remains at its previous value of 0.400 mA. If we assume a reasonable value for $\beta_{F(min)}$ of 30, the fan-out is given by $I_C = \beta_{F(min)} I_B$, or

$$I_C = 0.820N + 2.182 = 30(0.400) = 12.0 \text{ mA}$$

and $N = 11.97$. Since N must be an integer, a conservative choice is $N = 11$. Of course, the current rating of Q must not be exceeded.

The basic TTL NAND circuit is displayed in Fig. 6-37 and employs the topology of the DTL gate. The emitter junctions of the multiple-emitter transistor $Q1$ in Fig. 6-37 replaces the diodes D in the left-hand rectangle in Fig. 6-36. Also, $D1$ is replaced by the collector junction of $Q1$. The emitter junction and emitter resistance R_3 of $Q2$ in Fig. 6-37 replace $D2$ and R_2 in Fig. 6-36. Both circuits utilize an output inverter ($Q3$ or Q).

The explanation of the operation of the TTL gate parallels that of the DTL switch. Thus, if at least one input is at $V(0) = 0.2$ V, then

$$V_P = 0.2 + 0.7 = 0.9 \text{ V}$$

For the collector junction of $Q1$ to be forward-biased and for $Q2$ and $Q3$ to be ON requires V_P to be about $0.7 + 0.7 = 1.4$ V. Hence $Q2$ and $Q3$ are OFF; the output rises to $V_{CC} = 5$ V, and $Y = V(1)$. On the other hand, if all inputs are high (at 5 V), the input diodes (the emitter junctions) are reverse-biased and V_P rises toward V_{CC} and drives $Q2$ and $Q3$ into saturation. Then the output is $V_{CE(\text{sat})} = 0.2$ V, and $Y = V(0)$ (and V_P is clamped at about 1.6 V).

Input Transistor Action The explanation given in the preceding paragraph assumes that $Q1$ acts like isolated back-to-back diodes and not as a transistor. The preceding conclusions are also reached if the transistor behavior of $Q1$ is taken into consideration.

Condition 1. *At least one input is low*, $v_i = 0.2$ V. The emitter of $Q1$ is forward-biased and we assume that $Q2$ and $Q3$ are OFF. The current $I_{C1} (= I)$ into the collector P must be the current from emitter to base of $Q2$. Hence, I_{C1} equals the reverse saturation current of the emitter-junction diode of $Q2$. Since this current is very small (a few nanoamperes), $I_{B1} \gg I_{C1}/\beta_F$ and $Q1$ is in saturation. The voltage at P equals $V_{CE(\text{sat})} + v_i = 0.2 + 0.2 = 0.4$ V. This voltage is too small to put $Q2$ and $Q3$ ON. This argument justifies our assumptions that $Q2$ and $Q3$ are OFF and, therefore, $Y = V(1) = V_{CC}$.

FIGURE 6-37
The basic TTL NAND-gate configuration.

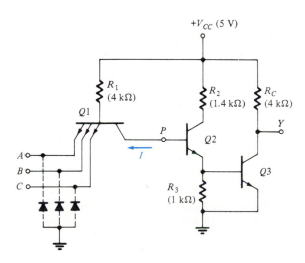

Condition 2. *All inputs are high.* The emitters of $Q1$ are reverse-biased, whereas the collector is forward-biased, because the p-type base is connected to the positive 5-V supply (through the 4-kΩ resistor). Hence $Q1$ is operating in the inverted mode (Sec. 3-3). The inverted-current gain β_R for an IC transistor is very small (<1). The input current (now the collector current of the inverted transistor) is $\beta_R I_{B1}$. The current I (now the emitter current of the inverted transistor) is $-(1 + \beta_R)I_{B1}$. This large current saturates $Q2$ and $Q3$ and $Y = V(0)$. This concludes the argument that Fig. 6-37 obeys NAND logic.

Low Storage Time   We now show that because of the transistor behavior of $Q1$ during turn-off, the storage time t_s (Sec. 3-14) is reduced considerably. Note that the base voltage of $Q2$, which equals the collector voltage of $Q1$, is at $0.8 + 0.8 = 1.6$ V during saturation of $Q2$ and $Q3$. If now any input drops to 0.2 V, then instantaneously the base voltage of $Q1$ is at 0.9 V. At this time the collector junction is reverse-biased by $1.6 - 0.9 = 0.7$ V, the emitter junction is forward-biased, and $Q1$ *is in its forward-active region.* The large collector current I of $Q1$ now quickly removes the stored charge in $Q2$ and $Q3$. It is this transistor action which gives TTL the highest speed of any saturated logic. It is not until all the charge is removed from $Q3$ and $Q2$ (so that these transistors go OFF) that $Q1$ saturates, as discussed in condition 1.

Input Clamping Diodes   These diodes (shown by dashed lines in Fig. 6-37) are often included from each input to ground, with the anode grounded. These diodes are effectively out of the circuit for positive input signals, but they limit negative voltage excursions at the input to a safe value. These negative signals may arise from ringing caused by parasitic lead inductance resonating with shunt capacitance.

6-12   TTL OUTPUT STAGES   In the discussion of fan-out in Sec. 6-11, two dc conditions are taken into account: (1) the output transistor must saturate when loaded by N gates, and (2) the current rating of this sink transistor must not be exceeded. Another (dynamic) condition is now considered.

At the output terminal of the TTL gate there is a capacitive load C_L, consisting of the capacitances of the reverse-biased diodes of the fan-out gates and any stray wiring capacitance. If the collector-circuit resistor of the inverter is R_C (called a *passive pull-up*), then, when the output changes from the low to the high state, the output transistor is cut off and the capacitance charges exponentially from $V_{CE(sat)}$ to V_{CC}. The time constant $R_C C_L$ of this waveform may introduce a prohibitively long delay time into the operation of these gates.

The Totem-Pole Stage   The output delay may be reduced by decreasing R_C, but this will increase the power dissipation when the output is in its low state as the voltage across R_C is $V_{CC} - V_{CE(sat)}$. A better solution to this problem is indicated in Fig. 6-38, where a transistor acts as an *active pull-up* circuit, replacing the

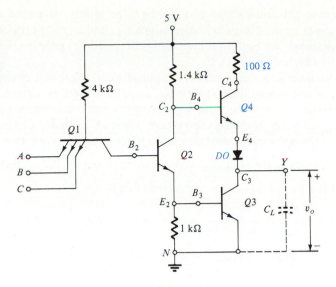

FIGURE 6-38
A TTL NAND gate with a totem-pole output. Active pull-up is provided by $Q4$, DO, and $100 \ \Omega$.

passive pull-up resistance R_C. This output configuration is called a *totem-pole* amplifier because the transistor $Q4$ "sits" atop $Q3$. It is also referred to as a *power-driver*, or *power-buffer*, output stage.

The transistor $Q2$ acts as a *phase splitter,* since the emitter voltage is out of phase with the collector voltage. (For an increase in base current, the emitter voltage increases and the collector voltage decreases.) We now explain the operation of this driver circuit in detail, with reference to the TTL gate in Fig. 6-38.

The output is in the low-voltage state when $Q2$ and $Q3$ are driven into saturation. For this state we should like $Q4$ to be OFF. Is it? Note that the collector voltage V_{CN2} of $Q2$ with respect to ground N is given by

$$V_{CN2} = V_{CE2(\text{sat})} + V_{BE3(\text{sat})} = 0.2 + 0.8 = 1.0 \text{ V}$$

Since the base of $Q4$ is tied to the collector of $Q2$, then $V_{BN4} = V_{CN2} = 1.0$ V. *If the output diode DO were missing,* the base-to-emitter voltage of $Q4$ would be

$$V_{BE4} = V_{BN4} - V_{CE3(\text{sat})} = 1.0 - 0.2 = 0.8 \text{ V}$$

which would put $Q4$ into saturation. Under these circumstances the steady current through $Q4$ would be

$$\frac{V_{CC} - V_{CE4(\text{sat})} - V_{CE3(\text{sat})}}{100} = \frac{5 - 0.2 - 0.2}{100} \text{ A} = 46 \text{ mA}$$

which is excessive and wasted current. The necessity for adding DO is now clear. With it in place, the sum of V_{BE4} and V_{DO} is 0.8 V. Hence both $Q4$ and DO are at cutoff. In summary, if C_L is at the high voltage $V(1)$ and the gate is excited, $Q4$ and DO go OFF, and $Q3$ conducts. Because of its large active-region current, $Q3$ quickly discharges C_L, and as v_o approaches $V(0)$, $Q3$ enters saturation. The bottom transistor $Q3$ of the totem pole is referred to as a *current sink*, which discharges C_L.

Assume now that with the output at $V(0)$, there is a change of state because one of the inputs drops to its low state. Then $Q2$ is turned OFF, which causes $Q3$ to go to cutoff because V_{BE3} drops to zero. The output v_o remains momentarily at 0.2 V because the voltage across C_L cannot change instantaneously. Now $Q4$ goes into saturation and DO conducts, as we can verify:

$$V_{BN4} = V_{BE4(\text{sat})} + V_{DO} + v_o = 0.8 + 0.7 + 0.2 = 1.7 \text{ V}$$

and the base and collector currents of $Q4$ are

$$I_{B4} = \frac{V_{CC} - V_{BN4}}{1.4} = \frac{5 - 1.7}{1.4} = 2.36 \text{ mA}$$

$$I_{C4} = \frac{V_{CC} - V_{CE4(\text{sat})} - V_{DO} - v_o}{0.1} = \frac{5 - 0.2 - 0.7 - 0.2}{0.1} = 39.0 \text{ mA}$$

Hence, if β_F exceeds $\beta_{F(\text{min})} = I_{C4}/I_{B4} = 39.0/2.36 = 16.5$, then $Q4$ is in saturation. The transistor $Q4$ is referred to as a *source*, supplying current to C_L. As long as $Q4$ remains in saturation, the output voltage rises exponentially toward V_{CC} with the very small time constant $(100 + R_{CS4} + R_f)C_L$, where R_{CS4} is the saturation resistance (Sec. 3-10) of $Q4$, and where R_f (a few ohms) is the diode forward resistance. As v_o increases, the currents in $Q4$ decrease, and $Q4$ comes out of saturation and finally v_o reaches a steady state when $Q4$ is at the cut-in condition. Hence the final value of the output voltage is

$$v_o = V_{CC} - V_{BE4(\text{cut-in})} - V_{DO(\text{cut-in})} \approx 5 - 0.5 - 0.6 = 3.9 \text{ V} = V(1)$$

If the 100-Ω resistor were omitted, there would result a faster change in output from $V(0)$ to $V(1)$. However, the 100-Ω resistor is needed to limit the current spikes during the turn-on and turn-off transients. In particular, $Q3$ does not turn off (because of storage time) as quickly as $Q4$ turns on. With both totem-pole transistors conducting at the same time, the supply voltage would be short-circuited if the 100-Ω resistor were missing. The peak current drawn from the supply during the transient is limited to $I_{C4} + I_{B4} = 39 + 2.4 \approx 41$ mA if the 100-Ω resistor is used. These current spikes generate noise in the power-supply distribution system, and also result in increased power consumption at high frequencies.

Three-State Output   It is often necessary to expand the capability of a digital system by combining a number of identical packages (Fig. 7-31). Consider such a design where the nth output Y_n corresponds to Y_{n1} from chip 1, to Y_{n2} from chip 2, to Y_{n3} from chip 3, etc. Depending on the specified logic, it is required that either Y_{n1}, Y_{n2}, or Y_{n3}, etc. (but only one of these) appear at an output Y_n. This result is obtained by connecting all leads Y_{n1}, Y_{n2}, Y_{n3}, etc., together (referred to as *wire* OR-*ing* or as the OR-*tied connection*) and by enabling only the *i*th chip during the interval when Y_{ni} is to appear at Y_n. The TTL totem-pole output stage (Fig. 6-38), modified to include such an enable, is indicated in Fig. 6-39*a* and the corresponding open-collector output circuit is shown in Fig. 6-39*b*.

In Fig. 6-39*a* if the *enable* or *chip-select* (*CS*) signal is low, *D*1 and *D*2 are OFF, and the output is either in state 1 or state 0, depending on whether the input data are 0 or 1. However, if *CS* is high, then *D*1 and *D*2 are ON, these diodes clamp *Q*3 and *Q*4 OFF, and the output *Y* is effectively an open circuit. This condition, referred to as the *high-impedance third state*, allows OR-ing of the outputs from the several packages. The circuit in Fig. 6-39*b* operates in a similar 3-state fashion. However, the manufacturers designate the configuration in Fig. 6-39*a* as the tristate (TS) output and that in Fig. 6-46*b* as the open-collector (OC) output.

The Transfer Characteristic The TTL NAND gate (Fig. 6-38) becomes an inverter when all the inputs are tied together. The piecewise linear approximation to the transfer characteristic of the resultant inverter, shown in Fig. 6-40, differs from that shown in Fig. 6-35 for the basic BJT inverter. The following qualitative argument justifies the shape of the characteristic in Fig. 6-40; the numerical evaluation of the critical voltage values is left to the student (Prob. 6-66).

For $v_i < V_{IL}$, both *Q*2 and *Q*3 are cut off, *Q*4 is in saturation, and the output is *V*(1). At point *A*, *Q*2 begins to conduct. However, the current produced in *Q*2 is insufficient to produce the voltage drop $V_{EN2} = V_{BE3}$ needed to turn on *Q*3. The accompanying decrease in V_{CN2} maintains *Q*4 in a conducting state, but it is no longer in saturation and accounts for the decrease in v_o. Increasing v_i to its value at *B* increases the emitter current in *Q*2 and hence turns on *Q*3. Between *B* and *C* in Fig. 6-40, *Q*3 is in its forward-active region and the output decreases with increasing v_i (similar to the region between *A* and *B* in Fig.

FIGURE 6-39
Three-state output stages controlled by an enable, strobe, or chip-select input (*CS*): (*a*) totem-pole state, (*b*) open-collector.

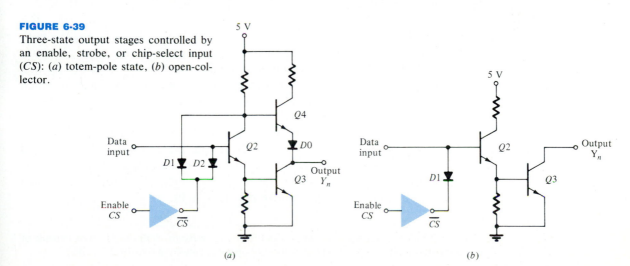

(*a*) (*b*)

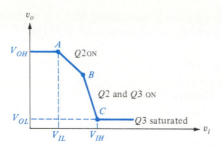

FIGURE 6-40
The straight-line (piecewise linear) representation of the TTL transfer characteristic.

6-35). As $Q2$ conducts more heavily, $Q3$ is driven into saturation and $Q4$ to cutoff (at point C) and the output is limited to $V_{CE(\text{sat})}$ for $v_i > V_{IH}$.

6-13   TTL LOGIC FAMILIES   The original TTL logic family was designated the 54/74 series[1] and used the NAND circuit in Fig. 6-38 (or incorporated the output stage in Fig. 6-39) as the basic building block. Schottky transistors were incorporated to improve speed, and this family is referred to as the 74S series. The propagation delay is reduced by a factor of 3 at the expense of doubling the power dissipation. The delay-power product, however, is improved over the 74 series. Both the 74 and 74S series have been supplanted by the Low-Power Schottky Series 74LS, the Advanced Schottky Series 74AS, and the Advanced Low-Power Schottky Series 74ALS.[2]

The 74LS series is, in current (1987) practice, the standard general-purpose TTL family for which the circuit in Fig. 6-41 is the basic NAND gate. Its topology is similar to that in Fig. 6-38 except that the BJTs are replaced by Schottky transistors. In Fig. 6-41 the transistor designations $Q1$ through $Q4$ correspond to the numbering in Fig. 6-38. Higher-numbered transistors indicate devices that have been added. Note that the multiple-emitter transistor $Q1$ in Fig. 6-38 is replaced by the Schottky diode AND circuit ($D1$, $D2$, and $D3$). The addition of $Q6$ virtually eliminates the second breakpoint B in Fig. 6-40 as both $Q6$ and $Q3$ must be conducting for current to exist in $Q2$. Furthermore, because both $Q6$ and $Q3$ turn on simultaneously, breakpoint A in Fig. 6-40 occurs at a higher input voltage. Breakpoint C corresponds to an input voltage at which the transistor $Q3$ is clamped because of its Schottky diode. As a result, the voltage difference between V_{IL} and V_{IH} is reduced.

The inclusion of $Q5$ helps provide larger load currents when the output is $V(1)$ than can be supplied by $Q4$ alone. Note that the emitters of both $Q4$ and $Q5$ (through the 4-kΩ resistance) are connected to the output. The diodes $D4$ and $D5$ are employed to increase the speed by which $Q4$ is turned off when the output must make a transition from $V(1)$ to $V(0)$.

---

[1]Both the 54 and 74 series have identical electrical characteristics, with the 54 series capable of operating from -55 to $+125°C$ and the 74 series operates only from 0 to 70°C.

[2]These circuits are also available in the 54 series.

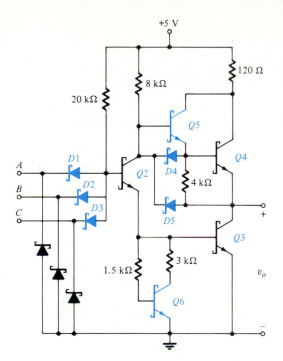

FIGURE 6-41
The circuit configuration of the low-power Schottky TTL NAND gate (54/74LS series). The devices shown in color replace corresponding devices in Fig. 6-38. Transistors $Q5$ and $Q6$ and diodes $D4$ and $D5$ are additional components. The device numbering corresponds to that used in Fig. 6-38.

The low-power operation is observed by comparing the resistance values associated with the transistor $Q2$ in Figs. 6-41 and 6-38. A reduction in power dissipation by a factor of 5 is obtained at comparable speed.

The 74AS NAND gate is the fastest of the TTL series. Design choices are made to minimize propagation delays at the expense of power dissipation. The circuit configuration is similar to that in Fig. 6-41 except that lower resistance values are used which account for both the speed improvements (lower time constants) and increased power consumption (higher currents).

The 74ALS series is a derivative of the 74LS series whose design is used to minimize power dissipation. Because propagation delay is also reduced, this series has the best delay-power product of any logic family. Power consumption is reduced because of the increase in resistance values and consequent current reduction. Speed is increased by the inclusion of additional active elements such as three *pnp emitter followers* to replace $D1$, $D2$, and $D3$. (In Sec. 6-14 we show that the output resistance of an emitter follower is low so that resultant time constants are small.) Improved processing techniques permit fabrication of smaller devices which also improve the speed characteristics of the 74ALS series.

6-14 EMITTER-COUPLED LOGIC (ECL) CIRCUITS   The fastest logic family available is *emitter-coupled logic* (ECL). Its speed results from the use of the nonsaturating current switch based on the emitter-coupled (differential) pair described in Sec.

3-12[1] and whose circuit is redrawn in Fig. 6-42a. In that section we showed that the currents and voltages I_{C1}, I_{C2}, v_{o1}, and v_{o2} each responded to the difference voltage $v_d = v_1 - v_2$ as illustrated in Figs. 3-38 and 3-39 and repeated in Figs. 6-42b and c for convenience. The sum of the currents $i_{C1} + i_{C2} = \alpha_F I_{EE} \simeq I_{EE}$ for all values of input voltages v_1 and v_2. For $v_d \geq 4V_T = 100\,\text{mV} = 0.1\,\text{V}$ (at room temperature), we observe that, for practical purposes, $i_{C1} \simeq I_{EE}$ and $i_{C2} \simeq 0$. The situation is reversed for negative v_d with $|v_d| \geq 4V_T$. Here, $Q1$ is virtually cut off and all the current is in $Q2$. If we select v_2 to be a fixed reference voltage V_R, then, as the input signal $v_i = v_1$ changes from $V_R + 0.1$ to $V_R - 0.1$, the current I_{EE} switches from $Q1$ to $Q2$. Excursions in v_1 from $V_R - 0.1$ to $V_R + 0.1$ cause the current to switch from $Q2$ to $Q1$. The

---

[1]The reader should review this section in preparation for the discussion of ECL circuits.

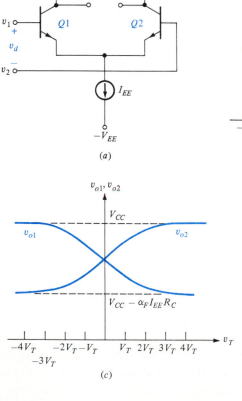

(a)

(c)

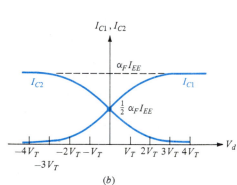

(b)

FIGURE 6-42

(a) An emitter-coupled pair, with transfer characteristics for (b) the collector currents and (c) the output voltages.

transistors $Q1$ and $Q2$, when conducting $(-0.1 < v_d < 0.1)$, are in the forward-active region and either $Q1$ or $Q2$ is virtually cut off when v_d is outside this range.

Note that for $|v_d| \geq 4V_T$, either v_{o1} is high and v_{o2} is low or vice versa. In boolean variable terms, the two outputs are complementary (if $v_{o1} = Y$, then $v_{o2} = \overline{Y}$). Emitter-coupled logic gates exploit this convenience of providing both outputs Y and $\overline{Y}$ simultaneously.

The Basic ECL OR/NOR Gate   The standard ECL OR/NOR gate topology is displayed in Fig. 6-43a. It is obtained from Fig. 6-42a by using a constant reference voltage V_R for v_2 of $Q2$ and by paralleling transistors which share a common collector resistance in place of $Q1$. A fan-in of 2 is depicted in Fig. 6-43a. If either input A or B is $V(1) > V_R + 0.1$, the outputs are $v_{o1} = V(0)$ and $v_{o2} = V(1)$. If both inputs are greater than $V_R + 0.1$, again $v_{o1} = V(0)$ and $v_{o2} = V(1)$. However, when both A and B are at $V(0) < V_R - 0.1$, $v_{o1} = V(1)$ and $v_{o2} = V(0)$. Consequently, v_{o2} is the OR output ($Y_2 = A + B \equiv Y$) and v_{o1} is the NOR output $[Y_1 = (\overline{A + B}) = \overline{Y}]$, as depicted in Fig. 6-43b.

The Reference Voltage V_R   The current source in Fig. 6-43 is realized in its simplest form by a resistor R_E placed between the common emitters and the negative supply voltage. In Fig. 6-43 (with R_E replacing I_{EE}), KVL applied to the loop consisting of V_R, V_{BE2}, R_E, and V_{EE} allows the current I in R_E to be expressed as $I = (V_R - V_{BE2} + V_{EE})/R_E$. Since the variation in V_{BE2} with current in the active region is small compared with $V_R + V_{EE}$, I remains essentially constant in the emitter-coupled pair. This constant current simulates the current source I_{EE} in Fig. 6-42a.

The circuit in Fig. 6-43a requires three power supplies (V_{CC}, V_{EE}, and V_R). A more practical single-supply (V_{EE}) ECL gate is obtained by setting V_{CC} to

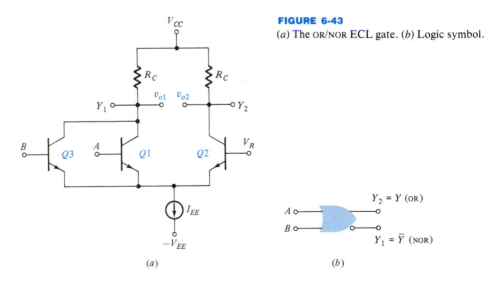

FIGURE 6-43
(a) The OR/NOR ECL gate. (b) Logic symbol.

(a)

(b)

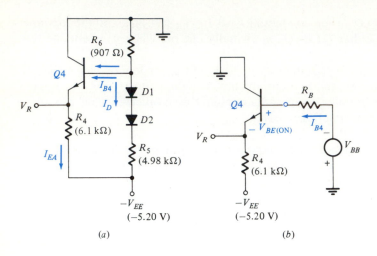

0 V (ground) and deriving V_R from the configuration in Fig. 6-44*a*. To evaluate V_R, note that the Thévenin equivalent of this circuit is shown in Fig. 6-44*b* for which

$$V_{BB} = (V_{EE} - 2V_D)\frac{R_6}{R_6 + R_5} \quad \text{and} \quad R_B = \frac{R_5 R_6}{R_5 + R_6} \quad (6\text{-}39)$$

where V_D is the voltage across each diode.

If we assume that I_{B4} produces a negligible voltage drop across R_B compared with V_{BB}, then

$$V_R = -V_{BB} - V_{BE(ON)4} \quad (6\text{-}40)$$

Using the numerical values shown in Fig. 6-44*a* and assuming $V_D = 0.75$ V, we find, from Eq. (6-39)

$$V_{BB} = \frac{(5.2 - 2 \times 0.75)(0.907)}{0.907 + 4.98} = 0.570 \text{ V}$$

and from Eq. (6-40)

$$V_R = -0.570 - 0.75 = -1.32 \text{ V}$$

The reason for selecting $V_{BE(ON)4} = 0.75$ V (rather than 0.7 V) is that for ECL speeds to be obtained, small-dimension devices must be employed. This results in the transistors being operated at higher current levels relative to I_{ES} than are generally encountered. This higher current requires that $V_{BE(ON)} > 0.7$ V, and it is common practice to assume a value of 0.75 V for the base-emitter voltage in ECL circuits. We now justify the assumption that $V_D = 0.75$ V. Since $I_D \gg I_{B4}$, then, from Fig. 6-44*a*, we obtain

$$I_D = \frac{V_{EE} - 2V_D}{R_B + R_5} = \frac{5.20 - 1.50}{0.907 + 4.98} = 0.63 \text{ mA} \quad (6\text{-}41)$$

From Fig. 5-19*a*, for a base-emitter diode with an open collector, we find $V_D \simeq 0.75$ V at 0.63 mA as assumed.

The Transfer Characteristic   The circuit used to determine the OR transfer function is obtained from Fig. 6-43 with R_E in place of the current source I_{EE} as depicted in Fig. 6-45*a*. Only one input transistor Q1 is indicated since all other parallel transistors (such as Q3 in Fig. 6-43) are considered to be nonconducting in this discussion. To obtain the straight-line approximation of the transfer characteristic for OR logic, two conditions must be satisfied: (1) if $v_i = V(0)$, Q1 is OFF, Q2 is ON, and $v_{o2} = V(0)$; and (2) if $v_i = V(1)$, Q1 is ON, Q2 is OFF, and $v_{o2} = V(1)$.

To calculate $V(0)$, consider Q2 conducting in Fig. 6-45*a*:

$$V_E = V_R - V_{BEON} = -1.32 - 0.75 = -2.07 \text{ V} \tag{6-42}$$

$$I_E = \frac{V_E - (-V_{EE})}{R_E} = \frac{V_E + V_{EE}}{R_E} \tag{6-43}$$

$$I_{C2} \simeq I_E = \frac{-2.07 + 5.20}{0.779} = 4.02 \text{ mA} \tag{6-44}$$

$$v_{o2} = -I_{C2}R_2 = -4.02 \times 0.245 = -0.98 \text{ V} = V(0)$$

Let us check condition 1 that Q1 is OFF for $v_i = V(0)$:

$$V_{BE1} = V(0) - V_E = -0.98 + 2.07 = 1.09 \text{ V}$$

Since this value is greater than the cut-in voltage $V_\gamma = 0.5$ V, Q1 is *not* cut off and *the circuit in Fig. 6-45a does not operate properly*. This difficulty is easily remedied. To cut off Q1, the value of $V(0)$ must be made more negative and is accomplished with the *level-shifting circuit* in Fig. 6-45*b*. By connecting

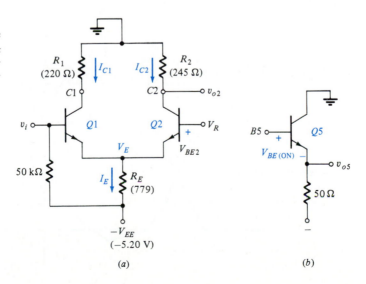

FIGURE 6-45
(*a*) The basic ECL gate with I_{EE} produced by R_E and V_{EE}. (*b*) Level-shifting (emitter-follower) circuit.

the base $B5$ of $Q5$ to the collector $C2$ of $Q2$ and taking the output at the emitter of $Q5$, we obtain

$$v_{o5} = v_{o2} - V_{BEON} = -0.98 - 0.75 = -1.73 \text{ V} = V(0)$$

Note that with the addition of $Q5$ and with $v_i = V(0)$, we have

$$V_{BE1} = v_i - V_E = -1.73 + 2.07 = 0.34 \text{ V} < V_\gamma = 0.5 \text{ V}$$

which confirms condition 1 that $Q1$ is OFF for an input of $V(0)$.

To calculate $V(1)$, assume condition 2, namely, that for $v_i = V(1)$, $Q1$ conducts and $Q2$ is OFF. Since $I_{C2} \simeq 0$, $v_{o2} \simeq 0$ and $v_{o5} = -V_{BE(ON)} = -0.75 \text{ V} = V(1)$. Let us now check the assumption that $Q2$ is indeed OFF. Since $V_E = v_i - V_{BE1} = V(1) - V_{BE(ON)} = -0.75 - 0.75 = -1.50 \text{ V}$ and $V_{BE2} = V_R - V_E = -1.32 + 1.50 = 0.18 \text{ V}$. Since this voltage is less than $V_\gamma = 0.5 \text{ V}$, we have verified that $Q2$ is OFF.

It should be noted that the values obtained for $V(0)$ and $V(1)$ are approximate in that the base currents have been neglected and the current in the cutoff transistor is not quite zero (it is about 2 percent of its ON value). Taking these terms into account (Prob. 6-75), $V(1) = -0.90 \text{ V}$ and $V(0) = -1.74 \text{ V}$. Note that these values are symmetrical about the reference voltage $V_R = -1.32 \text{ V}$ $(1.32 - 0.9 = 0.42 = 1.74 - 1.32)$. Both the actual and approximate transfer characteristics are displayed in Fig. 6-46.

The argument used here that I_E is essentially constant assumes that $Q2$ is always conducting with $V_{BE} = V_{BE(ON)}$. However, with $Q1$ ON, $Q2$ must be OFF, and I_E is determined by the input level $V(1)$. Thus $I_E = [V(1) - V_{BE1} + V_{EE}]/R_E \simeq I_{C1}$. It is found (Prob. 6-76) that I_{C1} is slightly larger than I_{C2} (with $Q2$ ON). However, to have symmetrical OR/NOR characteristics, the voltage $I_{C1}R_1$ when $Q1$ is conducting must equal the drop $(I_{C2}R_2)$ across R_2 when $Q2$ is ON. It follows that R_1 must be slightly smaller than R_2 as indicated in Fig. 6-47.

FIGURE 6-46
OR/NOR transfer characteristics.

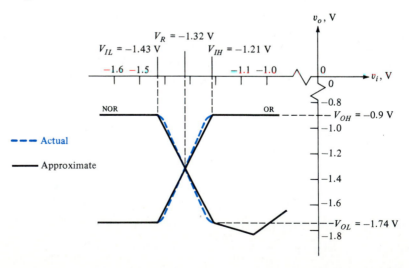

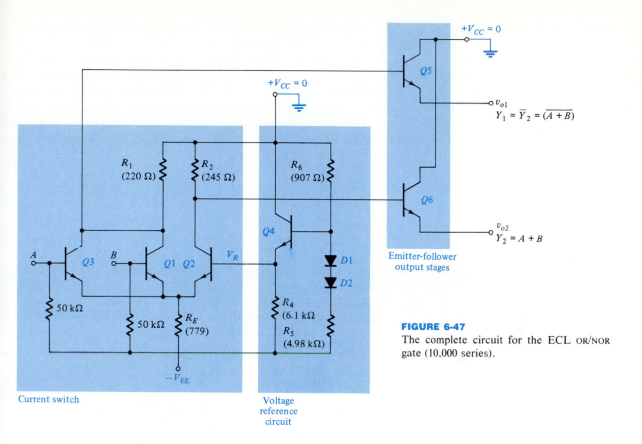

$+V_{CC} = 0$

$Q5$

v_{o1}
$Y_1 = \bar{Y}_2 = \overline{(A + B)}$

$+V_{CC} = 0$

R_1 (220 Ω)

R_2 (245 Ω)

R_6 (907 Ω)

$Q6$

v_{o2}
$Y_2 = A + B$

$Q4$

Emitter-follower output stages

A $Q3$ B $Q1$ $Q2$ V_R

$D1$

$D2$

50 kΩ

50 kΩ R_E (779)

R_4 (6.1 kΩ)

R_5 (4.98 kΩ)

FIGURE 6-47
The complete circuit for the ECL OR/NOR gate (10,000 series).

$-V_{EE}$

Current switch

Voltage reference circuit

Noise Margins

Determinations of V_{IL} and V_{IH} are usually made to correspond to differential input voltages that result in collector currents in the OFF transistor being 1 percent of the current in the ON transistor. The values of V_{IL} and V_{IH} so obtained differ only slightly from the values obtained at the unity-slope points (Prob. 6-74). With reference to Eq. (3-41), the change in v_d necessary to make $I_{C1}/I_{C2} = 100$ is evaluated as $v_d \approx 112$ mV. By symmetry, $V_{IH} = V_R + 0.112$ and $V_{IL} = V_R - 0.112$. Using $V_R = -1.32$ V, $V_{IH} = -1.21$ V and $V_{IL} = -1.43$ V.

The noise margins are calculated from Eq. (6-29) as

$$NM_H = V_{OH} - V_{IH} = -0.90 + 1.21 = 0.31 \text{ V}$$

$$NM_L = V_{IL} - V_{OL} = -1.43 + 1.74 = 0.31 \text{ V}$$

Note that the symmetry of the transfer function gives rise to equal noise margins, $NM_H = NM_L = 0.31$ V.[1]

---

[1]Use of the unity-slope points to evaluate NM_H and NM_L yields numerical results that differ by only several millivolts.

The NOR Characteristic The NOR output is obtained from the emitter of a level-shifter $Q6$ (identical with $Q5$ in Fig. 6-45b) connected to the collector of $Q1$. The NOR transfer characteristic is also displayed in Fig. 6-46. A notable difference between the OR and NOR characteristics occurs for large values of input voltages. As v_i increases beyond V_{IH}, I_{C1} continues to increase and the NOR output falls below V_{OL}. However, when v_i becomes large enough to saturate $Q1$, the increased emitter current causes V_E to increase. With V_{CE1} constant at $V_{CE(sat)} = 0.2$ V, the voltage V_{C1} now increases. Hence, as v_i continues to increase beyond the voltage at which saturation is reached in $Q1$, the NOR output rises again as illustrated in Fig. 6-46 (Prob. 6-73). Input voltages are restricted, however, in normal operation.

The Output Stage The circuit in Fig. 6-45b, in addition to being a level shifter, is also called an *emitter follower*. We now explain the meaning of this term. Since $V_{BE(ON)}$ is essentially constant, and independent of transistor current, any increase in base-to-ground voltage results in the same increase in emitter-to-ground voltage. Thus the emitter *follows* the base (in changes in voltage). The circuit is an amplifier with approximately unity voltage gain. This conclusion is valid even if the external emitter load is changed, and consequently the output resistance R_o of an emitter follower is extremely small. (See Sec. 10-12 for a quantitative calculation of R_o.) An emitter follower makes an excellent output stage as it provides the low output resistance necessary to obtain high speed when driving a capacitive load (the inputs to the gates that comprise the load).

ECL OR/NOR Topology The complete configuration is shown in Fig. 6-47 and consists of the three principal circuits already discussed: (1) the differential-pair current switch, (2) the voltage reference network, and (3) the emitter-follower output stages. The component values given in Fig. 6-47 are typical of those used in the 10,000-series ECL family discussed later in this section.

No on-chip load is provided for ECL gates, and, with no load, the emitters of $Q5$ and $Q6$ are open. The load resistance for each emitter follower is provided by the transmission path (often a 50-Ω transmission line) and the input resistance to the next stage. The 50-kΩ resistors, connected between the base and V_{EE} of the input transistors, can provide the load for the driver outputs. With no input signal provided (A and B remain open-circuited) note that these resistor connections ensure that the inputs are at $V(0)$. Thus ECL gates have active pull-up and passive pull-down.

The voltage reference circuit is designed so that V_R is essentially constant independent of the base current into $Q2$ by making I_{E2} much greater than I_{B2}. In part, this large value of I_{E2} accounts for the high power dissipation in ECL gates. The diodes $D1$ and $D2$ are used for temperature compensation of the base-emitter junction of $Q4$ and, in conjunction with R, maintain I equal to I_{E2}.

As indicated in Fig. 6-47, two different V_{CC} (ground) connections are shown, one for the emitter followers and one for the current switch and voltage reference. The reason for the separate connections is to isolate the switching transients (large current and voltage "spikes" caused by charging and discharging the external load and parasitic output capacitances) that appear at the output of the emitter followers from affecting the differential pair and reference circuit (the internal stages). Since ECL gates have low noise margins (0.3 V), this configuration helps to suppress noise levels.

Power Dissipation One consequence of using nonsaturating current switches is that a relatively high current must be supplied by V_{EE}. This results in high power dissipation relative to other logic families. The current I_{E4}, neglecting base currents, is

$$I_{E4} = \frac{V_R - (-V_{EE})}{R_4} = \frac{-1.32 + 5.2}{6.1} = 0.636 \text{ mA}$$

The diode current is $I_D = 0.63$ mA from Eq. (6-41) and $I_{EE} = 4.02$ mA from Eq. (6-44). The total current supplied by V_{EE} is $I_T = I_D + I_{EE} + I_{E4} = 0.63 + 4.02 + 0.64 = 5.29$ mA, and the power consumed is $P_D = 5.29 \times 5.2 = 27.5$ mW. Note that this value does not include the power supplied to the load by the emitter follower. (For typical power dissipation per gate, see Table 6-4.)

Fan-Out The fan-out in ECL gates is not limited by the dc current that can be supplied by the output stage but is determined by the load capacitance. Each load gate that is to be driven presents a capacitance C at the output of the driver gate and, for a fan-out of N, the total capacitance is NC. To preserve the speed (both propagation delay and rise time) of ECL gates only a few such gates (about 10) can be used.

The high speed of the ECL gates makes it imperative that the signal transmission between gates does not degrade the quality of the signal which has a rise time on the order of 1 ns. With the speed of light in a semiconductor approximately equal to 1.5×10^8 m/s (one-half that in a vacuum), a conducting path of more than 4 cm represents a delay of more than 0.25 ns, which is comparable with the rise time. Hence, transmission paths of this dimension must be treated as distributed-parameter systems (transmission lines). An inappropriately terminated transmission line will cause some of the signal transmitted to be reflected back to the sending end. At best, the reflected signal when combined with the original input, will deteriorate the signal quality (ringing). At worst, the reflected signal will be delayed sufficiently so that it appears as a second signal and introduce logic errors. To prevent any reflections from occurring, the transmission lines used to interconnect ECL gates must be terminated in their characteristic resistances (usually in the 50- to 100-Ω range).

ECL Families The most popular ECL logic family is the 10,000 (or 10K) series, which has a propagation delay of only about 2 ns. This is attributed to the small voltage swing between $V(0)$ and $V(1)$ and the nonsaturating switch driving a low-resistance circuit. There is also one other ECL logic family available commercially, the 100,000 (or 100K) series.[1]

The 10K series is designed so that the temperature-compensated reference voltage always lies midway between $V(0)$ and $V(1)$. However, these voltages vary with temperature, and the changes in logic levels can be detrimental. Circuit modification in the ECL 100K series makes the voltages $V(0)$ and $V(1)$ almost impervious to temperature.

The 100K series is newer and thus makes use of more advanced fabrication. Consequently, they afford the highest speeds available today with propagation delays of less than 1 ns. However, because of a more complex voltage-reference circuit, their power dissipation is increased over the 10K series.

Interfacing

The ECL logic family operates at different logic levels and with a negative supply voltage. This is different from the other logic families (TTL and CMOS) which are available in SSI and MSI circuits. Often, it is necessary and useful to construct systems in which different sections of the system utilize different logic families. Several manufacturers provide *translators* to facilitate the interconnection between ECL gates and the TTL or CMOS families. Both TTL to ECL and ECL to TTL interface circuits are available.

6-15 COMPARISON OF LOGIC FAMILIES Performance data for typical circuits in each of the logic families discussed in this chapter are displayed in Table 6-4. The specific characteristics in Table 6-4 are given for a normal operating temperature of 25°C. The term *logic swing,* the difference between V_{OH} and V_{OL} is based on the minimum and maximum values of these quantities, respectively. Similarly, $NM_H = V_{OH(\min)} - V_{IH(\max)}$ and $NM_L = V_{IL(\min)} - V_{OL(\max)}$ represent worst-case values. All other quantities in Table 6-4 are as defined previously in this chapter.

Comparison of the entries in Table 6-4 allows us to reach the following conclusions (previously stated in the chapter):

1. The ECL family has the lowest propagation delay and thus comprises the fastest logic circuits available.

2. The TTL 54/74 ALS family has the smallest delay-power product.

3. CMOS logic circuits have the lowest power dissipation.

The performance of NMOS gates is not included in Table 6-4 as these gates are not available as SSI or MSI circuits. However, NMOS circuits are widely used in LSI and VLSI implementations (Chap. 9) as they afford the highest

---

[1]The 10K and 100K series are designations introduced by Fairchild Semiconductor. Two similar series, MECLII and MECLIII, are used by the Motorola Company, whose engineers developed the original ECL series MECLI. The MECLII series is comparable with the 10K series but has a propagation delay of nearly twice that for the 10K family. The MECLIII series has virtually identical properties with the 100K series but has a faster rise time.

TABLE 6-4 Comparison of Logic Families

Family series	TTL			CMOS*		ECL	
Parameter	74 LS	74 AS	74 ALS	74 C	74 HC	10K	100K
Nominal supply voltage, V	5	5	5	5	5	−5.2	−4.5
Maximun V_{OL}, V	0.5	0.5	0.5	0.4	0.4	−1.7	−1.7
Minimum V_{OH}, V	2.7	2.7	2.7	4.2	4.2	−0.9	−0.9
Maximum V_{IL}, V	0.8	0.8	0.8	1.0	1.0	−1.4	−1.4
Minimum V_{IH}, V	2.0	2.0	2.0	3.5	3.5	−1.2	−1.2
NM_H, V	0.7	0.7	0.7	0.7	0.7	0.3	0.3
NM_L, V	0.3	0.3	0.3	0.6	0.6	0.3	0.3
Logic swing, V	2.0	2.0	2.0	3.8	3.8	0.8	0.8
Power dissipation per gate, mW	2	20	1	≈0	≈0	24	40
Delay-power product, pJ	10	1.5	4	30	10	2	0.75
Fan-out	100	10	100	>100	>100	10	10

*Measured at a load current $I_{OL} = 4$ mA. At $I_{OL} = 0.2$ mA, $V_{OL} = 0.1$ V and $V_{OH} = 4.8$ V.

component density on a chip of any technology. Thus it is interesting to note that each of the four major logic families described in this chapter is "best" with respect to one major performance criterion.

REFERENCES

1 Hodges, D. A., and H. G. Jackson: "Analysis and Design of Digital Integrated Circuits," McGraw-Hill Book Company, New York, 1983.

2 Taub, H., and D. Schilling: "Digital Integrated Electronics," McGraw-Hill Book Company, New York, 1977.

3 Sedra, A. S., and K. C. Smith: "Microelectronic Circuits," Holt, New York, 1982.

4 Ghausi, M. S.: "Electronic Devices and Circuits: Discrete and Integrated," Holt, New York, 1985.

5 Elmasry, M. I. (ed.): "Digital MOS Integrated Circuits," IEEE Press, New York, 1981.

6 Solomon, P. M.: A Comparison of Semiconductor Devices for High-Speed Logic, *Proceedings of IEEE,* vol. 70, no. 5, pp. 489–509, May 1982.

7 *IEEE Journal of Solid-State Circuits,* special issues on semiconductor logic and memory, October 1970 to present.

REVIEW QUESTIONS

6-1 What is meant by a binary number?

6-2 Define (*a*) positive logic; (*b*) negative logic.

6-3 What is meant by dynamic logic?

6-4   Define an OR gate and give its truth table.

6-5   Evaluate the following expressions: (*a*) $A + 1$; (*b*) $A + A$; (*c*) $A + 0$.

6-6   Define an AND gate and give its truth table.

6-7   Evaluate the following: (*a*) $A1$; (*b*) AA; (*c*) $A0$; (*d*) $A + AB$.

6-8   Define a NOT gate and give its truth table.

6-9   Evaluate the following expressions: (*a*) $\overline{\overline{A}}$; (*b*) $\overline{A}A$; (*c*) $\overline{A} + A$.

6-10   Define an inhibitor and give the truth table for $AB\overline{S}$.

6-11   Define an exclusive-OR and give its truth table.

6-12   Show two logic block diagrams for an exclusive-OR.

6-13   Verify that the following boolean expressions represent an exclusive-OR: (*a*) $+ \overline{AB}$; (*b*) $(A + B)(\overline{A} + \overline{B})$.

6-14   State the two forms of De Morgan's laws.

6-15   Show how to implement an AND with OR and NOT gates.

6-16   Show how to implement an OR with AND and NOT gates.

6-17   Define a NAND gate and give its truth table.

6-18   Define a NOR gate and give its truth table.

6-19   Define (*a*) fan-out; (*b*) fan-in.

6-20   Define noise margin.

6-21   Sketch an NMOS inverter with an enhancement load.

6-22   Repeat Rev. 6-21 for (*a*) a linear load; (*b*) a depletion load.

6-23   What effect does decreasing the aspect ratio W/L of the load transistor have on inverter performance?

6-24   Define (*a*) rise time; (*b*) fall time; (*c*) propagation delay.

6-25   Sketch a 2-input NMOS NOR gate and verify that it satisfies the boolean NOR equation.

6-26   Repeat Rev. 6-25 for a 2-input NAND gate.

6-27   Repeat Rev. 6-25 for an AND-OR-INVERT.

6-28   What is meant by wired logic?

6-29   (*a*)   Sketch the circuit of a CMOS inverter.
        (*b*)   Verify that this configuration satisfies the NOT operation.

6-30   Sketch the circuit of a 2-input NAND CMOS gate and verify that it satisfies the boolean NAND equation.

6-31   List five desirable properties of CMOS gates.

6-32   (*a*)   Sketch the circuit of a transmission gate by using CMOS transistors.
        (*b*)   Explain the operation of this switch.

6-33 Draw the circuit diagram of a TTL NAND gate and explain its operation.

6-34 Draw a totem-pole output buffer with a TTL gate. Explain its operation.

6-35 Explain the function of a 3-state TTL gate.

6-36 (*a*) Sketch a 2-input OR (and also NOR) ECL gate.
  (*b*) What parameters determine the noise margin?
  (*c*) Why are the two collector resistors unequal?
  (*d*) Explain why power line spikes are virtually nonexistent.

6-37 List and discuss at least four advantages and four disadvantages of the ECL gate.

6-38 Compare the relative merits of NMOS, CMOS, TTL, and ECL logic families.

Chapter 7
COMBINATORIAL DIGITAL CIRCUITS

Charles Babbage, in the mid-nineteenth century recognized that a digital system must contain a control (logic) unit, an arithmetic unit, and memory (the ability to store data) as well as appropriate input and output mechanisms. These units are also utilized in the organization of modern, electronic digital systems for computation, communication, and control.

Only a few types of basic networks are required to process the binary signals employed in digital systems. These elementary building blocks are used repeatedly in various topological configurations to realize specific functions. Control and binary arithmetic are realized by combinations of logic gates and, as emphasized in Sec. 6-3, all logic operations can be performed by a single type of gate (e.g., a NAND gate). A basic memory cell, the *FLIP-FLOP* (FF), can also be constructed from basic logic gates. Because it is a fundamental building block in sequential circuits, such as registers and counters, we treat FLIP-FLOPS in Chap. 8.

In this chapter we focus on circuits whose operation depends on combinatorial logic. The number of functions that must be performed is not large and includes binary addition, binary multiplication, data selection (multiplexing), and decoding (demultiplexing). Integrated-circuit (IC) manufacturers package single-chip circuits and subsystems which realize these functions (MSI) as well as packages containing several gates (SSI). These SSI and MSI circuits are usually available in all three technologies, TTL, CMOS, and ECL.

7-1 STANDARD GATE ASSEMBLIES The fundamental gates described in Chap. 6 are used in large numbers even in a relatively simple digital system. Consequently, it is more efficient and convenient to construct several (or many) gates on a single chip rather than package gates individually. The following list of standard digital (SSI) IC components is typical, but far from exhaustive:

Quad 2-input NAND	Quad 2-input NOR
Triple 3-input NAND	Quad 2-input OR
Dual 4-input NAND	Quad 2-input exclusive-NOR
Single 8-input NAND	Quad 2-input exclusive-OR
Dual 4-input AND	Triple 3-input NOR
Triple 3-input AND	Triple 3-input NOR
Hex inverter	Dual 4-input NOR
Dual 2-wide, 2-input AOI	Triple 3-input OR
Single 2-wide, 4-input AOI	Single 8-input NOR
Single, 4-wide, 4-2-3-2-input AOI	Hex 2-input OR
Single, 4-wide, 2-input AOI	Triple 4-input OR
Single 4-wide, 2-2-3-2 input AOI	Triple 4-input NOR

These combinations are available in most logic families (TTL, CMOS, etc.) listed in Sec. 6-15. The limitation on the number of gates per chip is usually set by the number of pins available. The most common package is the *dual-in-line* (plastic or ceramic) package (DIP), which has 14 leads, 7 brought out to each side of the IC (Fig. 7-1*c*). The dimensions of the assembly, which is much larger than the chip size, are approximately 2 by 0.75 by 0.5 cm. A schematic of the triple 3-input NAND is shown in Fig. 7-1*a*. Note that there are $3 \times 3 = 9$ input leads, 3 output leads, a power-supply lead, and a ground lead; a total of 14 leads is used.

In Fig. 7-1*b* is indicated the dual 2-wide, 2-input AOI (AND-OR INVERT). This combination needs 4 input leads and 1 output lead per AOI, or 10 for the dual array. If 1 power-supply lead and 1 ground lead are added, we see that 12 of

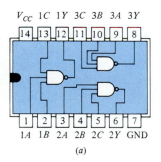

(*a*)

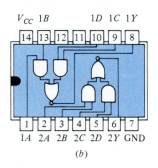

(*b*)

FIGURE 7-1

The load connections (top view) of (*a*) the triple 3-input NAND gate. (*b*) A dual 2-wide, 2-input, AND-OR-invert gate. (*c*) A dual-in-line package (DIP).

(*c*)

FIGURE 7-2
A TTL AND-OR-INVERT
gate.

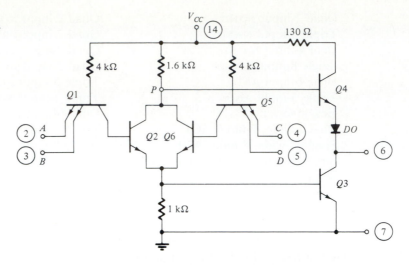

FIGURE 7-2
A TTL AND-OR-INVERT
gate.

the 14 available pins are used. The term "2-wide" indicates the number of AND gates that feed into the OR gate and refers to a two-level logic system.

The circuit diagram for this AOI gate is given in Fig. 7-2, implemented in TTL logic. The operation of this network should be clear from the discussion in Chap. 6. Thus $Q1$ and the input to $Q2$ (corresponding to the similarly numbered transistors in Fig. 6-37) constitute an AND gate. The identical arrangement of $Q5$ and $Q6$ constitutes a second AND gate. Since the collectors of $Q2$ and $Q6$ are tied together at P, the output at this node corresponds to either the inputs 2 AND 3 OR 4 AND 5. Also, because of the inversion through a transistor, the NOT operation is performed at P. The result is AND-OR-INVERT (AOI) logic $(\overline{AB + CD})$. Finally, note that $Q3$, DO, and $Q4$ form the totem-pole output stage of Fig. 6-38.

An alternative way of analyzing the circuit in Fig. 7-2 is to consider $Q1$ and $Q2$ (with the output at P) to constitute a NAND circuit. Similarly, $Q5$ and $Q6$ form a second NAND gate. The outputs of these two NAND configurations are short-circuited together by the lead connecting the collectors of $Q2$ and $Q6$ to form a wired-AND. Hence the output at P is, using De Morgan's law [Eq. (6-25)]

$$\overline{(AB)}\ \overline{(CD)} = \overline{AB + CD}$$

which confirms that AOI logic is performed.

Wired logic refers to the ability to perform additional logic operations by connecting the outputs of several gates together. The gate in Fig. 7-2 illustrates wired-AND logic. The NMOS AOI gate in Fig. 7-3 depicts a wired-OR configuration. The drivers $Q1$, $Q2$ and $Q3$, $Q4$ can be considered as performing the AND operation. Each of these configurations in conjunction with their depletion NMOS loads, $Q5$ and $Q6$, act as inverters. The connection between the respective outputs performs the OR operation. Note that only one depletion

FIGURE 7-3
An NMOS AND-OR-IN-VERT gate.

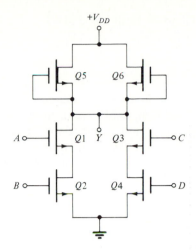

load transistor need be used as the interconnection between the gates places $Q5$ and $Q6$ in parallel.

Some of the more complicated functions to be described in this book require in excess of 14 pins, and these ICs are packaged with 16, 20, 24, and up to 64 leads.

The standard combinations considered in this section are examples of *small-scale integration* (SSI). Less than about 12 gates ($\sim$ 100 components) on a chip is considered SSI. The FLIP-FLOPS discussed in Sec. 8-3 are also SSI packages. Most other functions discussed in this chapter are examples of *medium-scale integration* (MSI), defined to have more than 12, but less than 100, gates per chip. The memories of Sec. 7-9 and the arrays in this and Chap. 9 may contain in excess of 100 gates ($>$ 1000 components) and are defined as *large-scale integration* (LSI). Many memory and signal-processing chips contain more than 10,000 components and are referred to as *very large-scale integration* (VLSI).

Design Philosophy   An electronics engineer should design a system so as to use standard ICs for as many subsystems as possible and must attempt to minimize the required number of packages (and hence the total cost). A single MSI is used in place of a number of SSI chips which could perform the same function. Similarly, an LSI package is used in the system wherever this IC can replace several MSI chips. In summary, in designing a digital system, it should be defined in terms of standard MSI and LSI packages. Discrete gates (SSI) should be used only for "interfaces" (also called the "glue") which may be required between the subsystem ICs.

A list of manufacturers of ICs is given in App. B-1. These companies have available data books, handbooks, and application notes which are invaluable to the system designer for keeping up to date on new packages and applications as they become available. The most important functions performed by MSI chips are given in this and Chap. 8. Both LSI and VLSI packages are discussed in Chap. 9. The most versatile LSI system is the microcomputer, a programmable *computer on a chip*.

7-2 BINARY ADDERS

A digital computer must obviously contain circuits which will perform arithmetic operations, i.e., addition, subtraction, multiplication, and division. The basic operations are addition and subtraction, since multiplication is essentially repeated addition, and division is essentially repeated subtraction.

Suppose we wish to sum two numbers in decimal arithmetic and obtain, say, the hundreds digit. We must add together not only the hundreds digit of each number but also a carry from the tens digit (if one exists). Similarly, in binary arithmetic we must add not only the digit of like significance of the two numbers to be summed, but also the carry bit (should one be present) of the next lower significant digit. This operation may be carried out in two steps: first, add the 2 bits corresponding to the 2^n digit, and then add the resultant to the carry from the 2^{n-1} digit. A 2-input adder is called a *half-adder,* because to complete an addition requires two such half-adders.

We show how a *half-adder* is constructed from the basic logic gates. A half-adder has two inputs—A and B—representing the bits to be added, and two outputs—D (for the digit of the same significance as A and B represent) and C (for the carry bit).

Half-Adder

The symbol for a half-adder is given in Fig. 7-4a, and the truth table in Fig. 7-4b. Note that the D column gives the sum of A and B as long as the sum can be represented by a single digit. When, however, the sum is larger than can be represented by a single digit, D gives the digit in the result which is of the same significance as the individual digits being added. Thus, in the first three rows of the truth table, D gives the sum of A and B directly. Since the decimal equation "1 plus 1 equals 2" is written in binary form as "01 plus 01 equals

FIGURE 7-4
(a) The symbol for a half-adder. (b) The truth table for the digit D and the carry C bits. (c) Implementation with standard logic blocks.

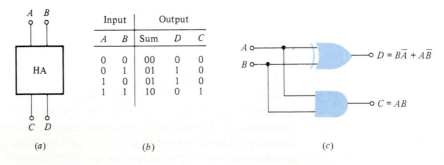

Input		Output		
A	B	Sum	D	C
0	0	00	0	0
0	1	01	1	0
1	0	01	1	0
1	1	10	0	1

$D = \overline{B}A + A\overline{B}$

$C = AB$

(a)   (b)   (c)

FIGURE 7-5
(a) The symbol for a full adder. (b) A 4-bit parallel binary adder using cascaded full adders.

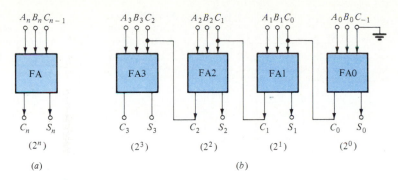

(a)

(b)

10," then in the last row $D = 0$. Because a 1 must now be carried to the place of next higher significance, $C = 1$.

From Fig. 7-4b we see that D obeys the exclusive-OR function and C follows the logic of an AND gate. These functions are indicated in Fig. 7-4c, and may be implemented in many different ways with the circuitry discussed in Chap. 6. For example, the exclusive-OR gate can be constructed with any of the four topologies of Sec. 6-15 and in any of the logic families in Table 6-4. The configuration in Fig. 6-8b ($Y = A\overline{B} + B\overline{A}$) is implemented in TTL logic with the AOI circuit of Fig. 7-2. The inverter for B (or A) is a single-input NAND gate. Since Y has an AND-OR (rather than an AND-OR-INVERT) topology, a transistor inverter is placed between node P and the base of $Q4$ in Fig. 7-2.

Parallel Operation   Two multidigit numbers may be added serially (one column at a time) or in parallel (all columns simultaneously). Consider parallel operation first. For an N-digit binary number there are (in addition to a common ground) N signal leads in the computer for each number. The nth line for number A (or B) is excited by A_n (or B_n), the bit for the 2^n digit ($n = 0, 1, \ldots, N - 1$).

Full Adder   In IC implementation, addition is performed using a complete adder, which (for reasons of economy of components) is not constructed from two half-adders. The symbol for the nth full adder (FA) is indicated in Fig. 7-5a. The circuit has three inputs: the addend A_n, the augend B_n, and the input carry C_{n-1} (from the next lower bit). The outputs are the sum S_n (sometimes designated Σ_n) and the output carry C_n. A parallel 4-bit adder is indicated in Fig. 7-5b. Since FA0 represents the least-significant bit (LSB), it has no input carry; hence $C_{-1} = 0$.

The circuitry within the block FA may be determined from Fig. 7-6, which is the truth table for adding 3 binary bits. From this table we can verify that the boolean expressions for S_n and C_n are given by

$$S_n = \overline{A}_n\overline{B}_nC_{n-1} + \overline{A}_nB_n\overline{C}_{n-1} + A_n\overline{B}_nC_{n-1} + A_nB_nC_{n-1} \tag{7-1}$$

$$C_n = \overline{A}_nB_nC_{n-1} + A_n\overline{B}_nC_{n-1} + A_nB_n\overline{C}_{n-1} + A_nB_nC_{n-1} \tag{7-2}$$

	Inputs			Outputs	
Line	A_n	B_n	C_{n-1}	S_n	C_n
0	0	0	0	0	0
1	0	0	1	1	0
2	0	1	0	1	0
3	0	1	1	0	1
4	1	0	0	1	0
5	1	0	1	0	1
6	1	1	0	0	1
7	1	1	1	1	1

FIGURE 7-6

Truth table for a 3-bit adder: C_{n-1} is the least-significant bit (LSB), and A_n is the most-significant bit (MSB).

Note that the first term of S_n corresponds to line 1 of the table, the second term to line 2, the third term to line 4, and the last term to line 7. (These are the only rows where $S_n = 1$.) Similarly, the first term of C_n corresponds to line 3 (where $C_n = 1$), the second term to line 5, and so on.

The AND operation ABC is sometimes called the *product of A and B and C*. Also, the OR operation $+$ is referred to as *summation*. Hence expressions such as those in Eqs. (7-1) and (7-2) represent a *boolean sum of products*. Such an equation is said to be in a *standard*, or *canonical, form*, and each term in the equation is called a *minterm*. A minterm contains the product of all boolean variables, or their complements.

The expression for C_n can be simplified considerably as follows: Since $Y + Y + Y = Y$, then Eq. (7-2), with $Y = A_n B_n C_{n-1}$, becomes

$$C_n = (\overline{A}_n B_n C_{n-1} + A_n B_n C_{n-1}) + (A_n \overline{B}_n C_{n-1} + A_n B_n C_{n-1})$$

$$+ (A_n B_n \overline{C}_{n-1} + A_n B_n C_{n-1}) \qquad (7\text{-}3)$$

Since $\overline{X} + X = 1$ where $X = A_n$ for the first parentheses, $X = B_n$ for the second parentheses, and $X = C_{n-1}$ for the third parentheses, then Eq. (7-3) reduces to

$$C_n = B_n C_{n-1} + C_{n-1} A_n + A_n B_n \qquad (7\text{-}4)$$

This expression could have been written down directly from the truth table in Fig. 7-6 by noting that $C_n = 1$ if and only if at least two out of the three inputs is 1.

It is interesting to note that if all 1s are changed to 0s and all 0s to 1s, then lines 0 and 7 are interchanged, as are 1 and 6, 2 and 5, and also 3 and 4. Because this switching of 1s and 0s leaves the truth table unchanged, whatever logic is represented by Fig. 7-6 is equally valid if all inputs and outputs are complemented. Therefore Eq. (7-3) is true if all variables are negated, or

$$\overline{C}_n = \overline{B}_n \overline{C}_{n-1} + \overline{C}_{n-1} \overline{A}_n + \overline{A}_n \overline{B}_n \qquad (7\text{-}5)$$

This same result is obtained (Prob. 7-2) by boolean manipulation of Eq. (7-4).

By evaluating $D_n \equiv (A_n + B_n + C_{n-1})\overline{C}_n$ and comparing the result with Eq. (7-1), we find that $S_n \equiv D_n + A_n B_n C_{n-1}$, or

$$S_n = A_n\overline{C}_n + B_n\overline{C}_n + C_{n-1}\overline{C}_n + A_n B_n C_{n-1} \qquad (7\text{-}6)$$

Equations (7-4) and (7-6) are implemented in Fig. 7-7 using AOI gates of the type shown in Figs. 7-2 and 7-3.

MSI Adders

There are commercially available 1-bit, 2-bit, and 4-bit full adders, each in one package. In Fig. 7-8 is indicated the logic topology for 2-bit addition. The inputs to the first stage are A_0 and B_0; the input marked C_{-1} is grounded. The output is the sum S_0. The carry C_0 is connected internally and is not brought to an output pin. This 2^0 stage (LSB) is identical with that in Fig. 7-7 with $n = 0$.

Since the carry from the first stage is C_0, it should be negated before it is fed to the 2^1 stage. However, the delay introduced by this inversion is undesirable, because the limitation upon the maximum speed of operation is the propagation delay (Sec. 6-15) of the carry through all the bits in the adder. The NOT-gate delay is eliminated completely in the carry by connecting $\overline{C}_0$ directly to the following stage and by complementing the inputs A_1 and B_1 before feeding these to this stage. This latter method is used in Fig. 7-8. Note that now the outputs S_1 and C_1 are obtained directly without requiring inverters. The logic followed by this second stage for the carry is given by Eq. (7-5), and for the sum by the modified form of Eq. (7-6), where each symbol is replaced by its complement.

FIGURE 7-7
Logic implementation of the nth stage of a full adder.

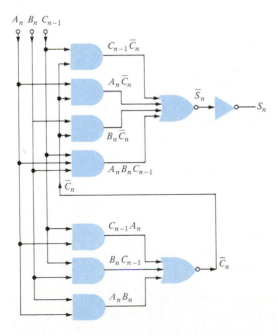

FIGURE 7-8
Logic diagram of an integrated 2-bit full adder.

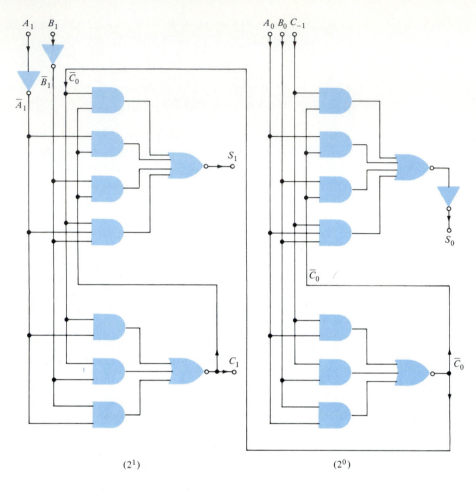

In a 4-bit adder C_1 is not brought out but is internally connected to the third stage, which is identical with the first stage. Similarly, the fourth and second stages have identical logic topologies. A 4-bit adder requires a 16-pin package: 8 inputs, 4 sum outputs, a carry output, a carry input, the power-supply input, and ground. The carry input is needed only if two arithmetic units are cascaded; for example, cascading a 2-bit with a 4-bit adder gives the sum of two 6-bit numbers. If the 2-bit unit is used for the 2^4 and 2^5 digits, then 4 must be added to all the subscripts in Fig. 7-8. For example, C_{-1} is now called C_3 and is obtained from the output carry of the 4-bit adder.

The MSI chip (54LS283) for a 4-bit binary full adder contains over 200 components (resistors, diodes, or transistors). The propagation delay time from data-in to data-out is typically 16 ns, and the power dissipation is 190 mW.

Serial Operation In a serial adder the inputs A and B are synchronous pulse trains on two lines in the computer. Figure 7-9a and b shows typical pulse trains representing, respectively, the decimal numbers 13 and 11. Pulse trains representing the sum

FIGURE 7-9
(*a*, *b*) Pulse waveforms
representing numbers B
and A. (*c*, *d*) Wave-
forms for sum and dif-
ference.

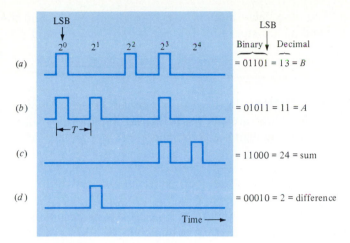

(24) and difference (2) are shown in Fig. 7-9*c* and *d*, respectively. A serial *adder* is a device which will take as inputs the two waveforms of Fig. 7-9*a* and *b* and deliver the output waveform in Fig. 7-9*c*. Similarly, a *subtractor* (Sec. 7-3) will yield the output shown in Fig. 7-9*d*.

We have already emphasized that the sum of two multidigit numbers may be formed by adding to the sum of the digits of like significance the carry (if any) which may have resulted from the next lower place. With respect to the pulse trains of Fig. 7-9, the above statement is equivalent to saying that, at any instant of time, we must add (in binary form) to the pulses A and B the carry pulse (if any) which comes from the resultant formed one period T earlier. The logic outlined above is performed by the full-adder circuit of Fig. 7-10. This circuit differs from the configuration in the parallel adder of Fig. 7-5 by the inclusion of a time delay TD which is equal to the time T between pulses. Hence the carry pulse is delayed a time T and added to the digit pulses in A and B, exactly as it should be.

A comparison of Figs. 7-5 and 7-10 indicates that parallel addition is faster than serial because all digits are added simultaneously in the former, but in sequence in the latter. However, whereas only one full adder is needed for serial arithmetic, we must use a full adder for each bit in parallel addition. Hence parallel addition is much more expensive than serial operation.

FIGURE 7-10
A serial full adder.

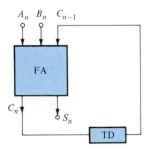

The time delay unit TD is a type D FLIP-FLOP, and the serial numbers A_n, B_n, and S_n are stored in *shift registers* (Secs. 8-4 and 8-5).

7-3 ARITHMETIC FUNCTIONS

In this and the next two sections other arithmetic units besides the adder are discussed. These include the subtractor, the ALU, the multiplier, the digital comparator, and the parity checker.

Binary Subtraction   The process of subtraction (B minus A) is equivalent to addition if the complement $\overline{A}$, called the *one's complement*, of the subtrahend is used. To justify this statement consider the following argument (applied specifically to a 4-bit number). The NOT function changes a 1 to a 0, and vice versa. Therefore[1]

$$A \text{ plus } \overline{A} = 1111$$

and

$$A \text{ plus } \overline{A} \text{ plus } 1 = 1111 \text{ plus } 0001 = 10000$$

so that

$$A = 10000 \text{ minus } \overline{A} \text{ minus } 1$$

Finally

$$B \text{ minus } A = (B \text{ plus } \overline{A} \text{ plus } 1) \text{ minus } 10000 \qquad (7\text{-}7)$$

This equation indicates that to subtract a 4-bit number A from a 4-bit number B it is only required to add B, $\overline{A}$, and 1 (a 2^0 bit). The operation B minus A must yield a 4-bit answer. The term "minus 10000" in Eq. (7-7) infers that the addition (B plus $\overline{A}$ plus 1) results in a fifth bit, which must be ignored. The addition of 1 to the one's complement of a binary number forms the two's complement of the number. Thus, for a binary number B, $\overline{B}$ plus 1 is the two's complement representation of B.

Example 7-1

Solution

Verify Eq. (7-7) for $B = 1100$ and $A = 1001$ (decimal 12 and 9).

$$B \text{ plus } \overline{A} \text{ plus } 1 = 1100 \text{ plus } 0110 \text{ plus } 0001 = 10011$$

The four (less significant) bits 0011 represent decimal 3 and the fifth bit 1 is a generated carry. Since, in decimal notation, B minus $A = 12 - 9 = 3$, then the correct answer is obtained by evaluating the sum in the parentheses of Eq. (7-7), provided that the carry is ignored.

---

[1]To avoid confusion with the OR operation, the word "plus" (minus) is used in place of $+(-)$ in the following equations.

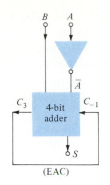

FIGURE 7-11
A simplified 4-bit parallel adder used as a subtractor.

In Eq. (7-7) the 1 in 10000 is the output carry $C_3 = 1$ from the 4-bit adder, and may be used to supply the 1 which must be added to $\overline{A}$. This bit is called the *end-around carry* (EAC) because this carry out is fed back to the carry input C_{-1} (Fig. 7-7) of the least-significant bit of A. This process of subtraction by means of a 4-bit parallel adder is indicated schematically in Fig. 7-11.

The two's complement method of subtraction just described is valid only if B is greater than A, so that a positive difference results and a carry is generated from (B plus $\overline{A}$ plus 1). If B is less than A, then the most-significant bit (MSB) of B (which differs from the corresponding bit of A) is 0 and that of A is 1. Since $\overline{A} = 0$, the MSB of (B plus $\overline{A}$) is 0. Hence no carry results from the sum (B plus $\overline{A}$ plus 0001), and the method indicated in Fig. 7-11 must be modified. We now demonstrate that if no carry results in the system of Fig. 7-11, the correct answer for B minus A is negative, and is obtained by forming the sum (B plus $\overline{A}$) and by complementing the sum digits S_0, S_1, S_2, and S_3. From Eq. (7-7)

$$B \text{ minus } A = (B \text{ plus } \overline{A}) \text{ minus } 1111$$

$$= \text{ minus } [1111 \text{ minus}(B \text{ plus } \overline{A})]$$

$$= \text{ minus}(\overline{B \text{ plus } \overline{A}}) \tag{7-8}$$

because 1111 minus a 4-bit binary number is the complement of the number. *In summary:* to subtract A from B form the sum (B plus $\overline{A}$) and check to see if a carry exists. If it does, the difference (B minus A) is positive and is given by S in Fig. 7-11. However, if the carry is zero, then the difference is negative and is given by $\overline{S}$. The circuits for testing for a carry and for obtaining the complement of S when the *EAC* is missing are not shown in Fig. 7-11.

An alternative method for representing negative binary numbers is to use an additional bit called the *sign bit*. To illustrate the use of the sign bit, the positive decimal number 46 is written in binary as $0_\wedge 0101110$, whereas negative 46 is $1_\wedge 0101110$. The sign bit is the digit to the left of the caret ($\wedge$); as indicated the sign bit is zero for positive numbers and is one for negative numbers. The magnitude of the number is represented by the binary digits to the right of the sign bit, with the most-significant bit (MSB) just to the right of the caret. The student should be cognizant of the fact that a caret is not always used to separate the sign bit from the magnitude of the binary number.

Arithmetic Logic Unit (ALU)/Function Generator Subtraction may be accomplished by using an ALU such as the 74AS-181A (or 74AS-881A). Using four (or three) function select lines, the following operations can be performed on two 4-bit numbers: B minus A, A minus B, A plus B, $A \oplus B$, $A + B$, AB, $A = B$, $A > B$, and a number of other arithmetic and logic operations. These MSI packages (24 pins) have the complexity of 85 equivalent gates ($\sim$ 800 components).

Binary Multipliers The 16-pin 74LS261 package is used to perform parallel multiplication of 4 × 4 bits and to produce a 4-bit output. Used in conjunction with the 74LS284 chip an 8-bit product is obtained in about 40 ns. The same operation can be obtained with a single high-complexity 20-pin 74AS274 chip. Note that a second chip is needed to obtain the 8 bits that result from a 4 × 4-bit multiplication. Without the second chip, only the four MSBs of the product are obtained.

7-4 DIGITAL COMPARATOR

It is sometimes necessary to know whether a binary number A is greater than, equal to, or less than another number B. The system for making this determination is called a *magnitude digital* (or *binary*) *comparator*. Consider single bit numbers first. As mentioned in Sec. 6-3, the exclusive-NOR gate is an *equality detector* because

$$E = \overline{A\overline{B} + \overline{A}B} = \begin{cases} 1 & A = B \\ 0 & A \neq B \end{cases} \tag{7-9}$$

The condition $A > B$ is given by

$$C = A\overline{B} = 1 \tag{7-10}$$

because if $A > B$, then $A = 1$ and $B = 0$, so that $C = 1$. On the other hand, if $A = B$ or $A < B$ ($A = 0$, $B = 1$), then $C = 0$.

Similarly, the restriction $A < B$ is determined from

$$D = \overline{A}B = 1 \tag{7-11}$$

The logic block diagram for the nth bit drawn in Fig. 7-12 has all three desired outputs C_n, D_n, and E_n. It consists of two inverters, two AND gates, and the AOI circuit of Fig. 7-2. Alternatively, Fig. 7-12 may be considered to consist of an exclusive-NOR and two AND gates. (Note that the outputs of the AND gates in the AOI block of Fig. 7-2 are not available, and hence additional AND gates must be fabricated to give C_n and D_n.)

FIGURE 7-12
A 1-bit digital comparator.

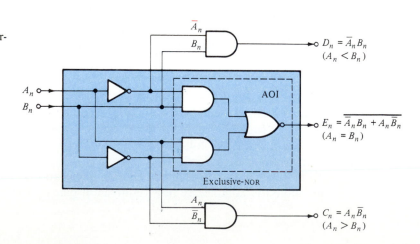

FIGURE 7-13
A 4-bit magnitude comparator. Assume $C' = 0$ and $E' = 1$. If $E = 1$, then $A = B$ and if $C = 1$, $A > B$. If $D = 1$, then $A < B$, where D has the same topology as C but with A and B interchanged.

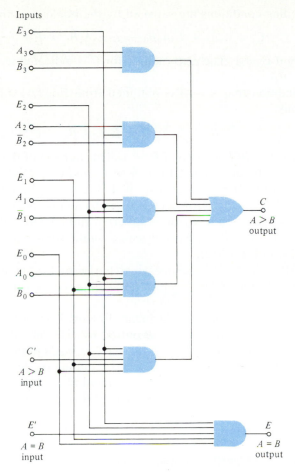

Consider now a 4-bit comparator. $A = B$ requires that

$$A_3 = B_3 \quad \text{and} \quad A_2 = B_2 \quad \text{and} \quad A_1 = B_1 \quad \text{and} \quad A_0 = B_0$$

Hence the AND gate E in Fig. 7-13 described by

$$E = E_3 E_2 E_1 E_0 \tag{7-12}$$

implies $A = B$ if $E = 1$ and $A \neq B$ if $E = 0$. (Assume that the input E' is held high; $E' = 1$.)

The inequality $A > B$ requires that

$$A_3 > B_3 \quad \text{(MSB)}$$

or $A_3 = B_3 \quad$ and $\quad A_2 > B_2$

or $A_3 = B_3 \quad$ and $\quad A_2 = B_2 \quad$ and $\quad A_1 > B_1$

or $A_3 = B_3 \quad$ and $\quad A_2 = B_2 \quad$ and $\quad A_1 = B_1 \quad$ and $\quad A_0 > B_0$

The preceding conditions are satisfied by the boolean expression

$$C = A_3\overline{B}_3 + E_3A_2\overline{B}_2 + E_3E_2A_1\overline{B}_1 + E_3E_2E_1A_0\overline{B}_0 \qquad (7\text{-}13)$$

if and only if $C = 1$. The AND-OR gate for C is indicated in Fig. 7-13. (Assume that $C' = 0$.)

The condition that $A > B$ is obtained from Eq. (7-13) by interchanging A and B. Thus

$$D = \overline{A}_3B_3 + E_3\overline{A}_2B_2 + E_3E_2\overline{A}_1B_1 + E_3E_2E_1\overline{A}_0B_0 \qquad (7\text{-}14)$$

implies that $A < B$ if and only if $D = 1$. This portion of the system is obtained from Fig. 7-13 by changing A to B, B to A, and C to D. Alternatively, D may be obtained from $D = \overline{EC}$ because, if $A \neq B$ ($E = 0$) and if $A \not> B$ ($C = 0$), then $A < B$ ($D = 1$). However, this implementation for D introduces the additional propagation delay of an inverter and an AND gate. Hence the logic indicated in Eq. (7-14) for D is fabricated on the same chip as that for C in Eq. (7-13) and E in Eq. (7-12).

The 74HC85 is an MSI package which performs 4-bit-magnitude comparison. If numbers of greater length are to be compared, several such units can be cascaded. Consider an 8-bit comparator. Designate the $A = B$ output terminal of the stage handling the less significant bits by E_L, the $A > B$ output terminal of this stage by C_L, and the $A < B$ output by D_L. Then the connections $E' = E_L$, $C' = C_L$, and $D' = D_L$ (Fig. 7-13) must be made to the stage with the more significant bits (Prob. 7-8). For the stage handling the less significant bits, the outputs C' and D' are grounded ($C' = 0$ and $D' = 0$) and the input E' is tied to the supply voltage ($E' = 1$). Why? The 74HC688 is one such 8-bit-magnitude comparator.

7-5 PARITY CHECKER-GENERATOR Another arithmetic operation that is often invoked in a digital system is determination of whether the sum of the binary bits in a word is odd (called *odd parity*) or even (designated *even parity*). The output of an exclusive-OR gate is 1 if and only if one input is 1 and the other is 0. Alternatively stated, the output is 1 if the sum of the digits is 1. An extension of this concept to the exclusive-OR tree in Fig. 7-14 leads to the conclusion that $Z = 1$ (or $Y = 0$) if the sum of the input bits A, B, C, and D is odd. Hence,

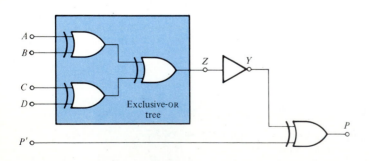

FIGURE 7-14

The odd-parity checker, or parity-bit generator for a 4-bit input word. Assume $P' = 0$; then $P = 0$ (1) represents odd (even) parity.

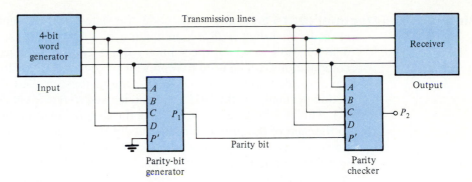

FIGURE 7-15
Binary data is tested by generating a parity bit at the input to a line and checking the parity of the transmitted bits plus the generated bit at the receiving end of the system.

if the input P' is grounded ($P' = 0$), then $P = 0$ for odd parity and $P = 1$ for even parity.

The system of Fig. 7-14 is not only a parity checker, but it may also be used to generate a parity bit P. Independently of the parity of the 4-bit input word, the parity of the 5-bit code A, B, C, D, and P is odd. This statement follows from the fact that if the sum of A, B, C, and D is odd (even), then P is 0 (1), and therefore the sum of A, B, C, D, and P is always odd.

The use of a parity code is an effective way of increasing the reliability of transmission of binary information. As indicated in Fig. 7-15, a parity bit P_1 is generated and transmitted along with the N-bit input word. At the receiver the parity of the augmented ($N + 1$)-bit signal is tested, and if the output P_2 of the checker is 0, it is assumed that no error has been made in transmitting the message, whereas $P_2 = 1$ is an indication that (say, due to noise) the received word is in error. Note that only errors in an odd number of digits can be detected with a single parity check.

An MSI 9-bit parity generator-checker is available (74HC280) with control inputs so that it may be used in either odd- or even-parity applications. For words of length greater than 8 bits, several such units may be cascaded (Prob. 7-14).

The MSI unit 74HC386 contains four 2-input exclusive-OR gates.

7-6 DECODER-DEMULTIPLEXER In a digital system, instructions as well as numbers are conveyed by means of binary levels or pulse trains. If, say, 4 bits of a character are set aside to convey instructions, then 16 different instructions are possible. This information is *coded* in binary form. Frequently a need arises for a multiposition switch which may be operated in accordance with this code. In other words, for each of the 16 codes, one and only one line is to be excited. This process of identifying a particular code is called *decoding*.

Binary-Coded-Decimal (BCD) System   This code translates decimal numbers by replacing each decimal digit with a combination of 4 binary digits. Since there are 16 distinct ways in which the 4 binary digits can be arranged in a row, any 10 combinations can be used to represent the decimal digits from 0 to 9. Thus we have a wide choice of BCD codes. One of these, called the "natural binary-coded-decimal," is the 8421 code and is used to represent decimal 264 in Table 7-1. This is a weighted code because the decimal digit in the 8421 code is equal to the sum of the products of the bits in the coded word times the successive powers of two starting from the right (LSB). We need N 4-bit sets to represent in BCD notation an N-digit decimal number. The first 4-bit set on the right represents units, the second represents tens, the third hundreds, and so on. For example, the decimal number 264 requires three 4-bit sets, as shown in Table 7-1. Note that this 3-decade BCD code can represent any number between 0 and 999; hence it has a resolution of 1 part in 1000, or 0.1 percent. It requires 12 bits, which in a straight binary code can resolve 1 part in $2^{12} = 4096$, or 0.025 percent.

TABLE 7-1   BCD Representation for the Decimal Number 264

Weighting factor	800	400	200	100	80	40	20	10	8	4	2	1
BCD code	0	0	1	0	0	1	1	0	0	1	0	0
Decimal digits			2				6				4	

BCD-to-Decimal Decoder   Suppose we wish to decode a BCD instruction representing one decimal digit, say 5. This operation may be carried out with a 4-input AND gate excited by the 4 BCD bits. For example, the output of the AND gate in Fig. 7-16 is 1 if and only if the BCD inputs are $A = 1$ (LSB), $B = 0$, $C = 1$, and $D = 0$. Since this code represents the decimal number 5, the output is labeled "line 5."

A BCD-to-decimal decoder is indicated in Fig. 7-17. This MSI unit (74HC42) has four inputs, A, B, C, and D, and 10 output lines. (Ignore the dashed lines, for the moment.) In addition, there must be a ground and a power-supply connection, and hence a 16-pin package is required. The complementary inputs $\overline{A}, \overline{B}, \overline{C}$, and $\overline{D}$ are obtained from inverters on the chip. Since NAND gates are used, an output is 0 (low) for the correct BCD code and is 1 (high) for any other (invalid) code. The system in Fig. 7-16 is also referred to as a "4-to-10-line decoder" designating that a 4-bit input code selects 1 of 10 output lines.

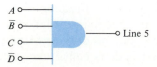

A
$\overline{B}$
C
$\overline{D}$
Line 5

FIGURE 7-16
The output is 1 if the BCD input is 0101 and is 0 for any other input instruction.

FIGURE 7-17
A BCD-to-decimal decoder.

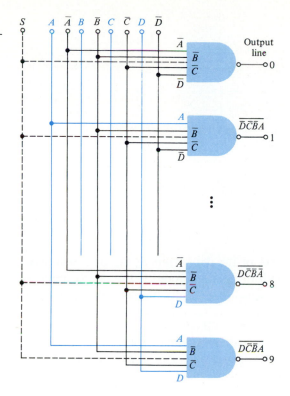

In other words, the decoder acts as a 10-position switch which responds to a BCD input instruction.

It is sometimes desired to decode only during certain intervals of time. In such applications an additional input, called a *strobe,* is added to each NAND gate. All strobe inputs are tied together and are excited by a binary signal S, as indicated by the dashed lines in Fig. 7-17. If $S = 1$, a gate is *enabled* and decoding takes place, whereas if $S = 0$, no coincidence is possible and decoding is inhibited. The strobe input can be used with a decoder having any number of inputs or outputs.

Demultiplexer

A *demultiplexer* is a system for transmitting a binary signal (serial data) on one of N lines, the particular line being selected by means of an address. A single-pole N-position rotary switch connected as in Fig. 7-18a is the mechanical analog of such a demultiplexer. The address determines the angle of rotation of the arm of the switch. A decoder is converted into a demultiplexer by means of the dashed connections in Fig. 7-17. If the data signal is applied at S, then the output will be the complement of this signal (because the output is 0 if all inputs are 1) and will appear only on the addressed line.

An enabling signal may be applied to a demultiplexer by cascading the system of Fig. 7-17 with that indicated in Fig. 7-19. If the *enable* input is 0, then S is

FIGURE 7-18

Mechanical analog of (a) demultiplexer and (b) multiplexer.

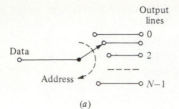

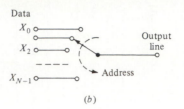

(a)

(b)

Data

Enable

o S

FIGURE 7-19

A decoder is converted to a demultiplexer (with an enabling input) if the S terminal in Fig. 7-17 is obtained from the AND gate.

the complement of the data. Hence, the data will appear (without inversion) on the line with the desired code. If the enable input is 1, $S = 0$, the data is inhibited from appearing on any line and all inputs remain at 1.

4-to-16-Line Decoder-Demultiplexer If an address corresponding to a decimal number in excess of 9 is applied to the inputs in Fig. 7-17, this instruction is rejected; that is, all 10 outputs remain at 1. If it is desired to select 1 of 16 output lines, the system is expanded by adding 6 more NAND gates and using all 16 codes possible with 4 binary bits.

The 74HC154 is a 4-to-16-line decoder-demultiplexer. It has 4 address lines, 16 output lines, two enable inputs, a ground pin, and a power-supply lead, so that a 24-pin package is required.

A dual 2-to-4-line (74HC139) and a 3-to-8-line (74HC138) decoder-demulti-plexer are also available in individual IC packages.

A 1-to-2-line demultiplexer is constructed from two 2-input NAND gates. The zero-output line comes from the NAND whose inputs are S and $\overline{A}$, whereas the one-output line is connected to the NAND whose inputs are S and A. The "address" A is called the *control* input because, if $A = 0(1)$, the complement of the data ($\overline{S}$) appear on line 0(1).

Decoder-Lamp Driver Some decoders are equipped with special output stages so that they can drive lamps such as the Burroughs Nixie tube. A Nixie indicator is a cold-cathode gas-discharge tube with a single anode and 10 cathodes, which are wires shaped in the form of numerals 0 to 9. These cathodes are connected to output lines 0 to 9, respectively, and the anode is tied to a fixed supply voltage. The decoder-lamp driver–Nixie indicator combination makes visible the dec-imal number corresponding to the BCD number applied. Thus, if the input is 0101, the numeral 5 will glow in the lamp.

A decoder for seven-segment numerals made visible by using light-emitting diodes is discussed in Sec. 7-11. These displays are quite common in calculators, clocks, and a variety of instruments.

Higher-Order Demultiplexers   If the number of output lines N exceeds 16, the demultiplexers discussed in the foregoing for $N = 16, 8, 4,$ or 2 are arranged in a "tree" formation to yield the desired number of output lines. For example, for $N = 32$, we can use a demultiplexer with the "trunk" $N_1 = 4$ and four "branches" $N_2 = 8$ as indicated in Fig. 7-20. Note that the total number of output lines is $N = N_1 N_2 = 32$. Lines 0 through 7 are decoded by demultiplexer N_{20}, whereas N_{21} decodes the next eight lines, and so on.

For $ED = 01$, lines 8 thru 15 are decoded in sequence as the address CBA changes from 000 to 001 to $\cdots$ to 111. For example, line 12 is decoded by the address $EDCBA = 01100$, which is the binary representation for decimal 12. Line 19 is decoded by $EDCBA = 10011$, etc. Since there are two 2-to-4-line decoders in one package, a total of $4\frac{1}{2}$ equivalent packages are needed for the system in Fig. 7-20. It is also possible to design this system with $N_1 = 8$, $N_2 = 4$ (Prob. 7-16) or with $N_1 = 2$, $N_2 = 16$, etc. The proper design choice is indicated by total cost.

A demultiplexer with 64 outputs can be designed with $N_1 = N_2 = 8$, for a total of 9 packages. Why? For very large values of N, higher-level branching is required (Prob. 7-17), where each output in Fig. 7-20 is an input to another demultiplexer.

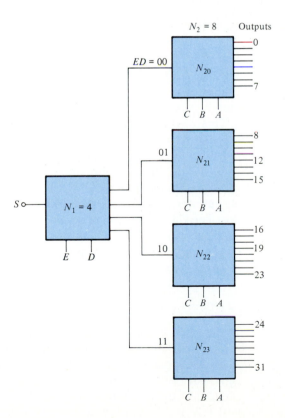

FIGURE 7-20
A 32-output demultiplexer tree where N_1 is a 4-output and N_2 is an 8-output demultiplexer.

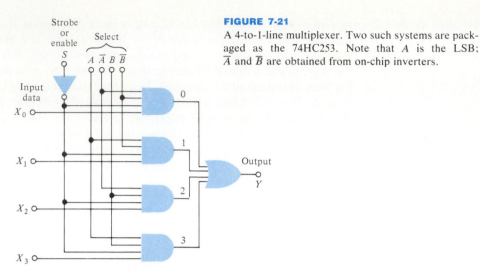

FIGURE 7-21
A 4-to-1-line multiplexer. Two such systems are packaged as the 74HC253. Note that A is the LSB;
$\overline{A}$ and $\overline{B}$ are obtained from on-chip inverters.

7-7 DATA SELECTOR-MULTIPLEXER The function performed by a *multiplexer* is to select 1 out of N input data sources and to transmit the selected data to a single information channel. The N-position switch connected as in Fig. 7-18*b* is the mechanical analog of a multiplexer. Compare Figs. 7-18*a* and 7-18*b*. Since in a demultiplexer there is only one input line and these data are caused to appear on 1 out of N output lines, a multiplexer performs the inverse process of a demultiplexer.

The demultiplexer in Fig. 7-17 is converted into a multiplexer by making the following two changes: (1) add a NAND gate whose inputs include all N outputs in Fig. 7-17, and (2) augment each NAND gate with an individual data input X_0, X_1, . . ., X_N. The logic system for a 4-to-1-line data selector-multiplexer is drawn in Fig. 7-21. This AND-OR logic is equivalent to the NAND-NAND logic as described in the above steps 1 and 2. Note that the same decoding config-uration is used in both the multiplexer and demultiplexer. If the select code is 01, then X_1 appears at the output Y, if the address is 11, then $Y = X_3$, etc., provided that the system is enabled ($S = 0$).

The following data selector-multiplexers are available: 16-to-1-line (74AS250), one per package; 8-to-1-line (74HC151A), one per package; 4-to-1-line (74HC253), two per package; and 2-to-1-line (74HC157), four per package. The 1-out-of-16 multiplexer is a 24-pin IC with 16 data inputs, a 4-bit select code, a strobe input, one output, a power-supply load, and a ground terminal. For this 16-to-1-line data selector, Fig. 7-21 is extended from four 4-input AND gates to sixteen 6-input AND gates.

Parallel-to-Serial Conversion Consider a 16-bit word available in parallel form so that X_0 represents the 2^0 bit, X_1 the 2^1 bit, etc. By means of a counter (Sec. 8-6), it is possible to change the select code so that it is 0000 for the first T seconds, 0001 for the next T seconds, 0010 for the third interval T, etc. With such excitation of the address, the output of the multiplexer will be X_0 for the first T seconds, X_1 for the next interval T, X_2 for the third period, etc. The output Y is a waveform which represents serially the binary data applied in parallel at the input. In other words, a parallel-to-serial conversion is accomplished of one 16-bit word. This process takes $16T$ seconds.

In a digital system, such as a computer, a data communication system, etc., a pulse train is often required for testing or control (gating) purposes. Such a *sequence generator* is obtained by means of the parallel-to-series converter. Any desired waveform may be obtained by properly choosing the input data X.

Sequential Data Selection   By changing the address with a counter in the manner indicated in the preceding paragraph, the operation of an electromechanical stepping switch is simulated. If the data inputs are pulse trains, this information will appear sequentially on the output channel: in other words, pulse train X_0 will appear for T seconds, followed by X_1 for the next T seconds, etc. If the number of data sources is M, then X_0 is again selected during the interval $MT < t < (M + 1)T$.

Higher-Order Multiplexers   If the number of input lines exceeds 16, then the logic block diagram assumes a topology which is the inverse to that shown in Fig. 7-20. For example, to select 1 out of 32 data inputs, the system in Fig. 7-22 may be used. Multiplexer N_{20} places the data inputs X_0 through X_7 in sequence onto line L_0 as the address

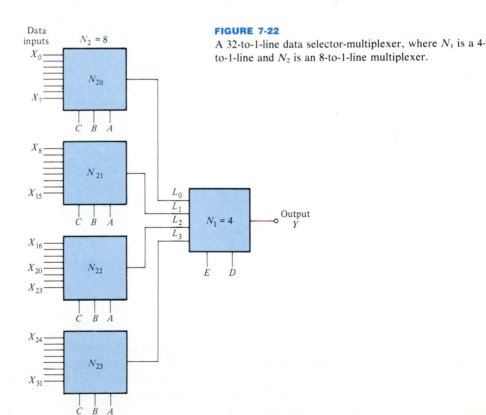

FIGURE 7-22
A 32-to-1-line data selector-multiplexer, where N_1 is a 4-to-1-line and N_2 is an 8-to-1-line multiplexer.

CBA changes from 000 to 001 to $\cdots$ to 111. Similarly N_{21} transmits the data X_8 through X_{15} onto line L_1 as CBA sequences from 000 through 111. Specifically, if the address is $CBA = 100$, then X_4 appears on L_0, X_{12} on L_1, X_{20} on L_2, and X_{28} on L_3. If it is desired that X_{20} be transmitted to the output, then ED must equal 10 so that N_1 will select the data on line L_2. In summary, for the address $EDCBA = 10100$, the multiplexer transfers the input data X_{20} to the output line Y. An alternative solution using two 16-data-input multiplexers to obtain 1 out of 32 data inputs is given in Prob. 7-20.

Note that the total number of input lines $N = N_2 N_1$. For very large values of N, a third level of multiplexers N_3 may be necessary. The outputs from N_3 are connected to the inputs of N_2, and the outputs from N_2 are the inputs to N_1. The system selects 1 out of $N = N_3 N_2 N_1$ inputs (Prob. 7-23).

Combinational Logic The boolean expression for the output Y of the multiplexer in Fig. 7-21 is

$$Y = X_0 \overline{BA} + X_1 \overline{B}A + X_2 B\overline{A} + X_3 BA \qquad (7\text{-}15)$$

As noted in Sec. 7-2, combinational logic of three variables is represented by a boolean sum of products of A, B, and C. Each minterm is of the form CBA or the complements of these variables. Hence, a multiplexer can satisfy any combinational-logic equation if the proper choices are made for the X inputs. Thus it may be required that $X = C$ or $X = \overline{C}$. If terms contain both C and $\overline{C}$, then $X = C + \overline{C} = 1$, and if a term is missing, then $X = 0$.

Example 7-2 Generate the following combinational-logic equation using a 4-input multiplexer:

$$Y = C\overline{BA} + \overline{C}\,\overline{B}A + C\overline{B}A + \overline{C}BA \qquad (7\text{-}16)$$

Solution Since $\overline{BA}$ represents decimal 0, the coefficient of $\overline{BA}$ is X_0. Hence $X_0 = C + \overline{C} = 1$. Since $\overline{B}A$ represents decimal 1, the factor multiplying $\overline{B}A$ is X_1. Hence $X_1 = C$. Since BA represents 3, $X_3 = \overline{C}$. Since $B\overline{A}$, which represents 2, is missing from the equation, $X_2 = 0$. In summary

$$X_0 = 1 \qquad X_1 = C \qquad X_2 = 0 \qquad \text{and} \qquad X_3 = \overline{C}$$

If these values of X are used in the multiplexer of Fig. 7-21, the output Y will equal the combinational logic in Eq. (7-16).

In the foregoing example, a standard boolean equation in three variables is generated with a 4-to-1-line multiplexer. In general, a combinational-logic equation in N variables is generated by using a 2^{N-1}-input data selector.

7-8 ENCODER A decoder is a system which accepts an M-bit word and establishes the state 1 on one (and only one) of 2^M output lines (Sec. 7-6). In other words, a decoder identifies (recognizes) a particular code. The inverse process is called *encoding*. An encoder has a number of inputs, only one of which is in the 1 state, and an N-bit code is *generated*, depending upon which of the inputs is excited.

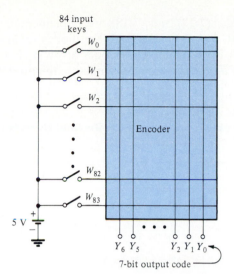

84 input keys

W_0
W_1
W_2

Encoder

W_{82}
W_{83}

5 V

$Y_6 \ Y_5$ $\quad$ $Y_2 \ Y_1 \ Y_0$

7-bit output code

FIGURE 7-23
A block diagram for an encoder which generates an output word for every character on a keyboard.

Consider, for example, that it is required that a binary code be transmitted with every stroke of an alphanumeric keyboard (a typewriter or teletype). There are 26 lowercase and 26 capital letters, 10 numerals, and about 22 special characters on such a keyboard so that the total number of codes necessary is approximately 84. This condition can be satisfied with a minimum of 7 bits ($2^7 = 128$, but $2^6 = 64$). Let us modify the keyboard so that, if a key is depressed, a switch is closed, thereby connecting a 5-V supply (corresponding to the 1 state) to an input line. A block diagram of such an encoder is indicated in Fig. 7-23. Inside the shaded block there is a rectangular array (or matrix) of wires, and we must determine how to interconnect these wires so as to generate the desired codes.

To illustrate the design procedure for constructing an encoder, let us simplify the above example by limiting the keyboard to only 10 keys, the numerals 0, 1, . . ., 9. A 4-bit output code is sufficient in this case, and let us choose BCD words for the output codes. The truth table defining this encoding is given in Table 7-2. Input W_n ($n = 0, 1, . . ., 9$) represents the nth key. When $W_n = 1$, key n is depressed. Since it is assumed that no more than one key is activated simultaneously, then in any row every input except one is a 0. From this truth table we conclude that $Y_0 = 1$ if $W_1 = 1$ or if $W_3 = 1$ or if $W_5 = 1$ or if $W_7 = 1$ or if $W_9 = 1$. Hence, in boolean notation,

$$Y_0 = W_1 + W_3 + W_5 + W_7 + W_9 \qquad (7\text{-}17)$$

Similarly

$$Y_1 = W_2 + W_3 + W_6 + W_7$$

$$Y_2 = W_4 + W_5 + W_6 + W_7 \qquad (7\text{-}18)$$

$$Y_3 = W_8 + W_9$$

The OR gates in Eqs. (7-17) and (7-18) are implemented with diodes in Fig. 7-24. (Compare with Fig. 2-13, but with the diodes reversed, because we are now considering positive logic.) An encoder array such as that in Fig. 7-24 is called a *rectangular diode matrix* and is similar to the programmable logic array (PLA) described in Sec. 7-15.

TABLE 7-2 Truth Table for Encoding the Decimal Numbers 0 to 9

Inputs										Outputs			
W_9	W_8	W_7	W_6	W_5	W_4	W_3	W_2	W_1	W_0	Y_3	Y_2	Y_1	Y_0
0	0	0	0	0	0	0	0	0	1	0	0	0	0
0	0	0	0	0	0	0	0	1	0	0	0	0	1
0	0	0	0	0	0	0	1	0	0	0	0	1	0
0	0	0	0	0	0	1	0	0	0	0	0	1	1
0	0	0	0	0	1	0	0	0	0	0	1	0	0
0	0	0	0	1	0	0	0	0	0	0	1	0	1
0	0	0	1	0	0	0	0	0	0	0	1	1	0
0	0	1	0	0	0	0	0	0	0	0	1	1	1
0	1	0	0	0	0	0	0	0	0	1	0	0	0
1	0	0	0	0	0	0	0	0	0	1	0	0	1

Incidentally, a decoder can also be constructed as a rectangular diode matrix (Prob. 7-29). This statement follows from the fact that a decoder consists of AND gates (Fig. 7-16), and it is possible to implement AND gates with diodes (Fig. 6-36).

Each diode of the encoder of Fig. 7-24 may be replaced by the base-emitter diode of a transistor. If the collector is tied to the supply voltage V_{CC}, then an emitter-follower OR gate results. Such a configuration is indicated in Fig. 7-25a for the output Y_2. Note that if either W_4 or W_5 or W_6 or W_7 is high, then the emitter-follower output is high, thus verifying that $Y_2 = W_4 + W_5 + W_6 + W_7$, as required by Eq. (7-18).

Only one transistor (with multiple emitters) is required for each encoder input. The base is tied to the input line, and each emitter is connected to a different output line, as dictated by the encoder logic. For example, since in Fig. 7-24 line W_7 is tied to three diodes whose cathodes go to Y_0, Y_1, and Y_2, then this combination may be replaced by the three-emitter transistor Q7 connected as in Fig. 7-25b. The maximum number of emitters that may be required equals the number of bits in the output code. For the particular encoder sketched in Fig. 7-24, Q1, Q2, Q4, and Q8 each have one emitter, Q3, Q5, Q6, and Q9 have two emitters each, and Q7 has three emitters.

Output Stages   A bipolar encoder uses the standard TTL output stages. If each output line from the encoder goes to the data input in Fig. 6-39a, a totem-pole output driver

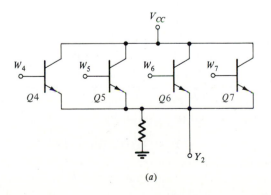

10 input
lines

W_0

W_1

W_2

W_3

W_4

W_5

W_6

W_7

W_8

W_9

5 V

Y_3 Y_2 Y_1 Y_0

4-bit output code

FIGURE 7-24
An encoding matrix to transform a decimal number into a BCD code. The key W_0 may be omitted since it is implied that the output is 0000 unless one of the other nine keys is activated.

FIGURE 7-25
(*a*) An emitter-follower OR gate, (*b*) the line W_7 in the encoder of Fig. 7-24 is connected to the base of the three-emitter transistor.

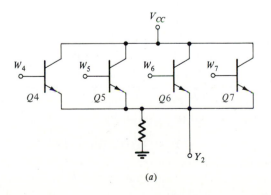

V_{CC}

W_4 W_5 W_6 W_7

$Q4$ $Q5$ $Q6$ $Q7$

Y_2

(*a*)

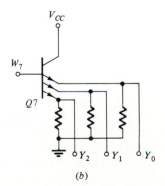

V_{CC}

W_7

$Q7$

Y_2 Y_1 Y_0

(*b*)

results. If an encoder output line goes to the data input in Fig. 6-39b an open-collector output is obtained.

Priority Encoder We now remove the restriction that only one key is depressed at any given time. If several keys are simultaneously pushed accidentally, then let us give priority to the highest-order data line and encode it. For example, if W_5 and W_6 are simultaneously activated, then it is desired that the output correspond to W_6. The truth table for a 10-to-4-line priority encoder is given in Table 7-3. An X in the table means that this entry is *irrelevant*. It may be either a 1 or a 0 and, hence, X is referred to as a *don't care state*. However, now a 0 in the table must be taken into account, whereas in Table 7-2 the zeros could be ignored because Table 7-2 is uniquely determined by the 1s along the diagonal.

The boolean expression for Y_1, obtained from Table 7-3, is

$$Y_1 = \overline{W}_9\overline{W}_8\overline{W}_7\overline{W}_6\overline{W}_5\overline{W}_4\overline{W}_3W_2 + \overline{W}_9\overline{W}_8\overline{W}_7\overline{W}_6\overline{W}_5\overline{W}_4W_3$$

$$+ \overline{W}_9\overline{W}_8\overline{W}_7W_6 + \overline{W}_9\overline{W}_8W_7 \quad (7\text{-}19)$$

This equation can be simplified considerably. Note that

$$Y_1 = \overline{W}_9\overline{W}_8(\overline{W}_7B + W_7) \quad (7\text{-}20)$$

where

$$B \equiv \overline{W}_6\overline{W}_5\overline{W}_4\overline{W}_3W_2 + \overline{W}_6\overline{W}_5\overline{W}_4W_3 + W_6 \quad (7\text{-}21)$$

TABLE 7-3   A Priority Encoder (10-line decimal to 4-line BCD)

				Inputs								Output	
W_9	W_8	W_7	W_6	W_5	W_4	W_3	W_2	W_1	W_0	Y_3	Y_2	Y_1	Y_0
0	0	0	0	0	0	0	0	0	1	0	0	0	0
0	0	0	0	0	0	0	0	1	X	0	0	0	1
0	0	0	0	0	0	0	1	X	X	0	0	1	0
0	0	0	0	0	0	1	X	X	X	0	0	1	1
0	0	0	0	0	1	X	X	X	X	0	1	0	0
0	0	0	0	1	X	X	X	X	X	0	1	0	1
0	0	0	1	X	X	X	X	X	X	0	1	1	0
0	0	1	X	X	X	X	X	X	X	0	1	1	1
0	1	X	X	X	X	X	X	X	X	1	0	0	0
1	X	X	X	X	X	X	X	X	X	1	0	0	1

From Eq. (6-18) with $A = W_7$, we obtain

$$Y_1 = \overline{W}_9\overline{W}_8(W_7 + B) \quad (7\text{-}22)$$

From Eq. (7-21),

$$B = \overline{W}_6 C + W_6 = W_6 + C \qquad (7\text{-}23)$$

where Eq. (6-18) is used again, and where

$$C \equiv \overline{W}_5 \overline{W}_4 \overline{W}_3 W_2 + \overline{W}_5 \overline{W}_4 W_3 = \overline{W}_5 \overline{W}_4 (\overline{W}_3 W_2 + W_3)$$

$$= \overline{W}_5 \overline{W}_4 (W_3 + W_2) \qquad (7\text{-}24)$$

From Eqs. (7-22), (7-23), and (7-24),

$$Y_1 = \overline{W}_9 \overline{W}_8 (W_7 + W_6 + \overline{W}_5 \overline{W}_4 W_3 + \overline{W}_5 \overline{W}_4 W_2) \qquad (7\text{-}25)$$

A NOR gate is used to generate $\overline{W}_9 \overline{W}_8 = \overline{W_9 + W_8}$ (De Morgan's law) and a 4-wide (2-2-4-4-input) AND-OR gate is required to generate Y_1. Proceeding in a similar manner, the combinational logic for Y_0, Y_2, and Y_3 is found (Probs. 7-30 and 7-31).

The above logic is fabricated on an MSI chip (74LS147) which priority encodes 10-line decimal to 4-line BCD. Applications include encoding of small keyboards, analog-to-digital conversion (Sec. 16-5), and controlling computer priority interrupts. The 74LS148 package encodes eight data lines to a three-line binary (octal code).

7-9 READ-ONLY MEMORY (ROM)

Consider the problem of converting one binary code into another. Such a code-conversion system (designated ROM and sketched in Fig. 7-26a) has M inputs (X_0, X_1, ..., X_{M-1}) and N outputs (Y_0, Y_1, ..., Y_{N-1}), where N may be greater than, equal to, or less than M. A definite M-bit code is to result in a specific output code of N bits. This code translation is achieved,

FIGURE 7-26
(a) Conversion of one code to another using a read-only memory (ROM). (b) A ROM may be considered to be a decoder for the input code followed by an encoder for the output code.

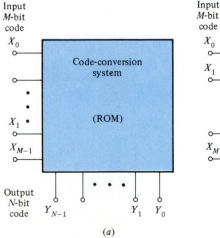

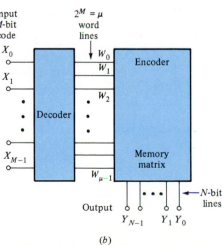

(a)

(b)

as indicated in Fig. 7-26b, by first decoding the M inputs onto $2^M \equiv \mu$ word lines ($W_0, W_1, \ldots, W_{\mu-1}$) and then encoding each line into the desired output word. If the inputs assume all possible combinations of 1s and 0s, then μ N-bit words are "read" at the output (not all these 2^M words need be unique, since it may be desirable to have the same output code for several different input words).

The functional relationship between output and input words is built into hardware in the encoder block of Fig. 7-26. Since this information is thus stored permanently, we say that the system has *nonvolatile memory*. The *memory elements* are the diodes in Fig. 7-24 or the emitters of transistors in Fig. 7-25. The output word for any input code may be read as often as desired. However, since the stored relationship between output and input codes cannot be modified without adding or subtracting memory elements (hardware), this system is called a *read-only memory*, abbreviated ROM.

Code Converters   The truth table for translating from a binary to a Gray code is given in Table 7-4. In the progression from one line to the next of the Gray code, 1 and only 1 bit is changed from 0 to 1, or vice versa. (This property does not uniquely define a code, and hence a number of Gray codes may be constructed.) The input bits [(1) in Table 7-4] are decoded in a ROM into the word lines $W_0, W_1, \ldots, W_{15}$, as indicated in Fig. 7-26b, and then are encoded [(2) in Table 7-4] into the desired Gray code $Y_3Y_2Y_1Y_0$. The W's are the minterm outputs of the decoder.

TABLE 7-4   Conversion from a Binary to a Gray Code [(1) to (2)] and from a Gray to a Binary Code [(1) to (3)]

(1) Inputs				Decoded word	(2) Gray code outputs				(3) Binary code outputs			
X_3	X_2	X_1	X_0	W_n	Y_3	Y_2	Y_1	Y_0	Y_3	Y_2	Y_1	Y_0
0	0	0	0	W_0	0	0	0	0	0	0	0	0
0	0	0	1	W_1	0	0	0	1	0	0	0	1
0	0	1	0	W_2	0	0	1	1	0	0	1	1
0	0	1	1	W_3	0	0	1	0	0	0	1	0
0	1	0	0	W_4	0	1	1	0	0	1	1	1
0	1	0	1	W_5	0	1	1	1	0	1	1	0
0	1	1	0	W_6	0	1	0	1	0	1	0	0
0	1	1	1	W_7	0	1	0	0	0	1	0	1
1	0	0	0	W_8	1	1	0	0	1	1	1	1
1	0	0	1	W_9	1	1	0	1	1	1	1	0
1	0	1	0	W_{10}	1	1	1	1	1	1	0	0
1	0	1	1	W_{11}	1	1	1	0	1	1	0	1
1	1	0	0	W_{12}	1	0	1	0	1	0	0	0
1	1	0	1	W_{13}	1	0	1	1	1	0	0	1
1	1	1	0	W_{14}	1	0	0	1	1	0	1	1
1	1	1	1	W_{15}	1	0	0	0	1	0	1	0

For example

$$W_0 = \bar{X}_3\bar{X}_2\bar{X}_1\bar{X}_0 \qquad W_5 = \bar{X}_3 X_2\bar{X}_1 X_0 \qquad W_9 = X_3\bar{X}_2\bar{X}_1 X_0 \qquad (7\text{-}26)$$

From the truth table (Table 7-4), we conclude that

$$Y_0 = W_1 + W_2 + W_5 + W_6 + W_9 + W_{10} + W_{13} + W_{14} \qquad (7\text{-}27)$$

This equation is implemented by connecting eight diodes with their cathodes all tied to Y_0 and their anodes connected to the decoder lines W_1, W_2, W_5, W_6, W_9, W_{10}, W_{13}, and W_{14}, respectively (or the base-emitter diodes of transistors may be used in an analogous manner to form an emitter-follower OR gate, as in Fig. 7-25a). Similarly, from Table 7-4, we may write the boolean expressions for the other output bits. For example,

$$Y_3 = W_8 + W_9 + W_{10} + W_{11} + W_{12} + W_{13} + W_{14} + W_{15} \qquad (7\text{-}28)$$

Consider the inverse code translation, from Gray to binary. The Gray-code inputs [(1) in Table 7-4] are arranged in the order W_0, W_1, . . . , W_{15} (corresponding to decimal numbers 0 to 15). The binary code corresponding to a given input word W_n is listed as the output code [(3) in Table 7-4] for that line. For example, from (1) and (2) of Table 7-4 at line W_{14}, we find that the Gray code 1001 corresponds to the binary code 1110, and this relationship is maintained in Table 7-4 [(1) and (3)] on line W_9. From this table we obtain the relationship between binary output (3) and Gray input (1) bits. For example,

$$Y_0 = W_1 + W_2 + W_4 + W_7 + W_8 + W_{11} + W_{13} + W_{14} \qquad (7\text{-}29)$$

This equation defines how the memory elements are to be arranged in the encoder. Note that the ROM for conversion from a binary to a Gray code uses the same decoding arrangement as that for conversion from a Gray to a binary code. However, the encoders are completely different. In other words, the IC chips for these two ROMs are quite distinct since individual masks must be used for the encoder matrix of memory elements.

Programming the ROM   Consider a 256-bit bipolar ROM 7488A arranged in 32 words of 8 bits each. The decoder input is a 5-bit binary select code, and its outputs are the 32 word lines. The encoder consists of 32 transistors (each base is tied to a different line) and with 8 emitters in each transistor. The customer fills out the truth table he wishes the ROM to satisfy, and then the vendor makes a mask for the metallization so as to connect one emitter of each transistor to the proper output line, or alternatively, to leave it floating. For example, for the Gray-to-binary-code conversion, Eq. (7-29) indicates that one emitter from each of transistors $Q1$, $Q2$, $Q4$, $Q7$, $Q8$, $Q11$, $Q13$, and $Q14$ is connected to line Y_0, whereas the corresponding emitter on each of the other transistors $Q0$, $Q3$, $Q5$, $Q6$ . . . is left unconnected. The process just described is called *custom programming* or *mask programming* of a ROM. Note that *hardware* (not *software*) programming is under consideration.

FIGURE 7-27
An NMOS ROM encoder. (Only 5 of the 1024 word lines are shown.)

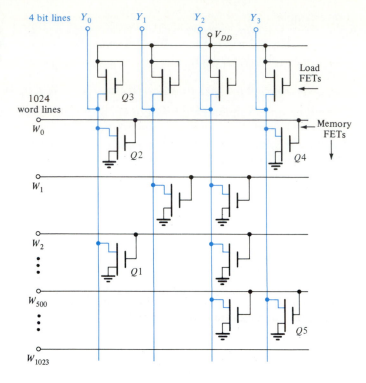

NMOS ROMS

Read-only memories are most widely implemented by means of NMOS technology, frequently as LSI or VLSI chips. Often, ROMs are fabricated as parts of larger systems on a single chip such as the microprocessor (Sec. 9-10). Consider, for example, a 10-bit input code resulting in $2^{10} = 1024$ word lines, and with 4 bits per output code. The memory matrix for this system consists of $1024 \times 4 = 4096$ intersections as indicated schematically in Fig. 7-27. This ROM is referred to as a *4-kilobit* (4-kb) ROM organized as 1 kb $\times$ 4. This colloquial designation stems from the fact that $2^{10} = 1024$ or approximately 10^3. Thus a 64-kb ROM contains $2^6 \times 2^{10} = 64 \times 2^{10}$ bits. The code conversion to be performed by the ROM is permanently programmed during the fabrication process by using a custom-designed mask so as to construct or omit a MOS transistor at each matrix intersection. Such an encoder is indicated in Fig. 7-27, which shows how the memory FETs are connected between *word* and *bit* lines.

Earlier in this section we demonstrated that the relationship between the output bits Y and the word lines W is satisfied by the logic OR function. Consider, for example, that it is required by the desired code conversion that

$$\bar{Y}_0 = W_0 + W_2 \qquad \bar{Y}_1 = W_1$$
$$\bar{Y}_2 = W_1 + W_2 + W_{500} \qquad \bar{Y}_3 = W_0 + W_{500}$$

(7-30)

These relationships are satisfied by the connections in Fig. 7-27. The NOR gate for Y_0 of Eq. (7-30) is precisely that drawn in Fig. 6-28a, with signals W_0 and W_2 applied as the inputs A and B, respectively.

The presence or absence of a MOS memory cell at a matrix intersection is determined during fabrication in the oxide-gate mask steps (Fig. 5-14). If the MOSFET has a normal thin-oxide gate, its threshold voltage V_T is low; if the gate oxide is thick, then V_T is high. In response to a positive pulse on the word line, the low-threshold device will conduct and a logic 0 (because of inverter action) will be detected on the bit line. On the other hand, if a positive pulse is applied to the thick-oxide gate (high-threshold device), it does not conduct; it is effectively missing from the circuit. In other words, growing a thick-oxide gate at a matrix location is equivalent to *not* constructing a MOSFET at this position, as shown in Fig. 7-28. The ROM is a *nonvolatile memory,* because if the power is interrupted and then restored, the relationship between input and output programmed into the ROM is not lost.

In a static ROM no clocks are needed and the output exists as long as the input address remains valid. Such ROMs are available in sizes from 1 to 64 kb (from Intel, Mostek, Texas Instruments, and other manufacturers), usually with four or eight output bits. Access times are in the range 0.1 to 1.0 μs (about half an order of magnitude longer than for bipolar ROMs) and power dissipations extend from 0.1 to 1 W.

An example of a static ROM is the 16-kb (2048 × 8) Intel 2316 in a 24-pin DIP. It uses n-channel MOSFETs (NMOS) and operates from a single 5-V supply so that inputs and outputs are TTL compatible. The two-dimensional addressing discussed in Sec. 7-10 is also used with LSI ROMs. The 2316 ROM is organized as indicated in Fig. 7-29. Eleven address bits are needed for 2048 words. Note that seven of these address inputs (A_4 through A_{10}) are used for the X decoder to obtain 128 rows. The memory matrix is square with 128 columns, and these must be multiplexed onto 8 outputs (O_0 through O_7). This

FIGURE 7-28

A MOS ROM matrix. (Only lines W_0 and W_{500} and bit lines Y_0 and Y_3 in Fig. 7-27 are shown.)

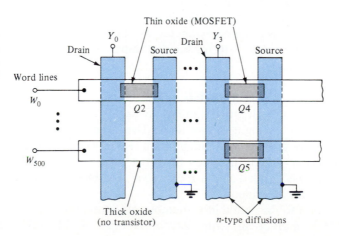

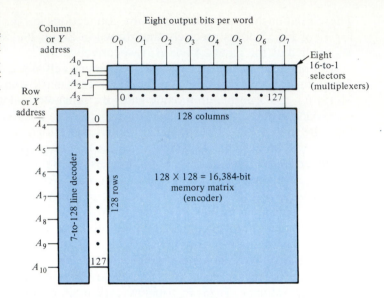

FIGURE 7-29
The organization of the
Intel 2316A 16-kb ROM
(2048 words × 8 bits).
Buffers and chip-select
circuitry are not shown.

is accomplished with eight (16-to-1) selectors, using four-column-address inputs (A_0 through A_3). This organization is an extension of that shown in Fig. 7-30 for a 2-kb ROM (512 × 4), although the two diagrams (Figs. 7-29 and 7-30) are drawn somewhat differently.

The decoder in a static MOS ROM (Fig. 7-29) contains NAND gates which are static. Power dissipation as a result is relatively high. A dynamic ROM uses clocked or dynamic inverters in the decoder and/or load FETs and requires a minimum clock rate, since otherwise the information is lost. However, its power dissipation is lower than for a static ROM. Most commercial ROMs are static because of the advantages of requiring no clocks and of giving an output which remains valid as long as the input address is applied.

Increasing the number of bits per word (called *word expansion*) or increasing the number of words, with the same number of bits per word (called *address expansion*), is attained by interconnecting several ROM packages in the manner discussed in Sec. 7-10 for bipolar chips.

7-10 TWO-DIMENSIONAL ADDRESSING OF A ROM Many manufacturers (listed in App. B-1) supply both bipolar and MOS ROMs in sizes from 256 bits to 64 kb, with either four or eight output lines. The larger ROMs are also available and are examples of large-scale integration (LSI). The time required for a valid output to appear on the bit lines from the time an input address is applied to the memory is defined as the *access time*. For bipolar ROMs access times of below 100 ns are obtained.

For a ROM with a large number of inputs, the decoding arrangement in Fig. 7-26 is impractical. Consider, for example, a 512 × 4 = 2048-bit ROM (M =

FIGURE 7-30

A 2-kb ROM (512 × 4 bits) with two-dimensional addressing. Note that the column address $A_8A_7A_6$ is applied to all four selectors-multiplexers. The chip-select input CS is used for enabling purposes.

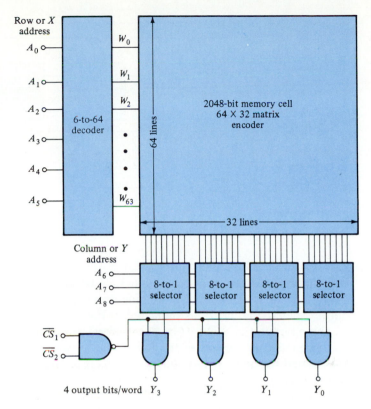

9 and $N = 4$). A total of 512 NAND gates are required in the decoder, one for each word line. Considerable economy results if the topology shown in Fig. 7-30 is used. A 6-bit X (row) address generates 64 horizontal lines. If 32 Y (vertical) lines are used in the memory matrix, the total number of bits is $64 \times 32 = 2048$ as required. However, since only four output lines are specified then four 8-to-1-line selectors are used. A 3-bit column address feeds each multiplexer. This arrangement is called X-Y or *two-dimensional addressing*. Note that now 64 NAND gates are needed for the decoder and $4 \times 9 = 36$ for the selectors from the NAND-NAND (AND-OR) configuration in Fig. 7-21 (it is clear that an 8-input selector requires nine gates). The total of $64 + 36 = 100$ NANDS for X-Y addressing is a tremendous saving over 512 NANDS required by the decoder arrangement of Fig. 7-26 for the same-size ROM. From Fig. 7-30, 64 transistors with 32 emitters each are needed for the encoder whereas in Fig. 7-26 there are 512 transistors, each with 4 emitters.

Word Expansion Increasing the number of bits per word is easily accomplished. For example, a 512×8-bit ROM is obtained by using two 512 × 4 ROMs. The addressing in Fig. 7-27 is applied to both chips simultaneously. The 4 bits of lower significance are obtained from one package and the 4 bits of higher significance are taken from the second chip.

FIGURE 7-31
Address expansion con-
verts four 256 × 4
ROMs into a 1024 × 4-
bit memory.

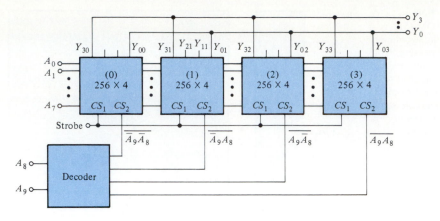

FIGURE 7-31
Address expansion con-
verts four 256 × 4
ROMs into a 1024 × 4-
bit memory.

Address Expansion   To obtain additional words (with no increase in the byte size) is more complicated. For example, to obtain 1024 words of 4 bits each requires four 256 × 4 ROMs, an external 2-to-4-line decoder, and the use of 3-state OR-tied output stages (Fig. 6-39), as indicated in Fig. 7-31.

The operation of this system is explained as follows. The addresses $A_7 \cdots A_0$ are applied in parallel to the four 256 × 4 ROMs, each of whose outputs are OR-tied together. A 2-bit address $A_9 A_8$ is applied to a decoder (Fig. 7-17), whose four outputs control the four chip selects CS_2. For example, if $A_9 = 0$ and $A_8 = 1$, then $CS_2 = 0$ for chip 1 and $CS_2 = 1$ for all other packages. Hence, chip 1 is enabled, whereas the other three packages present a high impedance at each output. Hence, only the outputs $Y_0 = Y_{01}$, $Y_1 = Y_{11}$, $Y_2 = Y_{21}$, and $Y_3 = Y_{31}$ from package 1 appear at the output. Each combination of 1s and 0s in an address $A_9 A_8 \cdots A_1 A_0$ yields one 4-bit output code, for a total of 1024 4-bit words in the memory.

7-11 ROM APPLICATIONS   As emphasized in the preceding section, a ROM is a code-conversion unit. However, many different practical systems represent a translation from one code to another. The most important of these ROM applications are discussed below.

Look-up Tables   Routine calculations such as trigonometric functions, logarithms, exponentials, square roots, etc., are sometimes required of a computer. If these are repeated often enough, it is more economical to include a ROM as a *look-up table,* rather than to use a subroutine or a software program to perform the calculation. A look-up table for $Y = \sin X$ is a code-conversion system between the input code representing the argument X in binary notation (to whatever accuracy is desired) and the output code giving the corresponding values of the sine function. Clearly, any calculation for which a truth table can be written may be implemented with a ROM—a different ROM for each truth table.

Sequence Generators If in a digital system, P pulse trains are required for testing or control purposes, these may be obtained by using P multiplexers connected for parallel-to-serial conversion (Sec. 7-7). A more economical way to supply these binary sequences is to use a ROM with P outputs and to change the address by means of a counter. As mentioned in Sec. 7-7, the input to the encoder changes from W_0 to W_1 to W_2, etc., every T seconds. Under this excitation the output Y_1 of the ROM represented by Table 7-4 (for Gray code to binary code conversion) is

$$Y_1 = 1100001100111100 \quad \text{(LSB)} \tag{7-31}$$

This equation is obtained by reading the digits in the Y_1 column from top to bottom. It indicates that for the first $2T$ seconds, Y_1 remains low; for the following $4T$ seconds, Y_1 is high; for the next $2T$ seconds, Y_1 is low; for the next $2T$ seconds, Y_1 is high; for the following $4T$ seconds, Y_1 is low; for the last $2T$ seconds, Y_1 is high; and after these $16T$ seconds, this sequence is repeated (as long as pulses are fed to the counter).

Simultaneously with Y_1, three other synchronous pulse trains, Y_0, Y_2, and Y_3, are created. In general, the number of sequences obtained equals the number of outputs from the ROM. Any desired serial binary waveforms are generated if the truth table is properly specified, i.e., if the ROM is correctly programmed.

Waveform Generator If the output of the digital sequence generator is converted into an analog voltage, then a *waveform generator* is obtained. Consider a 256×8-bit ROM sequenced by means of an 8-bit counter. Each step of the counter represents $\frac{360}{256} = 1.406°$ of the waveform. The ROM is programmed so the outputs Y_0 to Y_8 give the digital number corresponding to the analog amplitude at each step. The ROM outputs feed into a D/A (digital-to-analog converter, Sec. 16-4), and the output gives the analog waveform desired. This output changes in small discrete steps (each less than $\frac{1}{2}$ percent of full scale) and hence some simple filtering may be desirable.

Seven-Segment Visible Display It is common practice to make visible the reading of a digital instrument (a frequency meter, digital voltmeter, etc.) by means of the seven-segment numeric indicator sketched in Fig. 7-32a. A wide variety of readouts

FIGURE 7-32
(a) Identification of the segments in a seven-segment LED display. (b) The display which results from each of the sixteen 4-bit input codes.

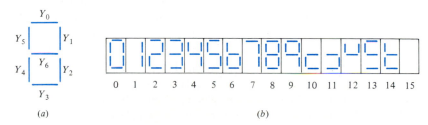

(a)

(b)

are commercially available. A solid-state indicator in which the segments obtain their luminosity from light-emitting gallium arsenide or phosphide diodes is operated at low voltage and low power and hence may be driven directly from IC logic gates.

The first 10 displays in Fig. 7-32b are the numerals 0 to 9, which, in the digital instrument, are represented in BCD form. Such a 4-bit code has 16 possible states, and the displays 10 to 15 of Fig. 7-32b are unique symbols used to identify a nonvalid BCD condition.

TABLE 7-5   Conversion from a BCD to a Seven-Segment-Indicator Code

Binary-coded-decimal inputs				Decoded word	Seven-segment-indicator code outputs						
$S_3 = D$	$X_2 = C$	$X_1 = B$	$X_0 = A$	W_n	Y_6	Y_5	Y_4	Y_3	Y_2	Y_1	Y_0
0	0	0	0	W_0	1	0	0	0	0	0	0
0	0	0	1	W_1	1	1	1	1	0	0	1
0	0	1	0	W_2	0	1	0	0	1	0	0
0	0	1	1	W_3	0	1	1	0	0	0	0
0	1	0	0	W_4	0	0	1	1	0	0	1
0	1	0	1	W_5	0	0	1	0	0	1	0
0	1	1	0	W_6	0	0	0	0	0	1	1
0	1	1	1	W_7	1	1	1	1	0	0	0
1	0	0	0	W_8	0	0	0	0	0	0	0
1	0	0	1	W_9	0	0	1	1	0	0	0
1	0	1	0	W_{10}	0	1	0	0	1	1	1
1	0	1	1	W_{11}	0	1	1	0	0	1	1
1	1	0	0	W_{12}	0	0	1	1	1	0	1
1	1	0	1	W_{13}	0	0	1	0	1	1	0
1	1	1	0	W_{14}	0	0	0	0	1	1	1
1	1	1	1	W_{15}	1	1	1	1	1	1	1

The problem of converting from a BCD input to the seven-segment outputs of Fig. 7-32 is easily solved using a ROM. If an excited (luminous) segment is identified as state 0 and a dark segment as the 1 state, the truth table (Table 7-5) is obtained. This table is verified as follows: For word W_0 (corresponding to numeral 0) we see from Fig. 7-32 that $Y_6 = 1$ and all other Y values are 0. For word W_4 (corresponding to the numeral 4) $Y_0 = Y_3 = Y_4 = 1$ and $Y_1 = Y_2 = Y_5 = Y_6 = 0$, and so forth. The ROM is programmed as explained in Sec. 7-9 to satisfy this truth table. For example,

$$Y_0 = W_1 + W_4 + W_6 + W_{10} + W_{11} + W_{12} + W_{14} + W_{15} \qquad (7\text{-}32)$$

It should be pointed out that a ROM may not use the smallest number of gates to carry out a particular code conversion. Consider Eq. (7-32) written as a sum of products. Replacing the minterm W_1 by $\bar{X}_3\bar{X}_2\bar{X}_1X_0 \equiv \bar{D}\bar{C}\bar{B}A$ and using analogous expressions for the outputs of the other decoders, Eq. (7-32) becomes

$$Y_0 = \bar{D}\bar{C}\bar{B}A + \bar{D}CB\bar{A} + \bar{D}CBA + D\bar{C}B\bar{A} + D\bar{C}BA$$

$$+ DC\bar{B}\bar{A} + DCB\bar{A} + DCBA \quad (7\text{-}33)$$

There are a number of algebraic and graphical techniques and computer programs for minimizing such boolean expressions. Note, for example, that the second and third minterms can be simplified to

$$\bar{D}C\bar{B}\bar{A} + \bar{D}CB\bar{A} = \bar{D}C\bar{A}$$

because $\bar{B} + B = 1$. Proceeding in this manner (Prob. 7-40), the following minimized form of Y_0 is obtained:

$$Y_0 = \bar{D}\bar{C}\bar{B}A + C\bar{A} + DB \quad (7\text{-}34)$$

Using the minimized expressions for $Y_0, Y_1, \ldots, Y_6$ results in some saving (about 20 percent) of components over those required in the ROM. A chip fabricated in this manner (e.g., 74HC4511) is designated a "BCD-to-seven-segment decoder/driver."

Minimization of boolean equations (particularly if the number of variables in each product exceeds five) is tedious and time-consuming. The engineering man-hours cost for minimization and for designing the special IC chip to realize the savings in components must be compared with that of simply programming an existing ROM. Unless a tremendous number of units are to be manufactured (and particularly if the matrix size is large), the ROM is the more economical procedure. The programmable logic arrays discussed in the next section afford a convenient method for designing complex logic functions by use of "fixed" hardware.

Combinational Logic If N logic equations of M variables are given in the sum-of-products canonical form, these equations may be implemented with an M-input, N-output ROM. As explained above, this is an economical solution if M and N are large (particularly if M is large). However, in the logic design of one stage of a full adder, where $M = 3$ and $N = 2$ (small numbers), and where this unit is sold in considerable quantities, using distinct gate combinations as in Fig. 7-7 is more economical than using a ROM.

Character Generator Alphanumeric characters may be "written" on the face of a cathode-ray tube (a television-type display) with the aid of a ROM.

Stored Programs Control programs (e.g., in a pocket calculator) are permanently stored in ROM.

7-12 PROGRAMMABLE ROMS (PROMS)

Many manufacturers supply *field-programmable* ROMs, called *PROMs* (see App. B-1). These IC chips can provide flexibility to the designer and can often reduce costs, particularly when only a small quantity of a particular ROM is required. The cost of making the connection mask, when amortized over a few units, is high; furthermore, delivery time may be too long to meet a desired product completion date.

A PROM contains an encoder matrix in which all possible connections that may be required are made. For example, the 256-bit ROM described in the previous section can be converted into a PROM containing 32 transistors, each having eight emitters (designated E_0, E_1, . . . , E_7). Each emitter E_0 is tied to output Y_0, E_1 emitters to Y_1, and so on. A narrow polysilicon strip is incorporated in series with each emitter and acts as a fuse. The fuse opens when a current in excess of a prescribed value is passed through the memory element. The *user* can easily "burn," "zap," or "blow" these memory-element fuses to open appropriate connections in order that the ROM provide the desired functional relationships between input and output.

An alternative view of a ROM is that it is a matrix consisting of an AND array and an OR array. This organization generates an input-output functional relationship in the sum-of-products form. Programmable read-only memories consist of a fixed AND array and a programmable OR array as illustrated in Fig. 7-33. The X's in Fig. 7-33 represent connections (fusible links) to the gate inputs. Note that only certain AND connections are present, whereas all OR input connections are made. Programming is accomplished by burning the undesired OR connections.

An instrument called a *PROM programmer* is used to "burn" in the program by providing the current necessary to open the fuses. Clearly, once a PROM is programmed by opening the fuses, the program is unalterable. However, it is possible to erase the program in certain MOS PROMs and to write a new program electrically.

7-13 ERASABLE PROMS   Two types of MOS PROMs in which the program can be erased are called the erasable PROM (EPROM) and the electrically erasable (alterable) PROM (E^2PROM or EAROM). Both types are discussed in this section.

EPROMS   Programmable read-only memories, programmed by blowing out fusible links, cannot have the program altered because the burned out fuse is not repairable. Erasable PROMs are based on the special MOS structure depicted in Fig. 7-34a. This double-gate NMOS transistor is sometimes referred to as a FAMOS (an acronym for floating-gate avalanche-injection metal oxide semiconductor). Gate 1 is a polysilicon gate which is left "floating" in that no contact is made to it. This gate is completely surrounded by S_1O_2, and thus no path for discharging and charge stored on the gate is present. By applying a large positive voltage to gate 2 and drain (≈ 25 V), the high electric field intensity in the depletion region of the drain-substrate pn junction causes avalanche breakdown. A large amount of additional current results from the junction breakdown. Consequently, high-energy electrons, accelerated by the electric field, penetrate the thin oxide layer and accumulate on gate 1. With no discharge path, the accumulated charge forces the potential on gate 1 to become negative when the voltages at gate 2 and drain are made zero. This negative potential

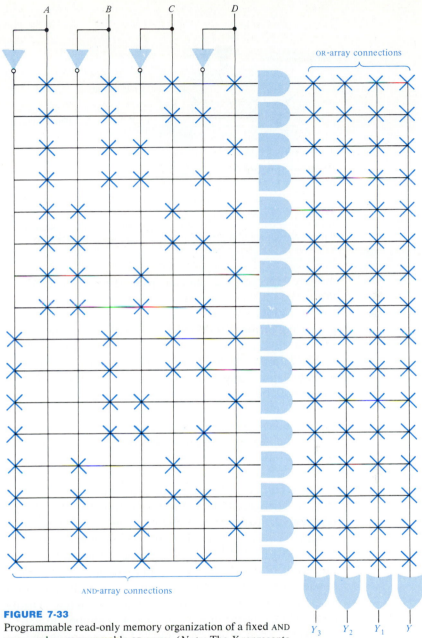

FIGURE 7-33
Programmable read-only memory organization of a fixed AND array and a programmable OR array. (*Note*: The *X* represents a fusible link.)

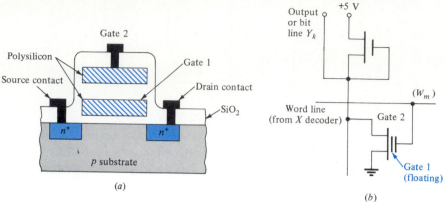

FIGURE 7-34
(a) Floating polysilicon gate structure (FAMOS) used to obtain an erasable PROM (EPROM). (b) The circuit diagram of a typical EPROM cell using the FAMOS.

on gate 1 prevents a channel from being induced between the n^+ source and drain regions when normal voltage levels for logic 1 (≈ 5 V) are applied to gate 2 (Fig. 7-34b). As a result, a logic 1 has been stored in this ROM cell. Programming of the PROM is accomplished by placing logic 1's in the appropriate cells in the array.

The superb insulating properties of SiO_2 can maintain the induced charge on gate 1 for many years. (It is estimated that greater then 70 percent of the charge will be present after *10 years* even if storage is at 125°C.) However, the memory cell can be erased by exposing it to UV light. The photoelectric current produced removes the charge on the gate because the SiO_2 becomes slightly conducting under direct illumination of UV light.

E²PROMS

A distinct disadvantage of EPROMs is the long exposure time required for UV erasure. Hence EPROMs cannot be employed in applications which need fast storage changes. The E²PROM overcomes this disadvantage by permitting relatively fast electrical erasure of the stored data. A structure similar to that in Fig. 7-34a is used in which the oxide thickness between gate 1 and the silicon regions is reduced by nearly an order of magnitude (to ≈ 100 Å). A voltage in the order of 10 V (again a higher voltage than normal positive logic levels) applied across the extremely narrow SiO_2 layer causes electrons to flow to gate 1 by a type of quantum-mechanical tunneling. The induced charge prevents a channel from forming when logic 1 is applied to gate 2 and consequently a 1 is stored. Erasure is accomplished by reversing the voltage needed to store logic 1.

7-14 PROGRAMMABLE ARRAY LOGIC The *programmable array logic* (PAL) is related to the PROM in that it, too, is a matrix comprising an AND array and an OR array. However, the OR array in a PAL is fixed and the AND is programmable.

FIGURE 7-35
Programmable array logic (PAL) formed from a programmable AND and fixed OR array.

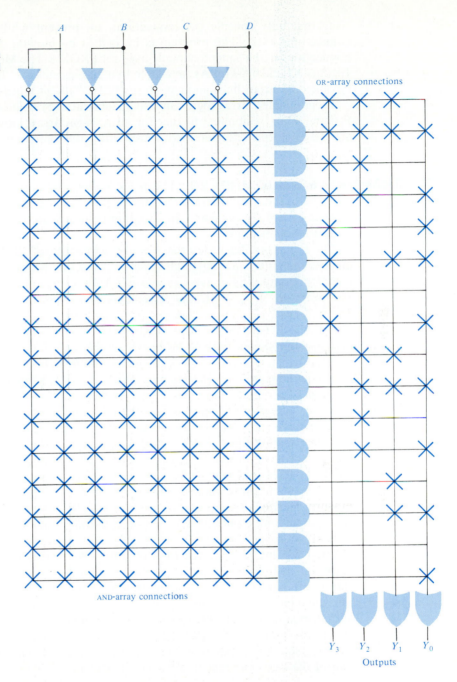

In many logic designs, simplified boolean expressions are implemented by the PAL arrangement compared with PROM realizations. A 16-word × 4-bit PAL is illustrated in Fig. 7-35, in which the X's indicate connections (fusible links).

Note that only specific OR connections are present while all possible AND connections exist. The program is burned in by blowing the undesired AND connections in the same technique used for PROMs. The Monolithic Memories 10H8 is a typical 8-kb PAL available in a 20-pin package. It contains 10 input lines and 8 output lines and can store 1024 (2^{10}) 8-bit words.

Programmable array logic can also be fabricated by using AOI arrays (10L8). Programming is identical to that for an AND-OR array; however, the outputs are the minterms associated with AND-NOR gate combinations.

7-15 PROGRAMMABLE LOGIC ARRAYS The *programmable logic array* (PLA) is the most versatile of the AND-OR array ICs. In the PLA, all connections to the AND- and OR-gate arrays are present. The program is burned in by blowing the fuses at undesired connections.

The following illustrates the use and versatility of PLAs. In Sec. 7-9 we discuss the (Intel 2316A) 16-kb (2048 $\times$ 8) ROM, which has $M = 11$ inputs and $N = 8$ outputs. For every increase in M by 1, the number of bits is doubled. For example, if $M = 16$ and N remains at 8, then the number of words is $2^{16} = 2^5 \times 2^{11} = 32 \times 2048 = 65,536$ and the number of bits is 65,536 $\times$ 8 = 524,288. This tremendous number of bits is not feasible in a single ROM chip and thirty-two 16-kb ROM packages are required, interconnected for address expansion (Sec. 7-10). This system implements $N = 8$ combinational logic equations in $M = 16$ variables (Sec. 7-11). Each equation is expressed in the sum-of-products canonical form. Each product contains 16 factors, and there are a total of 65,536 product terms (or words).

Consider a subset of the foregoing ROM system. The number of inputs and outputs are unchanged ($M = 16$ and $N = 8$), but each sum contains only, say, 48 product terms instead of 65,536. These are called *partial products* of the input variables, because each product does *not* contain all 16 inputs (or the complements of these inputs). Such a system of combinational logic is referred to as a 16 $\times$ 48 $\times$ 8 *programmable logic array* (PLA) and designates that there are 16 inputs, 8 outputs, and a total of 48 partial products, as indicated in Fig. 7-36.

The decoding array in the PLA of Fig. 7-36 consists of 48 AND gates. The output of each AND is a partial product term and the number of inputs to any gate is often small; the maximum equals the number of input data bits (16). The encoder matrix consists of eight OR gates whose outputs are the eight output functions of the PLA. The maximum number of inputs to any OR gate equals the number of product terms and is usually much smaller than this number (48). As an example, consider that two combinational logic equations (out of the eight implemented by the PLA of Fig. 7-36) are

$$O_0 = \overline{A}_3 A_0 + \overline{A}_9 A_4 \overline{A}_1 + A_{11}\overline{A}_{10}\overline{A}_7 A_3 + A_{13} + \overline{A}_{15}A_{14} \qquad (7\text{-}35)$$

$$O_1 = A_5 \overline{A}_4 A_0 + \overline{A}_9 A_4 \overline{A}_1 + A_{12}A_6 \qquad (7\text{-}36)$$

FIGURE 7-36

Programmable logic array (PLA). Both the AND array and the OR array are programmable.

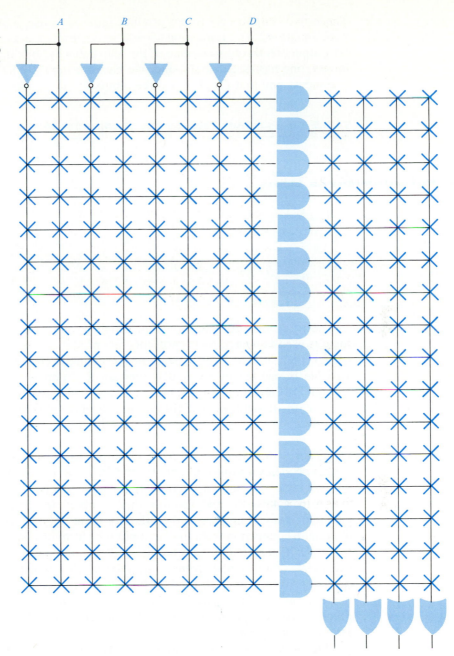

These two outputs use seven product terms because $\overline{A}_9A_4\overline{A}_1$ is common to both equations. The remaining $48 - 7 = 41$ terms are available for the other six outputs O_2 through O_7. One of the AND gates has one input, three have two inputs, one has three inputs, and one has four inputs. The OR gate for O_0 has five inputs and that for O_1 has three inputs.

TABLE 7-6   PLA Truth Table for Eqs. (7-35) and (7-36)

Term	15	14	13	12	11	10	9	8	7	6	5	4	3	2	1	0	1	0
					Inputs												*Outputs*	
0	X	X	X	X	X	X	X	X	X	X	X	X	0	X	X	1	0	1
1	X	X	X	X	X	X	0	X	X	X	X	1	X	X	0	X	1	1
2	X	X	X	X	1	0	X	X	0	X	X	X	1	X	X	X	0	1
3	X	X	1	X	X	X	X	X	X	X	X	X	X	X	X	X	0	1
4	0	1	X	X	X	X	X	X	X	X	X	X	X	X	X	X	0	1
5	X	X	X	X	X	X	X	X	X	X	1	0	X	X	X	1	1	0
6	X	X	X	1	X	X	X	X	X	1	X	X	X	X	X	X	1	0

The truth table for the foregoing equations is given in Table 7-6. Positive logic is used and each row represents one product term. If a data input is true (false), then a logic 1 (0) appears in the column representing this input. If a variable is missing in a product, then an X ("don't care") is placed in the column for the data input and product (row) under consideration. If the output O_k is 1 (0), it signifies that the product term represented by the row under consideration is present (absent) in the kth output function.

Programming a PLA   Table 7-6 (extended to cover 8 output functions and up to 48 product terms) becomes a *program table* for the $16 \times 48 \times 8$ PLA. The customer fills out this table to satisfy his combinational logic functions and the manufacturer makes a mask for the metallization so as to obtain the proper connections. For example, if an X appears at the pth data input and the rth product term, the aluminum connection from the pth input (nor from its complement) is *not* made to the rth AND gate. On the other hand, if the pth input is 1(0) for the rth term, the metallization is made from A_p ($\overline{A}_p$) to the rth AND gate. Similarly, if the kth output is 1(0) for the mth product line, aluminization connects (does not connect) the mth term to the input of the kth OR gate. An example of a mask-programmable logic array is the 6775 of Monolithic Memories, Inc. (or the DM 8575 of National Semiconductor) which has 14 inputs, 8 outputs, and 96 product terms ($14 \times 96 \times 8$). It is TTL compatible and the access time is approximately 50 ns.

Field-programmable logic arrays (FPLAs) are also available, such as the Signetics 82S100 ($16 \times 48 \times 8$), drawn in block-diagram form in Fig. 7-36. This bipolar chip uses (Schottky) diode AND gates and emitter-follower OR gates (Fig. 7-37). This FPLA is indicated in Fig. 7-37, where each symbol $\times$

FIGURE 7-37
The 82S100 FPLA (16 × 48 × 8). (*Courtesy of Signetics.*)

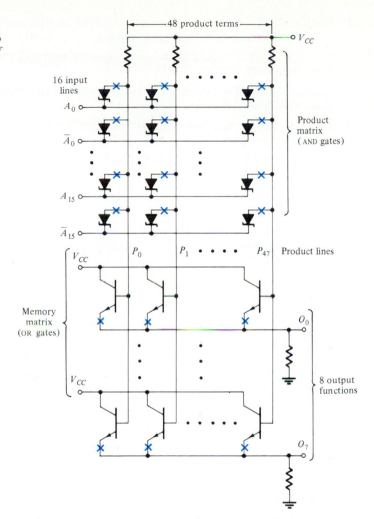

represents a fusible link. This system is programmed in the field by selectively blowing fuses so as to break the connections required in order to satisfy the program table, as explained in the preceding paragraph. The complement $\overline{A}_k$ of A_k is obtained by an inverter (not shown in Fig. 7-37). The complement of each output is also available on the chip, but this circuit is omitted from the figure for simplicity. The package is TTL compatible with either tristate or open-collector outputs (Fig. 6-39) and also includes chip-enable control.

The PLA (or FPLA) is designed for the implementation of complex logic functions. A PLA can handle more data inputs and is more economical than a ROM. It is useful for the same type of applications as the ROM (Sec. 7-11), provided that the number of product terms needed is a small fraction of the total input combinations possible.

REFERENCES

1 Hodges, D. A., and H. G. Jackson: "Analysis and Design of Digital Integrated Circuits," McGraw-Hill, New York, 1983.

2 Blakeslee, T. R.: "Digital Design with Standard MSI and LSI," John Wiley and Sons, New York, 1975.

3 Peatman, J. B.: "The Design of Digital Systems," 2d ed., McGraw-Hill Book Company, New York, 1981.

4 Taub, H., and D. Schilling: "Digital Integrated Electronics," McGraw-Hill Book Company, New York, 1975.

5 Mano, M. M.: "Computer System Architecture," 2d ed., Prentice-Hall, Englewood Cliffs, N.J., 1982.

6 Taub, H.: "Digital Circuits and Microprocessors," McGraw-Hill Book Company, New York, 1982.

7 *IEEE Journal of Solid-State Circuits,* annual special issue on semiconductor logic and memory, New York, October 1970 to present.

REVIEW QUESTIONS

7-1   (*a*)   How many input leads are needed for a chip containing quad 2-input NOR gates? Explain.
       (*b*)   Repeat part *a* for dual 2-wide, 2-input AOI gates.

7-2   Define SSI, MSI, LSI, and VLSI.

7-3   Draw the circuit configuration for an IC TTL AOI gate. Explain its operation.

7-4   Repeat Rev. 7-3 for an NMOS AOI gate.

7-5   What type of wired logic is used in Rev. 7-3? Rev. 7-4?

7-6   (*a*)   Find the truth table for the half-adder.
       (*b*)   Show the implementation for the digit D and the carry C.

7-7   Show the system of a 4-bit parallel binary adder, constructed from single-bit full adders.

7-8   (*a*)   Draw the truth table for a 3-input adder. Explain clearly the meaning of the input and output symbols in the table.
       (*b*)   Write the boolean expressions for the sum and the carry. (Do not simplify these.)

7-9   (*a*)   Show the system for a serial binary full adder.
       (*b*)   Explain the operation.

7-10   Define the one's complement and the two's complement of a binary number.

7-11   (*a*)   Consider two 4-bit numbers A and B with $B > A$. Verify that to subtract A from B it is only required to add B, $\overline{A}$, and 1.
       (*b*)   Indicate in simple form a 4-bit subtractor obtained from a full adder.

7-12   Consider two 1-bit numbers A and B. What are the logic gates required to test for (*a*) $A = B$; (*b*) $A > B$; and (*c*) $A < B$?

7-13 (a) Consider two 4-bit numbers A and B. If $E = 1$ represents the equality $A = B$, write the boolean expression for E. Explain.
(b) If $C = 1$ represents the inequality $A > B$, write the boolean expression for C. Explain.

7-14 Show the system for a 4-bit odd-parity checker.

7-15 (a) Show a system for increasing the reliability of transmission of binary information, using a parity checker and generator.
(b) Explain the operation of the system.

7-16 Write the decimal number 538 in the BCD system.

7-17 (a) Define a decoder.
(b) Show how to decode the 4-bit code 1011 (LSB).

7-18 (a) Define a demultiplexer.
(b) Show how to convert a decoder into a demultiplexer.
(c) Indicate how to add a strobe to this system.

7-19 Draw the logic block diagram for a 1-to-32-output demultiplexer tree using a "trunk" with four output lines. Indicate the correct addressing.

7-20 (a) Define a multiplexer.
(b) Draw a logic block diagram of a 4-to-1-line multiplexer.

7-21 Show how a multiplexer may be used as (a) a parallel-to-serial converter, and (b) a sequential data selector.

7-22 Draw the logic block diagram for a 32-to-1-line selector/multiplexer. Use selectors with a maximum of eight input lines. Indicate the correct addressing.

7-23 (a) Define an encoder.
(b) Indicate a diode matrix encoder to transform a decimal number into a binary code.

7-24 (a) Indicate an encoder matrix using emitter followers. In particular, for an encoder to transform a decimal number into a binary code, show the connections (b) to the output Y_2 and (c) to the line W_4.

7-25 (a) Define a priority encoder.
(b) Show the truth table for a 4-to-2-line priority encoder.

7-26 (a) Define a read-only memory.
(b) Show a block diagram of a ROM.
(c) What is stored in the memory?
(d) What hardware constitutes the memory elements?

7-27 Indicate a block diagram of a 624×4-bit ROM, using two-dimensional addressing. Use a 64×32 matrix encoder.

7-28 (a) Write the truth table for converting from a binary to a Gray code.
(b) Write the first six lines of the truth table for converting a Gray into a binary code.

7-29 Explain what is meant by mask-programming a ROM.

7-30 (a) Explain what is meant by a PROM.
(b) How is the programming done in the field?

7-31 List three ROM applications and explain these very briefly.

7-32 (*a*) What is a seven-segment visible display?

(*b*) Show the following two lines in the conversion table from BCD to seven-segment-indicator code: 0001 and 0101.

7-33 (*a*) What do the acronymns EPROM and E²PROM mean?

(*b*) Explain, briefly, the operation of each.

7-34 (*a*) Repeat Rev. 7-33*a* for the FAMOS device.

(*b*) Sketch the cross section of the FAMOS and briefly explain its operation.

7-35 (*a*) Compare and contrast a programmable logic array (PAL) with a programmable array logic (PAL).

(*b*) How are the PAL and PLA related to a ROM?

7-36 What is meant by a $16 \times 48 \times 8$ PLA?

Chapter 8
SEQUENTIAL CIRCUITS AND SYSTEMS

Many digital systems are required to operate in synchronism with a sequence of binary signals (a pulse train). The operation of a digital computer, for example, depends on first obtaining an instruction from memory (the fetch portion of a machine cycle) and storing it in a register until the instruction is executed. Moreover, the data to be processed must be obtained from its memory location. The third step is executing the instruction. Simultaneously, a program counter must be incremented to prepare for the next instruction. The timing and sequence of these steps is critical to successful operation of the system. Sequential circuits and systems are used to process binary signals synchronously. In addition, circuits capable of storing binary signals (memory) are needed.

The basic building block incorporated in sequential circuits is the FLIP-FLOP. In this chapter we treat FLIP-FLOPS and registers and counters, two classes of circuits based on FLIP-FLOPS. Several typical applications of these circuits are also described.

8-1 A 1-BIT MEMORY · All the systems discussed in Chap. 7 were based on combinational logic; the outputs at a given instant of time depend only on the values of the inputs at the same moment. Such a system is said to have no memory. Note that a ROM is a combinational circuit and, according to the above definition, it has no memory. *The memory of a ROM refers to the fact that it "memorizes" the functional relationship between the output variables and the input variables.* It does *not* store bits of information.

A 1-Bit Storage Cell   The basic digital memory circuit is obtained by cross-coupling two NOT circuits $N1$ and $N2$ (single-input NAND gates) in the manner shown in Fig. 8-1a. The output of each gate is connected to the input of the other, and this feedback combination is called a *latch*. The most important property of the latch is that it can exist in one of two stable states, either $Q = 1$ ($\overline{Q} = 0$), called the *1 state,* or $Q = 0$ ($\overline{Q} = 1$), referred to as the *0 state*. The existence

FIGURE 8-1

(a) A 1-bit memory or latch. (b) The latch provided with means for entering data into the cell.

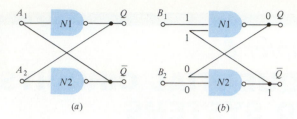

(a)   (b)

of these stable states is consistent with the interconnections shown in Fig. 8-1a. For example, if the output of $N1$ is $Q = 1$, so also is A_2, the input to $N2$. This inverter then has the state 0 at its output $\overline{Q}$. Since Q is tied to A_1, then the input of $N1$ is 0, and the corresponding output is $Q = 1$. This result confirms our original assumption that $Q = 1$. A similar argument leads to the conclusion that $Q = 0$; $\overline{Q} = 1$ is also a possible state. It is readily verified that the situation in which both outputs are in the same state (both 1 or both 0) is not consistent with the interconnection.

Since the configuration of Fig. 8-1a has two stable states, it is also called a *binary,* or *bistable circuit.* Since it may store one bit of information (either $Q = 1$ or $Q = 0$), it is a 1-*bit memory unit,* or a 1-*bit storage cell.* This information is locked, or latched, in place, which accounts for the name *latch.*

Suppose it is desired to store a specific state, say $Q = 1$, in the latch. Or conversely, we may wish to remember the state $Q = 0$. We may "write" a 1 or 0 into the memory cell by changing the NOT gates of Fig. 8-1a to 2-input NAND gates, $N1$ and $N2$, and by feeding this latch through two inputs B_1 and B_2 as in Fig. 8-1b. If we assume $B_1 = 1$ and $B_2 = 0$, then the state of each gate input and output is indicated on the diagram. Since $Q = 0$ we have verified that to enter a 0 into the memory required inputs $B_1 = 1$ and $B_2 = 0$. In a similar manner it can be demonstrated that to store a 1, it is necessary that the inputs be $B_1 = 0$ and $B_2 = 1$. If $B_1 = 1$ and $B_2 = 1$ these two leads may be removed from the NAND gates without affecting the logic. In other words, *the state of the latch is unaffected by the input combination $B_1 = B_2 = 1$; if $Q = 1$ (0) before this set of inputs is applied, the output will remain $Q = 1$ (0) after the inputs change to $B_1 = B_2 = 1$.* Note also that $B_1 = B_2 = 0$ is not allowed (Prob. 8-1).

A Chatterless Switch In a digital system it is often necessary to push a key in order to introduce a 1 or a 0 at a particular node. Most switches *chatter* or *bounce* several times before they settle down into the closed position. The storage cell just described may be used to give a single change of state when the key first closes, regardless of the chatter which follows. Consider the situation in Fig. 8-2, where the single-pole, double-throw switch is initially in the position to ground B_2 so that $B_2 = 0$. Neglecting the gate input current, the voltage at B_1 is 5 V, which is considered to be the 1-state voltage, so that initially $B_1 = 1$. At $t = t_1$ the key is pushed and the pole moves from 2 toward 1. The waveform for B_2 is shown in Fig. 8-2b. It takes an interval of time $t' = t_2 - t_1$ for the pole to reach contact 1 so that B_1 changes from 1 to 0 at $t = t_2$. However, as indicated in

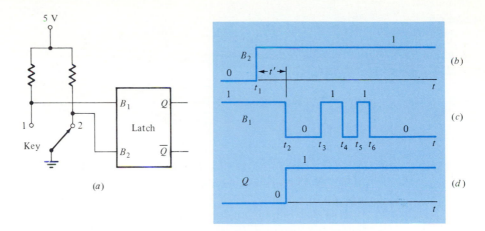

FIGURE 8-2
(*a*) The latch shown in Fig. 8-1*b* effectively causes the contact of a key to be without chatter. The waveforms (*b*) B_2, (*c*) B_1, and (*d*) Q.

Fig. 8-2*c* the contact chatters and the connection is broken in the intervals between t_3 and t_4 and between t_5 and t_6 (two bounces are assumed in drawing this figure). The output Q of the latch is indicated in Fig. 8-2*d* and is consistent with the logic discussed above; namely, $Q = 0$ if $B_1 = 1$ and $B_2 = 0$; $Q = 1$ if $B_1 = 0$ and $B_2 = 1$; Q is unaffected (does not change) if $B_1 = B_2 = 1$. Note that the latch has rendered the switch chatterless, since the output Q shows a single change of state from 0 to 1 at the first instant when B_1 is grounded. In a similar manner it can be shown that Q will show a single step from 1 to 0 if the switch blade is thrown from B_1 to B_2, even if there is bouncing at the contact B_2 (Prob. 8-2). For applications like the chatterless switch an IC containing four latches is available (74LS279A). Since B_1 is labeled $\overline{S}$ and B_2 is called $\overline{R}$ this package contains quadruple $\overline{SR}$ latches.

The Bistable Latch The addition of two NAND gates preceding $N1$ and $N2$ in Fig. 8-1*b* together with an inverter results in a system (Fig. 8-3) for the storage of one bit of binary information. When the enable input is high ($G = 1$), then the input data D are transferred to the output Q. This statement is readily verified, based on the logic which the latch $N1$-$N2$ satisfies. Thus, if $D = 0$, $S = 0$, $R = 1$, $B_1 = 1$, $B_2 = 0$, and $Q = 0$. Similarly, if $D = 1$, $S = 1$, $R = 0$, $B_1 = 0$, $B_2 = 1$, and $Q = 1$. As long as $G = 1$, any change in the data D appears at Q.

If the latch is inhibited ($G = 0$), then $B_1 = B_2 = 1$ independent of the value of D. Hence Q retains the binary value it had just before G changed from 1 to 0.

The memory cell in Fig. 8-3 may be built from the AOI configuration in Fig. 7-2 (Prob. 8-3). Four such bistable latches are fabricated (74LS375) on a 16-pin chip with complementary outputs (Q and $\overline{Q}$). There are also available eight latches (74ALS573) in a 20-pin package with a 3-state, buffered output.

FIGURE 8-3
A bistable latch trans-
fers data D to the output
Q if $G = 1$.

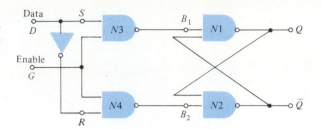

FIGURE 8-3
A bistable latch trans-
fers data D to the output
Q if $G = 1$.

8-2 THE CIRCUIT PROPERTIES OF A BISTABLE LATCH

The NAND-gate latch was described in terms of its logic relationships in Sec. 8-1. In this section we utilize the circuit properties of gates to demonstrate the bistable nature of the latch.

Latches can also be constructed from NOR gates as shown in Fig. 8-4. Consider the situation in which both inputs on each NOR gate are tied together, causing each to act as an inverter. The transfer characteristic of the inverter is displayed in Fig. 8-5a. With no feedback present (the colored path from the output of gate 2 to the input of gate 1 in Fig. 8-4), the two inverters are cascaded. The transfer function v_{o2} versus v_{i1} is depicted in Fig. 8-5b. The feedback path in Fig. 8-4 constrains v_{i1} to equal v_{o2}; this relation is plotted in color in Fig. 8-5b as a straight line with unity slope. Both the transfer characteristic and voltage constraint must be satisfied simultaneously. Circuit operation is consequently confined to be at the intersections of these curves. It appears that there are three possible states, A, B, and C in Fig. 8-5b, in the bistable latch. As explained in the following paragraphs, however, only the states corresponding to A and B can exist in the "real world."

The slope of the transfer characteristic (the gain of the circuit) between P and P' is greater than unity. Consequently, if the circuit state is at C, any extraneous signal (noise) drives the output to either A or B (depending on the polarity of the noise pulse). The situation is analogous to the possible outcomes obtained by flipping a coin. The coin can land on "heads," "tails," or on its edge (theoretically). However, any small perturbation will cause the coin standing on edge to flip over to either heads or tails. We conclude that two of the three states (A and B in Fig. 8-5b or heads and tails) are both *dynamically* and statically stable, whereas the third state (C and on edge) is unstable. Thus, with only two dynamically stable states the bistable nature of the latch is verified. Note that at A or B where the transfer characteristic has zero slope, any small perturbation has no effect.

Latches are available in all four IC technologies described in Chap. 6. Often, latches are used to *set* (S) and *reset* (R) circuits for synchronized operation

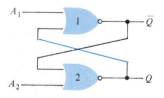

FIGURE 8-4
A latch formed by cross-coupling two NOR gates.

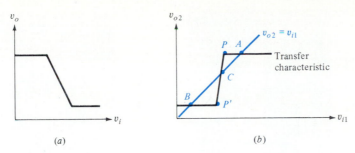

FIGURE 8-5

(*a*) Inverter transfer characteristic. (*b*) The voltage transfer characteristic of cascaded inverters (black). The feedback constraint (blue) indicates that points *A* and *B* are the two stable states of the latch. (Point *C* is dynamically unstable.)

with a pulse sequence. The *SR* latch is a common building block in sequential circuits, as is described in detail in Sec. 8-3. The topology of the 74L5279A is displayed in Fig. 8-6, in which the two NAND gates and feedback paths are identified. The ECL latch in Fig. 8-7 uses the NOR outputs of the basic OR/NOR gate. The circuits in Figs. 8-8*a* and 8-8*b* are NMOS and CMOS realizations, respectively, of *SR* latches which employ a coupled NOR-gate topology.

FIGURE 8-6

The 74LS5279 *SR* latch constructed from two TTL LS NAND gates.

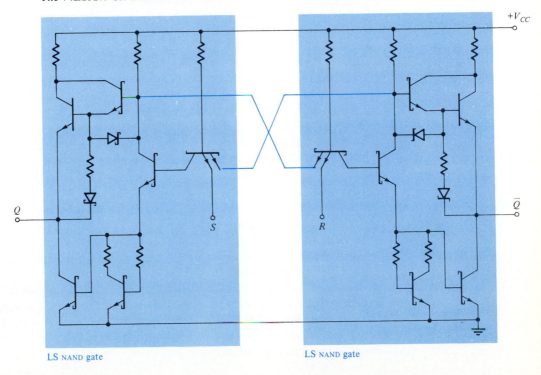

FIGURE 8-7
The ECL latch using the
NOR outputs.

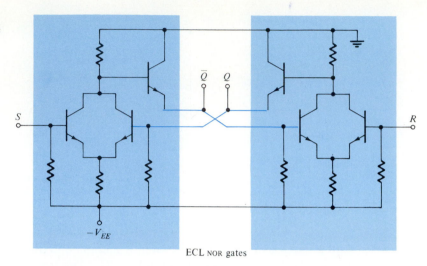

ECL NOR gates

8-3 THE CLOCKED *SR* FLIP-FLOP It is often required to set or reset a latch in synchronism with a pulse train. Such a triggered latch is called a FLIP-FLOP. One type of FLIP-FLOP is introduced in this section and the other commercially available types are discussed in Sec. 8-4.

A Sequential System Many digital systems are pulsed or clocked; i.e., they operate in synchronism with a pulse train of period T, called the system *clock* (abbreviated Ck), such as that indicated in Fig. 8-9. The pulse width t_p is assumed small compared with T. The binary values at each node in the system are assumed to remain constant in each interval between pulses. A transition from one state of the system to another may take place only with the application of a clock pulse. Let Q_n be the output (0 or 1) at a given node in the nth interval (bit time n) preceding the nth clock pulse (Fig. 8-9). Then Q_{n+1} is the corresponding output in the interval immediately after the nth pulse. Such a system where the values $Q_1, Q_2, Q_3, \ldots$, of Q_n are obtained in time sequence at intervals T is called a *sequential* (to distinguish it from a *combinational*) logic system. The value of Q_{n+1} may depend upon the nodal values during the previous (nth) bit time. Under these circumstances a sequential circuit possesses memory.

The *SR* FLIP-FLOP If the enable terminal in Fig. 8-3 is used for the clock (Ck) input and if the inverter is omitted to allow two data inputs S (*set*) and R (*reset*), the SR clocked FLIP-FLOP of Fig. 8-10 results. The gates $N1$ and $N2$ form a latch, whereas $N3$ and $N4$ are the *control*, or *steering*, gates which program the state of the FLIP-FLOP after the pulse appears.

Note that between clock pulses ($Ck = 0$), the outputs of $N3$ and $N4$ are 1 independently of the values of R or S. Hence the circuit is equivalent to the latch in Fig. 8-1a. If $Q = 1$, it remains 1, whereas if $Q = 0$, it remains 0. In other words, *the FLIP-FLOP does not change state between clock pulses;* it is invariant within a bit time.

FIGURE 8-8
(*a*) NMOS and (*b*) CMOS latch realizations using NOR gates.

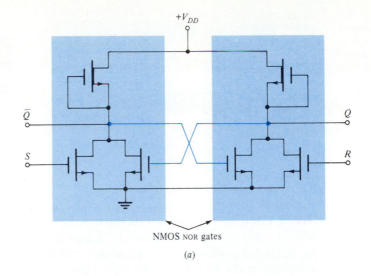

NMOS NOR gates

(*a*)

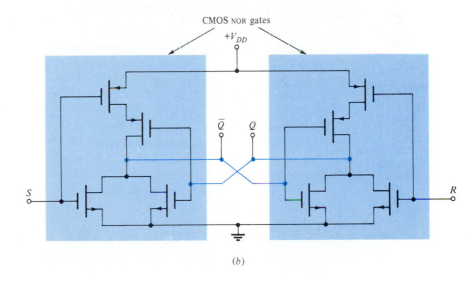

(*b*)

FIGURE 8-9
The output of a master oscillator used as a clock pulse train to synchronize a sequential system.

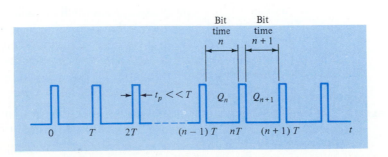

FIGURE 8-10

(a) An *SR* clocked FLIP-FLOP. (b) The truth table. (The question mark indicates that this state cannot be predicted.) (c) The circuit symbol.

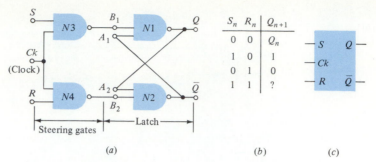

(a)   (b)   (c)

Now consider the time $t = nT (+)$ when a clock pulse is present ($Ck = 1$). If $S = 0$ and $R = 0$, then the outputs of N3 and N4 are 1. By the argument given in the preceding paragraph, the state Q_n of the FLIP-FLOP does not change. Hence, after the pulse passes (in the bit time $n + 1$), the state Q_{n+1} is identical with Q_n. If we denote the values of R and S in the interval just before $t = nT$ by R_n and S_n, then $Q_{n+1} = Q_n$ if $S_n = 0$ and $R_n = 0$. This relationship is indicated in the first row of the truth table of Fig. 8-10b.

If $Ck = 1$, $S_n = 0$ and $R_n = 1$, then $B_1 = 1$ and $B_2 = 0$, so that the situation is that pictured in Fig. 8-1b and the output state is 0. Hence, after the clock pulse passes (at bit time $n + 1$), we find $Q_{n+1} = 0$, confirming the third row of the truth table. If R and S are interchanged and if simultaneously Q is interchanged with $\overline{Q}$, then the logic diagram of Fig. 8-10a is unaltered. Hence the second row of Fig. 8-10b follows from the third row.

If $Ck = 1$, $S_n = 1$, and $R_n = 1$, then the outputs of the NAND gates N3 and N4 are both 0. Hence the input B_1 of N1 as well as B_2 of N2 is 0, so that the outputs of *both* N1 and N2 must be 1. This condition is logically inconsistent with our labeling the two outputs Q and $\overline{Q}$. We must conclude that the output transistor of each gate N1 and N2 is cut off, resulting in both outputs being high (1). At the end of the pulse the inputs at B_1 and B_2 rise from 0 toward 1. Depending upon which input increases faster and on circuit parameter asymmetries, either the stable state $Q = 1$ ($\overline{Q} = 0$) or $Q = 0$ ($\overline{Q} = 1$) will result. Therefore, we have indicated a question mark for Q_{n+1} in the fourth row of the truth table of Fig. 8-10b. This state is said to be *indeterminate, ambiguous,* or *undefined,* and the condition $S_n = 1$ and $R_n = 1$ is forbidden; it must be prevented from taking place.

8-4 J-K-, T-, AND D-TYPE FLIP-FLOPS In addition to the *SR* FLIP-FLOP, three other variations of this basic 1-bit memory are commercially available: the *J-K*, *T*, and *D* types. The *J-K* FLIP-FLOP removes the ambiguity in the truth table in Fig. 8-10b. The *T* FLIP-FLOP acts as a toggle switch and changes the output state with each clock pulse; $Q_{n+1} = \overline{Q}_n$. The *D* type acts as a delay unit which causes the output Q to follow the input D, but delayed by 1 bit time; $Q_{n+1} = D_n$. We now discuss each of these three FLIP-FLOP types.

The J-K FLIP-FLOP This building block is obtained by augmenting the *SR* FLIP-FLOP with two AND gates A1 and A2 (Fig. 8-11a). Data input J and the output $\overline{Q}$ are applied to A1. Since its output feeds S, then $S = J\overline{Q}$. Similarly, data input K and the output

FIGURE 8-11
(*a*) Conversion of an *SR* into a *J-K* FLIP-FLOP. (*b*) The truth table.

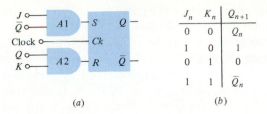

J_n	K_n	Q_{n+1}
0	0	Q_n
1	0	1
0	1	0
1	1	$\bar{Q}_n$

(*a*)   (*b*)

Q are applied to $A2$, and hence $R = KQ$. The logic followed by this system is given in the truth table in Fig. 8-11*b*. This logic can be verified by referring to Table 8-1. There are four possible combinations for the two data inputs J and K. For each of these there are two possible states for Q, and hence Table 8-1 has eight rows. From the J_n, K_n, Q_n, and $\bar{Q}_n$ bits in each row, $S_n = J_n\bar{Q}_n$ and $R_n = K_nQ_n$ are calculated and are entered into the fifth and sixth columns of the table. Using these values of S_n and R_n and referring to the *SR* FLIP-FLOP truth table of Fig. 8-10*b*, the seventh column is obtained. Finally, column 8 follows from column 7 because $Q_n = 1$ in row 4, $Q_n = 0$ in row 5, $\bar{Q}_n = 1$ in row 7, and $\bar{Q}_n = 0$ in row 8.

Columns 1, 2, and 8 of Table 8-1 form the *J-K* FLIP-FLOP truth table in Fig. 8-11*b*. Note that *the first three rows of a J-K truth table are identical with the corresponding row for an SR truth table* (Fig. 8-5*b*). However, the ambiguity of the state $S_n = 1 = R_n$ is now replaced by $Q_{n+1} = \bar{Q}_n$ for $J_n = 1 = K$. *If the two data inputs in the J-K* FLIP-FLOP *are high, the output will be complemented by the clock pulse.*

It is really not necessary to use the AND gates $A1$ and $A2$ of Fig. 8-11*a*, since the same function can be performed by adding an extra input terminal to each NAND gate $N3$ and $N4$ of Fig. 8-10*a*. This simplification is indicated in Fig. 8-12. (Ignore the dashed inputs; i.e., assume that they are both 1.) Now Q and $\bar{Q}$ at the inputs are obtained by the feedback connections (drawn in color) from the outputs.

Preset and Clear   The truth table in Fig. 8-11*b* tells us what happens to the output with the application of a clock pulse, as a function of the data inputs J and K. However,

TABLE 8-1   Truth Table for Fig. 8-11*a*

Column	*1*	*2*	*3*	*4*	*5*	*6*	*7*	*8*
Row	J_n	K_n	Q_n	$\bar{Q}_n$	S_n	R_n	Q_{n+1}	
1	0	0	0	1	0	0	Q_n	Q_n
2	0	0	1	0	0	0	Q_n	
3	1	0	0	1	1	0	1	1
4	1	0	1	0	0	0	Q_n	
5	0	1	0	1	0	0	Q_n	0
6	0	1	1	0	0	1	0	
7	1	1	0	1	1	0	1	$\bar{Q}_n$
8	1	1	1	0	0	1	0	

the value of the output before the pulse is applied is arbitrary. The addition of the dashed inputs in Fig. 8-12 allows the initial state of the FLIP-FLOP to be assigned. For example, it may be required to *clear* the latch, i.e., to specify that $Q = 0$ when $Ck = 0$.

The clear operation may be accomplished by programming the clear input to 0 and the *preset* input to 1; $Cr = 0$, $Pr = 1$, $Ck = 0$. Since $Cr = 0$, the output of N2 (Fig. 8-12) is $\overline{Q} = 1$. Since $Ck = 0$, the output of N3 is 1, and hence all inputs to N1 are 1 and $Q = 0$, as desired. Similarly, if it is required to preset the latch into the 1 state, it is necessary to choose $Pr = 0$, $Cr = 1$, $Ck = 0$. The preset and clear data are called *direct*, or *asynchronous*, inputs; i.e., they are not in synchronism with the clock, but may be applied at any time in between clock pulses. Once the state of the FLIP-FLOP is established asynchronously, the direct inputs must be maintained at $Pr = 1$, $Cr = 1$, before the next pulse arrives in order to *enable* the FLIP-FLOP. The data $Pr = 0$, $Cr = 0$, must not be used since they lead to an ambiguous state. Why?

The logic symbol for the *J-K* FLIP-FLOP is indicated in Fig. 8-12*b*, and the inputs for proper operation are given in Fig. 8-12*c*.

The Race-Around Condition   There is a possible physical difficulty with the *J-K* FLIP-FLOP constructed as in Fig. 8-12. Truth table 8-1 is based upon combinational logic, which assumes that the inputs are independent of the outputs. However, because of the feedback connection Q $(\overline{Q})$ at the input to K (J), the input will change during the clock pulse ($Ck = 1$) if the output changes state. Consider, for example, that the inputs to Fig. 8-12 are $J = 1$, $K = 1$, and $Q = 0$. When the pulse is applied, the output becomes $Q = 1$ (according to row 7 of Table 8-1), this change taking place after a time interval Δt equal to the propagation delay (Sec. 6-4) through two NAND gates in series in Fig. 8-7. Now $J = 1$, $K = 1$, and $Q = 1$, and from row 8 of Table 8-1, we find that the input changes back to $Q = 0$. Hence we must conclude that for the duration t_p (Fig. 8-9) of the pulse (while $Ck = 1$), the output will oscillate back and forth between 0 and 1. At the end of the pulse ($Ck = 0$), the value of Q is ambiguous.

The situation just described is called a *race-around condition*. It can be avoided if $t_p < \Delta t < T$. However, with IC components the propagation delay

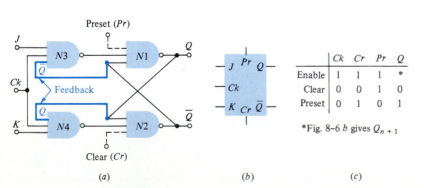

Preset (*Pr*)

Clear (*Cr*)

(*a*)

(*b*)

	Ck	Cr	Pr	Q
Enable	1	1	1	*
Clear	0	0	1	0
Preset	0	1	0	1

*Fig. 8-6 *b* gives Q_{n+1}

(*c*)

FIGURE 8-12

(*a*) A *J-K* FLIP-FLOP; (*b*) logic symbol; (*c*) the conditions necessary for synchronous operation (row 1), asynchronous clearing (row 2), or presetting (row 3).

FIGURE 8-13
A *J-K* master-slave FLIP-
FLOP.

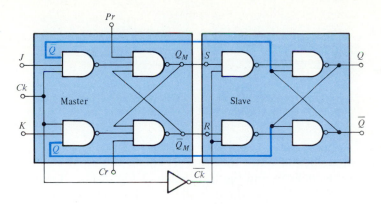

is very small, usually much less than the pulse width t_p. Hence the above inequality is *not* satisfied, and the output is indeterminate. Lumped delay lines can be used in series with the feedback connections of Fig. 8-12 in order to increase the loop delay beyond t_p, and hence to prevent the race-around difficulty. A more practical IC solution is now described.

The Master-Slave *J-K* FLIP-FLOP   In Fig. 8-13 is shown a cascade of two *SR* FLIP-FLOPS with feedback from the output of the second (called the *slave*) to the input of the first (called the *master*). Positive clock pulses are applied to the master, and these are inverted before being used to excite the slave. For $Pr = 1$, $Cr = 1$, and $Ck = 1$, the master is enabled and its operation follows the *J-K* truth table of Fig. 8-11*b*. Furthermore, since $\overline{Ck} = 0$, the slave *SR* FLIP-FLOP is inhibited (cannot change state), so that Q_n is invariant for the pulse duration t_p. Clearly, the race-around difficulty is circumvented with the master-slave topology. After the pulse passes, $Ck = 0$, so that the master is inhibited and $\overline{Ck} = 1$, which causes the slave to be enabled. The slave is an *SR* FLIP-FLOP, which follows the logic in Fig. 8-10*b*. If $S = Q_M = 1$ and $R = \overline{Q_M} = 0$, then $Q = 1$ and $\overline{Q} = 0$. Similarly, if $S = Q_M = 0$ and $R = \overline{Q_M} = 1$, then $Q = 0$ and $\overline{Q} = 1$. In other words, in the interval between clock pulses, the value of Q_M is transferred to the output Q. In summary, during a clock pulse the output Q does not change but Q_M follows *J-K* logic; at the end of the pulse, the value of Q_M is transferred to Q.

It should be emphasized that the data in *J* and *K* must remain constant for the pulse duration or an erroneous output may result (Prob. 8-12). The 16-pin MSI package (MC8-104135) contains two independent *J-K* master-slave FLIP-FLOPS. Some commercially available FLIP-FLOPS also have internal AND or AOI (54LS72) gates at the inputs to provide multiple *J* and *K* inputs, thereby avoiding the necessity of external gates in applications where these may be required.

The *D*-Type FLIP-FLOP   If a *J-K* FLIP-FLOP is modified by the addition of an inverter as in Fig. 8-14*a*, so that *K* is the complement of *J*, the unit is called a *D* (*delay*) FLIP-FLOP. From the *J-K* truth table of Fig. 8-11*b*, $Q_{n+1} = 1$ for $D_n = J_n = \overline{K}_n = 1$ and $Q_{n+1} = 0$ for $D_n = J_n = \overline{K}_n = 0$. Hence $Q_{n+1} = D_n$. The output Q_{n+1}

FIGURE 8-14
(a) Conversion of a J-K
FLIP-FLOP into a D-type
latch, (b) the logic sym-
bol, (c) the truth table.

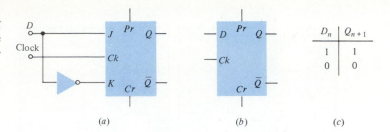

(a)   (b)   (c)

after the pulse (bit time $n + 1$) equals the input D_n before the pulse (bit time n), as indicated in the truth table of Fig. 8-14c. If the FLIP-FLOP in Fig. 8-14a is of the SR type, the unit also functions as a D-type latch with the clock (Ck) replaced by the enable (G) of Fig. 8-3. There is no ambiguous state because $J = K = 1$ is not possible.

The D-type FLIP-FLOP is a binary used to provide delay. The bit on the D line is transferred to the output at the next clock pulse, and hence this unit functions as a 1-bit delay device.

The *T*-Type FLIP-FLOP   This unit changes state with each clock pulse, and hence it acts as a toggle switch. If $J = K = 1$, then $Q_{n+1} = \bar{Q}_n$, so that the J-K FLIP-FLOP is converted into a T-type FLIP-FLOP. In Fig. 8-15a such a system is indicated with a data input T. The logic symbol is shown in Fig. 8-15b, and the truth table in Fig. 8-15c. The SR- and the D-type latches can also be converted into toggle, or complementing, FLIP-FLOPS (Prob. 8-14).

Summary   Four FLIP-FLOP configurations, SR, J-K, D, and T are important. The logic satisfied by each type is repeated for easy reference in Table 8-2. An IC FLIP-FLOP is driven synchronously by a clock, and in addition it may (or may not) have direct inputs for asynchronous operation, preset (Pr), and clear (Cr). A direct input can be 0 only in the interval between clock pulses when $Ck = 0$. When $Ck = 1$, both asynchronous inputs must be high; $Pr = 1$ and $Cr = 1$. The inputs must remain constant during a pulse width, $Ck = 1$. For a master-slave FLIP-FLOP the output Q remains constant for the pulse duration and

TABLE 8-2   FLIP-FLOP Truth Tables

SR			J-K			D		T		Direct inputs			
S_n	R_n	Q_{n+1}	J_n	K_n	Q_{n+1}	D_n	Q_{n+1}	T_n	Q_{n+1}	Ck	Cr	Pr	Q
0	0	Q_n	0	0	Q_n	1	1	1	$\bar{Q}_n$	0	1	0	1
1	0	1	1	0	1	0	0	0	Q_n	0	0	1	0
0	1	0	0	1	0					1	1	1	*
1	1	?	1	1	Q_n								
	Fig. 8-9			Fig. 8-11		Fig. 8-14		Fig. 8-15					

*Refer to truth table SR, J-K, D, or T for Q_{n+1} as a function of the inputs.

FIGURE 8-15
(*a*) A *J-K* FLIP-FLOP converted into a *T*-type FLIP-FLOP with data input *T*, (*b*) the logic symbol, (*c*) the truth table.

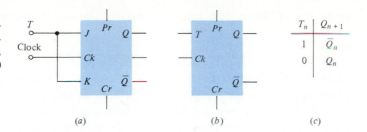

T_n	Q_{n+1}
1	$\bar{Q}_n$
0	Q_n

(*a*)　　　　　(*b*)　　　　　(*c*)

changes only after *Ck* changes from 1 to 0, at the *negative-going (trailing) edge* of the pulse. It is also possible to design a *J-K* FLIP-FLOP so that the output changes at the *positive-going (leading) edge* of the pulse. The 74LS109A chip is a dual positive-edge-triggered *J-K* FLIP-FLOP with preset and clear inputs. The MC10H176 contains six (hex) positive-edge-triggered *D*-type FLIP-FLOPS.

The toggle, or complementing, FLIP-FLOP is not available commercially because a *J-K* can be used as a *T* type by connecting the *J* and *K* inputs together (Fig. 8-15).

The FLIP-FLOP is available in all the IC digital families, and the maximum frequencies of operation are given in Table 6-4.

8-5 SHIFT REGISTERS

Since a binary is a 1-bit memory, then *n* FLIP-FLOPS can store an *n*-bit word. This combination is referred to as a *register*. To allow the data in the word to be read into the register serially, the output of one FLIP-FLOP is connected to the input of the following binary. Such a configuration, called a *shift register,* is indicated in Fig. 8-16. Each FLIP-FLOP is of the *SR* (or *J-K*) master-slave type. Note that the stage which is to store the most significant bit (MSB) is converted into a *D*-type latch (Fig. 8-14) by connecting *S* and *R* through an inverter. The 5-bit shift register indicated in Fig. 8-16 is available on a single chip in a 16-pin package (medium-scale integration). We shall now explain the

FIGURE 8-16
A 5-bit shift register.

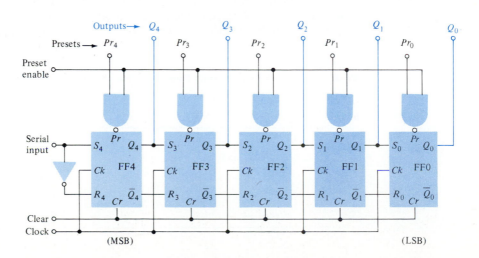

operation of this system by assuming that the serial data 01011 is to be registered. (The least significant bit is the rightmost digit, which in this case is a 1.)

Series-in, Parallel-out (SIPO) Register The FLIP-FLOPS are cleared by applying a 0 to the *clear* input (while the preset enable is low) so that every output $Q_0, Q_1, \ldots, Q_4$ is 0. Then Cr is set to 1 and Pr is held constant at 1 (by keeping the preset enable at 0). The serial data train and the synchronous clock are now applied. The least significant bit (LSB) is entered into the master latch of FF4 when Ck changes from a 0 to a 1 by the action of a D-type FLIP-FLOP. After the clock pulse, the 1 is transferred to the slave latch of FF4 and $Q_4 = 1$, while all other outputs remain at 0.

At the second clock pulse the state of Q_4 is transferred to the master latch of FF3 by the action of an SR FLIP-FLOP. Simultaneously, the next bit (a 1 in the 01011 word) enters the master of FF4. After the second clock pulse the bit in each master transfers to its slave and $Q_4 = 1$, $Q_3 = 1$, and the other outputs remain 0. The readings of the register *after* each pulse are given in Table 8-3. For example, after the third pulse, Q_3 has shifted to Q_2, Q_4 to Q_3, and the third input bit (0) has entered FF4, so that $Q_4 = 0$. We may easily follow this procedure and see that by registering each bit in the MSB FLIP-FLOP and then shifting to the right to make room for the next digit, the input word becomes installed in the register after the nth clock pulse (for an n-bit code). Of course, the clock pulses must stop at the moment the word is registered. Each output is available on a separate line, and they may be read simultaneously. Since the data entered the system serially and came out in parallel, this shift register is a *serial-to-parallel* converter. It is also referred to as a *series-in, parallel-out (SIPO) register*. A *temporal code* (a time arrangement of bits) has been changed to a *spatial code* (information stored in a static memory).

Master-slave FLIP-FLOPS are required because of the race problem between stages (Sec. 8-4). If all FLIP-FLOPS were to change states simultaneously, there would be an ambiguity as to what data would transfer from the preceding stage. For example, at the third clock pulse, Q_4 changes from 1 to 0, and it would be questionable as to whether Q_3 would become a 1 or a 0. Hence it is necessary that Q_4 remain a 1 until this bit is entered into FF3, and only then may it change

TABLE 8-3 Reading of Shift Register after Each Clock Pulse

Clock pulse	Word bit	Q_4	Q_3	Q_2	Q_1	Q_0
1	1	1	0	0	0	0
2	1	1	1	0	0	0
3	0	0	1	1	0	0
4	1	1	0	1	1	0
5	0	0	1	0	1	1

to 0. The master-slave configuration provides just this action. If in Fig. 8-13, the $J(K)$ input is called $S(R)$ and if the (blue) feedback connections are omitted an SR master-slave FLIP-FLOP results. The 74LS164 is an 8-bit SIPO shift register with gated (enable) inputs.

Series-in, Series-out (SISO) Register We may take the output at Q_0 and read the register serially if we apply n clock pulses, for an n-bit word. After the nth pulse each FLIP-FLOP reads 0. Note that the shift-out clock rate may be greater or smaller than the original pulse frequency. Hence here is a method for changing the spacing in time of a binary code, a process referred to as *buffering*.

The 74LS91 MSI package is an 8-bit SISO register with gated inputs and complementary outputs. Since a SISO chip requires only one data input pin and one data output pin, independent of the number of bits to be stored, a very long shift register is possible using LSI and VLSI technologies.

Parallel-in, Series-out (PISO) Register Consider the situation where the word bits are available in parallel, e.g., at the outputs from a ROM (Sec. 7-9). It is desired to present this code, say, 01011, in serial form.

The LSB is applied to Pr_0, the 2^1 bit to Pr_1, . . ., so that $Pr_0 = 1$, $Pr_1 = 1$, $Pr_2 = 0$, $Pr_3 = 1$, and $Pr_4 = 0$. The register is first cleared by $Cr = 0$, and then $Cr = 1$ is maintained. A 1 at the *preset enable* input activates all kth input NAND gates for which $Pr_k = 1$. The preset of the corresponding kth FLIP-FLOP is $Pr = 0$, and this stage is therefore preset to 1 (Table 8-2). In the present illustration FF0, FF1, and FF3 are preset and the input word 01011 is written into the register, all bits in parallel, by the preset enable pulse.

As explained above, the stored word may be read serially at Q_0 by applying five clock pulses. This is a *parallel-to-serial*, or a *spatial-to-temporal, converter*. The 74ALS166 package is a *parallel-in, series-out* (PISO), 8-bit register.

Parallel-in, Parallel-out (PIPO) Register The data is entered as explained above by applying a 1 at the preset enable, or *write,* terminal. It is then available in parallel form at the outputs Q_0, Q_1, ... If it is desired to *read* the register during a selected time, each output Q_k is applied to one input of a 2-input AND gate N_k, and the second input of each AND is excited by a read pulse. The output of N_k is 0 except for the pulse duration, when it reads 1 if $Q_k = 1$. (The gates N_k are not shown in Fig. 8-16.)

Note that in this application the system is not operating as a shift register since there is no clock required (and no serial input). Each FLIP-FLOP is simply used as an isolated 1-bit read-write memory.

Right-Shift, Left-Shift (Bidirectional) Register Some commercial shift registers are equipped with gates which allow shifting the data from right to left as well as in the reverse direction. One application for such a system is to perform multiplication or division by multiples of 2, as will now be explained. Consider first a right-shift register as in Fig. 8-16 and that the serial input is held low.

Assume that a binary number is stored in a shift register, with the least-significant bit stored in FF0. Now apply one clock pulse. Each bit then moves to the next lower significant place, and hence is divided by 2. The number now held in the register is half the original number, provided that FF0 was originally 0. Since the 2^0 bit is lost in the shift to the right, then if FF0 was originally in the 1 state, corresponding to the decimal number 1, after the shift the register is in error by the decimal number 0.5. The next clock pulse causes another division by 2, and so on.

Consider now that the system is wired so that each clock pulse causes a shift to the left. Each bit now moves to the next higher significant digit, and the number stored is multiplied by 2.

The logic diagram for the 74LS194A 4-bit bidirectional shift register is given in Fig. 8-17. This is a *universal* register because it can function in all the modes discussed in this section; SIPO, SISO, PISO, PIPO, and as a bidirectional register. It has two control inputs, S_0 and S_1, which allow the four operational modes listed in Table 8-4 to be realized. The verification of this table is considered in Prob. 8-17. The 8-bit 74ALS299 universal shift register has the same structure as that indicated in Fig. 8-17. It has the equivalent of 87 gates, and comes in a 24-pin package.

Digital Delay Line   A shift register may be used to introduce a time delay Δ into a system, where Δ is an integral multiple of the clock period T. Thus an input pulse train appears at the output of an n-stage register delayed by a time $(n - 1)T = \Delta$.

Sequence Generator   An important application of a shift register is to generate a binary sequence. This system is also called a *word, code,* or *character, generator.* The shift register FLIP-FLOPS are preset to give the desired code. Then the clock applies shift pulses, and the output of the shift register gives the temporal pattern corresponding to the specified sequence. Clearly, we have just described a parallel-in, series-out register. For test purposes it is often necessary that the code be repeated continuously. This mode of operation is easily obtained by feeding the output Q_0 of the register back into the serial input to form a "reentrant shift register." Such a configuration is called a *dynamic,* or *circulating, memory,* or a *shift-register read-only memory.*

TABLE 8-4   Modes of Operation of a Universal Register

S_0*	S_1*	Operational mode
0	0	Inhibit clock
1	1	Parallel input†
1	0	Shift right
0	1	Shift left

*S_0 and S_1 should be changed only while the clock input is high.

†Data is loaded into a FLIP-FLOP after a clock pulse. During loading, serial data flow is inhibited.

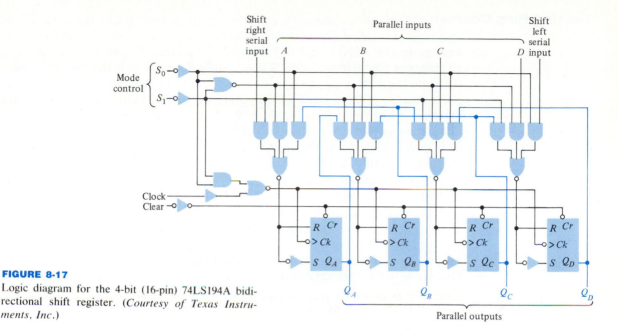

FIGURE 8-17

Logic diagram for the 4-bit (16-pin) 74LS194A bidirectional shift register. (*Courtesy of Texas Instruments, Inc.*)

A sequence generator may also be obtained from a multiplexer (Sec. 7-7) and a number of simultaneous sequences may be generated using a ROM (Sec. 7-9).

Shift-Register Ring Counter   Consider the 5-bit shift register (Fig. 8-16) with Q_0 connected to the serial input. Such a circulating memory forms a *ring counter*. Assume that all FLIP-FLOPS are cleared and then that FF0 is preset so that $Q_0 = 1$ and $Q_4 = Q_3 = Q_2 = Q_1 = 0$. The first clock pulse transfers the state of FF0 to FF4, so that after the pulse $Q_4 = 1$ and

$$Q_3 = Q_2 = Q_1 = Q_0 = 0$$

Succeeding pulses will transfer the state 1 progressively around the ring. The count is read by noting which FLIP-FLOP is in state 1; no decoding is necessary.

Consider a ring counter with N stages. If the interval between triggers is T, then the output from any binary stage is a pulse train of period NT, with each pulse of duration T. The output pulse of one stage is delayed by a time T from a pulse in the preceding stage. These pulses may be used where a set of sequential gating waveforms is required. Thus a ring counter is analogous to a stepping switch, where each triggering pulse causes an advance of the switch by one step.

Since there is one output pulse for each N clock pulses, the counter is also a *divide-by-N* unit, or an $N:1$ *scaler*. Typically, TTL shift-register counters operate at frequencies as high as 25 MHz.

Twisted-Ring Counter The topology where $\overline{Q}_0$ (rather than Q_0) is fed back to the input of the shift register is called a *twisted-ring, switched-tail, moebius,* or *Johnson, counter.* This system is a $2N:1$ scaler. To verify this statement consider that initially all stages in Fig. 8-16 are in the 0 state. Since $S_4 = \overline{Q}_0 = 1$, the first pulse puts FF4 into the 1 state; $Q_4 = 1$, and all other FLIP-FLOPS remain in the 0 state. Since now $S_3 = Q_4 = 1$ and S_4 remains in the 1 state, then after the next pulse there results $Q_4 = 1$, $Q_3 = 1$, $Q_2 = 0$, $Q_1 = 0$, and $Q_0 = 0$. In other words, pulse 1 causes only Q_4 to change state, and pulse 2 causes only Q_3 to change from 0 to 1. Continuing the analysis, we see that pulses 3, 4, and 5 cause Q_2, Q_1, and Q_0, in turn, to switch from the 0 to the 1 state. At the end of five pulses all FLIP-FLOPS are in the 1 state.

After pulse 5, $S_4 = \overline{Q}_0$ changes from 1 to 0. Hence the sixth pulse causes Q_4 to change to 0. The seventh pulse resets Q_3 to 0, and so on, until, at the tenth pulse, all stages have been returned to the 0 state, and the counting cycle is complete. We have demonstrated that this five-stage twisted-ring configuration is a 10:1 counter. To read the count requires a 5-to-10-line decoder, but because of the unique waveforms generated, only 2-input AND gates are required (Prob. 8-19).

Almost all the registers and counters are available in CMOS technology. The numbering of such IC packages is identical to that for the TTL family, except for the letters that indicate the technology. Thus a 74LS194A TTL bidirectional register has virtually the same characteristics as the CMOS 74HC194.

8-6 RIPPLE (ASYNCHRONOUS) COUNTERS

The ring counters discussed in the preceding section do not make efficient use of the FLIP-FLOPS. A 5:1 counter (or 10:1 with the Johnson ring) is obtained with five stages, whereas five FLIP-FLOPS define $2^5 = 32$ states. By modifying the interconnections between stages (*not* using the shift-register topology), we now demonstrate that n binaries can function as a $2^n:1$ counter.

Ripple Counter Consider a chain of four *J-K* master-slave FLIP-FLOPS with the output Q of each stage connected to the clock input of the following binary, as in Fig. 8-18. The pulses to be counted are applied to the clock input of FF0. For all stages J and K are tied to the supply voltage, so that $J = K = 1$. This connection converts each stage to a T-type FLIP-FLOP (Fig. 8-15), with $T = 1$.

It should be recalled that, for a T-type binary with $T = 1$, the master changes state every time the waveform at its clock input changes from 0 to 1 and that the new state of the master is transferred to the slave when the clock falls from 1 to 0. This operation requires that

1. Q_0 changes state at the *falling* edge of each pulse.
2. All other Q's make a transition when and only when the output of the preceding FLIP-FLOP changes from 1 to 0. This negative transition "ripples" through the counter from the LSB to the MSB.

FIGURE 8-18

FIGURE 8-18

A chain of FLIP-FLOPS connected as a ripple counter (74LS93). The 74LS393 is a dual 4- digit binary counter MSI package.

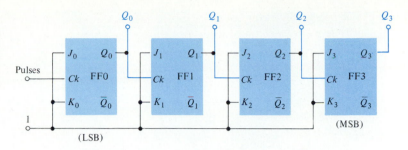

(LSB)

Following these two rules, the waveforms shown in Fig. 8-19 are obtained. Table 8-5 lists the states of all the binaries of the chain as a function of the number of externally applied pulses. This table may be verified directly by comparison with the waveform chart of Fig. 8-19. Note that in Table 8-5 the FLIP-FLOPS have been ordered in the reverse direction from their order in Fig. 8-18. We observe that the ordered array of states 0 and 1 in any row in Table 8-5 is precisely the binary representation of the decimal number of input pulses. Thus the chain of FLIP-FLOPS *counts* in the binary system.

A chain of n binaries will count up to the number 2^n before it resets itself into its original state. Such a chain is referred to as a counter *modulo* 2^n. To read the counter, the 4-bit words (numbers) in Table 8-5 are recognized with a decoder, which in turn drives visible numerical indicators (Sec. 7-11). Spikes are possible in any counter unless all FLIP-FLOPS change state simultaneously. To eliminate the spikes at the decoder output, a strobe pulse is used (S in Fig. 7-17) so that the counter is read only after the spikes have decayed and a steady state is reached.

Up-Down Counter A counter which can be made to count in either the forward or reverse direction is called an *up-down*, a *reversible*, or a *forward-backward*, counter.

TABLE 8-5 States of the FLIP-FLOPS Shown in Fig. 8-18

Number of input pulses	FLIP-FLOP outputs				Number of input pulses	FLIP-FLOP outputs			
	Q_3	Q_2	Q_1	Q_0		Q_3	Q_2	Q_1	Q_0
0	0	0	0	0	9	1	0	0	1
1	0	0	0	1	10	1	0	1	0
2	0	0	1	0	11	1	0	1	1
3	0	0	1	1	12	1	1	0	0
4	0	1	0	0	13	1	1	0	1
5	0	1	0	1	14	1	1	1	0
6	0	1	1	0	15	1	1	1	1
7	0	1	1	1	16	0	0	0	0
8	1	0	0	0					

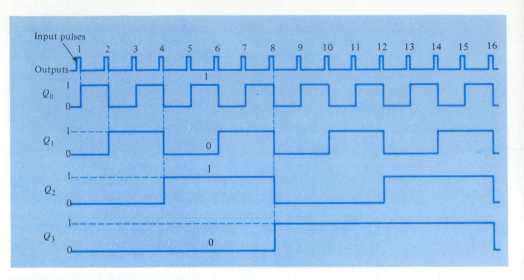

FIGURE 8-19
Waveforms for the 4-state ripple counter. Note that after pulse 5, $Q_0 = 1$, $Q_1 = 0$, $Q_2 = 1$, and $Q_3 = 0$. These binary outputs correspond to the decimal number 5.

Forward counting is accomplished, as we have seen, when the trigger input of a succeeding binary is coupled to the Q output of a preceding binary. The count will proceed in the reverse direction if the coupling is made instead to the $\overline{Q}$ output, as we shall now verify.

If a binary makes a transition from state 0 to 1, the output $\overline{Q}$ will make a transition from state 1 to 0. This negative-going transition in $\overline{Q}$ will induce a change in state in the succeeding binary. Hence, for the reversing connection, the following rules apply:

1. FLIP-FLOP FF0 makes a transition at each externally applied pulse.

2. Each of the other binaries makes a transition when and only when the preceding FLIP-FLOP goes from state 0 to state 1.

If these rules are applied to any of the numbers in Table 8-5, the next smaller number in the table results. For example, consider the number 12, which is 1100 in binary form. At the next pulse, the rightmost 0 (corresponding to Q_0) becomes 1. This change of state from 0 to 1 causes Q_1 to change state from 0 to 1, which in turn causes Q_2 to change state from 1 to 0. This last transition is in the direction not to affect the following binary, and hence Q_3 remains in state 1. The net result is that the counter reads 1011, which represents the number 11. Since we started with 12 and ended with 11, a reverse count has taken place.

The logic block diagram for an up-down counter is indicated in Fig. 8-20. For simplicity in drawing, no connections to J and K are indicated. For a ripple

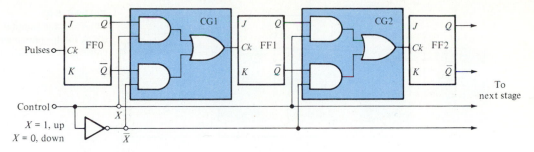

FIGURE 8-20
An up-down ripple counter. (It is understood that $J = K = 1$.)

counter it is always to be understood that $J = K = 1$ as in Fig. 8-18. The two-level AND-OR gates CG1 and CG2 between stages constitute a 2-input multiplexer which controls the direction of the counter. Note that this logic combination is equivalent to a NAND-NAND configuration. If the input X is a 1 (0), then Q ($\overline{Q}$) is effectively connected to the following FLIP-FLOP and pulses are added (subtracted). In other words, $X = 1$ converts the system to an *up* counter and $X = 0$ to a *down* counter. The control X may not be changed from 1 to 0 (or 0 to 1) between input pulses, because a spurious count may be introduced by this transition. (The synchronous counter of Fig. 8-22 does not have this difficulty and hence up-down counters are operated synchronously, Sec. 8-7).

Divide-by-*N* Counter   It may be desired to count to a base N which is not a power of 2. We may prefer, for example, to count to the base 10, since the decimal system is the one with which we are most familiar. To construct such a counter, start with a ripple chain of n FLIP-FLOPS such that n is the smallest number for which $2^n > N$. Add a feedback gate so that at count N all binaries are reset to zero. This feedback circuit is simply a NAND gate whose output feeds all *clear* inputs in parallel. Each input to the NAND gate is a FLIP-FLOP output Q which becomes 1 at the count N.

Let us illustrate the above procedure for a decade counter. Since the smallest value of n for which $2^n > 10$ is $n = 4$, then four FLIP-FLOPS are required. The decimal number 10 is the binary number 1010 (LSB), and hence $Q_0 = 0$, $Q_1 = 1$, $Q_2 = 0$, and $Q_3 = 1$. The inputs to the feedback NAND gate are therefore Q_1 and Q_3, and the complete circuit is shown in Fig. 8-21a. Note that after the tenth pulse Q_1 and Q_3 both go to 1, the output of the NAND gate becomes 0, and all FLIP-FLOPS are cleared (reset to 0). (Note that Q_1 and Q_3 first become 1 and then return to 0 after pulse 10, thereby generating a narrow spike.)

If the propagation delay from the clear input to the FLIP-FLOP output varies from stage to stage, the clear operation may not be reliable. In the above example, if FF3 takes an appreciably longer time to reset than FF1, then when Q_1 returns to 0, the output of the NAND gate goes to 1, so that $Cr = 1$ and Q_3 will not reset. Wide variations in reset propagation time may occur if the counter

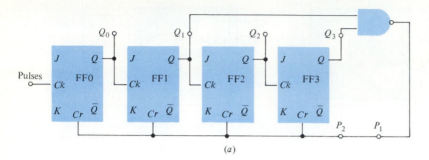

FIGURE 8-21
(*a*) A decade counter ($J = K = 1$).
(*b*) A latch eliminates resetting difficulties due to unequal internal delays.

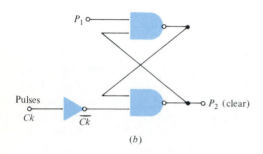

outputs are unevenly loaded. A method of eliminating the difficulty with resetting is to use a latch to memorize the output of the NAND gate at the Nth pulse. The lead in Fig. 8-21a between the NAND output P_1 and the clear input P_2 is opened, and the circuit drawn in Fig. 8-21b is inserted between these two points. The operation of the latch is considered in detail in Prob. 8-24. The 74LS90 decade counter, which does not require a latch, is indicated in Prob. 8-26. Two such counters are available in a single package (74LS390). A 12:1 counter (74LS92) is considered in Prob. 8-28.

A divide-by-6 counter is obtained using a 3-bit ripple counter, and since for $N = 6$, $Q_1 = 1 = Q_2$, then Q_1 and Q_2 are the inputs to the feedback NAND gate. Similarly, a divide-by-7 counter requires a 3-input NAND gate with inputs Q_0, Q_1, and Q_2.

In some applications it is important to be able to program the count (the value of N) of a divide-by-N counter, either by means of switches or through control data inputs at the preset terminals. Such a *programmable* or *presettable* counter is indicated in the figure of Prob. 8-29.

Consider that it is required to count up to 10,000 and to indicate the count visually in the decimal system. Since $10,000 = 10^4$, then four decade-counter units, such as in Fig. 8-21, are cascaded. A BCD-to-decimal decoder–lamp driver (Sec. 7-6) or a BCD-to-seven-segment display decoder (Sec. 7-11) is used with each unit to make visible the four decimal digits giving the count.

8-7 SYNCHRONOUS COUNTERS The *carry propagation delay* is the time required for a counter to complete its response to an input pulse. The carry time of a ripple counter is longest when each stage is in the 1 state. For in this situation, the next pulse must cause all previous FLIP-FLOPS to change state. Any particular binary will not respond until the preceding stage has nominally completed its transition. The clock pulse effectively "ripples" through the chain. Hence the carry time will be of the order of magnitude of the sum of the propagation delay times (Sec. 6-15) of all the binaries. If the chain is long, the carry time may well be longer than the interval between input pulses. In such a case, it will not be possible to read the counter between pulses.

If the asynchronous operation of a counter is changed so that all FLIP-FLOPS are clocked simultaneously (synchronously) by the input pulses, the propagation delay time may be reduced considerably. Repetition rate is limited by the delay of any one FLIP-FLOP plus the propagation times of any control gates required. Typically, the maximum frequency of operation of a 4-bit synchronous counter is greater than 100 MHz in the ECL family (MC10137). In the TTL family, this value is typically 75 MHz in the AS series. CMOS and the LS series TTL are limited to about 40-MHz maximum frequencies. The 100K series ECL family can operate at even higher frequencies than the 10K ECL family, which is about twice that of a ripple counter. Another advantage of the synchronous counter is that no decoding spikes appear at the output since all FLIP-FLOPS change state at the same time. Hence no strobe pulse is required when decoding a synchronous counter.

Series Carry

A 5-bit synchronous counter is indicated in Fig. 8-22. Each FLIP-FLOP is a T type, obtained by tying the J terminal to the K terminal of a J-K FLIP-FLOP (Fig. 8-15). If $T = 0$, there is no change of state when the binary is clocked, and if $T = 1$, the FLIP-FLOP output is complemented with each pulse.

The connections to be made to the T inputs are deduced from the waveform chart of Fig. 8-19.

Q_0 toggles with each pulse: $\qquad\qquad\qquad\qquad T_0 = 1$

Q_1 complements only if $Q_0 = 1$: $\qquad\qquad\qquad T_1 = Q_0$

Q_2 becomes $\overline{Q}_2$ only if $Q_0 = Q_1 = 1$: $\qquad\quad T_2 = Q_0 Q_1$

Q_3 toggles only if $Q_0 = Q_1 = Q_2 = 1$: $\qquad\quad T_3 = Q_0 Q_1 Q_2$

FIGURE 8-22
A 5-bit synchronous counter with series carry ($J = K = T$).

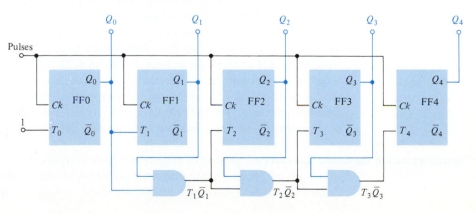

Extending this logic to Q_4, we conclude that $T_4 = Q_0Q_1Q_2Q_3$. Therefore, the T logic is given by

$$T_0 = 1 \qquad T_1 = Q_0 \qquad T_2 = T_1Q_1 \qquad T_3 = T_2Q_2 \qquad T_4 = T_3Q_3 \qquad (8\text{-}1)$$

Clearly, the 2-input AND gates of Fig. 8-22 perform this logic.

The minimum time T_{min} between pulses is the interval required for each J and K node to reach its steady-state value and is given by

$$T_{min} = T_F + (n - 2)T_G \qquad (8\text{-}2)$$

where T_F is the propagation delay of one FLIP-FLOP, and T_G is the propagation delay of one AND gate (actually, a NAND gate plus an inverter). The maximum pulse frequency for series carry is the reciprocal of T_{min}.

Parallel Carry   Since the carry passes through all the control gates in series in Fig. 8-22, this is a synchronous counter with *series*, or *ripple, carry*. The maximum frequency of operation can be improved by using parallel, or *look-ahead, carry*, where the toggle input to each binary comes from a multi-input AND gate excited by the outputs from every preceding FLIP-FLOP. From Eq. (8-1) it follows that

$$T_1 = Q_0 \qquad T_2 = Q_0Q_1 \qquad T_3 = Q_0Q_1Q_2 \qquad T_4 = Q_0Q_1Q_2Q_3 \qquad (8\text{-}3)$$

Hence T_4 is obtained from a 4-input AND gate fed by Q_0, Q_1, Q_2, and Q_3. Clearly, for parallel carry,

$$T_{min} = T_F + T_G \qquad (8\text{-}4)$$

which may be considerably smaller than the corresponding time for series carry given by Eq. (8-2), particularly if n is large (high division ratios).

The disadvantages of a parallel-carry counter are (1) the large fan-in of the gates; the gate feeding T_k requires k inputs; and (2) the heavy loading of the FLIP-FLOPS at the beginning of the chain; the fan-out of Q_0 is $n - 1$, since it must feed the carry gates of every succeeding stage.

Up-Down Synchronous Counter with Parallel Carry   As explained in the preceding section, a counter is reversed if $\overline{Q}$ is used in place of Q in the coupling from stage to stage. Hence a synchronous up-down counter is obtained if the control gates CG in Fig. 8-20 are interposed between the FLIP-FLOPS in Fig. 8-22. This change to an up-down synchronous counter is made in Fig. 8-23, where CG is now indicated as a NAND-NAND gate (equivalent to the AND-OR logic of Fig. 8-20). Note that CG1 is identical in Figs. 8-15 and 8-23. All control gates in the ripple counter are 2-input gates, whereas in the synchronous counter the fan-in for CG2 is 3, for CG3 is 4, etc. The extra input leads to the gates, as required by Eq. (8-3), are used for the parallel carry. In other words, the CG blocks in Fig. 8-23 perform both the up-down and the parallel-carry logic.

Synchronous Decade Counter   Design of a system which is to divide by a number that is not a multiple of 2 is much more difficult for a synchronous than for a ripple counter. Control matrices (Karnaugh maps) are used to simplify the procedure.

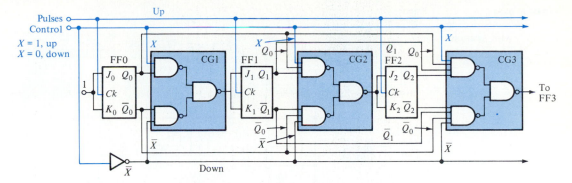

FIGURE 8-23
An up-down synchronous counter with parallel carry. The control X may be changed from up to down (or vice versa) between input pulses without introducing spurious counts because the counter responds only to the application of a clock pulse.

With a great deal of patience and intuition, the design may be carried out from direct observation of the waveform chart. Consider, for example, the synthesis of a synchronous decade counter with parallel carry. The waveform chart is that given in Fig. 8-19 except that *after the tenth pulse all waveforms return to 0*. Since $Q_0 = 0$ and $Q_2 = 0$ after the tenth pulse, FF0 and FF2 are excited as in the 16:1 synchronous counter. Hence, from Eq. (8-1)

$$T_0 = J_0 = K_0 = 1 \qquad T_2 = J_2 = K_2 = Q_0Q_1 \tag{8-5}$$

We note from Fig. 8-19 that FF1 toggles if $Q_0 = 1$. However, to prevent Q_1 from going to 1 after the tenth pulse, it is inhibited by Q_3. These statements are equivalent to the statement

$$T_1 = J_1 = K_1 = Q_0\overline{Q_3} \tag{8-6}$$

Finally, we wish FF3 to change state from 0 to 1 after the eighth pulse and to return to 0 after the tenth pulse. If

$$J_3 = Q_0Q_1Q_2 \qquad K_3 = Q_0 \tag{8-7}$$

then the desired logic is followed because $Q_0 = Q_1 = Q_2 = 1$, so that $J_3 = 1$, $K_3 = 1$, before pulse 8, whereas $Q_0 = 1$, $Q_1 = 0$, and $Q_2 = 0$, so that $J_3 = 0$, $K_3 = 1$, before pulse 10. The implementation of Eqs. (8-5) to (8-7) is given in the logic block diagrams of Fig. 8-24.

Synchronous up-down decade counters are available commercially (for example, MC10137 or 74ALS168) on a single MSI chip, as are 4-bit binary counters such as MC10154 or 74LS697. The FLIP-FLOPS are provided with *preset* (so that they are programmable) and *clear* inputs, not indicated in Fig. 8-23. Division by a number other than 2, 5, 6, 10, 12, or a power of 2 is not commercially available and must be designed as explained in the foregoing.

8-8 APPLICATIONS OF COUNTERS Many systems, including digital computers, data handling, and industrial control systems, use counters. We describe briefly some of the fundamental applications.

FIGURE 8-24
A synchronous decade counter with parallel carry.

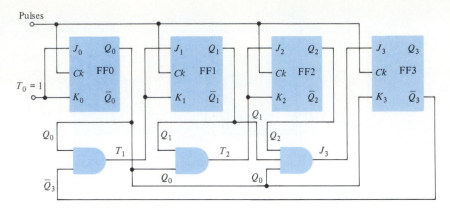

Direct Counting

Direct counting finds application in many industrial processes. Counters will operate with reliability where human counters fail because of fatigue or limitations of speed. It is required, of course, that the event which is to be counted first be converted into an electrical signal, but this requirement usually imposes no important limitation. For example, objects may be counted by passing them single-file on a conveyor belt between a photoelectric cell and a light source.

The *preset* input allows control of industrial processes. The counter may be preset so that it will deliver an output pulse when the count reaches a predetermined number. Such a counter may be used, for example, to count the number of pills dropped into a bottle. When the preset count is attained, the output pulse is used to divert the pills to the next container and at the same time to reset the counter for counting the next batch.

Divide-by-*N*

There are many applications where it is desired to change the frequency of a square wave from f to f/N, where N is some multiple of 2. From the waveforms in Fig. 8-19 it is seen that a counter performs this function.

If instead of square waves it is required to use narrow pulses or spikes for system synchronization, these may be obtained from the waveforms of Fig.

FIGURE 8-25
(*a*) An *N*:1 counter loaded by a network which converts the square-wave output (*b*) to pulses (*c*) or (*d*). If the input frequency is f, the spacing between positive pulses is $T = N/f$.

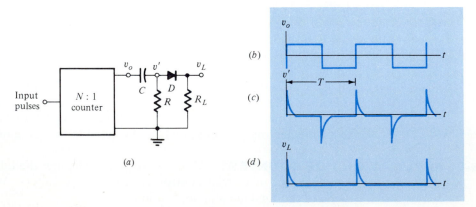

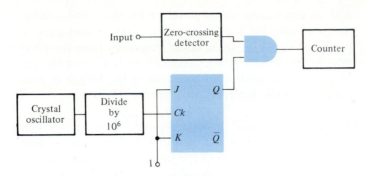

FIGURE 8-26
A system for measuring frequency by means of a counter.

8-19. A small RC coupling combination at the counter output, as in Fig. 8-25a, causes a positive pulse to appear at each transition from 0 to 1 and a negative pulse at each transition from 1 to 0, as in Fig. 8-25c. If now we count only the positive pulses, as in Fig. 8-25d (the negative pulses are eliminated, by using a diode, as in Fig. 8-25a), it appears that each binary divides by 2 the number of positive pulses applied to it. The four FLIP-FLOPS together accomplish a division by a factor $N = 2^4 = 16$. A single positive pulse will appear at the output for each 16 pulses applied at the input. A chain of n binaries used for this purpose of dividing or scaling down the number of pulses is referred to as a *scaler*. Thus a chain of four FLIP-FLOPS constitutes a scale-of-16 circuit, and so on.

Measurement of Frequency   The basic principle by which counters are used for the precise determination of frequency is illustrated in Fig. 8-26. The input signal whose frequency is to be measured is converted into pulses by means of the zero-crossing detector (Sec. 15-10) and applied through an AND gate to a counter. To determine the frequency, it is now only required to keep the gate open for transmission for a known time interval. If, say, the gating time is 1 s, the counter will yield the frequency directly in cycles per second (hertz). The *clock* for timing the gate interval is an accurate crystal oscillator whose frequency is, say, 1 MHz. The crystal oscillator drives a scale-of-10^6 circuit which divides the crystal frequency by a factor of 1 million. The divider output consists of a 1-Hz signal whose period is as accurately maintained as the crystal frequency. This divider output signal controls the gating time by setting a toggle FLIP-FLOP to the 1 state for 1 s. The system is susceptible to only slight errors. One source of error results from the fact that a variation of ± 1 count may be obtained, depending on the instant when the first and last pulses occur in relation to the sampling time. Beyond these, of course, the accuracy depends on the accuracy of the crystal oscillator.

Measurement of Time   The time interval between two pulses may also be measured with the circuit of Fig. 8-26. The FLIP-FLOP is now converted into set-reset type, with the first input pulse applied to the S terminal, the second pulse to the R terminal, and no connection made to Ck. With this configuration the first pulse opens the AND gate for transmission and the second pulse closes it. The crystal-oscillator signal (or some lower frequency from the divider chain) is converted into pulses, and these are passed through the gate into the counter. The number of counts recorded is proportional to the length of time the gate is open and hence gives the desired time interval.

Measurement of Distance   In radar or sonar systems a pulse is transmitted and a reflected pulse is received delayed by a time T. Since the velocity of light (or sound) is known, a measurement of the interval T, as outlined above, gives the distance from the transmitter to the object from which the reflection was received.

Measurement of Speed   A speed determination may be converted into a time measurement. For example, if two photocell-light-source combinations are set a fixed distance apart, the average speed of an object passing between these points is inversely proportional to the time interval between the generated pulses. Projectile velocities have been measured in this manner.

Digital Computer   In a digital computer a problem is solved by subjecting data to a sequence of operations in accordance with the program of instructions introduced into the computer. Counters may be used to count the operations as they are performed and to call forth the next operation from the memory when the preceding one has been completed.

REFERENCES

1   Mano, M. M.: "Computer System Architecture," 2d ed., John Wiley and Sons, New York, 1982.

2   Peatman, J. B.: "Design of Digital Systems," 2d ed., McGraw-Hill Book Company, New York, 1981.

3   Hodges, D. A., and H. G. Jackson: "Analysis and Design of Digital Integrated Circuits," McGraw-Hill Book Company, New York, 1983.

4   Taub, H.: "Digital Circuits and Microprocessors," McGraw-Hill Book Company, New York, 1982.

5   Chirlian, P. M.: "Digital Circuits," Matrix Press, Champaign, Ill., 1976.

6   Taub H., and D. Schilling: "Digital Integrated Electronics," McGraw-Hill Book Company, New York, 1977.

REVIEW QUESTIONS

8-1   (a)   Define a latch.
        (b)   Show how to construct a latch from inverters and verify that the circuit has two stable states.

8-2   Modify the latch described in Rev. 8-1 so that data may be entered into the latch, by means of an enable input.

8-3   (*a*) Define a sequential system.

       (*b*) How does it differ from a combinational system?

8-4   What is meant by a stable state?

8-5   (*a*) Draw the transfer characteristic of a bistable latch.

       (*b*) Explain why only two possible states exist in practice.

8-6   (*a*) Sketch the logic system for a clocked *SR* FLIP-FLOP.

       (*b*) Verify that the state of the system does not change in between clock pulses.

       (*c*) Give the truth table.

       (*d*) Justify the entries in the truth table.

8-7   (*a*) Augment an *SR* FLIP-FLOP with two AND gates to form a *J-K* FLIP-FLOP.

       (*b*) Give the truth table.

       (*c*) Verify part *b* by making a table of J_n, K_n, Q_n, $\overline{Q}_n$, S_n, R_n, and Q_{n+1}.

8-8   Explain what is meant by a race-around condition in connection with the *J-K* FLIP-FLOP in Rev. 8-7.

8-9   (*a*) Draw a clocked *J-K* FLIP-FLOP system and include *preset* (*Pr*) and *clear* (*Cr*) inputs.

       (*b*) Explain the clear operation.

8-10 (*a*) Draw a master-slave *J-K* FLIP-FLOP system.

       (*b*) Explain its operation and show that the race-around condition is eliminated.

8-11 (*a*) Show how to convert a *J-K* FLIP-FLOP into a delay (*D*-type) unit.

       (*b*) Give the truth table.

       (*c*) Verify this table.

8-12 Repeat Rev. 8-11 for a toggle (*T*-type) FLIP-FLOP.

8-13 Give the truth tables for each FLIP-FLOP type: (*a*) *SR*; (*b*) *J-K*; (*c*) *D*; and (*d*) *T*. What are the direct inputs *Pr* and *Cr* and the clock *Ck* for (*e*) presetting; (*f*) clearing; and (*g*) normal clocked operation?

8-14 (*a*) Define a register.

       (*b*) Construct a shift register from *SR* FLIP-FLOPS.

       (*c*) Explain its operation.

8-15 (*a*) Explain why there may be a race condition in a shift register.

       (*b*) How is this difficulty bypassed?

8-16 Explain how a shift register is used as a converter from (*a*) serial-to-parallel data and (*b*) parallel-to-serial data.

8-17 Explain how a shift register is used as a sequence generator.

8-18 Explain how a shift register is used as a circulating read-only memory.

       (*a*) Explain how a shift register is used as a ring counter.

       (*b*) Draw the output waveform from each FLIP-FLOP of a three-stage unit.

8-19 (*a*) Sketch the block diagram for a Johnson (twisted-ring) counter.

       (*b*) Draw the output waveform from each FLIP-FLOP of a three-stage unit.

       (*c*) By what number *N* does this system divide?

8-20 (*a*) Draw the block diagram of a ripple counter.

       (*b*) Sketch the waveform at the output of each FLIP-FLOP for a three-stage counter.

       (*c*) Explain how this waveform chart is obtained.

       (*d*) By what number *N* does this system divide?

8-21 (*a*) Draw the block diagram for an up-down counter.

(*b*) Explain its operation.

8-23 Explain how to modify a ripple counter so that it divides by N, where N is *not* a power of 2.

8-24 (*a*) Draw the block diagram of a decade ripple counter.

(*b*) Explain its operation.

8-25 Repeat Rev. 8-24 for a divide-by-6 ripple counter.

8-26 What is the advantage of a synchronous counter over a ripple counter?

8-27 (*a*) Draw the block diagram of a four-stage synchronous counter with series carry.

(*b*) Explain its operation.

(*c*) What is the maximum frequency of operation? Define the symbols in your equation.

8-28 (*a*) Repeat Rev. 8-27 if the counter uses parallel carry.

(*b*) What are the advantages and disadvantages of a parallel-carry counter?

8-29 Explain how to measure frequency by means of a counter.

8-30 List six applications of counters. Give no explanations.

Chapter 9
VERY LARGE SCALE INTEGRATED SYSTEMS

Chips containing more than 1000 components are referred to as *large-scale integrated* (LSI) systems and ones with over 10,000 components as very-large-scale integrated (VLSI) systems. In common terminology, however, the term "VLSI" is used to indicate systems with 100,000 or more components. By 1985, IC chips containing greater than 1,000,000 transistors were produced commercially. In this chapter we describe the most widely used digital LSI and VLSI chips. These and the PALs, PLAs, and PROMs described in Secs. 7-12 to 7-15 are used extensively in digital signal processing and control applications as well as in computer systems.

Memory chips are the most widely employed class of integrated circuits (ICs). Included are MOS shift-register serial memories and both *static* and *dynamic random-access memories* (RAMs), also known as *read-write memories*.

Component density, speed, and power consumption are three important design considerations in VLSI systems. We introduce two additional technologies, *charge-coupled devices* (CCDs), a MOS technology, and integrated-injection logic (I^2L), a bipolar technology, the use of which sometimes enhances circuit performance. Dynamic, or clocked, logic circuits in MOS and CMOS realizations are also used to increase component density or reduce power consumption. Very-large-scale-integration systems often employ dynamic logic circuits as basic building blocks and make use of the clock generator that is essential in establishing the timing for digital systems. The dynamic MOS shift register and *CMOS domino logic* are two such circuits and are described in the initial part of this chapter.

The chapter concludes with a brief introduction to VLSI system characteristics. The *microprocessor,* the most common single-chip system, is the basic component of personal computers (PCs), speech-synthesis chips, and a variety of control and instrumentation systems.

FIGURE 9-1
(*a*) A dynamic NMOS inverter. (*b*) A clock waveform.

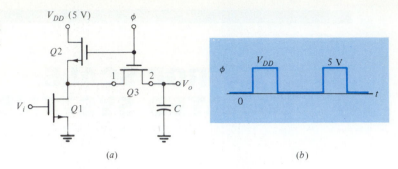

(*a*)   (*b*)

9-1 DYNAMIC MOS SHIFT REGISTERS

Very long shift registers (involving hundreds of bits) are impractical if constructed from FLIP-FLOPs as discussed in Sec. 8-5. Too much power is consumed and an excessive area of silicon is required. An alternative approach is to construct an LSI shift-register stage by cascading two dynamic MOS inverters. A bit is stored by charging the parasitic capacitance between gate and substrate of a MOSFET. A dynamic inverter is described first and then expanded into a 1-bit dynamic storage cell. While this technique is becoming obsolete in MOS memory systems because of the decrease in device dimensions, it is described as a means of introducing dynamic logic circuits.

A Dynamic MOS Inverter   The circuit in Fig. 9-1 shows a dynamic MOS[1] inverter which requires a clock waveform ϕ for proper operation. Positive logic using n-channel enhancement MOSFETs is assumed with a 0 state of 0 V and a 1 state of $V_{DD} = 5$ V. The capacitor C represents the parasitic capacitance (~ 0.1 pF) between the gate and substrate of the following MOS, fed by V_o.

When $\phi = 0$ V, then the gates of $Q2$ and $Q3$ are at 0 V and both these enhancement NMOSs are OFF. The supply voltage is disconnected from the circuit and delivers essentially no power. This differs from the standard NMOS inverter mentioned in Sec. 6-5 for which one MOS transistor is conducting and consequently the circuit always dissipates power. When the clock is at 5 V, both $Q2$ and $Q3$ are ON and inversion of V_i takes place. For example, if $V_i = 0$ V, $Q1$ is OFF, C charges to V_{DD}[†] through $Q2$ in series with $Q3$, and $V_o = 5$ V. If, however, $V_i = 5$ V, $Q1$ is ON, C discharges to ground through $Q3$ and $Q1$, and $V_o = 0$ V. Note that $Q3$ is a bidirectional switch: terminal 2 acts as the source when C charges to the supply voltage, whereas terminal 1 becomes the source when C discharges to ground.

The important features of MOSFETs for this dynamic inverter (and also for the shift register) are

---

[1]The following terms are used synonymously in this chapter: MOSFET, MOS, FET, NMOS, and transistor.

[†]It is assumed throughout this chapter that the threshold voltage V_T is smaller than the supply voltage V_{DD} and that $V_{ON} = 0$.

1. The MOS is a bidirectional switch.

2. The very high input resistance permits temporary data storage on the small gate-to-substrate capacitance of a MOS device.

3. The load FET may be turned off by a clock pulse to reduce static power dissipation.

The inverter discussed above is called a *ratioed inverter*. The name derives from the fact that when the input is high and the clock is high, transistors $Q1$ and $Q2$ form a voltage divider between V_{DD} and ground. Therefore, the output voltage V_o depends on the ratio of the ON resistance of $Q1$ and the effective load resistance of $Q2$ (typically, $< 1:5$). This ratio is related to the aspect ratios of $Q1$ and $Q2$.

Two-Phase Ratioed Memory Cell   Cascading two of the dynamic inverters in Fig. 9-1 allows each bit of information which is stored on the capacitance C of the first NOT gate to be transferred to the following inverter by applying a second clock pulse out of phase with the first waveform. A typical MOS dynamic-shift-register stage is shown in Fig. 9-2a and the required two-phase clock waveforms are indicated in Fig. 9-2b. The waveforms in Fig. 9-2b show a *nonoverlapping* two-phase clock as $t_3 > t_2$. When $t_3 < t_2$, the result is an *overlapping clock*. Each stage of the register requires six MOSFETs. The input V_i is the voltage on the gate capacitance C_1 of $Q1$, applied there by the previous stage (or by the input signal if this is the first stage of the shift register). When at $t = t_1$ the clock ϕ_1 goes positive (for NMOS devices), transistors $Q1$ and $Q2$ form an inverter and the bidirectional switch $Q3$ conducts. Hence, the complement of the level of C_1 is transferred to C_2. When ϕ_1 drops to 0 (at $t = t_2^+$), $Q2$ and $Q3$ are OFF and C_2 retains its charge as long as ϕ_1 remains at 0 V. However, at $t = t_3^+$, when $\phi_2 = V_{DD}$, then $Q4$ and $Q5$ act as an inverter and the switch $Q6$ is closed.

FIGURE 9-2

(a) A two-phase ratioed dynamic NMOS shift-register stage. (b) The two-phase clock waveforms ϕ_1 and ϕ_2.

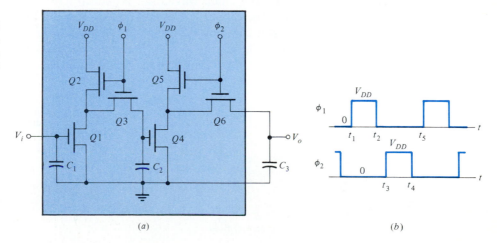

(a)   (b)

Hence, the data stored on C_2 is inverted and deposited on C_3. The bit (a 1 or a 0) transferred to the output V_o is identical with that which was at the input V_i but delayed by an amount determined by the clock period. In other words the register stage in Fig. 9-2a is a 1-bit delay line. The combination $Q1Q2Q3$ can be called the *master inverter,* and $Q4Q5Q6$ the *slave section.* To retain data stored in the register, the rate at which the data is clocked through the circuit must not fall below some minimum value. If the clock period is too long, the charge will leak off the parasitic capacitors and the data will be lost.

The load FETs in Fig. 9-2a are clocked because the gates are controlled by the clock waveform. Unclocked loads (the gates tied to fixed voltages) may also be used, but such circuits dissipate more power.

The Intel 2401 is a dual 1024-bit dynamic shift register fabricated with NMOS. It uses a single 5-V supply and is TTL-compatible. It operates at a minimum data rate of 25 kHz and a maximum rate of 1 MHz, with power dissipation of 0.12 mW/bit at 1 MHz. It is interesting to note that this chip contains $2 \times 1024 \times 6 = 12,288$ MOSFETs, exclusive of the control circuitry needed to convert it into a recirculating memory (Fig. 9-3).

Applications

Typical applications for MOS shift registers are as serial memories for calculators, cathode-ray tube displays, or communication equipment, as refresh or buffer memories, and as delay lines. A serial dynamic circulating shift-register memory is drawn in Fig. 9-3. The output of the register is returned to its input through an AND-OR combination as indicated. If the *write but not read (recirculate)* mode $W/\overline{R}$ is in the 1 state, the digital data at the *input* terminal is fed into the register. After a *clock* pulse cycle each bit is shifted to the right into the following stage, as explained in connection with Fig. 9-2. When the desired number of bits are entered sequentially into the register, the *recirculating* mode is commenced by changing $W/\overline{R}$ to the 0 state. In this mode further data is inhibited from entering the register and the bits stored in the memory are recirculated from the output back into the input of the shift register in synchronism with the clock waveform. Nondestructive reading of the data train is obtained at the output if the *read* input is excited by a logic 1.

If the register contains 1024 stages, then this recirculating memory may store one 1024-bit serial word. Consider that four systems, S_0, S_1, S_2, and S_3, of the type shown in Fig. 9-3, are used with *independent* data inputs and outputs.

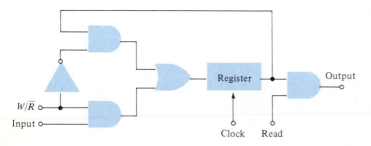

$W/\overline{R}$

Input

Clock   Read

Register

Output

FIGURE 9-3

A recirculating shift register ($W/\overline{R}$ is the abbreviation for *write but do not recirculate*). The AND, OR, and NOT gates are fabricated on the same chip with the register.

The $W/\overline{R}$ terminals are tied together as well as all the read terminals and the same clock synchronizes all systems. The resulting configuration is a serial memory which could be considered as storing 1024 words, each containing 4 bits. All 4 bits in a particular word appear simultaneously: the LSB at the output of S_0 and the MSB at the output of S_3. One clock period later another 4-bit word can be read. To expand the system to n-bit words, it is necessary to use n-recirculating shift registers. If more words are required, then longer shift registers must be used.

If the desired use has been made of the data circulating in the memory of Fig. 9-3, the $W/\overline{R}$ input is changed to a logic 1. This inhibits the bits from the last stage of the shift register from entering the first stage. In other words, the content of the memory is *erased* and new data may be simultaneously entered into the register.

Static MOS Shift Register   A "static" shift register is dc stable and can operate without a minimum clock rate. That is, it can store data indefinitely provided that power is supplied to the circuit. However, static shift-register cells are larger than the dynamic cells and consume more power. Consequently, their use is limited.

9-2   RATIOLESS SHIFT-REGISTER STAGES

It is pointed out in Sec. 9-1 that the load FET $Q2$ in Fig. 9-2 must have a much higher resistance than the driver $Q1$ in order for the low-state voltage V_{ON} to be close to zero. In Sec. 4-3 we emphasize that the FET resistance is proportional to L/W. Hence, $Q2$ must have a much larger channel length L and smaller width W than $Q1$. Consequently the inverter occupies more than the minimum possible area. Also, since the parasitic storage capacitance is charged through the load $Q2$ during a portion of the cycle, the

FIGURE 9-4

(a) A ratioless dynamic NMOS inverter, (b) $V_i = V(0)$ and $\phi = V_{DD}$, (c) $V_i = V(1)$ and $\phi = V_{DD}$ (during pulse), (d) input remains at $V(1)$ and $\phi = 0$ (after pulse terminates).

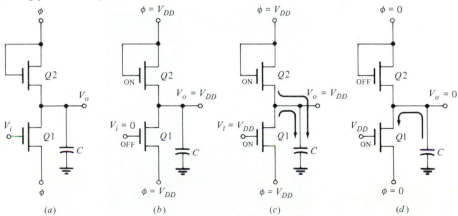

high resistance of $Q2$ limits the speed of operation of the register. Both of these difficulties may be avoided by using a dynamic *ratioless* inverter, as indicated in Fig. 9-4a (where $Q1$ and $Q2$ may have identical geometries). Note that no power supply (dc) voltage is used in this inverter. The clock pulse ϕ (Fig. 9-1b for NMOS devices) must supply the required energy to this circuit and the power dissipation is proportional to the clocking frequency.

To understand the operation of the ratioless inverter consider first the case where $V_i = 0$. Then during the pulse the situation is as pictured in Fig. 9-4b. Since the gate voltage of $Q1$ is 0 and that of $Q2$ is V_{DD}, then (for NMOS enhancement devices) $Q1$ is OFF and $Q2$ is ON. Therefore C charges through $Q2$ to V_{DD}. At the end of the pulse ϕ falls to 0 and both MOSFETs remain OFF. Hence, with $V_i = 0$ (logic 0), the output $V_o = V_{DD}$ (logic 1) and an inversion has taken place.

Consider now that $V_i = V_{DD}$ and that $\phi = V_{DD}$ as shown in Fig. 9-4c. Both MOSFETs are ON and deliver current to C as indicated. Hence C is quickly charged to V_{DD}. Since $V_o = V_i = V_{DD}$, there is no inversion during the pulse. However, at the termination of the pulse when the clock voltage returns to 0, we have the situation depicted in Fig. 9-4d. Now the gate G_2 of $Q2$ is at 0 and $Q2$ is OFF, whereas G_1 of $Q1$ is at V_{DD} and $Q1$ is ON. Consequently, C discharges to 0 V through $Q1$. Hence, shortly after the pulse ends, $V_o = 0$ while $V_i = V_{DD}$ indicating that a logical inversion has taken place.

Two-Phase Ratioless Dynamic Register Cell   If we cascade two inverters of the type shown in Fig. 9-4a through bidirectional transmission gates, the ratioless shift-register stage of Fig. 9-5 is obtained. The first inverter is powered by phase ϕ_1 and the second by phase ϕ_2, where these clock waveforms are drawn in Fig. 9-5b. At the beginning ($t = t_1^+$) of pulse ϕ_1 the switch $Q0$ closes and the voltage across C_0 (the input voltage of $Q1$) equals the input level V_i. By the inverter action described in connection with Fig. 9-4, the voltage across C_1 after the end of the pulse ϕ_1 (at $t = t_2^+$) corresponds to the complementary logic state of V_i.

FIGURE 9-5
(a) A two-phase ratioless dynamic NMOS shift-register stage and (b) the two-phase clock waveforms ϕ_1 and ϕ_2.

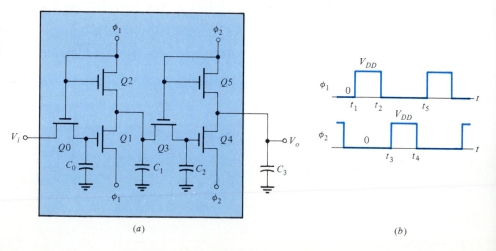

(a)   (b)

Since ϕ_1 is now at its low level, $Q0$ opens and V_i is retained on C_0 until the end of the period of ϕ_1 (at $t = t_5$).

At $t = t_3^+$ the second waveform ϕ_2 goes to its high level V_{DD} allowing transmission through $Q3$ and effectively placing C_1 and C_2 in parallel. If at $t = t_3^-$ the voltage on $C_1(C_2)$ is $V_1(V_2)$, then at $t = t_3^+$ the voltage V on C_2 (which must be the same as that on C_1) is found in Prob. 9-5 to be

$$V = \frac{C_1 V_1 + C_2 V_2}{C_1 + C_2} \tag{9-1}$$

If $C_1 \gg C_2$, note from Eq. (9-1) that $V \approx V_1$. In other words, pulse ϕ_2 causes the output voltage (across C_1) of the first inverter to appear at the input (across C_2) of the second NOT gate. Finally, by the inverter action described above, at the end of the pulse ϕ_2 (at $t = t_4^+$ and until $t = t_5$) the logic level V_0 across C_3 is the complement of that across C_2, which, in turn, is the complement of that across C_0. Clearly, in one period of the clock the input level V_i has shifted through the stage to the output V_o, as it should in a 1-bit delay line or 1-bit shift register.

No dc power supply is used in Fig. 9-5, but the clocking waveforms must be capable of furnishing the heavy capacitive currents. Also in order to ensure that C_1 be much larger than C_2, additional area must be added to the chip for C_1. We can reduce the loading on the clock drivers by adding another transistor to each inverter as in Prob. 9-6. This modification results in an eight-MOSFET stage. A number of four-phase ratioless shift registers capable of operation at high speed are described in the literature.[1] Because of the large amount of chip area required for two-phase ratioless shift registers and the additional complications of four-phase clock drivers these systems are seldom used.

A Dynamic CMOS Shift-Register Stage A dynamic CMOS shift-register stage, similar to the NMOS circuit in Fig. 9-5, can be constructed by interposing bidirectional CMOS transmission gates (Sec. 6-9) between CMOS static inverters (Sec. 6-8). Such a circuit, depicted in Fig. 9-6, utilizes the transmission gates $T1$ and $T2$ to fulfill the function of the MOS bidirectional switches in Fig. 9-5. The transmission gates are controlled by the complementary clocks ϕ and $\overline{\phi}$. When $\phi = V_{DD}$, $T1$ conducts whereas $T2$ acts as an open circuit. The CMOS inverters are marked $I1$ and $I2$.

The explanation of the operation of the register stage of Fig. 9-6 closely parallels that given in connection with Fig. 9-5. When $\phi = V_{DD}$ (logic 1), then $T1$ transmits and the input V_i appears across C_0. Because of the inverter action of $I1$, the complement of V_i appears across C_1 ($V_1 = \overline{V_i}$). On the next half-cycle $\phi = 0$, $T1$ opens, C_0 retains the voltage V_i, and V_1 remains at the $\overline{V_i}$. Also when $\phi = 0$, $T2$ closes, putting C_2 in parallel with C_1, and $I2$ causes the voltage across C_3 to be the complement of that across C_2. Consequently, at the end of a complete cycle $V_o = \overline{V_1} = V_i$ and we have demonstrated that this cell behaves as a 1-bit delay line or register.

---

[1] See Refs. 1, 2, 4, and 5 at the end of this chapter.

FIGURE 9-6

A dynamic CMOS shift-register cell.

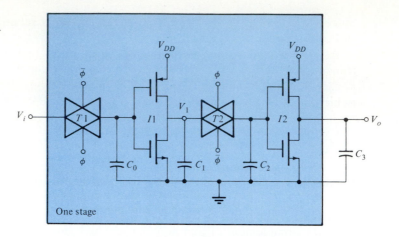

One stage

The CMOS stage consists of eight MOSFETs (or four complementary pairs). The power dissipation is very low since there are no dc current paths; power is used only for the transient charging of capacitors. From the explanation of the circuit given in the foregoing, it should be clear that the output voltage does not depend on the ratio of the resistances of any of the devices and, hence, ratioless operation is involved.

Another type of shift register (the charge-coupled device, CCD) is discussed in Sec. 9-8.

9-3 CMOS DOMINO LOGIC

Standard CMOS logic gates (Sec. 6-9) require one PMOS load transistor and one NMOS driver FET for each logic input. In Sec. 4-8 we indicated that PMOS devices require a larger area than NMOS transistors that handle the same current. To realize complex logic functions, that is, those containing many boolean variables, significant chip area is needed. (Compare this to NMOS realizations for which only one FET is added for each additional input.) Improved component density in CMOS circuits is attained by the use of a dynamic logic circuit, referred to as *domino logic*.

The circuit in Fig. 9-7 is a domino AND-OR gate used to realize the function $Y = AB + CDE$. The portion of the circuit containing $Q1$ to $Q7$ is an AOI gate used to drive the static CMOS inverter $Q8$ to $Q9$. Note that the AOI segment of the circuit is similar to the NMOS AOI gate in Fig. 7-3. The reduction in chip area of the 2- to 3-input AOI portion of the gate in Fig. 9-7 results from the fact that only seven FETs are needed, of which only one is a PMOS transistor, compared to the 10 (5 NMOS and 5 PMOS) transistors required in standard CMOS technology.

The action of the domino gate is controlled by the single-phase clock ϕ applied to PMOS load $Q7$ and the control NMOS transistor $Q1$. The parasitic load capacitance C_i acts as the load on the AOI portion of the circuit. When $\phi = 0$, $Q1$ is OFF and no current exists in the AND-OR branches of the AOI.

FIGURE 9-7

A CMOS domino AND-OR gate.

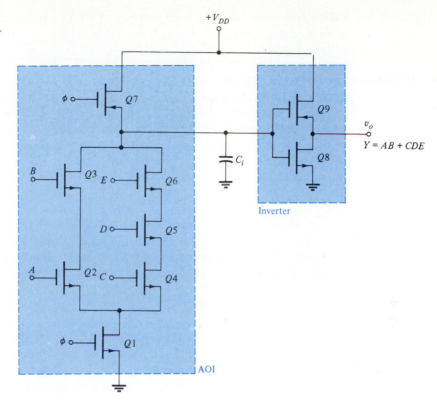

$+V_{DD}$

$Q7$

$Q9$

v_o

$Y = AB + CDE$

C_i

$Q8$

Inverter

B $Q3$ E $Q6$

D $Q5$

A $Q2$ C $Q4$

ϕ $Q1$

AOI

The PMOS load $Q7$ is ON and charges C_i to V_{DD}. With the input to the inverter high [$V(1)$], the output voltage $v_o = V(0)$.

Transistor $Q1$ turns on and $Q7$ turns off when $\phi = 1$. If either (or both) A and B or C and D and E is (are) at $V(1)$, C_i can discharge to ground through either $Q3 - Q2 - Q1$ or $Q6 - Q5 - Q4 - Q1$. The discharge of C_i causes the inverter input to be low [$V(0)$] and, consequently, $v_o = V(1)$. The capacitance C_i cannot discharge when $AB + CDE = V(0)$ because no path to ground exists, and hence no change in v_o occurs. It is important to recognize that logic inputs are permitted to change only when $\phi = 0$. Because a discharge path may exist, no input changes are permissible for $\phi = 1$.

Domino logic circuits result in improved component densities only when a large number of input variables are used. For domino gates to operate properly, $Q1$ and the inverter FETs $Q8$ and $Q9$ in Fig. 9-7 are required. Thus, for a simple 2-input OR gate, a domino circuit utilizes two input FETs and one PMOS load as well as $Q1$, $Q8$, and $Q9$ for a total of six transistors. This is the same number of devices used in a standard CMOS realization. As described previously in this section, however the 2- to 3-input AND-OR gate can be constructed by using both fewer PMOS and a fewer total number of transistors than does the static CMOS realization of the same gate. One application of CMOS domino logic is in the fabrication of PLAs (Sec. 7-15), where the reduced number of

PMOS transistors and the total number of FETs employed lead to significant savings in chip area.

9-4 RANDOM-ACCESS MEMORY (RAM) The operation of a digital system requires that data be stored and retrieved as desired. Semiconductor memories comprise an array of storage cells; each cell is capable of storing 1 bit of data. In such a memory, as contrasted with a shift register, information can be put *randomly* into, or taken out of, each storage element as required. Hence this system is called a *random-access memory,* abbreviated RAM. Since each bit can be read out of, or written into, each cell, the system is also referred to as a *read-write* (*R/W*) memory to distinguish it from the read-only memory (ROM) (see Sec. 7-9).

Both MOS and bipolar technologies are used to construct RAMs, and their supporting circuits with MOS technology are more prevalent. Random-access memories are available in MSI packages to store as few as 64 bits and as VLSI systems capable of storing 256 and 512 kb. Early in 1986 1-Mb RAMs (1 Mb $= 2^{10}$ kb $= 2^{20} = 1,048,576$ bits) became available commercially and it is expected that 4-Mb RAMs will be produced by 1990. Most large-storage RAMs (> 4 kb) are fabricated in polysilicon MOS technology. Most current computer systems utilize 64-kb and 256-kb RAMs for internal memory.

Static and dynamic circuits are used to construct RAMs with dynamic circuits employed almost exclusively for large storage capacity. Such circuits require fewer transistors and hence more cells can be fabricated on a given chip size (approximately 6×6 mm).

A major advantage of a RAM is that access time is the same for any bit in the matrix. In the shift-register serial memory, access time depends on the position of the bit at the moment of accessing. One disadvantage of the RAM is that they are *volatile,* which means that all stored information is lost if the power supply fails.[1] This differs from a ROM, in which the information is stored permanently during fabrication and is thus nonvolatile. (Recall that the data is stored during the masking operation.)

The remainder of this section deals with addressing and basic characteristics of RAMs. In the next section, static and dynamic memory-cell circuits are described.

Linear Selection To understand how the RAM operates, we examine the simple 1-bit *SR* FLIP-FLOP circuit shown in Fig. 9-8, with data input and output lines. From the figure we see that to read data out of or to write data into the cell, it is necessary to excite the *address line* ($X = 1$). To perform the write operation, the *write enable line* must also be excited. If the write input is a logic 1(0), then $S = 1(0)$ and $R = 0(1)$. Hence $Q = 1(0)$, and the data read out is 1(0), corresponding to that written in.

---

[1]This is the reason you should put important files and data on a disk or tape as these memory devices are permanent.

FIGURE 9-8

A 1-bit read/write memory.

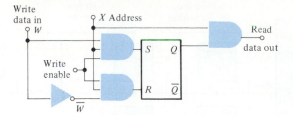

Suppose that we wish to have a 16-kb RAM organized as 1024 words of 16 bits each. This system requires 10 addresses, 16 data in and 16 data out lines. A total of $1024 \times 16 = 16,384$ storage cells must be used. Of this number, 16 cells are arranged in a horizontal line, all excited by the same address line. There are 1024 such lines, each excited by a different address. In other words, addressing is provided by exciting 1 of 1024 lines. This type of addressing is called *one-dimensional* or *linear selection* (Prob. 9-11). The number of pins on the package for addressing is reduced from the unreasonable number of 1024 to only 10 by including on the chip a 10-to-1024-line decoder.

Two-Dimensional Addressing A great economy of the number of NAND gates needed in the decoder mentioned above can be obtained (Prob. 9-12) by arranging the 1024 memory elements in a rectangular 32×32 array, each cell storing 1 bit of one word. Sixteen such packages are required, one for each of the 16 bits in each word.

Each word is identified by a matrix number X-Y in a memory cell of the rectangular matrix. To read (or write into) a specific cell (say, 1-3), an X decoder identifies row 1 (X_1) and a Y decoder locates column 3 (Y_3). Such *two-dimensional addressing* (also called *X-Y addressing* or *X-Y selection*) is indicated in Fig. 9-9 for a 16-kb (128×128) RAM.

Basic RAM Organization In the 1-bit memory of Fig. 9-8 separate read and write leads are required. For either the bipolar or the MOS RAM it is possible to construct a FLIP-FLOP (as we demonstrate in Figs. 9-12 and 9-14) which has a common terminal for both writing and reading, such as terminals 1 and 2 in Fig. 9-10. This configuration requires the use not only of the write data W (write 1), but also of its complement $\overline{W}$ (write 0). At the cell terminal to which $W(\overline{W})$ is applied, there is obtained the read $R(\overline{R})$ or sense $S(\overline{S})$ data output. Such a memory unit is indicated schematically in Fig. 9-10.

The basic elements of which a RAM is constructed are indicated in Fig. 9-9. These include the rectangular array of storage cells, the X and Y decoders, the write amplifiers to drive the memory, and the sense amplifiers to detect (read) the stored digital information. The amplifiers labeled R/W 0 and R/W 1 are not indicated explicitly in Fig. 9-9 (but are drawn in Fig. 9-12).

The organization (also called a *functional diagram*) for a 4096 word by 1-bit read/write memory is given in Fig. 9-11. Note that there are 64 rows and 64 columns in the memory array. Hence, each decoder has 6 inputs. The data

FIGURE 9-9

The organization of a 16,384-word × 1-bit static MOS RAM. Each colored square represents a 6-NMOS cell.

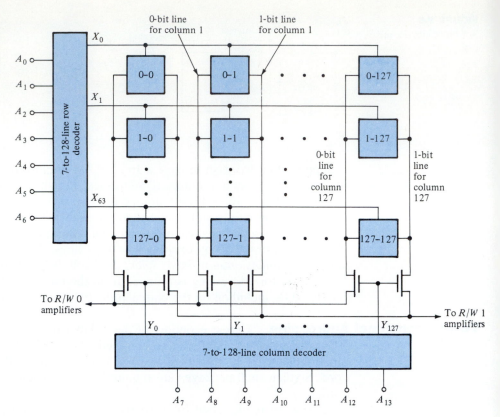

input D_{in} (output D_{out}) corresponds to $W(S)$ in Fig. 9-10. The complements of D_{in}, D_{out}, and the addresses $A_0 \cdots A_9$ are generated on the chip. Both decoders are also fabricated on the chip, referred to as *on-chip decoding*. The terminal labeled $\overline{CS}$ is the chip-select input (sometimes designated CE for *chip enable*). If $CS = 1$, the chip is selected. The complement of the write-enable input is

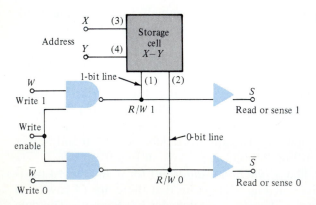

FIGURE 9-10

A basic storage cell can be constructed with complementary inputs and outputs and with the write and sense amplifiers meeting at a common node for (1) true data and (2) complementary data.

FIGURE 9-11

The organization of a 4096-word × 1-bit (4-kb × 1) static RAM. (Courtesy of *MOSTER Corporation*)

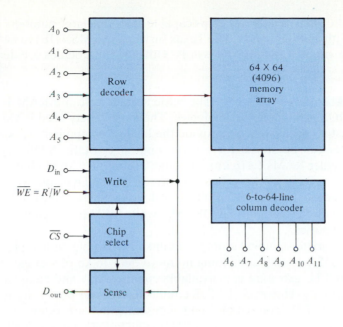

labeled $\overline{WE}$ or $R/\overline{W}$ (*read but not write*). Hence, reading takes place if $R/\overline{W} = 1$ and writing is done if $R/\overline{W} = 0$. The truth table giving the operating modes of this chip is indicated in Table 9-1.

TABLE 9-1   Truth Table for the RAM in Fig. 9-16

$\overline{CS}$	$R/\overline{W}$ or $\overline{WE}$	D_{in}	D_{out}	Mode
1	X^*	X	High impedance	Not selected
0	0	0	0	Write 0
0	0	1	1	Write 1
0	1	X	D_{out}	Read

*X = don't care.

To the 16 inputs indicated in Fig. 9-11 a ground and a power-supply lead must be added. Hence, this 4096-kb RAM comes in an 18-pin package.

Memory Expansion   Often words of more than 1 bit must be stored. If 4096 four-bit words are required, it is necessary to use four packages like the one indicated in Fig. 9-11. The 12 address lines are applied in parallel to all four packages, and with $CS = 1$ ($\overline{CS} = 0$), all chips are selected simultaneously. A specific address selects one of the 4096 words; 4 bits of data are written into (read out of) memory through four independent D_{in} (D_{out}) terminals with $R/\overline{W} = 0$ ($R/\overline{W} = 1$). Expansion of the number of words stored in RAM can be achieved by use of the address-expansion organization shown in Fig. 7-31, illustrated for ROM.

Commercial RAM chips capable of storing large numbers of bits are available with multibit words. The Texas Instrument TMS 4416 is a 64-kb RAM organized to store 16,384 four-bit words. Other RAM modules are described in Sec. 9-6.

9-5 READ-WRITE MEMORY CELLS

The basic storage cell in a RAM is fabricated in both MOS and bipolar technologies. The most widely used RAMs utilize MOS transistors because these provide the highest component density and hence, more bits can be stored for a given chip size. Static MOS cells are prevalent in smaller RAMs ($\leq$16 kb), although prototype CMOS chips with storage capacities of 256 kB have been reported (1985).[1] Dynamic MOS memory cells are used most frequently in 16kb to 1Mb RAMS. Dynamic random-access memories are often referred to as DRAMs, while SRAMs designate static random-access memories.

Emitter-coupled storage circuits, compatible with ECL, are used in bipolar RAMs, usually containing no more than 16 kb of storage. Static RAMs based on TTL gates are also available commercially and have 64-bit to 4-kb storage capacity. However, CMOS circuits have lower power dissipation, are designed to be TTL compatible, and, consequently, are often used in place of TTL RAMs. The various types of MOS and bipolar memory cells are discussed here and in Sec. 9-6.

Static MOS RAM The MOS FLIP-FLOP of Fig. 8-8 is a 1-bit memory and is the basic storage cell for the static MOS RAM. In Fig. 9-12 $Q1$ through $Q4$ form such a bistable unit and MOSFETs $Q5$-$Q6$ form the gating network through which the interior node N_1 (N_2) is connected to the 0-bit (1-bit) data line. Cell 1-3 is indicated in Fig. 9-12. This six-transistor cell is inserted into the memory array in the manner indicated in Fig. 9-9. The 0-bit and 1-bit lines are connected to every cell *in the same column*. To select a cell in a particular column (say, 3) it is necessary to address that column (Y_3). To select a cell in row 1, the row decoder must excite X_1. In other words, two-dimensional addressing is used to locate a specific cell 1-3.

Included in Fig. 9-12 are the read and write amplifiers for each of the data lines. Note that $Q17$ and $Q10$ ($Q9$) form an AND gate with inputs WE and W ($\overline{W}$), where WE = *write enable*, and W = *write* (also called *data input* D_{in}). The *sense S* or *read output* may also be labeled D_{out}.

It is desired to read cell 1-3. We must set X_1 and Y_3 to V_{DD} (logic 1 for NMOS). Assume that a 1 is stored in this cell ($Q2$ is ON and $Q1$ is OFF so that node N_2 is at 0 V and N_1 is at V_{DD}). In order to read, WE is set to 0. Then $Q17$ is OFF and hence $Q10$ ($Q9$) is nonconducting, so that the 1-bit (0-bit) data line is tied to V_{DD} *through the load $Q12$ ($Q11$).* Consequently current flows into $Q2$ through $Q12$, $Q8$, and $Q6$ from V_{DD} (as well as through $Q4$ from V_{DD}) and the 1-bit line is effectively grounded. Hence, $Q14$ is OFF and $S = D_{out} = V_{DD}$ (logic 1). Since $Q1$ is OFF, no current flows in $Q3$, $Q5$, $Q7$, and $Q11$ in series,

---

[1]See Ref. 5.

FIGURE 9-12

A storage cell (1–3) containing 6-NMOS transistors. The X_1 and Y_3 address lines and the write and read amplifiers are also shown. *A logic 1 is stored if Q2 conducts.*

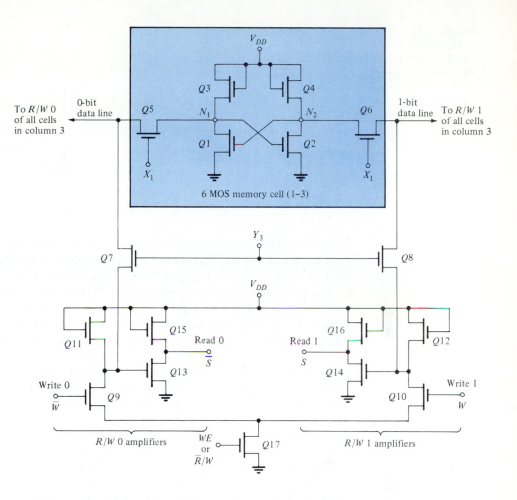

and the 0-bit data line is at V_{DD}, $Q13$ is ON, and $\overline{S} = 0$ V. We thus have correctly sensed that the 1-3 FLIP-FLOP stores a 1 (since $S = 1$ and $\overline{S} = 0$).

In order to write a 1 into the cell we address it ($X_1 = 1$ and $Y_3 = 1$), set $WE = 1$, $W = 1$, and $\overline{W} = 0$. Then $Q17$ and $Q10$ are ON and $Q9$ is OFF. Hence, the 1-bit line is grounded and the 0-bit line goes to V_{DD} through load $Q11$. Current now flows into the 1-bit line from V_{DD} through $Q4$, $Q6$, $Q8$, $Q10$, and $Q17$ to ground. Thus node N_2 is effectively grounded. This cuts off $Q1$, and N_1 rises to V_{DD}. Consequently $Q2$ is held ON and N_2 is maintained at 0. When the address is removed ($Q5$, $Q6$, $Q7$, and $Q8$ are OFF), $Q2$ is ON, $Q1$ is OFF, and a 1 has been written into the selected memory cell.

Static CMOS RAM Cell   The static CMOS RAM cell is similar in structure and operation to the NMOS cell depicted in Fig. 9-12. The CMOS circuit in Fig. 9-13 corresponds to the 6-MOS memory cell in Fig. 9-12 with identical transistor numbering. The sense amplifiers required to read and write data are not shown. Transistors $Q1$ to $Q4$, in Fig. 9-13, are the cross-coupled CMOS inverters which form the

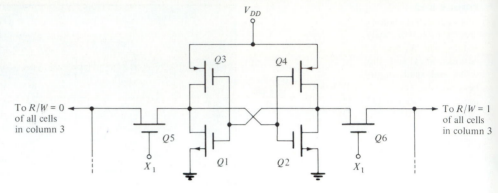

FIGURE 9-13
A static CMOS memory cell. The numbering of the transistors corresponds to that used in Fig. 9-12.

FLIP-FLOP. The NMOS transistors $Q5$ and $Q6$ form the transmission gates which provide the data path into (or out of) the memory cell. Reading and writing a 0 or a 1 is the same as in the NMOS circuit in Fig. 9-12.

Many IC manufacturers produce static MOS RAMs with memory capacities of 1 to 16 kb. Note that a 16-kb RAM using the six-transistor storage cell in either Fig. 9-12 or 9-13 has $6 \times 16{,}384 = 98{,}304$ MOSFETs in the memory array alone. With the required peripheral circuitry (sense amplifiers, etc.) such circuits contain over 100,000 devices. To increase memory capacity on chip areas that are conveniently fabricated, it is clearly important to reduce the number of transistors per cell. The following discussion of dynamic cells is the major approach used to accomplish this reduction.

Four-MOSFET Dynamic RAM Cell   The silicon area occupied by the six-transistor cell in Fig. 9-12 may be reduced by changing the load FETs $Q3$ and $Q4$ to clocked loads. In other words, the two cross-coupled inverters which form the latch are now dynamic inverters, as indicated in Fig. 9-14a. The gating excitation for each load is supplied by the word lines from the X decoder. The MOSFETs $Q3$ and $Q4$ act simultaneously as loads and as row-selection transistors, thus reducing the cell from a six- to a four-device memory unit. If $X = 0$, $Q3$ and $Q4$ are OFF, and no information can be written into (or read out of) the cell. However, if $X = 1$, then $Q3$ and $Q4$ are ON, and the four transistors form a latch which can store a 1 ($Q2$ ON) or a 0 ($Q1$ ON).

As with the dynamic MOS shift register of Fig. 9-2, the information in the dynamic memory cell in Fig. 9-14a is stored on the parasitic capacitances C_1 and C_2 between gate and source of $Q1$ and $Q2$, respectively. If a 1 is stored, then C_2 (C_1) is charged to V_{DD} (0), and if a 0 is written, the converse is true. Suppose that after the data are stored in the cell it is not accessed for a time T. The charge on the capacitors decreases during this interval because of the inevitable leakage currents. If T is too long the 1-state voltage may become small enough to be indistinguishable from the 0-V level and the information is lost. This same phenomenon is the reason why a dynamic shift register cannot be operated below a minimum operating frequency.

Clearly, some additional circuitry is required to refresh the stored data before the drop in capacitor voltage becomes excessive. Two transistors (Q and Q' in Fig. 9-14b) are added to restore *all the* FLIP-FLOPS *in a given column*. The refresh waveform v is a pulse of less than 1 μs occurring about every 2 ms. All cells in a given row are refreshed simultaneously by addressing that row while v is high. Note that during the refresh interval, $Q3$ in series with Q forms the load for $Q1$, and Q' in series with $Q4$ acts as the load for $Q2$. If at the beginning of the refresh cycle the voltage across C_2 exceeds that across C_1 (~ 0), then $Q1$ is OFF and C_2 charges toward V_{DD} by the current in Q and $Q3$. The current charging C_1 through Q' and $Q4$ is smaller than that of C_2 because $Q2$ is conducting. Hence, C_2 rises rapidly to V_{DD} and the voltage across $Q2$ falls to zero, maintaining the voltage across C_1 at 0 V. In other words, because of the regenerative feedback action in the FLIP-FLOP the cell is restored to its initial state (a logic 1 in this case).

Note that the organization of the four-transistor cell into a RAM is the same as that for the 6-MOS cell of Fig. 9-12. The number of transistors saved in going from a 6- to a 4-MOS cell in a 16-kb square RAM, taking into account the MOSFETs which must be added to generate the refresh voltage v, is 2 $\times$ 16,384 $-$ 2 $\times$ 128 $=$ 32,512. In addition to using much less chip area, the dynamic cell saves a great deal of power. The load devices conduct only during the refresh pulse and power is dissipated only during this very short interval.

FIGURE 9-14
(a) A 4-NMOS dynamic cell. A logic 1 is stored if the voltage across C_2 is V_{DD} so that $Q2$ is ON. (b) The memory cell as part of a RAM organized as in Fig. 9-9. All FLIP-FLOPS in column and are refreshed when a positive pulse is applied to turn on Q and Q' provided $X = 1$.

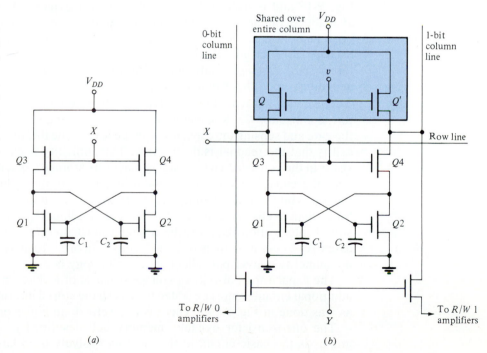

(a)   (b)

FIGURE 9-15
A 1-MOSFET dynamic
memory cell (in color).
It is organized in a RAM
as indicated in Fig. 9-9.
Refresh circuitry is not
shown.

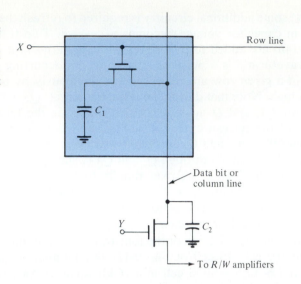

One-MOSFET Dynamic RAM Cell

In Fig. 9-14 the storage elements are capacitors, but there is no fundamental reason to use a FLIP-FLOP to charge or discharge the capacitors. It is possible to design a dynamic memory using a single capacitor and one transistor to act as a transmission gate to charge the capacitor or to remove the charge already stored there. This simplest of all RAM cells is indicated in Fig. 9-15 and is used in large (4-kb to 1-Mb) dynamic RAMs commercially available. Complementary inputs and outputs are not required, and hence the organization is that indicated in Fig. 9-9, except that only one bit (data) line is used to connect all cells in a column. As with the RAMs described in the foregoing discussions, only one cell in the memory is selected at a given time depending upon the X and Y addresses.

The cell is written into by applying the bit line voltage through the transistor to the capacitor C_1. Reading is done by connecting C_1 to the bit line through the gate and sensing the capacitor voltage level. One disadvantage of this simple cell is that the readout is destructive. This difficulty occurs because the transistor in the cell selected for reading places its storage capacitance C_1 in parallel with the capacitance C_2 of the data line. If V_1 is the voltage across C_1, the readout voltage V is given by Eq. (9-1) with $V_2 = 0$ or $V = C_1 V_1 / (C_1 + C_2)$. Since many cells are connected to the column line, $C_2 \gg C_1$ and $V \ll V_1$. The stored information that is to be retained must therefore be regenerated after every read operation to its original level V_1. In order to increase the ratio C_1/C_2, n-channel two-layer polysilicon gate technology is used.

The capacitor C_1 also loses voltage because of leakage currents and, hence, additional circuitry must be added to refresh the stored information periodically, as was done in Fig. 9-14b. There is one refresh amplifier on each data line.

The one-transistor dynamic memory cell described in the previous paragraphs is the basic circuit in the most extensively used large-capacity RAMs (64 kb to 1 Mb).

FIGURE 9-16
Typical chip layout of a 64-kb dynamic RAM (DRAM).

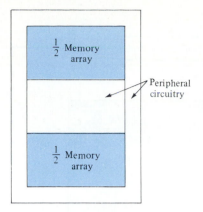

Peripheral circuitry

Dynamic RAM Chip Organization   Typical 16- to 64-kb DRAMs have chip layouts similar to that shown in Fig. 9-16. The memory array is divided into two equal segments and peripheral circuitry is placed along the borders of each array segment. The memory-support circuits include the row and column decoders and buffers, the sense and refresh amplifiers, input and output registers, and timing and control circuits. Dynamic RAMs generally come in the industry standard 16-pin DIP package.[1] For a 64-kb DRAM, such as the Mostek 4164 or Texas Instruments TMS 4164 (these are pin-compatible), the eight row and eight column address lines are multiplexed on eight pin connections. This is accomplished by adding two externally generated clock signals called the *row-address strobe* (RAS) and the *column-address strobe* (CAS). Actually, the complements of these signals, $\overline{RAS}$ and $\overline{CAS}$, are applied to the pins and latches the row and column addresses onto the chip. In addition to the eight address lines, 7 of the 8 pins remaining are used for the two strobe signals, the write enable signal ($\overline{WE}$), the data input and output lines, ground, and the 5-V supply. No connection is made to the last pin.

Larger-capacity DRAMs, such as the Texas Instruments 256-kb × 1-bit TMS4256, utilize chips organized as displayed in Fig. 9-17. This arrangement is easily identified in the photomicrograph of Fig. 9-18. The memory array is partitioned into four 64-kb arrays, each of which is organized as shown in Fig. 9-16. The nine row- and nine column-address lines needed to select 1 of 262,144 bits are latched onto the chip by means of the row-address and column-address strobes ($\overline{RAS}$ and $\overline{CAS}$). The 16 pin connections are the nine address lines, the two data lines, the two strobe lines, the write-enable signal ($\overline{WE}$), ground, and the 5-V supply. The timing and control signals and data lines are all TTL-compatible.

The basic organization in Fig. 9-17 can be modified so that the 256-kb capacity provides storage of 64K 4-bit words (64K × 4). Such a chip is shown (the Texas Instruments TMS 4464) in Fig. 9-19 and is available in an 18-pin DIP

[1]Many chips also are available in an 18-pin plastic chip-carrier package.

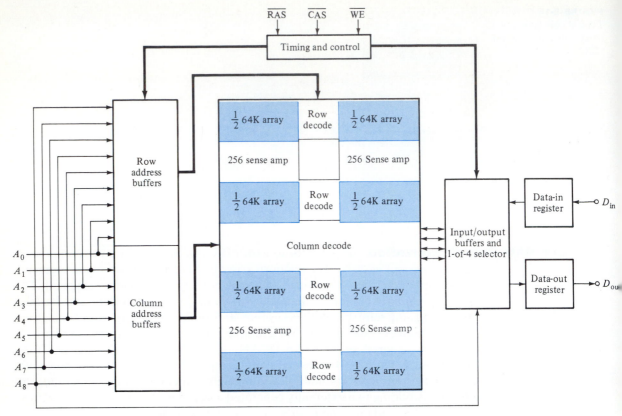

FIGURE 9-17
Chip layout and organization of a 256K × 1-bit DRAM (such as the Texas Instruments, Inc. TMS4256).

FIGURE 9-18
Photomicrograph of the TMS4256 256-kb DRAM. (*Courtesy of Texas Instruments, Inc.*)

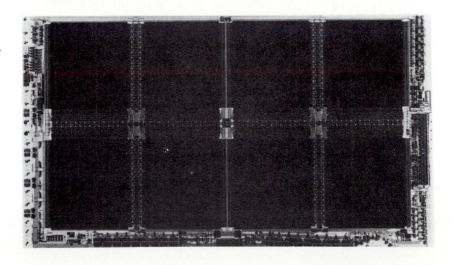

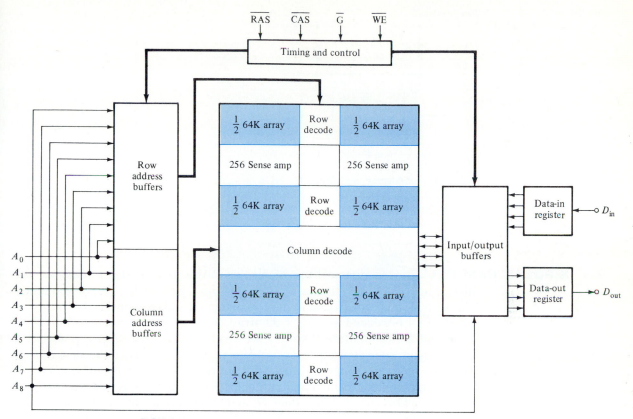

FIGURE 9-19

Organization of a 256-kb DRAM into a 64K × 4-bit memory (the TMS 4464).

package. The four data lines are used for both input and output. To provide for this multiplexing, an *output enable signal* ($\overline{G}$) is added to the timing and control portion of the system. Thus the 18 pin connections are used for the eight address lines, four data lines, four timing and control signals ($\overline{RAS}$, $\overline{CAS}$, $\overline{WE}$, $\overline{G}$), ground, and the 5-V supply.

Multiples of the 64-kb dynamic RAM are also packaged in modules containing several chips so that 64K words of several bits are stored. Texas Instruments packages several 4164 chips to provide 8- and 9-bit words (TMS 4164FM8 and TMS 4164EL9, respectively). Larger DRAM modules, based on the 256-kb TMS4256 chip, are available and provide 256K × 4-bit, 256K × 8-bit, and 256K × 9-bit memories. In addition, four TMS 4256 chips are packaged as a single module as a 1M × 1-bit DRAM (TMS 4256FC1).

The dynamic RAMs described have read or write cycles that are typically in the order of 250 ns with refresh times of less than 4 ms. Operating power of 300 mW and standby power as low as 12.5 mW are typical.

FIGURE 9-20
An emitter-coupled
memory cell.

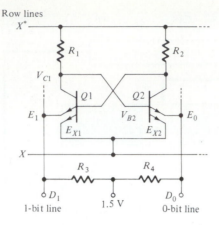

9-6 BIPOLAR RAM CELLS

The principal use of bipolar RAMs is in systems requiring the highest speeds of operation (as in ECL systems). Bipolar RAMs are frequently "word-oriented" arrays and require two-step decoding. Consider a 1-kb $\times$ 1-bit memory organized as a 32×32 array. The row decoder selects one of 32 rows and all 32 bits (the "word") are read out and placed in a register. A second 5-bit code is used to access the register and select the desired bit. Similarly, data are stored by writing an entire word simultaneously.

One common BJT memory circuit, depicted in Fig. 9-20, is called an *emitter-coupled cell* because the data lines D_1 and D_0 are connected to the emitters of the two transistors $Q1$ and $Q2$. Although both $Q1$ and $Q2$ have two emitters each, these BJTs operate in their normal modes and not in the inverted modes as in TTL gates. The transistor-resistor combinations, $Q1$-$R1$ and $Q2$-$R2$, are the two cross-coupled inverters that form the cell.

As seen in Fig. 9-20, two row-address lines, X and X^*, having different voltage levels are employed. The typical voltage levels for X^* are $V(0) = 1.3$ V and $V(1) = 4.3$ V, and those for X are $V(0) = 0.3$ V and $V(1) = 2.0$ V. Note that the voltage levels for X^* serve as the collector supply V_{CC}. The standby levels for X and X^* are their $V(0)$ values. Standby refers to those time intervals when the cell is not accessed for reading or writing. The data lines D_0 and D_1 are connected through R_4 and R_3, respectively, to a 1.5-V supply (typical). With no current in these resistors during standby, D_0 and D_1 have standby values of 1.5 V. The low standby voltage levels help to reduce the standby power dissipation of the memory cell. Only during reading and writing intervals do the voltage values at D_0 and D_1 differ from 1.5 V, as indicated in the ensuing discussion.

The operation of the cell is based on using the multiple-emitter transistors as current switches. Voltage levels are selected so that $Q1$ and $Q2$ never conduct simultaneously. Thus $Q1$ is used to read (write) a 1 and $Q2$ to read (write) a 0. The read or write operation is controlled by switching the current in the conducting BJT from the row line (X) to the appropriate data line (D_0 or D_1).

To write a 1, X and X^* are set to $V(1)$ and $D_1 = V(0)$. Thus the emitter E_1 of $Q1$ is forward-biased and current exists in $Q1$. The voltage $V_{C1} = V_{B2}$ decreases, and, with $D_0 = 1.5$ V and $X = V(1)$, both emitter junctions of $Q2$ are reverse-biased and $Q2$ is OFF. When voltage levels return to their standby levels, $X = V(0)$ and $D_0 = D_1 = 1.5$ V, $Q1$ remains ON as E_{X1} is forward-biased. Sufficient base current for $Q1$ is supplied through R_2. Although the emitter E_{X2} of $Q2$ is low, $X^* = V(0)$ causes $V_{C1} = V_{B2}$ to decrease from its value when $X^* = V(1)$, and the lower value of V_{B2} virtually eliminates any base current in $Q2$. Thus it is reasonable to assume that $Q2$ is OFF. A 1 is stored in the cell because $Q1$ is ON and a current path exists in the X line (through E_{X1}).

With $D_1 = V(1)$, reading the stored 1 is accomplished by making X and $X^* = V(1)$. These voltage values reverse-bias the emitters E_{X1} and E_{X2}. Because $X^* = V(1)$, sufficient base drive through R_2 is established to forward-bias the E_1 junction so that $Q1$ still conducts. The current path, however, is switched to the D_1 line and returns to ground through R_3 and the 1.5-V supply. The voltage at D_1 increases as a result of the voltage drop in R_3, and sensing this higher voltage indicates the presence of a 1.

To write a 0, $X = V(1)$ and $D_0 = V(0)$. The conditions that exist on E_1 and E_0 are reversed from those encountered for writing a 1. The operation of the circuit is the same except that it is $Q2$ that conducts with the current path through E_{X2} during standby. The stored 0 can be read from memory by making X and $X^* = V(1)$ with D_0 and D_1 remaining at 1.5 V. By analogy, the 0 is sensed because of the increased voltage drop in R_4 caused by the current switching to the D_0 line.

The circuit in Fig. 9-21 is a second BJT memory cell and is generally used when the bipolar process utilized permits fabrication of the Schottky diodes

FIGURE 9-21

A diode-coupled bipolar memory cell.

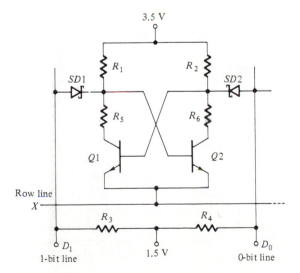

SD1 and SD2. This circuit is called a *diode-coupled cell* as data lines are connected to Q1 and Q2 through the Schottky diodes. Typical standby voltage values for the row and data lines are 2.5 and 1.5 V, respectively. The voltage values during writing intervals is 2.5 V; reading a stored bit is sensed by the decrease in the voltages that appear on the data lines. The behavior of the memory cell in Fig. 9-21 is similar to that for the emitter-coupled circuit in Fig. 9-20. Conduction in Q1 and Q2 determine whether a 1 or a 0 is stored, and switching the current in the conducting BJT from a row line to or from a data line allows a 0 or a 1 to be read out of memory.

9-7 CHARGE-COUPLED DEVICE (CCD)

A MOSFET designed with an extremely long channel and many ($\cong$ 1000) electrodes (gates) closely spaced between source and drain can function as a serial memory or a shift register. Each gate electrode and substrate forms an MOS capacitor (Sec. 5-9) which can store charge. For example, if logic 1 is applied to the source, charge is stored by the capacitor nearest the source provided, that an appropriate voltage is applied to the first gate E_1. If this voltage is removed from E_1 and simultaneously applied to E_2, this charge packet will move to E_2. Repetition of this process transfers the charge from capacitor to capacitor; hence this configuration is called a *charge-coupled device* (CCD). Extremely high density shift registers and serial memories can be constructed with CCDs. Because serial memories have more limited use than RAMs, CCDs are not widely used memory elements in digital systems. Applications of CCDs, however, are encountered in image processing and digital signal processing systems where the high-density serial nature of the device is often a valuable feature. Because image and digital signal processing are important areas in modern control (particularly robotics) and communication technology, we provide a brief introduction to CCD structures in this and the succeeding section.

Basic CCD Operation

To understand better the device operation discussed in the preceding paragraph consider a *p*-type substrate covered with a thin oxide layer on which there has been deposited a row of very closely spaced metallic electrodes, five of which are shown in Fig. 9-22. For simplicity of explanation, assume that the threshold voltage is zero and that no electrons are present. Consider the situation (Fig. 9-22) where the voltage on gate 3 is $+V$ and all other electrodes are grounded. This positive voltage repels holes in the substrate under E_3 and they are driven downward away from the SiO_2. Consequently, immobile negative ions are exposed and a depletion region is formed under E_3. Electric field lines extend from the positively charged electrode through the dielectric into the depletion region and onto these immobile negative charges. The magnitude of the potential profile (the potential variation with distance parallel to the oxide surface) is indicated in Fig. 9-22. This plot also represents the potential-energy barrier ("well") for electrons, the minority carriers. If a packet of electrons is introduced in the region under E_3, these charges can move freely within the

FIGURE 9-22
The simplest structure for an *n*-channel (*p*-substrate) CCD. The potential-energy "well" is formed under gate 3 if this electrode is at a positive voltage and all other gates are at the substrate potential (ground).

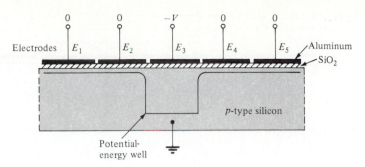

FIGURE 9-22
The simplest structure for an *n*-channel (*p*-substrate) CCD. The potential-energy "well" is formed under gate 3 if this electrode is at a positive voltage and all other gates are at the substrate potential (ground).

well, but cannot penetrate the potential-energy walls of the well (Sec. 1-2). In other words, as long as the voltage $+V$ is present, the negative charge cannot wander away but is trapped under E_3 near the channel surface.

We now consider how stored charge is moved from left to right down the channel, corresponding to shifting binary bits along this shift register. Consider the structure in Fig. 9-23a consisting of 10 plates where each third electrode is tied together. If at $t = t_1$ the applied voltages are $\phi_1 = +V$, $\phi_2 = \phi_3 = 0$, then as indicated in Fig. 9-23b, potential-energy wells (corresponding to Fig. 9-22) are formed under electrodes 1, 4, 7, and 10. The minus signs indicate schematically that charge is stored near the surface under E_1, E_7, and E_{10} but not under E_4, indicating that the digital word 1011 has been entered into the CCD. At a later time $t = t_2$, the voltage ϕ_2 changes to $+V$ but ϕ_1 and ϕ_2 maintain their previous values. The potential profile is thereby altered, as in Fig. 9-23c. The stored charge now is shared by two adjacent electrodes, owing to diffusion of the electrons from the original well into the newly created empty well.

Shortly after the situation in Fig. 9-23c is established, $|\phi_1|$ starts to decrease and, at $t = t_3$, $\phi_1 = +V/2$, whereas ϕ_2 and ϕ_3 remain unchanged. The potential profile at t_3 is shown in Fig. 9-23d. The fringing electric field caused by the potential difference between ϕ_1 and ϕ_2 causes the electrons to move into the deeper well as indicated. Finally at $t = t_4$, when $\phi_1 = 0$, $\phi_2 = +V$, and $\phi_3 = 0$, the potential profile is as shown in Fig. 9-23e. As a result of this sequence of voltage changes, the initial pattern of stored charge (1011) has transferred one electrode to the right, as is clear from a comparison of Fig. 9-23b and e.

The sequence just described represents transfer from one electrode to the next of the CCD shift register. Since three voltages are necessary, three-phase clocks are required. The waveforms ϕ_1, ϕ_2, and ϕ_3 necessary to conform with the profiles in Fig. 9-23 are given in Fig. 9-24. The times t_1, t_2, t_3, and t_4 in Fig. 9-23 are also indicated in Fig. 9-24. Note that at t_1 in Fig. 9-24, $\phi_1 = +V$, $\phi_2 = 0$, and $\phi_3 = 0$, as in Fig. 9-23; at t_2, $\phi_1 = +V$, $\phi_2 = +V$, and $\phi_3 = 0$ in both figures, etc. The first transfer takes place between t_1 and t_4, the second between t_5 and t_6, the third between t_7 and t_8, the fourth between t_9 and t_{10}. Clearly, for every input cycle of period T, three shifts take place. In the interval

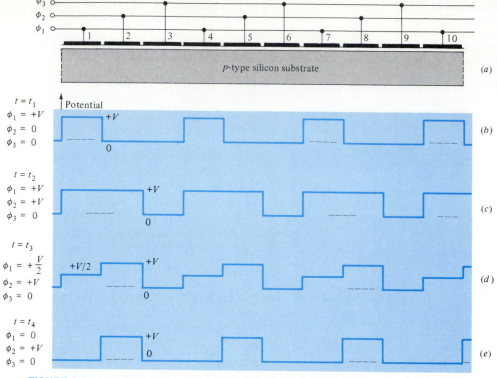

FIGURE 9-23
Illustration of charge transfer in a CCD. (a) Every third electrode is at the same potential and three-phase voltages, ϕ_1, ϕ_2, and ϕ_3 (Fig. 9-24), are applied; (b to e) the potential profile variations during one shift interval. The potential energy for negative charge is proportional to the magnitude of the potential; hence, these curves also represent the potential-energy wells for electrons.

between shifts (for example, between t_4 and t_5), the clock voltages remain constant and the potential profiles are unaltered.

Electrodes per Bit From Fig. 9-23b or e it is clear that if a logic bit is latched under one electrode, no information can be stored below the next two electrodes. In other words, a storage cell consists of three electrodes and 1 bit is stored in this cell. For this CCD the *electrodes per bit* is three ($E/B = 3$). The information is read at the output, say electrode 10, where at $t = t_1^-$ a 1 is observed. From Fig. 9-23 three shifts are required before the next bit (the 1 stored under electrode 7) can be sensed. Three transfers later the 0 under gate 4 appears at the output. Since three shifts take place in the period T, information must be read (or written) only once per cycle of the input waveform.

It has been assumed for simplicity that the threshold voltage V_T is negligible in the foregoing discussion. In reality, all levels marked 0 V in Figs. 9-23 and

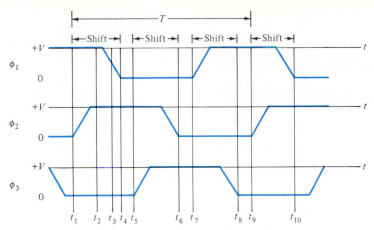

FIGURE 9-24
The three-phase excitation waveforms for the CCD in Fig. 9-23a. The potential profile in Fig. 9-23b corresponds to the instant t, of this figure; Fig. 9-23c corresponds to t_2; and so on.

9-24 should be at a voltage in excess of V_T for the electric field to penetrate into the channel and form the depletion region.

Minimum and Maximum Operating Frequencies   Steady-state (dc) operation of a CCD is not possible. Thermally generated carriers become trapped in empty potential-energy wells and, in time, change the logic state from a 0 to a 1. This phenomenon, called the *dark-current effect*, sets a lower limit to the clock frequency (10 kHz to 1 MHz).

No steady-state power is required by a CCD cell since power is dissipated only in charging the effective cell capacitances. Consequently the upper limit of clocking frequency (1 to 30 MHz) may be determined by the maximum allowable power dissipation. Also, an increase in frequency reduces the efficiency of transfer of charge from one cell to the next. Hence, the upper frequency may be limited by the point at which transfer losses become unacceptable.

9-8 CCD STRUCTURES   A CCD cannot be assembled from discrete components because a single continuous channel is required to provide the coupling between depletion regions. The gates (Fig. 9-23) must be separated by a small distance ($\sim$1 μm) to supply this coupling, and these narrow gaps are difficult to fabricate reliably. A number of alternative structures with both metallic and polysilicon gates have been developed to avoid difficulties in fabrication. One such polysilicon electrode structure is depicted in Fig. 9-25 for a three-phase, *n*-channel CCD. This planar electrode structure makes use of overlapping gates having different shapes. A large number of such CCDs are fabricated in rows parallel to one another to cover the chip area. With minimum separation between rows (2 to

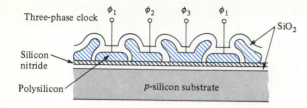

3 μm), a small area per bit is required for this three-electrode cell. For planar electrode structures (Figs. 9-23 and 9-25), it is necessary to use three-phase clocks to transfer charge longitudinally in one direction only. Construction of nonplanar electrodes permits the use of two-phase clocks needing only two electrodes per bit as described in the following.

Two-Phase CCD The nonplanar metallic-gate structure in Fig. 9-26*a* can utilize two-phase clocking. The right-hand half of each electrode is over a thinner oxide layer than the left-hand section, and, consequently, electric lines of force penetrate more deeply into the substrate at the right side of the metal. Hence the depletion region and the potential-energy profile has the same step shape as the two-level electrodes. Alternate electrodes are tied together, resulting in a two-phase system whose clock waveforms ϕ_1 and ϕ_2 are given in Fig. 9-27.

Let us again assume that the threshold voltage is 0. The potential-energy profiles are drawn in Fig. 9-26 for the values of time t_1, t_2, t_3, and t_4, indicated in Fig. 9-27. At $t = t_1$, $\phi_1 = 0$ and $\phi_2 = V$, so that there is no barrier under E_1, and the step barrier under E_2 is as drawn in Fig. 9-26*b*. We assume that a logic 1 is stored under E_2 and E_4, and we have indicated the minority carriers (electrons) by minus signs. It is convenient to place these minus signs near the bottom of the well, although, actually, the electrons are stored near the surface in the longitudinal position of the potential energy minimum.

At $t = t_2$, $\phi_1 = \phi_2 = \frac{1}{2}V$, and the profile under every electrode has the same shape, as indicated in Fig. 9-26*c*. The arrows on this sketch are intended to indicate that, as the time increases from t_1 to t_2 to t_3, the potential increases under the odd-numbered electrodes and decreases under the even-numbered plates. Hence, at $t = t_3$, the stair-step profile in Fig. 9-26*d* is obtained. The electrons stored under the right-hand side of E_2 and E_4 now are forced to the lowest potential energy and are trapped in sites under E_3 and E_5, respectively. Finally, at $t = t_4$, where $\phi_1 = V$ and $\phi_2 = 0$, the profile in Fig. 9-26*e* is obtained. In the interval $t_4 - t_1$ the information is shifted to the right by one electrode. Between t_5 and t_6 the second shift takes place. By the argument given in the preceding section, there are two electrodes per bit ($E/B = 2$). Hence, a shift-register cell contains two electrodes, and information must be read (or written) only once per clock period in the interval $t_5 - t_4$ or $t_7 - t_6$, which is also called the *input/output* (*I/O*) interval.

An excellent implementation of the electrode structure of the two-phase CCD is given in Fig. 9-28*a*. Polysilicon electrode E_1 is shaped like its metallic counterpart of Fig. 9-26*a*, with thick oxide at the left. The *p*-type ions implanted

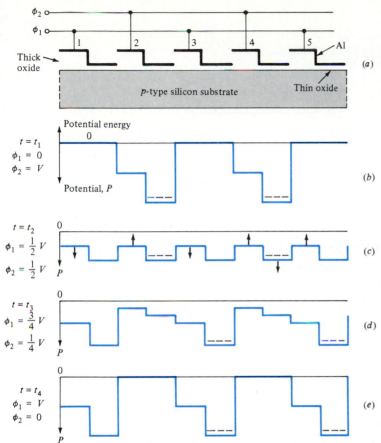

FIGURE 9-26
Illustrating charge transfer in a two-phase CCD: (*a*) electrode structure, (*b* to *e*) the potential-energy profiles corresponding to the times indicated in Fig. 9-27.

under the left-hand side of E_2 in the p-type substrate supply the desired potential offset under this electrode. When a positive voltage is applied to E_2, the holes are repelled, leaving the high concentration of immobile negative charges from the implantation. Consequently the lines of force from the left-hand side of E_2 terminate on these negative ions and do not penetrate deeply into the substrate. Hence, the potential-energy profile is much closer to the surface on the left side than on the right side of E_2, as required.

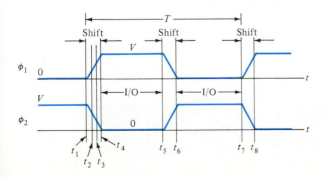

FIGURE 9-27
The two-phase clock waveforms for the CCD in Fig. 9-26.

FIGURE 9-28
Overlapping electrode structure of a two-phase CCD.

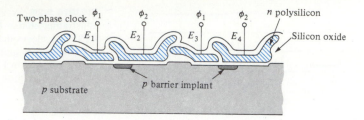

The waveforms ϕ_1 and ϕ_2 in Fig. 9-27 are essentially symmetrical square waves and, with nonzero rise and fall times, form an overlapping two-phase clock. Shorter, nonoverlapping positive clock pulses can also be utilized to shift the charge stored under one gate to the adjacent gate (Prob. 9-18). This arrangement can be adapted to single-phase operation by making ϕ_1, a constant potential, using a short, positive pulse ϕ_2 and appropriately adjusting their respective amplitudes (Prob. 9-20).

Four-phase CCDs are also fabricated in which the polysilicon gate arrangement combines the planar gate structure shown in Fig. 9-25 with the nonplanar structure in Fig. 9-28. Two data shifts occur in one clock cycle with four electrodes per bit required.

Input and Output Structures In Fig. 9-29 a source diffusion S and a gate G have been added to the input end of a CCD register. The potential well under the first electrode E_1 acts as a drain so that S, G, and E_1 form a MOSFET. Voltages are applied to S and to G so that current flows until the well fills with charge to the same potential as that of S.

The output is obtained from a drain diffusion D added at the end of the register as in Fig. 9-29, which senses the output current. Voltage or charge sensing is obtained by fabricating an output amplifier on the chip or adding one externally.

CCD Memory Organization Charge-coupled-device memories have slower access times than RAMs due to their serial operation. However, the CCD is an excellent display refresh memory for a CRT terminal and is a cost-efficient replacement for small shift-register memories.

Information must be shifted to the output part before it can be read in a CCD memory. The worst-case access time to any bit is called the *latency time*. For a given number of bits per chip, the latency time depends on how the chip is organized. Two commonly used organizations, called *serpentine* and *line-addressable random-access memory* (LARAM), are described in the following.

FIGURE 9-29
Structures for (a) injecting and (b) detecting charge in an n-channel CCD.

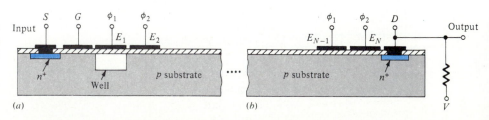

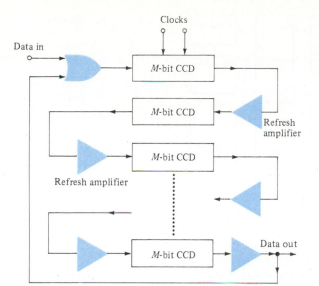

FIGURE 9-30
Serpentine or synchronous organization of a CCD memory. The multiphase clock is applied simultaneously to all CCD sections (between refresh amplifiers).

The serpentine is the simpler of the two CCD memories to fabricate and is illustrated in Fig. 9-30. It is a synchronous organization in which data is shifted from cell to cell in a long, snakelike manner (hence the name "serpentine") in a recirculating shift register.

The LARAM is designed for short access times and consists of a number of short recirculating memories operating in parallel. These parallel CCDs share common input and output lines. A decoder is used to address the registers at random and gives rise to the designation of this organization. The Intel 2464 is a 64-kb CCD organized as 256 independent registers of 256 bits each (Fig. 9-31). An 8-to-256-line decoder can select any register at random. The I/O operations are performed in a manner similar to that for a 256-bit RAM.

9-9 INTEGRATED-INJECTION LOGIC (I²L)

Integrated-injection logic or simply I²L, developed simultaneously in 1972 by engineers at Phillips Research in the Netherlands and IBM Laboratories in West Germany, is a VLSI technology that combines the higher component density of MOS with the higher speeds of BJTs. Advances in MOS technology, such as the reduction of minimum dimensions, have lead to the declining use of I²L. However, I²L RAMs, microprocessors, and A/D and D/A converters are currently available. In addition, research and development in this technology is proceeding, albeit at a diminished level compared to the decade since its introduction. For these reasons we include this section to introduce the principles underlying I²L operation.

Merging of Devices Increased component density in bipolar fabrication can come about by eliminating area-consuming resistors and significantly reducing (or eliminating) the isolation islands that separate devices. One such technique, already en-

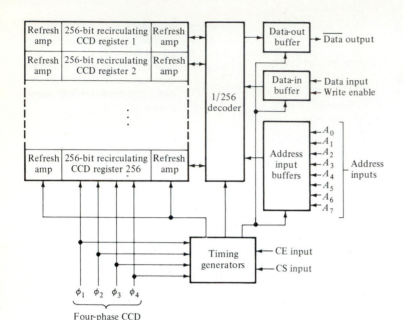

ϕ_1 ϕ_2 ϕ_3 ϕ_4

Four-phase CCD
clock inputs

FIGURE 9-31
The Intel 2464 65,536-bit (64-kb) CCD memory organized as 256 recirculating shift registers of 256 bits each. The registers can be addressed randomly. The four-phase clock is obtained from a separate chip (the Intel 5244). (*Courtesy of Intel Corporation*)

countered in the multiple-emitter BJTs used in TTL gates and emitter-coupled storage cells, is to merge devices. That is, when one semiconductor region is part of two or more components, these devices are said to be merged. This process saves considerable chip area.

In standard bipolar technology (Sec. 5-3), if isolation islands are not used, the collectors of all BJTs would be in a single n-epitaxial layer and would thus be at the same potential. In the bipolar DCTL NOR gate (Fig. 9-32), it is the emitters of all transistors that are at constant potential (ground). These emitters can be kept at constant potential if multiple-emitter transistors are fabricated and operated in the inverted mode. That is, each n^+ emitter region becomes a collector and the normal collector region becomes the common-emitter region for the merged device. This technique is used to form the three collectors and grounded emitters in Fig. 9-32 as illustrated in Fig. 9-33a. The n^+ substrate is used to improve the current gain of the inverted BJT.

FIGURE 9-32
Paralleled-BJT inverters.

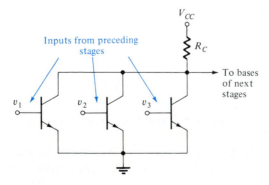

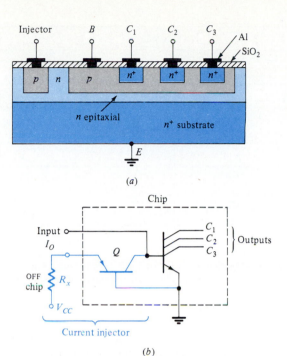

FIGURE 9-33
(a) The cross section of the multiple-collector transistor. Note that the n^+ regions that form a multiple-emitter transistor are used as the collectors. (b) Circuit configuration indicating pnp current injector.

Current Injection   Further savings in chip area are effected by removing R_C (Fig. 9-32) from the chip. Replacement of R_C by a current source having an output resistance (ideally infinite) of value much higher than can be fabricated for R_C, results in increased gain of the inverters in the NOR gate. As indicated in Sec. 6-5, higher gain results in improved gate performance. The current source, or *current injector*, is formed by using the grounded-base *pnp* transistor Q identified in Fig. 9-33b. Note that *the resistance R_x is external to the chip.* The value of the current source is clearly

$$I_O = \frac{V_{CC} - V_{BE}}{R_x} \qquad (9\text{-}2)$$

The collector current $\alpha_F I_O$ of Q is also the base current of the multicollector transistor. Implantation of an additional p region into the n-epitaxial layer results in formation of the lateral *pnp* transistor Q for the injector. The collector of Q is also the base of the multicollector BJT so that these regions are merged as seen in Fig. 9-33a.

It must be emphasized that *all* injector currents are obtained from V_{CC} through a *single* external resistor R_x. A chip usually is fabricated with long p-type diffusion lines called *injector rails*. Each rail delivers base current to all the *npn* transistors adjacent to it (within a diffusion length). A top view of a possible injector-logic-chip layout is shown in Fig. 9-34. A shaded rectangle represents an *n-p-n* transistor whose base (the transistor input) is shown as a

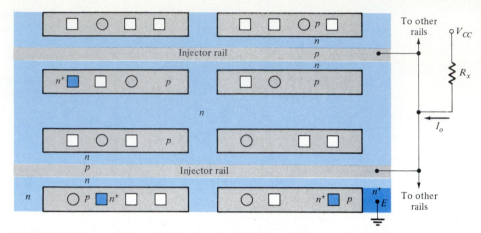

FIGURE 9-34
The top view of an integrated-injection logic (I²L) chip layout. Each circle represents an input (p base) connection of a vertical transistor and each square represents an output (n collector).

small circle and whose collectors (the inverter outputs) are indicated by small squares. All p regions are colored and all n regions are left white. We have arbitrarily chosen the number of collectors in each multicollector transistor. The positions of the collectors and of the base lead in each I²L gate is determined in such a way as to simplify the interconnections between gates to satisfy the desired logic. The bottom left-hand inverter in Fig. 9-34 corresponds to the electrode arrangement in Fig. 9-33. The emitter of each vertical transistor is the grounded n (unshaded) region. The organization in Fig. 9-34 shows only eight inverters, but it can, of course, be expanded both vertically and horizontally into a VLSI system.

With the background provided in the previous paragraphs, we now can describe the operation of the basic gates and FLIP-FLOPS fabricated in I²L technology.

Inverter

The inverter $Q1$ is loaded by $Q2$ in Fig. 9-35. Each transistor is biased by an injector current I_j. At the low logic level $V_i \approx 0$, $Q1$ is OFF, and the input signal V_i acts as a sink for I_j so that $I_{B1} = 0$ and $I_{C1} = 0$. Hence $I_{B2} = I_j$, and $Q2$ is ON so that $V_{BE2} \approx 0.75$ V $= V_{CE1}$. On the other hand, if the input is high, $V_i \approx 0.75$ V, the base current I_{B1} increases above I_j and $Q1$ tends to saturate. Consequently, V_{CE1} drops very low (≈ 0 V). Now $Q2$ is cut off because $Q1$ acts as a sink for I_j of $Q2$, so that $I_{C1} = I_j$, reducing I_{B2} to zero. Clearly, an

FIGURE 9-35
An I²L inverter: $Q1$ is directly coupled to the following stage $Q2$.

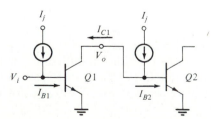

inversion has been performed by $Q1$ because $V_o = 0.75$ V for $V_i = 0$ and $V_o = 0$ V for $V_i = 0.75$ V. The logic swing is about 0.75 V, its exact value depending on the bias current I_j.

Note that in saturation the collector current is I_j and the base current has about the same magnitude. Hence, a value of the CE current gain β_F of only unity is required to cause saturation. A transistor operating in the inverse mode has a very much smaller value of β_F than in the normal mode (~ 100). Nevertheless, a CE current gain in excess of unity (approximately 2 to 10) is easily obtained for the upside-down transistor.

When an inverter is switched from one state to the other, the voltages of the transistor capacitances must change values, causing a propagation delay t_{pd}. The charging (and discharging) current of these capacitances is supplied by the injector. Large values of I_j result in small values of t_{pd}, but the penalty which must be paid for the decreased t_{pd} is an increased average-power dissipation P_{av}.

An advantage of the I²L configuration is that it can be operated over a wide range of speeds by simply altering the total injector current, by varying the one resistor R_x. The range of operation extends from about 1 nA to 1 mA. Thus after a chip is designed and built, the desired speed of operation may be selected by changing R_x.

NAND Gate

To obtain an AND gate in injector logic is extremely simple. In Fig. 9-36a, Y_1 is a logic variable at the output of an I²L inverter and Y_2 is another variable at the collector of a second I²L gate. Connecting Y_1 and Y_2 together yields $Y = Y_1Y_2$ at the common node in Fig. 9-36a. If Y is applied to the input of an inverter, the output is the NAND function $\overline{Y_1Y_2}$, as shown in Fig. 9-36a.

If $A(B)$ is an externally applied logic variable, then to obtain $A(B)$ *at the collector of an I²L gate*, two cascaded inverters must be used, as indicated by the solid lines in Fig. 9-36b. Proceeding in this manner, the NAND function $\overline{AB}$ is obtained as shown in the figure.

FIGURE 9-36
(a) A NAND gate using wired AND for internal (collector) logic variables. (b) A NAND gate for externally applied logic variables (solid connections). The dashed portion of the circuit is a NOR gate. (*Note:* The injectors are omitted for simplicity.)

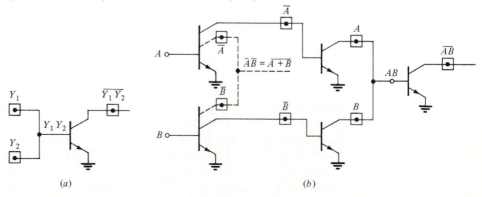

(a) (b)

FIGURE 9-37
(*a*) A bistable latch (a 1-
bit memory cell). (*b*) The
I²L interconnection dia-
gram for this FLIP-FLOP.

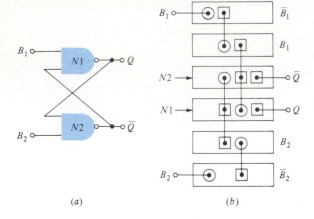

FIGURE 9-37
(*a*) A bistable latch (a 1-bit memory cell). (*b*) The I²L interconnection diagram for this FLIP-FLOP.

(*a*)   (*b*)

NOR Gate

In Chap. 6 it is verified that all combinational logic functions can be generated by using only NAND gates. From De Morgan's law (Sec. 6-3) $\overline{A + B} = \overline{A}\,\overline{B}$ and, hence, the NOR function is obtained from the wired-AND gate with inputs A and B. This procedure is illustrated by the dashed part of Fig. 9-36*b*. Note that the two input transistors have two collectors each, and the other three inverters each have one collector.

FLIP-FLOP

Sequential circuits such as registers and counters discussed in Chap. 8 are based upon FLIP-FLOPS, which are easily implemented with I²L gates. The 1-bit storage cell or latch in Fig. 8-1*b* is repeated in Fig. 9-37*a*. The interconnecting diagram, obtained by following the foregoing rules, is drawn in Fig. 9-37*b*.

Static RAMs, serial memories (such as those described earlier in this chapter), and a variety combinatoreal and sequential circuits (Chaps. 7 and 8) can be constructed from the basic I²L circuits of Figs. 9-35 to 9-37.

9-10 MICROPROCESSORS AND MICROCOMPUTERS A *microprocessor* is a single-chip system which contains, at a minimum, the arithmetic, logic, and control circuitry of a general-purpose data processing and/or computing system. Most modern microprocessors also contain a small amount of memory on the chip, and many incorporate an on-chip clock circuit as well. This combination of circuits (subsystems) is the central processing unit (CPU) of the system. The typical internal organization of a microprocessor shown in Fig. 9-38 displays the major subsystems incorporated.

Commercially available processors are available in word lengths of 4, 8, 16, and 32 bits. All the fabrication technologies discussed in this book are used to construct microprocessors. The 2-μm, polysilicon gate MOS process, both NMOS and CMOS, is the dominant technology used to fabricate the most recent microprocessors. For high-speed operation, bipolar technology, such as the TTL ALS family of circuits, is used. The photomicrograph in Fig. 9-39 is that of the Motorola MC68020, a 32-bit CMOS processor. This chip is approximately 9.5×8.9 mm (375×300 mils) and contains over 200,000 transistors.

FIGURE 9-38
Typical internal organization of a microprocessor.

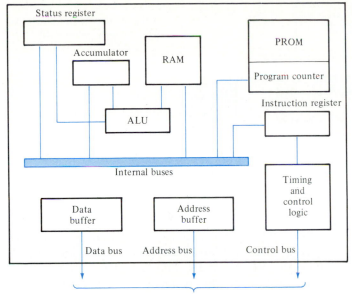

Status register

Accumulator

RAM

PROM

Program counter

Instruction register

ALU

Internal buses

Timing
and
control
logic

Data
buffer

Address
buffer

Data bus

Address bus

Control bus

To memory and I/O

FIGURE 9-39
Photomicrograph of the MC68020 microprocessor. (*Courtesy of Motorola, Inc.*)

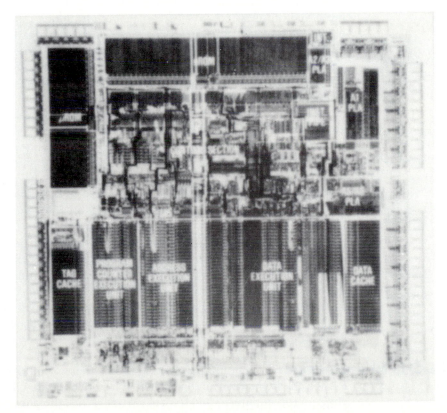

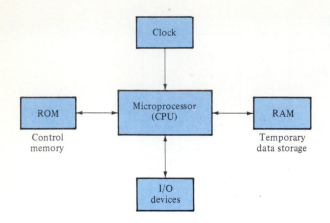

FIGURE 9-40
A microcomputer formed by augmenting a microprocessor with memory, input-output devices, and a clock.

The functional organization shown in Fig. 9-39 illustrates that the chip incorporates a number of sections (ROM, PLA, Data Execution, etc.) which are individually VLSI circuits. This technique of integrating several standard subsystems on a single chip is a major VLSI design approach.

Microcomputers   To be able to perform all the tasks required by a computer, the CPU must be augmented by additional memory, control circuits, and interface adapters for input and output (I/O) devices. An elementary block diagram of such a system is indicated in Fig. 9-40. In "dedicated" computers, that is, those designed to perform a specific set of tasks as in the case in electronic-fuel-injection systems, nonvolatile ROMs are used to store the program and tables. If the program is not completely fixed, PROMs, EPROMs and EEPROMs are used. Temporary storage resides in RAM chips, and, for large amounts of memory, magnetic disks (hard disks and/or flexible disks colloquially called "floppy" disks) are employed. The I/O devices include keyboards, CRT displays, and printers. Microprocessor-based communication and control systems often employ transducers and A/D and D/A converters which convert physical quantities to and from the electronic digital signals.

A variety of other control and processing subsystems are designed to operate in conjunction with the CPU (not shown in Fig. 9-40) and include coprocessors and memory management chips. Coprocessors, such as the Motorola MC68881 designed to operate with the MC68020, extend the floating-point numerical capability of the CPU. Control and direct-memory access (DMA) are provided by memory management and permit the efficient transfer of information to (from) RAM from (to) magnetic disks. The number of auxilliary chips surrounding the microprocessor may approach 100, and all chips are mounted on a printed-circuit (PC) board often no larger than the size of this page. The equivalent number of transistors on such a board often exceeds 1,000,000.

Single-Chip Microcomputer   Integration of the timing and control circuits, memory (ROM and RAM), the I/O access, and peripheral circuits with the CPU leads to the single-chip microcomputer. The Motorola MC68HC11, the photomicrograph of which

FIGURE 9-41
Photomicrograph of the MC68HC11, a single-chip microcomputer. (*Courtesy of Motorola, Inc.*)

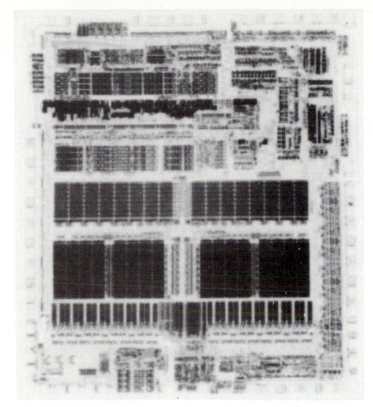

is displayed in Fig. 9-41, is one such chip. Containing over 100,000 transistors on a 6.5 × 7.4 mm (256 × 287 mil) die, this microcomputer comprises a CPU, timer, I/O, memory (RAM, ROM, EEPROM), and an A/D converter. Figure 9-41 also illustrates the relative size required to implement the three types of memory. Starting from the bottom of the chip, the first step contains 512 bytes of EEPROM. The 8192 bytes of ROM form the second strip, while the third strip is for the 256 bytes of RAM. The single-chip microcomputer can be augmented by other chips to both expand memory capacity and provide the interfaces with a variety of I/O and communications devices.

Applications   A detailed discussion of microprocessor applications is beyond the scope of this book. Indeed, today most electrical and computer engineering departments offer (and often require) courses in digital design, computer architecture, and microprocessor-based design where these applications are treated. However, the list on the following page indicates the range of applications which incorporate microprocessors.

Digital-signal processing is another area in which microprocessors are incorporated. Single-chip programmable signal processors, such as the Texas Instruments TMS 32010, are capable of performing the real-time functions of

Consumer electronics	Commercial and industrial	Instrumentation	Office and business
Calculators	Machine and process control	Self-calibrating equipment	Remote terminals for computer networks
Digital watches	Traffic control	Radiation-monitoring equipment	Automating banking equipment
Video games	Navigation	Laboratory analysis systems	Point-of-sales terminals
Appliance control	Automated aircraft landing systems	Data recorders	Reservations terminals
Home entertainment systems	Electronic fuel injection	Diagnostic tools	Stock and inventory control terminals
Home security systems	Mass-transit control	Medical instruments	Copying machines
Home energy management	Emissions (pollution) control	Security and fire control for buildings and manufacturing facilities	
	Data communications		

spectrum analysis, digital filtering, speech compression, data modems, and music synthesis.

These diverse applications indicate that the microprocessor has become a major *component* in electronic systems. Its programmable nature provides the flexibility of performing different functions with the *same* hardware. Indeed, the range of application is limited only by the ingenuity and imagination of the engineers who use them.

REFERENCES

1 Hodges, D. A. (ed.): "Semiconductor Memories," IEEE Press, New York, 1972.

2 Elmasry, M. I. (ed.): "Digital MOS Integrated Circuits," IEEE Press, New York, 1981.

3 Hodges, D. A.: Microelectronic Memories, *Scientific American,* vol. 237, no. 3, pp. 130–145, September 1977.

4 *IEEE Journal of Solid-State Circuits,* annual special issue on semiconductor logic and memory, October 1970 to present.

5 *Digest of Technical Papers,* International Solid-State Circuits Conference, annually 1954 to present.

6 Hodges, D. A., and H. G. Jackson: "Analysis and Design of Digital Integrated Circuits," McGraw-Hill Book Company, New York, 1983.

7 Mead, C., and L. Conway: "Introduction to VLSI Systems," Addison-Wesley Publishing Company, Reading, Mass., 1980.

8 J. E. Smith (ed.): "Integrated Injection Logic," IEEE Press, New York, 1980.

9 *IEEE Transactions Electron Devices,* vol. ED-23, no. 2, Special Issue on Charge-Transfer Devices, February 1976.

REVIEW QUESTIONS

9-1 (*a*) Draw the circuit for a single-phase dynamic MOS inverter.
      (*b*) Explain its operation.

9-2 List three important features of MOSFETs used in a dynamic inverter.

9-3 Explain what is meant by a ratioed inverter.

9-4 (*a*) Draw the circuit for one stage of a two-phase ratioed dynamic NMOS shift register.
(*b*) Draw the clocking waveforms.
(*c*) Explain the operation of the circuit.

9-5 (*a*) Draw the block diagram of a recirculating shift register.
(*b*) Explain its operation, including how to write into the register and how to read out nondestructively.

9-6 Explain how to obtain a serial memory which may store 512 words of 8 bits each.

9-7 (*a*) Sketch the circuit of a ratioless MOS inverter. Explain the operation (*b*) if $V_i = 0$ and (*c*) if $V_i = V_{DD}$.

9-8 Repeat Rev. 9-5 for one stage of a ratioless shift register.

9-9 (*a*) Draw one stage of a dynamic CMOS register.
(*b*) Explain its operation briefly.

9-10 (*a*) Draw the circuit diagram of a domino logic AOI gate.
(*b*) How can this be converted to an AND-OR gate?

9-11 (*a*) What is the principal advantage of CMOS domino logic?
(*b*) Explain.

9-12 Sketch the waveforms of both an overlapping and a nonoverlapping two-phase clock.

9-13 (*a*) List four advantages of semiconductor random-access memories over core memories.
(*b*) What advantage does a core memory have over a RAM?

9-14 (*a*) Draw the block diagram of a 1-bit read/write memory.
(*b*) Explain its operation.

9-15 Explain linear selection in a random-access memory (RAM).

9-16 Repeat Rev. 9-15 for two-dimensional addressing.

9-17 (*a*) Draw in block-diagram form the basic elements of a RAM with two-dimensional addressing used to store four words of 1 bit each.
(*b*) How is the system expanded to 3 bits per word?
(*c*) How is the system expanded to 25 words of 3 bits per word?

9-18 Indicate how to expand the memory of a 1024 × 1 RAM to one of 4096 × 1.

9-19 Explain how to expand a 1024 × 1 RAM to 1024 × 16.

9-20 (*a*) Sketch the circuit of a 6-MOSFET static RAM cell.
(*b*) Briefly explain its operation.

9-21 How many transistors are saved in going from a 6-MOSFET static RAM cell to a 4-MOSFET dynamic cell in a 64-kb RAM? Explain briefly.

9-22 (*a*) Draw a 4-MOSFET dynamic RAM cell.
(*b*) Why is additional circuitry required to refresh the data stored in the cell?

9-23 (*a*) Show how the 4-MOSFET cell in Rev. 9-22 is placed into a RAM.
(*b*) Which devices are associated with a given column?
(*c*) Explain the function served by each MOSFET.

9-24 Draw a 1-MOSFET dynamic memory cell. Briefly explain its operation.

9-25 Show the organization of a 64-kb DRAM.

9-26 How many transistors are saved in going from a 4-MOS cell to a 1-MOSFET dynamic cell in a 256-kb DRAM?

9-27 (a) Sketch the circuit diagram of an emitter-coupled RAM cell.
(b) Briefly explain its operation.

9-28 Why do bipolar RAM cells usually require two-step decoding?

9-29 List five important characteristics of RAMs.

9-30 (a) Explain how a potential-energy well is formed under an electrode of a CCD.
(b) If the substrate is p-type, are electrons or holes captured in the well?

9-31 Consider a CCD with planar electrodes and using three-phase excitation.
(a) How many shifts of a charge packet take place in one cycle?
(b) What is meant by electrodes per bit?
(c) What is the value of E/B for this CCD?

9-32 What determines (a) the minimum frequency of operation of a CCD; (b) the maximum frequency?

9-33 (a) Sketch schematically the shape of the electrodes in a two-phase charge-coupled memory.
(b) Draw the excitation waveforms.
(c) How many shifts of a charge packet take place in each cycle?

9-34 For the CCD memory in Rev. 9-33, illustrate how the information is shifted down the register.

9-35 Describe the serpentine organization of a CCD memory.

9-36 Repeat Rev. 9-35 for the LARAM organization.

9-37 Give three reasons why MOSFETs dominate in LSI over BJTs.

9-38 Define merging of devices.

9-39 (a) Explain why a collector resistor is not needed in DCTL and why it may be replaced by a current source.
(b) Indicate such a current injector.
(c) Which elements of the injector are external to the chip?

9-40 (a) Draw the cross section of an I²L inverter including the current source.
(b) Show a circuit model for an I²L unit.

9-41 Explain the operation of an I²L inverter.

9-42 (a) A and B are available at the collectors of two I²L inverters. Show how to obtain the NAND function $\overline{AB}$.
(b) Repeat part a if A and B are applied externally.

9-43 Show how to obtain the NOR function $\overline{A + B}$ in I²L.

9-44 Show the interconnection diagram for a bistable latch in I²L.

9-45 What functions are needed on a microprocessor chip?

9-46 How does a microcomputer differ from a microprocessor?

Part Three
AMPLIFIER CIRCUITS AND SYSTEMS

The availability of reliable devices to amplify signals is the foundation on which modern electronics rests. Without amplifying devices, nearly all of today's communication, control, instrumentation, and computer systems would be rendered impractical. In Secs. 3-1 and 4-1 we demonstrated that the controlled source is an amplifying element. Also, in those chapters it was seen that both BJTs and FETs, when appropriately biased, behave as controlled sources. In this part of the text we examine how transistors are utilized in amplifier stages and in turn, how such stages are interconnected in amplifier systems. When providing controlled-source characteristics, transistors behave linearly. Consequently, the analyses of amplifier circuits lend themselves to the use of small-signal, incremental models (Secs. 3-10 and 4-14).

The first two chapters in this part deal with the behavior of basic transistor amplifier configurations. In Chap. 10 we focus on low-frequency circuit performance and the biasing process. The frequency response, and the resultant limitations on performance of these basic circuits, is treated in Chap. 11. The very important topic of feedback and its use in controlling circuit performance is discussed in Chaps. 12 and 13. The feedback concept and low-frequency feedback amplifier behavior is treated in Chap. 12; stability and frequency response are discussed in Chap. 13. The objective in Chap. 14, where we describe in detail the operational amplifier, the predominant analog integrated circuit (IC), is twofold: (1) this single-chip amplifier system is an essential component in signal-processing and data-acquisition circuits; thus this treatment serves as a prelude to Part 4; and (2) amplifier design techniques are readily highlighted and the material in the four prior chapters is integrated by the discussion of the operational amplifier.

Chapter 10
BASIC AMPLIFIER STAGES AT LOW FREQUENCIES

The physical operation, volt-ampere characteristics, and behavior as a circuit element of bipolar and field-effect transistors (FETs) were described in Chaps. 3 and 4. We now utilize these discussions to investigate the performance of basic BJT and FET amplifier stages.

In Sec. 3-3 we demonstrated that a BJT behaved as a controlled source when biased in the forward-active region. Similarly, the FET functions as a controlled source when biased in the saturation region (Sec. 4-2). In Secs. 3-9 and 4-13 we demonstrated that transistors can be used as amplifying devices when so biased. Transistors are biased in their appropriate regions by means of externally applied direct voltages and currents. That is, these constant (with time) sources establish an operating (quiescent) point. Time-varying input signals (base current and gate-to-source voltage, for example) are superimposed on the quiescent levels to produce a time-varying output signal (collector current, drain voltage, etc.). It is the time-varying input signal we wish to amplify; the constant excitations establish the appropriate bias. In Chaps. 3 and 4 we showed that graphical methods and dc models are used to determine quiescent levels. Furthermore, the small-signal equivalent circuit was a convenient representation of the transistor that enabled us to obtain the output signal produced by the time-varying input signal. The use of small-signal models is predicated on the approximately linear behavior of BJTs operated in the forward-active region and of FETs biased in saturation.

Ideally, the output signal of an amplifier should, at a higher level of energy, faithfully reproduce the input waveform. As real devices have inherent limitations, the circuits in which they are employed cannot exhibit ideal behavior. The range of input amplitudes that can be processed effectively (referred to as the *dynamic range*) is restricted. Distortion, the degree to which the output signal does not resemble the input signal, is one consequence of dynamic-range limitations (Fig. 10-2b). Since transistors are not ideal controlled sources, as evidenced by their equivalent circuits, constraints exist on both the amplification, or gain, that can be realized and the frequency range over which effective amplification is obtained. The parameter values used in the equivalent circuits and the dynamic range both depend on the quiescent levels. In turn, dynamic performance is a factor in the selection of an operating point. Furthermore, the location of the operating point must be controlled because device characteristics vary with manufacturing tolerances and environmental changes such as temperature.

The previous paragraph underscores several aspects of transistor amplifiers with which we are concerned in this chapter. The initial sections focus on methods of obtaining a stable operating point in both integrated and discrete-component transistor circuits. A detailed discussion of the basic BJT amplifier configurations and the analogous FET stages follows. Only low-frequency performance, in which the internal device capacitances can be neglected, are treated in this chapter. The effects of these capacitances is investigated in Chap. 11.

Very often, a number of stages are used in cascade to amplify a signal from a source, such as a phonograph pickup, to a level which is suitable for the operation of another transducer, such as a loudspeaker. Consequently, we consider the low-frequency behavior of cascaded amplifiers. The operational amplifier, a widely used integrated-circuit (IC) cascaded-amplifier system, is also introduced.

Since both direct and time-varying voltages and currents exist in a transistor amplifier, the initial section of this chapter introduces the notation used to distinguish the various components.

10-1 WAVEFORMS FOR A SINUSOIDAL INPUT

Consider the elementary common-emitter stage depicted in Fig. 10-1a, where the sources I_{BB} and V_{CC} provide the bias and the current source $i_b(t)$ is the signal to be amplified. Two sets of collector characteristics, solid for $\beta_F = 50$ and dashed for $\beta_F = 125$, for the transistor in Fig. 10-1a are displayed in Fig. 10-1b. The two operating points, Q_1 for $\beta_F = 50$ and Q_2 for $\beta_F = 125$, are indicated on the load line, corresponding to $V_{CC} = 10$ V and $R_C = 1$ kΩ.

Let us first consider the situation for $\beta_F = 50$. The approximate transfer characteristic of the circuit in Fig. 10-1a, $v_o = v_C$ versus i_B, obtained as outlined in Sec. 3-8, is displayed in Fig. 10-2a. Note that Q_1 lies on the linear portion of the characteristic between cutoff and saturation. For a signal $i_b(t) = I_{bm} \sin \omega t = \sqrt{2}I_b \sin \omega t = 20 \sin \omega t$ μA, the instantaneous base current $i_B = I_{BB} + i_b$ varies sinusoidally about Q_1 and $40 \le i_B \le 80$ μA (between points A and B on the transfer characteristic). As seen in Fig. 10-2a, the BJT remains in the forward-active region for the entire excursion about Q_1 produced by $i_b(t)$. Hence the resultant output waveform is the faithful reproduction of the input sinusoid as illustrated by the solid curve in Fig. 10-2b.

In Fig. 10-1a, KVL requires that $v_C = -i_C R_C + V_{CC}$ or $i_C = (V_{CC} - v_C)/R_C$ and indicates that i_C contains dc and time-varying components. Similarly, each

FIGURE 10-1

(a) Elementary common-emitter stage. (b) Collector characteristics for transistor in (a); solid for $\beta_F = 50$, dashed for $\beta_F = 125$. On the load line, the operating point ($I_B = 60 \ \mu A$) is shown for both values of β_F.

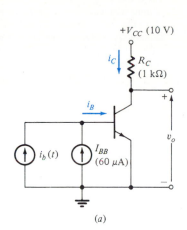

(a)

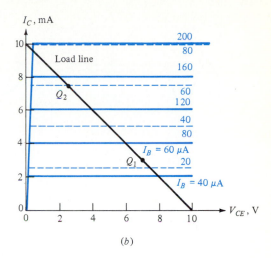

(b)

BJT current and voltage consists of a dc component and a time-varying component. When operation is confined to the linear portion of the transfer characteristic shown in Fig. 10-2a, the transistor behaves linearly. The waveforms in Fig. 10-3 display the BJT currents and voltages for a sinusoidal input and assumes linear operation. Each waveform in Fig. 10-3 comprises a distortion-free sinusoid superimposed on quiescent levels. A consequence of linear operation is that the dc component of the response is attributed to the dc excitation only and the time-varying component of the response is caused only by the input signal.

Notation

To avoid confusion, a standard terminology for the symbols representing specific current and voltage components has been adopted by the Institute for

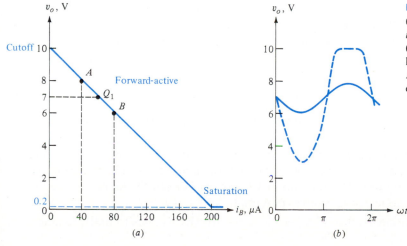

(a)

(b)

FIGURE 10-2

(a) The transfer characteristic (v_o versus i_B) of the circuit in Fig. 10-1a ($\beta_F = 50$). (b) The output waveform (solid) is a replica of the input sinusoid ($I_b = 20 \ \mu A$). A distorted output waveform (dashed) exists when $I_b = 80 \ \mu A$.

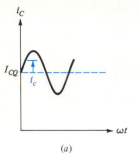

(a)

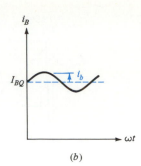

(b)

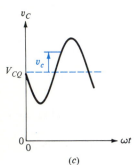

(c)

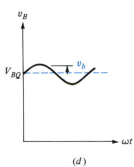

(d)

FIGURE 10-3

Sinusoidal waveforms for transistor currents and voltages: (a) collector current i_C, (b) base current i_B, (c) collector voltage v_C, (d) base voltage v_B. [*Note:* Each quantity comprises a constant (dc) term and a sinusoidally varying (ac) term.]

Electrical and Electronic Engineers (IEEE). These are given in Table 10-1 for the bipolar transistor and FET symbols are listed in Table 10-2.

The system underlying the notation is as follows:

1. Lowercase letters i and v designate instantaneous ("Inst." in Tables 10-1 and 10-2) current and voltage, respectively.

2. Uppercase V and I denote rms (or peak) values of ac components or average values (dc) of total quantities.

3. Lowercase subscripts are used only with the time-varying components of voltage and current.

4. Uppercase subscripts refer to total or dc quantities.

5. Repeated subscripts refer to the magnitudes of supply quantities.

6. Currents are taken as positive when entering a terminal from the external circuit.

7. Voltages are taken as positive when measured with respect to the reference node (usually ground). For voltages measured between a pair of terminals (not the reference), an additional subscript is used. Thus v_{CE} is the instantaneous collector-to-emitter potential.

8. The additional subscript Q is added to circuit variables to indicate quiescent values.

TABLE 10-1   Bipolar Transistor Voltage and Current Symbols

Item	Supply	Quiescent (static)	AC or time-varying component Inst.	AC or time-varying component RMS	Total (dc + ac) Inst.	Total (dc + ac) Avg.
Collector voltage	V_{CC}	V_{CQ}	v_c	V_c	v_C	V_C
Collector current	I_{CC}	I_{CQ}	i_c	I_c	i_C	I_C
Base voltage	V_{BB}	V_{BQ}	v_b	V_b	v_B	V_B
Base current	I_{BB}	I_{BQ}	i_b	I_b	i_B	I_B
Emitter voltage	V_{EE}	V_{EQ}	v_e	V_e	v_E	V_E
Emitter current	I_{EE}	I_{EQ}	i_e	I_e	i_E	I_E

TABLE 10-2   FET Voltage and Current Symbols

Item	Supply	Quiescent (static)	AC or time-varying component Inst.	AC or time-varying component RMS	Total (dc + ac) Inst.	Total (dc + ac) Avg.
Drain voltage	V_{DD}	V_{DQ}	v_d	V_d	v_D	V_D
Drain current	I_{DD}	I_{DQ}	i_d	I_d	i_D	I_D
Gate voltage	V_{GG}	V_{GQ}	v_g	V_g	v_G	V_G
Gate current	I_{GG}	I_{GQ}	i_g	I_g	i_G	I_G
Source voltage	V_{SS}	V_{SQ}	v_s	V_s	v_S	V_S
Source current	I_{SS}	I_{SQ}	i_s	I_s	i_S	I_S

Let us use the collector voltage as an example to illustrate the notation. The total quantity is the sum of the dc and ac components and can be expressed as

$$v_C = V_C + v_c \tag{10-1}$$

which for Fig. 10-3c becomes

$$v_C = V_{CQ} + \sqrt{2}\, V_c \sin \omega t \tag{10-2}$$

Note that the dc component in Fig. 10-3c is the quiescent voltage V_{CQ} as the average value of the sinusoid for one period is zero. This is a consequence of our assumption of linearity. If the instantaneous signal is distorted as in the dashed curve of Fig. 10-2b, the average value of v_C is not V_{CQ} and is a result of the nonlinear behavior of the transistor.

It may be useful to understand the notation if we consider how the quantities in Eqs. (10-1) and (10-2) are measured. The instantaneous voltage v_C is what is observed on the oscilloscope with the selector knob set to dc. Only v_c is displayed if the selector knob is set to ac.

Equation (10-1) can be rewritten as

$$v_c = v_C - V_C = \Delta v_C$$

and illustrates that v_c is the *incremental change* in v_C. Recall that incremental changes are the basis for the development of the small-signal equivalent circuits. These models allowed us to determine the time-varying component of the response analytically rather than graphically as in Fig. 10-2b. As a matter of fact, for very small signals the graphical technique requires interpolation between the printed characteristics of Fig. 10-1a and results in very poor accuracy. In subsequent sections of this chapter we assume small-signal (incremental) operation and utilize the equivalent circuits developed in Sec. 3-10 for the BJT and Sec. 4-14 for the FET. These models make it convenient to obtain the signal responses of amplifier stages using conventional network analysis techniques. In conjunction with the dc analysis in the next several sections, total voltages and currents can be evaluated.

10-2 THE OPERATING POINT OF A BJT The bipolar transistor functions most linearly when constrained to operate in the forward-active region. An operating point must be selected so that the transistor remains in the forward-active region at all times for which a time-varying signal is superimposed on quiescent levels. Distortion of the output signal results if an operating point is chosen that causes the BJT to be saturated or cut off (or both) when a signal is applied. The question now arises as to how to pick an operating point. We address this question by means of the common-emitter stage in Fig. 10-1a whose characteristics are displayed for $\beta_F = 50$ (solid) and $\beta_F = 125$ (dashed) in Fig. 10-1b.

The situation for $\beta_F = 50$ was described in the previous section. There we observed that for $i_b(t) = 20 \sin \omega t$ μA, operation is restricted to the linear portion of the transfer characteristic and an undistorted output signal v_o (solid curve in Fig. 10-2b) is obtained. However, if we increase the peak amplitude of $i_b(t)$ to be greater than 60 μA (say, 80 μA), the transistor is driven to cutoff during the negative half-cycle of $i_b(t)$. The dashed curve in Fig. 10-2b displays the clipped (distorted) output waveform for this situation. Because the transfer characteristic is approximately linear for $0 < i_B < 180$ μA, we can eliminate this distortion by changing the location of Q_1 to correspond to $I_{BB} = 90$ μA. With this value of I_{BB}, the signal causes $10 \leq i_B \leq 170$ μA and again operation is restricted to the linear segment of the transfer characteristic.

Figure 10-4a is the transfer characteristic of the circuit in Fig. 10-1a for $\beta_F = 125$ and is based on the dashed characteristics in Fig. 10-1b. Application of the signal $i_b(t) = 20 \sin \omega t$ μA causes the BJT to saturate during the positive

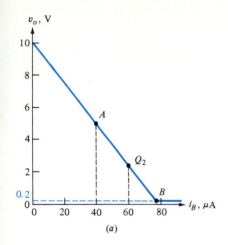

(a)

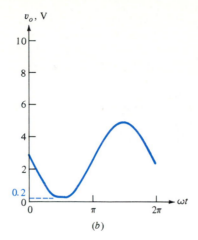

(b)

FIGURE 10-4
(a) Transfer characteristic of circuit in Fig. 10-1a for $\beta_F = 125$. (b) Output waveform resulting from an input signal $i_b(t) = 20 \sin \omega t \ \mu$A.

half-cycle of i_b and produces the clipped waveform in Fig. 10-4b. While a 20-μA peak-amplitude signal can be processed linearly by relocating Q_2 further up the transfer characteristic (reducing I_{BB}), no operating point can be selected to give a distortionless output if the peak amplitude of i_b is 80 μA. Indeed, on the basis of the transfer characteristic, the maximum sinusoidal signal that can be reproduced without distortion has an amplitude of about 39 μA. In reality, we must use a still lower value because the actual transfer characteristic is nonlinear near cutoff and saturation. Thus signal size determines the upper limit on dynamic range. The lower limit (the smallest signal that can be processed) is generally limited by the noise level present in the system.

We are not free to choose the operating point everywhere in the forward-active region because the various transistor ratings limit the range of useful operation. These ratings (App. B), which cannot be exceeded, are maximum collector dissipation $P_{C(\max)}$, maximum collector voltage $V_{CE(\max)}$, maximum collector current $I_{C(\max)}$, and maximum base-to-emitter voltage $V_{BE(\max)}$.

Bias Stability   The assumption in the previous section was that two different transistors were used, one with $\beta_F = 50$, the other with $\beta_F = 125$. This is typical, however, of the range of β_F values for a particular transistor type encountered in practice. (See manufacturers' specifications in App. B.) The variation in β_F represents the unit-to-unit variation in manufacture. Note, however, that this does not indicate poor manufacturing technique. The fabrication process controls α_F [not $\beta_F = \alpha_F/(1 - \alpha_F)$] since α_F is related to the geometry and doping levels employed. For a transistor having a β_F range from 50 to 125, the corresponding range of α_F is 0.980 to 0.992, a change of slightly more than 1 percent. For $\beta_F \geq 100$, control of β_F within 1 percent requires that α_F have a tolerance of less than 0.01 percent. (It is also commonplace for an individual BJT to exhibit such a range of β_F values over the extremes in operating temperatures.)

Now consider that the characteristics in Fig. 10-1b correspond to the min-

FIGURE 10-5

(*a*) Current mirror cir-
cuit. (*b*) Representation
of (*a*) with $Q2$ replaced
by equivalent diode. (*c*)
Norton equivalent of the
current mirror.

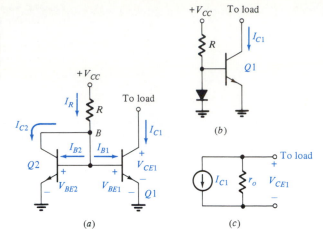

imum and maximum values of β_F for a given transistor over the operating
temperature range. For the circuit in Fig. 10-1*a*, the operating point can lie
anywhere between Q_1 and Q_2 (Fig. 10-1*b*) for $50 \leq \beta_F \leq 125$. Clearly, on the
basis of the output waveform in Fig. 10-4*b*, this bias arrangement is inappro-
priate. Proper circuit performance requires that the location of the operating
point be controlled; that is, biasing conditions must be stabilized. To achieve
bias stability, we must restrict the operation to a small region of the load line
over the specified range of β_F so that the signal (say, $i_b = 20 \sin \omega t \ \mu A$) is
accommodated. The import of the previous sentence is that *effective bias control occurs when the collector current, and consequently the emitter current,
remains essentially constant and independent of variations in β_F.* In Fig.
10-1*b* we see that maintaining the collector current at or slightly larger than
3 mA restricts the operation of the BJT to the forward-active region (Figs.
10-2*b* and 10-4*b*) for both $\beta_F = 50$ and $\beta_F = 125$. The following five sections describe circuits and techniques used to obtain bias stability.

10-3 BJT BIASING FOR INTEGRATED CIRCUITS The circuit in Fig. 10-5*a* is typical of
the biasing arrangement used in ICs. This and similar circuits are called *current
sources* or *current mirrors* and are designed to maintain the current I_{C1} at a
constant value.[1] Recall that we encountered this circuit in Sec. 3-11 and used
it in the differential amplifier stage (emitter-coupled stage) of Sec. 3-12 and
again in conjunction with ECL gates (Sec. 6-14). The differential amplifier is
one of the principal building blocks in IC amplifiers and is treated in detail in
Sec. 10-15. In Fig. 10-5*a* we recognize that $Q2$ is connected as a diode as shown
in Fig. 10-5*b*. The circuit in Fig. 10-5*c* is the equivalent representation of the
current source. The resistance r_o is the output resistance of the source and
reflects the fact that practical sources are nonideal.

---

[1]This circuit was investigated in Example 3-8.

The effectiveness of the current-source bias circuit is predicated on the features of IC technology. In particular, $Q1$ and $Q2$ are identical transistors, and R can be kept within the limits imposed by fabrication.

Since both transistors are identical, the collector currents I_{C1} and I_{C2} can be expressed as

$$I_{C1} = \alpha_F I_{ES}\, \epsilon^{V_{BE1}/V_T} \qquad I_{C2} = \alpha_F I_{ES}\, \epsilon^{V_{BE2}/V_T} \tag{10-3}$$

In Eq. (10-3) we neglect the Early effect (Sec. 3-5) and the current components due to the reverse-biased collector junctions. From Eq. (10-3) we obtain

$$\frac{I_{C1}}{I_{C2}} = \epsilon^{(V_{BE1} - V_{BE2})/V_T} \tag{10-4}$$

The KVL expression for the loop containing both emitter-base junctions shown in Fig. 10-5a requires

$$V_{BE1} = V_{BE2} = V_{BE} \tag{10-5}$$

and consequently, $I_{C1} = I_{C2} = I_C$ [Eq. (10-4)]. Since $\beta_{F1} = \beta_{F2} = \beta_F$, it follows that $I_{B1} = I_{B2} = I_B$.

The current I_R, called the *reference current,* is determined from the KVL relation for the loop containing V_{CC}, R, and V_{BE} and, after solving for I_R, yields

$$I_R = \frac{V_{CC} - V_{BE}}{R} \tag{10-6}$$

KCL at node B requires

$$I_C + 2I_B - I_R = 0 \tag{10-7}$$

Recalling that $I_C = \beta_F I_B$ and substituting Eq. (10-6), solving Eq. (10-7) for I_C, we obtain

$$I_C = \frac{\beta_F}{\beta_F + 2}\, I_R = \frac{\beta_F}{\beta_F + 2}\, \frac{V_{CC} - V_{BE}}{R} \tag{10-8}$$

The result expressed in Eq. (10-8) indicates that $I_C \approx I_R$ and is essentially constant over a wide range of β_F values. For $\beta_F \gg 1$, $\beta_F/(\beta_F + 2)$ is virtually unity, and, as an example, I_C varies by only 3 percent for $50 \le \beta_F \le 200$.

Note that if $\beta_F \gg 1$, $I_{C2} \approx I_R$ even if $Q1$ is disconnected. This is the basis of the mirror: V_{CC} and R set the value of I_{C2} (the "object"), and the connection of $Q1$ as shown in Fig. 10-5a makes the "image" $I_{C1} = I_{C2}$.

Output Resistance   Effective operation of the circuit requires that $Q1$ operate in the forward-active region. This fact is displayed in the volt-ampere characteristic of the current source, I_{C1} versus V_{CE1}, in Fig. 10-6. For $V_{CE1} < 0.3$ V, $Q1$ is saturated and behaves as a resistance $r_{CE(\text{sat})}$ (Sec. 3-6). In the forward-active region ($V_{CE1} > 0.3$ V), I_{C1} remains essentially constant. The slight increase in I_{C1} is attributed to the Early effect. The slope of the characteristic in this region is the reciprocal of the output resistance r_o of the current source. The value of r_o is

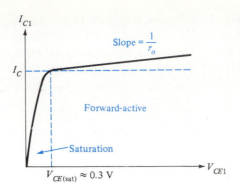

FIGURE 10-6
Volt-ampere characteristic of the current mirror. The horizontal dashed line is the characteristic of the source for an infinite Early voltage.

$$r_o = \frac{V_A}{I_C} \tag{10-9}$$

Note that if V_A, the Early voltage, is infinite, the volt-ampere characteristic is the dashed curve in Fig. 10-6 and $r_o \rightarrow \infty$ (an open circuit).

The operation of the current mirror is based in part on the fact that $I_{C1} = I_{C2}$. An additional consequence of the Early effect is to cause I_{C1}/I_{C2} to differ from unity. The operating values of V_{CE1} and V_{CE2} in Fig. 10-5a can differ significantly. The diode connection of $Q2$ makes $V_{CE2} = V_{BE} = 0.7$ V, whereas typical values of V_{CE1} can range from 1 to 30 V. As V_{CE1} increases, the Early effect causes I_{C1} to increase from I_C by a factor $(1 + V_{CE1}/V_A)$. Because $V_{CE2} = 0.7$ V $\ll V_A$, it follows that $I_{C2} \approx I_C$ and $I_{C1}/I_{C2} > 1$. For large values of V_{CE1}, I_{C1}/I_{C2} can differ from unity by 20 percent. Several of the circuits treated later in this section have output resistances higher than that given in Eq. (10-9). Consequently, the effective Early voltage increases appreciably and makes I_{C1}/I_{C2} more nearly unity.

Example 10-1

(a) For the circuit in Fig. 10-5a, determine R so that $I_C = 1$ mA. The transistor parameters are $V_{BE} = 0.7$ V, $\beta_F = 100$, and an infinite Early voltage is assumed. The supply voltage is 15 V. (b) For R obtained in part a, determine the percent change in I_C for $\beta_F = 200$. (c) Repeat part a for $I_C = 50$ μA.

Solution

(a) From Eq. (10-8), we have

$$1 = \frac{100}{100 + 2} \cdot \frac{15 - 0.7}{R}$$

from which $R = 14.0$ kΩ. Then (b)

$$I_C = \frac{200}{200 + 2} \cdot \frac{15 - 0.7}{14.0} = 1.0099 \text{ mA}$$

$$\text{Percent change} = \frac{1.0099 - 1}{1} \times 100 = 0.99 \text{ percent}$$

(c) Again, use of Eq. (10-8) gives

$$0.05 = \frac{100}{100 + 2} \cdot \frac{15 - 0.7}{R} \quad \text{and} \quad R = 280 \text{ k}\Omega$$

Current Repeaters   The diode-resistance combination in Fig. 10-5 that establishes the reference current I_R can be used to supply more than one load. Such a circuit, often called a *current repeater* or a *multiple current source,* is displayed in Fig. 10-7a. If all transistors are identical, then, neglecting the Early effect, the currents I_{C1}, I_{C2}, . . ., I_{CN} are identical and $\approx I_R$. Different values of I_{C1}, I_{C2}, . . ., I_{CN} can be achieved by scaling the emitter areas of $Q1, Q2, . . ., QN$. The ratio of maximum to minimum collector current obtained by this method is about 10 because of fabrication limitations.

In Fig. 10-7a, we observe that the bases of all BJTs are connected and all the emitters are grounded. Consequently, it is often convenient to fabricate the N transistors as a single multiple-collector merged device. The circuit in Fig. 10-7b is that shown in Fig. 10-7a, in which QM replaces $Q1, Q2, . . ., QN$. In the merged transistor, scaling the collector areas results in different values of collector current.

10-4   THE WIDLAR CURRENT SOURCE   The results in Example 10-1, parts *a* and *b* indicate that the circuit in Fig. 10-5a provides good bias stability with element values well within fabrication capabilities. However, this is not the case in Example 10-1, part *c,* where fabrication of $R = 280$ kΩ is virtually impossible. The *Widlar current source* illustrated in Fig. 10-8 is frequently used to realize low-value current sources. In the circuit in Fig. 10-8 we assume both transistors are identical. The function of the resistance R_E causes V_{BE1} and V_{BE2} to differ.

FIGURE 10-7
(*a*) A current repeater. If all the transistors are identical, the currents I_{C1}, $I_{C2} \cdots I_{CN}$ are equal.
(*b*) The current repeater using a multiple-collector transistor to conserve chip area.

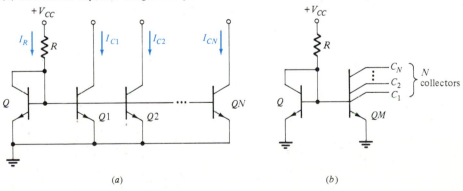

(a)   (b)

FIGURE 10-8
The Widlar current
source.

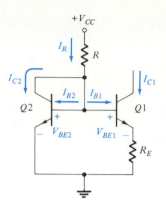

In this configuration V_{BE1} is less than V_{BE2}, and, consequently, I_{C1} is smaller than I_{C2} [Eq. (10-4)]. The asymmetric nature of the base-emitter loop causes this circuit to act as a "lens" rather than a "mirror." In effect, $Q2$, V_{CC}, and R establish the reference current I_R, and the value of R_E determines the degree to which I_{C1} is less than I_R.

The KVL equation for the emitter-base loop in Fig. 10-8 is

$$V_{BE2} = V_{BE1} + (I_{B1} + I_{C1})R_E$$

or

$$V_{BE2} - V_{BE1} = \Delta V_{BE} = (I_{B1} + I_{C1})R_E \qquad (10\text{-}10)$$

As indicated in Eq. (10-3), the collector current of a transistor depends heavily on the base-emitter voltage. For identical *npn* transistors, I_{C1} and I_{C2} are given by Eqs. (10-3) and their ratio in Eq. (10-4). Equation (10-4) can be rewritten by taking the natural logarithm of both sides as

$$V_{BE2} - V_{BE1} = \Delta V_{BE} = V_T \ln \frac{I_{C2}}{I_{C1}} \qquad (10\text{-}11)$$

Equating ΔV_{BE} in Eqs. (10-10) and (10-11) results in

$$R_E = \frac{V_T}{I_{C1}\left(1 + \dfrac{1}{\beta_F}\right)} \ln \frac{I_{C2}}{I_{C1}} \qquad (10\text{-}12)$$

The reference current I_R as given in Eq. (10-6) as the KVL expression for the loop containing V_{CC}, $Q2$, and R is the same in both Figs. 10-5*a* and 10-8. The KCL expression

$$I_R = I_{C2} + I_{B2} + I_{B1}$$

is rewritten as

$$I_R = I_{C2}\left(1 + \frac{1}{\beta_F}\right) + \frac{I_{C1}}{\beta_F} \qquad (10\text{-}13)$$

As we wish I_{C1} to be smaller than I_{C2}, the term I_{C1}/β_F can be neglected in Eq. (10-13). After combination with Eq. (10-6), Eq. (10-13) yields

$$I_{C2} = \frac{\beta_F}{\beta_F + 1} I_R = \frac{\beta_F}{\beta_F + 1} \frac{V_{CC} - V_{BE2}}{R} \approx \frac{V_{CC} - V_{BE2}}{R} \qquad (10\text{-}14)$$

for $\beta_F \gg 1$.

The value of I_{C2} is determined from Eq. (10-14) and, as I_{C1} is the specified current value, the resistance R_E is computed from Eq. (10-12). Example 10-2 illustrates the design.

Example 10-2   Determine R_E in the circuit in Fig. 10-8 for $V_{CC} = 15$ V, $R = 14.0$ kΩ, $V_{BE2} = 0.7$ V, $\beta_F = 100$, and the desired value of $I_{C1} = 50$ μA. Use $V_T = 25$ mV.

Solution   From Eq. (10-14)

$$I_{C2} = \frac{100}{100 + 1} \cdot \frac{15 - 0.7}{14.0} = 1.01 \text{ mA}$$

Use of Eq. (10-12) gives

$$R_E = \frac{0.025}{0.05(1 + \frac{1}{100})} \ln \frac{1.01}{0.05} = 1.49 \text{ kΩ}$$

Note that although two resistors are used in the circuit in Fig. 10-8, the total resistance obtained is $14.0 + 1.49 = 15.49$ kΩ. Each value is well within fabrication limits, and the total resistance is sufficiently small that excessive chip area is not consumed.

The output resistance R_o of the Widlar current source is evaluated using the small-signal equivalent circuit of the BJT. In Prob. 10-52 we show that $R_o \approx r_o(1 + g_m R_E)$. Clearly, this value is several times larger than r_o, the output resistance of a simple current mirror. Note that with $R_E = 0$, $R_o = r_o$; hence we can attribute the increase in R_o to the presence of R_E. This use of R_E is a form of feedback which, as described in Sec. 12-5, increases resistance levels in the circuit. Sometimes it is convenient to construct current sources in which both $Q1$ and $Q2$ have emitter resistances as illustrated in Fig. 10-9. If R_1 and R_2 are equal, the currents in $Q1$ and $Q2$ are also equal. Because of the emitter resistances, however, the output resistance of this current source is greater than r_o. The circuit in Fig. 10-9 is also used to provide different currents in $Q1$ and $Q2$ without the necessity of scaling emitter areas. Use of identical transistors makes the ratio I_{C1}/I_{C2} proportional to the ratio R_2/R_1 (Prob. 10-9).

Temperature Variation   Both the Widlar and simple current sources may be required to operate between temperatures as low as $-55°C$ to as high as $+150°C$. Consequently, the effect of temperature-induced changes in β_F and V_{BE} on the source current must be considered. We neglect the effects of I_{CO}, the collector reverse saturation curent, because of its extremely small value ($I_{CO} \approx 1$ pA at $T = 300°C$).

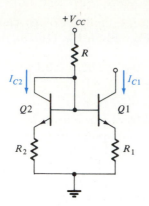

FIGURE 10-9
Current source with an emitter resistance in each transistor. If $R_1 = R_2$, the circuit acts as a mirror; for $R_1 > R_2$, behavior is that of the Widlar source in Fig. 10-8.

Both transistors in Figs. 10-5a and 10-8 are identical so that their β_F values "track"; that is, the β_F of each BJT varies in an identical manner. As the source current in each of these circuits is virtually independent of β_F, temperature variations in β_F with temperature usually produce changes in current values that are quite small. However, this need not be the case when we consider the effect of changes in V_{BE} which decreases by 2.2 mV for a 1°C rise in temperature.

The reference current I_R in Figs. 10-5a and 10-7 depends on the difference $V_{CC} - V_{BE}$. Frequently, $V_{CC} \gg V_{BE}$, so that variations in V_{BE} produce very small changes in I_R. In general, if ΔI_R is the change in I_R caused by a change ΔV_{BE} in V_{BE}, it can be shown (Prob. 10-10) that $\Delta I_R/I_R$ is inversely proportional to $V_{CC}/V_{BE} - 1$. Clearly, for small values of V_{CC}, the relative change $\Delta I_R/I_R$ can become significant.

10-5   THREE-TRANSISTOR CURRENT SOURCES
The ratio of the source and reference currents I_C/I_R differs from unity by $2/\beta_F$. For typical values of β_F, the difference of 1 to 2 percent is negligible. In addition, IC amplifiers also employ low-β_F

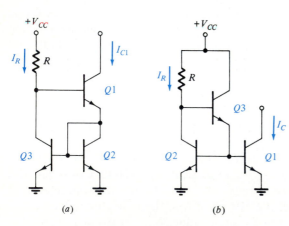

FIGURE 10-10
Three-transistor current sources: (a) Wilson circuit, (b) the current source with gain.

(a)   (b)

lateral *pnp* transistors (Sec. 5-3) as current sources for which the difference between I_C and I_R is a few percent. To make I_C more nearly equal to I_R, as required in some applications, current sources containing three transistors are often used. Two of the more common types are the *Wilson source* (Fig. 10-10a) and the *current source with gain* (Fig. 10-10b). Identical transistors are most often used in each of the circuits in Fig. 10-10.

The source current I_{C1} in Fig. 10-10a can be expressed as (Prob. 10-13)

$$I_{C1} = \frac{\beta_F^2 + 2\beta_F}{\beta_F^2 + 2\beta_F + 2} \cdot I_R = \frac{\beta_F^2 + 2\beta_F}{\beta_F^2 + 2\beta_F + 2} \frac{V_{CC} - 2V_{BE}}{R} \qquad (10\text{-}15)$$

The difference $I_{C1} - I_R = 2I_R/(\beta_F^2 + 2\beta_F + 2)$; clearly, this difference is extremely small for even modest values of β_F. For example, if $\beta_F = 20$, $I_{C1}/I_R = \frac{220}{221}$; and I_{C1} differs from I_R by less than 0.5 percent; if $\beta_F = 100$, $I_{C1}/I_R = \frac{5100}{5101}$. These values also indicate that variations in β_F have little effect on I_{C1}. Typically, changes in I_C of the order of a few hundredths of a percent occur for a 100 percent change in β_F.

The output resistance of the Wilson current source is substantially greater than r_o of Q1, since the diode-connected transistor Q3 acts an an emitter resistance.

The source current I_{C1} for the circuit in Fig. 10-10b can be derived (Prob. 10-14) as

$$I_{C1} = I_R \frac{\beta_F(\beta_F + 1)}{\beta_F^2 + \beta_F + 2} \qquad (10\text{-}16)$$

The similarity of Eqs. (10-15) and (10-16) indicates that I_{C1} for the circuit in Fig. 10-10b is essentially independent of β_F. The output resistance of this source is the r_o of Q1. However, this value can be increased by use of emitter resistances in Q1 and Q2 as is done in the circuit in Fig. 10-9. The two emitter resistors can also be used to make I_{C1} different from I_R.

Other circuits, notably the cascode current source in Fig. 10-11 are also employed. Such circuits are used to improve frequency response or to increase

FIGURE 10-11
Cascode current source.

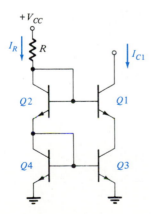

output resistance above that obtained in the Wilson source while ensuring that the source current is independent of β_F variations.

10-6 DISCRETE-COMPONENT BJT BIASING—ANALYSIS

Bias stabilization is as important a consideration in transistor stages constructed from discrete components as it is in ICs. However, as matched pairs of BJTs are costly and since no restrictions exist on resistance values, current sources are not used. The design objective still remains, keeping the collector current constant as β_F varies. To achieve bias stability, base current is allowed to vary with β_F. In Fig. 10-1b, to maintain $I_C \approx 3$ mA independent of β_F variations, requires that I_B decrease as β_F increases. The four-resistor circuit of Fig. 10-12 is the most widely used discrete-component biasing arrangement.

Capacitive Coupling   In Fig. 10-12, the capacitors C_{B1} and C_{B2} are called *blocking* or *coupling capacitors*. The capacitance C_{B1} is used to couple the signal from the input source v_s to the transistor, and C_{B2} couples the output signal from the BJT to the load R_L. Under quiescent conditions C_{B1} and C_{B2} act as open circuits because the reactance of a capacitor is infinite at zero frequency (dc). The values of these capacitors are chosen sufficiently large that, at the lowest signal frequency, their reactances are small enough that they may be considered short circuits. This serves to isolate v_s and R_L from the four resistances R_1, R_2, R_C, and R_E used to establish the bias. These capacitors block dc and freely pass signal voltages. For example, the quiescent collector voltage does not appear at the output, but v_o is an amplified replica of the input signal v_s. The output signal voltage often serves as the input to another amplifier stage (R_L is the input resistance of that stage) without affecting its bias because of the blocking effect of C_{B2}.

The capacitance C_E (called a *bypass* capacitor) is also selected so that it may be treated as a short circuit at the lowest signal frequency. Thus, for

FIGURE 10-12

A discrete-component common-emitter amplifier stage.

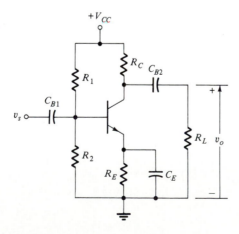

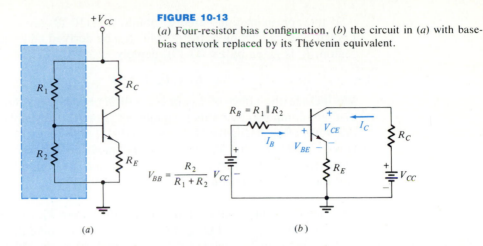

FIGURE 10-13
(a) Four-resistor bias configuration, (b) the circuit in (a) with base-bias network replaced by its Thévenin equivalent.

(a)

(b)

quiescent conditions, R_E is used to stabilize the bias, but at signal frequencies the emitter is grounded.

In this chapter we consider these capacitances so large that their reactance is zero at all signal frequencies. The effect of the finite size of blocking and bypass capacitors on the frequency response of an amplifier is considered in Sec. 11-13.

DC Analysis

The dc equivalent of the circuit in Fig. 10-12 is shown in Fig. 10-13a. This circuit was first analyzed in Example 3-4, where we found it convenient to replace the base-bias network V_{CC}, R_1, and R_2 by its Thévenin equivalent as depicted in Fig. 10-13b for which V_{BB} and R_B are repeated in Eq. (10-17).

$$V_{BB} = \frac{R_2}{R_1 + R_2} V_{CC} \qquad R_B = R_1 \| R_2 = \frac{R_1 R_2}{R_1 + R_2} \qquad (10\text{-}17)$$

The KVL expression for the base loop gives

$$V_{BB} = I_B R_B + V_{BE} + (I_B + I_C) R_E \qquad (10\text{-}18)$$

An approximate solution is readily obtained if $I_B \ll I_C$ ($\beta_F \gg 1$) and if $I_B R_B \ll V_{BB}$. Then

$$I_C = \frac{V_{BB} - V_{BE}}{R_E} \qquad (10\text{-}19)$$

Note that if V_{BE} is constant in Eq. (10-19), I_C is also constant. This circuit, in effect, behaves in a manner similar to a current source with V_{BB} and R_E serving the function of V_{CC} and R in Fig. 10-5a. Once I_C is evaluated, V_{CE} is found from the KVL expression for the collector loop in Eq. (10-20):

$$V_{CC} = I_C R_C + V_{CE} + (I_C + I_B) R_E \qquad (10\text{-}20)$$

If the approximations used to obtain Eq. (10-19) are not valid and if β_F is known, the calculation of the Q point can be derived analytically. In the active region, I_C is given by Eq. (3-19), namely

$$I_C = \beta_F I_B + (1 + \beta_F)I_{CO} \qquad (10\text{-}21)$$

We include the effect of I_{CO} in Eq. (10-21) because, sometimes, its effect in discrete-component circuits is significant. For IC transistors, the effects of I_{CO} are most often negligible. Equations (10-18), (10-20), and (10-21) can now be solved for I_B, I_C, and V_{CE} since V_{BE} is known in the active region. Note that the currents in the forward-active region are determined by the base circuit and the values of β_F and I_{CO}.

Example 10-3

The element values in the circuit in Fig. 10-13a are $V_{CC} = 28$ V, $R_C = 6.8$ kΩ, $R_E = 1.2$ kΩ, $R_1 = 90$ kΩ, and $R_2 = 10$ kΩ. Determine the Q point, assuming that I_{CO} is negligible, when (a) $\beta_F = 60$ and (b) $\beta_F = 150$.

Solution

(a) From Eq. (10-17), we obtain

$$V_{BB} = \frac{10}{90 + 10} \cdot 28 = 2.80 \text{ V} \qquad R_B = \frac{90 \times 10}{90 + 10} = 9.0 \text{ k}\Omega$$

Using $V_{BE} = 0.7$ V in the forward-active region, Eq. (10-18) is

$$2.80 = 9I_B + 0.7 + 1.2(I_B + I_C)$$

or

$$2.10 = 10.2I_B + 1.2I_C$$

Substitution of $I_C = \beta_F I_B = 60I_B$ or $I_B = I_C/60$ and solving for I_C yields

$$I_C = \frac{2.10}{0.17 + 1.2} = 1.53 \text{ mA} \qquad \text{and} \qquad I_B = \frac{1.53}{60} = 0.0255 \text{ mA}$$

Note that these values of current are obtained without reference to collector voltages and, hence, are independent of R_C and V_{CC}.

Substitution of values in Eq. (10-20) gives

$$28 = 1.53 \times 6.8 + V_{CE} + (1.53 + 0.0255) \times 1.2$$

from which $V_{CE} = 15.7$ V.

(b) The KVL relation for the base loop is that given in part a. Substitution of $I_B = I_C/150$ gives

$$I_C = \frac{2.10}{0.068 + 1.2} = 1.66 \text{ mA} \qquad \text{and} \qquad I_B = \frac{1.66}{150} = 0.011 \text{ mA}$$

Use of Eq. (10-20) yields

$$28 = 1.66 \times 6.8 + V_{CE} + (0.011 + 1.66) \times 1.2 \qquad \text{and} \qquad V_{CE} = 14.7 \text{ V}$$

Comparison of the results in parts a and b shows that the Q point changes by a small percentage for a 2.5:1 change in β_F, thus indicating the effectiveness of the circuit in Fig. 10-13a. We now analyze the current increments that result from changes in β_F and temperature variations. We have already indicated that β_F changes with transistor replacement and with temperature. In addition, V_{BE} decreases at the rate of 2.2 mV/°C and I_{CO} doubles for every 10°C rise in temperature.

We shall neglect changes of V_{CE} with temperature because this variation is very small and operation in the forward-active region makes I_C virtually independent of V_{CE}.

Solving Eq. (10-21) for I_B and substituting this relation into Eq. (10-18), we obtain after rearrangement of terms

$$I_C \frac{R_B + (1 + \beta_F)R_E}{\beta_F} = V_{BB} - V_{BE} + \frac{(R_B + R_E)(\beta_F + 1)}{\beta_F} I_{CO} \quad (10\text{-}22)$$

This expression indicates clearly that I_C will vary with changes in one or more of the parameters β_F, V_{BE}, or I_{CO}.

Current Increment for a Change in β_F Consider that a BJT whose current gain is β_{F1} is replaced by another having $\beta_{F2} > \beta_{F1}$. Let us find the resultant current change $\Delta I_C = I_{C2} - I_{C1}$, where I_{C2} (I_{C1}) corresponds to β_{F2} (β_{F1}). If $\beta_{F1} \gg 1$, the righthand side of Eq. (10-22) is essentially independent of β_F, and hence

$$I_{C2} \frac{R_B + R_E(1 + \beta_{F2})}{\beta_{F2}} = I_{C1} \frac{R_B + R_E(1 + \beta_{F1})}{B_{F1}} \quad (10\text{-}23)$$

Solving Eq. (10-23) for I_{C2}/I_{C1} and subtracting unity from the result yields

$$\frac{I_{C2} - I_{C1}}{I_{C1}} = \frac{\Delta I_C}{I_{C1}} = \left(1 + \frac{R_B}{R_E}\right) \frac{\Delta \beta M_2}{\beta_{F1}\beta_{F2}} \quad (10\text{-}24)$$

where $\Delta \beta = \beta_{F2} - \beta_{F1}$ and M is defined by

$$M \equiv \frac{1}{1 + R_B/[R_E(1 + \beta_F)]} \approx \frac{1}{1 + R_B/\beta_F R_E} \quad (10\text{-}25)$$

for $\beta_F \gg 1$. The parameter M_2 (M_1) corresponds to β_{F2} (β_{F1}). As we wish $\Delta I_C/I_{C1}$ to be small for good bias stability, it is clear that $R_B/\beta_F R_E$ should also be kept small. With $R_B \ll \beta_F R_E$, $M \approx 1$. Also, for a given spread in the value of β_F (say, $\beta_{F2}/\beta_{F1} = 3$), a circuit having a high value of β_{F1} will be more stable than one using a lower β_{F1} transistor.

Current Increment for a Change in I_{CO} From Eq. (10-22) with $\beta_F \gg 1$, it follows, if β_F and V_{BE} remain constant, that

$$\Delta I_C = \frac{R_B + R_E}{R_B/\beta_F + R_E} \Delta I_{CO} = \left(1 + \frac{R_B}{R_E}\right) M_1 \Delta I_{CO} \quad (10\text{-}26)$$

Current Increment for a Change in V_{BE} From Eq. (10-22) with $\beta_F \gg 1$, it follows, if β_F and I_{CO} remain constant, that

$$\Delta I_C = -\frac{\beta_F}{R_B + \beta_F R_E} \Delta V_{BE} = -\frac{M_1}{R_E} \Delta V_{BE} \qquad (10\text{-}27)$$

Note that in Eqs. (10-26) and (10-27) it is assumed that a BJT having β_{F1} is used.

Total Current Increment To obtain the total change in current over a specified temperature range due to simultaneous variations in β_F, I_{CO}, and V_{BE}, we add the individual increments found in Eqs. (10-25), (10-26), and (10-27). The fractional change in collector current is given by

$$\frac{\Delta I_C}{I_{C1}} = \left(1 + \frac{R_B}{R_E}\right) \frac{M_1 \Delta I_{CO}}{I_{C1}} - \frac{M_1 \Delta V_{BE}}{I_{C1} R_E} + \left(1 + \frac{R_B}{R_E}\right) \frac{M_2 \Delta \beta}{\beta_{F1} \beta_{F2}} \qquad (10\text{-}28)$$

where $M_1(M_2)$ corresponds to β_{F1} (β_{F2}). Note that as T increases, $\Delta I_{CO}/I_{C1}$ and $\Delta \beta$ increase, whereas $\Delta V_{BE}/I_{C1}$ decreases. Hence all terms in Eq. (10-28) are positive for increasing T and negative for decreasing T.

10-7 DISCRETE-COMPONENT BIASING—DESIGN

The following example illustrates the design technique for discrete-component biasing. It is used also to examine the relative sizes of the three components of $\Delta I_C/I_{C1}$ in Eq. (10-28). The transistor values used are typical of those encountered in practice. Many circuits for commercial applications are required to operate from 0 to 70°C; others, such as those used in aerospace, automotive, and military applications, often must operate from -55°C to between 100 and 150°C. Manufacturers frequently provide device data for the range -65 to 175°C to accommodate the wide variety of applications.

Example 10-4

A 12-V supply is used to bias the transistor in the circuit shown in Fig. 10-12. On the basis of the signal to be amplified, it is determined that $1.00 \le I_C \le 1.15$ mA and $5.0 \le V_{CE} \le 6.0$ V. The collector resistance $R_C = 1.5$ kΩ and is selected to achieve the desired gain. The BJT parameters are

$$
\begin{aligned}
T &= -55°C & \text{Lowest value of } \beta_F &= 40 & V_{BE} &= 0.88 \text{ V} \\
T &= +125°C & \text{Highest value of } \beta_F &= 400 & V_{BE} &= 0.48 \text{ V}
\end{aligned}
$$

(*a*) Design the circuit (R_1, R_2, and R_E) to meet the specifications. Consider that the effect of I_{CO} variations is negligible.

(*b*) Using values obtained in part *a*, determine the component of the total current increment ΔI_C attributed to changes in I_{CO} with temperature. Manufacturers' specifications indicate $I_{CO} = 2.0$ pA at $T = -55$°C and $I_{CO} = 525$ nA at $T = 125$°C.

(c) What is the range of values of I_C and V_{CE} that would be measured in the laboratory at $T = 25°C$ given that $75 \leq \beta_F \leq 200$ and $V_{BE} = 700 \pm 25$ mV? Use the component values obtained in part a.

Solution

(a) The approach we take is to consider that nominal circuit performance occurs at one temperature extreme ($-55°C$). Deviations from nominal behavior are measured at the other endpoint of the temperature range (125°C). The underlying principle of this "worst-case" design method is to satisfy circuit specifications for minimum β_F and to control deviations in performance when β_F is maximum. We already know that I_C increases with increasing β_F so that $I_C = 1.00$ mA must be achieved at $T = -55°C$. For $\beta_F = 400$ at $T = 125°C$, $I_C \leq 1.15$ mA or $\Delta I_C \leq 0.15$ mA. Similarly, as β_F increases, V_{CE} decreases (Example 10-3); consequently, at $T = -55°C$, we select $V_{CE} = 6.0$ V.

From Eq. (10-20), using data for $T = -55°C$, we obtain

$$12 = 1.00 \times 1.5 + 6.0 + \left(1.00 + \frac{1.00}{40}\right)R_E$$

Solving gives $R_E = 4.39$ kΩ. The ratio of R_B/R_E is obtained from Eq. (10-28), neglecting the ΔI_{CO} term. As R_E is known, this ratio specifies R_B. In our calculation, we assume $M_2 = 1$; that is, $R_B \ll \beta_F R_E$. The following data are needed in Eq. (10-28): $\Delta V_{BE} = 0.48 - 0.88 = -0.40$ V, $\beta_{F1} = 40$, $\beta_{F2} = 400$, and $\Delta\beta = 400 - 40 = 360$. Substitution gives

$$\frac{0.15}{1.00} = -\frac{1 \times (-0.40)}{4.39} + \left(1 + \frac{R_B}{R_E}\right)\frac{1 \times 360}{40 \times 400}$$

Solving gives $R_B/R_E = 1.62$ and $R_B = 1.62 \times 4.39 = 7.11$ kΩ. To obtain values of R_1 and R_2 in Eq. (10-17), V_{BB} must first be evaluated in Eq. (10-18) at $T = -55°C$:

$$V_{BB} = \frac{1.00}{40} \times 7.11 + 0.88 + \left(\frac{1.00}{40} + 1.00\right)4.39 = 5.56 \text{ V}$$

Solving Eq. (10-17) for R_1 and R_2 in terms of V_{CC}, V_{BB}, and R_B gives

$$R_1 = R_B \frac{V_{CC}}{V_{BB}} \quad \text{and} \quad R_2 = R_1 \frac{V_{BB}}{V_{CC} - V_{BB}} = R_B \frac{V_{CC}}{V_{CC} - V_{BB}}$$

Evaluation results in

$$R_1 = 7.11 \frac{12}{5.56} = 15.3 \text{ kΩ} \qquad R_2 = 7.11 \frac{12}{12 - 5.56} = 13.2 \text{ kΩ}$$

Let us check the value of V_{CE} at 125°C to see if it is within specification. From Eq. (10-20), we find

$$12 = 1.15 \times 1.5 + V_{CE} + \left(1.15 + \frac{1.15}{400}\right) = 4.39 \quad \text{and} \quad V_{CE} \times 5.22 \text{ V}$$

which is within specification.

The assumption that $M \approx 1$ should also be checked:

$$M_2 = \frac{1}{1 + R_B/\beta_{F2}R_E} = \frac{1}{1 + 7.11/(400 \times 4.39)} = 0.996$$

which is sufficiently close to unity so as to introduce virtually no error.

The circuit is now designed with $R_1 = 15.3$ kΩ, $R_2 = 13.2$ kΩ, $R_E = 4.39$ kΩ, and $R_C = 1.5$ kΩ.

(b) To obtain ΔI_C caused by variations in I_{CO}, we assume that β_F and V_{BE} are constant at their $T = -55°C$ values. Use of Eq. (10-26), with $\Delta I_{CO} \approx 525$ nA, yields

$$\Delta I_C = \left(1 + \frac{7.11}{4.39}\right) \times 1 \times 525 = 1376 \text{ nA} = 1.38 \text{ }\mu\text{A}$$

assuming $M_1 \approx 1$. The value of ΔI_C is very much smaller than I_C, so our assumption to neglect the effect of ΔI_{CO} is valid.

(c) In this part of the problem we investigate the effect of unit-to-unit variation at a given temperature. Solving Eq. (10-22) and neglecting the I_{CO} term, we can express I_C as

$$I_C = \frac{\beta_F(V_{BB} - V_{BE})}{R_B + (\beta_F + 1)R_E}$$

We observe that I_C will be maximum when β_F is maximum and V_{BE} is minimum. The opposite conditions result in the minimum value of I_C. Thus, using values obtained in part a, we obtain

$$I_{C(\text{max})} = \frac{200(5.56 - 0.675)}{7.11 + (200 + 1)4.39} = 1.098 \text{ mA}$$

$$I_{C(\text{min})} = \frac{75(5.56 - 0.725)}{7.11 + (75 + 1)4.39} = 1.064 \text{ mA}$$

The range of V_{CE} values is computed by using the values of $I_{C(\text{max})}$ and $I_{C(\text{min})}$ in Eq. (10-20). These are

$$12 = 1.064 \times 1.5 + V_{CE(\text{max})} + \left(1.064 + \frac{1.064}{75}\right)4.39$$

$$12 = 1.098 \times 1.5 + V_{CE(\text{min})} + \left(1.098 + \frac{1.098}{200}\right)4.39$$

from which $V_{CE(\text{min})} = 5.52$ V and $V_{CE(\text{max})} = 5.67$ V. The results in this part illustrate the effectiveness of the design method. At a single temperature we note that quiescent values shift by approximately 3 percent for an almost 3:1 spread of β_F values.

Design Considerations   Three observations concerning the results and methodology in Example 10-4 are worthy of note: (1) if the total deviation in quiescent values is to be determined, tolerances on resistance values and supply voltages must be taken into account;[1] (2) for operation over a wide temperature range the deviation produced by variations in V_{BE} are comparable to those resulting from β_F changes; and (3) most importantly, R_E and R_C cannot be specified independently once V_{CC} and Q are selected. From Eq. (10-20), assuming $I_C \gg I_B$, $(R_C + R_E) = (V_{CC} - V_{CE})/I_C$, and thus the sum of these resistances is fixed. Consequently, any increase in R_E must be accompanied by a decrease in R_C. The import of this statement can be discerned in Eq. (10-28). Assuming that ΔI_C is attributed equally to ΔV_{BE} and $\Delta\beta$, R_E is determined from Eq. (10-28) and hence R_C is specified. Increasing R_E for a given ΔV_{BE} decreases ΔI_C (provides better bias stability). Unfortunately, as we show in the next paragraph, the accompanying decrease in R_C reduces the gain of the stage. Thus static (bias) and dynamic (amplification) design requirements cannot be treated independently. The circuit designer is required to make choices based on trade-offs between static and dynamic performance. Of extreme importance, however, is that without a reasonable degree of bias stability, dynamic performance cannot satisfy design specifications (see Fig. 10-2b).

We noted earlier that under dc conditions C_B, C_C, and C_E in Fig. 10-12 act as open circuits. Quiescent conditions can be obtained by drawing a static (dc) load line corresponding to V_{CC} and the total resistance of the collector loop, namely, $R_C + R_E$ (assuming $\beta_F \gg 1$) as depicted in Fig. 10-14. Since we assume that C_E acts as a short circuit, the emitter is grounded at signal frequencies. Similarly, C_C acts as a short circuit, making the effective collector resistance $R'_L = R_C \| R_L$. Hence the equivalent ac resistance of the collector loop is R'_L. To determine the output signal, i.e., the variation about the Q point due to the input signal, we must draw a *dynamic load line*. This line, shown in Fig. 10-14, has a slope $-1/R'_L$ and passes through the Q point. Recall that when the input signal is zero ($\sin \omega t = n\pi$), only the bias is applied and the circuit is in its quiescent state. The projection on the V_{CE} axis of the segment of the dynamic

---

[1]Computer simulations, such as SPICE and MICROCAP II, are an invaluable aid in these calculations.

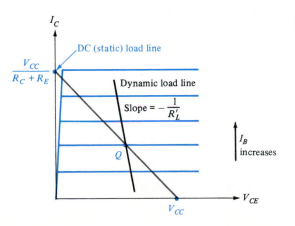

FIGURE 10-14
Static and dynamic load lines for a discrete-component BJT stage.

load line caused by the input signal Δi_B about I_{BQ} determines the output signal Δv_{CE}. If V_{CC} and the Q point are specified, the static load line is determined uniquely. However, a reduction in R_C and consequently R'_L (so R_E may be increased) causes the slope of the dynamic load line to increase (it becomes more vertical). For a given Δi_B, the projection of this line segment on the V_{CE} axis decreases (Δv_{CE} decreases). The reduced output signal for a given input signal is indicative of a decreased gain (amplification) in the stage.

10-8 FET BIASING

Biasing techniques which stabilize the Q point in IC and discrete-component FET circuits parallel those discussed in the previous two sections for the BJT. In MOS circuits, biasing schemes control deviations in the operating point caused by fabrication variations in the threshold voltage V_T and transconductance (processing) parameter k. Both IC and discrete-component JFET circuits are biased so that the effects of unit-to-unit variations in the pinch-off voltage V_P and zero-bias drain saturation current I_{DSS} are controlled. Both MOSFETs and JFETs are operated in their saturation regions at all times so that they provide controlled-source characteristics.

Current Sources

The circuit in Fig. 10-15a is a MOS IC current mirror. The transistor $Q1$ supplies the load with a current I_{D1}. The reference current $I_R = I_{D2}$ is provided by V_{DD}, R, and the enhancement-load transistor $Q2$ (Sec. 4-11). These components also determine the value of $V_{DS2} = V_{GS2}$, and because the gates are tied together, $V_{GS1} = V_{GS2}$. The values of V_T and k are identical for $Q1$ and $Q2$ since they are fabricated simultaneously. Hence, for MOSFETs having the same aspect ratio (W/L), the currents I_{D1} and I_{D2} are equal.

Example 10-5

Two identical transistors having characteristics given in Fig. 4-12 and repeated in Fig. 10-15b are used in the circuit in Fig. 10-15a. The supply voltage is 6 V, and $R = 20$ kΩ. Determine the source current.

Solution

The enhancement-load characteristic (shown in black) is drawn in Fig. 10-15b and is the locus of points for which $V_{DS2} = V_{GS2}$. The load line corresponding to $V_{DD} = 6$ V and $R = 20$ kΩ is also displayed in Fig. 10-15b. The value of $I_{D2} = I_R = 90$ μA is obtained at the intersection of the load line and the resistance characteristic. Because $Q1$ and $Q2$ are identical MOSFETs and $V_{GS1} = V_{GS2}$, $I_{D1} = 90$ μA.

Recalling that the drain current for specified values of V_T and k is scaled by W/L, making the aspect ratios of $Q1$ and $Q2$ different permits I_{D1} to differ from I_R as given in Eq. (10-29) (Prob. 10-34):

$$\frac{I_{D1}}{I_R} = \frac{(W/L)_1}{(W/L)_2} \qquad (10\text{-}29)$$

where $(W/L)_1$ and $(W/L)_2$ are the aspect ratios of $Q1$ and $Q2$, respectively.

FIGURE 10-15

(a) An NMOS current mirror. (b) The NMOS output characteristics, load line, and nonlinear resistance characteristic. The intersection determines the reference current for the mirror.

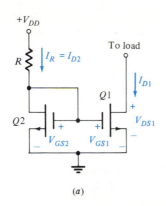

(a)

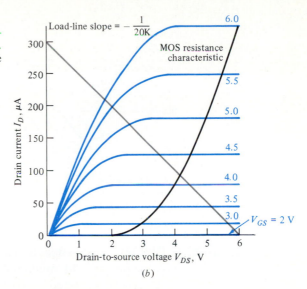

(b)

The volt-ampere characteristic of the current source in Fig. 10-15a is similar to that given in Fig. 10-6 for the BJT circuit. The low-voltage region in Fig. 10-6 indicating that the BJT is saturated corresponds to FET operation in the ohmic region. For the MOSFET, the nonzero slope of the V-I characteristic is attributed to channel-length modulation (Sec. 4-3). The reciprocal of this slope is the output resistance of the current mirror.

The resistance R, which must be large for low-value sources, is often replaced by an NMOS depletion load transistor (Fig. 4-26). Similarly, R can be replaced by a PMOS transistor connected as an enhancement-load resistance (Fig. 10-16). This type of circuit is fabricated by using standard CMOS technology. The source current can be scaled by adjustment of the aspect ratio of $Q3$ as well as for $Q1$ and $Q2$.

Improved current mirror characteristics can be obtained by using MOS analogs of the Wilson (Fig. 10-10a) and cascode (Fig. 10-11) current sources.

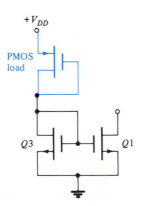

FIGURE 10-16

A MOSFET current source with a PMOS load. This circuit is constructed by use of CMOS fabrication techniques.

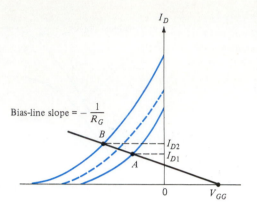

FIGURE 10-17
Maximum and minimum transfer characteristics for an *n*-channel JFET. The bias line
drawn through *A* and *B* ensures that the drain
current I_D is always between I_{D1} and I_{D2}.

Four-Resistor Bias Circuit Field-effect-transistor manufacturers usually supply information on the maximum and minimum values of I_{DSS} and V_P at room temperature. Data is also supplied to correct these quantities for temperature variations. The transfer characteristics for a given type of *n*-channel JFET (or depletion MOSFET) is displayed in Fig. 10-17. The top and bottom curves are for extreme values of device variation and temperature and the dashed curve is the characteristic at room temperature. Assume that, on the basis of considerations previously discussed, it is necessary to bias the device so that the drain current lies between I_{D1} (point *A*) and I_{D2} (point *B*) indicated on Fig. 10-17. The circuit in Fig. 10-18*a* can be designed to guarantee that the value of I_D is always between I_{D1} and I_{D2}. Note that this circuit was discussed in Sec. 4-10 (Fig. 4-21) for MOSFETs. Also note the similarity to the BJT circuit in Fig. 10-13. Not shown in Fig. 10-18*a* are the blocking capacitors used to couple the signal from the input to the transistor and from the FET to the load. These coupling capacitors as well as a bypass capacitor across R_S are employed in discrete-component FET stages. In this chapter we assume that, at the lowest signal frequencies, these capacitors have virtually zero reactance and can be considered short circuits.

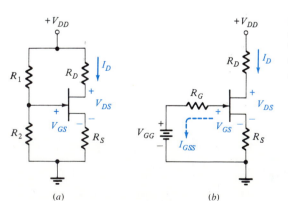

FIGURE 10-18
(*a*) Four-resistor FET bias network. (*b*) The circuit in (*a*) with gate-bias network replaced by its Thévenin equivalent.

Figure 10-18b is the same circuit as in Fig. 10-18a in which the gate bias network V_{DD}, R_1, and R_2 has been replaced by its Thévenin equivalent V_{GG} and R_G [Eq. (10-30)]:

$$V_{GG} = \frac{R_2}{R_1 + R_2} V_{DD} \qquad R_G = R_1 \| R_2 = \frac{R_1 R_2}{R_1 + R_2} \qquad (10\text{-}30)$$

For the circuit in Fig. 10-18b, KVL for the gate loop is

$$-V_{GG} + V_{GS} + I_D R_S = 0 \qquad \text{or} \qquad I_D = -\frac{1}{R_S} V_{GS} + \frac{V_{GG}}{R_S} \quad (10\text{-}31)$$

Equation (10-31) is the equation of the bias line (Sec. 4-10). Construction of the bias line on the transfer characteristics so that it passes through points A and B ensures that $I_{D1} \le I_D \le I_{D2}$. The slope of the bias line is $-1/R_S$, from which the value of R_S is determined. The x intercept specifies the value of V_{GG} needed.

In our analysis $I_G = 0$ was assumed. However, the small reverse saturation current I_{GSS} exists in the gate loop. The resistance R_G is selected to be as large as is feasible while maintaining the voltage drop $I_{GSS}R_G$ at a negligible value ($R_G I_{GSS} \ll V_{GG}$). Large values of R_G minimize loading effects on the previous stage and the current in the resistances R_1 and R_2. The following example illustrates the design method.

Example 10-6

The circuit in Fig. 10-18a is to be designed so that $5.0 \le I_D \le 6.0$ mA and $V_{DS} \ge 8.5$ V. The characteristics of the JFET used are displayed in Fig. 10-19. The supply voltage is 28 V, and $R_G \ge 100$ kΩ.

Solution

Identify points A and B in Fig. 10-19 corresponding to $I_D = 5$ and 6 mA, respectively. Construct the bias line through A and B as indicated and identify $V_{GG} = 3$ V. From the slope of the bias line

$$-\frac{1}{R_S} = \frac{4 - 0}{0 - 3} \qquad \text{and} \qquad R_S = 0.75 \text{ k}\Omega$$

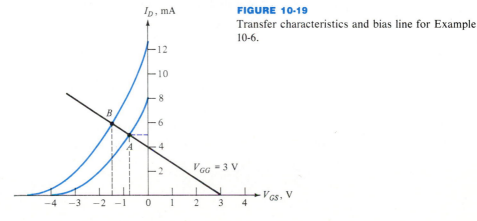

FIGURE 10-19

Transfer characteristics and bias line for Example 10-6.

The resistances R_1 and R_2 can be obtained from V_{GG} and R_G as

$$R_1 = R_G \frac{V_{DD}}{V_{GG}} = 100 \frac{28}{3} = 933 \text{ k}\Omega$$

$$R_2 = R_1 \frac{V_{GG}}{V_{DD} - V_{GG}}$$

$$= 933 \frac{3}{28 - 3} = 112 \text{ k}\Omega$$

Note that R_G is selected arbitrarily at its minimum value. The drain resistance is obtained from the KVL equation for the drain-source loop. The voltage V_{DS} is minimum when I_D is maximum; thus

$$-V_{DD} + I_D R_D + V_{DS} + I_D R_S = 0$$

$$-28 + 6R_D + 8.5 + 6 \times 0.75 = 0 \quad \text{and} \quad R_D = 2.5 \text{ k}\Omega$$

Note that R_D and R_S cannot be specified independently. This situation is the same as for the BJT circuit in Fig. 10-12 described in the previous section. (See also Example 10-4.)

10-9 LINEAR ANALYSIS OF TRANSISTOR CIRCUITS

In the previous sections of this chapter we were concerned with biasing a transistor to establish a stable operating point. We now consider the response of transistor circuits to time-varying applied signals. In particular, we treat small-signal operation for which the transistors are assumed to behave linearly. Under these conditions, the signal component of the response is best obtained by using the small-signal (incremental) equivalent circuits of BJTs and FETs.

The low-frequency, small-signal equivalent circuit of a BJT was initially depicted in Fig. 3-33 and is repeated as Fig. 10-20 for convenience. Similarly, the values of the parameters in the model, given in Eqs. (3-28), (3-29), (3-33), and (3-34), are restated in Eqs. (10-32) through (10-35):

$$\beta_o = \frac{\Delta i_c}{\Delta i_B}\bigg|_{V_{CEQ}} = \frac{i_c}{i_b}\bigg|_{v_{ce}=0} \qquad (10\text{-}32)$$

$$\beta_o = g_m r_\pi \quad \text{or} \quad r_\pi = \frac{\beta_o}{g_m} \qquad (10\text{-}33)$$

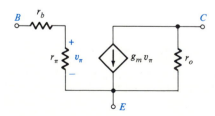

FIGURE 10-20
Low-frequency hybrid-π equivalent circuit.

$$g_m = \frac{|I_{CQ}|}{V_T} \approx \frac{|I_{CQ}| \; (\text{mA})}{25} \qquad \text{at room temperature} \qquad (10\text{-}34)$$

In addition, r_o is given in Eq. (10-9) and restated as Eq. (10-35):

$$r_o = \frac{V_A}{|I_{CQ}|} \qquad (10\text{-}35)$$

We observe in these equations that the small-signal parameters depend on I_{CQ}. Thus dc quantities (biasing) must be obtained *prior* to small-signal analysis. Device manufacturers generally provide data for β_o and the input resistance R_i at a specific operating point. For the *CE* configuration, β_o and R_i are most often listed on data sheets as h_{fe} and h_{ie}, respectively, as manufacturers usually measure the two-port h parameters (App. C). Note that if the circuit that you design is operated at different quiescent conditions, the parameter values in the model must be adjusted accordingly.

There are many transistor circuits which do not consist simply of the *CE*, *CB*, or *CC* configurations. For example, a *CE* amplifier may have a feedback resistor from collector to base, or it may have an emitter resistor. Furthermore, a circuit may consist of several transistors which are interconnected in some manner. An analytic determination of the small-signal behavior of even relatively complicated amplifier circuits may be made by following these simple rules:

1. Draw the actual wiring diagram of the circuit neatly.
2. Mark the points B (base), C (collector), and E (emitter) on this circuit diagram. Locate these points as the start of the equivalent circuit. Maintain the same relative positions as in the original circuit.
3. Replace each transistor by its model.
4. Transfer all components (resistors, capacitors, and signal sources) from the network to the equivalent circuit.
5. Since we are interested only in changes from the quiescent values, replace each independent dc source by its internal resistance. The ideal voltage source is replaced by a short circuit, and the ideal current source by an open circuit. This is an implication of linear operation which, in effect, allows the use of superposition.
6. Solve the resultant linear circuit for mesh (or branch) currents and node voltages by applying Kirchhoff's current and voltage laws.

It should be pointed out that the above procedure is not limited for use at low frequencies. A basic restriction is that the voltages and currents are small enough so that linear operation results. In other words, over the signal excursion, the parameters in the model must remain essentially constant. These rules apply equally to FET circuits.

10-10 THE COMMON-EMITTER AMPLIFIER This configuration is shown in Fig. 10-21a, where, for simplicity, we have omitted the biasing resistors, coupling, and bypass capacitors (if used). Figure 10-21b shows the common-emitter stage with the transistor replaced by the equivalent circuit (in color) in Fig. 10-20. We assume the voltages and currents vary sinusoidally (or, alternatively, are of the form $A\epsilon^{st}$) and can proceed with the analysis of this circuit by using the methods and notation given in App. C. The quantities of interest are *the current gain, the input resistance, the voltage gain,* and *the output resistance.*

The Current Gain or Current Amplification A_I[1] For the BJT amplifier stage, A_I is defined as the ratio of output to input currents, or

$$A_I \equiv \frac{I_o}{I_b} \tag{10-36}$$

For the circuit in Fig. 10-21 we see that $V_\pi = I_b r_\pi$ and, by use of the current-divider relation, $I_o = g_m V_\pi r_o/(r_o + R_C)$. Combination of these results and identification of $g_m r_\pi = \beta_o$ and $R_L = R_C \| r_o$ gives

$$A_I = \beta_o \frac{r_o}{r_o + R_C} = \beta_o \frac{R_L}{R_C} \tag{10-37}$$

For $r_o \gg R_C$, we find that $R_L = R_C$ and $A_I = \beta_o$. Subject to this approximation, A_I equals the short-circuit current gain of the BJT and is independent of the load R_C.

---

also used frequently to denote gain.

FIGURE 10-21
(a) Common-emitter stage and (b) its low-frequency, small-signal equivalent circuit. The small-signal model of the transistor is shown in blue. (*Note:* For simplicity, portions of the biasing network are omitted and are assumed to have negligible effect on the small-signal behavior of the circuit.)

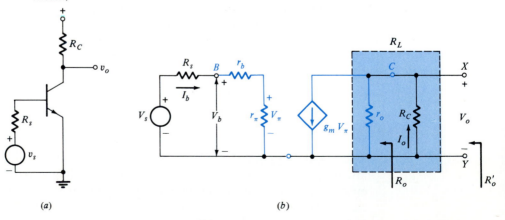

(a)   (b)

The Input Resistance R_i The resistance R_s in Fig. 10-21 represents the signal-source resistance. The resistance we see by looking into the transistor input terminals B and E is the amplifier *input resistance R_i* or

$$R_i \equiv \frac{V_b}{I_b} = r_b + r_\pi = h_{ie} \tag{10-38}$$

Note that R_i is also independent of the load and equals the short-circuit input resistance h_{ie}. If $r_\pi \gg r_b$, as is usually the situation, $R_i \approx r_\pi$.

The Voltage Gain or Voltage Amplification A_V The ratio of output voltage V_o to input voltage V_s is the voltage gain of the stage. Identifying R_L as the parallel combination of R_C and r_o, we obtain

$$A_V \equiv \frac{V_o}{V_s} = \frac{-g_m V_\pi R_L}{I_b(R_s + r_b + r_\pi)} \tag{10-39}$$

where $I_b(R_s + r_b + r_\pi)$ is KVL for the base loop. On substitution of $V_\pi = I_b r_\pi$ and Eqs. (10-38) and (10-33) in Eq. (10-39), we obtain

$$A_V = \frac{-\beta_o R_L}{R_s + R_i} = \frac{-\beta_o R_L}{R_s + r_b + r_\pi} \tag{10-40}$$

For $r_o \gg R_C$, $R_L \approx R_C$ and Eq. (10-40) reduces to

$$A_V = \frac{-\beta_o R_C}{R_s + r_b + r_\pi} \tag{10-41}$$

Note that the value of A_V cannot be increased by arbitrarily increasing R_C. If $R_C \gg r_o$, $R_L \approx r_o$, and A_V becomes

$$A_V = \frac{-\beta_o r_o}{R_s + r_o + r_\pi} \approx -g_m r_o \tag{10-42}$$

for $r_\pi \gg R_s + r_b$. Substitution of Eqs. (10-34) and (10-35) in (10-42) yields $|A_V| \approx V_A/V_T$; this is the maximum gain of the stage.

The situation given in Eq. (10-42) is often encountered in IC amplifier stages. As displayed in Fig. 10-22a, IC stages often employ *pnp* current sources in place of the resistance R_C in Fig. 10-21a. The current mirror ($Q2$ and $Q3$) both biases and provides the load resistance for $Q1$ as shown in the representation in Fig. 10-22b. If the current in R_0 is negligible in comparison to I_0, as is frequently true, $I_{C1} \approx I_0$ and the operating point of $Q1$ is stabilized. Figure 10-22c is the small-signal equivalent of Fig. 10-22b and is identical to Fig. 10-21b if R_0 is identified with R_C. Clearly, if $R_0 \gg r_o$, as is the case if a Widlar or Wilson source is used in place of the simple current mirror in Fig. 10-22a, A_V for the stage is given by Eq. (10-42).

The term ''active load'' is given to the current source used as described in the previous paragraph. A detailed examination of active loads is presented in Sec. 14-2.

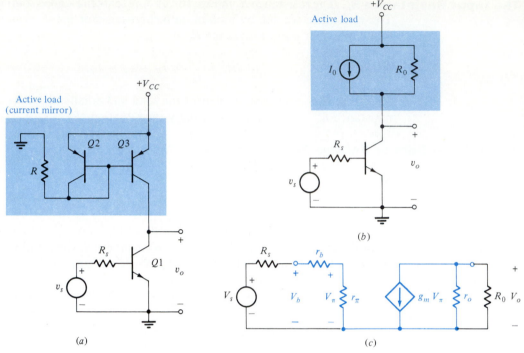

FIGURE 10-22
(*a*) A *pnp* current mirror used as a load in a common-emitter stage. (*b*) Equivalent representation of (*a*). (*c*) The small-signal equivalent circuit of the stage.

Sometimes, V_o/V_b is identified as a voltage gain in the professional literature. This quantity is the voltage transfer ratio from input to output of the transistor and is often called the *transducer gain*. The transducer gain and A_V and A_I are related by

$$A_V = \frac{R_i}{R_i + R_s} \times \frac{V_o}{V_b} \quad \text{or} \quad \frac{V_o}{V_b} = -A_I \frac{R_L}{R_i} \quad (10\text{-}42a)$$

Because the transducer gain does not include the effect of the signal-source resistance R_s, it is generally less useful in the design of practical amplifiers.

The Output Resistance   For the single-stage circuit, the output resistance R_o is the resistance seen by the load R_C. By definition, R_o is obtained by setting the source-voltage V_s to zero and $R_C \rightarrow \infty$, applying a source V_2 to the output terminals, and measuring the current I_2 produced. Then $R_o \equiv V_2/I_2$. With $V_s = 0$, I_b and V_π are zero. Thus $I_2 = V_2/r_o$, and

$$R_o = \frac{V_2}{I_2} = r_o \quad (10\text{-}43)$$

Amplifier systems usually consist of several stages. Consider the situation where terminals X-Y in Fig. 10-21b are connected to the input of another stage. Here it is useful to know the output resistance R'_o of the amplifier stage, that is, the output resistance that includes the effect of R_C. It is evident in Fig. 10-21b that R'_o is the parallel combination of r_o and R_C and

$$R'_o = \frac{r_o R_C}{r_o + R_C} = R_L \tag{10-44}$$

Note that for $r_o \gg R_C$, $R'_o = R_C$.

In practice, the situation where $r_o \gg R_C$ occurs with sufficient frequency that it is convenient to assume $r_o \to \infty$. Similarly, the values of r_b encountered are small enough so that assuming $r_b = 0$ introduces (almost always) negligible error. These assumptions greatly simplify pencil-and-paper calculations and permit the designer to quickly assess circuit performance. Where a high degree of accuracy is required, computer-aided circuit analysis programs (SPICE, MICROCAP II,etc.) are used. *The reader can assume $r_b = 0$ and $r_o \to \infty$ for the remainder of the text unless otherwise stated.*

The results for the common-emitter configuration are summarized in the first columns of Tables 10-3. The assumptions $r_o \to \infty$ and $r_b = 0$ are used in Table 10-3A; the results in Table 10-3B include these elements and assume only that $\beta_o r_o \gg r_b + r_\pi + R_s$.

10-11 THE EMITTER FOLLOWER

The circuit diagram of a common-collector (CC) transistor amplifier is given in Fig. 10-23a. This configuration is also called the *emitter follower,* because its voltage gain is close to unity [Eq. (10-50)], and hence a change in base voltage appears as an equal change across the load at the emitter. In other words, the emitter *follows* the input signal. It is shown below that the input resistance R_i of an emitter follower is very high (hundreds of kilohms) and the output resistance R_o is very low (tens of ohms). Hence the most common use for the CC circuit is as a buffer stage which performs the function of resistance transformation (from high to low resistance) over a wide range of frequencies, with voltage gain close to unity. In addition, the emitter follower increases the power level of the signal, that is, it provides power gain.

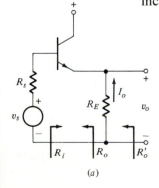

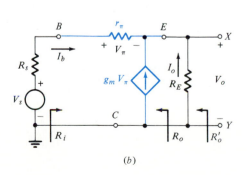

(a) (b)

FIGURE 10-23
(a) The common-collector (emitter-follower) stage and (b) the low-frequency equivalent circuit.

TABLE 10-3A Approximate Amplifier-Stage Equations ($r_o \to \infty$, $r_b = 0$)

Quantity*	CE	CE with R_E	CC	CB
			Configuration	
A_I	$\beta_o = h_{fe}$	β_o	$-(1 + \beta_o)$	$-\dfrac{\beta_o}{1 + \beta_o} \approx -1$
R_i	$r_\pi = h_{ie}$	$r_\pi + (\beta_o + 1)R_E$	$r_\pi + (\beta_o + 1)R_E$	$\dfrac{r_\pi}{1 + \beta_o} \approx \dfrac{1}{g_m}$
A_V	$\dfrac{-\beta_o R_C}{R_s + r_\pi}$	$\dfrac{-\beta_o R_C}{R_s + R_i}$	$\dfrac{(\beta_o + 1)R_E}{R_s + R_i}$	$-A_I\dfrac{R_C}{R_s + R_i} \approx \dfrac{R_C}{R_s}$
R_o	∞	∞	$\dfrac{R_s + r_\pi}{1 + \beta_o}$	∞
R'_o	R_C	R_C	$R_E\|R_o$	R_C

TABLE 10-3B Amplifier-Stage Equations

Quantity*	CE	CE with R_E	CC	CB
			Configuration	
A_I	$\dfrac{\beta_o r_o}{r_o + R_C} = \dfrac{\beta_o R_L}{R_C}$	$\dfrac{\beta_o r_o}{r_o + R_C + R_E}$	$\dfrac{-(\beta_o + 1)r_o}{r_o + R_E}$	$\dfrac{-\beta_o}{\beta_o + R_C/R_L}$
R_i	$r_b + r_\pi = h_{ie}$	$r_\pi + r_b + \dfrac{R_E[r_o(1 + \beta_o) + R_C]}{r_o + R_C + R_E}$	$r_b + r_\pi + R'_E(1 + \beta_o)$	$\dfrac{r_b + r_\pi}{(\beta_o R_L/R_C) + 1}$
A_V	$\dfrac{-\beta_o R_L}{R_s + r_b + r_\pi}$	$\dfrac{-\beta_o R_L}{R_s + r_b + r_\pi + R_E(1 + \beta_o R_L/R_C)}$	$\dfrac{(1 + \beta_o)R'_E}{R_s + R_i}$	$\dfrac{-R_L}{R_s + R_i}\dfrac{\beta_o}{1 + (\beta_o R_L/R_C)}$
R_o	r_o	$r_o\left[1 + \dfrac{\beta_o R_E}{R_s + r_b + r_\pi + R_E}\right]$	$r_o\|\dfrac{R_s + r_b + r_\pi}{1 + \beta_o}$	$r_o\left[1 + \dfrac{\beta_o R_s}{R_s + r_b + r_\pi}\right]$
R'_o	$R_L = r_o\|R_C$	$R_o\|R_C$	$R_E\|R_o$	$R_o\|R_C$

*The value of R_o can also be evaluated by determining the Thévenin resistance at the indicated terminals. We mention this now because the Thévenin equivalent is sometimes a more efficient method for obtaining R_o and is used throughout the text. The values $R_L = R_C\|r_o$ and $R'_E = R_E\|r_o$ are used in Table 10-3B.

The equivalent circuit of the emitter follower is given in Fig. 10-23b. Note that the collector is grounded with respect to the signal (because the supply V_{CC} is replaced by a short circuit in accordance with rule 5 in Sec. 10-9).

The Current Gain In Fig. 10-23b, the output current I_o, with the use of KCL at E, is given by

$$I_o = -I_b - g_m V_\pi \tag{10-45}$$

and

$$V_\pi = I_b r_\pi \tag{10-46}$$

Combination of Eqs. (10-45) and (10-46), identification of $\beta_o = g_m r_\pi$, and formation of the ratio I_o/I_b yields

$$A_I = \frac{I_o}{I_b} = -(\beta_o + 1) \tag{10-47}$$

The Input Resistance   The input resistance R_i is the ratio V_b/I_b. From KVL for the outer loop in Fig. 10-22b, we obtain

$$V_b = I_b r_\pi - I_o R_E \tag{10-48}$$

Substitution of I_o from Eq. (10-47) and division by I_b gives

$$R_i = \frac{V_b}{I_b} = r_\pi + (1 + \beta_o)R_E \tag{10-49}$$

In Eq. (10-49) we observe that R_i for the emitter follower is considerably greater than $R_i = r_\pi$ for the common-emitter stage, even for small values of R_E because $\beta_o \gg 1$.

The Voltage Gain   The output voltage $V_o = -I_o R_E$. Since $V_s = I_b R_s + V_b$, use of Eqs. (10-47) and (10-48) allows after some algebraic manipulation

$$A_V = \frac{V_o}{V_s} = \frac{(\beta_o + 1)R_E}{R_s + r_\pi + (\beta_o + 1)R_E} = \frac{(\beta_o + 1)R_E}{R_s + R_i} \tag{10-50}$$

For $(\beta_o + 1)R_E \gg R_s + r_\pi$, as is the usual case, A_V is approximately unity (but slightly less than unity).

The Output Resistance   The resistance R_o' is the Thévenin resistance seen at terminals X-Y. As the Thévenin voltage is simply $V_o = A_V V_s$, determination of the short-circuit current I_{sc} gives $R_o' = V_o/I_{sc}$. Note that $I_{sc} = -I_o$, and by letting $R_E = 0$ (short circuit), we can obtain (Prob. 10-46)

$$R_o' = \frac{(R_s + r_\pi)R_E/(1 + \beta_o)}{[(R_s + r_\pi)/(1 + \beta_o)] + R_E} \tag{10-51}$$

Equation (10-51) indicates that R_o' is the parallel combination of R_E and a resistance $(R_s + r_\pi)/(1 + \beta_o)$. In Fig. 10-23b we observe that $R_o' = R_o \| R_E$ and hence

$$R_o = \frac{R_s + r_\pi}{1 + \beta_o} \tag{10-52}$$

Note that *the output resistance is a function of the source resistance R_s.* Because $\beta_o \gg 1$, R_o of an emitter follower is small (ohms) in comparison with the input resistance, which is large (tens to hundreds of kilohms). The results for the common-collector stage are listed in column 3 in Table 10-3.

FIGURE 10-24
Low-frequency equiva-
lent circuit of the com-
mon-base stage.

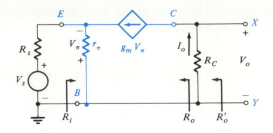

10-12 THE COMMON-BASE AMPLIFIER

The circuit in Fig. 3-8 is a common-base amplifier stage if V_{EE} and R_E are replaced by a signal source V_s having internal resistance R_s. The equivalent circuit is depicted in Fig. 10-24, and we observe that the use of the small-signal model and the results obtained are independent of whether an *npn* or a *pnp* transistor is under consideration. Applying the definitions of A_I, R_i, A_V, and R_o in Sec. 10-10 to this circuit, we obtain the results given in the fourth column in Table 10-3. The verification of these formulas is left to the reader (Prob. 10-42).

10-13 COMPARISON OF BJT AMPLIFIER CONFIGURATIONS

Numerical values for A_I, R_i, A_V, R_o, and R_o' for the three basic BJT amplifier configurations is given in Table 10-4 for $R_C = R_E = 1.5$ kΩ and $R_s = 0.60$ kΩ and a transistor having $\beta_o = 100$, $r_b = 50$ Ω, and $g_m = 0.10$ ℧. The value of g_m corresponds to a BJT biased at $I_{CQ} = 2.5$ mA [Eq. (10-34)]. The value $r_o = 50$ kΩ is obtained for an Early voltage $V_A = 125$ V. For each configuration, three values of A_I, R_i, A_V, R_o, and R_o' are given. The first column for each circuit is obtained by using the equations in Table 10-3A ($r_b = 0$, $r_o \rightarrow \infty$). Table 10-3B is applied to compute the remaining values, using $r_b = 50$ Ω, $r_o \rightarrow \infty$ in column 2 and $r_b = 0$, $r_o = 50$ kΩ in column 3 (for all three configurations). Comparison of the data for

TABLE 10-4 Comparison of BJT Configurations*

	CE				CC				CB		
Quantity	Approximate $r_b = 0$, $r_o \rightarrow \infty$		$r_b = 50$ Ω $r_o \rightarrow \infty$	$r_b = 0$ $r_o = 50$ kΩ	Approximate $r_b = 0$, $r_o \rightarrow \infty$		$r_b = 50$ Ω $r_o \rightarrow \infty$	$r_b = 0$ $r_o = 50$ kΩ	Approximate $r_b = 0$, $r_o \rightarrow \infty$		$r_b = 50$ Ω $r_o \rightarrow \infty$	$r_b = 0$ $r_o = 50$ kΩ
$|A_I|$	High	100	100	97.1	High	101	101	98.1	Low	0.990	0.990	0.990
R_i	Medium	1.00 kΩ	1.05 kΩ	1.00 kΩ	High	153 kΩ	153 kΩ	147 kΩ	Low	9.90 Ω	10.4 Ω	10.2 Ω
$|A_V|$	High	93.8	90.9	91.0	Low	0.990	0.989	0.989	Low	2.44	2.43	2.43
R_o	High	∞	∞	50 kΩ	Low	15.8 Ω	16.3 Ω	15.8 Ω	High	∞	∞	1.93 MΩ
R_o'	—	1.50 kΩ	1.50 kΩ	1.46 kΩ	—	15.6 Ω	16.1 Ω	15.6 Ω	—	1.50 kΩ	1.50 kΩ	1.50 kΩ

*Computed for $\beta_o = 100$, $g_m = 0.10$ ℧, $R_s = 0.60$ kΩ, and $R_C = R_E = 1.5$ kΩ.

each configuration readily demonstrates the utility of the approximate relations in Table 10-3A. With the exception of the values for R_o marked ∞, the results in column 1 differ by no more than 5 percent from the values in columns 2 and 3. In addition, the results in column 1 (for each circuit) agree, within 10 percent, with measured values for a transistor having the parameter values given.

The *CE* Configuration   From Table 10-4 we see that only the common-emitter stage is capable of both a voltage gain and a current gain greater than unity. This configuration is the most versatile and useful of the three connections. Note that the magnitudes of R_i and R_o lie between those for the *CB* and *CC* configurations.

The *CC* Configuration   For the common-collector stage, A_I is high (approximately equal to that of the common-emitter stage), A_V is less than unity (but close to unity), R_i is the highest, and R_o is the lowest of the three configurations. This circuit is widely applied as a buffer stage between a high-impedance source and a low-impedance load.

The *CB* Configuration   The common-base stage is rarely used alone (or as one of a number of cascaded common-base stages) because the severe mismatch in input and output resistances virtually precludes realization of any gain. In Table 10-3 we see that $A_I < 1$ is always true, and for R_C and R_s of the same order of magnitude, A_V can also be less than unity. In conjunction with other stages (such as a *CE-CB* cascade), the common-base stage is used because its extremely low input resistance helps to improve the frequency response of the combined stages. It is of interest to observe that the transducer gain V_o / V_e of the common-base stage is high, but because of the extremely low input resistance, the voltage gain of the stage is low.

10-14   THE COMMON-EMITTER AMPLIFIER WITH AN EMITTER RESISTOR   The

voltage gain of a common-emitter stage depends on β_o as indicated in Table 10-3. This BJT parameter depends on temperature, aging, fabrication processes, and other variables and displays the same degree of variability as does β_F. Consequently, it is often necessary to make the voltage gain A_V of the stage virtually insensitive to variations in β_o. (This is analogous to making I_C independent of β_F to stabilize the Q point.) A simple and effective way to obtain voltage-gain insensitivity is to add an emitter resistor R_E to a common-emitter stage, as indicated by Fig. 10-25a. The desensitivity obtained is a result of the feedback provided by R_E. (The general concept of feedback is discussed in Chap. 12.)

In this section we show that R_E has the following effects on the dynamic performance of the amplifier stage: it leaves the current gain essentially un-

FIGURE 10-25
(*a*) The common-emit-ter stage with emitter resistance. (*b*) The equivalent circuit of the stage in (*a*) valid at low frequencies.

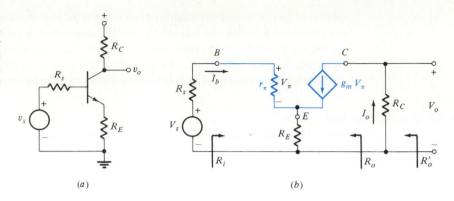

(*a*)   (*b*)

changed and increases the input resistance by $(1 + \beta_o)R_E$; the output resistance is also increased. For the condition that $(1 + \beta_o)R_E \gg R_s + r_\pi$, the voltage gain is essentially independent of β_o and approaches $-R_C/R_E$.

The low-frequency analysis of this circuit can be made by using the equivalent circuit in Fig. 10-25*b*. For the circuit in Fig. 10-25*b*, we find that

$$I_o = g_m V_\pi \tag{10-53}$$

$$V_o = -I_o R_C \tag{10-54}$$

$$V_\pi = I_b r_\pi \tag{10-55}$$

The KVL equation for the base loop gives

$$-V_s + I_b(R_s + r_\pi) + (I_b + g_m V_\pi)R_E = 0 \tag{10-56}$$

and

$$V_b = I_b r_\pi + (I_b + g_m V_\pi)R_E \tag{10-57}$$

Use of Eqs. (10-53) and (10-55) give

$$A_I = \frac{I_o}{I_b} = \beta_o \tag{10-58}$$

Substitution of Eq. (10-55) into Eq. (10-57) and division by I_b yields

$$R_i = \frac{V_b}{I_b} = r_\pi + (1 + \beta_o)R_E \tag{10-59}$$

Combination of Eqs. (10-53) to (10-56) and solving for V_o/V_s results in

$$A_V = \frac{V_o}{V_s} = \frac{-\beta_o R_C}{R_s + r_\pi + (1 + \beta_o)R_E} \tag{10-60}$$

For $(1 + \beta_o)R_E \gg R_s + r_\pi$, Eq. (10-60) reduces to

$$A_V \approx \frac{-R_C}{R_E} \tag{10-61}$$

and is independent of transistor parameters. The price we pay to make A_V independent of β_o variations is a significantly reduced gain compared to that obtained for a simple common-emitter stage. However, in many applications, the benefit derived outweighs the cost.

The output resistances R_o and R'_o in Fig. 10-25a are ∞ and R_C, respectively, as we assume $r_o \rightarrow \infty$. If the effect of r_o is included (Prob. 10-47), the output resistances are those given in Table 10-3B.

10-15 FET AMPLIFIER STAGES The principal FET amplifier configurations are analogous to the BJT stages discussed in the previous sections. The analysis of these stages is based on the small-signal FET model introduced in Sec. 4-14; the method employed parallels that for the BJT (Sec. 10-9). In this section we focus on the common-source (*CS*) and common-drain (*CD*) configurations and use the notation for FET currents and voltages as given in Table 10-2.

Small-Signal Equivalent Circuits The low-frequency, small-signal model valid for both JFETs and MOSFETs was given in Fig. 4-36 and is repeated in Fig. 10-26a for convenience. The parameters in the model, g_m and r_d, are defined in Eqs. (4-17), (4-19), and (4-23) and are repeated as Eqs. (10-62) and (10-63) for convenience:

$$g_m = \frac{2}{|V_p|} \sqrt{I_{DQ} I_{DSS}} \qquad \text{for JFETs} \qquad (10\text{-}62a)$$

$$g_m = 2\sqrt{k \frac{W}{L} I_{DQ}} \qquad \text{for MOSFETs} \qquad (10\text{-}62b)$$

$$r_d \approx \frac{1}{\lambda I_{DQ}} = \frac{V_A}{I_{DQ}} \qquad (10\text{-}63)$$

Note that the values of both g_m and r_d are bias-dependent. Comparison of Fig. 10-26a with the hybrid-π model of the BJT (Fig. 10-20) indicates that they are equivalent if $r_b = 0$ and $r_\pi \rightarrow \infty$ (open circuit).

The output resistance r_d is usually not sufficiently large, so it may not be neglected (as r_o is often neglected in the BJT model). Typically, r_d is in the order of a few tens of kilohms for MOSFETs; for JFETs, r_d can be as high as

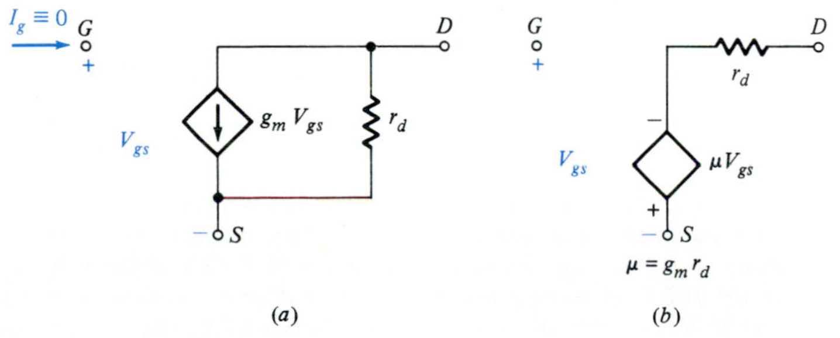

FIGURE 10-26
Two forms of the low-frequency small-signal equivalent circuit of the FET.

(a)   (b)

$\mu = g_m r_d$

several hundred kilohms. The value of g_m for a FET biased at I_{DQ} is lower than the value of g_m for a BJT biased at $I_{CQ} = I_{DQ}$. Hence, to achieve the same voltage gain as a BJT having a collector resistance R_C, the FET stage usually requires a drain resistance $R_D > R_C$. Because of the larger values of R_D often used, the incremental output resistance r_d cannot be neglected in the model. Because r_d must be included in FET amplifier analysis, it is often convenient to use the model in Fig. 10-26b. The voltage source-resistance $(\mu V_{gs} - r_d)$ combination in Fig. 10-26b is the voltage source equivalent of the $g_m V_{gs}$ current source in parallel with r_d in Fig. 10-26a. The quantity μ, called the *amplification factor*, is given by

$$\mu = g_m r_d \tag{10-64}$$

The open circuit between gate and source in the model[1] makes $I_g = 0$, so that it is meaningless to consider A_I and R_i for FET stages (usually). For most FET circuits, the voltage gain A_V and the output resistance R_o are the important quantities that describe amplifier performance.

The basic circuit we analyze is shown in Fig. 10-27a. Note that if the output voltage is measured from drain to ground (neutral, denoted by N), this stage is a *common-source stage with source resistance*. If $R_S = 0$, the stage is the standard *common-source (CS) stage* (Fig. 10-28a). Similarly, if the output is taken between source and ground, with $R_D = 0$, the circuit is a *common-drain (CD)* or *source-follower amplifier* in Fig. 10-28b. Inclusion of R_D converts this stage to a common-drain stage with drain resistance.

Analysis of the Generalized FET Amplifier Stage   The equivalent circuit of the generalized (basic) FET stage in Fig. 10-27a is given in Fig. 10-27b. Note that we use the equivalent circuit form in Fig. 10-26b. Application of KVL to the drain loop, assuming sinusoidal excitation, gives

$$I_d R_D + I_d r_d - \mu V_{gs} + I_d R_S = 0 \tag{10-65}$$

From Fig. 10-27b the voltage from G to S is given by

$$V_{gs} = V_i - I_d R_S \tag{10-66}$$

Combination of Eqs. (10-65) and (10-66) yields

$$I_d = \frac{\mu}{r_d + R_D + (1 + \mu)R_S} V_i \tag{10-67}$$

---

[1]A completely accurate model of the FET would include resistance r_{dg} and r_{sg} between drain and gate and source and gate, respectively. These resistors indicate current paths from source and drain to gate through the oxide layer in the MOSFET or through the reverse-biased pn junction in the JFET. Measurements (and theoretical analysis) indicate that r_{dg} and r_{sg} are greater than 10,000 MΩ ($\geq 10^{10}$ Ω), so it is reasonable to assume that these resistances are open circuits as shown.

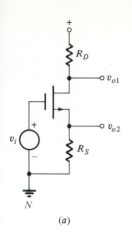

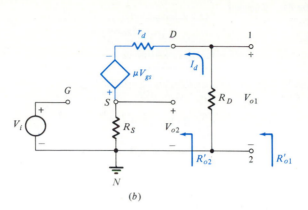

(a)

(b)

FIGURE 10-27
(a) The common-source stage with source resistance and (b) the small-signal equivalent circuit. The model in Fig. 10-26b (shown in blue) is used to represent the MOSFET in (a).

The output voltages V_{o1} and V_{o2}, taken between drain and ground and source and ground, respectively, are

$$V_{o1} = -I_d R_D = \frac{-\mu R_D}{r_d + R_D + (1 + \mu)R_S} V_i \qquad (10\text{-}68)$$

$$V_{o2} = +I_d R_S = \frac{\mu R_S}{r_d + R_D + (1 + \mu)R_S} V_i \qquad (10\text{-}69)$$

The voltage gain of the common-source stage with source resistance is

$$A_V = \frac{V_{o1}}{V_i} = \frac{-\mu R_D}{r_d + R_D + (1 + \mu)R_S} \qquad (10\text{-}70)$$

For $\mu \gg 1$ and use of Eq. (10-64), A_V can be rewritten as

$$A_V = \frac{-g_m R_L}{1 + g_m R_S R_L / R_D} \qquad (10\text{-}71)$$

The output resistance R'_{o1} is the Thévenin resistance seen "looking into" terminals 1–2. The short-circuit current I_{sc} is expressible as

$$-I_{sc} = I_d \Big|_{R_D = 0} = \frac{\mu}{r_d + (1 + \mu)R_S} V_i \qquad (10\text{-}72)$$

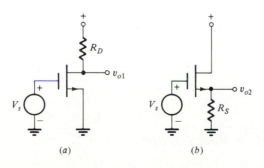

(a)

(b)

FIGURE 10-28
(a) The common-source stage and (b) the common-drain (source-follower) stage. These circuits can be derived from Fig. 10-27a by setting $R_S = 0$ [for (a)] or by setting $R_D = 0$ [for (b)].

TABLE 10-5 **Equations for FET Stages**

Quantity	CS		CS with source resistance		CD	
			Configuration			
A_V	$\dfrac{-\mu R_D}{r_d + R_D}$	$= \dfrac{-g_m R_D}{1 + R_D/r_d}$	$\dfrac{-\mu R_D}{r_d + R_D + (1 + \mu)R_S}$	$\approx \dfrac{-g_m R_L{}^*}{1 + g_m R_S R_L/R_D}$	$\dfrac{\mu R_S}{r_d + R_S(1 + \mu)}$	$\approx \dfrac{g_m R_S}{1 + g_m R_S}$
R_o		r_d	$r_d + R_S(1 + \mu) \approx r_d(1 + g_m R_S)$		$\dfrac{r_d}{1 + \mu} \approx \dfrac{1}{g_m}$	
R_o'		$R_D \parallel r_d$	$R_o \parallel R_D$		$R_S \parallel R_o$	

$*R_L = r_o \| R_D.$

Since V_{o1} is the Thévenin voltage, it follows that

$$R_{o1}' = \frac{V_{o1}}{I_{\text{sc}}} = \frac{R_D[r_d + R_S(1 + \mu)]}{R_D + r_d + R_S(1 + \mu)} = R_D \| [r_d + R_s(1 + \mu)] \quad (10\text{-}73)$$

The output resistance R_{o1}' of the common-source stage with source resistance is $R_{o1}' = R_{o1} \| R_D$ from which

$$R_{o1} = r_d + R_S(1 + \mu) \approx r_d(1 + g_m R_S) \quad (10\text{-}74)$$

for $\mu \gg 1$.

The results for the common-source stage with source resistance are tabulated in the second column in Table 10-5.

In similar fashion, obtaining the Thévenin resistance between S and N yields

$$R_{o2}' = \frac{V_{o2}}{I_d}\bigg|_{R_S=0} = \frac{R_S(r_d + R_D)}{R_D + r_d + (1 + \mu)R_S} = R_S \| \frac{R_D + r_d}{1 + \mu} \quad (10\text{-}75)$$

Again $R_{o2}' = R_{o2} \| R_S$ so that

$$R_{o2} = \frac{R_D + r_d}{1 + \mu} \quad (10\text{-}76)$$

For $\mu \gg 1$ and $r_d \gg R_D$, $R_{o2} \approx 1/g_m$.

The Common-Source Stage   By letting $R_S = 0$, we can reduce Eqs. (10-70) and (10-74) to

$$A_V = \frac{V_{o1}}{V_i} = \frac{-\mu R_D}{r_d + R_D} = \frac{-g_m R_D}{1 + R_D/r_d} \quad (10\text{-}77)$$

and

$$R_{o1} = r_d \quad (10\text{-}78)$$

From which $R_{o1}' = R_D \| R_{o1} = R_D \| r_d.$

The Common-Drain Configuration The common-drain or source-follower configuration is obtained from Fig. 10-27 by making $R_D = 0$ and using V_{o2} as the output as shown in Fig. 10-28b. For this situation, Eqs. (10-69) and (10-75) result in

$$A_V = \frac{V_{o2}}{V_i} = \frac{\mu R_S}{r_d + R_S(1 + \mu)} \tag{10-79}$$

$$R'_{o2} = R_S \| R_{o2} = R_S \| \frac{r_d}{1 + \mu} = \frac{R_S r_d}{r_d + (1 + \mu)R_S} \tag{10-80}$$

For $\mu \gg 1$ and $1/g_m \ll R_S$, $A_V \approx 1$ and $R'_{o2} \approx R_{o2} = 1/g_m$. These results indicate that the source follower has a gain of almost unity and a low output resistance, as does its BJT counterpart, the emitter follower. The results for the common-drain stage are listed in Table 10-5, column 3.

The equations in Table 10-5 can be obtained directly from Table 10-3B because of the similarities between the models of corresponding BJT and FET stages. The procedure is

1. Identify R_D, R_S, r_d, and g_m for the FET stage with R_C, R_E, r_o, and g_m, respectively, for the analogous BJT stage.

2. Set $r_b = 0$ and, after using $\beta_o = g_m r_\pi$, let $r_\pi \to \infty$. For example, using the value of R_o for the CC configuration in Table 10-3B, the identifications in step 1 give $R_o = r_d \| (R_S + r_b + r_\pi)/(1 + \beta_o)$. Use of step 2 results in $R_o = r_d \| (R_S + r_\pi)/(1 + g_m r_\pi)$, which, when $r_\pi \to \infty$, becomes $R_o = r_d \| 1/g_m = r_d/(1 + \mu)$. This is the value of R_o given in Table 10-5 for the CD stage.

The reader is advised not to draw misleading conclusions from the observation that the treatment of FET amplifiers is briefer by far than that for BJT stages. While BJT amplifiers are used more frequently, many modern IC amplifiers employ both FETs and BJTs on the same chip (BIMOS and BIFET technologies). In addition, a variety of signal-processing systems, which utilize both digital and analog circuits, are fabricated in NMOS or CMOS technology. Basic amplifier concepts are common to both types of transistor, and the BJT and FET configurations described are analogous. Since the results for BJT amplifiers can be applied directly to FET stages, we believe that a repetition for FET amplifiers of all the detailed analyses presented for BJT stages is unnecessary (and inefficient and boring).

10-16 CASCADED BJT AMPLIFIERS We observe in Table 10-3 that the voltage gain of a single-stage amplifier depends on the load resistance of the stage (R_C for common-emitter and common-base stages and R_E for the common-collector configuration). As described in Sec. 10-7, the size of the load resistance cannot be specified independently because of the constraints imposed by biasing. Consequently, the gain that is realized by a single-stage circuit may not be sufficient for the particular purpose. In addition, the input and output resistances may

FIGURE 10-29
Pictorial representation
of two cascaded stages.

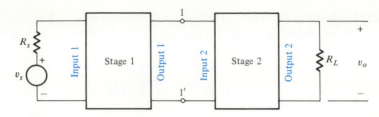

not be the correct magnitudes for the intended application. To overcome these limitations, two or more stages can be connected in cascade; thus, the output of one stage is connected to the input of the next stage as seen in Fig. 10-29.

The analysis of the cascade configuration is based on the results obtained for single stages and is described as follows. We first obtain a Thévenin equivalent of the first stage at terminals 1–1', that is, the portion of the circuit in Fig. 10-29 shown in Fig. 10-30a. For this single-stage amplifier, the output voltage (Thévenin voltage source) is $A_{V1}V_s$ and R'_{o1} is the output resistance. This combination acts as the signal source and source resistance of the second stage as illustrated in Fig. 10-30b. For the stage illustrated in Fig. 10-30b, the output voltage is A_{V2} times the input voltage or $V_o = A_{V1}A_{V2}V_s$ and

$$\frac{V_o}{V_s} = A_V = A_{V1}A_{V2} \tag{10-81}$$

The method leading to Eq. (10-81) is applicable to several stages by repeating the process. Note that if A_{V1} and A_{V2} are much greater than unity, A_V, the overall gain of the cascade, is much larger than the gain that can be achieved by either stage. We illustrate this in the following example.

Example 10-7

The cascaded amplifier in Fig. 10-31 consists of two common-emitter stages, one of which contains an emitter resistance and a common-collector stage. Note that biasing components are not shown and are assumed to have negligible effect. Transistor $Q1$ has $\beta_o = 100$ and $r_\pi = 1.0$ kΩ; transistors $Q2$ and $Q3$ have $\beta_o = 100$ and $r_\pi = 0.5$ kΩ. Determine the overall gain.

Solution

We first obtain the Thévenin equivalent of the first stage at terminals 1–1'. This is a common-emitter stage with an emitter resistance for which the voltage gain

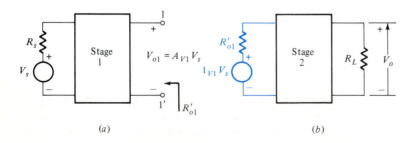

(a)

(b)

FIGURE 10-30

(a) The unloaded first stage of the amplifier in Fig. 10-29. (b) The second stage driven by the Thévenin equivalent of the first stage.

FIGURE 10-31
Three-stage cascaded amplifier for Example 10-7.

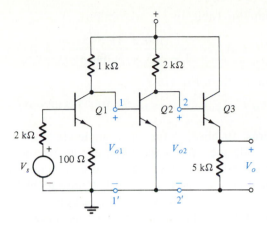

is given in the second column in Table 10-3A. Substitution of values gives

$$A_{V1} = \frac{-100 \times 1}{2 + 1 + (1 + 100)0.1} = -7.63$$

The value of $R'_{o1} = R_{C1} = 1$ kΩ is the Thévenin resistance. Now we obtain the Thévenin equivalent of the second stage at terminals 2–2'. As seen in Fig. 10-32b, the Thévenin equivalent of stage 1 is the signal source for stage 2. The voltage gain for this common-emitter stage, from column 1 in Table 10-3A, is

$$A_{V2} = \frac{-100 \times 2}{1 + 0.5} = -133$$

and the output voltage of stage 2 is $A_{V1}A_{V2}V_s = 1020\ V_s$. This voltage drives the common-collector stage and has a source resistance $R'_{o2} = 2$ kΩ as shown in Fig. 10-32c.

FIGURE 10-32
(a) The unloaded first stage of the amplifier in Fig. 10-31. (b) The second stage; the signal source V_{o1} and the source resistance R_{o1} form the Thévenin equivalent of the first stage. (c) The common-collector (third) stage; V_{o2} and R_{o2} are the Thévenin equivalent of (b).

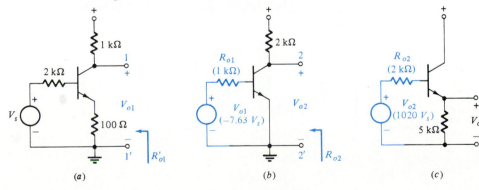

From column 3 in Table 10-3A, we obtain

$$A_{V3} = \frac{(100 + 1) \times 5}{2 + 0.5 + 5(100 + 1)} = 0.995$$

The overall gain is, from Eq. (10-81),

$$A_V = A_{V1}A_{V2}A_{V3} = (-7.63)(-133)(0.995) = 1010$$

It is of interest to observe that if a single-stage amplifier, say, $Q2$, were required to provide the same gain when driven by the signal source in Fig. 10-31, a collector resistance of 30 kΩ would be needed. At room temperature, the parameters of $Q2$ indicate $I_{CQ} = 5$ mA. Consequently, for this single stage, V_{CC} would be in excess of 150 V, a value which is evidently impractical for most transistor circuits. (The values given in this paragraph are verified in Prob. 10-65.)

The overall current gain A_I is *not* equal to the product of the current gains of the individual stages because the output current of one stage is not equal to the input current of the following stage. In Fig. 10-31 we can see that the output current of $Q1$ (in the 1-kΩ resistance) is not the input current (base current) of $Q2$. However, we can obtain A_I from the overall voltage gain A_V.

Consider the situation in Fig. 10-33a, in which the amplifier consists of a number of cascaded stages. For this circuit, $A_V = V_o/V_s$. Now let us convert the series combination of V_s and R_s to its current source equivalent as depicted in Fig. 10-33b. For the circuit in Fig. 10-33b, we can identify the current gain as $A_I = I_o/I_s$. However, $I_o = -V_o/R_L$, and, from the source conversion, $I_s = V_s/R_s$. Thus

$$A_I = \frac{I_o}{I_s} = \frac{-V_o/R_L}{V_s/R_s} = \frac{-R_s}{R_L}\frac{V_o}{V_s} = -\frac{R_s}{R_L}A_V \qquad (10\text{-}82)$$

Thus, by knowing R_s and R_L, we can always obtain A_I from A_V (or vice versa).

The input and output resistances of a cascade configuration are simply the input resistance of the first stage and the output resistance of the last stage. For the circuit in Fig. 10-31, the input resistance is that for a common-emitter stage with an emitter resistance. This is evidently higher than can be obtained with a simple common-emitter stage. Similarly, the common-collector output

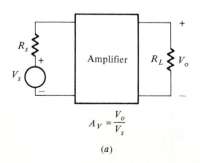

$$A_V = \frac{V_o}{V_s}$$

(a)

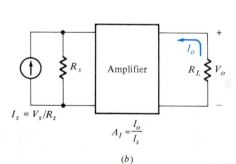

$$I_s = V_s/R_s$$

$$A_I = \frac{I_o}{I_s}$$

(b)

FIGURE 10-33
An amplifier driven by (*a*) a voltage source and (*b*) a current source.

stage provides a very low output resistance. Thus the amplifier in Fig. 10-31 has a high input and a low output resistance and a gain of about 1000. In Table 10-4 we see that this combination cannot be obtained from any single configuration.

Cascaded FET Stages Field-effect transistor stages are cascaded to achieve a larger voltage gain than can be realized by a single stage. The overall gain of cascaded amplifiers is given in Eq. (10-81) in which the A_{Vi} are the voltage gains of the individual stages (Table 10-5). Analogous configurations to most BJT cascaded amplifiers are used such as CS-CS, CS-CD (Example 10-7), and cascode (the CS, common-gate cascade in Prob. 10-72).

10-17 COMPOUND (COMPOSITE) TRANSISTOR STAGES

Three 2-stage cascaded amplifiers that are widely used as integrated circuits are shown in Figs. 10-34 and 10-35. Each of these circuits can be considered as a single equivalent BJT; hence the name "compound" or "composite transistor." These circuits combine or enhance the properties that can be obtained in the different BJT configurations. Behavior of compound transistor stages is similar to the cascade amplifier in Example 10-7. However, in IC design these composites are often used as individual stages in a cascaded amplifier.

The *CC-CC* (Darlington) Configuration The CC-CC cascade, shown in Fig. 10-34a, is often referred to as a *Darlington transistor* or a *Darlington pair*. The current source I_{EE} is used to provide the bias for the circuit. For the composite transistor (shaded), I_{b1} is the input current and $I_c = I_{c1} + I_{c2}$ is the output current. Note

FIGURE 10-34
Composite (compound) transistors: (*a*) the Darlington pair (*CC-CC* cascade), (*b*) the *CC-CE* cascade.

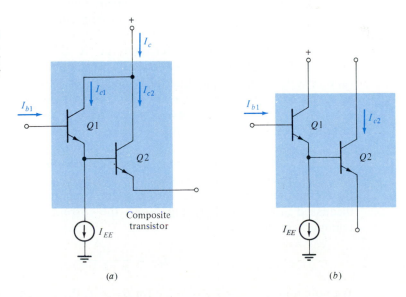

(*a*) (*b*)

FIGURE 10-35
The cascode or *CE-CB*
cascade.

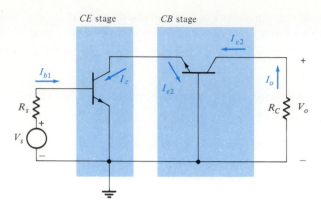

that the input signal current in $Q2$ is the emitter current of $Q1$. Thus

$$I_{c2} = \beta_o I_{b2} = \beta_o(\beta_o + 1)I_{b1} \quad \text{and} \quad I_c = I_{c1} + I_{c2} = \beta_o I_{b1} + \beta_o(\beta_o + 1)I_{b1}$$

from which the current gain of the composite transistor β_{oc} is

$$\beta_{oc} = \frac{I_c}{I_{b1}} = \beta_o(\beta_o + 2) \approx \beta_o^2 \tag{10-83}$$

for $\beta_o \gg 2$. For $\beta_o = 100$, $\beta_o^2 \approx 10^4$; clearly, the current gain is enhanced.

The Darlington transistor is most often used as an emitter follower because, as indicated in Table 10-3, the extremely high value of β_{oc} makes A_V virtually unity, R_i extremely large, and R_o extremely small.

The *CC-CE* Connection   The circuit displayed in Fig. 10-34b is a *CC-CE* circuit and has properties similar to those of the Darlington pair. The current gain of the composite transistor is approximately β_o^2, since the emitter current in $Q1$ is the base current in $Q2$.

The *CE-CC* composite is preferred to the *CC-CC* configuration in constructing equivalent common-emitter stages. The advantage of the *CC-CE* configuration results from the fact that the collectors of $Q1$ and $Q2$ are not connected as in the Darlington pair. Through connection of the collectors (essentially in parallel), the output resistance ($r_o < \infty$) is reduced. Also, the frequency response of the *CC-CE* connection is superior to that for the *CC-CC* circuit.

The *CE-CB* (Cascode) Connection   The primary use of the *cascode circuit* in Fig. 10-35 is to provide high voltage gain over a wider range of frequencies that can be achieved by a common-emitter stage. The frequency response of the *CE-CB* stage is discussed in Sec. 11-11. In Fig. 10-35, the current I_{b1} is $V_s/(R_s + R_i)$ for the common-emitter stage or

$$I_{b1} = \frac{V_s}{R_s + r_{\pi 1}} \tag{10-84}$$

We note that $\beta_{o1}I_{b1} = I_{c1} = -I_{e2}$; for $\beta_{o2} \gg 1$, the current gain of the common-base stage is unity. We can conclude that $I_o = I_{c2} \approx I_{c1} = \beta_o I_{b1}$. Thus the overall current gain $A_I = I_o/I_{b1} = \beta_o$.

The output voltage $V_o = -I_o R_C$, and by use of Eq. (10-84) we obtain

$$A_V = \frac{V_o}{V_s} = \frac{-\beta_o R_C}{R_s + r_{\pi 1}} \tag{10-85}$$

The value of A_V in Eq. (10-85) is that for a single common-emitter stage having a load resistance R_C. However, in the cascode circuit the load resistance of the common-emitter stage is R_i for the common-base stage. As indicated in Table 10-4, the value of R_i is significantly lower than the value of R_C needed to obtain the gain. It is this low load resistance on the common-emitter stage that gives the cascode circuit improved high-frequency performance.

10-18 THE DIFFERENTIAL AMPLIFIER

The *differential amplifier, emitter-coupled pair,* or *differential pair,* is an essential building block in modern IC amplifiers. This circuit, displayed in Fig. 10-36, was first introduced in Sec. 3-12, where we demonstrated that the operation of this circuit was predicated on the ability to fabricate matched components on the same chip. It was also shown that for a small difference voltage $V_d (4V_T > |V_d|$ in Fig. 3-33), the differential pair behaved as a linear amplifier. In this section we examine the behavior of this circuit at low frequencies in more detail.

Included in Fig. 10-36 is the output resistance R_E of the current-source bias network (Sec. 10-3). As we shall see shortly, this resistance has an important effect on performance. We assume that the current in R_E is negligible compared to I_{EE}. Note that no R_s is indicated in the base loop in Fig. 10-36. We will assume that $R_s = 0$ and that the base-spreading resistance $r_b = 0$ in our analysis of the differential amplifier. The effect of these elements is treated in the next section.

FIGURE 10-36
The emitter-coupled or differential pair stage.

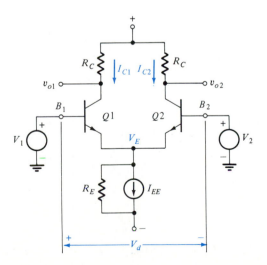

The Differential Mode For $V_1 = V_2$ and assuming $\beta_F \gg 1$, the collector and emitter currents in each stage are equal ($I_C \approx |I_E|$). All of these currents have magnitudes equal to $I_{EE}/2$ (approximately) because of the symmetry of the circuit and the negligible current in R_E.

Let us now incrementally increase V_1 by $\Delta v/2$ and simultaneously decrease V_2 by $\Delta v/2$. In effect, we are applying an incremental signal $\Delta v/2$ at B_1 and applying a signal $-\Delta v/2$ to B_2. The differential voltage $V_d = V_1 - V_2$ increases by Δv. For $\Delta v < 4V_T$, as indicated in the transfer characteristic for Fig. 3-33, the circuit behaves linearly. (This transfer characteristic is only a close approximation to the situation discussed here as Fig. 3-33 was developed for $R_E \rightarrow \infty$.) Thus I_{C1} increases by ΔI_C and I_{C2} decreases by an equal amount (the increment in $I_{C2} = -\Delta I_C$). As $I_C \approx |I_E|$, the changes in I_{C1} and I_{C2} also appear at the emitters. Consequently, the current in R_E remains unchanged (the incremental current in R_E is zero), causing the voltage V_E to remain constant. Recall that in small-signal analysis, constant potentials are replaced by short circuits. Thus, in our incremental model, each emitter is grounded.

The situation just described is referred to as the *differential mode* because the input signals ($\Delta v/2$) applied to $Q1$ and $Q2$ are equal and opposite and a difference signal V_d exists. For the differential mode, the incremental circuit can be drawn as shown in Fig. 10-37a. Observe that the model for the transistors is not explicity shown but is implied because of incremental operation. Since the two sides of the circuit are identical, only one side must be analyzed. This *half-circuit* concept is used in the next section to analyze the differential amplifier in detail.

The Common Mode Now let us consider that both V_1 and V_2 increase by $\Delta v/2$. The difference voltage V_d remains zero, and I_{C1} and I_{C2} remain equal. However, because R_E

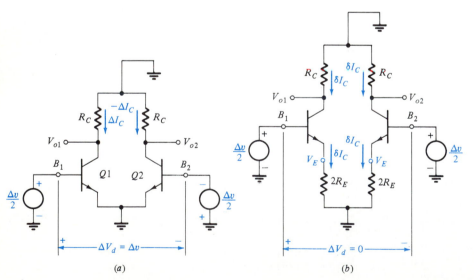

(a)　　　　　　　　　　　　(b)

FIGURE 10-37
The half-circuit concept: (a) the differential-mode small-signal equivalent and (b) the common-mode small-signal equivalent of the differential pair. (*Note:* It is implied that the transistors are replaced by their incremental equivalent circuit.)

is present, both I_{C1} and I_{C2} exhibit a small increase δI_C. Again, changes in I_C appear at the emitter, and hence the current in R_E increases by $2\delta I_C$. The voltage V_E is no longer constant but must increase by $2\delta I_C R_E$. This situation, where equal signals are applied to $Q1$ and $Q2$, is called the *common mode*. The incremental equivalent circuit is displayed in Fig. 10-37b, *in which it is implied that Q1 and Q2 are represented by their small-signal models.*

Two resistances, each of value $2R_E$, are shown in Fig. 10-37b. The voltage across each is $2\delta I_C R_E$ and equals the incremental change in V_E; thus the two resistances are in parallel and $2R_E \| 2R_E = R_E$. As shown in Fig. 10-37b, the two halves of the circuit are symmetrical and only one half must be analyzed. This common-mode equivalent is that of a common-emitter stage with an emitter resistance.

It is evident in Fig. 10-37 that, depending on the input signal, the differential amplifier behaves as either a common-emitter stage or a common-emitter stage with emitter resistance. Therefore, the gain of this stage is significantly higher for differential-mode operation than for common-mode operation. Usually, differential amplifiers are designed so that, for practical purposes, only difference signals are amplified.

As we have stated several times previously, large capacitors (for bypass and coupling) cannot be fabricated on a chip. Consequently, IC circuits are generally direct (dc)-coupled. However, the effect produced by bypass and coupling capacitors is present in differential amplifiers as a result of circuit symmetry. The voltage V_E remains constant in the differential mode, and, as shown in Fig. 10-37a, the emitter is at ground for small-signal analysis. Thus it appears that R_E is bypassed. Similarly, the voltage between the two collectors $V_{o1} - V_{o2}$ is zero in the common mode and is twice the change in V_{o1} (V_{o2}) for the differential mode. Since the applied signal Δv can be made positive or negative, the voltage $V_{o1} - V_{o2}$ can be positive or negative (about 0 V). This is simply the effect produced by a coupling capacitor.

Large resistances are also difficult to fabricate in any IC technology. Although R_C is shown as a resistor in Fig. 10-36, this resistance is usually realized by the output resistance of a current mirror (Fig. 10-22). These *active loads* are treated in Sec. 14-2.

10-19 ANALYSIS OF DIFFERENTIAL AMPLIFIERS

The analysis of the differential amplifier is based on the half-circuit concept referred to in Sec. 10-18. This method exploits the symmetry of the circuit in both differential and common modes.

Differential Mode Gain A_{DM} Consider that a signal V_{DM} is applied to the base of $Q1$ in Fig. 10-36 and that $-V_{DM}$ is applied to B_2. For this condition, the circuit in Fig. 10-37a is valid (with $\Delta v/2$ replaced by V_{DM}). Use of the half-circuit concept, that is, analysis of only one-half of the circuit, results in the small-signal model in Fig. 10-38a. This is the model for a common-emitter stage with $R_s = r_b = 0$ and from the A_V entry in column 1 in Table 10-3A:

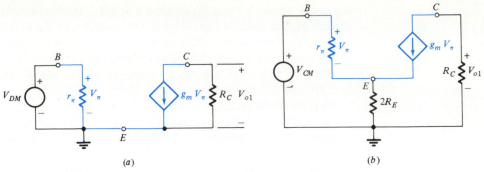

FIGURE 10-38
Small-signal models for (*a*) the differential mode and (*b*) the common mode.

$$A_{DM} = \frac{V_{o1}}{V_{DM}} = \frac{-\beta_o R_C}{r_\pi} = -g_m R_C \qquad (10\text{-}86)$$

For V_{DM} positive, $V_{o1} = A_{DM}V_{DM}$ and, as seen in Eq. (10-86), A_{DM} is negative, so that V_{o1} is 180° out of phase with V_{DM} (V_{o1} is inverted). Because Q_2 is driven by $-V_{DM}$, $V_{o2} = -A_{DM}V_{DM}$ and V_{o2} is in phase with V_{DM} (V_{o2} is noninverted).

Common-Mode Gain A_{CM} When a signal V_{CM} is applied to both bases in Fig. 10-36 (common mode), the circuit in Fig. 10-37*b* is valid and results in the equivalent circuit of Fig. 10-38*b*. For this circuit, the gain A_{CM} (from the A_V entry in column 2 of Table 10-3*A*) is

$$A_{CM} = \frac{V_{o1}}{V_{CM}} = \frac{-\beta_o R_C}{2(\beta_o + 1)R_E + r_\pi} \qquad (10\text{-}87)$$

Equation (10-87), with $\beta_o \gg 1$ and division by r_π, reduces to

$$A_{CM} = \frac{-g_m R_C}{1 + 2g_m R_E} \approx -\frac{R_C}{2R_E} \qquad (10\text{-}88)$$

for $2g_m R_E \gg 1$. Because the same signal is applied to $Q1$ and $Q2$, both V_{o1} and V_{o2} are 180° out of phase with V_{CM}.

The Common-Mode Rejection Ratio The differential amplifier is primarily designed to amplify differential signals; hence we require $A_{DM} \gg A_{CM}$. A convenient measure of differential amplifier performance is the *common-mode rejection ratio CMRR*, defined as

$$CMRR \equiv \frac{A_{DM}}{A_{CM}} \qquad (10\text{-}89)$$

Combination of Eqs. (10-86) and (10-88) yields

$$CMRR = 1 + 2g_m R_E \approx 2g_m R_E \qquad (10\text{-}90)$$

As seen in Eq. (10-90), large values of $CMRR$ require large values of R_E and often necessitate the use of current sources having high output resistances. Note that if $R_E \rightarrow \infty$, the $CMRR \rightarrow \infty$, $A_{CM} = 0$, and no common-mode component appears at the output. This is the condition for which the transfer characteristic in Fig. 3-33 was constructed.

Output for Arbitrary Input Signals   Our previous discussion assumed that either common-mode or differential-mode signals were present. This is unrealistic and rarely occurs in the "real world." However, arbitrary input signals can be decomposed into common-mode and differential-mode components. Consider signals V_1 and V_2 applied to $Q1$ and $Q2$, respectively. This pair of signals can be represented as the sum and difference of two other signals[1] V_{DM} and V_{CM}, or

$$V_1 = V_{CM} + V_{DM} \qquad V_2 = V_{CM} - V_{DM}$$

Solving these equations for V_{DM} and V_{CM} gives

$$V_{DM} = \frac{V_1 - V_2}{2} = \frac{V_d}{2} \qquad (10\text{-}91a)$$

$$V_{CM} = \frac{V_1 + V_2}{2} \qquad (10\text{-}91b)$$

The effect of this decomposition is illustrated in Fig. 10-39. Because the circuit behaves linearly, superposition applies. The output consists of two components, one attributed to the pair of sources V_{DM} and the other to the pair of sources V_{CM}. Thus one component of the output is due to the differential input signal, and the second is produced by the common-mode input. The output voltage V_{o1} is

$$V_{o1} = A_{DM}V_{DM} + A_{CM}V_{CM} \qquad (10\text{-}92a)$$

$$= A_{DM}\left(V_{DM} + \frac{V_{CM}}{CMRR}\right) \qquad (10\text{-}92b)$$

Equation (10-92b) demonstrates the importance of the $CMRR$ if we are to amplify only difference signals. As the $CMRR$ is increased, the common-mode output component has diminished significance.

The output voltage V_{o2} is expressed as

$$V_{o2} = -A_{DM}V_{DM} + A_{CM}V_{CM} \qquad (10\text{-}93a)$$

$$= -A_{DM}\left(V_{DM} - \frac{V_{CM}}{CMRR}\right) \qquad (10\text{-}93b)$$

---

[1]This is similar to the decomposition of a function into its odd and even parts used in calculus and Fourier series.

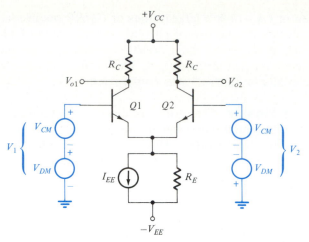

FIGURE 10-39

Representation of arbitrary signals V_1 and V_2 into differential (V_{DM})- and common (V_{CM})-mode components. These are applied to the bases of the differential pair.

Substitution of Eqs. (10-91) into Eqs. (10-92) and (10-93) gives

$$V_{o1} = \frac{A_{DM}}{2}\left(V_d + \frac{V_1 + V_2}{CMRR}\right) \tag{10-94a}$$

$$V_{o2} = \frac{-A_{DM}}{2}\left(V_d - \frac{V_1 + V_2}{CMRR}\right) \tag{10-94b}$$

Equations (10-94) are an alternative form for the output voltages that appear in the literature. Note that the difference signal V_d appears explicitly.

Example 10-8

The differential amplifier in Fig. 10-36 is designed with $R_C = 100$ kΩ and $R_E = 500$ kΩ. The transistor small-signal parameters are $\beta_o = 10^4$ and $g_m = 5.0$ m℧. A sinusoid having a 2-mV rms amplitude is applied to $Q1$ and no signal ($V_2 = 0$) is applied to $Q2$. (a) Determine the output voltage V_{o2}. (b) Repeat part a, assuming that $V_1 = 0$ and the sinusoid is applied to $Q2$.

Solution

We must evaluate A_{DM}, A_{CM} and decompose the input signal into its common- and differential-mode components. From Eqs. (10-86) and (10-88), we obtain

$$A_{DM} = -5 \times 100 = -500 \qquad A_{CM} = \frac{-5 \times 100}{1 + 2 \times 5 \times 500} = -0.10$$

(a) With $V_2 = 0$, Eqs. (10-91) give $V_{DM} = V_{CM} = V_1/2 = 1$ mV. Use of Eq. (10-93a) yields

$$V_{o2} = -(-500)10^{-3} + (-0.10)10^{-3} = 499 \text{ mV} \approx 500 \text{ mV}$$

(b) With $V_1 = 0$, Eqs. (10-91) give $V_{DM} = -(V_2/2) = -1$ mV and $V_{CM} = V_2/2 = 1$ mV. Again, use of Eq. (10-93a) results in

$$V_{o2} = -(-500)(-10^{-3}) + (-0.10)(10^{-3}) = -501 \text{ mV} \approx -500 \text{ mV}$$

From the results in parts a and b it is apparent that for practical purposes only the differential-mode signal has been amplified. This is a result of having a

$CMRR = A_{DM}/A_{CM} = 5000$. Manufacturers list the $CMRR$ in their data sheets and express it in decibels (dB). For this circuit, $CMRR$ in decibels = 20 log 5000 = 74 dB.

The common situation where only one output terminal is used (called a *single-ended output*) is illustrated in Example 10-8. Similarly, only one input signal is often applied. We observe in the results of Example 10-8 that the phase of the output relative to the input depends on which base is driven. When the signal is applied to B_2, V_{o2} is 180° out of phase with V_1; that is, it is inverted, and B_2 is called the *inverting input*. Input and output signals are in phase when the signal is applied to B_1 so that B_1 is referred to as the *noninverting input*.

The values used in Example 10-8 are typical of IC differential stages. The value of $\beta_o = 10^4$ can be realized by the *CC-CE* composite (Fig. 10-34b) and $R_C = 100$ kΩ can be achieved by an active load (current mirror).

Effect of the Source Resistance   The effect on A_{DM} and A_{CM} of the signal-source resistance R_s (and of r_b as these resistances appear in series) can be readily accounted for in the analysis of the emitter-coupled pair. Provided the circuit remains balanced—that is, a resistance R_s is connected to both B_1 and B_2—the analysis is identical to that described in this section. The values of A_{DM} and A_{CM} are modified to include R_s and r_b as given in Table 10-3b. Their effect, however, is most often negligible for typical values ($R_s \leq 10$ kΩ and $r_b \approx 100$ Ω). With use of the values given in Example 10-8, for which $r_\pi = \beta_o/g_m = 2$ MΩ, it is clear that $R_s + r_b \leq 10.1$ kΩ is negligible in comparison with r_π.

Input and Output Resistances   Input and output resistances can be identified for both common and differential modes. However, two are of particular interest; the differential-mode output resistance $R'_{o(DM)}$ and the differential mode input resistance $R_{i(DM)}$. For the differential mode $R'_{o(DM)}$ is just R'_o for the common-emitter stage, namely, R_C. A word of caution is in order: where R_C is derived from an active load, the output resistance r_o of the BJT cannot be ignored, but must be included as given in Table 10-3B.

The differential-mode input resistance $R_{i(DM)}$ is the resistance seen by the difference signal V_d; that is, it is the resistance measured between the base terminals of $Q1$ and $Q2$. Since both emitters are at ground in the differential mode, the input resistance is simply the sum of the input resistances of each transistor, namely

$$R_{i(DM)} = 2r_\pi \tag{10-95}$$

10-20 FET DIFFERENTIAL AMPLIFIERS   The *source-coupled pair* or *differential amplifier* is constructed with MOSFETs as shown in Fig. 10-40. (JFETs are also used in differential amplifiers). The balanced structure of this amplifier leads to an

FIGURE 10-40
A source-coupled pair.

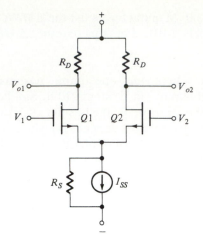

FIGURE 10-40
A source-coupled pair.

analysis that parallels that for the emitter-coupled pair described in Sec. 10-18. The differential-mode gain A_{DM} is the gain of a common-source stage (Table 10-5), and for the common-source stage with source resistance, column 2 in Table 10-5 gives the common-mode gain A_{CM}. The common-mode rejection ratio $CMRR$, defined in Eq. (10-89), is

$$CMRR = 1 + \frac{2R_S(1 + \mu)}{r_d + R_D} \qquad (10\text{-}96)$$

For $r_d \gg R_D$ and $\mu \gg 1$, Eq. (10-96) reduces to

$$CMRR = 1 + 2g_m R_S \approx 2g_m R_S \qquad (10\text{-}97)$$

Note that the resistance $2R_S$ is derived from the half-circuit concept used to describe the differential amplifier.

Active loads, often depletion MOSFETs, are used to realize R_D in Fig. 10-40. In fact, most MOSFET or JFET amplifiers fabricated on a chip employ active loads to both realize large resistance values and conserve chip area.

10-21 THE OPERATIONAL AMPLIFIER The *operational amplifier,* or Op-Amp, is a direct-coupled, high-gain amplifier used to perform a wide variety of functions. It is often referred to as the *basic linear* (or more accurately *analog*) *integrated circuit* (IC), and many manufacturers package one to four identical units on a single chip. Although many Op-Amps comprise the cascade of a differential pair, common-emitter (source) stage, and emitter (source) follower, it is used widely as a single-stage amplifier. Our purpose in this section is to provide an introduction to the basic Op-Amp configurations, which we will encounter again in Chaps. 11 to 16. Chapter 14 is devoted exclusively to a more detailed examination of the internal design and several applications of Op-Amps. Many other circuits in which the Op-Amp is an integral component are treated in Part 4.

FIGURE 10-41

The operational amplifier (Op-Amp) (a) circuit symbol and (b) equivalent circuit.

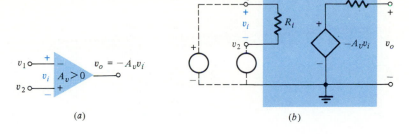

(a)

(b)

The schematic diagram of the Op-Amp is displayed in Fig. 10-41a and its equivalent circuit, in Fig. 10-41b. As seen in Fig. 10-41b, the Op-Amp is a voltage-controlled voltage source. The output voltage v_o is the amplified difference signal $v_i = v_1 - v_2$. The $-$ and $+$ symbols at the input of the Op-Amp refer to the inverting and noninverting input terminals. That is, if $v_2 = 0$, v_o is 180° out of phase (inverted) with respect to the input signal v_1. Similarly, when $v_1 = 0$, the output v_o and input v_2 are in phase (noninverting).

The Ideal Op-Amp   The ideal operational amplifier has the following characteristics:

1. *The input resistance $R_i \to \infty$ (open-circuited).* Consequently, *no current enters either input terminal.*

2. *The output resistance $R_o = 0$.*

3. *The voltage gain $A_v \to \infty$. The output voltage $v_o = -A_v v_i$ is finite ($|v_o| < \infty$); thus, as $A_v \to \infty$, it is required that $v_i = 0$.*

4. The amplifier responds equally at all frequencies. (The bandwidth is infinite.)

5. When $v_1 = v_2$, $v_o = 0$ and is independent of $|v_1|$. The converse is also true.

The same symbol is used for ideal and practical Op-Amps. To distinguish them, we indicate the finite gain A_v in the triangle for practical Op-Amps and omit it in the ideal case (see Fig. 10-42).

The circuit in Fig. 10-42a is an *inverting amplifier stage* utilizing an ideal Op-Amp. Because the input current is zero, the current I exists in both R_1 and R_2. Furthermore, since $V_i = 0$, it follows that

$$I = \frac{V_s}{R_1} = -\frac{V_o}{R_2} \tag{10-98}$$

from which

$$A_V = \frac{V_o}{V_s} = -\frac{R_2}{R_1} \tag{10-99}$$

In Eq. (10-99), we observe that A_V depends only on the resistor ratio. Recall that in the fabrication of ICs, resistor ratios can be controlled with much greater

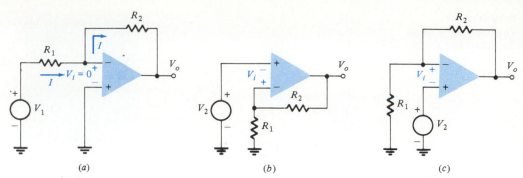

FIGURE 10-42
(*a*) Inverting Op-Amp stage. (*b*, *c*) Two commonly used circuit diagrams of the noninverting Op-Amp stage.

precision than the values of individual resistances. Furthermore, the circuit designer can obtain the gain required for a particular application by controlling resistance values *external to the amplifying device*. This is one property of feedback systems discussed in Sec. 12-3. (Observe that R_2 provides the feedback from output to input.)

The Op-Amp is used as a *noninverting amplifier stage* in the circuit shown in Fig. 10-42*b*. (An alternative schematic of this circuit is shown in Fig. 10-42*c*.) In Fig. 10-42*b*, making $V_i = 0$ requires that

$$V_i = V_1 - V_2 = \frac{R_1}{R_1 + R_2} V_o - V_s = 0 \qquad (10\text{-}100)$$

Solving for $A_V = V_o/V_s$ yields

$$A_V = \frac{R_1 + R_2}{R_1} = 1 + \frac{R_2}{R_1} \qquad (10\text{-}101)$$

Equation (10-101) once again indicates that the feedback provided by R_2 causes A_V to depend only on the resistance ratio R_2/R_1.

If we make $R_2 = 0$ in Fig. 10-42, $A_V = 1$ (and the resistance R_1 is unnecessary). This stage (shown in Fig. 10-43) is called a *unity-gain buffer* or *voltage follower* as it has infinite (high) input resistance, zero (low) output resistance, and unity gain. Note that this circuit has properties almost identical to the emitter- and source-follower circuits described previously.

We now demonstrate that practical Op-Amp stages closely approximate the stages.

Practical Inverting Op-Amp Stages   The equivalent circuit for an inverting stage using a practical Op-Amp is shown in Fig. 10-44*a*. To demonstrate the effect of the deviations from the ideal ($R_i < \infty$, $A_v < \infty$, $R_o \neq 0$), we will obtain the Thévenin equivalent of the stage seen by the load resistance R_L. First we replace the source V_1 and resistances R_1 and R_i by a Thévenin equivalent as displayed in

FIGURE 10-43
The voltage follower.

Fig. 10-44b. The equivalent source is $V_1R_i/(R_i + R_1)$, and the equivalent resistance is $R_1\|R_i$. Note that when $R_i \gg R_1$, as is the usual case, these quantities reduce to V_1 and R_1, respectively. Thus, the Op-Amp input resistance has negligible effect provided $R_i \gg R_1$. We assume this condition in the remainder of the analysis.

In Fig. 10-44b

$$V_o = -A_v V_i + IR_o \tag{10-102}$$

and

$$V_i = IR_2 + V_o \tag{10-103}$$

Application of KVL to the loop gives

$$-V_1 + I(R_1 + R_2 + R_o) - A_v V_i = 0 \tag{10-104}$$

Combination of Eqs. (10-102) to (10-104) and solving for V_o/V_s gives

$$A_V = \frac{V_o}{V_1} = \frac{-A_v R_2 + R_o}{R_1(1 + A_v) + R_2 + R_o} \tag{10-105}$$

Observe in Eq. (10-103) that if $A_v \gg 1$ and $A_v R_1 \gg R_2 + R_o$, A_V approaches $-R_2/R_1$. If A_v is sufficiently large, therefore, A_V has the same value as in the ideal case. The effect of R_o is negligible (A_V is independent of R_o), provided that $A_v R_1$ and $A_v R_2$ are each greater than R_o. We can then conclude that provided A_v is very large, the Thévenin voltage $A_V V_1 \approx -R_2 V_1/R_1$ and is virtually independent of deviations from the ideal in the Op-Amp.

To obtain the Thévenin resistance R_T (the output resistance of the stage seen by R_L), we evaluate the short-circuit current I_{sc}. For $R_L = 0$ (short circuit), the current I_A in the $R_1 - R_2$ branch is

$$I_A = \frac{V_1}{R_1 + R_2} \tag{10-106}$$

FIGURE 10-44
(a) Equivalent circuit of the nonideal inverting Op-Amp stage. (b) The circuit in a with V_1, R_1, and R_i replaced by their Thévenin equivalent, R_1'-V_1'.

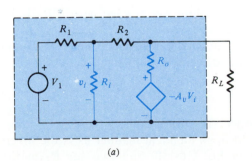

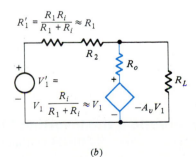

(a)   (b)

and the current I_B in R_o is

$$I_B = \frac{-A_v V_i}{R_o} \qquad (10\text{-}107)$$

Combining Eqs. (10-106), (10-107), and (10-103) with $V_o = 0$, we obtain

$$I_{sc} = I_A + I_B = \frac{-A_v(R_2 + R_o)V_1}{R_1(1 + A_v) + R_2 + R_o} \qquad (10\text{-}108)$$

from which

$$R_T = \frac{(R_1 + R_2)R_o}{R_1(1 + A_v) + R_2 + R_o} = \frac{[(R_1 + R_2)R_o]/(R_1 + R_2 + R_o)}{1 + [A_v R_1/(R_1 + R_2 + R_o)]} \qquad (10\text{-}109)$$

The value of R_T is significantly lower than R_o. In the right-hand form of Eq. (10-109), the numerator is $R_o \| (R_1 + R_2)$, which is smaller than R_o. This resistance is divided by a large (for $A_v \gg 1$) positive term, and, hence, $R_T \to 0$ as $A_v \to \infty$.

Example 10-9

An inverting Op-Amp stage is designed with $R_1 = 5$ kΩ, $R_2 = 10$ kΩ, and $R_L = 100\ \Omega$. The Op-Amp has $A_v = 5 \times 10^4$, $R_o = 500\ \Omega$, and $R_i \to \infty$. (*a*) Determine the voltage across R_L for an rms input signal of 1.5 V. (*b*) Repeat part *a*, considering that the Op-Amp is ideal.

Solution

(*a*) The Thévenin equivalent of the stage is given by Eqs. (10-105) and (10-109). Evaluation gives

$$A_v V_s = \frac{-5 \times 10^4 \times 10 + 0.5}{5(1 + 5 \times 10^4) + 10 + 0.5} \times 1.5 = -2.9998\ \text{V}$$

$$R_T = \frac{(5 + 10)0.5}{5(1 + 5 \times 10^4) + 10 + 0.5} = 0.027\ \Omega$$

The voltage across R_L, using the Thévenin equivalent, is

$$V_o = \frac{R_L}{R_L + R_o} A_V V_1 = \frac{100}{100 + 0.027} \times (-2.9998) = -2.999 \approx -3.0\ \text{V}$$

(*b*) For the ideal Op-Amp, the voltage across R_L is simply $A_V V_1$ or $V_o = -(R_2/R_1)\ V_1 = -\frac{10}{5} \times 1.5 = -3.0$ V.

The results indicate that for typical values of A_V and R_o, the difference between actual and ideal values is negligible.

The Practical Noninverting Stage   The practical noninverting Op-Amp stage (shown in Fig. 10-45) is analyzed in an identical manner as is the inverting stage. The value of A_V is given in Eq. (10-110) and that for R_T in Eq. (10-109), assuming $R_i \to \infty$:

FIGURE 10-45
The nonideal noninverting Op-Amp stage equivalent circuit.

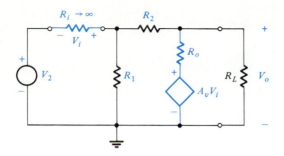

$$A_V = \frac{A_v(R_1 + R_2)}{R_1(1 + A_v) + R_2 + R_o}$$ (10-110)

Verification of these results is the object of Prob. 10-83. The conclusion that the noninverting stage closely approximates the ideal case when A_v is very large is easily discernible in Eqs. (10-110) and (10-109).

10-22 ELEMENTARY OP-AMP APPLICATIONS In this section, we introduce several basic Op-Amp circuits that are used extensively in amplifying systems and signal

Adder or Summing Amplifier The arrangement of Fig. 10-46 may be used to obtain an output which is a linear combination of a number of input signals. Since a virtual ground exists at the Op-Amp input, then

$$i = \frac{v_1}{R_1} + \frac{v_2}{R_2} + \cdots + \frac{v_n}{R_n}$$

and

$$v_o = -R'i = -\left(\frac{R'}{R_1} v_1 + \frac{R'}{R_2} v_2 + \cdots + \frac{R'}{R_n} v_n\right)$$ (10-111a)

If $R_1 = R_2 = \cdots = R_n$, then

$$v_o = -\frac{R'}{R_1} (v_1 + v_2 + \cdots + v_n)$$ (10-111b)

and the output is proportional to the sum of the inputs.

FIGURE 10-46
Inverting summing amplifier.

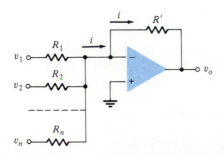

FIGURE 10-47
Noninverting summing
amplifier (adder).

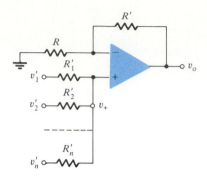

Many other methods may, of course, be used to combine signals. The present method has the advantage that it may be extended to a very large number of inputs requiring only one additional resistor for each additional input. The result depends, in the limiting case of large amplifier gain, only on the resistors involved, and because of the virtual ground, there is a minimum of interaction between input sources.

Noninverting Summing   An adder whose output is a linear combination of the inputs without a change of sign is obtained by using the noninverting amplifier. In Fig. 10-47 we show such a summer. From Eq. (10-101) the output is given by

$$v_o = \left(1 + \frac{R'}{R}\right)v_+ = \left(\frac{R + R'}{R}\right)v_+ \qquad (10\text{-}112)$$

where the voltage at the noninverting terminal v_+ is found by using superposition. For example, the contribution to v_+ due to v_2' is $v_2'R_{p2}'/(R_2' + R_{p2}')$, where R_{p2}' is the parallel combination of all the resistors tied to the noninverting node *with the exception of R_2'*; that is, $R_{p2}' = R_1' \| R_3' \| R_4' \cdots \| R_n'$.

For n equal resistors each of value R_2',

$$\frac{R_{p2}'}{R_2' + R_{p2}'} = \frac{R_2'/(n-1)}{R_2' + R_2'/(n-1)} = \frac{1}{n} \qquad (10\text{-}113)$$

and

$$v_+ = \frac{1}{n}(v_1' + v_2' + \cdots + v_n') \qquad (10\text{-}114)$$

The output is given by Eqs. (10-112) and (10-114).

It is possible to perform analog addition and subtraction simultaneously with a single Op-Amp by replacing the resistor R in Fig. 10-47 by the n input voltages and resistors in Fig. 10-46. Again, superposition is used to find the contribution to v_o from any of the input voltages. It should be emphasized that, when one of the voltages $v_1, v_2, \ldots, v_n$ is under consideration, then the positive input terminal is effectively grounded (if the bias current is negligible). Similarly when one of the voltages $v_1', v_2', \ldots, v_n'$ is under consideration, then R in Fig. 10-47 represents the parallel combination of $R_1, R_2, \ldots, R_n$.

Voltage-to-Current Converter (Transconductance Amplifier) Often it is desirable to convert a voltage signal to a proportional output current. This is required, for example, when we drive a deflection coil in a television tube. If the load impedance has neither side grounded (if it is floating), the simple circuit of Fig. 10-46 with R' replaced by the load impedance Z_L is an excellent *voltage-to-current converter*. For a single input $v_1 = v_s(t)$, the current in Z_L is

$$i_L = \frac{v_s(t)}{R_1} \tag{10-115}$$

Note that i is independent of the load Z_L, because of the virtual ground of the Op-Amp input. Since the same current flows through the signal source and the load, it is important that the signal source be capable of providing this load current. On the other hand, the amplifier of Fig. 10-48a requires very little current from the signal source due to the very large input resistance seen by the noninverting terminal.

If the load Z_L is grounded, the circuit of Fig. 10-48b can be used. In Prob. 10-88 we show that if $R_3/R_2 = R'/R_1$, then

$$i_L(t) = \frac{v_s(t)}{R_2} \tag{10-116}$$

Current-to-Voltage Converter (Transresistance Amplifer) Photocells and photomultiplier tubes give an output current which is independent of the load. The circuit in Fig. 10-49 shows an Op-Amp used as a current-to-voltage converter. Due to the virtual ground at the amplifier input, the current R_s is zero and i_s flows through the feedback resistor R'. Thus the output voltage v_o is $-i_s R'$. It must be pointed out that the lower limit on current measurement with this circuit is set by the bias current of the inverting input. It is common to parallel R' with a capacitance C' to reduce high-frequency noise and the possibility of oscillations. The current-to-voltage converter makes an excellent current-measuring instrument since it is an ammeter with zero voltage across the meter.

FIGURE 10-48
Voltage-to-current converter for (a) floating load and (b) grounded load.

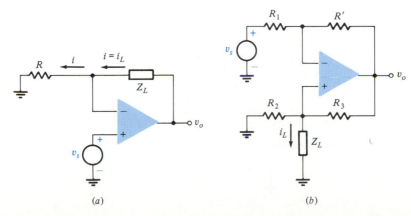

(a) (b)

FIGURE 10-49
Current-to-voltage con-
verter.

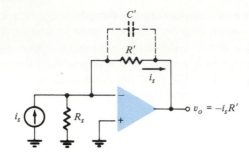

Integrators

If, in Fig. 10-42a, the resistance R_2 is replaced by a capacitor C, as displayed in Fig. 10-50, the circuit behaves as an *integrator*. In Fig. 10-50, $i_1 = v_s/R_1$ and $i_C = C(dv_o/dt)$; since no current enters the Op-Amp, $i_1 = -i_C$. Hence

$$\frac{v_s}{R_1} = -C\frac{dv_o}{dt}$$

or, on integrating and solving for v_o

$$v_o = -\frac{1}{R_1C}\int v_s\, dt \qquad (10\text{-}117)$$

The amplifier therefore provides an output voltage proportional to the integral of the input voltage.

If the input voltage is a constant, $v_s = V$, then the output will be a ramp, $v_o = -Vt/R_1C$. Such an integrator makes an excellent sweep circuit for a cathode-ray-tube oscilloscope, and is called a *Miller integrator,* or *Miller sweep.* The circuit in Fig. 10-50 is an ideal integrator. Practical integrators are treated in detail in Sec. 16-6.

REFERENCES

1 Ghausi, M. S.: "Electron Devices and Circuits: Discrete and Integrated," Holt, Rinehart and Winston, Inc., New York, 1985.

2 Sedra, A. S., and K. C. Smith: "Microelectronic Circuits," Holt, Rinehart and Winston, Inc., New York, 1981.

3 Schilling, D., and C. Belove: "Electronic Circuits—Discrete and Integrated," McGraw-Hill Book Company, New York, 1979.

FIGURE 10-50
Op-Amp integrator cir-
cuit.

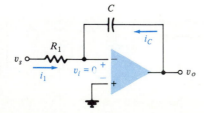

4 Gray, P. R., and R. G. Meyer: "Analysis and Design of Analog Integrated Circuits," 2d ed., John Wiley and Sons, New York, 1984.

5 Grebene, A. B.: "Bipolar and MOS Analog Integrated Circuit Design," John Wiley and Sons, New York, 1984.

6 Colclaser, R. A., D. N. Neaman, and C. F. Hawkins: "Electronic Circuit Analysis," John Wiley & Sons, New York, 1984.

7 Soclof, S.: "Analog Integrated Circuits," Prentice-Hall, Englewood Cliffs, N.J., 1985.

REVIEW QUESTIONS

10-1 Draw a fixed bias circuit and explain why this circuit is unsatisfactory if the transistor is replaced by another of the same type.

10-2 What is meant by dynamic range?

10-3 Explain why the quiescent collector current must be essentially independent of variations in β_F to achieve bias stabilization.

10-4 (a) Sketch the circuit diagram of a simple current mirror.
(b) Briefly explain how this circuit acts as a current source.

10-5 What effect does the Early voltage V_A have on the output current of a mirror?

10-6 (a) Draw the circuit of a Widlar current source.
(b) List two advantages of the Widlar circuit.

10-7 Repeat Rev. 10-6 for the Wilson current source.

10-8 (a) Draw the four-resistor bias network.
(b) Briefly explain how this configuration acts to maintain a constant I_{CQ} as β_F varies.

10-9 Explain the function of (a) a coupling capacitor and (b) a bypass capacitor.

10-10 In a discrete-component circuit, list three causes of variations in collector current.

10-11 A transistor is excited by a large sinusoidal base current whose magnitude exceeds the quiescent value I_{BQ} for $0 \le \omega t < \pi$ and is less than I_{BQ} for $\pi \le \omega t < 2\pi$. Is the magnitude of the collector current variation about I_{CQ} greater at $\omega t = \pi/2$ or $\omega t = 3\pi/2$? Explain your answer with the aid of a graphical construction.

10-12 Sketch the circuit diagram of a MOSFET current source and briefly explain its operation.

10-13 Explain how the four-resistor bias network tends to minimize variations in I_{DQ} caused by unit-to-unit changes in the FET.

10-14 Explain the differences between i_D, i_d, I_D, and I_d.

10-15 (a) For a capacitively coupled load, is the dc load larger or smaller than the ac load?
(b) Show the effect of the capacitively coupled load in terms of the static and dynamic load lines.

10-16 Relate (a) g_m, (b) r_π, and (c) r_o to the quiescent collector current I_{CQ}.

10-17 Evaluate the quantities in Rev. 10-16 for a transistor operating at room temperature with $I_{CQ} = 0.5$ mA and having $V_A = 120$ V.

10-18 Draw the low-frequency equivalent circuit of a common-emitter stage.

10-19 Repeat Rev. 10-18 for (a) a common-collector stage and (b) a common-base stage.

10-20 Which of the configurations (CE, CB, or CC) has the (a) highest R_i, (b) lowest R_i, (c) highest R_o, (d) lowest R_0, (e) lowest A_V, and (f) lowest A_I.

10-21 (a) Compare A_V, A_I, R_i, and R_o for a common-emitter stage with emitter resistance R_E with a simple common-emitter stage.
　　　 (b) What advantage is derived in using such a stage?

10-22 (a) Draw the cascade of a common-emitter stage and a common-collector stage.
　　　 (b) What are the overall voltage gain, input resistance, and output resistance of the cascade?

10-23 It is desired to have a high-gain amplifier with high input resistance and high output resistance. If a three-stage cascade is used, what configuration should be used for each stage? Explain.

10-24 (a) What is the effective load resistance of an interior stage of a cascaded amplifier?
　　　 (b) What is the effective source resistance of such a stage?

10-25 (a) Draw the circuit of an emitter follower using a Darlington connected pair.
　　　 (b) What is the advantage in using a Darlington pair?

10-26 (a) Sketch the circuit diagram of a cascode amplifier.
　　　 (b) How do A_V and A_I of this circuit compare to corresponding values for a common-emitter stage?

10-27 Draw the circuit diagram of an emitter-coupled (differential) pair.

10-28 (a) Define differential mode.
　　　 (b) Define common mode.
　　　 (c) Draw the half-circuit equivalent for parts a and b.

10-29 Define common-mode rejection ratio (CMRR).

10-30 Write an equation for the output voltage of a differential amplifier in terms of the CMRR, differential gain A_{DM}, and the common-mode and differential-mode input signals.

10-31 Over what range of difference voltage V_d does the emitter-coupled pair behave linearly?

10-32 Draw the low-frequency model of (a) the common-source stage and (b) the common-drain stage.

10-33 A FET amplifier is to be constructed having high gain and low output resistance. If four stages are used:
　　　 (a) What is the configuration of each stage?
　　　 (b) What is the overall gain?

10-34 List five properties of an ideal Op-Amp.

10-35 (*a*) Draw the circuit diagram of a noninverting Op-Amp stage and indicate its equivalent circuit.

(*b*) What is the voltage gain of this stage?

10-36 Repeat Rev. 10-35 for a noninverting stage.

10-37 (*a*) Show the circuit diagram of an integrator.

(*b*) Derive an expression which shows that the output is propertional to the integral of the input.

Chapter 11
FREQUENCY RESPONSE OF AMPLIFIERS

The signals utilized in many electronic systems require amplification with a minimum of distortion. Under these circumstances, the active devices involved must operate linearly; thus, small-signal conditions must apply. The first step in the analysis of such circuits is the use of a linear model to replace the actual circuit. Thereafter it is a matter of circuit analysis to determine the transmission characteristics of the linear network.

In Chap. 10, we focused on the low-frequency behavior of amplifier stages. To do so we considered that both the internal capacitances of the transistors and, where employed, the external coupling and bypass capacitors had negligible effect on performance. However, amplifiers are required to operate over a broad frequency range. The low end of this range may be dc (direct-coupled stages) or a few hertz, and the high end may extend to several tens of megahertz. The study of such wideband amplifiers was given impetus by the need to amplify the pulses that occur in communications systems such as television and radar. In the analysis of amplifiers over so wide a frequency band, the capacitances, heretofore neglected, must be included. Because capacitive reactance varies with frequency, the transmission characteristics of the linear models are frequency-dependent. Thus amplifier gain depends on the frequency of the input signal which can result in output signals that are frequency distorted.

In this chapter we consider how a low-level input signal that contains many frequency components from zero (dc) to a few megahertz can be amplified with a minimum of distortion.

In addressing this question, we first investigate the response of single-stage BJT and FET amplifiers. Multistage amplifiers are treated by relating the overall response to the frequency response of the component stages. Approximate methods for evaluating the frequency response, useful in amplifier design, are developed.

Integrated-circuit amplifiers, which are invariably direct-coupled, are limited only at high frequencies because of internal transistor capacitances (C_π and C_μ

in a BJT). Discrete-component stages are also limited at low frequencies because of the coupling and bypass capacitors used. Therefore, we treat single- and multistage behavior at high frequencies first and then discuss low-frequency limitations.

11-1 FREQUENCY-RESPONSE CHARACTERISTICS

The application of a low-level sinusoidal signal to the input of an amplifier results in a sinusoidal output waveform. For a nonsinusoidal excitation, however, the output waveshape is not an exact replica of the input signal because the input components at different frequencies are amplified differently. When the effects of internal device capacitances or when the external circuit (coupling capacitors or load impedances) has a reactive component, the gain $\mathbf{A} = A \angle \theta$ is a complex number. Both the magnitude A and the phase angle θ depend on the excitation frequency. The notation used in the previous sentence is used throughout the remainder of the text. Complex quantities are written in boldface type; the constituent parts of complex numbers such as magnitude and phase are not. The *frequency-response characteristic* of an amplifier is the plots of gain and phase versus frequency. Invariably, the Bode diagram is used to display the frequency response. The asymptotic Bode diagram is a convenient approximation of this characteristic.

Fidelity Considerations A criterion which may be used to compare one amplifier with another with respect to fidelity of reproduction of the input signal is suggested by the following considerations. Any arbitrary waveform of engineering importance may be resolved into a Fourier spectrum. If the waveform is periodic, the spectrum will consist of a series of sines and cosines whose frequencies are all integral multiples of a fundamental frequency. The fundamental frequency is the reciprocal of the time which must elapse before the waveform repeats itself. If the waveform is not periodic, the fundamental period extends in a sense from a time $-\infty$ to a time $+\infty$. The fundamental frequency is then infinitesimally small; the frequencies of successive terms in the spectrum differ by an infinitesimal amount rather than by a finite amount, and the Fourier series becomes instead a Fourier integral. In either case the spectrum includes terms whose frequencies extend, in the general case, from zero frequency to infinity.

Consider a sinusoidal signal of angular frequency ω represented by $V_m \sin(\omega t + \phi)$. If the voltage gain of the amplifier has a magnitude A and the signal suffers a phase change (lag angle)θ, the output will be

$$AV_m \sin(\omega t + \phi - \theta) = AV_m \sin\left[\omega\left(t - \frac{\theta}{\omega}\right) + \phi\right]$$

Therefore, *if the amplification A is independent of frequency and the phase shift θ is proportional to frequency (or is zero), the amplifier will preserve the form of the input signal, although the signal will be shifted in time (delayed) by an amount θ/ω.*

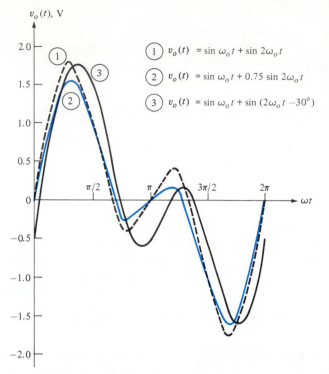

$v_o(t)$, V

① $v_o(t) = \sin \omega_o t + \sin 2\omega_o t$

② $v_o(t) = \sin \omega_o t + 0.75 \sin 2\omega_o t$

③ $v_o(t) = \sin \omega_o t + \sin (2\omega_o t - 30°)$

FIGURE 11-1
Illustrating amplitude and phase distortion. The amplitude of waveform 2 differs from that of waveform 1; also, the phase characteristics of waveforms 1 and 3 are different.

The waveforms in Fig. 11-1 illustrate the distortion that results when the amplification of different frequency components is not uniform. Each of the three curves in Fig. 11-1 is the output waveform of an amplifier excited by a voltage $0.1 \sin \omega_o t + 0.1 \sin 2\omega_o t$. The dashed curve corresponds to an amplifier having $A = 10$ and $\theta = 0°$ at both frequencies $f_o = \omega_o/2\pi$ and $2f_o$. The output voltage $v_o = 1.0(\sin \omega_o t + \sin 2\omega_o t)$ and is a replica of the input signal. The waveformshown in blue occurs when $A = 10$ at f_o and $A = 7.5$ at $2f_o$, with $\theta = 0°$ atboth frequencies. The output is $v_o = 1.0 \sin \omega_o t + 0.75 \sin 2\omega_o t$. In the third waveform, $v_o = 1.0 \sin \omega_o t + 1.0 \sin (2\omega_o t - 30°)$ and is the result of the amplifier's introduction of a $-30°$ phase shift at $2f_o$. The distortion due to changes in both A and θ with frequency are readily apparent. The situation where both A and θ vary with frequency, as is the case for most practical amplifiers, is the subject of Prob. 11-1.

This discussion suggests that the extent to which an amplifier's amplitude response is not uniform, and its time delay is not constant with frequency, may serve as a measure of the lack of fidelity to be anticipated in it. In principle, it is not necessary to specify both amplitude and delay response since, for most practical circuits, the two are related and, one having been specified, the other is uniquely determined. However, in particular cases it may well be that either the time-delay or amplitude response is the more sensitive indicator of frequency distortion.

High-Frequency Response Consider the circuit in Fig. 11-2a. The capacitor C_M, as is demonstrated in subsequent sections, is the effect of the internal capacitances of the amplifying device (C_π and C_μ for the BJT; C_{gs} and C_{gd} for the FET). The complex-frequency[1] (s-plane) representation of the circuit in Fig. 11-2a is displayed in Fig. 11-2b.

In Fig. 11-2b, $Z = R_i \parallel 1/sC_M = R_i/(1 + sR_iC_M)$ and, from the voltage-divider relation,

$$V_i = \frac{Z}{R_s + Z} V_s = \frac{R_i V_s}{R_s + R_i + sC_M R_i R_s} \tag{11-1}$$

As $V_o = -g_m R_L V_1$, substitution of Eq. (11-1) gives

$$\frac{V_o}{V_s} = A_{VH}(s) = \frac{-g_m R_i R_L}{R_s + R_i + sC_m R_i R_s} \tag{11-2}$$

Division by $R_i + R_s$ yields

$$A_{AH}(s) = \frac{-g_m R_L R_i/(R_s + R_i)}{1 + sC_M R_s R_i/(R_s + R_i)} = \frac{|A_{VO}|}{1 + s/\omega_H} \tag{11-3}$$

The numerator of Eq. (11-3), A_{VO}, is the gain of the circuit at $s = j\omega = 0$ (dc) for which the reactance of C_M is infinite (C_M is an open circuit). The angular frequency ω_H is the reciprocal of the time constant of the input circuit as $R_s R_i/(R_s + R_i)$ is the equivalent resistance seen by C_M.

For $s = j\omega$, the magnitude and phase of $A_{VH}(j\omega) = A_{VH}\angle\theta_H$ are

$$A_{VH} = \frac{|A_{VO}|}{\sqrt{1 + \left(\dfrac{f}{f_H}\right)^2}} \qquad \theta_H = -\tan^{-1}\frac{f}{f_H} \tag{11-4}$$

Observe in Eq. (11-4) that increase in the excitation frequency f causes A_{VH} to decrease; ultimately $A_{VH} \to 0$ as $f \to \infty$. Similarly, θ_H displays an increasing *phase lag* as f increases. At $f = f_H$, $A_{VH} = A_{VO}/\sqrt{2} = 0.707A_{VO}$. Expressed in decibels,[2] this corresponds to a reduction in A_{VH} of 3 db from the value A_{VO},

FIGURE 11-2

(a) Low-pass circuit.
(b) The frequency-domain (s-plane) representation of (a).

---

[1] A discussion of the complex frequency s is given in App. C-2 and App. C-5.
[2] A voltage gain A_V expressed in decibels is 20 log A_V.

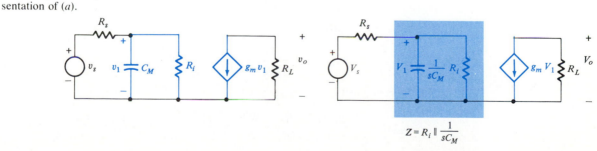

(a)   (b)

FIGURE 11-3
Normalized Bode diagram of the transfer function of the circuit in Fig. 11-2b.

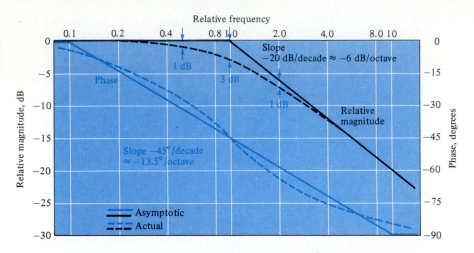

and f_H is often referred to as the *high*, or *upper 3-dB frequency*. As power is proportional to the square of the voltage, $A_{VH} = A_{VO}/\sqrt{2}$ corresponds to a power level at $f = f_H$ equal to one-half the power at $f = 0$. Thus f_H is also referred to as the *upper (high) half-power frequency*.

The frequency-response characteristic given in Eq. (11-5) is displayed in the Bode diagram in Fig. 11-3. The dashed curves indicate actual values, and the solid curves are for the asymptotic Bode diagram. Note that the frequency axis in Fig. 11-3 is normalized to f/f_H and that the gain is normalized to A_{VH}/A_{VO}. This circuit behaves as a *low-pass system*, for, as seen in the asymptotic Bode diagram, frequencies below f_H are transmitted with minimum attenuation and frequencies above f_H are attenuated.

For typical parameter values encountered ($C_M = 40$ pF, $R_i = R_s = 1$ kΩ) in practical amplifier stages, $f_H \approx 8$ MHz. Obviously, increasing (decreasing) R_i and R_s (or C_M) results in decreasing (increasing) f_H.

Low-Frequency Response   The circuit in Fig. 11-4 represents an amplifier stage in which $1/sC_C$ represents the impedance of the external coupling capacitor (C_B in Fig. 10-12 or C_G for an FET stage). For Fig. 11-4

$$V_1 = \frac{R_i V_s}{R_s + R_i + 1/sC_C}$$

and, from $V_o = -g_m R_L V_1$

$$A_{VL}(s) = \frac{V_o}{V_1} = \frac{-g_m R_L R_i}{R_i + R_s + 1/sC_C} \tag{11-5}$$

Equation (11-5) can be recast as

$$A_{VL}(s) = \frac{-g_m R_L R_i/(R_s + R_i)}{1 + 1/sC_C(R_s + R_i)} = \frac{|A_{vo}|}{1 + \omega_L/s} = |A_{vo}| \frac{s/\omega_L}{1 + s/\omega_L} \tag{11-6}$$

FIGURE 11-4
Frequency-domain representation of high-pass circuit.

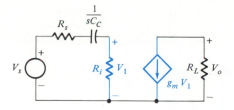

The angular frequency ω_L is the reciprocal of the time constant of the circuit $C_C(R_s + R_i)$ (this is the equivalent resistance seen by C_C), and, at $s = j\omega_L$, $1/\omega_L C_C = R_s + R_i$. Note that $A_{VL}(s) \rightarrow A_{Vo}$ for $s = j\omega \rightarrow \infty$. As $\omega \rightarrow \infty$, the reactance of $C_C \rightarrow 0$, the condition for which the effect of coupling capacitors is negligible. For $s = j\omega$, the magnitude and phase of $\mathbf{A}_{VL}(j\omega)$ are

$$A_{VL} = \frac{|A_{Vo}|}{\sqrt{1 + (f_L/f)^2}} \qquad \theta_L = \tan^{-1}\frac{f_L}{f} \qquad (11\text{-}7)$$

From Eq. (11-7) we observe that $A_{VL} \rightarrow 0$ as $f \rightarrow 0$ and indicates the low-frequency attenuation. This type of response is that of a *high-pass system* as seen in the Bode diagram in Fig. 11-5. The frequency f_L is identified as the *lower 3-dB* or *lower half-power frequency*. For typical values encountered ($C_C = 1\ \mu\text{F}$, $R_s = R_i = 1\ \text{k}\Omega$), $\omega_L = 500\ \text{rad/s}$ and $f_L \approx 80\ \text{Hz}$. It is clear that increasing (decreasing) C_C results in decreasing (increasing) f_L.

Total Response The circuit in Fig. 11-6 contains both the internal device capacitance C_M and the coupling capacitor C_C. It is evident from the preceding discussion that the response of this circuit is limited at both low (C_C) and high (C_M) frequencies. However, f_L and f_H are widely separated, as indicated by the typical values shown. Consequently, the frequencies at which C_C and C_M influence amplifier response are quite disparate. Table 11-1 lists the reactance of C_M and C_C at

FIGURE 11-5
Normalized Bode diagram of the transfer function of the circuit in Fig. 11-4.

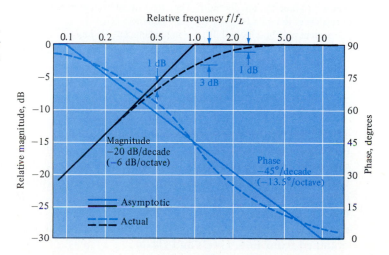

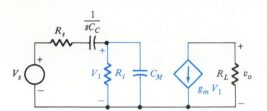

TABLE 11-1 Reactance of C_C and C_M at Various Frequencies

Angular frequency (rad/s)	500	5×10^3	5×10^6	50×10^6
Reactance of $C_C = 1 \ \mu\text{F} \ (\Omega)$	2K	200	0.20	0.02
Reactance of $C_M = 40 \ \text{pF} \ (\Omega)$	50M	5M	5K	500

various frequencies based on the numerical values given previously. The resistance in series with C_C is $R_s + R_i = 2 \ \text{k}\Omega$. As seen in Table 11-1, for $\omega \geq 10\omega_L = 5 \times 10^3$ rad/s, the reactance of C_C is negligible compared with $2 \ \text{k}\Omega$. (Recall $2000 + j200 \approx 2000$.) Thus, for $\omega \geq 5 \times 10^3$ rad/s, the effect of C_C can be neglected. Similarly, for $\omega \leq \omega_H/10 = 5$ Mrad/s, the reactance of C_M is much larger than $R_s \| R_i = 500 \ \Omega$. The effect of C_M can be considered negligible for $\omega \leq 5$ Mrad/s. The import of the five previous sentences is that we can conveniently divide the total response into three ranges: *midband, low,* and *high frequencies:*

Midband	High frequencies	Low frequencies
Neither C_C nor C_M affects the response; the gain is constant and no phase shift is introduced (this situation is considered throughout Chap. 10).	The effect of internal device capacitances (C_M) is important, but C_C has negligible reactance.	C_C affects the response and the reactance of C_M is so large as to render negligible effect.

To obtain the total response, the amplifier gain is determined in each of the three frequency ranges and these responses are combined. Thus the midband range is used to determine A_{VO}, the high-frequency equivalent circuit is used to obtain f_H, and f_L is evaluated from the low-frequency model for which C_C is included. On the basis of the parameter values, $R_s = R_i = 1 \ \text{k}\Omega$, $C_C = 1 \ \mu\text{F}$, $C_M = 100 \ \text{pF}$, $R_L = 2 \ \text{k}\Omega$, and $g_m = 0.1 \ \Omega$, the composite response of the circuit in Fig. 11-6 is displayed in Fig. 11-7. The values of A_{VO}, f_L, and f_H are obtained from Eqs. (11-3) and (11-6).

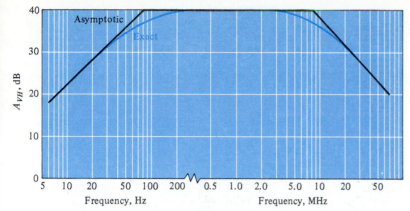

FIGURE 11-7
Bode diagram for system represented by Fig. 11-2*b* at high frequencies and by Fig. 11-4 at low frequencies. Note the break in the frequency axis.

Bandwidth

The frequency range from f_L to f_H is called the *bandwidth* (BW) of the amplifier stage. We may anticipate in a general way that a signal, all of whose frequency components of appreciable amplitude lie well within the range f_L to f_H, will pass through the stage without excessive distortion. This criterion must be applied, however, with caution. In the asymptotic Bode diagram in Fig. 11-7 we observe that between f_L and f_H, A_V is constant. However, the phase characteristic indicates a phase lead at f_L and a phase lag at f_H. In Fig. 11-1 we indicated that a phase shift introduces distortion even if the magnitude of the gain remains constant.

Many amplifiers have $f_H \gg f_L$; hence the bandwidth [Eq. (11-8)] is approximately f_H. In ICs which are invariably direct-coupled (no coupling capacitors), the low-frequency response extends to dc ($\omega = 0$) and the bandwidth is simply f_H. Thus only high-frequency limitations to amplifier response exist.

11-2 STEP RESPONSE OF AN AMPLIFIER

An alternative criterion of amplifier fidelity is the response of the amplifier to a particular input waveform. Of all possible available waveforms, the most generally useful is the step voltage. In terms of a circuit's response to a step, the response to an arbitrary waveform may be written in the form of the superposition (convolution) integral. Another feature which recommends the step voltage is the fact that this waveform is one which permits small distortions to stand out clearly. Additionally, from an experimental viewpoint, we note that excellent pulse (a short step) and square-wave (a repeated step) generators are available commercially.

As long as an amplifier can be represented by a single pole [Eq. (11-3)], the correlation between its frequency response and the output waveshape for a step input is that given in Fig. 11-8. Quite generally, even for more complicated amplifier circuits, there continues to be an intimate relationship between the distortion of the leading edge of a step and the high-frequency response. Similarly, there is a close relationship between the low-frequency response and

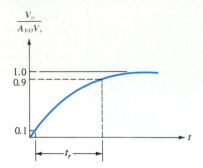

FIGURE 11-8
Normalized step response of the circuit in Fig. 11-2a.

the distortion of the flat portion of the step. We should, of course, expect such relationships, since the high-frequency response measures essentially the ability of the amplifier to respond faithfully to rapid variations in signal, whereas the low-frequency response measures the fidelity of the amplifier for slowly varying signals. An important feature of a step is that it is a combination of the most abrupt voltage change possible and the slowest possible voltage variation.

Rise Time

The response of the low-pass circuit in Fig. 11-2 to a step input of amplitude V_s is exponential with a time constant $1/\omega_H$. Since the capacitor voltage cannot change instantaneously, the output starts from zero and approaches its steady-state value $A_{VO}V_s$. The output is given by

$$v_o = A_{VO}V_s\,(1 - \epsilon^{-\omega_o t}) \tag{11-8}$$

The time required for v_o to reach one-tenth of its final value is readily found to be $0.1/\omega_H$, and the time needed to reach nine-tenths its final value is $2.3/\omega_H$. Note that the time constant is C_M multiplied by the equivalent resistance $R_s \parallel R_i$. The difference between these two values is called the *rise time* t_r of the circuit and is shown in Fig. 11-8. The time t_r is an indication of how fast the amplifier can respond to a discontinuity in the input voltage. We have

$$t_r = \frac{2.2}{\omega_H} = \frac{2.2}{2\pi f_H} = \frac{0.35}{f_H} \tag{11-9}$$

Note that the rise time is inversely proportional to the upper 3-dB frequency. For an amplifier with 1-MHz bandpass, $t_r = 0.35\ \mu s$.

The relationship between t_r and f_H in Eq. (11-9) is exact (to three significant figures) for a single-pole circuit. However, Eq. (11-9) is a good approximation (within 3 or 4 percent) for multiple-pole low-pass circuits.

Consider a pulse of width T_p. What must be the upper 3-dB frequency f_H of an amplifier if the signal is to be amplified without excessive distortion? A reasonable answer to this question is: *Choose f_H equal to or greater than the reciprocal of the pulse width T_p.* For $f_H = 1/T_p$, the output waveform (color) in Fig. 11-9 is the response to the input pulse indicated.

Normalized
amplitude

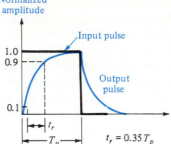

FIGURE 11-9
Normalized response of a low-pass circuit to a pulse.

Tilt or Sag

If a step of amplitude V_s is impressed on the high-pass circuit in Fig. 11-4, the output is

$$v_o = A_{VO}V_s\epsilon^{-t/(R_s + R_i)C_C} = V_o\epsilon^{-t/(R_s + R_i)C_C} \tag{11-10}$$

For times t which are small compared with the time constant $(R_s + R_i)C_C$, the response is given by

$$v_o \approx V_o\left[1 - \frac{t}{(R_s + R_i)C_C}\right] = V_o(1 - \omega_L t) \tag{11-11}$$

From Fig. 11-10 we see that the output is tilted, and the percent tilt, or sag, in time t_1 is given by

$$P \equiv \frac{V_o - V'_o}{V_o} \times 100\% = \frac{t_1}{(R_s + R_i)C_C} \times 100\% \tag{11-12}$$

It is found that this same expression is valid for the tilt of each half-cycle of a symmetrical square wave of peak-to-peak value V_o and period T, provided we set $t_1 = T/2$. If $f = 1/T$ is the frequency of the square wave, then we may express P in the form

$$P = \frac{T}{2(R_s + R_i)C_c} \times 100 = \frac{\omega_L}{2f(R_s + R_i)} \times 100 = \frac{\pi f_L}{f} \times 100\% \tag{11-13}$$

Note that the tilt is directly proportional to the lower 3-dB frequency. If we wish to pass a 50-Hz square wave with less than 10 percent sag, f_L must not exceed 1.6 Hz.

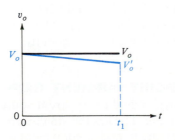

FIGURE 11-10
Illustrating tilt (sag) of a square wave.

Square-Wave Testing An important experimental procedure (called *square-wave testing*) is to observe with an oscilloscope the output of an amplifier excited by a square-wave generator. It is possible to improve the response of an amplifier by adding to it certain circuit elements, which then must be adjusted with precision. It is a great convenience to be able to adjust these elements and to see simultaneously the effect of such an adjustment on the amplifier ouput waveform. The alternative is to take data, after each successive adjustment, from which to plot the amplitude and phase responses. Aside from the extra time consumed in this latter procedure, we have the problem that it is usually not obvious which of the attainable amplitude and phase responses corresponds to optimum fidelity. On the other hand, the step response gives immediately useful information.

It is possible, by judicious selection of two square-wave frequencies, to examine individually the high-frequency and low-frequency distortion. For example, consider an amplifier which has a high-frequency time constant of 0.1 μs and a low-frequency time constant of 100 ms. A square wave of half period equal to several tenths of a microsecond, on an appropriately fast oscilloscope sweep, will display the rounding of the leading edge of the waveform and will not display the tilt. At the other extreme, a square wave of half period approximately 10 ms on an appropriately slow sweep will display the tilt, and not the distortion of the leading edge.

It should *not* be inferred from the above comparison between steady-state and transient response that the phase and amplitude responses are of no importance at all in the study of amplifiers. The frequency characteristics are useful for the following reasons. In the first place, much more is known generally about the analysis and synthesis of circuits in the frequency domain than in the time domain, and for this reason amplifier design is often done on a frequency-response basis. Second, it is often possible to arrive at least at a qualitative understanding of the properties of a circuit from a study of the steady-state response in circumstances where transient calculations are extremely cumbersome. Third, compensating an amplifier against unwanted oscillations (Chap. 13) is accomplished in the frequency domain. Finally, it happens occasionally that an amplifier is required whose characteristics are specified on a frequency basis, the principal emphasis being to amplify sinusoidal signals.

The high-frequency response of BJT and FET amplifier stages is treated in the next several sections. We concentrate on high-frequency behavior first because this is primary limitation in IC amplifiers. The results derived are applicable also to discrete-component stages. The low-frequency response characteristics of amplifiers is discussed subsequently.

11-3 THE COMMON-EMITTER SHORT-CIRCUIT CURRENT GAIN Consider a single-stage common-emitter amplifier excited by a current source I_b, and with $R_C = 0$ (short circuit). The circuit in Fig. 11-11 is the small-signal model of this stage with the transistor replaced by the hybrid-π equivalent of Fig. 3-32. Note that

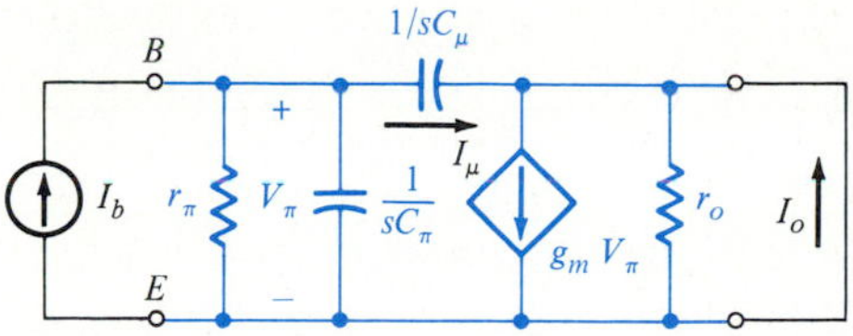

FIGURE 11-11
Circuit used to obtain the frequency-response characteristic for the common-emitter short-circuit current gain $\beta(s)$. The transistor is represented by its high-frequency hybrid-π model.

$r_b = 0$ is assumed in Fig. 11-11. The output current is $I_o = I_c$, so that the current gain of the stage $A_I = I_o/I_b$ is β of the transistor. It is evident in Fig. 11-11 that β varies with frequency because of the device capacitances C_π and C_μ. In Fig. 11-11, KCL requires

$$I_b = V_\pi\left(\frac{1}{r_\pi} + sC_\pi\right) + I_\mu = V_\pi\left[\frac{1}{r_\pi} + s(C_\pi + C_\mu)\right] \quad (11\text{-}14)$$

and

$$I_o = g_m V_\pi - I_\mu = V_\pi(g_m - sC_\mu) \quad (11\text{-}15)$$

Combination of Eqs. (11-14) and (11-15) and formation of I_o/I_b gives, after rearrangement of terms

$$\frac{I_o}{I_b} = A_I = \beta(s) = \frac{\beta_o(1 - sC_\mu/g_m)}{1 + sr_\pi(C_\pi + C_\mu)} = \frac{\beta_o(1 - s/\omega_z)}{1 + s/\omega_\beta} \quad (11\text{-}16)$$

In Eq. (11-16) we have used $g_m r_\pi = \beta_o$, and ω_z and ω_β can be readily identified. The following example, using typical parameter values for an IC transistor, illustrates the variation of β with frequency.

Example 11-1

(a) Sketch the magnitude of β as a function of frequency. Use the asymptotic Bode diagram. (b) Determine the (approximate) frequency f_T for which $|\beta(j\omega_T)| = 1$. The transistor parameters are $g_m = 0.05$ ℧, $r_\pi = 2$ kΩ, $C_\pi = 19.5$ pF, and $C_\mu = 0.5$ pF.

Solution

(a) In Eq. (11-16)

$$\beta_o = g_m r_\pi = 0.05 \times 2000 = 100 \qquad \omega_z = \frac{g_m}{C_\mu} = \frac{0.05}{5 \times 10^{-13}} = 10^{11} \text{ rad/s}$$

$$\omega_\beta = \frac{1}{r_\pi(C_\pi + C_\mu)} = \frac{1}{2 \times 10^3(19.5 + 0.5) \times 10^{-12}} = 2.5 \times 10^7 \text{ rad/s}$$

Thus,

$$\beta(s) = \frac{100(1 - s/10^{11})}{1 + s/(2.5 \times 10^7)}$$

for which the Bode diagram is displayed in Fig. 11-12. (b) From Fig. 11-12, we observe that the 0-dB crossing, corresponding to $\beta = 1$, occurs at $\omega_T = 2.5 \times 10^9$ rad/s or $f_T = \omega/2\pi = 395$ MHz.

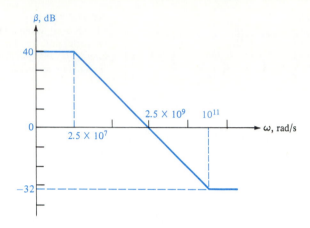

FIGURE 11-12
Asymptotic Bode diagram for $|\beta(j\omega)|$.

The Parameter f_T   In Example 11-1 the use of the asymptotic characteristic gives a highly accurate result for the unity gain frequency because of the wide separation of ω_β and ω_z. To build an amplifier with current gain greater than unity, it is evident that operating frequencies must be less than f_T. Consequently, for $\omega \leq \omega_T$, Eq. (11-16) can be approximated by the single-pole function

$$\beta(s) = \frac{\beta_o}{1 + s/\omega_\beta} \tag{11-17}$$

The approximation in Eq. (11-17) is equivalent to stating that the current I_μ in C_μ is a negligible component of I_o ($I_\mu \ll g_m V_\pi$). The asymptotic Bode diagram for this function is identical to that in Fig. 11-12 for the frequency range $\omega < \omega_z$.

To determine f_T, *the frequency at which the common-emitter short-circuit current gain has unit magnitude*, Eq. (11-17) is used. Hence

$$|\beta(j\omega_T)| = 1 = \frac{\beta_o}{\sqrt{1 + (\omega_T/\omega_\beta)^2}}$$

which, for $\beta_o^2 \gg 1$, gives

$$\omega_T = \beta_o \omega_\beta \qquad \text{or} \qquad f_T = \beta_o f_\beta \tag{11-18}$$

Substitution of ω_β in Eq. (11-16) into Eq. (11-18) yields

$$f_T = \frac{\beta_o}{2\pi\, r_\pi(C_\pi + C_\mu)} = \frac{g_m}{2\pi(C_\pi + C_\mu)} \tag{11-19}$$

Note that for $C_\pi \gg C_\mu$, $f_T \approx g_m/2\pi C_\pi$. The parameter f_T, as with other BJT parameters, depends on the operating conditions of the device. Typically, f_T varies with quiescent collector current as displayed in Fig. 11-13.

Since $f_T \approx \beta_o f_\beta$, this parameter may be given a second interpretation. It represents the *short-circuit current gain-bandwidth product*; that is, for the *CE* configuration with the output shorted, f_T is the product of the low-frequency

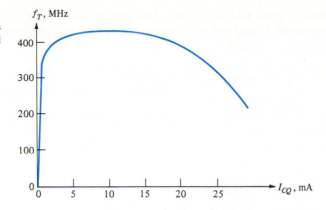

FIGURE 11-13
Variation of f_T with bias current I_{CQ} for a typical IC transistor.

current gain and the upper 3-dB frequency. It is to be noted that there is a sense in which gain may be sacrificed for bandwidth, and vice versa. Thus, if two transistors are available with equal f_T, the transistor with lower β_o will have a correspondingly larger bandwidth.

In practice, f_T is that it is used to determine the value of the capacitance C_π from measurement. The transconductance g_m is determined from the bias current I_{CQ} [Eq. (10-34)], f_T is determined from measurement (Prob. 11-8) or given in manufacturers' specifications, and C_μ is obtained by independent measurement. Manufacturers usually specify the common-base output capacitance $C_{ob} \simeq C_\mu$.

The frequency f_T also represents an upper bound on the frequency for which the hybrid-π model of the BJT is valid.[1] For excitation frequencies beyond f_T, the hybrid-π equivalent circuit does not accurately describe observed behavior. As transistors are used infrequently at $f > f_T$ (except in microwave circuits), a discussion of the models used is beyond the scope of this book. Note that the zero in Example 11-1 occurs at a frequency greater than f_T, so that the accuracy of the frequency of the zero ($z_1/2\pi$) is questionable.

The result in Eq. (11-18) is not peculiar to the evaluation of the current gain of the BJT. Many electronic circuits are often represented by single-pole functions over their useful range of operating frequencies. *For any single-pole system with "high gain," the unity gain frequency is the product of the zero-frequency (midband) gain and the 3-dB frequency.*

11-4 THE GENERALIZED GAIN FUNCTION Before we proceed to obtain the frequency-response characteristic $A_H(s)$ of both single- and multistage amplifiers, let us make a few general observations about the form of $A(s)$.

The high-frequency response of the amplifier in Fig. 11-2 is determined by a single time constant $C_M R_s R_i/(R_s + R_i)$. In reality, a multistage amplifier

---

[1] Often, $f_T/2$ is the limit used in the literature.

contains at least two capacitors and perhaps a third if the stage drives a capacitive load. Under these circumstances, the high-frequency transfer function is given by an equation of the form

$$A_H(s) = \frac{A_o(1 + s/z_1)(1 + s/z_2) \cdots (1 + s/z_m)}{(1 + s/p_1)(1 + s/p_2) \cdots (1 + s/p_n)} \tag{11-20}$$

In Eq. (11-20), A_o is the value of $A_H(s)$ evaluated at $s = 0$ and corresponds to the dc or midband gain. The values of s for which $A_H(s) \to \infty$ are called *poles* of the transfer function, where values of s which make $A_H(s) = 0$ are referred to as *zeros* of the transfer function. Assuming $n > m$, the transfer function in Eq. (11-20) has n poles at $-p_1, -p_2, \ldots, -p_n$, and m finite zeros $-z_1$, $-z_2, \ldots, -z_m$. As s becomes very large, $A_H(s) \to s^m/s^n = 1/s^{n-m}$, and as $s \to \infty$, $A_H(s) \to 0$. Thus $A_H(s)$ is said to have $n - m$ zeros at infinity. For the short-circuit current gain in Eq. (11-16), we note that $A_I(s)$ has one pole at $-\omega_\beta$ and one finite zero at $+\omega_z$. The transfer function in Eq. (11-3) shows that the stage in Fig. 11-2 has one pole at $-\omega_H$ and no finite zero. It does, however, have one zero at infinity [as $s \to \infty$, $A_{VH}(s) \to 0$]. Note that the frequency of a pole (zero) has a magnitude $p_i(z_i)/2\pi$, that is, $f_H = \omega_H/2\pi$.

Determination of the Number of Poles and Zeros   *The number of poles in a transfer function is equal to the number of independent energy-storing elements in the network.* In electronic amplifiers the storing components are almost exclusively capacitors. An *independent* capacitor is one to which you can assign an arbitrary voltage, independent of all other capacitor voltages. For example, two capacitors in parallel are *not* independent because the voltage across the first must be the same as that across the second. Similarly, two capacitors in series are *not* independent because the stored charge Q is the same in each component and the voltage across a capacitor C is Q/C. Also, if a network loop can be traversed by passing only through capacitors, then not all of these C values are independent (because the sum of the voltages around the closed path must be zero).

The number of zeros in a transfer function is determined from a knowledge of the number of poles and the behavior of the network as $s \to \infty$. In Eq. (11-20), $A_V(s) \to 0$, as $s \to \infty$, because there $n - m$ more poles than zeros. In general, if $A_V(s) \to 1/s^k$ as $s \to \infty$ *then the number of finite zeros is k less than the number of poles.*

The behavior of a network as $s \to \infty$ is usually obtained by inspection since the voltage across a capacitor is zero as $s \to \infty$. For example, in Fig. 11-2

$$A_{VH} = \frac{V_o}{V_s} \longrightarrow \frac{1}{s} \quad \text{as} \quad s \to \infty$$

Hence the number of zeros is one less than the number of poles. Since the circuit contains only one capacitor, A_{VH} must contain one pole and (by the preceding argument) no finite zero. This conclusion is confirmed in Eq. (11-3).

The Dominant-Pole Approximation The Bode diagram of $A_H(s)$ in Eq. (11-20) displays the frequency-response characteristic of the amplifier. The upper 3-dB frequency f_H is obtained from the Bode diagram. Note that to plot the frequency-response characteristic requires that all pole and zero locations be known; that is, z_1, z_2, . . ., z_m and p_1, p_2, . . ., p_n must be known.[1] If the smallest pole frequency $f_{p1} = p_1/2\pi$ in $A_H(s)$ is much smaller than all other pole and zero frequencies, however, the upper 3-dB frequency f_H of $A_H(s)$ is approximately f_{p1}. Thus, for excitation frequencies within the bandwidth of the amplifier, $A_H(s)$ behaves simply as a single-pole system having a transfer function $A_o/(1 + s/p_1)$. This approximation is referred to as the *dominant-pole approximation*. Note that the more widely separated are the other poles and zeros from p_1, the more accurate is the dominant-pole approximation.

Often, the high-frequency response of an amplifier has no finite zeros; that is, $A_H(s)$ contains only poles. For this situation an amplifier with three real poles[2] has a transfer function

$$A_H(s) = \frac{A_o}{(1 + s/p_1)(1 + s/p_2)(1 + s/p_3)} \tag{11-21}$$

Alternatively, Eq. (11-21) can, by performing the indicated multiplication, be rewritten as

$$A_H(s) = \frac{A_o}{1 + a_1 s + a_2 s^2 + a_3 s^3} \tag{11-22}$$

where

$$\left.\begin{aligned} a_1 &= \frac{1}{p_1} + \frac{1}{p_2} + \frac{1}{p_3} \\ a_2 &= \frac{1}{p_1 p_2} + \frac{1}{p_1 p_3} + \frac{1}{p_2 p_3} \\ a_3 &= \frac{1}{p_1 p_2 p_3} \end{aligned}\right\} \tag{11-23}$$

Consider the situation where $p_1 \ll p_2 < p_3$; in other words, where p_1 is the dominant pole. Then

$$\left.\begin{aligned} a_1 &\simeq \frac{1}{p_1} & \text{or} \quad p_1 &\simeq \frac{1}{a_1} \\ a_2 &\simeq \frac{1}{p_1 p_2} = \frac{a_1}{p_2} & \text{or} \quad p_2 &\simeq \frac{a_1}{a_2} \\ a_3 &\simeq \frac{1}{p_1 p_2 p_3} = \frac{a_2}{p_3} & \text{or} \quad p_3 &\simeq \frac{a_2}{a_3} \end{aligned}\right\} \tag{11-24}$$

---

[1] Many computer programs for evaluating the roots of a polynomial are available.

[2] That is, all the poles lie on the negative, real axis of the s plane.

The importance of Eq. (11-24) is that we can approximate the pole locations by knowing the coefficients a_1, a_2, and a_3 in $A_H(s)$. Furthermore, the dominant-pole approximation gives the value of the 3-dB frequency f_H as

$$f_H \simeq \frac{p_1}{2\pi} = \frac{1}{2\pi a_1}$$

(11-25)

The form of Eq. (11-24), $p_k = a_{k-1}/a_k$, applies to a transfer function of n real poles. In relation to amplifier response, however, we are interested in p_1 and p_2 (a_1 and a_2) only. First, p_1 determines the approximate value of f_H, and second, the separation between p_1 and p_2 indicates the degree to which the dominant-pole approximation is valid. For most of the transfer functions encountered, $p_2/p_1 = a_1^2/a_2 \geq 8$ gives f_H within 10 percent and p_1 within 20 percent of their actual values. As p_2/p_1 increases, the error between actual and approximate values decreases. Note that the dominant-pole approximation *always* results in values of f_H and p_1 that are less than the corresponding actual values.

Of greater importance than the numerical convenience afforded by Eqs. (11-24) and (11-25) is that the coefficients a_1, a_2 and so on can be determined by circuit time constants as described in Sec. 11-9. This permits the circuit designer to correlate overall response with the particular components (stages) which produce the response.

The student should note the following three limitations of the dominant-pole method:

1. It is valid only for transfer functions with real poles.

2. Any (all) zeros in the transfer function must be at least two octaves beyond the dominant pole.

3. The representation of the transfer function by a single dominant pole does not give accurate results for the phase characteristic.

11-5 THE HIGH-FREQUENCY RESPONSE OF A COMMON-EMITTER STAGE

The equivalent circuit used to evaluate the high-frequency performance of the basic common-emitter stage shown in Fig. 10-21 is displayed in Fig. 11-14. Typical numerical values of transistor parameters and circuit components are also indicated in Fig. 11-14. Since the circuit has two independent capacitors, the

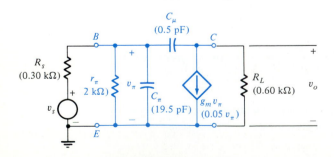

FIGURE 11-14

High-frequency equivalent circuit of the basic common-emitter stage (Fig. 10-21a). The hybrid-π model for the BJT is shown in blue.

transfer function has two poles. As $s \to \infty$, B and C are short-circuited (making $V_o = V_\pi$)[1] and the output falls to zero as $1/s$ as a result of the shunt capacitor C_π. From the discussion in the previous section, there must be one fewer zero than poles. Hence we expect a transfer function of two poles and one zero.

We can obtain the zero by inspection of Fig. 11-14. If, for $s = z_1$, $V_o = 0$, there is no current in R_L.[2] Thus the current $sC_\mu V_\pi$ in C_μ must equal the con-trolled-source current $g_m V_\pi$. Therefore, the zero is given by $sC_\mu V_\pi = g_m V_\pi$ or $s = g_m/C_\mu \equiv z_1$. Note that this is the same value found for the zero in the short-circuit current gain in Eq. (11-16). As expected, this zero arises in the ensuing evaluation of the transfer function by nodal analysis (App. C-2).

The Transfer Function   The node-voltage equations, taking V_π and V_o as variables, are

$$V_\pi \left(\frac{1}{R_s} + \frac{1}{r_\pi} + sC_\pi + sC_\mu \right) - sC_\mu V_o = \frac{V_s}{R_s} \tag{11-26}$$

$$V_\pi (g_m - sC_\mu) + V_o \left(sC_\mu + \frac{1}{R_L} \right) = 0 \tag{11-27}$$

Simultaneous solution of Eqs. (11-26) and (11-27), forming V_o/V_s and recasting the result in the form of Eq. (11-22), gives

$$A_{VH}(s) = \frac{V_o}{V_s} = \frac{[-\beta_o R_L/(R_s + r_\pi)] (1 - sC_\mu/g_m)}{1 + s[R_\pi^0 C_\pi + R_\pi^0 (1 + g_m R_L) C_\mu + R_L C_\mu] + s^2 R_\pi^0 R_L C_\pi C_\mu} \tag{11-28}$$

where

$$R_\pi^0 = \frac{r_\pi R_s}{r_\pi + R_s} = r_\pi \| R_s \tag{11-29}$$

The numerator term $-\beta_o R_L/(R_s + r_\pi)$ in Eq. (11-28) is identified as the dc (midband) gain A_{VO} of the stage [Eq. (10-40)]. The equivalent resistance R_π^0 can, by inspection of Fig. 11-14, be recognized as the zero-frequency resistance seen by C_π with $C_\mu = 0$. As is subsequently shown (Sec. 11-9), the equivalent resistance $R_\pi^0(1 + g_m R_L) + R_L$ which multiplies C_μ in the s coefficient in Eq. (11-28) is the resistance seen by C_μ with $C_\pi = 0$. Thus, as stated in the previous section, the s coefficient is related to circuit time constants.

Let us now see whether this transfer function, of the form $A_{VH}(s) = A_{VO}(1 - s/z_1)/(1 + a_1 s + a_2 s^2)$, has a dominant pole. Using the numerical values given, $a_1 = 9.43 \times 10^{-9}$ s and $a_2 = 1.53 \times 10^{-18}$ s². Use of Eq. (11-24) gives $p_1 \simeq 1/a_1 = 10.6 \times 10^7$ rad/s and $p_2 = a_1/a_2 = 6.16 \times 10^9$ rad/s. The separation is $p_2/p_1 = 58$, so that $-p_1$ is the dominant pole. The value of $z_1 = g_m/C_\mu = 10^{11}$ rad/s and is even further removed from p_1 than is p_2. Thus $-p_1$ is the dominant pole and the overall bandwidth, from Eq. (11-25), is

---

[1]Recall that V_o and V_π are the frequency-domain values of v_o and v_π, respectively.

[2]The resistance $R_L = R_C \| r_o$, which, for $R_C \ll r_o$, is approximately R_C. If the stage shown drives an external load, the load resistance also appears in parallel with R_C.

$f_H \simeq p_1/2\pi = 16.9$ MHz. Determination of the poles of $A_{VH}(s)$ by solving the quadratic equation in Eq. (11-28) gives $p_1 = 10.7 \times 10^7$ rad/s and $p_2 = 6.06 \times 10^9$ rad/s. Calculation of f_H from the transfer function gives $f_H = 17.1$ MHz. These results demonstrate the validity of the dominant-pole method.

The Unilateral Hybrid-π Equivalent The wide separation between p_1 and both p_2 and z_1 is a common occurrence in common-emitter stages. Thus a transfer function [Eq. (11-30)] containing a single dominant pole is a very good approximation of the frequency response.

$$A_{VH}(s) \simeq \frac{A_{VO}}{1 + s/p_1} = \frac{A_{VO}}{1 + a_1 s} = \frac{A_{VO}}{1 + s/2\pi f_H} \tag{11-30}$$

A simplified equivalent circuit which has the transfer function indicated in Eq. (11-30) can be obtained by applying Miller's theorem. Proceeding as in App. C-4, we obtain the circuit in Fig. 11-15, often called the *unilateral hybrid π*, with $K = V_o/V_\pi$. Assuming negligible current in C_μ (Sec. 11-3), $K = -g_m R_L$. The circuit in Fig. 11-15 has two independent time constants, one associated with the input capacitance $C_M = C_\pi + C_\mu(1 + g_m R_L)$ and the second with the output capacitance (shown by dashed line) $C_\mu(1 + g_m R_L)/g_m R_L \simeq C_\mu$ for $g_m R_L \gg 1$. The input time constant is C_M multiplied by the equivalent resistance seen by C_M; this equivalent resistance is $R_s \parallel r_\pi$ and is identified by R_π^0 in Eq. (11-29). Note that $C_M R_\pi^0$ corresponds to the first two terms of the s coefficient in Eq. (11-28). The output time constant, for $g_m R_L \gg 1$, is $C_\mu R_L$, the third term in the s coefficient in Eq. (11-28). Thus the circuit in Fig. 11-15 has the same coefficient a_1 as obtained in Eq. (11-28) and consequently gives the same dominant pole.

A further simplification results when we investigate typical numerical values for the two time constants. For values given[1]

$$R_\pi^0 C_M = \frac{2(0.3)}{2 + 0.3} [19.5 + 0.5(1 + 0.05 \times 600)] = 9.13 \text{ ns}$$
$$R_L C_\mu = 0.6 \times 0.5 = 0.30 \text{ ns}$$

Evidently $R_\pi^0 C_M \gg R_L C_\mu$, and it is often common practice to neglect the effect of the output time constant $R_L C_\mu$ in pencil-and-paper calculations. For the circuit in Fig. 11-15, a_1 in Eq. (11-30) becomes

$$a_1 = R_\pi^0 C_M = \frac{r_\pi R_s}{R_s + r_\pi} [C_\pi + C_\mu(1 + g_m R_L)] \tag{11-31}$$

With the use of numerical values, $f_H = 1/2\pi a_1 = 17.4$ MHz, a value within 2 percent of the actual value and certainly sufficiently accurate for pencil-and-paper calculations. If $R_L C_\mu$ were included in the a_1 coefficient, the resultant

---

[1] Note that resistances in kilohms (kΩ) and capacitances in picofarads (pF) result in time constants in nanoseconds (ns).

FIGURE 11-15
The unilateral hybrid-π equivalent circuit obtained by use of Miller's theorem.

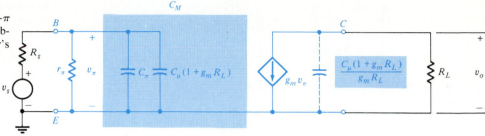

value of f_H would be identical with that obtained previously. Where highly accurate results are required, computer simulations (SPICE, Microcap II, etc.) are employed.

In the equivalent circuit in Fig. 11-15, no path exists between output and input (where C_μ is in Fig. 11-14). As a result, the unilateral hybrid-π model cannot be used to calculate the high-frequency output impedance Z_o. Here we must use the circuit in Fig. 11-14.

The Miller Input Impedance   The input impedance $Z_i(s)$ for the circuit in Fig. 11-15 is simply the parallel combination of r_π and C_M. In the derivation of Fig. 11-15, we assumed that the current I_μ in C_μ is negligible compared with $g_m V_\pi$ and $K = -g_m R_L$ was constant. Sometimes component values are such that this assumption introduces errors of several percent in $Z_i(s)$ and consequently in f_H. If the current I_μ is not neglected, $K = -g_m R_L/(1 + sR_L C_\mu)$ and Z_i must be modified. The resultant impedance is the parallel combination of r_π, C_π, R_x and C_x, where R_x and C_x arise from the application of Miller's theorem to C_μ. At a frequency $s = j\omega$, the values of R_x and C_x are

$$R_x = \frac{1}{g_m}\left(1 + \frac{1}{\omega^2 R_L^2 C_\mu^2}\right) \qquad C_x = C_\mu\left(1 + \frac{g_m R_L}{1 + \omega^2 R_L^2 C_\mu^2}\right) \qquad (11\text{-}32)$$

The derivation of Eq. (11-32) and the value of f_H that results are left to the student in Probs. 11-12 and 11-13.

11-6   THE GAIN-BANDWIDTH PRODUCT   By use of the single-pole transfer function in Eq. (11-30), it is found that the voltage gain-bandwidth product is

$$|A_{VO}f_H| = \frac{g_m}{2\pi C_M}\frac{R_L}{R_s} = \frac{f_T}{1 + 2\pi f_T R_L C_\mu}\frac{R_L}{R_s} \qquad (11\text{-}33)$$

The quantities f_H and A_{VO}, which characterize the transistor stage, depend on both R_L and R_s. The form of this dependence, as well as the order of magnitude of these quantities, is seen in Fig. 11-16. Here f_H has been plotted as a function of R_L for several values of R_s. The topmost f_H curve in Fig.

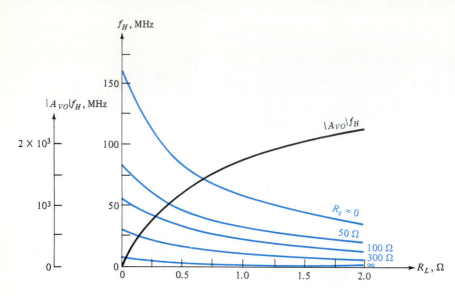

FIGURE 11-16
Bandwidth f_H as a function of R_L, with source resistance R_s as a parameter, for a single-stage common-emitter amplifier. Equation (11-31) is used to compute f_H using the numerical values given in Fig. 11-14.

11-16 for $R_s = 0$ corresponds to ideal-voltage-source drive.[1] We note that the bandwidth is highest for any value of R_L for the lowest R_s. The voltage gain-bandwidth product increases with increasing R_L and decreases with increasing R_s. Even if we know the gain-bandwidth product at a particular R_s and R_L, we cannot use the product to determine the improvement in bandwidth corresponding to a sacrifice in gain. If we change the gain by changing R_L or R_s or both, in general, the gain-bandwidth product will no longer be the same as it had been.

11-7 THE COMMON-SOURCE STAGE AT HIGH FREQUENCY

The high-frequency analysis of the common-source stage parallels that for the common-emitter stage discussed in the previous section. Note the similarity of the high-frequency model of the common-source stage in Fig. 11-17 with Fig. 11-14 for the common-emitter stage. If $r_b = 0$ and $r_\pi \to \infty$ in Fig. 11-14, the only difference between the two circuits is the existence of a third capacitance C_{ds} at the output in Fig. 11-17. Although there are three capacitances, these are not independent because they form a loop (Sec. 11-4). Thus the transfer function $A_{VH}(s) = V_o/V_s$ has only two poles and, by analogy with Fig. 11-14, one finite zero.

The nodal equations for the circuit in Fig. 11-17 are

$$\frac{V_s}{R_s} = V_{gs}\left(\frac{1}{R_s} + sC_{gs} + sC_{gd}\right) - sC_{gd}V_o \qquad (11\text{-}34a)$$

---

[1] Note that for $R_s = 0$, the value of f_H shown is in error because the time constant $C_\mu R_L$ is comparable to $R_\pi^0 C_M$. However, the form of the variation in f_H displayed is indicative of the physical situation that exists.

FIGURE 11-17
High-frequency equiv-
alent circuit of a com-
mon-source stage.

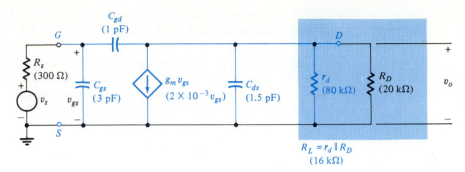

$$0 = V_{gs}(g_m - sC_{gd}) + V_o\left(\frac{1}{R_L} + sC_{gd} + sC_{ds}\right) \tag{11-34b}$$

Simultaneous solution of Eqs. (11-34), identification of $A_{VO} = -g_m R_L$, and some algebraic manipulation give

$$A_{VH}(s) = \frac{V_o}{V_s} = \frac{A_{VO}(1 - sC_{gd}/g_m)}{1 + a_1 s + a_2 s^2} \tag{11-35}$$

where

$$a_1 = R_s C_{gs} + R_s[(1 + g_m R_L) + R_L] C_{gd} + R_L C_{ds} \tag{11-36}$$
$$a_2 = R_s R_L(C_{gs}C_{gd} + C_{gs}C_{ds} + C_{ds}C_{gd})$$

Again we observe that a_1 comprises the sum of time constants; each of the equivalent resistances is equal to the zero-frequency (capacitors are open-circuited) resistance seen at each of the capacitor terminals. The coefficient a_2 can be considered as the product of time constants as described in Sec. 11-9.

Let us now examine Eq. (11-35) to determine whether the dominant-pole condition applies. Using the typical parameter values given in Fig. 11-17 we obtain, from Eq. (11-36)

$$a_1 = 0.3 \times 3 + [(1 + 2 \times 16) + 16] \times 1 + 16 \times 1.5$$
$$= 0.9 + 49 + 24 = 73.9 \text{ ns}$$
$$a_2 = 0.3 \times 16[3 \times 1 + 3 \times 1.5 + 1.5 \times 1]$$
$$= 43.2 \text{ (ns)}^2$$

From Eq. (11-24), we obtain

$$p_1 = \frac{1}{a_1} = \frac{1}{73.9 \times 10^{-9}} = 13.5 \times 10^6 \text{ rad/s}$$

$$p_2 = \frac{a_1}{a_2} = \frac{73.9 \times 10^{-9}}{43.2 \times 10^{-18}} = 1.71 \times 10^9 \text{ rad/s}$$

The separation between p_2 and p_1 is 126:1. The zero is at $s = g_m/C_{gd} = 2 \times 10^9$ rad/s. Inspection of these values indicates that p_1 is the dominant pole. Thus the transfer function can be approximated by the single-pole function of the form in Eq. (11-30) with $f_H = \omega_H/2\pi = 1/2\pi a_1 = 2.15$ MHz. Exact solution of Eq. (11-35) indicates that the values of f_H and p_1 obtained by the dominant-pole approximation deviate by less than 1 percent.

We can use Miller's theorem, with $K = V_o/V_{gs} = -g_m R_L$, to arrive at the approximate unilateral circuit in Fig. 11-18. Note that while C_{gs} is the smallest of the three capacitances, it has the most pronounced effect on a_1 and consequently on f_H. This is due to the Miller effect, which essentially multiplies C_{gd} by the gain of the stage. The input impedance of the stage, as seen in Fig. 11-18, is purely capacitave (approximately) and is

$$C_i = C_{gs} + C_{gd}(1 + g_m R_L) \qquad (11\text{-}37)$$

Note that the output impedance (Fig. 11-17) contains both resistive and capacitive components (Probs. 11-12 and 11-13).

11-8 EMITTER AND SOURCE FOLLOWERS AT HIGH FREQUENCIES In this section we examine the high-frequency response of emitter and source followers (common-collector and common-drain stages). The common-collector stage is treated first, and then by analogy we describe the common-drain stage. Because emitter and source followers often act as buffers, that is, unity (almost) gain, high input impedance, and low output impedance, we consider each of these quantities at high frequencies.

The Voltage Gain The high-frequency model of the emitter follower with a resistive load R_E is depicted in Fig. 11-19. Because of its low output impedance, the emitter follower is often used to drive capacitive loads (whose symbol is drawn dashed). We consider the load capacitor in conjunction with the discussion of the high-frequency output impedance.

The transfer function of the emitter follower has two poles (two independent capacitors) and one finite zero. As $s \to \infty$, the impedance $1/sC_\mu \to 0$ (a short circuit); hence $V_o \to 0$ as $1/s$ because of the short circuit at the input.

The transfer function is obtained from the nodal equations, using V_1 and V_o as the variables. These equations are

$$V_1\left(\frac{1}{R_s} + sC_\mu + \frac{1}{z_\pi}\right) - V_o\left(\frac{1}{z_\pi}\right) = \frac{V_s}{R_s} \qquad (11\text{-}38)$$

$$-V_1\left(\frac{1}{z_\pi}\right) + V_o\left(\frac{1}{z_\pi} + \frac{1}{R_E}\right) = g_m(V_1 - V_o) \qquad (11\text{-}39)$$

FIGURE 11-18
Unilateral representation of the model in Fig. 11-17.

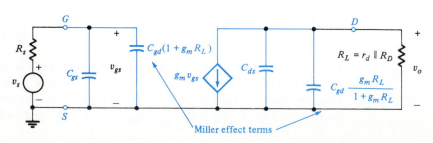

FIGURE 11-19
The high-frequency equivalent circuit of the emitter follower. The capacitor C_L forms part of the load.

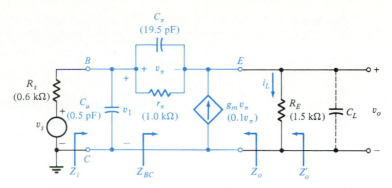

In Eq. (11-39), $V_\pi = V_1 - V_o$ is used. Simultaneous solution of Eqs. (11-38) and (11-39), after substitution of $z_\pi = r_\pi/(1 + sr_\pi C_\pi)$ and some algebraic manipulation, yields

$$A_{VH}(s) = \frac{V_o}{V_s} = \frac{(\beta_o + 1)R_E}{R_s + r_\pi + (\beta_o + 1)R_E} \frac{1 + sC_\pi r_\pi/(1 + \beta_o)}{1 + a_1 s + a_2 s^2} \quad (11\text{-}40)$$

where

$$a_1 = \frac{r_\pi(R_E + R_s)C_\pi}{R_s + r_\pi + (1 + \beta_o)R_E} + \frac{R_s[r_\pi + (1 + \beta_o)R_E]C_\mu}{R_s + r_\pi + (1 + \beta_o)R_E} \quad (11\text{-}41)$$

$$a_2 = \frac{R_E R_s r_\pi C_\pi C_\mu}{R_s + r_\pi + (1 + \beta_o)R_E} \quad (11\text{-}42)$$

We observe again that a_1 is the sum of time constants and that a_2 can be interpreted as the product of time constants. The zero z_1 at $s = -g_m/C_\pi$ in Eq. (11-40) can also be obtained from Fig. 11-19. With $V_o = 0$, no current exists in R_E, and hence the current in z_π must both be equal and opposite to $g_m V_\pi$. Thus

$$\frac{(1 + sr_\pi C_\pi)V_\pi}{r_\pi} = -g_m V_\pi$$

from which $z_1 = -(\beta_o + 1)/r_\pi C_\pi \approx -g_m/C_\pi$. Using the parameter values indicated in Fig. 11-19 (the resistances have the same values used to obtain the low-frequency behavior in Table 10-4), we find that

$$a_1 = 0.566 \text{ ns} \qquad a_2 = 0.0573 \times 10^{-18} \text{ s}$$

and
$$z_1 \approx \frac{g_m}{C_\pi} = 5.12 \times 10^9 \text{ rad/s}$$

The approximate dominant-pole angular frequency is

$$p_1 = \omega_H = \frac{1}{a_1} = \frac{1}{0.566 \times 10^{-9}} = 1.77 \times 10^9 \text{ rad/s}$$

$$f_H = \frac{\omega_H}{2\pi} = 281 \text{ MHz}$$

Similarly

$$p_2 = \frac{a_1}{a_2} = \frac{0.566}{0.0573} \times 10^9 = 9.87 \times 10^9 \text{ rad/s}$$

or $f_2 = p_2/2\pi = 1.57$ GHz.

The ratio $p_2/p_1 = 5.58$ and the dominant-pole approximation does not give particularly accurate results. Solving for the poles from Eq. (11-40) yields $p_1 = 2.30 \times 10^9$ rad/s and $p_2 = 7.57 \times 10^9$ rad/s. With these values, $f_H = 339$ MHz.

We observe that f_H, the pole frequencies, and the zero frequency are all of the order of magnitude of f_T. Because these values are at the frequency limit for which the equivalent circuit is valid, the numerical values are subject to question. However, they all display the approximate magnitudes obtained from computer simulations.

Comparison of f_H for the common-collector stage with $f_H = 16.9$ MHz for the common-emitter stage in Sec. 11-6 indicates that the common-collector stage has a considerably larger bandwidth than does the common-emitter stage. In reality, a common-emitter stage having $R_C = 1.5$ kΩ, driven from a source $R_s = 0.6$ kΩ and using the same transistor, has $f_H \approx 4.37$ MHz. Thus we conclude that when a common-collector stage is driven by (drives) a common-emitter stage, the value of f_H for the cascade is simply that of the common-emitter stage.

The Output Impedance Z_o   The high-frequency output impedances Z_o and Z_o' are obtained from the Thévenin equivalent of the stage. The open-circuit voltage, measured across R_E, is simply $V_o = A_{VH}V_s$. The current I_L is V_o/R_E and, for $R_E = 0$, gives the short-circuit current I_{sc} as

$$I_{sc} = \frac{(\beta_o + 1)V_s}{R_s + r_\pi} \frac{1 + sr_\pi C_\pi/(1 + \beta_o)}{1 + sR_s r_\pi (C_\pi + C_\mu)/(R_s + r_\pi)} \tag{11-43}$$

The Thévenin impedance $Z_o' = V_o/I_{sc}$ and is

$$Z_o' = \frac{R_E [(R_s + r_\pi)/(1 + \beta_o)]}{R_E + [(R_s + r_\pi)/(1 + \beta_o)]} \frac{1 + sr_\pi R_s (C_\pi + C_\mu)/(R_s + r_\pi)}{1 + a_1 s + a_2 s^2} \tag{11-44}$$

where a_1 and a_2 are as given in Eq. (11-41).

We observe in Fig. 11-19 that $Z_o' = Z_o \| R_E$ and if $R_E \to \infty$, $Z_o' = Z_o$. Hence from Eq. (11-44), we obtain

$$Z_o = \frac{R_s + r_\pi}{1 + \beta_o} \frac{1 + sr_\pi R_s (C_\pi + C_\mu)/(R_s + r_\pi)}{[1 + sr_\pi C_\pi/(1 + \beta_o)](1 + sR_s C_\mu)} \tag{11-45}$$

We can identify $(R_s + r_\pi)/(1 + \beta_o)$ as the low-frequency output resistance R_o (Table 10-3). Thus Z_o is of the form

$$Z_o = R_o \frac{(1 + s/z_1)}{(1 + s/p_1)(1 + s/p_2)} \tag{11-46}$$

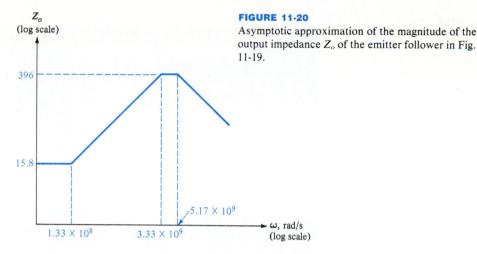

FIGURE 11-20

Asymptotic approximation of the magnitude of the output impedance Z_o of the emitter follower in Fig. 11-19.

With the use of values indicated in Fig. 11-19, $z_1 = 1.33 \times 10^8$ rad/s, $p_1 = 3.33 \times 10^9$ rad/s, and $p_2 = 5.17 \times 10^9$ rad/s. On the basis of these values, we display the log $|Z_o(j\omega)|$ versus log ω in Fig. 11-20. Note that $|Z_o(j\omega)|$ increases with increasing frequency for $\omega < p_1$. This is the behavior of an inductive impedance in this frequency range. When the emitter follower is used to drive capacitive loads (C_L in Fig. 11-19) at high frequencies, the circuit may behave as a resonant circuit. In systems excited by pulses, such as the ECL family high-speed logic (Sec. 6-13), the combination of the inductive output impedance and the load capacitance can cause excessive ringing on the output waveform. Usually, designers provide a sufficient amount of damping (resistive component of Z_o) to minimize this condition.

The Input Impedance Z_i   The input impedance Z_i of the emitter follower is, as seen in Fig. 11-19, the parallel combination of C_μ and $Z_{B'C}$. The derivation of the results for $Z_{B'C}(s)$ and $Z_i(s)$, given in Eqs. (11-47) and (11-48), are left to the student (Prob. 11-23):

$$Z_{B'C}(s) = \beta_o R_E \frac{1 + sC_\pi/g_m}{1 + sr_\pi C_\pi} \simeq \beta_o R_E \frac{1 + s/2\pi f_T}{1 + s/2\pi f_\beta} \tag{11-47}$$

$$Z_i(s) = \beta_o R_E \frac{1 + s(1/\omega_T)}{1 + s(1/\omega_\beta + \beta_o R_E C_\mu) + s^2(\beta_o R_E C_\mu/\omega_T)} \tag{11-48}$$

Note that in Eqs. (11-47) and (11-48) it is assumed that $\beta_o \gg 1$, $\beta_o R_E \gg r_\pi$ and $\omega_T \simeq g_m/C_\pi$.

For the situation where r_b is not assumed zero, $Z_i(s)$ becomes, using the approximations given for Eq. (11-48) and $\beta_o R_E \gg r_b$,

$$Z_i(s) = \beta_o R_E \frac{1 + s(1/\omega_T + r_b C_\mu) + s^2 r_b C_\mu/\omega_T}{1 + s(1/\omega_\beta + \beta_o R_E C_\mu) + s^2 (\beta_o R_E C_\mu/\omega_T)} \tag{11-49}$$

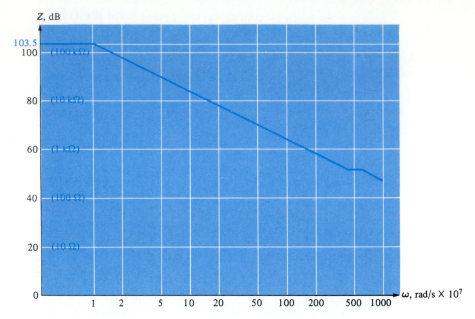

FIGURE 11-21
Asymptotic Bode diagram for the input impedance Z_i of the emitter follower in Fig. 11-19 and Eq. (11-50). *Note:* Pole and zero frequencies obtained by use of the dominant-pole approximation.

With the parameter values in Fig. 11-19 and $r_b = 50\ \Omega$, Eq. (11-49) becomes

$$Z_i(s) = 150\ \frac{[1\ +\ s/(4.44\ \times\ 10^9)]\ [1\ +\ s/(45\ \times\ 10^9)]}{[1\ +\ s/(1.05\ \times\ 10^7)]\ [1\ +\ s/(6.33\ \times\ 10^9)]}\quad k\Omega\quad (11\text{-}50)$$

The variation of the magnitude of this function with frequency (for $\omega < 10^{10}$ rad/s) is displayed in the asymptotic Bode diagram in Fig. 11-21. Note that the impedance decreases with frequency and this decrease occurs at frequencies below f_H for the stage. Indeed, at $f_H \simeq 300$ MHz ($\omega_H \simeq 1.90 \times 10^9$ rad/s), Z_i is less than 1 kΩ. The designer must be cognizant of this decrease in Z_i if the circuit is to provide adequate isolation.

For frequencies $f < f_T$, the impedance Z_{BC} can be approximated as

$$Z_{BC} = \frac{\beta_o R_E}{1\ +\ s/\omega_\beta}\qquad \text{or}\qquad Y_{BC} = \frac{1\ +\ s/\omega_\beta}{\beta_o R_E}\qquad (11\text{-}51)$$

The admittance in Eq. (11-51) represents the parallel combination of the capacitance and resistance given in Eq. (11-52):

$$R_{BC} = \beta_o R_E\qquad C_{BC} = \frac{1}{R_E \omega_T}\qquad (11\text{-}52)$$

The value of this capacitance is quite small (0.133 pF for the numerical values used). As C_{BC} appears in parallel with C_μ, the input capacitance over the useful range of operating frequencies is simply $C_\mu + 1/R_E\omega_T$. This value is considerably smaller than that for a common-emitter stage [$C_i = C_\pi + C_\mu(1 + g_m R_L)$]. Thus the emitter follower does not capacitively load the preceding stage, a factor of importance in high-speed, high-frequency systems. (Laboratory instruments often use emitter followers as input stages in order to minimize the effect of the instrument on the measurement.)

The small input capacitance can also be approximated by using Miller's theorem for which the effect of C_π on the input is $C_\pi(1 - K)$. As K is very near unity for an emitter follower, this term is approximately zero and leaves only C_μ at the input.

The Source Follower   The model of the source follower valid at high frequencies is displayed in Fig. 11-22. Observe that the circuit in Fig. 11-22 is similar to the emitter-follower equivalent circuit in Fig. 11-19 if we make the following identifications: $r_\pi \to \infty$, $C_{gd} = C_\mu$, $C_{gs} = C_\pi$, and $R_L = r_d \parallel R_S = R_E$. Only the capacitance C_{ds} cannot be identified, and its effect on the coefficients a_1 and a_2 must be included. Note that if the source follower drives a capacitive load C_L, then C_{ds} can be added to the load capacitance.

By making the identifications and adding the effect of C_{ds}, we can use the results for the emitter follower for the common-collector stage. These results for $A_{VH}(s)$, f_H, Z_i and Z_o are given as follows (the student is asked to derive these results in Probs. 11-24 and 11-25):

$$A_{VH}(s) = \frac{A_{Vo}(1 - g_m\,s/C_{gs})}{1 + a_1 s + a_2 s^2} \tag{11-53}$$

where
$$a_1 = R_s C_{gd} + \frac{(R_s + R_L)\,C_{gs}}{1 + g_m R_L} + \frac{R_L C_{ds}}{1 + g_m R_L}$$

$$a_2 = \frac{R_s R_L}{1 + g_m R_L}\,(C_{gd}C_{gs} + C_{gd}C_{ds} + C_{gs}C_{ds}) \tag{11-54}$$

The upper 3-dB frequency $f_H \approx 1/2\pi a_1$, which, for $g_m R_L \gg 1$ and $R_L > R_s$, becomes

$$f_H \approx \frac{1}{2\pi[R_s C_{gd} + (1/g_m)(C_{gs} + C_{ds})]} \tag{11-55}$$

$$Z_i(s) = \left(\frac{1}{sC_{gd}}\right) \Big\| \frac{(1 + g_m R_L)\,[1 + sR_L(C_{ds} + C_{gs})\,/\,(1 + g_m R_L)]}{sC_{gs}(1 + sR_L\,C_{ds})} \tag{11-56}$$

$$Z_o(s) = \frac{1}{g_m}\,\frac{1 + sR_s(C_{gd} + C_{gs})}{(1 + sC_{gs}/g_m)\,(1 + sR_s C_{gd})} \tag{11-57}$$

FIGURE 11-22
The high-frequency equivalent circuit of the source follower.

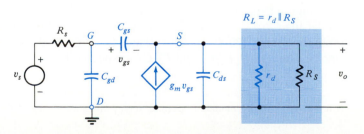

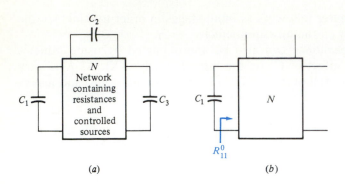

(a)

(b)

FIGURE 11-23
(a) Network containing three capacitors. (b) The network in (a) with C_2 and C_3 open-circuited. The zero-frequency resistance R^0_{11} is defined by this configuration.

11-9 THE TIME-CONSTANT METHOD OF OBTAINING THE RESPONSE   The form of Eqs. (11-28), (11-35), (11-40), and (11-53) in which the coefficient a_1 is the sum of time constants and a_2 is the product of time constants is not unique to simple common-emitter, common-source, common-collector, and common-drain stages. These coefficients in the characteristic polynomial of any linear system containing resistors, capacitors, and controlled sources can *always* be expressed in this form. In this section we describe a method for obtaining these coefficients by direct determination of the equivalent resistances needed to evaluate the time constants.[1]

The Coefficient a_1   Consider the network in Fig. 11-23a in which the portion of the circuit N within the rectangle contains only resistors and controlled sources. With three independent capacitors indicated, the transfer function has three poles, so that its denominator is expressed as given in Eq. (11-22). The coefficient a_1 can be written as

$$a_1 = R^0_{11}C_1 + R^0_{22}C_2 + R^0_{33}C_3 \qquad (11\text{-}58)$$

where R^0_{11}, R^0_{22}, R^0_{33} are the zero-frequency resistances seen by C_1, C_2, and C_3, respectively. Note that at zero frequency (dc), the capacitances are open-circuited and the RC products in Eq. (11-58) are often referred to as *open-circuit time constants*.

We can give the following heuristic argument to justify the form of a_1 in Eq. (11-58). Consider $C_2 = C_3 = 0$ (open circuits) so that the circuit contains only C_1 as shown in Fig. 11-23b. This transfer function of the circuit in Fig. 11-23b has only one pole whose angular frequency is simply the reciprocal of the circuit time constant τ. But τ for this case is C_1 multiplied by the equivalent resistance across its terminals, that is, R^0_{11} as indicated in Fig. 11-23b. Note that letting $C_2 = C_3 = 0$ in Eq. (11-58) gives the same result. Similar arguments apply if we consider $C_1 = C_3 = 0$ and $C_1 = C_2 = 0$. For these conditions the time constants are $R^0_{22}C_2$ and $R^0_{33}C_3$, respectively.

---

[1]A formal proof of this method is given in Refs. 3 and 4 at the end of the chapter.

The form of Eq. (11-58) can be extended by induction to a system which contains M capacitors as given in Eq. (11-59):

$$a_1 = \sum_{i=1}^{M} R_{ii}^0 C_i \qquad (11\text{-}59)$$

where R_{ii}^0 is the zero-frequency resistance seen by C_i. Alternatively, we can regard a_1 as the sum of open-circuit time constants, with $R_{ii}^0 C_i$ being the circuit time constant when all other capacitances are open-circuited. The application of the method is illustrated in the two examples that follow.

Example 11-2   Determine the coefficient a_1 in the transfer function of the common-emitter stage in Fig. 11-14.

Solution   With two capacitors, Eq. (11-59) gives a_1 as

$$a_1 = R_\pi^0 C_\pi + R_\mu^0 C_\mu$$

To evaluate R_π^0, the zero-frequency resistance seen by C_π, the capacitors are open-circuited and the independent voltage source v_s is suppressed (short-circuited) as illustrated in Fig. 11-24a. Observation of the circuit in Fig. 11-24a indicates

$$R_\pi^0 = R_s \| r_\pi = \frac{r_\pi R_s}{R_s + r_\pi}$$

In the circuit in Fig. 11-24b a test source I_t is supplied, and the voltage V_t is computed from which $R_\mu^0 = V_t/I_t$. In Fig. 11-24b, after identification of R_π^0 as shown, KVL yields

$$V_t = I_t R_\pi^0 + R_L(I_t + g_m v_\pi)$$

and substitution of $v_\pi = R_\pi^0 I_t$ gives

$$R_\mu^0 = \frac{V_t}{I_t} = R_\pi^0 + R_L(1 + g_m R_\pi^0) = R_\pi^0(1 + g_m R_L) + R_L$$

FIGURE 11-24
Circuits used to compute the zero-frequency (open-circuit) resistances (a) R_π^0 and (b) R_μ^0 for the common-emitter stage.

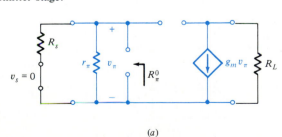

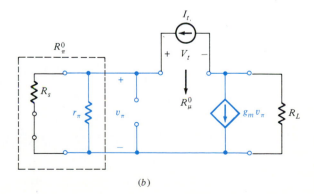

(a)   (b)

Thus

$$a_1 = R_\pi^0 C_\pi + [R_\pi^0(1 + g_m R_L) + R_L] C_\mu$$

which is the same expression as in Eq. (11-28).

Example 11-3

(a) Determine the a_1 coefficient in the transfer function of the amplifier stage whose equivalent circuit is displayed in Fig. 11-25. (b) Using the numerical values indicated, approximate the upper 3-dB frequency f_H. (c) Compare this value with that obtained in Sec. 11-7 for the common-source stage. (*Note:* This stage is a common-source amplifier with source resistance R_S. The device parameters and circuit components have the same values used in Sec. 11-7.)

Solution

(a) The circuit has three capacitors; hence, using Eq. (11-59), we have

$$a_1 = R_{gs}^0 C_{gs} + R_{gd}^0 C_{gd} + R_{ds}^0 C_{ds}$$

To evaluate the three time constants, R_{gs}^0, R_{gd}^0, and R_{ds}^0 must be found. By suppressing v_s and open-circuiting the capacitances, we can identify R_{gs}^0 as depicted in Fig. 11-26a. As the same current exists in R_D, r_d, and μV_{gs}, these elements can be converted to their current source $[\mu V_{gs}/(r_d + R_D)]$ − parallel resistance $(R_D + r_d)$ equivalent. This combination is in parallel with R_S and can be reconverted to its Thévenin equivalent as shown in Fig. 11-26b. The equivalent voltage source $\mu' V_{gs}$ and series resistance R_s are

$$\mu' = \frac{\mu R_A}{r_d + R_D} \qquad R_A = R_S \parallel (R_D + r_d)$$

Application of a test source V_t and computation of the current I_t as indicated in Fig. 11-26b gives $R_{gs}^0 = V_t/I_t$. With $V_{gs} = V_t$, use of KVL gives

$$I_t = \frac{V_t(1 + \mu')}{R_s + R_A}$$

from which

$$R_{gs}^0 = \frac{V_t}{I_t} = \frac{R_s + R_A}{1 + \mu'}$$

FIGURE 11-25
The high-frequency equivalent circuit of the common-source stage with source resistance.

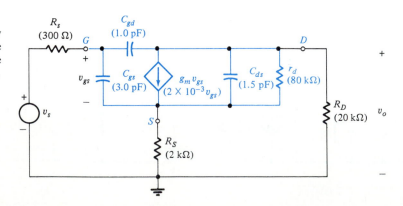

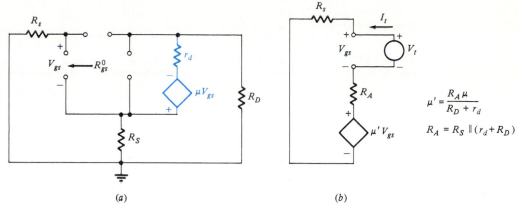

$$\mu' = \frac{R_A \mu}{R_D + r_d}$$

$$R_A = R_S \parallel (r_d + R_D)$$

(a)   (b)

FIGURE 11-26

(a) Circuit used to find R_{gs}^0. (b) Circuit in (a) with the drain-source portion (R_s, μV_{gs}, r_d, and R_D) replaced by a Thévenin equivalent $\mu' V_{gs} - R_A$.

The resistance R_{gd}^0 can be computed from the circuit in Fig. 11-27a. Application of KCL at D gives

$$I_1 = I_t - I_2 \tag{1}$$

$$V_{gs} = I_t R_s + I_1 R_S = I_t R_s + R_S(I_t - I_2) \tag{2}$$

Application of KVL for the right-hand loop gives

$$I_2 R_D = I_1(R_S + r_d) + \mu V_{gs} \tag{3}$$

Combination of (1) to (3) and solving for I_2 yields

$$I_2 = \frac{r_d + \mu R_s + (1 + \mu) R_s}{R_D + r_d + R_S(1 + \mu)} I_t \tag{4}$$

The voltage V_t, using KVL around the outside loop, is

$$V_t = I_t R_s + I_2 R_D \tag{5}$$

Substitution of (4) into (5) and formation of V_t/I_t gives

$$R_{gd}^0 = \frac{V_t}{I_t} = R_s + \frac{R_D[r_d + \mu R_s + (1 + \mu) R_s]}{R_D + r_d + R_S(1 + \mu)}$$

The third resistance R_{ds}^0 is obtained from the use of Fig. 11-27b. Note that no current exists in R_s, so that V_{gs} appears as indicated across R_S. Application of KCL requires $I_t = I_1 + I_2$, where

$$I_1 = \frac{V_t + \mu V_{gs}}{r_d} \quad \text{and} \quad I_2 = \frac{V_t - V_{gs}}{R_D}$$

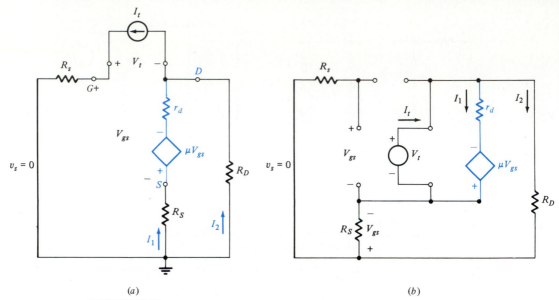

(a)   (b)

FIGURE 11-27
Circuit used to evaluate (a) R_{gd}^0 and (b) R_{ds}^0 for the stage in Fig. 11-25.

The control voltage $V_{gs} = I_2 R_S$. Combining these equations and solving for I_t, we obtain

$$R_{ds}^0 = \frac{V_t}{I_t} = \frac{r_d(R_S + R_D)}{R_D + r_d + (1 + \mu) R_S}$$

(b) Numerical evaluation of a_1 is as follows:

$$\mu = g_m r_d = 2 \times 80 = 160$$
$$R_A = R_S \parallel (R_D + r_d) = 2 \parallel (20 + 80) = 1.96 \text{ k}\Omega$$
$$\mu' = \frac{\mu R_A}{r_d + R_D} = \frac{160 \times 1.96}{80 + 20} = 3.14$$

and

$$R_{gs}^0 = \frac{0.3 + 1.96}{1 + 3.14} = 0.546 \text{ k}\Omega$$

$$R_{gd}^0 = 0.3 + \frac{20[80 + 160 \times 0.3 + (1 + 160)2]}{20 + 80 + 2(1 + 160)} = 21.6 \text{ k}\Omega$$

$$R_{ds}^0 = \frac{80(2 + 20)}{20 + 80 + (1 + 160)2} = 4.17 \text{ k}\Omega$$

Thus

$$a_1 = 0.546 \times 3 + 21.6 \times 1 + 4.17 \times 1.5 = 29.5 \text{ ns}$$

and

$$f_H = \frac{1}{2\pi a_1} = \frac{1}{2\pi \times 29.5 \times 10^{-9}} = 5.40 \text{ MHz}$$

(c) We observe that the value of f_H in part b is greater than the corresponding value (2.15 MHz) for the simple common-source stage. In Sec. 10-15 we noted that the feedback provided by R_S reduced the gain. Here we note that, concurrently, the bandwidth increases. The effect of feedback is treated in Chap. 12.

The Coefficient a_2   The location of the closest nondominant pole and, therefore, the separation between the dominant and nondominant poles, is determined by a_2. As seen in Eqs. (11-28) and (11-36), for example, a_2 comprises the product of time constants. As seen in Eq. (11-36), all possible pairs of capacitances form the time constants in a_2. Therefore, for the circuit in Fig. 11-23, a_2 is expressed as

$$a_2 = R_{11}^0 C_i R_{22}^1 C_2 + R_{11}^0 C_1 R_{33}^1 C_3 + R_{22}^0 C_2 R_{33}^2 C_3 \qquad (11\text{-}60)$$

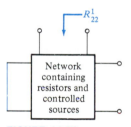

FIGURE 11-28
Network in Fig. 11-15 used to define R_{22}^1 (C_1 is short-circuited and C_3 is open-circuited).

where R_{ii}^i is the zero-frequency resistance seen by C_i when C_j is short-circuited. In Eq. (11-60), R_{22}^1 is the resistance seen by C_2 when C_1 is short-circuited and C_3 is open-circuited as depicted in Fig. 11-28. Similarly, R_{33}^1 is the resistance seen by C_3 with C_1 short-circuited and C_2 open-circuited; R_{33}^2 is obtained by short-circuiting C_2 and open-circuiting C_1. The rationale behind the notation is as follows: the subscript indicates the terminals at which the resistance is computed, and the superscript denotes the capacitance that is shorted. All capacitances not indicated in either subscript or superscript are open-circuited. Thus each term in Eq. (11-60) is the product of an open-circuit and short-circuit time constant.

The general form of each term in a_2 is of the form

$$R_{ii}^0 C_i R_{jj}^i C_j = R_{jj}^0 C_j R_{ii}^i C_i \qquad (11\text{-}61)$$

Equation (11-61) indicates that for any pair of capacitors C_i and C_j, we can find the open-circuit time constant for one and the short-circuit time constant for the second. Note that the capacitor that is shorted to determine R_{jj}^i (R_{ii}^i) is the one for which the open-circuit time constant is computed. Hence, the value of a_2 is unchanged if $R_{22}^0 C_2 R_{33}^2 C_3$ is replaced by $R_{33}^0 C_3 R_{22}^3 C_2$. Often, the choice of which form of Eq. (11-61) is used leads to a more convenient circuit analysis.

While there are four parameters in each term in Eq. (11-60), only the short-circuit resistances R_{ii}^i must be evaluated as the open-circuit resistances R_{ii}^0 are known from the calculation of a_1. The following example illustrates how the computation of the short-circuit time constants are made.

Example 11-4

Solution

Determine the a_2 coefficient for the circuit in Fig. 11-14.

For the two-capacitor system

$$a_2 = R_\pi^0 C_\pi R_\mu^\pi C_\mu = R_\mu^0 C_\mu R_\pi^\mu C_\pi$$

The open-circuit resistances R_π^0 and R_μ^0 are known (Example 11-2). Thus we may calculate either R_μ^π or R_π^μ. To illustrate the method, we shall do both using Fig. 11-29a for R_μ^π and Fig. 11-29b for R_π^μ.

In Fig. 11-29a, short-circuiting C_π makes $v_\pi = 0$. Consequently, R_π^0 is short-circuited and the current source $g_m v_\pi$ is open-circuited ($I = 0$). Thus $R_\mu^\pi = R_L$ and

$$a_2 = R_\pi^0 C_\pi R_L C_\mu$$

the result obtained in Eq. (11-28).

In Fig. 11-29b, short-circuiting C_μ places R_π^0, R_L, and the current source $g_m v_\pi$ in parallel. The voltage across the dependent source is v_π, and the current through it is $g_m v_\pi$, and hence this is a resistance $v_\pi/g_m v_\pi = 1/g_m$. Thus R_π^μ is simply the parallel combination of R_π^0, R_L, and $1/g_m$ or

$$R_\pi^\mu = R_\pi^0 \parallel R_L \parallel \frac{1}{g_m} = \frac{R_\pi^0 R_L}{R_\pi^0(1 + g_m R_L) + R_L}$$

Consequently, using the result for R_μ^0 in Example 11-2, we obtain

$$a_2 = [R_\pi^0(1 + g_m R_L) + R_L] C_\mu \times \frac{R_\pi^0 R_L}{R_\pi^0(1 + g_m R_L) + R_L} C_\pi = R_\pi^0 C_\pi R_L C_\mu$$

As expected, we obtain the same value of a_2.

The method described in this section is not an approximation, but gives the *exact* values of the coefficients a_1 and a_2. Thus, using the time-constant method, we obtain the same results that we would from the use of Cramer's rule or Gaussian elimination to solve, simultaneously, the nodal (mesh) equations for the circuit. For a two-node circuit, such as the common-emitter stage, the

FIGURE 11-29
Circuits for the common-emitter stage used to evaluate (a) R_μ^π and (b) R_π^μ.

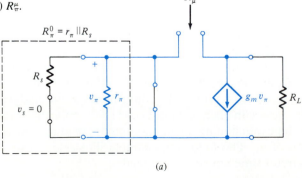

(a)

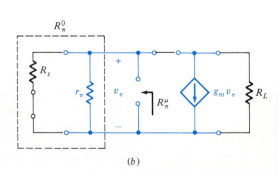

(b)

method does not provide any real advantage as the transfer function has only two poles and is readily obtained. However, when multistage amplifiers having four or more nodes are considered, the technique affords significant convenience as a_1 (and a_2) are obtained directly. By contrast, the use of nodal analysis, for example, requires that we obtain the entire characteristic polynomial (all n coefficients). Furthermore, as demonstrated in subsequent sections, open-circuiting and short-circuiting capacitors tend to decouple various portions (stages) of the circuit. Each time constant is associated with one part of the amplifier, and the impact of element values in this segment of the circuit on f_H is readily apparent. The circuit designer, consequently, can better evaluate the design trade-offs that result from changing component values. When used in conjunction with computer simulations, the engineer derives the advantages of insight into circuit behavior and accurate numerical data.

11-10   THE FREQUENCY RESPONSE OF CASCADED STAGES

Amplifiers are designed to provide gain over a specified frequency range. We demonstrated in Sec. 10-16 that by cascading stages, the overall gain, equal to the product of the gains of the individual stages, is increased markedly. In this section we examine the effect that cascading stages has on the frequency response of the overall amplifier. We demonstrate that the gain-bandwidth product of the cascade is increased in comparison to that for a single stage. For a specified value of f_H, therefore, the gain of the cascaded amplifier is greater than that achieved by a single stage having the same f_H.

The *CE-CE* and *CS-CS* Cascades at High Frequencies

A pair of cascaded common-emitter stages is shown in Fig. 11-30a, and its high-frequency equivalent circuit is given in Fig. 11-30b. Note that biasing components and coupling and bypass capacitors, where used, are not shown as these elements do not influence behavior at high frequencies.

As two stages are cascaded, we expect a transfer function that contains four poles and two zeros, that is, two poles and one zero from each stage. However, we cannot express the transfer function as the product of the individual high-frequency gains $A_{VH1}(s)$ and $A_{VH2}(s)$ because the stages are coupled to each other by means of $C_{\mu 1}$ and $C_{\mu 2}$. Thus, as we see in Fig. 11-30b, changing the value of R_{C2} affects the input of stage 2 (the Miller effect). As the input to stage 2 is part of the load on stage 1, this, too, is reflected into the input of stage 1 by the Miller effect. If we were to consider stage 1 (stage 2) separately, we would eliminate this coupling and the results obtained would be in error.

We can obtain the complete transfer function by writing the four nodal equations[1] for the circuit and solving the resultant set of simultaneous equations. In addition to being cumbersome, this approach makes it difficult to

---

[1] For a three-stage amplifier, six nodal equations are required. In general, two nodal equations are required per stage so that the number of simultaneous equations to be solved increases rapidly.

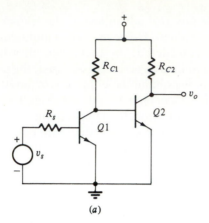

FIGURE 11-30
(a) A *CE-CE* cascaded amplifier and (b) its equivalent circuit at high frequencies.

(a)

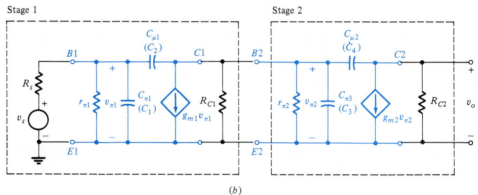

(b)

relate the performance of individual stages to the overall response. Instead, we make use of our previous discussion of the single-stage amplifier to approximate the high-frequency response. On the basis of this analysis we can reasonably assume that the zeros of the transfer function occur at sufficiently high frequencies that they can be neglected. As we are primarily interested in evaluating the upper 3-dB frequency f_H, we assume that dominant-pole conditions apply, and all that is needed is to compute the coefficient a_1. This approach is justifiable because many practical amplifiers are designed to have a dominant pole. Furthermore, as we show in the following development, the a_1 coefficient of the cascade is readily related to the a_1 coefficients of the individual stages, a factor essential in amplifier design. On the basis of the evaluation of a_1 we can subsequently determine the coefficient a_2 (if necessary) and examine the validity of the dominant-pole approximation. We make this calculation in Example 11-5.

The circuit in Fig. 11-31 is used to evaluate the open-circuit resistances needed in the expression for a_1 [Eq. (11-59)]. To prevent the notation from becoming too cumbersome, it is convenient to identify $C_1 = C_{\pi 1}$, $C_2 = C_{\mu 1}$, $C_3 = C_{\pi 3}$, and $C_4 = C_{\mu 2}$ as indicated in Fig. 11-30b. We note that one advantage of open-circuiting the capacitances in Fig. 11-31 is that the stages are decoupled. Thus, from the results obtained in the analysis of the single-stage common-

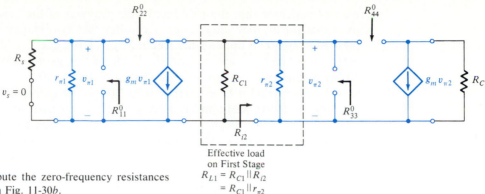

FIGURE 11-31

Circuit used to compute the zero-frequency resistances for *CE-CE* cascade in Fig. 11-30*b*.

Effective load
on First Stage
$R_{L1} = R_{C1} \| R_{i2}$
$= R_{C1} \| r_{\pi2}$

emitter amplifier, R_{11}^0 and R_{22}^0 for the first stage can be written as

$$R_{\pi1}^0 = R_{11}^0 = R_s \| r_{\pi1} \qquad R_{\mu1}^0 = R_{22}^0 = R_{11}^0(1 + g_{m1}R_{L1}) + R_{L1} \quad (11\text{-}62)$$

In Eq. (11-62), the resistance R_{L1}, identified in Fig. 11-31, is the effective load on the first stage, namely R_{C1} in parallel with the input resistance $R_{i2} = r_{\pi2}$ of the second stage.

The source resistance for the second stage is the output resistance $R_{01}' = R_{C1}$ of the first stage (Sec. 10-11), and in a fashion similar to the first stage, we obtain

$$R_{\pi2}^0 = R_{33}^0 = R_{C1} \| r_{\pi2} \qquad R_{\mu2}^0 = R_{44}^0 = R_{33}^0(1 + g_{m2}R_{C2}) + R_{C2} \quad (11\text{-}63)$$

Note that with $r_b = 0$, $R_{L1} = R_{\pi2}^0$. Using Eqs. (11-62) and (11-63) in conjunction with Eq. (11-59), we obtain

$$a_1 = R_{11}^0 C_1 + [R_{11}^0(1 + g_{m1}R_{L1}) + R_{L1}]C_2 + R_{33}^0 C_3 + [R_{33}^0(1 + g_{m2}R_{C2}) + R_{C2}]C_4 \quad (11\text{-}64)$$

We can identify the first two terms of Eq. (11-64) as the a_1 coefficient of the loaded first stage (let us call this a_{11}). The last pair of terms in Eq. (11-64) is the a_1 coefficient of the second stage (call this a_{12}). Using the dominant-pole approximation, the upper 3-dB frequency f_H of the cascade is

$$f_H = \frac{1}{2\pi a_1} = \frac{1}{2\pi a_{11} + 2\pi a_{12}} = \frac{1}{1/f_{H1} + 1/f_{H2}} \quad (11\text{-}65)$$

Clearly, Eq. (11-65) relates f_H to the upper 3-dB frequencies f_{H1} and f_{H2} of the individual stages. We also observe that f_H is smaller than either f_{H1} or f_{H2}, and hence we conclude that cascading stages decreases the bandwidth.

The reduction of the bandwidth is an "additive" process as seen in Eq. (11-65), whereas the midband gain increases multiplicatively. For numbers greater than unity, their product increases more rapidly than their sum.[1] Thus

---

[1] For n numbers $a_1, a_2, \ldots, a_n$, if each $a_i \geq n^{(1/n - 1)}$, the product $a_1 a_2 \cdots a_n \geq a_1 + a_2 + \cdots + a_n$. For $n = 2$, $a_i \geq 2$. For amplifier stages, usually both gain and bandwidth are greater than 2.

we can conclude that the gain-bandwidth product of the cascade is increased over that for an individual stage. Consider that two stages, characterized by midband gains and upper 3-dB frequencies of 100 and 0.1 MHz and 10 and 1 MHz, respectively, are cascaded. Each has a gain-bandwidth product of 10 MHz. The cascade has an overall gain of $100 \times 10 = 1000$ and, using Eq. (11-65), $f_H = 1/[(1/0.1) + (1/1)] = 0.91$ MHz. The gain-bandwidth product of the cascade is 910 MHz and is clearly much larger than that for either stage.

By analogy, the result given in Eq. (11-65) applies to a pair of cascaded common-source stages. The values of a_{11} and a_{12} are given by Eq. (11-36) in Sec. 11-7.

The result in Eq. (11-65) can be extended to an N-stage cascaded amplifier as follows: *The a_1 coefficient of the cascade is the sum of the a_1 coefficients of the individual stages. The effective load resistance on any stage is its collector (drain) resistance in parallel with the input resistance to the succeeding stage. The signal-source resistance of the stage is the output resistance of the previous stage. For the first stage the signal-source resistance is used.*

Example 11-5

The parameter values used in the *CE-CE* cascade in Fig. 11-30 are as follows: $R_s = 600\ \Omega$, $R_{C1} = 1.5\ \mathrm{k\Omega}$, $R_{C2} = 600\ \Omega$, $r_{\pi 1} = 1.2\ \mathrm{k\Omega}$, $g_{m1} = 0.1\ \mho$, $C_{\pi 1} = 24.5\ \mathrm{pF}$, $C_{\mu 1} = 0.5\ \mathrm{pF}$, $r_{\pi 2} = 2.4\ \mathrm{k\Omega}$, $g_{m2} = 0.05\ \mho$, $C_{\pi 2} = 19.5\ \mathrm{pF}$, and $C_{\mu 2} = 0.5\ \mathrm{pF}$. (*a*) Determine the approximate value of f_H and the approximate location of the dominant pole. (*b*) Determine the approximate location of the closest nondominant pole and comment on the validity of the dominant-pole approximation.

Solution

(*a*) To obtain f_H, we first evaluate the resistances in Eqs. (11-63) and (11-64). Thus

$$R_{11}^0 = 600 \parallel 1200 = 0.40\ \mathrm{k\Omega}$$
$$R_{L1} = R_{C1} \parallel r_{\pi 2} = 1.50 \parallel 2.4 = 0.923\ \mathrm{k\Omega}$$

and

$$R_{22}^0 = 0.40(1 + 100 \times 0.923) + 0.923 = 38.2\ \mathrm{k\Omega}^1$$
$$R_{33}^0 = 1.5 \parallel 2.4 = 0.923\ \mathrm{k\Omega}$$
$$R_{44}^0 = 0.923(1 + 50 \times 0.6) + 0.6 = 29.2\ \mathrm{k\Omega}$$

Use of Eq. (11-64) gives

$$a_1 = 0.40 \times 24.5 + 0.5 \times 38.2 + 0.923 \times 19.5 + 0.5 \times 29.2 = 61.5\ \mathrm{ns}$$

and, from Eq. (11-65),

$$f_H = \frac{1}{2\pi \times 61.5 \times 10^{-9}} = 2.59\ \mathrm{MHz}$$

---

[1]The value of $g_{m1} = 100\ \mathrm{m\mho}\ (0.1\ \mho)$; $1\ \mathrm{m\mho} \times 1\ \mathrm{k\Omega}$ is dimensionless and is a convenience in computation.

The dominant pole is located at $-p_1$, where

$$p_1 = \frac{1}{a_1} = \frac{1}{61.5 \times 10^{-9}} = 1.63 \times 10^7 \text{ rad/s}$$

It is of interest to determine the values of f_{H1} and f_{H2} of the amplifier. The first two terms in the calculation of a_1 give a_{11}, the last pair of terms a_{12}. These are

$$a_{11} = 28.9 \text{ ns} \qquad a_{12} = 32.6 \text{ ns}$$

from which

$$f_{H1} = \frac{1}{2\pi a_{11}} = \frac{1}{2\pi \times 28.9 \times 10^{-9}} = 5.51 \text{ MHz}$$

and

$$f_{H2} = \frac{1}{2\pi a_{12}} = \frac{1}{2\pi \times 32.6 \times 10^{-9}} = 4.88 \text{ MHz}$$

From these values we see the decrease in f_H caused by cascading the stages (to less than 60 percent of f_{H2}, the smaller of the two values). However, the midband gains of these stages are -100 and -18.5, respectively, and result in an overall gain of 1850. Again, it is clear that the gain has increased by a considerably larger percentage compared with the percentage decrease in f_H.

(b) To obtain the angular frequency p_2 of the second pole, we must evaluate the coefficient a_2. With four capacitors, there are six products of open- and short-circuit time constants; thus a_2 is expressible as

$$a_2 = R_{11}^0 C_1 R_{22}^1 C_1 + R_{11}^0 C_1 R_{33}^1 C_3 + R_{11}^0 C_1 R_{44}^1 C_4$$
$$+ R_{22}^0 C_2 R_{33}^2 C_3 + R_{22}^0 C_2 R_{44}^2 C_4 + R_{33}^0 C_3 R_{44}^3 C_4$$

Correlation of the preceding equation above with the circuit in Fig. 11-30b indicates that only two terms, previously unidentified, must be computed. In Fig. 11-30b we observe that with C_2 open-circuited, the portion of the circuit containing C_1 is decoupled from the second stage (C_3 and C_4). The resistances seen by C_3 and C_4 are thus independent of whether C_1 is open- or short-circuited. Hence $R_{33}^1 = R_{33}^0$ and $R_{44}^1 = R_{44}^0$, and the second and third terms in the expression for a_2 can be written as

$$R_{11}^0 C_1 [R_{33}^0 C_3 + R_{44}^0 C_4] = R_{11}^0 C_1 a_{12}$$

The resistance R_{22}^1 is computed from the circuit in Fig. 11-32a. With C_3 and C_4 open-circuited, $R_{22}^1 = R_{L1}$ is simply the value of R_{22}^1 for the loaded first stage. Similarly, short-circuiting C_3 completely decouples the first stage from the calculation of R_{44}^3, which, as seen in Fig. 11-32b, is R_{C2}, and the value of R_{44}^3 for the second stage. Thus the pair of terms containing R_{22}^1 and R_{44}^3 are just the a_2 coefficients of the individual stages.

FIGURE 11-32

Equivalent circuits used to calculate (a) R_{22}^1, (b) R_{44}^3, and (c) R_{44}^2.

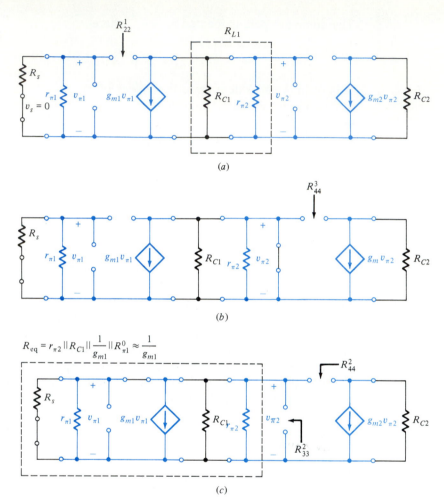

(a)

(b)

(c)

The remaining pair of short-circuit resistances is obtained by short-circuiting C_2. It is noted in Fig. 11-31 that short-circuiting C_2 causes the voltage across the source $g_{m1}v_{\pi1}$ to become $v_{\pi1}$. This v-i relationship represents a resistance $v_{\pi1}/g_{m1}v_{\pi1} = 1/g_{m1}$ as shown in the circuit in Fig. 11-32c. As noted in Fig. 11-32c, $1/g_{m1}$ is much smaller than either R_{C1} or R_{11}^0 and hence makes the parallel combination of these resistances approximately $1/g_{m1}$. The circuit in Fig. 11-32c is that used to compute R_{33}^0 and R_{44}^0, except that the effective source resistance is $1/g_{m1}$ instead of R_{C1} (see Fig. 11-30b). Thus

$$R_{33}^2 = r_{\pi2} \parallel R_{C1} \parallel \frac{1}{g_{m1}} \parallel R_{11}^0 \quad \text{and} \quad R_{44}^2 = R_{33}^2(1 + g_{m2}R_{C2}) + R_{C2}$$

Combination of all the terms in the three previous paragraphs gives

$$a_2 = R_{11}^0 C_1(R_{L1}C_2 + a_{12}) + R_{33}^0 C_3 R_{C2} C_4 + R_{22}^0 C_2(R_{33}^2 C_3 + R_{44}^2 C_4)$$

The resistances R_{33}^2 and R_{44}^2 are

$$R_{33}^2 = 2.4 \parallel 1.5 \parallel 0.01 \parallel 0.4 = 0.01 \text{ k}\Omega$$
$$R_{44}^2 = 0.01(1 + 50 \times 0.6) + 0.6 = 0.91 \text{ k}\Omega$$

Use of these and previously determined values in part *a* yields

$$a_2 = 0.40 \times 24.5(0.923 \times 0.5 + 32.6) + 0.923 \times 19.5 \times 0.6 \times 0.5$$
$$+ 38.2 \times 0.5(0.010 \times 19.5 + 0.91 \times 0.5) = 342 \times 10^{-18} \text{ s}$$

The angular frequency of the pole is

$$p_2 = \frac{a_1}{a_2} = \frac{61.5 \times 10^{-9}}{342 \times 10^{-18}} = 1.80 \times 10^8 \text{ rad/s}$$

and

$$f_2 = \frac{p_2}{2\pi} = \frac{1.80 \times 10^8}{2\pi} = 28.6 \text{ MHz}$$

As the poles are separated by more than a factor of 10, we can conclude that the dominant-pole approximation is valid. This is verified by computer simulation which gives $p_1 = 1.79 \times 10^7$ rad/s, $p_2 = 16.1 \times 10^7$ rad/s, and corresponding values $f_1 = 2.85$ MHz and $f_2 = 25.6$ MHz. The error in the dominant-pole calculation is nearly 10 percent, and the exact value of f_H is 2.71 MHz. Thus $f_H = 1/2\pi a_1$ gives a very close approximation of the actual value. Note also that $f_H = 1/2\pi a_1$ is less than the actual value.

One further observation is noteworthy. Stages 1 and 2, performing as single-stage amplifiers, would have pole frequencies of 5.51 and 4.88 MHz, respectively. As a result of the coupling between the stages when they are cascaded, however, the two corresponding pole frequencies of the cascade are approximately 2.59 and 28.6 MHz. Hence one pole has moved closer to the origin and the other, further away. This condition is often referred to as *pole splitting* and is often exploited in compensating operational amplifiers.

11-11 THE CASCODE (*CE-CB*) AMPLIFIER   The cascode amplifier (Fig. 10-35) comprises a common-emitter stage in cascade with a common-base stage. In Sec. 10-17 we demonstrated that the midband gain of this combination is virtually the same as the gain of a common-emitter stage having the same load resistance R_{C2} as does the common-base stage. Here we show that the frequency response of the composite is greater than that obtained for the corresponding common-emitter stage. To do so, we evaluate the a_1 coefficient of the cascode amplifier and compare the result with the a_1 coefficient of a common-emitter stage having a load R_{C2}.

The high-frequency equivalent circuit of the cascode amplifier is shown in Fig. 11-33*a*, and that used to evaluate the open-circuit resistances is given in Fig. 11-33*b*. Following the procedure outlined earlier in this section, the coefficient a_1 of the *CE-CB* cascade is the sum of the a_1 coefficients of the loaded common-emitter stage a_{11} and the common-base stage a_{12}. For the common-emitter stage

FIGURE 11-33
(a) The high-frequency
equivalent circuit of the
cascode (CB-CE cas-
cade) amplifier. (b) Cir-
cuit used to calculate
$R_{\pi 1}^0$ and $R_{\mu 1}^0$.

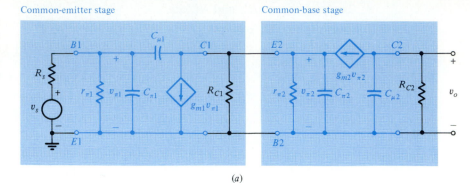

Common-emitter stage   Common-base stage

(a)

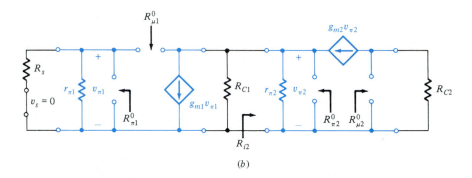

(b)

$$a_{11} = R_{\pi 1}^0 C_{\pi 1} + [R_{\pi 1}^0(1 + g_{m1}R_{L1}) + R_{L1}]C_{\mu 1} \qquad (11\text{-}66)$$

where $R_{\pi 1}^0$ is given in Eq. (11-62) as $R_s \parallel r_{\pi 1}$. The resistance $R_{L1} = R_{C1} \parallel R_{i2}$, where R_{i2} is the input resistance of the common-base stage. As described in Sec. 10-2 and listed in Table 10-4, the input resistance $R_i \approx r_\pi/(1 + \beta_o)$ of a common-base stage is extremely small; thus $R_{L1} \approx R_{i2}$ is also small. Equation (11-66) also applies to the corresponding common-emitter stage having R_{C2} as a load. As $R_{C2} \gg R_{i2}$, we see that the Miller effect multiplication of $C_{\mu 1}$ in the cascode amplifier is reduced significantly from that of the common-emitter stage. It is this reduction in the influence of $C_{\mu 1}$ that gives the *CE-CB* amplifier its improved frequency response.

The common-base stage has a higher f_H than does a common-emitter stage. As derived in Prob. 11-22a, the coefficient of a common-base stage can be approximated by

$$a_{12} \simeq \frac{1}{\omega_T} + C_{\mu 2}R_{C2} \qquad (11\text{-}67)$$

For typical parameter values, the value of a_{12} in Eq. (11-67) and hence f_{H2} is comparable to that of an emitter follower. Thus its effect on the overall response is minimal.

Example 11-6

A cascode amplifier has $R_{C1} = R_{C2} = 1.5$ kΩ and $R_s = 300$ Ω. The transistors are identical and have $r_\pi = 2$ kΩ, $g_m = 0.05$ ℧, $\beta_o = 100$, $C_\pi = 19.5$ pF, $C_\mu = 0.5$ pF, and $\omega_T = 2.5 \times 10^9$ rad/s. (a) Determine f_H for the circuit. (b) Determine f_H for a common-emitter stage having $R_C = 1.5$ kΩ, driven from a source having $R_s = 300$ Ω and using the transistor whose parameters are given above. Compare the result with that obtained in part a.

Solution

(a) For the common-emitter stage, from Eq. (11-62), we obtain

$$R_{\pi 1}^0 = 0.30 \parallel 2.0 = 0.261 \text{ kΩ}$$

Using Table 10-3 for the common-base amplifier, we have

$$R_{i2} = \frac{2.0}{1 + 100} = 0.0198 \text{ kΩ}$$

and

$$R_{L1} = R_{C1} \parallel R_{i2} = 1.5 \parallel 0.0198 = 0.0195 \text{ kΩ}$$

Hence

$$a_{11} = 0.261 \times 19.5 + 0.5[0.261(1 + 50 \times 0.0195) + 0.0195] = 5.36 \text{ ns}$$

Use of Eq. (11-67) for the common-base stage gives

$$a_{12} = \frac{1}{2.5} + 0.5 \times 1.5 = 1.15 \text{ ns}$$

Thus

$$a_1 = a_{11} + a_{12} = 5.36 + 1.15 = 6.51 \text{ ns}$$

and

$$f_H = \frac{1}{2\pi a_1} = \frac{1}{2\pi \times 6.51 \times 10^{-9}} = 24.4 \text{ MHz}$$

(b) For the equivalent common-emitter stage,

$$a_1 = 0.261 \times 19.5 + 0.5[0.261(1 + 50 \times 1.5) + 1.5] = 15.8 \text{ ns}$$

and

$$f_H = \frac{1}{2\pi a_1} = \frac{1}{2\pi \times 15.8 \times 10^{-9}} = 10.1 \text{ MHz}$$

Clearly, the cascode amplifier has a higher value of f_H than does the common-emitter stage. If the load resistance R_{C2} is made larger (say, 5 kΩ), the improvement in f_H is even more dramatic (18.7 MHz for the cascode and 3.82 MHz for the common-emitter stage).

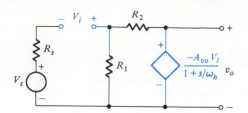

FIGURE 11-34
The noninverting stage using an Op-Amp represented as having a single dominant pole at $s = -\omega_h$.

11-12 THE OPERATIONAL AMPLIFIER AT HIGH FREQUENCIES

Practical Op-Amps, often the cascade of a differential amplifier, common-emitter stage, and emitter follower (Sec. 10-18), are designed so that their high-frequency response is characterized by a single dominant pole. Thus the gain A_v of the Op-Amp is

$$A_v(s) = \frac{A_{vo}}{1 + s/\omega_h} \tag{11-68}$$

where A_{vo} is the dc gain (midband) and ω_h is the angular frequency of the dominant pole. The gain-bandwidth product of the Op-Amp is $A_{vo}\omega_h$ (Sec. 11-3). We now determine the upper 3-dB frequency of the basic inverting and noninverting stages. To focus our attention on the effect of the dominant pole, we treat the Op-Amp as ideal in all respects except that the gain is given by Eq. (11-68).

The Noninverting Stage   The model of the noninverting stage is depicted in Fig. 11-34, in which the controlled source displays the frequency variation. The gain A_V of this stage is given in Eq. (10-10) and, with $R_o = 0$ and A_v given in Eq. (11-68), is

$$A_{VH}(s) = \frac{A_{vo}(R_1 + R_2)/(1 + s/\omega_h)}{R_1[1 + A_{vo}/(1 + s/\omega_h)] + R_2} \tag{11-69}$$

Clearing fractions and rearranging terms, we can write $A_{VH}(s)$ as

$$A_{VH}(s) = \frac{A_{vo}(R_1 + R_2)}{R_1(1 + A_{vo}) + R_2} \frac{1}{1 + \{s(R_1 + R_2)/\omega_h[R_1(1 + A_{vo}) + R_2]\}}$$

$$= \frac{A_{VO}}{1 + s/\omega_H} \tag{11-70}$$

where

$$A_{VO} = \frac{A_{vo}(R_1 + R_2)}{R_1(1 + A_{vo}) + R_2} \qquad \omega_H = \frac{\omega_h[R_1(1 + A_{vo}) + R_2]}{R_1 + R_2} \tag{11-71}$$

The angular gain-bandwidth product for the stage is $A_{VO}\omega_H$ and equals $A_{vo}\omega_h$ and can be seen in the Bode diagram in Fig. 11-35. The gain-bandwidth product is usually specified by manufacturers (as the unity gain 3-dB frequency). Con-

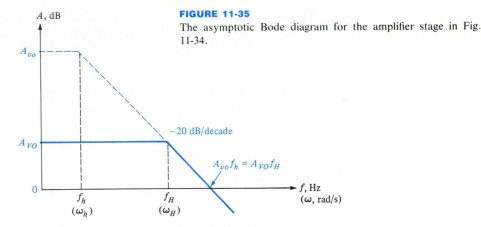

FIGURE 11-35

The asymptotic Bode diagram for the amplifier stage in Fig. 11-34.

sequently, for an Op-Amp with a 1-MHz gain-bandwidth product, a noninverting stage having a gain of 20 has a bandwidth of $f_H = \frac{1}{20}$ MHz = 50 kHz. As the gain A_{VO} is set by the resistor ratio R_2/R_1, both gain and bandwidth for a given Op-Amp are specified once the resistance ratio is selected.

Because the value of A_{vo} is large ($\simeq 10^5$), A_{VO} and f_H can be approximated as

$$A_{VO} \simeq 1 + \frac{R_2}{R_1} \qquad f_H \simeq f_h \frac{A_{vo}}{A_{VO}} \qquad (11\text{-}72)$$

Note the explicit relationship of f_H with the gain-bandwidth product and midband gain of the stage in Eq. (11-72). Typical Op-Amps have $A_{vo} \simeq 10^5$ and gain-bandwidth products of a few megahertz. Consequently, the dominant-pole frequency f_h is quite low (5 to 100 Hz is common).

The Inverting Stage   The analysis for the inverting stage parallels that for the noninverting stage just described. The results (Prob. 11-46) are given in Eqs. (11-73) and (11-74):

$$A_{VH}(s) = \frac{A_{vo}}{1 + s/\omega_H}$$

$$= \frac{-A_{vo}R_2}{R_1(1 + A_{vo}) + R_2} \frac{1}{1 + \{s(R_1 + R_2)/\omega_h[R_1(1 + A_{vo}) + R_2]\}}$$

$$(11\text{-}73)$$

Note that the value of ω_H in Eq. (11-73) is identical to that for the noninverting stage. Thus, for given values of R_1 and R_2, both inverting and noninverting stages using identical Op-Amps have the same bandwidth.

For $A_{vo} \gg 1$, Eq. (11-73) reduces to

$$A_{VO} = -\frac{R_2}{R_1} \qquad f_H = \frac{A_{vo}f_h}{1 + |A_{vo}|} \qquad (11\text{-}74)$$

Note that the gain-bandwidth product of the inverting stage is not equal to

$A_{vo}f_h$. For equal values of A_{VO}, the noninverting stage has a larger bandwidth. The reasons for this difference will become apparent in the discussion of the frequency response of feedback amplifiers (Chap. 13).

11-13 THE EFFECT OF COUPLING AND BYPASS CAPACITORS

The high-frequency response of amplifiers was treated in the eight previous sections. Almost call discrete-component circuits employ coupling and bypass capacitors. (Some IC amplifiers also utilize coupling capacitors.) Previously, we demonstrated that these capacitances affect amplifiers at low frequencies, and it is this frequency range with which we are concerned in this section.

The equivalent circuit of a common-emitter stage having both a bypass capacitor C_E and a coupling capacitor C_B is displayed in Fig. 11-36. Note that this is the model of the amplifier stage in Fig. 10-12 for which C_B in Fig. 11-36 is C_{B1}; C_{B2} is not considered here as it represents the coupling capacitor between the output of this stage and the input to a subsequent stage. Hence its effect is included in the low-frequency analysis of the second stage.

Each of the two capacitances C_B and C_E influences the low-frequency performance. First, let us assume that C_E bypasses R_E perfectly; that is, C_E acts as a short circuit (alternatively, C_E is assumed infinite). For this circuit we have one capacitor C_B, and consequently, the transfer function has one pole at the reciprocal of the circuit time constant. The gain is zero at dc ($s = j\omega = 0$) as C_B is open-circuited, making I_b, V_π, and hence $V_o = 0$. The equivalent resistance R_B^E in the base loop is $R_s + r_\pi$. Note that we are using the notation of the previous sections as R_B^E is the resistance seen by C_B when C_E is short-circuited. As $I_b = V_s/(R_B^E + 1/sC_B)$, $V_\pi = r_\pi I_b$ and $V_o = -g_m V_\pi R_C$, the transfer function is expressible as

$$A_{VL}(s) = \frac{A_{VO}sR_B^E C_B}{1 + sR_B^E C_B} = \frac{A_{VO}s/2\pi f_{LB}}{1 + s/2\pi f_{LB}} \tag{11-75}$$

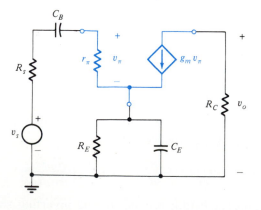

FIGURE 11-36
Low-frequency representation of a discrete-component common-emitter stage including coupling and bypass capacitors.

where

$$f_{LB} = \frac{1}{2\pi R_B^E C} = \frac{1}{2\pi(R_s + r_\pi)C_B} \qquad (11\text{-}76)$$

Note that A_{VO}, the midband gain, is obtained in Eq. (11-75) for $s = j\omega \to \infty$, that is, at frequencies where C_B acts as a short circuit and provides ideal coupling. The Bode diagram of Eq. (11-75) has the form displayed in Fig. 11-5.

Let us now investigate the effect of C_E when C_B provides ideal coupling. As seen in Fig. 11-36, the circuit has one capacitor C_E and the transfer function contains a single pole at $1/C_E R_E^B$, where R_E^B is the resistance seen by C_E with C_B short-circuited. We can expect one zero in the transfer function. This zero occurs at $s = -z_1$ for which $Z_E \to \infty$ (open-circuited), thus making I_b, V_π, and $V_o = 0$. The impedance Z_E is $R_E \parallel 1/sC_E = R_E/(1 + sR_EC_E)$, which becomes infinite when $sC_ER_E = -1$. Thus

$$z_1 = \frac{1}{R_E C_E} = \omega_E \qquad (11\text{-}77)$$

The equivalent resistance is identical to the output resistance R_o' of an emitter follower (see Fig. 10-23). Hence the transfer function is

$$A_{VL}(s) = \frac{A_{VO}\,(R_E^B/R_E)\,(1 + s/\omega_E)}{1 + sR_E^B C_E} \qquad (11\text{-}78)$$

where

$$R_E^B = \frac{R_E(R_s + r_\pi)/(1 + \beta_o)}{R_E + (R_s + r_\pi)/(1 + \beta_o)} \qquad (11\text{-}79)$$

The pole occurs at $s = -p_1 = -1/R_E^B C_E$, and as $R_E^B \ll R_E$ (recall that R_o' of the emitter follower is low), the pole-frequency p_1 is much greater than ω_E. Thus, for frequencies in the vicinity of p_1, the Bode diagram of Eq. (11-78) is that displayed in Fig. 11-5. The lower 3-dB frequency f_{LE} is

$$f_{LE} = \frac{p_1}{2\pi} = \frac{1}{2\pi C_E \left[\dfrac{R_E(R_s + r_\pi)/(1 + \beta_o)}{R_E + (R_s + r_\pi)/(1 + \beta_o)} \right]} \qquad (11\text{-}80)$$

The Complete Low-Frequency Response Both C_E and C_B affect the low-frequency response. When both capacitors are considered simultaneously, we expect a transfer function containing two poles and two zeros. The resultant transfer function (Prob. 11-51) is

$$A_{VL}(s) = \frac{A_{VO}\,(s/\omega_B)\,(1 + sR_EC_E)}{1 + a_1 s + a_2 s^2} = \frac{A_{VO}\,(s/\omega_B)\,(1 + s/z_1)}{(1 + s/p_1)\,(1 + s/p_2)} \qquad (11\text{-}81)$$

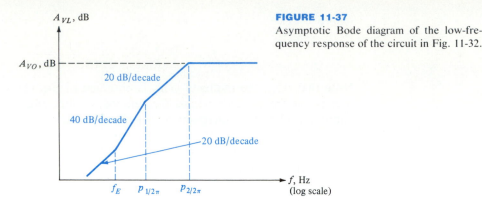

FIGURE 11-37
Asymptotic Bode diagram of the low-frequency response of the circuit in Fig. 11-32.

where

$$a_1 = R_E^0 C_E + R_B^0 C_B = R_E C_E + [R_s + r_\pi + (1 + \beta_o)R_E]C_B \atop a_2 = R_E^0 C_E R_B^E C_B = R_B^0 C_B R_E^E C_E = R_E C_E (R_s + r_\pi)C_B \Bigg\} \quad (11\text{-}82)$$

The form of the magnitude of the asymptotic Bode diagram of Eq. (11-81), assuming $z_1 < p_1 < p_2$, is displayed in Fig. 11-37. Again, we observe that midband occurs for large ω. Thus the lower 3-dB frequency is related to the pole farthest from the origin, that is, the largest pole-frequency (p_2). Using the approximation for p_2 in Eq. (11-24) gives

$$p_2 = \frac{a_1}{a_2} = \frac{R_E^0 C_E + R_B^0 C_B}{R_E^0 C_E R_B^E C_B}$$

which, after substitution for R_E^0, R_B^0, and R_B^E and division, yields

$$p_2 = \frac{1}{(R_s + r_\pi)C_B} + \frac{1}{\dfrac{C_E R_E (R_s + r_\pi)/(1 + \beta_o)}{R_E + (R_s + r_\pi)/(1 + \beta_o)}} \quad (11\text{-}83)$$

Comparison of Eqs. (11-76) and (11-80) with the terms in Eq. (11-83) results in

$$p_2 = \frac{1}{R_B^E C_B} + \frac{1}{R_E^B C_E} = 2\pi(f_{LB} + f_{LE}) \simeq 2\pi f_L \quad (11\text{-}84)$$

where f_{LB} and f_{LE} are the 3-dB frequencies associated with C_B and C_E, respectively [Eqs. (11-76) and (11-80)]. The frequency f_L is the lower 3-dB frequency (approximately) of the circuit.

The result in Eq. (11-84) can be extended to circuits containing more than two capacitances as indicated in Eq. (11-85):

$$2\pi f_L \simeq p_N = \sum_{K=1}^{N} \frac{1}{R_{KK}^\infty C_K} = 2\pi \sum_{K=1}^{N} f_{LK} \quad (11\text{-}85)$$

The resistance R_{KK}^∞ is defined as the resistance seen by C_K when *all* other capacitors are short circuits. Thus the lower 3-dB frequency can be approxi-

mated by simply summing the lower 3-dB frequencies attributed to each capacitor in the circuit.

Example 11-7   A transistor amplifier stage has $R_E = R_C = 1.5$ kΩ, $R_s = 600$ Ω, and transistor
parameters $\beta_o = 100$ and $r_\pi = 1.0$ kΩ. (a) Determine the values of C_B and C_E
needed to obtain $f_L = 50$ Hz. Assume that each capacitor contributes equally
to f_L. (b) Use the result in part a to determine the zero introduced by C_E.

Solution   (a) For an overall $f_L = 50$ Hz, $f_{LB} = f_{LE} = 25$ Hz. Using Eqs. (11-76) and
(11-80), we obtain

$$f_{LB} = 25 = \frac{1}{2\pi(0.6 + 1.0)C_B} \quad \text{or} \quad C_B = 3.98 \ \mu\text{F}$$

$$f_{LE} = 25$$

$$= \frac{1}{2\pi C_E\{[1.5(0.6 + 1.0)/ (1 + 100)]/[1.5 + (0.6 + 1.0)/(1 + 100)]\}}$$

or $C_E = 406 \ \mu\text{F}$

(b) From Eq. (11-66), $f_E = \omega_E/2\pi = 1/2\pi R_E C_E = 1/(2\pi \times 1.5 \times 0.406) = 0.261$ Hz.

The capacitance values obtained in Example 11-7 indicate the typical situation that exists in practice, namely, that the value of C_E required is much
larger than the value of C_B. In addition, the zero introduced by C_E occurs at
such a low frequency compared with the desired value of f_L that it has negligible
effect on the low-frequency response. Both size and cost of discrete capacitances increase with increasing capacitance value. Consequently, in amplifier
design, it is customary practice to select C_E to meet the specified value of f_L.
Then C_B is selected to make f_{LB} occur at a much lower frequency than f_L. A
good rule of thumb is to choose $f_{LB} \leq f_L/10$. For the values in Example 11-7,
this gives $C_E = 203 \ \mu\text{F}$ and $C_B \geq 19.9 \ \mu\text{F}$, thus making the total capacitance
about one-half of that found in Example 11-7.

Cascaded Stages at Low Frequencies   The lower 3-dB frequency f_L of cascaded amplifiers
is readily obtained by extension of the method described in Sec. 11-13 for the
single-stage amplifier. Originally given as Eq. (11-85) and repeated as Eq.
(11-86) in a somewhat altered form, f_L is expressible as

$$f_L = \sum_{K=1}^{N} f_{LK} = \frac{1}{2\pi} \sum_{K=1}^{N} \frac{1}{R_{KK}^\times C_K} \tag{11-86}$$

where $R_{KK}^\times C_K$ is the circuit time constant when all other capacitors are short-
circuited. It is clear in Eq. (11-86) that the overall lower 3-dB frequency f_L is
related to the lower 3-dB frequencies of the individual stages. Note that the
rule of thumb given immediately after Example 11-7 also applies to cascaded
stages at low frequencies.

Summary

The determination of the gain, upper and lower 3-dB frequencies of cascaded amplifiers can be summarized as follows:

1. Midband gain A_{VO} is the product of midband gains of individual stages.

2. The upper 3-dB frequency f_H is the sum of the reciprocals of the upper 3-dB frequencies f_{Hi} of the individual stages. The value of each f_{Hi} is the reciprocal of the sum of open-circuit time constants of the stage.

3. The lower 3-dB frequency f_L is the sum of the lower 3-dB frequencies f_{LK} of the individual stages. Each value of f_{LK} is the sum of the reciprocals of the short-circuit time constants of the stage.

The values of f_H and f_L are outlined by use of a dominant-pole approximation and generally give good correlation with measured values and computer simulations. Thus they are extremely useful in the "pencil-and-paper" calculations used in the initial phases of design.

REFERENCES

1 Gray, P. R., and R. G. Meyer: "Analysis and Design of Analog Integrated Circuits," 2d ed., John Wiley and Sons, New York, 1984.

2 Ghausi, M. S.: "Electronic Devices and Circuits: Discrete and Integrated," Holt, Rinehart and Winston, Inc., New York, 1985.

3 Moschytz, G. S.: "Linear Integrated Networks: Fundamentals," Van Nostrand, Reinhold Company, New York, 1974.

4 Cochrun, B. L., and A. Grabel: On the Determination of the Transfer Function of Electronic Circuits, *IEEE Trans. Circuit Theory*, vol. CT-20, pp. 16–20, January 1973.

5 Grebene, A. B.: "Bipolar and MOS Analog Integrated Circuit Design," John Wiley and Sons, New York, 1984.

6 Sedra, A. S., and K. C. Smith: "Microelectronic Circuits," Holt, Rinehart and Winston, Inc., New York, 1981.

7 Schilling, D., and C. Belove: "Electronic Circuits—Discrete and Integrated," McGraw-Hill Book Company, New York, 1979.

8 Soclof, S.: "Analog Integrated Circuits," Prentice-Hall, Englewood Cliffs, N.J., 1985.

REVIEW QUESTIONS

11-1 Define the frequency-response characteristic of an amplifier.

11-2 Sketch the high-frequency response of a single-pole transfer function.

11-3 Define f_H, the upper half-power frequency.

11-4 Repeat Rev. 11-2 for a single-pole high-pass system.

11-5 Define bandwidth.

11-6 (a) Sketch the step response of a single time constant low-pass system.
(b) Define the rise time t_r.

11-7 (a) Define tilt.
(b) What is the relationship between tilt and f_L?

11-8   (a)   Define f_β.
   (b)   Define f_T.
   (c)   Write an equation which relates f_β and f_T.

11-9   (a)   Write the transfer function of an amplifier having three poles and two finite zeros.
   (b)   Under what conditions does this amplifier have a dominant pole?

11-10   The three poles of an all-pole amplifier are widely separated.
   (a)   Write an expression for the approximate pole locations in terms of the coefficients of the transfer function.
   (b)   What is the approximate value of f_H?

11-11   Draw the unilateral hybrid-π equivalent circuit of a common-emitter stage.

11-12   For a common-emitter stage, use Miller's theorem to determine the input capacitance of a common-emitter stage.

11-13   Define the gain-bandwidth product (voltage).

11-14   Draw the equivalent circuit of a common-source stage valid at high frequencies.

11-15   Use Miller's theorem to obtain a unilateral model for a common-source stage at high frequencies.

11-16   (a)   Define open-circuit time constant.
   (b)   Write an equation for the a_1 coefficient in terms of the open-circuit time constants for a four-capacitor circuit.

11-17   What is meant by the zero-frequency resistance R_{ii}^0?

11-18   Express each term in the a_2 coefficient and explain the significance of each.

11-19   (a)   Approximate the first two pole frequencies in terms of the a_1 and a_2 coefficients.
   (b)   What is the approximate 3-dB bandwidth?

11-20   Under what conditions is the approximation in Rev. 11-19 valid?

11-21   Briefly explain how to compute R_{11}^0 and R_{22}^1.

11-22   (a)   What is larger, ω_H for a common-emitter stage or ω_H for a common-collector stage? Explain.
   (b)   A common-emitter stage is cascaded with a common-collector stage. What is the overall 3-dB frequency?

11-23   How do coupling and bypass capacitors affect the frequency response of an amplifier stage?

11-24   Write an expression by which the lower 3-dB frequency can be approximated. Identify each term.

11-25   (a)   Why does cascading of stages increase the gain-bandwidth product of an amplifier?
   (b)   Is the answer in part a always true?

11-26   Discuss the advantages of a cascode amplifier.

11-27   Relate the midband gain A_{VO} and half-power frequency f_H of a noninverting Op-Amp stage to the dc gain A_{vo} and bandwidth f_h of the Op-Amp.

11-28   Repeat Rev. 11-27 for an inverting stage.

Chapter 12
FEEDBACK AMPLIFIERS

Feedback is one of the fundamental processes in nature. It is the mechanism of the hand-eye coordination you used to turn to this page, of the control we use to drive an automobile at constant speed, of maintaining constant internal body temperature, and of natural population control in an ecosystem. What we mean by "feedback" is the process whereby a portion of the output is returned to the input to form part of the system excitation. This action, appropriately applied, tends to make systems self-regulating.

We have seen instances of feedback as it applies to electronic circuits in previous chapters. Feedback was used to make the operating point of a transistor insensitive to both manufacturing variations in β_F and temperature (Sec. 10-7). In a common-emitter stage containing an emitter resistance (Sec. 10-14), the feedback provided by R_E helped to maintain an essentially constant gain as β_o varied. The bandwidth of the operational-amplifier (Op-Amp) stage in Sec. 11-12 was shown to be greater than the bandwidth of the Op-Amp, the improvement attributed to the feedback resistance(s) R_2 (and R_1). The low-input and high-output resistances of the emitter (source) follower are another example of the effects of feedback.

The aforementioned examples highlight some of the advantages that can be derived by the appropriate use of feedback, namely, control of impedance levels, bandwidth improvement, and rendering circuit performance relatively insensitive to manufacturing and environmental changes. The latter is of principal importance in modern electronics because controlled, precise circuit performance can be attained without resorting to costly, precision components (or by minimizing their number). The instances cited are examples of negative feedback; that is, the signal fed back from output to input is 180° out of phase with the applied excitation. Thus the input signal to the amplifier is proportional to the difference between the excitation and output signals. As we shall see in subsequent sections, this differencing mechanism is fundamental to deriving the benefits of negative feedback.

The advantages of feedback, however, are accompanied by corresponding disadvantages. To achieve the desensitivity in the common-emitter stage containing an emitter resistance and the improved bandwidth in the Op-Amp stage, the gain is reduced from its value prior to introducing the feedback. Since the magnitude and phase of the gain vary with frequency, it is possible to introduce sufficient phase shift to cause positive feedback. Under these circumstances, the amplifier may become unstable and generate an output signal independent of the input (or with no input signal); in other words, it oscillates. Although positive feedback is useful in establishing the two stable states in a FLIP-FLOP (Sec. 8-1) and is exploited in constructing oscillator circuits, unwanted oscillations render an amplifier useless. Furthermore, electronic circuits often contain unwanted but unavoidable feedback paths. The signals fed back through such paths can deteriorate performance. The Miller effect multiplication of C_μ (C_{gd}) in transistor amplifiers, which reduces the upper 3-dB frequency, is one example of feedback that is unavoidable (and often undesirable). Similar effects caused by parasitic elements (such as the capacitance between input and output leads of an IC package) are sometimes observed.

The effects of feedback on amplifier gain, distortion, impedance levels, and sensitivity to parameter changes are investigated in this chapter. In particular, the performance of the four basic single-loop feedback amplifiers is examined. The chapter concludes with a brief introduction to multiloop feedback amplifiers. In Chap. 13 we treat the stability and frequency response of feedback amplifiers.

12-1   CLASSIFICATION AND REPRESENTATION OF AMPLIFIERS

Before proceeding with the concept of feedback, it is useful to classify practical amplifiers based on the controlled sources they are designed to approximate. The four broad categories of this classification correspond to the four types of ideal controlled sources. Each of the two voltage and two current sources, dependent on either a voltage or a current, has zero or infinite input impedance and zero or infinite output impedance. Consequently, source and load impedances have no effect on the input-output relationships of these ideal sources. Practical amplifiers, however, have finite, nonzero input and output impedances. Thus the size of amplifier impedance levels relative to load and source impedances is a necessary consideration in amplifier classification.

The Voltage Amplifier   The two-port network representation of an amplifier having one stage or a number of stages in cascade is displayed in Fig. 12-1. Note the similarity of this circuit with that for the practical Op-Amp in Sec. 10-21. The output portion of Fig. 12-1 (shown in color) represents the Thévenin equivalent of the amplifier, and R_i is the input resistance[1] of the amplifier (also shown in color). The resistance R_L is the load, and R_s is the resistance of the source V_s. If the

---

[1] The figures in this section show resistive input and output impedances. The discussion applies equally to generalized impedances Z_i and Z_o as displayed in Table 12-1.

FIGURE 12-1
Equivalent circuit of a
voltage amplifier.

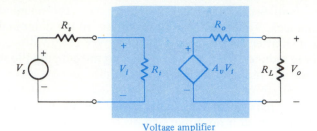

Voltage amplifier

input resistance is much greater then R_s, then $V_i \approx V_s$. Similarly, if $R_L \gg |R_o|$, then $V_o \approx A_v V_i = A_v V_s$. This amplifier provides an output voltage proportional to the input voltage, and *the proportionality factor is independent of the magnitudes of the source and lead resistances*. Such a circuit is called a *voltage amplifier* or a *voltage-voltage converter* and its behavior is that of the voltage-controlled voltage source. The symbol A_v in Fig. 12-1 represents V_o/V_i with $R_L \rightarrow \infty$ and hence is the *open-circuit voltage gain*.

The Current Amplifier An ideal current-controlled current source is unilateral, has zero input resistance R_i, and has infinite output resistance R_o. The practical *current amplifier*, or *current-current converter*, shown in Fig. 12-2 approaches ideal behavior when $|R_i| \ll R_s$, so that $I_i \approx I_s$ and when $R_L \ll |R_o|$, making $I_o = A_i I_i \approx A_i I_s$. Thus the output current is proportional to the input current and independent of either R_s or R_L. Note that $A_i \equiv I_o/I_i$, with $R_L = 0$, represents the *short-circuit current gain*. The circuit in Fig. 12-2 is analogous to the simplest BJT model based on the Ebers-Moll equations in which $I_c = \beta I_b$.

The Voltage-Current Converter or Transconductance Amplifier The *voltage-current converter* or *transconductance amplifier* is based on the ideal voltage-controlled current source. Note the similarity of this amplifier type, shown in Fig. 12-3, to the unilateral hybrid-π model of the BJT. To approximate ideal characteristics, $R_s \ll |R_i|$ and $R_L \ll |R_o|$ in the practical voltage-current converter. These conditions make $V_i \approx V_s$ and $I_o \approx G_m V_i \approx G_m V_s$, so that G_m, the proportionality factor, is independent of both load and source resistances. The parameter $G_m \equiv I_o/V_s$ with $R_L = 0$ is referred to as the *short-circuit transfer conductance* (or simply the *transconductance*). Note the similarity of G_m for the overall amplifier with the definition of g_m for the transistor.

FIGURE 12-2
Current amplifier equiv-
alent circuit.

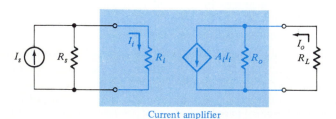

Current amplifier

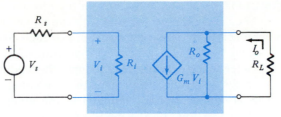

FIGURE 12-3
Equivalent circuit of a
voltage-to-current(trans-
conductance) amplifier.

Voltage-to-current converter

The Current-Voltage Converter or Transimpedance Amplifier The fourth amplifier
type, depicted in Fig. 12-4, approximates the behavior of the ideal current-
controlled voltage source. Because the output voltage is proportional to the
input current, this amplifier category is called a *transimpedance amplifier* or
current-voltage converter. The practical amplifier must have $|R_i| \ll R_s$ and
$|R_o| \ll R_L$ to approach ideal behavior, that is, $I_i \approx I_s$ and $V_o \approx Z_m I_i \approx Z_m I_s$.
The parameter $Z_m \equiv V_o/I_i$ with $R_L \to \infty$ is referred to as the *open-circuit transfer
impedance* or simply the *transimpedance*. The characteristics of the four amplifier types are summarized in Table 12-1.

TABLE 12-1 Basic Amplifier Characteristics

| | Amplifier type | | | | | | | |
| | Voltage | | Current | | Transconductance | | Transimpedance | |
Parameter	Ideal	Practical	Ideal	Practical	Ideal	Practical	Ideal	Practical
Z_i	∞	High; $|Z_i| \gg R_s$	0	Low; $|Z_i| \ll R_s$	∞	High; $|Z_i| \gg R_s$	0	Low; $|Z_i| \ll R_s$
Z_o	0	Low; $|Z_i| \ll R_L$	∞	High; $|Z_o| \gg R_L$	∞	High; $|Z_o| \gg R_L$	0	Low; $|Z_o| \ll R_L$
Gain or transfer ratio	$V_o = A_v V_s$	$V_o \approx A_v V_s$	$I_o = A_i I_s$	$I_o \approx A_i I_s$	$I_o = G_m V_s$	$I_o \approx G_m V_s$	$V_o = Z_m I_s$	$V_o \approx Z_m I_s$
Circuit model	Fig. 12-1		Fig. 12-2		Fig. 12-3		Fig. 12-4	

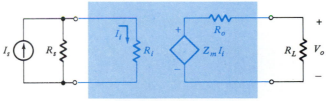

Current-to-voltage converter

12-2 THE FEEDBACK CONCEPT The characteristics of the four basic amplifier types were described in Sec. 12-1. The use of feedback can make practical amplifier characteristics approach those of ideal amplifiers. For each amplifier we may sample[1] the output signal by a suitable network and transmit this signal back to the input through a feedback network. At the input the feedback signal is combined with the external signal source by means of a summing (comparison) or mixer network. This combined signal is applied to the input of the basic practical amplifier as shown in Fig. 12-5. Embodied in the basic single-loop amplifier topology in Fig. 12-5 are the five constituent elements of a feedback system. These are the input and output signals, the measure (sample) of the output, comparison, and the processing of the compared signal by the basic amplifier.

The Signal Source (Input) This block in Fig. 12-5 represents the signal to be amplified. The signal source is modeled by either a voltage source V_s in series with R_s or a current source I_s in parallel with R_s.

The Output Signal The output can be either the voltage across or the current in the load resistance R_L (or impedance Z_L). It is the output signal that we desire to be independent of the load and insensitive to parameter variations in the basic amplifier.

The Sampling Network The function of the sampling network is to provide a measure of the output signal, that is, a signal that is proportional to the output. Two sampling networks are shown in Fig. 12-6. In Fig. 12-6a, the output voltage is sampled by connecting the output port of the feedback network in parallel with the load. This configuration is referred to as a *shunt* connection. The output current is sampled in Fig. 12-6b, in which the output port of the feedback network is connected in *series* with the load.

---

[1]In this context, "sampling" refers to obtaining a continuous signal proportional to the output and *not* the periodic time-sampling used to generate a discrete-time (sampled-data) signal.

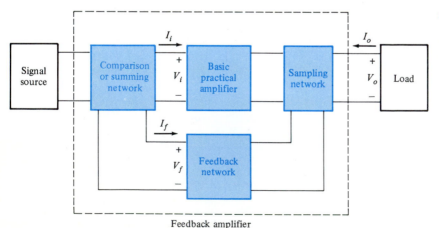

FIGURE 12-5
The basic structure of a single-loop feedback amplifier. The basic amplifier may be any of the four circuits shown in Figs. 12-1 to 12-4.

Feedback amplifier

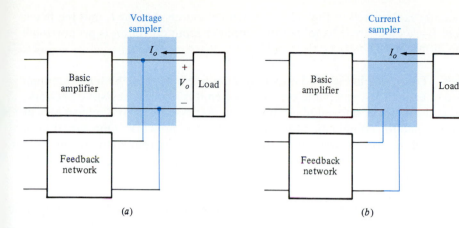

(a)   (b)

FIGURE 12-6
Feedback connections at the output of a basic amplifier which provides a measure (sample) of the output (a) voltage and (b) current.

Both circuits in Fig. 12-6 show equal output and sampled signals. Although this situation is prevalent, it is not required for proper performance. All that is necessary is for the sampled signal to be directly proportional to the output signal.

The Comparison or Summing Network   Two very common networks used for the comparison or summing of the input and feedback signals are displayed in Fig. 12-7. The circuit in Fig. 12-7a is a *series* connection and is used to compare the signal voltage V_s and feedback signal V_f. The amplifier input signal V_i is proportional to the voltage difference $V_s - V_f$ that results from comparison. A differential amplifier (Sec. 10-19) is often used for comparison, as its output voltage is proportional to the difference between the signals at the two inputs. A *shunt*

FIGURE 12-7
Feedback connections at the input of a basic amplifier: (a) voltage summing (series comparison), (b) current summing (shunt comparison).

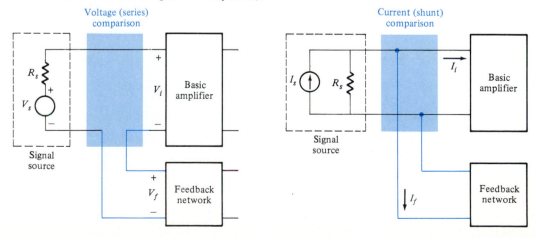

connection is depicted in Fig. 12-7b, in which source current I_s and feedback current I_f are compared. Note that the amplifier input current I_i is proportional to the difference $I_s - I_f$.

The Feedback Network   This block in Fig. 12-5 is usually a passive network which may contain resistors, capacitors, and inductors. Most often it is a purely resistive network. In several classes of integrated circuits (ICs) discussed in Part 4, capacitors or resistor-capacitor combinations form the feedback network. One function of the feedback network is to convert the sample of the output signal to the form of the signal appropriate for comparison. For example, consider a feedback amplifier in which the output is a voltage and currents are compared at the input. The transmission from the output port to the input port of the feedback network must convert the output voltage to a proportionally valued current at the input.

The Basic Amplifier   The basic amplifier in Fig. 12-5 is one of the four configurations given in Figs. 12-1 to 12-4. This circuit amplifies the difference signal that results from comparison, and it is this process that is responsible for the desensitivity and control of the output in a feedback system.

Consider the feedback amplifier in Fig. 12-8, in which the basic amplifier is the current amplifier in Fig. 12-2. Let us assume that A_i increases as a result, perhaps, to an increase in β_o, in one of the transistors which comprise A_i. An increase in A_i tends to increase the load current I_o, and hence the feedback current I_f also increases. Neglecting the current in R_s, the control current $I_i = I_s - I_f$ decreases. With reduced input drive, the amplifier output tends to decrease and thus offsets the effect of the increased A_i. This action is the basis of *negative feedback*. Because the summing network provides a difference signal, the amplifier input changes in a direction opposite to the change in the output. The net effect is to maintain a constant output signal that is independent

FIGURE 12-8
Feedback amplifier with current summing and current sampling.

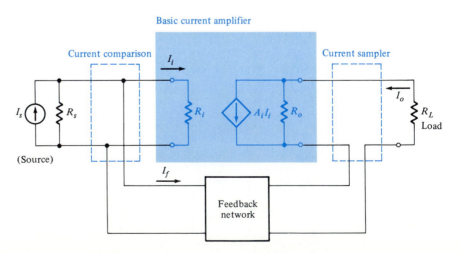

of variations in A_i. A quantitative discussion of the desensitivity provided by feedback amplifiers is given in the next section.

Feedback Amplifier Topologies   There are four basic amplifier types, each approximating the characteristics of an ideal controlled source. As expected, there are four basic single-loop feedback amplifier topologies, all of which have the structure given in Fig. 12-5. The four feedback configurations are called the *shunt-shunt* (or simply *shunt*), *shunt-series*, *series-shunt*, and *series-series* (simply *series*) amplifiers. These designations correspond to the input- and output-port connections, respectively, between the feedback network and the basic amplifier. For example, in the shunt-series amplifier, the input ports of the feedback network and amplifier are parallel-connected and the output ports are series-connected. Hence currents are compared and the output current is sampled. An alternative nomenclature is based on the quantity sampled and the input connection used. Thus a *current-shunt* topology corresponds to the shunt-series connection. Similarly, *current-series*, *voltage-series*, and *voltage-shunt* topologies correspond to the series-series, series-shunt, and shunt-shunt topologies, respectively. We use the initial designations of the feedback structures as these names are more prevalent in the literature.

Impedance levels decrease when networks are connected in parallel and increase when they are series-connected. Thus we expect the shunt-series configuration to have a low input impedance and a high output impedance. These impedance levels correspond to the current amplifier in Fig. 12-2 and Table 12-1. We treat in detail the characteristics and properties of the four single-loop feedback amplifier topologies in Sec. 12-6.

12-3 THE IDEAL FEEDBACK AMPLIFIER   Each of the four feedback amplifier topologies has several common characteristics and each is represented by the configuration in Fig. 12-5. In this section we examine the effects of feedback on overall amplifier performance (gain, desensitivity, distortion, etc.). The impedance levels in a feedback amplifier are discussed in Sec. 12-5.

As a first step toward a method of analysis which emphasizes the benefits of feedback, consider the representation of the ideal feedback amplifier in Fig. 12-9. The basic amplifier in Fig. 12-9a may be any of the four amplifiers listed in Table 12-1 connected in one of the four feedback topologies described in the previous section. The input signal X_s, the output signal X_o, the feedback signal X_f, and the difference (comparison) signal X_i each represents either a voltage or a current. These signals and the transfer functions A and β are summarized in Table 12-2 for the different feedback topologies. The symbol indicated by the circle with the summation sign Σ enclosed represents the summing network whose output is the algebraic sum of the inputs. Thus

$$X_i = X_s + X_f \tag{12-1}$$

The signal X_i, representing the output of the summing network, is the amplifier input $\hat{X}_i$. The term[1] $\hat{X}_i$ is introduced for convenience; in subsequent sections

---

[1] The term $\hat{X}_i$ is read X_i, "hat."

TABLE 12-2   Signals and Transfer Ratios in Feedback Amplifiers

Signal or Ratio	Feedback topology			
	Shunt-shunt	Shunt-series	Series-series	Series-shunt
X_o	Voltage	Current	Current	Voltage
X_s, X_i, X_f	Current	Current	Voltage	Voltage
A	V_o/I_i	I_o/I_i	I_o/V_i	V_o/V_i
β	I_f/V_o	I_f/I_o	V_f/I_o	V_f/V_o

it is beneficial to distinguish between the comparison signal and the input to the amplifier. If the feedback signal X_f is 180° out of phase with the input X_s, as is true in negative-feedback systems, then X_i is a difference signal. That is, X_i decreases as $|X_f|$ increases.

The reverse transmission of the feedback network β is defined by

$$\beta \equiv \frac{X_f}{X_o} \tag{12-2}$$

The transfer ratio β is often a real number but, in general, is a function of frequency. (The student should not confuse this symbol with that used for the common-emitter short-circuit current gain.)

The gain of the amplifier A is defined by

$$A \equiv \frac{X_o}{\hat{X}_i} = \frac{X_o}{X_i} \tag{12-3}$$

FIGURE 12-9
Ideal feedback amplifier model: (*a*) block diagram; (*b*) signal-flow graph.

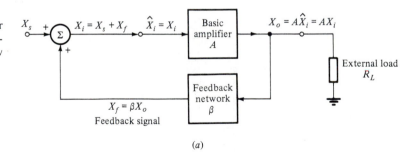

(*a*)

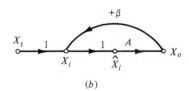

(*b*)

The gain with feedback A_F is obtained by substituting Eqs. (12-1) and (12-2) into (12-3) and is

$$A_F \equiv \frac{X_o}{X_s} = \frac{A}{1 - A\beta} \tag{12-4}$$

The gain A in Eqs. (12-3) and (12-4) represents the transfer function without feedback. If $\beta = 0$, eliminating the fed-back signal, no feedback exists and Eq. (12-4) reduces to Eq. (12-3). Frequently, A is referred to as the *open-loop gain* ($\beta = 0$) and designated by A_{OL}. When $\beta \neq 0$, a feedback loop exists and A_F is often called the *closed-loop gain*.

If $|A_F| < |A|$, the feedback is termed *negative*; if $|A_F| > |A|$, the feedback is *positive* (*regenerative*). We see that in the case of negative feedback $|1 - A\beta| > 1$ [Eq. (12-4)].

The Return Ratio or Loop Gain   The signal $\hat{X}_i$ in Fig. 12-9a is multiplied by A in passing through the amplifier and by β in transmission through the feedback network. Such a path takes us from the amplifier input around the loop consisting of the amplifier and feedback network. The product $-A\beta$ is called the *loop gain* or *return ratio* T. Equation (12-4) can be rewritten in terms of A_{OL} and T as

$$A_F = \frac{A}{1 - A\beta} = \frac{A_{\text{OL}}}{1 + T} \tag{12-5}$$

For negative feedback, $-A\beta = T > 0$.

We can give physical interpretation to the return ratio by considering the input signal $X_s = 0$, and the path between X_i and $\hat{X}_i$ is open. If a signal $\hat{X}_i$ is now applied to the amplifier input, then $X_i = X_f = A\beta$ or

$$T = -A\beta = -\left. \frac{X_i}{\hat{X}_i} \right|_{X_s = 0} \tag{12-6}$$

The return ratio is then the negative of the ratio of the fed-back signal to the amplifier input. Often, the quantity $F = 1 - A\beta = 1 + T$ is referred to as the *return difference*. If negative feedback is considered, both F and T are greater than zero (positive numbers).

The signal-flow graph (Sec. C-6) in Fig. 12-9b describes the same relationships given in Eqs. (12-2) and (12-3). The transmittance A represents the amplifier in Fig. 12-9a, and the branch β expresses the reverse transmission through the feedback network. Evaluation of $A_F = X_o/X_s$ by reduction of the graph gives Eq. (12-4). With $X_s = 0$, the loop consisting of A and β is readily discerned. In subsequent sections, the flow-graph representation of the system is used as an aid in conceptualizing practical circuits.

Fundamental Assumptions   Three conditions are implicit in the representation of the ideal feedback amplifier in Fig. 12-9 that lead to the expression for A_F in Eqs. (12-4) and (12-5):

1. The input signal is transmitted to the output through the amplifier A and *not* through the feedback network β. Thus, if the amplifier is deactivated by making $A = 0$ (by reducing g_m of a transistor to zero, for example), the output signal must become zero. This assumption is equivalent to stating that the *feedback network is unilateral.*

2. The feedback signal is transmitted from output to input through the feedback network *only.* That is, the *amplifier A is unilateral and transmits only from input to output.*

3. *The transfer ratio β is independent of the source and load resistances R_s and R_L* (Fig. 12-8).

Practical feedback amplifiers only approximate these conditions. For example, the feedback network usually consists of passive elements (R, C, L) and thus transmits a signal from input to output. Similarly, both A and β are affected by the load and source resistances. These deviations from the ideal can be included in an approximate analysis which is valid for most practical circuits. We point out the approximations made in each of the amplifier configurations studied. In addition, in Sec. 12-8, we formulate a more general approach to analyzing practical feedback amplifiers.

12-4 PROPERTIES OF NEGATIVE-FEEDBACK AMPLIFIERS
Since negative feedback results in reduced gain, why is it used? The answer lies in the several benefits that are obtained for the price of gain reduction. We now examine some of the advantages of negative feedback.

Desensitivity The amplifier transfer ratio A_F changes with variations due to tolerances, replacement, temperature, aging, and other variables of the transistor characteristics and circuit components. The sensitivity function S_x^G, the ratio of the fractional change in G to the fractional change in x, as defined in Eq. (12-7), is a convenient method of expressing the effect that variations in x have on system performance G.

$$S_x^G \equiv \frac{\Delta G/G}{\Delta x/x} \tag{12-7}$$

When $\Delta x/x \ll 1$, $\Delta G/G$ is dG/G and Eq. (12-7) becomes[1]

$$S_x^G \approx \frac{x}{G}\frac{dG}{dx} = \frac{dG/G}{dx/x} \tag{12-8}$$

A value of $S_x^G \ll 1$ indicates that G is insensitive to variations in x as $\Delta G/G \ll \Delta x/x$. Alternatively, $S_x^G \gg 1$ signifies that G is highly sensitive to

---

[1]When G depends on more than one variable, the derivative dG/dx is replaced by the partial derivative $\partial G/\partial x$.

changes in x. A unity value (approximately) of S_x^G reflects the fact that G is directly proportional to x and, hence, fractional changes in G and x are virtually equal. This situation exists for the variation of the return ratio T with respect to gain changes in the internal amplifier (A_{OL}). Thus $S_T^{A_F}$ also indicates the sensitivity of A_F with respect to variations in the gain of the basic amplifier. This is demonstrated by rewriting Eq. (12-4) as

$$A_F = \frac{A}{1 - A\beta} \frac{-\beta}{-\beta} = -\frac{1}{\beta} \frac{-A\beta}{1 - A\beta} = K \frac{T}{1 + T} \tag{12-9}$$

where $K = -1/\beta$. In Eq. (12-9), if T changes by ΔT, A_F changes by ΔA_F, which can be expressed as

$$\Delta A_F = \frac{K(T + \Delta T)}{1 + T + \Delta T} - \frac{KT}{1 + T} = \frac{K \, \Delta T}{(1 + T + \Delta T)(1 + T)}$$

Forming $\Delta A_F / A_F$ and using Eq. (12-7), we obtain

$$S_T^{A_F} = \frac{1}{1 + T + \Delta T} \tag{12-10}$$

which, if $T \gg |\Delta T|$, becomes

$$S_T^{A_F} \approx \frac{1}{1 + T} \tag{12-11}$$

The result in Eq. (12-11) is exactly that which is obtained if Eq. (12-8) is used to evaluate $S_T^{A_F}$.

Equation (12-10) demonstrates that the closed-loop gain can be rendered insensitive to changes in basic amplifier gain by increasing the value of T. For example, in an amplifier having a value of $T = 49$, a change in T of $\Delta T = +25$ (about a 50 percent increase in the basic amplifier gain), gives $S_T^{A_F} = 1/(1 + 49 + 25) = \frac{1}{75}$. Note that Eq. (12-10) is used because of the large value of ΔT. The corresponding fractional change in A_F is, from Eq. (12-7),

$$\frac{\Delta A_F}{A_F} = \frac{\Delta T}{T} S_T^{A_F} = \frac{25}{49} \frac{1}{75} = 0.0068$$

or A_F changes by about $\frac{2}{3}$ of 1 percent. Similarly, a change $\Delta T = -25$ (a 50 percent decrease in gain) gives $S_T^{A_F} = \frac{1}{25}$ and produces a 2 percent decrease in A_F. Further increase in the value of T diminishes the variation in A_F. These values illustrate the effectiveness of negative feedback. The closed-loop gain A_F can be controlled precisely even when the internal-amplifier gain varies substantially. The insensitivity of the closed-loop gain to open-loop gain variations that results from increasing T can be seen in Eq. (12-9). For $T \gg 1$, $T/(1 + T) \approx 1$ and

$$A_F \approx K = -\frac{1}{\beta} \tag{12-12}$$

FIGURE 12-10
Ideal voltage amplifier.

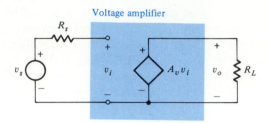

Voltage amplifier

Since β is the transfer function of the usually passive feedback network, A_F is essentially *independent of the gain of the basic amplifier and depends only on the ratio of passive components*. This is the situation we encountered in the inverting and noninverting Op-Amp stages (Sec. 10-21) for which the gain of these stages was proportional to the resistance ratio R_2/R_1.

Although it is exact only for small variations in T, Eq. (12-11) provides an estimate of the improvement in sensitivity; that is, the percentage change in A_F is the percentage change in T divided by $(1 + T)$. For the numerical values given above, the percentage change in A_F is 1 percent for 50 percent changes in T. Note that while Eq. (12-11) indicates that $\Delta A_F/A_F$ is the same for positive and negative changes in T, this is not the case for large variations in T.

Nonlinear Distortion The analysis in this and Chaps. 10 and 11 assumed that the amplifier stages behaved linearly; that is small-signal conditions applied. However, if a large signal is applied, the amplifier characteristic exhibits nonlinearity and the output waveform becomes distorted (Fig. 10-2b). We illustrate the effect of feedback on nonlinear distortion in the following discussion.

The voltage amplifier in Fig. 12-10 is ideal in all respects except that its dynamic range is limited. That is, the range of input-signal amplitudes that can be accommodated for linear operation is restricted. This is depicted in the voltage transfer characteristic in Fig. 12-11, which displays the relationship

FIGURE 12-11
Voltage transfer characteristic for amplifier in Fig. 12-10.

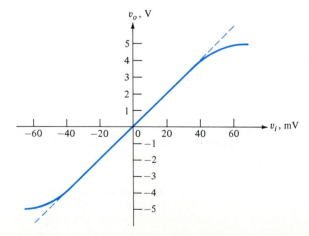

between the output signal v_o and the input signal v_i. The origin in Fig. 12-11 represents the operating point, and the slope of the curve is the voltage gain. The dashed line in Fig. 12-11 is the extension of the linear portion of the transfer characteristic ($0 \le |v_i| \le 40$ mV) and indicates a voltage gain $A_v = 100$. In the nonlinear region of the characteristic, $A_v < 100$ and becomes zero for $|v_i| \ge 60$ mV. (For a BJT amplifier, the horizontal segments in Fig. 12-11, for which $A_v = 0$, correspond to cutoff or saturation.) From the transfer characteristic, it is evident that input signals $|v_i| \ge 40$ mV result in distorted output waveforms.

The voltage transfer characteristic in Fig. 12-11 can be expressed analytically as follows:

$$\left. \begin{array}{ll} |v_o| = 100\,|v_i|; & 0 \le |v_i| \le 40 \text{ mV} \\ |v_o| = 100\,(|v_i| - 0.04) - 2500\,(|v_i| - 0.04)^2; & 40 \le v_i \le 60 \text{ mV} \\ |v_o| = 5; & |v_i| > 60 \text{ mV} \end{array} \right\} \quad (12\text{-}13)$$

For convenience, representative points on the transfer characteristic are listed in Table 12-2.

TABLE 12-2   Values of $|v_o|$ vs. $|v_i|$ from Eq. (12-13)

$|v_o|$, V	1.0	2.0	4.0	4.44	4.75	4.94	5.0
$|v_s| = |v_i|$, mV	10	20	40	45	50	55	60

The feedback amplifier in Fig. 12-12 utilizes the amplifier described by Eq. (12-13) and Fig. 12-11. The feedback network is designed to make $v_f = 0.09v_o$. We now construct the transfer characteristic of the feedback amplifier v_o/v_s by computing the values of v_s that correspond to the values of v_o and v_i in Table 12-2. In Fig. 12-12, KVL for the input loop gives

$$v_s = v_i + v_f = v_i + 0.09v_o$$

Substitution of values yields the data in Table 12-3, from which the transfer characteristic, shown in color in Fig. 12-13, is obtained.

TABLE 12-3   Values of $|v_o|$ vs. $|v_s|$ for Fig. 12-13

$|v_o|$ (V)	1.0	2.0	4.0	4.44	4.75	4.94	5.0
$|v_s|$ (mV)	100	200	400	444	478	500	510
$|v_i|$ (mV)	10	20	40	45	50	55	60

As observed in Fig. 12-13, the transfer characteristic is essentially linear over the entire range of input voltages. Hence we expect almost no nonlinear distortion for $|v_s| \le 500$ mV.

Also plotted in Fig. 12-13 is the transfer characteristic shown in Fig. 12-11. It is clearly evident from the curves that the feedback amplifier has the lower gain ($A_F = 10$). By cascading of two of these feedback amplifier stages, how-

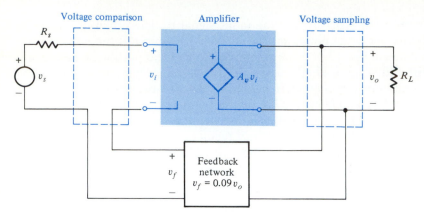

Voltage comparison   Amplifier   Voltage sampling

FIGURE 12-12
Feedback amplifier with voltage summing and voltage sampling. The amplifier has the transfer characteristic in Fig. 12-11.

ever, the overall gain is once again 100, and, for $|v_i| \leq 50$ mV, a minimum of distortion is introduced.

In the discussion just concluded, we assumed that the internal amplifier was ideal and limited only by the amplitude of the signal it could process. For practical amplifiers, however, the essential feature we just illustrated remains valid; specifically, the transfer characteristic of the feedback amplifier is much more nearly linear than is that for the basic amplifier.

Suppose that the signal applied to the amplifier in Fig. 12-10 is a sinusoid whose amplitude extends operation slightly beyond the linear range (a sinusoid with a 50-mV amplitude for the characteristic in Fig. 12-11, perhaps). The relatively small amount of distortion consists simply of a second harmonic signal generated within the device. In Eq. (12-13), for $|v_i| \geq 40$ mV, the output voltage v_o depends on v_i^2 and, recalling that $\sin^2 \omega t = \frac{1}{2} - \frac{1}{2} \cos 2\omega t$, the output contains the second harmonic. This distortion voltage can be considered as an external signal source v_d applied at the output of the amplifier. Negative feedback is now introduced and the signal-source amplitude is increased (preamplification) by the same amount the gain is reduced. Hence the distortion voltage v_d introduced at the basic amplifier output has the same value as in the nonfeedback amplifier. The signal-flow graph for this system is depicted in Fig. 12-14a; the equivalent block diagram representation is displayed in Fig. 12-14b. Using superposition, we obtain

$$v_o = A_F v_s + \frac{v_d}{1 + T} \tag{12-14}$$

and we observe that the distortion at the output is reduced by $(1 + T)$. Since T, in general, is a function of frequency, it must be evaluated at the second harmonic frequency. Note that the decrease in distortion by $(1 + T)$ corresponds to the linearization of the transfer characteristic in Fig. 12-13.

The signal applied to the feedback amplifier may be the actual signal externally available, or it may be the output of an amplifier preceding the feedback stage or stages under consideration. To multiply the input to the feedback amplifier by the factor $|1 + T|$, it is necessary either to increase the nominal

FIGURE 12-13
Transfer characteristic
of feedback amplifier in
Fig. 12-12.

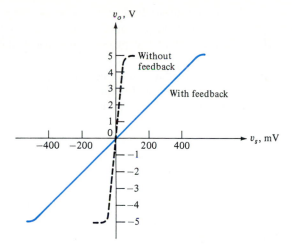

FIGURE 12-13
Transfer characteristic
of feedback amplifier in
Fig. 12-12.

gain of the preamplifying stages or to add a new stage. If the full benefit of the feedback amplifier in reducing nonlinear distortion is to be obtained, these preamplifying stages must not introduce additional distortion because of the increased output demanded of them. Since, however, appreciable harmonics are introduced only when the output swing is large, most of the distortion arises in the last stage. The preamplifying stages are of lesser importance in considerations of harmonic generation.

It has been assumed in the derivation of Eq. (12-14) that the small amount of additional distortion that might arise from the second-harmonic component fed back from the output to the input is negligible. This assumption leads to little error. Further, it must be noted that the result given by Eq. (12-14) applies only in the case of small distortion. The principle of superposition has been used in the derivation, and for this reason it is required that the device operate approximately linearly.

Reduction of Noise   By employing the same reasoning as that in the discussion of nonlinear distortion, it can be shown that the noise introduced at the output of an amplifier is divided by the factor $1 + T$ if feedback is employed. Noise introduced at the input is equivalent to applying a second signal and is thus unaffected by

FIGURE 12-14
Representation of the
distortion signal v_d applied to a feedback am-
plifier by (a) signal-flow
graph and (b) block dia-
gram.

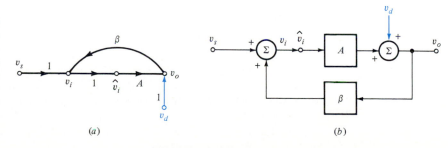

feedback. If $1 + T$ is much larger than unity, this would seem to represent a considerable reduction in the output noise. However, as noted above, for a given output the amplification of the preamplifier for a specified overall gain must be increased by the factor $1 + T$. Since the noise generated is independent of the signal amplitude, there may be as much noise generated in the pream- plifying stage as in the output stage. Furthermore, this additional noise will be amplified, as well as the signal, by the feedback amplifier, so that the complete system may actually be noisier than the original amplifier without feedback. Special, low-noise preamplifiers are used in several applications, such as high- quality stereo systems, to exploit the benefits of feedback and improve the signal-to-noise rates. If the additional gain required to compensate what is lost because of the presence of negative feedback can be obtained by a readjustment of the circuit parameters rather than by the addition of an extra stage, a definite reduction will result from the presence of the feedback. In particular, the hum introduced into the circuit by a poorly filtered power supply may be decreased appreciably.

12-5   IMPEDANCE IN FEEDBACK AMPLIFIERS   Earlier in this chapter we stated that feedback is used to make practical amplifier characteristics approximate those of ideal amplifiers. To do so requires that the input and output resistances (impedances) of the feedback amplifier have appropriate values (Table 12-1). We now examine the effect of the topology of a feedback on the amplifier on these impedance levels.

Input Resistance   *If the feedback signal is returned to the input in series with the applied voltage, the input impedance increases.*[1] The preceding is valid independent of the output connection. Thus the increase in input resistance is exhibited in both the series-shunt and series-series configurations.

Consider the circuit in Fig. 12-15a, which depicts the series-connected input circuit of a feedback amplifier. The KVL for the loop gives

$$V_i = V_s + V_f$$

The feedback signal is $V_f = \beta X_o$, where X_o is the output signal and $X_o = AV_i$. Combination of these relationships gives

$$V_i = IR_i = \frac{V_s}{1 - A\beta}$$

from which the resistance with feedback R_{IF} is

$$R_{IF} \equiv \frac{V_s}{I} = R_i (1 - A\beta) = R_i (1 + T) \qquad (12\text{-}15)$$

---

[1]Although we treat circuits only at midband frequencies in this chapter, the relationships developed apply at all signal frequencies.

FIGURE 12-15

Circuits for the calculation of the input resistance to a feedback amplifier: (*a*) series connection; (*b*) shunt connection.

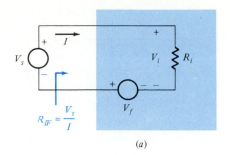

(*a*)

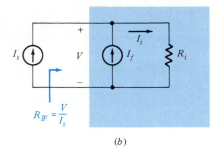

(*b*)

With $V_f = 0$ (no feedback and $\beta = 0$), the input resistance is simply R_i; clearly, the feedback has increased the input resistance. Qualitatively, we can justify this result as follows: Since V_f is 180° out of phase with V_s in a negative-feedback amplifier, V_i is less than it would be if V_f were absent. Hence $I = V_i/R_i$ decreases and causes the ratio V_s/I to increase.

When a shunt connection is used at the input of a negative-feedback amplifier, the input impedance decreases (and is independent of the output connection). The shunt-input connection of a feedback amplifier is displayed in Fig. 12-15*b*. For this circuit

$$I_i = I_s + I_f \qquad I_f = \beta X_o; \qquad X_o = AI_i$$

Combination of these equations and use of $V = I_i R_i$ yields

$$R_{IF} \equiv \frac{V}{I_s} = \frac{R_i}{1 - A\beta} = \frac{R_i}{1 + T} \tag{12-16}$$

Equation (12-16) clearly indicates that the resistance with feedback is less than the resistance without feedback when negative feedback ($T > 0$) is used. Both the shunt-series and shunt-shunt topologies exhibit this decrease in input impedance.

Output Impedance   *When the output of a feedback amplifier employs a shunt connection, negative feedback reduces the output resistance* (and is independent of the input connection). Consider the circuit in Fig. 12-16*a*, which displays the shunt output connection of a feedback amplifier. Since we are considering an ideal feedback amplifier, the fundamental assumptions stated in Sec. 12-3 apply. Thus the output voltage V_o is attributed to the basic amplifier AX_i and β is independent of the load resistance. We can evaluate the output resistance R_{OF} by the use of Thévenin's theorem. (Recall that the Thévenin resistance is the output resistance and equals the ratio of open-circuit voltage to short-circuit current.) The open-circuit voltage is V_o and equals

$$V_o = \frac{A}{1 - A\beta} X_s$$

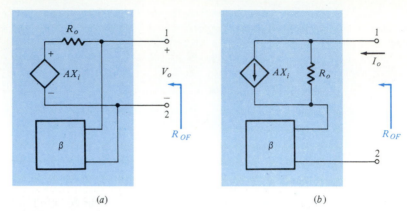

FIGURE 12-16
Pertaining to the calculation of the input resistance to a feedback amplifier: (a) shunt connection; (b) series connection.

Note that the input signal X_s is not shown but implied. The short-circuit current is obtained by short-circuiting terminals 1 and 2 in Fig. 12-16a and is

$$I_{\text{sc}} = \frac{AX_i}{R_o} = \frac{AX_s}{R_o}$$

With $V_o = 0$ (short circuit), no signal is fed back, $X_f = 0$, and $X_i = X_s$. Formation of the ratio V_o/I_{sc} yields

$$R_{OF} \equiv \frac{V_o}{I_{\text{sc}}} = \frac{R_o}{1 - A\beta} = \frac{R_o}{1 + T} \tag{12-17}$$

Equation (12-17) demonstrates that the resistance without feedback R_o is reduced by the addition of negative feedback ($T > 0$).

The output impedance is increased when a negative-feedback amplifier employs a series-connected output (and is independent of the input configuration). The circuit in Fig. 12-16b shows a series-connected output. In a manner similar to that used for the shunt-connected output, we obtain

$$I_o = -I_{\text{sc}} = \frac{A}{1 - A\beta} X_s$$

With terminals 1 and 2 open-circuited, no signal is fed back ($X_f = 0$) and $X_i = X_s$ and the open-circuit voltage $V_{\text{oc}} = -AX_iR_o$. Combining these relations and forming $V_{\text{oc}}/I_{\text{sc}}$, we obtain

$$R_{OF} = R_o(1 - A\beta) = R_o(1 + T) \tag{12-18}$$

Equations (12-15) to (12-18) for R_{IF} and R_{OF} are special cases of *Blackman's impedance formula* (given in Eq. (12-21) and derived in Sec. 12-9). These equations apply to the ideal feedback system and satisfy the fundamental assumptions given in Sec. 12-3. Practical amplifiers only approximate this behavior. The values of R_o, R_i, A, and β (and hence T) must be modified to include the source and load resistances R_s and R_L and the nonunilateral nature of the feedback network before Eqs. (12-15) to (12-18) are used. We illustrate this in subsequent sections and indicate the approximations used.

Blackman's Impedance Formula The input and output resistances given in Eqs. (12-15) to (12-18) can be obtained by calculating R_i (or R_o) and T independently. For example, in Eq. (12-15), $R_{IF} = R_i$ when $T = 0$. Since $T = 0$ can be achieved by making $A = 0$, that is, reducing the gain of the basic amplifier to zero, R_i is simply the input resistance of the resultant passive network. Bode[1] referred to this situation as the "dead system" because $A = 0$ corresponds to suppressing the controlled source in the system. Let us call this dead-system input resistance R_{ID}.

The return ratio T can be evaluated by using Eq. (12-6) as described in Sec. 12-3. In Eq. (12-6), T is computed by setting $X_s = 0$, that is, by suppressing the signal source. In Fig. 12-15a suppressing V_s short-circuits the input terminals. Thus T is measured with the input short-circuited and, using Bode's nomenclature, identify this value of return ratio by T_{SC}. Thus Eq. (12-15) can be rewritten as

$$R_{IF} = R_{ID} (1 + T_{SC}) \qquad (12\text{-}19)$$

Similarly, for the circuit in Fig. 12-15b, setting $X_s = I_s = 0$ is equivalent to open-circuiting the input terminals. The return ratio measured is identified as T_{OC}; Eq. (12-16) becomes

$$R_{IF} = \frac{R_{ID}}{1 + T_{OC}} \qquad (12\text{-}20)$$

where R_{ID} is again the input resistance with $A = 0$.

In Eqs. (12-17) and (12-18), $R_{OF} = R_o$ when $T = 0$ ($A = 0$) and is the dead-system output resistance R_{OD}. The value of T (Fig. 12-16a) is measured when terminals 1 and 2 are open-circuited and T_{OC} is identified. (If terminals 1 and 2 are short-circuited, no feedback exists.) Thus Eq. (12-17) reduces to Eq. (12-20) with R_{ID} replaced by R_{OD}. The reverse situation exists in the circuit in Fig. 12-16b, where open-circuiting terminals 1 and 2 eliminates the feedback. The return ratio $T = T_{SC}$ is measured when terminals 1 and 2 are short-circuited. Thus Eq. (12-18) can be rewritten as in Eq. (12-19) with R_{OD} substituted for R_{ID}.

Combination of the ideas in the previous paragraphs serves as the justification for Eq. (12-21), the general form of Blackman's impedance formula:

$$Z_F = Z_D \frac{1 + T_{SC}}{1 + T_{OC}} \qquad (12\text{-}21)$$

---

[1]Many of the terms used to describe feedback amplifiers were introduced by Bode.

In Eq. (12-21), we have the following conditions:

1. Z_F is the impedance seen looking into any pair of terminals A and B of a feedback amplifier

2. Z_D is the dead-system value of Z_F, that is, it is the impedance seen looking in to terminals A and B when the gain of the amplifier is made zero.

3. T_{SC} is the return ratio measured when terminals A and B are short-circuited.

4. T_{OC} is the return ratio measured when terminals A and B are open-circuited.

Note that Eq. (12-21) reduces to Eq. (12-19) if $T_{OC} = 0$ and to Eq. (12-20) when $T_{SC} = 0$.

Blackman's impedance formula applies to all feedback amplifiers and not simply to the ideal situation described in Figs. 12-15 and 12-16. Embodied in the quantities T_{OC}, T_{SC}, and Z_D are the loading effects of the feedback network β on the basic amplifier A and the effects of the load and source resistances R_s and R_L, respectively, on the values of A and β. In the analysis of the four basic single-loop topologies in the next several sections, we include these loading effects.

12-6 PROPERTIES OF FEEDBACK AMPLIFIER TOPOLOGIES

Some of the general characteristics of single-loop feedback amplifiers were described in the four previous sections. Each of the four topologies introduced in Sec. 12-2 approximates one of the four amplifier types (Sec. 12-1). In this section, we investigate specific characteristics of the four topologies. Transistor realizations that approximate these circuits are discussed in the remainder of the chapter.

The Shunt-Shunt Amplifier The two-port network representation of a shunt-shunt amplifier is shown in Fig. 12-17. The parallel connection at the output signifies that the output voltage is sampled, whereas the parallel-connected input results in current comparison. Thus the feedback network provides voltage-to-current transfer.

The use of Thévenin or Norton's theorem allows representation of the internal amplifier by any of the four amplifier types discussed in Sec. 12-1. Note that $V_i = z_i I_i$ and that the current source $g V_i$ in parallel with r_o can be converted to its voltage-source equivalent. The representation in Fig. 12-17 is based on the two-port y parameters.[1]

We expect networks in parallel to have low impedance levels; this is demonstrated by using Blackman's impedance formula [Eq. (12-21)] as follows. When terminals 1–1′ in Fig. 12-17 are short-circuited, I_i and V_i are zero, and

---

[1]Two 2-port networks in parallel can be represented by an equivalent network having y parameters equal to the sum of the y parameters of the constituent networks. In practical circuits, however, it is often difficult to identify the individual networks.

FIGURE 12-17

The shunt-shunt (volt-age-shunt) feedback amplifier topology.

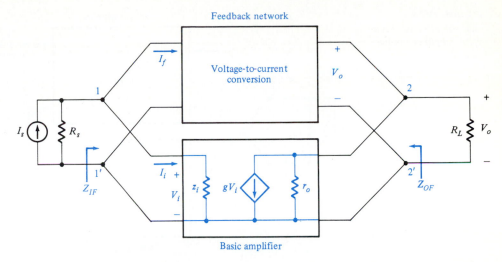

hence $T_{SC} = 0$. If terminals 1–1′ are open-circuited, however, $I_i = -I_f$ and $T_{OC} \neq 0$. The resultant input impedance Z_{IF} is low as Z_{ID} is divided by $1 + T_{OC}$. Similarly, in the calculation of Z_{OF}, short-circuiting terminals 2–2′ makes $V_o = 0$, and consequently, $I_f = I_i = 0$, so that $T_{SC} = 0$. Open-circuiting terminals 2–2′ permits a signal to be fed back to the input, and $T_{OC} \neq 0$. Again, Z_{OF} is low because of the division Z_{OD} by $1 + T_{OC}$. The low values of Z_{IF} and Z_{OF} obtained make the transfer function independent of R_s and R_L. As indicated in Table 12-1, the shunt-shunt amplifier realizes the current-voltage converter or transimpedance amplifier. The properties of the four feedback amplifiers are listed in Table 12-4 at the end of this section.

The Series-Series Amplifier The series-series connection is illustrated in Fig. 12-18, in which we identify that the output current I_o is sampled and the series input connection

FIGURE 12-18

Two-port representation of series-series feedback amplifier.

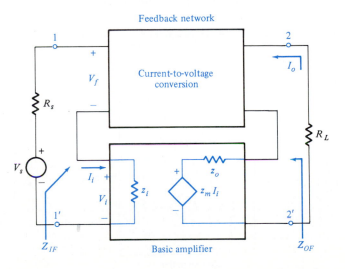

FIGURE 12-19

The shunt-series feed-back amplifier configuration.

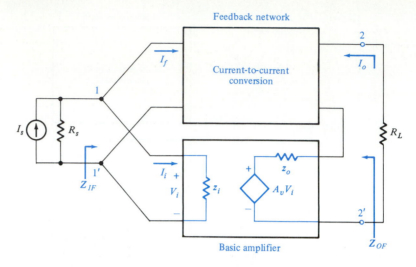

requires that voltages are compared. (The current is the same in all elements in a series circuit.)

To determine Z_{IF}, we must evaluate T_{OC} and T_{SC}. Open-circuiting terminals 1–1' makes both I_i and V_i zero. Thus open-circuiting a series connection results in $T_{OC} = 0$. With 1–1' short-circuited, $I_i \neq 0$, and hence $T_{SC} \neq 0$. From Eq. (12-21) we conclude that Z_{ID} is increased by the feedback by a factor $1 + T_{SC}$. By analogy, Z_{OF} is large as $T_{OC} = 0$ for a series connection and $T_{SC} \neq 0$. With high input and output impedances, this series-series amplifier behaves as the voltage-current converter or transconductance amplifier (Table 12-1).

The Shunt-Series Amplifier This configuration is displayed in Fig. 12-19. On the basis of the prior discussion, this amplifier has low input and high output impedances. The output current I_o is essentially independent of the load resistance R_L, and

FIGURE 12-20

The two-port representation of the series-shunt feedback amplifier.

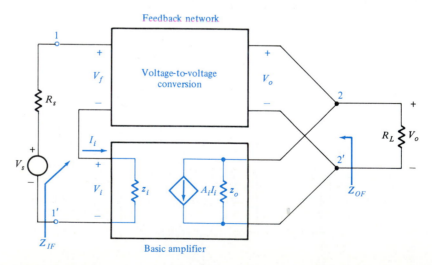

TABLE 12-4   Properties of Feedback Amplifier Structures

Topology	Amplifier classification	Comparison signal	Output signal (sample)	Input impedance	Output impedance
Shunt-shunt	Current-voltage converter	Current	Voltage	Low	Low
Shunt-series	Current	Current	Current	Low	High
Series-series	Voltage-current converter	Voltage	Current	High	High
Series-shunt	Voltage	Voltage	Voltage	High	Low

current comparison exists at the input. From Table 12-4 we recognize that the shunt-series topology is that of a current amplifier.

The Series-Shunt Amplifier   The topology depicted in Fig. 12-20 is that of the series-shunt amplifier. Again, using the previous discussion, we conclude that the series input connection results in high input impedance and voltage comparison. Similarly, low output impedance and voltage sampling are the characteristics of the parallel-connected output port. Thus the voltage amplifier (Table 12-4) is approximated by the series-shunt feedback configuration.

12-7   APPROXIMATE ANALYSIS OF A FEEDBACK AMPLIFIER   Practical feedback amplifiers are usually designed to approximate the characteristics of one of the four basic topologies. Therefore, it is convenient to analyze these circuits in a fashion similar to that used for the ideal amplifiers previously discussed. The method of analysis is based on the following assumptions:

1. The basic amplifier is unilateral, but its gain reflects the loading of the feedback network, and source and load resistances. The gain of this block is the gain of the amplifier without feedback; we designate it by $A_{\rm OL}$ and thus distinguish it from the ideal situation.

2. The feedback network is unilateral (Sec. 12-3). This assumption is equivalent to the statement that the feedforward transmission through the β network is negligible compared to that through the amplifier.

The first step in the analysis is to identify the topology. The *input loop* is defined as the mesh containing the applied signal voltage V_s and either (*a*) the base-to-emitter region of the input bipolar transistor, or (*b*) the gate-to-source region of the first FET in the amplifier, or (*c*) the section between the two inputs of a differential or operational amplifier. The input connection is identified as *series* if, in the input circuit, there is a circuit component W in series with V_s and if W is connected to the output (the portion of the system containing the load). If this condition is true, the voltage across W is the feedback signal $X_f = V_f$.

If the previous condition is not satisfied, we must test for a shunt connection. The *input node* is defined as either (*a*) the base of the first BJT, or (*b*) the gate of the first FET, or (*c*) the inverting terminal of a differential or operational amplifier. A current source is now used for the external excitation so that the current signal I_s enters the input node. The configuration is identified as *shunt* if there is a connection between the input node and the output circuit. The current in this connection is the feedback signal $X_f = I_f$.

The sampled output quantity can be either a voltage or a current. The output node at which the output voltage V_o (with respect to ground) is taken must be specified. This voltage V_o appears across the load resistor (often designated by R_L) and the output current I_o is the current in R_L. Tests for the type of sampling are the following:

1. Set $V_o = 0$ (that is, set $R_L = 0$, thereby short-circuiting the output). If X_f becomes zero, the original system exhibited *voltage sampling* and a shunt connection exists.

2. Set $I_o = 0$ (that is, set $R_L = \infty$ and open-circuit the output). If X_f becomes zero, *current sampling* is present in the original amplifier and a series connection exists.

The Amplifier without Feedback It is desirable to separate the feedback amplifier into two blocks, the basic amplifier A_{OL} and the feedback network β, because with a knowledge of A_{OL} and β, we can calculate the important characteristics of the feedback system. The basic amplifier configuration *without feedback but taking the loading of the β network into account* is obtained by applying the following rules:

To find the input circuit:

1. Set $V_o = 0$ for a shunt output connection. In other words, short-circuit the output node.

2. Set $I_o = 0$ for a series-connected output. In other words, open-circuit the output loop.

To find the output circuit:

1. Set $V_i = 0$ for current comparison. In other words, short-circuit the input node (so that none of the feedback current enters the amplifier input).

2. Set $I_i = 0$ for series (voltage) comparison. In other words, open-circuit the input loop (so that none of the feedback voltage reaches the amplifier input).

These procedures ensure that the feedback is reduced to zero without altering the loading on the basic amplifier.

Outline of Analysis   To find A_F, R_{IF}, and R_{OF} the following steps are carried out:

1. Identify the topology as indicated previously. These tests determine whether X_f is a voltage or a current.
2. Draw the basic amplifier circuit without feedback, following the rules listed.
3. Replace each active device by its proper model.
4. Identify X_f and X_o on the circuit obtained.
5. Evaluate $\beta = X_f / X_o$.
6. Evaluate A_{OL} by applying KVL and KCL to the equivalent circuit obtained.
7. From A_{OL} and β, find T, and A_F.
8. From the equivalent circuit find R_{ID} and R_{OD}. Apply Blackman's impedance formula to obtain R_{IF} and R_{OF}.

We illustrate the approximate analysis procedure in the following two examples.

Example 12-1

Determine A_F, T, R_{IF}, and R_{OF} for the emitter follower in Fig. 12-21a.

Solution

Since the input contains a component R_E which is connected to the output (V_o is taken across R_E), the input is series-connected and voltages are compared. The feedback voltage V_f is measured across R_E as indicated. The polarity used

FIGURE 12-21
(a) Emitter follower. (b) Representation of the amplifier without feedback. (c) The equivalent circuit.

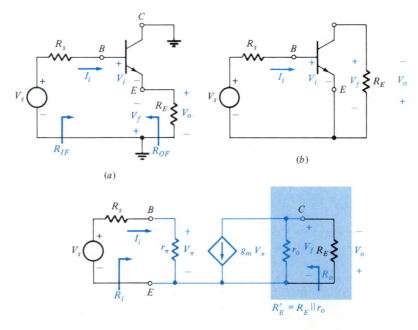

is consistent with the summing network in Fig. 12-9a for which $X_i = X_s + X_f$. Clearly, V_f is negative; thus an increase in V_f causes V_i to decrease, as is required in negative feedback.

The output connection of the amplifier is determined by setting $V_o = 0$ ($R_E = 0$). With $R_E = 0$, the feedback is eliminated and $V_f = 0$. Thus the output is shunt-connected and the emitter-follower topology is series-shunt.

We must now draw the amplifier without feedback; to do so, we follow the steps listed earlier in this section. To find the input circuit, set $V_o = 0$, and hence V_s appears directly across B and E. The output circuit is obtained by setting $I_i = 0$ (the input is opened), and R_E appears only in the output loop. Following these rules, the circuit in Fig. 12-21b is obtained and for which Fig. 12-21c is the equivalent circuit. To be consistent with Fig. 12-21a, V_o is measured from emitter to collector.

In Fig. 12-21c we observe that $V_f = -V_o$ and $\beta \equiv V_f/V_o = -1$. Also in Fig. 12-21c, $V_o = g_m V_\pi R'_E$, where $R'_E = R_E \| r_o$ and $V_\pi = r_\pi V_s/(R_s + r_\pi)$.

Combining these relations, we obtain

$$A_{\text{OL}} = \frac{V_o}{V_s} = \frac{g_m r_\pi R'_E}{R_s + r_\pi}$$

and $T = -\beta A_{\text{OL}}$ is

$$T = \frac{g_m r_\pi R'_E}{R_s + r_\pi}$$

Using these values and after clearing fractions, we have

$$A_F = \frac{A_{\text{OL}}}{1 + T} = \frac{\beta_o R'_E}{R_s + r_\pi + \beta_o R'_E}$$

(The relationship $g_m r_\pi = \beta_o$ is used in determining A_F.) If $R_E \ll r_o$, as is usually the case, $R'_E = R_E$ and $A_F = \beta_o R_E/(R_s + r_\pi + \beta_o R_E)$. Comparison of this result with the A_v entry for the common-collector stage in Table 10-3A shows that they are the same if $\beta_o \gg 1$. The slight difference in the two results is attributed to neglecting the forward transmission of the feedback network.

To determine R_{IF} and R_{OF}, we use Blackman's impedance formula [Eq. (12-21)]. Inspection of Fig. 12-21c gives $R_{ID} = R_i = r_\pi$. To evaluate T_{OC}, we must open-circuit the input. This is accomplished by letting $R_s \to \infty$ in the expression for T. If $R_s \to \infty$, $T_{\text{OC}} = 0$ and is consistent with our previous discussions for series-connected inputs. Setting $R_s = 0$ is equivalent to short-circuiting the input; hence $T_{\text{SC}} = T|_{R_s=0} = \beta_o R'_E/r_\pi = g_m R'_E$. Use of Eq. (12-21) gives

$$R_{IF} = r_\pi (1 + g_m R'_E) = r_\pi + \beta_o R'_E$$

Again, for $\beta_o \gg 1$, this result is the same as given in Table 10-3.

The "dead" system output resistance $R_{OD} = r_o$. To evaluate T_{SC} and T_{OC} from T, we set $R_E = 0$ and let $R_E \to \infty$, respectively. Thus $T_{\text{SC}} = 0$ and $T_{\text{OC}} = \beta_o r_o/(R_s + r_\pi)$ from which

$$R_{OF} = \frac{r_o}{1 + [\beta_o r_o/(R_s + r_\pi)]} = \frac{r_o(R_s + r_\pi)/\beta_o}{r_o + (R_s + r_\pi)/\beta_o} \approx \frac{R_s + r_\pi}{\beta_o}$$

for $r_o \gg (R_s + r_\pi)/\beta_o$. For $\beta_o \gg 1$, R_{OF} is the entry given for the output resistance in Table 10-3A.

Example 12-2

Determine A_F, T, and R_{OF} for the common-source stage with source resistance in Fig. 12-22a.

Solution

The input circuit is analogous to that for the emitter follower and is thus series-connected. Making $V_o = 0$ does not eliminate the feedback because I_o and, hence, V_f do not become zero. When $I_o = 0$, $V_f = 0$ and the output is also series connected. (This amplifier is a series-series type.)

To determine the input circuit of the amplifier without feedback, open-circuit the output ($I_o = 0$). The feedback resistance R_S appears in series with V_s as shown in Fig. 12-22b. To find the output circuit, set $I_i = 0$. Again, as indicated in Fig. 12-22b, R_S appears in the output loop. The equivalent circuit in Fig. 12-22b is depicted in Fig. 12-22c.

In Fig. 12-22c, $V_o = -I_o R_D$ and $V_f = -I_o R_S$; hence $\beta = V_f/I_o = -R_S$. Since no current exists in the gate loop, $V_{gs} = V_s$. Use of KVL for the drain loop gives $I_o = \mu V_s/(r_d + R_D + R_S)$. It follows that

$$A_{OL} = \frac{I_o}{V_s} = \frac{\mu}{r_d + R_D + R_S}$$

The return ratio is

$$T = -\beta A_{OL} = \frac{\mu R_S}{r_d + R_D + R_S}$$

FIGURE 12-22
(a) Common-source amplifier with source resistance. (b) Schematic diagram and (c) equivalent circuit of the amplifier without feedback.

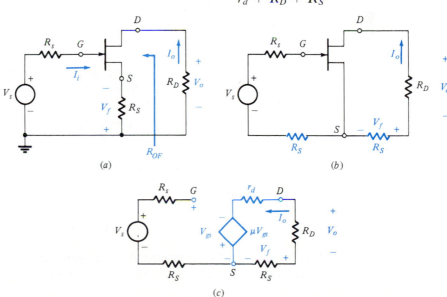

Combination of these equations and clearing fractions yields

$$A_F = \frac{A_{OL}}{1 + T} = \frac{\mu}{r_d + R_D + (1 + \mu)R_D}$$

Since $V_o = - I_o R_D$, V_o/V_s is the same result given in Eq. (10-70).

Inspection of Fig. 12-22c shows that R_{OD}, the output resistance with the controlled source suppressed, is $R_S + r_d$. The return ratios T_{OC} and T_{SC} are obtained by setting $R_D = \infty$ and $R_D = 0$, respectively. Thus $T_{OC} = T\,|_{R_D \to \infty} = 0$ and $T_{SC} = T\,|_{R_D = 0} = \mu R_S/(r_d + R_S)$. The output resistance with feedback $R_{OF} = R_{OD}\,(1 + T_{SC}) = (R_s + r_d)\,[1 + \mu R_S/(r_d + R_S)]$. Clearing fractions and rearranging terms, we obtain $R_{OF} = r_d + R_S\,(1 + \mu) \approx r_d\,(1 + g_m R_S)$ for $\mu \gg 1$. This result is identical to that given in Eq. (10-74) and Table 10-5.

The approximate analysis used in this example and the actual values are identical because no feedforward path exists in this stage. Because the gate circuit is open-circuited at low frequencies, setting $\mu = 0$ causes I_o and V_o to be zero. In contrast, in Example 12-1, setting β_o of the transistor to zero does not make $V_o = 0$ because a path consisting of V_s, R_s, r_π and R_E exists.

12-8 GENERAL ANALYSIS OF FEEDBACK AMPLIFIERS

In Example 12-1 we showed that the approximate analysis gives results that deviate from actual values because the feedback network is assumed to be unilateral (from output to input). Before proceeding with multistage feedback amplifiers, let us develop an analysis method which accounts for the feedforward in the β network. The analysis is based on the block diagram in Fig. 12-5. However, no approximations concerning the amplifier block or feedback network are made.

Two sources are present in a single-loop feedback amplifier: the signal source X_s and the controlled source (the basic amplifier) whose control variable is X_i. Initially, let us treat the controlled source as an independent source, that is, assume that $\hat{X}_i$ is an independent variable. This is the same technique used in writing a set of nodal or mesh equations. The controlled source is treated initially as an independent source and an equation, called a constraint equation, relating the control variable to the nodal (mesh) variables, is then written. Since the feedback amplifier is assumed to behave linearly, superposition applies. Hence *any* voltage or current in the system has two components, one attributed to X_s, and the second to $\hat{X}_i$. The output X_o is expressable as

$$X_o = t_{11}X_s + t_{12}\,\hat{X}_i \tag{12-22}$$

where $t_{11}X_s$ and $t_{12}\hat{X}_i$ are the output components due to X_s and $\hat{X}_i$, respectively. The compared signal X_i can also be written in the form of Eq. (12-22), namely,

$X_i = AX_s + B\hat{X}_i$. By substitution of the value of $\hat{X}_i$ obtained from Eq. (12-22), however, X_i can be written as

$$X_i = t_{21}X_s + t_{22}X_o \tag{12-23}$$

On the basis of Eqs. (12-22) and (12-23), the signal-flow graph in Fig. 12-23 is constructed. Note that no feedback is present in Fig. 12-23, as both sources are considered independent variables. The constraint equation is simply $\hat{X}_i = X_i$. This relationship introduces the unity transmittance between X_i and $\hat{X}_i$ shown dashed in Fig. 12-23 and corresponds to closing the loop. *The signal-flow graph in Fig. 12-23, applicable to all single-loop feedback amplifiers independent of their topology, is simply a consequence of the superposition principle.*

On the basis of Eqs. (12-22) and (12-23), the various t parameters are defined as

$$t_{11} \equiv \frac{X_o}{X_s}\bigg|_{\hat{X}_i=0} \qquad t_{12} \equiv \frac{X_o}{\hat{X}_i}\bigg|_{X_s=0} \tag{12-24}$$

$$t_{21} \equiv \frac{X_i}{X_s}\bigg|_{X_o=0} \qquad t_{22} \equiv \frac{X_i}{X_o}\bigg|_{X_s=0}$$

Observe that the t parameters are defined in a manner analogous to the definitions of the various two-port parameters.

In Eq. (12-24), note that making $\hat{X}_i = 0$ is equivalent to suppressing the controlled source; that is, it makes the gain of the basic amplifier zero.

With $\hat{X}_i = 0$, however, the output signal $X_o \neq 0$ [Eq. (12-22)] but is $t_{11}X_s$. It is the branch t_{11} that accounts for the transmission from input to output through the feedback network. With the controlled source—represented by the branch t_{12}—suppressed, the system is passive; that is, it is the dead system. Thus, setting $t_{12} = 0$ has the same effect as making $\hat{X}_i = 0$.

The branch transmittance t_{21} represents the component of X_i produced by the signal source X_s and reflects the fact that practical sources and amplifier inputs have finite, nonzero impedances. The transmission through the feedback network from the output to input is represented by the branch t_{22}.

The Gain (Transfer Ratio) with Feedback   The *closed-loop gain* A_F is defined as the ratio X_o/X_s. In Fig. 12-23, the branch t_{11} is in parallel with the path containing t_{21}, t_{12}, and the feedback branch t_{22}. Hence

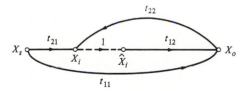

FIGURE 12-23
Signal-flow graph with the amplifier input (X_i) disconnected from the summing network output X_i. Since S_i is now an independent variable, X_o and X_i each depend on X_s and $\hat{X}_i$ can be determined by use of superposition.

$$A_F \equiv \frac{X_o}{X_s} = \frac{t_{11} + t_{12}\, t_{21}}{1 - t_{12}\, t_{22}} \tag{12-25}$$

As previously noted, t_{12} is the only term in Eq. (12-25) that depends on the gain of the basic amplifier block. If this gain is reduced to zero, A_F becomes

$$A_F = \frac{X_o}{X_s}\bigg|_{t_{12}=0} \equiv A_D = t_{11} \tag{12-26}$$

The quantity A_D in Eq. (12-26) is called the *dead-system gain*. Observe that $A_D = t_{11}$, the parameter associated with the feedforward path through the feedback network.

The return ratio T indicates the transmission around the feedback loop formed by the amplifier t_{12} and the feedback network t_{22}. If the branch between X_i and $\hat{X}_i$ in Fig. 12-23 is removed and the signal source X_s is suppressed, then use of Eq. (12-6) gives

$$T \equiv -\frac{X_i}{\hat{X}_i}\bigg|_{X_s=0} = -t_{12}\, t_{22} \tag{12-27}$$

Note that by removing the branch between X_i and $\hat{X}_i$, we no longer have a feedback system and, as seen in Fig. 12-23, T is simply the cascade of the basic amplifier and feedback network.

When the transmittance t_{22} between the amplifier output and the summing circuit is zero, the feedback path is eliminated. The transfer ratio of the resultant nonfeedback amplifier, the open-loop gain A_{OL}, is, from Eq. (12-25),

$$A_{OL} \equiv \frac{X_o}{X_s}\bigg|_{t_{22}=0} = t_{11} + t_{12}t_{21} = A_D + t_{12}\, t_{21} \tag{12-28}$$

As we are interested in constructing an amplifier, the magnitude of A_{OL} is generally much larger than unity. The magnitude of A_D is most often less than unity since it is the transfer ratio of a passive (resistive) network. Thus $A_{OL} \approx t_{12}\, t_{21}$.

Using the terms defined in Eqs. (12-26) through (12-28), we can rewrite A_F as

$$A_F = \frac{A_{OL}}{1 + T} \tag{12-29}$$

An alternate form of A_F is expressable as

$$A_F = \frac{t_{11} + (t_{21})(-t_{12}t_{22})/(-t_{22})}{1 - t_{12}t_{22}} = \frac{A_D + KT}{1 + T} \tag{12-30}$$

where

$$K = \frac{-t_{21}}{t_{22}} \tag{12-31}$$

The parameter K depends only on the passive elements which comprise t_{21} and t_{22}.

If $|A_D| \ll |KT|$, then $A_{OL} = KT$ and

$$A_F \approx \frac{KT}{1 + T} \qquad (12\text{-}32)$$

For $T \gg 1$, $A_F \approx K = -1/\beta$ and illustrates that the closed-loop gain is essentially independent of the basic amplifier gain and depends only on the ratio of passive components. This is the situation we encountered in the Op-Amp stages of Sec. 10-21 for which the gain was proportional to the resistance ratio R_2/R_1.

Observe that Eqs. (12-32) and (12-9) are identical because in Eq. (12-32) we neglect A_D, the feedforward transmission.

We can give additional meaning to $K = -1/\beta$ [Eq. (12-7)] by examining Eq. (12-22). If the gain of the basic amplifier is infinite (t_{12} and hence A_{OL} and T become infinite) and X_o *remains finite*, the input to the amplifier $X_i = \hat{X}_i = 0$. Setting $X_i = 0$ in Eq. (12-22) indicates that $X_o/X_s = -(t_{21}/t_{22}) = K$ and is the gain of the feedback amplifier when $T \to \infty$. Making $X_i = 0$ is equivalent to stating that the feedback signal $t_{22}X_o$ and the component of the input signal $t_{21}X_s$ are equal in magnitude and opposite in phase. We have encountered this situation in Sec. 10-21 in the discussion of the inverting Op-Amp stage. There we showed that when $A_v \to \infty$, $V_i \to 0$ and the current produced by the signal source V_s/R_1 was balanced by the feedback signal V_o/R_2.

The Analysis Procedure   For a specific circuit, the t parameters are determined by first assuming the controlled source in the device model is an independent source; thus we must first identify $\hat{X}_i$ and then apply Eqs. (12-24). This is illustrated in the two examples that follow. We have deliberately selected two circuits previously analyzed so that we may focus on the techniques used to obtain the t parameters and compare the results with prior analyses.

Example 12-3   (a) Evaluate the t parameters for the emitter follower in Fig. 12-24a. (b) Use the results in part a to obtain A_F, T, A_D, and K.

Solution   (a) The equivalent circuit of the emitter follower is given in Fig. 12-24. The parameters $X_s = V_s$, $X_i = V_\pi$, $X_o = V_o$, and $\hat{X}_i = \hat{V}_\pi$ are identified in the equivalent circuit. Note that $\hat{V}_\pi$ is associated with the controlled source, thus making $g_m V_\pi$ behave as an independent current source.

To determine t_{11} by use of Eq. (12-17), $\hat{X}_i = \hat{V}_\pi$ is suppressed and Fig. 12-24 reduces to the circuit in Fig. 12-25a. In Fig. 12-25a, use of the voltage-divider relationship gives

$$t_{11} = \left.\frac{V_o}{V_s}\right|_{\hat{v}_\pi = 0} = \frac{R_E}{R_s + r_\pi + R_E}$$

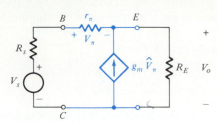

FIGURE 12-24
Small-signal model of emitter follower used in Example 12-3. *Note:* For the purposes of analysis the controlled source is identified as $g_m \hat{V}_\pi$ and is distinguished from V_π across r_π.

Setting $V_o = 0$, as shown in Fig. 12-25*b*, gives

$$t_{21} = \left.\frac{V_\pi}{V_s}\right|_{V_o=0} = \frac{r_\pi}{R_s + r_\pi}$$

In the circuit of Fig. 12-25*c*, $V_s = 0$ and $R_E \parallel (R_s + r_\pi)$. Thus

$$t_{12} = \left.\frac{V_o}{\hat{V}_\pi}\right|_{V_s=0} = g_m \frac{R_E (R_s + r_\pi)}{R_s + r_\pi + R_E}$$

For evaluation of t_{22}, the circuit in Fig. 12-25*d* is used. Recall that t_{22} signifies the transmission from output to the comparison circuit through the feedback network. Consequently, V_o is treated as the independent variable, as we are interested in what fraction of V_o contributes to V_π rather than how V_o is obtained (the parameters t_{12} and t_{11} indicate this). In Fig. 12-25*d*, use of the voltage-divider technique gives

$$t_{22} = \frac{-r_\pi}{R_s + r_\pi}$$

FIGURE 12-25
Circuits used to evaluate the *t* parameters in Example 12-3: (*a*) t_{11}, the dead system; (*b*) t_{21}; (*c*) t_{12}; (*d*) t_{22}.

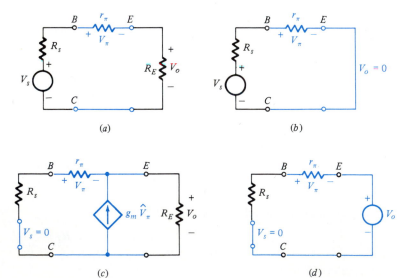

(*a*) (*b*)

(*c*) (*d*)

(b) From Eqs. (12-26), (12-27), (12-31), and (12-25), respectively

$$A_D = t_{11} = \frac{R_E}{R_s + r_\pi + R_E}$$

$$T = -t_{11}t_{22} = -\frac{g_m R_E (R_s + r_\pi)}{R_s + r_\pi + R_E} \frac{-r_\pi}{R_s + r_\pi}$$

and, recalling $g_m r_\pi = \beta_o$

$$T = \frac{\beta_o R_E}{R_s + r_\pi + R_E}$$

$$K = -\frac{t_{21}}{t_{22}} = -\frac{r_\pi/(R_s + r_\pi)}{-r_\pi/(R_s + r_\pi)} = 1$$

$$A_F = \frac{V_o}{V_s} = \frac{A_D + KT}{1 + T} = \frac{[R_E/(R_s + r_\pi + R_E)] + (1) [\beta_o R_E/(R_s + r_\pi + R_E)]}{1 + [\beta_o R_E/(R_s + r_\pi + R_E)]}$$

Clearing fractions gives

$$A_F = \frac{(\beta_o + 1) R_E}{R_s + r_\pi + (\beta_o + 1) R_E}$$

which is the value given in Table 10-3.

Alternatively, T is computed directly by using the circuit in Fig. 12-17c. Using current-divider techniques, we obtain

$$I_b = -g_m \hat{V}_\pi \frac{R_E}{R_E + R_s + r_\pi}$$

and $V_\pi = I_b r_\pi$. Combination of these relations and use of Eq. (12-27) gives

$$T = \frac{V_\pi}{\hat{V}_\pi}\bigg|_{V_s = 0} = \frac{\beta_o R_E}{R_E + R_s + r_\pi}$$

as before.

Note that only t_{12} depends on the controlled-source parameter g_m (or β_o). Each of the other t parameters depends on the resistive elements in the circuit.

Example 12-4 (a) Determine the voltage gain V_o/V_s for the noninverting Op-Amp stage in Fig. 12-26a by first evaluating A_D, A_{OL}, T, and K.

Solution The equivalent circuit of the stage, assuming $R_i \to \infty$ and $R_o = 0$ for the Op-Amp, is given in Fig. 12-26b. In Fig. 12-26b we identify $X_s = V_s$, $X_o = V_o$, and $X_i = V_i$. The controlled source is made independent by considering its value to be $A_v \hat{V}_i$ as shown.

FIGURE 12-26
(*a*) Noninverting Op-Amp stage, (*b*) equivalent circuit for Example 12-4.

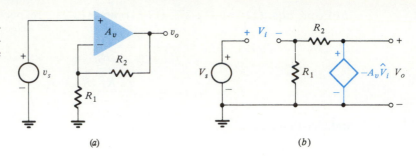

(*a*)   (*b*)

In Fig. 12-26*b*, $V_o = -A_v \hat{V}_i$; setting $\hat{V}_i = 0$ makes $V_o = 0$ and hence $t_{11} = A_D = 0$. Also, suppression of V_s gives

$$t_{12} = \left. \frac{V_o}{\hat{V}_i} \right|_{V_s=0} = -A_v$$

With $V_s = 0$, the voltage-divider relation gives

$$V_i = -A_v \hat{V}_i \frac{R_1}{R_1 + R_2}$$

Then, from Eq. (12-27)

$$T = \left. -\frac{V_i}{\hat{V}_i} \right|_{V_s=0} = \frac{A_v R_1}{R_1 + R_2}$$

and

$$t_{22} = -\frac{T}{t_{12}} = \frac{R_1}{R_1 + R_2}$$

When $V_o = 0$, no voltage drop exists across R_1 and $V_i = -V_s$. Thus $t_{21} = -1$.

From Eq. (12-28)

$$A_{OL} = A_D + t_{12}t_{21} = 0 + (-A_v)(-1) = A_v$$

$$K = -\frac{t_{21}}{t_{22}} = -\frac{-1}{R_1/(R_1 + R_2)} = \frac{R_1 + R_2}{R_1} = 1 + \frac{R_2}{R_1}$$

The closed-loop gain A_F, using Eq. (12-29), is

$$A_F = \frac{A_{OL}}{1 + T} = \frac{A_v}{1 + A_v R_1/(R_1 + R_2)} = \frac{A_v(R_1 + R_2)}{R_1(1 + A_v) + R_2}$$

and is the result obtained in Eq. (10-110) with $R_o = 0$. Again we observe that for large values of A_v, $A_F \approx K = 1 + R_2/R_1$.

In each of the examples we note that A_F becomes independent of the amplifying device (β_o, A_v, etc.) for large values of T. Consequently, $A_F \approx K$ and depends only on the resistive elements external to the amplifying device, and the behavior of the feedback amplifier closely approximates the ideal amplifiers discussed in Sec. 12-1. Similarly, in Sec. 12-3 we showed that large values of T are required to improve performance, that is, to reduce distortion, improve desensitivity, and so on. Thus, in the design of feedback amplifiers, external circuit element values are selected to meet the overall gain requirement A_F and the gain of the internal amplifier is chosen to yield the required value of return ratio T.

The analysis in the two examples in this section assumed that the amplifying devices operated at midband frequencies. This was a convenience used to illustrate the technique. The method of analysis is applicable at all frequencies, provided that appropriate device models are used. Here the t parameters are functions of frequency, and their use gives the values of $A_F(s)$, $T(s)$, $A_{OL}(s)$, $A_D(s)$, and $K(s)$.

12-9 IMPEDANCE IN FEEDBACK AMPLIFIERS REVISITED

The effect of feedback on the input and output resistances was discussed in Sec. 12-5. In that section, Blackman's impedance formula [Eq. (12-21)] was introduced. We now use the general analysis of the previous section to derive this equation.

Suppose we wish to determine the input impedance of a feedback amplifier. We can apply a voltage V_s and measure the current I_s as shown in Fig. 12-27a. Then the input impedance $Z_{IF} = V_s/I_s$. Alternatively, we may apply a current I_s and measure the voltage V_s as indicated in Fig. 12-27b for which $Z_{IF} = V_s/I_s$. Both methods must give the same result as the input impedance is independent of the method used to calculate it. Let us now evaluate Z_{IF} for each circuit in Fig. 12-27.

The circuit in Fig. 12-27a is a feedback amplifier and by identifying $X_s = V_s$, $X_o = I_s$, we can write a pair of equations analogous to Eqs. (12-22) and (12-23) or

$$I_s = t_{11}V_s + t_{12}\hat{X}_i$$
$$X_i = t_{21}V_s + t_{22}I_s \tag{12-33}$$

Observe that Eq. (12-33) is independent of both the type of controlled source used and whether the signal X_i is a voltage or a current. The transfer function

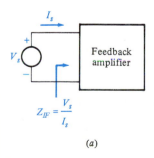

$$Z_{IF} = \frac{V_s}{I_s}$$

(a)

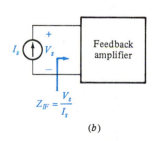

$$Z_{IF} = \frac{V_s}{I_s}$$

(b)

FIGURE 12-27
Two possible circuits to determine the input impedance: (a) apply a voltage and measure the current; (b) apply a current and measure the voltage.

of this system, of the form given in Eq. (12-25), is

$$\frac{I_s}{V_s} = \frac{1}{Z_{IF}} = \frac{t_{11} + t_{12}\,t_{21}}{1 - t_{12}\,t_{22}} \tag{12-34}$$

Alternatively,

$$Z_{IF} = \frac{1}{t_{11}}\,\frac{1 - t_{12}\,t_{22}}{1 + t_{12}\,t_{21}/t_{11}} \tag{12-35}$$

In Eq. (12-34) we can identify $-t_{12}\,t_{22}$ as the return ratio T. Recall that T is measured with the applied source suppressed ($X_s = 0$). Thus the return ratio is measured with the input terminals short-circuited ($V_s = 0$) and is T_{SC}.

The term t_{11} in Eq. (12-34) is I_s/V_s when the gain of the internal amplifier is made zero ($\hat{X}_i = 0$). Hence $1/t_{11}$ is the input impedance for the dead system and is denoted by Z_{ID} and

$$Z_{IF} = Z_{ID}\,\frac{1 + T_{SC}}{1 + t_{12}\,t_{21}/t_{11}} \tag{12-36}$$

Now, let us evaluate Z_{IF} for the circuit in Fig. 12-27b. For this feedback amplifier, $X_s = I_s$ and $X_o = V_s$, so that

$$V_s = t'_{11}\,I_s + t'_{12}\hat{X}_i \qquad X_i = t'_{21}I_s + t'_{22}V_s \tag{12-37}$$

The transfer function for the system described by Eq. (12-37) is

$$\frac{V_s}{I_s} = Z_{IF} = \frac{t'_{11} + t'_{12}t'_{21}}{1 - t'_{12}t'_{22}} = t'_{11}\,\frac{1 + t'_{12}\,t'_{21}/t'_{11}}{1 - t'_{12}t'_{22}} \tag{12-38}$$

Making $\hat{X}_i = 0$ reduces the circuit to the "dead" system; hence $t'_{11} = Z_{ID}$. The return ratio is evaluated with $I_s = 0$, that is, an open circuit, so that $T_{OC} = -t'_{12}t'_{22}$, and

$$Z_{IF} = Z_{ID}\,\frac{1 + t'_{12}\,t'_{21}/t'_{11}}{1 + T_{OC}} \tag{12-39}$$

Equating Eqs. (12-22) and (12-23), we obtain

$$Z_{IF} = Z_{ID}\,\frac{1 + T_{SC}}{1 + T_{OC}} \tag{12-40}$$

which is Blackman's impedance formula.

Example 12-5

(a) Determine the input resistance R_{IF} of an inverting Op-Amp stage. Include the Op-Amp input resistance R_i in the model for the stage. (b) Evaluate R_{IF} for $R_i \to \infty$.

Solution

(a) The equivalent circuit of the stage is displayed in Fig. 12-28. The dead-system input resistance R_{ID} obtained by setting $\hat{V}_i = 0$, is

$$R_{ID} = R_1 + R_i\|R_2$$

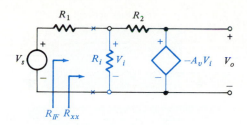

With the input terminals open-circuited, no current exists in R_1, and consequently

$$T_{OC} = \frac{-V_i}{\hat{V}_i} = \frac{R_i A_v}{R_i + R_2}$$

The resistances R_1 and R_i are in parallel when the input terminals are short-circuited. Hence

$$T_{SC} = \frac{R_1 \| R_i}{(R_1 \| R_i) + R_2} A_v$$

Observe that neither T_{OC} nor T_{SC} is zero in this amplifier. Substitution of these values into Eq. (12-40) gives

$$R_{IF} = [R_1 + R_i \| R_2] \frac{1 + (R_i \| R_1) A_v/[(R_1 \| R_i) + R_2]}{1 + [R_i A_v/(R_i + R_2)]} = R_1 + \frac{R_2 R_i}{R_i (1 + A_v) + R_2}$$

In Fig. 12-28, it is clear that $R_{IF} = R_1 + R_{xx}$; therefore

$$R_{xx} = \frac{R_2 R_i}{R_i (1 + A_v) + R_2} = \frac{R_i [R_2/(1 + A_v)]}{R_i + [R_2/(1 + A_v)]}$$

The resistance R_{xx} can be recognized as $R_i \| R_2/(1 + A_v)$; the resistance $R_2/(1 + A_v)$ is exactly that which is reflected across the X–X terminals by using Miller's theorem. (b) From part a, allowing $R_i \rightarrow \infty$, gives

$$R_{IF} = R_1 + \frac{R_2}{1 + A_v} = R_1 + R_{xx}$$

Note that for large values of A_v ($A_v \rightarrow \infty$), the input resistance is simply R_1 as $R_{xx} \rightarrow 0$. However, this is the expected result, for when $A_v \rightarrow \infty$, the inverting terminal is a virtual ground (Sec. 10-21).

12-10 THE SHUNT-FEEDBACK TRIPLE The feedback amplifiers analyzed in the three previous sections each contained only one active element. In general, practical amplifiers have two or more stages, so that both high closed-loop gain and large return ratios can be realized simultaneously. In this and the three succeeding sections, we introduce four commonly used multistage feedback amplifiers. Each amplifier, using "real-world" transistors, approximates one of the basic

FIGURE 12-29
A bipolar shunt-triple
feedback amplifier.

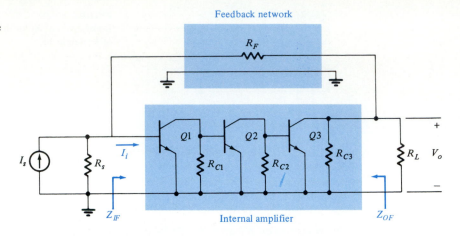

FIGURE 12-29
A bipolar shunt-triple
feedback amplifier.

single-loop feedback topologies. For example, transistor stages have inherent feedback provided by C_μ (C_{gd}) and, therefore, are not truly single-loop amplifiers at high frequencies. Also, when common-emitter (common-source) stages containing R_E (R_S) are used as one stage of the basic amplifier, even at low frequencies, they are feedback amplifiers. The type of feedback described in the two prior sentences is called *local feedback* because the feedback loop is around only one stage. In multistage circuits, we are concerned with *global feedback*, that is, the overall feedback around a number of stages in cascade.

The three-stage amplifier in Fig. 12-29 has the shunt-shunt structure of Fig. 12-17 and is commonly called a *shunt triple*. The internal three-stage amplifier identified in Fig. 12-29 can be modeled as a single equivalent amplifier as depicted in Fig. 12-30. The controlled source $Z_m I_i$ and series resistance r_o are the Thévenin equivalent of the amplifier block in Fig. 12-29. The resistance r_i^1

---

[1]Note that r_b is included. Because the shunt connection has a very low input resistance, neglecting of r_b *may* sometimes introduce an error.

FIGURE 12-30
The equivalent-circuit
representation of the
shunt triple.

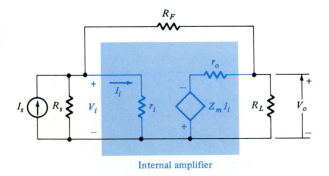

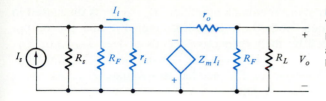

FIGURE 12-31
The approximate model of the basic amplifier (without feedback) of the shunt triple. The resistances R_F at both input and output are the effects of the loading of the basic amplifier by the feedback network.

is the input resistance of the three-stage cascaded amplifier. These quantities are (Prob. 12-23)

$$r_i = r_{b1} + r_{\pi 1} \qquad r_o = R_{C3} \tag{12-41}$$

$$Z_m = \beta_{o1} R_{C1} A_{v2} A_{v3} \tag{12-42}$$

where A_{v2} and A_{v3} are the voltage gains of the second and third stages, respectively.

The amplifier without feedback is shown in the circuit in Fig. 12-31, in which the feedback provided by R_F is eliminated. *Note that the loading effect of R_F on the input and output circuits is included.*

The feedback network in the shunt-shunt configuration transmits a current I_f to the input that is proportional to the output voltage V_o. Thus, when $V_o = 0$, no current is fed back to the input and R_F is connected from input to ground as shown in Fig. 12-31. Similarly, when I_s is suppressed, elimination of the feedback makes $V_i = 0$ as no current exists in r_i. Consequently, connection of R_F from output to ground indicates the output loading effect of R_F.

For the circuit in Fig. 12-31, calculation of $A_{OL} = V_o/I_s$ (Prob. 12-26) gives

$$A_{OL} = \frac{-Z_m R_s'}{R_s' + r_i} \frac{R_L'}{R_L' + r_o} \tag{12-43}$$

where

$$R_s' = R_F \| R_s \qquad \text{and} \qquad R_L' = R_F \| R_L$$

The value of $K = -1/\beta$ can be obtained directly by letting the comparison signal $X_i = 0$ and solving for X_o/X_s. In the circuit in Fig. 12-30, $I_i = 0$ implies that $V_i = 0$; use of KCL at the input node gives $V_o/R_F = -I_s$ and $K = -R_F$ or $\beta = 1/R_F$.

The return ratio, when A_D is assumed negligible, is

$$T = \frac{A_{OL}}{K} = -\beta A_{OL} = \frac{A_{OL}}{R_F} \tag{12-44}$$

Hence Eq. (12-44) is used to evaluate T from the approximate value of A_{OL}.

For the shunt triple, substitution of Eq. (12-43) into Eq. (12-44) yields

$$T = \frac{Z_m}{R_F + R_L} \frac{R_L}{R_L' + r_o} \frac{R_s}{R_s' + r_i} \frac{R_F}{R_F + R_s} \tag{12-45}$$

The Input and Output Impedances In a shunt feedback amplifier, $T_{SC} = 0$, as discussed in Secs. 12-5 and 12-6. Hence, both the input impedance Z_{IF} and the output impedance Z_{OF} are reduced by their respective values of $1 + T_{OC}$. For Z_{IF}, the value of T_{OC} is obtained from T by allowing $R_s \to \infty$ in Eq. (12-45). Similarly, allowing $R_L \to \infty$ gives the value of T_{OC} needed to evaluate Z_{OF}. Applying Blackman's impedance formula, we have

$$Z_{IF} = \frac{r_i \| R_F}{1 + T \mid_{R_s \to \infty}} \tag{12-46}$$

$$Z_{OF} = \frac{r_o \| R_F}{1 + T \mid_{R_L \to \infty}} \tag{12-47}$$

where Z_{ID} and Z_{OD} are obtained from Fig. 12-31.[1]

Example 12-6

The shunt triple in Fig. 12-29 is designed to be supplied from a 600-Ω source and to drive a 600-Ω load. The transistor and component values used are listed in Table 12-5. The feedback resistance $R_F = 20$ kΩ. (*a*) Determine the open-loop gain, return ratio, and closed-loop gain at midband frequencies. (*b*) Determine the input and output impedances.

TABLE 12-5 Parameter Values for Shunt Triple in Fig. 12-29 and Example 12-6*

Parameter	Stage 1	Stage 2	Stage 3
g_m (m$\mho$)	4.0	10	40
r_π	25	10	2.5
β_o	100	100	100
R_C (kΩ)	30	10	0.60

*Note: $r_b = 0$ and $r_o \to \infty$ is assumed for all three transistors

Solution

(*a*) We must first characterize the basic amplifier. From Eqs. (12-41), we obtain

$$r_i = 0 + 25 = 25 \text{ k}\Omega \qquad r_o = 0.60 \text{ k}\Omega$$

As described in Sec. 10-16

$$A_{V2} = \frac{-\beta_{o2}R_{C2}}{R_{C1} + r_{\pi2}} = \frac{-100 \times 10}{30 + 10} = -25.0$$

$$A_{V3} = \frac{-\beta_{o3}R_{C3}}{R_{C2} + r_{\pi3}} = \frac{-100 \times 0.6}{10 + 2.5} = -4.8$$

Substitution of these values into Eq. (12-42) yields

$$Z_m = 100 \times 30 \times (-25.0)(-4.8) = 3.60 \times 10^5 \text{ k}\Omega$$

---

[1]Evaluation of Z_{ID} from Fig. 12-30 gives $Z_{ID} = r_i\|(R_F + R_L\|r_o)$ which, for $R_F \gg R_L$ (as in Example 12-6), reduces to Eq. (12-46). Similarly, for $R_F \gg R_s$, evaluation of Z_{OD} from Fig. 12-30 reduces to Eq. (12-47).

The return ratio and the open-loop gain are evaluated by use of Eqs. (12-44) and (12-43), respectively.

In these equations

$$R'_s = R_s \parallel R_F = 0.6 \parallel 20 = 0.582 \text{ k}\Omega$$

and

$$R'_L = R_F \parallel R_L = 20 \parallel 0.6 = 0.582 \text{ k}\Omega$$

$$A_{\text{OL}} = \frac{-3.6 \times 10^5 \times 0.582}{0.582 + 25} \frac{0.582}{0.582 + 0.6} = -4040$$

$$T = \frac{-4040}{-20} = 202$$

$$A_F = \frac{A_{\text{OL}}}{1 + T} = \frac{-4040}{1 + 202} = -19.9 \text{ k}\Omega \approx -20 \text{ k}\Omega = -R_F$$

(b) We use Eqs. (12-46) and (12-47) to determine Z_{IF} and Z_{OF}. For Z_{ID}, we obtain

$$Z_{ID} = 25 \parallel 20 = 11.1 \text{ k}\Omega$$

The open-circuit return ratio T_{OC} is obtained from T by letting $R_s \rightarrow \infty$. For this condition, $R'_s \rightarrow R_F$ and from Eq. (12-45)

$$T_{\text{OC}} = \frac{3.6 \times 10^5}{20 + 0.6} \frac{0.6}{0.582 + 0.6} \frac{20}{20 + 25} = 3.94 \times 10^3$$

Hence

$$Z_{IF} = \frac{11.1}{1 + 3.94 \times 10^3} \approx \frac{11.1}{3.94 \times 10^3} = 0.00282 \text{ k}\Omega = 2.82 \text{ }\Omega$$

The dead-system output impedance Z_{OD} is

$$Z_{OD} = r_o \parallel R_F = 0.60 \parallel 20 = 0.582 \text{ k}\Omega$$

When $R_L \rightarrow \infty$, $R'_L \rightarrow R_F$, and $R_L/(R_L + R_F) \rightarrow 1$; therefore, from Eq. (12-45),

$$T_{\text{OC}} = 3.6 \times 10^5 \,(1) \frac{1}{0.60 + 20 + 0.6} \frac{0.582}{0.582 + 25} = 398$$

and

$$Z_{OF} = \frac{0.582}{1 + 398} = 0.00146 \text{ k}\Omega = 1.46 \text{ }\Omega$$

The results obtained clearly indicate that $Z_{IF} \ll R_s$ and $Z_{OF} \ll R_L$. Thus we have satisfied the conditions given in Table 12-1, and this feedback amplifier closely approximates the ideal current-voltage converter (transimpedance amplifier).

Design Choices

The pertinent characteristics of the shunt triple are given in Eqs. (12-43), (12-45), (12-46), and (12-47). Let us qualitatively describe some of the design choices that must be made. We focus on midband performance, as the frequency response and amplifier stability are related and treated in subsequent sections.

Obviously, the amplifier has a gain requirement A_F; that is, the range of output signals corresponding to the input-signal range is known. The source resistance R_s and load resistance R_L that the amplifier must drive are also prescribed. For the amplifier to approach ideal behavior, within specified limits, the given values of R_s and R_L determine the values of Z_{IF} and Z_{OF}, respectively. In addition, variation in gain (desensitivity) and nonlinear distortion of the output signal are design requirements.

For $|T| \gg 1$, the gain of the amplifier in Fig. 12-29 is $A_F \approx K = -R_F$. Thus the gain requirement determines R_F, the feedback-network resistance. Each of the other design specifications given in the previous paragraph depends on a particular value of the return ratio T (Secs. 12-3 to 12-5). The circuit designer must select the largest value of T computed to meet each specification independently. That is, a value of T is determined to satisfy the distortion requirement, a second value of T (T_{OC}) is computed to meet input impedance specification, and so on, and the largest value is selected. Note that an increase in T reduces Z_{IF}, Z_{OF}, the amount of nonlinear distortion and the sensitivity. Thus, with selection of the largest value of T, all other design requirements are satisfied. As T depends on the transfer ratio of the internal amplifier Z_m (and R_F), the value of Z_m is prescribed. Thus the basic design requirements have been "translated" into specifications on the individual networks of the feedback amplifier.

The student should be aware that the design process, while applicable to all topologies, is not as simple as outlined in the previous paragraph. The frequency response and design choices relating to dc performance (bias stabilization), among others, have not been considered. In general, the design process is interactive; choices affecting signal-processing characteristics influence dc performance and vice-versa.

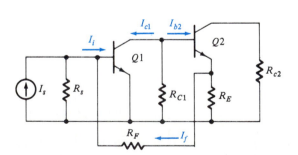

FIGURE 12-32
The representation of the BJT shunt-series pair.

12-11 THE SHUNT-SERIES PAIR

The two-stage amplifier in Fig. 12-32 is called the *shunt-series pair*. Clearly, current comparison takes place at the input. We now show that current sampling occurs at the output by demonstrating that the feedback current I_f is proportional to the output current I_o. Assuming that $\beta_o \gg 1$, the emitter current in $Q2$ is I_o. The paths through the resistances R_E and R_F form a current divider and, hence I_f is proportional to I_o. In addition, setting $I_o = 0$ causes I_f to be zero.

The circuit in Fig. 12-32 only approximates the shunt-series amplifier of Fig. 12-19 because of the local feedback in $Q2$ provided by R_E.

The amplifier without feedback shown in Fig. 12-33 is used to determine the approximate value of $A_{OL} = I_o/I_s$. As seen in Fig. 12-33, the loading effect of the feedback network, obtained by using the rules in Sec. 12-7, is included. We also note that only the global feedback loop is eliminated; the local feedback due to the emitter resistance in $Q2$ is included as part of A_{OL}. Analysis of the circuit in Fig. 12-33 (Prob. 12-30), making the following assumptions listed, gives the results stated in Eqs. (12-48).

$$A_{OL} \approx \frac{-\beta_o R_{C1}}{R_E} \qquad T \approx \frac{\beta_o R_{C1}}{R_F} \qquad K = -\frac{1}{\beta} = \frac{-R_F}{R_E} \qquad (12\text{-}48)$$

where it is assumed that $\beta_o \gg 1$, $\beta_o R_E \gg$ both $r_{\pi 2}$ and R_{C1}, $R_s \gg r_{\pi 1}$, and $R_F \gg$ both R_E and $r_{\pi 1}$.

The Input and Output Resistances

The input impedance Z_{IF} is reduced because of the shunt connection at the input. Similar to the discussion in the previous section, we obtain

$$Z_{IF} = \frac{r_{\pi 1}}{1 + T}\Big|_{R_s \to \infty} \qquad (12\text{-}49)$$

where it is assumed that $R_F \gg r_{\pi 1}$.

If $r_{o2} \to \infty$, then the dead-system impedance is also infinite. When r_{o2} is included in the model (Prob. 12-29) Z_{OD}, assuming $R_F \gg R_E$, is

FIGURE 12-33

The model of the basic amplifier of the shunt-series pair including the loading effect of the feedback network at both input (R_F in series with R_E) and output ($R_F \| R_E$).

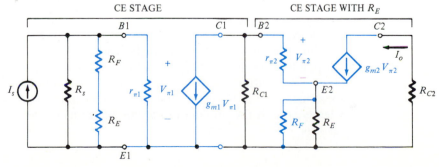

$$Z_{OD} \approx r_{o2} \left(1 + \frac{\beta_{o2} R_E}{R_E + r_{\pi 2} + R_{C1}} \right) \tag{12-50}$$

Note that R_{C1} is the effective source resistance for this stage. Equation (12-50) indicates that the local series feedback provided by R_E increases the output resistance.

For a series connection, $T_{OC} = 0$ and T_{SC} is evaluated by setting $R_{C2} = 0$ in the expression for T. According to the same approximations used to derive Eq. (12-48), $T_{SC} \approx T$, and

$$Z_{OF} = r_{o2} \left(1 + \frac{\beta_o R_E}{R_E + r_{\pi 2} + R_{C1}} \right) \left(1 + \frac{\beta_o R_{C1}}{R_F} \right) \tag{12-51}$$

Example 12-7

A shunt-series amplifier is designed with the following parameter values: $R_s = 10 \text{ k}\Omega$, $R_E = 0.50 \text{ k}\Omega$, $R_F = 10 \text{ k}\Omega$, $R_{C1} = R_{C2} = 2.0 \text{ k}\Omega$, $r_{\pi 1} = 0.5 \text{ k}\Omega$, $\beta_{o1} = \beta_{o2} = \beta_o = 100$, and $r_{\pi 2} = 0.50 \text{ k}\Omega$. The output resistances r_{o1} and r_{o2} are sufficiently large that they can be neglected. Determine the approximate values of A_{OL}, T, and A_F.

Solution

The assumptions that $R_F \gg R_E$, $R_F \gg r_{\pi 1}$, $\beta_o R_E \gg R_{C1}$ and that negligible current exists in R_s are used. Hence Eqs. (12-48) are employed and give

$$A_{\text{OL}} \approx \frac{-100 \times 2}{0.5} = -400 \qquad K \approx -\frac{10}{0.5} = -20 \qquad T \approx \frac{-400}{-20} = 20$$

Then, $A_F = -400/(1 + 20) = -19.0$.

Calculation of these quantities, but making none of the stated assumptions, yields $A_{\text{OL}} = -358$, $T = 17.1$, $K = -(R_F + R_E)/R_E = -21$, and $A_F = -19.7$. The approximate values, based on the assumptions stated, are reasonably close to the more accurate ones. We point this out to highlight the fact that by making such approximations the design engineer can estimate quickly the performance of the circuit.

12-12 THE SERIES-SHUNT PAIR A BJT realization of the *series-shunt pair* is depicted in Fig. 12-34a. The parallel-connected output is clearly evident. At the input, the comparison signal $V_{\pi 1} = V_f$ is approximately the difference between V_s and the voltage across R_E.

The equivalent circuit of the two-stage cascaded amplifier in Fig. 12-34b is the amplifier in Fig. 12-34a with the global feedback network eliminated according to the rules given in Sec. 12-7. The two stages are a common-emitter stage with emitter resistance and a simple common-emitter stage. Hence, the local feedback loop in the first stage is included in the calculation of A_{OL}. Clearly, A_{OL} is the product of the voltage gains of each stage and, from Table 10-3, is

$$A_{\text{OL}} = A_{v1} A_{v2} = \frac{-\beta_{o1} R_{C1}}{R_s + r_{\pi 1} + (1 + \beta_{o1})R_E'} \frac{-\beta_{o2} R_{L2}}{R_{C1} + r_{\pi 2}} \tag{12-52}$$

FIGURE 12-34
(a) Series-shunt feedback pair and (b) the equivalent circuit of the amplifier without feedback.

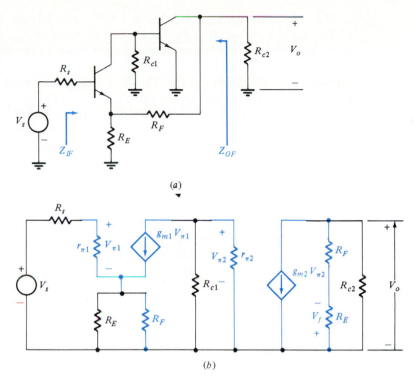

(a)

(b)

where

$$R'_E = R_E \| R_F \qquad \text{and} \qquad R_{L2} = R_{C2} \| (R_F + R_E)$$

The value of β is determined following the procedure given in Sec. 12-7. In Fig. 12-34b, the voltage-divider network gives $V_f = -R_E V_o/(R_E + R_F)$ and

$$\beta = \frac{V_f}{V_o} = -\frac{R_E}{R_E + R_F} \tag{12-53}$$

Assuming $\beta_{o1} \gg 1$ and $\beta_{o1}R'_E \gg R_s + r_{\pi1}$, the approximate value of T is

$$T \approx \frac{\beta_{o2}R_{C1}R_{L2}}{R_{C1} + r_{\pi2}} \tag{12-54}$$

The Input and Output Impedance   In the series-shunt amplifier we expect an increase in the input impedance Z_{IF} and a decrease in the output impedance Z_{OF}. Assuming $R_F \gg R_E \| (R_s + r_{\pi1})/(1 + \beta_{o1})$,

$$Z_{OF} \approx \frac{R_F}{1 + T \mid_{R_{C2} \to \infty}} \tag{12-55}$$

The input impedance is given by

$$Z_{IF} = [r_{\pi1} + R_E(1 + \beta_{o1})](1 + T \mid_{R_s = 0}) \tag{12-56}$$

The derivation of Eqs. (12-55) and (12-56) is left to the student (Prob. 12-34). Note, however, that Z_{ID} (and the exact value of Z_{OD}) contains the effect of the local feedback loop on the input impedance of the first stage.

Example 12-8

The series-shunt amplifier in Fig. 12-34a is designed by using transistors having the following parameters: $r_{\pi 1} = 5.0$ kΩ, $\beta_{o1} = 125$ (for $Q1$); $r_{\pi 2} = 2.50$ kΩ, $\beta_{o2} = 125$ (for $Q2$). The circuit elements used are as follows: $R_{C1} = 9.0$ kΩ, $R_{C2} = 3.0$ kΩ, $R_E = 0.20$ kΩ, and $R_F = 6.0$ kΩ. The amplifier is driven by a source having an internal resistance $R_s = 2.5$ kΩ. Determine A_{OL}, T, and A_F.

Solution

We use the results of the approximate analysis [Eq. (12-52)] to evaluate A_{OL} as

$$A_{OL} = \frac{-125 \times 9.0}{2.5 + 5.0 + (125 + 1) \, 0.194} \frac{-125 \times 2.02}{9.0 + 2.5} = 773$$

in which $R'_E = R_E \parallel R_F = 0.20 \parallel 6.0 \approx 0.194$ kΩ and $R_{L2} = R_{C2} \parallel (R_F + R_E) = 3.0 \parallel (6.0 + 0.20) = 2.02$ kΩ have been used.

Use of Eq. (12-53) gives

$$\beta = -\frac{0.2}{0.2 + 6.0} \approx -\frac{1}{31} \quad \text{and} \quad T \approx -A_{OL}\beta = \frac{773}{31} = 24.9$$

On the basis of these values,

$$A_F = \frac{773}{1 + 24.9} = 29.8$$

Examination of the results in Example 12-8 leads to the following two observations: (1) the gain $A_F = 29.8$ is typical of what can be achieved by a single common-emitter stage (Sec. 10-10), and (2) the desensitivity provided by $T = 24.9$ can also be obtained with a single common-emitter stage by using the local feedback provided by R_E (Prob. 12-27). However, the common-emitter stage containing R_E cannot realize both the gain and the desensitivity simultaneously, as we now demonstrate. In Prob. 12-11, we show that $T = \beta_o R_E/(R_s + r_\pi + R_E)$ for a common-emitter stage with an emitter resistance. Using values in Example 12-8 for the first stage in Fig. 12-34a, we find that $T = 3.25$ and $|A_V| = \beta_o R_{C1}/[R_s + r_{\pi 1} + (\beta_o + 1)R_E] = 34.4$. Clearly, $|A_V|$ is comparable to A_F, but the return ratio for this stage is considerably lower than that obtained for the series-shunt pair. To achieve the same desensitivity in both amplifiers, R_E in the common-emitter stage must be increased. This, however, then causes $|A_V|$ to decrease. Consequently, both gain and desensitivity requirements cannot be realized simultaneously by the single stage with local feedback.

A similar situation exists when the shunt-series pair in Example 12-8 is compared with a common-emitter stage containing local feedback from collector to base (Prob. 12-28). We can conclude from this discussion that, almost always, *the use of global feedback is more effective than is the use of local feedback*. Hence global feedback is preferred in circuit design.

12-13   THE SERIES TRIPLE   The FET realization of the series-series amplifier in Fig. 12-18 is shown in Fig. 12-35a and is commonly called a *series triple*. The output current is fed back through R_F and contributes to the voltage drop V_S across R_S. Voltage comparison occurs in the input loop. The total current in R_S is $I_1 + I_o$ and indicates that R_S is part of both the global feedback network and the local feedback network of the first stage. Similarly, R_F and R_S act as a resistance in the source of the third stage and thus provide local feedback for this stage.

The equivalent circuit of the amplifier without feedback is displayed in Fig. 12-35b and includes the loading effects of the feedback network. This three-stage cascaded amplifier consists of two common-source stages containing local feedback (stages 1 and 3) and one simple common-source amplifier (stage 2). Assuming $R_F \gg R_S$ so that $R_F \parallel R_S \approx R_S$ and $R_F + R_S \approx R_F$, A_{OL} is the product of the gains of these stages, or

$$A_{OL} = \frac{I_o}{V_s} = \frac{-\mu_1 R_{D1}}{R_{D1} + r_{d1} + (1 + \mu_1)R_S}(-g_{m2}R_{L2})\frac{\mu_3}{R_{D3} + r_{d3} + (1 + \mu_3)R_F}$$

$$(12\text{-}57)$$

FIGURE 12-35
(a) A FET series triple.
(b) The equivalent circuit of the amplifier without feedback.

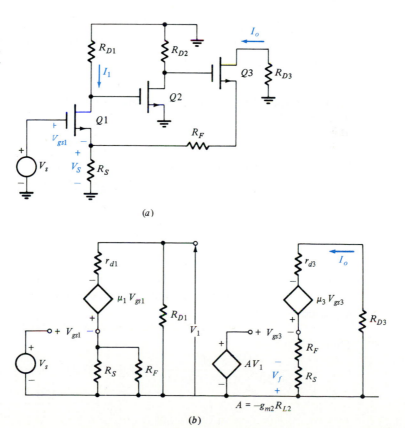

(a)

(b)

Note that the last term in Eq. (12-57) is the voltage-to-current transfer ratio of the third stage.

In Fig. 12-35b letting $V_{gs1} = 0$ results in $V_f = I_o R_S$. Hence $\beta = V_f/I_o = R_S$ and the return ratio can be approximated by

$$T = -A_{OL}\beta \qquad (12\text{-}58)$$

$$= \frac{\mu_1 \mu_3 g_{m2} R_{D1} R_{L2} R_S}{[R_{D1} + r_{d1} + (1 + \mu_1)R_S][R_{D3} + r_{d3} + (1 + \mu_3)R_F]}$$

The Input and Output Impedances The input resistance of this stage is virtually infinite at midband frequencies. The gate-to-source resistance of the MOSFET is in the order of 10^{12} Ω, and this value is multiplied by $1 + T_{SC}$ because of the series feedback. Consequently, Z_{IF} for this circuit is usually considered as an open circuit. In discrete-transistor stages, the input impedance is simply the equivalent resistance of the gate-bias network.

The dead-system output impedance Z_{OD} is the output resistance of a common-source stage containing a source resistance as determined in Table 10-5. Assuming $R_F \gg R_S$, we have

$$Z_{OF} = [r_{d3} + (1 + \mu_3)R_F](1 + T \mid _{R_{D3} = 0}) \qquad (12\text{-}61)$$

where T_{SC} is obtained from T by setting $R_{D3} = 0$.

Example 12-9

The FET parameters and drain resistances used in the series triple in Fig. 12-35 are listed in Table 12-6. The feedback resistor $R_F = 10$ kΩ, and $R_S = 0.50$ kΩ. Find A_{OL}, T, and A_F for this stage.

TABLE 12-6 FET Parameters and Component Values for Example 12-9 and Fig. 12-35

Transistor	g_m, m℧	r_d, kΩ	μ	R_D, kΩ
		Parameter		
$Q1$	3.0	50	150	50
$Q2$	1.5	100	150	50
$Q3$	1.0	130	130	10

Solution

We use the results of the approximate analysis in Eqs. (12-57) and (12-58) to evaluate A_{OL} and T, respectively. Note that the assumption that $R_F \gg R_s$ is reasonable and will introduce only a small error. In Eq. (12-57), the value of R_{L2} used is

$$R_{L2} = R_{D2} \parallel r_{d2} = 50 \parallel 100 = 33.3 \text{ k}\Omega$$

Then

$$A_{OL} = \frac{-150 \times 50}{50 + 50 + (1 + 150)\,0.5}(-1.5 \times 33.3)$$

$$\times \frac{130}{10 + 130 + (1 + 130)\,10} = 191.6 \text{ m℧}$$

Since $\beta = -R_S = -0.5$ kΩ, we find that

$$T = -A_{OL}\beta = -191.6\,(-0.5) = 95.8$$

The closed-loop gain A_F is

$$A_F = \frac{A_{OL}}{1 + T} = \frac{191.6}{1 + 95.8} = 1.98 \text{ m}\mho$$

and is very nearly the value of $-1/\beta$ since $T \gg 1$. When the loading of R_F on R_s is considered, that is, $R_F \parallel R_s$ and $R_F + R_s$ are computed and used in Eq. (12-57), the value of $A_{OL} = 187.2$ m$\mho$.

12-14   GENERAL ANALYSIS OF MULTISTAGE FEEDBACK AMPLIFIERS

Multistage feedback amplifiers can be analyzed to include the transmission from input to output of the feedback network. The method used is based on the single-stage analysis discussed in Sec. 12-8. The procedure is as follows:

1. Draw the equivalent circuit of the amplifier.
2. Identify $\hat{X}_i$. It is convenient to identify $\hat{X}_i$ for a common-emitter (source) stage since V_π (V_{gs}) is measured with respect to ground.
3. Apply the definitions of Eqs. (12-24) and evaluate the t parameters. The t parameters are then used to determine A_{OL}, T, and A_F.

The example that follows illustrates this procedure.

Example 12-10   (a) Determine the t parameters for the series-shunt amplifier in Fig. 12-34a. (b) Use the element values in Example 12-8 and evaluate A_D, A_{OL}, T, and A_F.

Solution   (a) The equivalent circuit is given in Fig. 12-36 in which $\hat{X}_i = \hat{V}_{\pi 2}$ is identified in the common-emitter stage. To evaluate t_{11}, Fig. 12-37a is used. When viewed from terminals $X - X'$, the first stage is an emitter follower; use of Thévenin's theorem at $X - X'$ results in the circuit shown in Fig. 12-37b. From Table 10-3A,

$$A_v = \frac{(\beta_{o1} + 1)R_E}{R_s + r_{\pi 1} + (\beta_o + 1)R_E} \qquad R_o' = R_E \parallel \frac{R_s + r_{\pi 1}}{1 + \beta_{o1}} = R_E \parallel R_o$$

FIGURE 12-36
The equivalent circuit of the series-shunt pair. The controlled source of the CE stage is identified by $g_{m2} \hat{V}_{\pi 2}$ for the purposes of analysis.

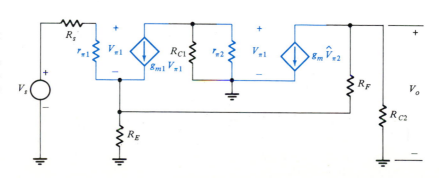

FIGURE 12-37
Equivalent circuits used
to compute the *t* param-
eters in Example 12-10.
(*a*) Circuit for t_{11} and (*b*)
Thévenin equivalent. (*c*)
Circuit for t_{21}; (*d*) circuit
for t_{12} and *T*.

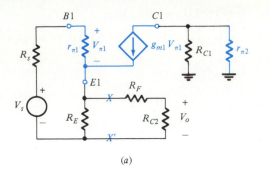
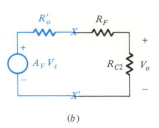

(*a*)

(*b*)

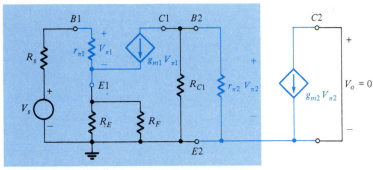

CE stage with $R'_E = R_E \| R_F$

(*c*)

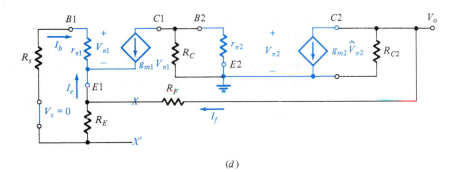

(*d*)

In Fig. 12-37*b*, the voltage-divider relation gives

$$t_{11} = A_D = \frac{V_o}{V_s} = \frac{R_{C2}}{R'_o + R_F + R_{C2}} A_V$$

To compute t_{21}, $V_o = 0$ and the circuit used is shown in Fig. 12-37*c*. The
portion of the circuit enclosed by the blue rectangle is a common-emitter stage
with emitter resistance $R'_E = R_E \| R_F$. Identifying $R_{L1} = r_{\pi 2} \| R_{C1}$, use of Table
10-3*A* yields

$$t_{21} = \frac{V_{\pi2}}{V_s} = \frac{-\beta_{o1}R_{L1}}{R_s + r_{\pi1} + (\beta_{o1} + 1)R_E'}$$

Both t_{12} and the return ratio T are evaluated by using Fig. 12-37d. The effective load on the common-emitter stage is $R_{L2} = R_{C2} \| (R_F + R_o')$. Hence $V_o = -g_{m2}R_{L2}\hat{V}_{\pi2}$ and

$$t_{12} = \frac{V_o}{\hat{V}_{\pi2}} = -g_{m2}R_{L2}$$

The current $I_f = V_o/(R_F + R_o') = -g_{m2}R_{L2}\hat{V}_{\pi2}/(R_F + R_o')$ splits between R_E and R_o of the emitter follower. Use of the current-divider relationship gives

$$I_e = \frac{I_f R_E}{R_E + R_o}$$

Since $I_e = -(\beta_{o1} + 1)I_b$ and $V_{\pi1} = I_b r_{\pi1}$, then $V_{\pi1} = -r_{\pi1}I_e/(\beta_{o1} + 1)$. Combination of the relations for $V_{\pi1}$, I_e, I_f, V_o, and $V_{\pi2}$ and formation of $T = -V_{\pi2}/\hat{V}_{\pi2}$ yields

$$T = \frac{\beta_{o1}R_{L1}}{\beta_{o1} + 1}\frac{R_E}{R_E + R_o}\frac{g_{m2}R_{L2}}{R_F + R_o'}$$

Evaluation of $t_{22} = -T/t_{12}$ gives

$$t_{22} = \frac{\beta_{o1}R_{L1}}{\beta_{o1} + 1}\frac{R_E}{R_E + R_o}\frac{1}{R_F + R_o'}$$

(b) Using the values given in Example 12-8, we obtain

$$R_o' = 0.20 \| \frac{2.5 + 5.0}{1 + 125} = 0.0459 \text{ k}\Omega, \qquad R_o = \frac{2.5 + 5.0}{1 + 125} = 0.0595 \text{ k}\Omega$$

$$R_{L1} = 2.5 \| 9.0 = 1.96 \text{ k}\Omega, \qquad R_{L2} = 3.0 \| 6.0 = 2.0 \text{ k}\Omega$$

$$R_E' = 0.20 \| 6.0 = 0.194 \text{ k}\Omega$$

$$A_D = t_{11} = \frac{3.0}{0.0459 + 6.0 + 3.0} = 0.332$$

$$T = \frac{125\,(1.96)}{125 + 1}\frac{0.20}{0.20 + 0.0595}\frac{50\,(2)}{6.0 + 0.0459} = 24.8$$

where $g_{m2} = \beta_{o2}/r_{\pi2} = 125/2.5 = 50 \text{ m}\mho$ is used.

$$A_{OL} = t_{11} + t_{12}\,t_{21} = 0.332 + (-50)\,(2.0)\left(\frac{-125 \times 1.96}{2.5 + 5.0 + 126 \times 0.194}\right)$$

$$= 767$$

Thus

$$A_F = \frac{A_{OL}}{1 + T} = \frac{767}{1 + 24.8} = 29.7$$

The values obtained in this example are all within 1 percent of those obtained in Example 12-8 by use of the approximate analysis.

12-15 MULTILOOP FEEDBACK AMPLIFIERS

The decrease in sensitivity provided by single-loop feedback amplifiers is limited by the value of the return ratio. Since real-world devices have finite gain, the sensitivity cannot be made zero; that is, the transfer function A_F cannot be made completely independent of the gain of the basic amplifier (without feedback). In addition, variations in the values of the passive components which comprise the feedback network also cause A_F to change. Note that these element variations are unaffected by the amount of feedback. For example, in the noninverting Op-Amp stage shown in Fig. 12-26, for A_v and hence $T \gg 1$, $A_F = 1 + R_2/R_1$, and the precision with which we can specify A_F is directly proportional to the precision with which the resistor ratio can be fabricated.

Single-loop amplifiers fail when there is a catastrophic failure in one of the amplifying elements. If the gain of the internal amplifier becomes zero, the gain with feedback also approaches zero (it becomes A_D). Protection against catastrophic failure is important in several applications such as medical instrumentation (cardiac-care monitoring equipment), manned space flight, and satellite and undersea cable communications (where repair is costly, time-consuming, and extremely difficult).

Multiloop feedback amplifiers are often used to overcome these limitations. As used in this section, a multiloop feedback amplifier is a multistage circuit that contains two or more global feedback loops.[1] The following discussion is a brief introduction of the properties of several classes of multiloop feedback amplifiers.

Positive-Negative-Feedback Amplifier The three-stage amplifier in Fig. 12-38 contains two feedback loops. The interior loop formed by A_2, A_3, and f_2 provides positive feedback as the fed-back signal and applied signal (from A_1) at X_2 are in phase. The feedback introduced by f_1 around all three stages is negative as the signal fed back is 180° out of phase (at midband) with the input signal X_s. For convenience, we assume that all feedforward paths have negligible effect.

This amplifier is designed to make the output signal invariant with respect to changes in the gain A_1. The transfer ratio can be expressed as

$$\frac{X_o}{X_s} = A_F = \frac{-A_1 \left[A_2 A_3/(1 - A_2 A_3 f_2) \right]}{1 + \left[A_1 A_2 A_3 f 1/(1 - A_2 A_2 f_2) \right]}$$

$$= \frac{-A_1 A_2 A_3}{1 - A_2 A_3 f_2 + A_1 A_2 A_3 f_1} \tag{12-60}$$

---

[1]Cascaded single-loop amplifiers and single-loop circuits that contain local feedback are not, in general, treated as multiloop circuits.

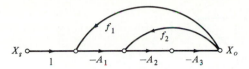

FIGURE 12-38
Signal-flow graph of a positive-negative multiloop feedback amplifier.

In Eq. (12-60) we observe that if $A_2A_3f_2 = 1$, then $A_F = 1/f_1$ and is independent of A_1.

An alternative view of this amplifier is to consider that f_1 is a global feedback about an amplifier comprising A_1 in cascade with the feedback amplifier of gain $A_2A_3/(1 - A_2A_3f_2)$. For $A_2A_3f_2 = 1$, this interior feedback amplifier has infinite gain. Consequently, A_{OL} and T are both infinite, and with $T \to \infty$, the sensitivity is zero.

The circuit in Fig. 12-39 is a BJT realization of this amplifier. The interior positive-feedback loop consists of the series-series connection of R_1 and R_2 about $Q2$ and $Q3$. The shunt feedback connection from $Q3$ to $Q1$ through R_F provides the overall negative feedback.

The first of two potential drawbacks of this circuit is its sensitivity to the components in the positive-feedback loop. For proper operation, $A_2A_3f_2 = 1$, and any deviation in this value results in a dependence of A_F and A_1. The second drawback is the stability of the amplifier; almost any use of positive feedback has the potential to cause oscillation.

The McMillan Structure   The amplifier in Fig. 12-40, originally proposed by McMillan for re-liable underwater cable transmission, utilizes both feedback and feedforward paths. The amplifiers A_1 and A_2 in the cross-coupled, parallel-channel config-uration are usually single-loop feedback amplifiers. This topology is used to protect against a catastrophic failure in one channel without disrupting signal transmission from input to output. We now demonstrate this.

The transfer function of this amplifier (Prob. 12-44) can be expressed as

$$A_F = \frac{X_o}{X_s} = \frac{A_1(1 - A_2f_{21}) + A_2(1 - A_1f_{12})}{1 - A_1A_2f_{12}f_{21}} \tag{12-61}$$

FIGURE 12-39

Bipolar implementation of the amplifier in Fig. 12-38.

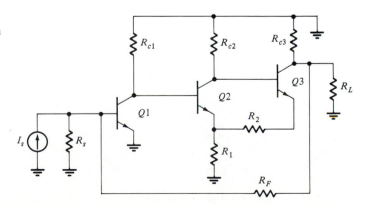

FIGURE 12-40
A two-channel Mc-
Millan feedback-feed-
forward amplifier. Only
one channel must op-
erate normally to pro-
vide the desired output.

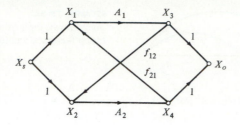

Nominally, the two channels are designed to be identical with $A_1 = A_2 = A$ and $f_{12} = f_{21} = f$. Under these circumstances

$$A_F = \frac{2A(1 - Af)}{1 - A^2f^2} = \frac{2A(1 - Af)}{(1 + Af)(1 - Af)} = \frac{2A}{1 + Af} \qquad (12\text{-}62)$$

If $Af = 1$, then, from Eq. (12-62), $A_F = A$.

Suppose that $A_1 = 0$, that is, a catastrophic failure occurs in one channel. Use of Eq. (12-61) indicates that $A_F = A_2 = A$ and the same transmission between input and output exists. Recall that A_2 is a negative-feedback amplifier, so that even with $A_1 = 0$, the circuit performance displays the benefits of a single-loop amplifier.

The amplifier structure in Fig. 12-40 is also insensitive to any changes that occur in one channel. Thus, if A changes so that $A_1f_{12} \neq 1$ while $A_2f_{21} = 1$, from Eq. (12-61) we have

$$A_F = \frac{A_2(1 - A_1f)}{1 - A_1f} = A_2 = A$$

and no variation in A_F exists. Similarly, if f_{12} changes so that $Af_{12} \neq 1$, then

$$A_F = \frac{A(1 - Af_{12})}{1 - Af_{12}} = A$$

We conclude that this amplifier is invariant to changes in both active and passive elements. Again, one drawback is that Af must be exactly unity, and deviation in this value (in both channels) causes A_F to change.

The topology in Fig. 12-40 can be extended to more than two loops as shown in the three-channel circuit in Fig. 12-41. In this circuit, two catastrophic failures

FIGURE 12-41
A three-channel Mc-
Millan feedback-feed-
forward amplifier.

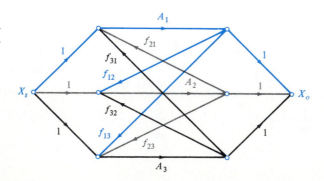

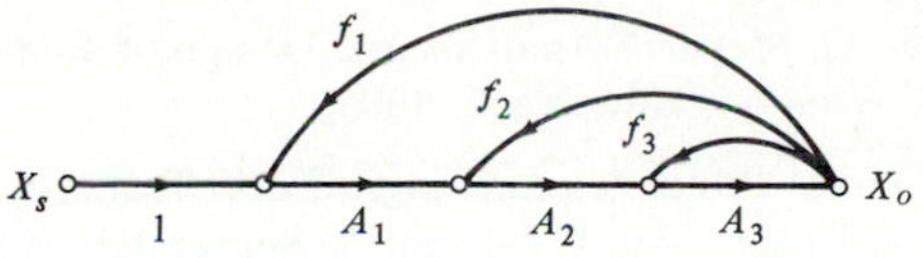

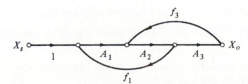

FIGURE 12-42
The signal-flow graph of a nested-loop (follow-the-leader) multiloop feedback amplifier.

can be tolerated without disrupting performances. In effect, the redundancy provided by the parallel channels allows for signal transmission as long as one channel is operating.

Follow-the-Leader Feedback   The signal-flow graph for a three-stage "follow-the-leader feedback" (FLF) amplifier is depicted in Fig. 12-42. In this topology, also referred to as a "nested" amplifier, each feedback loop originates at the same node (the output). The terminus of any one loop is at the input of one of the amplifiers. This structure is often used in active filters because each term in the transfer characteristic can be set by one feedback loop. This is demonstrated in Eq. (12-63), derived from Fig. 12-42 (Prob. 12-40):

$$A_F = \frac{A_1 A_2 A_3}{1 - A_3 f_3 - A_3 A_2 f_2 - A_3 A_2 A_1 f_1} \tag{12-63}$$

Leap-Frog Feedback   The "leap-frog" feedback amplifier topology, shown in Fig. 12-43, is also used in active-filter realizations. As seen in Fig. 12-43, the overlapping feedback loops result in dependence of the input signal to each amplifier on the output signal of each amplifier. This is exploited in active filters to make each term in the transfer function [Eq. (12-64)] depend on the loop gain of one feedback loop (Prob. 12-41):

$$A_F = \frac{A_1 A_2 A_3}{1 - A_2 A_1 f_1 - A_2 A_3 f_3} \tag{12-64}$$

Both the leap-frog and the follow-the-leader topologies can be extended to contain more feedback loops than are indicated in Figs. 12-42 and 12-43.

REFERENCES

1   Bode, H. W.: "Network Analysis and Feedback Amplifier Design," D. Van Nostrand Company, Princeton, N.J., 1945.

2   Blackman, R. B.: Effect of Feedback on Impedance, *Bell System Tech. J.*, vol. 22, no. 3, p. 2, 1943.

3   Sedra, A. S., and K. C. Smith: "Microelectronic Circuits," Holt, Rinehart & Winston, New York, 1981.

FIGURE 12-43
The signal-flow graph of the leap-frog multiloop feedback amplifier.

4 Gray, P. R., and R. G. Meyer: "Analysis and Design of Analog Integrated Circuits," Holt, Rinehart & Winston, New York, 1985.

5 Blecher, F. H.: Design Principles for Single Loop Transistor Feedback Amplifiers, *IRE Trans. Circuit Theory*, vol. CT-4, no. 5, September 1957.

6 Ghausi, M. S.: "Electronic Devices and Circuits: Discrete and Integrated," Holt, Rinehart & Winston, New York, 1985.

7 Schilling, D., and H. Belove: "Electronic Circuits—Discrete and Integrated," McGraw-Hill Book Company, New York, 1979.

8 Soclof, S.: "Analog Integrated Circuits," Prentice-Hall, Englewood Cliffs, N. J., 1985.

9 Black, H. S.: Stabilized Feedback Amplifiers, *Bell System Tech. J.*, vol. 14, pp. 1–18, January 1934.

REVIEW QUESTIONS

12-1 (a) Draw the equivalent circuit of a voltage amplifier.
(b) What conditions must be satisfied if this amplifier is to behave ideally?

12-2 Repeat Rev. 12-1 for a current amplifier.

12-3 Repeat Rev. 12-1 for a voltage-to-current converter.

12-4 Repeat Rev. 12-1 for a current-to-voltage converter.

12-5 List the five constituent parts of a single-loop feedback amplifier.

12-6 Draw a block diagram of a single-loop feedback amplifier and describe the function of each block.

12-7 List the four basic single-loop feedback amplifier topologies.

12-8 (a) Draw the block diagram of a single-loop feedback amplifier.
(b) Define A and β.
(c) What relationship exists between A_F and A?

12-9 Identify X_s, X_o, X_f, and X_i as either a current or a voltage for each feedback topology.

12-10 For each of the four topologies, identify A and β.

12-11 Define (a) negative feedback and (b) positive feedback.

12-12 Define in words and by an equation (a) the open-loop gain A_{OL} and (b) the return ratio T.

12-13 Express A_F in terms of A_{OL} and T.

12-14 (a) Define sensitivity.
(b) For large values of T, what is the value of A_F?
(c) What is the significance of the result in (b)?

12-15 List five characteristics of an amplifier that are modified by negative feedback.

12-16 State the three fundamental assumptions used in the approximate method of analysis.

12-17 To obtain the amplifier without feedback, describe how to obtain the (a) input circuit and (b) output circuit.

12-18 For each of the four topologies, indicate whether the (a) input impedance and (b) output impedance increases or decreases as a result of feedback.

12-19 (a) State Blackman's impedance formula.
(b) Define the terms T_{SC} and T_{OC}.
(c) What is meant by the dead system?

12-20 (a) Draw the two-port representation of a shunt-shunt amplifier.
(b) In terms of Blackman's impedance formula, explain whether the input and output resistances increase or decrease.

12-21 Repeat Rev. 12-20 for a shunt-series amplifier.

12-22 Repeat Rev. 12-20 for the series-series topology.

12-23 Repeat Rev. 12-20 for the series-shunt topology.

12-24 (a) Express X_o and X_i as the superposition of two terms.
(b) Define the t parameters.

12-25 (a) In terms of the t parameters, identify A_D, T, K, and A_{OL}.
(b) What is the significance of each term?

12-26 What is the difference between global and local feedback?

12-27 What four common circuits are used to approximate the four basic feedback amplifiers?

12-28 (a) Which of the circuits in Rev. 12-27 contain local feedback loops?
(b) Which do not?

12-29 List four types of multiloop feedback amplifiers.

12-30 What properties can multiloop topologies exhibit that cannot be achieved in single-loop amplifiers?

12-31 (a) Draw the signal-flow graph for a two-channel McMillan amplifier.
(b) What characteristic does this amplifier have that is not present in the other configurations discussed?

Chapter 13
STABILITY AND RESPONSE OF FEEDBACK AMPLIFIERS

$\mathbf{N}$egative-feedback amplifiers are required to operate over a specified band of frequencies or, alternatively, to produce a desired response to a step-function excitation. In this chapter, we examine the effect of feedback on amplifier response. Methods for testing whether the closed-loop response is stable are developed. Included in our discussion are the compensation techniques employed that ensure that feedback amplifier operation is stable and produces the desired response.

13-1 EFFECT OF FEEDBACK ON BANDWIDTH Consider a feedback amplifier in which the feedback network, source impedance, and load impedance are all resistive. For these conditions, the open-loop gain A_{OL} (the amplifier without feedback) and the return ratio T have the same poles. Initially, let us consider that $A_{\mathrm{OL}}(s)$ has a single dominant pole so that

$$A_{\mathrm{OL}}(s) = \frac{A_O}{1 + s/\omega_h} \qquad \text{and} \qquad T(s) = \frac{T_O}{1 + s/\omega_h} \qquad (13\text{-}1)$$

where A_O and T_O are the midband values of A_{OL} and T, respectively, and ω_h is the angular frequency of the dominant pole. Use of Eq. (12-5) gives

$$A_F(s) = \frac{A_O/(1 + s/\omega_h)}{1 + T_O/(1 + s/\omega_h)} = \frac{A_O/(1 + T_O)}{1 + s/\omega_h(1 + T_O)} = \frac{A_{FO}}{1 + s/\omega_H} \qquad (13\text{-}2)$$

In Eq. (13-2), we identify the midband closed-loop gain $A_{FO} = A_{\mathrm{OL}}/(1 + T_O)$, and ω_H, the 3-dB angular frequency as

$$\omega_H = (1 + T_O)\omega_h \qquad (13\text{-}3)$$

It is clear in Eq. (13-3) that negative feedback has increased the bandwidth by a factor $(1 + T_O)$, the same factor by which A_{OL} is reduced. Thus, for a single dominant-pole function

$$A_O \omega_h = A_{FO} \omega_H \qquad (13\text{-}4)$$

and indicates that the gain-bandwidth product of the amplifier without feedback equals the gain-bandwidth product with feedback. Note that we have seen this before in the discussion of Op-Amp stages in Sec. 11-12.

The Two-Pole Function Let us now assume that $A_{OL}(s)$ and $T(s)$ are represented by a function having two poles on the negative, real axis at $s_1 = -\omega_1$ and $s_2 = -\omega_2$. Thus

$$A_{OL}(s) = \frac{A_O}{1 + s(1/\omega_1 + 1/\omega_2) + s^2/\omega_1\omega_2} = \frac{A_O}{1 + a_1 s + a_2 s^2} \qquad (13\text{-}5)$$

$$T(s) = \frac{T_O}{1 + s\left[(1/\omega_1) + (1/\omega_2)\right] + s^2/\omega_1\omega_2} = \frac{T_O}{1 + a_1 s + a_2 s^2} \qquad (13\text{-}6)$$

Recall that if ω_1 and ω_2 are widely separated, then $\omega_1 \approx 1/a_1$ and $\omega_2 \approx a_1/a_2$. Substitution of Eqs. (13-5) and (13-6) into Eq. (12-5) gives

$$A_F(s) = \frac{A_{FO}}{1 + \dfrac{a_1 s}{1 + T_O} + \dfrac{a_2 s^2}{1 + T_O}} = \frac{A_{FO}}{1 + a_1' s + a_2' s^2} \qquad (13\text{-}7)$$

where $A_{FO} = A_O/(1 + T_O)$.

Application of the dominant-pole approximation to Eq. (13-7) is generally invalid. We observe that the approximate pole frequency $\omega_1' = 1/a_1' = (1 + T_O)\omega_1$ is increased by $(1 + T_O)$, whereas the second pole at $\omega_2' = a_1'/a_2'$ remains at ω_2. Clearly, it is possible to select T_O so that $(1 + T_O)\omega_1 > \omega_2$. The closed-loop poles for this situation are complex (see Prob. 11-26), and the bandwidth ω_H is computed from $|\mathbf{A}_F(j\omega_H)| = 0.707\,A_{FO}$. However, the reduction in the $a_1' = a_1/(1 + T_O)$ is indicative of improved bandwidth. Hence we conclude that negative feedback decreases the frequency and phase distortion. The response of the two-pole system is treated in detail in Sec. 13-5.

Equation (13-7) indicates that the poles of A_F are functions of T_O, the amount of feedback. The movement of the closed-loop poles as T_O is increased is displayed in Fig. 13-1. These poles start at $-\omega_1$ and $-\omega_2$, the poles of $T(s)$,

FIGURE 13-1
Root locus of a two-pole transfer function.

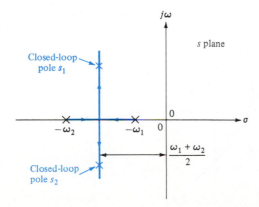

and move toward each other along the negative real axis as T_O is increased from zero. The poles coincide at $-(\omega_1 + \omega_2)/2$; thus, A_F has two equal poles at $-(\omega_1 + \omega_2)/2$. The poles become complex as T_O is increased further, with the real part remaining at $-(\omega_1 + \omega_2)/2$. We note that the closed-loop poles always remain in the left half plane and, as described in Sec. 13-2, indicate a stable system.

The Three-Pole Function   When $T(s)$ and $A_{OL}(s)$ are three-pole functions, as is generally true for a three-stage amplifier, the problem of determining the bandwidth is still more complex. The closed-loop gain is expressible as

$$A_F(s) = \frac{A_{FO}}{1 + \dfrac{s}{1 + T_O}\left(\dfrac{1}{\omega_1} + \dfrac{1}{\omega_2} + \dfrac{1}{\omega_3}\right) + \dfrac{s^2}{1 + T_O}\left(\dfrac{1}{\omega_1\omega_2} + \dfrac{1}{\omega_1\omega_3} + \dfrac{1}{\omega_2\omega_3}\right) + \dfrac{s^3}{(1 + T_O)\omega_1\omega_2\omega_3}}$$

or

$$A_F(s) = \frac{A_{FO}}{1 + \dfrac{a_1 s}{1 + T_O} + \dfrac{a_2 s^2}{1 + T_O} + \dfrac{a_3 s^3}{1 + T_O}} \tag{13-9}$$

The angular frequencies of the open-loop poles are ω_1, ω_2, and ω_3 and are all on the negative real axis as displayed in Fig. 13-2. As is evident in the root locus in Fig. 13-2, increasing T_O can drive two of the poles into the right half plane. These right-half-plane poles introduce terms having positive exponents in the transient response thus making the amplifier unstable. However, observe in Fig. 13-2 that when the poles remain in the left half plane so that the amplifier is stable, the decrease in the a_1 coefficient by $1 + T_O$ illustrates that the bandwidth is improved. In subsequent sections, we describe in detail the degree of bandwidth improvement and the costs which must be borne to obtain it.

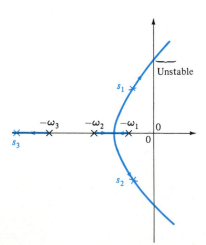

FIGURE 13-2
Root locus of a three-pole transfer function. The poles without feedback ($T = 0$) are $-\omega_1$, $-\omega_2$, and $-\omega_3$, whereas the poles after feedback is added ($T > 0$) are s_1, s_2, and s_3.

If the feedback network contains reactive elements, additional poles (and zeros) are introduced in $T(s)$ and perhaps in $A_{\mathrm{OL}}(s)$. These must be included in determination of the frequency response. In addition, the poles and zeros of the feedback network can cause the amplifier to become unstable. Because the question of stability is of paramount importance, we treat it first. Subsequently, we return and examine in detail the response of multipole feedback systems.

13-2 STABILITY Negative feedback, for which the return ratio $T > 0$, has been considered in some detail in Chap. 12. If $T < 0$, the feedback is termed *positive* or *regenerative*. Under these circumstances, the resultant gain $|A_F|$ may be greater than the open-loop gain $|A_{\mathrm{OL}}|$ (the gain without feedback). Consider $-1 < T < 0$; referring to Eq. (12-5), we conclude that $|A_F| < |A_{\mathrm{OL}}|$. In the earliest days of electronics, the amplifying devices (vacuum triodes) available were incapable of achieving even moderate gains. Regenerative amplifiers, first proposed by Armstrong, were used to increase the effective gain of these devices. However, the development of new devices obviated the need for positive feedback. This fact, coupled with poor stability, has resulted in the rare use of positive feedback.

To illustrate the instability in an amplifier with positive feedback, consider the following situation. No signal is applied, but because of some transient disturbance, a signal X_o appears at the output terminals. Referring to Fig. 12-9, a portion of this signal, $t_{22}X_o$ (βX_o) will be fed back to the input circuit and will appear in the output as an increased signal $t_{12}t_{22}X_o$ ($-A\beta X_o$). If this term just equals X_o, the spurious output has regenerated itself. In other words, if $-TX_o = X_o$ (that is, if $T = -1$), the amplifier will oscillate. Hence, if an attempt is made to obtain large gain by making $|T|$ almost equal to unity, there is the possibility that the amplifier may break out into spontaneous oscillation. This would occur if, because of variation in supply voltages, aging of transistors, etc., $-T$ becomes equal to unity. There is little point in attempting to achieve amplification at the expense of stability. In fact, because of all the advantages enumerated in Sec. 12-3, feedback in amplifiers is almost always negative. However, combinations of positive and negative feedback are used (Sec. 12-15).

Even amplifiers designed to have negative feedback at midband or over a range of frequencies can oscillate. For the comparison signal $X_i = t_{21}X_s + t_{22}X_o$ (Fig. 12-9) to be a difference signal, $t_{22}X_o$ must be 180° out of phase with $t_{21}X_s$. In Chap. 11 we demonstrated that phase shift is introduced in the amplifier response for frequencies outside the midband range. When an additional 180° of phase shift is introduced in the feedback loop, the feedback signal $t_{22}X_o$ is now in phase with $t_{21}X_s$, resulting in positive feedback.

The Definition of Stability If an amplifier is designed to have negative feedback in a particular frequency range but breaks out into oscillation at some higher or lower frequency, it is useless as an amplifier. Hence, in the design of a feedback amplifier, it must be ascertained that the circuit is stable at *all* frequencies, and not merely over the frequency range of interest. In the sense used here, the system is stable if a transient disturbance of finite duration results in a response which dies out. A system is unstable if such a transient disturbance produces an output which persists indefinitely or increases until it is limited only by some nonlinearity in the circuit.

The ideas in the preceding paragraph provide a physical picture of what is meant by stability. Mathematically, the definition of stability is as follows: *A system is stable if and only if all bounded input signals produce bounded output signals*. A signal $x(t)$ is bounded if $|x(t)| \leq$ constant for all t. For example, $\sin \omega t$ is bounded by unity. Similarly, if $x(t) = 0$ for $t < 0$ and ϵ^{-t} for $t \geq 0$, $x(t)$ is also bounded by unity whereas ϵ^{+2t} is not bounded for $t \geq 0$. Usually we are concerned with response defined for $t \geq 0$ as it is customary to assume that excitations are applied at $t = 0$.

The question of stability involves study of the transfer function of the circuit since this determines the transient behavior of the network. If a pole exists with a positive real part, this results in a component of the output that increases exponentially with time and hence is unbounded. Thus the consequence of the definition of stability for linear systems is that all poles of the transfer function must lie in the open left half of the complex-frequency $(s-)$ plane. The term open in the previous sentence refers to the fact that the j axis is excluded from the left half plane.

Consider a system having a transfer function $1/s$, that is, a pole exists at the origin (which is a point on the j axis). If this system is excited by a unit step function, the system response is the ramp function t. The ramp function is unbounded, whereas the input step is bounded. Hence, according to the definition, this system is unstable. Thus poles in the left half of the s plane indicate that the real parts of the poles must be negative. (Sinusoidal oscillators discussed in Sec. 15-1 are examples of systems deliberately designed to have poles on the j axis.)

Stability in Feedback Amplifiers   The closed-loop gain $A_F(s)$ of a feedback amplifier, given in Eq. (12-5), is repeated for convenience as Eq. (13-10):

$$A_F(s) = \frac{A_{OL}(s)}{1 + T(s)} \tag{13-10}$$

The method for determining A_{OL} and T is described in Sec. 12-7, and the frequency response of these quantities can be evaluated by use of the time-constant techniques in Sec. 11-9. We employ this technique in Sec. 13-8.

As seen in Eq. (13-10), the poles of A_F are the zeros of $1 + T(s)$ and any poles of $A_{OL}(s)$ that are not common to $T(s)$. If we assume that the amplifier without feedback is stable, all the poles of A_{OL} lie in the left half plane. Then *the feedback amplifier is stable when the zeros of $1 + T(s)$ all lie in the left half plane*. Alternatively, no zeros of $1 + T = F$ can lie in the right half plane if the amplifier is to be stable. Methods for testing the stability of a feedback system are treated in the next section.

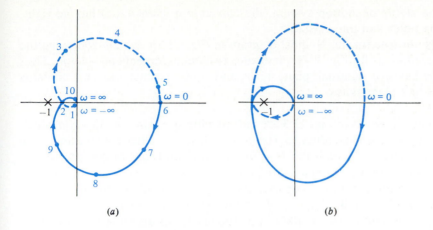

FIGURE 13-3
Nyquist diagrams for (*a*) a stable and (*b*) an unstable system. [*Note*: In (*a*), the point $-1 + j0$ is not encircled, whereas in (*b*) this point is encircled.]

13-3 TESTS FOR STABILITY

In 1931 Nyquist developed a graphical construction, subsequently named the *Nyquist diagram*, to determine whether a feedback amplifier had any right-half-plane poles. The Nyquist diagram is a plot of $T(j\omega) = T(j\omega)\underline{/\theta(j\omega)}$ in polar coordinates. That is, at each angular frequency $-\infty < \omega < +\infty$, $T(j\omega)$ and $\theta(j\omega)$ are evaluated, and each of these values is the coordinate of a single point on the curve.

Two Nyquist diagrams are shown in Fig. 13-3; in each the solid curve corresponds to $\omega \geq 0$ and the dashed curve, to $\omega < 0$. Only values of $T(j\omega)$ for $\omega \geq 0$ must be evaluated. Because the coefficients of $T(s)$ are real, as must be true in a physical system, $\mathbf{T}(-j\omega) = \mathbf{T}^*(j\omega)$; that is, $\mathbf{T}(-j\omega)$ is the conjugate of $\mathbf{T}(j\omega)$. Thus the dashed curves in Fig. 13-3 are the mirror images of the solid curves obtained for $\omega \geq 0$.

The Nyquist Criterion The *Nyquist criterion* states that the number of clockwise encirclements of the point $-1 + j0$ equals the difference between the number of zeros and the number of poles of $F(s) = 1 + T(s)$ in the right half plane. For stability, we must ascertain that $F(s)$ has no right-half-plane zeros, that is, $A_F(s)$ has no right-half-plane poles. Since $F = 1 + T$, the poles of F are identical to the poles of T, and if the amplifier without feedback is stable, $F(s)$ has no poles in the right half plane. Thus, for these conditions, the number of encirclements of $-1+j0$ must be zero for the feedback amplifier to be stable.

The clockwise encirclements of $-1+j0$ are determined by drawing a radius vector from $-1+j0$ to the Nyquist diagram and tracing the locus of points from $-\infty < \omega < +\infty$ as indicated in Fig. 13-3.[1] In Fig. 13-3*a*, the process begins at point 1 ($\omega = -\infty$) and proceeds consecutively through the numbered points so that at point 10 we arrive at $\omega = +\infty$. If the radius vector rotates by 360° about $-1+j0$, an encirclement exists. For Fig. 13-3*a*, no encirclement is present

---

[1]The student might find a paper clip useful for this procedure.

and, for a stable open-loop system, the closed-loop gain $A_{F1}(s)$ has no right-half-plane poles and thus is stable.

This is not true for the Nyquist diagram in Fig. 13-3b, which, following the same procedure as for Fig. 13-3a, indicates two encirclements of $-1+j0$. Thus $1 + T_2(s)$ has two right-half-plane zeros and $A_{F2}(s)$ is unstable as it contains two right-half-plane poles.

Phase Margin

The Nyquist diagrams in Fig. 13-3 are redrawn for $\omega \geq 0$ in Fig. 13-4, on which the unit circle, corresponding to $T(j\omega) = 1$ (0 dB) is also constructed. The angular frequency at which the Nyquist diagram and unit circle intersect is called the *gain-crossover angular frequency* ω_G because $T(j\omega) > 1$ for $\omega < \omega_G$ and $T(j\omega) < 1$ for $\omega > \omega_G$. Comparing the two Nyquist diagrams, we observe that at ω_{G1}, $\angle T_1 > -180°$, that is, $|\angle T_1| < 180°$ and the system is stable, whereas $\angle T_2(j\omega_{G2}) < -180°$ ($|\angle T_2| > 180°$) corresponds to an unstable system. It is, therefore, convenient to introduce the *phase margin* ϕ_M, defined as

$$\phi_M \equiv \angle T(j\omega_G) + 180° \tag{13-11}$$

Note that $\angle T(j\omega_G)$ is, in general, a negative number. An alternative to the statement of Nyquist's criterion, extremely useful in design, is: *The closed-loop system is stable when the phase margin is positive* ($\phi_M > 0$). Hence $\angle T(j\omega_G)$ must be less negative than $-180°$.

Gain Margin

In Fig. 13-4 the angular frequency at which the Nyquist diagram intersects the negative real axis, corresponding to $\angle T = -180°$, is defined as the *phase-crossover angular frequency* ω_ϕ. For $\omega > \omega_\phi$, $\angle T < -180°$ and $\angle T > -180°$ for $\omega < \omega_\phi$. The magnitude $T(j\omega_\phi)$ is used to define the *gain margin GM* as

$$GM = -20 \log T(j\omega_\phi) = -T(j\omega_\phi) \quad \text{dB} \tag{13-12}$$

FIGURE 13-4
The portion of the Nyquist diagram in Fig. 13-3 for $\omega \geq 0$ used to define phase margin ϕ_M and gain margin GM. In (a) $\phi_M > 0$ and $GM > 0$ and the system is stable, whereas for the unstable system in (b) both ϕ_M and GM are negative.

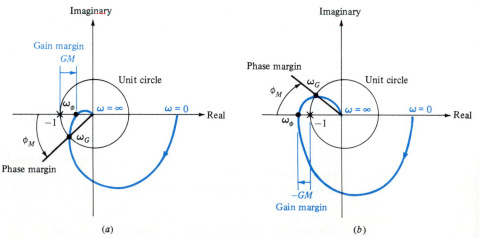

(a)   (b)

FIGURE 13-5
Asymptotic Bode diagrams corresponding to (*a*) Fig. 13-4*a* and (*b*) Fig. 13-4*b*.

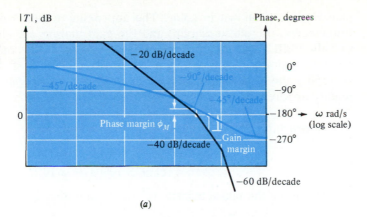

(*a*)

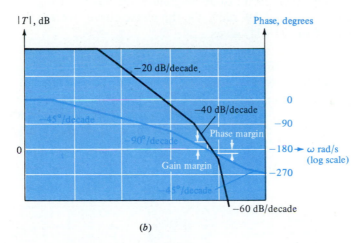

(*b*)

We observe in Fig 13-4 that for the stable system, $T_1(j\omega_{\phi 1}) < 1$ and, as the logarithm of a number less than unity is negative, $GM_1 > 0$. For the unstable system, $T_2(j\omega_{\phi 2}) > 1$ and consequently $GM_2 < 0$. We can conclude that $GM > 0$ is an alternative statement of stability. Its use is more limited in design, however, than the phase margin.

The Bode Diagram The information contained in the Nyquist diagram is often more conveniently displayed in a Bode diagram since we can readily alter the Bode diagram if a pole (zero) location or the midband value of T is changed. In particular, the asymptotic Bode diagram is extremely useful for the pencil-and-paper calculations made by circuit designers. Again, computer simulations are used to obtain the accuracy required for the component values in the final design.

The asymptotic Bode diagrams in Fig. 13-5 correspond[1] to the Nyquist diagrams in Fig. 13-4. In Fig. 13-5, $T(j\omega)$ (in decibels) is drawn in black and the

---

[1]Since no numerical values are given, the Bode diagrams in Fig. 13-5 illustrate one of several possible situations.

phase curve is shown in color. The approximate gain- and phase-crossover frequencies, the phase margin, and the gain margin are also indicated. Note that the 0-dB value for the magnitude curve and $-180°$ on the phase characteristic are drawn at the same vertical position. This is often a convenience in sketching the Bode diagram; for most systems, since $\omega_G < \omega_\phi$, stability is readily apparent.

Example 13-1

The return ratio of a two-pole amplifier is

$$T(s) = \frac{100}{(1 + s/10^6)(1 + s/10^7)}$$

(a) Determine the phase margin. (b) Is the amplifier stable?

Solution

(a) The asymptotic Bode diagram is displayed in Fig. 13-6, from which $\omega_G = 10^{7.5} = 3.16 \times 10^7$ rad/s. On the phase curve, we see that $\angle T = -157.5°$, and use of Eq. (13-11) gives

$$\phi_M = -157.5 + 180 = 22.5°$$

as indicated on Fig. 13-6. (b) As $\phi_M > 0$, the amplifier is stable. Calculation of ω_G and ϕ_M using the actual Bode diagram, and verified by MICROCAP II, gives $\omega_G = 3.09 \times 10^7$ rad/s and $\phi_M = 20.2°$. These are in good agreement with the values obtained from the asymptotic Bode diagram.

FIGURE 13-6
Asymptotic Bode diagram for Example 13-1.

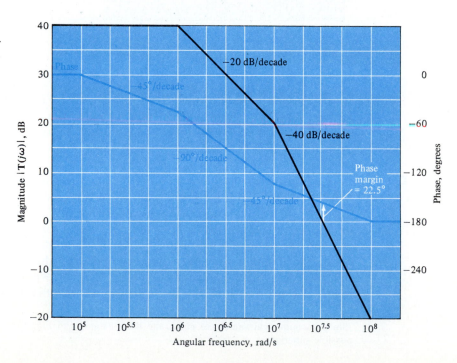

Note that in Example 13-1 we cannot identify the phase-crossover frequency and, consequently, the gain margin. This results from the fact that for a two-pole system, the angle is never $-180°$ but approaches it asymptotically as $\omega \to \infty$. Thus we conclude that *a two-pole feedback amplifier is always stable*.[1] This is also verified by the root locus in Fig. 13-1.

Example 13-2 The return ratio of a three-pole amplifier is

$$T(s) = \frac{T_O}{(1 + s/\omega_1)(1 + s/10^7)(1 + s/10^8)}$$

(*a*) Determine the gain and phase margins for $T_O = 10^4$ when (1) $\omega_1 = 10^6$ rad/s and (2) $\omega_1 = 100$ rad/s. (*b*) Is the closed-loop amplifier stable for each case in part *a*? (*c*) Repeat parts *a* and *b* for $\omega_1 = 10^6$ rad/s, but when T_O is reduced to 10.

Solution (*a*) (1). The asymptotic Bode diagram for $T(s)$ with $\omega_1 = 10^6$ rad/s is displayed as the black curves in Fig. 13-7. For $\omega > 10^8$ rad/s, the slope of $T(j\omega)$ is -60 dB/decade and $T(j10^8) = 20$ dB. Thus ω_G occurs at -20 dB/-60 dB/decade $= \frac{1}{3}$ decade $(10^{1/3})$ beyond 10^8 rad/s or $\omega_G = 10^8 \times 10^{1/3} = 10^{8.33}$ rad/s $= 2.15 \times 10^8$ rad/s. The slope of the phase curve is $-45°$/decade for $10^8 \le \omega \le 10^9$ rad/s. Thus in one-third of a decade the change in phase is $-15°$ and makes $\angle T(j10^{8.33}) = -240°$. The value of ϕ_M is then

$$\phi_M = -240 + 180 = -60°$$

as indicated in Fig. 13-7. In similar fashion the phase-crossover angular frequency is obtained. For $10^6 < \omega < 10^8$ rad/s, the slope of the phase characteristics is $-90°$/decade. At $\omega = 10^7$ rad/s, $\angle T = -135°$, so that ω_ϕ occurs at $-45°/-90°$/decade $= +0.5$ decade, or $10^{0.5}$ beyond 10^7 rad/s. Thus $\omega_\phi = 10^7 \cdot 10^{0.5} = 10^{7.5} = 3.16 \times 10^7$ rad/s, the same value as is read on a carefully constructed curve. On the magnitude characteristic, $T(j10^{7.5}) = 40$ dB, making $GM = -40$ dB. (2) The Bode diagram for $\omega_1 = 100$ rad/s, displayed in Fig. 13-8, gives $\omega_G = 10^6$ rad/s and $\omega_\phi = 10^{7.5}$ rad/s. The corresponding gain and phase margins are $GM = 40$ dB and $\phi_M = 90°$.

(*b*) The amplifier in part *a*, part 1 is unstable ($\phi_M = -60°$), and that in part 2 is stable ($\phi_M = 90°$). Again, the effectiveness of the asymptotic Bode diagram is seen by comparing the results in part *a* with a computer simulation. For $\omega_1 = 10^6$ rad/s, exact values are $\omega_G = 2.09 \times 10^8$ rad/s, $\omega_\phi = 3.30 \times 10^7$ rad/s, $\phi_M = -61.5°$, and $GM = -38.4$ dB. When $\omega_1 = 100$ rad/s, $\omega_G = 0.996 \times 10^6$ rad/s, $\omega_\phi = 3.16 \times 10^7$ rad/s, $GM = 40.8$ dB, and $\phi_M = 83.8°$.

(*c*) The magnitude of the Bode diagram for $T_O = 10$ is shown in Fig. 13-7 as the lower black curve. As the phase is unaffected by changes in magnitude, the phase curve is the same as in part *a*, part 1 as displayed in Fig. 13-7. From these characteristics, the approximate values are $\omega_G = 10^7$ rad/s, $\omega_\phi = 10^{7.5}$ rad/s, $\phi_M = 45°$, and $GM = 20$ dB. Thus the amplifier is stable.

---

[1]The student must be wary, however, that additional poles caused by parasitic elements (package capacitances, lead inductances, etc.) do not affect circuit performance and stability.

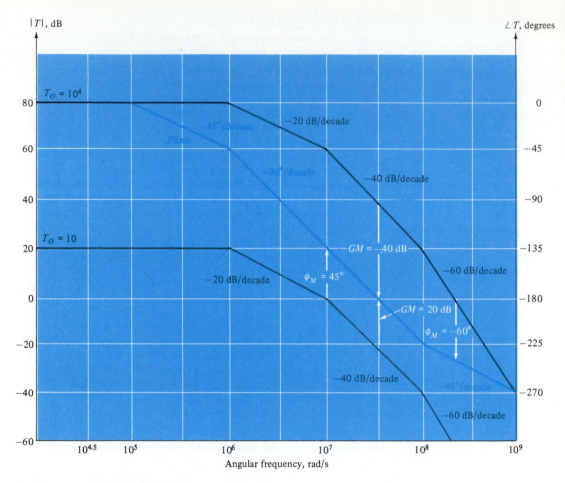

FIGURE 13-7
The asymptotic Bode diagram for the three-pole amplifier in Example 13-2. Note that the phase characteristic is the same for both values of T_O.

Several conclusions can be reached by comparing the results obtained in Example 13-2. The unstable amplifier, having $T_O = 10^4$ and $\omega_1 = 10^6$ rad/s, was stabilized by reducing either T_O or ω_1. The decrease in T_O was accompanied by a corresponding decrease in ω_G without altering ω_ϕ and thus made the amplifier stable. This, however, is an ineffective method for stabilizing the amplifier because decrease in T_O also results in decreased benefits (sensitivity, distortion, etc.) derived from negative feedback.

Decrease in ω_1 also results in decreased value of ω_G without change in T_O. For this situation $\omega_1 \ll \omega_2$ and no change in ω_ϕ occurred. This is because the dominant pole (ω_1) of $T(s)$ can introduce a maximum of only 90° of phase lag. Hence ω_ϕ must then be due to the 90° of phase lag provided by the nondominant

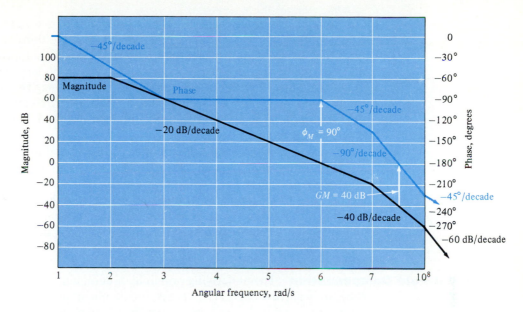

FIGURE 13-8

Asymptotic Bode diagram for a three-pole amplifier illustrating compensation by narrow-banding the amplifier. By moving the dominant pole of $T(s)$ nearer the origin, the feedback amplifier is stabilized.

poles (ω_2 and ω_3). If, as is the case in Example 13-2, $10\omega_1 \leq \omega_2/10$, the asymptotic Bode diagram indicates that $-90°$ of phase shift is provided by the pole at $-\omega_1$ prior to the introduction of any phase contribution by the pole at $-\omega_2$. In almost all practical amplifiers, the phase margin ϕ_M is at least 45°. Consequently, the phase margin is determined by the phase of the pole at $-\omega_2$ (and perhaps the remaining nondominant poles). Furthermore, for $\phi_M \geq 45°$, $\omega_2 \geq \omega_G$ and the slope of the magnitude characteristic is -20 dB/decade for $\omega \leq \omega_G$. Under these circumstances, ω_1 and ω_G are related by T_O and specification of ω_G and T_O determines the value of ω_1 needed to stabilize the amplifier. This is the basis of the compensation technique discussed in the next section.

13-4 COMPENSATION We can consider the procedure used to design a feedback amplifier to comprise three steps:

1. Design the amplifier at midband frequencies to satisfy the gain, desensitivity, distortion, impedance level, and other specifications.

2. Test for amplifier stability.

3. Make it work, anyway! That is, ensure that the amplifier is stable while satisfying nominal design specifications.

In Chap. 12 we focused on the first step, and step 2 was treated in earlier sections of this chapter. In this section, we address the third and crucial step.

In Chap. 11 we demonstrated that feedback amplifiers approach ideal controlled-source behavior for large values of T_O. Similarly, large values of T_O are needed to reduce distortion and control gain variations. However, feedback amplifiers are likely to be unstable for large values of T_O. Even for an inherently stable, two-pole system, the response obtained (Sec. 13-5) when T_O is large may be unsatisfactory.

The response of the feedback amplifier is determined by the poles of $A_F(s)$; these poles are related to the return ratio $T(s)$. As T_O cannot be changed because of the midband requirements (step 1), we must alter the poles of $T(s)$ to ensure that the closed-loop amplifier will be stable and have the desired response. To accomplish this (step 3), the feedback amplifier is *compensated*; that is, additional components are inserted in the circuit which alter the location of one or more poles of $T(s)$ without changing T_O.

Dominant-Pole Compensation   The underlying principle of dominant-pole compensation is to deliberately narrow-band $T(s)$. That is, the dominant pole in the uncompensated return ratio is moved closer to the origin in a manner similar to the decrease of ω_1 in Example 13-2a. The question arises as to where this dominant pole should be located to obtain stability and the desired closed-loop response. In the next section we show that the poles of A_F, and hence the closed-loop response, depend on the phase margin. Consequently, the location of the dominant pole in the compensated return ratio is related to the desired phase margin.

To illustrate the method, consider the uncompensated return ratio to be of the form

$$T(s) = \frac{T_O}{[1 + (s/\omega_1)] \, [1 + (s/\omega_2)] \, [1 + (s/\omega_3)]} \qquad (13\text{-}13)$$

To stabilize the amplifier, ω_1 is moved closer to the origin[1] so that $10\omega_1 \leq \omega_2/10$. In addition, if we assume that $\omega_3 \geq 10\omega_2$, then Fig. 13-9a is the asymptotic phase characteristic in the vicinity of ω_2. The horizontal portion of the curve at $-90°$ for $\omega < \omega_2/10$ is the phase contribution due to the dominant pole at $-\omega_1$. Selection of the phase margin ($90° \geq \phi_M \geq 45°$) specifies ω_G as indicated in Fig. 13-9a. As stated in Sec. 13-3, for $\phi_M \geq 45°$, $\omega_G \leq \omega_2$ and for $\omega \leq \omega_G$, the slope of $T(j\omega)$ is -20 dB/decade. The location of ω_1 can now be determined by use of the asymptotic magnitude characteristic illustrated in Fig. 13-9b. Since ω_G is known, a line of slope -20 dB/decade is drawn through ω_G and extended back (lower frequency) until it intersects the horizontal line corresonding to T_O. The angular frequency of the dominant pole ω_1 is the first corner frequency of $T(s)$, and hence this intersection defines the value of ω_1. Note that ϕ_M is determined solely by the portion of the phase characteristic due to the nondominant pole $-\omega_2$. As ϕ_M is decreased, both ω_G and ω_1 are increased.

---

[1] Specific circuit techniques for compensation are discussed in Sec. 13-9.

FIGURE 13-9

Illustrating the compensation technique. Specification of ϕ_M in (a) allows determination of ω_G. In (b), identification of ω_G allows ω_1 to be evaluated.

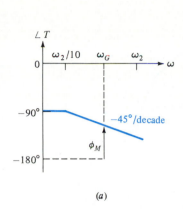

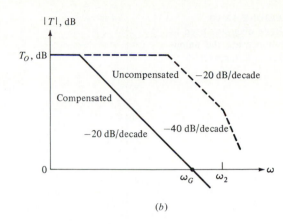

(a)   (b)

Example 13-3

The compensated return ratio of a single-loop amplifier is

$$T(s) = \frac{10^4}{[1 + (s/\omega_1)] [1 + (s/10^7)] [1 + (s/10^8)]}$$

Determine ω_1 so that the phase margin is approximately 67.5°.

Solution

We assume that compensation results in $10\omega_1 \le \omega_2/10$ or $\omega_1 \le 10^4$ rad/s. Consequently, the phase characteristic is the solid curve displayed in Fig. 13-10a. For $\phi_M \approx 67.5°$, $\angle T(j\omega) = -112.5°$, and, as indicated on Fig. 13-10a, $\omega_G = 10^{6.5}$ rad/s. On the magnitude characteristic in Fig. 13-10b, a line of slope -20 dB/decade is constructed to pass through ω_G and extended back until it intersects the horizontal line at 80 dB, corresponding to the value of T_O. From this intersection $\omega_1 = 10^{2.5} = 316$ rad/s. Note that the values of ω_1 and ω_G are approximate as they are based on asymptotic characteristics. They are, however, quite close to actual values. Actual values of ω_G and ϕ_M for $\omega_1 = 316$ rad/s are 3.03×10^6 rad/s and 71.4°, respectively.

The return ratio in this example is the same as that in Example 13-2. The dashed curves in Fig. 13-10 correspond to $\omega_1 = 10^6$ rad/s (Example 13-2a, part 1) and illustrate the effect on $T(s)$ produced by compensation. We observe that for $\omega_1 < \omega \le \omega_G$, $T(j\omega)$ is decreased significantly from uncompensated values. Recall that the reduction in nonlinear distortion, desensitivity, and so on depends on $|1 + T|$. Thus we do not derive the benefits of negative feedback over as broad a frequency band as in the uncompensated amplifier. This is the cost in performance that is borne in order to stabilize the amplifier.

Pole-Zero Cancellation   An alternative compensation method which produces the same results as the dominant-pole method is *pole-zero cancellation*. In this technique a network having a one-pole one-zero transfer function is inserted into the amplifier so that the compensated return ratio $T_C(s)$ is

$$T_C(s) = T(s) \frac{1 + s/z_C}{1 + s/\omega_C}$$

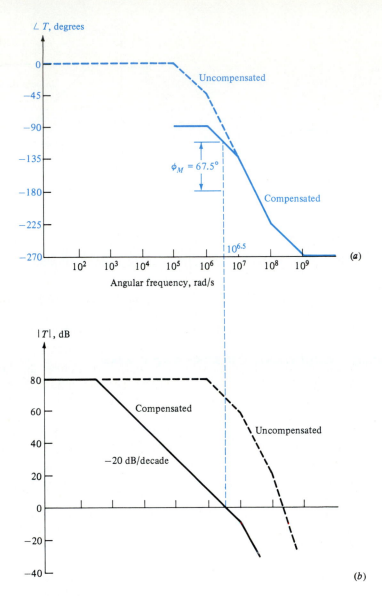

FIGURE 13-10
Bode diagrams used to compensate the amplifier in Example 13-3.

In this expression, $T(s)$ is the uncompensated return ratio and z_C and ω_C are the angular frequencies of the zero and pole, respectively, of the compensating network. The value of z_C is selected to cancel the pole closest to the origin in $T(s)$, and ω_C is chosen to obtain the desired phase margin. With use of the value of $T(s)$ in Example 13-2a, part 1, pole-zero cancellation requires $z_C = 10^6$ rad/s; that is, it must cancel the pole at $-\omega_1 = -10^6$ rad/s. To achieve $\phi_M = 67.5°$, as in Example 13-3, we choose $\omega_C = 10^{2.5} = 316$ rad/s. Note that the compensated return ratio, for a given ϕ_M, is the same when either dominant-pole or pole-zero cancellation compensation is used.

One drawback of pole-zero cancellation is that of component sensitivity. That is, the values of z_C and ω_C are not necessarily determined by the same circuit elements. Thus, if these components have different tolerances, age differently, vary differently with temperature, and so on, the desired cancellation may not be obtained.

The asymptotic Bode diagram is of invaluable aid to the circuit designer. Its use in compensating a feedback amplifier provides the initial design value of ω_1 and, hence, the component values needed. However, the asymptotic characteristic is an approximation of $T(s)$, and thus its use gives only approximate design values. The analysis presented in subsequent sections in conjunction with computer analyses are used to obtain final design values.

A Note to the Student In the three prior sections we concentrated on the question of stability by considering the high-frequency response. This frequency range is the major concern in integrated-circuit (IC) amplifiers as these are usually direct-coupled. In discrete-component circuits employing coupling and bypass capacitors, however, sufficient phase shift can be introduced at low frequencies to make the feedback positive. Consequently, we must test for stability at low frequencies and, where necessary, provide adequate compensation.

13-5 FREQUENCY RESPONSE OF FEEDBACK AMPLIFIERS—THE DOUBLE-POLE TRANSFER FUNCTION

Each of the four feedback amplifier configurations frequently employs resistive feedback networks. If we assume that the feedforward or dead-system gain A_D is negligible, then the open-loop gain A_{OL} and return ratio T have the same poles. This condition was utilized in Sec. 13-1 to demonstrate that the bandwidth of a one-pole system increased by a factor $(1 + T_O)$. Also, we alluded to the fact that negative feedback resulted in bandwidth improvement in multipole systems. We investigate the response of a two-pole transfer function in this section and treat higher-order systems in the next section.

Consider $A_{OL}(s)$ and $T(s)$ to be the two-pole functions given in Eqs. (13-5) and (13-6). The closed-loop gain $A_F(s)$ is stated in Eq. (13-7) and repeated as Eq. (13-14) for convenience:

$$A_F(s) = \frac{A_{FO}}{1 + \dfrac{a_1 s}{1 + T_O} + \dfrac{a_2 s^2}{1 + T_O}}$$

$$= \frac{A_{FO}}{1 + \dfrac{s}{1 + T_O}\left(\dfrac{1}{\omega_1} + \dfrac{1}{\omega_2}\right) + \dfrac{s^2}{(1 + T_O)\omega_1\omega_2}} \tag{13-14}$$

Alternatively

$$A_F(s) = \frac{A_{FO}}{1 + (s/\omega_o)(1/Q) + (s/\omega_o)^2} \tag{13-15}$$

FIGURE 13-11
Root locus of a two-pole
feedback amplifier.

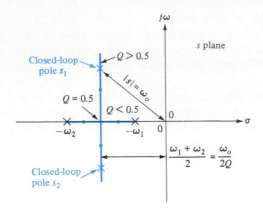

where $A_{FO} = A_O/(1 + T_O)$ is the midband value of A_F and ω_o and Q are defined by

$$\omega_o \equiv \sqrt{\omega_1\omega_2(1 + T_O)} \qquad Q \equiv \frac{\omega_o}{\omega_1 + \omega_2} \qquad (13\text{-}16)$$

The poles of A_F are given by

$$\frac{s}{\omega_o} = -\frac{1}{2Q} \pm \frac{1}{2Q}\sqrt{1 - 4Q^2} \qquad (13\text{-}17)$$

or

$$s = -\frac{\omega_1 + \omega_2}{2} \pm \frac{\omega_1 + \omega_2}{2}\sqrt{1 - 4Q^2} \qquad (13\text{-}18)$$

Note that when $T_O = 0$ (no feedback), $\omega_o = \sqrt{\omega_1\omega_2}$, $Q_{\min} = \sqrt{\omega_1\omega_2}/(\omega_1 + \omega_2)$, and the poles of A_F are at $-\omega_1$ and $-\omega_2$, the poles of A_{OL}. Clearly, this is the correct result; with no feedback, the gain of the system must be $A_{OL}(s)$.

We showed the motion of the poles of A_F as T_O is increased in the root locus of Fig. 13-1 and repeat this in Fig. 13-11. From Eq. (13-17) we observe that the poles of A_F are real, negative, and unequal for $Q < 0.5$; negative, real, and equal to $-(\omega_1 + \omega_2)/2$ for $Q = 0.5$; and complex for $Q > 0.5$.

Circuit Model We now demonstrate that the network in Fig. 13-12 is the analog of the two-pole feedback amplifier. The transfer function of the circuit in Fig. 13-12 can be expressed as

$$\frac{V_o(s)}{V_s(s)} = \frac{1}{1 + s(L/R) + s^2LC} \qquad (13\text{-}19)$$

Introduction[1] of

$$\omega_o \equiv \frac{1}{\sqrt{LC}} \qquad Q \equiv R\sqrt{\frac{C}{L}} = \frac{R}{\omega_o L} = \omega_o RC \qquad (13\text{-}20)$$

---

[1]Note that conversion of V_s and sL to their Norton equivalents demonstrates that this circuit behaves as a parallel-resonant circuit.

FIGURE 13-12

An *RLC* circuit equivalent of a two-pole amplifier.

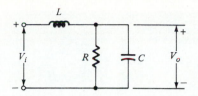

leads to

$$\frac{V_o(s)}{V_i(s)} = \frac{1}{1 + (s/\omega_o)(1/Q) + (s^2/\omega_o^2)} = \frac{A_F(s)}{A_{FO}} \qquad (13\text{-}21)$$

where the second equality follows from Eq. (13-15). Clearly, Fig. 13-12 is a circuit model of a two-pole amplifier in the sense that both have the same frequency and transient response. Physical meanings can now be given to ω_o and Q, introduced in connection with the feedback amplifier. By analogy with the resonance phenomena, from Eqs. (13-20) we observe that

ω_o = undamped $(R \rightarrow \infty)$ resonant angular frequency of oscillation

Q = quality factor at the resonant frequency

One important consequence of the analogy just drawn is that *the response of networks containing resistors, capacitors, and inductors (RLC networks) can be obtained by the use of feedback with circuits that contain only resistors, capacitors, and controlled sources (transistor amplifiers).* This is extremely important as inductors cannot be fabricated on an IC. In discrete-component circuits, the values of L needed are often so large that their use is precluded by the physical size of the inductor. The ability to obtain *RLC* behavior using amplifiers, resistors, and capacitors is the basis for active filters (Sec. 16-8).

Frequency Response   If s in Eq. (13-21) is replaced by $j\omega$, then this expression gives the frequency response of the two-pole amplifier with feedback. It is convenient to use the *damping factor k* in place of Q. These are related by

$$k \equiv \frac{1}{2Q} \qquad (13\text{-}22)$$

Thus, from Eqs. (13-21) and (13-22), we obtain

$$\left| \frac{A_F}{A_{FO}} \right| = \frac{1}{\sqrt{[1 - (\omega/\omega_o)^2]^2 + 4k^2(\omega/\omega_o)^2}} \qquad (13\text{-}23)$$

and

$$\angle \frac{A_F}{A_{FO}} = -\tan^{-1}\frac{2k(\omega/\omega_o)}{1 - (\omega/\omega_o)^2} \qquad (13\text{-}24)$$

Equation (13-23) is the normalized magnitude or amplitude characteristic, and the phase characteristic is given by Eq. (13-24). The peaks of the amplitude

FIGURE 13-13
Normalized plot of the
amplitude-frequency
response of a two-pole
feedback amplifier.

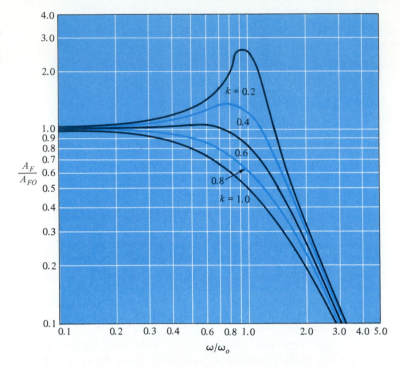

response are obtained by setting the derivative of the quantity under the square-root sign equal to zero. We find that a peak occurs at

$$\omega = \omega_o\sqrt{1 - 2k^2} \tag{13-25}$$

and the magnitude of the peak is given by

$$\left|\frac{A_F}{A_{FO}}\right|_{peak} = \frac{1}{2k\sqrt{1 - k^2}} \tag{13-26}$$

Note that if $2k^2 > 1$ or $k > 0.707$ or $Q < 0.707$, the magnitude response will not exhibit a peak. A plot of the normalized magnitude response is given in Fig. 13-13.

For $(\omega/\omega_o) \ll 1$, Eq. (13-24) shows that the phase characteristic is approximately linear and its slope determined by the value of k. The phase characteristic is $-90°$ at $\omega = \omega_o$ for all values of k and asymptotically approaches $-180°$ when $(\omega/\omega_o) \gg 1$.

Step Response   It has been proved in this section that regardless of how much negative feedback is employed, a two-pole amplifier remains stable (its poles are always in the left-half s plane). However, if the loop gain T_O is too large, the transient response of the amplifier may be entirely unsatisfactory.

For example, in Fig. 13-14 there is indicated one possible response to a voltage step. Note that the output overshoots its final value by 37 percent and oscillates before settling down to the steady-state voltage. For almost all applications such a violent response is not acceptable.

The important parameters of the waveform are indicated in Fig. 13-14 and are defined as follows:

Rise time = time for waveform to rise from 0.1 to 0.9 of its steady-state value

Delay time = time for waveform to rise from 0 to 0.5 of its steady-state value

Overshoot = peak excursion above the steady-state value

Damped period = time interval for one cycle of oscillation

Settling time = time for response to settle to within $\pm P$ percent of the steady-state value (P specified for a particular application, say $P = 0.1$)

Analytical expressions for the response of the amplifier to a step of amplitude V is obtained by setting $V_i(s) = V/s$ into Eq. (13-21) and solving for the inverse Laplace transform. Recalling that $Q = 1/2k$, the poles, given in Eq. (13-18), can be put into the form

$$s = -k\omega_o \pm \omega_o \sqrt{k^2 - 1} \tag{13-27}$$

FIGURE 13-14

The step response of a two-pole feedback amplifier for a damping factor $k = 0.3$.

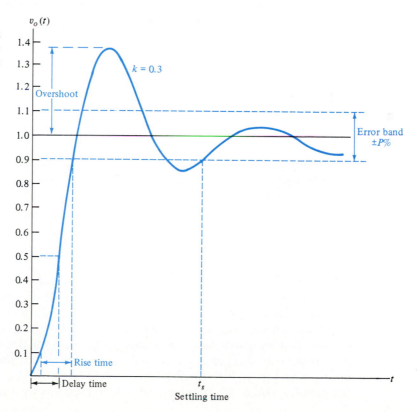

If $k = 1$, the two poles coincide, corresponding to the *critically damped* case. If $k < 1$, the poles are complex conjugates, corresponding to an *underdamped* condition, where the response is a sinusoid whose amplitude decays with time. If $k > 1$, both poles are real and negative, corresponding to an *overdamped* circuit where the response approaches its final value monotonically. For the underdamped case it is convenient to introduce the damped frequency

$$\omega_d \equiv \sqrt{1 - k^2}\, \omega_o \qquad (13\text{-}28)$$

and the response $v_o(t)$ *to a step of magnitude V* into an amplifier of midband gain A_{FO} is given by the following equations:

Critically damped, $k = 1$:

$$\frac{v_o(t)}{VA_{FO}} = 1 - (1 + \omega_o t)\epsilon^{-\omega_o t} \qquad (13\text{-}29)$$

Overdamped, $k > 1$:

$$\frac{v_o(t)}{VA_{FO}} = 1 - \frac{1}{2\sqrt{k^2 - 1}} \left(\frac{1}{k_1} \epsilon^{-k_1 \omega_o t} - \frac{1}{k_2} \epsilon^{-k_2 \omega_o t} \right) \qquad (13\text{-}30)$$

where

$$k_1 \equiv k - \sqrt{k^2 - 1} \quad \text{and} \quad k_2 \equiv k + \sqrt{k^2 - 1}$$

If $4k^2 \gg 1$, the response may be approximated by

$$\frac{v_o(t)}{VA_{FO}} \approx 1 - \epsilon^{-\omega_o t/2k} \qquad (13\text{-}31)$$

Underdamped, $k < 1$:

$$\frac{v_o(t)}{VA_{FO}} = 1 - \left(\frac{k\omega_o}{\omega_d} \sin \omega_d t + \cos \omega_d t \right) \epsilon^{-k\omega_o t} \qquad (13\text{-}32)$$

These equations are plotted in Fig. 13-15 using the normalized coordinates $x \equiv t/T_o$ and $y \equiv v_o(t)/VA_{FO}$ where $T_o \equiv 2\pi/\omega_o$ is the undamped period. If the derivative of Eq. (13-32) is set equal to zero, the positions $x = x_m$ and magnitudes $y = y_m$ of the maxima and minima are obtained. The results are

$$x_m = \frac{\omega_o t_m}{2\pi} = \frac{m}{2(1 - k^2)^{1/2}} \qquad y_m = \frac{v_o(t_m)}{VA_{FO}} = 1 - (-1)^m \epsilon^{-2\pi k x_m} \qquad (13\text{-}33)$$

where m is an integer. The maxima occur for odd values of m, and the minima are obtained for even values of m. By using Eq. (13-33) the waveshape of the underdamped output may be sketched very rapidly. From Eq. (13-33) it follows that the *overshoot* is given by $\exp[-\pi k m/(1 - k^2)^{1/2}]$.

Note that for heavy damping (k large or Q small) the rise time t_r is very long. As k is decreased (Q or T_O increased), t_r decreases. For the critically damped case we find from Fig. 13-15 that $t_r = 0.53T_o = 3.33/\omega_o$. If the feedback is increased so that $k < 1$, the rise time is decreased further, but this improvement

FIGURE 13-15
The normalized step re-
sponse of a two-pole
feedback amplifier.

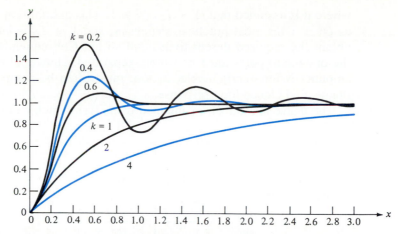

is obtained at the expense of a ringing (oscillatory) response which may be unacceptable for some applications. Often $k \geq 0.707$ ($Q \leq 0.707$) is specified as a satisfactory response and corresponds to an overshoot of 4.3 percent or less. In general, the overshoot rarely exceeds 10 percent so that $k \geq 0.6$ ($Q \leq 0.83$).

13-6 PHASE MARGIN OF THE TWO-POLE FEEDBACK AMPLIFIER

The design of a two-pole feedback amplifier requires that the open-loop pole frequencies be selected to give the desired closed-loop response. Our objective in this section is to relate the closed-loop response to the phase margin and the design values of the two open-loop pole frequencies.

It is convenient to introduce the pole-separation factor $n = \omega_2/\omega_1$. Then Eqs. (13-16) become

$$\omega_o = \omega_1\sqrt{n(1 + T_O)} \qquad Q = \frac{\sqrt{n(1 + T_O)}}{n + 1} \qquad (13\text{-}34)$$

and the closed-loop poles, given in Eq. (13-18), can be expressed as

$$s = -\frac{\omega_1(1 + n)}{2}(1 \pm \sqrt{1 - 4Q^2}) \qquad (13\text{-}35)$$

The desired closed-loop response, displayed in the frequency-response characteristic shown in Fig. 13-13 or the step response depicted in Fig. 13-15, is used to specify the value of Q (or k). That is, on the basis of the amount of frequency peaking, if any, or the overshoot in the step response, if any, these curves indicate the value of Q of the circuit. Hence Eq. (13-34) can now be used to obtain the pole-separation factor n. Solution of the resultant quadratic equation for n gives

$$n \approx \frac{1 + T_O}{Q^2} \qquad (13\text{-}36)$$

where it is assumed that $(1 + T_O)/Q^2 \gg 1$. This assumption is quite reasonable as $Q^2 < 1$ for most practical responses, and $1 + T_O$ is large (at least 10) to obtain the required desensitivity, and so on. Consequently, we observe that the open-loop poles must be widely separated. Hence, although the two-pole amplifier is inherently stable, it, too, must often be compensated to achieve the desired closed-loop response.

The phase margin ϕ_M is obtained from $T(s)$, which can be expressed as

$$T(s) = \frac{T_O}{(1 + s/\omega_1)(1 + s/n\omega_1)} \tag{13-37}$$

The angular gain-crossover frequency ω_G, obtained from Eq. (13-37) by forming $T(j\omega_G) = 1$, is

$$\frac{\omega_G}{\omega_1} = \sqrt{\frac{n^2 + 1}{2}} \left[\sqrt{\frac{4n^2(T_O^2 - 1)}{(n^2 + 1)^2} + 1} - 1 \right]^{1/2} \tag{13-38}$$

For $n^2 \gg 1$ and $T_O^2 \gg 1$, which is the usual case, and use of Eq. (13-36), we can write Eq. (13-38) as

$$\frac{\omega_G}{\omega_1} = \frac{T_O}{Q^2\sqrt{2}} \left(\sqrt{4Q^4 + 1} - 1 \right)^{1/2} = \frac{n}{\sqrt{2}} \left(\sqrt{4Q^4 + 1} - 1 \right)^{1/2} \tag{13-39}$$

In Eq. (13-39) we note that ω_G is also widely separated from ω_1 for normally encountered values of T_O and Q.

The phase margin ϕ_M [given in Eq. (13-11)] for $T(s)$ in Eq. (13-37) is

$$\phi_M = -\tan\frac{\omega_G}{\omega_1} - \tan^{-1}\frac{\omega_G}{n\omega_1} + 180°$$

and can be rewritten as

$$\phi_M = \left(90 - \tan^{-1}\frac{\omega_G}{\omega_1} \right) + \left(90 - \tan^{-1}\frac{\omega_G}{n\omega_1} \right)$$

$$= \tan^{-1}\frac{\omega_1}{\omega_G} + \tan^{-1}\frac{n\omega_1}{\omega_G} \tag{13-40}$$

Since $\omega_1 \ll \omega_G$, $\tan^{-1}(\omega_1/\omega_G)$ is a very small angle and often can be neglected. Then

$$\phi_M \approx \tan^{-1}\frac{n\omega_1}{\omega_G} = \tan^{-1}\frac{\omega_2}{\omega_G} \tag{13-41}$$

and substitution of Eq. (13-39) yields

$$\phi_M \approx \tan^{-1}\sqrt{2} \, (\sqrt{4Q^4 + 1} - 1)^{-1/2} \tag{13-42}$$

Observe that Eq. (13-41) expresses the same relation as is discussed in Sec. 13-4, namely, that ϕ_M is determined by ω_2 when the open-loop poles are widely separated.

Equation (13-15) is used to obtain the closed-loop bandwidth ω_H by solving for ω_H when $A_F(j\omega_H) = A_{FO}/\sqrt{2}$ and is

$$\omega_H = \frac{\omega_o}{Q} \sqrt{\frac{2Q^2 - 1}{2}} \left[1 + \sqrt{1 + \frac{4Q^4}{(2Q^2 - 1)^2}} \right]^{1/2} \qquad Q^2 > 0.5$$

$$= \frac{\omega_o}{Q} \sqrt{\frac{1 - 2Q^2}{2}} \left[\sqrt{1 + \frac{4Q^4}{(1 - 2Q^2)^2}} - 1 \right]^{1/2} \qquad Q^2 < 0.5 \qquad \left.\begin{matrix} \\ \\ \\ \end{matrix}\right\} \text{(13-43)}$$

$$= \omega_o \qquad Q^2 = 0.5$$

Thus specification of ω_H and Q determines the value of ω_o needed, and, by use of Eq. (13-34), ω_1 is evaluated. The use of the separation factor n gives the required value of ω_2. We illustrate the design procedure in the following example.

Example 13-4 A two-pole feedback amplifier is to be designed with $T_O = 99$ and $\omega_H = 10^7$ rad/s. Determine ω_1, ω_2, and ϕ_M for the following values of Q: (a) 0.316, (b) 0.500, (c) 0.707, (d) 0.833, and (e) 1.00.

Solution (a) The pole-separation factor n is obtained from Eq. (13-36) as

$$n = \frac{1 + T_O}{Q^2} = \frac{1 + 99}{(0.316)^2} = 1000$$

The value of ω_o, from Eq. (13-43), is

$$10^7 = \frac{\omega_o}{0.316} \sqrt{\frac{1 - 2(0.316)^2}{2}} \left[\sqrt{1 + \frac{4(0.316)^4}{[1 - 2(0.316)^2]^2}} - 1 \right]^{1/2}$$

$$\omega_o = 2.85 \times 10^7 \text{ rad/s}$$

Use of Eq. (13-34) yields

$$2.85 \times 10^7 = \omega_1 \sqrt{1000(1 + 99)} \qquad \text{and} \qquad \omega_1 = 9.01 \times 10^4 \text{ rad/s}$$

Hence $\omega_2 = n\omega_1 = 1000 \times 9.01 \times 10^4 = 9.01 \times 10^7$ rad/s. The phase margin is, from Eq. (13-42)

$$\phi_M \approx \tan^{-1} \sqrt{2} \left[\sqrt{4(0.316)^4 + 1} - 1 \right]^{-1/2} = 84.3°$$

The values of ω_1, ω_2, and ϕ_M for each of the Q values given are obtained in the same manner. The results are listed in Table 13-1.

The values in Table 13-1 indicate that for a given value of T_O, ω_1 does not

TABLE 13-1   Open-Loop Pole Frequencies and Phase Margins for Example 13-4

Q	n	ω_o, Mrad/s	ω_o/ω_H	ω_1, krad/s	ω_2, Mrad/s	ϕ_M, degrees
0.316	1000	28.5	2.85	90.1	90.1	84.3
0.500	400	15.5	1.55	77.7	31.1	75.3
0.707	200	10.0	1.00	70.7	14.14	65.5
0.833	144	8.71	0.871	72.6	10.4	59.2
1.000	100	7.86	0.786	78.6	7.86	51.8

change appreciably with Q whereas ω_2 and the pole-separation factor change markedly. In Example 13-5 we show the variation in performance when T_O and ω_2 remain constant.

Example 13-5

A feedback amplifier is designed having $T_O = 99$ and $\omega_2 = 10^7$ rad/s. Determine ω_1 and ω_H for the following values of Q: (a) 0.316, (b) 0.500, (c) 0.707, (d) 0.833, and (e) 1.00.

Solution

The pole-separation factor depends only on T_O and Q. Hence the values of n in Table 13-1 apply to this problem. For $Q = 0.316$, $n = 1000$ and

$$\omega_1 = \frac{\omega_2}{n} = \frac{10^7}{1000} = 10^4 \text{ rad/s}$$

The value of ω_o is, using Eq. (13-34),

$$\omega_o = 10^4 \sqrt{1000(1 + 99)} = 3.16 \times 10^6 \text{ rad/s}$$

The ratio ω_H/ω_o depends only on the value of Q [Eq. (13-43)]. Hence the ratio ω_o/ω_H given in Table 13-1 also applies to this problem, and, for $Q = 0.316$

$$\frac{\omega_o}{\omega_H} = 2.85 \quad \text{or} \quad \omega_H = \frac{\omega_o}{2.85} = \frac{3.16 \times 10^6}{2.85} = 1.11 \times 10^6 \text{ rad/s}$$

The remaining values are listed in Table 13-2 and are determined in similar fashion.

The results listed in Table 13-2 indicate that when ω_2 is fixed, ω_1 and the closed-loop bandwidth increase with increasing Q (decreasing k). This is graphically illustrated in Fig. 13-13, for which ω_H, the frequency at which $A_F/A_{FO} = 0.707$, increases with decreasing k.

Also note that the values of ω_o are simply $Q\omega_2$. This is verified from Eq. (13-34) as

$$\omega_o = \omega_1 \sqrt{n(1 + T_O)} = \frac{\omega_2}{n} \sqrt{n(1 + T_O)} = \omega_2 \sqrt{(1 + T_O)/n} = Q\omega_2$$

TABLE 13-2   Values of ω_1 and ω_H for Example 13-5

Q	ω_1, krad/s	ω_o, Mrad/s	ω_H, Mrad/s
0.316	10.0	3.16	1.11
0.500	25.0	5.00	3.23
0.707	50.0	7.07	7.07
0.833	69.4	8.33	9.56
1.00	100.00	10.00	12.7

13-7 THREE-POLE FEEDBACK AMPLIFIER RESPONSE If A_{OL} and T are three-pole functions, the closed-loop gain $A_F(s)$ is given by Eq. (13-8). The root locus in Fig. 13-2 shows the motion of the poles of A_F as T_O increases from zero. We observe in Fig. 13-2 that the two poles nearest the origin ($-\omega_1$ and $-\omega_2$) move toward one another along the negative real axis, coincide, and then become complex. The third pole ($-\omega_3$) remains on the negative real axis but moves further from the origin.

The movement of the three poles indicates that, for stable systems, the response is due primarily to the closed-loop poles nearest the origin, that is, the closed-loop poles corresponding to the open-loop poles at $-\omega_1$ and $-\omega_2$. In Sec. 13-4 we showed that a stable system is obtained for moderate and high values of T_O, when ω_1 is widely separated from both ω_2 and ω_3. If, in addition, ω_2 and ω_3 are separated by at least two octaves ($\omega_3 \geq 4\omega_2$), the feedback causes the third pole to be separated from ω_1 and ω_2 to an even greater degree. Consequently, the three-pole system can be approximated quite well by a two-pole function corresponding to open-loop poles at $-\omega_1$ and $-\omega_2$. Hence the results obtained in Sec. 13-5 for the two-pole transfer function are applicable to the three-pole amplifier. The accuracy of this approximation[1] is usually sufficient for the pencil-and-paper calculations needed to obtain initial design values. As is almost always true, final design values are based on computer analyses.

The two-pole approximation is also employed in the analysis and design of amplifiers having more than three poles. Compensation by narrow-banding $T(s)$ is required if these amplifiers are to be stable. Consequently, their response is dominated by the two poles nearest the origin. Recall that we have already made use of the two-pole approximation in characterizing the frequency response of a cascaded amplifier. When we considered a two-stage common-emitter or common-source amplifier (Sec. 11-11), we observed that it contained four poles and two zeros. However, the zero and one pole introduced by each stage were far removed from the dominant poles, and we concluded that the two-pole representation adequately represented the transfer function of the amplifier.

13-8 APPROXIMATE ANALYSIS OF A MULTIPOLE FEEDBACK AMPLIFIER In the general case, the determination of the exact response of a feedback amplifier is so complicated that computer simulations must be used. If the open-loop poles are widely separated, however, a simple approximate method can be used. We justify and describe this technique in the following discussion.

It has been shown[2] that, for a three-pole amplifier, if the pole frequencies of $T(s)$ are $\omega_1 = 10^7$, $\omega_2 = 7 \times 10^7$, and $\omega_3 = 1.8 \times 10^8$ rad/s, the closed-loop

---

[1] For $Q \leq 0.83$, the two-pole approximation usually results in an error of less than 12 percent in closed-loop pole locations.

[2] See Ref. 10.

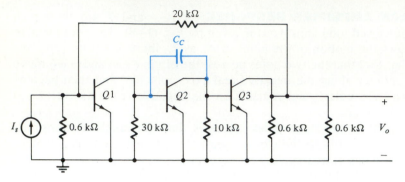

FIGURE 13-16
The shunt triple for Example 13-6. The capacitance C_C is used to compensate the amplifier.

system is unstable for $T_O \geq 31$. The value of T_O must be made significantly lower than 31 for this amplifier if it is to have an acceptable response (Figs. 13-13 and 13-16). Often, we must use values of T_O that are fairly large (often $T_O > 31$) if the feedback amplifier is to have the requisite desensitivity and reduction of nonlinear distortion and so on. Consequently, as described in Sec. 13-4, the dominant pole in $T(s)$ must be moved nearer the origin, causing ω_1 to be widely separated from the remaining poles of the transfer function. If, in addition, the first nondominant pole at $s = -\omega_2$ is separated from the remaining poles by at least two octaves ($4\omega_2 \leq \omega_3$), both $T(s)$ and $A_{OL}(s)$ can be approximated by two-pole transfer functions. Thus

$$T(s) \approx \frac{T_O}{1 + a_1 s + a_2 s^2} \qquad A_{OL} \approx \frac{A_O}{1 + a_1 s + a_2 s^2} \qquad (13\text{-}44)$$

Because the two poles nearest the origin are widely separated, dominant-pole conditions apply and

$$\omega_1 \approx \frac{1}{a_1} \qquad \omega_2 \approx \frac{a_1}{a_2} \qquad (13\text{-}45)$$

Hence the pole-separation factor $n = \omega_2/\omega_1$ is

$$n = \frac{a_1^2}{a_2} \qquad (13\text{-}46)$$

and Eqs. (13-34) can be rewritten as

$$\omega_o = \sqrt{\frac{1 + T_O}{a_2}} \qquad Q = \frac{\sqrt{a_2(1 + T_O)}}{a_1} \qquad (13\text{-}47)$$

Equations (13-45) to (13-47), in conjunction with the results for the two-pole system discussed in Secs. 13-5 and 13-6, are used to approximate the response of a multipole feedback amplifier. The reader should note that in Eq. (13-44) it is assumed that the feedback network is resistive and all zeros of the transfer function are sufficiently far removed from ω_2 to render negligible effect. In the next section and in Sec. 13-9, we show how a_1 and a_2, and hence the response of the feedback amplifier, are related to circuit elements.

Phase Margin   Substitution of Eq. (13-45) into Eq. (13-41) permits the gain-crossover angular frequency to be written as

$$\omega_G \approx \frac{a_1}{a_2 \tan \phi_M} \tag{13-48}$$

Similarly, with $T_O \gg 1$ and $\omega_G \gg 1/a_1$, $\omega_1 \approx \omega_G/T_O \sin \phi_M$ and

$$n = \frac{\omega_2}{\omega_1} \approx T_O \sin \phi_M \tan \phi_M \tag{13-49}$$

Since $n \approx T_O/Q^2$, we obtain, from Eq. (13-49)

$$Q \approx \frac{1}{\sqrt{\sin \phi_M \tan \phi_M}} \tag{13-50}$$

and clearly demonstrate the relation between Q and phase margin. Note that these expressions are all approximations. An accurate calculation of the phase margin must account for the small amount of phase shift introduced by the remaining nondominant poles (ω_3, ω_4, etc.).

In Table 13-1 we indicate the relationship between ϕ_M and Q. Recall that $Q = 0.5$ (critical damping) results in two identical poles on the real axis. Thus, for $\phi_M > 76.3°$, we expect the closed-loop poles to lie on the negative real axis, whereas $\phi_M < 76.3°$ results in complex poles. When $Q = 0.707$, we observed in Fig. 13-13 that the amplitude response exhibits no peaking. This situation corresponds to $\phi_M = 65.5°$.

Earlier in this chapter we noted that $\phi_M > 45°$ is most common in practical amplifiers. From both Eqs. (13-50) and (13-42), we find that $Q \approx 1.18$ for $\phi_M = 45°$. The curve in Fig. 13-16 indicates that $k = 1/2Q = 0.42$ results in an overshoot of nearly 20 percent and the amplitude response (Fig. 13-13) displays a moderate amount of peaking. Clearly, any further increase in Q (decrease in k) results in an unacceptable response.

The Dominant Pole   A feedback amplifier can be designed so that the closed-loop transfer function exhibits a dominant pole. This is often the case in commercial Op-Amps (see Sec. 14-8). In Table 13-1 we note that $\phi_M = 84.3°$ corresponds to $Q = 0.316 = 1/\sqrt{10}$, or $Q^2 = 0.1$. Substitution of this value in Eq. (13-35) gives closed-loop poles $s_1 = -0.113\omega_1(1 + n)$ and $s_2 = -0.887\omega_1(1 + n)$. These poles are separated by almost three octaves, and we can conclude that dominant-pole conditions exist for $Q \leq 0.316$ ($Q^2 \leq 0.1$). We also can state that as $\phi_M \rightarrow 90°$, the closed-loop response can be approximated by a single dominant pole.

A phase margin of approximately 90° can be achieved only when the open-loop poles are very widely separated (n is very large). Under these circumstances, both $A_{OL}(s)$ and $T(s)$ can be represented by a single-pole transfer function, the pole located at $s \approx -1/a_1 = -1/\omega_1$. Hence the closed-loop gain is expressable as the one-pole function in Eq. (13-2) and $\omega_H \approx \omega_1 T_O = T_O/a_1$ for $T_O \gg 1$.

Now, let us determine the location of the nondominant poles of the closed-loop response assuming that dominant-pole conditions apply. Examination of Eqs. (13-5) and (13-7) for a two-pole system indicates that if $A_F(s)$ has a dominant pole, A_{OL} must also have a dominant pole ($\omega_2 \gg \omega_1$). Hence $\omega_1 \approx 1/a_1$ and $\omega_2 \approx a_2/a_1$ for the open-loop system. The angular frequency of the closed-loop poles are (approximately) $1/a_1' = (1 + T_O)/a_1 = (1 + T_O)\omega_1$ and $a_1'/a_2' = a_1/a_2 = \omega_2$. From this analysis we can conclude that when the closed-loop response exhibits a dominant pole, the nondominant poles are located, approximately, at the same frequencies as the nondominant poles of the open-loop amplifier.

13-9   APPROXIMATE DETERMINATION OF THE OPEN-LOOP POLES   The discussion in the previous sections of this chapter was based on the assumption that the pole frequencies of both the open-loop gain A_{OL} and the return ratio T are known. As described in Sec. 11-9, the precise evaluation of the poles of a multistage amplifier is both difficult and cumbersome. Indeed, the accurate determination of the open-loop and closed-loop poles of a multistage feedback amplifier can be accomplished only with the use of computer simulations.

As indicated previously, the information regarding the open-loop response must be known prior to fabrication (to test for stability and to control closed-loop performance). However, the designer of a feedback amplifier requires more information than simply the accurate values of the open-loop poles. The circuit designer must be able to relate the selection of specific element values to performance in order to assess the design trade-offs that accompany the choice of our changes in component values. For example, in our discussion of the single-stage common-emitter circuit in Chaps. 10 and 11 we recognized that the voltage gain A_{VO} can be increased by increasing the collector resistance R_C. To maintain the same operating point, and consequently the same values of small-signal BJT parameters, however, an increase in R_C of ΔR_C must be accompanied by an increase in the supply voltage V_{CC} of $\Delta R_C I_{CQ}$. Furthermore, the bandwidth decreases as R_C is increased. Hence the designer is faced with the choice between gain, bandwidth, and power consumption. Similarly, in a feedback amplifier we must know which stages introduce poles nearest the origin so that effective compensation is achieved and the closed-loop response can be predicted. This is the objective in this section: to approximate the poles of the open-loop response and relate them to specific element values.

The approach we take to achieve the objective is based on the following:

1. The poles of A_{OL} and T are identical when resistive feedback is used.

2. The approximate analysis introduced in Sec. 12-7 is used to determine the poles of A_{OL} and hence T.

3. Only the two smallest angular pole frequencies, ω_1 and ω_2, are needed to approximate the closed-loop response.

4. These pole frequencies can be approximated by evaluating the coefficients a_1 and a_2 in the transfer function using the method described in Sec. 11-9.

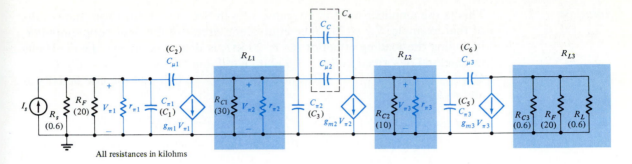

All resistances in kilohms

FIGURE 13-17
The high-frequency equivalent circuit of the amplifier without feedback for the shunt triple of Example 13-6. The loading effect of R_F is included at both input and output.

5. On the basis of estimated values of ω_1 and ω_2, the closed-loop performance is predicted.

6. The approximate results are compared with computer analyses.

In performing the analysis, we assume that midband design requirements are satisfied. Consequently, all device parameters and resistor values used are known. Also, the zeros of the transfer function, such as those introduced by C_μ, for example, are assumed to be sufficiently far removed from ω_1 and ω_2 that their effect is negligible. The two examples that follow illustrate the method of analysis.

Example 13-6   The shunt triple in Fig. 13-16 is designed with $C_C = 55$ pF and transistor parameters given in Table 13-3. Determine (a) the approximate open-loop angular pole frequencies ω_1 and ω_2 and (b) the approximate closed-loop poles. Using the results in part a, (c) estimate the phase margin from the asymptotic Bode diagram, (d) estimate the phase margin from the value of Q, and (e) compare the results in parts c and d.
    It is assumed that $r_b = 0$ and $r_o \to \infty$ for all transistors.

TABLE 13-3 Transistor Parameters for Shunt Triple in Fig. 13-17

			Parameter		
Transistor	r_π, $k\Omega$	g_m, $m\mho$	β_o	C_π, pF	C_μ, pF
Q1	25	4.0	100	1.6	0.5
Q2	10	10	100	4.0	0.5
Q3	2.5	40	100	10.0	0.5

Solution

This is the amplifier used in Example 12-6 in Sec. 12-10. (*a*) From the results of that example, $T_O = 202$. The equivalent circuit of the open-loop amplifier, including the loading effect of $R_F = 20$ kΩ, is displayed in Fig. 13-17. Using the method described in Sec. 11-9 and identifying $C_{\pi 1} = C_1$, $C_{\mu 1} = C_2$, $C_{\pi 2} = C_3$, $C_{\mu 2} + C_C = C_4$, $C_{\pi 3} = C_5$, and $C_{\mu 3} = C_6$, we find that the coefficients a_1 and a_2 are

$$a_1 = R_{11}^0 C_1 + R_{22}^0 C_2 + R_{33}^0 C_3 + R_{44}^0 C_4 + R_{55}^0 C_5 + R_{66}^0 C_6$$

$$a_2 = R_{11}^0 C_1 (R_{22}^1 C_2 + R_{33}^1 C_3 + R_{44}^1 C_4 + R_{55}^1 C_5 + R_{66}^1 C_6)$$

$$+ R_{22}^0 C_2 (R_{33}^2 C_3 + R_{44}^2 C_4 + R_{55}^2 C_5 + R_{66}^2 C_6)$$

$$+ R_{33}^0 C_3 (R_{44}^3 C_4 + R_{55}^3 C_5 + R_{66}^3 C_6)$$

$$+ R_{44}^0 C_4 (R_{55}^4 C_5 + R_{66}^4 C_6) + R_{55}^0 C_5 R_{66}^5 C_6$$

The resistance values (in kilohms) needed in the calculation of a_1 and a_2 are

$$R_{L1} = R_{C1} \| r_{\pi 2} = 30 \| 10 = 7.5$$

$$R_{L2} = R_{C2} \| r_{\pi 3} = 10 \| 2.5 = 2.0$$

$$R_{L3} = R_{C3} \| R_F \| R_L = 0.60 \| 20 \| 0.60 = 0.296$$

$$R_{11}^0 = R_s \| R_F \| r_{\pi 1} = 0.60 \| 20 \| 25 = 0.569$$

$$R_{22}^0 = R_{11}^0 (1 + g_{m1} R_{L1}) + R_{L1}$$

$$= 0.569(1 + 4.0 \times 7.5) + 7.5 = 25.1$$

$$R_{33}^0 = R_{L1} = 7.5$$

$$R_{44}^0 = R_{33}^0 (1 + g_{m2} R_{L2}) + R_{L2} = 7.5(1 + 10 \times 2.0) + 2.0 = 159.5$$

$$R_{55}^0 = R_{L2} = 2.0$$

$$R_{66}^0 = R_{55}^0 (1 + g_{m3} R_{L3}) + R_{L3} = 2.0(1 + 40 \times 0.296) + 0.296 = 25.9$$

$$R_{22}^1 = R_{L1} = 7.5 \qquad R_{33}^1 = R_{33}^0 = 7.5 \qquad R_{44}^1 = R_{44}^0 = 159.5$$

$$R_{55}^1 = R_{55}^0 = 2.0 \qquad R_{66}^1 = R_{66}^0 = 25.9$$

$$R_{33}^2 = R_{L1} \| \frac{1}{g_{m1}} \| R_{11}^0 = 7.5 \| 0.25 \| 0.569 = 0.174$$

$$R_{44}^2 = R_{33}^2 (1 + g_{m2} R_{L2}) + R_{L2} = 0.174(1 + 10 \times 2.0) + 2.0 = 5.65$$

$$R_{55}^2 = R_{55}^0 = 2.0 \qquad R_{66}^2 = R_{66}^0 = 25.9$$

$$R_{44}^3 = R_{L2} = 2.0 \qquad R_{55}^3 = R_{55}^0 = 2.0$$

$$R_{66}^3 = R_{66}^0 = 25.9$$

$$R_{55}^4 = R_{L2}\|\frac{1}{g_{m2}}\|R_{33}^0 = 2.0\|0.10\|7.5 = 0.094$$

$$R_{66}^4 = R_{55}^4 (1 + g_{m3}R_{L3}) + R_{L3} = 0.094(1 + 40 \times 0.296) + 0.296 = 1.50$$

$$R_{66}^5 = R_{L3} = 0.296$$

The student is encouraged to verify these values.

Substitution of the resistance and capacitance values into the expressions for a_1 and a_2 yields

$$a_1 = 156.0 + 159.5C_C = 156.0 + 159.5 \times 55 = 8928 \text{ ns}$$

$$a_2 = 1757 + 657.1C_C = 1794 + 657.1 \times 55 = 37,900 \text{ (ns)}^2$$

from which

$$\omega_1 = \frac{1}{a_1} = \frac{1}{8928} = 0.1120 \times 10^6 \text{ rad/s};$$

$$f_1 = \frac{\omega_1}{2\pi} = \frac{0.112 \times 10^6}{2\pi} = 17.83 \text{ kHz}$$

$$\omega_2 = \frac{a_1}{a_2} = \frac{8928}{37,900} = 235.5 \times 10^6 \text{ rad/s};$$

$$f_2 = \frac{\omega_2}{2\pi} = \frac{235.5 \times 10^6}{2\pi} = 37.47 \text{ MHz}$$

(b) The separation of the open-loop poles is

$$n = \frac{\omega_2}{\omega_1} = \frac{a_1^2}{a_2} = \frac{(8928)^2}{37,900} = 2103$$

Thus, from Eqs. (13-47)

$$Q = \frac{\sqrt{2103(1 + 202)}}{2103 + 1} = 0.3105$$

and from Eq. (13-35)

$$s = \frac{-0.112(2103 + 1)}{2} [1 + \sqrt{1 - 4(0.3105)^2}]$$

or

$$s_1 = -2.555 \times 10^7 \text{ rad/s} \qquad s_2 = -2.102 \times 10^8 \text{ rad/s}$$

The separation of the closed-loop poles is

$$\left|\frac{s_2}{s_1}\right| = \frac{2.102 \times 10^8}{2.552 \times 10^7} = 8.24$$

As this is more than three octaves, dominant-pole conditions apply. For these conditions, we showed in Eq. (13-3) that the closed-loop dominant pole was

$$\omega_H = |s_1| = (1 + T_O)\omega_h = (1 + T_O)\omega_1$$

or

$$|s_1| = (1 + 202) \times 0.112 \times 10^6 = 2.274 \times 10^7 \text{ rad/s}$$

Clearly, the two values are approximately equal.

Also in Sec. 13-1, it was observed that the nondominant pole was essentially unaffected by feedback when dominant-pole conditions are satisfied. Comparison of $|s_2|$ and ω_2 supports this conclusion.

(c) From the results in part a, we can write

$$T(s) = \frac{T_O}{[1 + (s/\omega_1)][1 + (s/\omega_2)]}$$

$$= \frac{202}{\{1 + [s/(0.112 \times 10^6)]\}\{1 + [s/(2.354 \times 10^8)]\}}$$

The asymptotic Bode diagram for $T(j\omega)$ is displayed in Fig. 13-18a, from which we observe that $\phi_M = 90°$.

The asymptotic Bode diagram for $\mathbf{T}(j\omega)$ in Fig. 13-18b includes the effect of the third pole of the system. Using the extension of the method in Sec. 11-9, it can be shown that

$$a_3 = 1704 + 417.5C_C = 1704 + 417.5 \times 55 = 24{,}670 \text{ (ns)}^3$$

and

$$\omega_3 = \frac{a_2}{a_3} = \frac{37{,}930}{24{,}670} = 1.538 \times 10^9 \text{ rad/s}$$

Note that the inclusion of the third pole does not change ϕ_M. (d) The phase margin is determined by the use of Eq. (13-42):

$$\phi_M = \tan^{-1} \sqrt{2} \, [\sqrt{4(0.3106)^4 + 1} - 1]^{-1/2} = 84.5°$$

(e) The results in parts c and d compare favorably. The difference is attributed to the small error introduced by the asymptotic approximation of the phase characteristic. Note that at $\omega_G = 2.263 \times 10^7$ the phase contribution of ω_2 is

$$\theta_2 = -\tan^{-1} \frac{\omega_G}{\omega_2} = \frac{2.354 \times 10^6}{2.263 \times 10^7} = -5.5°$$

and this is exactly the difference between the results in parts c and d.

Computer analysis of the shunt triple in this example yields the following results: $T_O = 202$, $\omega_1 = 0.1120 \times 10^6$ rad/s, $\omega_2 = 2.406 \times 10^8$ rad/s, and closed-loop poles $s_1 = -2.87 \times 10^7$ rad/s, $s_2 = -1.76 \times 10^8$ rad/s. We observe that the open-loop poles are virtually identical to those obtained by the ap-

FIGURE 13-18
Asymptotic Bode diagrams for Example 13-6: (*a*) the two-pole approximation, (*b*) inclusion of the third pole.

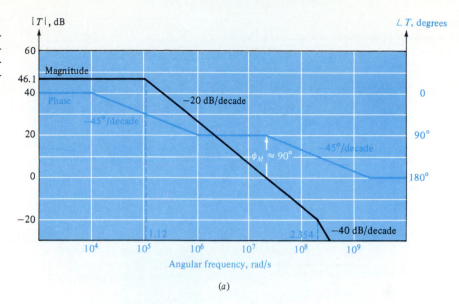

(*a*)

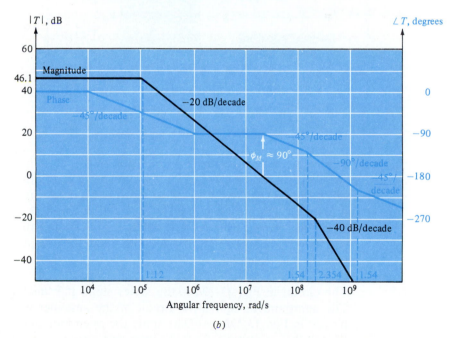

(*b*)

proximate calculation in part *a*. However, the closed-loop poles are not as widely separated as calculated. This is attributed to our assumption that the zeros of the transfer function have negligible effect. With $C_4 = C_{\mu 2} + C_C = 0.5 + 55 = 55.5$ pF, the second stage introduces a zero at $s = +g_{m2}/C_4 = +10/55.5 = 1.80 \times 10^8$ rad/s (Sec. 11-5). Clearly, this zero has an angular

FIGURE 13-19

(*a*) Ac schematic of a series-shunt pair. (*b*) The high-frequency model of the amplifier without feedback and including the loading effect of the feedback network.

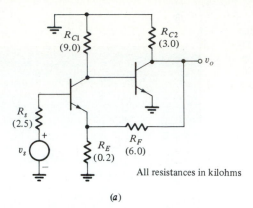

All resistances in kilohms

(*a*)

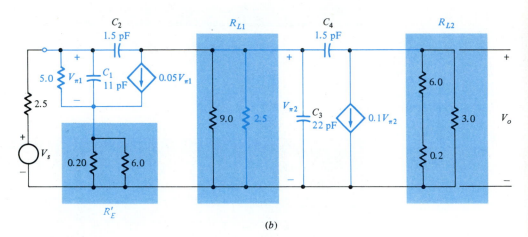

(*b*)

frequency that is less than ω_2. Inclusion of this zero in the expression for $T(s)$ results in a phase margin of 77.5° because the zero introduces an additional 7° of phase shift. As we showed in the root locus of Fig. 13-1 and in our prior discussions, reduction in ϕ_M results in increase in Q and, hence the closed-loop poles move closer to one another as indicated by the computer simulation. In the next section we discuss in more detail the compensation technique (C_C) used in this example.

Example 13-7

The circuit in Fig. 13-19*a* is the series-shunt pair analyzed in Example 12-8. The approximate equivalent circuit of the amplifier without feedback is displayed in Fig. 13-19*b*. (*a*) Determine the open-loop and closed-loop poles. (*b*) Sketch the asymptotic Bode diagram and estimate the phase margin.

Solution

(*a*) The resistances R'_E, R_{L1}, and R_{L2} identified in Fig. 13-19*b* are

$$R'_E = R_E \| R_F = 0.20 \| 6.0 = 0.194 \text{ k}\Omega$$

$$R_{L1} = R_{C1} \| r_{\pi2} = 9.0 \| 2.5 = 1.96 \text{ k}\Omega$$

$$R_{L2} = R_{C2} \| (R_F + R_E) = 3.0 \| (6.0 + 0.20) = 2.02 \text{ k}\Omega$$

To obtain the open-loop pole frequencies, we evaluate the coefficients a_1 and a_2 by use of the method in Sec. 11-9. Thus

$$a_1 = R_{11}^0 C_1 + R_{22}^0 C_2 + R_{33}^0 C_3 + R_{44}^0 C_4$$

$$a_2 = R_{11}^0 C_1 (R_{22}^1 C_1 + R_{33}^1 C_3 + R_{44}^1 C_4) + R_{22}^0 C_2 (R_{33}^2 C_3 + R_{44}^2 C_4) + R_{33}^0 C_3 R_{44}^3 C_4$$

The resistances R_{11}^0 and R_{22}^0 are the equivalent resistances seen by the capacitors in the stage containing local feedback. Hence, we will use Blackman's imped-ance formula to calculate these values.

For R_{11}^0: The dead system is shown in Fig. 13-20a, from which

$$R_{11D}^0 = r_{\pi1}\|(R_s + R_E') = 5.0\|(2.5 + 0.194) = 1.75 \text{ k}\Omega$$

Clearly, short-circuiting the terminals of C_1 makes $T_{SC} = 0$. The circuit in Fig. 13-20b is used to evaluate T_{OC}. The current divider relation gives

$$I_\pi = -g_{m1} \hat{V}_\pi \times \frac{R_E'}{R_E' + R_s + r_{\pi1}}$$

$$V_\pi = I_\pi r_{\pi1} = \frac{-g_{m1} r_{\pi1} R_E' \hat{V}_\pi}{R_E' + R_s + r_{\pi1})}$$

from which

$$T_{OC} = \frac{V_\pi}{\hat{V}_\pi} = \frac{\beta_o R_E'}{R_E' + R_s + r_{\pi1}} = \frac{125 \times 0.194}{0.194 + 2.5 + 5.0}$$

$$= 3.15$$

Thus

$$R_{11}^0 = \frac{R_{11D}^0}{1 + T_{OC}} = \frac{1.75}{1 + 3.15} = 0.422 \text{ k}\Omega$$

FIGURE 13-20

Circuits used to compute R_{22}^0 by the use of Blackman's impedance relation: (a) For the dead system, (b) for the calculation of T_{OC}, (c) for the calculation of T_{SC}.

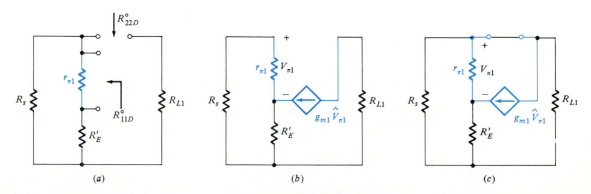

(a)   (b)   (c)

For R_{22}^0: in Fig. 13-21a, we find that

$$R_{22D}^0 = R_{L1} + R_s\|(r_{\pi 1} + R_E') = 1.96 + 2.5\|(5.0 + 0.194) = 3.65 \text{ k}\Omega$$

The value of T_{OC}, obtained from the circuit in Fig. 13-20b, is

$$T_{OC} = 3.15$$

To evaluate T_{SC} when C_2 is short-circuited, we use the circuit in Fig. 13-20c. Recognizing that R_s and R_{L1} are in parallel, we find that

$$I_\pi = -g_{m1} \hat{V}_\pi \frac{R_E' + R_s\|R_{L1}}{R_E' + (R_s\|R_{L1})} \qquad \text{and} \qquad V_\pi = I_\pi r_{\pi 1}$$

Combination of these relationships gives

$$T_{SC} = -\frac{V_\pi}{\hat{V}_\pi} = \frac{\beta_o[R_E' + R_s\|R_{L1}]}{r_\pi + R_E' + (R_s\|R_{L1})} = \frac{125[1.94 + 2.5\|1.96]}{5.0 + 0.194 + (2.5\|1.96)} = 25.7$$

Hence

$$R_{22}^0 = R_{22D}^0 \frac{1 + T_{SC}}{1 + T_{OC}} = 3.65 \frac{1 + 25.7}{1 + 3.15} = 23.4 \text{ k}\Omega$$

In the common-emitter stage containing C_3 and C_4, the resistances are

$$R_{33}^0 = R_{L1} = 1.96 \text{ k}\Omega \qquad \text{and} \qquad R_{44}^0 = R_{33}^0(1 + g_{m2}R_{L2}) + R_{L2}$$
$$= 1.96(1 + 50 \times 2.02) + 2.02 = 202 \text{ k}\Omega$$

The remaining resistances are found as follows. Short-circuiting C_1 gives

$$R_{22}^1 = R_{L1} + R_E'\|R_s = 1.96 + 0.194\|2.5 = 2.14 \text{ k}\Omega$$

$$R_{33}^1 = R_{33}^0 = 1.96 \text{ k}\Omega \qquad R_{44} = R_{44} = 202 \text{ k}\Omega$$

With C_2 short-circuited

$$R_{33}^2 = R_{L1}\|R_s\|\left(R_E' + \frac{1}{g_m}\right) = 1.96\|2.5\|(0.194 + 0.020) = 0.179 \text{ k}\Omega$$

The calculation for R_{44}^2 is similar to that for R_{44}^0 except that R_{33}^0 is replaced by R_{33}^2. Then

$$R_{44}^2 = R_{33}^2(1 + g_{m2}R_{L2}) + R_{L2} = 0.179(1 + 50 \times 2.02) + 2.02 = 20.3 \text{ k}\Omega$$

The entire left-hand portion of the circuit is eliminated when C_3 is short-circuited; hence

$$R_{44}^3 = R_{L2} = 2.02 \text{ k}\Omega$$

Using the capacitance values given in Fig. 13-19b and the computed values of resistance, we obtain

$$a_1 = 0.422 \times 11 + 23.4 \times 1.5 + 1.96 \times 22 + 202 \times 1.5 = 385.9 \text{ ns}$$

$$a_2 = 0.422 \times 11(2.14 \times 1.5 + 1.96 \times 22 + 202 \times 1.5) +$$

$$23.4 \times 1.5(0.179 \times 22 + 20.3 \times 1.5) + 1.96 \times 22 \times 2.02 \times 1.5$$

$$= 2959 \text{ (ns)}^2$$

Then the open-loop poles have angular frequencies of $\omega_1 = 1/a_1 = 2.59 \times 10^6$ rad/s, $\omega_2 = a_1/a_2 = 1.30 \times 10^8$ rad/s. In Example 12-8, the value of $T = 24.9$ was obtained. Use of Eqs. (13-47) and (13-35) results in

$$Q = \frac{\sqrt{2959(1 + 24.9)}}{385.9} = 0.717$$

$$s = \frac{-2.59 \times 10^6(50.3 + 1)}{2}(1 \pm \sqrt{1 - 4Q^2})$$

$$= -6.64 \times 10^7 (1 \pm j1.03) \text{ rad/s}$$

The value of $n = \omega_2/\omega_1 = 1.30 \times 10^8/2.59 \times 10^6 = 50.3$ is used in the preceding equations.

Note that the value of $Q = 0.717$ is nearly 0.707 for which no peaking in the amplitude response exists. Thus we can conclude that the amplitude response of this series-shunt pair exhibits virtually no peaking. Similarly, the step response has little overshoot (<3 percent).

(b) The asymptotic Bode diagram for

$$T(s) = \frac{24.9}{(1 + s/2.59 \times 10^6)(1 + s/1.30 \times 10^8)}$$

is displayed in Fig. 13-21. The phase margin, indicated in Fig. 13-21, is $\phi_M = 60.5°$. If Eq. (13-42) is used, $\phi_M = 65.0°$. The difference between the two results is attributed to the errors in the asymptotic approximation of the phase characteristic in Fig. 13-21. The error in phase due to the pole at $-\omega_2$ is approximately 5° (about the difference between the two values).

Computer analysis of the circuit in Fig. 13-19 gives

$$\omega_1 = 2.64 \times 10^6 \text{ rad/s} \qquad \omega_2 = 1.27 \times 10^8 \text{ rad/s}$$

$$s = -6.59 \times 10^7 (1 \pm j\, 1.035) \text{ rad/s}$$

The approximate values are all within 3 percent of actual (computer-simulated) values and demonstrate the effectiveness of the approximate calculations as a design tool.

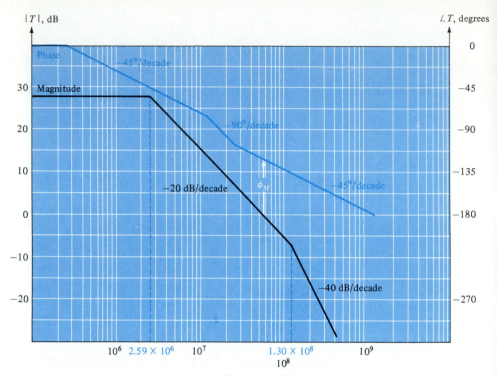

FIGURE 13-21
Asymptotic Bode diagram of $T(j\omega)$ for Example 13-7.

Angular frequency, rad/s

13-10 COMPENSATION REVISITED We have demonstrated in prior sections that the dominant pole $-\omega_1$ of $T(s)$ must be widely separated from the next pole at $-\omega_2$ to achieve a particular closed-loop response. Often the basic amplifier must be compensated if the required pole separation is to be realized. In Sec. 13-4 we demonstrated that effective compensation occurs when the amplifier without feedback is deliberately narrow-banded. In this section we show that the method for approximating the open-loop poles discussed in Sec. 13-8 can also be used to obtain the initial design values of the compensating circuit elements.

The insertion of a compensating capacitor C_C into the open-loop amplifier is the simplest technique by which narrow-banding is achieved. Clearly, the addition of C_C increases the value of the a_1 coefficient in the transfer function and, hence, reduces the value of $\omega_1 \approx 1/a_1$. If a_{1C} is the s coefficient of the compensated amplifier, then

$$a_{1C} = a_1 + R^0_{CC}C_C \qquad (13\text{-}51)$$

where a_1 is the s coefficient of the uncompensated amplifier and R^0_{CC} is the equivalent open-circuit resistance seen by C_C. [Equation (13-51) is obtained in a fashion similar to that used in Example 13-6.]

Since each term in the coefficient of s^2 is expressible as the product of an open-circuit and short-circuit time constant, a_{2C}, the compensated value of this coefficient, may be written as

$$a_{2C} = a_2 + R_{CC}^0 C_C \left(\sum_{\substack{i=1 \\ i \neq C}}^N R_{ii}^C C_i \right) \tag{13-52}$$

In Eq. (13-52), $R_{ii}^C C_i$ is the time constant for capacitance C_i when C_C is shorted. Note that we expressed a_{1C} and a_{2C} in the form of Eqs. (13-51) and (13-52) in Example 13-6.

Use of Eq. (13-46) gives the pole-separation factor n as

$$n = \frac{T_O}{Q^2} = \frac{a_1^2}{a_2} = \frac{(a_1 + R_{CC}^0 C_C)^2}{a_2 + R_{CC}^0 C_C \left(\sum_{\substack{i=1 \\ i \neq C}}^N R_{ii}^C C_i \right)} \tag{13-53}$$

Since n, a_1, a_2, and the open- and short-circuit resistances are known, solving Eq. (13-53) yields the initial design value of C_C needed to obtain the requisite pole separation.

Let us examine the two extremes that result from simple capacitive compensation. To narrow-band the amplifier significantly, it is clear that $R_{CC}^0 C_C$ in Eq. (13-51) must be much larger than a_1. (If $a_{1C} \geq 10a_1$, as is required to move $\omega_{1C} \approx 1/a_{1C}$ one decade nearer the origin than ω_1, $R_{CC}^0 C_C \geq 9a_1$.) Hence we may employ the approximation $a_{1C} \approx R_{CC}^0 C_C$. In the first extreme, consider that

$$R_{CC}^0 C_C \left(\sum_{\substack{R=1 \\ i \neq C}}^N R_{ii}^C C_i \right) \gg a_2$$

Then

$$\omega_{1C} \approx \frac{1}{R_{CC}^0 C_C} \qquad \omega_{2C} = \frac{a_{1C}}{a_{2C}} \approx \frac{1}{\displaystyle\sum_{\substack{i=1 \\ i \neq C}}^N R_{ii}^C C_i} \tag{13-54}$$

Note that Eq. (13-54) indicates that ω_{2C} is independent of C_C and thus is constant. Hence C_C can be obtained directly from the relation for ω_{1C} in Eq. (13-54). The value of ω_{1C} used can be evaluated from the pole-separation factor n or by use of the asymptotic Bode diagram technique presented in Sec. 13-4.

We assume that $a_2 \gg R_{CC}^0 C_C \left(\sum_{i=1}^N R_{ii}^C C_i \right)$ in the second extreme case. Consequently

$$\omega_{1C} \approx \frac{1}{R_{CC}^0 C_C} \qquad \omega_{2C} \approx \frac{R_{CC}^0 C_C}{a_2} \tag{13-55}$$

Some practical amplifiers exhibit the extreme conditions described in Eqs. (13-54) and (13-55). However, this is not always the situation.

Pole-Splitting

The technique, used in Example 13-6, of adding a capacitor C_C between base and collector of the interior stage is commonly called *pole-splitting*. Use of C_C in this manner takes advantage of the Miller effect multiplication (Sec. 11-5) and results in capacitance values readily fabricated on a chip. It can be shown that this type of compensation causes $\omega_{1C} < \omega_1$ and $\omega_{2C} > \omega_2$. Thus ω_{1C} moves from ω_1 toward the origin (narrow-banding), and ω_{2C} moves from ω_2 further from the origin; hence the term "pole-splitting" is used. Note that pole-splitting is evident in the situation described by Eq. (13-55).

Example 13-8

Determine C_C when $Q^2 = 0.1$ for the amplifier in Example 13-6.

Solution

In Example 13-6 we evaluated a_{1C} and a_{2C} as

$$a_{1C} = 156.0 + 159.5C_C \qquad a_{2C} = 1794 + 657.1C_C$$

To achieve $Q^2 = 0.1$ with $T_O = 202$, Eq. (13-36) states that

$$n = \frac{T_O + 1}{Q^2} = \frac{202 + 1}{0.1} = 2030$$

Use of Eq. (13-53) gives

$$2030 = \frac{(156.0 + 159.5C_C)^2}{1794 + 657.1C_C}$$

and solving the resultant quadratic equation yields $C_C = 54.8$ pF. In Example 13-6 we found that $Q^2 = 0.0966$ for $C_C = 55$ pF. As n increases and Q decreases with increasing C_C, we expect a slight reduction in Q^2 for a small increase in C_C.

Observe that in the expression for a_{1C}, a value of $C_C > 30$ pF gives $a_{1C} \approx 159.5C_C$ and $a_{2C} \approx 651.7C_C$. Hence

$$\omega_{2C} \approx \frac{a_{1C}}{a_{2C}} = \frac{159.5C_C}{651.7C_C} = 2.447 \times 10^8 \text{ rad/s}$$

and

$$\omega_{1C} = \frac{\omega_{2C}}{n} = \frac{2.447 \times 10^8}{2030} = 0.120 \times 10^6 \text{ rad/s}$$

Solving for C_C yields

$$C_C = \frac{a_{1C}}{159.5} = \frac{1}{159.5\omega_{1C}} = \frac{1}{159.5 \times 0.120 \times 10^6} = 52.2 \text{ pF}$$

These results are all very nearly equal to the actual values obtained. Therefore, we can conclude that this situation approximates the first extreme case previously discussed.

FIGURE 13-22
Shunt triple with shunt-capacitance compensation.

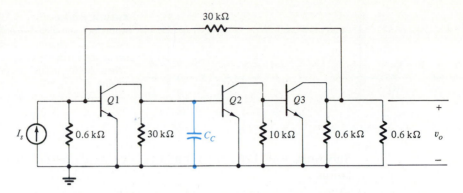

Shunt-Capacitance Compensation

One potential disadvantage of pole-splitting is that the zero at $s = g_{m2}/(C_C + C_{\mu2})$ also moves nearer to the origin and affects the phase margin and closed-loop poles (Example 13-6). An alternative compensation technique is to add a capacitance C_C at the input of the second stage as shown in Fig. 13-22. Because C_C is shunted to ground, the zero, introduced by the second stage, remains at $g_{m2}/C_{\mu2}$ and has a negligible effect on the phase margin. With the use of component values in Example 13-6, Eqs. (13-51) and (13-5) become

$$a_{1C} = 156.0 + 7.5C_C \quad \text{ns} \qquad a_{2C} = 1794 + 262.46C_C \quad \text{(ns)}^2$$

For $n = 2030$, solving Eq. (13-53) gives $C_C = 9430$ pF. Clearly, this value is much larger than the value obtained in Example 13-6. Furthermore, a 9430-pF capacitance cannot be fabricated on a chip rendering this technique impractical in IC design.

A second disadvantage of this method is that while it provides the appropriate pole separation, the resultant closed-loop bandwidth decreases. Using $C_C = 9430$ pF, we obtain

$$a_{1C} = 70.89 \ \mu s \qquad a_{2C} = 2.477 \ (\mu s)^2$$

from which

$$\omega_{1C} = 1.410 \times 10^4 \ \text{rad/s} \qquad \omega_{2C} = 2.864 \times 10^7 \ \text{rad/s}$$

Both of these values are lower than the corresponding values in Example 13-6. Consequently, $\omega_o = \sqrt{(1 + T_O)}\omega_1\omega_2$ is lower for shunt-capacitance compensation, and, as given by Eq. (13-43), the bandwidth is reduced. We also observe that $\omega_{2C} < \omega_2$ when this technique is used and pole-splitting does not occur.

Root-Locus Analysis (Optional)

The poles of the compensated open-loop amplifier are given by (approximately)

$$1 + a_{1C}s + a_{2C}s^2 = 0$$

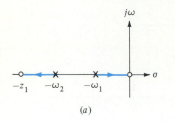

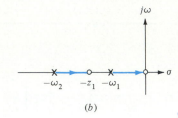

FIGURE 13-23
Root loci showing (*a*) pole-splitting and (*b*) with both poles being narrow-banded. The situation in (*a*) arises when Miller effect compensation is used. When shunt-capacitance compensation is used, (*b*) results.

Substitution of Eqs. (13-51) and (13-52) yields, after some algebraic manipulation

$$\frac{R_{CC}^0 \, C_C s \left(1 + s \sum\limits_{\substack{i=i \\ i \neq C}}^{N} R_{ii}^C \, C_i \right)}{1 + a_1 s + a_2 s^2} = \frac{R_{CC}^0 C_C \, s(1 + s/z_1)}{(1 + s/\omega_1)(1 + s/\omega_2)} = -1 \quad (13\text{-}56)$$

Let us assume that $z_1 > \omega_2$. The locus of poles as C_C varies is illustrated in Fig. 13-23*a*. Clearly, pole-splitting occurs, and this is the situation that exists in Examples 13-6 and 13-8. Alternatively, let us now assume that $\omega_1 < z_1 < \omega_2$, for which Fig. 13-23*b* is the root locus. Note that both ω_{1C} and ω_{2C} are less than their corresponding values in the uncompensated amplifier. Shunt-capacitance compensation often leads to this condition and results in decreased bandwidth.

Summary

The steps in the analysis and design of a feedback amplifier can be summarized as follows:

1. Using component values required to satisfy midband specifications, approximate $A_{OL}(s)$ and $T(s)$, employing the methods described in Sec. 13-8.

2. Test for stability as outlined in Sec. 13-3.

3. Compensate the amplifier to give the approximate closed-loop response desired. The two-pole approximation in Secs. 13-5 and 13-6 is used to predict the closed-loop response, and the methods described here and in Sec. 13-4 provide the basis for compensation.

4. Use computer simulations to obtain the actual closed-loop response and open-loop transfer function.

5. Compare the response in step 4 with predicted values.

6. Adjust component values to reduce the difference between the actual and predicted responses.

7. Repeat steps 4, 5, and 6 until final design values are obtained.

REFERENCES

1 Sedra, A. S., and K. C. Smith: "Microelectronic Circuits," Holt, New York, 1981.

2 Gray, P. R., and R. G. Meyer: "Analysis and Design of Analog Integrated Circuits," John Wiley and Sons, New York, 1984.

3 Blecher, F. H.: Design Principles in Single Loop Transistor Feedback Amplifiers, *IRE Trans. Circuit Theory,* vol. CT-4, no. 5, September 1957.

4 Ghausi, M. S.: "Electronic Devices and Circuits: Discrete and Integrated," Holt, New York, 1985.

5 Grebene, A. B.: "Bipolar and MOS Analog Integrated Circuits," John Wiley and Sons, New York, 1984.

6 Bode, H. W.: "Network Analysis and Feedback Amplifier Design," D. Van Nostrand Company, Princeton, N.J., 1945.

7 Schilling, D., and C. Belove: "Electronic Circuits—Discrete and Integrated," McGraw-Hill Book Company, New York, 1985.

8 Soclof, S.: "Analog Integrated Circuits," Prentice-Hall, Englewood Cliffs, N.J., 1985.

9 Nyquist, H.: Regeneration Theory, *Bell System Tech. J.,* vol. 11, pp. 126–147, January 1932.

10 Thornton, R. D., C. L. Searle, D. O. Pederson, R. B. Adler, and E. J. Angelo, Jr: "Multistage Transistor Circuits," SEEC Committee Series, vol. 5, pp. 108–118, John Wiley and Sons, New York, 1965.

REVIEW QUESTIONS

13-1 Consider a feedback amplifier with a single-pole transfer function.
 (a) What is the relationship between the high 3-dB frequency with and without feedback?
 (b) Repeat part *a* for the low 3-dB frequency.
 (c) Repeat part *a* for the gain-bandwidth product.

13-2 Consider a feedback amplifier with a double-pole transfer function.
 (a) Without proof sketch the locus of the poles in the *s* plane after feedback.
 (b) Why is this amplifier stable, independent of the amount of negative feedback?

13-3 (a) Indicate (without proof) a circuit having the same transfer function as the double-pole feedback amplifier.
 (b) Sketch the step response of this amplifier for both the underdamped and overdamped condition.

13-4 For an underdamped two-pole amplifier response, define (a) rise time; (b) delay time; (c) overshoot; (d) damped period; (e) settling time.

13-5 (a) Sketch (without proof) the root locus of the poles of a three-pole amplifier after feedback is added.
 (b) Indicate where the amplifier becomes unstable.

13-6 Consider a multipole amplifier with $|s_1| < |s_2| < |s_3| < \cdots < |s_n|$. Under what circumstances is the response with feedback determined by (a) s_1 and s_2 and (b) s_1 alone?

13-7 (*a*) Define stability.

(*b*) For stability where must the poles of $A_F(s)$ lie?

13-8 (*a*) State Nyquist's criterion.

(*b*) Draw a Nyquist diagram for a stable system.

(*c*) Repeat part *b* for an unstable system.

13-9 (*a*) Define phase margin ϕ_M.

(*b*) Indicate ϕ_M on the diagrams in Revs. 13-8*b* and 13-8*c*.

13-10 (*a*) Define gain margin.

(*b*) Indicate the gain margin on the diagrams in Revs. 13-8*b* and 13-8*c*.

13-11 (*a*) Draw the Bode diagrams corresponding to Revs. 13-8*b* and 13-8*c*.

(*b*) Identify the gain and phase margins on the Bode diagrams in part *a*.

13-12 What is meant by compensation?

13-13 With the aid of a Bode diagram, explain how an amplifier may be compensated.

13-14 Describe the method by which the first two dominant poles of the open-loop amplifier can be determined.

13-15 (*a*) If the closed-loop response exhibits a dominant pole, must the open-loop response have a dominant pole? Explain.

(*b*) Comment on the converse of part *a*.

13-16 Describe pole-splitting.

13-17 Compare the Miller effect and shunt-capacitance methods of dominant-pole compensation.

13-18 (*a*) What is meant by positive feedback?

(*b*) How are A_F and A related in a positive-feedback amplifier?

(*c*) If $T = -1$, what is the gain A_F?

Chapter 14

OPERATIONAL-AMPLIFIER CHARACTERISTICS

The operational amplifier (Op-Amp) is the most extensively used analog integrated circuit (IC). The objective of this chapter is to describe the properties of practical Op-Amps and relate these characteristics to analog IC design techniques. Since the Op-Amp is a multistage circuit that almost always employs feedback, the material treated in this chapter brings together many of the concepts discussed in Chaps. 10 to 13.

14-1 OPERATIONAL-AMPLIFIER ARCHITECTURES The Op-Amp, introduced in Sec. 10-21, is a two-input voltage-controlled voltage source whose output voltage is proportional to the difference between the two input voltages. The characteristics of Op-Amps and their use in the basic inverting and noninverting amplifiers were described in Sec. 10-21. The frequency response of these amplifiers is discussed in Sec. 11-13. We summarize the performance of ideal and practical Op-Amps in Table 14-1.

TABLE 14-1   Characteristics of Op-Amps

Property	Ideal	Practical (Typical)
Open-loop gain	Infinite	Very high ($\geq 10^4$)
Open-loop bandwidth	Infinite	Dominant pole (≈ 10 Hz)
Common-mode rejection ratio	Infinite	High (≥ 70 dB)
Input resistance	Infinite	High (≥ 10 MΩ)
Output resistance	Zero	Low (< 500 Ω)
Input currents	Zero	Low (< 0.5 μA)
Offset voltages and currents	Zero	Low (< 10 mV, < 0.2 nA)

Now let us examine some of the reasons underlying the values listed in Table 14-1. Since the basic Op-Amp stages are feedback amplifiers, high open-loop gain—and hence large return ratios—is desirable to ensure exclusive dependence of the closed-loop gain on the feedback resistances R_1 and R_2 (see Fig. 10-42). Similarly, most Op-Amps are designed to have a transfer function that contains a dominant pole. Under these conditions, both the open- and closed-loop gain-bandwidth products are equal. Thus, for a specified closed-loop gain, the closed-loop bandwidth is readily determined.

A high common-mode rejection ratio $CMRR$ is needed to ensure that the output signal is proportional to the difference between the input voltages. With high $CMRR$ values, common-mode signals, often containing dc components, have only a small effect on the amplifier output.

To approximate the characteristics of an ideal voltage amplifier, the Op-Amp must possess high-input and low-output resistances. The current in each of the two inputs is ideally zero. These dc currents are part of the bias currents of the input stage and, because ICs are invariably direct-coupled, must be small to prevent adverse interaction with the signal source.

In an "ideal" IC, we can exactly match transistor characteristics and component values. Regardless of how sophisticated the fabrication technology is, exact matching is impossible in the "real world." The offset voltage and current are measures of the degree of mismatch in the circuit, and clearly these should be small.

The Two-Stage Architecture   A majority of commercially available operational amplifiers employ the structure displayed in Fig. 14-1. This cascade configuration is commonly referred to as a *two-stage operational amplifier* because only the differential amplifier and gain stage contribute to the overall voltage gain. The differential amplifier is used as the input stage to provide the inverting and noninverting inputs, the high $CMRR$, and the high-input resistance, as well as voltage gain. The low-output resistance of the Op-Amp is achieved by the emitter-follower output stage. The level shifter adjusts the dc voltages so that the output voltage signal is referenced to ground. The adjustment of dc levels is required because the gain stages are direct-coupled. Since large-valued capacitors cannot be fabricated on a chip, virtually all ICs are direct-coupled. The interior gain stage is a high-gain voltage amplifier used to obtain a large open-loop gain.

In the previous paragraph we note that the input and output stages are required to match the Op-Amp with the "external world." That is, these stages serve as the interfaces between the amplifier and the input signal sources and

FIGURE 14-1
The architecture of a two-stage Op-Amp.

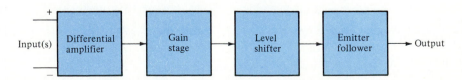

between the amplifier and the load. In the design of the input and output stages, gain may sometimes be sacrificed to achieve the appropriate interface with the external world. Under these circumstances, the gain of the internal amplifier stage is increased so that the overall amplification satisfies design requirements.

We devote the next four sections to describing each of the four stages in the architecture depicted in Fig. 14-1. The focus is on BJT stages; FET Op-Amps are discussed in Sec. 14-10. The illustrative problems in these sections use numerical data for the 741-type Op-Amp. Introduced in 1966 by Fairchild Semiconductor, Inc., this Op-Amp is now fabricated by many manufacturers. The 741 employs the two-stage architecture of Fig. 14-1 and, at present, is probably the most widely used Op-Amp.

14-2 THE GAIN STAGE WITH ACTIVE LOAD

The interior stage of the Op-Amp is required to have a large voltage gain. In Sec. 10-10 we showed that the gain of a BJT stage depends on both the collector resistance used and the value of β_o of the transistor. Often, high-β_o composite transistors such as the Darlington pair and *CC-CE* cascade (Sec. 10-17) are used in the gain stage. Large values of collector resistance, however, cannot be fabricated conveniently on a chip. Even if this were not the case, the resultant voltage levels are impractical. For example, there is a 100-V drop across a 100-kΩ resistance carrying a dc current of 1 mA. Consequently, power supplies in excess of 120 V are required to obtain output signals of 20 V peak to peak. Clearly this is undesirable.

To overcome this limitation, *active loads* are used. An active load is a current source whose output resistance is used in place of the collector resistance as illustrated in Fig. 14-2. The current mirror (active load) in Fig. 14-2 is formed from a pair of lateral *pnp* transistors $Q3$ and $Q4$ (Sec. 5-3). The use of *pnp* transistors provides the appropriate direction of collector current in the *npn* transistor $Q1$ and the high-output resistance. Since nearly all of the current

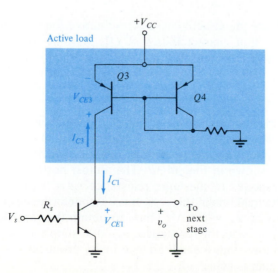

FIGURE 14-2

A common-emitter stage in which the current mirror is used as an active load.

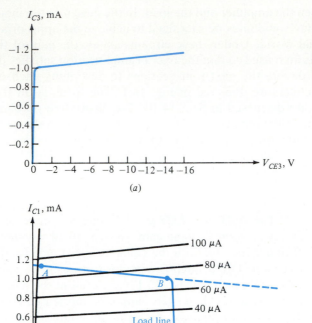

exists in the source and not in the output resistance, the need for high-voltage supplies is obviated. It is also evident that the output resistance of the source is the collector resistance in the small-signal model.

The Load Line

To demonstrate the effectiveness of the active load, let us construct the load line on the output characteristics of $Q1$. Assume that the current source is designed to give a current of 1 mA and that the *pnp* transistors have an Early voltage $V_A = 100$ V. The supply voltage is 15 V. Kirchhoff's laws indicate that

$$-I_{C3} = I_{C1} \quad \text{and} \quad V_{CE1} = V_{CC} - V_{CE3} = 15 - V_{CE3} \quad (14\text{-}1)$$

The volt-ampere characteristic of the current mirror is displayed in Fig. 14-3*a* and reflects the Early voltage. The curve in Fig. 14-3*a* is that of a nonlinear resistance, and, following the method described in Sec. 4-11, we can construct the load line shown in Fig. 14-3*b*. The almost horizontal load line between A and B corresponds to the large resistance $r_o = V_A/I_{C3} = 100/1 = 100$ kΩ. Indeed, if the load line in this region is extended (the dashed line), it intersects the V_{CE1} axis at $V_A = 100$ V. Thus, to obtain the same load line (between A and B) using a collector resistance, a 100-V supply is required. Inspection of Fig. 14-3*b* shows that a small change in I_{B1} produces a large change in V_{CE1}; hence high gain is achieved.

FIGURE 14-4
The equivalent circuit at low frequencies of the circuit in Fig. 14-2.

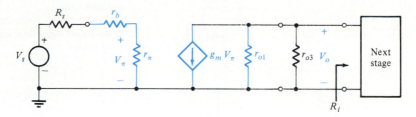

The Small-Signal Model The small-signal model of the stage containing $Q1$ and the active load is shown in Fig. 14-4. Since the collector resistance r_{o3} of the *pnp* transistor is often comparable in value to the output resistance r_{o1} of the *npn* transistor, both must be included in the model. The effective load on this stage is the parallel combination of r_{o1}, r_{o3}, and the input resistance R_i of the next stage. Observe that R_i must also be a large resistance to minimize loading. If this is not the case, the advantage of the active load is counteracted and the gain of the stage reduced. In Fig. 14-1, the load on the gain stage is the input resistance of the emitter follower. This high-input resistance helps to minimize the loading effect on the gain stage.

Example 14-1 A simplified schematic of the gain stage of the 741 Op-Amp is displayed in Fig. 14-5. The signal source and source resistance shown comprise the Thévenin equivalent of the differential amplifier which serves as the input to the gain stage. The *pnp* transistor $Q13B$ is part of the current-source active load on the stage. The transistors are numbered to correspond to the circuit diagram of the entire amplifier in Fig. 14-19. The transistors are biased at $I_{C16} = 16\ \mu A$ and $I_{C17} = I_{C13B} = 550\ \mu A$. All transistors have $\beta_o = 250$, and the Early voltages are 100 and 50 V for *npn* and *pnp* devices, respectively. Assume $r_b = 0$ for all BJTs.

Determine the voltage gain V_o/V_s, the input resistance R_{i2}, and the output resistance R'_{o2} for this stage.

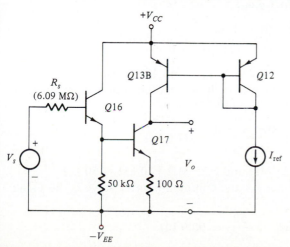

FIGURE 14-5
The gain stage of a 741-type Op-Amp. The transistors $Q16$ and $Q17$ form a *CC-CE* composite, and $Q13B$ and $Q12$ are the active load. The series combination $V_s - R_s$ represents the Thévenin equivalent of the differential stage.

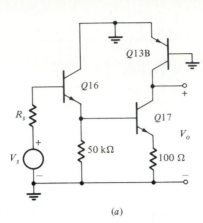

FIGURE 14-6
(*a*) The ac representation of Fig. 14-5. (*b*) The small-signal equivalent circuit in (*a*) valid at low frequencies.

(*a*)

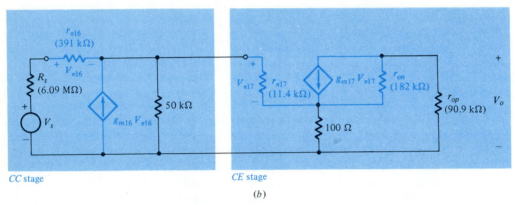

CC stage   CE stage

(*b*)

Solution

The gain stage is a *CC-CE* cascade. The small-signal schematic circuit diagram of this amplifier is shown in Fig. 14-6*a*. The incremental model of the circuit in Fig. 14-6*a* is displayed in Fig. 14-6*b*. Following the procedure described in Sec. 10-16, we first obtain the gain A_{V1} of the emitter follower (common-collector stage). Use of Table 10-3 gives

$$A_{V1} = \frac{(250 + 1)50}{6090 + 391 + (250 + 1)50} = 0.659$$

in which $r_\pi = \beta_o/g_m = \beta_o V_T/I_{C16} = 250 \times \frac{25}{16} = 391$ kΩ is used.

The output resistance of this stage (which acts as the source resistance of the common-emitter stage) is, from Eq. (10-52),

$$R_o = 50 \,\| \, \frac{6090 + 391}{250 + 1} = 17.0 \text{ k}\Omega$$

Note that r_o of $Q16$ is $\frac{100}{16} = 6.25$ MΩ and 6.25 M$\Omega \,\|\, 50$ k$\Omega \approx 50$ kΩ.

The effective collector resistance for the common-emitter stage is the value of r_o of the *pnp* load transistor, or

$$R_C = r_{op} = \frac{50}{0.55} = 90.9 \text{ k}\Omega$$

The output resistance r_{on} of the *npn* transistor $Q17$ is

$$r_{on} = \frac{100}{0.55} = 181.8 \text{ k}\Omega$$

Use of the gain equation given in Table 10-3 for a *CE* stage with an emitter resistance yields

$$A_{V2} = \frac{-250 \times 90.9\|181.8}{17.0 + 11.4 + 0.1 + 250 \times 181.8 \times 0.1/(181.8 + 90.9)} = -334$$

The overall gain of the stage is

$$A_2 = A_{V1}A_{V2} = 0.659 \,(-334) = -220$$

The output resistance of the *CC-CE* cascade is the output resistance of the common-emitter stage. Using the result in Table 10-3B, we obtain

$$R'_{o2} = 90.9 \,\|\, \left[181.8 \left(1 + \frac{250 \times 0.1}{17.0 + 11.4 + 0.1} \right) \right] = 71.8 \text{ k}\Omega$$

The input resistance R_{i2} is the resistance seen looking into the common collector stage. The effective emitter resistance of this stage is 50 kΩ in parallel with the input resistance R_x of the common-emitter stage. Thus, use of Table 10-3 gives

$$R_x = 11.4 + (250 + 1)0.1 = 36.5 \text{ k}\Omega$$

and

$$R_E = 50 \,\|\, 36.5 = 21.1 \text{ k}\Omega$$

Hence

$$R_{i2} = 391 + (250 + 1)21.1 = 5.69 \text{ M}\Omega$$

Limitations of *pnp* Current Sources   Lateral *pnp* transistors have lower values of β_F, smaller current-carrying capacity, and lower Early voltages V_A than do *npn* transistors. Consequently, the performance of *pnp* current mirrors is somewhat inferior to that of *npn* sources. The lower value of V_A results in a lower output resistance. This limitation can be overcome by using a Widlar, Wilson, or cascode current source, each of which has an output resistance higher than that in a simple mirror.

Circuit techniques are usually employed to compensate for the two other limitations. One such circuit is the composite *pnp-npn* current source depicted in Fig. 14-7. The basic mirror is formed by the *pnp* transistors $Q3$ and $Q4$, and the *npn* transistors $Q1$ and $Q2$ act as a current amplifier. If A_2/A_1 is the ratio of the emitter areas of $Q2$ and $Q1$, then

$$I_o \approx I_R \left(1 + \frac{A_2}{A_1} \right) \tag{14-2}$$

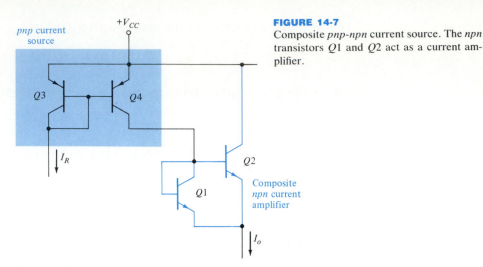

FIGURE 14-7
Composite *pnp-npn* current source. The *npn* transistors $Q1$ and $Q2$ act as a current amplifier.

provided the *npn* transistors have $\beta_F \gg A_2/A_1$ (Prob. 14-4). The significance of Eq. (14-2) is that the output current can be increased but only $Q1$ and $Q2$ carry the large current. For example, suppose $I_R \approx 100\ \mu A$, a practical limit for a lateral *pnp* transistor, then making $A_2/A_1 = 4$ results in $I_o = 500\ \mu A$. However, $Q2$, the larger area *npn* transistor, carries most of this current ($\approx 400\ \mu A$).

Recent advances in fabrication technology have made it possible to construct *npn* and *pnp* transistors that have complementary characteristics. Fabrication of such devices is more expensive as several additional processing steps are required. Analog Devices, Inc., employs this fabrication technology in some of their Op-Amps and other analog IC products.

14-3   THE DIFFERENTIAL STAGE

The basic structure of an IC differential-amplifier stage is shown in Fig. 14-8. The active element in Fig. 14-8 is a BJT (or FET) or a compound stage such as a cascode configuration or a Darlington pair (Sec. 10-17). Three important characteristics of the differential input stage are the common-mode rejection ratio *CMRR*, the input differential resistance R_{id}, and the differential-mode gain A_{DM}.

The Common-Mode Rejection Ratio

The common-mode rejection ratio of a BJT differential stage was derived in Eq. (10-90) and is repeated as Eq. (14-3) for convenience.

$$CMRR = 1 + 2g_m R_E \qquad (14\text{-}3)$$

where it is assumed $r_\pi \gg R_s$ and $\beta_o \gg 1$. Inspection of Eq. (14-3) clearly indicates that R_E must be made large if a high *CMRR* is to be achieved. In Fig. 14-8, R_E is identified as the output resistance of the current source used to bias the active elements. If a simple current mirror is used and the active device is a BJT, then $R_E = V_A/I_O$ and $g_m = I_O/2V_T$. (Recall that each half of the differential pair draws one-half of the source current.) Use of these values in Eq. (14-3) yields

FIGURE 14-8

(*a*) Basic topology of a differential amplifier. (*b*) An emitter-coupled implementation of (*a*).

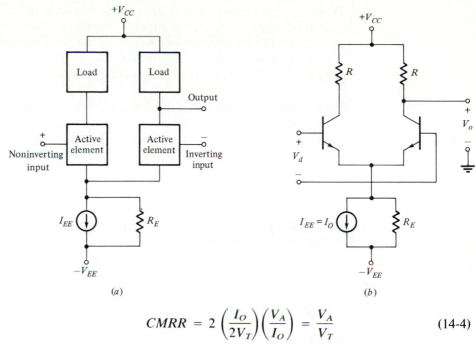

(*a*)

(*b*)

$$CMRR = 2 \left(\frac{I_O}{2V_T} \right) \left(\frac{V_A}{I_O} \right) = \frac{V_A}{V_T} \tag{14-4}$$

For an *npn* transistor with $V_A = 100$ V, $CMRR = 100$ V/25 mV $= 4000$, or $CMRR = 72$ dB. This is (approximately) the minimum acceptable common-mode rejection ratio listed in Table 14-1. To increase the common-mode rejection ratio, the effective Early voltage, that is, the output resistance of the current source, must be increased. Operational amplifiers with common-mode rejection ratios of 80 to 90 dB usually employ Wilson, Widlar, or cascode current sources.

Input Resistance R_{id} The differential-input resistance R_{id} of the differential stage is the input resistance of the Op-Amp. For approximation of the input of an ideal voltage-controlled voltage source, R_{id} must be large. The differential-input resistance is

$$R_{id} \approx 2r_\pi = \frac{2\beta_o}{g_m} = \frac{2\beta_o V_T}{I_C} \tag{14-5}$$

In Eq. (14-5) we observe that a large R_{id} requires bias currents that are quite small. For example, for a transistor having $\beta_o = 250$, a collector current $I_C = 12.5$ μA is needed for $R_{id} = 1$ MΩ.

Two techniques commonly employed to increase R_{id} are the use of FET input stages and the use of high-β transistors in the differential pair. The use of FET differential stages, discussed in Sec. 14-10 (BIFET and BIMOS processes), result in input resistance values greater than 10^{12} Ω. This technique is employed in the Analog Devices AD5449 Op-Amp.

When the effective β_o of the active elements is increased by use of a Darlington pair composite transistor, the input impedance increases significantly.

However, limitations such as frequency response often preclude Darlington pair input stages. The use of super-β transistors (Sec. 5-3) in the differential pair results in high input resistances at current levels usually encountered. For a super-β transistor ($\beta_o = 5000$) biased at $I_C = 12.5~\mu A$, $R_{id} = 20~M\Omega$. Clearly, reduction in the current results in increased R_{id}.

The Differential-Mode Gain A_{DM}   Since the input stage of the Op-Amp is one of the two gain stages, it is desirable to make A_{DM}, the differential-mode gain, large. Consequently, active loads are also employed in these stages. From Eq. (10-86), we obtain

$$|A_{DM}| = \left|\frac{v_o}{v_{DM}}\right| = \frac{\beta_o R_L}{r_\pi} = g_m R_L \qquad (14\text{-}6)$$

Recalling that $v_{DM} = V_d/2$ [Eq. (10-91)], we can express the differential-output voltage of the amplifier as

$$V_o = \frac{g_m R_L}{2} V_d \qquad (14\text{-}7)$$

where R_L is the parallel combination of the active-load resistance R_C and the output resistance r_o of the active element. Expression of g_m in terms of the bias current yields

$$|V_o| = \frac{I_C}{2V_T}|V_d| R_L = \frac{I_O}{4V_T}|V_d| R_L \qquad (14\text{-}8)$$

Equation (14-8) indicates that the effective transconductance of a differential stage is one-fourth that of a single BJT biased by a collector current of I_O.

Example 14-2

The input stage of the 741 Op-Amp is depicted in Fig. 14-9. The *npn-pnp* transistor combinations $Q1$ and $Q3$ and $Q2$ and $Q4$ form the active element of the differential pair. The active loads are provided by the three-transistor current source $Q5$, $Q6$, and $Q7$. Transistors $Q8$ and $Q9$ form a *pnp* current mirror used for base-biasing and ensure that the transistors remain in the active region when no input signal is applied. Transistors $Q1$ to $Q6$ are biased at $I_C = 9.5~\mu A$, and all have $\beta_o = 250$. The Early voltages are 100 and 50 V for the *npn* and *pnp* transistors, respectively.

Determine the gain V_o/V_d, the differential-input resistance R_{id}, and the output resistance R_o. Use the small-signal schematic representation of the differential stage in Fig. 14-10.

Solution

A convenient method for analyzing this circuit is to obtain the Norton equivalent of the stage. Conversion of the Norton equivalent to a Thévenin equivalent gives both the output resistance and the voltage gain. In Fig. 14-10, KCL requires that $I_o = I_{c4} + I_{c6}$. The composite active element $Q2$ and $Q4$ can be considered as an emitter follower ($Q2$) driving $Q4$, connected as a common-base stage, that is, a cascode circuit. This is illustrated in the equivalent circuit

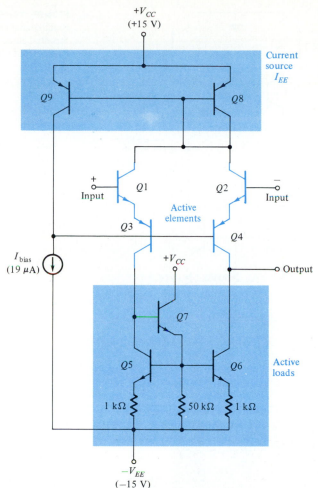

FIGURE 14-9
The input stage of a 741-type Op-Amp. The active element in the differential pair is formed by the CC-CB connected transistors $Q1$-$Q3$ and $Q2$-$Q4$.

in Fig. 14-11a. The input resistance of the common-base stage is $r_{\pi 4}/(1 + \beta_o) \approx 1/g_{m4}$, and the emitter follower can be represented by its Thévenin equivalent $-V_d/2$ in series with $1/g_{m2}$ as shown in Fig. 14-11b. Note that under open-circuit conditions ($R_E \rightarrow \infty$), the gain of the common-collector stage is unity and with $R_s = 0$, its output resistance is $r_{\pi 2}/(1 + \beta_o) \approx 1/g_{m2}$. The current I_{e4}, from KVL for the loop, is

$$I_{e4} = \frac{-V_d/2}{1/g_{m2} + 1/g_{m4}} = \frac{-g_m V_d}{4}$$

since g_{m2} and g_{m4} are equal because $|I_{c2}| = |I_{c4}|$. Assuming that $\beta_o \gg 1$, $I_{c4} = -I_{e4} = g_m V_d/4$. By the symmetry of the circuit $I_{c3} = g_m V_d/4$, and, as I_{c3} is the current in the active load, $I_{c6} = I_{c3} = g_m V_d/4$. Hence the output current I_o is

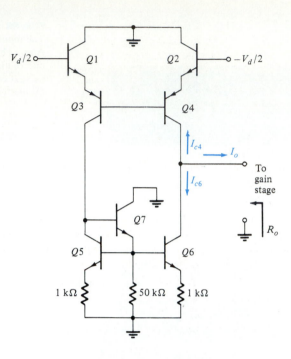

FIGURE 14-10
Small-signal representation of the differential stage of a 741-type Op-Amp. The current source with gain ($Q5$, $Q6$, and $Q7$) is the active load.

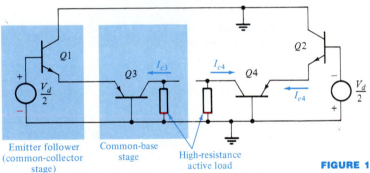

Emitter follower
(common-collector
stage)

Common-base
stage

High-resistance
active load

(a)

$1/g_{m2}$

$\dfrac{V_d}{2}$

$V_{\pi 4}$ $r_{\pi 4}$

$g_{m4}V_{\pi 4}$

Active
load

Thévenin equivalent of
common-collector stage

(b)

FIGURE 14-11
(a) The circuit in Fig. 14-10 redrawn to show the *CC-CB* connection. (b) The equivalent circuit of the common-base portion of the amplifier. The common-collector stage is replaced by its Thévenin equivalent $V_d/2$ in series with $1/g_{m2}$. (*Note*: The half-circuit concept presented in Sec. 10-19 is used.)

$$I_o = \frac{g_m V_d}{4} + \frac{g_m V_d}{4} = \frac{g_m V_d}{2}$$

Evaluation gives

$$g_m = \frac{9.5}{25} = 0.38 \text{ m℧} \quad \text{and} \quad I_o = 0.19 V_d \quad \text{mA}$$

The output resistance of the stage, as seen in Fig. 14-10, is the parallel combination of the output resistances R_{o4} and R_{o6} of $Q4$ and $Q6$, respectively. The output resistance of the common-collector stage ($Q2$) acts as the emitter resistance of $Q4$ as displayed in Fig. 14-12. Both R_{o4} and R_{o6} are given by

$$R_o = r_o \left(1 + \frac{\beta_o R_E}{r_\pi + R_E} \right)$$

as $R_s = r_b = 0$. The parameter values needed to compute R_{o4} and R_{o6} are

$$g_{m2} = g_{m4} = g_{m6} = \frac{I_c}{V_T} = \frac{9.5}{25} = 0.38 \text{ m℧} \qquad \frac{1}{g_{m2}} = \frac{1}{0.38} = 2.63 \text{ k}\Omega$$

$$r_{o4} = \frac{V_A}{I_C} = \frac{50}{9.5} = 5.26 \text{ M}\Omega \qquad r_{o6} = \frac{100}{9.5} = 10.5 \text{ M}\Omega$$

$$r_{\pi4} = r_{\pi6} = \frac{\beta_o}{g_{m2}} = \frac{250}{0.38} = 658 \text{ k}\Omega$$

Thus

$$R_{o4} = 5.26 \left(1 + \frac{250 \times 2.63}{658 + 2.63} \right) = 10.5 \text{ M}\Omega$$

$$R_{o6} = 10.5 \left(1 + \frac{250 \times 1}{658 + 1} \right) = 14.5 \text{ M}\Omega$$

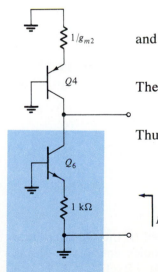

and

$$R_o = R_{o4} \| R_{o6} = 10.5 \| 14.5 = 6.09 \text{ M}\Omega$$

The Thévenin equivalent voltage V_o is

$$V_o = I_o R_o = 0.19 V_d \times 6090 = 1157 V_d$$

Thus the gain of the differential-input stage is

$$A_1 = \frac{V_o}{V_d} = 1157$$

FIGURE 14-12

The ac representation of the differential stage used to compute the output resistance R_o.

Active load

The differential-mode input resistance R_{id} of this stage is twice the input resistance of the emitter-follower stage $Q2$ [see Eq. (14-5)]. The emitter resistance of this stage is $1/g_{m4}$, the input to the common-base stage ($Q4$). Hence

$$R_{id} = 2\left(r_{\pi 2} + \frac{\beta_{o1} + 1}{g_{m4}}\right) \approx 2(r_{\pi 2} + r_{\pi 4})$$

Evaluation gives

$$R_{id} = 2\ (658 + 658) = 2.63\ \text{M}\Omega$$

The overall gain of the first two stages is $A_1 A_2 = 220 \times 1157 = 2.54 \times 10^5$. This value is effectively the open-loop gain of the Op-Amp since both the level-shifting stage and the emitter-follower output stage have virtually unity voltage gain. Manufacturers usually specify the minimum value of open-loop gain as 2×10^5. Differences are attributed to fabrication tolerances in bias currents, β_o, and the Early voltages of the transistor. In addition, parasitic effects associated with the substrate tend to decrease the gain of the amplifier. Performance data for 741-type and other Op-Amps discussed in this chapter are summarized in Table 14-2.

TABLE 14-2 Typical Performance of Selected Op-Amp Types

	Type 741 (two-stage architecture)	LM 118 (three-stage architecture)	LM 108 (super-beta)	AD 611 (BIFET)	AD 507 K (wide-band)
Input offset voltage (mV)	≤5	≤4	≤2	≤0.5	≤5
Bias current (nA)	≤500	≤250	≤2	≤0.025	≤15
Offset current (nA)	≤200	≤50	≤0.4	≤0.010	≤15
Open-loop gain (dB)	106	100	95	98	100
Common-mode rejection ratio (dB)	80	90	95	80	100
Input resistance (MΩ)	2	5	100	10^6	300
Slew rate (V/μs)	0.5	≥50	0.2	13	35
Unity-gain bandwidth (MHz)	1	15	1	2	35
Full-power bandwidth (kHz)	10	1000	4	200	600
Settling time (μs)	1.5	4	1	3	0.9

14-4 DC LEVEL SHIFTING Since no coupling capacitors can be used (if the Op-Amp is to operate down to dc), it may be necessary to shift the quiescent voltage of one stage before applying its output to the following stage. Level shifting is also

FIGURE 14-13
Level shifters using an emitter follower.

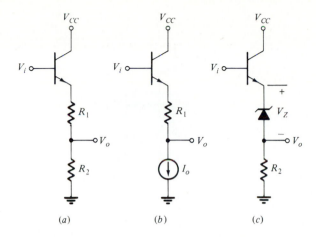

(a) (b) (c)

required in order for the output to be close to zero in the quiescent state (no input signal). The input resistance of the level-shifting stage should be high to prevent loading of the gain stage. Similarly, it is desirable that the output resistance be low to effectively drive the output stage. An emitter follower (Fig. 14-13) can serve as such a buffer and, simultaneously, as a voltage shifter. If the output V_o is taken at the emitter then the change in level is $V_o - V_i = -V_{BE} \approx -0.7$ V. If this shift is not sufficient, the output may be taken at the junction of two resistors in the emitter leg, as shown in Fig. 14-13a. The voltage shift is then increased by the drop across R_1. The disadvantage with this arrangement is that the signal voltage suffers an attenuation $R_2/(R_1 + R_2)$. This difficulty is avoided by replacing R_2 by a current source I_o, as shown in Fig. 14-13b. The level shift is now $V_o - V_i = -(V_{BE} + I_oR_1)$, and there is no ac attenuation for a very high resistance current source.

Another voltage translator is indicated in Fig. 14-13c, where an avalanche diode is used. Then $V_o - V_i = -(V_{BE} + V_Z)$. A number of forward-biased pn diodes may also be used in place of the Zener diode. If the dynamic resistance of the Zener diode (or of the string of diodes) is small compared with R_2, the attenuation of the signal may be neglected.

The V_{BE} Multiplier   An interesting voltage source easily fabricated in monolithic form is displayed in Fig. 14-14a. If the base current is negligible compared with the current in R_3 and R_4, the circuit acts as a "V_{BE} multiplier" because

$$V = \frac{V_{BE}}{R_4}(R_3 + R_4) = V_{BE}\left(1 + \frac{R_3}{R_4}\right) \tag{14-9}$$

This voltage source is used in place of R_1 in Fig. 14-13a as illustrated in Fig. 14-14b. The change in dc level $V_i - V_o$ is expressible as

$$V_i - V_o = V_{BE}\left(2 + \frac{R_3}{R_4}\right) \tag{14-10}$$

FIGURE 14-14
(*a*) A V_{BE} multiplier is a voltage source V. (*b*) The circuit in (*a*) used as a level shifter.

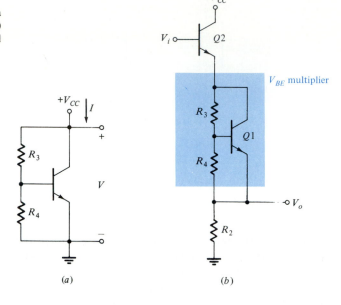

(*a*)   (*b*)

The small-signal voltage gain, assuming $\beta_o \gg 1$, is

$$A_V = \frac{v_o}{v_i} \approx \frac{g_{m2}R_2}{1 + g_{m2}R_2 + g_{m2}(R_3 + R_4)/(1 + g_{m1}R_4)} \approx 1 \qquad (14\text{-}11)$$

for $g_{m2}R_2 \gg 1$ and $R_2 \gg (R_3 + R_4)/(1 + g_{m1}R_4)$.

One advantage of the circuit in Fig. 14-14*b* is that the dc level shift depends on the accurately controlled resistor ratio R_3/R_4 and is achieved at unity gain. The major disadvantage of the circuit is that the temperature dependence of $V_o - V_i$ is the same as that for V_{BE} (-2.2 mV/°C). The level-shifting stage of a 741-type Op-Amp is usually a simple emitter follower.

14-5 OUTPUT STAGES The output stage of an Op-Amp must be capable of supplying the external load current and must have a low output resistance. This stage must also provide a large output voltage swing; ideally, the peak-to-peak output voltage should approach the total supply voltage $V_{CC} + V_{EE}$. A common configuration for the output stage that possesses these features is the *complementary emitter follower* shown in Fig. 14-15*a*. If the input signal V_i goes positive, the *npn* transistor $Q1$ acts as a source of supply current to the load R_L and the *pnp* transistor $Q2$ is cut off. Alternatively, if V_i becomes negative, $Q1$ is cut off and $Q2$ acts as sink to remove current from the load, that is, to decrease I_L. Hence, if V_i is a sinusoid, $Q1$ drives the load during positive half-cycles and $Q2$ acts during the negative half-cycles. As each transistor is ON for only half the time, the output voltage swing is twice the value that can be achieved by a single-stage emitter follower.

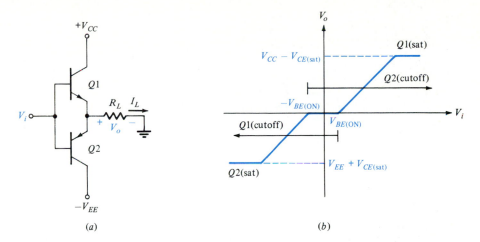

FIGURE 14-15
Complementary emitter-follower output stage: (*a*) Circuit diagram, (*b*) voltage transfer characteristic. The horizontal portion of the characteristic in the vicinity of the origin introduces crossover distortion.

There is a fundamental difficulty with the circuit in Fig. 14-15*a* because the output voltage remains virtually zero until $V_i = V_{BE(ON)}$. This phenomenon is known as *crossover distortion* and is readily observed in the transfer characteristic illustrated in Fig. 14-15*b*. [Actually, the output becomes nonzero for $V_i \approx V_{BE(cut\ in)} = V_\gamma \approx 0.5$ V. However, the current in the transistor is so small that the output voltage is negligible.]

Crossover distortion can be virtually eliminated by applying a bias voltage $V > 2V_\gamma$ between the two bases so that a small current exists in the transistors in the quiescent state. One common technique is to employ a pair of series-connected *pn* diodes as displayed in Fig. 14-16. It is common to fabricate diodes *D*1 and *D*2 as diode-connected BJTs (Sec. 5-6). The transfer characteristic of the circuit in Fig. 14-16*a* is displayed in Fig. 14-16*b*, in which we see that the crossover distortion is virtually eliminated. However, the characteristic does not pass through the origin, and, with $V_i = 0$, $V_o \neq 0$. Recalling that V_i is obtained from the level-shift stage, we obtain $V_o = 0$ with zero input signal by making the quiescent value of $V_i \approx -V_{BE2}$.

The circuit in Fig. 14-17 is also used to eliminate crossover distortion. The block labeled V is the V_{BE}-multiplier circuit in Fig. 14-14*b* and is used in place of diodes *D*1 and *D*2 in Fig. 14-16*a*. The output voltage of this block is designed to apply approximately 1.1 V between the two bases. Hence both *Q*1 and *Q*2 are conducting slightly under quiescent conditions.

The output stages in Figs. 14-16*a* and 14-17 are both employed in commercially fabricated 741-type Op-Amps. The basic output-stage configuration is displayed in Fig. 14-18. Transistors *Q*14 and *Q*20 form the complementary emitter follower. The small resistances R_6 and R_7 provide for current-limiting

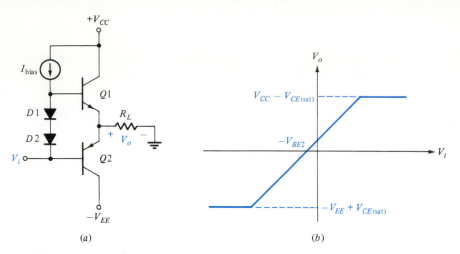

(a)   (b)

FIGURE 14-16
(a) The use of series-connected diodes in the complementary emitter follower. (b) The voltage transfer characteristic exhibits no crossover distortion. However, when $V_i = 0$, V_o is not zero but equals $-V_{BE2}$.

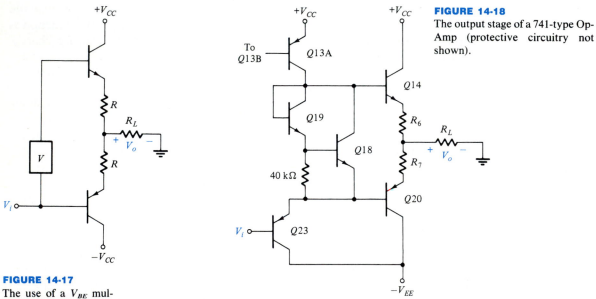

FIGURE 14-18
The output stage of a 741-type Op-Amp (protective circuitry not shown).

FIGURE 14-17
The use of a V_{BE} multiplier (the box identified by V) to eliminate crossover distortion in the complementary emitter follower.

at the output. The Darlington pair $Q18$ and $Q19$ are used in place of diodes $D1$ and $D2$. This arrangement is favored over two diode-connected BJTs in series since the Darlington pair can be fabricated in a smaller area. The current source in Fig. 14-16a is realized, in part, by $Q13B$. The complete circuit diagram of the 741-type Op-Amp is displayed in Fig. 14-19.

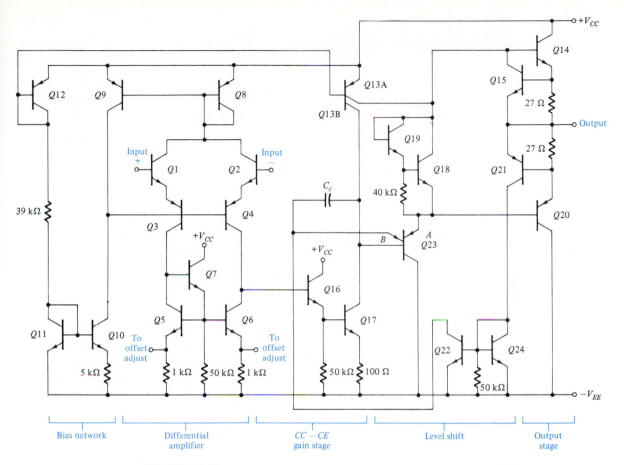

FIGURE 14-19
The schematic diagram of the 741-type Op-Amp.

14-6 OFFSET VOLTAGES AND CURRENTS In previous sections we observed that the ideal Op-Amp is perfectly balanced, that is, $V_o = 0$ when $V_1 = V_2 = 0$. A real Op-Amp exhibits an imbalance caused by a mismatch of the input transistors. This mismatch results in unequal bias currents in the input terminals and unequal base-emitter voltages (Fig. 14-20). Often, an input offset voltage applied between the two input terminals is required to balance the amplifier.

We are concerned in this section with the dc error voltages and currents, representing deviations from the ideal, that can be measured. In addition, we describe the important specifications of Op-Amp performance. The idealized model of the Op-Amp (Fig. 10-41) must be modified to include the offset voltage and bias currents as depicted in Fig. 14-20b. The main specifications used to describe Op-Amp performance are as follows:

FIGURE 14-20
(a) The input bias currents I_{B1} and I_{B2} and the offset voltage V_{io}. (b) The Op-Amp equivalent circuit showing bias currents and offset voltage.

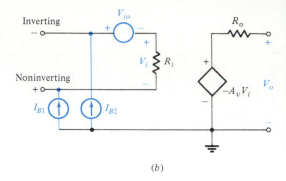

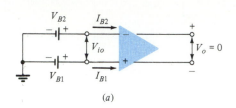

(a)

(b)

Input bias current The input bias current is one-half the sum of the separate currents entering the two input terminals of a balanced amplifier, as shown in Fig. 14-20. Since the input stage is of the type shown in Fig. 14-9, the input bias current is $I_B \equiv (I_{B1} + I_{B2})/2$ when $V_o = 0$.

Input offset current The input offset current I_{io} is the difference between the separate currents entering the input terminals of a balanced amplifier. As shown in Fig. 14-20, we have $I_{io} \equiv I_{B1} - I_{B2}$ when $V_o = 0$.

Input offset current drift The input offset current drift $\Delta I_{io}/\Delta T$ is the ratio of the change of input offset current to the change of temperature.

Input offset voltage The input offset voltage V_{io} is that voltage which must be applied between the input terminals to balance the amplifier.

Input offset voltage drift The input offset voltage drift $\Delta V_{io}/\Delta T$ is the ratio of the change of input offset voltage to the change in temperature.

Output offset voltage The output offset voltage is the difference between the dc voltages present at the two output terminals (or at the output terminal and ground for an amplifier with one output) when the two input terminals are grounded.

Input common-mode range The common-mode input-signal range is that range within which a differential amplifier remains linear.

Input differential range This is the maximum difference signal that can be applied safely to the Op-Amp input terminals.

Output voltage range This is maximum output swing that can be obtained without significant distortion (at a given load resistance).

Full-power bandwidth This is the maximum frequency at which a sinusoid whose size is the output voltage range is obtained.

Power-supply rejection ratio The power-supply rejection ratio $PSRR$ is the ratio of the change in input offset voltage to the corresponding change in one power-supply voltage, with all remaining power voltages held constant.

Slew rate The slew rate is the time rate of change of the closed-loop amplifier output voltage under large-signal conditions.

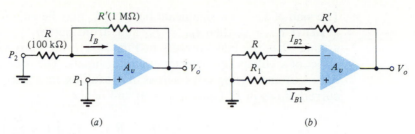

FIGURE 14-21
Illustrative example. If P_1 in (a) is disconnected from ground and a signal voltage is applied to P_1, a noninverting Op-Amp is under consideration. Alternatively, if P_2 is removed from ground and a signal is applied to P_2, the circuit becomes an inverting Op-Amp stage.

Example 14-3

(a) Inverting and noninverting Op-Amp stages have the same configuration (Fig. 14-21a) with no input signal voltages applied. Assuming the input offset voltage $V_{io} = 0$, find the output voltage due to the input bias current when $I_{B1} = I_{B2} = I_B = 100$ nA. (b) How can the effect of the bias current be eliminated so that $V_o = 0$? (c) Using the result in part b, calculate V_o, assuming $I_{B1} - I_{B2} = I_{io} = 20$ nA. (d) Assuming $I_{io} = 0$, determine V_o when $V_{io} = 5$ mV. (e) Find the value of V_o when $I_{io} = 20$ nA and $V_{io} = 5$ mV.

Solution

(a) As mentioned in Sec. 10-21 for very large values of A_v there exists a short circuit between the two input terminals. Hence there is no current in R. The current I_B must exist in R' and hence $V_o = I_B R'$.

Using the value $I_B = 100$ nA, we obtain

$$V_o = 100 \times 10^{-9} \times 10^6 = 0.1 \text{ V} = 100 \text{ mV}$$

(b) Add a resistor R_1 between the noninverting terminal and ground, as indicated in Fig. 14-21b. If $V_o = 0$, then R and R' are in parallel ($R \parallel R' = R_p$) and the voltage from the inverting terminal to ground is $-I_{B2}R_p$. Since there is zero voltage between input terminals, $-I_{B2}R_p$ must equal $-I_{B1}R_1$ or (for $I_{B1} = I_{B2}$)

$$R_1 = R_p = \frac{RR'}{R + R'} = \frac{100 \times 1000}{1100} = 90.9 \text{ k}\Omega$$

If $I_{B1} \neq I_{B2}$, we must choose $I_{B1}R_{p1} = I_{B2}R_p$.

(c) In Fig. 14-21b set $I_{B2} = I_{B1} - I_{io}$. In part b it is demonstrated that due to I_{B1} entering both the inverting and noninverting terminals, the output is $V_o = 0$. Applying superposition to the two current sources I_{B1} and I_{io}, we may now set $I_{B1} = 0$ and find the effect of I_{io}. Since the drop across R_1 is $I_{B1}R_1 = 0$ and the two input terminals are at the same potential, the drop across R is 0 and the current R is also 0. Hence I_{io} flows in R' and

$$V_o = -I_{io}R'$$

For the numerical values given,

$$V_o = -20 \times 10^{-9} \times 10^6 \text{ V} = -20 \text{ mV}$$

The sign of V_o is not significant because I_{io} may be either positive or negative.

(d) If $I_{io} = 0$, then $I_{B1} = I_{B2}$ and, from part b, $V_o = 0$. Hence we may assume that the bias currents in Fig. 14-21b are zero and consider only the effect of a voltage V_{io} between input terminals. The drop across R_1 is zero (for $I_{B1} = 0$) and V_{io} appears across R, resulting in a current V_{io}/R. This same current flows in R' (since $I_{B2} = 0$) and hence

$$V_o = \frac{V_{io}}{R}(R + R') = V_{io}\left(1 + \frac{R'}{R}\right)$$

Evaluation yields, $V_o = \pm(5)(1 + 10) = \pm 55$ mV. Note that (for the indicated parameter values) the effect of V_{io} is comparable to that due to I_{io}.

(e) Use of superposition gives

$$V_o = -I_{io}R' + V_{io}\left(1 + \frac{R'}{R}\right)$$

If all resistance values are divided by a factor M, the output due to V_{io} is not altered, whereas the component of V_o caused by I_{io} is divided by M. The inverting and also the noninverting gains depend only on resistance ratios and, hence, are independent of the factor M.

Universal Balancing Techniques When we use an operational amplifier, it is often necessary to balance the offset voltage. This means that we must apply a small dc voltage in the input so as to cause the dc output voltage to become zero. The techniques shown here allow offset-voltage balancing with regard to the internal circuitry of the amplifier. The circuit shown in Fig. 14-22a supplies a small voltage effectively in series with the noninverting input terminal in the range $\pm V[R_2/(R_3 + R_2)] = \pm 15$ mV if ± 15-V supplies are used and $R_3 = 100$ kΩ, $R_2 = 100$ Ω. This circuit is useful for balancing inverting amplifiers even when the feedback element R' is a capacitor or a nonlinear element. If the Op-Amp is used as a noninverting amplifier, the circuit in Fig. 14-22 is used for balancing the offset voltage.

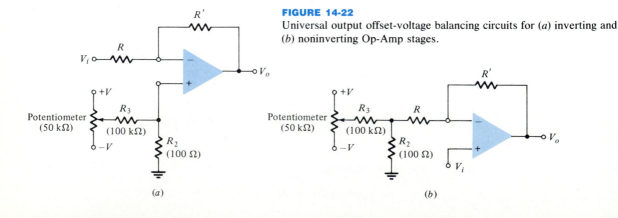

FIGURE 14-22
Universal output offset-voltage balancing circuits for (a) inverting and (b) noninverting Op-Amp stages.

(a)   (b)

FIGURE 14-23
System for measuring V_{io}, I_{B1}, and I_{B2}.

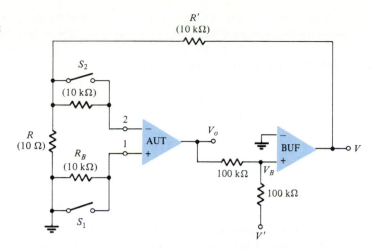

FIGURE 14-23
System for measuring V_{io}, I_{B1}, and I_{B2}.

14-7 MEASUREMENT OF OPERATIONAL-AMPLIFIER PARAMETERS
In this section we describe practical methods of measuring some of the important parameters of Op-Amps. Specifically, we examine (1) input offset voltage V_{io}, (2) input bias current I_B and input offset current I_{io}, (3) open-loop voltage gain A_{OL}, (4) common-mode rejection ratio, and (5) slew rate. In the circuits discussed in this section the Op-Amp whose parameters are to be determined is labeled AUT (*amplifier under test*). The AUT is cascaded with another Op-Amp labeled BUF (*buffer*) which increases the open-loop gain and also allows the output voltage of the AUT to be adjusted to any desired value. The input offset voltage of the buffer in Fig. 14-23 is balanced out by means of the arrangement of Fig. 14-22 applied to the inverting terminal. Since the BUF is within the feedback loop, the potential difference between its input terminals is zero. By neglecting the bias current of the BUF, it follows that $V_o = -V'$ since $V_B = 0$. Hence the output of the AUT is always equal to the negative of V', which can be set at any desired value from an external voltage supply.

The system in Fig. 14-23 may oscillate if not properly compensated (Sec. 13-4). A capacitor across R' will normally stabilize the loop (Sec. 14-8).

Input Offset Voltage V_{io}   For this measurement, set $V' = 0$ so that $V_o = 0$. Both switches S_1 and S_2 are closed. From the circuit model in Fig. 14-20b, if $V_o = 0$, then $V_i = 0$ and V_{io} appears between the inverting and noninverting terminals. In other words, V_{io} of the AUT is across R and the corresponding current V_{io}/R (which is much larger than the bias current) also passes through the feedback resistor R'. Hence

$$V = \frac{V_{io}}{R}(R + R') = 1001 V_{io} \approx 10^3 V_{io} \equiv V_3$$

From the meter reading V_3 in volts, we obtain V_{io} in millivolts. Note that V_{io} is measured with the output of the AUT set at zero, as required by the definition of the *input offset voltage*.

The power-supply rejection ratio is obtained by repeating the V_{io} measurement for two values of the supply voltage V_{CC}. Then the power-supply rejection ratio is calculated from $\Delta V_{io}/\Delta V_{CC}$, where ΔV_{io} (ΔV_{CC}) represents the difference in the two input offset (power-supply) voltages.

Input Bias Current Switch S_1 in Fig. 14-23 is opened, S_2 is closed, and $V' = 0$ for this measurement. The voltage across R is now, from Fig. 14-20b, $V_{io} - R_B I_{B1}$ and

$$V = \frac{R + R'}{R} (V_{io} - R_B I_B) \approx 10^3 (V_{io} - 10^4 I_{B1}) \equiv V_4 \qquad (14\text{-}13)$$

From Eqs. (14-12) and (14-13), it follows that

$$-I_{B1} = (V_4 - V_3)10^{-7} \qquad A = 100(V_4 - V_3) \qquad nA \qquad (14\text{-}14)$$

If S_2 is left open but S_1 is closed and $V' = 0$, then I_{B2} is obtained by proceeding as above and $+I_{B2}$ is given by Eq. (14-14). The bias current is given by $I_B = \frac{1}{2}(I_{B1} + I_{B2})$ and the offset current by $I_{io} = I_{B1} - I_{B2}$.

Open-Loop Differential Voltage Gain $A_v = A_{DM}$ The open-loop gain is defined as the ratio of the output voltage to the differential voltage input signal. A direct measurement of A_{DM} based on this definition is extremely difficult. It is essential that the effect of the input offset voltage and current in the *open-loop* amplifier be canceled almost exactly, since otherwise the high amplification of the unbalanced input would result in output saturation of the amplifier (whereas it should be operating in its linear region). If an output of, say, 10 V is desired, then, with $A_{DM} = 100,000$, an accurately calibrated input signal of 0.1 mV is required. With such small signals, noise voltages may be troublesome. All of these difficulties are circumvented by using the AUT in the closed-loop arrangement shown in Fig. 14-23.

Switches S_1 and S_2 are closed and V' is set to the recommended output voltage, say, -10 V. Then $V_o = +10$ V. Since the output resistance of the AUT is very small compared with its load of 100 kΩ, then from Fig. 14-19b, $A_v V_i = V_o$. The voltage across the resistor R between the input terminals of the AUT is $V_{io} + V_i$. Hence

$$V = \frac{R + R'}{R} (V_{io} + V_i) \approx 10^3 \left(V_{io} + \frac{V_o}{A_v} \right) \equiv V_5 \qquad (14\text{-}15)$$

Subtracting Eq. (14-12) from (14-15) yields (for $V_o = 10$ V)

$$A_{DM} = A_v = \frac{10^3 \, V_o}{V_5 - V_3} = \frac{10^4}{V_5 - V_3} \qquad (14\text{-}16)$$

If V' is adjusted to $+10$ V and the above procedure is repeated, the A_v for an output of $V_o = -10$ V is obtained. If the voltage gain A_V under load is desired, it is only necessary to place the required load resistor R_L from V_o to ground while carrying out the above measurements.

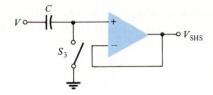

FIGURE 14-24
Sample-and-hold circuit.

For $A_v = 100,000$, $V_5 - V_3 = 0.1$ V, and very poor accuracy is obtained since two large and almost equal numbers must be subtracted. This difficulty is overcome as follows. The subtractions required in Eq. (14-16), and also in Eq. (14-17), may be performed electronically by use of the circuit in Fig. 14-24. The Op-Amp is a noninverting unity gain follower, with very high input resistance. The capacitor C will store the measured voltage V_3. The input to this circuit [called a *sample-hold-subtract* (SHS) configuration] is the output V of the BUF in Fig. 14-23. The experimental procedure is as follows: S_3, S_2, and S_1 are closed and $V' = 0$, so that $V = V_3$ is stored on the high-quality capacitor C. Now S_3 is opened and the procedure outlined above for measuring A_v (or I_{B1}) is followed. Then V is V_5 (or V_4) and $V_{SHS} = V_5 - V_3$ (or $V_4 - V_3$).

Common-Mode Rejection Ratio   This ratio is defined in Eq. (10-89) as $CMRR \equiv |A_{DM}/A_{CM}|$ where A_{DM} is the differential gain and A_{CM} is the common-mode gain. The circuit for measuring the common-mode rejection ratio is that of Fig. 14-23 with switches S_1 and S_2 closed, $V' = 0$, and a signal voltage V_s inserted between the noninverting terminal and ground. These modifications result in the network shown in Fig. 14-25. From Eq. (10-92a) applied to the AUT, with $V_o = 0$,

$$V_o = A_{DM}V_{DM} + A_{CM}V_{CM} = 0 \qquad (14\text{-}17)$$

To obtain V_{DM} and V_{CM} we first obtain V_1 and V_2 in Fig. 14-24. Clearly, $V_1 = V_s$. Using superposition, we have

$$V_2 = V_s \frac{R'}{R + R'} + V \frac{R}{R + R'} \approx V_s + V \frac{R}{R'} \qquad (14\text{-}18)$$

FIGURE 14-25
Measurement of the common-mode rejection ratio *CMRR*.

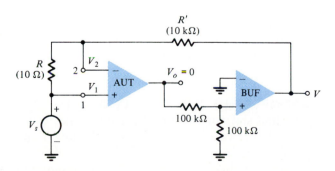

since $R' \gg R$. The difference voltage V_d is the voltage $-V_i$ across R_i. If we take the input offset voltage into account (Fig. 14-20b) and use Eqs. (14-17) and (14-12) with $R' \gg R$, we obtain

$$V_d = 2V_{DM} = V_1 - V_2 - V_{io} = -\frac{VR}{R'} - V_{io} = -\frac{R}{R'}(V + V_3) \quad (14\text{-}19)$$

and

$$V_{CM} = \tfrac{1}{2}(V_1 + V_2) = V_s + \frac{VR}{2R'} \quad (14\text{-}20)$$

Substitution of Eqs. (14-19) and (14-20) into Eq. (14-17) yields

$$-A_{DM}\frac{R}{2R'}(V + V_3) + A_{CM}\left(V_s + \frac{VR}{2R'}\right) = 0 \quad (14\text{-}21)$$

Since $A_{DM} \gg A_{CM}$ the fourth term in this equation may be neglected compared with the first term. Hence, if the measured value of V is designated by V_6, we obtain

$$CMRR\frac{R}{2R'}(V_6 + V_3) = V_s \quad (14\text{-}22)$$

For $CMRR = 10^5$, $R/R' = 2 \times 10^{-3}$, and $V_s = 10$ V we find that $V_6 + V_3 = 0.1$ V. For $V_{io} = 5$ mV, $V_1 = 5$ V. Hence $V_6 = -4.9$ V, and very poor accuracy is obtained from this measurement since two large and almost equal voltages $|V_6|$ and $|V_3|$ must be subtracted. This difficulty is overcome by changing the input to a new value V_s' and measuring the new value of V, called V_6'. Then, corresponding to Eq. (14-22), we have

$$CMRR\frac{R}{2R'}(V_6' + V_3) = V_6' \quad (14\text{-}23)$$

Subtraction of Eq. (14-22) from Eq. (14-23) results in elimination of V_3 and yields

$$CMRR = \frac{2R'}{R}\frac{V_s' - V_s}{V_6' - V_6} \quad (14\text{-}24)$$

If $V_s' = 5$ V, $V_s = -5$ V, $CMRR = 10^5$, and $R'/R = 500$, we obtain $V_6' - V_6 = 0.1$ V. However, this subtraction can now be done electronically with the SHS subtract circuit in Fig. 14-24. The switch S_3 is closed for the V_6 measurement and opened for the V_6' measurement.

14.8 FREQUENCY RESPONSE AND COMPENSATION

The closed-loop response of the basic inverting and noninverting Op-Amp stages often is required to exhibit dominant-pole behavior for all values of the low-frequency closed-loop gain. Thus the Op-Amp can be represented by a one-pole transfer function. In previous sections we demonstrated that the open-loop poles must be widely separated and the phase margin $\phi_M \approx 90°$ to achieve this kind of closed-loop response. For example, for the closed-loop poles to be separated by at least three octaves ($Q^2 \leq 0.1$), the required pole separation [Eq. (13-36)] is $n \geq T_O/Q^2 \geq 10T_O$. One decade separation in the closed-loop response requires $Q^2 \leq \frac{10}{121}$ and $n \geq 12.1T_O$. Since $T \approx A_{OL}/A_F$, the maximum value of T occurs for $A_F = 1$; that is, the Op-Amp stage is used as a unity gain buffer. Hence dominant-pole behavior requires that $n \geq 12.1A_O$, where A_O is the low-frequency value of A_{OL}; for typical values of A_O ($\approx 10^5$), the open-loop poles must be separated by more than six decades. This separation can be obtained only by narrow-banding (compensating) the open-loop amplifier. Note that for increasing Q (decreasing ϕ_M), compensation is still required because of the large value of A_O. Indeed, without compensation, the closed-loop response of a typical two-stage Op-Amp is unstable. We can demonstrate this easily for a 741-type Op-Amp. Analysis of this amplifier gives $A_O = 2 \times 10^5$, $a_1 = 8.86\ \mu s$, and $a_2 = 4.10\ ps$, from which the two dominant poles are

$$f_1 = \frac{1}{2\pi a_1} = \frac{1}{2\pi \times 8.86} = 18.0\ \text{kHz}$$

$$f_2 = \frac{a_1}{2\pi a_2} = \frac{8.86}{2\pi \times 4.10} = 344\ \text{kHz}$$

The asymptotic Bode diagram in Fig. 14-26 is constructed on the basis of these values.[1] Observe that even if $A_{OL}(s)$ contained only the two poles at $-2\pi f_1$

---

[1] Computer simulation gives $f_1 = 18.9$ kHz and $f_2 = 328$ kHz. Note that the a_1 and a_2 coefficients give very nearly the same values.

FIGURE 14-26

Asymptotic Bode diagram of the open-loop gain of a 741-type Op-Amp (*Note*: A two-pole approximation is used.)

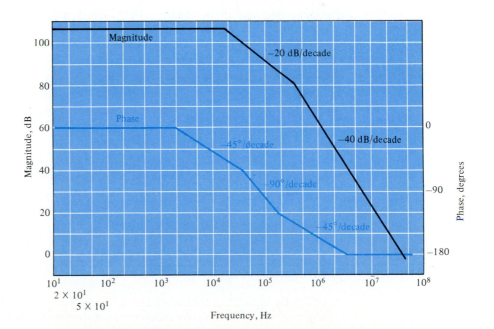

and $-2\pi f_2$, ϕ_M approaches zero and leads to an unacceptable closed-loop response. In reality, the phase shift introduced by the nondominant poles causes $\phi_M < 0$, resulting in instability. Hence the open-loop amplifier must be compensated.

The compensation of most commercial Op-Amps can be classified as internal or custom.

Internal Compensation The compensation network is fabricated on the chip, and, usually, no external access to the compensation network is provided. The manufacturer specifies the phase margin for a closed-loop gain of unity. The 741-type Op-Amp is designed in this manner. Most often, Miller-effect compensation is employed.

Custom Compensation The Op-Amp is not compensated by the manufacturer. The IC package contains terminals which provide access to the internal amplifier so that an external compensation network can be connected. The user is responsible for compensating the amplifier to match the particular application. The LM 108 is one such Op-Amp.

Sometimes a combination of the two methods is used. The manufacturer compensates the amplifier to have a prescribed phase margin at a given closed-loop gain (usually greater than unity). In addition, access is provided so the circuit designer can modify ϕ_M or compensate the amplifier for unity gain.

Miller-Effect Compensation The simplest and most common form of compensation employed is to connect a capacitor between the output and input of the gain stage. This method is similar to the compensation of the shunt triple in Example 13-5. The effective value of the compensating capacitance C_C is increased by the gain of the stage because of the Miller effect. Thus the large values of capacitance required to narrow-band the amplifier are realized by small capacitors that can be fabricated on the chip. Used extensively for internal compensation, the Miller-effect technique is also used in custom compensation.

The 741-type Op-Amp employs a 30-pF compensating capacitor, connected as shown in Fig. 14-27a. The small-signal equivalent circuit of this stage is shown in Fig. 14-27b, in which the numerical values are those obtained in Examples 14-1 and 14-2. The model in Fig. 14-27b can be modified to make the controlled source depend on V_1 as shown in Fig. 14-28 since $V_1 = (R_o + R_{i2})V_s/R_{i2}$. The form of the circuit in Fig. 14-27 is analogous to that used to compute the resistance associated with C_μ in the BJT (Example 11-2). Hence

$$R_{CC}^0 = [6.09 \parallel 5.69](1 + 455) + 0.0718 = 1340 \text{ M}\Omega$$

and, using Eq. (13-51)

$$a_{1C} = a_1 + R_{CC}^0 C_C \approx R_{CC}^0 C_C = 1340 \times 30 = 40.2 \text{ ms}$$

The frequency of the dominant pole is

$$f_1 = \frac{1}{2\pi a_{1C}} = \frac{1}{2\pi R_{CC}^0 C_C} = \frac{1}{2\pi \times 40.2 \times 10^3} = 4.0 \text{ Hz} \qquad (14\text{-}25)$$

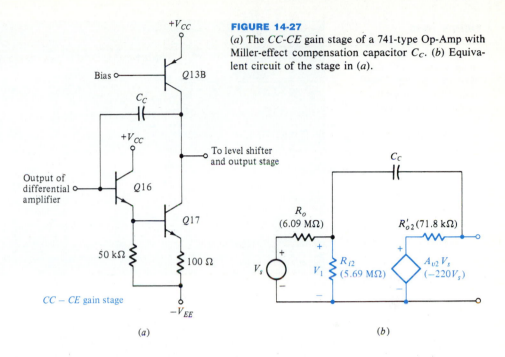

FIGURE 14-27
(a) The *CC-CE* gain stage of a 741-type Op-Amp with Miller-effect compensation capacitor C_C. (b) Equivalent circuit of the stage in (a).

The nearest nondominant pole must be separated from f_1 by T_O/Q^2. Assuming that the closed-loop poles are separated by one decade, $n = 2 \times 10^5/\frac{10}{121} = 2.42 \times 10^6$ and $f_2 = nf_1 = 9.68$ MHz. The asymptotic Bode diagram for the compensated amplifier is depicted in Fig. 14-29 from which $f_G \approx 0.8$ MHz and $\phi_M \approx 90°$.[1] The dashed curves in Fig. 14-29 are the magnitude and phase characteristics of the uncompensated amplifier. Computer simulation of the 741-type Op-Amp yields $f_1 = 5$ Hz, $f_2 = 15$ MHz, $f_G = 1$ MHz, and $\phi_M = 84°$. These values confirm the approximate values computed and correlate with measured values.[2] The difference in the phase margin is due to the phase shift introduced by the nondominant poles.

---

[1] If the calculated value of $A_O = 2.5 \times 10^5$ is used, $f_G \approx 1$ MHz.

[2] Use of $A_O = 2 \times 10^5$, the manufacturer's specified value, corresponds to $R^0_{CC} = 1070$ MΩ and results in $f_1 = 4.95$ Hz and $f_G = 0.99$ MHz.

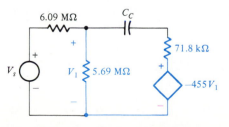

FIGURE 14-28
Equivalent representation of Fig. 14-27b in which the controlled source is dependent on V_1.

FIGURE 14-29
The asymptotic Bode diagram showing the compensation of the 741-type Op-Amp. The dashed curves are for the uncompensated amplifier.

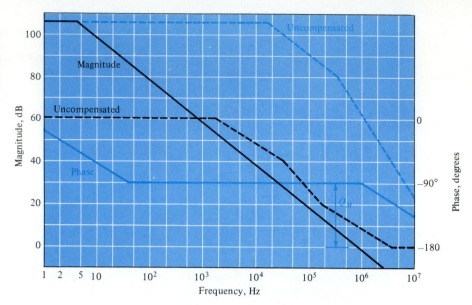

In Sec. 13-4 we observed that for frequencies below f_G, $T(s)$ could be represented by a one-pole function $T_O/(1 + s/2\pi f_1)$. Thus the gain-crossover frequency $f_G \approx T_O f_1$ is determined by the pole of the gain stage. An alternative method to compute the value of the compensating capacitor is based on the circuit in Fig. 14-30. The signal source $g_m V_d/2$ is the output current of the differential amplifier (Example 14-2), and the amplifier shown in Fig. 14-30 is the gain stage. Since the gain is large, $|V_o| \gg |V_i|$, and, in a manner analogous to the analysis of Op-Amp stages (Sec. 10-21), KCL requires

$$\frac{g_m V_d}{2} = I = -j\omega C_C V_o \qquad (14\text{-}26)$$

In Eq. (14-26), sinusoidal excitation is assumed. The frequency f_G is determined when $|V_o/V_d| = 1$; solving Eq. (14-25) gives

$$f_G = \frac{g_m}{2\pi \times 2C_C} = \frac{g_m}{4\pi C_C} \qquad (14\text{-}27)$$

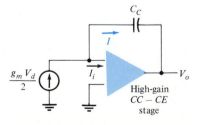

FIGURE 14-30

Alternative representation of gain stage used to determine the value of C_C needed for compensation.

Evaluation of Eq. (14-27) for the 741-type Op-Amp with $g_m = 0.38$ m℧ (Example 14-1) and $C_C = 30$ pF gives

$$f_G = \frac{0.38 \times 10^{-3}}{4\pi \times 30 \times 10^{-12}} = 1.01 \text{ MHz}$$

Since $f_1 = f_G/T_O$, $f_1 = 1.01 \times 10^6/2 \times 10^5 = 5$ Hz as indicated earlier. This simple method is useful since f_G is the approximate gain-bandwidth product of the closed-loop amplifier. That is, f_G is the 3-dB bandwidth of the unity gain buffer and is often a design specification. By knowing f_G and T_O, we can compute the value of the compensating capacitor C_C needed from Eq. (14-26).

Example 14-3

A noninverting stage using a 741-type Op-Amp is designed to have a closed-loop gain of 10. Determine ϕ_M and Q for this amplifier.

Solution

To obtain a closed-loop gain $A_{FO} = 10$, we use Eq. (10-101) to yield

$$10 = 1 + \frac{R_2}{R_1} \quad \text{or} \quad \frac{R_2}{R_1} = 9$$

The low-frequency return ratio of the stage is

$$T_O = A_O \frac{R_1}{R_1 + R_2} = \frac{A_O}{1 + R_2/R_1} = \frac{A_O}{A_{FO}} = \frac{2 \times 10^5}{10} = 2 \times 10^4$$

The gain-crossover frequency is

$$f_G = T_O f_1 = 2 \times 10^4 \times 5 = 100 \text{ kHz}$$

Since the phase of $\mathbf{T}(j\omega)$ is unchanged as T_O varies, the phase margin is obtained from Fig. 14-28 as $\phi_M = 90°$. Use of Eq. (13-36) gives

$$Q = \frac{\sqrt{nT_O}}{n+1} \approx \sqrt{\frac{T_O}{n}} = \sqrt{\frac{2 \times 10^4}{2.42 \times 10^6}} = 0.0909$$

Observe that this value of Q is considerably smaller than $Q = \sqrt{\frac{10}{121}} = 0.287$, the value needed to obtain a one-decade separation of the closed-loop poles. For $Q = 0.287$ and $T_O = 2 \times 10^4$, the open-loop pole separation required is $n = T_O/Q^2 = 2.42 \times 10^5$. The internal compensation for $A_{FO} = 1$ in the 741 results in overcompensation when $A_F > 1$. That is, the open-loop amplifier has a lower bandwidth than is required to obtain the closed-loop response.

Custom compensation allows the designer to select the value of C_C to meet the requirements of the particular circuit being constructed. In addition, the circuit designer is not restricted to using Miller-effect compensation but can employ other circuit configurations to achieve the desired phase margin.

Consider an uncompensated 741-type Op-Amp in which the terminals at which C_C is connected (Fig. 14-27) are externally accessible. Assume that we wish a gain-crossover frequency of 1 MHz (the same value as in the internally compensated unity gain buffer) for $A_{FO} = 10$. Then, as determined in Example 14-3, $T_O = 2 \times 10^4$. Since the slope of the magnitude characteristics of $T(s)$

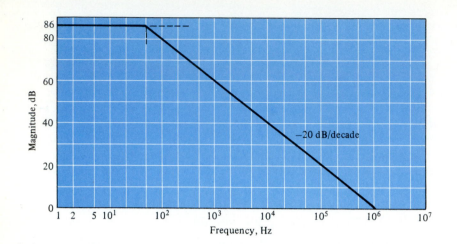

is -20 dB/decade for $f < f_G$, the asymptotic Bode diagram in Fig. 14-31 is used to obtain $f_1 = 50$ Hz (Sec. 13-4). The value of C_C required is, from Eq. (14-25),

$$C_C = \frac{1}{2\pi \times 1.34 \times 10^9 \times 50} = 2.4 \text{ pF}$$

Note that this value of C_C is much smaller than that used in the internally compensated Op-Amp, and with $f_1 = 50$ Hz, the bandwidth of $T(s)$ is increased (by a factor of 10). The advantages of negative feedback are provided over a wider range of frequencies in this stage than if an internally compensated 741 were used to achieve $A_{FO} = 10$.

Pole-Zero Cancellation   Narrow-banding of the open-loop amplifier can also be achieved by means of pole-zero cancellation (Sec. 13-4). This technique is used in situations where the designer can adjust the pole and zero frequencies. The circuits in Fig. 14-32 display two methods by which a zero can be inserted in the open-

FIGURE 14-32
Two circuits used for compensation by pole-zero cancellation.

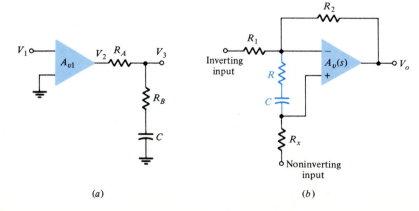

(a) (b)

loop transfer function. The analysis of these circuits is the subject of Probs. 14-32 and 14-33.

Many other circuit configurations can be used to compensate a feedback amplifier. To increase the bandwidth of $T(s)$, one class of compensating networks is used to make $T(s)$ have the form

$$T(s) = \frac{T_O (1 + s/\omega_Z)}{(1 + s/\omega_A)^2 (1 + s/\omega_2)}$$ (14-28)

The asymptotic Bode diagram for the magnitude of Eq. (14-28) is displayed in Fig. 14-33. The dashed curve in Fig. 14-33 shows the degree to which $T(s)$ is narrow-banded if simple compensation is used to give the same gain-crossover frequency. Inspection of the Bode diagrams clearly indicates that the bandwidth of $T(s)$ in Eq. (14-28) is the larger.

14-9 SLEW RATE

The value of capacitance C_C used to stabilize the Op-Amp and provide the desired closed-loop response is computed by means of small-signal analysis. The behavior of the Op-Amp when a large input signal is applied is also of importance. The slew rate, defined in Sec. 14-6 as the maximum time rate of change of the output voltage $dV_o/dt|_{max}$ describes the large-signal limitation of the Op-Amp.

In most two-stage architectures, the slew rate is directly proportional to the time required to charge the compensating capacitor. The Op-Amp model useful in approximating the slew rate is depicted in Fig. 14-30. The current I driving the gain stage is the output of the differential stage. Application of KCL gives

$$I = I_i - C_C \frac{d}{dt} (V_o - V_i)$$ (14-29)

Since the gain of the *CC-CE* stage is large, $|V_o| \gg |V_i|$ and I_i is negligible compared to I. Hence

$$I \approx -C_C \frac{dV_o}{dt}$$ (14-30)

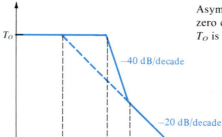

FIGURE 14-33
Asymptotic Bode diagram (magnitude) for two-pole one-zero compensation. The dashed curve extending back to T_O is for Miller-effect compensation.

The maximum current that can be delivered by the differential stage is twice the quiescent collector current observed in the transfer characteristic of the emitter-coupled pair (Fig. 3-38). Thus

$$\frac{dV_o}{dt} = \text{slew rate} = \frac{2I_C}{C_C} \tag{14-31}$$

For a type 741 Op-Amp ($I_C \approx 9.5~\mu\text{A}$ and $C_C = 30~\text{pF}$), the slew rate is $2 \times 9.5/30 = 0.63~\text{V}/\mu\text{s}$.

Substitution of Eq. (14-27) into Eq. (14-31) results in

$$\text{Slew rate} = \frac{8\pi I_C}{g_m} f_G \tag{14-32}$$

and, as $g_m = I_C/V_T$, Eq. (14-32) becomes

$$\text{Slew rate} = 8\pi V_T f_G \tag{14-33}$$

In Eq. (14-33) we observe that the slew rate is increased by increasing f_G, the unity gain bandwidth of the Op-Amp. However, increases in f_G are limited by the frequency response of the transistors used. For a two-stage Op-Amp, such as a type 741, f_G can be increased only marginally since f_T of the lateral *pnp* transistors is of the order of 5 to 10 MHz. Significant improvement in f_G is usually achieved by employing three-stage architectures. The National Semiconductor LM118 is one such amplifier and has $f_G = 15$ MHz and a slew rate of 50 V/μs.

Improvement in the slew rate can also be achieved by decreasing g_m for a given f_G. We have seen previously that the use of an emitter resistance (emitter degeneration) decreases the gain of a common-emitter stage (Sec. 10-11). Consequently, the effective value of g_m for the stage is reduced for a given bias current and results in an increased slew rate. Employment of emitter degeneration in the input stage of a type 741 Op-Amp improves the slew rate by almost an order of magnitude.

Effect of Slew Rate on an Input Signal   Let us consider a unity gain noninverting Op-Amp stage biased by ± 15-V supplies to which a 15-V step input voltage is applied. The Op-Amp has a slew rate of 0.5 V/μs. Since the output voltage V_o cannot change by more than 0.5 V/μs, the output waveform is as shown in Fig. 14-34. As indicated in Fig. 14-34, V_o does not reach 15 V, the expected output, until 30 μs have elapsed.

Now, let us consider that the input voltage V_s to this Op-Amp stage is $V_s = V_m \sin \omega t$. With no slew-rate limitation, $V_o = V_m \sin \omega t$ and

$$\frac{dV_o}{dt} = \omega V_m \cos \omega t \tag{14-34}$$

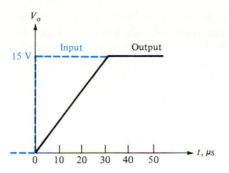

FIGURE 14-34
The response of the Op-Amp to a large step input
voltage illustrating slew rate.

The maximum value of dV_o/dt occurs at the zero crossings of the input signal,
that is, when $\omega t = n\pi$, $n = 0, 1, 2, \ldots$. Thus

$$\left.\frac{dV_o}{dt}\right|_{max} = \omega V_m \qquad (14\text{-}35)$$

Faithful reproduction of this sinusoid requires that $\omega V_m \leq$ slew rate. Use
of $V_m = 15$ V and slew rate $= 0.7$ V/μs gives $\omega =$ slew rate/$V_m = 0.5 \times$
$10^6/15 = 3.33 \times 10^4$ rad/s or $f = \omega/2\pi = 5.31$ kHz as the maximum frequency
of the input signal that can be amplified without distortion. The waveform in
Fig. 14-35 displays the effect of applying a 15-V sinusoidal input signal whose
maximum slope is greater than the slew rate. Note, in Fig. 14-35, the distortion
in the vicinity of the zero crossings of the input waveform.

14-10 BIFET AND BIMOS CIRCUITS

Ion implantation (Sec. 5-2) permits the compatible fabrication of JFETs (MOSFETs) and BJTs on the same chip. The term "BIFET
(BIMOS) technology" is commonly used for IC circuits fabricated by this
process.

All BIFET (BIMOS) Op-Amps employ FET input stages; the remaining
stages use BJTs. Such amplifiers, with JFET input stages, were introduced

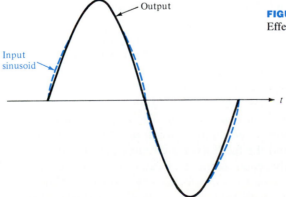

FIGURE 14-35
Effect of slew rate on a sinusoidal signal.

initially in the 1970s. By the mid-1980s, BIMOS circuits were produced commercially. Field-effect transistor differential stages have several advantages compared with BJT input stages: higher differential-mode input resistance, lower input current and (hence) decreased input offset currents, and higher slew rates.

The gate-to-source resistance of a FET (very nearly an open circuit) compared with r_π of a BJT accounts for the much higher input resistance. Often, FET differential stages have input resistances of more than four orders of magnitude higher than can be realized with BJTs.

The input bias current of a JFET is the reverse saturation current I_{GSS} of the reverse-biased gate-to-channel junction. This current is usually much smaller than the base current of a BJT biased to have $I_C = I_D$. Because $I_{in} = I_{GSS}$ is very small, the offset current caused by device mismatch is also much smaller than those which arise in BJT circuits. The use of MOSFET differential stages further reduce these quantities as leakage current through the gate oxide is much smaller than I_{GSS} of the JFET.

For a given drain current I_D, the value of g_m in a FET is smaller than the g_m of a BJT biased at $I_C = I_D$. Thus, as seen in Eq. (14-32), a reduction of g_m for a specified value of f_G results in an increased slew rate. This lower value of g_m usually results in lower differential-mode gain A_{DM} in a FET stage compared with values of A_{DM} realized with BJT circuits. To overcome this limitation, BIFET and BIMOS Op-Amps often employ three-stage architectures such as those described in the next section. The characteristics of the Analog Devices AD611 are listed in Table 14-2.

An additional advantage of FET input stages is lower noise and is a consequence of the fact that FETs are inherently lower noise devices than are BJTs.

14-11 THREE-STAGE OPERATIONAL AMPLIFIERS Most high-frequency and BIFET (BIMOS) Op-Amps employ three gain stages (a differential amplifier input stage and two gain stages) in addition to the level-shifting stage and the emitter-follower output stage. A typical structure is shown in Fig. 14-36a; the signal-flow graph for this amplifier is displayed in Fig. 14-36b. Note that this architecture is similar to the nested-loop (FLF) multiloop structure of Fig. 12-42. With three stages that provide the open-loop gain, g_m of the differential stage can be reduced to improve the slew rate [Eq. (14-32)]. Also, the three-stage multiloop feedback architecture can be designed to have higher values of the gain-crossover frequency f_G than are obtainable in two-stage Op-Amps. Thus both the slew-rate and unity gain bandwidth are increased.

Since each stage contributes a dominant pole in the open-loop amplifier, stabilization and compensation are more difficult. Both feedback networks f_1 and f_2 and the feedforward circuit a_d are required to compensate the amplifier. Typically, each of these circuits is an RC network rather than the single capacitor used to compensate a two-stage amplifier. The overall feedback loop f_1 around both gain stages is used to obtain a dominant pole in the open-loop

FIGURE 14-36
(a) Three-stage Op-Amp architecture. (b) A signal-flow graph representation of (a).

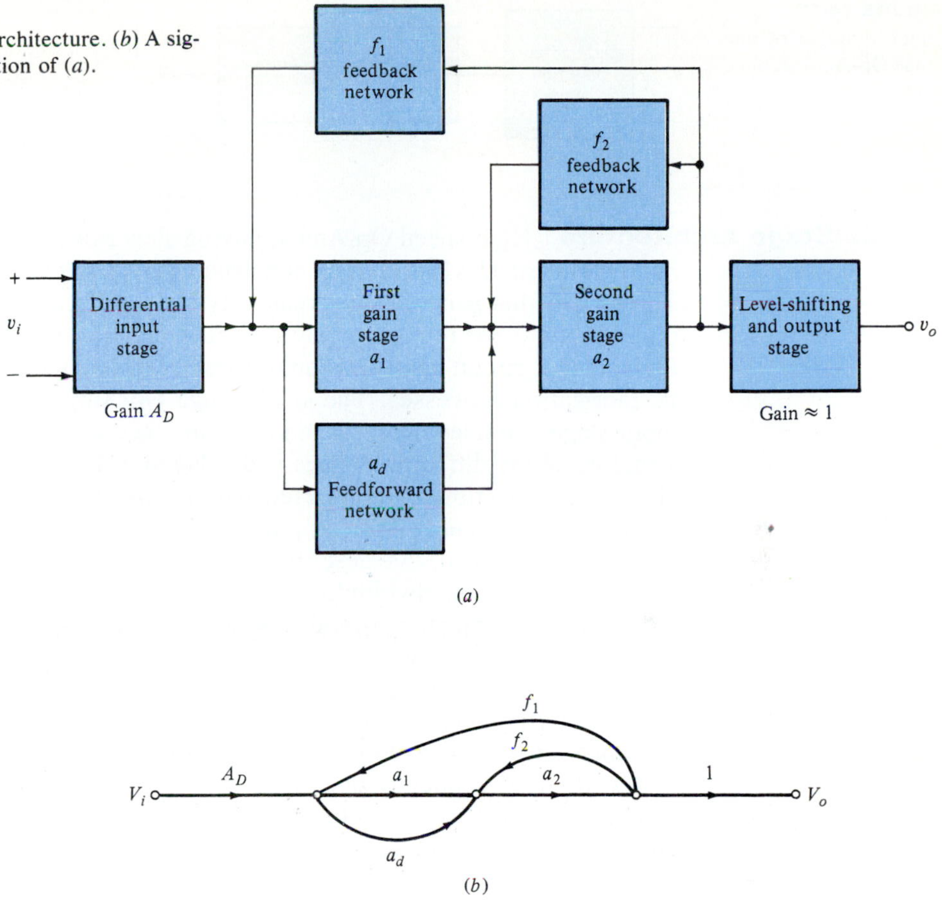

(a)

(b)

transfer function. The interior feedback loop f_2 is designed to make the pole of the second gain stage be the nondominant pole of the amplifier. A zero in the amplifier transfer function is introduced by the feedforward network. The positive phase shift of this zero improves the phase margin and helps to stabilize the amplifier. In addition, the positive phase shift tends to increase the gain-crossover frequency f_G (Prob. 14-33). The National Semiconductor LM118 is a three-stage Op-Amp and has a unity gain bandwidth of 15 MHz and a slew rate of 50 V/μs. Table 14-2 lists typical performance data of the Op-Amps described thus far in this chapter.

14-12 OTHER TYPES OF OPERATIONAL AMPLIFIERS Several other amplifier configurations are used in commercially available Op-Amps. In this section, we describe briefly three of these: the single-stage architecture, the instrumentation amplifier, and the operational-transconductance amplifier (OTA).

FIGURE 14-37

Block diagram of one-stage Op-Amp architecture.

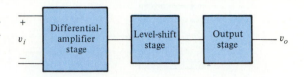

Single-Stage Architecture

High-speed Op-Amps, having slew rates of 50 V/μs and unity gain bandwidths of 15 MHz, can be constructed by using single-stage architecture. To achieve this performance, relatively simple circuit configurations are employed, but these require highly complex fabrication processes. In contrast, the complex circuit configuration of three-stage architecture relies on standard IC fabrication processes. The single-stage Op-Amp consists of a differential-input stage and a level-shifting and output stage as depicted in Fig. 14-37. The structure of the differential stage is displayed in Fig. 14-38 in which the active element is a Darlington-connected pair of *npn* transistors. The load usually consists of cascode-connected *pnp* level-shift transistors driving a cascode-connected *npn* load. As described in Sec. 11-11, the cascode configuration results in a larger bandwidth for a given gain than does a common-emitter stage. It is this arrangement that yields the superior high-frequency performance. To achieve this performance, high-frequency *pnp* transistors are needed, thus precluding the use of lateral *pnp* devices. The increased processing complexity arises from the need to fabricate *pnp* transistors having values of β_F, β_o, and f_T comparable to those of *npn* devices. A disadvantage of single-stage Op-Amps is that the low-frequency open-loop gain is usually a factor of 10 lower than can be achieved in two- and three-stage circuits (typically 80 dB compared with 100 dB as indicated in Table 14-2).

Instrumentation Amplifiers

Transducers are devices which convert a physical quantity and its variations into an electrical signal. Strain gages, thermocouples (temperature measurement), and hot-wire anemometers (fluid flow) are examples of such transducers. Each of these transducers generates a small-difference signal which usually must be amplified. *Instrumentation amplifiers* provide an output that is a precise multiple of the difference between two input signals.

A simple instrumentation amplifier can be constructed by using one Op-Amp as shown in Fig. 14-39. Use of superposition, and assumption that the input current to the Op-Amp is negligible, gives

$$V_o = \frac{R_4}{R_3 + R_4}\left(1 + \frac{R_2}{R_1}\right) V_2 - \frac{R_2}{R_1} V_1 \tag{14-36}$$

If $R_3/R_4 = R_1/R_2$ it follows that

$$V_o = \frac{R_2}{R_1}(V_2 - V_1) \tag{14-37}$$

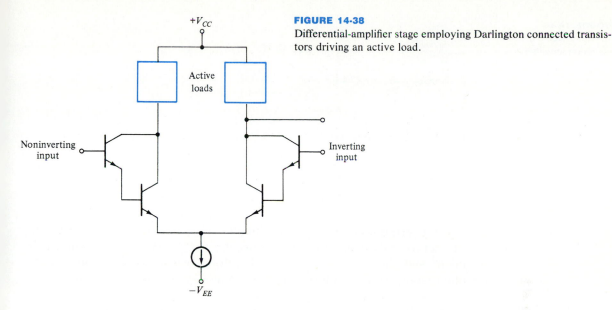

FIGURE 14-38
Differential-amplifier stage employing Darlington connected transistors driving an active load.

If the signals V_1 and V_2 have source resistances R_{s1} and R_{s2}, then these resistances add to R_3 and R_1, respectively. Note that the signal source V_1 sees a resistance $R_3 + R_4 = 101$ kΩ. If $V_2 = 0$, the inverting input is at ground potential and hence V_1 is loaded by R_1. If this is too heavy a load for the transducer, a high-resistance buffer may be used preceding each input in Fig. 14.39. The resulting system in Fig. 14-40 of three Op-Amps represents a dc instrumentation amplifier with very high input resistance and improved common-mode rejection ratio. (Since two, three, or four Op-Amps are available on a single chip, the cost of this configuration is low.)

It is easy to demonstrate that the gain of each buffer $A1$ and $A2$ is unity for a common-mode voltage but is high for a difference signal. Because there is almost zero voltage between amplifier input terminals, the top node of R is at a voltage V_1 and the bottom node of this resistor is at V_2. If a common-mode signal is under consideration, $V_1 = V_2$ and the voltage across R is zero. Hence,

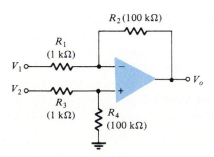

FIGURE 14-39
An Op-Amp used as an instrumentation amplifier. Setting $R_1/R_2 = R_3/R_4$ causes V_o to be proportional to $V_2 - V_1$.

FIGURE 14-40
An improved instrumentation amplifier.

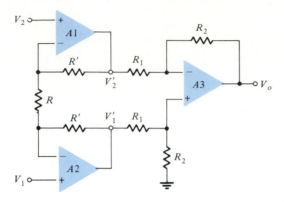

there is no current in R or R'. Consequently, $V'_2 = V_2$ and $V'_1 = V_1$, and the buffers act as unity gain amplifiers. However, if $V_1 \neq V_2$, there is current in R and R' and $V'_1 - V'_2 > V_1 - V_2$. Therefore, the differential gain and the common-mode rejection ratio of the two-stage system have been increased over the single-stage circuit of Fig. 14-39. Continuing this analysis (Prob. 14-38), we obtain

$$V_o = \left(1 + \frac{2R'}{R}\right) \frac{R_2}{R_1} (V_2 - V_1) \tag{14-38}$$

Note that the difference gain may be varied by using an adjustable resistance for R.

The system consisting of only $A1$, $A2$, R, and R' is an amplifier with a double-ended output (*a differential-output amplifier*). Clearly, $V'_1 - V'_2 = (1 + 2R'/R)(V_1 - V_2)$.

Monolithic (single-chip) instrumentation amplifiers are also available commercially. These are designed to have very high differential input resistances (>100 MΩ) and extremely high common-mode rejection ratios (in the order of 120 dB). High differential-input resistance is required to minimize loading effects on both the amplifier and the measurement system. Because the amplification of very small difference signals (≈ 10 μV) in the presence of relatively high common-mode signals (in the order of 1 V) is often the case, it is essential that instrumentation amplifiers have extremely large values of common-mode rejection ratio.

In contrast to the circuits in Figs. 14-39 and 14-40, monolithic instrumentation amplifiers are designed to operate in an open-loop condition, that is, without global feedback. The gain of the amplifier is determined by the ratio of two resistances called the *gain* and *sense* resistances, R_G and R_S, respectively. These external precision resistors are connected to lie within the amplifier and are thus isolated from the input circuit. Consequently, R_S and R_G do not load the input-signal sources and can be adjusted to provide gains typically between 1 and 1000.

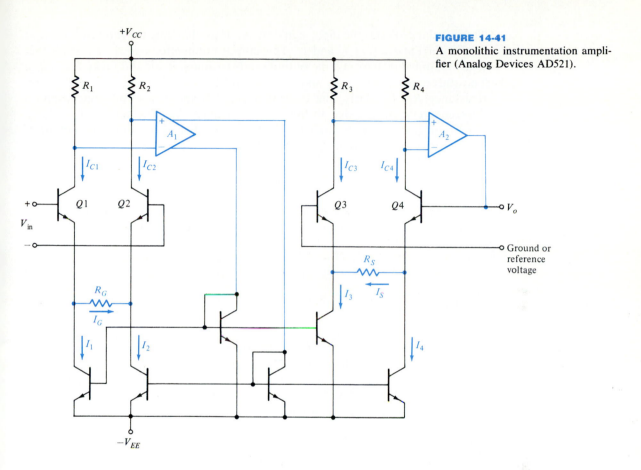

FIGURE 14-41

A monolithic instrumentation amplifier (Analog Devices AD521).

The Analog Devices AD 521 is a typical monolithic instrumentation amplifier whose circuit configuration has the form shown in Fig. 14-41. The reference currents for the matched current sources I_1, I_2, I_3, and I_4 are generated by the two outputs of the differential amplifier A_1. With $V_{in} = 0$, that is, when the difference signal is zero and common-mode conditions apply, the circuit is balanced with $I_1 = I_2 = I_3 = I_4$. The inputs to A_2 are equal because of the balanced conditions and result in $V_o = 0$. Application of a difference signal $V_{in} \neq 0$ causes the emitter currents in $Q1$ and $Q2$ to become imbalanced. Use of KVL for the loop containing V_{in}, R_G, and the emitter-base junctions of $Q1$ and $Q2$ gives

$$-V_{in} + V_{BE1} + I_G R_G - V_{BE2} = 0 \qquad (14\text{-}39)$$

Since $V_{BE1} = V_{BE2}$, solving Eq. (14.39) for I_G yields

$$I_G = \frac{V_{in}}{R_G} \qquad (14\text{-}40)$$

The imbalance varies the input signals to A_1, thus changing the reference current to the current sources I_1 and I_2. The output connections of A_1 are such that when an imbalance exists, the output tends to adjust itself and make $I_{C1} = I_{C2}$. The difference between I_1 and I_2, from KCL, is the current I_G in the gain resistance R_G. Similarly, the current sources I_3 and I_4 are imbalanced and a difference current I_s exists in the sense resistance R_S. The action of A_2 when the circuit becomes imbalanced is analogous to that of A_1; thus, the output of A_2 tends to render $I_{C3} = I_C$. Since V_o is the difference between the base voltages of $Q3$ and $Q4$

$$V_o = I_S R_S \tag{14-41}$$

The currents I_S and I_G are equal, a consequence of using matched differential current sources. Combination of Eqs. (14-40) and (14-41) gives

$$\frac{V_o}{V_{\text{in}}} = A_V = \frac{R_S}{R_G} \tag{14-42}$$

Clearly, adjustment of the resistance ratio permits the realization of different values of A_V.

The Operational Transconductance Amplifier   A voltage-to-current converter (Sec. 12-1) is an amplifier that produces an output current that is proportional to an input voltage. The constant of proportionality is referred to as the *transconductance* of the amplifier. An OTA is a single-chip amplifier in which the transconductance is controlled by an externally connected resistance.

FIGURE 14-42
An operational trans-
conductance amplifier.

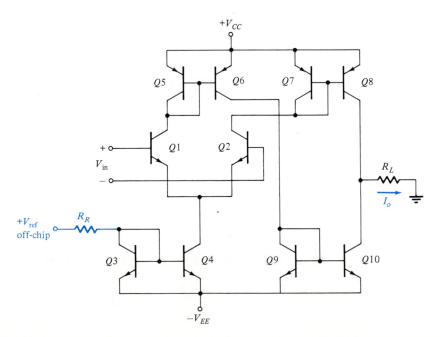

Figure 14-42 displays a simple OTA circuit in which transistors $Q1$ and $Q2$ form a differential pair. The collector currents in $Q1$ and $Q2$ are the reference currents for the complementary current sources ($Q7$-$Q8$ and $Q9$-$Q10$) which drive the load. The variable transconductance of the stage is controlled by the external resistance R_R the supply voltage V_{ref}. Adjustment of these values determines the reference current I_R for the current source $Q3$ and $Q4$. Since $I_{c4} = I_R$, the collector currents $I_{c1} = I_{c2} = I_R/2$ and make $g_{m1} = g_{m2} = g_m = I_R/2V_T$. The collector currents I_{c1} and I_{c2} are $g_m V_{in}/2$ and $-g_m V_{in}/2$, respectively. These reference currents make $I_{c10} = I_{c8} = g_m V_{in}/2$ and, hence, $I_o = g_m V_{in} = I_R V_{in}/2V_T$. Clearly, by varying either (both) R_1 or (and) V_1, we can alter the gain of the stage. Furthermore, if $R_L \ll R \leftrightarrow$ of the output current sources, then

$$A_V = \frac{V_o}{V_{in}} = \frac{I_o R_L}{V_{in}} = \frac{I_R R_L}{2V_T} \tag{14-43}$$

signifying that the voltage gain of the circuit is controlled by the bias current I_R.

14-13 MOS OPERATIONAL AMPLIFIERS

Metal oxide semiconductor Op-Amps are utilized in LSI and VLSI applications in which both analog and digital circuit functions are to be realized on the same chip. Typical uses of MOS Op-Amps are in analog-to-digital (A/D) and digital-to-analog (D/A) converters (Sec. 16-5) and active filters (Sec. 16-6) employed in digital signal processing. Packaged MOS Op-Amps are not available at present because their performance is generally inferior to that of bipolar circuits. The reduced performance is, however, adequate in many applications and the advantage of high-component density achieved in MOS technology is exploited.

NMOS Circuits

The basic structure of an NMOS Op-Amp, displayed in Fig. 14-43, is a derivative of the two-stage architecture described in Sec. 14-1. In part, the modification in the circuit configuration is due to the fact that complementary devices are not available. Hence circuits analogous to the complementary emitter-follower output stage and *pnp* active loads in bipolar Op-Amps cannot be employed.

The input differential stage, shown in Fig. 14-44, uses NMOS current sources as active loads. As seen in Figs. 14-43 and 14-44, a differential output is pro-

FIGURE 14-43

Typical architecture of an NMOS Op-Amp.

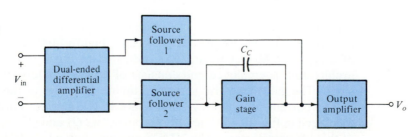

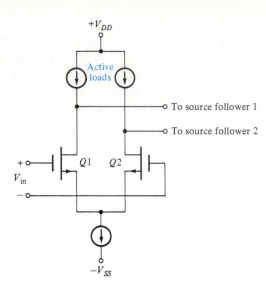

FIGURE 14-44
An NMOS differential stage with active load. Note that a double-ended output is provided, each driving the source follower shown in Fig. 14-43.

vided, each output driving a source follower. Level shifting is provided by the source followers; the outputs of these stages form the input signals to both the gain and output stages.

The voltage gains that can be realized by both the differential and gain stages is low because of the reduced value of g_m in an NMOS transistor (compared to that in a BJT at the same quiescent current). Consequently, the output stage provides some voltage gain so that the open-loop gain of the Op-Amp is adequate (typically 1000 to 10,000). Note that these values are at least a factor of 10 less than are obtained in BJT circuits.

Compensation is provided by the pole-splitting capacitor C_C (Miller effect), connected from the output to the input of the gain stage. The feedforward path through source follower 2 helps to stabilize the amplifier and increase the phase margin in much the same manner as in the three-stage Op-Amp (Sec. 14-11). The feedforward is needed to minimize the effects of the phase shift introduced by the gain stage at $s \approx g_m/C_C$. Because of the low value of g_m, the location of the zero is in the vicinity of the gain-crossover frequency; therefore, its phase shift cannot be neglected. (For $g_m = 0.5$ m℧ and $C_C = 20$ pF, the frequency of the zero is approximately 4 MHz. If the unity gain bandwidth is to be 1 to 5 MHz, then clearly, the phase shift of the zero must be taken into account.)

CMOS Op-Amps The disadvantage of the lack of complementary devices in NMOS technology is overcome by the use of CMOS circuits. Figure 14-45 shows a simplified form of a typical CMOS Op-Amp. As indicated in Fig. 14-45, a simple two-stage architecture is employed and consists of an input differential stage and a gain stage. The gain stage also serves as the output stage of the Op-Amp.

The current sources I_1, I_2, and I_3 are PMOS current mirrors similar to that displayed in Fig. 10-15. The PMOS transistors $Q1$ and $Q2$ are the active devices of the differential stage. The NMOS current source formed by $Q3$ and $Q4$ serve

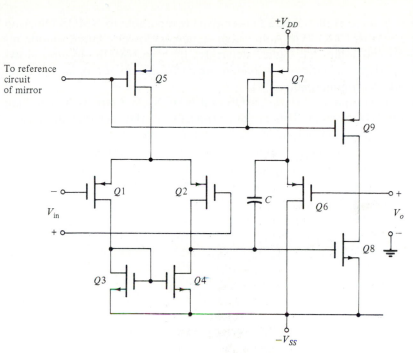

FIGURE 14-45
A two-stage CMOS Op-Amp circuit. Note that the gain stage is also the output stage. Since the output drives the input of another CMOS stage (assumed), the low output resistance of a source follower is not needed.

as the active load of the stage. The gain stage comprises $Q5$ and its active load provided by the current source I_3.

The amplifier is compensated by the pole-splitting capacitance C. This capacitor is connected to the output by means of the source follower ($Q6$ and its active load I_2). Since the gain of the source follower is nearly unity, C is, in effect, connected between the output and the input of the gain stage. Hence, at the output of the differential stage, the effect of C is the Miller input capacitance of the gain stage. However, the effect of the source follower is to isolate C from the output as represented in Fig. 14-46. Because of the unilateral nature of the source follower (unity gain buffer), the zero in the transfer function occurs at $-g_m/C_{gd}$ rather than $g_m/(C_{gs} + C)$. Since $C > C_{gd}$, the frequency of the zero is beyond the gain-crossover frequency and the effect of the phase shift introduced by the zero is minimal.

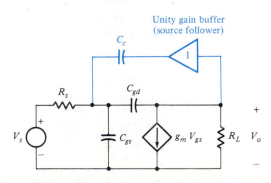

Unity gain buffer (source follower)

FIGURE 14-46
Compensation circuit for a MOS Op-Amp. The unity gain buffer allows Miller-effect compensation without accompanying movement of the zero in the transfer function.

The performance of the CMOS Op-Amps is comparable to NMOS Op-Amp performance. Most CMOS Op-Amps have somewhat lower common-mode rejection ratios and somewhat higher slew rates than do NMOS circuits. However, neither MOS technology can match the performance of bipolar Op-Amp circuits available commercially.

The output resistance of both CMOS and NMOS Op-Amps is higher than is obtained in BJT circuits. This is due, primarily, to the fact that the output in MOS Op-Amps is the output of the common-source gain stage rather than the emitter-follower output in bipolar circuits. Since the principal use of MOS Op-Amps is to drive other very high-input resistance MOS circuits, the moderate output resistance has minimal effect on performance.

REFERENCES

1. Grebene, A. B.: "Bipolar and MOS Analog Integrated Circuit Design," John Wiley and Sons, New York, 1984.

2. Gray, P. R., and R. G. Meyer: "Analysis and Design of Analog Integrated Circuits," 2d ed., John Wiley and Sons, New York, 1984.

3. Soclof, S.: "Analog Integrated Circuits," Prentice-Hall, Englewood Cliffs, N. J., 1985.

4. Ghausi, M.S.: "Electronic Devices and Circuits: Discrete and Integrated," Holt, New York, 1985.

5. Hamilton, D. J., and W. G. Howard: "Basic Integrated Circuit Engineering," McGraw-Hill Book Company, New York, 1975.

6. Sedra, A. S., and K. C. Smith: "Microelectronic Circuits," Holt, New York, 1981.

7. Solomon, J. E.: The Monolithic Op-Amp: A Tutorial Study, *IEEE Journal of Solid-State Circuits*, vol. SC-9, pp. 314–332, December 1974.

8. Gray, P. R., D. A. Hodges, and R. W. Broderson (eds.): "Analog MOS Integrated Circuits," IEEE Press, New York, 1980.

9. Brokaw, A. P., and M. P. Timko: An Improved Monolithic Instrumentation Amplifier, *IEEE J. Solid-State Circuits*, vol. SC-10, pp. 417–423, December 1975.

10. Roberge, J. K.: "Operational Amplifiers: Theory and Practice," John Wiley and Sons, New York, 1975.

REVIEW QUESTIONS

14-1 Why is it desirable for an Op-Amp to have a high *CMRR*?

14-2 (*a*) Draw a block diagram of a two-stage Op-Amp.
(*b*) Explain the function of each block.

14-3 (*a*) Sketch the circuit diagram of a differential pair with an active load.
(*b*) What advantages accrue with the use an active load?

14-4 (*a*) What is the differential-input resistance of an emitter-coupled pair?
(*b*) How does this resistance depend on quiescent current?

14-5 Indicate two methods by which very high Op-Amp input resistances can be obtained.

14-6 (*a*) Show two forms of level-shifting circuits using an emitter follower.
(*b*) What is the expression for the shift in each circuit?

14-7 Draw the circuit of a V_{BE} multiplier and explain its operation.

14-8 Why is a complementary emitter follower used as an output stage?

14-9 (*a*) Draw a simple complementary emitter-follower circuit.
(*b*) Explain why this circuit exhibits crossover distortion.
(*c*) How can the circuit in part *a* be modified to eliminate this distortion?

14-10 Define (*a*) input bias current, (*b*) input offset current, (*c*) input offset voltage, (*d*) output offset voltage, and (*e*) input offset drift voltage.

14-11 What are the relative advantages and disadvantages of internal compensation and custom compensation?

14-12 Why is Miller-effect compensation often employed to compensate an Op-Amp?

14-13 Show two circuits which can provide pole-zero cancellation.

14-14 (*a*) What is meant by pole-zero compensation?
(*b*) What are the advantages and disadvantages of this technique?

14-15 (*a*) Define slew rate.
(*b*) How does this limit the response of an Op-Amp?

14-16 Describe how the slew rate of an Op-Amp can be improved.

14-17 Show the block diagram of an Op-Amp employing three-stage architecture.

14-18 Explain why many three-stage Op-Amps have two feedback loops and one feedforward path.

14-19 What are the advantages and disadvantages of three-stage architecture?

14-20 Repeat Rev. 14-19 for single-stage architecture.

14-21 (*a*) Draw the circuit diagram of a simple instrumentation amplifier.
(*b*) Write an expression for the output voltage of this circuit in terms of the input voltages and circuit resistances.

14-22 Compare the performance and operation of a monolithic Op-Amp with that in Rev. 14-21*a*.

14-23 (*a*) What is meant by an operational transconductance amplifier?
(*b*) How is the performance of this amplifier controlled?

14-24 (*a*) Sketch the architecture of an NMOS Op-Amp.
(*b*) List three reasons why this architecture is employed.

14-26 Repeat Rev. 14-25 for a CMOS Op-Amp.

14-27 What is meant by a BIMOS or BIFET amplifier?

Part Four
SIGNAL PROCESSING AND DATA ACQUISITION

The transmission, reception, and processing of information in the form of electrical signals is the basis of modern electronic systems for control, communication, and computation. Many of these systems utilize both analog and digital signals to perform their function. Clearly, a variety of different signal waveforms are required. Furthermore, the form of these signals (amplitude, phase, frequency, duration, rise time, etc.) must be appropriate for the particular application for effective processing. In the two chapters in this part of the book we treat a number of different circuits used for signal generation and processing. Chapter 15 deals with waveform generation and waveshaping. Included are sinusoidal oscillators, clock (square-wave) generators, and time-base generators. The conversion of data from analog-to-digital (A/D) and digital-to-analog (D/A) form is described in the second chapter of this part. In addition, signal-conditioning circuits such as logarithmic amplifiers, integrators, multipliers, and active filters are discussed. The circuits treated in this section utilize the basic building blocks (logic gates, Op-Amps, switches, etc.) described in the previous parts of the book.

Chapter 15
WAVEFORM GENERATORS AND WAVESHAPING

Three basic waveforms extensively used are the sinusoid (frequency generation), square-wave (clock function), and ramp (time-base generation). In this chapter, the oscillator, multivibrator, and sawtooth circuits used to generate these signals are discussed.

The comparator is introduced as a basic building block, and the Schmitt trigger (regenerative comparator) is used to generate a variety of waveforms.

15-1 SINUSOIDAL OSCILLATORS

In Sec. 13-2 we observed that if sufficient phase shift was introduced into the feedback loop when the loop gain was greater than unity, the feedback amplifier became unstable; that is, it oscillated. Under these circumstances the closed-loop poles were drawn into the right half-plane and no excitation was needed to produce an output. If the closed-loop poles can be made to lie on the j axis, the natural response of the system is a sinusoid whose frequency is the pole frequency. This idea is the basis of sinusoidal oscillator circuits. That is, a sinusoidal oscillator is a feedback amplifier designed to have closed-loop poles on the j axis at the frequency of the desired output.

To illustrate the concept of an oscillator, consider the signal-flow graph in Fig. 15-1a. This is the basic flow graph of a single-loop feedback amplifier (Sec. 12–3) prior to closing the loop. (The connection between X_c, the output of the summing network, and $\hat{X}_c$, the input to the amplifier network, is open.) If

FIGURE 15-1
(a) The signal-flow graph for a single-loop amplifier prior to closing the loop. (b) The system in (a) with a signal applied directly to the amplifier.

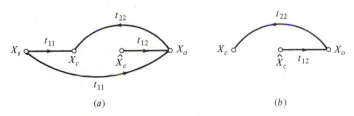

(a) (b)

$X_s = 0$, then Fig. 15-1b is the flow graph for the system. Now let us apply a signal $X_i = \hat{X}_c$ directly into the amplifier. As a consequence of this signal, the amplifier output $X_o = t_{12}\hat{X}_c$. The output of the feedback network is $X_c = t_{22}X_o = t_{12}t_{22}\hat{X}_c = -T\hat{X}_c$, where T is the return ratio of the amplifier. Suppose it should happen that matters are adjusted in such a way that the signal X_c is identical to the externally applied input signal $\hat{X}_c$. Since the amplifier has no means of distinguishing the source of the input signal applied to it, it would appear that, if the external source were removed and if node X_c were connected to node $\hat{X}_c$, the amplifier would continue to provide the same output signal X_o as before. Note, of course, that the statement $X_c = \hat{X}_c$ means that the instantaneous values of X_o and $X_c = \hat{X}_c$ are exactly equal at all times. The condition $X_c = \hat{X}_c$ is equivalent to $-T = 1$; the *return ratio must equal minus one*.

The Barkhausen Criterion For a sinusoidal output waveform, the relation $X_c = \hat{X}_c$ is equivalent to the condition that the *amplitude*, *phase*, and *frequency* of X_c and $\hat{X}_c$ must be identical. Hence we have the following important principle: *The frequency at which a sinusoidal oscillator will operate is the frequency f_o for which*

$$\mathbf{T}(j2\pi f_o) = -1 \tag{15-1}$$

Alternatively, the condition for sinusoidal oscillation can be expressed as

$$T(j2\pi f_o) = 1 \qquad \angle\ T(j2\pi f_o) = -180° \tag{15-2}$$

or

$$\text{Real part } \mathbf{T}(j2\pi f_o) = -1 \qquad \text{Imaginary part } \mathbf{T}(j2\pi f_o) = 0 \tag{15-3}$$

Equations (15-1)–(15-3) indicate that two conditions must be satisfied by a circuit to sustain oscillations:

1. The phase shift through the amplifier and feedback network must be 360° (or $2\pi n$ radians). Recall that the definition of T includes a minus sign which is equivalent to 180° of phase shift.
2. The magnitude of the gain of the amplifier and feedback network must be unity.

The condition that $\mathbf{T}(j\omega) = -1$ is called the *Barkhausen criterion*. This condition is consistent with our analysis of feedback amplifiers for which $A_F = A_{\text{OL}}/(1 + T)$. For $T = -1$, $A_F \to \infty$ and may be interpreted to mean that an output exists even in the absence of an externally applied signal. In Chap. 13 we described compensation techniques to prevent oscillation and to ensure $T(j\omega) < 1$ when $\angle T(j\omega) = -180°$. Thus the Barkhausen criterion is equivalent to stating that both the phase margin and gain margin are zero. Consequently, the phase-crossover and gain-crossover frequencies are equal. The frequency oscillation is the frequency for which $\phi_M = 0$.

Practical Considerations   Referring to Fig. 15-1, it appears that if $|T|$ at the oscillator frequency is precisely unity, then, with the feedback signal connected to the input terminals, the removal of the external generator will make no difference. If $|T|$ is less than unity (the gain margin is positive), the removal of the external generator will result in a cessation of oscillations. But now suppose that $|T|$ is greater than unity. Then, for example, a 1-V signal appearing initially at the input terminals will, after a trip around the loop and back to the input terminals, appear there with an amplitude larger than 1 V. This larger voltage will then reappear as a still larger voltage, and so on. It seems, then, that if $|T|$ is larger than unity, the amplitude of the oscillations will continue to increase without limit. But of course, such an increase in the amplitude can continue only as long as it is not limited by the onset of nonlinearity of operation in the active devices associated with the amplifier. Such a nonlinearity becomes more marked as the amplitude of oscillation increases. This onset of nonlinearity to limit the amplitude of oscillation is an essential feature of the operation of all practical oscillators, as the following considerations will show. The condition $|T| = 1$ does not give a range of acceptable values of $|T|$, but rather a single and precise value. Now suppose that initially it were even possible to satisfy this condition. Then, because circuit components and, more importantly, transistors change characteristics (drift) with age, temperature, voltage, etc., it is clear that if the entire oscillator is left to itself, in a very short time $|T|$ will become either less or larger than unity. In the former case the oscillation simply stops, and in the latter case we are back to the point of requiring nonlinearity to limit the amplitude. An oscillator in which the loop gain is exactly unity is an abstraction completely unrealizable in practice. It is accordingly necessary, in the adjustment of a practical oscillator, always to arrange to have $|T|$ somewhat larger (say, 5 percent) than unity in order to ensure that, with incidental variations in transistor and circuit parameters, $|T|$ shall not fall below unity. While the first two principles stated above must be satisfied on purely theoretical grounds, we may add a third general principle dictated by practical considerations, that is: *In every practical oscillator the loop gain is slightly larger than unity, and the amplitude of the oscillations is limited by the onset of nonlinearity.*

15-2   THE PHASE-SHIFT OSCILLATOR   We select the *phase-shift oscillator* (Fig. 15-2) as a first example because it exemplifies very simply the principles set forth above. Here a discrete-component JFET amplifier is followed by three cascaded arrangements of a capacitor C and a resistor R, the output of the last RC combination being returned to the gate. If the loading of the phase-shift network on the amplifier can be neglected that is, $R \gg R_L$, the amplifier shifts by 180° the phase of any voltage which appears on the gate, and the network of resistors and capacitors shifts the phase by an additional amount. At some frequency the phase shift introduced by the RC network will be precisely 180°, and at this frequency the total phase shift from the gate around the circuit and back to the gate will be exactly zero. This particular frequency will be the one at

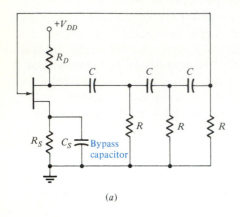

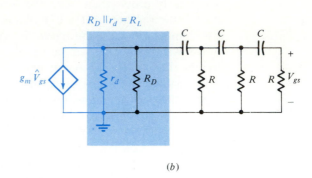

(a)

(b)

FIGURE 15-2
(a) A JFET RC phase-shift oscillator and (b) its equivalent circuit. (c) An Op-Amp version of the phase-shift oscillator.

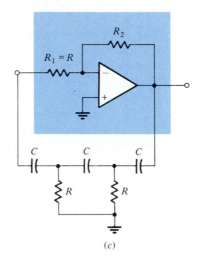

(c)

which the circuit will oscillate, provided that the magnitude of the amplification is sufficiently large.

Determination of $\mathbf{T}(j\omega)$ by the method described in Sec. 12-7 (Prob. 15-1) gives

$$\mathbf{T}(j\omega_N) = \frac{jg_m R_L \omega_N^3}{(1 - 6\omega_N^2) + j\omega_N(5 - \omega_N^2)} \tag{15-4}$$

where $\omega_N = \omega RC$. Application of Eq. (15-3) yields $\omega_N^2 = \frac{1}{6}$, from which the frequency of oscillation f_o becomes

$$f_o = \frac{1}{2\pi RC\sqrt{6}} \tag{15-5}$$

At $\omega_N = 1/\sqrt{6}$,

$$\mathbf{T}(j\omega_N) = \frac{jg_m R_L/6\sqrt{6}}{j(5 - \frac{1}{6})/\sqrt{6}}$$

from which $g_m R_L = 29$. Thus, to sustain oscillations, the gain of the JFET amplifier stage must be at least 29.

If the loading effect of the RC phase-shift network is included (Prob. 15-3), the frequency of oscillation decreases and the gain of the JFET amplifier stage must be increased.

The FET in Fig. 15-2a may be replaced by an Op-Amp as indicated in Fig. 15-2c. Because of the virtual ground, the resistance from the input node P to ground is $R_1 = R$ and, hence, the phase-shift network in Fig. 15-2c is identical with that in Fig. 15-2a. Therefore, the oscillation frequency is given by Eq. (15-5). Since the Op-Amp gain $A_V = -R_1/R$ and $|A_V|$ must be at least 29, then R_2/R must be greater than 29 (by about 5 percent).

It is possible to replace the Op-Amp in Fig. 15-2c by a single transistor stage with $R_2 = \infty$ and $R_1 = R - R_i$ (Prob. 15-4).

It should be pointed out that it is not always necessary to make use of an amplifier with transfer gain $|A| > 1$ to satisfy the Barkhausen criterion. It is necessary only that $|T| > 1$. Passive network structures exist for which the transfer function of the feedback network is greater than unity at some particular frequency. In Prob. 15-6 we show an oscillator circuit consisting of a source follower and the RC circuit in Fig. 15-2 appropriately connected.

Variable-Frequency Operation   The phase-shift oscillator is particularly suited to the range of frequencies from several hertz to several hundred kilohertz, and so includes the range of audio frequencies. The frequency of oscillation may be varied by changing any of the impedance elements in the phase-shifting network. For variations of frequency over a large range, the three capacitors are usually varied simultaneously. Such a variation keeps the input impedance to the phase-shifting network constant (Prob. 15-2) and also keeps constant the magnitude of T. Hence the amplitude of oscillation will not be affected as the frequency is adjusted. The phase-shift oscillator is operated in class A in order to keep distortion to a minimum.

Two active phase shifters may be used in place of the passive feedback network in Fig. 15-2c to obtain a sinusoidal oscillator with quadrature outputs, sine, and cosine waveforms.

15-3   THE WIEN BRIDGE OSCILLATOR   An oscillator circuit in which a balanced bridge is used as the feedback network is the Wien bridge oscillator shown in Fig. 15-3a. The "bridge" is clearly indicated in Fig. 15-3b. The four arms of the bridge are Z_1, Z_2, R_1, and R_2. The input to the bridge is the output V_o of the Op-Amp, and the output of the bridge between nodes 1 and 2 supplies the differential input to the Op-Amp.

There are two feedback paths in Fig. 15-3a: *positive* feedback through Z_1 and Z_2, whose components determine the frequency of oscillation, and *negative* feedback through R_1 and R_2, whose elements affect the amplitude of oscillation and set the gain of the Op-Amp stage. The loop gain is given by

$$T(s) = -\left(1 + \frac{R_2}{R_1}\right)\frac{Z_2}{Z_1 + Z_2} \qquad (15-6)$$

FIGURE 15-3
(*a*) A Wien bridge oscillator. (*b*) The bridge network.

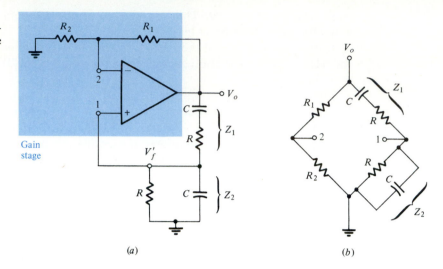

(*a*)   (*b*)

For $Z_1 = (RCs + 1)/Cs$ and $Z_2 = R/(RCs + 1)$, application of the Barkhausen criterion gives

$$f_o = \frac{1}{2\pi RC} \quad \text{and} \quad R_1 = 2R_2 \tag{15-7}$$

Thus the gain of the noninverting Op-Amp stage is 3 (or about 5 percent greater than 3) to sustain oscillation. The maximum frequency of oscillation is limited by the slew rate of the amplifier. Continuous variation of frequency is accomplished by varying simultaneously the two capacitors (ganged variable-air capacitors). Changes in frequency range are accomplished by switching in different values for the two identical resistors R.

Amplitude Stabilization We consider modification of the circuit in Fig. 15-3, which serves to stabilize the amplitude against variations due to fluctuations occasioned by the aging of transistors, components, etc. One modification consists simply in replacing the resistor R_2 by a sensistor (a resistor which has a positive thermal coefficient).

The amplitude of oscillation is determined by the extent to which the loop gain is greater than unity. If the output V_o increases (for any reason), the current in R_2 increases and A decreases. The regulation mechanism introduced by the sensistor operates by automatically changing A so as to keep the return ratio more constant. The temperature of R_2 is determined by the root-mean-square (rms) value of the current which passes through it. If the rms value of the current changes, then, because of the thermal lag of the sensistor, the temperature will be determined by the average value over a large number of cycles of the rms value of the current. An important fact to keep in mind about the mechanism just described is that, because of the thermal lag of the sensistor,

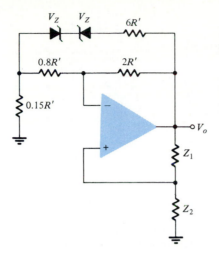

FIGURE 15-4
Zener diodes are used to automatically control the
gain of the oscillator and, hence, stabilize the amplitude of the sinusoid.

the resistance of the sensistor during the course of a single cycle is very nearly absolutely constant. Therefore, at any fixed amplitude of oscillation, the sensistor behaves entirely like an ordinary linear resistor.

A thermistor which has a negative temperature coefficient can also be used but it must replace R_1 rather than R_2.

Another method of stabilizing the amplitude is indicated in Fig. 15-4. Initially, both Zener diodes are nonconducting, and the loop gain is

$$\frac{1}{3}\left(1 + \frac{R_1}{R_2}\right) = \frac{1}{3}\left(1 + \frac{2R'}{0.15R' + 0.8R'}\right) = 1.04 > 1$$

and, hence, oscillations start. Because the loop gain exceeds unity, the amplitude of these oscillations grows until the peaks exceed the diode breakdown voltage V_Z. When this happens, the shunting action of the resistor $6R'$ reduces the gain and limits the amplitude to approximately V_Z. Distortion can be reduced to approximately 0.5 percent with this circuit.

The two methods of amplitude stabilization described in the foregoing are examples of automatic gain control (AGC). An active AGC loop may also be used with a FET as a voltage-controlled resistor.

15-4   A GENERAL FORM OF OSCILLATOR CONFIGURATION   Many oscillator cir-
cuits fall into the general form shown in Fig. 15-5a. In the analysis that follows we assume an active device with extremely high input resistance such as an Op-Amp or a FET. Figure 15-5b shows the linear equivalent circuit in Fig. 15-5a, using an amplifier with an open-circuit negative gain $-A_v$ and output resistance R_o. Clearly, the topology shown in Fig. 15-5 is that of shunt-series feedback.

FIGURE 15-5
(*a*) General form of an oscillator circuit. (*b*) The equivalent circuit of (*a*).

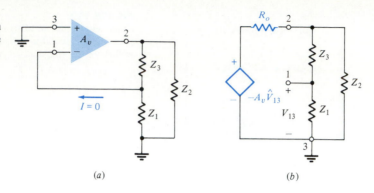

(*a*) (*b*)

The Return Ratio The value of T will be obtained by considering the circuit of Fig. 15-5*a* to be a feedback amplifier with output taken from terminals 2 and 3 and with input terminals 1 and 3. The load impedance Z_L consists of Z_2 in parallel with the series combination of Z_1 and Z_3. Then

$$V_o = \frac{A_v \hat{V}_{13} Z_L}{Z_L + R_o} \qquad \text{and} \qquad V_{13} = \frac{Z_2}{Z_1 + Z_2} V_o \qquad (15\text{-}8)$$

Combination of the relationships in Eq. (15-8) and substitution for Z_L results in a return ratio of

$$T = -\frac{V_{13}}{\hat{V}_{13}} = \frac{A_v Z_1 Z_2}{R_o(Z_1 + Z_2 + Z_3) + Z_2(Z_1 + Z_3)} \qquad (15\text{-}9)$$

LC-Tunable Oscillators The oscillators described in the two previous sections are *RC*-tunable circuits. That is, the frequency of oscillation is determined by the resistance and capacitance values used. Often, the frequency obtainable in such circuits is limited to a few hundred kilohertz. Where higher frequencies of oscillation are required, such as those used in AM and FM receivers, tuning is accomplished by varying a capacitance or inductance. In the general oscillator configuration shown in Fig. 15-5, making Z_1, Z_2, and Z_3 pure reactances (either inductive or capacitive), an *LC*-tunable oscillator results. If we let $\mathbf{Z}_1 = jX_1$, $\mathbf{Z}_2 = jX_2$, and $\mathbf{Z}_3 = jX_3$, where $X = \omega L$ for an inductance and $-1/\omega C$ for a capacitance, Eq. (15-9) becomes

$$\mathbf{T} = \frac{+ A_v X_1 X_2}{jR_o(X_1 + X_2 + X_3) - X_2(X_1 + X_3)} \qquad (15\text{-}10)$$

For **T** to be real

$$X_1 + X_2 + X_3 = 0 \qquad (15\text{-}11)$$

and

$$T = \frac{A_v X_1 X_2}{-X_2(X_1 + X_3)} = \frac{-A_v X_1}{X_1 + X_3} \qquad (15\text{-}12)$$

From Eq. (15-11) we see that the circuit will oscillate at the resonant frequency of the series combination of X_1, X_2, and X_3.

Use of Eq. (15-11) in Eq. (15-12) yields

$$T = \frac{+A_v X_1}{X_2} \qquad (15\text{-}13)$$

Since T must be positive and at least unity in magnitude, then X_1 and X_2 must have the same sign (A_v is positive). In other words, they must be the same kind of reactance, either both inductive or both capacitive. Then, from Eq. (15-11), $X_3 = -(X_1 + X_2)$ must be inductive if X_1 and X_2 are capacitive, or vice versa.

If X_1 and X_2 are capacitors and X_3 is an inductor, the circuit is called a *Colpitts oscillator*. If X_1 and X_2 are inductors and X_3 is a capacitor, the circuit is called a *Hartley oscillator*. In this latter case, there may be mutual coupling between X_1 and X_2 (and the above equations will then not apply).

Transistor versions of the types of LC oscillators described above are possible. As an example, a transistor Colpitts oscillator is indicated in Fig. 15-6a. Qualitatively, this circuit operates in the manner described above. However, the detailed analysis of a transistor oscillator circuit is more difficult, for two fundamental reasons. First, the low input impedance of the transistor shunts Z_1 in Fig. 15-5a, and hence complicates the expressions for the loop gain given above. Second, if the oscillation frequency is beyond the audio range, the simple low-frequency model is no longer valid. Under these circumstances the high-frequency hybrid-π model in Fig. 3-32 must be used. A transistor Hartley oscillator is shown in Fig. 15-6b.

FIGURE 15-6
Inductance-capacitance LC oscillators: (a) Colpitts, (b) Hartley oscillators.

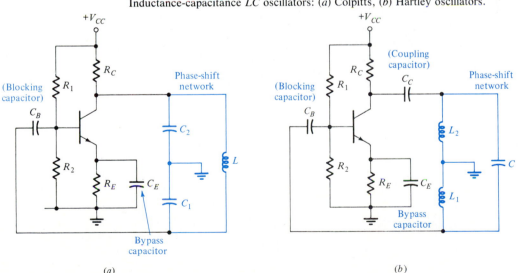

(a)

(b)

15-5 CRYSTAL OSCILLATORS

If a piezoelectric crystal, usually quartz, has electrodes plated on opposite faces and a potential is applied between these electrodes, forces will be exerted on the bound charges within the crystal. If this device is properly mounted, deformations take place within the crystal, and an electromechanical system is formed which will vibrate when properly excited. The resonant frequency and the Q depend on the crystal dimensions, how the surfaces are oriented with respect to its axes, and how the device is mounted. Frequencies ranging from a few kilohertz to a few hundred megahertz, and Q values in the range from several thousand to several hundred thousand, are commercially available. These extraordinarily high values of Q and the fact that the characteristics of quartz are extremely stable with respect to time and temperature account for the exceptional frequency stability of oscillators incorporating crystals.

The electrical equivalent circuit of a crystal is indicated in Fig. 15-7. The inductor L, the capacitor C, and the resistor R are the analogs of the mass, the compliance (the reciprocal of the spring constant), and the viscous-damping factor of the mechanical system. Typical values for a 90-kHz crystal are $L = 137$ H, $C = 0.0235$ pF, and $R = 15$ kΩ, corresponding to $Q = 5500$. The dimensions of such a crystal are $30 \times 4 \times 1.5$ mm. Since C' represents the electrostatic capacitance between electrodes with the crystal as a dielectric, its magnitude (~ 3.5 pF) is very much larger than C.

If we neglect the resistance R, the impedance of the crystal is a reactance jX whose dependence upon frequency is given by

$$jX = -\frac{j}{\omega C'}\frac{\omega^2 - \omega_s^2}{\omega^2 - \omega_p^2} \tag{15-14}$$

where $\omega_s^2 = 1/LC$ is the series resonant frequency (the zero impedance frequency), and $\omega_p^2 = (1/L)(1/C + 1/C')$ is the parallel resonant frequency (the infinite impedance frequency). Since $C' \gg C$, then $\omega_p \approx \omega_s$. For the crystal whose parameters are specified above, the parallel frequency is only three-tenths of 1 percent higher than the series frequency. For $\omega_s < \omega < \omega_p$, the

FIGURE 15-7
A piezoelectric crystal: (a) symbol, (b) circuit model, (c) the reactance as a function of frequency assuming $R = 0$.

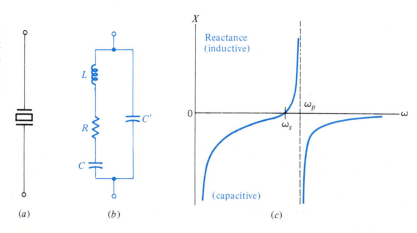

(a) (b) (c)

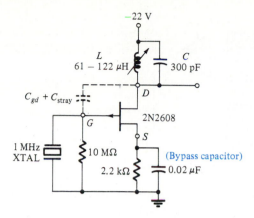

FIGURE 15-8
A 1-MHz FET crystal oscillator. (*Courtesy of Siliconix Co.*)

reactance is inductive, and outside this range it is capacitive, as indicated in Fig. 15-7.

A variety of crystal-oscillator circuits is possible. If a crystal is used for Z_1 in the basic configuration in Fig. 15-5a, a tuned LC combination for Z_2, and the capacitance C_{gd} between gate and drain for Z_3, the resulting circuit is as indicated in Fig. 15-8. From the theory given in the preceding section, the crystal reactance, as well as that of the LC network, must be inductive. For the loop gain to be greater than unity, we see from Eq. (15-13) that X_1 cannot be too small. Hence the circuit will oscillate at a frequency which lies between ω_s and ω_p but close to the parallel-resonance value. Since $\omega_p \approx \omega_s$, the oscillator frequency is essentially determined by the crystal, and not by the rest of the circuit.

15-6 MULTIVIBRATORS

The oscillators described in the previous sections are one part of the class of *regenerative circuits*. We observed that at the frequency of oscillation, sinusoidal oscillators are positive-feedback amplifiers. Multivibrators are another important group of regenerative circuits that are used extensively in timing applications. Multivibrators are conveniently classified as (1) bistable circuits, (2) monostable circuits, or (3) astable circuits.

The bistable latches and flip-flops described in Secs. 7-1 to 7-3 can all be considered as *bistable multivibrators*. In Sec. 15-11 we describe the regenerative comparator (Schmitt trigger), another bistable circuit. One important characteristic of a bistable circuit is that it maintains a given output state (level) unless an external signal (trigger) is applied. Application of an appropriate external signal causes a change of state, and this output level is maintained indefinitely until a second trigger is applied. Thus a bistable circuit requires two external triggers before it returns to its initial state.

The *monostable*, or "one-shot," multivibrator generates a single pulse of specified duration in response to each external trigger signal. As its name implies, only one stable state exists. Application of a trigger causes a change

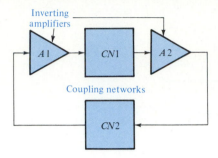

FIGURE 15-9
Basic block diagram of a multivibrator.

to the *quasistable state*. The circuit remains in the quasistable state for a fixed interval of time and then reverts to its original stable state. In effect, an internal trigger signal is generated which produces the transition to the stable state. Usually, the charging and discharging of a capacitor provides this internal trigger signal.

Astable, or *"free-running," multivibrators* have two quasistable states (no stable states), and circuit conditions oscillate between them. Note that no external signals are required to produce the changes in state. The duration the circuit remains in each state is determined by the component values. Because of the oscillation between states, astable circuits are used to generate square waves. Precise control of the period of the square wave, often by means of a crystal, enables such circuits to be used as the clock generators in digital systems.

Figure 15-9 shows a common multivibrator configuration. As indicated in Fig. 15-9, the two inverting amplifiers $A1$ and $A2$ form a positive-feedback amplifier. Often, logic gates are used in place of the amplifiers. As described in Sec. 6-2, the magnitude of the slope of the transfer characteristic between logic states is greater than unity and denotes amplification. The nature of the coupling networks between stages determines the type of multivibrator. When both $CN1$ and $CN2$ are resistive, bistable operation exists. A signal applied to $A1$ that produces a transition is transmitted through $CN1$ and causes $A2$ to change state. Similar action in $CN2$ forces $A1$ to remain in its new state until another trigger signal is applied.

However, if either $CN1$ or $CN2$ (or both) contains a series capacitor, dc signals cannot be transmitted indefinitely. Because capacitance voltage cannot change instantaneously, transitions in $A1$ ($A2$) are transmitted for a short duration. The action of the capacitors charging (discharging) as a result of the initial transition generates an internal "trigger," causing circuit operation to revert to its initial state. Monostable circuits employ one such coupling network, whereas both $CN1$ and $CN2$ are capacitive in astable multivibrators.

Monostable Multivibrators A simple monostable circuit which employs positive-logic CMOS NOR gates as the amplifying element is depicted in Fig. 15-10. Note that the coupling network between NOR gate 1 and NOR gate 2 is capacitive and between NOR gate 2 and NOR gate 1, resistive ($R = 0$). Let us assume that the NOR gates

FIGURE 15-10
A monostable multivi-
brator using NOR gates.

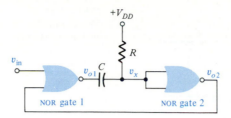

have $V(1) = V_{DD}$, $V(0) = 0$ and that $V_T > 0$ is the threshold voltage of the NMOS driver transistors. Furthermore, we assume for simplicity that the transition between states in the gates is instantaneous; thus the switching speed of the NOR gates is small compared to the duration of the desired output pulse.

Let us consider that v_{in}, the trigger signal, is as shown in Fig. 15-11a. For $t < 0$, no current exists in R and $v_x = V_{DD} = V(1)$. Consequently, the output of inverter-connected NOR gate 2 is $v_{o2} = V(0) = 0$. With both inputs to NOR gate 1 at $V(0)$, its output voltage $v_{o1} = V(1) = V_{DD}$. Hence the capacitor voltage v_C is

$$v_C = v_{o1} - v_x \quad \text{and} \quad v_C = V_{DD} - V_{DD} = 0 \tag{15-15}$$

Application of the trigger signal $v_{in} > V_T$ at $t = 0$ causes a transition in NOR gate 1, and v_{o1} becomes $V(0) = 0$ as seen in Fig. 15-11b. Because v_C cannot change instantaneously $[v_C(0^+) = 0]$, from Eq. (15-15), it follows that $v_x(0^+) = 0$. The application of $V(0)$ at the input of NOR gate 2 makes $v_{o2}(0^+) = V_{DD}$ (Fig. 15-11c). Consequently, $V(1)$ is transmitted to the input of NOR gate 1 and maintains $v_{o1} = 0$. The capacitor voltage v_C, however, tends to charge to $-V_{DD}$ through the resistance R as indicated in Fig. 15-12a. Hence v_x tends to increase from zero toward V_{DD} (Fig. 15-11d) as given by Eq. (15-16):

$$v_x(t) = V_{DD}(1 - \epsilon^{-t/RC}) \tag{15-16}$$

When $v_x = V_T$, NOR gate 2 changes state, $v_{o2} = 0$ and produces a transition in NOR gate 1 from $V(0)$ to $V(1)$. The time T_1 at which the transition occurs is obtained from Eq. (15-16) as follows:

$$v_x(T_1) = V_T = V_{DD}(1 - \epsilon^{-T_1/RC}) \quad \text{or} \quad \epsilon^{-T_1/RC} = \frac{V_{DD} - V_T}{V_{DD}}$$

Taking the logarithm of both sides and solving for T_1 yields

$$T_1 = RC \ln \frac{V_{DD}}{V_{DD} - V_T} \tag{15-17}$$

If $V_T = V_{DD}/2$, as is often the case in double-buffered CMOS gates, Eq. (15-17) reduces to

$$T_1 = RC \ln 2 \simeq 0.693RC \tag{15-18}$$

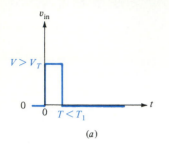

(a)

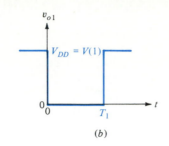

(b)

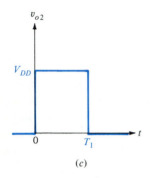

(c)

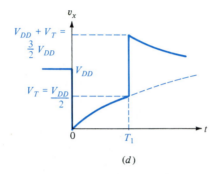

(d)

FIGURE 15-11
Waveforms for the monostable multivibrator in Fig. 15-10: (a) the trigger pulse, (b, c) the NOR-gate output voltages, (d) the input voltage to NOR gate 2.

Just prior to the transitions in the NOR gates, at $t = T_1$, $v_C (T_1) = -V_T$ [Eq. (15-15)]. At $t = T_1^+$, $v_{o1} = V_{DD}$ and to maintain $v_C (T_1^+) = -V_T$, $v_x(T_1^+) = -V_{DD} + V_T$. The capacitor voltage discharges to zero (Fig. 15-12b), and the circuit conditions return to those that existed for $t < 0$. The waveforms for various voltages are displayed in Fig. 15-11 for the condition that $V_T = V_{DD}/2$. Note that the discharge of the capacitor ($t > T_1$) has the same time constant RC as during the interval $0 \le t \le T_1$.

The input voltage v_x to NOR gate 2, as displayed in Fig. 15-11d, rises to $3V_{DD}/2$ at $t = T_1$. Often, this voltage level is excessive for the MOS devices in the NOR gate. To preclude this, *a catching diode*, as depicted in Fig. 15-13a, is used. The diode D is open-circuited during most of the cycle. However, while NOR gate 2 is turned on at $t = T_1$, D conducts and ensures that v_x does not exceed V_{DD} by more than the turn-on voltage V_γ of the diode. Actually, v_x is slightly higher than $V_{DD} + V_\gamma$ because of the small forward resistance R_f of the diode. Since $R_f \ll R$, their parallel combination is approximately R_f. Thus

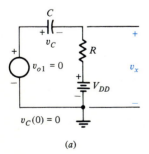

(a)

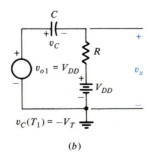

(b)

FIGURE 15-12
Equivalent circuits for the circuit in Fig. 15-10. Referring to Fig. 15-11: (a) $0 \le t \le T_1$, (b) $t \ge T_1$.

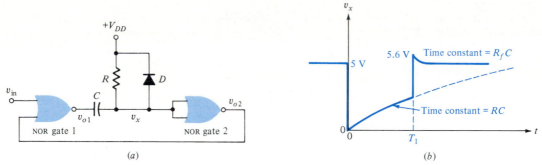

FIGURE 15-13

(a) The use of a catching diode D in a monostable multivibrator. (b) The waveform for the input voltage to NOR gate 2. Note that the peak voltage at $t - T_1$ is 5.6 V ($V_{DD} = V$ of the diode) rather than $3V_{DD}/2$ as in Fig. 15-11d.

the discharge of the capacitor from $V_{DD} + V_\gamma$ to V_{DD} has a time constant equal to R_fC and occurs quickly. The resultant waveform for v_x is displayed in Fig. 15-13b for $V_T = V_{DD}/2$ and $V_\gamma = 0.6$ V.

Monostable multivibrators of the type displayed in Fig. 15-10 are readily constructed by use of commercial CMOS gates. Since CMOS gates are double-buffered, the use of a 5-V supply renders $V_T = V_{DD}/2 = 2.5$ V. In addition, most CMOS gates have diode-protected inputs to prevent excessive voltages from being applied. Hence the catching diodes are fabricated on the chip. The only components that are externally connected are the timing elements R and C.

The basic 54/74 TTL family of logic gates can also be used to construct single-chip monostable circuits. The TI 9600 family employs a TTL NAND-gate input and a standard totem-pole output stage. The interior stages comprise the cascade of a bistable latch, monostable circuit, and Schmitt trigger (Sec. 15-9) as shown in Fig. 15-14. All five circuits are fabricated on a single chip. The pulse duration is established by the externally connected resistance R and capacitance C. Typically, $R \geq 5$ kΩ and $C \geq 1000$ pF; these values result in pulse widths greater than 1 μs. However, use of lower values of R and C results in reduced pulse duration by one order of magnitude ($\approx$ 100 ns). If the totem-pole stage is eliminated and the output taken at the Schmitt trigger, both positive and negative pulses are available (dashed-line outputs in Fig. 15-14).

FIGURE 15-14

Block diagram of a TTL monostable multivibrator.

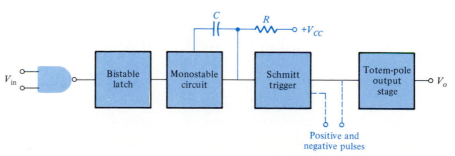

FIGURE 15-15
(a) A NOR-gate astable (free-running) multivibrator. Both NOR gates are connected as inverters. (b) The crystal converts the circuit in (a) into a simple clock generator.

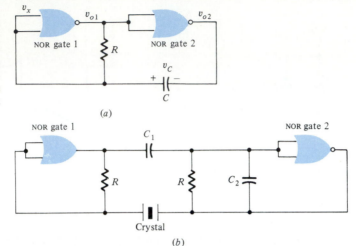

(a)

(b)

Astable Multivibrators

The circuit in Fig. 15-10 can be modified to form an astable multivibrator as displayed in Fig. 15-15a. Note that both NOR gates are connected as inverters. Let us assume that the NOR gates have $V(0) = 0$, $V(1) = V_{DD}$, the supply voltage, and a threshold voltage $V_T = V_{DD}/2$. Consider that, as depicted in Fig. 15-16, NOR gate 1 undergoes a transition from $V(1)$ to $V(0)$ at $t = 0$. Thus, at $t = 0^-$, $v_{o2} = V_{DD}$ and $v_{o2} = 0$. The input voltage to NOR gate 1 is $v_x = V_T$ as it is about to switch states. Since $v_C = v_x - v_{o2}$, $v_C(0^-) = V_T$. At $t = 0^+$, just after NOR gate 1 changes state, $v_{o1}(0^+) = 0$, which causes NOR gate 2 to undergo a transition, making $v_{o2}(0^+) = V_{DD}$. Because v_C cannot change instantaneously, v_x tends toward $V_T + V_{DD}$. (The use of a catching diode, however, limits v_x to $V_{DD} + V_\gamma$.) With $v_{o1} = 0$ and $v_{o2} = V_{DD}$, the capacitor charges exponentially to V_{DD} with a time constant $\tau = RC$. (The voltage v_C tends toward $-V_{DD}$.) As the capacitor charges, v_x decreases toward zero. At $t = T_1$, $v_x = V_T$, NOR gate 1 is turned off, $v_{o1} = V_{DD}$ causing NOR gate 2 to make a transition. The output voltage v_{o2} of NOR gate 2 drops to zero, and, to maintain v_C constant during switching, v_x decreases by V_{DD}. The capacitor voltage v_C charges exponentially to $+V_{DD}$ ($v_{o1} = V_{DD}$ and $v_{o2} = 0$) with a time constant

FIGURE 15-16
Waveforms for the circuit in Fig. 15-15: (a, b) the output voltages of the NOR gates; (c) the input voltage to NOR gate 1; (d) the capacitor voltage.

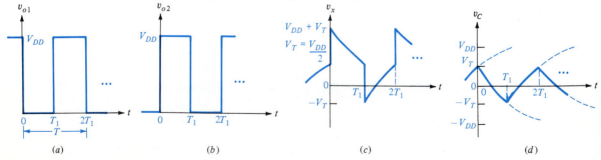

(a) (b) (c) (d)

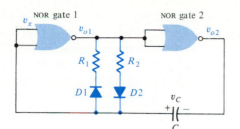

FIGURE 15-17
Astable circuit with catching diodes. Choosing different resistance values ($R_1 \neq R_2$) results in an asymmetric square-wave output voltage.

$\tau = RC$. As v_C tends toward V_{DD}, so does v_x and, at $t = T_2$, $v_x = V_T$, completing one cycle. This process is repeated in each cycle as indicated by the waveforms in Fig. 15-16. The output square wave is symmetrical, with each gate having an output voltage V_{DD} for one half-cycle and $V(0) = 0$ for the other halfcycle. The time for half the period is $RC \ln 3$. With ideal catching diodes, $T = 2T_1 = 2RC \ln 2 = 1.39RC$; the frequency of oscillation $f_0 = 1/T$ is

$$f_o = \frac{1}{T} = \frac{1}{1.39RC} = \frac{0.721}{RC} \qquad (15\text{-}19)$$

The circuit can be designed to produce an asymmetrical square wave by changing either V_{DD} or V_T so that $V_T \neq V_{DD}/2$ (Prob. 15-26). The diode-resistor combinations in Fig. 15-17 are also used to generate an asymmetric output signal (Prob. 15-27). In this circuit the diodes control the charging path; selection of $R_1 \neq R_2$ results in a different time constant for each transition period leading to an asymmetric square wave.

The circuit in Fig. 15-15a is modified as shown in Fig. 15-15b to form a simple clock generator in which the crystal frequency f_0 precisely controls the period of the square wave. The capacitor C_2 is chosen so that $2\pi C_2 R f_0 \simeq 1$ (that is, the impedance of $R \parallel sC_2$ has a pole at $s = -1/2\pi f_0$ and is therefore open-circuited at f_0). The capacitor C_1 is selected to have a negligible reactance at f_0. Thus, at the oscillation frequency the circuits in Fig. 15-15a and b are identical once we recognize that the crystal replaces C in Fig. 15-15a. The two capacitors help to suppress the higher harmonics generated and provide a stable output frequency. This simple clock generator can operate at high frequencies (< 30 MHz) if the CMOS series 74HC or TTL series 74LS or 74ALS inverters are used.

15-7 COMPARATORS

The mono- and astable circuits discussed in Sec. 15-6 exploited the voltage-controlled switching action of CMOS digital logic gates. That is, the input voltage level determined the binary state of the output, $V(0)$ or $V(1)$. In Secs. 3-12 and 10-15 we showed that the differential amplifier (emitter-coupled pair) exhibited a binary output voltage for input signals $|v_i| > 4V_T$. Operated in this fashion, the differential amplifier acts as an analog comparator and thus is useful in waveform generation.

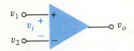

FIGURE 15-18
The Op-Amp operated open-loop becomes a comparator.

An *analog comparator,* or simply a *comparator,* has two input voltages v_1 and v_2 and one output voltage v_o. Often, one input (v_2) is a constant reference voltage V_R, and the other is a time-varying signal. Recall that this arrangement is used in the input stage of the ECL OR/NOR gate described in Sec. 6-14. The ideal comparator in Fig. 15-18, having the voltage transfer characteristic shown in Fig. 15-19a, has a constant output voltage $v_o = V(0)$ if $v_1 - v_2 = v_i < 0$ and a different constant voltage $v_o = V(1)$ if $v_i > 0$. Hence, if $v_2 = V_R$, a reference voltage, $v_o = V(0)$ when $v_1 < V_R$ and $v_o = V(1)$ when $v_1 > V_R$. Clearly, the input is *compared* with the reference and the output is digitized into one of two states: a 0 level of voltage $V(0)$ and a 1 level of $V(1)$. Voltages $V(0)$ and $V(1)$ compatible with TTL, ECL, or MOS logic levels may be obtained. Other limiting voltages, such as ± 10 V, are also available.

As stated previously, the voltage transfer characteristic of a differential pair approximates that of the ideal comparator. However, the total input swing between the two extreme output levels is $\sim 8V_T = 200$ mV. This range may be reduced drastically by cascading the differential amplifier with other high-gain stages. Since this configuration corresponds to the Op-Amp topology shown in Fig. 14-1, an Op-Amp may be used (open-loop) as a comparator. A typical Op-Amp transfer characteristic is given by the solid curve in Fig. 15-19b. It is now observed that the change in the output state takes place with an increment in input Δv_i of only 2 mV. Note that the input offset voltage contributes an error in the point of comparison between v_1 and V_R of the order of 1 mV. For some applications this offset may be too large and it will be necessary to balance it out, as indicated in Fig. 14-22.

A number of operational amplifiers have been designed specifically for comparator applications and are designated on the manufacturer's specification sheets as *voltage comparator/buffer* instead of Op-Amp. Since a comparator is not intended to be used with negative feedback, frequency-compensation components may be omitted, so that greater bandwidth (higher speed) is attainable than with Op-Amps. The designation "buffer" denotes that the com-

FIGURE 15-19
The transfer characteristic of (a) an ideal and (b) a practical (commercial) comparator.

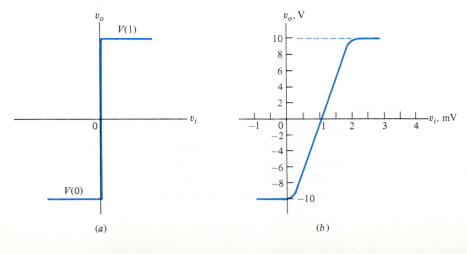

parator will not load down the signal source because of its very high input resistance. Among the many comparator chips available are the Fairchild μA710, the National LM111, the Analog Devices AD 604, and the Harris HA 2111. The uncertainty region Δv_i may be as small as 15 μV and the *response time* (the interval necessary for the comparator to change state) ranges from about 20 to 200 ns. Packages with two or four independent comparators are also available. Some chips are designed with a digital-signal strobe input so that the comparator may be disabled during input transients.

To obtain limiting output voltages, which are independent of the power-supply voltages, a resistor R and two back-to-back Zener diodes are added to clamp the output of the comparator, as indicated in Fig. 15-20a. The resistance value is chosen so that the avalanche diodes operate at the recommended Zener current. The solid curves give the output v'_o across the diodes, whereas the dashed curves represent the output v_o from the comparator. If the input signal is applied at the noninverting terminal and the reference V_R is tied to the inverting terminal, the noninverting comparator is obtained. If the positions of v_i and v_R are interchanged, the inverting-comparator characteristic results. The limiting voltages of v'_o are $V_{Z1} + V_D \equiv V_o$ and $-(V_{Z2} + V_D) \equiv -V_o$, where V_D ($\sim$0.7 V) is a pn diode forward voltage. A second advantage of adding the Zener diodes is that the limiting may be much sharper for v'_o than for v_o. A disadvantage is the poor transient response of the avalanche diodes.

The comparator has many uses in signal-processing and waveform-generating circuits. Each application exploits the ability of the comparator to detect whether one signal is greater or less than a second signal (or reference voltage). The circuits discussed in this chapter deal with waveform generation; other signal-processing applications are treated in Chap. 16.

15-8 SQUARE-WAVE GENERATION FROM A SINUSOID
The comparator performs highly nonlinear waveshaping because the output bears no resemblance to the input waveform. It is often used to transform a signal which varies slowly with time to another which exhibits an abrupt change. One such application is the generation of a square wave from a sinusoidal signal.

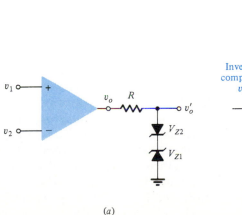

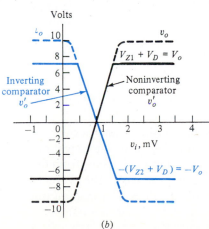

(a)

(b)

FIGURE 15-20

(a) A comparator cascaded with a resistor–Zener diode combination. (b) The transfer characteristics for v_o and v'_o. If $v_1 = v_i$ and $v_2 = V_R$, the comparator is noninverting (black curve). An inverting comparator is obtained (blue curve) for $v_1 = V_R$ and $v_2 = v_i$.

FIGURE 15-21

A zero-crossing detector converts a sinusoid v_1 into a square wave v_o. The pulse waveforms v' and v_L result from v_o being fed into a short time-constant RC circuit in cascade with a diode clipper.

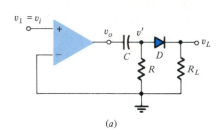

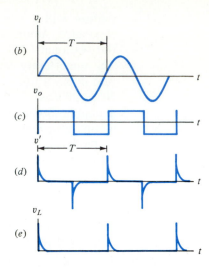

Timing-Markers Generator from a Sine Wave

If V_R is set equal to zero, the output will change from one state to the other very rapidly (limited by the slew rate) every time that the input passes through zero. Such a configuration is called a *zero-crossing detector*. Among the many applications of the zero-crossing detector are the following.

If the input to a comparator is a sine wave, the output is a square wave. If a zero-crossing detector is used (Fig. 15-21a), a symmetrical square wave results, as shown in Fig. 15-21c. This idealized waveform is shown with vertical sides which, in reality, should extend over a range of a fraction of a millivolt of input voltage v_i.

Timing-Markers Generator from a Sine Wave The square-wave output v_o of the preceding application is applied to the input of an RC series circuit (Fig. 15-21a). If the time constant RC is very small compared with the period T of the sine-wave input, the voltage v' across R is a series of positive and negative pulses, as indicated in Fig. 15-21. If v' is applied to a clipper with an ideal diode (Fig. 15-21a), the load voltage v_L contains only positive pulses (Fig. 15-21e). Thus the sinusoid has been converted into a train of positive pulses whose spacing is T. These may be used for timing markers (on the sweep voltage of a CRT, for example).

Note that the waveshaping performed by the configuration in Fig. 15-21a is very drastic—a sinusoid having been converted into either a square wave or a pulse train.

Spurious positive- and negative-voltage spikes, called *noise,* superimposed on the input signal in the neighborhood of the amplitude V_R, may cause the output to "chatter" (change from one binary voltage to the other) several times before settling down to the correct level. This difficulty can be avoided, and also reduced values of transition time can be obtained, if *positive feedback* or *regeneration* is added to a comparator, as discussed in the following section.

15-9 REGENERATIVE COMPARATOR (SCHMITT TRIGGER)

The transfer characteristic in Fig. 15-20b makes the change in output from -7 to $+7$ V for a swing in input of about 1.0 mV. Hence, the voltage gain is 14,000. By employing positive feedback the gain may be increased greatly. Consequently the total output excursion takes place in a time interval during which the input is changing by much less than 1 mV. Theoretically, if the return ratio T is adjusted to be -1, then the gain with feedback A_V becomes infinite [Eq. (12-4)]. Such an idealized situation results in an abrupt (zero rise time) transition between the extreme values of output voltage. If a loop gain in excess of unity is chosen, the output waveform continues to be virtually discontinuous at the comparison voltage. However, the circuit now exhibits a phenomenon called *hysteresis,* or *backlash,* which is explained in the following.

The regenerative comparator of Fig. 15-22a is commonly referred to as a *Schmitt trigger* (after the inventor of a vacuum-tube version of this circuit). The input voltage is applied to the inverting terminal 2 and the feedback voltage to the noninverting terminal 1. Assuming that the output resistance of the comparator is negligible compared with $R_1 + R$, we obtain

$$v_1 = \frac{R_2}{R_1 + R_2} v_o$$

FIGURE 15-22

(a) The regenerative comparator or Schmitt trigger. The output waveforms showing a transition (b) from $+V_o$ to $-V_o$ and (c) from $-V_o$ to $+V_o$. (d) The output voltage for one cycle showing the hysteresis ($V_1 - V_2$).

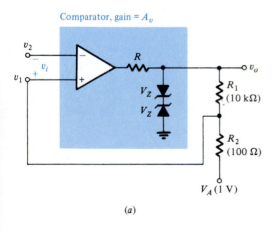

(a)

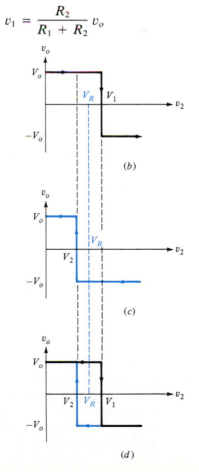

(b)

(c)

(d)

Since $v_1 = v_i$ with $v_2 = 0$, $v_o = A_v v_i$ and use of small-signal analysis gives the return ratio as

$$T = \frac{-R_2 A_v}{R_1 + R_2}$$

Clearly, with $A_v > 0$, $T < 0$ and the feedback is positive (regenerative). For $R_1 = 10 \text{ k}\Omega$, $R_2 = 100 \Omega$ and $A_v = 14{,}000$:

$$T = -\frac{0.1 \times 14{,}000}{10 + 0.1} = -139$$

It is easily verified that the feedback is regenerative. If the output *increases* by Δv_o, the signal fed back to v_1, the noninverting terminal, is $\Delta v_o R_2/(R_1 + R_2)$. Hence v_o will *increase* further by $\Delta v_o R_2 A_v/(R_1 + R_2) = -T \Delta v_o$, indicating positive feedback.

Let $V_o \equiv V_Z + V_D$ and assume that $v_2 < v_1$ so that $v_o = +V_o$. From Fig. 15-22 we find that the voltage at the noninverting terminal is given by

$$v_1 = V_A + \frac{R_2}{R_1 + R_2}(V_o - V_A) \equiv V_1 \qquad (15\text{-}20)$$

If v_2 is now increased, the v_o remains constant at V_o, and $v_1 = V_1 = $ constant until $v_2 = V_1$. At this *threshold, critical,* or *triggering voltage,* the output regeneratively switches to $v_o = -V_o$ and remains at this value as long as $v_2 > V_1$. This transfer characteristic is indicated in Fig. 15-22b.

The voltage at the noninverting terminal for $v_2 > V_1$ is

$$v_1 = V_A - \frac{R_2}{R_1 + R_2}(V_o + V_A) \equiv V_2 \qquad (15\text{-}21)$$

For the parameter values given in Fig. 15-22 and with $V_o = 7$ V, we obtain

$$V_1 = 1 + \frac{0.1 \times 6}{10.1} = 1 + 0.059 = 1.059 \text{ V}$$

$$V_2 = 1 - \frac{0.1 \times 8}{10.1} = 1 - 0.079 = 0.921 \text{ V}$$

Note that $V_2 < V_1$, and the difference between these two values is called the *hysteresis* V_H:

$$V_H = V_1 - V_2 = \frac{2R_2 V_o}{R_1 + R_2} = 0.138 \text{ V} \qquad (15\text{-}22)$$

If we now decrease v_2, the output remains at $-V_o$ until v_2 equals the voltage at terminal 1 or until $v_2 = V_2$. At this voltage a regenerative transition takes place and, as indicated in Fig. 15-22c, the output returns to $+V_o$ almost instantaneously. The complete transfer function is indicated in Fig. 15-22d, where the portions without arrows may be traversed in either direction, but the other segments can only be obtained if v_2 varies as indicated by the arrows. Note that because of the hysteresis, the circuit triggers at a higher voltage for increasing than for decreasing signals.

We note above that transfer gain increases from 14,000 toward infinity as the return ratio decreases from zero to -1, and that there is no hysteresis as long as $-T \leq 1$. However, adjusting the gain precisely to -1 is not feasible. The comparator parameters and, hence, the gain A_v are variable over the signal excursion. Hence an adjustment which ensures that the maximum $|T|$ is unity would result in voltage ranges where this amplification is less than unity, with a consequent loss in speed of response of the circuit. Furthermore, the circuit may not be stable enough to maintain T precisely -1 for a long period of time without frequent readjustment. In practice, therefore, $|T|$ in excess of unity is chosen and a small amount of hysteresis is tolerated. In some applications a large backlash range will not allow the circuit to function properly. Thus if the peak-to-peak signal were smaller than V_H, then the Schmitt circuit, having responded at a threshold voltage by a transition in one direction, would never reset itself. In other words, once the output has jumped to, say, V_o, it would remain at this level and never return to $-V_o$.

By the same argument given in the preceding paragraph it follows that, if v_2 just exceeds V_1, an output transition takes place and v_o remains at $-V_o$, even if there is some noise superimposed on the input signal. As long as the peak-to-peak noise voltage does not exceed the hysteresis V_H, v_2 cannot fall below V_2 and, therefore, a change of state back to $+V_o$ is avoided. In other words, the noise chattering mentioned in Sec. 15-8 is eliminated.

The output offset voltage for the Schmitt trigger in Fig. 15-22a is $I_B R_B$, where I_B is the input bias current and $R_B \equiv R_1 \parallel R_2$. The addition of a resistance R_B in series with the input signal v_2 results in an output offset voltage $I_{io}R_B$, where I_{io} is the input offset current. Since $I_{io} < I_B$, the resultant offset voltage is reduced.

One of the most important uses made of the Schmitt trigger is to convert a slowly varying input voltage into an output waveform displaying an abrupt, almost discontinuous, change that occurs at a precise value of input voltage. For example, the use of the Schmitt trigger as a squaring circuit is illustrated in Fig. 15-23. The input signal is arbitrary except that it has a sufficiently large excursion to carry the input beyond the limits of the hysteresis range V_H. The

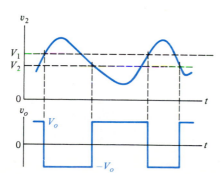

FIGURE 15-23

Response of the inverting Schmitt trigger to an arbitrary input signal.

output is usually an asymmetric square wave (as indicated in Fig. 15-23), the amplitude of which is independent of the peak-to-peak value of the input signal. Clearly, the output waveform has much faster leading and trailing edges than does the input.

Symmetrical square waves can be generated by setting $V_A = 0$, and, from Eqs. (15-20) and (15-21), it follows that $V_2 = -V_1 = -R_2V_o/(R_1 + R_2)$. Application of an input sinusoid of frequency $f = 1/T$ and peak amplitude V_M to such a comparator results in a symmetrical output waveform of half-period $T/2$. The leading and trailing edges of the square wave do not occur at the times the sine wave passes through zero as in the circuit in Fig. 15-21a. These edges are shifted in phase by 0, where $0 = \sin^{-1} V_1/V_m$.

Special-purpose Schmitt triggers are commercially available. The TTL family TI132 chip behaves as a positive-logic NAND gate with a totem-pole output and hysteresis of 0.8 V ($V_1 = 1.7$ V and $V_2 = 0.9$ V). Four 2-input NAND Schmitt triggers are contained in the TI132 package. This regenerative comparator topology is the basis of the TI9600 monostable multivibrator (Sec. 15-6).

The Emitter-Coupled Schmitt Trigger   The basic emitter-coupled pair can be converted into a regenerative comparator as illustrated in Fig. 15-24. The resistances R_1 and R_2 are unequal ($R_1 > R_2$) and, hence, Q1 and Q2 have different currents when saturated. This difference results in the hysteresis as different input voltages are required to saturate and cut off Q1 and Q2.

Consider that V_{in} is sufficiently small so as to cut off Q1. The current in R_1 is sufficient to cause Q2 to just barely saturate, making $V_o = V_{CC} - I_{C2(sat)}R_2$. For $\beta_F \gg 1$, the voltage $V_E \approx I_{C2(sat)}R_E$; thus to turn Q1 on, V_{in} must be increased to at least $V_{BE(ON)} + V_E$. As V_{in} is increased above this threshold, Q1 conducts and the voltage V_{C1} decreases, cuts off Q2, and causes $V_o \approx V_{CC}$. If V_{in} is sufficiently large to just barely saturate Q1, the voltage $V_E \approx I_{C1(sat)}R_E$. Since $R_1 > R_2$, $I_{C1(sat)} < I_{C2(sat)}$. Consequently, V_{in} must now be decreased below $I_{C1(sat)}R_E + V_{BE(ON)}$ to turn off Q1. This threshold level is below that required to turn Q1 on, and the difference between these levels is the hysteresis voltage V_H. Note that V_H depends on the degree of mismatch in R_1 and R_2.

FIGURE 15-24
An emitter-coupled
Schmitt trigger.

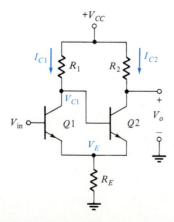

FIGURE 15-25

(*a*) A square-wave generator. (*b*) Output and capacitor voltage waveforms.

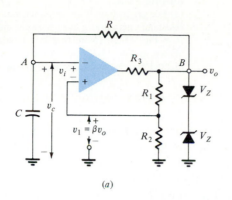

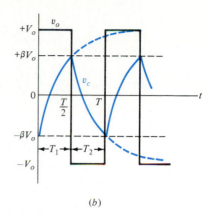

(*a*)

(*b*)

The qualitative analysis in the preceding paragraph assumed that $Q1$ and $Q2$ are just barely saturated. This is not necessary for circuit operation as $Q1$ and $Q2$ can be driven into saturation. Typically, emitter-coupled Schmitt triggers are designed with V_H in the order of several tenths of a volt (see Probs. 15-43 and 15-44.)

15-10   SQUARE-WAVE AND TRIANGLE-WAVE GENERATORS

The inverting Schmitt trigger can be used to obtain a free-running square-wave generator (astable multivibrator) by connecting an RC feedback network between the output and the inverting input. The circuit is displayed in Fig. 15-25*a* and indicates that the external signal is replaced by the RC feedback network. In Fig. 15-25*a*, the differential input voltage v_i is given by

$$v_i = v_c - v_1 = v_c - \frac{R_2}{R_1 + R_2}v_o = v_c - \beta v_o \qquad (15\text{-}23)$$

From the ideal-comparator characteristic $v_o = V_Z + V_D = V_o$ if $v_i < 0$ and $v_o = -V_o$ if $v_i > 0$. Consider an instant of time when $v_i < 0$ or $v_c < \beta v_o = \beta V_o$. The capacitor C now charges exponentially toward V_o through the RC combination. The output remains constant at V_o until v_c equals $+\beta V_o$, at which time the comparator output reverses to $-V_o$. Now v_c charges exponentially toward $-V_o$. The output voltage v_o and capacitor voltage v_c waveforms are shown in Fig. 15-24. If we let $t = 0$ when $v_c = -\beta V_o$ for the first half-cycle, we have (since v_c approaches V_o exponentially with a time constant RC)

$$v_c(t) = V_o [1 - (1 + \beta)\epsilon^{-t/RC}] \qquad (15\text{-}24)$$

Since at $t = T/2$, $v_c(t) = +\beta V_o$, we find T, solving Eq (15-24), to be given by

$$T = 2RC \ln \frac{1 + \beta}{1 - \beta} = 2RC \ln \left(1 + \frac{2R_1}{R_2}\right) \qquad (15\text{-}25)$$

Note that T is independent of V_o.

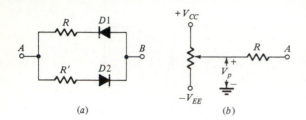

FIGURE 15-26

(a) For generation of an asymmetrical square wave, the diode-resistance network is used between nodes A and B to replace R in Fig. 15-25. (Note that this is equivalent to the network shown in Fig. 15-17.) (b) Alternatively, the configuration indicated may be tied to node A in Fig. 15-25 so that $T_1 \neq T_2$.

This square-wave generator is particularly useful in the frequency range of 10 Hz to 10 kHz. At higher frequencies the slew rate of the Op-Amp limits the slope of the output square wave. Waveform symmetry depends on the matching of the two Zener diodes (Prob. 15-39).

If it is desired that the output be $\pm V_o$ but that $T_1 \neq T_2$ in Fig. 15-25, the resistor R between nodes A and B is replaced by the network shown in Fig. 15-26a. This is the same technique used in the CMOS astable circuit in Fig. 15-18. During the interval when the output is positive, $D1$ conducts but $D2$ is OFF. Hence, the circuit reduces to that in Fig. 15-25 except that V_o is reduced by the diode drop. Since the period is independent of V_o, then T_1 is given by $T/2$ in Eq. (15.25). During the interval when the output is negative, $D1$ is OFF and $D2$ conducts. Hence, the discharge-time constant is now $R'C$, and, therefore, T_2 is given by $T/2$ in Eq. (15-25) with R replaced by R'. If $R' = 2R$, clearly, $T_2 = 2T$.

An alternative method for obtaining an unsymmetrical square wave is to connect the network in Fig. 15-26b to node A of Fig. 15-25a. Assume that the potentiometer resistance is small compared with R and that the voltage from the potentiometer resistance is small compared with R and that the voltage from the potentiometer arm to ground is V_p. Then the capacitor charges with a time constant $RC/2$ toward $(V_p + V_o)/2$, but C discharges toward $(V_p - V_o)/2$ (with the same time constant). Hence $T_1 \neq T_2$.

Triangle-Wave Generators   We observe from Fig. 15-25b that the exponential charge and discharge of the capacitor C makes v_C an almost triangular waveform. To linearize the triangles, it is required that C be charged with a constant current, thus making v_C vary linearly with time (a ramp) rather than the exponential current supplied through R in Fig. 15-25. Recall that the compensating capacitance connected between the ouput and input of the high-gain stage of an Op-Amp is supplied by a constant current and resulted in the slew-rate limitation. An Op-Amp with a feedback capacitor C (an integrator) is used to supply a constant current to C in the triangle-wave generating circuit in Fig. 15-27. Because of the phase reversal in the Op-Amp integrator, the output of this stage is fed back to the noninverting terminal of the comparator rather than to the inverting terminal as displayed in Fig. 15-25. Thus the comparator behaves as a noninverting Schmitt trigger. In effect, the output of the Op-Amp stage is used in place of the reference voltage V_A in the Schmitt trigger.

To find the maximum value of the triangular waveform assume that the output voltage v_o of the Schmitt trigger is at its negative value, $-(V_Z + V_D) = -V_o$.

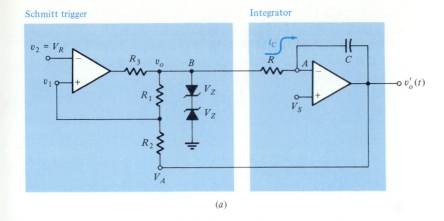

(a)

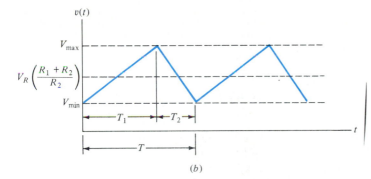

(b)

FIGURE 15-27

(a) A triangle-wave generator. (b) Output waveform. [*Note*: $T_1 = T_2$ if $V_S = 0$. Also, $V_{max} = V_o R_2/R_1 = -V_{o(min)}$ if $V_R = 0$. The square-wave output is $-V_o$ during the interval T_1 and $+V_o$ for the interval $T_2 - T_1$.]

With a negative input, the output $v'(t)$ of the integrator is an *increasing* ramp. The voltage at the noninverting comparator input v_1 is obtained by the use of superposition and is

$$v_1 = -\frac{V_o R_2}{R_1 + R_2} + \frac{v_o' R_1}{R_1 + R_2} \qquad (15\text{-}26)$$

When v_1 rises to V_R, the comparator changes state, $v_o = +V_o$, and $v_o'(t)$ starts *decreasing* linearly. Hence the peak V_{max} of the triangular waveform occurs for $v_1 = V_R$. From Eq. (15-26),

$$V_{max} = V_R \frac{R_1 + R_2}{R_1} + V_o \frac{R_2}{R_1} \qquad (15\text{-}27)$$

By a similar argument it is found that

$$V_{min} = V_R \frac{R_1 + R_2}{R_1} - V_o \frac{R_2}{R_1} \qquad (15\text{-}28)$$

The peak-to-peak swing is

$$V_{max} - V_{min} = 2V_o \frac{R_2}{R_1} \qquad (15\text{-}29)$$

The triangular waveform is indicated in Fig. 15-27b. From Eqs. (15-27) and (15-28) it should be clear that the average value is $V_R(R_1 + R_2)/R_1$. Note that, if $V_R = 0$, the waveform extends between $- V_o R_2/R_1$ and $+ V_o R_2/R_1$. Its displacement in voltage is controlled by an adjustment of V_R and the peak-to-peak swing is varied by changing the ratio R_2/R_1.

We now calculate the sweep times T_1 and T_2 for $V_s = 0$. The capacitor-charging current is

$$i_c = C \frac{dv_c}{dt} = - C \frac{dv'_o}{dt} \tag{15-30}$$

where $v_c = - v'_o$ is the capacitor voltage. For $v_o = - V_o$, $i = - V_o/R$, and the positive-sweep speed is $dv'_o/dt = V_o/RC$. Hence

$$T_1 = \frac{V_{max} - V_{min}}{V_o/RC} = \frac{2R_2 RC}{R_1} \tag{15-31}$$

where use was made of Eq. (15-29). since the negative-sweep speed has the same magnitude as that calculated above, $T_2 = T_1 = T/2 = 1/2f$, where the frequency f is given by

$$f = \frac{R_1}{4R_2 RC} \tag{15-32}$$

Note that the frequency is independent of V_o. The maximum frequency is limited by either the slew rate of the integrator or its maximum output current, which determines the charging rate of C. The slowest sweep is limited by the bias current of the Op-Amp. Decade changes in frequency are obtained by switching capacitance values by factors of 10, and increments of frequency within a decade result from continuous variations of resistance R.

Duty-Cycle Modulation   If unequal sweep intervals $T_1 \neq T_2$ are desired, then R in Fig. 15-27a may be replaced by the network in Fig. 15-26a. An alternative method is to apply an adjustable voltage $V_s \neq 0$ to the noninverting terminal of the integrator, as indicated in Fig. 15-27a. The positive-sweep speed is now $(V_o + V_S)/RC$ and the negative-ramp slope is $(V_o - V_S)/RC$. (Why?) The peak-to-peak triangular amplitude is unaffected by the symmetry control voltage V_S. Hence

$$\frac{T_1}{T_2} = \frac{V_o - V_S}{V_o + V_S} \tag{15-33}$$

The oscillation frequency can be shown (Prob. 15-40) to be given by Eq. (15-32) multiplied by $[1 - (V_S/V_o)^2]$. The frequency is lowered for $V_S \neq 0$.

The *duty cycle* δ of a square- or triangular-wave oscillator is defined as T_1/T, where $T = T_1 + T_2$. From Eq. (15-33) it follows that

$$\delta \equiv \frac{T_1}{T} = \frac{1}{2}\left(1 - \frac{V_S}{V_o}\right) \tag{15-34}$$

The system in Fig. 15-27 with the addition of V_S is a *duty-cycle modulator*. The duty cycle varies linearly with V_S and extends from 0 for $V_S = V_o$, to $\frac{1}{2}$ for $V_S = 0$, and to 1 for $V_S = -V_o$.

Voltage-Controlled Oscillator (VCO)

Note that V_S in Fig. 15-27 not only modifies the duty cycle but also affects the period $T = 1/f$. This is an example of a VCO or of a *voltage-to-frequency converter*. However, f is a nonlinear function of V_S, since the frequency depends upon $1 - (V_S/V_o)^2$.

A system for obtaining a square- or triangular-waveform generator whose frequency depends *linearly* on a modulation voltage v_m is indicated in Fig. 15-28. The CMOS inverter formed by $Q1$ and $Q2$ acts as a single-pole, double-throw switch (SPDT). The buffer stage shown in Fig. 15-28 is a voltage follower which drives the integrator from a low impedance. This system differs functionally from that of Fig. 15-27 in that now the sweep speed is determined by v_m but the waveform amplitude continues to be fixed by the comparator parameters, as in Fig. 15-25, namely, $\pm \beta V_o$. The negative voltage $-v_m$ is obtained from an Op-Amp unity gain inverter.

Assume that the Schmitt comparator output is $v_o = V_o$, where V_o exceeds the maximum value of v_m. Then, for the CMOS inverter, switch $Q1$ is OFF and

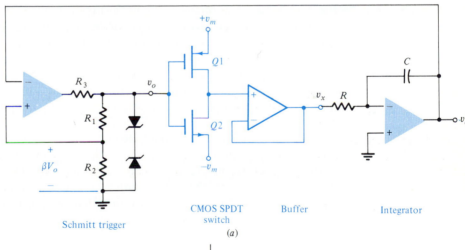

Schmitt trigger CMOS SPDT switch Buffer Integrator

(a)

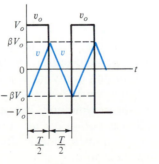

(b)

FIGURE 15-28

(a) A voltage-controlled oscillator (VCO) whose frequency varies linearly with the modulating voltage v_m. *(b)* The square wave v_o and triangular wave v_o'.

$Q2$ is ON. The input v_x to the integrator (the output of the voltage follower) is $-v_m$. Hence $v_o'(t)$ increase linearly with a sweep speed of v_m/RC V/s until v_o' reaches the comparator threshold level $\beta V_o = V_o R_2/(R_1 + R_2)$. Then the Schmitt output changes state to $v_o = -V_o$, as depicted in Fig. 15-28b. Now $Q1$ is ON, $Q2$ is OFF, and the CMOS switch output becomes $+v_m$, resulting in a linear negative ramp $v_o' = -v_m t/RC$ until the negative threshold $-\beta V_o$ is reached. Clearly, the two half-cycles are identical and

$$\frac{v_m}{RC}\frac{T}{2} = \beta V_o - (-\beta V_o) = 2\frac{R_2}{R_1 + R_2}V_o \tag{15-35}$$

The frequency of the oscillator is given by $f = 1/T$, or

$$f = \frac{R_1 + R_2}{4RCR_2}\frac{v_m}{V_o} \tag{15-36}$$

clearly indicating that this VCO frequency varies linearly with the modulation voltage v_m. Experimentally, it is found that this linearity extends over more than three decades (from below 2 mV to above 2 V). The system of Fig. 15-28 is that of a *frequency-modulated* square or triangular waveform.

15-11 PULSE GENERATORS

The square-wave generator in Fig. 15-25 is modified as shown in Fig. 15-29a to operate as a monostable multivibrator by adding a clamping diode ($D1$) across C. A narrow negative triggering pulse v_t is applied to the noninverting terminal through diode $D2$. To see how the circuit operates, assume that it is in its stable state with the output at $v_o = +V_o$ and with the capacitor clamped at the diode $D1$ ON voltage $V_1 \approx 0.7$ V (with $\beta V_o > V_1$). If the trigger amplitude is greater than $\beta V_o - V_1$, it will cause the comparator to switch to an output $v_o = -V_o$. As indicated in Fig. 15-29b, the capacitor will now charge exponentially with a time constant $\tau = RC$ through R toward $-V_o$ because $D1$ becomes reverse-biased. When v_c becomes more negative than $-\beta V_o$, the comparator output swings back to $+V_o$. The capacitor now starts charging toward $+V_o$ through R until v_c reaches V_1 and C becomes clamped again at $v_c = V_1$. In Prob. 15-41 we find that the pulse width T is given by

$$T = RC \ln\frac{1 + V_1/V_o}{1 - \beta} \tag{15-37}$$

If $V_o \gg V_1$ and $R_2 = R_1$ so that $\beta = \frac{1}{2}$, then $T = 0.69RC$.

The triggering pulse width T_p must be much smaller than the duration T of the generated pulse. The diode $D2$ is not essential but it serves to avoid malfunctioning if any positive noise spikes are present in the triggering line.

Since the one-shot generates a rectangular waveform which starts at a definite instant of time and, hence, can be used to gate other parts of a system, it is called a *gating circuit*. Furthermore, since it generates a fast transition at a predetermined time T after the input trigger, it is also referred to as a *time-delay circuit*.

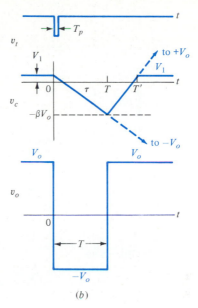

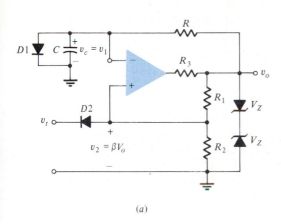

FIGURE 15-29
(a) A monostable multivibrator. (b) The waveforms for the negative, short-duration trigger pulse v_t, the capacitor voltage v_c, and the negative-output pulse v_o. (It is assumed that $T > T_p$.)

Note that the capacitor voltage v_c in Fig. 15-29b does not reach its quiescent value $v_c = V_1$ until time $T' > T$. Hence, there is a *recovery time* $T' - T$ during which the circuit may not be triggered again. In other words, the next synchronizing trigger must be delayed from the previous input pulse by at least T' seconds. An alternative monostable circuit with a faster recovery time is given in Prob. 15-42.

A Retriggerable Monostable Multivibrator Consider the configuration in Fig. 15-30a. In the quiescent state (before a trigger is applied) the JFET is cut off by the reverse-biased gate-to-source voltage $-V_{GG}$ of Q. The capacitor is charged to the supply voltage V_{CC} so that the voltage at the inverting terminal of the comparator is $v_c = V_{CC}$. The noninverting input voltage is constant and equals $\beta V_\alpha = V_\alpha R_2/(R_1 + R_2)$. Since $v_c > \beta V_\alpha$, the comparator output is at its low level, $v_o = -V_o$.

Assume that at $t = 0$, a narrow, positive triggering signal v_t is applied, with the pulse amplitude approximately equal to V'. The JFET conducts with a large constant current and rapidly discharges C linearly toward ground. For small voltages, v_c no longer falls linearly, but it approaches zero exponentially with a time constant $r_{DS(ON)}C$ (Sec. 4-3). The waveforms for v_c and v_o are plotted in Fig. 15-30b. As soon as v_c falls below βV_{CC}, the comparator output changes to its high level, $v_o = +V_o$.

We assume that the pulse width T_p is large enough so that $v_c \approx 0$ at the end of the input signal. Then, at $t = T_p$, the capacitor charges exponentially with a time constant RC toward V. When $v_c = \beta V_{CC}$, the comparator switches again and for $v_c > \beta V_{CC}$, v_o remains at $-V_o$, thereby generating the positive gating waveform of width T shown in Fig. 15-30b. It can be shown that

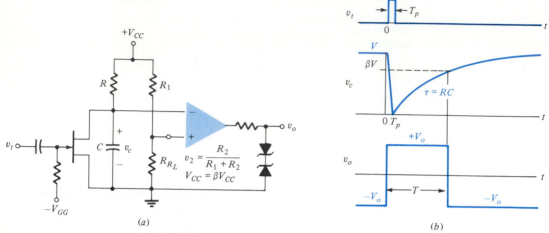

FIGURE 15-30
(a) A retriggerable monostable multivibrator. (b) The waveforms for the trigger pulse v_t, the capacitor voltage v_c, and the output pulse v_o (with $T > T_p$).

$$T = RC \ln \left(1 + \frac{R_2}{R_1} \right) \tag{15-38}$$

In deriving this equation it is assumed that $T \gg T_p$. A better approximation is to add T_p to the right-hand side of Eq. (15-38).

Note that, unlike most monostable configurations (for example, that in Fig. 15-29), no recovery time is required before the system in Fig. 15-30 can be triggered again. If a second positive input pulse appears at any time t' (less than or greater than T), the JFET reduces the voltage on C to zero and the waveforms indicated in Fig. 15-30b are generated at $t = t'$ instead of $t = 0$. Therefore a new gating interval T is initiated at $t = t'$. Such a circuit is called a *retriggerable monostable multi*.

15-12 THE 555 IC TIMER The 555 IC timer chip is widely used as both a monostable and astable multivibrator. Originally introduced by Signetics using bipolar technology, the 555 is now available from several manufacturers in both bipolar and CMOS technology. The basic configuration of the 555 is illustrated in Fig. 15-31 and is seen to consist of two comparators, an SR latch, a discharge transistor $Q1$, and a totem-pole output stage. Used with $V_{CC} = 5$ V, the timer is compatible with the 54/74 series TTL and CMOS logic families.

The circuit in Fig. 15-31a is connected as a monostable multivibrator; the resistance R and the capacitance C are external to the chip, and their values determine the output pulse width. (Note that external elements and connections are drawn in black; circuits that are on the chip are shown in color.) The three equal resistances R_1 establish the reference voltages V_1 and V_2 for comparators

1 and 2, respectively, as $V_1 = 2V_{CC}/3$ and $V_2 = V_{CC}/3$. The value of R_1 cannot be controlled precisely. However, IC fabrication techniques control resistance ratios accurately so that V_1 and V_2 are precise.

Prior to the application of the trigger voltage v_t, the SR latch is reset with $Q = V(0)$ and $\overline{Q} = V(1)$. These levels render $v_o = V(0) \approx 0$ and cause $Q1$ to saturate, making the threshold voltage $v_x \approx 0$. Since $v_x < V_1$, the output of comparator 1 is $V(0)$. The output of comparator 2 is also $V(0)$ as $v_t > V_2$.

At $t = 0$, application of a trigger pulse $v_t < V_2$ causes the output of comparator 2 to be $V(1)$, thus setting the latch. Hence $Q = V(1)$, $\overline{Q} = V(0)$, make $v_o = V(1)$ and turn off $Q1$. The timing capacitor C charges toward V_{CC} with a time constant $\tau = RC$. When v_x reaches V_1 at $t = T_1$, comparator 1 switches and its output becomes $V(1)$. This transition resets the latch and returns the output v_o to is original level $V(0)$. The low saturation resistance of $Q1$ discharges C quickly. The waveforms for v_t, v_x, and v_o are displayed in Figs. 15-31b, 15-31c, and 15-31d.

The pulse width T_1 is determined by the time required for the capacitance voltage v_x to charge to V_1. For $0 \leq t \leq T_1$

$$v_x = V_{CC} - [V_{CC} - V(0)]\, \epsilon^{-t/RC} \tag{15-39}$$

FIGURE 15-31
(a) The basic configuration of the 555 IC timer connected as a monostable multivibrator. The waveforms for (b) the trigger pulse v_t, (c) the threshold voltage v_x and (d) the output pulse v_o.

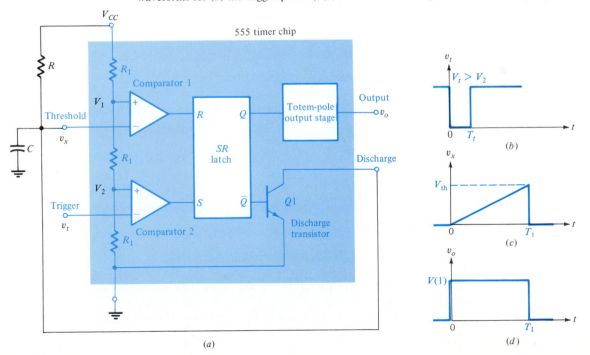

FIGURE 15-32
The 555 IC timer con-
nected as an astable
multivibrator.

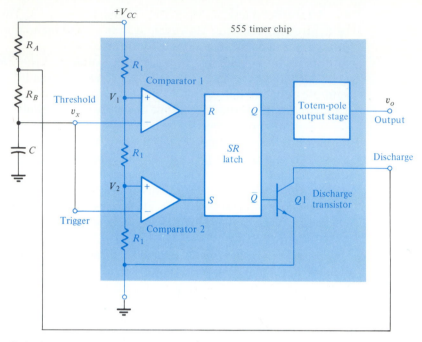

Solution of Eq. (15-39), at $t = T_1$ when $v_x(T_1) = V_1 = 2V_{CC}/3$, yields

$$T_1 = RC \ln \frac{V_{CC} - V(0)}{V_{CC}/3} \qquad (15\text{-}40)$$

If $V(0) = 0$, Eq. (15-41) reduces to

$$T_1 = RC \ln 3 = 1.1RC \qquad (15\text{-}41)$$

The Astable Multivibrator Figure 15-32 shows the connection of the 555 timer as an astable multivibrator. As in Fig. 15-31, external components and connections are shown in black. Assume that at $t = 0$, $v_x = V_2$, causing comparator 1 to switch and making its output $V(1)$. The latch is now reset, thus saturating $Q1$ and discharging C through R_B. At $t = T_1$, the threshold voltage is $V_1 = V_{CC}/3$ and the output of comparator 2 becomes $V(1)$ and sets the latch. Transistor $Q1$ is cut off, and C charges toward V_{CC} through $R_A + R_B$. At time T_2, $v_x = V_2$, causing a transition in comparator 1, and thus completing the cycle. The pulse durations (Prob. 15-45) are given by

$$T_1 = R_B C \ln 2 \qquad T_2 - T_1 = (R_A + R_B)C \ln 2 \qquad (15\text{-}42)$$

In Eq. (15-42) we assume that $V(0) = 0$. Note that the square wave is not symmetrical and is attributed to the different time constants during charging and discharging.

The period of the square wave is T_2, and hence the frequency f_o of oscillation is

$$f_o = \frac{1}{T_2} = \frac{1}{(R_A + 2R_B)C \ln 2} \qquad (15\text{-}43)$$

15-13 VOLTAGE TIME-BASE GENERATORS   A linear time-base generator is one that provides an output waveform, a portion of which exhibits a linear variation of voltage or current with time. A very important application of such a waveform is in connection with a cathode-ray oscilloscope ("scope"). The display on the screen of a scope of the variation with respect to time of an arbitrary waveform requires that there be applied to one set of deflecting plates a voltage which varies linearly wtih time. Since the waveform is used to *sweep* the electron beam horizontally across the screen, it is called a *sweep voltage*. There are, in addition, many other applications for time-base circuits, such as in radar and television indicators, in precise time measurements, and in time modulation.

The typical form of a time-base voltage is shown in Fig. 15-33a. The voltage, starting from some initial value, increases linearly with time to a maximum amplitude V_s, after which it drops to its initial value. The time T_r required for the return to the initial value is called the *restoration time*, the *return time*, or the *flyback time*. Very frequently the shape of the waveform during the restoration time and the interval T_r are unimportant.

In some cases, however, a restoration time is desired which is very short in comparison with the time occupied by the linear portion of the waveform. If it should happen that the restoration time is extremely short and that a new linear voltage is initiated at the instant the previous one is terminated, then the waveform will appear as in Fig. 15-33b. This figure suggests the designation *sawtooth generator* or *ramp generator*. It is customary to refer to waveforms of the type indicated in Fig. 15-33 as *sweep* waveforms even in applications not involving the deflection of an electron beam.

Clearly, the triangular voltage of Fig. 15-27b is a sweep waveform with a sweep time T_1 and a return time T_2. A sawtooth waveshape is obtained by making $T_2 \ll T_1$. The flyback time cannot be reduced to zero because of the limitations set by the slew rate of the integrator or by its maximum output current I (since the sweep speed is $dv/dt = I/C$).

The Triggered Sweep   A waveform may not be periodic but may occur rather at irregular intervals. In such a case it is desirable that the sweep circuit, instead of running continuously, should remain quiescent and wait to be initiated by the waveform itself. Even if it should happen that the waveform does recur regularly, it may be that the interesting part of the waveform is short in time duration in comparison with the period of the waveform. For example, the waveform might

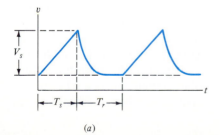

(a)

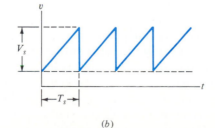

(b)

FIGURE 15-33
(a) A general sweep voltage. The sweep time is T_s and the return time, T_r. The sweep amplitude is V_s. (b) A sawtooth voltage waveform of period T_s.

FIGURE 15-34

A block diagram of the time-base generating system for a cathode-ray tube (CRT).

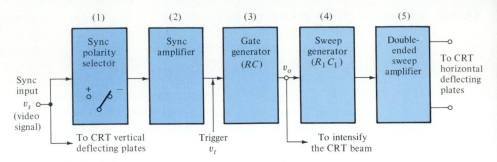

consist of 1-μs pulses with a time interval of 100 μs between pulses. In this case the fastest recurrent sweep which will provide a synchronized pattern will have a period of 100 μs. If, typically, the time base is spread out over 10 cm, the pulse will occupy 1 mm and none of the detail of form of the pulse will be apparent. If, on the other hand, a sweep of period 1 μs or somewhat larger could be used, the pulse would be spread across the entire screen. Therefore, what is required here is a sweep set for, say, a 1.5-μs interval which remains quiescent until it is initiated by the pulse. Such a monostable circuit is known as a *driven* sweep or a *triggered* sweep.

A block diagram for a time-base system for a CRT is indicated in Fig. 15-34. The waveform v_s to be observed is applied through a high-quality video amplifier (not indicated in Fig. 15-34) to the vertical deflecting plates of the CRT. This signal is simultaneously applied to the sweep system as the synchronizing input. In block 1 the sync polarity selection is made by taking the output across either a collector or an emitter resistor. The sync amplifier (block 2) need not operate linearly, since all that is required is that the output v_t be large and fast enough to be able to trigger the (monostable) gate generator. In some scopes a Schmitt trigger is used to obtain a sharp pulse on either the rising or falling portion of the signal, as desired. Since this trigger is used to start the sweep, a selected portion of the input signal appears on the scope face.

The third block in Fig. 15-34 is a monostable multivibrator whose gate width is determined by the time constant RC (Fig. 15-29a). A negative gating waveform (v_o in Fig. 15-29b) is applied to the sweep generator (4), whose sweep speed depends upon a resistor R_1 and a capacitor C_1 (Fig. 15-35). The sweep generator output is amplified linearly (5) and applied to the horizontal deflecting plate of the CRT.

In a case in which the sweep time is short in comparison with the time between sweeps the CRT beam will remain in one place most of the time. If the intensity is reduced to prevent screen burns, the fast trace will be very faint. To intensify the trace during the sweep, a positive gate which is derived from the outputs of the multi is applied to the CRT grid. As a matter of fact, in the presence of this "unblanking" or "intensifier signal" the beam brightness may be adjusted so that the spot is initially invisible but the trace will become visible as soon as the sweep starts.

Sweep Generators   The simplest sweep is obtained by charging a capacitor C_1 through a resistor R_1 from a supply voltage V_{CC}, as indicated in Fig. 15-35a. At $t = 0$ the switch S is opened and the sweep $v'_o(t)$ is given by

$$v'_o = V_{CC}(1 - \epsilon^{-t/R_1C_1}) \tag{15-44}$$

For the present discussion the physical form of the switch S is unimportant. After an interval T_s, when the sweep amplitude reaches V_s, the switch again closes. The resulting sweep waveform is indicated in Fig. 15-35b (assuming zero switch resistance).

Note that the sweep voltage is exponential and not linear. In the case of a cathode-ray oscilloscope, an important requirement of the sweep is that the sweep speed be constant. Hence, a reasonable definition of the deviation from linearity is given by the *slope* or *sweep-speed error e_s*:

$$e_s \equiv \frac{\text{Difference in slope at beginning and end of sweep}}{\text{Initial value of slope}} \tag{15-45}$$

If this definition is applied to Eq. (15-44), we find (Prob. 15-50) that, *independent of the time constant*, for a fixed sweep amplitude V_s and power-supply voltage V_{CC},

$$e_s = \frac{V_s}{V} \tag{15-46}$$

The linearity improves as the ratio V_s/V decreases. Hence the simple circuit in Fig. 15-35a is useful only in applications requiring sweep voltages of the order of volts or tens of volts. For example, a 20-V sweep can be obtained with a sweep-speed error of less than 10 percent by using a supply voltage of at least 200 V. Time-base voltages of hundreds of volts require power supplies of thousands of volts, which are inconveniently large.

A tremendous improvement in linearity is obtained by using the Op-Amp (Miller) integrator in Fig. 15-35c instead of the simple circuit in Fig. 15-35a. If the magnitude of the amplifier voltage gain is A_v, if the input resistance is $R_i = \infty$, and if the output resistance is $R_o = 0$, then $v' = A_v v_i$. The input v_i is V_s/A_v when the sweep amplitude at the amplifier output is V_s. Hence, from Eq. (15-46), $e_s = V_s/A_v V$ which indicates that

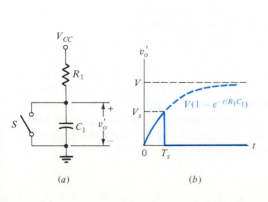

(a)   (b)

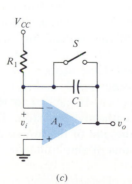

(c)

FIGURE 15-35
(a) Charging a capacitor through a resistor from a fixed voltage. (b) The resultant exponential waveform. (c) A Miller integration sweep circuit.

FIGURE 15-36

(*a*) A triggered sweep generator. The switch S in Fig. 15-35*a* or 15-35*c* is implemented by the diode-resistor-FET circuit. (*b*) A video pulse waveform v_s to be observed on the scope. (*c*) The output voltage v_o of the monostable multivibrator. (*d*) The generated sweep is synchronized with the input signal.

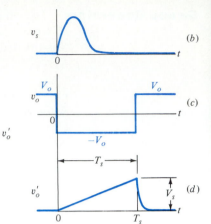

$$e_s(\text{Fig. 15-35}c) = \frac{1}{A_v} e_s(\text{Fig. 15-35}a) \qquad (15\text{-}47)$$

Since $A_v \approx 100,000$ the IC circuit generates extremely linear ramp voltages.

An approximately linear sweep may also be obtained by using the *bootstrap* configuration in Prob. 15-53.

The switch S may be a JFET driven by a gate generator, as indicated in Fig. 15-36*a* (corresponding to Fig. 15-35*a*). The video signal v_s to be observed is indicated in Fig. 15-36*b*. As shown in Fig. 15-34, this signal is amplified to form a trigger v_t for the monostable multi, whose output v_o is pictured in Fig. 15-36*c*. There is a small delay (not shown) between the beginning of the pulse in Fig. 15-36*b* and the beginning of the gate in Fig. 15-36*c*. In the quiescent state Q is ON because $v_o = +V_o$, and the capacitor voltage is held close to zero since $r_{DS(ON)} \ll R_1$. During the gating interval T_s the FET is cut off by the gate voltage $-V_o$ and the capacitor charges, thus generating the sweep v_o in Fig. 15-36*d*. At the end of the interval T_s, v_o returns to $+V_o$, which drives the FET ON, discharging C rapidly for a short retrace time, as explained in connection with the waveform v_c in Fig. 15-30*b*. (The diode D prevents the gate of Q from drawing appreciable current.)

We have already noted that the sweep speed is determined by R_1C_1 in the sweep generator, whereas the gate width is determined by RC in the gate generator. If the sweep amplitude is to remain nominally constant, the gate controls R and C must be adjusted whenever the sweep speed controls R_1 and C_1 are varied. Capacitors C_1 and C are switched simultaneously to change the range of sweep speed, and resistor R_1, which is used for continuous variation of sweep speed, is ganged to R. No attempt is made to maintain constant amplitude with any precision. The sweep amplitude is deliberately made so large that the end of the sweep occurs at a point well off the CRT screen, so that variations of amplitude are not observed.

15-14 STEP (STAIRCASE) GENERATORS   The simple implemention in Fig. 15-37a is used to generate the staircase waveform displayed in Fig. 15-37b. A train of negative clock pulses v_p (Fig 15-37b) is applied to an Op-Amp integrator. The output v_o of the integrator rises linearly for the short duration T_p of each pulse and remains constant in between pulses (Fig. 15-37c). If $T_p \ll T$ = the clock period, the waveform v approaches the ideal staircase. Note that the counter and switch S in Fig. 15-37a play no part in forming the step waveform; they are required only for resetting v_o to zero after the desired number of steps (as explained in the following paragraphs). If V is the pulse amplitude, the sweep speed is V/RC, and the size V' of each step is given by

$$V' = \frac{VT_p}{RC} \tag{15-48}$$

If it is desired to terminate the staircase after, say, seven steps, a three-stage ripple counter is used. The output of each of the three FLIP-FLOPS is applied to an AND gate (Fig. 15-37a). After the seventh pulse there is a coincidence and the AND gate output v_A goes high and remains high until after the eighth pulse (refer to the diagram in Fig. 15-34). The resulting waveform is used to control the switch S in Fig. 15-37a, which discharges C rapidly to zero, as shown in Fig. 15-37c. Resetting at any desired step can be accomplished by modifying a ripple counter with an appropriate feedback gate, as explained in connection with Fig. 8-16.

A Storage Counter   The change from one step to the next in Fig. 17-37c takes place in the time T_p (one pulse width). A much more abrupt rise may be obtained with the *storage-counter* configuration of Fig. 17-38. To understand the operation, assume that capacitor C_1 is uncharged and C_2 is charged to a voltage v. An input pulse will cause the capacitor C_1 to charge through the diode $D1$. The time constant with which C_1 charges is the product of C_1 times the sum of the diode and the voltage-follower resistances. This time constant can be very small in comparison with the duration of the pulse, and C_1 will charge fully to the value $v_1 = V$, with the polarity indicated. During the charging time of C_1, the diode $D2$ does not conduct and the voltage across C_2 remains at v_o. At the termination of the input pulse, the capacitor C_1 is left with the voltage $v_1 = V$, which now appears across $D1$. The polarity of this voltage is such that $D1$ will not conduct. The capacitor C_1 will, however, discharge through $D2$ and the amplifier output resistance into C_2. The virtual ground at the input terminals of the operational amplifier takes no current. Hence, all the charge C_1V which leaves C_1 must transfer to C_2. The increase in voltage across C_2 is, therefore,

$$V' = \frac{C_1V}{C_2} \tag{15-49}$$

and the voltage across C_1 is reduced to zero. By the foregoing argument, the next pulse again charges C_1 to the voltage V during T_p and abruptly transfers

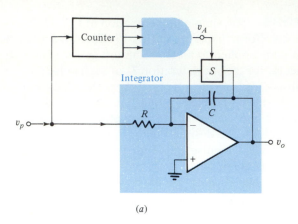

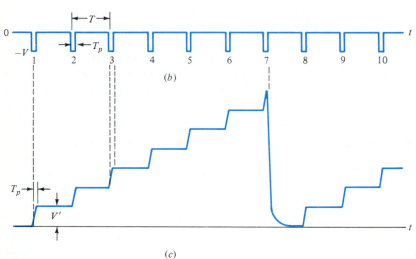

FIGURE 15-37
(*a*) A staircase waveform v_o is obtained by applying a train of narrow pulses v_p to a Miller integrator. The counter, the AND gate, and the controlled switch S perform the resetting operation. Waveforms v_p and v_o are displayed in (*b*) and (*c*), respectively.

the charge C_1V to C_2 at the end of the pulse, so that v decreases by another step of the same size V' given by Eq. (15-49).

Applications   The staircase waveshape is frequently useful to vary some voltage in a step fashion. A (very-high-frequency) sampling scope uses such a step generator. The staircase waveform may also be used to trace out a family of BJT or FET volt-ampere characteristics on a CRT. In this application each step of the staircase corresponds to a particular constant value of base current or gate voltage.

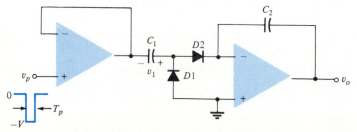

FIGURE 15-38
A storage-counter staircase generator. The resetting circuitry is identical with that used in Fig. 15-37.

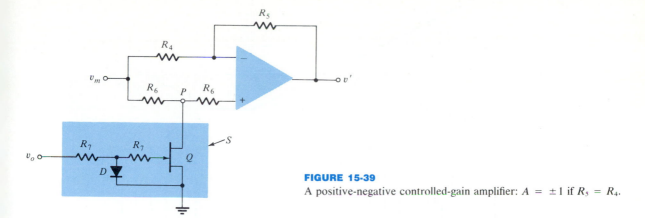

FIGURE 15-39

A positive-negative controlled-gain amplifier: $A = \pm 1$ if $R_5 = R_4$.

15-15 MODULATION OF A SQUARE WAVE

The variation of a high-frequency-*carrier* characteristic proportional to a lower-frequency signal is called *modulation*. The parameter being modulated may be frequency, amplitude, or pulse width. The VCO system in Fig. 15-28 is an example of a frequency-modulated (FM) square waveform. Equation (15-36) shows that the frequency f is proportional to the modulating-signal magnitude v_m.

Amplitude Modulation By multiplying any carrier waveform by a modulating signal v_m an amplitude-modulated (AM) signal is obtained because the instantaneous value of the carrier is proportional to v_m. An analog multiplier (Sec. 16-14) can be used in this application for a sinusoidal carrier.

If the carrier is a square wave, the multiplication can be performed very simply with the *biphase amplifier* (Fig. 15-39). If $v_o = +V_o$, the output voltage $v' = -v_m$; with $v_o = -V_o$, $v' = +v_m$. The analysis is left to the student in Prob. 15-41. In Fig. 15-40a the modulating signal v_m is shown as a piecewise linear signal (for ease of drawing), and the carrier is the square wave v_o in Fig. 15-40b. The AM wave is sketched in Fig. 15-40c. Note that, when $-v_o$ is positive, $v' = v_m$ and when $-v_o$ is negative, $v' = -v_m$. In other words, the square wave is multiplied by the modulating signal. This system is often referred to as a *pulse-height modulator* or a *pulse-amplitude modulator* (PAM).

A Chopper Modulator A very simple amplitude modulator is obtained by "chopping" the signal with a switch which is controlled synchronously by the square wave. In Fig. 15-41 the switch S_1 is controlled by the negative of the square waveform in Fig. 15-40b. An excellent implementation for S_1 is the JFET switch S in Fig. 15-36 or the CMOS analog switch in Fig. 6-32. During T_2 when v_o (in Figs. 15-40 and 15-41) is negative, S_1 is open and $v = v_m$. During T_1 when v_o is positive, S_1 is closed, and $v = 0$, assuming that the closed resistance of S_1 is much smaller than R. For the modulating signal v_m and the chopping signal v_o in Figs. 15-40a and 15-40b, respectively, the waveform v is as indicated in Fig. 15-41b. Observe that the waveform v is a *chopped* or *sampled* version of the waveform v_m. It is for this reason that the circuit of Fig. 15-41a is called a *chopper*.

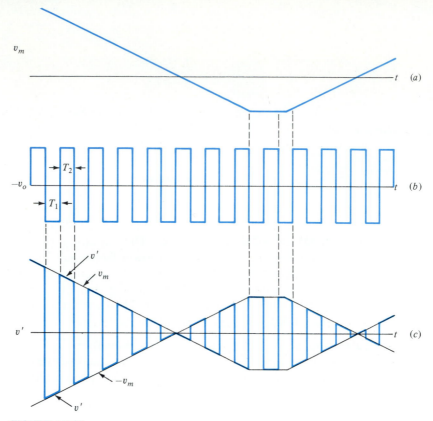

FIGURE 15-40
(*a*) A moduling signal. (*b*) A constant-frequency square-wave carrier. (*c*) The amplitude-modulated waveform.

FIGURE 15-41
(*a*) A chopper modulator. (*b*) A chopped reproduction of the modulating signal of Fig. 15-40*a*. (*c*) The amplitude-modulated waveform.

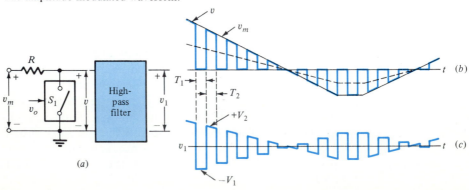

We note that when S_1 is open, the signal v reproduces the input signal v_m. As we have drawn the figure, a perceptible voltage change takes place in v_m during any interval when S_1 is open. Thus, when v_m is positive, the positive extremities of the waveform v_m are not at a constant voltage and, similarly, for the negative extremities when v_m is negative. More customarily, the frequency of operation of the switch is very large (typically 100 times), in comparison with the frequency of the signal v_m. Therefore, no appreciable change takes place in v_m during the interval when S_1 is open. Accordingly, it is proper to describe the waveform v_m as a square wave of amplitude proportional to v_m and having an average value (shown dashed) that is also proportional to the signal v_m. Alternatively stated, the waveform v is a square wave at the switching frequency, amplitude-modulated by the input signal, and superimposed on a signal which is proportional to the input signal v_m itself.

The low-frequency cutoff of the high-pass filter is such that the relatively high frequency square wave passes with small distortion while the signal frequency is well below the cutoff point. Consequently, at the output of the filter, we obtain the waveform v_1 in Fig. 15-41c, which corresponds to v but with the average value subtracted. Note that v_1 is an attenuated replica of the amplitude-modulated waveform v' obtained in Fig. 15-40.

Demodulators The process of recovering the modulating signal v_m from the PAM waveform is called *demodulation*. The positive-negative gain amplifier, which was used as a modulator, functions equally well as a demodulator. This statement is justified by the following argument. If the modulated waveform v' of Fig. 15-40c is used as the input v_m to Fig. 15-39, then, in the interval T_1 (Fig. 15-40b), when $v' = -v_m$, the gain A is -1, and in the next half-period T_2, when $v' = v_m$, $A = +1$. Hence the output v' (in Fig. 15-39) in any interval is v_m (in Fig. 15-40). Clearly, we have reconstituted the original signal v_m.

An alternative demodulator, corresponding to the chopper modulator of Fig. 15-41 is indicated in Fig. 15-42, where switch S_2 is controlled by $+v_o$ and, hence, operates in synchronism with switch S_1 in Fig. 15-41. For example, in the interval T_1 of Fig. 15-41c, S_2 is closed and the output is zero. Hence during T_1, the negative extremity of v_1 is clamped to ground, and the voltage across C is $-V_1$, as indicated in Fig. 15-42. In the next half-cycle T_2 of the square wave, S_2 is open, $v_1 = +V_2$, and $v_2 = V_2 + V_1$, which is the amplitude of v (Fig. 15-41) during T_2. As a consequence of the clamping action of C and the controlled switch S_2, the waveform v_1 is reconverted into the chopped mod-

FIGURE 15-42
A synchronous demodulator.

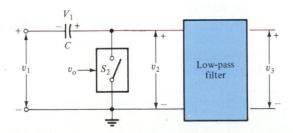

ulated signal v in Fig. 15-41b. If this waveform v is passed through the low-pass filter in Fig. 15-42, which rejects the high-frequency square wave and transmits the low-frequency signal, the resulting waveform v_3 is the modulating waveform v_m in Fig. 15-40a. The combination of the capacitor C, the switch S_2, and the low-pass filter constitutes a *synchronous demodulator*.

A Chopper-Stabilized Amplifier   A modulator-demodulator system which has a particularly interesting application is now discussed. Let us assume that it is required to amplify a small signal v_m (t) (say, of the order of millivolts) and that dv_m/dt is extremely small. For example, if the signal is periodic, the period may be minutes or even hours in duration. An ac amplifier with the customary coupling between stages would not be feasible, since these blocking capacitances would be impractically large. Instead, it would be necessary to use direct coupling between stages. With such a dc amplifier we would not be able to distinguish between a change in output voltage as a result of a variation in input voltage or as a consequence of a drift in some active device or some component, perhaps because of a temperature change. If the amplifier has high gain, even a tiny shift in the operating point of the first stage, amplified by the following stages, might cause a large variation in output. In summary: for this application an extremely stable (i.e., drift-free) dc amplifier is required.

A method of circumventing the above difficulty is to use an ac amplifier, but to precede it by a modulator and to follow it by a demodulator. This system is indicated in Fig. 15-43. Since the slowly varying input signal v_m is chopped, it is easily handled by a conventional ac amplifier (which is a high-pass system). The amplified waveform is then demodulated to reconstitute an enlarged replica of the input v_m. This system is called a chopper-stabilized amplifier. Note, however, that the amplifier is not stabilized by the choppers, but rather that the synchronous modulator-demodulator combination eliminates the necessity for an extremely drift-free direct-coupled amplifier.

The frequency response of a chopper-stabilized amplifier is very low. However, high-frequency-stabilized amplifiers are available from several manufacturers, because they augment the chopper with a high-frequency ac-coupled Op-Amp so that the overall response extends down to zero frequency. For example, the Harris Semiconductor HA 2900 or the Burr-Brown 3292 have the following excellent characteristics: offset voltage drift of ± 0.3 μV/°C, offset current drift of ± 1 pA/°C, a unity gain bandwidth of 3 MHz, and a minimum open-loop gain of 140 dB.

Pulse-Width Modulation   If a triangular waveform $v(t)$ is applied to a comparator whose reference voltage V_R is not constant but rather is an audio signal $v_m(t)$, a succession of pulses is obtained. The width of these pulses reflects the audio information. Such a *pulse-width modulation* system is indicated in Fig. 15-44a.

FIGURE 15-43
A chopper-stabilized amplifier.

v_m — | Modulator | — | Ac amplifier | — | Demodulator | — Av_m

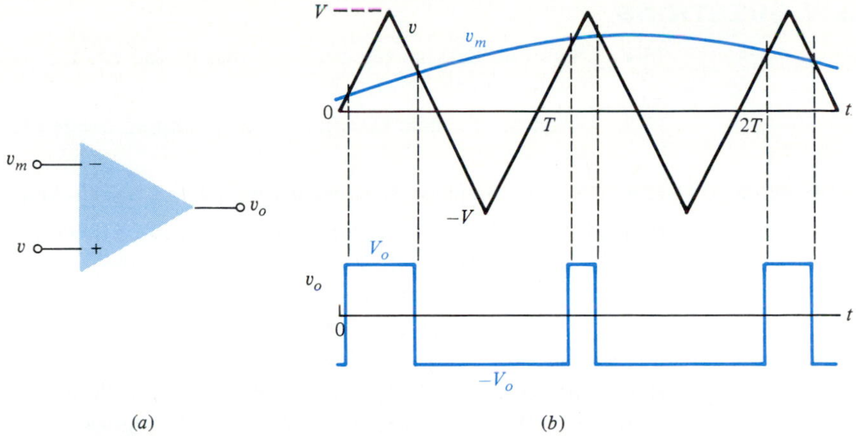

FIGURE 15-44

(*a*) A comparator used as a pulse-width modulator. (*b*) A triangular waveform v_r is used for the reference, v_m is the modulating signal, and v_o is the output pulse train.

If $v > v_m$, the comparator output $v_o = V_o$, and if $v < v_m$ then $v_o = -V_o$, as indicated in Fig. 15-44*b*. When $v_m = 0$ the pulse width is $T/2$, where T is the period of the triangular wave. As v_m increases, it linearly reduces the width of the output pulses v_o. the pulse train has an average value which is proportional to the modulating signal. Hence, an average-value detector may be used as a demodulator. Note that equal negative and positive switching delays in the comparator cancel out and do not affect the pulse width.

The system just described is also a *linear duty-cycle modulator* (Sec. 15-10). The duty cycle is given by $\delta = 0.5(1 - v_m/V)$, where V is the peak value of the triangular wave.

REFERENCES

1 Hodges, D. A., and H. G. Jackson: "Analysis and Design of Digital Integrated Circuits," McGraw-Hill Book Company, New York, 1983.

2 Grebene, A. B.: "Bipolar and MOS Analog Integrated Circuit Design," John Wiley and Sons, New York, 1984.

3 Ghausi, M. S.: "Electronic Devices and Circuits: Discrete and Integrated," Holt, New York, 1985.

4 Schilling, D., and C. Belove: "Electronic Circuits: Discrete and Integrated," McGraw-Hill Book Company, New York, 1979.

5 Soclof, S.: "Applications of Analog Integrated Circuits," Prentice-Hall, Englewood Cliffs, N. J., 1985.

6 Taub, H., and D. Schilling: "Digital Integrated Electronics," McGraw-Hill Book Company, New York, 1977.

7 Sedra, A. S., and K. C. Smith: "Microelectronic Circuits," Holt, New York, 1981.

8 Millman, J., and H. Taub: "Pulse, Digital and Switching Waveforms," McGraw-Hill Book Company, New York, 1965.

REVIEW QUESTIONS

15-1 State the Barkhausen criterion, that is, the conditions necessary for sinusoidal oscillations to be sustained.

15-2 What are the gain margin and phase margin needed to sustain sinusoidal oscillations?

15-3 Sketch the phase-shift oscillator using (*a*) an Op-Amp and (*b*) a JFET.

15-4 (*a*) Sketch the topology for a generalized resonant-circuit oscillator, using impedances Z_1, Z_2, Z_3.
(*b*) At what frequency will the circuit oscillate?
(*c*) Under what conditions does the configuration reduce to a Colpitts oscillator? A Hartley oscillator?

15-5 (*a*) Sketch the circuit of a Wien bridge oscillator.
(*b*) Which components determine the frequency of oscillation?
(*c*) Which elements determine the amplitude of oscillation?

15-6 (*a*) Draw the electrical model of a piezoelectric crystal.
(*b*) Sketch the reactance versus frequency function.
(*c*) Over what portion of the reactance curve do we desire oscillations to take place when the crystal is used as part of a sinusoidal oscillator? Explain.

15-7 Sketch a circuit of a crystal-controlled oscillator.

15-8 Compare and contrast the three types of multivibrators.

15-9 Draw a NOR-gate monostable multivibrator and explain its operation.

15-10 What is the function of a catching diode in a CMOS multivibrator?

15-11 Repeat Rev. 15-9 for an astable circuit.

15-12 (*a*) Sketch the characteristic of an ideal comparator with a reference voltage V_R.
(*b*) Repeat part *a* for a commercially available comparator.

15-13 (*a*) List two improvements in comparator characteristics which may be obtained by cascading the Op-Amp with a series combination of a resistor R and two back-to-back Zener diodes.
(*b*) What determines the magnitude of the resistance R?

15-14 (*a*) Sketch the system indicated in Rev. 15-13 for an inverting comparator with a reference V_R.
(*b*) Draw the transfer characteristics realistically if the output voltage is taken at the Op-Amp output terminal and also if it is taken across the two Zener diodes.

15-15 Sketch the circuit for converting a sinusoid into (*a*) a square wave and (*b*) a series of positive pulses, one per cycle.

15-16 Explain how to measure the phase difference between two sinusoids.

15-17 If noise spikes are present on the input signal to a comparator in the neighborhood of the amplitude V_R, why might the output ''chatter''?

15-18 (*a*) Sketch a regenerative comparator (Schmitt trigger) and explain its operation.
(*b*) What parameters determine the loop gain?
(*c*) What parameters determine the hysteresis?
(*d*) Sketch the transfer characteristic and indicate the hysteresis.

15-19 Repeat Rev. 15-18 for an emitter-coupled Schmitt trigger.

15-20 (*a*) Draw the system of a square-wave generator using one comparator.

(*b*) Explain its operation by drawing the capacitor and output voltage waveforms.

(*c*) Indicate one method for obtaining a nonsymmetrical square wave ($T_1 \neq T_2$).

15-21 (*a*) Using a comparator and an integrator, draw the system of a triangular-waveform generator with $T_1 = T_2$.

(*b*) Explain its operation by drawing the capacitor voltage waveform.

15-22 (*a*) Draw the configuration for a positive-negative controlled-gain amplifier.

(*b*) Explain its operation.

15-23 (*a*) In a VCO, what oscillator characteristic is controlled by the externally applied voltage?

(*b*) What is meant by *duty-cycle modulation*?

15-24 (*a*) Draw the configuration of a pulse generator (a one-shot) using a comparator.

(*b*) Explain its operation by referring to the capacitor and output waveforms.

15-25 (*a*) A capacitor C is charged through a resistor R from a supply V. An n-channel JFET is used as a switch across C and is biased so that the transistor Q is OFF. The capacitor voltage v_c is applied to the inverting terminal of a comparator having a reference voltage $V_R < V$. At $t = 0$ a triggering pulse v_t turns Q ON. Sketch the waveforms v_t, v_c, and v_o (the comparator output).

(*b*) Explain the operation and show that this configuration functions as a re-triggerable monostable multi.

15-26 Draw a block diagram of a time-base system for a CRT.

15-27 Sketch the configuration for a triggered-sweep generator with an output waveform which is (*a*) exponential and (*b*) linear. (*c*) Indicate one form for the reset switch.

15-28 (*a*) Sketch the system using an integrator for generating a staircase waveform v, starting with a pulse train v_p.

(*b*) Sketch v_p and v and explain the operation.

(*c*) Explain how to reset the system after N pulses.

15-29 Repeat Rev. 15-28 for a storage-counter step generator.

15-30 Explain how to amplitude-modulate a sinusoidal carrier v_c with a lower-frequency waveform v_m.

15-31 (*a*) Explain how to modulate the amplitude of a square-wave carrier v_c with a lower-frequency waveform v_m, using a positive-negative controlled-gain amplifier A.

(*b*) Explain why A may also be used as an amplitude demodulator.

15-32 Sketch the system of a chopper modulator and explain its operation.

15-33 What is a chopper-stabilized amplifier? Explain carefully.

15-34 (*a*) Explain how a comparator is used as a pulse-width modulator.

(*b*) Draw the modulating waveform v_m and the corresponding output waveform v_o.

Chapter 16
SIGNAL CONDITIONING AND DATA CONVERSION

Electronic control, communication, computation, and instrumentation systems can be considered to process information contained in the electrical signals present within the system. In previous chapters we indicated that the information exists in the characteristics of the signal waveform. For example, data can be contained by the frequency, phase, amplitude, pulse duration, or presence or absence of a pulse at a specific time. Circuits for generating a variety of these waveforms are treated in Chap. 15. In this chapter we describe a number of circuits used for signal conditioning and data conversion, specifically, circuits which provide appropriate signal characteristics for the particular application. Included are analog-to-digital (A/D) and digital-to-analog (D/A) converters (also referred to as ADCs and DACs, respectively) active-*RC* filters including switched-capacitor circuits, and a variety of circuits for analog computation.

16-1 SIGNALS AND SIGNAL PROCESSING

Continuous signals and discrete signals are convenient classifications used to describe electrical waveforms. Continuous signals are displayed in Fig. 16-1, and the waveforms shown in Fig. 16-2 are discrete signals. As illustrated in Fig.16-1, continuous signals are described by time functions which are defined for all values of t; that is, t is a continuous variable. The discrete signals exist only at specific instances of time; their functional description is valid only for the discrete time intervals.

Another feature of the waveforms in Figs. 16-1 and 16-2 can be discerned if we consider the signal in Fig. 16-1*a* to be a voltage representing a physical quantity. (Perhaps it can be the output voltage of a microphone.) The discrete signal in Fig. 16-2*a* has the same amplitude at times $t = 0$, T_1, T_2, and T_3 as does the continuous signal in Fig. 16-1*a*. Both voltage waveforms have a one-to-one correspondence in time and amplitude with the physical quantity represented. The waveform in Fig. 10-2*a* is referred to as a *sampled-data signal*

FIGURE 16·1

Two waveforms of continuous signals.

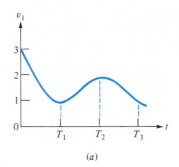

(a)

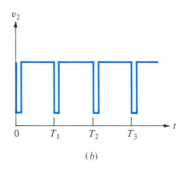

(b)

or simply a *sampled signal*. Systems which utilize such signals care called *sampled-data systems*.

In the context of this discussion, the sequence of pulses in each time interval of Fig. 16-2b is a numeric, or digital, representation of the corresponding voltage samples shown in Fig. 16-2a. The waveforms in Fig. 16-1b may be the clock signal which sets the timing sequence used in the generation of the pulses in Fig. 16-2. Neither the amplitude or time of the signals in Figs. 16-1b and 16-2b correspond to the physical quantity v_1. Essentially, these are signals in which the information is contained by the presence or absence of a pulse during a given time interval.

The waveforms in Figs. 16-1a and 16-2a are analog signals, and those depicted in Figs. 16-1b and 16-2b are digital signals. Both types of signals are often present in modern electronic systems. Clearly, circuits which process these signals and convert one type to the other are required. The following qualitative descriptions help to indicate the several circuit functions that must be performed.

The pictorial representation of a commercial amplitude-modulation radio system is shown in Fig. 16-3. The primary purpose of the system is to transfer the audio information at the transmitting end to the receiving end. The first step in the process is to convert the acoustic energy into an electrical signal. The conversion is effected by a transducer, usually a microphone. As the output of the transducer is a low-level signal, amplification is necessary. Radio-frequency (rf) signals (signals whose frequencies are greater than 500 kHz) are

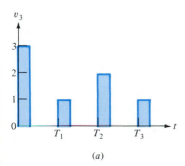

(a)

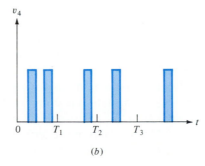

(b)

FIGURE 16-2

Two discrete signals. The waveform in (a) represents pulses whose amplitudes are those in Fig. 16-1a at times (b) T_1, T_2, and T_3, respectively. The waveform in (b) is the 2-bit binary representation of the amplitudes in (a) at T_1, T_2, and T_3.

FIGURE 16-3
Pictorial representation
of a commercial AM ra-
dio system.

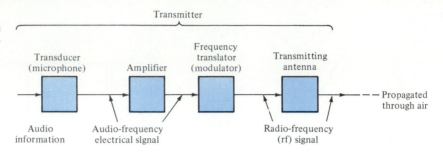

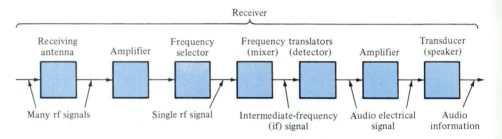

much more readily propagated through the atmosphere than are audio frequencies (20 to 20,000 Hz). Therefore, the audio information is frequency
translated to radio frequencies. The frequency translation is achieved by a
process called *modulation*.

At the receiver, the process of extracting the information is nearly the reverse
of transmission. The received signal is weak and must be amplified. In addition,
because many signals (stations) are present at the receiving antenna, the desired
signal must be identified and extracted. This function is referred to as *frequency
selection*. Practical considerations dictate two frequency translations (demodulation), the mixer and the detector, before the desired audio signal is extracted.
The final transducer, usually a loudspeaker, reconverts the electrical signal into
the audible acoustic wave.

In an AM system, the audio information is contained in the amplitude of the
rf carrier (Fig. 16-4*a*). Variation of the frequency of the rf wave (Fig. 16-4*b*)
contains the information in a frequency-modulated (FM) system. Voltage-to-
frequency and frequency-to-voltage converters are used in the modulation and
demodulation processes in FM systems.

If the audio information to be broadcast is the music stored on a digital-disk
recording, the stored digital signal is converted to an analog waveform by means
of a D/A converter. [Recall that a phonograph record can be considered a read-
only memory (ROM).]

Modern telephone transmission employs pulse-code-modulated (PCM) signals, where the vocal information is converted to a digital signal at the transmitter and then reconstructed as an analog signal at the receiver. Long-distance
transmission of PCM signals is more effective than transmission of analog
signals because digital data are affected less by noise than is an analog wave-

form. To account for losses in transmission, analog signals must be amplified periodically (approximately every 50 km). The noise introduced by these repeaters (amplifiers) is cumulative and can lead to serious signal degradation. The repeaters in a PCM system detect the incoming signal and regenerate a "clean" signal for transmission to the next repeater (or station). Consequently, only the noise introduced in a single transmission link is added to the signal. Note that digital circuits having large noise margins (Chap. 6) make such transmission effective.

A simplified pictorial representation of a telephone system is displayed in Fig. 16-5. The audio signal, generated by a microphone, is sampled (sample-and-hold circuit) and converted to a digital signal by an A/D converter. The digital signal is used to modulate the transmitted PCM signal. At the receiver, the carrier is demodulated and reconverted to an analog signal. In data transmission (e.g., from a terminal to a mainframe computer), a modulator-demodulator (MODEM) performs this function.

Two filters (frequency selection) are also shown in Fig. 16-5. The *antialiasing* filter is used at the transmitter and eliminates ambiguity in the sampled signal. To illustrate the origin of this ambiguity (aliasing), consider two sinusoidal signals $v_1 = 2 \sin \pi \times 10^3 t$ and $v_2 = -2 \sin 7\pi \times 10^3 t$ as shown in Fig. 16-6. The frequencies of these signals are 0.5 and 3.5 kHz, respectively. If v_1 and v_2 are sampled at a 4-kHz rate (i.e., once every 0.25 ms), the sampled values are those indicated by the black dots. As seen in Fig. 16-6, both v_1 and v_2 have the same value at these times. Consequently, an ambiguity exists and

FIGURE 16-4

(*a*) Amplitude-modulated (AM) and (*b*) frequency-modulated (FM) waveforms.

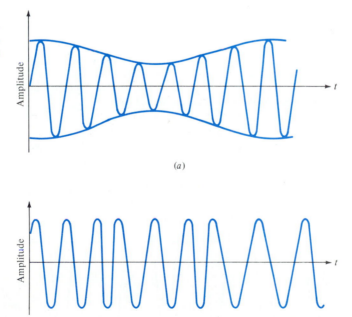

(*a*)

(*b*)

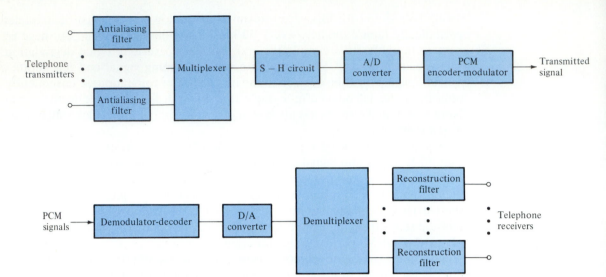

FIGURE 16-5

A pictorial representation of a telephone system. Usually, 24 signals are applied to the multiplexer and a 24:1 demultiplexer is used to separate the different conversations.

renders unique reconstruction of the original signal impossible. If, however, v_1 and v_2 are sampled at an 8-kHz rate (one every 0.125 ms), no ambiguity exists, as indicated by the blue dots. To prevent aliasing, the sampling frequency must be at least twice the highest frequency contained in the analog signal. The purpose of the antialiasing filter is to limit the maximum frequency of the

FIGURE 16-6

Illustration of aliasing. The 0.5-kHz sinusoid (black) and the 3.5-kHz sinusoid (blue) have the same amplitudes if these waves are sampled every 0.25 ms (blue dots) (a 4-kHz sampling frequency). However, sampling every 0.125 ms (an 8-kHz sampling frequency and indicated by blue dots) gives different values for the amplitudes of the two signals.

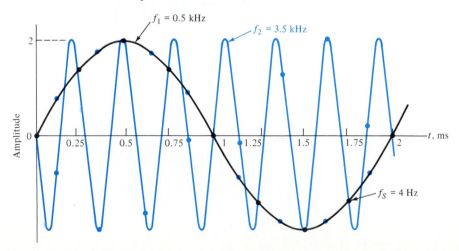

analog signal to be sampled to one-half the sampling frequency. The reconstruction filter is often required to "smooth" the output waveform of the D/A converter.

The representation in Fig. 16-5 is applicable to many systems which employ digital processing. For example, the input signal may be proportional to the velocity of the conveyor used to transport the wafer carrier in an IC fabrication facility. For utilization of the power of a digital computer (or microprocessor) in the control process, the input signal is sampled and converted to its digital equivalent. The computer operates on this information and provides an output signal (digital) which indicates whether the velocity of the conveyor must be corrected. The digital output is converted to an analog signal which is then amplified (usually) and applied to the drive system. If, in addition, the position of the conveyor is required for control, the velocity signal may be integrated and also converted to a digital signal. Multiplexing allows for the sampling of both the velocity and position (displacement) signals with the same circuitry.

The qualitative descriptions in the previous paragraphs indicate the variety of circuit functions that need to be performed to process signals effectively. The next several sections deal with the interconversion of analog and digital signals. The remainder of the chapter treats a variety of circuits useful in signal conditioning and data acquisition.

16-2 SAMPLE-AND-HOLD SYSTEMS

A typical data-acquisition system receives signals from a number of different sources and transmits these signals in suitable form to a computer or a communication channel. A multiplexer (Sec. 16-3) selects each signal in sequence, and then the analog information is converted into a constant voltage over the gating-time interval by means of a *sample-and-hold system*. The constant output of the sample-and-hold circuit may then be converted to a digital signal by means of an analog-to-digital (A/D) converter (Sec. 16-13) for digital transmission.

A sample-and-hold circuit in its simplest form is a switch S in series with a capacitor, as in Fig. 16-7a. The voltage across the capacitor tracks the input

FIGURE 16-7

(a) A simple sample-and-hold circuit. (b) A practical sample-and-hold system. The MOSFET switch replaces the switch S in (a). The low output resistance of the voltage follower $A1$ charges C quickly when the MOSFET switch is closed. The high input resistance of $A2$ maintains the charge on C when the switch is open.

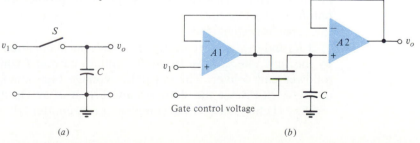

signal during the time T_g when a logic control gate closes S and holds the instantaneous value attained at the end of the interval T_g when the control gate opens S. The switch may be a bipolar transistor switch, a MOSFET controlled by a gating-signal voltage, or a CMOS transmission gate (Fig. 6-32).

The configuration shown in Fig. 16-7b is one of the simplest practical sample-and-hold systems. A positive pulse at the gate of the NMOS will turn the switch ON, and the holding capacitor C will charge to the instantaneous value of the input voltage with a time constant $(R_o + r_{DS(ON)})C$, where R_o is the very small output resistance of the input Op-Amp voltage follower $A1$ and $r_{DS(ON)}$ is the ON resistance of the FET (Sec. 4-2). In the absence of a positive pulse, the switch is turned OFF and the capacitor is isolated from any load through the Op-Amp $A2$. Thus it will hold the voltage impressed on it. It is recommended, in discrete-component or hybrid circuits, that a capacitor with polycarbonate, polyethylene, polystyrene, Mylar, or Teflon dielectric be used. Most other capacitors do not retain the stored voltage, as the result of a polarization phenomenon which causes the stored voltage to decay with a time constant of several seconds. In effect, this is the leakage resistance of the capacitor.

Recall that the basic cell in a MOS dynamic RAM (Sec. 9-5) had to be refreshed once every few milliseconds. Since a MOSFET is a capacitor, the decrease in charge in an IC is analogous to that described above. However, if an IC sample and hold is sampled frequently (at least once every few tenths of a millisecond, as is the usual practice), there is little leakage since the time between samples is considerably shorter than the time constant.

Dielectrics other than those mentioned above also exhibit a phenomenon called *dielectric absorption*, which causes a capacitor to "remember" a fraction of its previous charge (if there is a change in capacitor voltage). Even if the polarization and absorption effects do not occur, the OFF current of the switch (< 1 nA) and the bias current of the Op-Amp will flow through C. Since the maximum input bias current is < 1 nA, it follows that with a 0.5-μF capacitance the drift rate during the HOLD period will be less than 2 mV/s.

Two additional factors influence the operation of the circuit: the *aperture time* (typically less than 100 ns) is the delay between the time that the pulse is applied to the switch and the actual time the switch closes, and the *acquisition time* is the time it takes for the capacitor to change from one level of holding voltage to the new value of input voltage after the switch has closed.

When the hold capacitor is larger than 0.05 μF, an isolation resistor of approximately 10 kΩ should be included between the capacitor and the + input of the Op-Amp. This resistor is required to protect the amplifier in case the output is short-circuited or the power supplies are abruptly shut down while the capacitor is charged.

If R_o and $r_{DS(ON)}$ were negligibly small, the acquisition time would be limited by the slew rate, that is, by the maximum current I which the input Op-Amp follower can deliver. The capacitor voltage then changes at a peak rate of $dv_c/dt = I/C$. Since the short-circuit current of an Op-Amp is limited (25 mA for the 741 chip), an external complementary emitter follower is used to increase

FIGURE 16-8
An improved sample-
and-hold system. (The
complementary emitter
follower is discussed in
Sec. 14-6.)

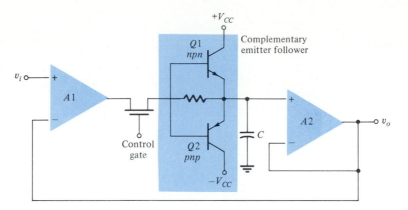

the current available to charge (or discharge) C extremely rapidly. Such an arrangement is indicated in Fig. 16-8 between the sampling switch and the capacitor. Note that $A1$ is no longer operated as a follower, but its negative-input terminal now is connected to the output v_o. This connection ensures that $v_o = v_i$ during the *sample interval*. In the *hold interval* v_o remains at the value which v_i attained at the end of the sampling time, except for the very small changes in voltage across C as a result of the bias current of the output Op-Amp and the leakage current of the switch and emitter follower. The larger the value of the capacitance C, the smaller is the drift in voltage during the hold mode. However, the smaller the capacitance C, the smaller is the acqui-sition time and, hence, the greater the fidelity with which the output follows the input during the sampling mode. Furthermore, the holding capacitor creates an additional pole which one must account for when considering loop trans-mission and stability. Hence the value of C must be chosen as a compromise between these three conflicting requirements, depending on the application.

A sample-and-hold (S-H) system is available on a single monolithic chip (e.g., Harris Semiconductor, HA 2420 or National Semiconductor, LF 198), with the storage capacitor added externally. The inverting terminal of $A1$ is available at an external pin, and hence, this chip may be used to build either a noninverting or an inverting S-H system which exhibits gain, if the usual external resistors are added (Prob. 16-5).

16-3 ANALOG MULTIPLEXER AND DEMULTIPLEXER As indicated in Fig. 7-17b, a *multiplexer* selects one out of N sources and transmits the (analog) signals to a single transmission line. Of all the switches (mentioned in the preceding section) which are available to feed the input signals to the output channel, the best performance is obtained with the CMOS transmission gate (Fig. 6-32). If dielectric isolation is used in the fabrication of this gate, then typically a leakage current of only 1 nA at $+125°C$ with a switching time of 250 ns is obtainable. Large arrays of such CMOS gates are available for this application.

FIGURE 16-9
A 16-input analog mul-
tiplexer using CMOS
transmission gates.

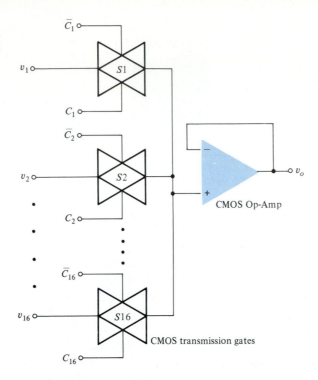

A block diagram of a 16-input analog commutator is indicated in Fig. 16-9. Time-division multiplexing results if the complementary MOSFET switch $S1$ closes (i.e., it is in its low-resistance state) for a time T, switch $S2$ closes for the second interval T, $S3$ transmits for the third period T, and so forth. In Fig. 16-9 the symbol C_k ($k = 1, 2, \ldots, 16$) represents the digital control voltage and $\bar{C}_k$ is its complementary value obtained from an inverter (not shown). If C_k equals binary 1, the CMOS gate transmits the analog signal v_k to the output, but if C_k is binary 0, no transmission is allowed.

The block diagram for obtaining the required digital control voltages for the analog multiplexer in Fig. 16-9 is indicated in Fig. 16-10. The control C_k is the output of the kth line of a 4-to-16-line decoder (Sec. 7-6). The four address lines A, B, C, and D are the outputs from a binary counter which is excited by a pulse generator. If the time interval between pulses is T, time-division multiplexing is obtained with the system shown in Figs. 16-10 and 16-9 (cor-

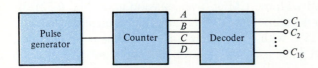

FIGURE 16-10
A system for generating the digital control voltages C_k for the multiplexer.

responding to parallel-to-serial conversion of the digital data, discussed in Sec. 7-7).

Analog Demultiplexer The multiplexer described above has entered the analog data on a single channel, each analog signal occupying its own time slot. At the end of the transmission line, each signal must be separated from the others and placed into an individual channel. This reverse process is called *demodulation* and is represented schematically in Fig. 7-17a. The multiposition switch in this figure is replaced by N CMOS transmission gates, and the serial data are applied to the input of all these gates. The control signals C_k are obtained in the manner indicated in Fig. 16-10. These systems for C_k must be synchronized at the sending and receiving ends of the channel. Such a multiplexer-demultiplexer system saves the size, weight, and cost of $N - 1$ transmission channels since all the analog signals have been transmitted on a single channel (N may be as large as several hundred). The National Semiconductor CD4051M is an eight-channel analog multiplexer-demultiplexer.

16-4 DIGITAL-TO-ANALOG (D/A) CONVERTERS

Many systems accept a digital word as an input signal and translate or convert it to an analog voltage or current. These systems are called *digital-to-analog*, or *D/A*, *converters* (or *DACs*). The digital word is presented in a variety of codes, the most common being pure binary or binary-coded-decimal (BCD).

The output V_o of an N-bit D/A converter is given by the following equation:

$$V_o = (2^{N-1}a_{N-1} + 2^{N-2}a_{N-2} + \cdots + 2^2 a_2 + 2^1 a_1 + a_o)V$$

$$= \left(a_{N-1} + \frac{1}{2}a_{N-2} + \frac{1}{4}a_{N-3} + \cdots + \frac{1}{2^{N-2}}a_1 + \frac{1}{2^{N-1}}a_o\right)2^{N-1}V \quad (16\text{-}1)$$

where V is a proportionality factor determined by the system parameters and where the coefficients a_n represent the binary word and $a_n = 1(0)$ if the nth bit is 1(0). A stable reference voltage V_R, from which V is derived, is used in this circuit. The most-significant bit (MSB) is that corresponding to a_{N-1}, and its weight is $2^{N-1}V$, while the least significant bit (LSB) corresponds to a_o, and its weight is $2^0 V = V$.

Consider, for example, a 5-bit word ($N = 5$) so that Eq. (16-1) becomes

$$V_o = (16a_4 + 8a_3 + 4a_2 + 2a_1 + a_o)V \quad (16\text{-}2)$$

For simplicity, assume $V = 1$. Then, if $a_o = 1$ and all other a are zero, we have $V_o = 1$. If $a_1 = 1$ and all other a values are zero, we obtain $V_o = 2$. If $a_0 = a_1 = 1$ and all other a values are zero, $V_o = 2 + 1 = 3$ V, etc. Clearly, V_o is an analog voltage proportional to the digital input.

A *binary-weighted D/A converter* is indicated schematically in Fig. 16-11. The blocks $S_0, S_1, S_2, \ldots, S_{N-1}$ in Fig. 16-11 are electronic switches which are digitally controlled. For example, when a 1 is present on the MSB line, switch S_{N-1} connects the resistor R to the reference voltage $-V_R$; conversely,

FIGURE 16-11
A D/A converter (DAC)
with binary-weighted
resistors.

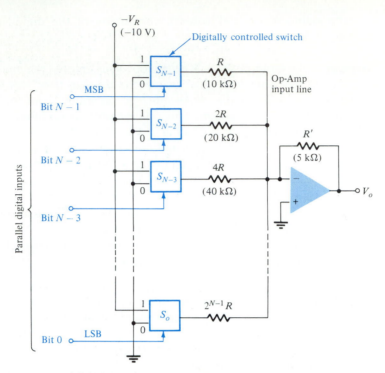

when a 0 is present on the MSB line, the switch connects the resistor to the ground line. Thus the switch is a single-pole double-throw (SPDT) electronic switch. The operational amplifier acts as a current-to-voltage converter (Sec. 10-22). We see that if the MSB is 1 and all other bits are 0, the current through the resistor R is $-V_R/R$ and the output is $V_R R'/R$. Similarly, the output of the LSB (if $N = 5$) becomes $V_o = V_R R'/16R$. If all five bits are 1, the output becomes

$$V_o = (1 + \tfrac{1}{2} + \tfrac{1}{4} + \tfrac{1}{8} + \tfrac{1}{16}) \frac{V_R R'}{R} = (16 + 8 + 4 + 2 + 1) \frac{V_R R'}{16R} \quad (16\text{-}3)$$

which agrees with Eq. (16-1) if $V = V_R R'/16R$. This argument confirms that the analog voltage V_o is proportional to the digital input.

Many implementations are possible for the digitally controlled switches shown in Fig. 16-11, two of which are indicated in Fig. 16-12. A totem-pole MOSFET driver in Fig. 16-12a feeds each resistor connected to the Op-Amp input. The two complementary gate inputs Q and $\overline{Q}$ come from a MOSFET S-R FLIP-FLOP or register which holds the digital information to be converted to an analog number. Let us assume that logic 1 corresponds to -10 V and logic 0 corresponds to 0 V (negative logic). A 1 on the bit line sets the FLIP-FLOP at $Q = 1$ and $\overline{Q} = 0$; and thus transistor $Q1$ is ON, connecting the resistor R_1 to the reference voltage $-V_R$, while transistor $Q2$ is kept OFF. Similarly, a 0 at the input bit line will connect the resistor to the ground terminal.

An excellent alternative single-pole double-throw electronic switch is that shown in Fig. 16-12b. This configuration consists of a CMOS inverter feeding

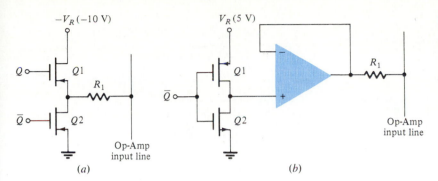

FIGURE 16-12
Two implementations of the digitally controlled switch in Fig. 16-10. (*a*) A totem-pole and (*b*) a CMOS inverter configuration. The resistance R_1 depends on the bit under consideration, thus, for the $N - 3$ bit in Fig. 16-10, $R_1 = 4R$.

an Op-Amp follower which drives R_1 from a very low output resistance. A positive logic system is indicated with $V(1) = V_R = +5$ V and $V(0) = 0$ V. The complement $\overline{Q}$ of the bit $Q = a_n$ under consideration is applied to the input. Hence, if $a_n = 1$, then $\overline{Q}$ of the bit $Q = a_n$ under consideration is applied to the input. Hence, if $a_n = 1$, then $\overline{Q} = 0$, the output of the inverter is logic 1, and 5 V is applied to R_1. On the other hand, if the nth is a binary 0, $\overline{Q} = 1$ and the output of the inverter is 0 V, so that R_2 is connected to ground. This confirms the proper operation of the circuit in Fig. 16-12*b* as an SPDT switch.

The accuracy and stability of the DAC in Fig. 16-11 depend primarily on the absolute accuracy of the resistors and the tracking of each other with temperature. Since all resistors are different and the largest is $2^{N-1}R$, where R is the smallest resistor, their values become excessively large, and it is very difficult and expensive to obtain stable, precise resistances of such values. For example, for a 12-bit DAC, the largest resistance is 5.12 MΩ if the smallest is 2.5 kΩ. The voltage drop across such a large resistance due to the bias current would affect the accuracy. Also, practical fabrication of such large resistance values is precluded. On the other hand, if the largest resistance is a reasonable value (51.2 kΩ), the smallest (25 Ω) may become comparable to the output resistance of the switch, again affecting the accuracy. Consequently, this type of DAC is seldom used where more than 4 bits are required. The ladder-type converter described in the following avoids these difficulties of extreme resistance values and is frequently used in high-resolution data-conversion systems.

A Ladder-Type D/A Converter A circuit utilizing twice the number of resistors in Fig. 16-11 for the same number of bits (N) but of values R and $2R$ only is shown in Fig. 16-13. The ladder used in this circuit is a current-splitting device, and thus the ratio of the resistors is more critical than their absolute value. We observe from the figure that at any of the ladder nodes the resistance is $2R$ looking to the left or the right or toward the switch.

For example, to the left of node 0 there is $2R$ to ground; to the left of node 1 there is the parallel combination of two $2R$ resistors to ground in series with R, for a total resistance of $R + R = 2R$, and so forth. Hence, if any switch, say, $N - 2$, is connected to V_R; the resistance seen by V_R is $2R + 2R \parallel 2R = 3R$ and the voltage at node $N - 2$ is $(V_R/3R)R = V_R/3$.

FIGURE 16-13

An $R - 2R$ ladder D/A converter.

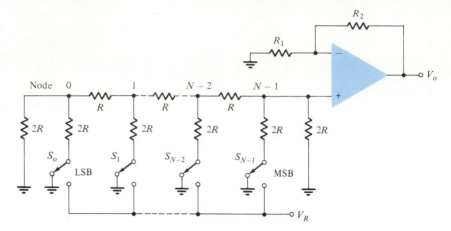

Consider now that MSB is logic 1 so that the voltage at node $N - 1$ is $\frac{1}{3}V_R$, the output is

$$V_o = \frac{V_R}{3}\frac{R_1 + R_2}{R_1} \equiv V' \tag{16-4}$$

Similarly, when the second MSB bit $(N - 2)$ is binary 1 and all other bits are logic 0, the output voltage at node $N - 2$ is $V_R/3$, but at node $N - 1$ the voltage is half this value, because of the attenuation due to the resistance R between the nodes and the resistance R from node $N - 1$ to ground. Hence $V_o = \frac{1}{2}V'$ for the second MSB $(N - 2)$. In a similar manner (Prob. 16-6) it can be shown that the third MSB gives an output $\frac{1}{4}V'$, and so forth. Clearly, the output is of the form of Eq. (16-1) with $V' = 2^{N-1}V$.

Because of the stray capacitance from the nodes to ground, there is a propagation delay time from left to right down the ladder network. When switch S_0 closes, the propagation delay is much longer than when the MSB switch closes. Hence, when the digital voltage changes, a transient waveform will appear at the output before V_o settles down to its proper value. These transients are avoided by using an inverted-ladder DAC (Prob. 16-7).

Multiplying D/A Converter   A D/A converter which may use a varying analog signal V_a instead of a fixed reference voltage is called a *multiplying D/A converter*. From Eq. (16-1) we see that the output is the product of the digital word and the analog voltage $V_a(= 2^{N-1}V)$ and its value depends on the binary word (which represents a number smaller than unity). This arrangement is often referred to as a *programmable attenuator* because the output V_o is a fraction of the input V_a and the attenuator setting can be controlled by digital logic. This type of DAC is sometimes used to control the center frequency or bandwidth of a state-variable filter (Sec. 16-11).

The basic DACs described in this section must be augmented by additional circuitry. Such circuitry includes the reference voltage, the Op-Amp, and the latches and logic circuits needed to input the data. Monolithic systems containing all these circuits are commercially available. The Analog Devices AD558 is an 8-bit bipolar DAC in which the digital circuits are realized in I^2L technology. The R-$2R$ ladder is also used in the AD7541, a 12-bit DAC (which can function as a multiplying D/A converter).

A two-stage segmented architecture is often used to achieve 16-bit resolution. The 4 MSBs are digitally decoded to select a voltage from a resistor chain similar to that used in the flash A/D converter (Fig. 16-16). This voltage becomes the reference V_R in an R-$2R$ ladder DAC which converts the 12 LSBs. In effect, the 4 MSBs divide V_R into 16 segments from 0 to $15V_R/16$ V. The 12 LSBs further divide the appropriate fraction of V_R into 4096 (2^{12}) parts. Thus, for V_R = 10 V, the 16-bit DAC provides an analog output from 0 to 10 V in 153-μV increments. This architecture is used in the AD7546, which incorporates the basic ladder of the AD7541.

Sixteen-bit DACs are also fabricated in the BIMOS technology (Sec. 14-10) to utilize the low power of CMOS logic with high-speed bipolar analog circuits (AD569). Recently introduced D/As employing switched-capacitor CMOS circuitry have been effective in reducing power consumption as most resistors are eliminated. (See Sec. 16-12.)

16-5 ANALOG-TO-DIGITAL (A/D) CONVERTERS

It is often required that data taken in a physical system be converted into digital form. Such data would normally appear in electrical analog form. For example, a temperature difference would be represented by the output of a thermocouple, the strain of a mechanical member would be represented by the electrical unbalance of a strain-gauge bridge, etc. The need therefore arises for a device that converts analog information into digital form. A very large number of such devices have been invented. We shall consider the four most popular systems: (1) the counting analog-to-digital converter (ADC), (2) the successive-approximation ADC, (3) the parallel-comparator ADC, and (4) the dual-slope or ratiometric ADC.

The Counting A/D Converter

This system will be explained with reference to Fig. 16-14a. The *clear* pulse resets the counter to the zero count. The counter then records in binary form the number of pulses from the clock line. The clock is a source of pulses equally spaced in time. Since the number of pulses counted increases linearly with time, the binary word representing this count is used as the input of a D/A converter whose output is the staircase waveform shown in Fig. 16-14b. As long as the analog input V_a is greater than V_d, the comparator (which is a high-gain differential amplifier; see Sec. 15-7) has an output which is high and the AND gate is open for the transmission of the clock pulses to the counter. When V_d exceeds V_a, the comparator output changes to the low value and the AND gate is disabled. This stops the counting at the time when $V_a \approx V_d$ and the counter can be read out as the digital word representing the analog input voltage.

FIGURE 16-14
(*a*) A counting A/D converter (ADC) and (*b*) D/A output staircase waveform.

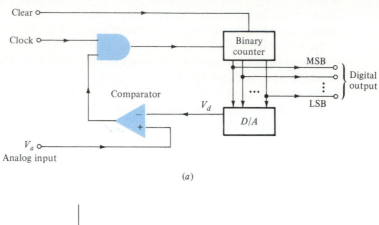

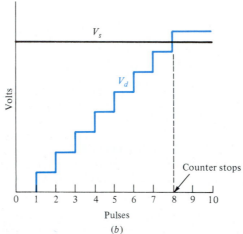

If the analog voltage varies with time, it is not possible to convert the analog data continuously, but it will be necessary that the input signal be sampled at fixed intervals. If the maximum value of the analog voltage is represented by n pulses and if the period of the clock is T seconds, the minimum interval between samples (the conversion time) is nT seconds.

An improved version of the counting ADC, called a *tracking* or *servo converter*, is obtained by using an up-down counter (Sec. 8-6, Fig. 8-18). This modification of the system of Fig. 16-14*a* is indicated in Fig. 16-15. Neither a START command (a clear pulse) nor an AND gate is now used. However, an up-down counter is now required and the comparator output feeds the UP-DOWN control of the counter. To understand the operation of the system, assume initially that the output of the DAC is less than the analog input V_a. Then the positive comparator output causes the counter to read UP. The D/A converter output increases with each clock pulse until it exceeds V_a. The UP-DOWN control line changes state so that it now counts DOWN (but by only one count, LSB). This causes the control to change to UP and the count to increase by 1 LSB. This process keeps repeating so that the digital output bounces back and forth

FIGURE 16-15
A tracking A/D converter.

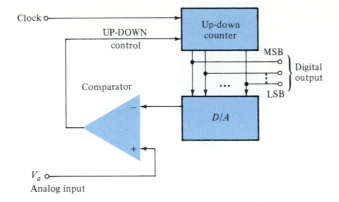

by ± 1 LSB around the correct value. The conversion time is small for small changes in the sampled analog signal, and hence this system can be used effectively as a tracking A/D converter.

Successive-Approximation A/D Converter   Instead of a binary counter, as shown in Fig. 16-15, this system uses a programmer. The programmer sets the MSB to 1, with all other bits to 0, and the comparator compares the D/A output with the analog signal. If the D/A output is larger, the 1 is removed from the MSB, and it is tried in the next MSB. If the analog input is larger, the 1 remains in that bit. Thus a 1 is tried in each bit of the D/A decoder until the binary equivalent of the analog signal is obtained at the end of the process. For an N-bit system, the conversion time is N clock periods as opposed to a worst case of 2^N pulse intervals for the counting-type A/D converter. The AD7582 (Analog Devices Co.), which is a 28-pin dual-in-line CMOS package, is a 12-bit A/D converter which makes use of the successive-approximation technique.

The Parallel-Comparator (Flash) A/D Converter   This system is by far the fastest of all converters. Its operation is easily understood if reference is made to the 3-bit A/D converter in Fig. 16-16. The analog voltage v_a is applied simultaneously to a bank of comparators with equally spaced thresholds (reference voltages $V_{R1} = V/8$, $V_{R2} = 2V/8$, etc.). This type of processing is called *bin conversion*, because the analog input is sorted into a given voltage range or "voltage bin" determined by the thresholds to two adjacent comparators. Note that the comparator outputs W take on a very distinctive pattern: low output (logic 0) for all comparators with thresholds *above* the input voltage and high output (logic 1) for each comparator whose threshold is *below* the analog input. For example, if $\frac{2}{8}V < v_a < \frac{3}{8}V$, then $W_1 = 1$, $W_2 = 1$, and all other W values are 0. For this situation the digital output should be 2 ($Y_2 = 0$, $Y_1 = 1$, $Y_0 = 0$), which is interpreted to mean an input analog voltage between $\frac{2}{8}V$ and $\frac{3}{8}V$.

The truth table with inputs W and outputs Y is given in Table 16-1. A comparison with Table 7-3 shows that the logic is that of a 3-bit priority encoder. The X values in Table 7-3 are all replaced by 1s. The column labeled W_0 in

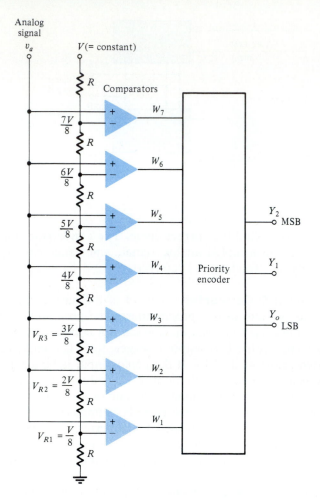

FIGURE 16-16

A 3-bit parallel-comparator (flash) A/D converter.

TABLE 16-1   Truth Table for the A/D Converter in Fig. 16-15

		Inputs						Outputs	
W_7	W_6	W_5	W_4	W_3	W_2	W_1	Y_2	Y_1	Y_0
0	0	0	0	0	0	0	0	0	0
0	0	0	0	0	0	1	0	0	1
0	0	0	0	0	1	1	0	1	0
0	0	0	0	1	1	1	0	1	1
0	0	0	1	1	1	1	1	0	0
0	0	1	1	1	1	1	1	0	1
0	1	1	1	1	1	1	1	1	0
1	1	1	1	1	1	1	1	1	1

Table 7-3 is missing in Table 16-1 because, if $v_a < \frac{1}{8}V$ then W_1 through W_7 are all 0, and the output is zero ($Y_2 = 0$, $Y_1 = 0$, $Y_0 = 0$).

Conversion time is limited only by the speed of the comparator and of the priority encoder. By using an Advanced Micro Devices AMD 686A comparator and a TI147 priority-encoder conversion, delays of the order of 20 ns can be obtained.

An obvious drawback of this technique is the complexity of the hardware. The number of comparators needed is $2^N - 1$, where N is the desired number of bits (seven comparators for the 3-bit converter in Fig. 16-16). Hence the number of comparators approximately doubles for each added bit. Also the larger the N, the more complex is the priority encoder.

Dual-Slope or Ratiometric A/D Converter This widely used system is depicted in Fig. 16-17. Consider unipolar operation with $V_a > 0$ and $V_R < 0$. Initally S_1 is open, S_2 is closed, and the counter is cleared. Then at $t = t_1$, S_1 connects V_a to the integrator and S_2 opens. The sampled (and hence constant) analog voltage V_a is now integrated for a fixed number n_1 of clock pulses. If the clock period is T, the integration takes place for a definite known time $T_1 = n_1T$, and the waveform v at the output of the integrator (Sec. 16-7) is indicated in Fig. 16-18.

If an N-stage ripple counter is used and if $n_1 = 2^N$, then at time t_2 (the end of the integration of V_a) all FLIP-FLOPS in the counter read 0. This is clearly indicated in the waveform chart of Fig. 8-14 for a four-stage ripple counter where, after $n_1 = 2^4 = 16$ counts, $Q_0 = 0$, $Q_1 = 0$, $Q_2 = 0$, and $Q_3 = 0$. In other words, the counter automatically resets itself to zero at the end of the interval T_1. Note also from Fig. 8-14 that at the 2^N the pulse, the state of Q_{N1} (MSB), changes from 1 to 0 for the first time. This change of state can be used as the control signal for the analog switch or transmission gate (Fig. 6-32).

Because of the counter operation described in the preceding paragraph, the reference voltage V_R is automatically connected to the input of the integrator at $t = t_2$, at which time the counter reads zero. Since V_R is negative, the waveform v has the positive slope shown in Fig. 16-18. We have assumed that $|V_R| > V_a$, so that the integration time T_2 is less than T_1, as indicated. As

FIGURE 16-17
Schematic representation of a dual-slope ADC.

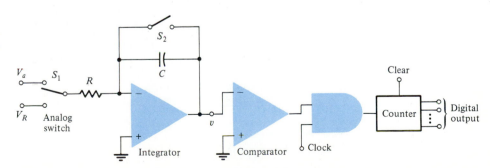

FIGURE 16-18
The output waveform of
the integrator shown in
Fig. 16-17.

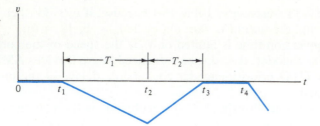

long as v is negative, the output of the comparator is positive and the AND gate allows clock pulses to be counted. When v falls to zero, at $t = t_3$, the AND gate is inhibited and no further clock pulses enter the counter.

We now show that the reading of the counter at time t_3 is proportional to the analog input voltage. The value of v at t_3 is given by

$$v = -\frac{1}{RC} \int_{t_1}^{t_2} V_a \, dt - \frac{1}{RC} \int_{t_2}^{t_3} V_R \, dt = 0$$

With V_a and V_R constant,

$$V_a(t_2 - t_1) + V_R(t_3 - t_2) = 0 \qquad \text{or} \qquad V_a = |V_R| T_2/T_1$$

If the number of pulses accumulated in the interval T_2 is n_2, then $T_2 = n_2 T$. Since $T_1 = n_1 T = 2^N T$, then

$$V_a = \frac{T_2 |V_R|}{T_1} = \frac{n_2 |V_R|}{n_1} = n_2 \frac{|V_R|}{2^N} \tag{16-5}$$

Since $|V_R|$ and N are constant, we have verified that V_a is proportional to the counter reading n_2. Note that this result is independent of the time constant RC.

The system includes automatic logic sequencing (not shown in Fig. 16-17), which clears the counter between t_3 and t_4, takes a new sample of the analog voltage, and moves S_1 back to V_a at t_4, so that the process is repeated; thus a new reading of V_a is obtained each $t_3 = t_1 + T_1 + T_2$ seconds. This technique can be very accurate; six-digit digital voltmeters employ such signal processing. The counter feeds a decoder/lamp driver so that the output is visible. For each cycle of operation a new voltage reading is obtained.

The dual-slope system is inherently noise-immune because of input-signal integration, i.e., the ubiquitous 60-Hz interference can be all but eliminated by choosing the integration time to be an integral number of power line periods. This statement also brings to light the obvious disadvantage of the system,

FIGURE 16-19
A Miller integrator.

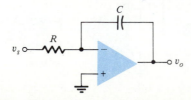

namely, the conversion time is long since $\frac{1}{60}$ s $\approx$ 16 ms. Such a dual-slope A/D converter can be obtained in various degrees of user complexity. The Datel Intersil ICL7109 is a monolithic 12-bit dual-slope A/D with microprocessor compatibility.

16-6 INTEGRATOR AND DIFFERENTIATOR CIRCUITS

The analog integrator is very useful in many signal-processing applications. The ideal integrator, introduced in Sec. 10-22 and repeated for convenience in Fig. 16-19, employs an ideal Op-Amp. Several waveform generation circuits which incorporated the circuit shown in Fig. 16-19 were described in Chap. 15. Our objective in this section is to treat practical integrator circuits in which the nonideal behavior of the Op-Amp is included.

DC Offset and Bias Current The input stage of the Op-Amp used in Fig. 16-19 is usually a differential amplifier. The dc input offset voltage V_{io} appears across the amplifier input, and this voltage will be integrated and will appear at the output as a linearly increasing voltage. The input bias current will also flow through the feedback capacitor, charging it and producing an additional linearly increasing component of the output voltage. These two effects (error sources) cause a continually increasing output until the amplifier reaches its saturation point. We see then that a limit is set on the feasible integration time by the above error components. The effect of the bias current can be minimized by increasing the feedback capacitor C while simultaneously decreasing the value of R for a given value of the time constant RC.

Finite Gain and Bandwidth The integrator supplies an output voltage proportional to the integral of the input voltage, provided the operational amplifier shown in Fig. 16-19 has infinite gain $|A_v| \to \infty$ and infinite bandwidth. The voltage gain as a function of the complex variable s is, upon transforming Eq. (10-117),

$$A_v(s) = \frac{V_o(s)}{V_s(s)} = -\frac{Z_2'}{Z_1} = -\frac{1}{RCs} \tag{16-6}$$

and it is clear that the ideal integrator has a pole at the origin.

Let us assume that in the absence of C the Op-Amp has a dominant pole at f_1, or $s_1 \equiv -2\pi f_1$. Hence its voltage gain A_v is approximated by

$$A_v = \frac{A_{vo}}{1 + j\,(f/f_1)} = \frac{A_{vo}}{1 - s/s_1} \tag{16-7}$$

If we assume that the Op-Amp output resistance $R_o = 0$ and the input resistance $R_i \to \infty$, then, for $A_{vo} \gg 1$ and $A_{vo}\,RC \gg 1/\,|\,s_1\,|$, we obtain

$$A_v(s) = \frac{-A_{vo}}{(1 + s/A_{vo}|s_1|)(1 + sRCA_{vo})} \tag{16-8}$$

where A_{vo} is the low-frequency voltage gain of the Op-Amp.

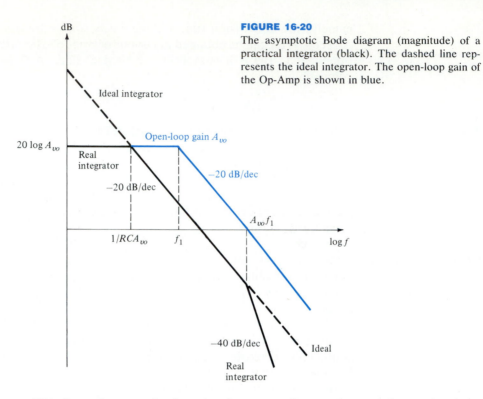

FIGURE 16-20

The asymptotic Bode diagram (magnitude) of a practical integrator (black). The dashed line represents the ideal integrator. The open-loop gain of the Op-Amp is shown in blue.

The foregoing transfer function has two poles on the negative real axis as compared with one pole at the origin for the ideal integrator. In Fig. 16-20 we show the Bode plots of the magnitudes of Eqs. (16-6) to (16-8). We note that the response of the real integrator departs from the ideal at both low and high frequencies. At high frequencies the integrator performance is affected by the finite bandwidth ($-s_1/2\pi$) of the operational amplifier, while at low frequencies the integration is limited by the finite gain of the Op-Amp.

Practical Circuit A practical integrator can be provided with an external circuit to introduce initial conditions, as shown in Fig. 16-21. When switch S is in position 1, the

FIGURE 16-21

A commercial integrator. (*Courtesy of National Semiconductor Corporation*)

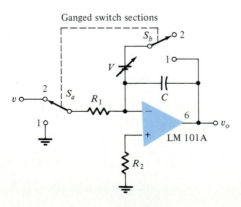

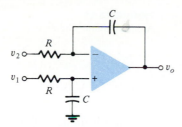

FIGURE 16-22
A differential integrator. The output voltage v_o is proportional to the integral of the difference voltage $v_1 - v_2$.

input is zero and capacitor C is charged to the voltage V, setting an initial condition of $v_o = V$. When switch S is in position 2, the amplifier is connected as an integrator and its output will be V plus a constant times the time integral of the input voltage v. If $R_2 = R_1$, the bias current through C is I_{io} (why?) rather than I_B, thus minimizing the error due to bias current.

The capacitor C must have very low leakage and it usually has a Teflon, polystyrene, or Mylar dielectric with typical capacitance values ranging from 0.001 to 10 μF.

The Differential Integrator The circuit shown in Fig. 16-22 is frequently referred to as a *differential integrator* because the output voltage v_o can be expressed as (Prob. 16-15)

$$v_o = \frac{1}{RC} \int (v_1 - v_2) \, dt \qquad (16\text{-}9)$$

or in the frequency domain as

$$V_o = \frac{V_1 - V_2}{RCs} \qquad (16\text{-}10)$$

Note that this circuit is the integrator equivalent of the instrumentation (difference) amplifier in Fig. 14-39. The circuit in Fig. 16-22 is used in several active-filter structures (Secs. 16-10 and 16-11).

Differentiator In the circuit shown in Fig. 16-19, if the positions of R and C are interchanged, as shown in Fig. 16-23, the resultant circuit is a *differentiator*. With a virtual ground at the inverting input of the Op-Amp (shown in blue), we find that

$$i_C = C \frac{dv_s}{dt} \qquad \text{and} \qquad i_R = -\frac{v_o}{R}$$

Since $i_C = i_R$, solving for v_o yields

$$v_o = -Ri = -RC \frac{dv_s}{dt} \qquad (16\text{-}11)$$

Hence the output is proportional to the time derivative of the input. If the input signal is $v = \sin \omega t$, then the output will be $v_o = -RC\omega \cos \omega t$. Thus the magnitude of the output increases linearly with increasing frequency, and the differentiator circuit has high gain at high frequencies. This results in amplification of the high-frequency components of amplifier noise, and the noise output may completely obscure the differentiated signal. At lower frequencies, how-

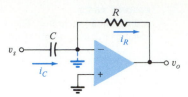

FIGURE 16-23
A differentiator circuit.

ever, the linear variation of v_o with the frequency of the sinusoidal excitation makes the differentiator behave as a simple *frequency-to-voltage* converter.

16-7 ELECTRONIC ANALOG COMPUTATION

The Op-Amp is the fundamental building block in an electronic analog computer. As an illustration, let us consider how to program the differential equation

$$\frac{d^2v}{dt^2} + K_1\frac{dv}{dt} + K_2v - v_1 = 0 \tag{16-12}$$

where v_1 is a given function of time, and K_1 and K_2 are real positive constants.

We begin by assuming that d^2v/dt^2 is available in the form of a voltage. Then, a voltage proportional to dv/dt is obtained by means of an integrator. A second integrator gives a voltage proportional to v. Then an adder (and scale changer) gives $-K_1(dv/dt) - K_2v + v_1$. From the differential equation [Eq. (16-12)], this equals d^2v/dt^2, and hence the output of this summing amplifier is fed to the input terminal, where we had assumed that d^2v/dt^2 was available in the first place.

The procedure outlined above is carried out in Fig. 16-24. The voltage d^2v/dt^2 is assumed to be available at an input terminal. The integrator (1) has a time constant $RC = 1$ s, and hence its output at terminal 1 is $-dv/dt$. This voltage is fed to a similar integrator (2), and the voltage at terminal 2 is $+v$. The voltage at terminal 1 is fed to the inverter and scale changer (3), and its output at

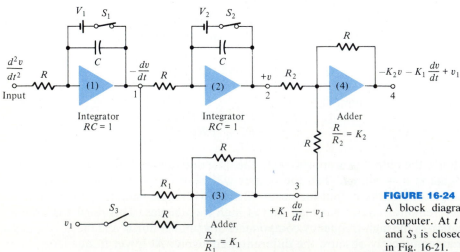

FIGURE 16-24
A block diagram of an electronic analog computer. At $t = 0$, S_1 and S_2 are opened and S_3 is closed. Each Op-Amp input is as in Fig. 16-21.

terminal 3 is $+K_1(dv/dt)$. This same Op-Amp (3) is used as an adder. Hence, if the given voltage $v_1(t)$ is also fed into it as shown, the output at terminal 3 also contains the term $-v_1$, or the net output is $+K_1(dv/dt) - v_1$. Scale changer-adder (4) is fed from terminals 2 and 3 and hence delivers a resultant voltage $-K_2v - K_1(dv/dt) + v_1$ at terminal 4. According to Eq. (16-12), this must equal d^2v/dt^2, which is the voltage that was assumed to exist at the input terminal. Hence the computer is completed by connecting terminal 4 to the input terminal. (This last step is omitted from Fig. 16-24 for the sake of clarity of explanation.)

The specified initial conditions (the value of dv/dt and v at $t = 0$) must now be inserted into the computer. We note that the voltages at terminals 1 and 2 in Fig. 16-24 are proportional to dv/dt and v, respectively. Hence initial conditions are taken care of (as in Fig. 16-21) by applying the correct voltages V_1 and V_2 across the capacitors in integrators 1 and 2, respectively.

The solution is obtained by opening switches S_1 and S_2 and simultaneously closing S_3 (by means of relays) at $t = 0$ and observing the waveform at terminal 2. If the derivative dv/dt is also desired, its waveform is available at terminal 1. The indicator may be a cathode-ray tube (CRT) (with a triggered sweep) or a recorder or, for qualitative analysis with slowly varying quantities, a high-impedance voltmeter.

The solution of Eq. (16-12) can also be obtained with a computer which contains differentiators instead of integrators. However, integrators are almost invariably preferred over differentiators in analog computer applications because the gain of an integrator decreases with frequency whereas the gain of a differentiator increases nominally linearly with frequency, and thus it is easier to stabilize the former than the latter with respect to spurious oscillations. As a result of its limited bandwidth, an integrator is less sensitive to noise voltages than a differentiator. Furthermore, if the input waveform changes rapidly, the amplifier of a differentiator may overload. Finally, as a matter of practice, it is convenient to introduce initial conditions in an integrator.

16-8 ACTIVE-*RC* FILTERS   The systems pictorially represented in Figs. 16-3 and 16-5 and described in Sec. 16-1 illustrate the need for frequency selection in signal processing. *Active-RC filters* are a class of frequency-selective circuits in which resistances, capacitances and Op-Amps (active elements) are the only components used. The fact that no inductance is required is an important advantage as modern IC fabrication precludes the use of inductors. Even in discrete-component circuits, the use of inductors should be avoided, if possible, because they are bulky, heavy, and nonlinear. In addition, they generate stray magnetic fields and may dissipate considerable power. For example, at $\omega = 2\pi \times 10^2$ rad/s, a 10-kΩ reactance requires an inductance of 1.6 henrys (H). Many turns are needed to construct a 1.6-H coil. Hence it is physically large and its resistance may dissipate considerable energy.

Ideal Characteristics   Consider the system depicted in Fig. 16-25*a*, in which the input signal $v_1(t)$ contains many (several) components at different frequencies. The filter is

FIGURE 16-25
(*a*) Time-domain and (*b*) frequency-domain representations of a filter.

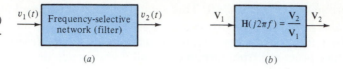

(*a*)   (*b*)

used to separate one band of frequencies from those present. That is, the filter output signal $v_2(t)$ contains only some of the frequency components of $v_1(t)$. It is convenient to describe the frequency-selective properties of the filter in terms of the transfer function $\mathbf{H}(j\omega) = \mathbf{V}_2/\mathbf{V}_1$ as indicated in Fig. 16-25*b*. The four ideal frequency-response characteristics are displayed in Fig. 16-26 and are useful in classifying filters.

The *low-pass* characteristic in Fig. 16-26*a* illustrates that all frequencies from zero (dc) to the cutoff frequency f_H are transmitted without loss. Inputs with frequency components $f > f_H$ give zero output. That is, $|\,\mathbf{H}(j2\pi f)\,| = H(j2\pi f) = H_o$ for $f < f_H$ and $H(j2\pi f) = 0$ for $f > f_H$. The high-frequency behavior of common-emitter and common-source stages described in Chap. 11 approximates this response. Figure 16-26*b* is a *band-pass* characteristic and indicates transmission between f_1 and f_2 and rejection of all other frequency components. Thus $H(j2\pi f) = 0$ for $f < f_1$ and $f > f_2$, and $H(j2\pi f) = H_o$ if $f_1 \leq f \leq f_2$. The *high-pass* response shown in Fig. 16-26*c* reveals that $H(j2\pi f) = 0$ for $f < f_L$ and uniform signal transmission [$H(j2\pi f) = H_o$] for $f > f_L$. The effect of the coupling and bypass capacitors in Sec. 11-10 approximates the high-pass response. To reject a band of frequencies between f_1 and f_2, one uses a *band-reject* or *band-elimination* filter whose characteristic is displayed in Fig. 16-26*d*. For this situation, $H(j2\pi f) = 0$ if $f_1 < f < f_2$; for all other values of f, $H(j2\pi f) = H_o$.

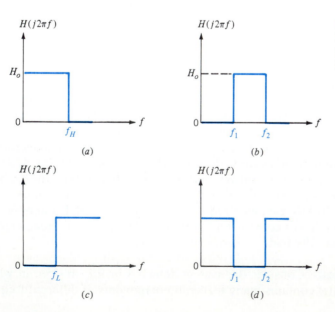

(*a*)   (*b*)

(*c*)   (*d*)

FIGURE 16-26
Ideal filter responses: (*a*) low-pass, (*b*) band-pass, (*c*) high-pass, and (*d*) band-reject.

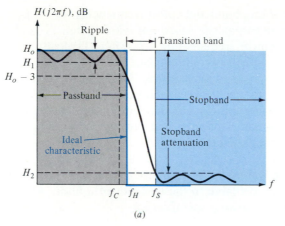

$H(j2\pi f)$, dB

Ripple

Transition band

H_o
H_1
$H_o - 3$

Passband

Stopband

Ideal characteristic

Stopband attenuation

H_2

f_C f_H f_S

f

(a)

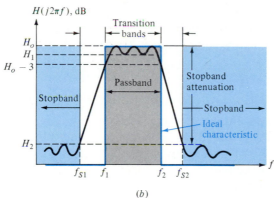

$H(j2\pi f)$, dB

Transition bands

H_o
H_1
$H_o - 3$

Stopband

Passband

Stopband attenuation

Stopband

Ideal characteristic

H_2

f_{S1} f_1 f_2 f_{S2}

f

(b)

FIGURE 16-27
Real frequency characteristics of (*a*) a low-pass and (*b*) a band-pass filter.

Realistic Frequency-Response Characteristics   The ideal characteristics shown in Fig. 16-26 are only approximated by practical circuits. Figure 16-27*a* shows a realistic low-pass response. The *passband* indicated in Fig. 16-27*a* is the range of frequencies that is transmitted without excessive attenuation. Note that $H(j2\pi f)$ need not be constant in the passband; the difference $H_o - H_1$ is called the *ripple* γ. Typically, the ripple is no greater than 1 dB with ≈ 0.5 dB common. The frequency f_H for which $H(j2\pi f_H) = H_o - 3$ (in decibels) is often used to indicate the edge of the passband, that is, the cutoff frequency. Alternatively, it is sometimes convenient to use f_C in Fig. 16-27*a* for the cutoff frequency. Observe that for the low-pass characteristic, $H(j2\pi f) = H_o - \gamma$ for all $f \leq f_C$. Because f_C is the maximum frequency for which $H(j2\pi f) = H_1$, it is sometimes called the "ripple" bandwidth.

The *stopband* denotes the range of frequencies that are attenuated. The attenuation is the difference (expressed in decibels), $H_o - H_2$. The stopband frequency f_S is the minimum frequency at which the attenuation is achieved. Note that stopband ripple may be present.

The change between the passband and the stopband in a practical filter is not abrupt as it is in the ideal response illustrated in Fig. 16-26. The difference between the stopband and cutoff frequencies ($f_S - f_H$ or, alternatively, $f_S - f_C$) is referred to as the *transition band*. Often f_S is chosen to be one octave above the cutoff frequency ($f_S = 2f_H$). Typically, attenuation in excess of 24 dB at $f_S = 2f_H$ is commonplace.

The band-pass response in Fig. 16-27b shows two stopbands and two transition bands, one each above and below the cutoff frequencies f_2 and f_1 which define the passband. Similar characteristics for realistic high-pass and band-reject filters can be drawn and corresponding identifications of the various frequency bands made.

Filter Specification   We use the low-pass response in Fig. 16-27a to illustrate the information that must be available to the designer of a filter. At a minimum, the designer requires the following specifications:

1. The cutoff frequency f_H (or f_C), that is, the range of passband frequencies.

2. The stopband attenuation $H_o - H_2$.

3. The stopband frequency range. That is, f_S is specified.

4. The allowable passband ripple $\gamma = H_o - H_1$. If no ripple is permitted, $\gamma = 0$ and $H_1 = H_o$.

An other factor that is usually specified is the impedance level at both the filter input and output (the interfaces with the signal source and load). The characteristic illustrated in Fig. 16-27a is the magnitude of the transfer function $\mathbf{H}(j\omega)$. Often, the phase response (delay) and transient response (rise time, overshoot) of the filter are also specified.

Since commercially available Op-Amps have unity gain bandwidths in excess of 100 MHz, it is possible to design filters up to frequencies of several megahertz.[1] Because of slew rate limitations and unit-to-unit variation in Op-Amp gain-bandwidth product and open-loop gain, however, most IC active filters are used at audio frequencies. Clearly, decrease in the unity gain bandwidth of the Op-Amp results in decrease of the maximum filter frequency.

16-9 BUTTERWORTH AND CHEBYSHEV FILTER FUNCTIONS   The frequency responses displayed in Fig. 16-27 are approximations of the ideal low-pass and band-pass characteristics in Figs. 16-26a and 16-26b, respectively. Similar approximations for the high-pass and band-elimination characteristics in Figs. 16-26c and 16-26d can be made. To design a "real" filter, the characteristics in Fig. 16-27 must be expressed mathematically. The general form of the transfer function $H(s)$ can be expressed as

$$H(s) = \frac{A(s)}{B(s)} \tag{16-13}$$

---

[1]At these frequencies, Op-Amps are usually costly and alternative filter realizations are often used.

TABLE 16-2 Biquadratic Transfer Functions

Type of characteristic	Form of transfer function $H(s) = A(s)/B(s)$	
Low-pass	$\dfrac{K}{s^2 + (\omega_o/Q)s + \omega_o^2}$	$\dfrac{K(s + z)}{s^2 + (\omega_o/Q)s + \omega_o^2}$
High-pass	$\dfrac{Ks^2}{s^2 + (\omega_o/Q)s + \omega_o^2}$	$\dfrac{Ks(s + z)}{s^2 + (\omega_o/Q)s + \omega_o^2}$
Band-pass	$\dfrac{Ks}{s^2 + (\omega_o/Q)s + \omega_o^2}$	
Band-elimination	$\dfrac{K(s^2 + \omega_r^2)}{s^2 + (\omega_o/Q)s + \omega_o^2}$	

where $A(s)$ and $B(s)$ are polynomials in the frequency variable s. Clearly, for stability, the zeros of $B(s)$ lie in the left half plane. The zero locations of $A(s)$ are unrestricted. However, the number of finite zeros of $N(s)$ is assumed to be equal to or less than the number of zeros in $B(s)$, that is, the poles of $H(s)$.

The Biquadratic Function Consider $H(s)$ to be of the form

$$H(s) = \frac{a_2 s^2 + a_1 s + a_0}{s^2 + b_1 s + b_0} \tag{16-14}$$

The expression for $H(s)$ in Eq. (16-14) is called a *biquadratic function* or simply a *biquad* because both numerator and denominator are quadratics in s. All four responses in Fig. 16-26 can be approximated by Eq. (16-14) by appropriate adjustment of the coefficient values. For the band-pass case depicted in Fig. 16-26b, $H(s) = 0$ for $s = j2\pi f$ equal to zero and infinity. The coefficient $a_0 = 0$ if $H(0) = 0$; similarly, if $H(j2\pi f) \to 0$ as $f \to \infty$, it is necessary that $a_2 = 0$. Analogous reasoning for the three remaining cases leads to the results given in Table 16-2.

The low-pass function in the left-hand column in Table 16-2 can be rewritten as

$$H(s) = \frac{H_o}{(s^2/\omega_o^2) + (1/Q)(s/\omega_o) + 1} \tag{16-15}$$

where $H_o = K/\omega_o^2$. Equation (16-15) has the same form as Eq. (13-13) for the two-pole feedback amplifier. The frequency response of this function is plotted in Fig. 13-13 as the damping factor $k = 1/2Q$ varies. As seen in Fig. 13-13, the two-pole function does not provide much attenuation at $s/\omega_o = 2$. For example, if $k > 0.6$ is assumed so that peaking is minimized, the attenuation at $s/\omega_o = 2$ is less than 14 dB and provides selectivity insufficient for most filter applications. In general, higher-order functions are required to provide typical attenuation levels encountered in practice.

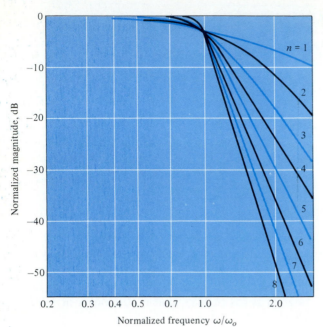

Butterworth Polynomials The use of Butterworth polynomials is a common all-pole approximation of the low-pass characteristic. Hence $H(s) = H_o/B(s)$, where $B(s)$ is a Butterworth polynomial whose magnitude is given by

$$B^2(\omega) = 1 + \left(\frac{\omega}{\omega_o}\right)^{2n} \qquad (16\text{-}16)$$

Filters that use Butterworth polynomials are called *Butterworth filters*. The normalized frequency response for various values of n is displayed in Fig. 16-28. Note that the magnitude of $H(j\omega)/H_o$ is down 3 dB at $\omega = \omega_o$ for all n and is montonically decreasing. The larger the value of n, the more closely the curve approximates the ideal response shown in Fig. 16-26a. We also observe that no passband ripple exists and the response is essentially constant for $\omega < \omega_o$. Butterworth polynomials are part of a class of *maximally-flat-magnitude* (*MFM*) filters; that is, the first $n - 1$ derivatives of $H(j\omega)$ evaluated at $\omega = 0$ are identically zero.

If we normalize the frequency by assuming $\omega_o = 1$ rad/s, we see that Table 16-3 gives the Butterworth polynomials $B_n(s)$ for n up to 8 (up to eighth-order). Note that for n even, the polynomials are products of quadratic factors of the form of the denominator of Eq. (16-15). Odd-order polynomials each contain a factor $(s + 1)$. An interesting property of the Butterworth polynomials is that their roots lie on the unit circle.

The value of n—that is, the order of the filter—is determined from the required stopband attenuation as illustrated in the following example.

TABLE 16-3 Normalized Butterworth Polynomials

n	Factors of polynomial $B_n(s)$
1	$(s + 1)$
2	$(s^2 + 1.414s + 1)$
3	$(s + 1)(s^2 + s + 1)$
4	$(s^2 + 0.765s + 1)(s^2 + 1.848s + 1)$
5	$(s + 1)(s^2 + 0.618s + 1)(s^2 + 1.618s + 1)$
6	$(s^2 + 0.518s + 1)(s^2 + 1.414s + 1)(s^2 + 1.932s + 1)$
7	$(s + 1)(s^2 + 0.445s + 1)(s^2 + 1.247s + 1)(s^2 + 1.802s + 1)$
8	$(s^2 + 0.390s + 1)(s^2 + 1.111s + 1)(s^2 + 1.663s + 1)(s^2 + 1.962s + 1)$

Example 16-1

Determine the order of a low-pass Butterworth filter that is to provide 40-dB attenuation at $\omega/\omega_o = 2$.

Solution

From Eq. (16-16), the normalized magnitude of the filter transfer function is

$$\left| \frac{H(j\omega)}{H_o} \right|^2 = \frac{1}{1 + (\omega/\omega_o)^{2n}}$$

An attenuation of 40 dB corresponds to $H(j\omega)/H_o = 0.01$, and hence

$$(0.010)^2 = \frac{1}{1 + 2^{2n}} \quad \text{or} \quad 2^{2n} = 10^4 - 1$$

Solving for n by taking the logarithm of both sides gives

$$2n = \frac{\log (10^4 - 1)}{\log 2} \quad \text{and} \quad n = 6.64$$

Since the order of the filter must be an integer, $n = 7$.

Chebyshev Filters   When specifications permit a small amount of passband ripple, a frequently used all-pole approximation is the Chebyshev filter. The transfer function is of the form

$$H^2(j\omega) = \frac{H_o^2}{1 + \epsilon^2 C_n^2(\omega/\omega_C)} \tag{16-17}$$

where $C_n(\omega/\omega_C)$ are Chebyshev polynomials defined by

$$C_n\left(\frac{\omega}{\omega_C}\right) = \cos\left(n \cos^{-1} \frac{\omega}{\omega_C}\right) \quad 0 \le \frac{\omega}{\omega_C} \le 1$$

$$= \cosh\left(n \cosh^{-1} \frac{\omega}{\omega_C}\right) \quad \frac{\omega}{\omega_C} > 1 \tag{16-18}$$

The parameter ϵ is related to the passband ripple γ in decibels by

TABLE 16-4   Normalized Polynomials for Chebyshev Filters

n	Factors of Chebyshev filter polynomials
	0.5-dB ripple ($\epsilon = 0.3493$)
1	$s + 2.863$
2	$s^2 + 1.425 s + 1.516$
3	$(s + 0.626) (s^2 + 0.626s + 1.142)$
4	$(s^2 + 0.351s + 1.064) (s^2 + 0.845s + 0.356)$
5	$(s + 0.362) (s^2 + 0.224s + 1.036) (s^2 + 0.586s + 0.477)$
6	$(s^2 + 0.1554s + 1.024) (s^2 + 0.4142s + 0.5475) (s^2 + 0.5796s + 0.157)$
7	$(s + 0.2562) (s^2 + 0.1014s + 1.015) (s^2 + 0.3194s + 0.6657) (s^2 + 0.4616s + 0.2539)$
8	$(s^2 + 0.0872s + 1.012)(s^2 + 0.2484s + 0.7413)(s^2 + 0.3718s + 0.3872)(s^2 + 0.4386s + 0.08805)$
	1.0-dB ripple ($\epsilon = 0.5089$)
1	$s + 1.965$
2	$(s^2 + 1.098s + 1.103)$
3	$(s + 0.494) (s^2 + 0.494s + 0.994)$
4	$(s^2 + 0.279s + 0.987) (s^2 + 0.674s + 0.279)$
5	$(s + 0.289) (s^2 + 0.179s + 0.988)(s^2 + 0.468s + 0.429)$
6	$(s^2 + 0.1244s + 0.9907) (s^2 + 0.3398s + 0.5577) (s^2 + 0.4642s + 0.1247)$
7	$(s + 0.2054) (s^2 + 0.0914s + 0.9927) (s^2 + 0.2562s + 0.6535) (s^2 + 0.3702s + 0.2304)$
8	$(s^2 + 0.07s + 0.9942) (s^2 + 0.1994s + 0.7236) (s^2 + 0.2994s + 0.3408) (s^2 + 0.3518s + 0.0702)$

$$\epsilon^2 = 10^{\gamma/10} - 1 \tag{16-19}$$

For 0.5-dB ripple, $\epsilon = 0.3493$ and $\epsilon = 0.5089$ if $\gamma = 1$ dB. The frequency $f_C = \omega_C/2\pi$ is the ripple bandwidth (Fig. 16-27). The 3-dB frequency f_H is related to f_C by

$$f_H = f_C \cosh \left(\frac{1}{n} \cosh^{-1} \frac{1}{\epsilon} \right) \tag{16-20}$$

The first eight Chebyshev filter polynomials are given in Table 16-4 for 0.5- and 1.0-dB ripple. Each polynomial is normalized for 0.5- and 1.0-dB ripple and to $\omega_C = 1$ rad/s. It can be shown that the roots of the functions presented in Table 16-4 lie on an ellipse. The eccentricity of the ellipse depends on the ripple.

The normalized frequency response of a 1-dB Chebyshev filter for different values of n is shown in Fig. 16-29a. The ripple in the passband is displayed in Fig. 16-29b for $n = 3$ and $\gamma = 1$ dB. Note again that as n increases the response more closely approximates the ideal characteristic.

Example 16-2   (a) Determine the order of a 1-dB ripple Chebyshev filter that gives a 40-dB attenuation at $\omega/\omega_C = 2$.(b) Determine the 3-dB bandwidth of the filter.

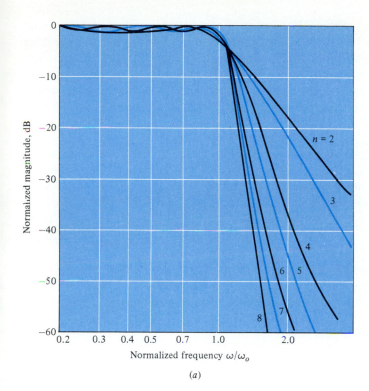

FIGURE 16-29
Chebyshev filter characteristics for different values of n. Each characteristic has 1-dB passband ripple.

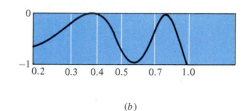

(a)

(b)

Solution

(a) The normalized response is, from Eq. (16-17),

$$\frac{H^2(j\omega)}{H_o^2} = \frac{1}{1 + \epsilon^2 C_n^2(\omega/\omega_C)}$$

A 40-dB attenuation corresponds to $H(j\omega)/H_o = 0.01$. Thus

$$(0.01)^2 = \frac{1}{1 + (0.5089)^2 C_n^2(2)}$$

and

$$C_n^2 (2) = \frac{10^4 - 1}{(0.5089)^2} = 3.861 \times 10^4$$

or

$$C_n(2) = 196.5$$

Use of Eq. (16-18) yields $196.5 = \cosh (n \cosh^{-1}2)$. Solution for n gives $n = 4.536$ and $n = 5$ is selected

(b) The value of f_H/f_C is obtained from Eq. (16-20) with $n = 5$ as

$$\frac{f_H}{f_C} = \cosh \frac{1}{5} \cosh^{-1} \frac{1}{0.5089} = 1.034$$

Comparison of the results in Examples 16-1 and 16-2 reveals that if passband ripple is permitted, a lower-order filter can be used. That is, a Chebyshev filter requiring fewer stages (lower order) is needed to obtain the same attenuation as a Butterworth filter.

This observation can also be seen by inspection of the response characteristics in Figs. 16-28 and 16-29a. Many other functions are used to approximate filter characteristics.[1]

Frequency Transformation The Butterworth and Chebyshev filter functions are also used to approximate band-pass, high-pass, and band-reject responses. The following transformations are used to convert low-pass functions to each of the others.

To transform a low-pass function to a high-pass one, we substitute

$$\frac{p}{\omega_o} = \frac{1}{s/\omega_o} \qquad \text{or} \qquad s = \frac{\omega_o^2}{p} \qquad (16\text{-}21)$$

into the expression for $H(s)$. Thus the low-pass function $H(s) = 1/[1 + (s/\omega_o)]$ becomes $H(p) = (p/\omega_o)/[1 + (p/\omega_o)]$, which is a high-pass function.

The low-pass–band-pass and low-pass–band-elimination transformations are given in Eqs. (16-22) and (16-23), respectively. Verification of these transformations is left to the student (Prob. 16-18):

$$p = \frac{s^2 + \omega_o^2}{\omega_H s} = \frac{Q\,[(s/\omega_o)^2 + 1]}{s/\omega_o} \qquad (16\text{-}22)$$

$$p = \frac{\omega_H s}{s^2 + \omega_o^2} = \frac{s/\omega_o}{Q[(s/\omega_o)^2 + 1]} \qquad (16\text{-}23)$$

where $Q = \omega_o/\omega_H$, $f_o = \omega_o/2\pi$ is the center frequency, and $f_H = \omega_H/2\pi$ is the 3-dB frequency.

Equations (16-21) to (16-23) are used to convert the low-pass Butterworth and Chebyshev filter functions in Tables 16-3 and 16-4 to their high-pass, band-pass, and band-reject equivalents. Note that the 5-pole low-pass function becomes a 10-pole band-pass or band-reject function. In the band-pass case, five poles provide the attenuation at frequencies $f < f_o$ and five poles give the attenuation for $f > f_o$.

16-10 SINGLE-AMPLIFIER BIQUAD SECTIONS Often, filters are constructed by cascading a number of sections, each of which realizes one of the biquadratic transfer functions listed in Table 16-2. Thus a six-pole 0.5-dB ripple, low-pass Chebyshev filter has three sections; each section is used to provide one of the quadratic factors listed in Table 16-4. In this section we examine a number of widely used biquad sections that employ only one Op-Amp.

---

[1]Many of them are tabulated in Refs. 3 and 8 given at the end of the chapter.

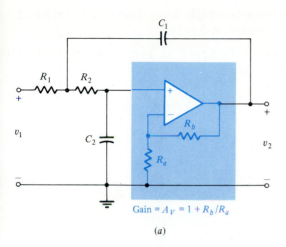
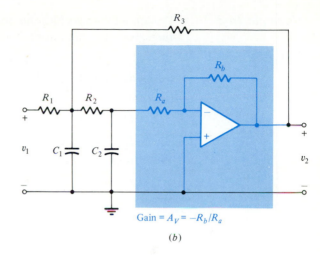

Gain $= A_V = 1 + R_b/R_a$

(a)

Gain $= A_V = -R_b/R_a$

(b)

FIGURE 16-30
Sallen and Key low-pass sections using (a) a noninverting amplifier (positive-feedback) and (b) an inverting amplifier (negative feedback).

Low-Pass Sections   The Sallen and Key circuit in Fig. 16-30a uses a noninverting Op-Amp stage to provide positive feedback. The transfer function of this circuit can be expressed as (Prob. 16-21)

$$H(s) = \frac{A_V}{R_1 R_2 C_1 C_2 s^2 + s\,[C_2(R_1 + R_2) + R_1 C_1(1 - A_V)] + 1} \tag{16-24}$$

where $A_V = 1 + R_b/R_a$ is the gain of the Op-Amp stage. Comparison of Eq. (16-24) with the low-pass function in Eq. (16-15) gives

$$\omega_o = \frac{1}{\sqrt{R_1 R_2 C_1 C_2}} \qquad Q = \frac{\sqrt{R_1 R_2 C_1 C_2}}{R_1 C_1\,(1 - A_V) + C_2(R_1 + R_2)} \tag{16-25}$$

Clearly, the five circuit parameters R_1, R_2, C_1, C_2, and A_V provide more degrees of freedom than are needed to specify ω_o and Q. Often, to simplify fabrication, $C_1 = C_2 = C$ and $R_1 = R_2 = R$. Equations (16-24) and (16-25) for these conditions reduce to

$$H(s) = \frac{A_V}{R^2 C^2 s^2 + RCs(3 - A_V) + 1} \tag{16-26}$$

$$\omega_o = \frac{1}{RC} \qquad Q = \frac{1}{3 - A_V} \tag{16-27}$$

In Eq. (16-27), observe that making both resistors and both capacitors equal results in the time constant RC establishing ω_o and the gain A_V determining Q. Furthermore, note $A_V < 3$ for stability. If $A_V \geq 3$, then from Eq. (16-26), the s-coefficient is ≤ 0 indicating $H(s)$ has right-half-plane poles.

Example 16-3

Design a low-pass filter having no more than 1-dB ripple from dc to 1 kHz and that gives a minimum of 40-dB attenuation at 2 kHz.

Solution

In Example 16-2 we showed that with 1-dB ripple, an attenuation of 40 dB is obtained with a fifth-order Chebyshev filter. The normalized characteristic polynomial, obtained from Table 16-4, is

$$B(s) = (s + 0.289)(s^2 + 0.179s + 0.988)(s^2 + 0.468s + 0.429)$$

Since $H(s) = H_o/B(s)$ has three factors, a three-section filter is used. These sections are

$$H_1(s) = \frac{H'_{o1}}{s + 0.289} \qquad H_2(s) = \frac{H'_{o2}}{s^2 + 0.179s + 0.988}$$

$$H_3(s) = \frac{H'_{o3}}{s^2 + 0.468s + 0.429}$$

Each function given above is normalized to the cutoff angular frequency $\omega_c = 1$ rad/s. The cutoff frequency desired is $f_C = 1$ kHz or $\omega_C = 2\pi \times 10^3$ rad/s. The unnormalized functions, written in the form of Eq. (16-15), are

$$H_1(s) = \frac{H_{o1}}{(s/0.289\omega_C) + 1} \qquad H_2(s) = \frac{H_{o2}}{(s^2/0.988\omega_C^2) + (0.179s/0.988\omega_C) + 1}$$

$$H_2(s) = \frac{H_{o3}}{(s^2/0.429\omega_C^2) + (0.468s/0.429\omega_C) + 1}$$

Both H_2 and H_3 can be realized with Sallen and Key low-pass sections (Fig. 16-30a) with $R_1 = R_2$ and $C_1 = C_2$. Comparison of H_2 and H_3 with Eqs. (16-26) and (16-27) gives

$$\omega_{o2} = \sqrt{0.988\omega_C^2} = 0.994\omega_C = 0.994 \times 2\pi \times 10^3 \text{ rad/s}$$

$$\frac{1}{Q_2} = \frac{0.179}{\sqrt{0.988}} \qquad \text{and} \qquad Q_2 = 5.55$$

$$\omega_{o3} = \sqrt{0.429\omega_C^2} = 0.655\omega_C = 0.655 \times 2\pi \times 10^3 \text{ rad/s}$$

$$\frac{1}{Q_3} = \frac{0.468}{\sqrt{0.429}} \qquad \text{and} \qquad Q_3 = 1.4$$

Equations (16-27) relate Q and ω_o to the circuit parameters; note, however, that specification of ω_o does not permit unique determination of R and C but only their product. In filter fabrication, particularly in hybrid technology, it is sometimes convenient to use the same value of capacitance in all sections. We choose $C = 0.05 \ \mu$F. Hence

$$R_2 = \frac{1}{\omega_{o2}C} = \frac{1}{0.994 \times 2\pi \times 10^3 \times 0.05 \times 10^{-6}} = 3.20 \text{ k}\Omega$$

$$R_3 = \frac{1}{\omega_{o3}C} = \frac{1}{0.655 \times 2\pi \times 10^3 \times 0.05 \times 10^{-6}} = 4.86 \text{ k}\Omega$$

FIGURE 16-31
Two circuits which re-
alize a zero on the neg-
ative real axis: (a) non-
inverting, (b) inverting.

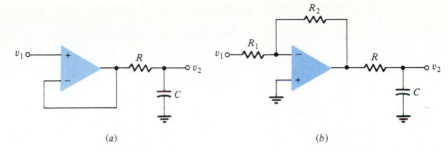

(a) (b)

Solution of Eq. (16-27) for A_V gives $A_V = 3 - 1/Q$. Hence

$$A_{V2} = 3 - \frac{1}{5.55} = 2.82 \qquad A_{V3} = 3 - \frac{1}{1.4} = 2.285$$

Since $A_v \gg 1$ for a commercial Op-Amp, $A_V = 1 + R_b/R_a$. To minimize both the number of resistance values used and the spread of element values (the ratio of the largest to the smallest resistance), we choose $R_a = 4.86$ kΩ for both sections. Then

$$R_{b2} = (A_{V2} - 1)R_{a2} = (2.82 - 1)4.86 = 8.84 \text{ k}\Omega$$

$$R_{b3} = (A_{V3} - 1)R_{a3} = (2.285 - 1)4.86 = 6.4 \text{ k}\Omega$$

The remaining section, $H_1(s)$, contains a real pole of $s = -0.289\omega_C = 0.289 \times 2\pi \times 10^3$ rad/s. The simple voltage follower driving an RC circuit as shown in Fig. 16-31a has a transfer function

$$\frac{V_2}{V_1} = \frac{1}{RCs + 1}$$

and is used to realize the real pole in $H_1(s)$. Comparison of the transfer functions yields

$$RC = \frac{1}{0.289\omega_C}$$

Choice of $C = 0.05$ μF results in

$$R = \frac{1}{0.289} \times 2\pi \times 10^3 \times 0.05 \times 10^{-6} = 11.0 \text{ k}\Omega$$

The final circuit is displayed in Fig. 16-32.

Negative-feedback circuits employing inverting Op-Amp stages are also used to obtain low-pass transfer functions. The Sallen and Key circuit in Fig. 16-30b is a low-pass biquad section. Real poles can be realized by using the circuit shown in Fig. 16-31b. The transfer functions for these circuits are

$$H(s) = \frac{H_o}{a_2s^2 + a_1s + 1} \tag{16-28}$$

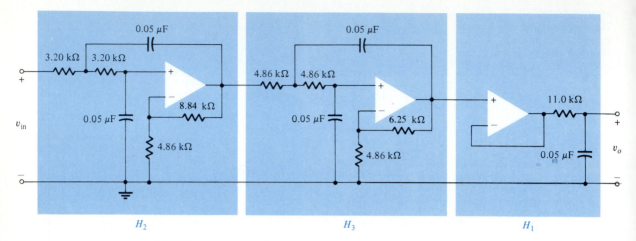

FIGURE 16-32
Circuit diagram for 5-pole 1-dB Chebyshev low-pass filter described in Example 16-3.

where
$$a_1 = \frac{R_3(R_1C_1 + R_1C_2 + R_2C_2)}{R_3 + R_1(1 + A_V)}$$

$$a_2 = \frac{R_1R_2R_3C_1C_2}{R_3 + R_1(1 + A_V)} \qquad H_o = \frac{A_V R_3}{R_3 + R_1(1 + A_V)}$$

$$\frac{V_2}{V_1} = -\frac{R_2}{R_1} \times \frac{1}{RCs + 1} \tag{16-29}$$

where $A_V = -R_b/R_a$.

High-Pass Sections   The Sallen and Key circuits in Figs. 16-30 are easily converted to high-pass sections by interchanging the resistors and capacitors as displayed in Fig. 16-33. Similarly, interchanging the position of R and C in Fig. 16-31 results in circuits having transfer functions of the form $H(s) = A_V RCs/(RCs + 1)$.

Band-Pass Sections   Two biquad sections used to obtain band pass response are shown in Fig. 16-34. The circuit in Fig. 16-34a uses a noninverting Op-Amp (positive feedback), and negative feedback is employed in the realization of Fig. 16-34b. Note that an ideal Op-Amp (infinite gain) is assumed in Fig. 16-34b, whereas a finite-gain stage is used in the Sallen and Key circuit in Fig. 16-34a. The voltage transfer ratios are given in Eqs. (16-30) and (16-31) for Figs. 16-34a and 16-34b, respectively:

$$H(s) = \frac{V_2}{V_1}$$

$$H(s) = \frac{A_V R_2 R_3 C_2 s/(R_1 + R_3)}{\dfrac{s^2 R_1 R_2 R_3 C_1 C_2}{R_1 + R_3} + \dfrac{s[C_2 R_3(R_1 + R_2) + C_2 R_1 R_2(1 - A_V) + C_1 R_3 R_1]}{R_1 + R_3} + 1} \tag{16-30}$$

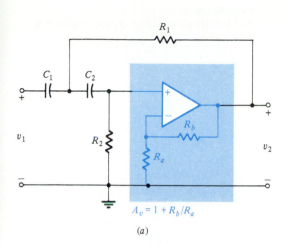

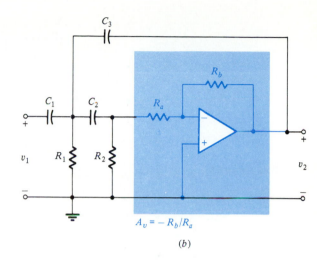

$A_v = 1 + R_b/R_a$

(a)

$A_v = - R_b/R_a$

(b)

FIGURE 16-33
Sallen and Key high-pass sections using (a) positive and (b) negative feedback.

$$H(s) = \frac{V_2}{V_1} = \frac{-R_2C_1s}{R_1R_2C_1C_2s^2 + sR_1(C_1 + C_2) + 1} \qquad (16\text{-}31)$$

Verification of these equations is left to the student in Probs. 16-28 and 16-29.

The addition of positive feedback (R_a and R_b in Fig. 16-35) to the circuit in Fig. 16-34b results in improved performance. The combination of positive and negative feedback allows practical fabrication of the higher-Q circuits required in high-order filters. It can be shown (Prob. 16-31) that the transfer function of this circuit is given by

$$H(s) = \frac{V_2}{V_1} = \frac{-A_V R_2 C_1 s/(A_V - 1)}{R_1R_2C_1C_2s^2 + s[R_1(C_1 + C_2) - R_2C_1/(A_V - 1)] + 1} \qquad (16\text{-}32)$$

where $A_V = 1 + R_b/R_a$

FIGURE 16-34
(a) Sallen and Key non-inverting band-pass section. (b) A noninverting band-pass circuit.

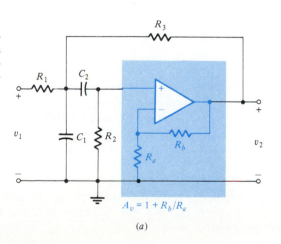

$A_v = 1 + R_b/R_a$

(a)

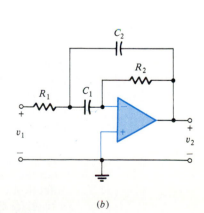

(b)

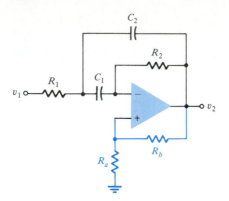

FIGURE 16-35
The Delyiannis band-pass circuit using both positive and negative feedback.

The band-pass sections in Figs. 16-34 and 16-35 are useful in realizing narrow-band circuits; that is, the passband is a fraction of the center frequency. The response of narrow-band circuits is similar to that obtained with a simple series or parallel resonant circuit having moderate Q. For this situation, the upper and lower 3-dB frequencies f_2 and f_1, respectively (Fig. 16-26b), are quite close so that the center frequency $f_o = \sqrt{f_1 f_2} \approx (f_1 + f_2)/2$ and the bandwidth $f_2 - f_1 = f_o/Q$. In some applications $f_2 - f_1 > f_o$ and broad-band band-pass circuits are required. This characteristic can be obtained by cascading low-pass and high-pass sections as shown in Fig. 16-36a. If $\omega_H > \omega_L$ (Fig. 16-36b), the band-pass response shown in Fig. 16-36c results. Note in Fig. 16-36c that attenuation at low frequencies ($\omega < \omega_L$) is provided by the high-pass network, whereas for $\omega > \omega_H$, attenuation results from the low-pass section. Both networks transmit the signal in the passband ($\omega_L \le \omega \le \omega_H$).

The cascade structure cannot be used in the narrow-band case because of component variations (sensitivity). If ω_L and ω_H are nearly equal, a small change in either (both) causes a significant error in the passband (Prob. 16-32).

Band-Reject Sections   The parallel-channel configuration in Fig. 16-37a is used to realize a broad-band band-reject filter. If the low-pass and high-pass networks have frequency responses depicted in Fig. 16-36b and $\omega_H < \omega_L$, then Fig. 16-37b is the response of the circuit in Fig. 16-37a. Both the high-pass and low-pass sections provide attenuation in the stopband between ω_H and ω_L. For $\omega < \omega_H$, transmission is through the low-pass section and for $\omega > \omega_L$, the signal is transmitted through the high-pass section.

Narrow-band band-reject circuits are often called *notch filters*. This can be discerned from the band-elimination entry in Table 16-2 for which $H(j\omega_r) = 0$ when $s = j\omega_r$. Note that $H(j\omega) \ne 0$ for all $\omega \ne \omega_r$ as displayed in the frequency response shown in Fig. 16-38. The circuit in Fig. 16-39 is commonly used to obtain the response shown in Fig. 16-38. The passive elements form a twin-tee network which provides the j-axis zeros. As depicted in Fig. 16-39, selection of $R_1 = R_2 = R$, $C_1 = C_2 = C$, $R_3 = R/2$, and $C_3 = 2C$ results in a transfer function (with $Y = 0$)

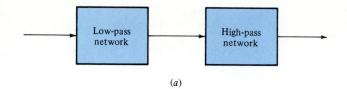

FIGURE 16-36
(*a*) A low-pass–high-pass cascade to form a broad-band band-pass circuit. (*b*) The frequency-response characteristics of the low-pass and high-pass net-works result in the (*c*) band-pass response of the cascade.

(*a*)

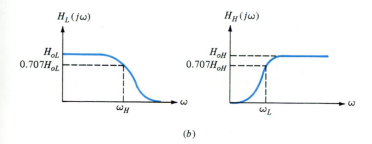

(*b*)

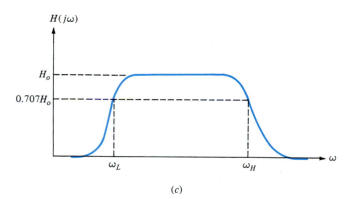

(*c*)

$$H(s) = \frac{V_2}{V_1} = \frac{A_V\,(R^2C^2s^2 + 1)}{R^2C^2s^2 + 2RC\,(2 - A_V)s + 1} \qquad (16\text{-}33)$$

In Eq. (16-33) we observe that $\omega_o = \omega_r = 1/RC$ and the Q of the notch is determined by the gain $A_V = 1 + R_b/R_a$ of the Op-Amp stage. Inspection of Eq. (16-33) also indicates that stability requires $A_V < 2$.

Often, it is desirable to have $\omega \neq \omega_o$. Circuits for which $\omega_r > \omega_o$ are called *high-pass notches*, and when $\omega_r < \omega_o$, the circuits are referred to as *low-pass notches*. The addition of the admittance Y (shown as the blue dashed-line element in Fig. 16-39) converts the circuit to either a high-pass or low-pass notch network. Choice of $Y = 1/R_4$ yields a low-pass notch; a high-pass notch is obtained for $Y = sC_4$. In both circuits, $C_3 = C_1 + C_2$ and $1/R_3 = 1/R_1 + 1/R_2$.

All-Pass Networks   The selection of one of the Butterworth or Chebyshev filter functions (Sec. 16-8) to approximate the magnitude of the filter response also specifies the phase characteristic $\angle H(j\omega)$ of the filter. It is often desirable to also control

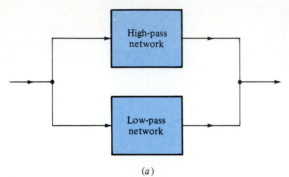

FIGURE 16-37
(*a*) A parallel-channel band-reject circuit. (*b*) If $\omega_L < \omega_H$ for the high-pass and low-pass characteristics in Fig. 16-36*b*, the band-reject response is obtained.

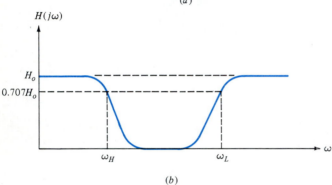

FIGURE 16-38
A notch response.

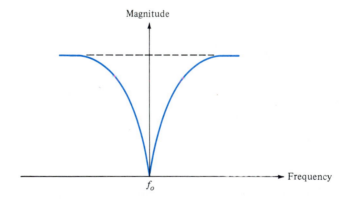

the phase response of the filter as well. One method is to cascade an *all-pass section* with the filter. An all-pass network has a magnitude $H(j\omega) = 1$ at all frequencies; its phase response, however, varies with frequency. The circuit in Fig. 16-40 is a one-pole all-pass network whose transfer function is

$$H(s) = \frac{V_2}{V_1} = \frac{1 - RCs}{1 + RCs} \tag{16-34}$$

Note that the pole and zero frequencies in Eq. (16-34) are equal in magnitude. However, the zero lies in the right half plane. Inspection of Eq. (16-34) indicates that $H(j\omega) = 1$ for all ω but

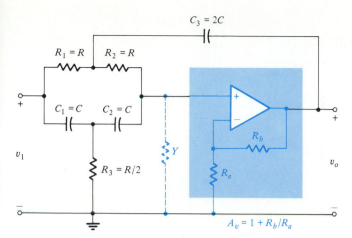

FIGURE 16-39
A twin-tee network used to obtain a notch. If $Y = 0$, the notch frequency and pole frequencies are equal. Inclusion of $Y \neq 0$ converts the circuit to a high-pass or low-pass notch.

FIGURE 16-40
A one-pole all-pass network.

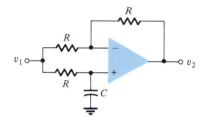

$$\angle H(j\omega) = -2\tan^{-1} RC \qquad (16\text{-}35)$$

The phase characteristic is displayed in Fig. 16-41; the black curve is the straight-line (Bode) approximation. Adjustment of the RC time constant of the circuit can be used to add 0 to $-180°$ of phase shift over the approximate frequency range $1/10RC < \omega < 10/RC$ to the filter characteristic.

Biquadratic all-pass networks, having the form in Eq. (16-36), can be realized by using the circuit in Fig. 16-42.

$$\frac{V_2}{V_1} = H(s) = \frac{s^2 - (\omega_o/Q)s + \omega_o^2}{s^2 + (\omega_o/Q)s + \omega_o^2} \qquad (16\text{-}36)$$

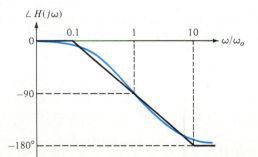

FIGURE 16-41
The phase characteristic of a one-pole all-pass network. The asymptotic Bode diagram (phase) is also shown.

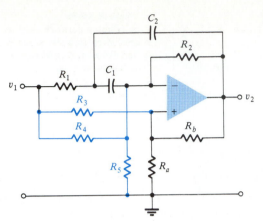

FIGURE 16-42
Friend's general biquad filter section.

Examination of Eq. (16-36) indicates that when $\omega < \omega_o$, $\angle H(j\omega) < 0$ and for $\omega > \omega_o$, $\angle H(j\omega) > 0$. Thus both positive and negative phase shift can be added to the phase response of the filter.

Friend's General Biquad Section Modification of the circuit in Fig. 16-35 (the elements shown in blue in Fig. 16-42) results in a general biquad section developed by Friend at Bell Telephone Laboratories. This circuit, called the *standard tantalum active resonator* (STAR), can be used to realize all biquadratic transfer functions except low-pass by eliminating (open-circuiting) appropriate resistances. Its advantage in large-scale communication systems is that the same topology can be used in all filter sections and in all filters in a multiplexed system. The additional elements introduce feedforward paths between input and output. These paths account for s^2 and constant terms added to the numerator of Eq. (16-32). Consequently, $H(s)$ for the circuit in Fig. 16-42 is of the form given in Eq. (16-14).

16-11 MULTIPLE OP-AMP BIQUAD SECTIONS

In the fabrication of practical filters, the performance of biquad sections is often improved by the introduction of additional Op-Amp stages. The cost of increased power consumption of additional Op-Amps is frequently more than offset by reduced sensitivity to component variations, ease in tuning (adjustment of ω_o and Q for each section), and the use of a standard topology to realize three or four of the basic frequency responses.

One such circuit, depicted in Fig. 16-43, can be used as a low-pass, band-pass, or high-pass filter. The stages A_{V1} and A_{V2} in Fig. 16-43 are finite-gain ideal voltage amplifiers and are constructed by using the basic Op-Amp stages (Fig. 10-42). The transfer function of this circuit (Prob. 16-43) is shown to be

$$H(s) = \frac{A_{V1}A_{V2}Z_BZ_D}{(Z_A + Z_B)(Z_C + Z_D) - Z_AZ_DA_{V1}A_{V2}} \tag{16-37}$$

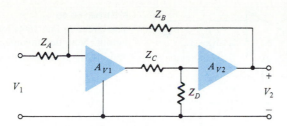

FIGURE 16-43
Basic structure of a 2-Op-Amp biquad section. If each impedance is chosen as either R or C, the response can be either low-pass, high-pass, or band-pass.

Table 16-5   Impedance Selection for the Circuit in Fig. 16-43

Response	Z_A	Z_B	Z_C	Z_D
Low-pass	R_1	$1/sC_1$	R_2	$1/sC_2$
Band-pass	R_1	$1/sC_1$	$1/sC_2$	R_2
High-pass	$1/sC_1$	R_1	$1/sC_2$	R_2

Two of the four impedances, Z_A, Z_B, Z_C, and Z_D, are resistors and two are capacitors. The particular choice determines the nature of the response as shown in Table 16-5.

The Universal or State-Variable Filter Section   The universal or state-variable biquad section in Fig. 16-44 is capable of providing low-pass, band-pass, and high-pass outputs simultaneously. As seen in Fig. 16-44, the circuit consists of two integrators and an inverting gain stage. Negative feedback around all three stages is provided by R_5 where R_6 and R_7 form a positive-feedback loop around the first two stages. The voltages v_A, v_B, and v_C are the high-pass, band-pass, and low-pass outputs, respectively.

FIGURE 16-44
The universal or state-variable filter section. Low-pass, high-pass, and band-pass responses are available at v_C, v_A, and v_B, respectively.

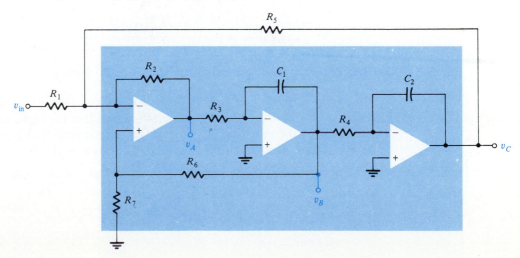

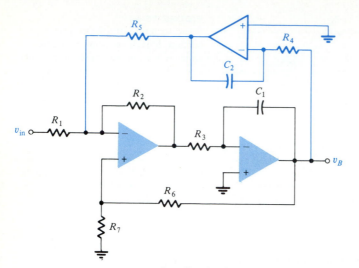

FIGURE 16-45

The circuit in Fig. 16-44 redrawn to show v_B as the output. Note that the integrator formed by R_4, C_2, and the Op-Amp are part of the feedback loop.

Qualitatively, we can demonstrate that v_C is the low-pass output as follows. Consider the circuit in Fig. 16-44 as a feedback amplifier in which R_5 and R_1 form a resistive feedback loop around the amplifier contained in the blue rectangle. The gain without feedback $A(s) = A_o/D(s)$, where $D(s)$ has two roots, one attributed to each of the integrators in the amplifier. The return ratio $T = -\beta A(s) = -\beta A_o/D(s)$ contains the same two poles as $A(s)$ since β is real because the feedback network is resistive. Hence, $A_F(s) = A(s)/[1 + T(s)] = A_o/[D(s) - \beta A_o]$. Clearly, A_F has two poles and no finite zeros and is a low-pass biquadratic function.

The circuit in Fig. 16-45 is the same as that in Fig. 16-44 redrawn. In this feedback amplifier, one integrator is part of the feedback network (shown in color). Thus β is not real but contains a pole at $s = 0$ due to the integrator, that is, $\beta = K/s$, where K is a real constant. The amplifier includes only one integrator and thus has only one pole $A(s) = A'_o/(1 + s/p_1)$. Again, use of Eq. (12-5) yields

$$A_F(s) = \frac{A'_o/(1 + s/p_1)}{1 + [A'_o/(1 + s/p_1)] \times (K/s)} = \frac{A'_o s}{D(s)}$$

where $D(s)$ is a quadratic. Comparison of $A_F(s)$ with the band-pass entry in Table 16-2 shows that they are equivalent.

In similar fashion, we can show that v_A is the high-pass output as both integrators are part of the feedback network and that the basic amplifier, consisting only of the inverting Op-Amp stage, is frequency-independent.

The various transfer functions can be expressed as (Prob. 16-44)

$$H_L = \frac{V_C}{V_{in}} = \frac{-R_s/R_1}{D(s)} \qquad H_B = \frac{V_B}{V_{in}} = \frac{R_4 R_5 C_2 s/R_1}{D(s)}$$

$$H_H = \frac{V_A}{V_{in}} = \frac{-R_3 R_4 R_5 C_1 C_2 s^2/R_1}{D(s)}$$

$$(16\text{-}38)$$

where

$$D(s) = \frac{C_1 C_2 R_3 R_4 R_5}{R_2} s^2 + \frac{R_4 R_7 C_2 (R_1 R_2 + R_1 R_5 + R_2 R_5)}{R_1 R_2 (R_6 + R_7)} s + 1 \quad (16\text{-}39)$$

The biquad section in Fig. 16-44 is commercially available from several manufacturers, including Burr-Brown, Inc., and General Instrument, Inc. These sections typically contain two precision 1000-pF capacitors (C_1 and C_2) and four precision resistances (usually R_2, R_3, R_4, and R_6). The remaining three resistances are selected by the designer to realize the desired values of ω_o, Q, and the maximum gain.

16-12 SWITCHED-CAPACITOR FILTERS

The active-RC circuits described in the three previous sections are all continuous-time filters; that is, the input and output signals exist for all time. Most often, hybrid technology consisting of monolithic Op-Amps and thin-film resistors and capacitors are used to implement the biquad sections which comprise these filters. *Switched-capacitor filters* are analog sampled-data systems (Sec. 16-1) containing only capacitors, Op-Amps, and analog switches. If the signal frequencies are much smaller than the switching frequency of the analog switches, these sampled-data filters provide an alternative but equivalent substitute for active-RC filters. Among the advantages that may accrue from this replacement are

1. The entire filter may be fabricated in monolithic form.

2. High-component-density MOS technology can be employed. This often results in single-chip implementation of systems requiring both analog and digital signal processing.

3. In the biquad sections described previously, the angular frequency ω_o (Table 16-2) usually depended on the RC time constants. With switched-capacitor realizations, ω_o can be made to depend on capacitor ratios. Since component ratios are more precisely controlled than are individual component values, more accuracy is obtained in the transfer ratio of the fabricated filter.

4. Elimination of resistors diminishes power consumption.

The topology of many switched-capacitor filters are derivatives of the continuous-time filters described in the two previous sections. Our objective in this section is to describe the fundamentals of switched-capacitor operation and their use in basic gain and integrating Op-Amp stages.

FIGURE 16-46
(a) A switched capacitor and (b) its equivalent resistance.

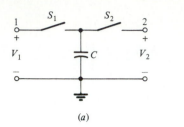
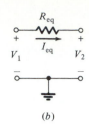

(a)   (b)

Resistance Simulation   Consider the circuit in Fig. 16-46a in which switches S_1 and S_2 are complementary. Switch S_1 (S_2) is closed (open) for T_1 seconds and open (closed) for T_2 seconds. The period of one switching cycle $T = T_1 + T_2$ and $f_s = 1/T$ is the switching frequency. Voltages V_1 and V_2 are ideal voltage sources, and we assume $V_1 > V_2$ in our discussion. With S_1 closed and S_2 open, C is charged to V_1. At $t = T_1$, S_1 is open, S_2 is closed, and C discharges to V_2. The cycle is repeated at $t = T = T_1 + T_2$. During one cycle the charge Q, transported from node 1 to node 2, is $Q = C(V_1 - V_2)$. Since this action occurs in T seconds, this is equivalent to a current I_{eq}

$$I_{eq} = \frac{Q}{T} = \frac{C}{T}(V_1 - V_2) = Cf_s(V_1 - V_2) \tag{16-40}$$

The current I_{eq} is the same that exists in R_{eq} in Fig. 16-46b. Thus

$$\frac{V_1 - V_2}{R_{eq}} = Cf_s(V_1 - V_2) \tag{16-41}$$

or

$$R_{eq} = \frac{1}{Cf_s} \tag{16-42}$$

Equation (16-42) demonstrates that periodically switching the capacitor is equivalent to a resistance connected as shown in Fig. 16-46b. Note that this is true if the switching frequency f_S is much greater than the frequency of the signals V_1 and V_2.

Integrators   The circuit in Fig. 16-47a is the switched-capacitor realization of the continuous-time integrator in Fig. 16-47b. The transfer function of the circuit in Fig. 16-47b is given in Eq. (16-6) and repeated as Eq. (16-43) for convenience:

$$A_V = \frac{V_o}{V_s} = \frac{-1}{RC_1 s} \tag{16-43}$$

Substitution of Eq. (16-42) for R in Eq. (16-43) gives the transfer function of the switched-capacitor integrator as

$$A_V = \frac{V_o}{V_s} = -\frac{C_2 f_s}{C_1 s} \tag{16-44}$$

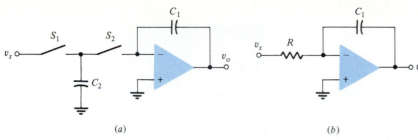

(a)   (b)

FIGURE 16-47
(a) Switched-capacitor integrator and (b) its continuous-time equivalent.

Clearly, Eq. (16-44) demonstrates that A_V depends on the *capacitor ratio C_2/C_1*.

The clocked MOS transistors in the integrator in Fig. 16-48a implement switches S_1 and S_2 in Fig. 16-47a The signals ϕ and $\bar{\phi}$ are derived from a nonoverlapping two-phase clock of frequency f_S (Fig. 16-48b). The value $V(1)$ of the clock pulse must be larger than the threshold voltage V_T of the NMOS transistor. Similarly, $V(0) < V_T$ so that the MOSFET is an open switch.

Gain Stages

An inverting Op-Amp stage is converted to its switched-capacitor equivalent by replacing each resistor in the stage by the configuration in Fig. 16-46a. This is illustrated in Fig. 16-49 with MOSFET switches. Since both switch pairs are operated at the same frequency f_S, the transfer function of the circuit is

$$A_V = \frac{V_o}{V_s} = -\frac{C_1}{C_2} \tag{16-45}$$

The gain A_V can be controlled accurately as it depends on the ratio of component values.

The circuit in Fig. 16-49a however, is impractical. Since both switches used to realize $R_{2(\text{eq})}$ are never closed simultaneously, no feedback around the Op-Amp is provided. The arrangement depicted in Fig. 16-49b is used to overcome this difficulty. When $\phi = 1$, C_1 is charged to the input signal and C_2 is discharged (so that the charge previously stored is not retained). When $\phi = 0$, the voltage on C_1 is applied to the Op-Amp and C_2 forms the feedback path. The gain is given by Eq. (16-45).

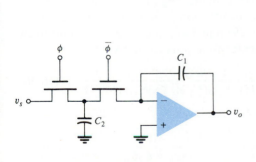

(a)

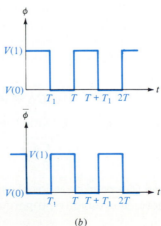

(b)

FIGURE 16-48
(a) The circuit in Fig. 16-47a with clocked MOSFETs used as the switches. (b) The nonoverlapping two-phase clock waveforms.

FIGURE 16-49
(*a*) An impractical switched-capacitor inverting Op-Amp gain stage and (*b*) a practical circuit.

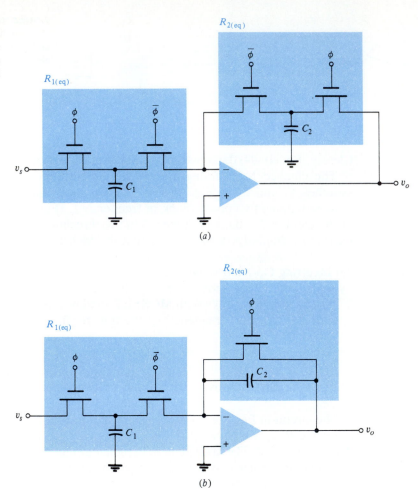

(*a*)

(*b*)

One-Pole Sections Tables 16-4 and 16-5 illustrate that any odd-order filter contains a pole on the real axis. The circuits in Fig. 16-31 provide negative real axis poles in continuous-time filters. The switched-capacitor implementations of Fig. 16-31 are shown in Fig. 16-50. From Eq. (16-29) and Example 16-3, substitution of Eqs. (16-45) and (16-42) allows the transfer functions to be expressed as

$$H(s) = \frac{V_2}{V_1} = \frac{-C_1 s/C_4 f_s}{(C_1 C_2 s^2/C_3 C_4 f_s^2) + [s(C_1 + C_2)/C_3 f_s] + 1} \qquad (16\text{-}46)$$

Comparison of Eq. (16-46) with Table 16-1 yields

$$\omega_o = f_s \sqrt{\frac{C_3 C_4}{C_1 C_2}} \qquad Q = \frac{1}{1 + C_1/C_2} \sqrt{\frac{C_1 C_3}{C_2 C_4}} \qquad (16\text{-}47)$$

Both ω_o and Q depend on ratios of capacitors which can be controlled precisely. Furthermore, ω is directly proportional to f_s, which, in turn, is the frequency

FIGURE 16-50

Switched-capacitor realizations of the circuits in Fig. 16-31 which realize real poles.

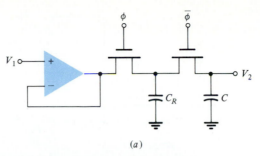

(a)

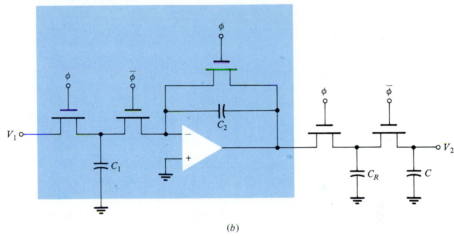

Inverting Op-Amp stage

(b)

of the two-phase clock. Since the clock frequency can be controlled to high accuracy, both ω_o and Q can be manufactured with high precision.

The circuit in Fig. 16-51 can be tuned by varying the clock frequency f_S. Specification of the capacitor ratios determines Q; hence, increasing (decreasing) ω_o by adjustment of the clock frequency causes the bandwidth ω_o/Q to increase (decrease).

Low-order switched-capacitor filters are available from several manufacturers. The MF6-100 (National Semiconductor) is a six-pole low-pass Butterworth filter fabricated by means of CMOS technology. The cutoff frequency is variable from 0.1 to 20 kHz and requires a clock frequency 100 times the cutoff frequency. The passband gain is unity, so that a 12-pole filter is readily obtained by cascading two circuits.

EG&G Reticon has a series (R56XX) of six-pole Chebyshev band-pass filters whose center frequency can be varied from 0.5 to 20 kHz. Also available in this series are high-pass and notch filters.

A monolithic switched-capacitor universal filter (National Semiconcuctor MF10) can be designed to provide low-pass, high-pass, band-pass, band-reject (notch), and all-pass characteristics. A four-pole filter can be designed and requires an external clock and as many as eight external resistors.

FIGURE 16-51
Switched-capacitor re-
alization of the band-
pass circuit in Fig. 16-
34*b*.

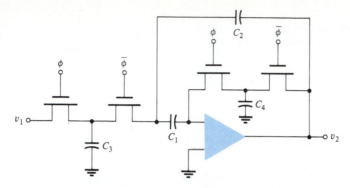

16-13 LOGARITHMIC AND EXPONENTIAL AMPLIFIERS

In Fig. 16-52 an Op-Amp is shown with the feedback resistor R_2' replaced by the diode $D1$. This amplifier is used when it is desired to have the output voltage proportional to the logarithm of the input voltage.

From Eq. (2-3) the volt-ampere diode characteristic is

$$i_f = I_S(\epsilon^{v_f/\eta V_T} - 1) \approx I_S\epsilon^{v_f/\eta V_T}$$

provided $v_f/\eta V_T \gg 1$. Hence

$$v_f = \eta V_T(\ln i_f - \ln I_S) \tag{16-48}$$

Since $i_s = v_s/R$ due to the virtual ground at the amplifier input, then

$$v_o = -v_f = -\eta V_T\left(\ln \frac{v_s}{R} - \ln I_S\right) \tag{16-49}$$

Logarithmic Amplifier Using Matched Transistors

We note from Eq. (16-49) that v_o is temperature-dependent as a result of the scale factor ηV_T and the saturation current I_S. The factor η, whose value normally depends on the diode current, can be eliminated by replacing the diode with a grounded-base transistor. Another important advantage of using a transistor in place of a diode is that the exponential relationship between current and voltage extends over a much wider voltage range for a transistor than a diode. By augmenting Fig. 16-52

FIGURE 16-52
An elementary logarith-
mic amplifier.

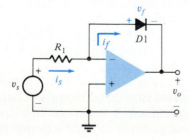

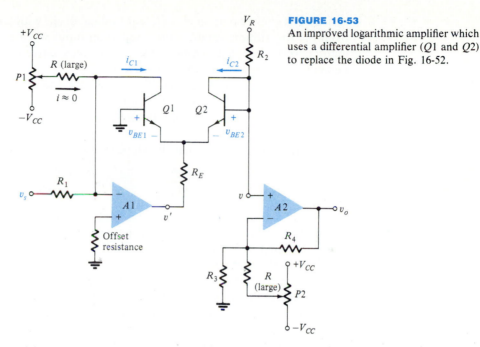

FIGURE 16-53
An improved logarithmic amplifier which uses a differential amplifier ($Q1$ and $Q2$) to replace the diode in Fig. 16-52.

with a second matched transistor, it is possible to eliminate from the expression for v_o the reverse saturation current I_S (which doubles for every 10°C rise in temperature). The final system shown in Fig. 16-53 includes an output noninverting Op-Amp stage with a gain $A_V = 1 + R_4/R_3$.

We now derive the logarithmic expression for v_o. For the present discussion, ignore *the high-resistance-balancing-potentiometer arrangements*. For matched transistors and with $i_B \ll i_C$, the positive input to $A2$ is at a voltage

$$v \equiv V_{BE2} - V_{BE1} = V_T \ln i_{C2} - V_T \ln i_{C1} = -V_T \ln \frac{i_{C1}}{i_{C2}} \quad (16\text{-}50)$$

Since v equals the small difference in the base-emitter voltages of $Q2$ and $Q1$, we neglect v compared with the reference voltage V_R. Then, since $i_{B2} \ll i_{C2}$ and because of the virtual ground at the input of $A1$, it follows that

$$i_{C2} = \frac{V_R}{R_2} \quad \text{and} \quad i_{C1} = \frac{v_s}{R_1} \quad (16\text{-}51)$$

Since $A2$ is a noninverting Op-Amp, $v_o = v(R_3 + R_4)/R_3$. Combination of this equation with Eqs. (16-50) and (16-51) gives

$$v_o = -V_T \frac{R_3 + R_4}{R_3} \ln \left(\frac{v_s}{R_1} \frac{R_2}{V_R} \right) \quad (16\text{-}52)$$

Experimentally, it is found that Eq. (16-52) is satisfied over a dynamic range of four decades, from input voltages of 2 mV to 20 V. Beyond 20 V the higher values of transistor currents passing through the ohmic collector and base resistances give a component of linear voltage drop, which leads to a departure from the logarithmic relationship. Below an input voltage of about 2 mV, the input current becomes comparable to the bias current and the logarithmic dependence of v_o on v_s is no longer valid.

The potentiometer $P1$ is used to balance out the offset voltage of $A1$; that is, with $v_s = 0$, $P1$ is varied until $v' \approx 0$ (less than 50 μV). The system is nulled as follows. With $v_s = V_R R_1/R_2$, potentiometer $P2$ is varied until $v_o = 0$, thus satisfying Eq. (16-52).

Note from Eq. (16-52) that the slope of the characteristic

$$\frac{dv_o}{d(\ln v_s)} = -V_T \frac{R_3 + R_4}{R_3} \tag{16-53}$$

This result has been verified experimentally. Since V_T is proportional to temperature, R_3 should be chosen as a temperature-sensitive resistance. If R_3 increases linearly with T, the slope in Eq. (16-53) can be made to be quite constant as the temperature changes.

Log amplifiers having dynamic ranges of five orders of magnitude can be constructed with Op-Amps having low bias currents. The use of such log amplifiers in signal processing can be illustrated as follows. Consider an analog input signal whose dynamic range is five orders of magnitude that is to be converted to a digital signal. A 20-bit A/D converter is required if the resolution is to be 10 percent of the smallest signal. A flash A/D (Fig. 16-16) needs 2^{20} Op-Amps and is clearly impractical. The slower A/D converter in Fig. 16-15 and with a 20-MHz clock requires 1 μs per conversion. The log amplifier compresses the dynamic range of the output signal v_o so that an 8-bit A/D suffices.

Exponential (Antilog) Amplifier   After processing, the compressed dynamic range of the log amplifier output must often exhibit the same dynamic range as the original input. That is, the output of the D/A converter used in reconstructing an analog signal must also display the large dynamic range of the input analog signal. An *antilog* or *exponential amplifier* is used for this purpose. This system is depicted in Fig. 16-54 and should be compared with that of Fig. 16-53. In the exponential amplifier the feedback current i_{C1} is constant and is derived from the reference voltage V_R, whereas i_{C2} depends on the input signal. In the logarithmic amplifier the converse is true.

Because of the virtual ground at the inputs to $A1$ and $A2$, the collector and base of $Q1$ are at the same potential $-v = V_{BE1} - V_{BE2}$. Neglecting v relative to V_R, we obtain

$$i_{C1} = \frac{V_R}{R_2} \quad \text{and} \quad i_{C2} = \frac{v_o}{R_1} \tag{16-54}$$

From the input attenuator it is clear that

$$-v = \frac{R_3 v_2}{R_3 + R_4} = V_T \ln \frac{i_{C1}}{i_{C2}} \tag{16-55}$$

FIGURE 16-54

An exponential ampli-
fier.

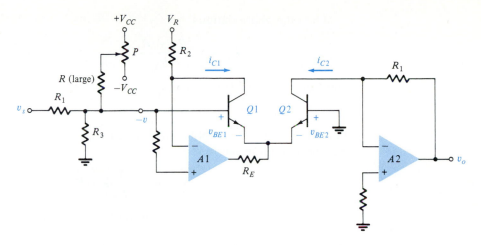

where use is made of Eq. (16-50). Substituting the currents I_{C1} and I_{C2} from
Eq. (16-54) into Eq. (16-55) we obtain

$$v_2 = -V_T \frac{R_3 + R_4}{R_3} \ln\left(\frac{v_o}{R_1} \frac{R_2}{V_R}\right) \tag{16-56}$$

Note that this equation becomes identical with Eq. (16-52) if v_s and v_o are
interchanged. Hence v_o is proportional to the antilog or exponential of v_s. From
Eq. (16-56), we obtain

$$v_o = \frac{R_1 V_R}{R_2} \exp\left(-\frac{v_s}{V_T} \frac{R_3}{R_3 + R_4}\right) \tag{16-57}$$

The system is calibrated for mismatch and offset voltages by setting the input
$v_s = 0$ and then adjusting the potentiometer P until $v_o = R_1 V_R/R_2$.

Logarithmic Multiplier The log and antilog amplifiers can be used for multiplication or division
of two analog voltages v_{s1} and v_{s2}. In Fig. 16-55 the logarithm of each input is
taken, then the two logarithms are added, and finally the antilog of the sum is
taken. We now verify that the output is proportional to the product of the two
inputs.

FIGURE 16-55

Logarithmic multiplier
of two analog signals.

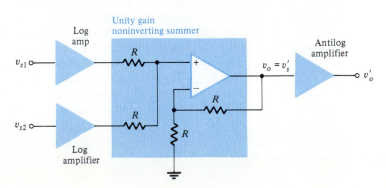

When the abbreviations

$$K_1 \equiv V_T \frac{R_3 + R_4}{R_3} \qquad \text{and} \qquad K_2 \equiv \frac{R_2}{R_1 V_R} \qquad (16\text{-}58)$$

are used, Eq. (16-52) becomes

$$v_o = -K_1 \ln K_2 v_s \qquad (16\text{-}59)$$

For the exponential amplifier with an input v_s', and an output v_o', Eq. (16-57) may be written in the form

$$v_o' = \frac{1}{K_2} \epsilon^{-v_s/K_1} \qquad (16\text{-}60)$$

According to this notation, the output v_o of the summing Op-Amp in Fig. 16-55 is

$$v_o = -K_1 \ln K_2 v_{s1} - K_1 \ln K_2 v_{s2} = -K_1 \ln K_2^2 v_{s1} v_{s2} \qquad (16\text{-}61)$$

Since v_o is the input to the antilog amplifier, $v_o = v_s'$, and from Eqs. (16-60) and (16-61)

$$v_o' = \frac{1}{K_2} \exp(\ln K_2^2 v_{s1} v_{s2}) = K_2 v_{s1} v_{s2} \qquad (16\text{-}62)$$

We show in Prob. 16-88 that it is possible to raise the input v_s to an arbitrary power by cascading log and antilog amplifiers.

The input signals can be divided if we subtract the logarithm of v_{s1} from that of v_{s2} and then take the antilog. We must point out that the logarithmic multiplier or divider is useful for unipolar inputs only. This is often called *one-quadrant operation*. Other techniques are available for the accurate multiplication of two signals, one of which is described in the next section.

16-14   ANALOG MULTIPLIERS   From Eqs. (10-86) and (10-88), we observe that the output voltage of a differential amplifier depends on the current source I_{EE}; that is, g_m is directly proportional to I_{EE}. The differential amplifier can function as a multiplier by varying the source current as shown in Fig. 16-56a. Application of a signal v_{s2} causes the reference current and hence i_{EE} to vary directly with v_{s2}. If, in addition, a signal v_{s1} is applied to the differential amplifier, the output is proportional to the product of the two signals, v_{s1}, v_{s2}. Both the inverting and noninverting outputs of the differential amplifier drive a differential amplifier (Sec. 14-12). This tends to eliminate common-mode output components. The circuit symbol for a multiplier is depicted in Fig. 16-56b. The constant K is a scale factor which affects the dynamic range of the input signals.

The circuit in Fig. 16-56a is a two-quadrant multiplier since $v_{s2} > V_{BE(ON)}$ of $Q3$ and $Q4$. This limitation is overcome by using the *Gilbert multiplier cell* shown in Fig. 16-57 (which replaces the differential stage and current source

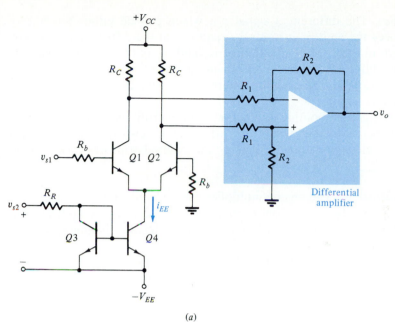

FIGURE 16-56
(a) A differential amplifier used as a two-quadrant analog multiplier. (b) Multiplier symbol.

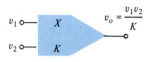

(a)

(b)

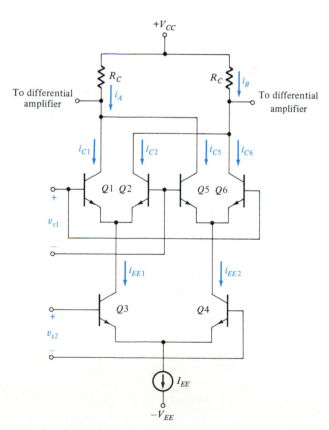

FIGURE 16-57
The Gilbert cell used to obtain a four-quadrant multiplier.

in Fig. 16-56*a*). The differential signal v_{s2}, which can be either positive or negative, causes the emitter currents i_{EE1} and i_{EE2} in the differential pairs $Q1$-$Q2$ and $Q5$-$Q6$ to vary. Multiplication of v_{s1} and v_{s2} is thus accomplished in each of these differential pairs. Four-quadrant operation is obtained as the output currents i_A and i_B are each the difference between the collector currents of the differential amplifiers.

The AD534L is a monolithic 4-quadrant multiplier-divider with a basic accuracy of 0.25 percent, a bandwidth of 1 MHz, and a slew rate of 20 V/μs. The circuit is completely pretrimmed; that is, no external trim networks are required.

Squaring and Square-Root Circuits The analog multiplier can be used to obtain the square and square root of a function. Connnection of the two inputs, as shown in Fig. 16-58*a*, renders the output $v_o = v_s^2/K$. If the input signal v_s is amplified by K before driving the multiplier, $v_o = v_s^2$.

The square-root circuit in Fig. 16-58*b* employs the multiplier in the feedback loop of an inverting Op-Amp stage. The virtual ground at the input of the Op-Amp results in

$$i_1 = \frac{v_s}{R_1} \qquad i_2 = \frac{v_2}{R_2} \tag{16-63}$$

the multiplier output $v_2 = v_o^2/K$, and since $i_1 = -i_2$, combination of these relations and solving for v_o results in

$$v_o = \sqrt{\frac{KR_2}{R_1}} \, |v_s| \tag{16-64}$$

The quantity $|v_s|$ is needed in Eq. (16-64) to ensure that the terms contained in the radical are positive. Note that making $R_2/R_1 = 1/K$ yields $v_o = \sqrt{|v_s|}$.

Balanced Modulator The analog multiplier can be used to generate an amplitude-modulated (AM) signal. If $v_{s1} = V_1 \cos \omega_C t$ and $v_{s2} = V_2 \cos \omega_S t$, where ω_C is the carrier angular frequency and ω_S is the angular frequency of the signal, the output v_o of the multiplier in Fig. 16-56*b* is

$$v_o = V_1 V_2 \cos \omega_C t \cos \omega_S t = (V_2 V_2 \cos \omega_S t) \cos \omega_C t \tag{16-65}$$

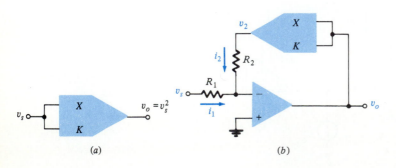

FIGURE 16-58

The use of a multiplier as (*a*) a squaring circuit and (*b*) a square-root circuit.

(*a*) (*b*)

Equation (16-65) demonstrates that the amplitude of the carrier varies directly with the signal. Use of the identity cos $(x + y)$ = cos x cos $y \mp$ sin x sin y in Eq. (16-65) allows v_o to be expressed as

$$v_o = \frac{V_1 V_2}{2} \left[\cos (\omega_C - \omega_S)t + \cos (\omega_C + \omega_S)t \right] \qquad (16\text{-}66)$$

Since the carrier frequency $f_C = \omega_C/2\pi$ does not explicitly appear in Eq. (16-66), the circuit is referred to as a *balanced modulator*.

16-15   PRECISION AC/DC CONVERTERS

If a sinusoid whose peak value is less than the threshold or cut-in voltage V_γ ($\sim$ 0.6 V) is applied to the rectifier circuit in Fig. 2-13, we see that the output is zero for all times. In order to be able to rectify millivolt signals, it is clearly necessary to reduce V_γ. By placing the diode in the feedback loop of an Op-Amp, the cut-in voltage is divided by the open-loop gain A_v of the amplifier. Hence V_γ is virtually eliminated and the diode approaches the ideal rectifying component. If the input v_s in Fig. 16-59a goes positive by at least V_γ/A_v, then v' exceeds V_γ and D conducts. Because of the virtual connection between the noninverting and inverting inputs (due to the feedback with D ON), $v_o \approx v_s$. Therefore, the circuit acts as a voltage follower for positive signals (in excess of approximately $0.6/10^5$ V = 60 μV). When v_s swings negatively, D is OFF and no current is delivered to the external load except for the small bias current of the Op-Amp and the diode reverse saturation current.

Precision Limiting   By modifying the circuit of Fig. 16-59a, as indicated in Fig. 16-59b, we can obtain an almost ideal limiter. If $v_s < V_R$, then v' is positive and D conducts. As explained above, under these conditions the output equals the voltage at the noninverting terminal, or $v_o = V_R$. If $v_s > V_R$, then v' is negative, D is OFF, and $v_o = v_s R_L/(R_L + R) \approx v_s$ if $R \ll R_L$. In summary, the output follows the input for $v_s > V_R$ and v_o is clamped to V_R if v_i is less than V_R by about 60 μV. When D is reverse-biased in Fig. 16-29a or 16-29b a large differential voltage may appear between the inputs, and the Op-Amp must be able to withstand this voltage. Also note that when $v_s > V_R$, the Op-Amp saturates because the feedback through D is missing.

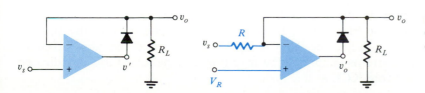

(a)   (b)

FIGURE 16-59
(a) A precision rectifier. (b) A precision clamp (limiting circuit).

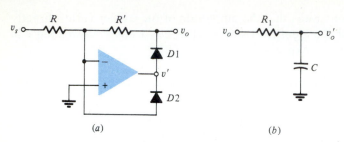

FIGURE 16-60
(a) A precision half-wave rectifier. (b) A low-pass RC filter which can be cascaded with the circuit in (a) to obtain an average detector.

Fast Half-Wave Rectifier   By adding R' and $D2$ to Fig. 16-59b and setting $V_R = 0$, we obtain the circuit shown in Fig. 16-60a. If v_s goes negative, $D1$ is ON, $D2$ is OFF, and the circuit behaves as an inverting Op-Amp, so that $v_o = -(R'/R)v_i$. If v_s is positive, $D1$ is OFF and $D2$ is ON. Because of the feedback through $D2$, a virtual ground exists at the input and $v_o = 0$. If v_s is a sinusoid, the circuit performs half-wave rectification.

The principal limitation of this circuit is the slew rate of the Op-Amp. As the input passes through zero, the Op-Amp output v' must change as quickly as possible from $+0.6$ to -0.6 V (or vice versa) in order for the conduction to switch very rapidly from one diode to the other. If the slew rate is 1 V/μs, this switching time is 1.2 μs. Hence 1.2 μs must be a small fraction of the period of the input sinusoid.

An alternative noniverting configuration to that in Fig. 16-60a is to ground the left-hand side of R and to impress v_i at the noninverting terminal. The output now has a value of $(R + R')/R$ times the input for positive voltages and $v_o = v_s$ for negative inputs if $R_L \gg R'$. Hence half-wave rectification is obtained if $R' \gg R$. For either the inverting or noninverting half-wave rectifier, diodes $D1$ and $D2$ may *both* be reversed.

Full-Wave Rectifier   The system indicated in Fig. 16-61a gives full-wave rectification without inversion and with a gain R/R_1, controllable by the one resistor R_1. Consider first the half-cycle where v_i is positive. Then $D1$ is ON and $D2$ is OFF. Since $D1$ conducts, there is a virtual ground at the input to $A1$. Because $D2$ is non-conducting and there is no current in the R which is connected to the noninverting input to $A2$, it follows that $v_1 = 0$. Hence, the system consists of two Op-Amps in cascade, with the gain of $A1$ equal to $-R/R_1$ and the gain of $A2$ equal to $-R/R = -1$. The result is

$$v_o = +\frac{R}{R_1} v_i > 0 \qquad \text{for} \qquad v_i > 0 \qquad (16\text{-}67)$$

Consider now the half-cycle where v_i is negative. Then $D1$ is OFF and $D2$ is ON, as indicated in Fig. 16-61b. Because of the virtual ground at the input to $A2$, $v_2 = v_1 \equiv v$. Since the input terminals of $A1$ are at the same (ground) potential, the currents coming to the inverting terminal of $A1$ are as indicated in the figure. From KCL at this node

$$\frac{v_s}{R_1} + \frac{v}{2R} + \frac{v}{R} = 0 \qquad \text{or} \qquad v = -\frac{2}{3}\frac{R}{R_1} v_s \qquad (16\text{-}68)$$

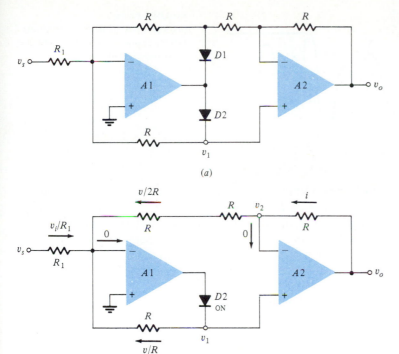

FIGURE 16-61
(a) A full-wave rectifier system. (b) During the half-cycle when v_s is negative, $D1$ is OFF and $D2$ is ON as indicated. Note that $v_1 = v_2 = v$ and $i = v/2R$.

The output voltage is $v_o = iR + v$, where the current i equals $v/2R$ because the inverting terminal of $A2$ takes no current. Hence

$$v_o = \frac{v}{2R}R + v = \frac{3}{2}v = -\frac{R}{R_1}v_s > 0 \qquad \text{for} \qquad v_i < 0 \qquad (16\text{-}69)$$

where use is made of Eq. (16-58). Note that the sign of v_o is positive in Eq. (16-69) because v_s is negative in this half-cycle. Since v_o in Eq. (16-69) equals v_o in Eq. (16-68), the outputs for the two half-cycles are identical, thus verifying that the system performs full-wave rectification (with a gain of R/R_1). Note that for any input waveform, v_o is proportional to the absolute value of the input $|v_s|$.

Active Average Detector   Consider the circuit in Fig. 16-60a to be cascaded with the low-pass filter in Fig. 16-60b. If v_s is an amplitude-modulated carrier, the R_1C filter removes the carrier and v_o' is proportional to the average value of the audio signal. In other words, this configuration represents an *average detector*.

Active Peak Detector   If a capacitor is added at the output of the precision diode in Fig. 16-59a, with $R_L = \infty$, a peak detector results. The capacitor in Fig. 16-62a will hold the output at $t = t'$ to the most positive value attained by the input v_s prior to t', as indicated in Fig. 16.62b. This operation follows from the fact that if $v_s > v_o$, the voltage at the noninverting terminal exceeds that at the inverting

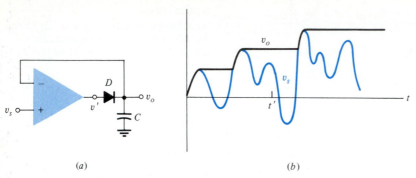

(a)   (b)

terminal and the Op-Amp output v' is positive, so that D conducts. The capacitor is then charged through D (by the output current of the amplifier) to the value of the input because the circuit is a voltage follower. When v_s falls below the capacitor voltage, the Op-Amp output goes negative and the diode becomes reverse-biased. Thus the capacitor charges until its voltage equals the most positive value of the input. To reset the circuit, a low-leakage switch such as a MOSFET gate must be placed across the capacitor.

The bias current of the Op-Amp is also integrated by the capacitor. Also, if the output is loaded, C discharges through the load. Both of these difficulties are avoided by modifying the system with a source follower, as indicated in Fig. 16-63. When the inverting terminal is connected to the load at the output, v_o is forced to equal the peak value of v_s, as desired (but the capacitor voltage differs from v_o by the gate-to-source voltage of the FET). This network is a special case of a sample-and-hold circuit, and the capacitor-leakage-current considerations given in the following section also apply to this configuration. For an ideal capacitor, the voltage across C in the hold position changes only because of the very small FET input current and the diode reverse current.

If the input v_s falls below the output v_o, the Op-Amp will saturate (the maximum input differential range may also be exceeded). To prevent this difficulty, another diode is added to the circuit, as indicated in Fig. 16-63b. If now $v_s < v_o$, then $D2$ conducts and the Op-Amp is a voltage follower, so that there is a virtual short circuit between the input terminals. If $v_s > v_o$, then $D2$ is cut off and the circuit reduces to the peak detector of Fig. 16-63a.

To obtain a peak detector that measures the most negative value of the input voltage, it is necessary only to reverse the diode D in Fig. 16-62 or 16-63. Why?

FIGURE 16-63
An improved version of the peak detector.

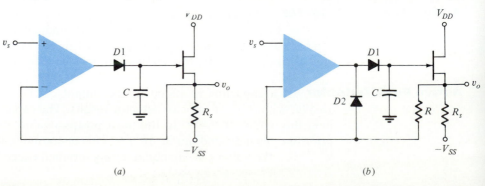

(a)   (b)

REFERENCES

1 Soclof, S: "Applications of Analog Integrated Circuits," Prentice-Hall, Englewood Cliffs, N.J., 1985.

2 Grebene, A. B.: "Bipolar and MOS Analog Integrated Circuit Design," John Wiley and Sons, New York, 1984.

3 Ghausi, M. S., and K. R. Laker: "Modern Filter Design," Prentice-Hall, Englewood Cliffs, N.J., 1981.

4 Allen, P. E., and E. Sanchez-Sinencio: "Switched-Capacitor Circuits," Van Nostrand Reinhold Company, New York, 1984.

5 Schaumann, R., M. A. Soderstrand, and K. R. Laker (Eds.): "Modern Active Filter Design," IEEE Press, New York, 1981.

6 Sallen, P. R., and E. L. Key: A Practical Method of Designing *RC* Active Filters, *IRE Trans. Circuit Theory*, vol. CT-2, pp. 74–85, March 1955.

7 Butterworth, S.: On the Theory of Filter Amplifiers, *Wireless Engineer*, vol. 7, pp. 536–541, October 1930.

8 Temes, G. C., and J. W. LaPatra: "Circuit Synthesis and Design," McGraw-Hill Book Company, New York, 1977.

9 Gray, P. R., and R. G. Meyer: "Analysis and Design of Analog Integrated Circuits," 2d ed., John Wiley and Sons, New York, 1984.

10 Dooley, D. J.: "Data Conversion Integrated Circuits," IEEE Press, New York, 1980.

11 Sheingold, D. H. (Ed.): "Analog-Digital Conversion Handbook," 3d ed., Analog Devices, Inc., Norwood, Mass., 1986.

12 Hamilton, D. J., and W. G. Howard: "Basic Integrated Circuit Engineering," McGraw-Hill Book Company, New York, 1975.

13 Gilbert, B.: A Precise Four-Quadrant Multiplier with Subnanosecond Response, *IEEE J. Solid-State Circuits*, vol. SC-3, pp. 365 – 373, December 1968.

14 Gray, P. R., D. A. Hodges, and R. W. Broderson (Eds.): "Analog MOS Integrated Circuits," IEEE Press, New York, 1980.

15 Friend, J.: STAR: An Active Biquadratic Filter Section, *IEEE Trans. Circuits and Syst.*, vol. CAS-22, February 1975.

REVIEW QUESTIONS

16-1 What functions must be performed in an AM system?

16-2 (*a*) Qualitatively describe the operation of a PCM system.
(*b*) Why must an antialiasing filter be used?

16-3 (*a*) Sketch a sample-and-hold system with very high input resistance and very low output resistance.
(*b*) Explain the operation of this system.

16-4 (*a*) What limits the acquisition time in a sample-and-hold configuration?
(*b*) Sketch a system for minimizing the acquisition time.

16-5 (*a*) Sketch an analog multiplexer system.
(*b*) How are the switches implemented?

16-6 Draw the block diagram from which to obtain the gating signals for time-division multiplexing.

16-7 (*a*) Draw a schematic diagram of a D/A converter. Use resistance values whose ratios are multiples of 2.

 (*b*) Explain the operation of the converter.

16-8 Indicate two possible implementations for the digitally controlled switch of a D/A converter.

16-9 Repeat Rev. 16-7 for a ladder network whose resistances have one of two values, R or $2R$.

16-10 Explain how a DAC may function as a programmable attenuator for an analog signal.

16-11 (*a*) Draw the block diagram for a counting A/D converter.

 (*b*) Explain the operation for this system.

16-12 Repeat Rev. 16-11 for a servo ADC.

16-13 Repeat Rev. 16-11 for a 2-bit parallel-comparator A/D converter.

16-14 Explain by means of an asymptotic Bode diagram why a practical integrator deviates from an ideal one at both high and low frequencies.

16-15 Sketch the circuit of a differential integrator and explain its operation.

16-16 Show how an integrator can be modified to become a differentiator.

16-17 Sketch the ideal frequency-response characteristics of low-pass, high-pass, band-pass, and band-reject systems.

16-18 (*a*) Write the transfer function for a general biquadratic function.

 (*b*) What coefficients in part *a* must be zero in order to obtain a low-pass characteristic?

 (*c*) Repeat part *b* for high-pass and band-pass characteristcs.

 (*d*) Repeat part *b* for band-reject response.

16-19 Define by means of a diagram (*a*) the passband, (*b*) the stopband, (*c*) the transition band, (*d*) passband ripple.

16-20 What differences exist in the frequency response of Butterworth and Chebyshev filters (low-pass) of the same degree?

16-21 Draw the circuit diagram of a low-pass Sallen and Key positive-feedback section.

16-22 How can the circuit in Rev. 16-21 be modified to become a high-pass section?

16-23 Repeat Rev. 16-21 for a negative-feedback section.

16-24 Draw a block diagram of a broad-band band-pass filter and explain its operation.

16-25 Repeat Rev. 16-24 for a band-reject filter.

16-26 Draw the circuit diagram of a band-pass section using an ideal Op-Amp.

16-27 (*a*) What is meant by a notch network?

 (*b*) Define high-pass and low-pass notches.

16-28 (*a*) What is meant by an all-pass network?

 (*b*) Of what use is such a network?

16-29 (*a*) Write the expression for the transfer function of a one-pole all-pass system.
(*b*) Repeat part *a* for a two-pole system.
(*c*) What is the maximum range of the phase shift that can be achieved in parts *a* and *b*?

16-30 Draw the circuit diagram of a universal biquad filter and show how low-pass, band-pass and high-pass outputs can be achieved simultaneously.

16-31 Show how a switched capacitor behaves as a resistance.

16-32 What are three advantages of switched-capacitor filters?

16-33 Sketch the circuit of a logarithmic amplifier using one Op-Amp and explain its operation.

16-34 Of what use is a logarithmic amplifier?

16-35 Sketch the circuit of a two-quadrant analog multiplier and explain its operation.

16-36 Repeat Rev. 16-35 for the Gilbert multiplier cell.

16-37 Briefly describe three uses of an analog multiplier.

16-38 (*a*) Sketch the circuit of a precision half-wave rectifier and explain its operation.
(*b*) How can this circuit be used as an average detector?

16-39 Sketch the circuit of a peak detector in which the external load does not discharge the capacitor.

Part Five

LARGE-SIGNAL ELECTRONICS

The one chapter in this, the last part of the book, treats electronic circuits used in high-voltage, high-current, and/or high-power applications. The two major subject areas investigated are the conversion of alternating current to direct current (power supplies) and large-signal amplifiers of the type needed to drive loudspeakers and cathode-ray-tube (CRT) displays.

Chapter 17
POWER CIRCUITS AND SYSTEMS

Almost all electronic circuits require a dc source of power. For portable low-power systems, batteries may be used. More frequently, however, electronic equipment is energized by a power supply, a circuit which converts the ac waveform of the power lines to direct voltage of constant amplitude. The process of ac-to-dc conversion is examined and is based on the simple rectifier circuits introduced in Chap. 2. In this chapter we consider the regulator circuits used to control the amplitude of a dc supply voltage. These circuits are a special class of feedback amplifiers. We also introduce dc-to-dc conversion (switching regulators).

An amplifying system usually consists of several stages in cascade. The input and intermediate stages operate in a small-signal class A mode. Their function is to amplify the small-input excitation to a value large enough to drive the final device. This output stage feeds a transducer such as a cathode-ray tube (CRT), a loudspeaker, a servomotor, etc., and hence must be capable of delivering a large voltage or current swing or an appreciable amount of power. In this chapter we study such large-signal amplifiers. Thermal considerations are very important with power amplifiers and are discussed here. Both bipolar and FET power transistors are introduced.

17-1 AC TO DC CONVERSION

Fixed-amplitude, fixed-frequency alternating current is the primary source of electric energy. (In the United States it is a 110- to 220-V-rms 60-Hz sinusoid; in many parts of Europe 220-V-rms 50-Hz sinusoids are available.) The overwhelming majority of electronic circuits require constant voltages to ensure appropriate operation. For example, many minicomputers require 5-V sources capable of providing a current of 100 A. Other signal-processing systems often require 12- and 15-V supplies in which the current produced varies with load conditions. In addition, many motor drives and control systems require dc supplies whose voltage levels can be adjusted to meet desired operating conditions.

FIGURE 17-1
Block diagram of a
power supply.

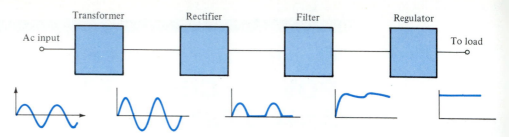

The block diagram for a dc power supply obtained from a primary ac source is depicted in Fig. 17-1. With the exception of the rectifier, whether each of the remaining circuit functions is employed depends on the application. As indicated by the waveforms in Fig. 17-1, the functions of the various circuits are as listed.

1. Transformer: Adjusts the ac level so that the appropriate dc amplitude is achieved.
2. Rectifier: Converts the sinusoidal voltage to a pulsating dc signal.
3. Filter: "Smooths" the waveform by eliminating the ac components from the rectifier output.
4. Regulator: Maintains a constant voltage level independent of load conditions or variations in the amplitude of the ac supply.

The transformer can be of the step-up or step-down type and its power-handling capacity must be sufficient to supply the load and account for losses in the rectifier, filter, and regulator. The turns ratio is determined by the output level required relative to the ac input amplitude. The remaining circuits are discussed in the next several sections.

17-2 RECTIFIERS

Almost all electronic circuits require a dc source of power. For portable low-power systems, batteries may be used. More frequently, however, electronic equipment is energized by a *power supply*, a piece of equipment which converts the alternating waveform from the power lines into an essentially direct voltage. The study of the ac-to-dc conversion is initiated in this section.

A Half-Wave Rectifier A device, such as the semiconductor diode, which is capable of converting a sinusoidal input waveform (whose average value is zero) into a unidirectional (though not constant) waveform, with a nonzero average component, is called a *rectifier*. The basic circuit for half-wave rectification is shown in Fig. 17-2. Since in a rectifier circuit the input $v_i = V_m \sin \omega t$ has a peak value V_m which is most often very large compared with the cut-in voltage V_γ of the diode, we assume in the following discussion that $V_\gamma = 0$. With the diode

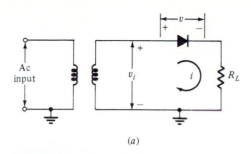

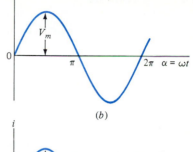

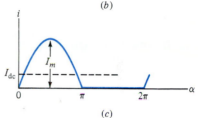

FIGURE 17-2
(a) Basic circuit of a half-wave rectifier. (b) Transformer output sinusoidal voltage v_i. (c) Diode and load current i.

idealized to be a resistance R_f in the ON state and an open circuit in the OFF state, the current i in the diode or load R_L is given by

$$i = I_m \sin \alpha \qquad \text{if } 0 \le \alpha \le \pi$$
$$i = 0 \qquad \text{if } \pi \le \alpha \le 2\pi \tag{17-1}$$

where $\alpha \equiv \omega t$ and

$$I_m \equiv \frac{V_m}{R_f + R_L} \tag{17-2}$$

The transformer secondary voltage v_i is shown in Fig. 17-2b, and the rectified current in Fig. 17-2c. Note that the output current is unidirectional. We now calculate this nonzero value of the average current.

A dc ammeter is constructed so that the needle deflection indicates the average value of the current passing through it. By definition, the average value of a periodic function is given by the area of one cycle of the curve divided by the base. Expressed mathematically,

$$I_{dc} = \frac{1}{2\pi} \int_0^{2\pi} i \, d\alpha \tag{17-3}$$

For the half-wave circuit under consideration, it follows from Eqs. (17-1) that

$$I_{dc} = \frac{1}{2\pi} \int_0^{\pi} I_m \sin \alpha \, d\alpha = \frac{I_m}{\pi} \tag{17-4}$$

Note that the upper limit of the integral has been changed from 2π to π since the instantaneous current in the interval from π to 2π is zero and so contributes nothing to the integral.

FIGURE 17-3
The voltage across the
diode in Fig. 17-2.

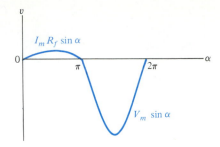

FIGURE 17-3
The voltage across the
diode in Fig. 17-2.

The Diode Voltage The dc (average) output voltage is clearly given as

$$V_{dc} = I_{dc}R_L = \frac{I_m R_L}{\pi} \tag{17-5}$$

However, the reading of a dc voltmeter placed across the diode is *not* given by $I_{dc}R_f$ because the diode cannot be modeled as a constant resistance, but rather it has two values: R_f in the ON state and ∞ in the OFF state.

A dc voltmeter reads the average value of the voltage across its terminals. Hence, to obtain V'_{dc} across the diode, the instantaneous voltage must be plotted as in Fig. 17-3 and the average value obtained by integration. Thus

$$V'_{dc} = \frac{1}{2\pi} \left(\int_0^\pi I_m R_f \sin \alpha \, d\alpha + \int_\pi^{2\pi} V_m \sin \alpha \, d\alpha \right)$$

$$= \frac{1}{\pi}(I_m R_f - V_m) = \frac{1}{\pi}[I_m R_f - I_m(R_f + R_L)]$$

where use has been made of Eq. (17-2). Hence

$$V'_{dc} = - \frac{I_m R_L}{\pi} \tag{17-6}$$

This result is negative, which means that if the voltmeter is to read upscale, its positive terminal must be connected to the cathode of the diode. From Eq. (17-5) the dc diode voltage is seen to be equal to the negative of the average voltage across the load resistor. This result is evidently correct because the sum of the dc voltages around the complete circuit must add up to zero.

The AC Current (Voltage) *A root-mean-square ammeter (voltmeter) is constructed so that the needle deflection indicates the effective, or rms, current (voltage).* Such a "square-law" instrument may be of the thermocouple type. By definiton, the effective or rms value squared of a periodic function of time is given by the area of one cycle of the curve, which represents the square of the function, divided by the base. Expressed mathematically, this is

$$I_{rms} = \left(\frac{1}{2\pi} \int_0^{2\pi} i^2 \, d\alpha \right)^{1/2} \tag{17-7}$$

By use of Eqs. (17-1), it follows that

$$I_{rms} = \left(\frac{1}{2\pi} \int_0^\pi I_m^2 \sin^2\alpha \, d\alpha\right)^{1/2} = \frac{I_m}{2} \tag{17-8}$$

The rms output voltage is given by $I_m R_L/2$.
Applying Eq. (17-7) to the *sinusoidal input voltage*, we obtain

$$V_{rms} = \frac{V_m}{\sqrt{2}} \tag{17-9}$$

Regulation

The variation of dc output voltage as a function of dc load current is called *regulation*. The percentage regulation is defined as

$$\% \text{ regulation} \equiv \frac{V_{\text{no load}} - V_{\text{load}}}{V_{\text{load}}} \times 100\% \tag{17-10}$$

where *no load* refers to zero current and *load* indicates the normal load current. For an ideal power supply the output voltage is independent of the load (the output current) and the percentage regulation is zero.

The variation of V_{dc} with I_{dc} for the half-wave rectifier is obtained as follows: From Eqs (17-4) and (17-2),

$$I_{dc} = \frac{I_m}{\pi} = \frac{V_m/\pi}{R_f + R_L} \tag{17-11}$$

Solving Eq. (17-11) for $V_{dc} = I_{dc}R_L$, we obtain

$$V_{dc} = \frac{V_m}{\pi} - I_{dc}R_f \tag{17-12}$$

This result is consistent with the circuit model given in Fig. 17-4 for the dc voltage and current. Note that the rectifier circuit functions as if it were a constant (open-circuit) voltage source $V = V_m/\pi$ in series with an effective internal resistance (the *output resistance*) $R_o = R_f$. This model shows that V_{dc} equals V_m/π at no load and that the dc voltage decreases linearly with an increase in dc output current. In practice, the resistance R_s of the transformer secondary is in series with the diode, and in Eq. (17-12) R_s should be added

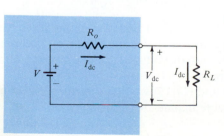

FIGURE 17-4
The Thévenin equivalent of a power supply used to determine the load voltage and current.

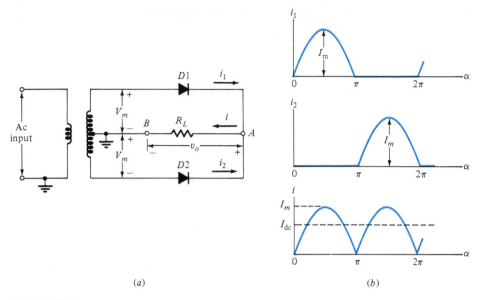

(a)   (b)

FIGURE 17-5
(a) A full-wave rectifier circuit. (b) The diode currents i_1 and i_2 and the load current i. The output voltage is $v_o = iR_L$.

to R_f. The best method of estimating the diode resistance is to obtain a regulation plot of V_{dc} versus I_{dc} in the laboratory. The negative slope of the resulting straight line gives $R_f + R_s$. Clearly, Fig. 17-4 represents a Thévenin model, and hence a rectifier behaves as a linear circuit with respect to average current and voltage.

A Full-Wave Rectifier   The circuit of a full-wave rectifier is shown in Fig. 17-5a. This circuit is seen to comprise two half-wave circuits connected so that conduction takes place through one diode during one half of the power cycle and through the other diode during the second half of the cycle.

The current to the load, which is the sum of these two currents, $i = i_1 + i_2$, has the form shown in Fig. 17-5b. The dc and rms values of the load current and voltage in such a system are readily found to be

$$I_{dc} = \frac{2I_m}{\pi} \quad I_{rms} = \frac{I_m}{\sqrt{2}} \quad V_{dc} = \frac{2I_m R_L}{\pi} \tag{17-13}$$

where I_m is given by Eq. (17-2) and V_m is the peak transformer secondary voltage from one end to the center tap. Note by comparing Eq. (17-13) with Eq. (17-5) that the dc output voltage for the full-wave connection is twice that for the half-wave circuit.

From Eqs. (17-2) and (17-13) we find that the dc output voltage varies with current in the following manner:

$$V_{dc} = \frac{2V_m}{\pi} - I_{dc}R_f \tag{17-14}$$

This expression leads to the Thévenin dc model of Fig. 17-4, except that the internal (open-circuit) supply is $V = 2V_m/\pi$ instead of V_m/π.

When the turn-on voltage V_γ of the diode is included in the analysis (the model in Fig. 17-6a), diode current exists for less than one half-cycle (half-wave rectifier) as shown in Fig. 17-6b. For this situation, referring to Fig. 17-6, we obtain

$$\theta_1 = \sin^{-1}\frac{V_\gamma}{V_m} \qquad \theta_2 = \pi - \theta_1 \tag{17-15}$$

The average current I_{dc} (Prob. 17-2) is given by

$$I_{dc} = \frac{V_m}{\pi(R_f + R_L)}\cos\theta_1 - \frac{\pi - 2\theta_1}{2\pi}\frac{V_\gamma}{R_f + R_L} \quad \text{(half-wave rectifier)} \tag{17-16}$$

$$I_{dc} = \frac{2V_m}{\pi(R_f + R_L)}\cos\theta_1 - \frac{\pi - 2\theta_1}{2\pi}\frac{V_\gamma}{R_f + R_L} \quad \text{(full-wave rectifier)} \tag{17-17}$$

Peak Inverse Voltage   For each rectifier circuit there is a maximum voltage to which the diode can be subjected. This potential is called the *peak inverse voltage* because it occurs during that part of the cycle when the diode is nonconducting. From Fig. 17-2 it is clear that, for the half-wave rectifier, the peak inverse voltage is V_m. We now show that, for a full-wave circuit, twice this value is obtained. At the instant of time when the transformer secondary voltage to midpoint is at its peak value V_m, diode $D1$ is conducting and $D2$ is nonconducting. If we apply KVL around the outside loop and neglect the small voltage drop across $D1$, we obtain $2V_m$ for the peak inverse voltage across $D2$. Note that this result is obtained without reference to the nature of the load, which can be a pure resistance R_L or a combination of R_L and some reactive elements which may be introduced to "filter" the ripple. We conclude that, *in a full-wave circuit, independently of the filter used, the peak inverse voltage across each diode is twice the maximum transformer voltage measured from midpoint to either end.*

Rectification of a sinusoid whose peak value is less that V_γ is discussed in Sec. 16-15.

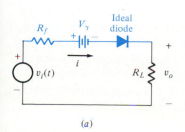

(a)

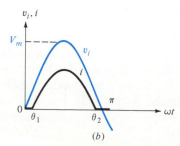

(b)

FIGURE 17-6

(a) Equivalent circuit of a rectifier. The diode is represented by its large-signal model R_f, V_γ, and an ideal diode. (b) The waveforms for the input voltage and load current. (*Note:* The diode does not conduct until v_i exceeds V_γ; this accounts for the ignition and extinction angles θ_1 and θ_2, respectively.)

FIGURE 17-7
A full-wave bridge rectifier.

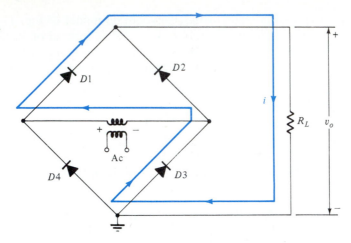

17-3 OTHER FULL-WAVE CIRCUITS

A variety of other rectifier circuits find extensive use. Among these are the bridge circuit, several voltage-doubling circuits, and a number of voltage-multiplying circuits. The bridge circuit finds application not only for power circuits, but also as a rectifying system in rectifier ac meters for use over a fairly wide range of frequencies.

The Bridge Rectifier   The essentials of the bridge circuit are shown in Fig. 17-7. To understand the action of this circuit, it is necessary only to note that two diodes conduct simultaneously. For example, during the portion of the cycle when the transformer polarity is that indicated in Fig. 17-7, diodes 1 and 3 are conducting, and current passes from the positive to the negative end of the load. The conduction path is shown in the figure. During the next half-cycle, the transformer voltage reverses its polarity, and diodes 2 and 4 send current through the load in the same direction as during the previous half-cycle.

The principal features of the bridge circuit are the following: the currents drawn in both the primary and the secondary of the supply transformer are sinusoidal, and therefore a smaller transformer may be used than for the full-wave circuit of the same output; a transformer without a center tap is used; and each diode has only transformer voltage across it on the inverse cycle. The bridge circuit is thus suitable for high-voltage applications.

The Rectifier Meter   This instrument, illustrated in Fig. 17-8, is essentially a bridge-rectifier system, except that no transformer is required. Instead, the voltage to be measured is applied through a multiplier resistor R to two corners of the bridge, a dc milliammeter being used as an indicating instrument across the other two corners. Since the dc milliammeter reads average values of current, the meter scale is calibrated to give rms values when a sinusoidal voltage is applied to the input terminals. As a result, this instrument will not read correctly when used with waveforms which contain appreciable harmonics.

FIGURE 17-8
The rectifier voltmeter.

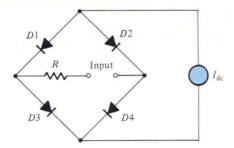

Voltage Multipliers   A common voltage-doubling circuit which delivers a dc voltage approximately equal to twice the transformer maximum voltage at no load is shown in Fig. 17-9. This circuit is operated by alternately charging each of the two capacitors to the transformer peak voltage V_m, current being continually drained from the capacitors through the load. The capacitors also act to smooth out the ripple in the output.

17-4   CAPACITOR FILTERS   Filtering is frequently effected by shunting the load with a capacitor. The action of this system depends upon the fact that the capacitor stores energy during the conduction period and delivers this energy to the load during the inverse, or nonconducting, period. In this way, the time during which the current passes through the load is prolonged, and the ripple is considerably decreased. The ripple voltage is defined as the deviation of the load voltage from its average dc value.

Consider the half-wave capacitive rectifier of Fig. 17-10. Suppose, first, that the load resistance $R_L = \infty$. The capacitor will charge to the potential V_m, the transformer maximum value. Further, the capacitor will maintain this potential, for no path exists by which this charge is permitted to leak off, since the diode will not pass a negative current. The diode resistance is infinite in the inverse direction, and no charge can flow during this portion of the cycle. Consequently, the filtering action is perfect, and the capacitor voltage v_o remains constant at its peak value, as is seen in Fig. 17-11.

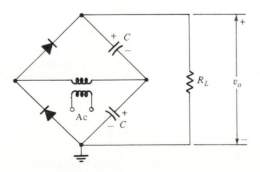

FIGURE 17-9
The bridge rectifier as a voltage-doubling circuit. The two capacitors in this circuit replace two diodes in Fig. 17-7.

FIGURE 17-10

Half-wave rectifier with
a capacitor filter.

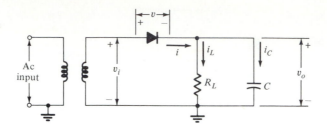

FIGURE 17-10

Half-wave rectifier with
a capacitor filter.

The voltage v_o across the capacitor is, of course, the same as the voltage across the load resistor, since the two elements are in parallel. The diode voltage v is given by

$$v = v_1 - v_o \qquad (17\text{-}18)$$

We see from Fig. 17-11 that the diode voltage is always negative and that the peak inverse voltage is twice the transformer maximum. Hence the presence of the capacitor causes the peak inverse voltage to increase from a value equal to twice the transformer maximum value when the filter is used.

Suppose now, that the load resistor R_L is finite. Without the capacitor input filter, the load current and the load voltage during the conduction period will be sinusoidal functions of time. The inclusion of a capacitor in the circuit results in the capacitor charging in step with the applied voltage. Also, the capacitor must discharge through the load resistor, since the diode will prevent a current in the negative direction. Clearly, the diode acts as a switch which permits charge to flow into the capacitor when the transformer voltage exceeds the capacitor voltage, and then acts to disconnect the power source when the transformer voltage falls below that of the capacitor.

Output Voltage under Load During the time interval when the diode in Fig. 17-10 is conducting, the transformer voltage is impressed directly across the load (assuming that the diode drop can be neglected). Hence, the output voltage is $v_o = V_m \sin \omega t$. During the interval when D is nonconducting, the capacitor discharges through the load with a time constant CR_L. The output waveform in Fig. 17-12 consists of portions of sinusoids (when D is ON) joined to exponential segments (when D is OFF). The point at which the diode starts to conduct is

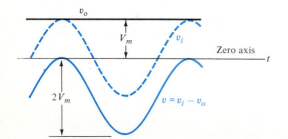

FIGURE 17-11

Voltages in a capacitor-filtered half-wave rectifier at no load. The output voltage v_o is constant (perfect filtering). The diode voltage is negative for all values of time, and the peak inverse voltage is $2V_m$.

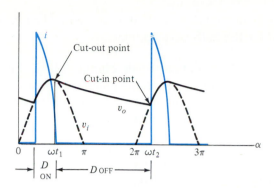

FIGURE 17-12
Theoretical sketch of the diode current
and output voltage waveforms in a capacitor-filtered half-wave rectifier.

called the *cut-in* point t_2, and that at which it stops conducting is called the
cut-out point t_1. These times are indicated in Fig. 17-13.

The cut-out time is obtained from the expression (Prob. 17-12) for the current
i in Fig. 17-10 when $v_o = V_m \sin \omega t$. Then the time for which $i = 0$ gives the
cut-out angle ωt_1. The cut-in point t_2 is obtained graphically by finding the time
when the exponential portion of v_o in Fig. 17-12 intersects the curve $V_m \sin \omega t$
(in the following cycle). The validity of this statement follows from the fact
that at an instant of time greater than t_2, the transformer voltage v_1 (the sine
curve) is greater than the capacitor voltage v_o (the exponential curve). Since
the diode voltage is $v = v_1 - v_o$, the v will be positive beyond t_2 and the diode
will become conducting. Thus t_2 is the cut-in point.

The use of a large capacitance to improve the filtering at a given load R_L is
accompanied by a high-peak diode current I_m. for a specified average load
current, i becomes more peaked and the conduction period decreases as C is
made larger. It is to be emphasized that the use of a capacitor filter may impose
serious restrictions on the diode, since the average current may be well within
the current rating of the diode, and yet the peak current may be excessive.

Full-Wave Circuit Consider a full-wave rectifier with a capacitor filter obtained by placing a
capacitor C across R_L in Fig. 17-5. The analysis of this circuit requires a simple
extension of that just made for the half-wave circuit. If in Fig. 17-12 a dashed
half-sinusoid is added between π and 2π, the result is the dashed full-wave
voltage in Fig. 17-13. The cut-in point now lies between π and 2π, where the

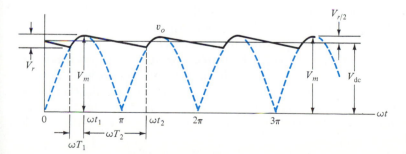

FIGURE 17-13
Approximate load-voltage waveform for a
full-wave rectifier with a capacitor filter.

exponential portion of v_o intersects this sinusoid. The cut-out point is the same as that found for the half-wave rectifier.

Approximate Analysis It is possible to obtain the dc output voltage for given values of the parameters ω, R_L, C, and V_m from the graphical construction indicated in Fig. 17-13. Such an analysis is involved and tedious. Hence we now present an approximate solution which is simple and yet sufficiently accurate for most engineering applications.

We assume that the output-voltage waveform of a full-wave circuit with a capacitor filter may be represented by the approximately piecewise linear curve shown in Fig. 17-13. For large values of C (so that $\omega C R_L \gg 1$) we note that $\omega t_1 \to \pi/2$ and $v_o \to V_m$ at $t = t_1$. Also, with C very large, the exponential decay can be replaced by a linear fall. If the total capacitor discharge voltage (the ripple voltage) is denoted by V_r, then from Fig. 17-13, the average value of the voltage is approximately

$$V_{dc} = V_m - \frac{V_r}{2} \tag{17-19}$$

It is necessary, however, to express V_r as a function of the load current and the capacitance. If T_2 represents the total nonconducting time, the capacitor, when discharging at the constant rate I_{dc}, will lose an amount of charge $I_{dc}T_2$. Hence the change in capacitor voltage is $I_{dc}T_2/C$, or

$$V_r = \frac{I_{dc}T_2}{C} \tag{17-20}$$

The better the filtering action, the smaller will be the conduction time T_1 and the closer T_2 will approach the time of half a cycle. Hence we assume that $T_1 = T/2 = 1/2f$, where f is the fundamental power-line frequency. Then

$$V_r = \frac{I_{dc}}{2fC} \tag{17-21}$$

and from Eq. (17-19),

$$V_{dc} = V_m - \frac{I_{dc}}{4fC} \tag{17-22}$$

This result is consistent with Thévenin's model of Fig. 17-4, with the open-circuit voltage $V = V_m$ and the effective output resistance $R_o = 1/4fC$.

The ripple is seen to vary directly with the load current I_{dc} and also inversely with the capacitance. Hence, to keep the ripple low and to ensure good regulation, very large capacitances (of the order of tens of microfarads) must be used. The most common type of capacitor for this rectifier application is the electrolytic capacitor. These capacitors are polarized, and care must be taken to insert them into the circuit with the terminal marked $+$ to the positive side of the output.

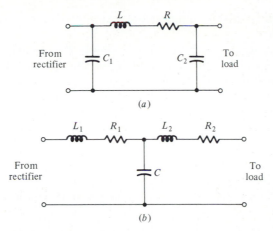

FIGURE 17-14
(*a*) Capacitor-input and (*b*) choke-input filters.

The desirable features of rectifiers employing capacitor-input filters are the small ripple and the high voltage at light load. The no-load voltage is equal, theoretically, to the maximum transformer voltage. The disadvantages of this system are the relatively poor regulation, the high ripple at large load currents, and the peaked currents that the diodes must pass.

An approximate analysis similar to that given above applied to the half-wave circuit shows that the ripple, and also the drop from no load to a given load, are double the values calculated for the full-wave rectifier.

Capacitor-Input and Choke-Input Filters More efficient filtering of the rectifier output waveform is accomplished by using more than one energy storage element. The circuit in Fig. 17-14*a* is referred to as a *capacitor-input* filter, and that in Fig. 17-14*b* is called a *choke (inductor)-input filter*. In both circuits the resistances R, R_1, and R_2 indicate the coil resistance associated with each inductor. The reactances of the inductances L_1, L_2, and L in Fig. 17-14 are chosen to be high at the alternating frequency. Hence they act to attenuate the ripple voltage but, because they have zero reactance at dc ($\omega = 0$), do not affect the dc output. The analysis of these filters is the subject of Prob. 17-13 and 17-14.

17-5 REGULATED POWER SUPPLIES An ideal *regulated power supply* is an electronic circuit designed to provide a predetermined dc voltage V_O which is independent of the current I_L drawn from V_O, of the temperature, and also of any variations in the ac line voltage. An unregulated power supply consists of a transformer, a rectifier, and a filter, as shown in Figs. 17-5 and 17-10.

There are three reasons why an unregulated power supply is not sufficient for many applications. The first is its poor regulation; the output voltage is not constant as the load varies. The second is that the dc output voltage varies with the ac input. In some locations the line voltage (of nominal value 115 V) may vary over as wide a range as 90 to 130 V, and yet it is necessary that the dc voltage remain essentially constant. The third reason is that the dc output voltage varies with the temperature, particularly because semiconductor devices are used.

The Zener diode can be used as a simple regulator in the circuit shown in Fig. 2-32. This circuit, as described in Sec. 2-11, is limited by the current (and power) capability of the Zener diode used. Typically, the Zener diode must be able to handle a current that is greater than that supplied to the load.

The feedback circuit shown in Fig. 17-15 is used to overcome the three shortcomings described previously and the current limitation of the Zener diode. Such a system is called a *regulated power supply*. From Fig. 17-15 we see that the regulated power supply represents a case of series-shunt (voltage-series feedback). If we assume that the voltage gain of the emitter follower $Q1$ ($Q1$ is also called the *pass transistor* or *element*) is approximately unity, then $V'_O \approx V_O$ and

$$V'_O = A_V V_i = A_V(\beta V_O - V_R) \approx V_O \qquad (17\text{-}23)$$

where

$$\beta \equiv \frac{R_2}{R_1 + R_2} \qquad (17\text{-}24)$$

From Eq. (17-23) it follows that

$$V_O = V_R \frac{A_V}{1 + \beta A_V} \qquad (17\text{-}24)$$

If $\beta A_V \gg 1$, $V_O \approx V_R/\beta$; note, however, that V_O must be less than the unregulated voltage source V_{dc}.

The output voltage V_O can be changed by varying β, by changing the fraction of V_O that is fed back. The emitter follower $Q1$ is used to provide current gain, because the current delivered by Op-Amp A_V usually is not sufficient. Also, the pass element must absorb the difference between the unregulated input

FIGURE 17-15
A regulated power-supply system.

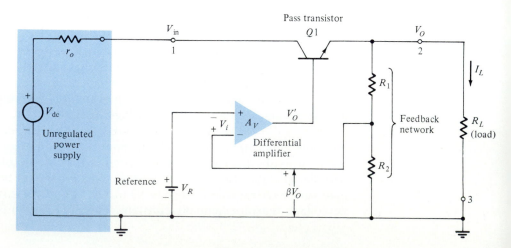

voltage V_{in} and the regulated output voltage V_O. The dc collector voltage required by the error amplifier A_V is obtained from the unregulated voltage.

STABILIZATION   Since the output dc voltage V_O depends on the input supply dc voltage V_{dc}, load current I_L, and temperature T, the change ΔV_O in output voltage of a power supply can be expressed as follows:

$$\Delta V_O = \frac{\partial V_O}{\partial V_{\text{dc}}}\Delta V_{\text{dc}} + \frac{\partial V_O}{\partial I_L}\Delta I_L + \frac{\partial V_O}{\partial T}\Delta T$$

or
$$\Delta V_O = S_V\Delta V_{\text{dc}} + R_o\Delta I_L + S_T\Delta T \qquad (17\text{-}26)$$

where the three coefficients are defined as
    Input regulation factor:

$$S_V = \left.\frac{\Delta V_O}{\Delta V_{\text{dc}}}\right|_{\substack{\Delta I_L=0 \\ \Delta T=0}} \qquad (17\text{-}27)$$

Output resistance:

$$R_o = \left.\frac{\Delta V_O}{\Delta I_L}\right|_{\substack{\Delta V_{\text{dc}}=0 \\ \Delta T=0}} \qquad (17\text{-}28)$$

Temperature coefficient:

$$S_T = \left.\frac{\Delta V_O}{\Delta T}\right|_{\substack{\Delta V_{\text{dc}}=0 \\ \Delta I_L=0}} \qquad (17\text{-}29)$$

The smaller the value of the three coefficients, the better the regulation of the power supply. The input-voltage change ΔV_{dc} may be due to a change in ac line voltage or may be ripple because of inadequate filtering.

17-6 MONOLITHIC REGULATORS   It is interesting to note that if we were to construct a discrete-component regulator, it would resemble Fig. 17-15 topologically: the amplifier A_V would be an Op-Amp (such as the μA741 or LM301A) and the battery V_R would be replaced by a reference diode (an LM103, LM199, or a Zener). With the advent of microelectronics it has become technically and economically feasible to incorporate all components in monolithic form. All the benefits of ICs are thus obtained: excellent performance, small size, ease of use, low cost, and high reliability.

An example of a monolithic regulator is the Motorola MC7800C series of three-terminal, positive, fixed-voltage regulators. Figure 17-16 is the standard application, and shows the degree to which user complexity has been all but eliminated. Input capacitor C_i is required to cancel inductive effects associated with long power-distribution leads. Output capacitor C_o improves the transient

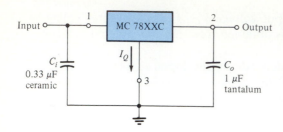

FIGURE 17-16
A standard three-terminal, positive, fixed-voltage, monolithic regulator. The quiescent current is I_Q.

response. These devices, requiring no adjustment, have an output preset by the manufacturer to an industry standard voltage of 5, 6, 8, 12, 15, 18, or 24 V. (An MC7824C represents a 24-V regulator.) There must be a minimum of 2 V between input and output. Such regulators are capable of output currents in excess of 1.0 A. They have internal short-circuit protection which limits the maximum current the circuit will pass, thermal shutdown, and output-transistor safe-operating-area protection. Typical values for the stabilization coefficients are

$$S_V = 3 \times 10^{-3} \qquad R_O = 30 \text{ m}\Omega \qquad S_T = 1 \text{ mV/°C}$$

The level of complexity afforded by monolithic IC techniques can be appreciated by examining Fig. 17-17, the circuit diagram of the MC7800C. To the left of the shaded block is the reference voltage V_R in Fig. 17-15. This is the level shifter in Fig. 14-11a with a Zener diode input to the emitter-follower buffer. The shaded circuit in Fig. 17-17 is the difference amplifier A_V in Fig. 17-15. The design similarity with the 741 Op-Amp configuration in Fig. 14-19 should be noted. The resistor divider R_1 and R_2 in Fig. 17-17 corresponds to the same feedback network in Fig. 17-15. The Darlington pair Q' and Q'' in Fig. 17-17 constitutes the pass element $Q1$ in Fig. 17-15.

The protection circuitry is shown in heavy outline and merits explanation. Current limiting is performed by R_3, R_4, and Q_2. Safe-operating protection is accomplished in the following way. If the output is pulled low by an overload, thus increasing the collector-emitter voltage of Q'', Zener $D1$ (which under normal loads is OFF) will conduct. Under these circumstances sufficient base current is supplied to $Q2$ so that it conducts, which in turn, "robs" base drive from the $Q'Q''$ Darlington combination. In this manner the volt-ampere product of the pass element is limited to a reasonable power dissipation.

Consider next the thermal overload protection. A fraction of the reference voltage appearing across R_5 is applied to the base-emitter junction of $Q3$. For a fixed value of V_{BE3}, the collector current I_3 increases rapidly with increasing temperature. Hence, at sufficiently elevated temperatures (caused by either power dissipation or a high ambient), transistor $Q3$ will conduct heavily and once again starve the pass transistors $Q'Q''$ of base drive, thereby providing thermal shutdown.

Using monolithic regulators, it is possible to distribute unregulated voltage through electronic equipment and provide regulation locally, for example, on individual printed-circuit (PC) boards. Among the advantages of this approach

FIGURE 17-17

The circuit diagram of the MC7800C series monolithic regulator. (*Courtesy of Motorola Semiconductor, Inc.*)

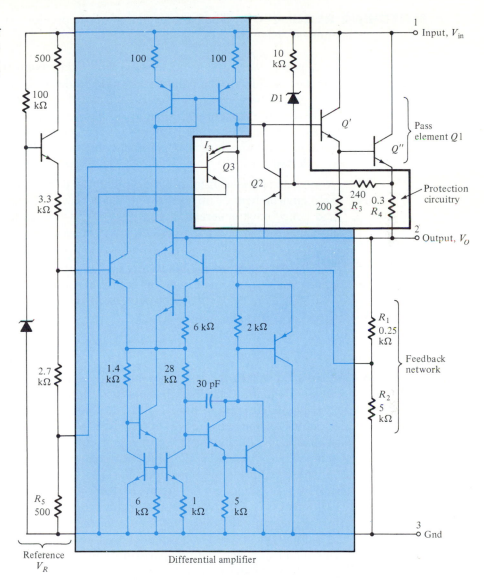

Reference V_R

Differential amplifier

are greater flexibility in voltage levels, regulation for individual stages, and improved isolation and decoupling of these stages.

Monolithic regulators are available in a multitude of performance levels: fixed or variable, positive or negative output voltage, high output current (> 1 A), high output voltage (> 24 V), and single or dual ($\pm$) outputs. The engineer can also use the standard three-terminal regulator (Fig. 17-16) as a basic building block to tailor its performance to specific needs. Such techniques are considered in Probs. 17-19 and 17-20.

17-7 A SWITCHING REGULATOR   The pass regulators of the previous section, despite their usefulness, have three drawbacks.

1. In a power supply that includes ac-to-dc conversion (transformer, bridge rectifier, and filter) the polarity and magnitude of the raw (unregulated) dc voltage can be a design parameter, and thus no inherent problem exists. If, however, in a system with one dc supply voltage (such as +5 V for TTL gates) there exists a need for ±15 V for Op-Amp operation, it may be economically (or physically) impracticable to add the facility for additional raw dc voltages.

2. A system operated from a battery, such as a communication system in a field environment or on a satellite in deep space, has no ac source available and hence must generate all voltages (positive or negative) from the single dc voltage source. Such a system is a *dc-to-dc converter*.

3. The input-voltage magnitude must be greater than the output magnitude, and series pass regulators are inherently inefficient. The greater the input-output differential for a given current, the greater the losses. A TTL system regulator operating from 10 V is at best 50 percent efficient, and from 20 V the efficiency drops to 25 percent.

Basic Switching Regulator Topology   All three difficulties can be avoided with the use of a *switching regulator*. The basic regulating control loop is shown in Fig. 17-18. The unregulated input voltage is V_{in} and the regulated output voltage is V_O. The output current delivered to the load R_L is to be large (say, several amperes). The shaded block contains low-power circuits which are fabricated on a single IC chip. The *reference regulator* is the series pass regulator described in Sec. 17-6 whose output is the regulated reference voltage V_{ref} which serves as the power-supply voltage for all circuits on the chip. Since the current drawn from V_{ref} is small (say, 10 mA), the small power loss in the pass regulator does not affect appreciably the overall efficiency of the system.

The topology in Fig. 17-18 is that of a series-shunt feedback system (voltage-series feedback), and the comparison of the fixed input V_{ref} with a fraction $R_1/(R_1 + R_2)$ of the output V_O is made with the differential amplifier (error amplifier). A triangular waveform generator of period T (circuit not indicated in Fig. 17-18) is also on the chip, and its output v is applied to the noninverting terminal of a comparator which functions as a *pulse-width modulator* (PWM). The error amplifier output voltage v_m is applied to the inverting terminal of the PWM, as shown in Fig. 17-18. This modulator operates as described in Sec. 15-15, producing a square wave v_A of period T, whose duty cycle δ varies linearly with v_m. The output v_A of the PWM drives a power switch (indicated by the SPDT block of Fig. 17-18), creating a square wave (of period T and duty cycle δ), whose minimum value is 0 and whose maximum is V_{in}. This square wave is filtered by the LC combination, which acts as a low-pass filter. If the reactance of C is much smaller than that of L at the fundamental frequency,

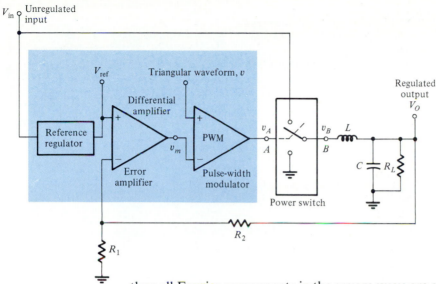

FIGURE 17-18
Basic switching regulator to-pology. The circuits in the shaded rectangle are fabri-cated on a single chip. All other components are dis-crete elements externally connected to the chip.

then all Fourier components in the square wave are greatly attenuated. In other words, if $T/2\pi C \ll 2\pi L/T$ or if $\sqrt{LC} \gg T/2\pi$, then V_O will be a constant, equal to the average value of the square wave.

The Regulated Output Voltage   Since there is a virtual short circuit between the input terminals of the error amplifier, $V_{\text{ref}} = R_1 V_O/(R_1 + R_2)$ and the output is given by

$$V_O = V_{\text{ref}} \left(1 + \frac{R_2}{R_1} \right) \qquad (17\text{-}30)$$

Note that this regulated voltage is independent of variations in the raw input voltage V_{in} and of changes in load current. It depends only on the constancy of the regulated voltage V_{ref} and the ratio R_2/R_1. If, for example, the reference voltage is the supply for TTL logic gates, so that $V_{\text{ref}} = 5$ V, and if an output voltage $V_O = 15$ V is desired, it is necessary only to select $R_2 = 2R_1$. As noted above, V_O is the dc value of the power-switch square-wave output voltage v_B, whose peak value is V_{in}. Hence this configuration can be used only if $V_{\text{in}} > V_O$. This control system operates in such a manner that an error voltage v_m is generated automatically, so that the PWM has the correct duty cycle δ to cause v_B to have a dc value V_O, given by Eq. (17-30).

Efficiency   An inspection of Fig. 17-18 reveals that the output current passes from V_{in} through the power switch and the inductor through the load. Hence, using a switch with low losses (a transistor switch with small $V_{CE(\text{sat})}$ and high switching speed) and a filter with high Q (an inductor with low resistance), the conversion efficiency often exceeds 90 percent.

The Power Switch   The action of the SPDT switch in Fig. 17-18 may be obtained by the com-bination in Fig. 17-19a of a diode and SPST switch (to be replaced by a transistor

FIGURE 17-19

(a) The SPDT switch of Fig. 17-18 is replaced by a SPST switch and flyback diode. (b) A practical implementation of the switch using transistors. The output is positive and smaller than the input.

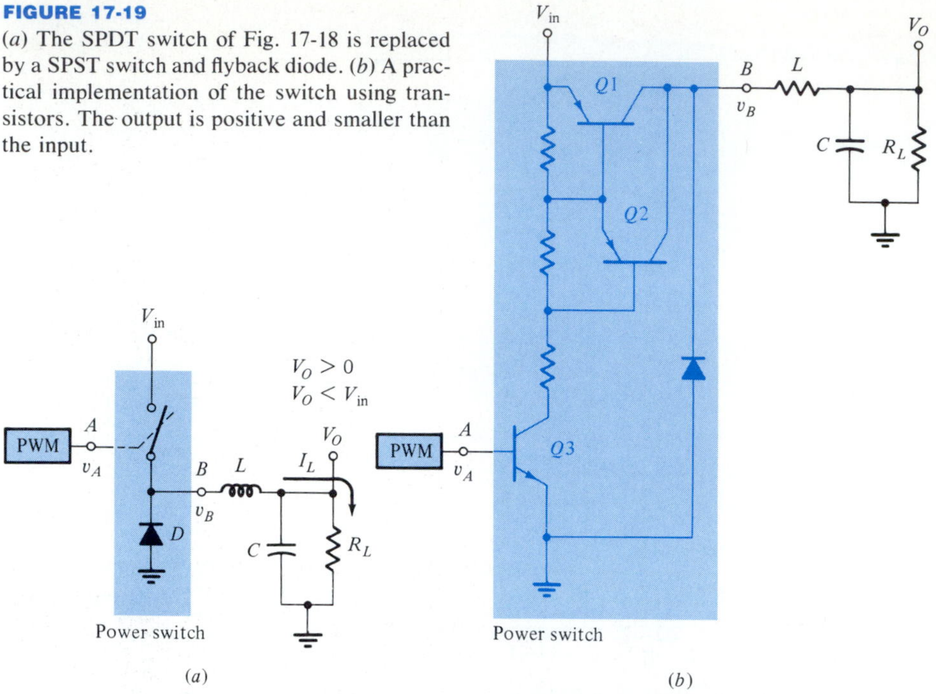

(a)

(b)

in Fig. 17-19b). The LC filter, the load R_L, and the PWM block driving the switch are also included, but R_2, R_1, and the IC block are omitted for simplicity.

The circuit operates as follows: When the switch is closed the diode is reverse-biased by V_{in} and the load current I_L is supplied from $v_B = V_{in}$ through L. In the second portion of the cycle when the switch is opened, the inductor current cannot decrease instantaneously. (If it did, then the inductor voltage $L \, di/dt$ would be negative infinity.) Hence, at the instant that the switch opens, i_L remains constant and the current path must be from ground through the diode and the inductor into the load. Neglecting the drop across the diode, $v_B = 0$. Hence v_B is a square wave of period T and duty cycle δ, with a minimum value of 0 and a maximum value of V_{in}. This waveform is identical with that for v_B in Fig. 17-18. Therefore the circuit in Fig. 17-19a operates exactly as that in Fig. 17-19. Since, when the switch opens, v_B flies back from V_{in} to zero, the configuration in Fig. 17-19 gives rise to the name *flyback converter*.

The SPST switch in Fig. 17-19a may be simulated with a *pnp* power transistor $Q1$, as indicated in Fig. 17-19b. If the load current is 1 A, then the collector current of $Q1$ is 1 A and, for $\beta_F = 100$, the base current is 10 mA. The transistor $Q2$ is used to supply this large base current. Note that $Q1$ and $Q2$ form a Darlington pair (Sec. 10-14). To drive these transistors with the proper voltage polarity the PWM output voltage v_A must be inverted and, hence, the transistor $Q3$ is needed to complete the switch in Fig. 17-19b.

For v_A positive, $Q3$ conducts and its collector current (through the switch resistors) biases $Q1$ and $Q2$ ON, so that $v_B \approx V_{in}$. On the other hand, for v_A negative or 0, $Q3$ is nonconducting and there is no current in the biasing resistors. Hence $Q1$ and $Q2$ are OFF and the switch is open. By the action described in the preceding paragraph, the flyback diode goes ON and $v_B = 0$. This behavior indicates that the power switch in Fig. 17-19b is the practical implememtation of the idealized switch in Fig. 17-19a. Incidentally, the low-power transistor $Q3$ is fabricated as part of the IC chip shown in the shaded block of Fig. 17-18.

17-8 ADDITIONAL SWITCHING REGULATOR TOPOLOGIES
For the configuration in Fig. 17-19 the output voltage is positive and less than the input voltage ($V_O < V_{in}$), as verified in the preceding section. This restriction is removed by using the configuration in Fig. 17-20, as will now be demonstrated. Consider the interval T_1 when the switch is closed. The diode is reverse-biased by the positive voltage V_O, the feedback loop is open, and C discharges through R_L. By choosing $CR_L \gg T_1$, the drop in V_O (the ripple voltage) is small. During this interval the input voltage is across L and the inductor i_L increases by $di_L = V_{in}\, dt/L = V_{in}T_1/L$.

Consider now the interval T_2 during which the switch is open. Since the current in an inductor cannot change instantaneously, $i_L(T_1 -) = i_L(T_1 +)$ and, hence, the diode goes ON and i_L passes through the diode and into C. In the steady state the voltage across C must be the same at the end of the period $T = T_1 + T_2$ as it was at the beginning, $t = 0$. Similarly, the current must *decrease* ($di_L/dt < 0$) during T_2 by the amount $V_{in}T_1/L$ by which it increased during T_1. Neglecting the diode voltage we obtain from Fig. 17-20 that v_o (the instantaneous output voltage) is given by

$$v_o = V_{in} - L\, di/dt > V_{in}$$

because di_L/dt is negative. This argument verifies that the output V_O exceeds the input for this configuration. Incidentally, the switch action is obtained by using the Darlington pair $Q1$-$Q2$ driven by $Q3$ in a manner similar to that indicated in Fig. 17-19b.

Negative Output Voltages To obtain a negative supply from a raw dc voltage, the configuration seen in Fig. 17-21 is used for the power components. We assume that $V_O < 0$ and then justify this assumption. The argument is similar to that used

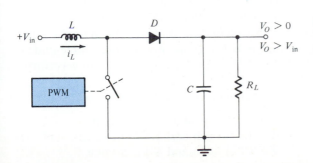

$V_O > 0$
$V_O > V_{in}$

FIGURE 17-20
The output is positive and greater than the input for this arrangement of the power components in a switching regulator.

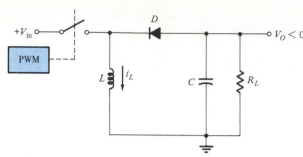

FIGURE 17-21
This topology in a switching regulator results in a negative output voltage.

in the preceding paragraph. During the interval T_1 when the switch is closed, the diode is OFF because the cathode voltage is $+V_{in}$ and the anode voltage is negative. The capacitor discharges slightly through the load and the inductor current increases by $V_{in}T_1/L$. At the instant the switch opens, i_L cannot change, and the diode is forced ON so that i_L flows in the loop formed by L, C, and D. Since i_L enters the bottom plate of C, this plate is charged positively and the output voltage is negative. An alternative proof that $V_O < 0$ is that i_L must decrease in the interval T_2 by the amount it increased during T_1, so that $di_L/dt < 0$ and, therefore, $v_O \approx L \, di_L/dt$ is negative. No restriction is placed on the magnitude of V_O; it may be larger or smaller that V_{in}. Its value is determined by the control loop of Fig. 17-18. If V_O is negative, level shifting must be used in order for the effective feedback voltage to be positive. This configuration is indicated in Prob. 17-22.

Transformer-Coupled Push-Pull DC-to-DC Converter   This switching regulator configuration has the most flexibility because the output V_O may be greater than or less than the raw dc input V_{in}, and the sign of V_O may be the same as, or opposite to, that of V_{in}. The topology of the power components is indicated in Fig. 17-22, which uses an iron-core transformer with a center-tapped primary ($v_{P1} = v_{P2}$) as well as a center-tapped secondary ($v_{S1} = v_{S2}$). The number of turns in the secondary is n times that in the primary, so that $v_{S1} = nv_{P1}$ and $v_{S2} = nv_{P2}$. If $n > 1$, then it is possible to obtain $V_O > V_{in}$, whereas for $n \leqslant 1$, $V_{in} \gg V_O$.

The two switches $SW1$ and $SW2$ are controlled by waveforms v_{A1} and v_{A2}, which are obtained from the PWM output v_A (as explained in Fig. 17-24). The waveforms v_A, v_{A1}, and v_{A2} are sketched in Fig. 17-23a, b, and c, respectively. The waveform v_A is obtained from the PWM in the shaded block of Fig. 17-18. Note that $SW1$ and $SW2$ are closed for the same duty cycle, but each is operated only once for every other period of the PWM waveform v_A. In other words, each switch operates at one half the frequency of the single-ended converter of Fig. 17-18. Switch $SW1$ ($SW2$) is a transistor whose base waveform is v_A (v_{A2}).

From Fig. 17-22 it follows that the primary voltages are given by

$$v_{P1} = v_{P2} = \begin{cases} -V_{in} & \text{if } SW1 \text{ is closed and } SW2 \text{ is open} \\ +V_{in} & \text{if } SW1 \text{ is open and } SW2 \text{ is closed} \\ 0 & \text{if } SW1 \text{ is open and } SW2 \text{ is open} \end{cases} \quad (17\text{-}31)$$

FIGURE 17-22

A push-pull trans-former-coupled switch-ing regulator.

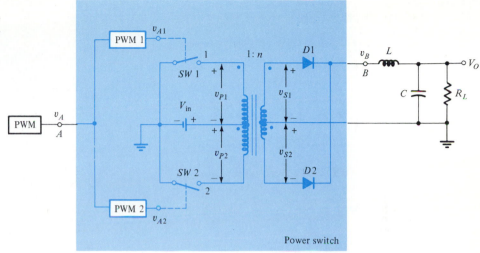

FIGURE 17-22

A push-pull trans-former-coupled switch-ing regulator.

This waveform is indicated in Fig. 17-23d. The secondary voltages $v_{S1} = v_{S2}$ have this same waveshape but are n times as large. During the intervals when $v_{S1} = v_{S2}$ is positive $D1$ conducts, $D2$ is OFF, and $v_B = nV_{IN}$. If $v_{S1} = v_{S2}$ is negative $D2$ conducts, $D1$ is OFF, and $v_B = nV_{in}$ again. When $v_{S1} = v_{S2} = 0$ the two diodes are connected in parallel from point B to ground and, hence, act as a flyback diode, as shown in Fig. 17-19, so that $v_B = 0$ during this interval. Consequently, the waveform v_B is as indicated in Fig. 17-23. Note that v_B is proportional to v_A. Because of the LC filter, the dc output voltage V_O equals the average value of the v_B waveform and may be greater (less) than V_{in}, depending on whether n exceeds (is smaller than) unity. If the diodes are reversed, the sign of V_O is negative. The block between A and B in Fig. 17-22 replaces the power-switch block in the feedback loop of Fig. 17-19b. The regulated output is given by Eq. (17-19).

Generating the Switching Waveforms   We now indicate how to obtain the two switching waveforms v_{A1} and v_{A2} from the PWM waveform v_A. The block diagram is shown in Fig. 17-24 and the waveforms in Fig. 17-25. The square-wave oscillator waveform v_{osc} in Fig. 17-25a is used to generate the triangular voltage needed

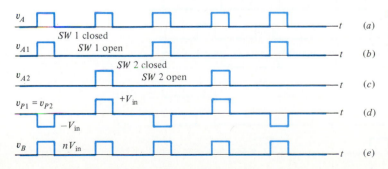

FIGURE 17-23

The waveforms in Fig. 17-22. (For simplicity, all amplitudes are drawn equal.)

FIGURE 17-24

FIGURE 17-24

The block diagram of the system used to generate the waveforms v_{A1} and v_{A2} in Fig. 17-23 from the pulse-width modulator output v_A. The power switch transistors $Q1$ and $Q2$ and the base-drive transistors $Q3$ and $Q4$ are also shown.

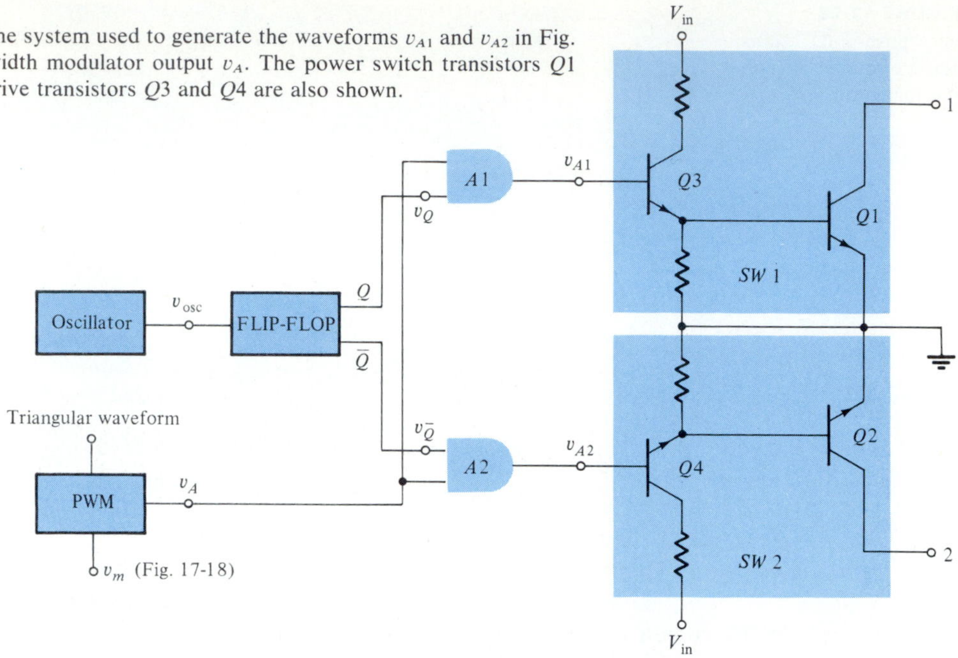

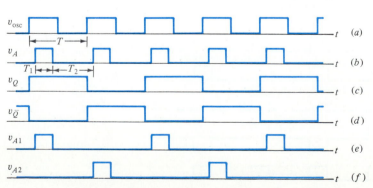

for the pulse-width modulator. This PWM waveshape v_A is given in Fig. 17-25b. The duty cycle δ of v_A is $T_1/(T_1 + T_2)$. The FLIP-FLOP is used as a divide-by-2 circuit, whose input is v_{osc}, and the two complementary FLIP-FLOP outputs v_Q and $v_{\bar{Q}}$ are shown in Fig. 17-25c and d, respectively. The inputs to AND gate $A1$ ($A2$) are v_Q ($v_{\bar{Q}}$) and v_A, and the outputs v_{A1} and v_{A2} are drawn in Fig. 17-23b and f respectively. These are the waveforms used in Figs. 17-23b and 17-23c.

The switch $SW1$ ($SW2$) in Fig. 17-22 is replaced by the power transistor $Q1$ ($Q2$) in Fig. 17-24. The base currents for $Q1$ and $Q2$ are supplied by transistors $Q3$ and $Q4$, which are driven by waveforms v_{A1} and v_{A2} respectively. The complexity of the switching regulator system would preclude its usefulness were it not for the increased level of sophistication attainable in modern microelectronics. The Silicon General SG1524 package is such an example.

FIGURE 17-25

The waveforms shown in Fig. 17-24. (All amplitudes are drawn equal for simplicity.)

Contained on this chip are all of the following circuits: reference regulator, pulse-width modulator (consisting of the sawtooth oscillator and comparator), error amplifier, two uncommitted transistors (for $Q3$ and $Q4$ in Fig. 17-24 or $Q3$ in Fig. 17-19), steering FLIP-FLOP and two AND gates (Fig. 17-24), and provisions for current-limiting and shutdown. The waveforms in Fig. 17-24 are shown in Fig. 17-25.

The SG1524 is placed in the feedback loop of Fig. 17-18 to form a switching regulator by adding the feedback resistors R_1 and R_2 and the discrete power-switch components of Fig. 17-19b or Fig. 17-22. For $V_{in} = 28$ V, it is possible to obtain a regulated output voltage V_O of 5 V at 1 A for the single-ended system and 5 V at 5 A for the push-pull system. For the $L \approx 1$ mH and $C \approx 1000$ μF as filter components with the SG1524, line and load regulation of 0.2 percent with less than 1 percent maximum variation is achieved. The control circuitry operates at a reference voltage of 5 V, draws less than 10 mA, and is capable of operation beyond 100 kHz (external resistors R_T and C_T set the frequency). The output transistors on the chip are rated at 100 mA and are short-circuit protected. The feedback loop is stabilized by adding an RC lag network.

17-9 LARGE-SIGNAL AMPLIFIERS

A simple transistor amplifier that supplies power to a pure resistance load R_L is indicated in Fig. 17-26. Using the notation of Table 10-1, i_C represents the total instantaneous collector current, i_c designates the instantaneous variation from the quiescent value I_C of the collector current. Similarly, i_B, i_b, and I_B represent corresponding base currents. The total instantaneous collector-to-emitter voltage is given by v_C, and the instantaneous variation from the quiescent value V_C is represented by v_c.

Let us assume that the static output characteristics are equidistant for equal increments of input base current i_b, as indicated in Fig. 17-27. Then, if the input signal i_b is a sinusoid, the output current and voltage are also sinusoidal, as shown. Under these circumstances the nonlinear distortion is negligible, and the power output may be found as follows:

$$P = V_c I_c = I_c^2 R_L \qquad (17\text{-}32)$$

FIGURE 17-26
A simple discrete-component transistor stage.

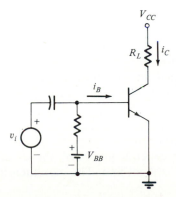

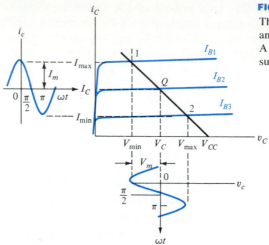

FIGURE 17-27

The output characteristics and the current and voltage waveforms shown in Fig. 17-26. A sinusoidal base-current excitation is assumed.

where V_c and I_c are the rms values of the signal output voltage v_c and current i_c, respectively, and R_L is the load resistance. The numerical values of V_c and I_c can be determined graphically in terms of the maximum and minimum voltage and current swings, as indicated in Fig. 17-27. If I_m (V_m) represents the peak sinusoidal current (voltage) swing, it is seen that

$$I_c = \frac{I_m}{\sqrt{2}} = \frac{I_{max} - I_{min}}{2\sqrt{2}} \tag{17-33}$$

and

$$V_c = \frac{V_m}{\sqrt{2}} = \frac{V_{max} - V_{min}}{2\sqrt{2}} \tag{17-34}$$

so that the power becomes

$$P = \frac{V_m I_m}{2} = \frac{I_m^2 R_L}{2} = \frac{V_m^2}{2R_L} \tag{17-35}$$

which may also be written in the form

$$P = \frac{(V_{max} - V_{min})(I_{max} - I_{min})}{8} \tag{17-36}$$

This equation allows the output power to be calculated very simply. All that is necessary is to plot the load line on the volt-ampere characteristics of the device and to read off the values of V_{max}, V_{min}, I_{max}, and I_{min}.

17-10 HARMONIC DISTORTION

In the preceding section the active device is idealized as a perfectly linear device. In general, however, the dynamic transfer characteristic (i_c versus i_b) is not a straight line. This nonlinearity arises because the static output characteristics are not equidistant straight lines for constant increments of input excitation. Referring to Fig. 10-3, we see that the waveform of the output voltage differs from that of the input signal. Distortion of this type is called *nonlinear*, or *amplitude*, *distortion*.

The reader may wonder why the question of distortion has not been addressed in the earlier chapters on amplification. The answer is signal magnitude. The underlying precept of Chap. 10 is that any device, independent of the transfer characteristic, can be treated analytically in a linear fashion for sufficiently small excursions about a quiescent operating point. This is not the case with power amplifiers. By its very nature a power amplifier must generate a large output signal, and the entire transfer curve, linear or nonlinear, must therefore be examined.

Second-Harmonic Distortion   To investigate the magnitude of this distortion, we assume that the dynamic curve with respect to the quiescent point Q can be represented by a parabola rather than a straight line. Thus, instead of relating the alternating output current i_c with the input excitation i_b by the equation $i_c = Gi_b$ resulting from a linear circuit, we assume that the relationship between i_c and i_b is given more accurately by the expression

$$i_c = G_1 i_b + G_2 i_b^2 \tag{17-37}$$

where the G values are constants. Actually these two terms are the beginning of a power-series expansion of i_c as a function of i_b.

If the input waveform is sinusoidal and of the form

$$i_b = I_{bm} \cos \omega t \tag{17-38}$$

the substitution of this expression in Eq. (17-37) leads to

$$i_c = G_1 I_{bm} \cos \omega t + G_2 I_{bm}^2 \cos^2 \omega t$$

Since $\cos^2 \omega t = \frac{1}{2} + \frac{1}{2} \cos 2\omega t$, the expression for the instantaneous total current i_c reduces to the form

$$i_C = I_C + i_c = I_C + B_0 + B_1 \cos \omega t + B_2 \cos 2\omega t \tag{17-39}$$

where the B values are constants which may be evaluated in terms of the G values. The physical meaning of this equation is evident. It shows that the application of a sinusoidal signal on a parabolic dynamic characteristic results in an output current which contains, in addition to a term of the same frequency as the input, a second-harmonic term, and also a constant current. This constant term B_0 adds to the original dc value I_C to yield a total dc component of current $I_C + B_0$. *Parabolic nonlinear distortion introduces into the output a component whose frequency is twice that of the sinusoidal input excitation.* This was also observed in Sec. 12-3 in the discussion of the effect of negative feedback on distortion. Also, *since a sinusoidal input signal changes the average value of the output current rectification takes place.*

The amplitudes B_0, B_1, and B_2 for a given load resistor are readily determined from the static characteristics. We observe from Fig. 17-27 that

When $\omega t = 0$: $\qquad\qquad\qquad\qquad i_C = I_{max}$

When $\omega t = \dfrac{\pi}{2}$: $\qquad\qquad\qquad i_C = I_C$ $\qquad\qquad\qquad\qquad$ (17-40)

When $\omega t = \pi$: $\qquad\qquad\qquad i_C = I_{min}$

By substituting these values in Eq. (17-39), there results

$$I_{max} = I_C + B_0 + B_1 + B_2$$

$$I_C = I_C + B_0 - B_2 \qquad\qquad\qquad (17\text{-}41)$$

$$I_{min} = I_C + B_0 - B_1 + B_2$$

This set of three equations determines the three unknowns B_0, B_1, and B_2. It follows from the second of this group that

$$B_0 = B_2 \qquad\qquad\qquad (17\text{-}42)$$

By subtracting the third equation from the first, there results

$$B_1 = \frac{I_{max} - I_{min}}{2} \qquad\qquad\qquad (17\text{-}43)$$

With this value of B_1, the value for B_2 may be evaluated from either the first or the last of Eqs. (17-41) as

$$B_2 = B_0 = \frac{I_{max} + I_{min} - 2I_C}{4} \qquad\qquad\qquad (17\text{-}44)$$

The second-harmonic distortion D_2 is defined as

$$D_2 \equiv \frac{|B_2|}{|B_1|} \qquad\qquad\qquad (17\text{-}45)$$

(To find the percent second-harmonic distortion, D_2 is multiplied by 100.) The quantities I_{max}, I_{min}, and I_C appearing in these equations are obtained directly from the characteristic curves of the transistor and from the load line.

If the dynamic characteristic is given by the parabolic form (17-37) and if the input contains two frequencies ω_1 and ω_2, then the output will consist of a dc term and sinusoidal components of frequencies ω_1, ω_2, $2\omega_1$, $2\omega_2$, $\omega_1 + \omega_2$, and $\omega_1 - \omega_2$ (Prob. 17-22). The sum and difference frequencies are called *intermodulation* frequencies.

Higher-Order Harmonic Generation The preceding analysis assumes a parabolic dynamic characteristic. This approximation is usually valid for amplifiers where the swing is small. For a power amplifier with a large input swing, however, it is necessary to express the dynamic transfer curve with respect to the Q point by a power series of the form

$$i_c = G_1 i_b + G_2 i_b^2 + G_3 i_b^3 + G_4 i_b^4 + \cdots \qquad (17\text{-}46)$$

If we assume that the input wave is a simple cosine function of time, of the form in Eq. (17-38), the output current will be given by

$$i_C = I_C + B_0 + B_1 \cos \omega t + B_2 \cos 2\omega t + B_3 \cos 3\omega t + \cdots \qquad (17\text{-}47)$$

This equation results when Eq. (17-38) is inserted in Eq. (17-46) and the proper trigonometric transformations are made.

Note that now a third harmonic and higher-order harmonics are present. The Fourier coefficients B_0, B_1, B_2, B_3, . . . may be obtained by an extension of the foregoing procedure used with Eq. (17-47) instead of Eq. (17-39).

The harmonic distortion is defined as

$$D_2 \equiv \frac{|B_2|}{|B_1|} \qquad D_3 \equiv \frac{|B_3|}{|B_1|} \qquad D_4 \equiv \frac{|B_4|}{|B_1|} \qquad (17\text{-}48)$$

where D_s $(s = 2, 3, 4, \ldots)$ represents the distortion of the sth harmonic.

Power Output   If the distortion is not negligible, the power delivered at the fundamental frequency is

$$P_1 = \frac{B_1^2 R_L}{2} \qquad (17\text{-}49)$$

However, the total power output is

$$P = (B_1^2 + B_2^2 + B_3^2 + \cdots) \frac{R_L}{2} = (1 + D_2^2 + D_3^2 + \cdots) P_1$$

or

$$P = (1 + D^2) P_1 \qquad (17\text{-}50)$$

where *the total harmonic distortion* (THD), or *distortion factor*, is defined as

$$D \equiv \sqrt{D_2^2 + D_3^2 + D_4^2 + \cdots} \qquad (17\text{-}51)$$

If the total distortion is 10 percent of the fundamental, then

$$P = [1 + (0.1)^2] P_1 = 1.01 P_1$$

The total power output is only 1 percent higher than the fundamental power when the distortion is 10 percent. Hence little error is made in using only the fundamental term P_1 in calculating the power output.

In passing, it should be noted that the total harmonic distortion is not necessarily indicative of the discomfort to someone listening to music. Usually, the same amount of distortion is more irritating, the higher the order of the harmonic frequency.

17-11 AMPLIFIER CLASSIFICATION It has been tacitly assumed in all previous amplifier design and analysis that the transistor is biased in the middle of its operating range, as indicated in Fig. 17-27 (note the location of the point Q in the i_C-v_C plane). This is not always the case with power circuits, and a classification (A, B, AB, and C) has evolved to describe amplifier operation, dependent on the type of biasing employed. The significance of this classification is discussed in the following sections.

Class A

A class A amplifier is one in which the operating point and the input signal are such that the current in the output circuit (in the collector, or drain electrode) flows at all times. A class A amplifier operates essentially over a linear portion of its characteristic.

Class B

A class B amplifier is one in which the operating point is at an extreme end of its characteristic, so that the quiescent power is very small. Hence either the quiescent current or the quiescent voltage is approximately zero. If the signal excitation is sinusoidal, amplification takes place for only one half of a cycle. For example, if the quiescent output circuit current is zero, this current will remain zero for one half of a cycle.

Class AB

A class AB amplifier is one operating between the two extremes defined for class A and class B. Hence the output signal is zero for part but less than one half of an input sinusoidal signal cycle.

Class C

A class C amplifier is one in which the operating point is chosen so that the output current (or voltage) is zero for more than one half of an input sinusoidal signal cycle.

17-12 EFFICIENCY OF A CLASS A AMPLIFIER If a power amplifier design is constrained either by a limited source of power (as would be the case for a satellite) or by maximum power-dissipation consideration, as in Sec. 17-13, attention must be focused on the issue of power conversion.

Conversion Efficiency A measure of the ability of an active device to convert the dc power of the supply into the ac (signal) power delivered to the load is called the *conversion efficiency*, or *theoretical efficiency*. This figure of merit, designated η, is also called the *collector-circuit efficiency* for a transistor amplifier. By definition, the percentage efficiency is

$$\eta \equiv \frac{\text{signal power delivered to load}}{\text{dc power supplied to output circuit}} \times 100 \text{ percent} \qquad (17\text{-}52)$$

In general,

$$\eta = \frac{\frac{1}{2}B_1^2 R_L}{V_{CC}(I_C + B_0)} \times 100 \text{ percent} \qquad (17\text{-}53)$$

If the distortion components are negligible, then

$$\eta = \frac{\frac{1}{2}V_m I_m}{V_{CC} I_C} \times 100 = 50 \frac{V_m I_m}{V_{CC} I_C} \qquad \text{percent} \qquad (17\text{-}54)$$

where V_m (I_m) represents the peak sinusoidal voltage (current) swing. The collector-circuit efficiency differs from the overall efficiency because the power taken by the base is not included in the denominator of Eq. (17-53).

From the definitions in Sec. 17-11, the amplifier of Sec. 17-9 operates in class A. Let us now qualitatively examine its efficiency using two limiting cases.

1. *Small signal.* With a small output signal the output power is correspondingly small. However, the power consumed by the class A biasing remains at $V_{CC} I_C$, which may be substantial, resulting in an extremely small conversion efficiency. Note also that the load must dissipate a large fraction of the dc power $V_{CC} I_C$ even under zero excitation.

2. *Maximum signal.* With careful selection of the bias point, the transistor may be driven from the edge of saturation to cutoff. It can be shown (Prob. 17-25) that under this condition $I_m = I_C$ and $V_m = \frac{1}{2}V_{CC}$, yielding $\eta = 25$ percent. For every 1 W of output power, 3 W are being consumed internally. Clearly, from an efficiency standpoint, class A operation is a poor choice for power amplification.

17-13 CLASS B PUSH-PULL AMPLIFIERS

If $V_{BB} = 0$ in Fig. 17-26, the quiescent current is $I_C = 0$. From the definitions given in Sec. 17-11, this zero-bias circuit is a class B amplifier. Similarly, the emitter follower in Fig. 17-28a operates in class B. Let us assume that the transistor output characteristics are equally spaced for equal intervals of excitation. For such an idealized transistor the dynamic transfer curve (i_C versus i_B) is a straight line passing through the origin (Fig. 17-28b). The graphical construction from which to determine the collector-current waveshape is indicated. Note that for this class B circuit the load current $i_L \approx i_C$ is sinusoidal during one half of each period and is zero during the second half-cycle. In other words, this circuit behaves as a rectifier rather than as a power amplifier.

The foregoing difficulty is overcome by using the complementary emitter-follower Op-Amp output stage shown in Fig. 14-13, which is repeated in Fig. 17-29 for convenience. This configuration is called a class B *push-pull amplifier*. For positive values of the sinusoidal input v_i, $Q1$ conducts and $Q2$ is OFF ($i_2 = 0$), so that i_1 is the positive half sine wave in Fig. 17-28b. For negative values of v_i, $Q1$ is nonconducting ($i_1 = 0$), and $Q2$ conducts, resulting in a positive half sinusoid for i_2 which is 180° out of phase with that shown in Fig. 17-28b. Since the load current is the difference between the two transistor emitter currents, it follows that

$$i_L = i_1 - i_2 \qquad (17\text{-}55)$$

FIGURE 17-28
(a) The emitter follower with zero bias operating as a class B amplifier.
(b) Graphical construction for determining the output current waveform.

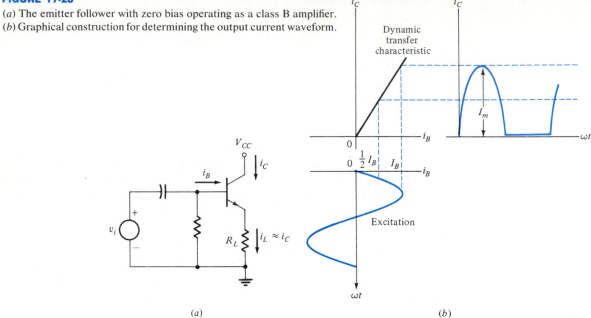

(a)   (b)

Consequently, for the idealized transfer characteristic of Fig. 17-28b, the load current is a perfect sinusoid.

The advantages of class B as compared with class A operation are the following: It is possible to obtain greater power output, the efficiency is higher, and there is negligible power loss at no signal. For these reasons, in systems where the power supply is limited, such as those operating from solar cells or batteries, the output power is usually delivered through a push-pull class B transistor circuit. The disadvantages are that the harmonic distortion may be higher and the supply voltages must have good regulation. The power output circuit in most modern IC amplifiers is the complementary emitter-follower push-pull stage.

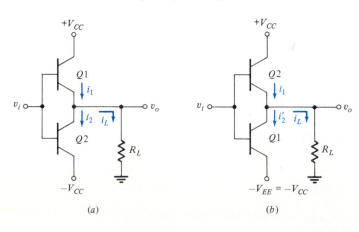

(a)   (b)

FIGURE 17-29

(a) A complementary emitter follower and (b) a complementary common-emitter push-pull amplifier.

Efficiency

In Fig. 17-29 the peak load voltage is $V_m = I_m R_L$. The power output is

$$P = \frac{I_m V_m}{2} \tag{17-56}$$

The corresponding direct collector current in each transistor under load is the average value of the half sine loop of Fig. 17-28b. Since $I_{dc} = I_m/\pi$ for this waveform, the dc input power from the supply is

$$P_i = 2\frac{I_m V_{CC}}{\pi} \tag{17-57}$$

The factor 2 in this expression arises because two transistors are used in the push-pull system.

Taking the ratio of Eqs. (17-56) and (17-57), we obtain for the collector-circuit efficiency

$$\eta \equiv \frac{P}{P_i} \times 100 = \frac{\pi}{4}\frac{V_m}{V_{CC}} \times 100 \text{ percent} \tag{17-58}$$

If the drop across a transistor is negligible compared with the supply voltage, then $V_m \approx V_{CC}$. Under these conditions, Eq. (17-58) shows that the maximum possible conversion efficiency is $25\pi = 78.5$ percent for a class B system compared with 25 percent for class A operation. This large value of η results from the fact that there is no current in a class B system if there is no excitation, whereas there is a drain from the power supply in a class A system even at zero signal. We also note that in a class B amplifier the dissipation at the collectors is zero in the quiescent state and increases with excitation, whereas the heating of the collectors of a class A system is a maximum at zero input and decreases as the signal increases. Since the direct current increases with signal in a class B amplifier, the power supply must have good regulation.

Dissipation

The dissipation P_C (in both transistors) is the difference between the power input to the collector circuit and the power delivered to the load. Since $I_m = V_m/R_L$, it follows that

$$P_C = P_i - P = \frac{2}{\pi}\frac{V_{CC}V_m}{R_L} - \frac{V_m^2}{2R_L} \tag{17-59}$$

This equation shows that the collector dissipation is zero at no signal ($V_m = 0$), rises as V_m increases, and passes through a maximum at $V_m = 2V_{CC}/\pi$. The peak dissipation is found to be

$$P_{C(\max)} = \frac{2V_{CC}^2}{\pi^2 R_L} \tag{17-60}$$

The maximum power which can be delivered is obtained for $V_m = V_{CC}$ or

$$P_{\max} = \frac{V_{CC}^2}{2R_L} \tag{17-61}$$

Hence

$$P_{C(max)} = \frac{4}{\pi^2} P_{max} \approx 0.4 P_{max} \qquad (17\text{-}62)$$

If, for example, we wish to deliver 10 W from a class B push-pull amplifier, then $P_{C(max)} = 4$ W, or we must select transistors which have collector dissipations of approximately 2 W each. In other words, we can obtain a push-pull output of five times the specified power dissipation of a single transistor. On the other hand, if we paralleled two transistors and operated them class A to obtain 10 W out, the collector dissipation of each transistor would have to be at least 20 W (assuming 25 percent efficiency). This statement follows from the fact that $P_i = P/\eta = 10/0.25 = 40$ W. This input power must all be dissipated in the two collectors at no signal, or $P_C = 20$ W per transistor. Hence at no excitation there would be a steady loss of 20 W in each transistor, whereas in class B the standby (no-signal) dissipation is zero. This example clearly indicates the superiority of the push-pull over the parallel configuration.

Distortion

The distortion properties of a push-pull system are rather unique. Consider the operation shown in Fig. 17-29 when the transfer characteristic is not linear. Either $Q1$ or $Q2$ is conducting, depending upon the polarity of the input signal. If the devices are matched, then the current i_2 is identical with i_1, except shifted in phase by 180°. The current of $Q1$ is given by Eq. (17-47) and is repeated here for convenience:

$$i_1 = I_C + B_0 + B_1 \cos \omega t + B_2 \cos 2\omega t + B_3 \cos 3\omega t + \cdots \qquad (17\text{-}63)$$

The output current of transistor $Q2$ is obtained by replacing ωt by $\omega t + \pi$ in the expression for i_1; that is

$$i_2(\omega t) = i_1(\omega t + \pi) \qquad (17\text{-}64)$$

whence

$$i_2 = I_C + B_0 + B_1 \cos (\omega t + \pi) + B_2 \cos \cos 2(\omega t + \pi) + \cdots$$

or

$$i_2 = I_C + B_0 - B_1 \cos \omega t + B_2 \cos 2\omega t - B_3 \cos 3\omega t + \cdots \qquad (17\text{-}65)$$

From Eq. (17-55)

$$i_L = i_1 - i_2 = 2(B_1 \cos \omega t + B_3 \cos 3\omega t + \cdots) \qquad (17\text{-}66)$$

This expression shows that a push-pull circuit will balance out all even harmonics in the output and will leave the third-harmonic term as the principal source of distortion. This conclusion was reached on the assumption that the two transistors are identical. If their characteristics differ appreciably, the appearance of even harmonics must be expected.

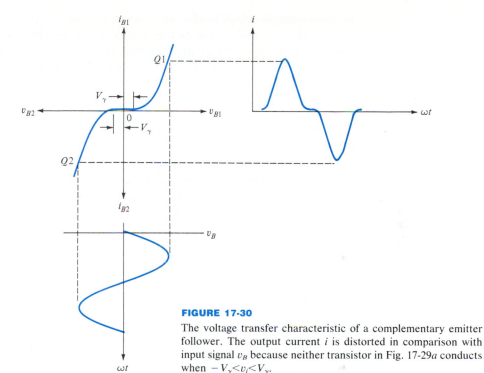

FIGURE 17-30

The voltage transfer characteristic of a complementary emitter follower. The output current i is distorted in comparison with input signal v_B because neither transistor in Fig. 17-29a conducts when $-V_\gamma < v_i < V_\gamma$.

17-14   CLASS AB OPERATION

In addition to the distortion introduced by not using matched transistors and that due to the nonlinearity of the collector characteristics, there is one more source of distortion, that caused by nonlinearity of the input characteristic. As pointed out in Sec. 3-3 and Fig. 3-9, no appreciable base current flows until the emitter junction is forward-biased by at least the cut-in voltage V_γ, which is 0.5 V for silicon. Under these circumstances a sinusoidal base-voltage excitation will not result in a sinusoidal output current. Although already mentioned briefly in Sec. 14-5, the significance of the nonlinear input characteristic merits further discussion.

The distortion caused by this nonlinear curve is indicated in Fig. 17-30. The i_B-v_B curve for each transistor is drawn, and the construction used to obtain the output current (assumed proportional to the base current) is shown. In the region of small currents (for $v_B < V_\gamma$) the output is much smaller than it would be if the response were linear. This effect is called *crossover distortion*. Such distortion would not occur if the driver were a true current generator, in other words, if the base current (rather than the base voltage) were sinusoidal.

To minimize crossover distortion, the transistors must operate in a class AB mode, where a small standby current flows at zero excitation. For example, in the circuit of Fig. 14-14 the difference between the base voltages of the two transistors is adjusted to be approximately equal to $2V_\gamma$. Class AB operation results in less distortion than class B, but the price which must be paid for this improvement is a loss in efficiency and waste of standby power. The calculations of the distortion components in a class AB or class A push-pull amplifier due to the nonlinearity of the collector characteristics is somewhat involved since it requires the construction of composite output curves for the pair of transistors.

17-15 INTEGRATED-CIRCUIT POWER AMPLIFIERS Manufacturers (App. B-1) have available a wide range of IC power amplifiers. An industry standard Op-Amp, such as the 741 (at a cost of under 50 cents), is capable of delivering about 100 mW of power with no additional external components. Two examples of IC audio amplifiers with ratings of 4 and 20 W, respectively, are indicated in the following.

The LM384 amplifier shown in Fig. 17-31 is designed to provide 34 dB of amplification for signals as high as 300 kHz and to deliver 5 W of power to a capacitively coupled load. The component values shown result in total harmonic distortion at 1 kHz of less than 1 percent at 5-W output power into an 8-Ω load. When this device is used, care must be exercised to lay out the circuit properly and to avoid stray coupling or feedback from the output to the input, which may result in oscillations. To avoid oscillations, the input cable must be shielded and the lag compensating network R_1C_2 must be connected from the output pin to ground. Capacitor C_3 is used to cancel the effects of inductance in the power-supply leads, while C_1 acts as a low-frequency bypass.

The 20-W amplifier of Fig. 17-32 is another example of the state of linear, monolithic, power amplifier technology. Connected as shown with 260-mV input, the SGS TDA2020 will typically produce 20 W into 4 Ω at less than 1 percent distortion and 57 percent efficiency. The frequency response (-3 dB) is 10 Hz to 160 kHz for a gain of 30 dB. Furthermore, the device has short-circuit protection for current overloads and thermal shutdown if the recommended maximum power dissipation limit is exceeded.

FIGURE 17-31

A 5-W audio amplifier. (*Courtesy of National Semiconductor*)

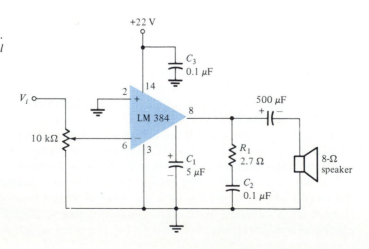

FIGURE 17-32
A split-supply 20-W audio power amplifier. (*Courtesy of SGS/ATES Corporation*)

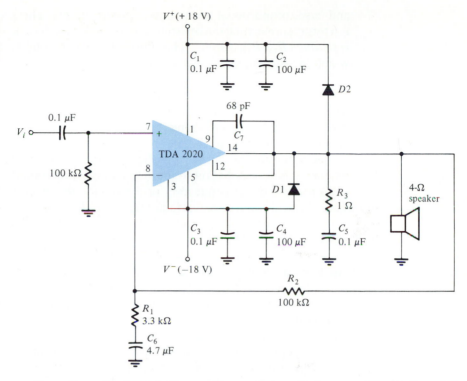

Capacitors C_1 through C_4 provide power-supply bypass. The R_3C_5 and R_1C_6 networks produce output and input lag compensation, respectively. Further compensation is controlled by C_7. Since the output dc level is set at $(V^+ + V^-)/2$, split power-supply operation yields 0 V dc output, and the load can be direct-coupled, eliminating the need for a very large coupling capacitor. Diodes $D1$ and $D2$ clamp (and thus protect) the output from inductive excursions greater than the supply voltages.

17-16 THERMAL DESIGN CONSIDERATIONS

The power amplifier in Fig. 17-32 raises a very important question. At 20-W output and 57 percent efficiency, the input power is $20/0.57 = 35.1$ W. Hence 15.1 W must be dissipated by the transistors. We will now indicate how this heat is removed and what factors must be considered in maintaining proper device operation.

Maximum Junction Temperature All semiconductor devices have a maximum operating junction temperature $T_{J(\text{max})}$, ranging typically from 125 to 200°C for silicon. Above this temperature catastrophic irreversible failure will occur.

Thermal Resistance The heat generated within the device will conduct away from its source (the collector junction) to the case, causing a gradient in temperature. Therefore, there will be a steady-state temperature difference ΔT_{JC} between junction and case, proportional to the power dissipated P_D. The proportionality factor is a term representing the resistance to the heat transfer and is called the *thermal resistance* R_{th}, $R_{\theta JC}$, or θ_{JC}. The subscripts on θ denote the two points between which the measurement is taken. Hence

$$T_J - T_C = \Delta T_{JC} = P_D\theta_{JC} \qquad (17\text{-}67)$$

where P_D is in watts and θ is in degrees Celsius per watt. The electrical analog is obvious: if P_D (θ_{JC}) is likened to current I (resistance R), then ΔT_{JC} must be analogous to voltage drop ΔV.

The value of the thermal resistance depends on the size of the transistor, on convection or radiation to the surroundings, on forced-air cooling (if used), and on the thermal connection of the device to a metal chassis or to a heat sink. Typical values for various transistor designs vary from 0.2°C/W for a high-power transistor with an efficient heat sink to 1000°C/W for a low-power transistor in free air.

Dissipation Derating Curve Manufacturers usually present a power-temperature derating curve, such as shown in Fig. 17-33. The maximum junction temperature $T_{J(\text{max})}$ can be deduced by noting that no power (0 W) can be dissipated at 200°C. Zero power dissipation implies no temperature gradient and, hence, the junction must also be at 200°C [Eq.(17-67)].

The specifications for the 2N5671 silicon *npn* transistor are given in App. B-8. This transistor has high power (140 W), high current (I_C = 30 A, I_B = 10 A), and high speed (switching time ~ 1 μs).[1] Since the maximum ordinate in Fig. 17-33 corresponds to $P_{D(\text{max})}$ = 140 W, then from Eq. (17-67), θ_{JC} = (200 − 25)/140 = 1.25°C/W. The reciprocal of the thermal resistance is the slope of the line in Fig. 17-33, and is called the *power derating factor* (1/1.25 = 0.8 W/°C). The value of the thermal resistance is inversely related to the surface area of the case. The 2N2222A which has a much smaller case than the 2N5671 has θ_{JC} = 83°C/W.

To carry the heat away from the case into the ambient (the surrounding air), a *heat sink* is used with a power transistor. The heat sink is a metallic structure

---

[1]The ratings of the 2N5671 power transistor should be compared with those of the 2N2222A small-signal transistor (App. B-3).

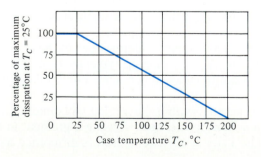

FIGURE 17-33
Dissipation derating curve for the 2N5671 power transistor. (*Courtesy of RCA Solid-State Division.*)

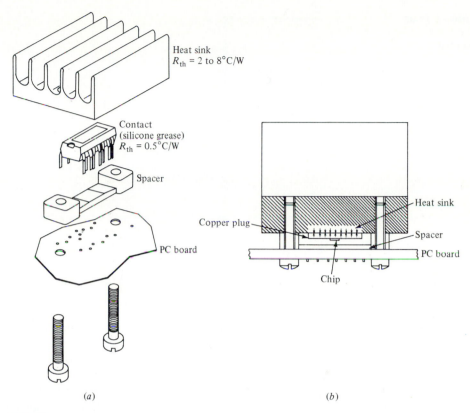

Heat sink
R_{th} = 2 to 8°C/W

Contact
(silicone grease)
R_{th} = 0.5°C/W

Spacer

PC board

Heat sink

Copper plug

Spacer

PC board

Chip

(a)   (b)

FIGURE 17-34

Heat sink mounting of TDA2020. A number of heat sinks are available with thermal resistances in the range 2 to 8°C/W. (b) Cross section of assembled system. (*Courtesy of SGS/AETA*)

with a relatively large heat-radiating surface to which the transistor case is attached. Figure 17-34 depicts the mounting system for the TDA2020 chip in Fig. 17-32.

Example 17-1

In the foregoing discussion we observed that the 20-W TDA2020 amplifier must dissipate 15.1 W of internal power. The ambient temperature is $T_A = 30°C$. If the maximum allowable junction temperature is $T_{J(max)} = 150°C$ and if $\theta_{JC} = 3°C/W$, what is the maximum heat-sink thermal resistance θ_{SA} (sink-to-ambient) that can be tolerated?

Solution

Using the electrical analog of Eq. (17-67), we obtain the series circuit model given in Fig. 17-35 for the power flow.

$$T_J = \Delta T_{JC} + \Delta T_{CS} + \Delta T_{SA} + T_A$$
$$= P_D(\theta_{JC} + \theta_{CS} + \theta_{SA}) + T_A \qquad (17\text{-}68)$$

FIGURE 17-35
Electrical analog of thermal system.

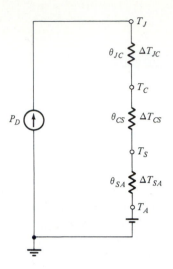

With $\theta_{CS} = 0.5°C/W$, as indicated in Fig. 17-34a, Eq. (17-68) becomes

$$150 = 15.1(3 + 0.5 + \theta_{SA}) + 30$$

which yields $\theta_{SA} = 4.5°C/W$ maximum. The heat sink in Fig. 17-34 is satisfactory, since its maximum thermal resistance may be chosen to be less than $4.5°C/W$.

17-17 POWER FIELD-EFFECT TRANSISTOR (VMOS)

In 1976 Siliconix Inc. introduced a new type of FET power transistor which overcomes many of the limitations of the bipolar power transistor. This new device is an *n*-channel enhancement MOSFET, but it is fabricated so that the current flows vertically. Hence this transistor is designated VMOS.[1] This construction distinguishes the VMOS from the low-power MOSFETs described in Chap. 4, where the carriers flow horizontally from source to drain.

The fabrication of the power FET starts with a silicon n^+ substrate onto which is grown an n^- epitaxial layer. Two successive diffusions then take place, the first with *p*-type and the second with *n*-type impurities, as indicated in Fig. 17-36. The structure obtained at this stage of the construction is identical with the discrete bipolar transistor shown in Fig. 5-7*d*. In the BJT the top (bottom) n^+ region is the emitter (collector), whereas in Fig. 17-36 the top (bottom) n^+ section becomes the source (drain). In Fig 5-7*d* the *p* region is the base, but in the VMOS the *p* section is the *n* channel. In order to be able to place a control gate over the channel, extending from source to drain, a novel fabrication step is introduced; an isosceles V-shaped groove is anisotropically etched into the silicon, as indicated in Fig. 17-36. Continuing with the standard fabrication processes described in Chap. 5, a thin silicon dioxide layer is grown and then metallization is used to form the gate electrode and the source contact.

---

[1] VMOS is the designation used by Siliconix, Inc. At present, several other manufacturers fabricate this type of power MOSFET.

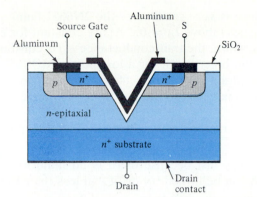

FIGURE 17-36
Cross section of a vertical FET.

Note that the drain area (at the bottom of Fig. 17-36) is large and can be placed in contact with a heat sink for optimum removal of the power dissipated within the device. The channel length L (the vertical extent of the p region) is determined by the difference in the depths of the p and n^+ (source) diffusions. Hence, L can be made reliably quite small; for example, $L \approx 1.5$ μm. It should be recalled that in the standard (horizontal) MOSFET the channel length is determined by masking, etching, and the *lateral* diffusion of the source and drain, so that L is much longer than for the VMOS. The V-shaped gate controls two vertical MOSFETs, one on each side of the notch. Hence, by paralleling the two S terminals in Fig. 17-36 the current capacity is doubled.

The low-power MOSFETs described in Chap. 4 are symmetrical devices, between source and drain. Clearly, from Fig. 17-36, the VMOS is built unsymmetrically so that S and D may not be interchanged.

VMOS Characteristics   The volt-ampere curves of a vertical FET are indicated in Fig. 17-37 and should be compared with the low-power n-channel enhancement MOSFET in Fig. 4-12. The peak VMOS current is 2 A (contrasted with 50 mA for the horizontal MOSFET). Also, note that the characteristics in the saturation region of Fig. 17-37 are much flatter than in Fig. 4-12 (I_D = constant and hence the

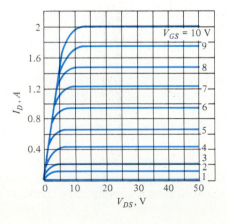

FIGURE 17-37
The output characteristics of an n-channel enhancement transistor (*Courtesy of Texas Instruments, Inc.*)

output conductance is very small). For the 2N6657 family shown the spacing of the characteristics (above $I_D = 0.4$ A) are constant for equal increments of gate voltage. Therefore, the transconductance g_m is constant (≈ 0.25 A/V) for $I_D \geqslant 0.4$ A. On the other hand, for a low-power MOSFET, g_m varies as the square root of the drain current [Eq. (4-18)], rather than remaining constant.

The VMOS has many advantageous properties, including the following:

1. The transfer characteristic I_D versus V_{GS} is linear ($g_m = $ constant) for $I_D \geqslant 0.4$ A.

2. Switching is very fast because there is no minority carrier storage. For example, 2 A can be turned ON or OFF in less than 10 ns.

3. Thermal runaway (Sec. 10-3) is not possible because the drain-source resistance has a positive temperature coefficient and the current becomes limited as the device heats up. (No hot spots develop and secondary breakdown does not occur.)

4. There is no "current hogging" when VMOS devices are operated in parallel to increase the current capacity. If one transistor tries to take more than its share of the current the positive drain-to-source temperature coefficient increases V_{DS}, thereby limiting I_D.

5. Because of its very high input resistance the VMOS requires extremely small input power and may be driven from CMOS logic gates. The power gain is extremely high.

6. The ON resistance is very low. From the slope of the curves at the origin in Fig. 17-37, we see that $r_{DS(ON)} \approx 3 \; \Omega$.

7. Power FETs have extremely low noise figures.

8. The threshold voltage V_T ranges from 0.8 to 2 V, so that VMOS devices are compatible with TTL logic.

9. From Fig. 17-36 it is seen that the overlap of the gate and drain (and therefore the capacitance between these electrodes) is quite small. Hence the capacitive feedback between output and input is minimized and VMOS devices may be used for high-frequency (broadband) circuits ($f_T \approx 600$ MH$_Z$).

10. The VMOS breakdown voltage between drain and source is high. This feature results from the fact that the epitaxial layer absorbs the depletion region from the reverse-biased body-drain pn diode.

Applications

The VMOS may be used as the output stage of an audio or RF power amplifier or of a switching regulator power supply. Other applications include industrial process control, motor control, solenoid or relay driver, plasma display driver, ultrasonic transducer driver, and so on.

REFERENCES

1 Millman, J., and C. C. Halkias: "Integrated Electronics: Analog and Digital Circuits and Systems," McGraw-Hill Book Company, New York, 1972.

2 Grebene, A. B.: "Bipolar and MOS Analog Integrated Circuit Design," John Wiley and Sons, New York, 1984.

3 Ghausi, M. S.: "Electronic Devices and Circuits: Discrete and Integrated," Holt, New York, 1985.

4 Bohn, D. (Ed.): "Audio Handbook," National Semiconductor Company, Santa Clara, Calif., 1976.

5 Mammamo, R.: "Simplifying Converter Design with a New Integrated Regulating Pulse-Width Modulator," Application Note, Silicon General, Inc., Irvine, Calif., 1980.

National Semiconductor Corporation, Fairchild Semiconductor Company, Texas Instruments, Inc., Silicon General, Inc., and Unitrode Corporation all publish Voltage Regulator Handbooks.

REVIEW QUESTIONS

17-1   List four components of an ac-to-dc converter and explain the function of each.

17-2   (*a*)   Sketch the circuit of a half-wave rectifier.
 (*b*)   Derive the expression for the (1) dc current and (2) the rms load current.

17-3   Repeat Rev. 17-2 for a full-wave rectifier.

17-4   (*a*)   Define regulaton.
 (*b*)   Derive the regulation equation for a full-wave circuit.

17-5   Draw the Thévenin model for a full-wave rectifier.

17-6   (*a*)   Define peak inverse voltage.
 (*b*)   What is the peak inverse voltage for a full-wave circuit using ideal diodes?
 (*c*)   Repeat part *b* for a half-wave rectifier.

17-7   Sketch the circuit of a bridge rectifier and explain its operation.

17-8   Repeat Rev. 17-7 for a rectifier meter circuit.

17-9   Repeat Rev. 17-7 for a voltage-doubler circuit.

17-10   (*a*)   Draw the circuit of a half-wave capacitive rectifier.
 (*b*)   At no load draw the steady-state voltage across the capacitor and also across the diode.

17-11   (*a*)   Draw the circuit of a full-wave capacitive rectifier.
 (*b*)   Sketch the load voltage for this circuit.

17-12   Draw the circuit of a choke-input filter and explain how this circuit reduces ripple.

17-13   Repeat Rev. 17-12 for a capacitor-input filter.

17-14   Give three reasons why an unregulated supply is inadequate for some applications.

17-15   Define input regulation factor, output resistance, and temperature coefficient for a voltage regulator.

17-16   (*a*)   Draw a simplified circuit diagram of a regulated power supply.
 (*b*)   What type of feedback is employed by this regulator?

17-17 List three disadvantages of pass regulators which may be overcome with a switching regulator.

17-18 (*a*) Draw the basic switching-regulator topology.
(*b*) Explain how the dc output voltage is determined by this feedback system.

17-19 Explain why a switching regulator is capable of very high conversion efficiency.

17-20 (*a*) Draw the power switch of a switching regulator as an SPDT switch. How is the switch controlled, and what is the switch output waveform?
(*b*) Verify that the switch in part (*a*) is equivalent to an SPST switch in series with a diode to ground. Why is the diode referred to as a flyback diode?

17-21 (*a*) Indicate the SPDT power switch of the basic regulator as a combination of three transistors and a diode.
(*b*) Explain the function of each transistor and the diode.

17-22 (*a*) Draw the power components in a switching regulator for which V_O is positive and greater than V_{in}.
(*b*) Verify that for this configuration $V_O > V_{\text{in}}$.
(*c*) What determines the numerical value of V_O?

17-23 (*a*) Repeat Rev. 17-22 for a regulator for which V_O is negative.
(*b*) Give an argument to justify that $V_O < 0$.

17-24 (*a*) Draw the power switch for a push-pull transformer-coupled switching regulator.
(*b*) Indicate the pulse-width-modulator waveform v_A, and also the waveforms v_{A1} and v_{A2} controlling the two SPST switches in series with the transformer primaries.
(*c*) Sketch the transformer secondary waveforms.
(*d*) Draw the waveform from the output switch (the input voltage to the filter).

17-25 (*a*) Draw in block-diagram form the system for obtaining the waveforms v_{A1} and v_{A2} in Rev. 17-24, part *b*.
(*b*) Explain the operation of the system with the aid of a waveform chart.
(*c*) Show the switches controlled by v_{A1} and v_{A2} simulated by transistors.

17-26 List all the low-power control circuits which are fabricated on a single IC chip and used with a switching regulator.

17-27 Derive an expression for the output power of a class A large-signal amplifier in terms of V_{max}, V_{min}, I_{max}, and I_{min}.

17-28 Discuss how rectification may take place in a power amplifier.

17-29 Define intermodulation distortion.

17-30 Define total harmonic distortion.

17-31 Define a (*a*) class A, (*b*) class B, and (*c*) class AB amplifier.

17-32 (*a*) Define the conversion efficiency η of a power stage.
(*b*) Derive a simple expression for η for a class A amplifier.
(*c*) What is the theoretical maximum efficiency for a class A amplifier?

17-33 (*a*) Draw the circuit of a class B power stage.
(*b*) For a sinusoidal input, what is the output waveform?

17-34 (*a*) Draw the circuit of a class B push-pull power amplifier.
(*b*) State three advantages of class B over class A.

17-35 Derive a simple expression for the output power of an idealized class B push-pull power amplifier.

17-36 Show that the maximum conversion efficiency of the idealized class B push-pull circuit is 78.5 percent.

17-37 Obtain the expression for the collector dissipation of a class B push-pull stage in terms of V_m and R_L.

17-38 Demonstrate that even harmonics are eliminated in a balanced push-pull amplifier.

17-39 (*a*) Explain the origin of crossover distortion.
(*b*) Describe a method to minimize this distortion.

17-40 (*a*) Define thermal resistance θ.
(*b*) Sketch a dissipation derating curve for a power amplifier.
(*c*) How is θ related to the curve in part *b*?

17-41 (*a*) What is a heat sink?
(*b*) Explain why a heat sink must be used with a power amplifier.

17-42 (*a*) Sketch the cross-sectional view of a power FET.
(*b*) Explain briefly how this device is fabricated.

17-43 Give two important differences in the output characteristics of a VMOS and a low-power MOSFET.

17-44 List six advantages of a VMOS.

Appendix A
CONSTANTS AND CONVERSION FACTORS

A-1 PROBABLE VALUES OF GENERAL PHYSICAL CONSTANTS*

Constant	Symbol	Value
Electronic charge	q	1.602×10^{-19} C
Electronic mass	m	9.109×10^{-31} kg
Ratio of charge to mass of an electron	q/m	1.759×10^{11} C/kg
Mass of atom of unit atomic weight (hypothetical)	$\cdots$	1.660×10^{-27} kg
Mass of proton	m_p	1.673×10^{-27} kg
Ratio of proton to electron mass	m_p/m	1.837×10^3
Planck constant	h	6.626×10^{-34} J $\cdot$ s
Boltzmann constant	$\overline{k}$	1.381×10^{-23} J/K
	k	8.620×10^{-5} eV/K
Stefan-Boltzmann constant	σ	5.670×10^{-8} W/(m$^2 \cdot$ K^4)
Avogadro number	N_A	6.023×10^{23} molecules/mol
Gas constant	R	8.314 J/(deg $\cdot$ mol)
Velocity of light	c	2.998×10^8 m/s
Faraday constant	F	9.649×10^3 C/mol
Volume per mole	V_o	2.241×10^{-2} m^3
Acceleration due to gravity	g	9.807 m/s^2
Permeability of free space	μ_o	1.257×10^{-6} H/m
Permittivity of free space	ϵ_o	8.849×10^{-12} F/m

*E. A. Mechtly, ''The International System of Units: Physical Constants and Conversion Factors,'' National Aeronautics and Space Administration, NASA SP-7012, Washington, D.C., 1964.

A-2   CONVERSION FACTORS AND PREFIXES

1 ampere (A)	$= 1$ C/s	1 lumen per square	$= 1$ footcandle
1 angstrom unit (Å)	$= 10^{-10}$ m	foot	(fc)
	$= 10^{-4}$ μm	mega (M)	$= \times 10^{6}$
1 atmosphere	$= 760$ mmHg	1 meter (m)	$= 39.37$ in
pressure		micro (μ)	$= \times 10^{-6}$
1 coulomb (C)	$= 1$ A $\cdot$ s	1 micron	$= 10^{-6}$ m
1 electron volt (eV)	$= 1.60 \times 10^{-19}$ J		$= 1$ μm
1 farad (F)	$= 1$ C/V	1 mil	$= 10^{-3}$ in
1 foot (ft)	$= 0.305$ m		$= 25$ μm
1 gram-calorie	$= 4.185$ J	1 mile	$= 5280$ ft
giga (G)	$= \times 10^{9}$		$= 1.609$ km
1 henry (H)	$= 1$ V $\cdot$ s/A	milli (m)	$= \times 10^{-3}$
1 hertz (Hz)	$= 1$ cycle/s	nano (n)	$= \times 10^{-9}$
1 inch (in)	$= 2.54$ cm	1 newton (N)	$= 1$ kg $\cdot$ m/s^2
1 joule (J)	$= 10^{7}$ ergs	pico (p)	$= \times 10^{-12}$
	$= 1$ W $\cdot$ s	1 pound (lb)	$= 453.6$ g
	$= 6.25 \times 10^{18}$ eV	1 tesla (T)	$= 1$ Wb/m^2
	$= 1$ N $\cdot$ m	1 volt (V)	$= 1$ W/A
	$= 1$ C $\cdot$ V	1 watt (W)	$= 1$ J/s
kilo (k)	$= \times 10^{3}$	1 weber (Wb)	$= 1$ V $\cdot$ s
1 kilogram (kg)	$= 2.205$ lb	1 weber per square	
1 kilometer (km)	$= 0.622$ mi	meter (Wb/m^2)	$= 10^{4}$ G (Gauss)
1 lumen	$= 0.0016$ W		$= 1$ T
	(at 0.55 μm)		

Appendix B
SEMICONDUCTOR MANUFACTURERS AND DEVICE SPECIFICATIONS

B-1 ELECTRONIC DEVICE MANUFACTURERS Databooks and applications information may be obtained from the following semiconductor companies:

Advanced Micro Devices   901 Thompson Pl., Sunnyvale, CA 94086
American Microsystems Inc.   3800 Homestead Road, Santa Clara, CA 95051
Analog Devices   2 Technology Way, Norwood, MA 02062
Burr-Brown Research Corp.   6730 S. Tucson Blvd., Tucson, AZ 85734
Fairchild Semiconductor   464 Ellis St., Mountain View, CA 94042
Ferranti Electric   E. Bethpage Rd., Plainview, NY 11803
General Electric Co.   Schenectady, NY 13201
General Instrument Corp.   600 West John St., Hicksville, NY 11802
Harris Semiconductor   Box 833, Melbourne, FL 32901
Hitachi America, Ltd.   111 E. Wacker Dr., Chicago, IL 60601
Imsai   14860 Wicks Blvd., San Leandro, CA 94577
Intel Corp.   3065 Bowers Ave., Santa Clara, CA 95051
Intersil Inc.   10900 N. Tantau Ave., Cupertino, CA 95014
ITT Semiconductors   74 Commerce Way, Woburn, MA 01801
Monolithic Memories, Inc.   1165 E. Argues Ave., Sunnyvale, CA 94086
Mostek Corp.   1215 W. Crosby Rd., Carollton, TX 75006
Motorola Semiconductor Products   Box 20912, Phoenix, AZ 85036
National Semiconductor, Inc.   2900 Semiconductor Dr., Santa Clara, CA 95051
Plessey Semiconductors   1674 McGraw Ave., Santa Ana, CA 92705
Raytheon Semiconductor   350 Ellis St., Mountain View, CA 94042
RCA Solid State Division   Box 3200, Somerville, NJ 08876
SGS/ATES Semiconductor Corp.   796 Massasoit Street, Waltham, MA 03254
Signetics Corp.   811 E. Argues Ave., Sunnyvale, CA 94086
Silicon General   73826 Bolsoo Ave., Westminster, CA 92683
Siliconix, Inc.   2201 Laurelwood Road, Santa Clara, CA 95054
Stewart-Warner Microcircuits   730 E. Evelyn Ave., Sunnyvale, CA 94086

Teledyne Semiconductor   1300 Terra Bella Ave., Mountain View, CA 94043
Texas Instruments Semiconductor Group   Box 5012, Dallas, TX 75222
Toshiba America   280 Park Ave., New York, NY 10017
TRW Microelectronics Center   One Space Park, Redondo Beach, CA 90278
Unitrode Corporation   580 Pleasant St., Watertown, MA 02172

B-2 SPECIFICATIONS FOR 1N4153 SILICON DIODE (Courtesy of Texas Instruments, Inc.)

High-speed switching diodes for computer and general-purpose applications.

TABLE B2-1   Absolute Maximum Ratings (25 °C)

		1N4151	1N4152	1N4153	1N4154	Unit
V_{RM}	Peak reverse voltage	75	40	75		V
$V_{RM(wkg)}$	Working peak reverse voltage	50	30	50	25	V
P	Continuous power dissipation at (or below) 25°C free-air temperature*			500		mW
T_{stg}	Storage temperature range			−65 to 200		°C
T_L	Lead temperature $\frac{1}{16}$ in from case for 10 s			300		°C

*Derate linearly to 200°C at the rate of 2.85 mW/°C.

TABLE B2-2   Electrical Characteristics (25°C unless Otherwise Noted)

Parameter		Test conditions	1N4153 Min	1N4153 Max	Unit
V_{br}	Reverse breakdown voltage	$I_R = 5\ \mu A$	75		V
I_R	Static reverse current	$V_R = $ rated $V_{RM(wkg)}$		0.05	μA
		$V_R = $ rated $V_{RM(wkg)}$ $T_A = 150°C$		50	μA
V_F	Static forward voltage	$I_F = 0.1$ mA	0.49	0.55	V
		$I_F = 0.25$ mA	0.53	0.59	V
		$I_F = 1$ mA	0.59	0.67	V
		$I_F = 2$ mA	0.62	0.70	V
		$I_F = 10$ mA	0.70	0.81	V
		$I_F = 20$ mA	0.74	0.88	V
C_T	Total capacitance	$V_R = 0$ $f = 1$ MHz		2	pF
t_{rr}	Reverse recovery time	$I_F = 10$ mA, $I_{RM} = 10$ mA $R_L = 100\ \Omega$		4	ns
		$I_F = 10$ mA, $V_R = 6$ V $R_L = 100\ \Omega$		2	ns

B-3 SPECIFICATIONS FOR 2N2222A NPN SILICON BIPOLAR JUNCTION TRANSISTOR (Courtesy of Motorola, Inc.)

Widely used "Industry Standard" transistor for applications as medium-speed switches and as amplifiers from audio to VHF frequencies. Complements to *pnp* transistor 2N2907A.

TABLE B3-1 Absolute Maximum Ratings*

Characteristic	Symbol	Rating	Unit
Collector-emitter voltage	V_{CEO}	40	V
Collector-base voltage	V_{CB}	75	V
Emitter-base voltage	V_{EB}	6.0	V
Collector current—continuous	I_C	800	mA
Total device dissipation at $T_A = 25°C$ Derate above 25°C	P_D	0.5 3.33	W mW/°C
Total device dissipation at $T_C = 25°C$ Derate above 25°C	P_D	1.8 12	W mW/°C
Operating and storage junction Temperature range	T_J, T_{stg}	65 to +200	°C

*T_A = ambient, T_C = case, and T_J = junction temperature.

TABLE B3-2 Electrical Characteristics (T_A = 25°C unless Otherwise Noted)

OFF characteristics	Symbol	Min	Max	Unit
Collector-emitter breakdown voltage ($I_C = 10$ mA, $I_B = 0$)	BV_{CEO}	40		V
Collector-base breakdown voltage ($I_C = 10$ μA, $I_E = 0$)	BV_{CBO}	75		V
Emitter-base breakdown voltage ($I_E = 10$ μA, $I_C = 0$)	BV_{EBO}	60		V
Collector cutoff current ($V_{CE} = 60$ V, $V_{EB(OFF)} = 3.0$ V)	I_{CEX}		10	nA
Collector cutoff current ($V_{CB} = 60$ V, $I_E = 0$, $T_A = 150°C$)	I_{CBO}		10	μA
Emitter cutoff current ($V_{EB} = 3.0$ V, $I_C = 0$)	I_{EBO}		10	nA
Base cutoff current ($V_{CE} = 60$ V, $V_{EB(OFF)} = 3.0$ V)	I_{BL}		20	nA

ON characteristics

See Fig. 3-13 for $V_{CE(sat)}$ and $V_{BE(sat)}$. See Fig. 3-14 for $h_{FE} \approx \beta_F$.

TABLE B3-3 Electrical Characteristics (continued)

Small-signal characteristics	Symbol	Min	Max	Unit
Current-gain–bandwidth product ($I_C = 20$ mA, $V_{CE} = 20$ V, $f = 100$ MHz)	f_T	300		MHz
Output capacitance ($V_{CB} = 10$ V, $I_E = 0$, $f = 100$ kHz)	C_{ob}		8.0	pF
Input capacitance ($V_{EB} = 0.5$ V, $I_C = 0$, $f = 100$ kHz)	C_{ib}		25	pF
Input impedance ($I_C = 1.0$ mA, $V_{CE} = 10$ V, $f = 1.0$ kHz) ($I_C = 10$ mA, $V_{CE} = 10$ V, $f = 1.0$ kHz)	h_{ie}	2.0 0.25	8.0 1.25	kΩ
Voltage feedback ratio ($I_C = 1.0$ mA, $V_{CE} = 10$ V, $f = 1.0$ kHz) ($I_C = 10$ mA, $V_{CE} = 10$ V, $f = 1.0$ kHz)	h_{re}		8.0 4.0	$\times 10^{-4}$
Small-signal current gain ($I_C = 1.0$ mA, $V_{CE} = 10$ V, $f = 1.0$ kHz) ($I_C = 10$ mA, $V_{CE} = 10$ V, $f = 1.0$ kHz)	h_{fe}	50 75	300 375	
Output admittance ($I_C = 1.0$ mA, $V_{CE} = 10$ V, $f = 1.0$ kHz) ($I_C = 10$ mA, $V_{CE} = 10$ V, $f = 1.0$ kHz)	h_{oe}	5.0 25	35 200	µmhos
Collector-base time constant ($I_E = 20$ mA, $V_{CB} = 20$ V, $f = 31.8$ MHz)	$r_b'C_c$		150	ps
Noise figure ($I_C = 100$ µA, $V_{CE} = 10$ V, $R_S = 1.0$ kΩ, $f = 1.0$ kHz)	NF		4.0	dB

Switching characteristics	Symbol	Min	Max	Unit
Delay time ($V_{CC} = 30$ V, $V_{BE(OFF)} = 0.5$ V $I_C = 150$ mA,	t_d		10	ns
Rise time $I_{B1} = 15$ mA)	t_r		25	ns
Storage time ($V_{CC} = 30$ V, $I_C = 150$ mA, $I_{B1} = 15$ mA	t_x		225	ns
Fall time $I_{B2} = 15$ mA)	t_f		60	ns
Active-region time constant ($I_C = 150$ mA, $V_{CE} = 30$ V)	t_A		2.5	ns

B-4 SPECIFICATIONS FOR 2N4869 DEPLETION-MODE *N*-CHANNEL SILICON JUNCTION FIELD-EFFECT TRANSISTOR (Courtesy of Siliconix, Inc.)

Specifically designed for audio or subaudio frequency applications where noise must be at an absolute minimum.

TABLE B4-1   Absolute Maximum Ratings (25°C)

Gate-drain or gate-source voltage*	-40 V
Gate current or drain current	50 mA
Total device dissipation (derate 1.7 mW/°C)	300 mW
Storage temperature range	-65 to $+200$°C

*As a result of symmetrical geometry, these units may be operated with source and drain leads interchanged.

TABLE B4-2   Electrical Characteristics (25°C unless Otherwise Noted)

		Characteristic	Min	Max	Unit	Test conditions	
S T A T I C	I_{GSS}	Gate reverse current		-0.25	nA	$V_{GS} = -30$ V, $V_{DS} = 0$	
				-0.25	μA		150°C
	BV_{GSS}	Gate-source breakdown voltage	-40		V	$I_G = -1$ μA, $V_{DS} = 0$	
	$V_{GS(OFF)}$	Gate-source cutoff voltage	-1.8	-5	V	$V_{DS} = 20$ V, $I_D = 1$ μA	
	I_{DSS}	Saturation drain current*	2.5	7.5	mA	$V_{DS} = 20$ V, $V_{GS} = 0$	
D Y N A M I C	g_{fs}	Common-source forward transconductance*	1300	4000	μ℧		$f = 1$ kHz
	g_{os}	Common-source output conductance		10	μ℧	$V_{DS} = 20$ V, $V_{GS} = 0$	$f = 1$ kHz
	C_{rss}	Common-source reverse transfer capacitance		5	pF		$f = 1$ MHz
	C_{iss}	Common-source input capacitance		25	pF		$f = 1$ MHz

*Pulse test duration = 2 ms.

B-5 SPECIFICATIONS FOR 3N163 P-CHANNEL ENHANCEMENT-TYPE SILICON MOS FIELD-EFFECT TRANSISTOR (Courtesy of Siliconix, Inc.) Normally OFF MOSFET for analog and digital switching general-purpose amplifiers.

TABLE B5-1 Absolute Maximum Ratings (25°C)

Drain-source or gate-source voltage	-40 V
Transient gate-source voltage	± 150 V
Drain current	-50 mA
Storage temperature	-65 to $+200$°C
Operating junction temperature	-55 to $+150$°C
Total device dissipation (derate 3.0 mW/°C to 150°C)	375 mW
Lead temperature $\frac{1}{16}$ in from case for 10 s	265°C

TABLE B5-2 Electrical Characteristics (25°C and $V_{BS} = 0$ unless Otherwise Noted)

		Characteristic	Min	Max	Unit	Test conditions
S T A T I C	I_{GSS}	Gate-body leakage current		-10 -25	pA pA	$V_{GS} = -40$ V, $V_{DS} = 0$, 125°C
	BV_{DSS}	Drain-source breakdown voltage	-40		V	$I_D = -10$ μA, $V_{GS} = 0$
	BV_{SDS}	Source-drain breakdown voltage	-40		V	$I_S = -10$ μA, $V_{GD} = V_{BD} = 0$
	V_{GS}	Gate-source voltage	-3	-6.5	V	$V_{DS} = -15$ V, $I_D = -0.5$ mA
	$V_{GS(th)}$	Gate-source threshold voltage	-2	-5	V	$V_{DS} = V_{GS}$, $I_D = -10$ μA
	I_{DSS}	Drain cutoff current		-200	pA	$V_{DS} = -15$ V, $V_{GS} = 0$
	I_{SDS}	Source cutoff current		-400	pA	$V_{SD} = -20$ V, $V_{GD} = 0$, $V_{DB} = 0$
	$I_{D(ON)}$	ON drain current	-5	-30	mA	$V_{DS} = -15$ V, $V_{GS} = -10$ V
	$r_{DS(ON)}$	Drain-source ON resistance		250	Ω	$V_{GS} = -20$ V, $I_D = -100$ μA
D Y N A M I C	g_{fs}	Common-source forward transconductance	2000	4000	μ℧	$V_{DS} = -15$ V, $I_D = -10$ mA, $f = 1$ kHz
	g_{oss}	Common-source output conductance		250	μ℧	
	C_{iss}	Common-source input capacitance		2.5	pF	
	C_{rss}	Common-source reverse transfer capacitance		0.7	pF	$V_{DS} = -15$ V, $I_D = -10$ mA, $f = 1$ MHz
	C_{oss}	Common-source output capacitance		3	pF	
S W	t_d	Turn-on delay time		12	ns	$V_{DD} = -15$ V
	t_r	Rise time		24	ns	$I_{D(ON)} = -10$ mA
	t_{OFF}	Turn-off time		50	ns	$R_G = R_L = 1.5$ kΩ

B-6 SPECIFICATIONS FOR SCHOTTKY LOW-POWER TTL POSITIVE NAND GATE (LS5410 OR LS7410) WITH TOTEM-POLE OUTPUT (Courtesy of Texas Instruments, Inc.)

The other TTL families have very similar characteristics for NAND gates or inverters.

TABLE B6-1 Recommended Operating Conditions

Parameter	Family	Min	Nom	Max	Unit
Supply voltage, V_{CC}	54	4.5	5	5.5	V
	74	4.75	5	5.25	
High-level output current, I_{OH}	54			−400	μA
	74			−400	
Low-level output current, I_{OL}	54			4	mA
	74			8	
Operating free-air temperature, T_A	54	−55		125	°C
	74	0		70	

TABLE B6-2 Electrical Characteristics over Recommended Operating Free-Air Temperature Range (unless Otherwise Noted)

Parameter		Test conditions*	Family	Min	Typ†	Max	Unit
V_{IH}	High-level input voltage			2			V
V_{IL}	Low-level input voltage		54			0.7	V
			74			0.8	
V_I	Input clamp voltage	V_{CC} = min, I_I = −18 mA				−1.5	V
V_{OH}	High-level output voltage	V_{CC} = min, V_{IL} = V_{IL} max I_{OH} = max	54	2.5	3.4		V
			74	2.7	3.4		
V_{OL}	Low-level output voltage	V_{CC} = min, V_{IH} = 2 V I_{OL} = max	54		0.25	0.4	V
			74		0.35	0.5	
I_I	Input current at maximum input voltage	V_{CC} = max, V_I = 5.5 V				0.1	mA
I_{IH}	High-level input current	V_{CC} = max, V_{HH} = 2.7 V			20		μA
I_{IL}	Low-level input current	V_{CC} = max, V_{IL} = 0.4 V				−0.36	mA
I_{OS}	Short-circuit output current‡	V_{CC} = max	54	−6		−40	mA
			75	−5		−42	

*For conditions shown as min or max, use the appropriate value specified under recommended operating conditions.

†All typical values are at V_{CC} = 5 V, T_A = 25°C.

‡Not more than one output should be short-circuited at a time.

B-7 SPECIFICATIONS FOR LM741 OPERATIONAL AMPLIFIER (Courtesy of National Semiconductor, Inc.)

A high-performance monolithic Op-Amp intended for a wide range of analog applications. It is short-circuit protected and requires no external components for frequency compensation. The LM741C is identical to the LM741, except that the performance of the former is guaranteed over a 0 to 70°C temperature range, instead of −55 to 125°C.

TABLE B7-1   Absolute Maximum Ratings

Supply voltage LM741	±22 V
LM741C	±18 V
Power dissipation*	500 mW
Differential input voltage	±30 V
Input voltage†	±15 V
Output short-circuit duration	Indefinite
Storage temperature range	−65 to 150°C
Lead temperature (soldering, 10 s)	300°C

*The maximum junction temperature of the LM741 is 150°C, while that of the LM741C is 100°C. For operating at elevated temperatures, devices in the TO-5 package must be derated based on a thermal resistance of 150°C/W, junction to case.

†For supply voltages less than ±15 V, the absolute maximum input voltage is equal to the supply voltage.

TABLE B7-2   Electrical Characteristics*

Parameter	Conditions	LM741 Min	LM741 Typ	LM741 Max	LM741C Min	LM741C Typ	LM741C Max	Unit
Input offset voltage	$T_A = 25°C$, $R_s \leqslant 10$ kΩ		1.0	5.0		1.0	6.0	mV
Input offset current	$T_A = 25°C$		30	200		30	200	nA
Input bias current	$T_A = 25°C$		200	500		200	500	nA
Input resistance	$T_A = 25°C$	0.3	1.0		0.3	1.0		MΩ
Supply current	$T_A = 25°C$, $V_S = \pm15$ V		1.7	2.8		1.7	2.8	mA
Large-signal voltage gain	$T_A = 25°C$, $V_S = \pm15$ V $V_{OUT} = \pm10$ V, $R_L \geqslant 2$ kΩ	50	160		25	160		V/mV
Input offset voltage	$R_s < 10$ kΩ			6.0			7.5	mV
Input offset current				500			300	nA
Input bias current				1.5			0.8	μA
Large-signal voltage gain	$V_S = \pm15$ V, $V_{OUT} = \pm10$ V $R_L \geqslant 2$ kΩ	25			15			V/mV
Output voltage swing	$V_S = \pm15$ V, $R_L = 10$ kΩ $R_L = 2$ kΩ	±12 ±10	±14 ±13		±12 ±10	±14 ±13		V V
Input voltage range	$V_S = \pm15$ V	±12			±12			V
Common-mode rejection ratio	$R_s \leqslant 10$ kΩ	60	90		70	90		dB
Supply voltage rejection ratio	$R_s \leqslant 10$ kΩ	77	96		77	96		dB

*These specifications apply for $V_S = \pm15$ V and $-55°C \leqslant T_A \leqslant 125°C$, unless otherwise specified. With the LM741C, however, all specifications are limited to $0°C \leqslant T_A \leqslant 70°C$ and $V_S = \pm15$ V.

B-8 SPECIFICATIONS FOR 2N5671 *npn* SILICON POWER TRANSISTOR (Courtesy of RCA Solid State Division.)

This transistor has high-current- and high-power-handling capability and fast-switching speed. It is especially suitable for switching-control amplifiers, power gates, switching regulators, power-switching circuits, converters, inverters, control circuits, dc-rf amplifiers, and power oscillators.

TABLE B8-1   Absolute Maximum Ratings

Characteristic	Symbol	Rating	Unit
Collector-base voltage	V_{CBO}	120	V
Collector-emitter sustaining voltage:			
With base open	$V_{CEO(sus)}$	90	V
With external base-emitter resistance (R_{BE}) $\leq 50\ \Omega$	$V_{CER(sus)}$	110	V
With external base-emitter resistance $< 50\ \Omega$ and $V_{BE} = -1.5$ V	$V_{CEX(sus)}$	120	V
Emitter-base voltage	V_{EBO}	7	V
Collector current	I_C	30	A
Base current	I_B	10	A
Transistor dissipation at case temperatures up to 25°C and V_{CE} up to 24 V	P_D	140	W
Temperature range	—	≤ 200	°C

TABLE B8-2   Electrical Characteristics, Case Temperature (T_C) = 25°C

Characteristic	Symbol	Test conditions DC collector voltage (V) V_{CB}	V_{CE}	DC emitter or base voltage (V) V_{EB}	V_{BE}	DC current (A) I_C	I_E	I_B	2N5671 Min	Max	Unit
	I_{CEO}		80				0			10	mA
Collector cutoff current	I_{CEV}		110			−1.5				12	mA
Emitter cutoff current	I_{EBO}			7		0				10	mA
Collector-emitter sustaining voltage: With base open	$V_{CEO(sus)}$					0.2	0		90*		V
With external base-emitter resistance (R_{BE}) ⩽ 50 Ω	$V_{CER(sus)}$					0.2	0		110*		V
With base-emitter junction reverse biased and R_{BE} ⩽ 50 Ω	$V_{CEX(sus)}$				−1.5	0.2			120*		V
Base-emitter saturation voltage	$V_{BE(sat)}$					15	1.2			1.5	V
Base-emitter voltage	V_{BE}		5			15				1.6	V
Collector-emitter saturation voltage	$V_{CE(sat)}$					15	1.2			0.75	V
Dc forward-current transfer ratio	h_{FE}		2			15			20	100	
			5			20			20		
Gain-bandwidth product	f_T		10			2			50		MHz
Output capacitance (at 1 MHz)	C_{ob}	10					0			900	pF
Saturated switching turn-on time (delay time + rise time)	t_{ON}	V_{CC} = 30 V				15		$I_{B1} = I_{B2}$ = 1.2		0.5	μs
Saturated switching storage time	t_s	V_{CC} = 30 V				15		$I_{B1} = I_{B2}$ = 1.2		1.5	μs
Saturated switching fall time	t_f	V_{CC} = 30 V				15		$I_{B1} = I_{B2}$ = 1.2		0.5	μs
Thermal resistance (junction to case)	θ_{J-C}		40			0.5				1.25	°C/W

Caution: The sustaining voltages $V_{CEO(sus)}$, $V_{CER(sus)}$, and $V_{CEX(sus)}$ must not be measured on a curve tracer.

B9 SPECIFICATIONS FOR CA3045, CA3046 GENERAL-PURPOSE TRANSISTOR ARRAYS (Courtesy of RCA)

Five general-purpose transistors on a common substrate; two transistors are internally connected to form a differential pair. Intended for a variety of applications from dc to VHF, the transistors may be used as discrete devices. In addition, they provide significant integrated-circuit advantages of close electrical and thermal matching. The CA3046 is electrically identical to the CA3045 but is supplied in a dual-in-line plastic package for applications over a limited temperature range.

TABLE B9-1 Absolute Maximum Ratings at 25°C

	CA3045		CA3046		
	Each transistor	Total package	Each transistor	Total package	Unit
Power dissipation:					
T_A up to 55°C			300	750	mW
$T_A > 55$°C			Derate at 6.67		mW/°C
T_A up to 75°C	300	750			mW
$T_A > 75$°C	Derate at 8				mW/°C
Collector-to-emitter voltage, V_{CEO}	15		15		V
Collector-to-base voltage, V_{CBO}	20		20		V
Collector-to-substrate voltage, V_{CIO}*	20		20		V
Emitter-to-base voltage, V_{EBO}	5		5		V
Temperature range:					
Operating	-55 to $+125$		-55 to $+125$		°C
Storage	-65 to $+150$		-65 to $+150$		°C
Lead temperature (during soldering): At distance $\frac{1}{16} \pm \frac{1}{32}$ in (1.59 ± 0.79 mm) from case for 10 s max	$+265$		$+265$		°C

*The collector of each transistor of the CA3045 and CA3046 is isolated from the substrate by an integral diode. *The substrate (terminal 13) must be connected to the most negative point in the external circuit to maintain isolation between transistors and to provide for normal transistor action.*

TABLE B9-2 DC Characteristics at 25°C

Characteristics	Symbols	Special test conditions	Limits: Type CA3045, Type CA3046			Unit
			Min	Typical	Max	
Collector-to-base breakdown voltage	$V_{(BR)CBO}$	$I_C = 10\ \mu A,\ I_E = 0$	20	60		V
Collector-to-emitter breakdown voltage	$V_{(BR)CEO}$	$I_C = 1\ mA,\ I_B = 0$	15	24		V
Collector-to-substrate breakdown voltage	$V_{(BR)CIO}$	$I_C = 10\ \mu A,\ I_{CI} = 0$	20	60		V
Emitter-to-base breakdown voltage	$V_{(BR)EBO}$	$I_E = 10\ \mu A,\ I_C = 0$	5	7		V
Collector-cutoff current	I_{CBO}	$V_{CB} = 10\ V,\ I_E = 0$		0.002	40	nA
Collector-cutoff current	I_{CEO}	$V_{CE} = 10\ V,\ I_B = 0$			0.5	μA
Static forward current-transfer ratio (static-beta)	h_{FE}	$V_{CE} = 3\ V \begin{cases} I_C = 10\ mA \\ I_C = 1\ mA \\ I_C = 10\ \mu A \end{cases}$	40	100 100 54		
Input offset current for matched pair Q_1 and Q_2.		$V_{CE} = 3\ V,\ I_C = 1\ mA$		0.3	2	μA
Base-to-emitter voltage	V_{BE}	$V_{CE} = 3\ V \begin{cases} I_E = 1\ mA \\ I_E = 10\ mA \end{cases}$		0.715 0.800		V V
Magnitude of input offset voltage for differential pair $\lvert V_{BE_1} - V_{BE_2} \rvert$		$V_{CE} = 3\ V,\ I_C = 1\ mA$		0.45	5	mV
Magnitude of input offset voltage for isolated transistors $\lvert V_{BE3} - V_{BE4} \rvert,\ \lvert V_{BE4} - V_{BE5} \rvert,$ $\lvert V_{BE5} - V_{BE3} \rvert$		$V_{CE} = 3\ V,\ I_C = 1\ mA$		0.45	5	mV
Temperature coefficient of base-to-emitter voltage	$\dfrac{\Delta V_{BE}}{\Delta T}$	$V_{CE} = 3\ V,\ I_C = 1\ mA$		-1.9		mV/°C
Collector-to-emitter saturation voltage	V_{CES}	$I_B = 1\ mA,\ I_C = 10\ mA$		0.23		V
Temperature coefficient: Magnitude of input-offset voltage	$\dfrac{\lvert \Delta V_{IO} \rvert}{\Delta T}$	$V_{CE} = 3\ V,\ I_C = 1\ mA$		1.1		μV/°C

TABLE B9-3 Small-Signal Characteristics at 25°C

Characteristics	Symbols	Special test conditions	Limits: Type CA3045, Type CA3046			Unit
			Min	Typical	Max	
Low-frequency noise figure	NF	$f = 1\ kHz,\ V_{CE} = 3\ V,$ $I_C = 100\ \mu A$ Source resistance $= 1\ k\Omega$		3.25		dB
Low-frequency small-signal equivalent circuit characteristics:						
Forward current-transistor ratio	h_{fe}			110		
Short-circuit input impedance	h_{ie}			3.5		$k\Omega$
Open-circuit output impedence	h_{oe}	$f = 1\ kHz,\ V_{CE} = 3\ V,$ $I_C = 1\ mA$		15.6		$\mu\mho$
Open-circuit reverse voltage-transfer ratio	h_{re}			1.8×10^{-4}		
Admittance characteristics:						
Forward transfer admittance	$\mathbf{Y}_{fe}$			$31 + j1.5$		
Input admittance	$\mathbf{Y}_{ie}$	$f = 1\ MHz,\ V_{CE} = 3\ V,$ $I_C = 1\ mA$		$0.3 + j0.04$		
Output admittance	$\mathbf{Y}_{oe}$			$0.0001 + j0.03$		
Gain-bandwidth product	f_T	$V_{CE} = 3\ V,\ I_C = 3\ mA$	300	550		MHz
Emitter-to-base capacitance	C_{EB}	$V_{EB} = 3\ V,\ I_E = 0$		0.6		pF
Collector-to-base capacitance	C_{CB}	$V_{CB} = 3\ V,\ I_C = 0$		0.58		pF
Collector-to-substrate capacitance	C_{CI}	$V_{CS} = 3\ V,\ I_C = 0$		2.8		pF

Appendix C
SUMMARY OF
NETWORK THEORY

In this book we use linear passive elements such as resistors, capacitors, and inductors in combination with voltage and/or current sources and solid-state devices to form networks. The theorems discussed in this appendix are used frequently in the analysis of these electronic circuits.

C-1 RESISTIVE NETWORKS

Voltage and Current Sources   In this section we review some basic concepts and theorems in connection with resistive networks containing voltage and current sources. The circuit symbols and reference directions of independent voltage and current sources are shown in Fig. C-1. An ideal voltage source is defined as a voltage generator whose output voltage $v = v_s$ is independent of the current delivered by the generator. The output voltage is usually specified as a function of time such as, for example, $v_s = V_m \cos \omega t$ or a constant dc voltage. Similarly, an ideal current source delivers an arbitrary current $i = i_s$ independent of the voltage between the two terminals of the current source. The reference polarity for the voltage source v_s means that 1 coulomb (C) of positive charge moving from the negative to the positive terminal through the voltage source acquires v_s joules of energy. In the same way the arrow reference for the current source i_s indicates that i_s coulombs of positive charge per second pass through the source in the direction of the arrow. In a practical voltage or current source there is always some energy converted to heat in an irreversible energy-conversion process. This energy loss can be represented by the loss in a series or parallel source resistance R_s, as shown in Figs. C-1c and C-1d.

FIGURE C-1
(a, b) Ideal and (c, d) practical voltage and current sources. A circle with a + and − sign is the symbol for the ideal voltage generator. An arrow inside a circle is the symbol for an ideal current generator. The source resistance is designated by R_s, drawn either in series with a voltage source v_s or in parallel with a current source i_s.

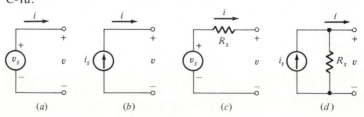

(a)   (b)   (c)   (d)

FIGURE C-2
(*a*) A bipolar transistor model contains a voltage-controlled current source $g_m v_\pi$. (*b*) The Op-Amp equivalent circuit contains the voltage-controlled voltage source $A_v v_i$.

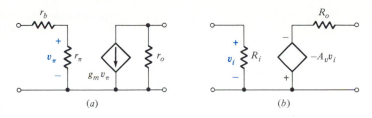

(a) $\qquad\qquad\qquad$ (b)

A *controlled* or *dependent* source is one whose voltage or current is a function of the voltage or current elsewhere in the circuit. For example, Fig. C-2*a* represents a small-signal circuit model of a transistor at low frequencies. At the output there is a dependent current generator $g_m v_\pi$ whose *current* is proportional to the *voltage* v_π and the proportionality factor is g_m.

Another active device studied in this book is the Op-Amp, and its equivalent low-frequency small-signal model is indicated in Fig. C-2*b*. Note that at the output there is a dependent voltage source $A_v v_i$ controlled by the input *voltage* and the proportionality factor is A_v.

Resistance

Ohm's law states that the voltage V across a conductor is proportional to the current I in this circuit element. The proportionality factor V/I is called the *resistance* and is expressed in ohms (abbreviated Ω) if V is in volts and I in amperes.

$$V = IR \qquad\qquad\qquad (C\text{-}1)$$

In most electronic circuits it is convenient to express the resistance values in kilohms (abbreviated kΩ). Then Eq. (C-1) continues to be valid if I is expressed in milliamperes (mA) and V in volts (V). If a conductor does not obey Eq. (C-1), it is said to be a nonlinear (or nonohmic) resistor.

To find the resistance R seen between two points in a network, an external voltage source V is considered to be applied between these two points and the current I drawn from the source V is determined. The effective resistance is $R = V/I$, provided that in the above procedure each *independent* source in the network is replaced by its internal source resistance R_s; an ideal voltage source by a short circuit, and an ideal current source by an open circuit (Fig. C-1). *All dependent sources, however, must be retained in the network.*

The two basic laws which allow us to analyze electric networks (linear or nonlinear) are known as *Kirchhoff's current law* (KCL) and *Kirchhoff's voltage law* (KVL).

Kirchhoff's Current Law (KCL)   *The sum of all currents leaving a node must be zero at any instant of time.* A *node* is a point where two or more circuit components meet such as points 1 and 2 in Fig. C-3*a*. When we apply this law, currents directed away from a node are usually taken as positive and those directed toward are taken as negative. The opposite convention can also be used as long as we are consistent for all nodes of the network. The positive reference direction of the current through any resistor of the network can be assigned arbitrarily with the understanding that, if the computed current is determined to be negative, the actual current direction is opposite to that assumed. The physical principle on which KCL is based is the law of the conservation of charge, since a violation of KCL would require that some electric charge be "lost" or be "created" at the node.

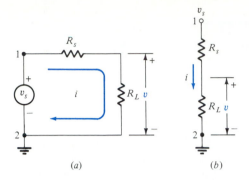

FIGURE C-3
(*a*) A load R_L is placed across a voltage source whose internal resistance is R_s. (*b*) The *same* circuit is redrawn in a different way. The tiny circle at node 1 is used to indicate that a power supply v_s exists between this node and node 2 which has been designated as the reference node. Since one terminal of a generator is usually connected to the metal chassis on which it is built, this terminal is called *ground*. The standard symbol for a ground is shown at node 2.

Kirchhoff's Voltage Law (KVL)   *The sum of all voltage drops around a loop must be zero at all times.* A closed path in a circuit is called a *loop* or a *mesh*. A voltage *drop* V_{12} between two nodes 1 and 2 in a circuit (the potential of point 1 with respect to point 2) is defined as the energy in joules (J) removed from the circuit when a positive charge q of 1 C moves from point 1 to point 2. For example, a voltage drop of $+5$ V across the terminal nodes 1 and 2 of a resistor means that 5 J of energy are removed from the circuit and dissipated as heat in the resistor when a positive charge of 1 C moves *from point 1 to point 2*. If the voltage is -5 V, then point 2 is at a higher voltage than point 1 ($V_{12} = -5$ V represents a voltage *rise*), and a positive charge of 1 C moving from point 1 to point 2 gains 5 J of energy. This, of course, is impossible when a resistor is connected between the two nodes 1 and 2, and is possible only if the negative terminal of a battery is connected to node 1 and the positive terminal to node 2.

It should be clear that KVL is a consequence of the law of conservation of energy. In writing the KVL equations, we go completely around a loop, add all voltage drops, and set the sum equal to zero. Remember these two rules:

1. There is a positive drop in the direction of the current in a resistor.

2. There is a positive drop through a battery (or dc source) in the direction from the $+$ to the $-$ terminal, independent of the direction of the current.

The two fundamental laws (KCL and KVL) are illustrated in the following examples. Consider first the situation where a resistor R_L is placed directly across the output terminals of a real (nonidealized) voltage source (Fig. C-1*c*). This added component is called the *load resistor* or, simply, the *load*. The result is that a single mesh is formed as indicated in Fig. C-3. We wish to find the voltage v across R_L.

The current i around the loop flows through R_s and R_L. Traversing this mesh in the assumed direction of current starting at node 2, adding all voltage drops, and setting the sum to zero (as required by KVL) yields

$$-v_s + iR_s + iR_L = 0$$

or $\qquad\qquad i = \dfrac{v_s}{R_s + R_L} \qquad$ and $\qquad v = iR_L = \dfrac{R_L v_s}{R_s + R_L} \qquad\qquad$ (C-2)

Note that under *open-circuit* conditions (defined as $R_L \to \infty$), $v = v_s$. This result is obviously correct since no current can flow in an open circuit so that $i = 0$, $iR_s = 0$, and $v = v_s = $ *open-circuit voltage*. Also note that under *short-circuit* conditions (defined as $R_L = 0$; an ideal zero-resistance wire), the output voltage drops to zero, $v = 0$. Now the current is a maximum (with respect to variations in the value of R_L) and $i = v_s/R_s = $ *short-circuit current*. The voltage v_s may be a function of time, and then v will also be a function of time.

An equivalent alternative way to draw the circuit of Fig. C-3a is shown in Fig. C-3b. The caption explains the meaning of the symbols at nodes 1 and 2. This configuration is referred to as a *voltage divider*. Note that v is less than v_s (for any finite R_L).

$$\frac{v}{v_s} = \frac{R_L}{R_s + R_L} \tag{C-3}$$

Example C-1

(a) Find the currents I_1, I_2, and I_3 in the circuit shown in Fig. C-4. (b) Find the voltage drop V_{24}.

Solution

(a) We assign the arbitrary reference directions of positive currents as shown in the figure. We must sum the voltage drops in each loop by going around each mesh in the arbitrary direction shown by the loop arrows. Note that the current in R_1 is the mesh current I_1 and that in R_2 is the mesh current I_2. However, the current in R_3 is the sum of I_1 and I_2. Applying KVL, we obtain the following equations:

Loop 1 $\qquad\qquad\qquad V_{12} + V_{24} + V_{41} = 0 \qquad\qquad\qquad$ (C-4)

Loop 2 $\qquad\qquad\qquad V_{32} + V_{24} + V_{43} = 0 \qquad\qquad\qquad$ (C-5)

where the individual voltage drops are given below:[1]

$$V_{12} = I_1R_1 = I_1 \qquad V_{24} = -I_3R_3 = -2I_3 \qquad V_{41} = -6$$

$$V_{32} = I_2R_2 = 9I_2 \qquad V_{43} = 14$$

Substituting these values into Eqs. (C-4) and (C-5) gives

$$I_1 - 2I_3 - 6 = 0$$

$$9I_2 - 2I_3 + 14 = 0$$

Since we have only two equations for our three unknowns, we must use the KCL equation at node 2 to obtain the additional equation

---

[1] We express R in kilohms and I in milliamperes ($1\ \text{k}\Omega \cdot 1\ \text{mA} = 1\ \text{V}$).

FIGURE C-4

(a) Two-loop resistive network. (b) The same network with the voltages (with respect to ground) at nodes 1 and 3 indicated, but with the battery symbols omitted.

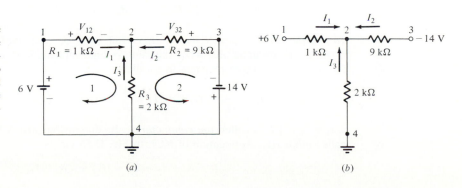

$$I_1 + I_2 + I_3 = 0 \quad \text{or} \quad I_3 = -(I_1 + I_2)$$

Substituting this value of I_3 into the equations for I_1 and I_2 gives

$$3I_1 + 2I_2 = 6$$

$$2I_1 + 11I_2 = -14$$

Solving these simultaneous algebraic equations, we find

$$I_1 = 3.242 \quad I_2 = -1.862 \quad \text{and} \quad I_3 = -1.379 \text{ mA}$$

(*b*) The voltage drop V_{24} is

$$V_{24} = -I_3 R_3 = 1.379 \times 2 = 2.758 \text{ V}$$

The voltage drop between any two nodes in a network is independent of the path chosen between the nodes. For example, V_{24} may be found by going from 2 to 1 to 4 and adding all voltage drops along this path. Thus

$$V_{24} = -I_1 R_1 + 6 = -3.242 + 6 = 2.758 \text{ V}$$

which agrees with the value found by going directly from 2 to 4 through R_3.

In solving the above illustrative problem, we chose the two internal meshes 1 and 2. There is a third mesh in this network; the one around the outside loop 4-1-2-3-4. However, this outside mesh is not independent of the other two meshes. *An independent loop is one whose KVL equation includes at least one voltage not included in the other equations.* The number of independent KVL equations is equal to the number of independent loops.

A *junction* is defined as a point where three or more circuit elements meet. Of the four nodes in Fig. C-4, nodes 2 and 4 are junctions. *The number of independent KCL equations is equal to one less than the number of junctions.* Hence, in solving the above problem only one KCL equation is required.

Series and Parallel Combinations of Resistors   The circuit of Fig. C-5*a* consists of three resistors in *series*, which means that the same current flows in each resistor. From KVL, we obtain

$$-V + IR_1 + IR_2 + IR_3 = 0$$

The equivalent resistance R between nodes 1 and 2 is, by definition, given by

$$R \equiv \frac{V}{I} = R_1 + R_2 + R_3 \tag{C-6}$$

To find the total resistance in a series circuit, add the individual values of resistance.

Resistors are in parallel when the same voltage appears across each resistor. Hence, Fig. C-5*b* shows three resistors in parallel:

$$I_1 = \frac{V}{R_1} = G_1 V \quad I_2 = \frac{V}{R_2} = G_2 V \quad I_3 = \frac{V}{R_3} = G_3 V$$

where $G \equiv 1/R$ is called the *conductance*. Its dimensions are A/V or reciprocal ohms, called *mhos* (℧). Application of KCL to Fig. C-5*b* yields

$$I = I_1 + I_2 + I_3 = (G_1 + G_2 + G_3)V$$

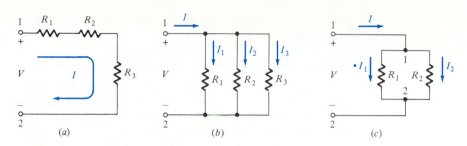

FIGURE C-5
(a) Resistors in series.
(b) Resistors in parallel.
(c) A current divider.

The equivalent conductance between nodes 1 and 2 is, by definition

$$G \equiv \frac{I}{V} = G_1 + G_2 + G_3 \tag{C-7}$$

To find the total conductance in a parallel circuit, add the individual values of conductance. Equation (C-7) is equivalent to

$$\frac{1}{R} = \frac{1}{R_1} + \frac{1}{R_2} + \frac{1}{R_3} \tag{C-8}$$

Of course, the number of resistors in the series or parallel circuits of Fig. C-5 is not limited to three; it can be any number, two or more. For the special case of two resistors, Eq. (C-8) is equivalent to

$$R = R_1 \| R_2 = \frac{R_1 R_2}{R_1 + R_2} \tag{C-9}$$

where the symbol $\|$ is to be read "in parallel with." It follows from this equation that two resistors in parallel have an effective resistance which is *smaller* than either resistor.

Just as a series circuit gives voltage attenuation [Fig. C-3b and Eq. (C-3)] so a parallel circuit gives current attenuation. In Fig. C-5c the current I_1 in R_1 (or I_2 in R_2) is less than the current I entering node 1. Thus, using Eq. (C-9), we have

$$V = IR = \frac{I R_1 R_2}{R_1 + R_2} = I_1 R_1$$

or
$$I_1 = \frac{R_2 I}{R_1 + R_2} \tag{C-10}$$

Note that if $R_1 = 0$, $I_1 = I$. This result is intuitively correct since all the current should flow in the short circuit. On the other hand, if $R_1 \to \infty$, then $I_1 = 0$, which is certainly true because no current can flow in an open circuit.

C-2 NETWORK THEOREMS The currents and voltages in any network regardless of its complexity may be obtained by a systematic application of KCL and KVL. However, the analysis may often be simplified by using one or more of the additional network theorems discussed in this section.

Superposition Theorem *The response of a linear network containing several independent sources is found by considering each generator separately and then adding the individual responses.* When evaluating the response due to one source, replace each of the other independent generators by its internal resistance, that is, set $v_s = 0$ for a voltage source and $i_s = 0$ for a current generator.

Example C-2

Find the currents I_1, I_2, and I_3 in the circuit of Fig. C-4, using the superposition theorem.

Solution

First consider the currents I_1', I_2', and I_3' due to the 6-V supply. Then node 3 must be short-circuited to node 4, so as to eliminate the response due to the -14-V source. This connection puts R_2 and R_3 in parallel, as indicated in Fig. C-6a. From Eq. (C-9) this parallel combination has a resistance of

$$\frac{R_2 R_3}{R_2 + R_3} = \frac{9 \times 2}{9 + 2} = 1.636 \text{ k}\Omega$$

The resistance seen by the 6-V supply is R_1 plus the above value; hence

$$I_1' = \frac{6}{1 + 1.636} = 2.276 \text{ mA}$$

From the current attenuation formula, Eq. (C-10), we obtain

$$I_2' = \frac{-I_1' R_3}{R_2 + R_3} = \frac{-2.276 \times 2}{9 + 2} = -0.414 \text{ mA}$$

Similarly

$$I_3' = \frac{-I_1' R_2}{R_2 + R_3} = \frac{-2.276 \times 9}{2 + 9} = -1.862 \text{ mA}$$

We now find the currents I_1'', I_2'', and I_3'' due to the -14-V supply. To eliminate the effect of the 6-V source, connect nodes 1 and 4 together as shown in Fig. C-6b. Proceeding as above, we find

$$I_2'' = \frac{-14}{9 + (1 \times 2)/3} = -1.448 \text{ mA}$$

$$I_1'' = +1.448 \times \tfrac{2}{3} = 0.9655 \text{ mA}$$

$$I_3'' = +1.448 \times \tfrac{1}{3} = 0.4826 \text{ mA}$$

The net current is the algebraic sum of the currents due to each excitation. Thus

$$I_1 = I_1' + I_1'' = 2.276 + 0.966 = 3.242 \text{ mA}$$

$$I_2 = I_2' + I_2'' = -0.414 - 1.448 = -1.862 \text{ mA}$$

$$I_3 = I_3' + I_3'' = -1.862 + 0.483 = -1.379 \text{ mA}$$

FIGURE C-6

Superposition is applied to the network shown in Fig. C-4. The circuit from which to calculate the response due to (a) the 6-V supply and (b) the -14-V supply.

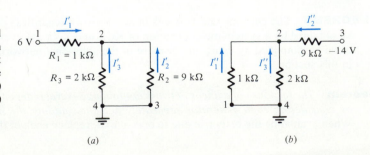

(a)   (b)

These are the same values obtained in the previous section. Note that for this particular network the analysis in Sec. C-1 using KVL and KCL is simpler than that given here using superposition.

Thévenin's Theorem *Any linear network may, with respect to a pair of terminals, be replaced by a voltage generator* V_{Th} *(equal to the open-circuit voltage) in series with the resistance* R_{Th} *seen between these terminals.* To find R_{Th} all *independent* voltage sources are short-circuited and all *independent* current sources are open-circuited. This theorem is often used to reduce the number of meshes in a network. For example, the two-mesh circuit in Fig. C-4 may be reduced to a single loop by replacing the components to the left of terminals 2 and 4 (including R_3) by the Thévenin equivalent. For convenience, the circuit in Fig. C-4 is redrawn in Fig. C-7a. The components in the shaded box are those to the right of nodes 2 and 4, and they are redrawn unaltered in Fig. C-7b. The other circuit elements do not appear in Fig. C-7b but are replaced by V_{Th} and R_{Th}. Thévenin's theorem states that I_2 and V_{24} calculated from this reduced circuit are identical to the corresponding values in Fig. C-4.

The open-circuit voltage V_{Th} is obtained by disconnecting the components in the box from Fig. C-7a. From the voltage attenuator formula Eq. (C-2), we obtain

$$V_{Th} = \frac{6 \times 2}{1 + 2} = 4 \text{ V}$$

To find the resistance seen to the left of nodes 2 and 4, the 6-V supply is imagined reduced to zero, which is equivalent to connecting the top of the 1-kΩ resistor to ground. Hence this resistor is now placed in parallel with the 2-kΩ resistor, and

$$R_{Th} = \frac{1 \times 2}{1 + 2} = 0.667 \text{ kΩ}$$

From the equivalent circuit of Fig. C-7b we obtain

$$I_2 = \frac{-(14 + V_{Th})}{9 + R_{Th}} = \frac{-18}{9.667} = -1.862 \text{ mA}$$

and
$$V_{24} = -9I_2 - 14 = 9 \times 1.862 - 14 = 2.758 \text{ V}$$

FIGURE C-7
The circuit in Fig. C-4 redrawn. (*b*) Thévenin's theorem applied to the circuit in (*a*) looking to the left of nodes 2 and 4.

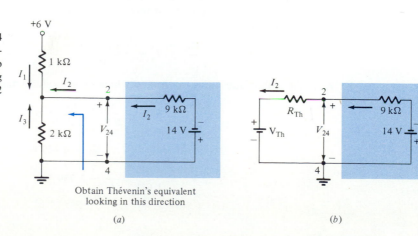

Obtain Thévenin's equivalent looking in this direction

(*a*)   (*b*)

These two values agree with the numerical values found in Sec. C-1. The currents I_3 and I_1 do not appear in Fig. C-7b and must be found from Fig. C-7a. Thus

$$I_3 = \frac{-V_{24}}{2} = \frac{-2.758}{2} = -1.379 \text{ mA}$$

and

$$I_1 = \frac{6 - V_{24}}{1} = 6 - 2.758 = 3.242 \text{ mA}$$

which are the same currents found previously.

Norton's Theorem *Any linear network may, with respect to a pair of terminals, be replaced by a current generator (equal to the short-circuit current) in parallel with the resistance seen between the two terminals.*

From Thévenin's and Norton's theorems it follows that a voltage source V in series with a resistance R is equivalent to a current source I in parallel with R, provided that $I = V/R$. These equivalent circuits are indicated in Figs. C-1c and d with $v_s = V$, $R_s = R$, and $i_s = I = V/R_s$.

As corollaries to Thévenin's and Norton's theorems we have the following relationships. If V represents the *open-circuit voltage,* I the *short-circuit current,* and R (G) the resistance (conductance) between two terminals in a network, then

$$V = IR = \frac{I}{G} \qquad I = \frac{V}{R} = GV \qquad R = \frac{V}{I} \tag{C-11}$$

In spite of their disarming simplicity, these equations (reminiscent of Ohm's law) should not be overlooked because they are most useful in analysis. For example, the first equation, which states "open-circuit voltage equals short-circuit current divided by conductance," is often the simplest way to find the voltage between two points in a network.

Nodal Method of Analysis When the number of junction voltages (with respect to the reference, or ground, node) is less than the number of independent meshes, then the choice of nodal voltages as the unknowns leads to a simpler solution than considering the mesh currents as the unknowns. For example, the circuit of Fig. C-4 has two independent meshes but only one independent junction voltage. In terms of the one unknown independent voltage V_{24}, the currents are

$$I_1 = \frac{6 - V_{24}}{1} \qquad I_2 = \frac{-14 - V_{24}}{9} \qquad I_3 = \frac{-V_{24}}{2} \tag{C-12}$$

By KCL the sum of these three currents (which enter node 2) must equal zero. Hence

$$\frac{6}{1} - \frac{V_{24}}{1} - \frac{14}{9} - \frac{V_{24}}{9} - \frac{V_{24}}{2} = 0$$

$$V_{24}(\tfrac{1}{1} + \tfrac{1}{9} + \tfrac{1}{2}) = \tfrac{6}{1} - \tfrac{14}{9} = 4.444 \text{ mA}$$

and

$$V_{24} = \frac{4.444}{1.611} = \frac{I}{G} = 2.759 \text{ V}$$

The formal procedure for writing nodal equations is outlined in the following steps:

1. Convert all voltage sources in series with resistances to current sources in parallel with conductances, as outlined in Eq. (C-11). The circuit is then redrawn.

2. Select a reference node O and identify the node-voltage variables $V_A, V_B, \ldots, V_N$ as the voltage drops from nodes $A, B, \ldots, N$ to node O. The choice of a reference is arbitrary; its selection is often based on convenience.

3. Write the KCL equations at nodes $A, B, \ldots, N$ in terms of the node-voltage variables. For circuits which contain no dependent sources, the resultant set of equations is of the form

A:
$$G_{AA}V_A - G_{AB}V_B - \cdots - G_{AN}V_N = I_A$$

B:
$$-G_{AB}V_A + G_{BB}V_B - \cdots - G_{BN}V_N = I_B$$

$$\cdots \cdots \cdots \cdots \cdots \cdots \cdots \cdots$$

N:
$$-G_{AN}V_A - G_{BN}V_B - \cdots + G_{NN}V_N = I_N$$

where G_{JJ} = sum of all conductances connected to node J
G_{JK} = sum of all conductances connected between nodes J and K
I_J = sum of all current sources entering node J

4. Solve the equations for the desired node voltages. Other voltages and currents in the circuit are determined by application of Kirchhoff's voltage law and Ohm's law.

If the circuit contains controlled sources, the control variable (v_π and v_i in Figs. C-2a and C-2b, respectively) must be expressed in terms of the node-voltage variables prior to solution. (That is, v_π and v_i must be expressed in terms of $V_A, V_B, \ldots, V_N$.) The form of the equations in step 3 (above), is as shown *except* $G_{KJ} \neq G_{JK}$.

Mesh Analysis

Mesh analysis is analogous to the node-voltage method except that the KVL equations are formulated in terms of mesh-current variables. A mesh current, indicated by 1 and 2 in Fig. C-4a, is assumed to exist in each element in the loop. Hence the current in any branch (component) is the algebraic sum of the mesh currents that exist in it. For example, in Fig. C-4a, if I_A is the current in mesh 1 and I_B is the current in mesh 2, KVL for these loops become

$$-6 + I_A \cdot 1 + (I_A + I_B) \cdot 2 = 0$$

$$14 + 9I_B + (I_A + I_B) \cdot 2 = 0$$

or

$$3I_A + 2I_B = 6$$

$$2I_A + 11I_B = -14$$

Solution of these equations yields $I_A = I_1 = 3.242$ mA and $I_B = I_2 = -1.862$ mA. The current I_3 in R_3 is $-(I_A + I_B) = -1.379$ mA. Clearly, these values are the same as those obtained in Example C-1.

The formal procedure for writing mesh equations is outlined in the following steps:

1. Convert each current-source parallel-conductance combination to a voltage source in series with resistance. For convenience, the circuit is redrawn.
2. Select a mesh current variable for each loop.
3. Write a KVL equation for each loop in the direction of the mesh current for the loop. The resultant form, for circuits which contain no dependent sources, is

1:
$$R_{11}I_1 - R_{12}I_2 - \cdots - R_{1N}I_N = V_1$$

2:
$$-R_{12}I_1 + R_{22}I_2 - \cdots - R_{2N}I_N = V_2$$

$$\cdots\cdots\cdots\cdots\cdots\cdots\cdots\cdots\cdots\cdots$$

N:
$$-R_{1N}I_1 - R_{2N}I_2 - \cdots - R_{NN}I_N = V_N$$

where R_{JJ} = sum of all resistances contained in mesh J
  R_{JK} = sum of all resistances common to both meshes J and K
  V_J = sum of all the source-voltage rises in mesh J, taken in the direction of I_J

4. Solve the set of equations for the desired mesh currents. Other currents and voltages in the circuit can be obtained by use of Ohm's law and Kirchhoff's current law.

For circuits containing controlled sources, the control variable is expressed in terms of the mesh currents $I_1, I_2, \ldots, I_N$. In these cases, the form of the equations in step 3 is as indicated *except $R_{IJ} \neq R_{JI}$.*

C-3 THE SINUSOIDAL STEADY STATE

If a sinusoidal excitation (a voltage or current) is applied to a linear network, then the response (the voltage between any two nodes or the current in any branch of the network) will also be sinusoidal. (It is assumed that all transients have died down so that a steady state is reached.) Let us verify this general statement for the simple parallel combination of resistor R and capacitor C in Fig. C-8 to which has been applied the sinusoidal source voltage

$$v = V_m \cos \omega t = V_m \cos 2\pi ft \tag{C-13}$$

where f is the *frequency* of the source in hertz (Hz), $\omega = 2\pi f$ is called the *angular frequency,* and V_m is the *maximum* or *peak* value of voltage. We shall now prove that the generator current i is also a sinusoidal waveform.

A capacitor C is a component (say, two metals separated by a dielectric) which stores charge q (coulombs) proportional to the applied voltage v (volts) so that

$$q = Cv \tag{C-14}$$

where the proportionality factor C is called the *capacitance*. The dimensions of C are

FIGURE C-8
A parallel *RC* combination excited by a sinusoidal voltage.

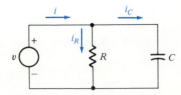

coulombs per volt, which is abbreviated as *farads* (F). The capacitor current i_C is therefore

$$i_C = \frac{dq}{dt} = C\,\frac{dv}{dt} \tag{C-15}$$

or, using Eq. (C-13),

$$i_C = -\omega C V_m \sin \omega t \tag{C-16}$$

From Ohm's law, the resistor current i_R is given by

$$i_R = \frac{v}{R} = \frac{V_m}{R} \cos \omega t \tag{C-17}$$

From Kirchhoff's current law (KCL), $i = i_R + i_C$, or

$$i = \frac{V_m}{R} \cos \omega t - \omega C V_m \sin \omega t \tag{C-18}$$

which has the form

$$i = I_m \cos \theta \cos \omega t - I_m \sin \theta \sin \omega t \tag{C-19}$$

where

$$I_m \cos \theta \equiv \frac{V_m}{R} \quad \text{and} \quad I_m \sin \theta \equiv \omega C V_m \tag{C-20}$$

From the trigonometric identity, we obtain

$$\cos(\theta + \alpha) = \cos \theta \cos \alpha - \sin \theta \sin \alpha \tag{C-21}$$

then Eq. (C-19), with $\alpha \equiv \omega t$, is equivalent to

$$i = I_m \cos(\omega t + \theta) \tag{C-22}$$

We have thus verified that the generator current is indeed a sinusoid. The peak current is then I_m, and i is shifted in phase by the angle θ with respect to the source voltage $V_m \cos \omega t$. We say that "the generator current *leads* its voltage by the phase angle θ."

The peak current I_m and the phase θ are obtained from Eqs. (C-20). If the two equations are each squared and then added, we obtain

$$I_m^2 \cos^2 \theta + I_m^2 \sin^2 \theta = \frac{V_m^2}{R^2} + \omega^2 C^2 V_m^2 \tag{C-23}$$

Since $\cos^2 \theta + \sin^2 \theta = 1$, then

$$I_m = V_m \sqrt{\frac{1}{R^2} + \omega^2 C^2} \tag{C-24}$$

Division of the second equation in Eq. (C-20) by the first yields

$$\frac{I_m \sin \theta}{I_m \cos \theta} = \frac{\omega C V_m}{V_m/R}$$

or

$$\tan \theta = \omega C R \tag{C-25}$$

FIGURE C-9
(*a*) The current repre-
sented as a phasor of
magnitude *I* and phase
θ. (*b*) Phasor addition,
representing $\mathbf{I} = \mathbf{I}_R +$
$\mathbf{I}_C$.

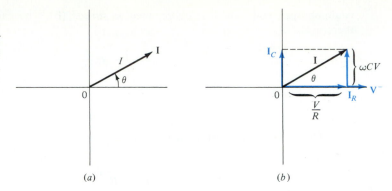

(*a*)   (*b*)

For a more complicated network than the one in Fig. C-8, the analysis would involve a prohibitive amount of trigonometric manipulation. Hence we now present a simpler alternative general method for solving sinusoidal networks in the steady state. Some important concepts (such as phasors, complex plane, and impedance) are first intro-duced.

Phasors

Each current (or voltage) in a network is a sinusoid, which has a peak value and a phase angle. Hence it can be represented by a directed line segment having a length and direction. For a sinusoid this directed line segment is called a *phasor*. Its magnitude represents the effective or rms value and is given by the peak value divided by $\sqrt{2}$. The direction of the phasor is the phase θ in the sinusoidal waveform $I_m \cos(\omega t + \theta)$, and the angle θ is measured counterclockwise with respect to the horizontal axis. In this section we use boldface $\mathbf{I}$ ($\mathbf{V}$) to denote a phase current (voltage). In phasor notation, the current in Eq. (C-22) is written

$$\mathbf{I} = I \angle \theta \tag{C-26}$$

where $I = I_m/\sqrt{2}$. This phasor is indicated in Fig. C-9*a*.

The applied voltage phasor is, from Eq. (C-13), $\mathbf{V} = V \angle 0°$, where $V = V_m/\sqrt{2}$, and the current in the resistor is, from Eq. (C-17), $\mathbf{I}_R = V/R \angle 0°$. These phasors are indicated in Fig. C-9*b*. Note that *the current in a resistor is in phase with the voltage across the resistor.*

Since Eq. (C-16) may be written $i_C = \omega C V_m \cos(\omega t + 90°)$, the phasor representing the capacitor current is

$$\mathbf{I}_C = \omega C V \angle 90° \tag{C-27}$$

where $V = V_m/\sqrt{2}$ is the rms voltage. Note that *the current in a capacitor leads the voltage across the capacitor by* 90°. The phase $\mathbf{I}_C$ is plotted in Fig. C-9*b*. The generator current is the sum of the resistor current and the capacitor current, or in phasor notation

$$\mathbf{I} = \mathbf{I}_R + \mathbf{I}_C = \frac{\mathbf{V}}{R} \angle 0° + \omega C \mathbf{V} \angle 90° \tag{C-28}$$

This phasor sum is indicated in Fig. C-9*b*, where it is found that

$$|\mathbf{I}|^2 = \frac{V^2}{R^2} + \omega^2 C^2 V^2 \quad \text{and} \quad \tan\theta = \omega CR$$

in agreement with Eqs. (C-24) and (C-25). Note how simple the phasor method of analysis is compared with the above solution, with the use of instantaneous values of current and voltage and manipulating the equations with the aid of trigonometric identities. By introducing the concept of the complex plane, the analysis may be further simplified. An essentially algebraic, rather than a trigonometric, solution is now obtained.

The j Operator

A useful convention is to take the symbol j to represent a *phase lead* of 90°. In place of Eq. (C-27) we now write $\mathbf{I}_C = j\omega C\mathbf{V}$, and for the total current in Eq. (C-28) we have

$$\mathbf{I} = \frac{\mathbf{V}}{R} + j\omega C\mathbf{V} \tag{C-29}$$

This equation is interpreted to mean that $\mathbf{I}$ is a phasor formed by combining the phasor $\mathbf{V}/R$ horizontally (at zero phase) with $\omega C\mathbf{V}$ plotted vertically (at a phase of 90°). Hence the vertical axis is also called the j axis. The current $\mathbf{I}$, shown in Fig. C-9b, is identical with that found above.

From the definition of j it follows that $j\mathbf{I}$ is a phasor whose magnitude is that of $\mathbf{I}$ but whose phase is 90° greater than the phase of $\mathbf{I}$. In other words, j "multiplying" a phasor $\mathbf{I}$ is an operator which rotates $\mathbf{I}$ in the counterclockwise direction by 90°. Consider $\mathbf{I} = 1$, a phasor of magnitude 1 and phase 0. Then $j\mathbf{I} = j1$ has a magnitude 1 and phase 90°, as indicated in Fig. C-10. Then $j(j1)$ represents a rotation of $j1$ by 90°, which results in a phasor of unit magnitude pointing along the negative horizontal axis, as indicated in Fig. C-10. In a purely formal manner we may write

$$j(j1) = j^2 1 = -1 \qquad \text{or} \qquad j = \sqrt{-1} \tag{C-30}$$

Because of this formalism the vertical axis is called the j or *imaginary axis* and the horizontal axis is designated as the *real axis*. The plane of Fig. C-10 is now called the *complex plane*.

Note that higher powers of j are easily found. Thus

$$j^3 = j(j^2) = j(-1) = -j \tag{C-31}$$

which represents a phasor of magnitude 1 and phase $-90°$. The reciprocal of j is $-j$, as is easily verified. Thus

$$\frac{1}{j} = \frac{1}{j}\frac{j}{j} = \frac{j}{j^2} = -j \tag{C-32}$$

FIGURE C-10
(*a*) Concerning the operator j. (*b*) A phasor current plotted in the complex plane.

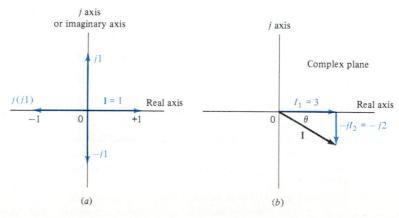

(a) (b)

because $j^2 = -1$ from Eq. (C-30). A point in the complex plane is called a *complex number*, and it is evident that a phasor is a complex number. Hence the analysis of sinusoidal circuits is carried out most simply by treating currents and voltages as complex numbers representing phasors.

Assume that a complicated circuit is analyzed (by the general method outlined in Sec. C-4) and that the following complex current is obtained

$$\mathbf{I} = I_1 - jI_2 = 3 - j2 \quad \text{mA} \tag{C-33}$$

This phasor is indicated in the complex plane in Fig. C-10b. From this diagram it follows that the rms current $|\mathbf{I}|$ and the phase angle θ are given by

$$|\mathbf{I}| = \sqrt{I_1^2 + I_2^2} = \sqrt{13} = 3.61 \text{ mA}$$

and

$$\theta = -\arctan\frac{I_2}{I_1} = -\arctan\frac{2}{3} = -33.7° = -0.588 \text{ rad}$$

If the frequency f is 1 kHz, then the instantaneous current is, from Eq. (C-22), $i = 3.61\sqrt{2}\cos(6280t - 0.588)$ mA.

C-4 SIMPLIFIED SINUSOIDAL NETWORK ANALYSIS

Consider a linear network containing resistors, capacitors, inductors, and sinusoidal sources. The steady-state response is desired. A straightforward method of solution is possible which is analogous to that used with networks containing only resistive components and constant (dc) supply voltages (or currents). The analysis consists of writing the KVL and KCL equations for the network and then solving for the complex (phasor) currents and voltages. In order to carry out such an analysis, it is first necessary to introduce the concept of *complex resistance* or *reactance*. After defining reactance, a number of specific circuits are solved by using this simple method of analysis.

Reactance

The ratio of the voltage $\mathbf{V}$ across a passive circuit component to the current $\mathbf{I}$ through the element for each of the three basic components is as follows:

Resistance:
$$\frac{\mathbf{V}}{\mathbf{I}} = R$$

Capacitance:
$$\frac{\mathbf{V}}{\mathbf{I}} = \frac{1}{j\omega C} = \frac{-j}{\omega C} = +j\left(\frac{-1}{\omega C}\right) \tag{C-34}$$

Inductance
$$\frac{\mathbf{V}}{\mathbf{I}} = j\omega L$$

The first equation of Eqs. (C-34) is Ohm's law. The second equation follows from Eq. (C-27). An inductor is a component (say, a coil of wire) whose terminal voltage v is proportional to the rate of change of current. The proportionality factor L (henrys, H) is called the *inductance*. From $v = L \, di/dt$, the third equation of Eqs. (C-34) can be obtained in a manner analogous to that used in the preceding section to obtain Eq. (C-27).

From Eqs. (C-34) it follows that a capacitor behaves as a "complex resistance" $-j/\omega C$ and an inductor acts like a "complex resistance" $j\omega L$. A more commonly used phrase for complex resistance is *reactance*, denoted by the real positive symbol X:

$$\text{Capacitive reactance} = +jX_C \quad \text{where } X_C \equiv \frac{-1}{\omega C}$$

and $\qquad\qquad$ Inductive reactance $= +jX_L \quad \text{where } X_L \equiv \omega L$

In applying KVL to a circuit containing reactive elements, it must be remembered that the drop across a capacitor is $-jX_C I = -jI/\omega C$ and the drop across an inductor is $jX_L I = j\omega L I$. It follows from the above considerations that KVL applied to the series circuit of Fig. C-11 yields

$$V = RI + j\omega L I - \frac{j}{\omega C} I \tag{C-35}$$

or $$I = \frac{V}{R + j(\omega L - 1/\omega C)} = \frac{V}{R + jX} \tag{C-36}$$

where the total series reactance is $X \equiv \omega L - 1/\omega C$. The current may be expressed in standard complex number form $I = I_1 + jI_2$ by multiplying both the numerator and the denominator by the complex conjugate (change j to $-j$) of the denominator. Thus

$$I = \frac{V}{R + jX}\frac{R - jX}{R - jX} = \frac{V}{R^2 + X^2}(R - jX) \tag{C-37}$$

From this equation we find that the magnitude and phase of I are given by

$$|I| = \frac{V}{\sqrt{R^2 + X^2}} \quad \text{and} \quad \tan\theta = -\frac{X}{R} \tag{C-38}$$

Complex (phasor) quantities can thus be expressed in three ways: rectangular form ($I = I_1 + jI_2$), polar form ($I = I\angle\theta$); and exponential form ($I = Ie^{j\theta}$). Conversion between these forms is given by

$$I = \sqrt{I_1^2 + I_2^2} \qquad \theta = \tan^{-1}\frac{I_2}{I_1}$$
$$I_1 = I\cos\theta \qquad I_2 = I\sin\theta \tag{C-39}$$

Impedance

The ratio of the phasor voltage between any two points A and B of a network to the phasor current in this portion of the circuit is called the *impedance* **Z** between A and B. For the circuit in Fig. C-11

$$Z \equiv \frac{V}{I} = R + j\left(\omega L - \frac{1}{\omega C}\right) \tag{C-40}$$

from Eq. (C-35). Since the generator **V** is placed directly between A and B, then **Z** is

FIGURE C-11
An *RLC* series circuit: (a) in the time domain and (b) in the frequency domain.

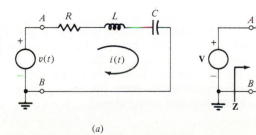

(a) $\qquad\qquad\qquad\qquad\qquad\qquad$ (b)

the impedance "seen" by the source $\mathbf{V}$. Note that for a series circuit the impedance equals the sum of the resistances plus reactances in the loop. This statement is analogous to the law for a dc series circuit, which states that the total resistance is the sum of the resistances in series. It should be emphasized that, whereas $\mathbf{Z}$ is a complex quantity, it is not a phasor, since it does *not* represent a current or a voltage varying sinusoidally with time.

Two impedances $\mathbf{Z}_1$ and $\mathbf{Z}_2$ in parallel represent an equivalent impedance $\mathbf{Z}$ given by

$$\mathbf{Z} = \frac{\mathbf{Z}_1 \mathbf{Z}_2}{\mathbf{Z}_1 + \mathbf{Z}_2} \tag{C-41}$$

corresponding to Eq. (C-9) for two resistors in parallel. For the parallel combination of a resistor R and a capacitor C as in Fig. C-8, $\mathbf{Z}_1 = R$, $\mathbf{Z}_2 = -j/\omega C = 1/j\omega C$, and from Eq. (C-41),

$$\mathbf{Z} = \frac{R(1/j\omega C)}{R + 1/j\omega C} = \frac{R}{1 + j\omega CR} \tag{C-42}$$

This same result is obtained by applying KCL to Fig. C-8. Using phasor notation, we have

$$\mathbf{I} = \mathbf{I}_R + \mathbf{I}_C = \frac{\mathbf{V}}{R} + \frac{\mathbf{V}}{1/j\omega C} = \frac{\mathbf{V}}{R} + j\omega C \mathbf{V} \tag{C-43}$$

and $\mathbf{Z} = \mathbf{V}/\mathbf{I}$ gives the result in Eq. (C-42).

Admittance

The reciprocal of the impedance is called the *admittance* and is designated by $\mathbf{Y}$, so that

$$\mathbf{Y} \equiv \frac{1}{\mathbf{Z}} = G + jB \tag{C-44}$$

The real part of $\mathbf{Y}$ is the *conductance* G and the imaginary part is the *susceptance* B. If a resistor is under consideration then $\mathbf{Z} = R$, $G = 1/R$, and $B = 0$. On the other hand, if the circuit element is a capacitor, $\mathbf{Z} = 1/j\omega C$ and $\mathbf{Y} = j\omega C$ so that $B = \omega C$ and $G = 0$.

Since $\mathbf{I} = \mathbf{V}/\mathbf{Z}$, then $\mathbf{I} = \mathbf{Y}\mathbf{V}$. For a resistor $\mathbf{I}_R = G\mathbf{V}_R$ and for a capacitor $\mathbf{I}_C = j\omega C \mathbf{V}_C$. For the circuit in Fig. C-8, with R and C in parallel, $\mathbf{V}_R = \mathbf{V}_C = \mathbf{V}$, and the total current is

$$\mathbf{I} = \mathbf{I}_R + \mathbf{I}_C = (G + j\omega C)\mathbf{V}$$

The admittance of this combination is $\mathbf{Y} = \mathbf{I}/\mathbf{V} = G + j\omega C$, which agrees with Eq. (C-43) if $G = 1/R$.

Network Analysis

The network theorems developed for resistive networks in Sec. C-2 are also applicable to circuits excited by sinusoids. For example, the Thévenin equivalent is the open-circuit voltage *phasor* $\mathbf{V}_{\text{Th}}$ in series with the Thévenin *impedance* $\mathbf{Z}_{\text{Th}}$. Similarly, superposition denotes the fact that the phasor voltage (current) response to a number of excitations of the same frequency is simply the sum of the phasor voltage (current) components due to each excitation acting alone.

FIGURE C-12
Pertaining to Miller's theorem. By definition, $K \equiv V_2/V_1$. The networks in (a) and (b) have identical node voltages. Note that $I_1 = -I_2$.

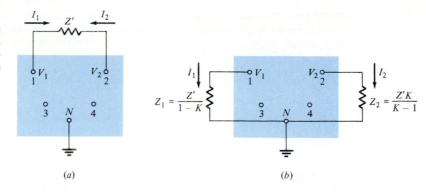

(a)   (b)

The procedures for mesh and nodal analysis are also analogous to the resistive case. The voltage and current variables are phasor quantities in the sinusoidal case and the resistances are replaced by impedances.

Miller's Theorem This theorem is particularly useful in connection with transistor high-frequency amplifiers. Consider an arbitrary circuit configuration with N distinct nodes 1, 2, 3, . . ., N, as indicated in Fig. C-12. Let the node voltages be $\mathbf{V}_1$, $\mathbf{V}_2$, $\mathbf{V}_3$, . . ., $\mathbf{V}_N$, where $\mathbf{V}_N = 0$, since N is the reference, or ground, node. Nodes 1 and 2 (referred to as N_1 and N_2) are interconnected with an impedance $\mathbf{Z}'$. We postulate that we know the ratio $\mathbf{V}_2/\mathbf{V}_1$. Designate this ratio $\mathbf{V}_2/\mathbf{V}_1$ by $\mathbf{K}$. We shall now show that the current $\mathbf{I}_1$ drawn from N_1 through $\mathbf{Z}'$ can be obtained by disconnecting $\mathbf{Z}'$ from terminal 1 and by bridging an impedance $\mathbf{Z}'/(1 - \mathbf{K})$ from N_1 to ground, as indicated in Fig. C-12b.
The current $\mathbf{I}_1$ is given by

$$\mathbf{I}_1 = \frac{\mathbf{V}_1 - \mathbf{V}_2}{\mathbf{Z}'} = \frac{\mathbf{V}_1(1 - \mathbf{K})}{\mathbf{Z}'} = \frac{\mathbf{V}_1}{\mathbf{Z}'/(1 - \mathbf{K})} = \frac{\mathbf{V}_1}{\mathbf{Z}_1} \tag{C-45}$$

Therefore, if $\mathbf{Z}_1 \equiv \mathbf{Z}'/(1 - \mathbf{K})$ were shunted across terminals N_1-N, the current $\mathbf{I}_1$ drawn from N_1 would be the same as that from the original circuit. Hence the same expression is obtained for $\mathbf{I}_1$ in terms of the node voltages for the two configurations (Fig. C-12a and b).
In a similar way, it may be established that the correct current $\mathbf{I}_2$ drawn from N_2 may be calculated by removing $\mathbf{Z}'$ and by connecting between N_2 and ground an impedance $\mathbf{Z}_2$, given by

$$\mathbf{Z}_2 \equiv \frac{\mathbf{Z}'}{1 - 1/\mathbf{K}} = \frac{\mathbf{Z}'\mathbf{K}}{\mathbf{K} - 1} \tag{C-46}$$

Since identical nodal equations (KCL) are obtained from the configurations of Fig. C-12a and b, these two networks are equivalent. It must be emphasized that this theorem will be useful in making calculations only if it is possible to find the value of $\mathbf{K}$ by some independent means.

C-5 EXPONENTIAL EXCITATION Sinusoidal excitation can be considered as a special case of exponential excitation for which $s = j\omega$. For this situation, the impedance and admittance of the various elements are given in Table C-1b. (Note that making $s = j\omega$ in Table C-1b yields the results listed in Table C-1c).

TABLE C-1 Element Volt-Ampere Relationship with (*a*) Time-Varying Excitation, (*b*) Exponential Excitation, and (*c*) Sinusoidal Excitation

	Resistance (conductance)	*Inductance*	*Capacitance*	*Passive networks*

a Time-varying excitation in which $v = v(t)$ and $i = i(t)$

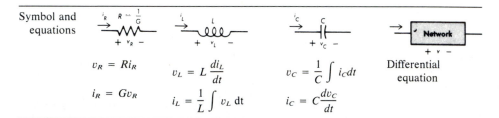

| Symbol and equations | $v_R = Ri_R$ $i_R = Gv_R$ | $v_L = L\dfrac{di_L}{dt}$ $i_L = \dfrac{1}{L}\int v_L\,dt$ | $v_C = \dfrac{1}{C}\int i_C dt$ $i_C = C\dfrac{dv_C}{dt}$ | Differential equation |

b Exponential excitation in which $v = V\epsilon^{st}$ and $i = I\epsilon^{st}$

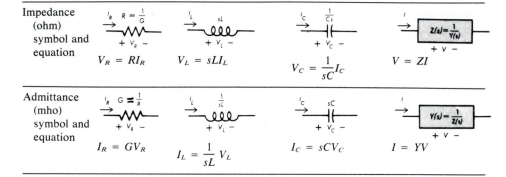

| Impedance (ohm) symbol and equation | $V_R = RI_R$ | $V_L = sLI_L$ | $V_C = \dfrac{1}{sC}I_C$ | $V = ZI$ |
| Admittance (mho) symbol and equation | $I_R = GV_R$ | $I_L = \dfrac{1}{sL}V_L$ | $I_C = sCV_C$ | $I = YV$ |

c Sinusoidal excitation in which $v = V_m \cos(\omega t + \theta)$ and $i = I_m \cos(\omega t + \alpha)$

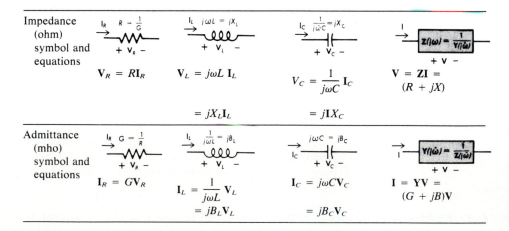

| Impedance (ohm) symbol and equations | $V_R = RI_R$ | $V_L = j\omega L\,I_L$ $= jX_L I_L$ | $V_C = \dfrac{1}{j\omega C}I_C$ $= jIX_C$ | $V = ZI =$ $(R + jX)$ |
| Admittance (mho) symbol and equations | $I_R = GV_R$ | $I_L = \dfrac{1}{j\omega L}V_L$ $= jB_L V_L$ | $I_C = j\omega CV_C$ $= jB_C V_C$ | $I = YV =$ $(G + jB)V$ |

For the generalized frequency variable $s = \sigma + j\omega$, the methods of circuit analysis are identical to the algorithms developed in Secs. C-2 to C-4; that is, resistors, capacitors, and inductors (time-domain elements) are replaced by their impedances (admittances) in the frequency (*s*-plane) domain. Once the network is so transformed, all the network theorems and analysis techniques outlined in Sec. C-2 are directly applicable. Thus, if $j\omega$ is replaced by s in Fig. C-11b, the impedance $Z(s)$ is

$$Z(s) = \frac{V}{I} = R + sL + \frac{1}{sC} = \frac{LCs^2 + RCs + 1}{sC} \tag{C-47}$$

and $I = V/Z$ becomes

$$I = \frac{sC}{LCs^2 + RCs + 1} V \tag{C-48}$$

Observe, in Eq. (C-48), that setting $s = 0$ (dc) makes $I = 0$; similarly, for values of s which make $LCs^2 + RCs + 1 = 0$, $I \to \infty$.

In general, the ratio of the response to the excitation (in the frequency domain) is called the *transfer function* of the circuit. Thus, if V_o is the response to an excitation V_i,

$$A(s) = \frac{V_o}{V_i} = \frac{N(s)}{D(s)} \tag{C-49}$$

where $A(s)$ is the transfer function. Values of s which make $A(s) = 0$ are called *zeros* of the transfer function; values of s for which $A(s) \to \infty$ are called *poles*. In Eq. (C-47), values of s, which cause $N(s) = 0$ are zeros of $A(s)$, whereas s values for which $D(s) = 0$ are poles of $A(s)$.

The use of poles and zeros (and their significance) becomes more evident in the following sections.

C-6 STEP RESPONSE OF AN *RC* CIRCUIT

The most common transient problem encountered in electronic circuits is that resulting from a step change in dc excitation applied to a series combination of a resistor and a capacitor. Consider the high-pass *RC* circuit in Fig. C-13 to which is applied a step of voltage v_i. The output voltage of v_o is taken across the resistor.

The High-Pass *RC* Circuit   A *step voltage* is one which maintains the value zero for all times $t < 0$ and maintains the value V for all times $t > 0$. The transition between the two voltage levels takes place at $t = 0$ and is accomplished in an arbitrarily short time interval. Thus in Fig. C-14, $v_i = 0$ immediately before $t = 0$ (to be referred to as time $t = 0^-$), and $v_i = V$ immediately after $t = 0$ (to be referred to as time $t = 0^+$).

FIGURE C-13
The high-pass *RC* circuit.

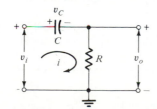

FIGURE C-14
Step-voltage response
of the high-pass RC circuit. The dashed line is
tangent to the exponen-
tial at $t = 0^+$.

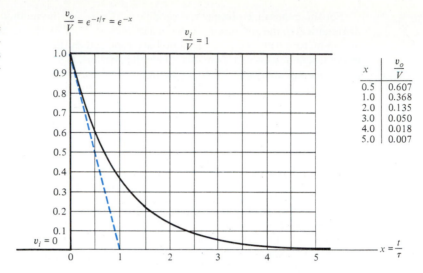

FIGURE C-14
Step-voltage response
of the high-pass RC circuit. The dashed line is
tangent to the exponential at $t = 0^+$.

x	$\dfrac{v_o}{V}$
0.5	0.607
1.0	0.368
2.0	0.135
3.0	0.050
4.0	0.018
5.0	0.007

From elementary considerations, the response of the network is exponential, with a time constant $RC \equiv \tau$, and the output voltage is of the form

$$v_o = B_1 + B_2 \epsilon^{-t/\tau} \tag{C-50}$$

The constant B_1 is equal to the steady-state value of the output voltage because as $t \to \infty$, $v_o \to B_1$. If this final value of output voltage is called V_f, then $B_1 = V_f$. The constant B_2 is determined by the initial output voltage, say, V_i, because at $t = 0$, $v_o = V_i = B_1 + B_2$ or $B_2 = V_i - V_f$. Hence the general solution for a single-time-constant circuit having initial and final values V_i and V_f, respectively, is

$$v_o = V_f + (V_i - V_f)\epsilon^{-t/\tau} \tag{C-51}$$

This basic equation is used many times throughout this text.

The constants V_f and V_i must now be determined for the circuit in Fig. C-13. The input is a constant ($v_i = V$) for $t > 0$. Since $i = C(dv_C/dt)$, then in the steady state $i = 0$, and the final output voltage iR is zero, or $V_f = 0$.

The above result may also be obtained by the following argument: We have already emphasized that a capacitor C behaves as an open circuit at zero frequency (because the reactance of C varies inversely with f). Hence any constant (dc) input voltage is "blocked" and cannot reach the output, and thus $V_f = 0$.

The value of V_i is determined from the following basic considerations. If the instantaneous current through a capacitor is i, then the change in voltage across the capacitor in time t_1 is $(1/C) \int_0^{t_1} i \, dt$. Since the current is always of finite magnitude, the above integral approaches zero as $t_1 \to 0$. Hence, it follows that *the voltage across a capacitor cannot change instantaneously.*

Applying the above principle to the network of Fig. C-13, we must conclude that since the input voltage changes discontinuously by an amount V at $t = 0$, the output must also change abruptly by this same amount. If we assume that the capacitor is initially uncharged, then the output at $t = 0^+$ must jump to V. Hence $V_i = V$, and since $V_f = 0$, Eq. (C-51) becomes

$$v_o = V\epsilon^{-t/\tau} \tag{C-52}$$

FIGURE C-15
The low-pass *RC* circuit.

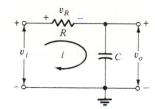

Input and output are shown in Fig. C-14. Note that the output is 0.61 of its initial value at 0.5τ, 0.37 at 1τ, and 0.14 at 2τ. The output has completed more than 95 percent of its total change after 3τ and more than 99 percent of its swing if $t > 5\tau$. Hence, although the steady state is approached asymptotically, we may assume for most applications that the final value has been reached after 5τ.

Discharge of a Capacitor through a Resistor Consider a capacitor C charged to a voltage V. At $t = 0$, a resistor R is placed across C. We wish to obtain the capacitor voltage v_o as a function of time. Since the action of shunting C by R cannot instantaneously change the voltage, $v_o = V$ at $t = 0^+$. Hence $V_i = V$. Clearly, at $t = \infty$, the capacitor will be completely discharged by the resistor and, therefore, $V_f = 0$. Substituting these values of V_i and V_f into Eq. (C-51), we obtain Eq. (C-52), and the capacitor discharge is indicated in Fig. C-14.

The Low-Pass *RC* Circuit The response of the circuit of Fig. C-15 to a step input is exponential with a time constant RC. Since the capacitor voltage cannot change instantaneously, the output starts from zero and rises toward the steady-state value V. The output is given by Eq. (C-51) as shown in Fig. C-16, or

$$v_o = V(1 - \epsilon^{-t/RC}) \tag{C-53}$$

Note that the circuits in Figs. C-13 and C-15 are identical except that the output $v_o = v_R$ is taken across R in Fig. C-13, whereas the output in Fig. C-15 is $v_o = v_C$. From Fig. C-15, we obtain

$$v_C = v_i - v_R = V - V\epsilon^{-t/RC}$$

where v_R is given by Eq. (C-52). This result for v_C agrees with Eq. (C-53).

Note that the impedance $Z(s)$ in the circuits in Figs. C-13 and C-15 is

FIGURE C-16
Step-voltage response of the low-pass *RC* circuit.

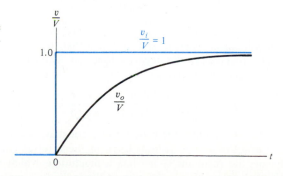

$$Z(s) = R + \frac{1}{Cs} = \frac{RCs + 1}{Cs}$$

Clearly $Z(s) = 0$ when $s = -1/RC = -1/\tau$. Since $I = V_i/Z$ and, in Fig. C-15, $V_o = I/sC$, we obtain

$$\frac{V_o}{V_i} = \frac{1}{1 + RCs} \tag{C-54}$$

In Eq. (C-54) we observe that V_o/V_i has a pole at $s = -1/RC$, and it is this value of s (the pole of the transfer function) that determines the time constant of the circuit.

C-7 THE ASYMPTOTIC BODE DIAGRAM   The frequencies of the sinusoids applied to electronic circuits display wide variation. For example, the excitation in an audio system can be as low as 20 Hz and as high as 20 kHz. Hence, to evaluate the response of the network, we must know the magnitude and the phase of the network transfer function $G(s)$ at each frequency. A convenient method by which this information is obtained is the frequency-response characteristic. This characteristic is the plot of the magnitude of $\mathbf{G}(j\omega)$ versus ω and the $\angle \mathbf{G}(j\omega)$ versus ω. Usually $|\mathbf{G}(j\omega)|$ is expressed in decibels (dB), given by

$$G(j\omega) \text{ in dB} = 20 \log |\mathbf{G}(j\omega)| \tag{C-55}$$

When $G(j\omega)$ in dB is plotted (along with the phase), the frequency-response characteristic is called the *Bode diagram*.

Determining the frequency-response characteristics of a network or system by algebraic manipulation is a moderate chore. For many purposes, an approximate frequency-response characteristic is adequate. The nature of the Bode diagram leads to a simply drawn approximate characteristic, called the *asymptotic Bode diagram*.

In general, a network function can be expressed as a quotient of two polynomials in s or $j\omega$. If the network function is put in the form

$$G(s) = K \frac{1 + a_1 s + a_2 s^2 + \cdots + a_m s^m}{1 + b_1 s + b_2 s^2 + \cdots + b_n s^n} \tag{C-56}$$

the numerator and denominator polynomials can be factored and the function represented by

$$G(s) = K \frac{(1 + s/z_1)(1 + s/z_2) \cdots (1 + s/z_m)}{(1 + s/p_1)(1 + s/p_2) \cdots (1 + s/p_n)} \tag{C-57}$$

Note that $-z_1$, $-z_2$, etc., and $-p_1$, $-p_2$, etc., are the roots of the numerator and denominator polynomials, respectively, and also that the $-z$ terms are the zeros and the $-p$ terms are the poles of the network function. The frequency-response curve is found by letting s become $j\omega$, giving

$$\mathbf{G}(j\omega) = K \frac{(1 + j\omega/z_1)(1 + j\omega/z_2) \cdots (1 + j\omega/z_m)}{(1 + j\omega/p_1)(1 + j\omega/p_2) \cdots (1 + j\omega/p_n)} \tag{C-58}$$

The value of $\mathbf{G}(j\omega)$ is evidently the product of a constant and a group of terms having the form $(1 + j\omega/\omega_o)$ or $1/(1 + j\omega/\omega_o)$. Each of these terms can be thought of as an individual phasor; the resultant $\mathbf{G}(j\omega)$ has a magnitude which is the product of the magnitudes and an angle which is the sum of the individual angles.

The magnitude-curve portion of the Bode diagram is plotted in decibels and, from Eq. (C-55), is a logarithmic function. Thus the product

$$\left(1 + \frac{j\omega}{z_1}\right)\left(1 + \frac{j\omega}{z_2}\right)\cdots\left(1 + \frac{j\omega}{z_m}\right)$$

becomes the sum

$$\left(1 + \frac{j\omega}{z_1}\right)_{dB} + \left(1 + \frac{j\omega}{z_2}\right)_{dB} + \cdots + \left(1 + \frac{j\omega}{z_m}\right)_{dB}$$

when the individual terms are expressed in decibels. As a result, the Bode diagram magnitude and phase curves can both be considered to be composed of sums produced by the individual factors. The behavior of the $(1 + j\omega/\omega_o)$ and $1/(1 + j\omega/\omega_o)$ terms is then seen to be of importance in the construction of Bode diagrams. The development of their characteristics will show certain simplifying approximations which are useful for the rapid sketching of these diagrams.

Consider the functions

$$\mathbf{G}_1(j\omega) = 1 + \frac{j\omega}{\omega_o} \quad \text{and} \quad \mathbf{G}_2 = \frac{1}{1 + j\omega/\omega_o} \tag{C-59}$$

At low frequencies ($\omega/\omega_o \ll 1$), the magnitude of both functions is approximately

$$G_1(j\omega) = G_2(j\omega) = 1$$

or

$$G_1(j\omega)_{dB} = G_2(j\omega)_{dB} = 20 \log 1 = 0 \tag{C-60}$$

At high frequencies ($\omega/\omega_o \gg 1$), the functions become

$$G_1(j\omega) = \frac{\omega}{\omega_o} \quad \text{and} \quad G_2(j\omega) = \frac{1}{\omega/\omega_o}$$

or

$$G_1(j\omega)_{dB} = 20 \log \frac{\omega}{\omega_o} \quad \text{and} \quad G_2(j\omega)_{dB} = -20 \log \frac{\omega}{\omega_o} \tag{C-61}$$

The low-frequency magnitudes are seen to be 0 dB (unity magnitude). The high-frequency magnitudes are $G_1 = G_2 = 0$ dB at $\omega/\omega_o = 1$; $G_1 = 20$ dB, $G_2 = -20$ dB at $\omega/\omega_o = 10$; $G_1 = 40$ dB, $G_2 = -40$ dB at $\omega/\omega_o = 100$; etc. The value of G_1 increases (and G_2 decreases) by a factor of 20 dB for each factor of 10 (decade) increase in ω/ω_o. Since factors of 10 are linear increments on the logarithmic frequency scale, Eqs. (C-61) are straight lines on the Bode plots; their slopes are $+20$ dB/decade for G_1 and -20 dB/decade for G_2. Often the slopes of the straight lines are expressed in units of decibels per octave, where an octave represents a factor of 2 in frequency. For $\omega/\omega_o = 2$, the values of G_1 and G_2 from Eq. (C-61) are 6 dB and -6 dB, respectively. Thus in one octave, from $\omega/\omega_o = 1$ to $\omega/\omega_o = 2$, G_1 has changed by 6 dB and G_2 by -6 dB. The corresponding slopes are 6 dB/octave and -6 dB/octave. It may be noted that 6 dB/octave and 20 dB/decade define identical slopes.

The angles associated with G_1 and G_2 are

$$\text{Angle } G_1 = \tan^{-1}\frac{\omega}{\omega_o} \quad \text{and} \quad \text{angle } G_2 = -\tan^{-1}\frac{\omega}{\omega_o} \tag{C-62}$$

FIGURE C-17
Asymptotic Bode diagrams for (a) $1 + j\omega/\omega_o$ and (b) $1/(1 + j\omega/\omega_o)$.

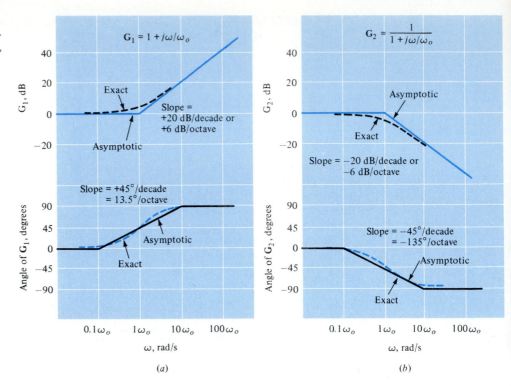

FIGURE C-17
Asymptotic Bode diagrams for (a) $1 + j\omega/\omega_o$ and (b) $1/(1 + j\omega/\omega_o)$.

At $\omega = \omega_o$, the angles of G_1 and G_2 are $+45°$ and $-45°$, respectively. For frequencies where $\omega \geq 10\omega_o$, the angle of G_1 is nearly $90°$, that for the angle of G_2 nearly $-90°$. At low frequencies ($\omega \leq 0.1\omega_o$), both angles are nearly zero. These results lead to the straight-line approximations for the angles of G_1 and G_2 shown in Fig. C-17. Also depicted in Fig. C-17 are the straight-line (asymptotic) magnitude characteristics. The dashed curves in the figure indicate the exact magnitude and phase responses. The exact and approximate curves are reasonably close. The greatest error in the asymptotic curve occurs at $\omega = \omega_o$ and is $+3$ dB for G_1 and -3 dB for G_2. One octave away from the corner frequency ($\omega = \omega_o/2$ and $\omega = 2\omega_o$), the error is $+1$ dB for G_1 and -1 dB for G_2. For angular frequencies more than one octave from the break frequency, errors are less than 1 dB and are generally neglected. The maximum error in the phase characteristic occurs at one decade away from the corner and is nearly $6°$. At the break frequency there is zero error; one octave away the error is nearly $5°$. The curves in Fig. C-17 indicate the algebraic sign of the errors for the angles of G_1 and G_2.

The process of drawing asymptotic Bode plots then becomes one of expressing the function in the form of Eq. (C-58), locating the break frequencies, drawing the component asymptotic curves, and adding these curves to get the resultant.

Example C-3

(a) Sketch the asymptotic Bode diagram for

$$G(s) = \frac{10^4(s + 40)}{s^2 + 410s + 4000}$$

(b) Determine the value of $G(j800)$.

Solution

(a) The equation for $G(s)$ is factored and put in the form of Eq. C-58:

$$G(s) = \frac{100(1 + s/40)}{(1 + s/10)(1 + s/400)}$$

or

$$\mathbf{G}(j\omega) = \frac{100(1 + j\omega/40)}{(1 + j\omega/10)(1 + j\omega/400)}$$

The break frequencies are 10 and 400 rad/s for the denominator and 40 rad/s for the numerator. The component curves are drawn in Fig. C-18. Note the constant value of 40 dB = 20 log 100 represents the constant multiplier in $\mathbf{G}(j\omega)$. The resultant magnitude and angle characteristics are indicated by the dashed lines in Fig. C-18.

FIGURE C-18
Asymptotic Bode diagram for Example C-3.

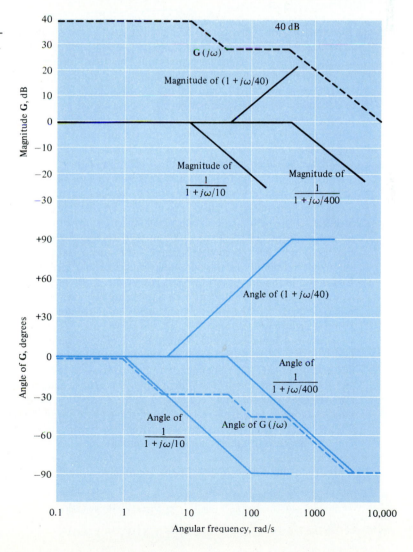

(*b*) From the resultant curves

$$G(j800) = 22 \text{ dB}$$

so that, from $20 \log G = G(\text{dB})$, we obtain

$$G(j800) = \log^{-1} \frac{22}{20} = 12.6$$

and

$$\angle G(j800) = -58.5°$$

The frequency response, given by the asymptotic Bode diagram in Fig. C-18, is determined with far less computation than is necessary in order to obtain the exact characteristic. With little additional effort, incorporating the errors at a few frequencies within the asymptotic plot produces a result of sufficient accuracy for most engineering analyses.

C-8 TWO-PORT NETWORKS Many networks can be considered as having two pairs of terminals: a pair of input terminals at which the excitation is usually applied, and a pair of output terminals at which the desired signals are extracted. The use of frequency-response characteristics and transfer functions highlights the importance of input-output relationships in systems. Indeed, as systems are composed of interconnected networks, overall system response is dependent on individual network responses. Just as the Thévenin equivalent is effective in representing the behavior of networks at a pair of terminals, equivalent circuits which focus on input-output characteristics are convenient representations of complex networks.

Networks which contain two pairs of terminals, one input pair and one output pair, are referred to as *two-terminal-pair networks* or *two-port networks*. A *port* refers to a pair of terminals at which energy can be supplied or extracted and at which measurements can be made. It is, therefore, customary to represent two-port networks as shown in Fig. C-19, which also indicates the standard convention for positive directions and polarities of the port currents and voltages. In Fig. C-19, terminals 1 and 1' represent the input port and terminals 2 and 2', the output port. Certain two-port and most electronic circuits have the additional property that terminals 1' and 2' are common. Embodied in the definition of a port is that the currents leaving terminals 1' and 2' are exactly equal to the currents entering terminals 1 and 2, respectively. In addition, measurements can be made only at the ports but are not permissible between terminals 1 and 2 or terminals 1' and 2'.

A two-port network can be described by four variables which are the port currents and voltages. Two of the variables may be considered to be the independent variables and the other two, the dependent variables. Because the system behaves linearly, the variables are related by a set of linear equations. These equations relate the port currents and voltages and define a set of *two-port parameters*. There are six combinations by which two of the four variables can be expressed in terms of the remaining two variables. Of the six possible parameter sets, three are used extensively in electronic circuit analysis because of their ease of measurement.

The *two-port admittance parameters,* or *y parameters,* are used to relate the port currents to the port voltages. In the frequency domain, the defining set of equations are

$$I_1 = y_{11}(s)V_1 + y_{12}(s)V_2 \tag{C-63}$$

$$I_2 = y_{21}(s)V_1 + y_{22}(s)V_2 \tag{C-64}$$

FIGURE C-19
Representation of a two-
port network indicating
reference voltage polar-
ities and current direc-
tions.

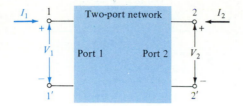

The elements $y_{11}(s)$, $y_{12}(s)$, $y_{21}(s)$, and $y_{22}(s)$ have the dimensions of mhos and are called the y parameters. Often the functional dependence of the parameters on the complex-frequency variable s is assumed to be implicit, and the parameters written simply as y_{11}, y_{12}, y_{21}, and y_{22}. Both notations will be used hereafter.

The specific name given each parameter is determined from its volt-ampere relationship. If terminals 2 and 2′ in Fig. C-19 are short-circuited, the voltage V_2 is constrained to be zero. Under these conditions, Eqs. (C-63) and (C-64) yield

$$y_{11} = \left.\frac{I_1}{V_1}\right|_{V_2=0} = \text{short-circuit input admittance} \tag{C-65}$$

$$y_{21} = \left.\frac{I_2}{V_1}\right|_{V_2=0} = \text{short-circuit forward transfer admittance} \tag{C-66}$$

The term "forward transfer" in Eq. (C-66) indicates the network is being used in its normal fashion with the excitation applied at the input port and the response measured at the output port. If the excitation is applied at port 2 and port 1 is short-circuited, Eqs. (C-63) and (C-64) give

$$y_{12} = \left.\frac{I_1}{V_2}\right|_{V_1=0} = \text{short-circuit reverse transfer admittance} \tag{C-67}$$

$$y_{22} = \left.\frac{I_2}{V_2}\right|_{V_1=0} = \text{short-circuit output admittance} \tag{C-68}$$

"Reverse transfer" indicates excitation applied at the output port and the response measured at the input port.

Another terminology is also used to identify y parameters, particularly when it is employed to describe electronic devices. With this terminology Eqs. (C-63) and (C-64) become

$$I_1 = y_i V_1 + y_r V_2 \tag{C-69}$$

$$I_2 = y_f V_1 + y_o V_2 \tag{C-70}$$

The subscripts i, r, f, and o indicate that the parameters with which they are associated are the *i*nput, *r*everse transfer, *f*orward transfer, and *o*utput, respectively. The y parameters in Eqs. (C-69) and (C-70) are defined by Eqs. (C-65) to (C-68).

It is often convenient to represent a two-port network by an equivalent circuit which exhibits the same terminal relations as those expressed in the defining equations. Figure C-20 depicts the *equivalent y-parameter circuit*.

FIGURE C-20

The y-parameter equivalent circuit.

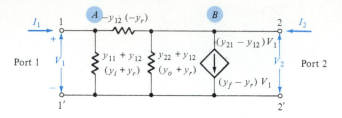

z Parameters

A second set of parameters can be established from the simultaneous solution of Eqs. (C-63) and (C-64) for V_1 and V_2. The results are

$$V_1 = \frac{y_{22}}{y_{11}y_{22} - y_{12}y_{21}} I_1 + \frac{-y_{12}}{y_{11}y_{22} - y_{12}y_{21}} I_2 \qquad \text{(C-71)}$$

$$V_2 = \frac{y_{21}}{y_{11}y_{22} - y_{12}y_{21}} I_1 + \frac{y_{22}}{y_{11}y_{22} - y_{12}y_{21}} I_2 \qquad \text{(C-72)}$$

In these equations the port voltages V_1 and V_2 are functions of the port currents I_1 and I_2. The general forms of Eqs. (C-71) and (C-72) are

$$V_1 = z_{11}I_1 + z_{12}I_2 = z_i I_1 + z_r I_2 \qquad \text{(C-73)}$$

and

$$V_2 = z_{21}I_1 + z_{22}I_2 = z_f I_1 + z_o I_2 \qquad \text{(C-74)}$$

The parameters z_{11}, z_{12}, z_{21}, and z_{22} are called the *impedance* or *z parameters*. Specific parameters are defined by first open-circuiting port 2, which makes $I_2 = 0$, and exciting port 1, and then repeating the process by open-circuiting port 1 and exciting port 2. The results are

$$z_{11} = z_i = \frac{V_1}{I_1}\bigg|_{I_2=0} = \text{open-circuit input impedance}$$

$$z_{21} = z_f = \frac{V_2}{I_1}\bigg|_{I_2=0} = \text{open-circuit forward transfer impedance}$$

$$\qquad \text{(C-75)}$$

$$z_{22} = z_o = \frac{V_2}{I_2}\bigg|_{I_1=0} = \text{open-circuit output impedance}$$

$$z_{12} = z_r = \frac{V_1}{I_2}\bigg|_{I_1=0} = \text{open-circuit reverse transfer impedance}$$

The equivalent circuit most often used to represent the z parameters is depicted in Fig. C-21.

h Parameters

A third set of two-port parameters is called the *hybrid* or *h parameters*. They are defined by the equations

$$V_1 = h_{11}I_1 + h_{12}V_2 = h_i I_1 + h_r V_2 \qquad \text{(C-76)}$$

$$I_2 = h_{21}I_1 + h_{22}V_2 = h_f I_1 + h_o V_2 \qquad \text{(C-77)}$$

The specific h parameters may be defined by first exciting the input port and short-circuiting the output port ($V_2 = 0$), and then exciting the output port and open-circuiting

FIGURE C-21
The z-parameter equivalent circuit.

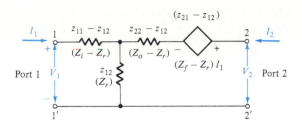

the input port ($I_1 = 0$). The results are expressed in Eqs. (C-78).

$$h_{11} = h_i = \left.\frac{V_1}{I_1}\right|_{V_2=0} = \text{short-circuit input impedance}$$

$$h_{21} = h_f = \left.\frac{I_2}{I_1}\right|_{V_2=0} = \text{short-circuit forward current gain}$$

$$\text{(C-78)}$$

$$h_{22} = h_o = \left.\frac{I_2}{V_2}\right|_{I_1=0} = \text{open-circuit output admittance}$$

$$h_{12} = h_r = \left.\frac{V_1}{V_2}\right|_{I_1=0} = \text{open-circuit reverse voltage gain}$$

The quantities h_r and h_f are dimensionless and are each represented by a controlled source in the circuit model shown in Fig. C-22. When used with transistors, an additional subscript is often used with the h parameters to indicate the transistor connection. Thus h_{fe} is the short-circuit forward current gain in the common-emitter configuration.

Example C-4

The circuit shown in Fig. C-23 is the equivalent circuit of a common-emitter amplifier stage.
(*a*) Determine h_{ie} and h_{fe}.
(*b*) Sketch an asymptotic Bode diagram for h_{ie} and h_{fe} for angular frequencies below 10^{10} rad/s.
(*c*) Determine the angular frequency at which $|h_{fe}(j\omega)|$ is unity.

Solution

(*a*) From the definitions of Eqs. (C-78), both h_{ie} and h_{fe} are computed by short-circuiting port 2 and exciting port 1 with a current source I_1. The resultant circuits used to determine h_{ie} and h_{fe} are shown in Figs. C-24a and C-24b, respectively. For the circuit in Fig. C-24a, we obtain

$$V_1 = I_1 \times 50 + I_1 Z$$

FIGURE C-22
The h-parameter equivalent circuit.

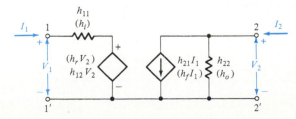

FIGURE C-23
Circuit for Example C-4.

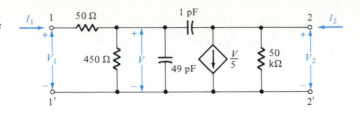

where

$$Z = \frac{450 \times 1/s(49 + 1) \times 10^{-12}}{450 + 1/s(40 + 1) \times 10^{-12}} = \frac{450}{1 + s \times 2.25 \times 10^{-8}}$$

Substitution of Z into the equation for V_1 and formation of the ratio V_1/I_1 gives

$$\frac{V_1}{I_1} = h_{ie} = 50 + \frac{450}{1 + s \times 2.25 \times 10^{-8}}$$

Rearrangement gives

$$h_i = \frac{500(1 + s/4.44 \times 10^8)}{(1 + s/4.44 \times 10^7)}$$

The circuit in Fig. C-24b is used to determine h_f. The current I_2, by use of KCL, is

$$I_2 = \frac{V}{5} - I_\mu = \frac{V}{5} - s \times 1 \times 10^{-12} V$$

as the current in the 50-kΩ resistance is zero because $V_2 = 0$. The voltage V is $I_1 Z$ and is given in the calculation for h_i as

$$V = I_1 Z = I_1 \frac{450}{1 + s \times 2.25 \times 10^{-8}}$$

FIGURE C-24
(a) Circuit to determine h_i and (b) circuit to determine h_f in Example C-4.

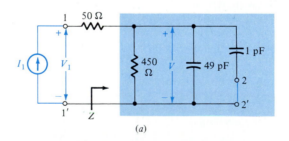

(a)

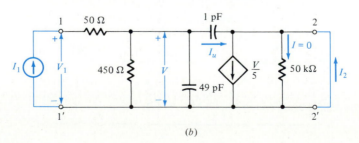

(b)

FIGURE C-25
Asymptotic Bode diagrams for (a) h_{ie} and (b) h_{fe} in Example C-4.

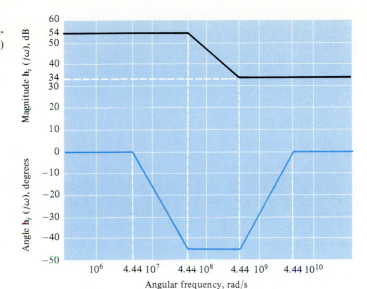

(a)

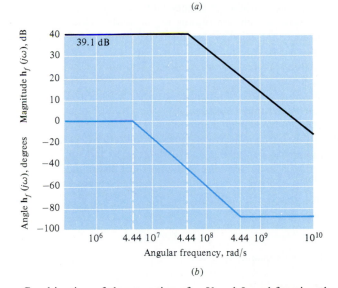

(b)

Combination of the equations for V and I_2 and forming the ratio I_2/I_1 yields

$$\frac{I_2}{I_1} = h_{fe} = \frac{90(1 - s/2 \times 10^{11})}{1 + s/4.44 \times 10^7}$$

(b) The asymptotic Bode diagrams are plotted in Fig. C-25.

(c) From the Bode diagram, the angular frequency at which $h_{fe}(j\omega) = 0$ dB is 4×10^9 rad/s.

Note that the break frequency at 2×10^{11} rad/s occurs at a frequency far removed from all other critical frequencies of interest. As a result, h_{fe} is often approximated by

$$h_{fe} = \frac{90}{1 + s/4.44 \times 10^7}$$

FIGURE C-26

(a) Signal-flow graph element and (b) a simple flow graph.

 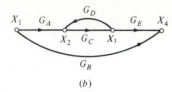

(a)   (b)

C-9 SIGNAL-FLOW GRAPHS

Simply stated, a signal-flow graph is a diagrammatic representation of a linear system of equations. As such, a signal-flow graph is often used to schematically describe a system in terms of its constituent parts. The two basic elements in a signal-flow graph are *nodes* and *branches*. A node is used to indicate a variable, a branch to indicate the relationship between a pair of variables. Figure C-26a shows a typical component in the graph. The variables X_1 and X_2 are represented by nodes. The directed arrow is the branch whose branch transmittance G defines the functional relationship $X_2 = GX_1$. The significance of the arrow is that it denotes the unilateral nature of the relation between X_2 and X_1. The arrow, being directed from X_1 to X_2, indicates that X_2 depends on X_1. Thus, in the graph of Fig. C-26b, the dependence of X_3 on X_2 is described by the branch transmittance G_C and that of X_2 on X_3 by the branch transmittance G_D.

In Fig. C-26b, the node variable X_2 is $G_D X_3 + G_A X_1$ and highlights the fact that the value of a node variable is defined only by the branches entering (incident upon) it. Each incident branch contributes to the node value an amount equal to the branch transmittance multiplied by the value of the node variable from which the branch leaves. Nodes having only incident branches are called *sink nodes,* and those having only branches that leave are *source nodes*. The nodes X_1 and X_4 in Fig. C-26b are source and sink nodes, respectively.

Because a signal-flow graph describes a set of linear equations, the elements of the graph can be combined algebraically. This process of graph reduction permits the transfer function to be evaluated and, in essence, is a method for solving the set of equations for one of the variables. Two elementary reductions are shown in Fig. C-27. The parallel-branch configuration in Fig. C-27a reduces to the sum of branch transmittances, and that for the cascade structure is the product of the individual branch transmittances.

Two other commonly encountered configurations are the self-loop and the feedback structure, which are illustrated in Figs. C-28a and C-28b, respectively. For the self-loop

$$X_2 = G_1 X_1 + H_1 X_2$$

and, after rearrangement, gives

$$X_2 = \frac{G_1}{1 - H_1} X_1$$

FIGURE C-27

Signal-flow graph equivalents for (a) parallel configuration and (b) cascade configuration.

$$X_1 \underset{G_2}{\overset{G_1}{\rightleftarrows}} X_2 \equiv X_1 \xrightarrow{G_1 + G_2} X_2$$

(a)

$$X_1 \xrightarrow{G_1} X_2 \xrightarrow{G_2} X_3 \equiv X_1 \xrightarrow{G_1 G_2} X_3$$

(b)

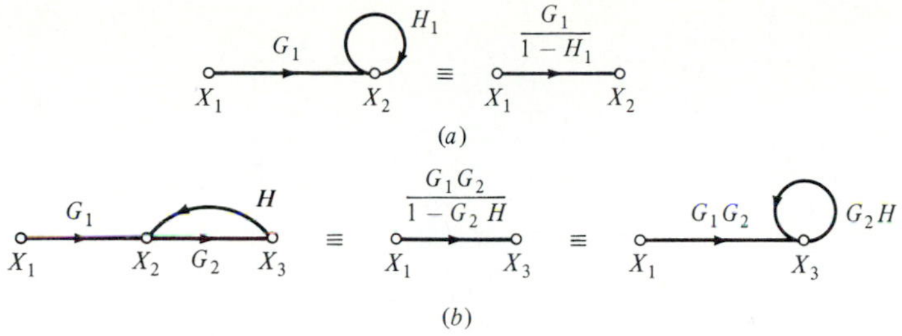

FIGURE C-28

Signal-flow graph equivalents for (*a*) self-loop and (*b*) feedback loop.

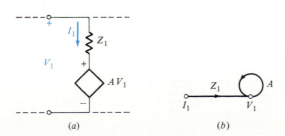

The relation for the equivalent branch in Fig. C-28*a* leads to the general rule for eliminating self-loops from a graph. Stated without further proof, this rule is that all branches incident on a node containing a self-loop have their branch transmittances divided by 1 minus the self-loop transmittance.

Figure C-29*a* depicts a portion of a circuit and demonstrates how a self-loop can arise in the formulation of network equations. The KVL expression for the circuit segment is

$$V_1 = I_1 Z_1 + A V_1$$

for which the graph in Fig. C-29*b* can be drawn. Both using the graph reduction and algebraically solving for V_1 in the KVL expression result in

$$V_1 = \frac{I_1 Z_1}{1 - A}$$

The reduction of the feedback loop in Fig. C-28*b* proceeds from

$$X_2 = G_1 X_1 + H_1 X_3 \qquad \text{and} \qquad X_3 = G_2 X_2$$

Substitution of X_2 into the equation for X_3 and combination of terms yields

$$X_3 = \frac{G_1 G_2}{1 - G_2 H}$$

which is the relation for the equivalent branch in the reduced graph in Fig. C-28*b*. Note that the same result is obtained for the flow graph containing the self-loop in Fig. C-28*b*.

Example C-5

The two-port network in Fig. C-30 is characterized by its y parameters.
(*a*) Construct a signal-flow graph for the circuit shown using V_s, V_1, I_1, I_2, and V_2 as nodes.
(*b*) Use the flow graph to evaluate the transfer function V_2/V_s.

FIGURE C-29

(*a*) Circuit segment; (*b*) flow graph for (*a*) showing self-loop.

FIGURE C-30
Circuit for Example
C-5.

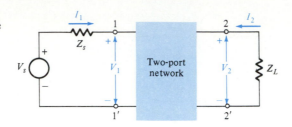

Solution

(a) The equations which relate the variables must be obtained first. Note, however, that V_s is the excitation and is represented by a source node. No other variable can be indicated by a source node. In general, all other node variables have branches both leaving and entering, except sink nodes. The basic equations which relate I_1, I_2, V_1, and V_2 are the two-port y parameters given in Eqs. (C-63) and (C-64) and restated here:

$$I_1 = y_{11}V_1 + y_{12}V_2$$

$$I_2 = y_{21}V_1 + y_{22}V_2$$

These equations are identified as branches A through D in the signal-flow graph in Fig. C-31. At port 1 in Fig. C-30, the KVL relation is expressible as

$$V_1 = V_s - I_1 Z_s$$

and indicated by branches E and F.

The last relation required is the Ohm's law equation at port 2:

$$V_2 = -I_2 Z_L$$

which is drawn as branch G. The seven branches constitute one possible signal-flow graph which characterizes the system.

(b) The transfer function is obtained by reducing the graph. Branches C, G, and D form a feedback loop. These are replaced by the equivalent branch H (see Fig. C-28b) whose transmittance is

$$H = \frac{-y_{21}Z_L}{1 + y_{22}Z_L}$$

and shown in Fig. C-32a. In Fig. C-32a, a feedback loop is formed by branches B, F, and A. This reduction is depicted in Fig. C-32b in which the feedback loop is replaced by branch J and self-loop K. The respective branch transmittances are

$$J = -y_{12}Z_s \qquad K = -y_{11}Z_s$$

The flow graph in Fig. C-32c results when the self-loop is eliminated. The branch transmittances affected by this reduction are E and J whose values are divided by

FIGURE C-31
Signal-flow graph for
circuit in Fig. C-30.

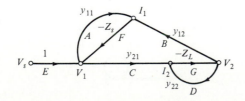

FIGURE C-32
Reduction of signal-flow
graph of Fig. C-31.

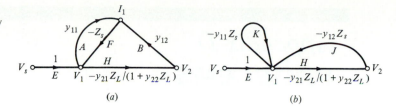

(a) (b)

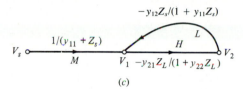

(c)

$(1 + y_{11}Z_s)$, the self-loop transmittance. The equivalent branches for J and E are L and M, respectively.

Figure C-32c indicates that a feedback loop is constituted by branches M, L, and H and their reduction gives

$$\frac{V_2}{V_s} = \frac{[1/(1 + y_{11}Z_s)] \times [-y_{21}Z_L/(1 + y_{22}Z_L)]}{1 - [-y_{21}Z_L/(1 + y_{22}Z_L)] \times [-y_{12}Z_s/(1 + y_{11}Z_s)]}$$

Clearing fractions and combining terms yields

$$\frac{V_2}{V_s} = \frac{-y_{21}Z_L}{1 + y_{11}Z_s + y_{22}Z_L + Z_sZ_L(y_{11}y_{22} - y_{12}y_{21})}$$

Appendix D
PROBLEMS

CHAPTER 1

1-1 An electron is emitted from an electrode with a negligible initial velocity and is *accelerated* by a potential V. Find the value of V, given that the final velocity of the particle is 9.4×10^6 m/s.

1-2 An electron having an initial kinetic energy of 10^{-17} J at the surface of one of two parallel-plane electrodes and moving normal to the surface is slowed down by the retarding field caused by a potential V_x applied between the electrodes. What value of V_s is required for the electron to reach the second electrode with zero velocity?

1-3 The essential features of the display tube of an oscilloscope are shown in the accompanying figure. The voltage difference between K and A is V_a and between P_1 and P_2 is V_p. Neither electric field affects the other one. The electrons are emitted from the electrode K with initial zero velocity, and they pass through a hole in the middle of electrode A. Because of the field between P_1 and P_2 they change direction while they pass through these plates and, after that, move with constant velocity toward the screen S. The distance between plates is d.
(a) Find the velocity v_x of the electrons as a function of V_a as they cross A.
(b) Find the Y component of velocity v_y of the electrons as they come out of the field of plates P_1 and P_2 as a function of V_p, l_d, d, and v_x.
(c) Find the distance from the middle of the screen (d_s), when the electrons reach the screen, as a function of tube distances and applied voltages.
(d) For $V_a = 2.0$ kV, and $V_p = 100$ V, $l_d = 1.27$ cm, $d = 0.5$ cm, and $l_s = 20$ cm, find the numerical values of v_x, v_y, and d_s.
(e) If we want to have a deflection of $d_s = 1.0$ cm of the electron beam, what must be the value of V_a? All other values are given in (d).

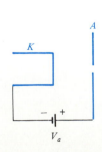
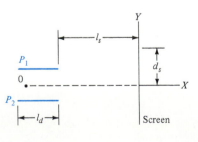

1-4 A flat aluminum strip has a resistivity of $3.44 \times 10^{-8}\ \Omega \cdot$ m, a cross-sectional area of 2×10^{-4} mm^2, and a length of 5 mm. What is the voltage drop across the strip for a current of 50 mA?

1-5 For the aluminum strip described in Prob. 1-4, what current exists if the voltage across the strip is 30 μV?

1-6 (*a*) Calculate the electric field required to give an electron in silicon (Si) an average energy of 1.1 eV.
  (*b*) Is it practical to generate electron-hole pairs by applying a voltage across a bar of silicon? Explain.

1-7 Repeat Prob. 1-5 for an intrinsic silicon strip at 800 K.

1-8 Compute the mobility of the free electrons in aluminum (Al) for which the density is 2.70×10^3 kg/m^3 and the resistivity is $3.44 \times 10^{-8}\ \Omega \cdot$ m. Assume that Al has three valence electrons per atom and an atomic weight of 26.98.

1-9 (*a*) Determine the concentration of free electrons and holes at 300 K for a silicon sample which has a donor atom concentration of $N_D = 2 \times 10^{14}$ atoms/cm^3 and an acceptor atom concentration of $N_A = 3 \times 10^{14}$ atoms/cm^3.
  (*b*) Is the sample in (*a*) *p*- or *n*-type silicon?

1-10 Repeat Prob. 1-9 for $N_A = N_D = 10^{15}$ atoms/cm^3.

1-11 Repeat Prob. 1-9 for $N_D = 10^{16}$ atoms/cm^3 and $N_A = 10^{14}$ atoms/cm^3.

1-12 (*a*) Find the concentration of holes and electrons in a *p*-type silicon sample at 300 K, assuming that the resistivity is $0.02\ \Omega \cdot$ cm.
  (*b*) Repeat (*a*) for *n*-type silicon.

1-13 Repeat Prob. 1-12 for a resistivity of $5\ \Omega \cdot$ cm.

1-14 Donor impurities are added to intrinsic silicon and the resistivity decreases to $1\ \Omega \cdot$ cm. Compute the ratio of donor atoms to Si atoms per unit volume.

1-15 If silicon (Si) were a monovalent metal, what would the ratio of its conductivity be to that of intrinsic Si at 300 K?

1-16 The electron concentration in a semiconductor is shown.
  (*a*) Derive an expression and sketch the electron current density $J_n(x)$, assuming that there is no externally applied electric field.
  (*b*) Sketch and derive an expression for the built-in electric field that must exist if the net electron current is to be zero.
  (*c*) Determine the potential between points $x = 0$ and $x = W$, given $n(0)/n_o = 10^3$.

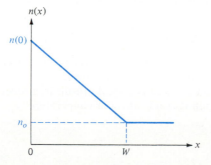

Prob. 1-16

1-17 Verify Eq. (1-40) for an open-circuited graded semiconductor.

1-18 Verify the expression for the contact potential V_o given in Eq. (1-42) for the step-graded junction shown in Fig. 1-10b by considering the electron current density $J_n = 0$.

1-19 The junction in Fig. 1-10b is doped with N_A corresponding to 1 acceptor atom per 10^6 Si atoms. Calculate the contact difference of potential V_o at room temperature.

1-20 Determine the *change* in the contact difference of potential in an open-circuited pn junction at 300 K, assuming that N_D is changed by a factor of 2500 and N_A remains unchanged.

1-21 (*a*) Repeat Prob. 1-20, assuming that N_D remains unchanged and N_A is changed by a factor of 8000.
(*b*) Does the answer in (*a*) depend on whether N_A is increased or decreased? Explain briefly.

1-22 The resistivities of the two sides of a step-graded Si junction are 5 $\Omega \cdot$ cm (p side) and 2.5 $\Omega \cdot$ cm (n side). Calculate the height of the potential barrier V_0.

1-23 Repeat Prob. 1-22, assuming that the resistivities of the two sides are interchanged.

CHAPTER 2

2-1 Sketch logarithmic plots of carrier concentration versus distance for an abrupt silicon junction if $N_A = 5 \times 10^{14}$ atoms/cm³ and $N_D = 5 \times 10^{16}$ atoms/cm³. Give numerical values for ordinates. Label the n, p, and depletion regions.

2-2 The resistivities of the two sides of an abrupt silicon junction are 2.4 $\Omega \cdot$ cm (p side) and 25 $\Omega \cdot$ cm (n side). Sketch the logarithmic plots of carrier concentration versus distance. Give numerical values for the ordinates. Label the n, p, and depletion regions.

2-3 (*a*) For what voltage will the reverse current in a pn junction silicon diode reach 95 percent of its saturation value at room temperature?
(*b*) What is the ratio of the current for a forward bias of 0.2 V to the current for the same magnitude of reverse bias?
(*c*) If the reverse saturation current is 10 pA, what are the forward currents for voltages of 0.5, 0.6, and 0.7 V, respectively?

2-4 If the reverse saturation current in a pn junction silicon diode is 1 nA, what is the applied voltage for a forward current of 2.5 μA?

2-5 (*a*) A silicon diode at room temperature (300 K) conducts 1 mA at 0.7 V. Given that the voltage increases to 0.8 V, calculate the diode current. Assume $\eta = 2$.
(*b*) Calculate the reverse saturation current.
(*c*) Repeat (*a*) for $\eta = 1$.

2-6 (*a*) What increase in temperature would result in a reverse saturation current which is 60 times its value at room temperature?
(*b*) What decrease in temperature would result in a reverse saturation current which is one-tenth its value at room temperature?

2-7   A diode is mounted on a chassis in such a manner that, for each degree of temperature rise above ambient, 0.1 mW is thermally transferred from the diode to its surroundings. (The "thermal resistance" of the mechanical contact between the diode and its surroundings is 0.1 mW/°C.) The ambient temperature is 25°C. The diode temperature is not to be allowed to increase by more than 10°C above ambient. If the reverse saturation current is 5 nA at 25°C and increases at the rate 0.07°C^{-1}, what is the maximum reverse-biasing voltage that can be maintained across the diode?

2-8   A silicon diode is operated at a constant forward voltage of 0.7 V. What is the ratio of the maximum to minimum current in the diode over a temperature range -55 to $+100$°C?

2-9   The silicon diode described in Fig. 2-5 is used in the circuit in Fig. 2-8a with $V_{AA} = 6$ V and $R = 100$ Ω.
   (a)   Determine the diode current and the voltage.
   (b)   If V_{AA} is decreased to 3 V, what must the new value of R be if the diode current is to remain at the value in (a)?

2-10   A silicon diode whose characteristics are shown is used in the circuit in Fig. 2-8a with $V_{AA} = 5$ V and $R = 1$ kΩ.
   (a)   Determine the current in and voltage across R.
   (b)   What is the power dissipated by the diode?
   (c)   What is the diode current if R is successively changed to 2 and 5 kΩ?

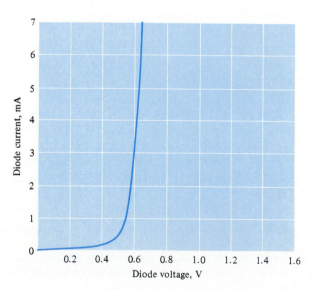

Diode current, mA

Diode voltage, V

Prob. 2-10

2-11   (a)   Repeat Prob. 2-10, parts (a) and (b) for $V_{AA} = 10$ V and $R = 2$ kΩ.
   (b)   What is the load current if V_{AA} is reduced to 5.0 V?
   (c)   What is the diode current if V_{AA} is increased to 20 V?

2-12 The circuit shown uses the diode in Prob. 2-10. Find V_o, given $V_{BB} = 9$ V.

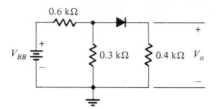

0.6 kΩ

V_{BB}

0.3 kΩ

0.4 kΩ V_o

+

−

Prob. 2-12

2-13 (*a*) The Si diode whose characteristics are given in Fig. 2-5 is used in the circuit from Prob. 2-12 with $V_{BB} = 60$ V. Determine the power dissipated by the 0.4-kΩ resistance.

2-14 A constant current $I = 70$ mA is supplied to the diode-resistance combination shown. The resistance R is a precision 1.00-kΩ resistor. At a temperature of 25°C, the diode voltage is 700 mV.
(*a*) Plot a graph of I_R versus temperature T from -55 to 125°C.
(*b*) Comment on the use of the circuit as a thermometer. Assume that the values of R and the diode current have negligible variation over the temperature range.

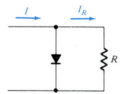

I I_R

R

Prob. 2-14

2-15 Determine the current in the circuit in Fig. 2-8*a* for $V_{AA} = 12$ V and $R = 4$ kΩ, assuming the diode
(*a*) Is ideal
(*b*) Can be represented as in Fig. 2-11 with $V_\gamma = 0.6$ V and $R_f = 20$ Ω

2-16 In the circuit in Prob. 2-12, determine V_o, assuming that the diode
(*a*) Is ideal
(*b*) Is represented as in Fig. 2-11 with $V_\gamma = 0.6$ V and $R_f = 30$ Ω

2-17 (*a*) Represent the silicon diode in Fig. 2-5 by the model given in Fig. 2-11; that is, estimate V_γ and R_f.
(*b*) Use this representation to solve Prob. 2-9, part (*a*).
(*c*) Compare your answer to (*b*) with the answer to Prob. 2-9, part (*a*).

2-18 (*a*) Repeat Prob. 2-17, part (*a*) for the diode characteristic in Prob. 2-10.
(*b*) Use this representation to solve Prob. 2-12.
(*c*) Compare your answer with the answer for Prob. 2-12.

2-19 The current in the circuit of Fig. 2-8 is to be 10 mA with $V_{AA} = 1.5$ V. Determine the value of R_L, given that the diode is
(*a*) Ideal
(*b*) Represented by $V_\gamma = 0.5$ V and $R_f = 50$ Ω

2-20 Sketch the output voltage $v_o(t)$ in the circuit shown for $0 \le t \le 5$ ms, assuming that the diode is ideal.

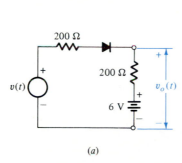

(a)

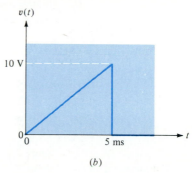

(b)

2-21 Repeat Prob. 2-20, given that the diode is represented by $V_\gamma = 0.5$ V and $R_f = 50\ \Omega$.

2-22 Sketch the output voltage for the circuit shown for $0 \le t \le 10$ ms, assuming that the diode is
(a) Ideal
(b) Represented by $V_\gamma = 0.6$ V and $R_f = 20\ \Omega$

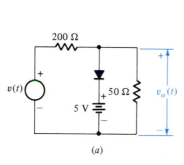

(a)

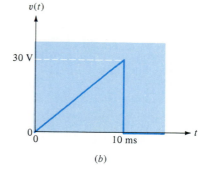

(b)

2-23 Sketch the voltage transfer characteristic (v_o versus v_i) for the circuit in Prob. 2-20, assuming that the diode is
(a) Ideal
(b) Represented by $V_\gamma = 0.6$ V and $R_f = 25\ \Omega$

2-24 Sketch the voltage transfer characteristic (v_o versus v) for the circuit in Prob. 2-22, assuming that the diode is
(a) Ideal
(b) Represented by $V_\gamma = 0.5$ V and $R_f = 40\ \Omega$

2-25 Obtain the voltage transfer characteristic for the circuit shown, assuming that the diodes are identical and have $V_\gamma = 0.6$ V and $R_f = 0$.

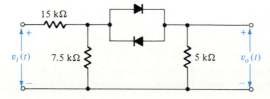

2-26 (a) Obtain the voltage transfer characteristic of the circuit shown, assuming that the diodes are ideal.

(b) Sketch one cycle of the output voltage, assuming that the input voltage is $v_i(t) = 20 \sin \omega t$.

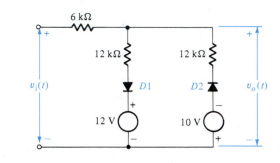

Prob. 2-26

2-27 The input voltage to the network whose voltage transfer characteristic is shown is $v_i = 2 + 2 \sin \omega t$. Sketch the output voltage $v_o(t)$ for one cycle of the input.

2-28 (a) The voltage transfer characteristic of a diode network is shown. Sketch the output voltage for $v_i(t) = 2.0 + 3 \sin \omega t$.

(b) Design a simple diode network, using ideal diodes, that has the transfer function given.

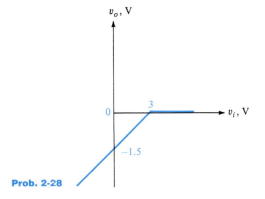

Prob. 2-27

Prob. 2-28

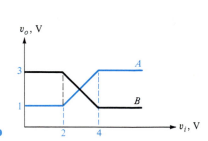

Prob. 2-29

2-29 (a) A sinusoid $v_i(t) = 3 + 2 \sin \omega t$ is applied to a diode network whose voltage transfer characteristic is denoted by A. Sketch the output waveform $v_o(t)$ for one cycle.

(b) What changes would you expect in the output waveform if B is the voltage transfer characteristic of the network?

(c) Using ideal diodes, design a circuit having the A characteristic.

2-30 (a) Obtain the voltage transfer characteristic of the circuit in Fig. 2-23.

(b) Using (a), verify that the circuit is a two-level clipper.

2-31 (*a*) The current I in Prob. 2-14 changes by $\Delta I \ll I$. Use small-signal analysis to determine ΔI_R.

(*b*) For $R = 1\ \mathrm{k\Omega}$, what is the minimum value of I for which $(\Delta I_R/\Delta I) \le 0.01$ at room temperature? Neglect the effect of C_D.

2-32 In the circuit in Prob. 2-12, V_{BB} increases from 6.0 to 6.25 V. Determine

(*a*) The change ΔV_o in V_o

(*b*) The new value of V_o

2-33 In the circuit in Prob. 2-20, part (*a*), $v(t) = 8 + 0.02 \sin \omega t$. Neglecting the effect of the diffusion capacitance, and assuming that the dc model of the diode is $V_\gamma = 0.6$ and $R_f = 0$, determine and sketch the output voltage that would appear on an oscilloscope if the selector is set to

(*a*) Ac

(*b*) Dc

2-34 The waveforms for v_1 and v_2 are applied to the diode-resistance circuit shown. Sketch $v_o(t)$ for $0 \le t \le 4$ ms. Assume that the silicon diode switches instantaneously and that $V_\gamma = 0.6$ V and $R_f = 20\ \Omega$.

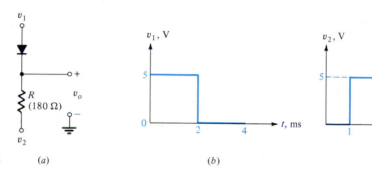

Prob. 2-34   (*a*)   (*b*)   (*c*)

2-35 (*a*) The transfer characteristic of a diode circuit is shown. Sketch the output voltage for one cycle, assuming that $v_i = 6 + V_m \sin \omega t$.

(*b*) If $v_i = 6 + \Delta V_i$, what is the change in v_o, ΔV_o for both a positive and a negative ΔV_i?

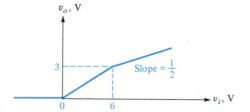

Prob. 2-35

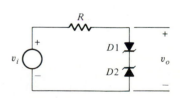

Prob. 2-36

2-36 Sketch the transfer characteristic for the Zener diode circuit shown, assuming that $D1$ and $D2$ are identical and have parameters V_Z, V_γ, and R_f.

2-37 Repeat Prob. 2-36 for $V_{Z1} = 5$ V and $V_{Z2} = 10$ V. Assume $V_\gamma = 0.6$ V, $R = 10\ \mathrm{k\Omega}$, $R_f = 20\ \Omega$, and the saturation current is negligible.

2-38 In the circuit in Fig. 2-32, a 5-V Zener diode is used which provides regulation for 50 mA $\leq I_A \leq$ 1.0 A. Determine the range of load currents for which regulation is achieved if the unregulated voltage V_S varies between 7.5 and 10 V. The resistance R_S = 4.75 Ω.

2-39 The regulator in Fig. 2-32 is to provide a 6-V load voltage for all load currents $I_L \leq$ 0.5 A. The unregulated supply varies between 8 and 10 V, and the Zener diode provides regulation for $I_Z >$ 0. Determine
(a) The series resistance R_S needed
(b) The power dissipation rating of the Zener diode

2-40 The circuit in Fig. 2-32 is designed with R_S = 20 Ω. The 5.6-V Zener diode provides regulation for 1 mA $\leq I_Z \leq$ 300 mA and for a load current of 0 $\leq I_L \leq$ 200 mA. Determine the range of amplitudes of the unregulated supply for which the load remains regulated.

2-41 Reverse-biased diodes are frequently employed as electrically controllable variable capacitors. The depletion capacitance of an abrupt junction diode is 4 pF at 4 V. Compute the change in capacitance for
(a) A 0.5-V increase in bias
(b) A 0.5-V decrease in bias

2-42 The derivation of Eq. (2-40) for the diffusion capacitance C_D assumes that the p side is more heavily doped than the n side, so that the current at the junction is essentially the hole current. Derive an expression for C_D when this approximation is not made.

2-43 For the circuit shown the cut-in voltage of a diode is 0.6 V and the drop across a conducting diode is V' = 0.7 V. Calculate v_o for the following input voltages and indicate the state of each diode (ON or OFF). Justify your assumptions about the state of each diode:
(a) v_1 = 10 V, v_2 = 0 V
(b) v_1 = 5 V, v_2 = 0 V
(c) v_1 = 10 V, v_2 = 5 V
(d) v_1 = 5 V, v_2 = 5 V

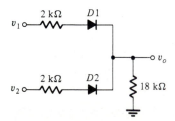

Prob. 2-43

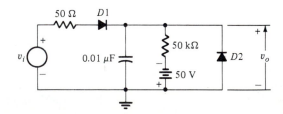

Prob. 2-45

2-44 Repeat Prob. 2-43, assuming that the 18-kΩ resistor is returned to ground through a 5-V supply.

2-45 In the circuit shown, v_i is a 5-V pulse whose duration is 10 to 40 ns. Sketch the output pulse for input pulse widths of 10, 20, 30, and 40 ns. Assume that the diodes are ideal. (*Hint:* For $x \ll 1$, $\epsilon^{-x} \approx 1 - x$.)

CHAPTER 3

3-1 A current-controlled current source is used in the circuit in Fig. 3-3. The controlled source is defined by

$$i_2 = 100 \, i_1 \text{ mA} \quad \text{for } i_1 \geq 0$$

$$i_2 = 0 \quad \text{for } i_1 < 0$$

and is otherwise ideal. The parameter values are $R_s = 100 \, \Omega$, $R_2 = 1 \, k\Omega$, and $V_{22} = 10 \, V$.
(a) Draw a set of output characteristics (i_2 versus v_2) for $0 \leq i_1 \leq 200 \, \mu A$.
(b) Plot a transfer characteristic (v_2 versus v_s) for $v_s \geq 0$.
(c) What value of v_s is needed to make $v_2 \leq 0.5 \, V$?

3-2 A nonideal controlled source is used in the circuit shown. Sketch the transfer characteristic (v_2 versus v_s) for $r_o \rightarrow \infty$ (open circuit). Which segments of the characteristic would you use if the circuit were to act as a switch? As an amplifier? (*Hint:* First draw a set of output characteristics for the device. Assume that the cut-in and ON voltages of the diode are both 0.5 V and $R_f = 50 \, \Omega$.)

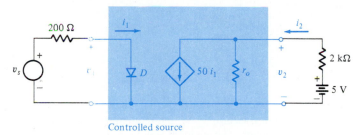

Controlled source
Prob. 3-2

3-3 Repeat Prob. 3-2, given $r_o = 20 \, k\Omega$.

3-4 The device in Fig. 3-2 is an ideal *current-controlled voltage source*. Draw a set of typical output characteristics and then explain how this device can be used as a controlled switch or as an amplifier.

3-5 (a) With the base open-circuited ($I_B = 0$), determine whether the emitter-base and collector-base junctions are forward- or reverse-biased.
(b) Evaluate the current that exists for *pnp* transistor having $I_{ES} = 1 \, pA$, $\alpha_F = 0.99$, and $\alpha_R = 0.5$. Assume room temperature.

3-6 Derive Eq. (3-16); modify Eq. (3-16) for an *npn* transistor.

3-7 Derive Eq. (3-24) from the Ebers-Moll equations.

3-8 The circuit in Fig. 3-24a is used to bias a 2N2222A transistor, whose characteristics are given in Fig. 3-16, at $V_{CEQ} = 5 \, V$ and $I_{CQ} = 15 \, mA$. The supply voltage is $V_{CC} = 12 \, V$.
(a) Determine the values of R_B and R_C needed.
(b) Estimate the value of β_F at these bias conditions.

3-9 Draw the circuit, analogous to Fig. 3-24a, used to bias the *pnp* 2N2907A transistor. (This transistor and the 2N2222A are complementary.) For $V_{CC} = 15 \, V$, determine the values of R_B and R_C needed to establish $V_{CE} = -10 \, V$ and $I_C = -20 \, mA$.

3-10 A 2N2222A transistor is used in the circuit in Fig. 3-24a with $R_C = 225 \, \Omega$, $R_B = 100 \, k\Omega$, and $V_{CC} = 9 \, V$. Determine I_C and V_{CE}.

3-11 Use the circuit in Fig. 3-25a to bias the 2N2222A at $V_{CE} = 5$ V and $I_C = 15$ mA with $V_{EE} = 10$ V.

3-12 A transistor having $\beta_F = 99$ and negligible reverse saturation current is used in the circuit in Fig. 3-25a with $R_C = 2$ kΩ, $R_E = 1$ kΩ, $R_B = 200$ kΩ, and $V_{EE} = 6$ V.
(a) Determine I_C and V_{CE}.
(b) Repeat (a) for β_F changed to 199.

3-13 The transistor used in the circuit shown has $\beta_F = 150$ and negligible reverse saturation current.
(a) Determine I_C and V_{CE}.
(b) Repeat (a) for β_F reduced to 50.

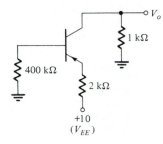

Prob. 3-13

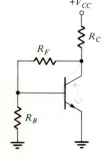

Prob. 3-14

3-14 A transistor with $\beta_F = 99$ and negligible reverse saturation current is used in the circuit shown. The element values are $V_{CC} = 10$ V, $R_C = 2.7$ kΩ, and $R_F = 180$ kΩ, and R_B is open-circuited.
(a) Determine the values of V_{CE} and I_C.
(b) For β_F increased to 199, repeat (a).

3-15 The circuit in Prob. 3-14 is used to establish $V_{CE} = 5$ V and $I_C = 5$ mA with $V_{CC} = 9$ V. The transistor in Prob. 3-14 is used and R_B is open-circuited.
(a) Determine R_C and R_F.
(b) Find the new values of I_C and V_{CE} for β_F changed to 49.

3-16 The circuit in Prob. 3-14 is used with the following values: $R_C = 2$ kΩ, $R_B = 25$ kΩ, and $V_{CC} = 12$ V. The transistor has $\beta_F = 49$ and negligible reverse saturation current.
(a) Determine R_F so that $I_E = -2$ mA.
(b) Using the value of R_F in (a), determine I_E for β_F changed to 150.

3-17 The circuit shown uses a transistor having $\beta_F = 100$ and parameter values $R_C = 0.5$ kΩ, $R_E = 1.0$ kΩ, $R_B = 44$ kΩ, $V_{CC} = 15$ V, $V_{EE} = -15$ V, and $V_{BB} = 0$.
(a) Determine V_{O1} and V_{O2}.
(b) What new value of R_C makes $V_{O1} = 0$?
(c) What new value of R_C makes $V_{O2} = 0$?

Neglect the reverse saturation current.

3-18 For the circuit in Prob. 3-17, the supply voltages V_{BB}, V_{CC}, and V_{EE} can each be 10, -10, or 0 V. List all the possible combinations of supply voltages for which it is possible for the transistor to be biased in the forward-active region.

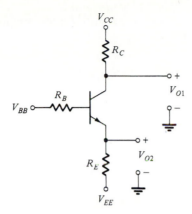

Prob. 3-17

3-19 Repeat Prob. 3-18 for the reverse-active region.

3-20 A transistor having $\beta_F = 125$ and $\beta_R = 1$ is used in the circuit in Fig. 3-27a. For $V_{EE} = 6$ V, $R_E = R_C = 1$ kΩ, determine R_B so that the current in R_E is 1 mA.

3-21 The element values in the circuit of Fig. 3-26a are $R_1 = 150$ kΩ, $R_2 = 37.5$ kΩ, $R_C = 7$ kΩ, and $R_E = 3$ kΩ. The transistor has $\beta_F = 100$ and negligible reverse saturation current. For $V_{CC} = 9$ V
(*a*) Determine V_{CE} and I_C.
(*b*) Repeat (*a*) for $\beta_F = 50$.

3-22 The circuit in Fig. 3-26a uses the transistor given in Prob. 3-21. The element values are $R_1 = 90$ kΩ, $R_2 = 10$ kΩ, $R_C = 10$ kΩ, $R_E = 0.9$ kΩ, and $V_{CC} = 12$ V.
(*a*) Determine V_{CE} and I_C.
(*b*) Repeat (*a*) for $\beta_F = 200$.

3-23 The circuit in Fig. 3-26a is to be used with a *pnp* transistor having $\beta_F = 50$ and negligible reverse saturation current. A positive supply voltage of 12 V is available. The emitter and collector resistors are each 2 kΩ. Determine the values of R_1 and R_2 which make $V_{CE} = -6$ V.

3-24 In the circuit from Prob. 3-17, determine the value of V_{BB} which
(*a*) Just barely saturates the transistor
(*b*) Makes $\beta_{\text{forced}} = 10$

The element values are those given in Prob. 3-17.

3-25 The element values in the circuit from Prob. 3-17 are $V_{CC} = 0$, $V_{EE} = -10$ V, $R_E = 0$, $R_C = 2$ kΩ, and $R_B = 50$ kΩ. The transistor has $\beta_F = 125$ and negligible reverse saturation current. Sketch the transfer characteristic V_{O1} versus V_{BB} clearly indicating the region of operation of the transistor.

3-26 Repeat Prob. 3-25, assuming $V_{CC} = 10$ V and that all other parameters are those given.

3-27 (*a*) Repeat Prob. 3-25 for the following parameters: $R_C = 5$ kΩ, $R_B = 100$ kΩ, $R_E = 2$ kΩ, $V_{CC} = 9$ V, and $V_{EE} = 0$ V. The transistor has $\beta_F = 150$ and negligible reverse saturation current.
(*b*) Sketch the transfer characteristic V_{O2} versus V_{BB}.

3-28 Sketch the transfer characteristic V_o versus V_i for the circuit shown. The transistor has $\beta_F = 75$ and $I_{CO} \approx 0$.

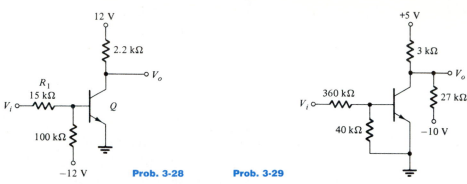

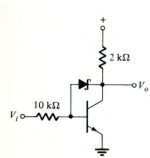

Prob. 3-30

Prob. 3-28   Prob. 3-29

3-29 Sketch the transfer characteristic V_o versus V_i for the circuit shown. The transistor used is described in Prob. 3-28.

3-30 Sketch the transfer characteristic V_o versus V_i for the circuit shown. The transistor has $\beta_F = 150$ and negligible reverse saturation current. The Schottky diode has a 0.4-V drop when conducting.

3-31 The transistors $Q1$ and $Q2$ in the circuit shown are identical and have $\beta_F = 100$ and negligible reverse saturation current.
  (a) Find V_o when $V_i = 0$. Assume that $Q1$ is OFF and justify the assumption.
  (b) With $V_i = 6$ V, find V_o. Assume $Q2$ is OFF and justify the assumption.
  (c) Sketch the voltage transfer characteristic V_o versus V_i as V_i increases from 0 to 6 V.
  (d) Repeat (c) for V_i decreasing from 6 to 0 V.

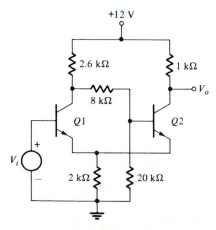

Prob. 3-31

3-32 The input voltage v_i in the circuit is $v_i(t) = 2.0 + 1.0 \sin 2\pi \times 10^3 t$. The transistor used is described in Prob. 3-31, and $v_E(t)$ is as shown. Sketch $v_o(t)$ for one cycle.

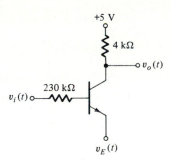

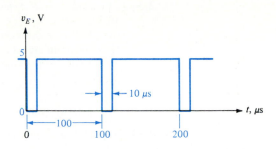

Prob. 3-32

3-33 The transistor in Prob. 3-31 is used in the circuit shown.
  (a) Determine R_B so that the transistor is barely saturated for $V_i = 5$ V.
  (b) If V_i is the rectangular pulse displayed, sketch $v_o(t)$. Assume that the *transistor* responds instantaneously.

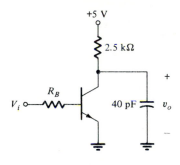

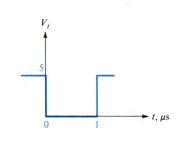

Prob. 3-33

3-34 (a) Sketch the voltage transfer characteristic for the circuit shown. The transistor has $\beta_F = 120$, $\beta_R = 2$, and $I_{CO} \approx 0$. The Zener diode is rated at $V_Z = 5.6$ V.
  (b) Sketch I_Z versus V_i.

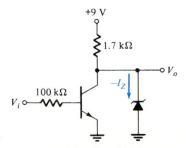

Prob. 3-34

3-35 A transistor is biased at $I_C = 0.5$ mA and has $\beta_o = 150$.
  (a) Determine g_m and r_π at room temperature.
  (b) The input resistance $h_{ie} = 7.6$ kΩ. Find r_b.
  (c) A load resistance $R_C = 2$ kΩ is used and the transistor is driven from a 300-Ω source. Estimate the voltage gain.

3-36 The 2N2222A is biased at I_{CQ} = 20 mA and V_{CEQ} = 5 V. The supply voltage is 10 V.

  (a) Estimate the value of β_o for the transistor.

  (b) An input signal $i_b(t)$ = 20 sin ωt μA is superimposed on the quiescent level. Estimate the signal component of the collector current.

3-37 (a) Draw the small-signal equivalent, valid at low frequencies, of the circuit in Fig. 3-25a.

  (b) Derive an expression for the resistance seen between base and ground.

3-38 Repeat Prob. 3-37 for the circuit in Fig. 3-26a.

3-39 In the circuit of Prob. 3-28, the transistor has β_F = 100 and β_o = 100. The reverse saturation current is negligible, the Early voltage $V_A \rightarrow \infty$, and the base-spreading resistance r_b = 0. The voltage V_i = 3.75 + ΔV_i V.

  (a) Draw the small-signal model of the circuit at low frequencies; include numerical values for the transistor parameters.

  (b) Use (a) to evaluate the change ΔV_o in V_o caused by ΔV_i.

  (c) Evaluate ΔV_o for ΔV_i = 0.25 V.

  (d) Compare the result in (c) with a dc analysis of the circuit for V_i = 4.0 V. Explain any differences.

3-40 (a) Draw the small-signal equivalent circuit of the common-base stage in Fig. 3-13.

  (b) At low frequencies, calculate the resistance seen between emitter and base (looking into the transistor).

3-41 (a) Draw the small-signal equivalent circuit, valid at low frequencies, of the circuit in Fig. 3-37.

  (b) Determine V_{o2}, given V_1 = $-V_2$ = 25 V. The transistor parameter values are β_o = 125, r_b = 0, and r_o = 1 MΩ. The current source I_{EE} = 0.2 mA and R_C = 250 kΩ.

3-42 Repeat Prob. 3-41, part (b) for V_1 = 25 μV and V_2 = 0.

3-43 Repeat Prob. 3-41, part (b) for V_1 = 0 and V_2 = 25 μV.

3-44 The low-frequency small-signal parameters for the transistor in the common-collector circuit are g_m = 40 m$\mho$, β_o = 150, $r_o \rightarrow \infty$, and $r_b \approx 0$.

  (a) Draw the small-signal equivalent of this stage.

  (b) Determine R_{in} and R_o.

  (c) Evaluate the transfer function V_o/V_s.

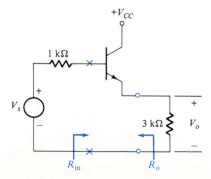

Prob. 3-44

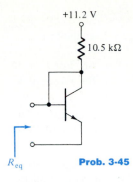

+11.2 V

10.5 kΩ

R_{eq} **Prob. 3-45**

3-45 The transistor in the circuit shown is described in Prob. 3-44. Determine the small-signal equivalent resistance R_{eq} for the diode-connected transistor.

3-46 A transistor having $\beta_F = 100$ is used in the circuit of Fig. 3-36a. With $V_{CC} = 15$ V, determine the value of R that makes $I_C = 0.2$ mA.

3-47 The parameters in Fig. 3-32 are $r_b = 50\ \Omega$, $r_\pi = 950\ \Omega$, $C_\pi = 50$ pF, $C_\mu = 1$ pF, $r_o = 50$ kΩ, and $g_m = 0.1\ \mho$. With terminals c and e short-circuited determine
(a) The ratio I_c/I_b as a function of frequency
(b) At what frequency the magnitude of the ratio in (a) is unity
(c) The impedance $Z_{in}(s)$ seen looking between terminals b and e

CHAPTER 4

4-1 The device in the circuit shown is an ideal voltage-controlled current source defined by $I_2 = 3 \times 10^{-3} V_1$ mA.
(a) Sketch the output characteristics (I_2 versus V_1) for $0 \le V_1 \le 3$ V in 0.5-V increments.
(b) For $V_i = 1.5$ V, determine I_2 and V_2.
(c) If V_i is a positive pulse, what must its amplitude be for the circuit to behave as a controlled switch?

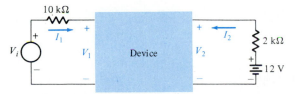

Prob. 4-1

4-2 For the device and circuit of Prob. 4-1
(a) Sketch the transfer characteristic V_2 versus V_i.
(b) For $V_i = 1.5 + 1.0 \sin \omega t$, sketch one cycle of the waveform for V_2.
(c) For the conditions in (b), sketch one cycle of the voltage across the 2-kΩ resistor.
(d) If V_2 were observed on an oscilloscope with the selector knob set to AC, sketch one cycle of the waveform you would see.

4-3 A voltage-controlled current source defined by $I_2 = 2.5 \times 10^{-3} V_1 + 5 \times 10^{-5} V_2$ is used in the circuit in Prob. 4-1. Repeat Prob. 4-2.

4-4 Consider an n-channel device with donor concentration of N_D atoms/cm^3 and a heavily doped gate with acceptor concentration of N_A atoms/cm^3, such that $N_A \gg N_D$ and an abrupt channel-gate junction. Assume that $V_{DS} = 0$ and that the junction contact potential is much smaller than $|V_P|$. Prove that, for the geometry of Fig. 4-6,

$$|V_P| = \frac{qN_D}{2\epsilon} a^2$$

where ϵ = dielectric constant of the channel material and q = magnitude of the electronic charge. Find V_P for a silicon n-channel JFET with $a = 2\ \mu$m, $N_D = 7 \times 10^{14}$ atoms/cm^3, and $\epsilon_r = 12$.

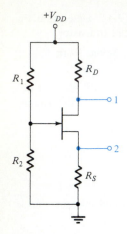

Prob. 4-10

4-5 Derive Eq. (4-1).

4-6 (*a*) For the JFET whose characteristics are given in Fig. 4-7, evaluate $r_{DS(ON)}$ at $V_{GS} = 0$ V.

(*b*) An *n*-channel silicon JFET has the structure shown in Fig. 4-6. For $L = 10$ μm, $a = 2$ μm, $W = 8$ μm, and $V_p = -4$ V, find $r_{DS(ON)}$ for $V_{GS} = 0$ V. (*Hint:* Use the expression for V_p in Prob. 4-4.)

4-7 The JFET whose characteristics are given in Fig. 4-7 is used in the circuit in Fig. 4-19. The element values are $V_{DD} = 24$ V, $R_D = 4$ kΩ, $R_S = 1$ kΩ, and $R_G \geq 100$ kΩ. Determine V_{DS}, I_D, and V_{GS}.

4-8 The circuit in Fig. 4-19 uses the JFET in Fig. 4-7. The supply voltage is 30 V, and it is desired to have $V_{DS} = 17.5$ V and $I_D = 2.5$ mA. Determine R_D and R_S.

4-9 A *p*-channel JFET has $V_p = 5$ V and $I_{DSS} = -12$ mA. The supply voltage available is 12 V. Using a circuit analogous to that in Fig. 4-19 for a *p*-channel device, determine R_D and R_S so that $I_D = -4$ mA and $V_{DS} = -6$ V.

4-10 An *n*-channel JFET has $V_p = -5$ V and $I_{DSS} = 12$ mA and is used in the circuit shown. The parameter values are $V_{DD} = 18$ V, $R_S = 2$ kΩ, $R_D = 2$ kΩ, $R_1 = 400$ kΩ, and $R_2 = 90$ kΩ. Determine V_{DS} and I_D.

4-11 (*a*) The resistance R_2 is changed in Prob. 4-10. What must the new value of R_2 be if $I_D = 8$ mA?

(*b*) Using values given in Prob. 4-10 but changing V_{DD}, find the new value of V_{DD} for which $I_D = 8$ mA.

(*c*) For the conditions in (*b*), what is the new value of V_{DS}?

4-12 The circuit from Prob. 4-10 is used to obtain $I_D = 2.5$ mA and $V_{DS} = 17.5$ V with a 30-V supply for the JFET in Fig. 4-7. The resistance $R_G = R_1 \| R_2 \geq 100$ kΩ and $R_S = 1.2$ kΩ. Determine R_1, R_2, and R_D.

4-13 The MOSFET described by the transfer characteristic in Fig. 4-13 is used in the circuit in Fig. 4-21a. The parameter values are $V_{DD} = 18$ V, $R_D = 50$ kΩ, and $R_S = 10$ kΩ.

(*a*) Determine the ratio R_1/R_2 which makes $I_D = 0.1$ mA.

(*b*) What is the value of V_{DS}?

4-14 An NMOS enhancement transistor having $k = 1$ mA/V^2, $W/L = 2$, and $V_T = 4$ V is used in the circuit in Fig. 4-21a. The supply voltage is 12 V, $R_D = R_S = 2$ kΩ, $R_1 = 100$ kΩ, and $R_2 = 300$ kΩ. Determine

(*a*) I_D and V_{DS}

(*b*) The new value of R_S needed to maintain the value of I_D in (*a*) if $W/L = 4$

4-15 A PMOS enhancement transistor has $V_T = -1$ V, $W/L = 1$, and $k = 0.2$ mA/V^2 and is used in a circuit analogous to that in Fig. 4-21a. The supply voltage is 9 V, $R_1 = 240$ kΩ, and $R_2 = 120$ kΩ.

(*a*) Determine R_S so that $V_{GS} = -2$ V.

(*b*) Determine the value of R_D needed to make $V_{DS} = 3$ V.

(*c*) Find the new value of R_2 that maintains the value of I_D in (*a*) and (*b*) for V_T changed to -1.5 V. Assume that all other parameter values remain at the values given or computed in (*a*) and (*b*).

4-16 The transistors Q1 and Q2, used in the circuit shown, are identical and have the characteristics given in Fig. 4-24b.

(*a*) Determine the drain current in Q1 and the voltage V_o.

(*b*) What is the value of V_{DS2}?

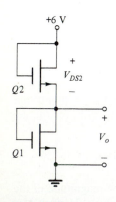

Prob. 4-16

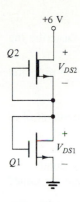

+6 V

$Q2$

+
V_{DS2}
−

+
V_{DS1}
−

$Q1$

Prob. 4-18

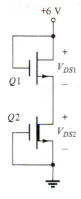

+6 V

$Q1$

+
V_{DS1}
−

$Q2$

+
V_{DS2}
−

Prob. 4-21

4-17 (a) Repeat Prob. 4-16, part (a), assuming that the aspect ratio W/L of $Q2$ is decreased by a factor of 4 and $Q1$ is unchanged.
(b) Repeat Prob. 4-16, part (a), assuming that the aspect ratio of $Q1$ is decreased by a factor of 4 and $Q2$ is unchanged.

4-18 The characteristics for $Q1$ and $Q2$, used in the circuit shown, are given in Figs. 4-24b and 4-26, respectively. Determine V_{DS1} and V_{DS2}.

4-19 Repeat Prob. 4-18, assuming that the aspect ratio of $Q2$ is decreased by a factor of 5 and $Q1$ is unchanged.

4-20 Repeat Prob. 4-18, assuming that the aspect ratio of $Q1$ is decreased by a factor of 5 and $Q2$ is unchanged.

4-21 Repeat Prob. 4-18 for the circuit shown.

4-22 Repeat Prob. 4-21, assuming that
(a) The aspect ratio of $Q1$ is decreased by a factor of 5 and $Q2$ is unchanged.
(b) The aspect ratio of $Q2$ is decreased by a factor of 5 and $Q1$ is unchanged.
(c) The aspect ratios of both $Q1$ and $Q2$ are increased by a factor of 3.

4-23 In the circuit shown, $Q1$, $Q2$, and $Q3$ are identical transistors having the characteristics given in Fig. 4-12. Determine I_o and V_o.

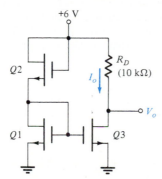

Prob. 4-23

4-24 Repeat Prob. 4-23, assuming that the connections of the 10-kΩ resistance and $Q2$ are interchanged.

4-25 Repeat Prob. 4-23, assuming that $Q2$ is replaced by a depletion transistor connected as a resistance and having the characteristics given in Fig. 4-26.

4-26 In the circuit in Prob. 4-23, $Q1$, $Q2$, and $Q3$ are identical transistors having $k = 40$ μA/V², $W/L = 5$, and $V_T = 1.0$ V. Determine R_D so that $V_o = 3.5$ V.

4-27 Sketch the transfer function V_o versus V_i for the circuit in Fig. 4-24a for $V_{DD} = 6$ V. Transistor $Q1$ has the characteristics given in Fig. 4-24b, and $Q2$ is an identical transistor having an aspect ratio of 0.4 times that of $Q1$.

4-28 The transistors used in the circuit in Fig. 4-24a have $k = 50$ μA/V² and $V_T = 1$ V. The gate dimensions of $Q1$ are $W = 50$ μm and $L = 5$ μm; $Q2$ has $W = 10$ μm and $L = 5$ μm. Sketch the transfer function V_o versus V_i for $V_{DD} = 5$ V.

4-29 (a) An NMOS circuit has the transfer characteristic given in Fig. 4-30. For $v_i = 3.0 + 0.25 \sin \omega t$, sketch the waveform of v_o for one cycle when display is seen on an oscilloscope with the selector set to DC.
(b) Repeat (a) for the oscilloscope selector knob set to AC.

4-30 (a) Repeat Prob. 4-29 for the transfer characteristic in Fig. 4-25.

(b) The amplitude of the sinusoidal input is increased to 1.25 V. Describe the output waveform.

4-31 The JFET used in the circuit of Fig. 4-31 has $V_p = -6$ V, $I_{DSS} = 15$ mA, and $\lambda = 0.02$ V^{-1} and is biased at $I_D = 6$ mA and $V_{DS} = 10$ V.

(a) Draw the small-signal model of the circuit at low frequencies.

(b) What value of R_D is required if the amplitude of the signal component of v_o is to be 10 times the amplitude of v_s?

4-32 (a) Draw the small-signal equivalent of the circuit in Prob. 4-10.

(b) Determine the output resistance seen between terminal 1 and ground.

(c) If $R_S = 0$, does the resistance in (a) increase, decrease, or remain the same?

4-33 In the circuit in Prob. 4-10

(a) Determine the resistance seen looking into the circuit between terminal 2 and ground at low frequencies

(b) Evaluate the resistance in (a) for $R_D = 5$ kΩ, $R_S = 3$ kΩ, $R_1 = 240$ kΩ, $R_2 = 80$ kΩ, $g_m = 2$ m$\mho$, and $r_d = 50$ kΩ.

(c) Repeat (b) for $R_D = 0$.

4-34 The JFET in the circuit shown has the characteristics displayed in Fig. 4-32. For $I_{DD} = 2.5$ mA, determine the signal component of v_o produced by an input signal $v_s = 2 \sin \omega t$ mV. The relevant parameter values are $R_D = 100$ kΩ and $r_d = 100$ kΩ. You may assume that R_D draws negligible dc current and the signal frequency is sufficiently small so that the low-frequency FET model is valid.

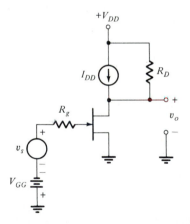

Prob. 4-34

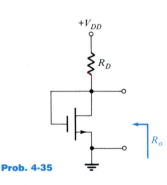

Prob. 4-35

Prob. 4-36

4-35 (a) Draw the low-frequency small-signal model of the circuit shown.

(b) Determine R_o.

(c) Evaluate R_o for $g_m = 1.0$ m$\mho$, $r_d = 50$ kΩ, and $R_D = 10$ kΩ.

4-36 Repeat Prob. 4-35, parts (a) and (b) for the depletion resistance shown.

4-37 (a) Draw the incremental model of the circuit in Prob. 4-35 valid at high frequencies.

(b) What is the equivalent capacitance seen between drain and ground?

4-38 (a) Draw the high-frequency model of the circuit in Prob. 4-36.

(b) Determine the equivalent capacitance seen between source and ground.

4-39 (a) Draw the small-signal equivalent of the MOSFET stage shown in Fig. 4-29a.
(b) Derive Eq. (4-19).
(c) Derive an expression, valid at low frequencies, which relates the output and input signal amplitudes. (*Hint:* The results in Prob. 4-36 may be useful.)

4-40 Sketch the transfer characteristic of the circuit shown for $V_{DD} = 6$ V and where $Q1$ and $Q2$ are identical transistors described by Fig. 4-12.

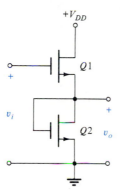

Prob. 4-40

4-41 (a) Draw the low-frequency small-signal model of the circuit in Prob. 4-40.
(b) Derive an expression for the signal component of v_o produced by the signal input v_i.

4-42 The NMOS and PMOS transistors in Fig. 4-38 are complementary and have $k = 20$ μA/V^2, $W/L = 1$, and $V_T = 1.0$ V. For $V_{DD} = 5$ V, sketch the transfer characteristic v_o versus v_i.

4-43 In Fig. 4-38 the NMOS transistor has $k = 15$ μA/V^2, $W/L = 10$, and $V_T = 2$ V. The PMOS transistor has $V_T = -1.0$ V, $W/L = 10$, and $k = 15$ μA/V^2. For $V_{DD} = 6$ V, sketch the transfer characteristic v_o versus v_i.

4-44 The transistors in Fig. 4-38 are complementary devices whose parameters are given in Prob. 4-42. The aspect ratio W/L of the PMOS transistor is changed to 2. Sketch the transfer characteristic of the circuit.

CHAPTER 5

5-1 List in order the steps required in fabricating a silicon IC transistor by the epitaxial-diffused method. Sketch the cross section after each oxide growth.

5-2 (a) Consider an IC *npn* transistor $Q1$ built upon a *p*-type substrate S. Show that between the four terminals E, B, C, and S there exists a *pnp* transistor $Q2$ in addition to $Q1$.
(b) If $Q1$ is in its active region, in what mode is $Q2$ operating? Explain.
(c) Repeat part (b) for $Q1$ in saturation.
(d) Repeat part (b) for $Q1$ in cutoff.

5-3 Sketch the five basic diode connections (in circuit form) for the monolithic ICs. Which will have the lowest forward-voltage drop? Highest breakdown voltage?

5-4 A 1-mil-thick silicon wafer has been doped uniformly with phosphorus to a concentration of 10^{17} cm^{-3}, plus boron to a concentration of 5×10^{16} cm^{-3}. Find its sheet resistance.

5-5 (*a*) What is the total length required to fabricate a 20-kΩ resistor whose width is 25 μm if $R_S = 200$ Ω/square.

(*b*) What is the width required to fabricate a 5-kΩ resistor whose length is 25 μm?

5-6 A thin-film capacitor has a capacitance of 0.4 pF/$(\mu m)^2$. The thickness of the SiO_2 layer is 500 Å. Calculate the relative dielectric constant ϵ_r of silicon dioxide.

5-7 A MOS capacitor is fabricated with an oxide thickness of 500 Å. How much chip area is required to obtain a capacitance of 200 pF? The relative dielectric constant ϵ_r of silicon dioxide is 3.5.

5-8 For the circuit shown, find (*a*) the *minimum* number and (*b*) the *maximum* number, of isolation regions.

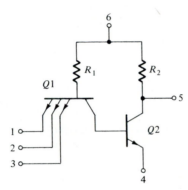

Prob. 5-8

5-9 (*a*) What is the minimum number of isolation regions required to realize in monolithic form the logic gate shown?

(*b*) Draw a monolithic layout of the gate in the fashion of Fig. 5-1.

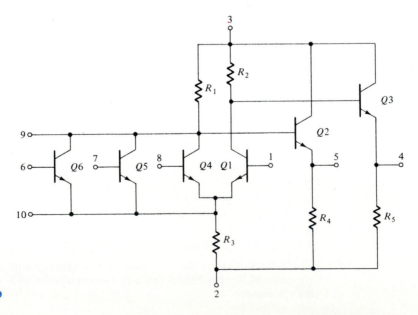

5-10 Repeat Prob. 5-9 for the difference amplifier shown.

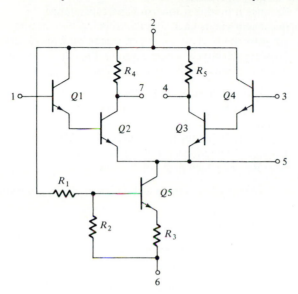

Prob. 5-10

5-11 For the circuit shown, (*a*) find the minimum number of isolation regions and (*b*) draw a monolithic layout.

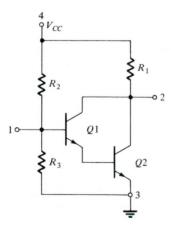

Prob. 5-11

(*Note*: In Probs. 5-12 to 5-22, indicate your answer by giving the letter of the statement you consider correct.)

5-12 The typical number of diffusions used in making epitaxial-diffused silicon ICs is (*a*) 6, (*b*) 3, (*c*) 4, (*d*) 5, (*e*) 2.

5-13 The "buried layer" in an *npn* transistor fabricated on a *p*-type substrate in an IC is
(*a*) Used to reduce the parasitic capacitance
(*b*) p^+-doped
(*c*) Located in the emitter region
(*d*) n^+-doped

5-14 Epitaxial growth is used in ICs
 (a) Because it produces low parasitic capacitance
 (b) Because it yields back-to-back isolating *pn* junctions
 (c) To grow single-crystal *n*-doped silicon on a single-crystal *p*-type substrate
 (d) To grow selectively single-crystal *p*-doped silicon of one resistivity on a *p*-type substrate of a different resistivity

5-15 Silicon dioxide (SiO_2) is used in ICs
 (a) To control the location of diffusion and to protect and insulate the silicon surface
 (b) Because it facilitates the penetration of diffusants
 (c) To control the concentration of diffusants
 (d) Because of its high heat conduction

5-16 When a hole is opened in the SiO_2 and impurities are introduced, they will diffuse vertically
 (a) A greater distance than laterally
 (b) The same distance as laterally
 (c) A shorter distance than laterally
 (d) Twice the lateral distance

5-17 The *p*-type substrate in a monolithic circuit should be connected to
 (a) Any dc ground point
 (b) Nowhere, i.e., be left floating
 (c) The most positive voltage available in the circuit
 (d) The most negative voltage available in the circuit

5-18 The sheet resistance of a semiconductor is
 (a) A parameter whose value is important in a thin-film resistance
 (b) A characteristic whose value determines the required area for a given value of integrated capacitance
 (c) An important characteristic of a diffused region, especially when used to form diffused resistors
 (d) An undesirable parasitic element

5-19 Isolation in ICs is required to
 (a) Minimize electrical interaction between circuit components
 (b) Simplify interconnections between devices
 (c) Protect the components from mechanical damage
 (d) Protect the transistor from possible "thermal runaway"

5-20 Most resistors are made in a monolithic IC
 (a) During metallization
 (b) During the emitter diffusion
 (c) While growing the epitaxial layer
 (d) During the base diffusion

5-21 In a monolithic-type IC
 (a) Each transistor is diffused into a separate isolation region
 (b) Resistors and capacitors of any value may be made
 (c) All isolation problems are eliminated
 (d) All components are fabricated into a single crystal of silicon

5-22 Repeat Prob. 5-16, assuming that the impurities are introduced by ion implantation.

5-23 In order, list the steps required in fabricating an NMOS enhancement transistor. Sketch a cross section following each oxidation step.

5-24 Repeat Prob. 5-23 for a depletion transistor.

5-25 Draw the layout of the circuits shown.

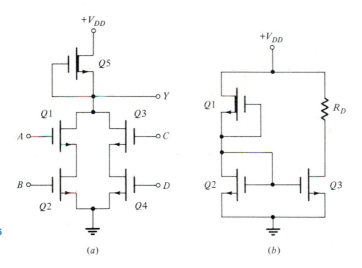

Prob. 5-25

(a)   (b)

5-26 In order, list the steps required in fabricating the CMOS circuit shown.

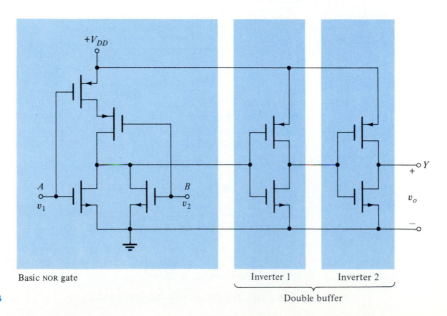

Basic NOR gate   Inverter 1   Inverter 2

Prob. 5-26   Double buffer

CHAPTER 6

6-1 Convert the following decimal numbers to binary numbers: (*a*) 127, (*b*) 360, (*c*) 1066.

6-2 Repeat Prob. 6-1 for (*a*) 222, (*b*) 302, (*c*) 1776.

6-3 Convert the decimal numbers in Prob. 6-1 to octal (base 8) numbers.

6-4 Convert the decimal numbers in Prob. 6-2 into hexadecimal (base 16) numbers.

6-5 For the waveforms shown, express *A*, *B*, and *C* as 8-bit binary numbers, assuming that
    (*a*)   Positive-logic system is used
    (*b*)   Negative-logic system is used

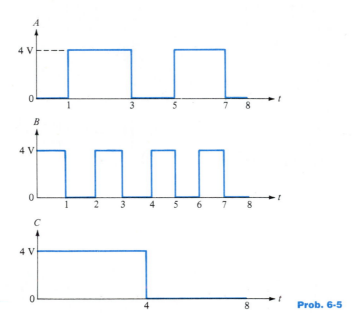

Prob. 6-5

6-6 Consider that the switch in Fig. 6-1 is controlled by a voltage v and is closed if $v = V(1)$ and open if $v = V(0)$. When closed, the switch has $R_{ON} = 50\ \Omega$, and when open, the switch can be represented by $R_{OFF} = 50\ k\Omega$. Determine the range of values for R which guarantee that $V(0) \leq 0.2\ V$ and $V(1) \geq 4.5\ V$.

6-7 Repeat Prob. 6-6 for $V(0) \leq 0.3\ V$ and $V(1) \geq 4.7\ V$.

6-8 The circuit in Fig. 6-1 is used as described in Prob. 6-6. The value of R used is 5 kΩ. Determine
    (*a*)   The minimum value of R_{OFF} for which $V(1) \geq 4.8\ V$
    (*b*)   The maximum value of R_{ON} for which $V(0) \leq 0.2\ V$

6-9 The waveforms in Prob. 6-5 are the three inputs to a positive-logic OR gate.
    (*a*)   Sketch the waveform for the output voltage of the gate.
    (*b*)   Write the truth table for the gate.

6-10 Repeat Prob. 6-9 for a negative-logic gate.

6-11 The waveforms in Prob. 6-5 are the three inputs to a positive-logic AND gate.
(a) Write the truth table for a 3-input AND gate.
(b) Sketch the output voltage waveform for the given inputs.

6-12 Repeat Prob. 6-11 for a negative-logic AND gate.

6-13 Each of the three inputs shown in Prob. 6-5 are fed into an inverter (NOT gate). The inverter outputs are used as the inputs to a positive-logic AND gate.
(a) Sketch the output waveform of the AND gate.
(b) What logical operation is performed on inputs A, B, and C?

6-14 Repeat Prob. 6-13, assuming that the inverter outputs are the inputs to an OR gate.

6-15 The waveform C in Prob. 6-5 is fed into an inverter. The output of the inverter and A and B are the three inputs of an AND gate.
(a) Sketch the output waveform of the AND gate.
(b) What logical operation is performed?

6-16 The waveforms given in Prob. 6-5 are applied to a 3-input NOR gate. Sketch the output waveform of the gate.

6-17 Repeat Prob. 6-16, assuming that the waveforms are applied to a NAND gate.

6-18 Using only 2-input NAND gates, construct AND, OR, and NOT gates.

6-19 Use boolean algebra to verify
(a) $(A + B)(A + C)(B + C) = AB + AC + BC$
(b) $(A + B)(\overline{A} + C) = AC + \overline{A}B$
(c) $(AB + \overline{BC} + AC = AB + \overline{BC}$

6-20 (a) Using only NOR gates, construct a logic circuit which realizes each side of the boolean equation in Prob. 6-19, part (b).
(b) Repeat (a) using only NAND gates.
(c) Which of the circuits in (a) and (b) uses the fewest number of gates?

6-21 Repeat Prob. 6-20 for the boolean equation in Prob. 6-19, part (c).

6-22 (a) Using only NOR gates, construct an exclusive-OR circuit.
(b) Repeat (a) using NAND gates.

6-23 A half-adder is a 2-input, 2-output logic circuit that has the following truth table:

Input 1	Input 2	Output 1	Output 2
0	0	0	0
0	1	1	0
1	0	1	0
1	1	0	1

Implement this circuit using
(a) NAND gates
(b) NOR gates

6-24 The circuit shown is a positive-logic inverter driving N identical circuits in parallel. The controlled switch has $R_{ON} = 100 \ \Omega$, $R_{OFF} = 50 \ k\Omega$, and $R_{in} = 200 \ k\Omega$. Determine the fan-out. The logic levels are $V(0) \le 0.5 \ V$ and $V(1) \ge 3.0 \ V$.

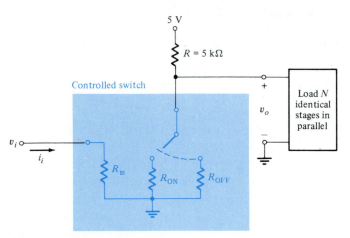

Prob. 6-24

6-25 In the circuit from Prob. 6-24, $R_{ON} = 0.5 \ k\Omega$ and $R_{OFF} = 100 \ k\Omega$ and the logic levels are $V(0) \le 0.5 \ V$ and $V(1) \ge 2.5 \ V$.
(a) What is the minimum value of R_{in} if the fan-out is to be 10?
(b) Given the value of R_{in} in (a), what effect does decreasing R_{OFF} have on the fan-out and logic levels?
(c) Repeat (b) for the situation where R increases.

6-26 The controlled switch in the circuit shown is closed for $v_i = V(1)$ and open for $v_i = V(0)$. When closed, the switch is characterized by R_{ON}; the open switch is represented by R_{OFF}. The input voltage v_i has been at $V(1)$ for a long time. At $t = 0$, v_i becomes $V(0)$. Derive an expression for the propagation delay t_{pLH}.

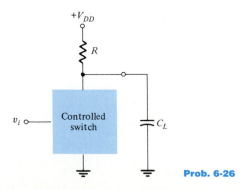

Prob. 6-26

6-27 The input voltage v_i in the circuit from Prob. 6-26 has been $V(0)$ for a long time. At $t = 0$, v_i becomes $V(1)$. Derive an expression for t_{pHL}.

6-28   The parameter values for the circuit described in Prob. 6-26 are $V_{DD} = 5$ V, $R = 10$ kΩ, $C = 50$ pF, $R_{ON} = 417$ Ω, and $R_{OFF} = 40$ kΩ. At $t = 0$, v_i goes from $V(0)$ to $V(1)$; at $t = 0.2$ μs, v_i becomes $V(0)$ again.
   (a)   Determine the propagation delay (average).
   (b)   What is the maximum instantaneous current the switch must be capable of handling?
   (c)   What is the minimum cycle time for the circuit?

6-29   Assume that the switch in Prob. 6-28 is ON one-half of the time and OFF one-half of the time.
   (a)   Determine the average power dissipated by the circuit in one cycle.
   (b)   Evaluate the delay-power product.

6-30   In the circuit of Fig. 6-20b, both $Q1$ and $Q2$ have $k = 25$ μA/V^2 and $V_T = 1.5$ V. The aspect ratios are $W/L = 5$ for $Q1$ and $W/L = 1$ for $Q2$. The supply voltage $V_{DD} = 5$ V.
   (a)   Sketch the transfer characteristic of the gate.
   (b)   Determine V_{OL}, V_{OH}, V_{IL}, V_{IH}, and the noise margins.

6-31   The inverter in Prob. 6-30 is subject to variations in fabrication. Repeat Prob. 6-31, given that the value of k varies by ± 20 percent. Assess performance changes.

6-32   The transistors in Fig. 6-20b are identical with $V_T = 1.25$ V. For $Q1$, $kW/L = 100$ μA/V^2; for $Q2$, $kW/L = 50$ μA/V^2. The supply voltage $V_{DD} = 5$ V.
   (a)   Sketch the voltage transfer characteristic.
   (b)   Evaluate the noise margin.

6-33   Fabrication variations cause V_T to vary by ± 0.25 V. Repeat Prob. 6-32 for these variations in V_T and assess the changes in circuit performance.

6-34   The supply voltages in Fig. 6-23a are $V_{DD} = 5$ V and $V_{GG} = 10$ V. Both $Q1$ and $Q2$ are identical and have $kW/L = 1$ mA/V^2 and $V_T = 1.5$ V. Sketch the transfer characteristics of the circuit and evaluate the noise margins.

6-35   (a)   The aspect ratio of $Q2$ in the circuit in Prob. 6-34 increases by 10 percent. What is the percentage change in the noise margins?
   (b)   Repeat (a), assuming that only the aspect ratio of $Q1$ increases by 10 percent.

6-36   The gate-bias supply V_{GG} in the circuit described in Fig. 6-23 is varied between 7 and 12 V. Plot curves of the noise margins as a function of V_{GG}.

6-37   In the circuit in Fig. 6-24a, the enhancement transistor has $kW/L = 0.1$ mA/V^2 and $V_T = 1.5$ V and the depletion MOSFET has $kW/L = 20$ μA/V^2 and $V_T = -1.5$ V. For $V_{DD} = 5$ V
   (a)   Sketch the voltage transfer characteristic.
   (b)   Evaluate the noise margins.

6-38   The circuit in Fig. 6-24a uses $Q1$ described in Prob. 6-37 and has $V_{DD} = 5$ V. The depletion transistor has $kW/L = 25$ μA/V^2, and its threshold voltage V_T is varied between -0.5 and -2.5 V. Plot curves of the noise margins as a function of V_T of Q_2. Use 0.5-V increments.

6-39   The circuit in Fig. 6-24a uses $Q1$ described in Prob. 6-37 and a supply $V_{DD} = 5$ V. The depletion MOSFET has $V_T = -1.5$ V and $k = 10$ μA/V^2.
   (a)   Sketch the transfer characteristic for aspect ratio of $Q2$, of 1, 2.5, 5, 7.5, and 10.
   (b)   Plot curves of the noise margins as a function of the ratio of kW/L for $Q1$ to kW/L of $Q2$.

6-40 Determine t_{pLH} for the circuit in Example 6-5.

6-41 Evaluate the average propagation delay for the circuit described in Prob. 6-37.

6-42 For the circuit in Prob. 6-38, with $V_T = -1.0 \, \text{V}$, evaluate the delay-power product. Assume that the output is high 50 percent of the time.

6-43 Evaluate the delay-power product of the circuit in Prob. 6-30. Assume that the output is high 25 percent of the time.

6-44 In the NOR gate in Fig. 6-28a, consider that both drivers have $V(1) = 5 \, \text{V}$ applied. In effect, the driver MOSFETs are in parallel. The MOSFETs used are described in Prob. 6-37 and $V_{DD} = 5 \, \text{V}$. Determine the drain current in each transistor. (*Hint*: Since the drivers are in parallel, draw a composite characteristic of the driver and then construct the load line.)

6-45 (*a*) Consider two isolated NMOS inverters. The input to one is *A*, and the input to the other is *B*. The two outputs are now wired together, and the common output is *Y*. What is the logical relation among *Y*, *A*, and *B*?
(*b*) Draw the circuit and show that the logic in (*a*) is satisfied. Omit one load FET since the loads are in parallel.

6-46 Repeat Prob. 6-45 using 2-input NAND gates in place of the inverters.

6-47 (*a*) Three inverters are cascaded. Each has the voltage transfer characteristic shown. Sketch the transfer characteristic of the cascade. (*Hint*: You will need to take several values in the range $2.45 < v_i < 2.55 \, \text{V}$.)
(*b*) Compare the slope of the transfer characteristic in the linear region of the cascade with that for a single inverter.
(*c*) Compare the noise margins of the single inverter and of the cascade.

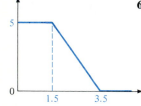

Prob. 6-47

6-48 Draw the circuit diagram of a 2-input CMOS NAND gate.

6-49 If the inverters in Prob. 6-45 are fabricated using CMOS technology, can the outputs be wired together to obtain the same logical relation among *A*, *B*, and *Y*? Explain.

6-50 Consider the circuit in Fig. 6-30a for which the MOSFETs are described in Sec. 6-8. The input voltage v_i varies linearly with time and reaches 5 V in 100 μs.
(*a*) Sketch the current in the circuit as a function of time.
(*b*) What is the average power dissipated during each 100-μs interval?

6-51 The input to the CMOS inverter in Fig. 6-30a and described in Sec. 6-8 is shown.
(*a*) Determine the average power dissipated in one cycle.
(*b*) As *T* decreases (the frequency increases), does the result in (*a*) increase, decrease, or remain unchanged.

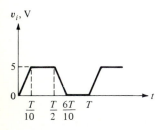

Prob. 6-51

6-52 Draw the CMOS circuit that performs the logical operation applied in Prob. 6-46.

6-53 Consider the transmission gate shown in Fig. 6-32 with control voltages $V(0) = -5 \, \text{V}$, $V(1) = 5 \, \text{V}$, and a sinusoid of 5-V peak amplitude. Assume that the threshold voltage $V_T = 0$.
(*a*) Verify that the entire sinusoid appears at the output of the control $C = V(1)$.
(*b*) Show that transmission is inhibited if $C = V(0)$.
(*c*) Repeat (*a*) and (*b*), given $V_T = 2.5 \, \text{V}$. Indicate the range of input voltage for which both *Q*1 and *Q*2 conduct.

(d) Assume a peak sinusoidal input voltage of 7.5 V. Assuming that the control voltage is $V(1)$, sketch the output voltage.

(e) Repeat (d), given that the control voltage is $V(0)$ and $V_T = 2.5$ V.

6-54 The BJT inverter in Fig. 6-34a is designed with $R_B = 12$ kΩ, $R_C = 3$ kΩ, and $V_{CC} = 6$ V. Reverse saturation currents are negligible.

(a) Determine the minimum value of β_F needed to just barely saturate the transistor when $v_s = V(1) = 6$ V.

(b) Assuming that the transistor output is $V(1)$ 50 percent of the time, calculate the average power dissipated.

6-55 The transistor used in Fig. 6-34a has $50 \le \beta_F \le 150$. The supply voltage is 5 V and $V(0) = 0.3$ V and $V(1) = 4.8$ V. An output current pulse is to be 10 mA.

(a) Determine R_C and R_B so that the transistor is just barely saturated at the minimum β_F.

(b) Assuming that the transistor is ON 50 percent of the time, determine the average power dissipated by the gate. Use $\beta_F = 150$.

(c) Is the answer to (b) significantly different if $\beta_F = 50$? Justify your answer (without solving the problem again).

6-56 The inverter shown is required to drive N identical gates.

(a) For $\beta_F = 40$, what value of $v_o = V(1)$ just barely saturates the transistor?

(b) Given $v_i = V(0) = 0.3$ V, evaluate N, assuming that each of these stages is barely saturated.

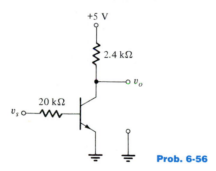

+5 V

2.4 kΩ

v_o

20 kΩ

v_s

Prob. 6-56

6-57 The inverter in Prob. 6-55 is required to drive N identical gates.

(a) Compute the minimum value of β_F of the transistor if an ON transistor is just barely saturated.

(b) Determine the value of $v_o = V(1)$.

(c) What are the approximate noise margins?

6-58 A 2N2222A transistor is used in the inverter of Fig. 6-34a with $V_{CC} = 10$ V, $R_C = 500$ Ω, and $R_B = 50$ kΩ. Plot the voltage transfer characteristic of the gate for $0 \le v_i \le 10$ V.

6-59 The circuit shown is sometimes used as an inverter in TTL logic chips. The transistors used are identical and have $\beta_F = 25$ and $\beta_R = 0.5$. For $V(0) = 0.2$ V and $V(1) = 3.5$ V

(a) Verify that the circuit behaves as an inverter.

(b) Determine the base and collector current in each transistor for $v_s = V(0)$ and $v_s = V(1)$.

(c) What is the fan-out of the circuit?

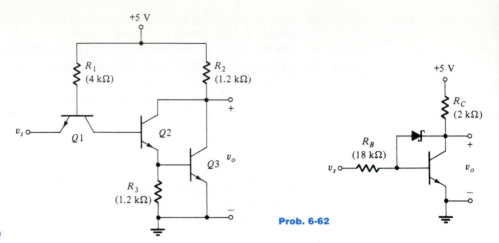

Prob. 6-59

Prob. 6-62

6-60 Obtain the voltage transfer characteristic of the circuit in Prob. 6-59.

6-61 A Schottky diode is connected between collector and base in the circuit in Prob. 6-56. Sketch the voltage transfer characteristic for $0 \leq v_s \leq 5$ V and estimate the noise margins.

6-62 The transistor in the circuit shown has $\beta_F = 50$. Determine v_o and the collector, base, and diode currents for $v_s = V(1) = 4$ V.

6-63 Given $V(0) = 0.3$ V in the circuit in Prob. 6-62, determine the fan-out.

6-64 (a) For the circuit shown, verify that $Y = \overline{ABC}$.
(b) If $\beta_F = 25$, what is the fan-out?
(c) What is the average power dissipated by the gate assuming $Y = V(1)$ 50 percent of the time?

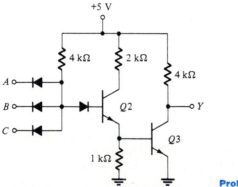

Prob. 6-64

6-65 (a) For the TTL NAND gate in Fig. 6-37, calculate $\beta_{F(min)}$ for proper operation. Assume that $Q2$ and $Q3$ saturate if all inputs are $V(1)$ and for $Q1$, $\beta_R = 0.1$.
(b) Repeat (a), assuming that $Q2$ remains in the active region and $Q3$ saturates when all inputs are $V(1)$.

6-66 (a) Sketch the voltage transfer function for the TTL NAND gate in Fig. 6-37. Carefully indicate the state of each transistor for each segment of the characteristic. Assume $\beta_F = 25$ and $\beta_R = 0.2$ for all transistors.

(b) Determine the noise margins.

(c) Determine the fan-out.

6-67 The TTL NAND gate shown uses a modified totem-pole stage. Assume that the inputs are derived from the outputs of identical gates and $\beta_F = 20$ and $\beta_R = 0.5$.

(a) Given $A = B = C = V(1)$, determine the current in each resistor, each collector, and each base; evaluate the voltage at each base and each collector with respect to ground. Verify that $Q5$ is in the forward-active region.

(b) Repeat (a) for the situation where at least one logic level is $V(0)$. Verify that $Q5$ is in saturation.

(c) Determine the logic levels.

(d) Determine the fan-out.

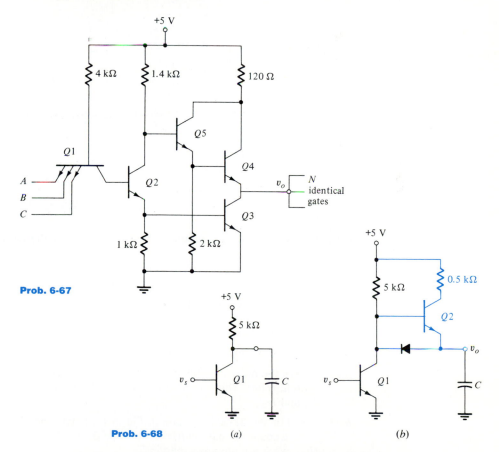

Prob. 6-67

Prob. 6-68 (a) (b)

6-68 (a) In the circuit shown in (a) $v_s = V(1) = 5$ V for a long time. At $t = 0$, $v_s = V(0) = 0.2$ V. Determine the rise time for v_o.

(b) To reduce the rise time in (a), the active pull-up circuit, indicated in color in (b), is added across the 5-kΩ resistor. Explain how the circuit works and why it is desirable to reduce the rise time.

(c)   Why is the simple replacement of the 5-kΩ resistor by a 0.5-kΩ resistor an effective method for decreasing the rise time?

6-69   The output of the TTL gate in Fig. 6-38 is accidentally short-circuited to ground. Determine the short-circuit current, given $\beta_F = 20$ and that
(a)   All inputs are at $V(1)$
(b)   At least one input is at $V(0)$

6-70   Both inputs of the TTL gate are tied together as shown. The transistors are identical and have $\beta_R = 0.5$.
(a)   Determine $\beta_{F(min)}$ for proper operation. Assume that $Q2$ and $Q3$ saturate for $v_s = V(1)$.
(b)   Repeat (a), assuming that $Q2$ does not saturate.

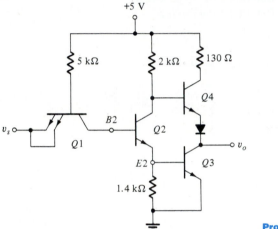

Prob. 6-70

6-71   (a)   Sketch the voltage transfer characteristic for the circuit in Prob. 6-70. Assume $\beta_F = 25$.
(b)   What is the fan-out?
(c)   What are the noise margins (approximate)?

6-72   (a)   Derive the voltage transfer characteristic for the OR output of the ECL gate in Fig. 6-47.
(b)   Evaluate the noise margins.

6-73   Repeat Prob. 6-72 for the NOR output.

6-74   Derive the noise margins corresponding to the unity-slope points of the current switch. (*Hint:* This can be done analytically using the exponential relations for I_{C1} and I_{C2}.)

6-75   (a)   For the basic ECL gate in Fig. 6-47, determine $V(0)$ and $V(1)$, taking into account the base currents. Assume $\beta_F = 50$.
(b)   What are the noise margins?

6-76   Show that when $Q2$ is ON, I_{C2} in Fig. 6-45a is greater than I_{C1} (with $Q1$ ON).

6-77   For the circuit shown, $V_{EE} = 5.0$ V, $V_R = -1.2$ V, and v_s is the output of an identical gate. The logic levels are $V(1) = -0.8$ V and $V(0) = -1.6$ V, and the

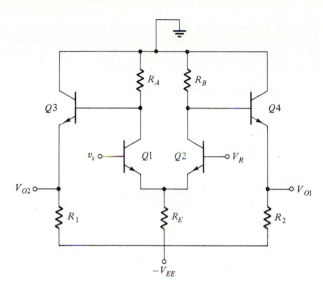

Prob. 6-77

maximum current in any transistor is 6 mA. Assume that $\beta_F \gg 1$ so that base currents may be neglected. Determine the values of the resistors R_1, R_2, R_A, R_B, and R_E.

CHAPTER 7

7-1 (a) Indicate how to implement S_n in Eq. (7-1) with AND, OR, and NOT gates.

(b) Verify that the sum S_n in Eq. (7-1) for a full adder can be put in the form

$$S_n = A_n \oplus B_n \oplus C_{n-1}$$

7-2 (a) For convenience, let $A_n = A$, $B_n = B$, $C_{n-1} = C$, and $C_n = C'$. Using Eq. (7-4) for C', prove that

$$\overline{C'} = \overline{B}\,\overline{C} + \overline{C}\,\overline{A} + \overline{A}\,\overline{B}$$

(b) Evaluate $D \equiv (A + B + C)\overline{C'}$ and prove that S_n in Eq. (7-1) is given by

$$S_n = D + ABC$$

7-3 Consider a digital system for majority logic. There are three inputs A, B, and C. The output Y is to equal 1 if two or three inputs are 1.

(a) Write the truth table.

(b) From the truth table obtain the boolean expression for Y.

(c) Minimize Y and show the logic block diagram.

7-4 The time to add two numbers in parallel is limited by the time it takes to propagate the carry through the word. At the expense of more logic, this carry propagation time can be avoided by generating a carry-look-ahead signal. Prove that if two 4-bit words ($A_3A_2A_1A_0$ and $B_3B_2B_1B_0$ with A_3 as the MSB) are added, then the carry-out C_3 is given by

$$\overline{C_3} = \overline{C_{-1}}\overline{(B_0A_0)}\overline{(B_1A_1)}\overline{(B_2A_2)}\overline{(B_3A_3)} + \overline{(A_0 + B_0)}\overline{(B_1A_1)}\overline{(B_2A_2)}\overline{(B_3A_3)}$$

$$+ \overline{(A_1 + B_1)}\overline{(B_2A_2)}\overline{(B_3A_3)} + \overline{(A_2 + B_2)}\overline{(B_3A_3)} + \overline{(A_3 + B_3)}$$

where C_{-1} is the input carry. Note that the carry out is given as a function of only input variables and does not involve intermediate carries. [*Hint:* Apply Eq. (7-5) recursively four times (for $n = 0, 1, 2,$ and 3). Start with Eq. (7-5) in the form $\overline{C}_n = \overline{C}_{n-1}(\overline{B_n A_n}) + (A_n + B_n)$.]

7-5 The system shown is called a true-complement–zero-one element. Verify the truth table.

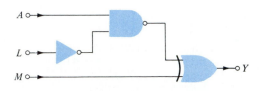

A		
L		
M		

Control inputs		Output
L	M	Y
0	0	$\overline{A}$
0	1	A
1	0	1
1	1	0

Prob. 7-5

7-6 (*a*) Verify that an exclusive-OR gate is a true-complement unit.
(*b*) One input is A, the other (control) input is C, and the output is Y. Is $Y = A$ for $C = 1$ or $C = 0$?

7-7 (*a*) Make a truth table for a binary half subtractor A minus B (corresponding to the half-adder in Fig. 7-4). Instead of a carry C, introduce a *borrow* P.
(*b*) Verify that the digit D is satisfied by an exclusive-OR gate and that P follows the logic "*B* but not *A*."

7-8 Consider an 8-bit comparator. Justify the connections $C' = C_L$, $D' = D_L$, and $E' = E_L$ for the chip handling the more significant bits. [*Hint:* Add 4 to each subscript in Fig. 7-13. Extend Eq. (7-12) for E and Eq. (7-13) for C to take all 8 bits into account.]

7-9 Consider a comparator which has as input two n-bit words, and has as outputs E, C, and D as in Fig. 7-13, but input leads E', C', and D' are not available. What additional logic is required to compare two $2n$-bit numbers using two n-bit comparators?

7-10 Consider two 5-bit words $S_A A_3 A_2 A_1 A_0$ and $S_B B_3 B_2 B_1 B_0$, where S_A and S_B are the sign bits, while the remaining bits indicate the magnitude of the word. S_A (S_B) $= 0$ means that the corresponding word is positive and S_A (S_B) $= 1$ means that the corresponding word is negative. Design a system to compare the two words, using a 4-bit comparator to compare the magnitudes and a 1-bit comparator to compare the sign bits.

7-11 (*a*) By means of a truth table verify the boolean identity

$$Y = (A \oplus B) \oplus C = A \oplus (B \oplus C)$$

(*b*) Verify that $Y = 1$ (0) if an odd (even) number of variables equals 1. This result is *not* limited to three inputs, but is true for any number of inputs. It is used in Sec. 7-5 to construct a parity checker.

7-12 Construct the truth table for the exclusive-OR tree in Fig. 7-14 for all possible inputs A, B, C, and D. Include $A \oplus B$ and $C \oplus D$ as well as the output Z. Verify that $Z = 1$ (0) for odd (even) parity.

7-13 (*a*) Draw the logic circuit diagram for an 8-bit parity check-generator system.
(*b*) Verify that the output is 0 (1) for odd (even) parity.

7-14 (a) Indicate an 8-bit parity checker as a block having 8 input bits (collectively designated A_1), an output P_1, and an input control P_1'. Consider a second 8-bit unit with inputs A_2, output P_2, and control P_2'. Show how to cascade the two packages in order to check for odd parity of a 16-bit word. Verify that the system operates properly if $P_1' = 1$. Consider the four possible parity combinations of A_1 and A_2.

(b) Show how to cascade three units to obtain the parity of a 24-bit word. Should $P_1' = 0$ or 1 for odd parity?

(c) Show how to cascade units to obtain the parity of a 10-bit word.

7-15 (a) Draw a 4-to-10-line decoder.

(b) Show how to convert this to a 3-to-8-line decoder.

7-16 Draw a block diagram of a demultiplexer tree with 32 outputs using $N_1 = 8$ and $N_2 = 4$. Explain the operation with reference to line 25.

7-17 (a) Draw a block diagram of a demultiplexer tree with 1024 outputs. Note that $1024 = 16$ times 8 times 8. Hence two levels of branching are required.

(b) How many equivalent packages are used?

(c) Assuming that 1024 were broken up into the product of 16 times 16 times 4, indicate the system and state the number of equivalent packages required.

7-18 (a) How many NAND-gate inputs does a 1-to-16 demultiplexer have?

(b) How many gate inputs does a 1-to-16-tree demultiplexer (using only 1-to-4 demultiplexers) have?

7-19 (a) Draw a logic diagram for a 6-to-1-line multiplexer.

(b) How is the network in (a) augmented to become an 8-to-1-line multiplexer?

7-20 Design a system to convert two 1-out-of-16 data-selector chips to form a 1-out-of-32 data selector. Explain the operation of this system. (*Hint:* The enable input S_2 to the higher-order chip is the complement of S_1 to the lower-order chip. Also, the outputs Y_1 and Y_2 from the two chips are the inputs to an OR gate whose output Y is the output of the system.)

7-21 (a) Draw a block diagram of a 32-to-1-line selector as in Fig. 7-22, but with $N_2 = 4$ and $N_1 = 8$. Explain the operation with respect to the input X_{25}.

(b) How many equivalent packages are needed?

7-22 Repeat Prob. 7-21 for a 64-to-1 multiplexer, using identical chips.

7-23 (a) Draw a block diagram of a multiplexer using 2048 inputs, noting the fact that $2048 = 16$ times 16 times 8.

(b) How many packages are needed?

7-24 (a) Generate Eq. (7-1) for the sum S_n of a full adder using a multiplexer. Find the X values in terms of C, $\overline{C}$, 0, and 1. (*Note:* For simplicity, drop the subscripts on A, B, and C and let $Y \equiv S_n$).

(b) Generate Eq. (7-2) for the carry C_n using a multiplexer. (*Note:* Let $C_{n-1} \equiv C$ and $C_n \equiv Y$.)

(c) Can the same multiplexer be used for S_n and C_n? Explain.

7-25 Use a multiplexer to generate the combinational-logic equation

$$Y = \overline{DCBA} + D\overline{CBA} + \overline{D}C\overline{BA} + \overline{DC}B\overline{A}$$
$$+ DC\overline{BA} + \overline{D}C\overline{B}A + DCB\overline{A} + \overline{D}CB\overline{A}$$

How many data inputs are needed? Find the values of the data inputs X.

7-26 Consider a digital system for majority logic. There are four inputs A, B, C, and D. The output Y is equal to 1, if three or four inputs are 1.
(a) Write the boolean expression for Y.
(b) Use a multiplexer-selector to satisfy this majority logic. What are the values of the data inputs X?

7-27 Design an encoder satisfying the following truth table, using a diode matrix.

Inputs				Outputs			
W_3	W_2	W_1	W_0	Y_3	Y_2	Y_1	Y_0
0	0	0	1	0	1	1	1
0	0	1	0	1	1	0	0
0	1	0	0	1	1	0	1
1	0	0	0	0	0	1	0

7-28 (a) Design an encoder, using multiple-emitter transistors, to satisfy the following truth table.
(b) How many transistors are needed, and how many emitters are there in each transistor?

Inputs			Outputs				
W_2	W_1	W_0	Y_4	Y_3	Y_2	Y_1	Y_0
0	0	1	1	0	1	1	0
0	1	0	1	1	1	0	0
1	0	0	0	1	0	1	1

7-29 A block diagram of a 3-input (A, B, and C) and 8-output (Y_0 to Y_7) decoder matrix is indicated. The bit Y_5 is to be 1 (5 V) if the input code is 101 corresponding to decimal 5.

(a) Indicate how diodes are to be connected to line Y_5.
(b) Repeat for Y_2, Y_3, and Y_4.

5 V

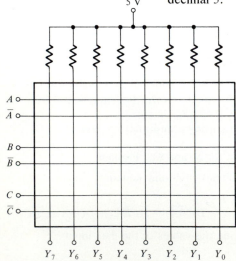

Y_7 Y_6 Y_5 Y_4 Y_3 Y_2 Y_1 Y_0 **Prob. 7-29**

7-30 For the priority encoder in Table 7-3 verify that:

(a) $Y_3 = W_9 + W_8$

(b) $Y_2 = (\overline{W_9 + W_8})(W_7 + W_6 + W_5 + W_4)$

7-31 For the 10-line-decimal-to-4-line BCD priority encoder verify that

$$Y_0 = W_9 + \overline{W_8}(W_7 + \overline{W_6}W_5 + \overline{W_6}\,\overline{W_4}W_3 + \overline{W_6}\,\overline{W_4}\,\overline{W_2}W_1)$$

7-32 (a) Fill in the truth table for an 8-data-line-to-3-line binary (octal) priority encoder, using X to indicate a don't care state.

(b) Obtain the expression for Y_0.

7-33 Repeat Prob. 7-32 for Y_1.

7-34 (a) Implement the code conversion indicated below using a read-only memory (ROM). Indicate *all* connections between the X inputs and the Y outputs. Use the standard symbols for inverters, AND gates, and OR gates.

Inputs		Outputs			
X_1	X_0	Y_3	Y_2	Y_1	Y_0
0	0	1	0	1	1
0	1	0	1	0	1
1	0	0	1	1	1
1	1	1	1	0	0

(b) Draw the OR gates as multiple-emitter transistors.

7-35 (a) Draw a block diagram of a 1024 × 4-bit ROM using two-dimensional addressing.

(b) How many NAND gates are required?

(c) How many transistors must be used in the memory matrix, and how many emitters must each transistor have?

7-36 Consider a 1024 × 8-bit ROM using two-dimensional addressing with 8-to-1 selectors.

(a) How many bits are needed to address the ROM?

(b) How many bits are needed for the X address?

(c) How many NAND gates are required?

(d) Specify the number of transistors in the memory matrix and the number of emitters in each transistor.

7-37 (a) Write the expressions for Y_0 and Y_2 in the binary-to-Gray-code converter.

(b) Indicate how to implement the relationship for Y_0 with diodes.

7-38 (a) Give the relationships between the output and input bits for the Gray-to-binary-code translator for Y_3 and Y_2.

(b) Indicate how to implement the equation for Y_3 with transistors.

7-39 (a) Write the sum-of-products canonical form for Y_5 in Table 7-5 for the seven-segment indicator code.

(b) Verify that this expression can be minimized to

$$Y_5 = \overline{DC}\overline{A} + \overline{CB}\overline{A} + BA$$

7-40 Minimize the number of terms in Eq. (7-33) to obtain Eq. (7-34).

7-41 Consider a 4-kb ROM with 4 output bits. If the encoder is square, how many bits are needed for the (a) X address? (b) The Y address? (c) Sketch the block diagram of the system.

7-42 Consider the 8-kb ROM with 8 output bits. If the memory matrix has 128 rows, how many bits are needed for (a) the X address? (b) The Y address?
(c) Repeat (a) and (b) assuming that there are 64 rows in the encoder.
(d) How many words does this ROM have and how many bits are needed to decode these words? Check your answer against the sum of the bits in the X and Y addresses for each of the two ROMs considered in this problem.

7-43 Two 16-kb (2048 $\times$ 8) ROMs are available. Show how to connect these so as to obtain (a) a 32-kb (2048 $\times$ 16) ROM and (b) a 32-kb (4096 $\times$ 8) ROM.

7-44 Indicate in block diagram form how to assemble thirty-two 16-kb (2048 $\times$ 8) ROMs to obtain an equivalent ROM with 16 address lines and 8 output lines.

7-45 (a) A 32 $\times$ 8 ROM is to be converted into a 64 $\times$ 4 ROM. The eight outputs are $0_0 \cdots 0_7$ and the addresses are $A_0 \cdots A_4$. Add one more address $X = A_5$ to control AND-OR gates so that, with $X = 1$, the four outputs $0_0 \cdots 0_3$ are used and for $X = 0$ the four outputs $0_4 \cdots 0_7$ are used. Indicate this 64 $\times$ 4 ROM system.
(b) Indicate how to convert two 32 $\times$ 8 ROM chips into a 128 $\times$ 4 ROM.

7-46 (a) Show the block diagram of a system for converting a 64 $\times$ 8 ROM into a 512 $\times$ 1 ROM, using a selector-multiplexer.
(b) Repeat (a) for converting a 64 $\times$ 8 ROM into a 256 $\times$ 2 ROM.

CHAPTER 8

8-1 (a) Verify that it is not possible for both outputs in Fig. 8-1 to be in the same state.
(b) Verify that $B_1 = B_2 = 0$ in Fig. 9-1b is not allowed.

8-2 Consider the chatterless switch in Fig. 8-2. At time $t_1' > t_6$ the key is depressed so that the pole is moved from 1 to 2. It reaches 2 at time t_2' and then bounces three times. Indicate the waveforms B_2, B_1, and Q and explain your reasoning.

8-3 (a) Verify that the AOI topology shown gives the same logic as the latch of Fig. 8-3.
(b) Transform the block diagram so that it becomes equivalent to that in Fig. 8-3.

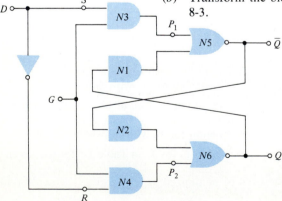

Prob. 8-3

8-4 The NOR gates in Fig. 8-4 are fabricated in NMOS technology. The enhancement drivers have $kW/L = 400$ μA/V^2 and $V_T = 1.0$ V; the depletion load has $kW/L = 100$ μA/V^2. Using $V_{DD} = 5$ V, determine the output levels of the bistable circuit.

8-5 The NAND gates in Fig. 8-1 are fabricated in TTL technology with $V_{CC} = 5$ V. The TTL gates have $V(1) = 2.7$ V, $V(0) = 0.3$ V, and $NM_H = NM_L = 0.2$ V. Assuming that the TTL NAND gate has a voltage transfer chracteristic as given in Fig. 8-5a, determine the output levels of the latch and the minimum trigger signal needed to switch states.

8-6 The NOR gates in Fig. 8-4 are fabricated in CMOS technology. The NMOS device has $kW/L = 200$ μA/V^2 and $V_T = 2$ V; the PMOS device has $kW/L = 200$ μA/V^2 and $V_T = -2$ V. Using $V_{DD} = 5$ V, determine (a) the output levels of the latch and (b) the minimum input signal needed to cause the output to change state.

8-7 Show how to build the bistable latch of Fig. 8-3 by using the AOI configuration.

8-8 The excitation table for a J-K FLIP-FLOP is shown. An X in the table is to be interpreted to mean that it does not matter whether this entry is a 1 or a 0. It is referred to as a don't care condition. Thus the second row indicates that if the output is to change from 0 to 1, the J input must be 1, whereas K can be either 1 or 0. Verify this excitation table by referring to the truth table of Fig. 8-11.

Q_n	Q_{n+1}	J_n	K_n
0	0	0	X
0	1	1	X
1	0	X	1
1	1	X	0

8-9 Verify that the J-K FLIP-FLOP truth table is satisfied by the difference equation $Q_{n+1} = J_n\overline{Q}_n + \overline{K}_nQ_n$.

8-10 (a) Show that the J-K FLIP-FLOP in Fig. 7-7 will preset correctly ($Pr = 0$, $Cr = 1$) only if $\overline{K} + \overline{Ck} = 1$.
  (b) Show that the J-K FLIP-FLOP will clear correctly ($Pr = 1$, $Cr = 0$) only if $\overline{J} + \overline{Ck} = 1$.
  (c) Verify that $Cr = Pr = Ck = 0$ leads to an indeterminate state.
  (d) Show that $Pr = 1$, $Cr = 1$ will enable the FLIP-FLOP.

8-11 (a) Verify that there is no race-around difficulty in the J-K circuit in Fig. 8-12 for any data input combination except $J = K = 1$.
  (b) Explain why the race-around condition does not exist (even for $J = K = 1$) provided that $t_p < \Delta t < T$.

8-12 (a) For the master-slave J-K FLIP-FLOP in Fig. 8-13, assume $Q = 0$, $\overline{Q} = 1$, $Ck = 1$, $J = 0$, and K arbitrary. What is Q_M?
  (b) If J changes to 1, what is Q_M?
  (c) If J returns to 0, what is Q_M? Note that Q_M does not return to its initial value. Hence J (and K) must not vary during the pulse.

8-13 The indicated waveforms J, K, and Ck are applied to a J-K FLIP-FLOP. Plot the output waveform for Q and $\overline{Q}$ lined up with respect to clock pulses. (*Note:* Assume that the output $Q = 0$ when the first clock pulse is applied and that $Pr = Cr = 1$.)

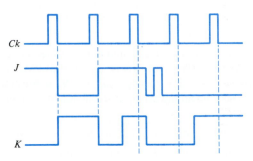

Prob. 8-13

8-14 (*a*) Verify that an SR FLIP-FLOP is converted to a T type if S is connected to $\overline{Q}$ and R to Q.

(*b*) Verify that a D-type FLIP-FLOP becomes a T type if D is tied to $\overline{Q}$.

8-15 The truth table for an A-B FLIP-FLOP is as shown. Show how to build this FLIP-FLOP using a J-K FLIP-FLOP and any additional logic required.

A_n	B_n	Q_{n+1}
0	0	$\overline{Q}_n$
1	0	Q_n
0	1	1
1	1	0

8-16 A 4-bit cascadable priority register consisting of D-type latches is shown.

(*a*) Set $P_0 = 0$, $D_0 = D_1 = D_3 = 0$, and $D_2 = 1$. Verify that $Y_2 = 1$ and all other outputs are 0.

(*b*) Set $P_0 = 0$, $D_0 = D_1 = 0$, and $D_2 = D_3 = 1$. Verify that only $Y_2 = 1$.

(*c*) Generalize the above results to show that the lowest-order D_n among those in the high (1) state is transferred to make the corresponding Y_n high.

(*d*) Cascade two such 4-bit packages. Put $P_0 = 0$ for the lower-order chip. For the higher-order chip tie P_0 to the complement of the P_1 output of the lower-order package. Demonstrate that this cascaded system functions as an 8-bit priority register.

8-17 For the bidirectional shift register in Fig. 8-17 verify the mode of operation indicated in Table 8-4 for (*a*) the second row, (*b*) the third row, and (*c*) the fourth row.

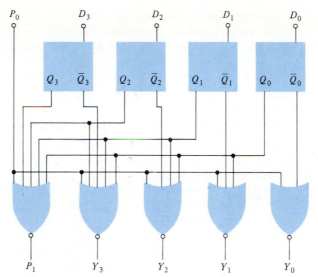

P_1 Y_3 Y_2 Y_1 Y_0 **Prob. 8-16**

8-18 Augment the shift register in Fig. 8-16 with a 4-input NOR gate whose output is connected to the *serial input* terminal. The NOR-gate inputs are Q_4, Q_3, Q_2, and Q_1.

(a) Verify that regardless of the initial state of each FLIP-FLOP, when power is applied, the register will assume correct operation as a ring counter after P clock pulses, where $P \le 4$.

(b) Assuming that initially $Q_4 = 0$, $Q_3 = 1$, $Q_2 = 1$, $Q_1 = 0$, and $Q_0 = 1$, sketch the waveform at Q_0 for the first 16 pulses.

(c) Repeat (b) for $Q_4 = 1$, $Q_3 = 1$, $Q_2 = 0$, $Q_1 = 1$, and $Q_0 = 0$.

8-19 (a) Draw a waveform chart for the twisted-ring counter; i.e., indicate the waveforms Q_4, Q_3, Q_2, Q_1, and Q_0 for, say, 12 pulses. Assume that initially $Q_0 = Q_1 = Q_2 = Q_3 = Q_4 = 0$.

(b) Write the truth table after each pulse.

(c) By inspection of the table show that 2-input AND gates can be used for decoding. For example, pulse 1 is decoded by $Q_4\overline{Q}_3$. Why?

8-20 (a) For the modified ring counter shown, assume that initially $Q_0 = 0$, $Q_1 = 0$, and $Q_2 = 1$. Make a table of the readings Q_0, Q_1, Q_2, J_2, and K_2 after each clock pulse. How many pulses are required before the system begins to operate as a divide-by-N counter? What is N?

(b) Repeat (a) given initially $Q_0 = 0$, $Q_1 = 1$, and $Q_2 = 0$.

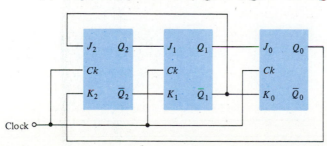

Prob. 8-20

8-21 A 25 : 1 ripple counter is desired.
    (*a*)   How many FLIP-FLOPS are required?
    (*b*)   If 4-bit FLIP-FLOPS are available on a chip, how many chips are needed? How are these interconnected?
    (*c*)   Indicate the feedback connections to the clear terminals.

8-22 (*a*)   Indicate a divide-by-20 ripple-counter block diagram. Include a latch in the clear input.
    (*b*)   What are the inputs to the feedback NAND gate for a 125 : 1 ripple counter?

8-23 Consider the operation of the latch in Fig. 8-21. Make a table of the quantities Ck, Q_1, Q_3, P_1, $\overline{Ck}$, and $P_2 = Cr$ for the following conditions:
    (*a*)   Immediately after the tenth pulse.
    (*b*)   After the tenth pulse and assuming Q_1 has reset before Q_3.
    (*c*)   During the eleventh pulse.
    (*d*)   After the eleventh pulse.
This table should demonstrate that
    (*a*)   The tenth pulse sets the latch to clear the counter.
    (*b*)   The latch remains set until all FLIP-FLOPS are cleared.
    (*c*)   The positive edge of the eleventh pulse resets the latch so that $Cr = 1$.
    (*d*)   The negative edge of the eleventh pulse initiates the new counting cycle.

8-24 Draw a block diagram of a counter system that can be used to generate signals from 0.1 s to 1 h in 0.1-s increments. A 1.8-MHz frequency clock signal is available.

8-25 (*a*)   Indicate a divide-by-11 ripple counter in the block-diagram form. Indicate the connections to J, K, and Ck for each FLIP-FLOP and the inputs to the feedback gate to the clear inputs. (You may omit the latch.) The preset inputs are held at the 1 level.
    (*b*)   There is a second method of obtaining an 11-to-1 ripple counter. The clear inputs are now held at the 1 level and the feedback gate excites the preset inputs. Draw the block diagram for such a programmable ripple counter. Indicate all connections carefully.

8-26 (*a*)   For the logic diagram of the decade counter shown write the truth table for Q_0, Q_1, Q_2, and Q_3 (starting with 0000) after each pulse. If no connection is shown to a J or K input, then this terminal is understood to be high (a 1). Verify that this system is a 10 : 1 counter.
    (*b*)   How can this system be used as a 5 : 1 counter?

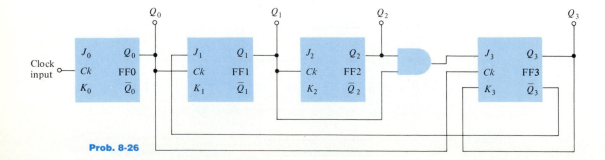

Prob. 8-26

8-27 Modify the logic diagram to Prob 8-26 as follows. Remove the clock from the input to FF0 and apply the Q_3 output to this input. Remove Q_0 from Ck of FF1. Apply the clock input to Ck of FF1. Change no other connections. Write the truth table for Q_0, Q_1, Q_2, and Q_3 (starting with 0000) after each pulse. Verify that this system is a 10 : 1 counter. This is called a *biquinary* counter because a symmetrical square wave is obtained at Q_0. Your truth table should show that this statement is true.

8-28 (a) For the block diagram shown write the truth table for Q_0, Q_1, Q_2, and Q_3 (starting with 0000) after each pulse. Verify that this is a 12 : 1 counter.

　　(b) How can this system be used as a 6 : 1 counter?

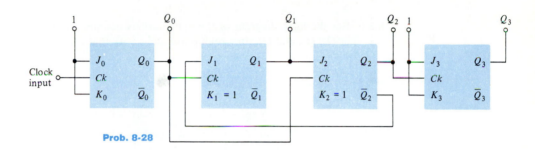

Prob. 8-28

8-29 (a) The circuit shown is a *programmable* ripple counter. Initially $Ck = 0$ and the counter is cleared by momentarily setting $Cr = 0$. Thereafter it is understood that $J = K = Cr = 1$, and that the latch in Fig. 8-21 exists between P_1 and P_2. If $Pr_0 = Pr_1 = 0$ and $Pr_2 = Pr_3 = 1$, and *if a pulse from an external source* (not shown) is applied to the preset input, to what state is each FLIP-FLOP set? If a clock-pulse train is now applied to the counter input, what is the count N? Explain the operation of the system carefully.

　　(b) Why is the latch required?

　　(c) Generalize the result of (a) as follows. The counter has n stages and is to divide by N, where $2^n > N > 2^{n-1}$. How must the preset inputs be programmed?

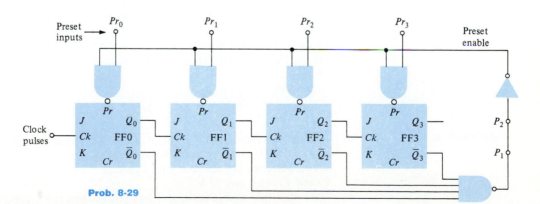

Prob. 8-29

8-30 Draw the logic diagram of a 5-bit up-down synchronous counter with series carry.

8-31 Verify that the system shown is a 3 : 1 synchronous counter. Start with $Q_0 = Q_1 = 0$ and show the state of Q_0 and Q_1 after each pulse.

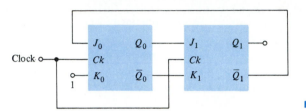

Prob. 8-31

8-32 For the logic diagram of the synchronous counter shown, write the truth table of Q_0, Q_1, and Q_2 after each pulse and verify that this is a 5 : 1 counter.

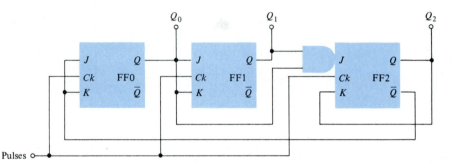

Prob. 8-32

8-33 Consider a two-stage synchronous counter (both stages receive the pulses at the Ck input). In each counter $K = 1$. Given $J_0 = \overline{Q}_1$ and $J_1 = Q_0$, draw the circuit. From a truth table of Q_0 and Q_1 after each pulse, demonstrate that this is a 3 : 1 counter.

8-34 Draw the waveform chart for a 6 : 1 divider from Fig. 8-19 and deduce the connections for a synchronous counter. Draw the logic block diagram.

8-35 Solve Prob. 8-34 for 5 : 1 divider.

8-36 Assume you have a crystal-oscillator circuit which provides a series of clock pulses at a frequency of 131.0 kHz. Construct a system whose output is a light-emitting diode flickering approximately once every second, and use the crystal pulses as input. How many seconds does the system "miss" over a 1-h period? (*Hint:* $2^{17} = 131,072$.)

CHAPTER 9

9-1 (*a*) Modify the dynamic MOS inverter of Fig. 9-1 by adding another FET $Q4$ in series with $Q1$. Designate the input to $Q4$ ($Q1$) by V_4 (V_1). Verify that this circuit performs the function of a dynamic NAND gate. The input levels of V_1 and V_4 are 0 and 10 V.

(*b*) Show that this circuit dissipates less power than the corresponding static NMOS NAND gate in Fig. 6-29*a*.

9-2 Modify the circuit in Fig. 9-1 by adding another FET Q_4 in parallel with $Q1$. Repeat Prob. 9-1 (with the word NAND replaced by NOR and Fig. 6-29a replaced by Fig. 6-28a).

9-3 (a) Consider the shift-register stage of Fig. 9-2 but with unclocked loads; that is, the gates of $Q2$ and $Q5$ are tied to V_{DD} instead of being excited by the clock waveforms. Explain the operation of the circuit.

(b) Show that more power is dissipated in this cell than in the clocked-load version of Fig. 9-2.

9-4 (a) An NMOS dynamic shift-register stage is shown. The two-phase waveforms ϕ_1 and ϕ_2 are sketched in Fig. 9-2b. Carefully explain the operation of this circuit. Assume $C_1 \gg C_2$.

(b) Are the inverters ratioed or ratioless? Explain.

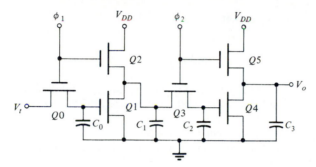

Prob. 9-4

9-5 Verify Eq. (9-1). (*Hint:* When the transmission gate $Q3$ closes, the same charge which leaves C_1 must be added to C_2.)

9-6 (a) Consider the two-phase NMOS inverter shown which uses the clock waveforms pictured in Fig. 9-5b. Explain the circuit operation by considering first the interval $t_1 - t_2$, then $t_2 - t_3$, and so on.

(b) Is this a ratioed or ratioless inverter? Explain.

(c) Using two such inverters and two bidirectional gates, sketch one stage of a shift register. (*Hint:* Interchange ϕ_1 and ϕ_2 in the second inverter and read the output during ϕ_1.)

(d) Explain the operation of this shift-register cell.

9-7 Consider the four-phase dynamic NMOS shift-register cell shown. Note that the four clock pulses are nonoverlapping so that, if one of the phases is high, the

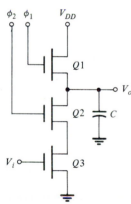

Prob. 9-6

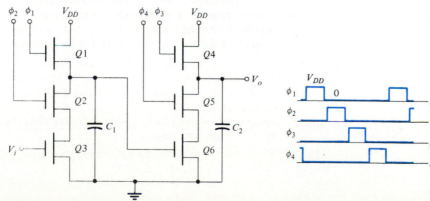

Prob. 9-7

other three are low. Explain the operation and verify that V_o equals the value which V_i had one period earlier.

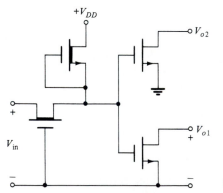

9-8 The circuit shown is sometimes referred to as a "push-pull" NMOS NOR gate and is used to improve the delay-power product over that of conventional NMOS gates.

(a) Verify that NOR logic is performed.

(b) Compare the voltages V_{GS3} and V_{GS4} with V_{GS} of the load transistor in a standard NMOS NOR gate during an input transition from $V(0)$ to $V(1)$.

(c) How do these levels affect the charge and discharge of C_L during a transition?

Prob. 9-8

9-9 (a) For the circuit shown, what is the logic relation between V_{in} and V_{o1}?

(b) What change in input level is required to cause a transition from $V(0)$ to $V(1)$?

(c) Repeat (b) from $V(1)$ to $V(0)$.

(d) Can the results in (b) and (c) be used to improve the delay-power product?

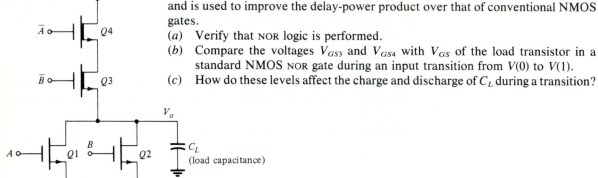

Prob. 9-9

9-10 A 1024-bit RAM consists of 128 words of 8 bits each. If linear selection is used, show a block diagram of the system organization. *Note:* Use one rectangle to represent the 1-bit read-write cell in Fig. 9-8, with three terminals: X for the *address input, W* for the *write input,* and R for the *read output.*

9-11 (a) How many NAND gates and how many inputs to each gate are there in the decoder (or decoders) for a 4096×1 RAM, if linear selection is used?

(b) Repeat (a), assuming that two-dimensional addressing is used to give a square memory array.

(c) Repeat (a), assuming that two-dimensional addressing is used to produce a 256×16 memory array.

9-12 In Fig. 9-19 chip (0) contains words 0 to 1023, chip (1) has words 1024 to 2047, and so on. What word is decoded by (a) $A_{11} \cdots A_0 = 011100101011$? (b) 111000010110? (c) What address must be applied to obtain word 2600?

9-13 Draw a block diagram of a 4096 × 16 RAM system built from 1024 × 1 RAMs.

9-14 Draw a block diagram of a 128K × 4-bit read-write system assembled from 16-kb × 1-bit RAMs.

9-15 Consider the CCD structure in Fig. 9-23a operated by the two-phase waveforms of Fig. 9-27. All odd-numbered electrodes are tied to ϕ_1, and all even-numbered electrodes are excited by ϕ_2. Draw potential profiles as in Fig. 9-23, and demonstrate that this system is unsatisfactory because the direction of charge transfer is indeterminate.

9-16 Consider a two-phase CCD. The effective length of each electrode is 8 μm and its width is 8 μm. The separation between rows of electrodes is also 8 μm.
(a) Calculate the area in mm² occupied by a memory cell.
(b) Mnemonics, Inc. has built a 64-kb (65,536-bit) memory using the cell described in (a). The chip size is 218 × 235 mils. What fraction of the chip area is occupied by the auxiliary circuits (input, output, clocks, etc.)?

9-17 Show the organization of the RAM in Prob. 9-10, given the cell used as shown in Fig. 9-20.

9-18 Consider the two-phase CCD structure of Fig. 9-26a excited by the positive clock pulses shown. Assume that $V_2 = V$ and $V_1 = \frac{1}{2}V$. Draw the potential-energy profiles under the first four electrodes for the five times $t_1 \cdots t_5$ indicated. Start with charge under E_1 at $t = t_1$ and demonstrate that it is shifted to E_2 at $t = t_5$. Use quadrilled paper.

9-19 Consider the two-phase CCD structure of Fig. 9-26a excited by the negative clock pulses shown. Assume that $V_2 = V$ and $V_1 = 0$. Draw the potential-energy profiles under the first four electrodes for the five times $t_1 \cdots t_5$ indicated. Start with a bit stored under E_1 at $t = t_1$ and show that the information is transferred to the well under E_2 at $t = t_5$. Use quadrilled paper.

9-20 (a) Consider a single-phase CCD structure. The odd electrodes are biased to a constant voltage $\frac{1}{2}V$. The even electrodes are excited by the positive-pulse waveform ϕ_2 shown in Fig. 9-27 with $V_1 = 0$ and $V_2 = V$. Draw the potential-energy profiles under the first four electrodes for the times t_2, t_3, and t_4. Start with electrons stored under E_1 at time t_2 and show that the charge is retained in the well under E_2 at $t = t_4$. Use quadrilled paper.
(b) Draw the potential-energy profile for a time t_4' (where $\phi_2 = \frac{1}{4}V$), for t_5 or t_6 (where $\phi_2 = 0$), and for t_7 (where $\phi_2 = \frac{1}{2}V$). Demonstrate that the information under E_1 has been transferred to E_3 in one clock period.

9-21 Consider three logic variables A, B, and C at the collectors of three I²L inverters. Connect these three outputs together. Show by a physical argument that at the common node the logic variable is $Y = ABC$. In other words, justify the wired-AND operation for injection logic.

9-22 Given the four external variables A, B, C, and D, draw an I²L connection diagram for the AOI output $Y = AB + CD$.

9-23 The three inputs to a decoder are A, B, and C. Draw an I²L connection diagram to obtain the eight outputs.

9-24 Consider a 2-to-1-line multiplexer without a strobe. Draw an I²L connection diagram for this data selector.

9-25 The carry in a full adder is of the form $C' = AB + BC + CA$. Draw an I²L connection diagram for C'.

9-26 Draw an I²L connection diagram for the J-K clocked FLIP-FLOP.

CHAPTER 10   The following transistors are used throughout the problems in this chapter:

	Transistor				
Quantity	A	B	C	D	E
Type	*npn*	*npn*	*npn*	*pnp*	*pnp*
β_F	125	150	200	150	50
β_o	125	150	200	150	50
V_A, V	∞	100	∞	∞	50

It is assumed that $r_b = 0$ for these transistors and operation is at $T = 25°C$ unless otherwise stated.

10-1 The current mirror in Fig. 10-5a is designed by using transistor A and is to provide a 0.5-mA current with $V_{CC} = 10$ V.
   (a) Determine the value of R.
   (b) Assuming that all other parameters remain constant and the change in V_{BE} is -2.2 mV/°C, what is the permissible temperature change if I_{C1} is to remain within 1 percent of its nominal design value?

10-2 Transistor C is used in the circuit in Fig. 10-5a with $V_{CC} = 5$ V and $R = 5$ kΩ.
   (a) Determine I_{C1}.
   (b) What are the minimum and maximum values of β_F if the change in I_{C1} is to be no greater than 1 percent of the value in (a)?

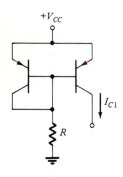

$+V_{CC}$

I_{C1}

R

Prob. 10-3

10-3 The *pnp* current mirror shown uses transistor D. For $I_{C1} = 1$ mA and $V_{CC} = 15$ V
   (a) Determine R.
   (b) Determine the percentage change in I_{C1} for a 50°C change in temperature if V_{BE} changes by 2.2 mV/°C and all other parameters are unchanged.

10-4 The transistors in the circuit shown are identical. What are the minimum and maximum values of I_C if $75 \leq \beta_F \leq 175$?

10-5 Transistor C is used in the circuit shown.
   (a) Determine I_{C1} and I_{C2}.
   (b) Find R_C so that $V_o = 6$ V.

10-6 Transistor B is used in the circuit shown.
   (a) Consider $V_A \rightarrow \infty$ and find R_C so that $V_o = 0$.
   (b) Using the value of R_C in (a), determine V_o when $V_A = 100$ V. [*Hint:* $I_C = (I_{CS} \epsilon^{V_{BE}/V_T})(1 + V_{CE}/V_A)$ accounts for the Early voltage.]

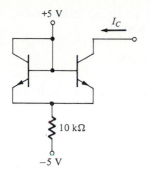

Prob. 10-4

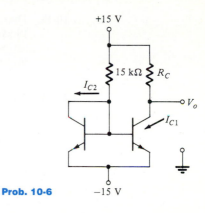

Prob. 10-5

Prob. 10-6

10-7 The current mirror in Prob. 10-1 is to be converted to the circuit in Fig. 10-8 and provide $I_{C1} = 50$ μA. Determine R_E.

10-8 The circuit in Fig. 10-8 is designed with $V_{CC} = 15$ V, $R = 30$ kΩ, and $R_E = 1$ kΩ. Transistor C is used.
(a) Determine I_{C1}.
(b) What is the percentage change in I_{C1} if V_{CC} increases by 0.3 V?

10-9 The circuit in Fig. 10-9 is designed with identical transistors. Derive an equation for I_{C1}/I_{C2} and show that this ratio is proportional to R_2/R_1.

10-10 For the circuit in Fig. 10-8 show that $\Delta I_R/I_R$ is inversely proportional to $[(V_{CC}/ V_{BE}) - 1]$ if V_{BE} changes by ΔV_{BE}. Assume that all other parameter values remain constant.

10-11 A Widlar source is to be designed using *pnp* transistors (transistor D). A negative supply voltage of magnitude 9 V is available and $R = 25$ kΩ is used. Determine R_E so that the source current has a magnitude of 40 μA.

10-12 (a) Repeat Prob. 10-11 using a positive supply voltage of 9 V.
(b) Draw the circuit diagram of the current source.

10-13 Derive Eq. (10-15).

10-14 Derive Eq. (10-16).

10-15 The circuit in Fig. 10-10a is used to obtain a current of 1 mA using a 12-V supply and transistor A.
(a) Determine the value of R.
(b) If β_F decreases by 60 percent, what is the percent change in I_{C1}?

10-16 Repeat Prob. 10-15 for the circuit in Fig. 10-10b.

10-17 The transistors used in Fig. 10-11 are identical and have $V_A \to \infty$.
(a) Derive the expression for I_{C1} in terms of β_F, V_{BE}, R, and V_{CC}.
(b) For $V_{CC} = 15$ V and $\beta_F = 150$, determine R so that $I_{C1} = 300$ μA.
(c) Assuming that all other parameters remain unchanged, what is the permissible change in temperature if $| \Delta I_{C1}| \leq 30$ μA and V_{BE} changes by -2.2 mV/°C?

10-18 The circuit in Fig. 10-12 uses transistor C. The parameter values used are $V_{CC} = 11.2$ V, $R_C = 1.2$ kΩ, $R_E = 0.3$ kΩ, $R_1 = 90$ kΩ, and $R_2 = 10$ kΩ.
(a) Determine the operating point.
(b) If β_F decreases by 50 percent, what are the new values of I_{CQ} and V_{CEQ}?

10-19 The circuit in Fig. 10-12 is designed with $V_{CC} = 15$ V, $R_1 = 72$ kΩ, $R_2 = 18$ kΩ, $R_E = 1.4$ kΩ, and $R_C = 4.0$ kΩ. Transistor A is used.
  (a) Determine the operating point.
  (b) Given β_F doubled, determine the new operating point.
  (c) Comment on the effectiveness of the circuit.

10-20 The circuit shown uses transistor C and is designed to make $V_o = 0$ and $V_{CEQ} = 3$ V.
  (a) Determine R_C and R_E.
  (b) Using the values obtained in (a), find the change in V_o, given that β_F is halved.
  (c) The supply voltages each change by 5 percent. Determine the maximum change in V_o. Use the parameter values obtained in (a).

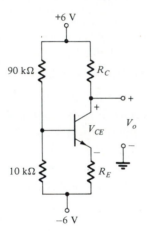

Prob. 10-20

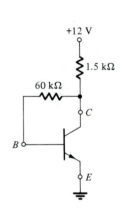

Prob. 10-21

10-21 The circuit shown uses transistor B (consider $V_A \to \infty$).
  (a) Determine I_{CQ} and V_{CEQ}.
  (b) A resistance R is added to the circuit from base to ground. What value of R is needed to make $V_{CEQ} = 6.7$ V?
  (c) If β_F changes by ± 100, what is the range of values of V_{CEQ} in the circuit in (b)?

10-22 The circuit in Fig. 10-12 is to be designed by using a 28-V supply. The transistor exhibits $50 \le \beta_F \le 200$ and is to be used from $T = 0°C$ to $T = 100°C$. The nominal quiescent point is $I_{CQ} = 1.5$ mA and $V_{CEQ} = 13$ V. The worst-case condition requires that I_C be with 150 μA. Neglect I_{CO} and assume that both β_F and V_{BE} variations produce equal deviations. Determine R_1, R_2, R_C, and R_E.

10-23 The circuit shown is an emitter-coupled pair in which $Q3$ and $Q4$ are used to bias $Q1$ and $Q2$. Transistors $Q5$, $Q6$, and $Q7$ form a current repeater, and $Q6$ and $Q7$ form the loads for $Q1$ and $Q2$. All *pnp* transistors have $\beta_F = 50$; the *npn* transistors have $\beta_F = 150$. Assume $V_A \to \infty$. Find R so that the current relationships are satisfied.

10-24 Transistor A is used in the current repeater shown. Determine I_{C1}, I_{C2}, and I_{C3}.

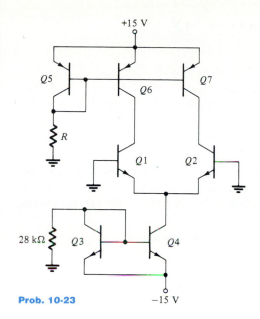

Prob. 10-23

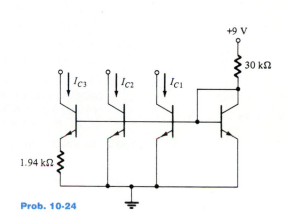

Prob. 10-24

10-25 The MOSFETs in Fig. 10-15a are described in saturation by

$$I_D = 25 \left(\frac{W}{L}\right)(V_{GS} - 1.5)^2 \; \mu A$$

(a) For $W/L = 4$, find R so that $I_D = 400 \; \mu A$ with $V_{DD} = 9$ V.

(b) Using the value of R found in (a), determine the change in I_{D1} if W/L for $Q1$ is made 2.

(c) Repeat (b) for $W/L = 8$.

10-26 The circuit in Fig. 10-15a uses MOSFETs whose characteristics are given in Fig. 10-15b. The supply voltage is 5 V. Determine R so that $I_D = 100 \; \mu A$.

10-27 The MOSFETs in Fig. 10-15a have $k(W/L) = 200 \; \mu A$ and $V_T = 2$ V. They are supplied from a 12-V source. Determine R so that $I_D = 0.5$ mA.

10-28 The enhancement transistors in the circuit shown have $I_D = 100(V_{GS} - 3)^2 \; \mu A$. The depletion transistor has $I_D = 100(V_{GS} + 1)^2 \; \mu A$. Determine I_{D1}.

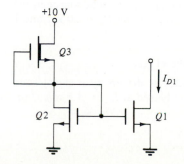

Prob. 10-28

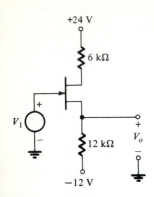

+24 V

6 kΩ

V_1

12 kΩ

V_o

−12 V

10-29 The FET shown has $I_{DSS} = 4$ mA and $V_P = -4$ V.
    (*a*) For $V_1 = 0$, find V_o.
    (*b*) For $V_1 = 15$ V, find V_o.
    (*c*) For $V_o = 0$, find V_i.

10-30 The FET in the circuit of Fig. 10-18*a* has $I_{DSS} = 3$ mA and $V_P = -3$ V. For $R_1 = 1.5$ mΩ, $R_2 = 0.3$ MΩ, $R_D = 20$ kΩ, $R_S = 5$ kΩ, and $V_{DD} = 60$ V, determine I_{DQ}, V_{DSQ}, and V_{GSQ}.

10-31 The change in V_P in the FET in Prob. 10-30 is ± 0.5 V. Determine the range of values of I_{DQ} and V_{DSQ}.

10-32 The FET whose transfer characteristic is shown in Fig. 10-19 is used in the circuit in Fig. 10-18*a*. The drain current is restricted to be between 4.0 and 5.0 mA and $V_{DS} \geq 6$ V. The supply voltage is 24 V and $R_G \geq 100$ kΩ. Determine R_1, R_2, R_C, and R_S.

10-33 The manufacturer of a *p*-channel JFET provides the following data:

	Minimum	Maximum
V_P	5 V	6 V
I_{DSS}	−2.5 mA	−4.5 mA

A circuit analogous to that shown in Fig. 10-18*a* for *p*-channel devices is to be designed so that I_{EQ} lies between −1.6 and −2.0 mA. For $V_{DD} = -30$ V and $R_G \geq 100$ kΩ determine
    (*a*) Determine R_1, R_2, and R_S.
    (*b*) Given $R_L = 10$ kΩ, what are the minimum and maximum values of V_{DSQ}?

10-34 Derive Eq. (10-29).

10-35 Transistor *A* is used in the circuit in Fig. 10-21*a* and is biased at $I_{CQ} = 1$ mA. For $R_s = 300$ Ω and $R_C = 1.2$ kΩ, determine A_V and R_i.

10-36 Transistor *C*, biased at $I_C = 0.5$ mA, is used in the circuit in Fig. 10-21*a*.

10-37 Transistor *B* is used in the circuit in Fig. 10-22*b*; I_o is obtained from a current mirror using transistor *E* and is 50 μA. For $R_s = 5$ kΩ, determine A_V, R_i, and R_o.

10-38 Transistor *E* is used to obtain the current source in Fig. 10-22*b*. Transistor *B* is driven from a signal source having $R_s = 20$ kΩ and is to be biased such that $R_s = R_i$.
    (*a*) Find I_o.
    (*b*) Determine the value of A_V.

10-39 The circuit in Fig. 10-22*a* is driven by a signal source having $R_s = 10$ kΩ. Transistor *C* is used and biased at $I_{CQ} = 1.5$ mA. For $R_E = 2$ kΩ, determine
    (*a*) A_V
    (*b*) R_i
    (*c*) R_o and R'_o

10-40 A common-collector stage uses transistor D biased at $I_{CQ} = -0.25$ mA and driven from a 3-kΩ source.
   (a)   What value of R_E is needed to make $R_o' = 110$ Ω?
   (b)   Using the value of R_E in (a), determine A_V and R_i.

10-41 An emitter-follower uses transistor A biased at $I_{CQ} = 2$ mA and is required to have $R_i \geq 500$ kΩ.
   (a)   Find R_E.
   (b)   For $R_s = 5$ kΩ, determine A_V, R_o, and R_o'.

10-42 Verify the approximate equations for the common-base stage in Table 10-3A.

10-43 The circuit in Fig. 10-25a uses transistor A biased at 0.2 mA. For $R_s = 2$ kΩ, $R_E = 100$ Ω, and $R_C = 5$ kΩ, determine
   (a) A_V and R_i
   (b)   The range of values for A_V for β_o varying by 60 percent.

10-44 (a)   Repeat Prob. 10-43, assuming that transistor B is used.
   (b)   Determine R_o and R_o' for the amplifier.

10-45 (a)   Assuming $\beta_o \gg 1$, derive an expression for $\Delta A_V/A_V$ for changes in β_o of $\Delta \beta$ in the circuit in Fig. 10-25a.
   (b)   Using transistor C and considering a 50 percent change in β_o, derive an equation for R_E which restricts $|\Delta A_V/A_V| \leq 0.1$.
   (c)   For $R_s = 0.6$ kΩ and the transistor biased at $I_{CQ} = 0.5$ mA, find R_E.
   (d)   Assuming that the nominal value of A_V is 10, find R_C.

10-46 Verify the results in Table 10-3B for the common-collector stage.

10-47 Verify the results in Table 10-3B for the common-emitter stage with an emitter resistance.

10-48 Verify the results in Table 10-3b for the common-base stage.

10-49 Verify the approximate numerical results in Table 10-4.

10-50 Repeat Prob. 10-49 for the case where $r_o = 50$ kΩ.

10-51 For each of the configurations in Table 10-4, determine A_V, assuming $r_b = 50$ Ω and $r_o = 50$ kΩ.

10-52 Derive an expression for the output resistance of a Widlar source.

10-53 A JFET having $I_{DSS} = 5$ mA and $V_P = -4$ V is biased at $V_{GSQ} = -1$ V. It is used in the circuit in Fig. 10-27a for which the parameter values are $R_D = 16$ kΩ and $R_S = 1$ kΩ.
   (a)   Find $A_V = V_{o1}/V_s$.
   (b)   Find R_o and R_o' seen between v_{o1} and ground. Use $1/\lambda = 90$ V.

10-54 For the JFET circuit in Prob. 10-53
   (a)   Find $A_V = V_{o2}/V_s$.
   (b)   Find R_o and R_o' seen between v_{o2} and ground.

10-55 The JFET in Prob. 10-53 is biased at $V_{GSQ} = -2$ V.
   (a)   Determine R_D so that $|A_V| = 20$ in a common-source stage.
   (b)   Assuming that I_{DSS} remains constant and using R_D in (a), find the new value of A_V for $V_P = -5$ V.
   (c)   Repeat (b) for $V_P = -3$ V.

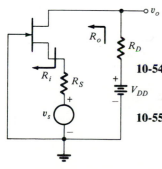

Prob. 10-56

10-56 The circuit shown is a grounded- or common-gate stage. Derive the expressions for A_V, R_i, and R_o.

10-57 A JFET is used in the circuit of Fig. 10-27a with $R_D = 20$ kΩ and $R_S = 1.5$ kΩ. The parameter values of the JFET are $g_m = 1$ m℧ and $r_d = 40$ kΩ.
   (a)   Determine $A_V = V_{o1}/V_s$.
   (b)   Suppose that I_{DSS} changes by 20 percent. Determine the new value of A_V, assuming that V_P and V_{GS} remain constant.

10-58 The JFET in Prob. 10-57 is used as a source follower.
   (a)   Find R_S so that $A_V = 0.95$.
   (b)   For R_S in (a), find R_o and R'_o.

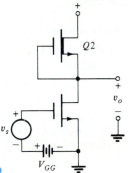

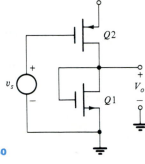

Prob. 10-59

Prob. 10-60

10-59 The circuit shown is a MOSFET common-source stage for which $Q1$ has parameters g_{m1} and r_{d1} and $Q2$ has g_{m2} and r_{d2}. Find $A_V = V_o/V_s$ and comment on the effectiveness of the stage.

10-60 Repeat Prob. 10-59 for the CMOS circuit shown.

10-61 Each stage of a *CE-CE* cascade uses transistor A and is biased at $I_{CQ} = 1$ mA. The component values are $R_s = 0.6$ kΩ and $R_{C1} = R_{C2} = 1.2$ kΩ. Determine A_{V1}, A_{V2}, and A_V.

10-62 A third stage is added to the cascade in Prob. 10-61 and uses transistor A biased at $I_{CQ} = 2$ mA. The collector resistance is 0.6 kΩ.

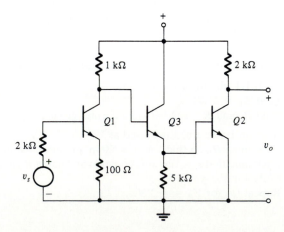

Prob. 10-62

(a) Determine A_V of the amplifier, assuming that this stage follows the two previous stages.

(b) Repeat (a), assuming that this stage precedes the two stages in Prob. 10-61.

(c) Repeat (a), assuming that this stage is placed between the two stages in Prob. 10-61.

10-63 The transistor amplifier stages in Example 10-7 are connected as shown.
(a) Determine the overall gain A_V.
(b) Compare the answer in (a) with A_V in Example 10-7 and justify this comparison in words.

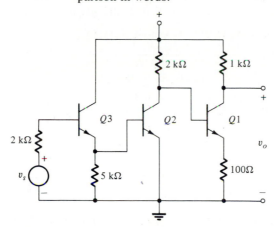

Prob. 10-64

10-64 Repeat Prob. 10-63 for the circuit shown.

10-65 The signal source and source resistance in Example 10-7 drive a single common-emitter stage using $Q2$.
(a) Determine the value of R_C needed to obtain the same overall gain as in Example 10-7.
(b) What is the minimum supply voltage required if the value of R_C in (a) is used?

10-66 The CC-CE composite of Fig. 10-34b is used in the amplifier shown. Both $Q1$ and $Q2$ have $\beta_F = 150$ and $V_A = 130$ V and are biased at $I_{C1Q} = 100$ μA and $I_{C2Q} = 100$ μA. For $R_s = 50$ kΩ and $R_C = 250$ kΩ, determine the gain $A_V = V_o/V_s$.

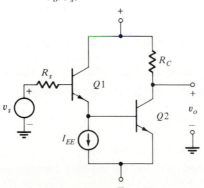

Prob. 10-66

10-67 Two common-base stages are cascaded, and each uses transistor C biased at $I_{CQ} = 0.5$ mA. The circuit is driven from a 50-Ω signal source and $R_{C1} = R_{C2} = 5$ kΩ. Determine
(a) The gains of the individual stages
(b) The gain of the cascade

10-68 Transistor C is used for each transistor in a cascode amplifier. Each is biased at 0.2 mA, and the combination is driven from a source with $R_s = 1$ kΩ. For $R_C = 5$ kΩ
(a) Determine A_V.
(b) What is the percentage change in A_V if R_C changes by ± 20 percent?
(c) Repeat (b), assuming that R_s changes by ± 10 percent.

10-69 In the circuit shown, verify that

(a) $A_{V1} = \dfrac{V_{o1}}{V_s} \simeq \dfrac{g_{m1} \beta_o R_S}{1 + g_{m1} \beta_o R_S}$

(b) $A_{V2} = \dfrac{V_{o2}}{V_s} \simeq \dfrac{g_{m1} \beta_o (R_S + R_C)}{1 + g_m \beta_o R_S}$

Assume $R_D \gg r_\pi$, $r_d \gg r_\pi$, $\beta_o \gg 1$, and $\mu \gg 1$.

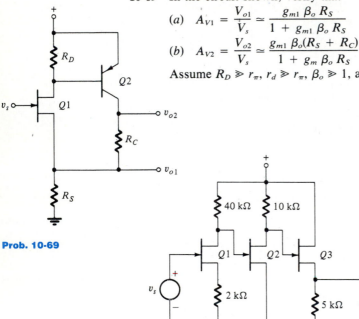

Prob. 10-69

Prob. 10-71

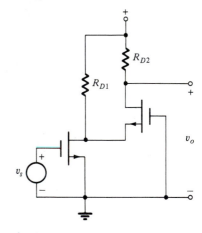

Prob. 10-72

10-70 A Darlington pair (Fig. 10-34a) is used as an emitter follower with $R_E = 500$ Ω and is driven from a 50-kΩ source. Transistor B is used, and $Q2$ is biased at 1.0 mA and $Q1$ is biased at 15 μA. Find A_V, R_o, and R_i.

10-71 Junction FETs $Q1$, $Q2$, and $Q3$ are identical and have the parameters given in Prob. 10-57. Determine
(a) The gain of each stage
(b) The overall gain V_o/V_s
(c) The output resistances R_o and R_o'

10-72 Derive an expression for V_o/V_s for the FET cascode amplifier shown.

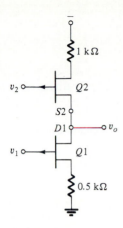

10-73 Transistor $Q1$ has $r_{d1} = 10$ kΩ and $g_{m1} = 3$ m$\mho$; $Q2$ has $r_{d2} = 15$ kΩ and $g_{m2} = 2$ m$\mho$.
(a) Find the gain V_o/V_2 for $V_1 = 0$.
(b) Find the gain V_o/V_1 for $V_2 = 0$.
(c) For $v_1 = 5 \sin \omega t$ and $v_2 = -2.5 \sin \omega t$, find v_o.

10-74 The differential amplifier in Fig. 10-36 uses transistor C biased at $I_{CQ} = 100$ μA. Determine R_C and R_E so that $|A_{DM}| = 500$ and the $CMRR = 80$ dB.

10-75 The inputs to the differential amplifier in Prob. 10-74 are:

$$v_1 = 15 \sin 120 \pi t + 5 \sin 2\pi \times 10^3 t \text{ mV}$$

$$v_2 = 15 \sin 120 \pi t - 5 \sin 2\pi \times 10^3 t \text{ mV}$$

The signal at 60 Hz represents an interference signal, and that at 1 kHz is the signal to be processed.
(a) Determine $v_{o1}(t)$.
(b) Determine $v_{o2}(t)$.

10-76 In the circuit shown, transistor C is used for $Q1$ and $Q2$.
(a) With $v_1 = v_2 = 0$, determine the bias currents I_{CQ} and I_{BQ}.
(b) Find v_{o1} and v_{o2} for the conditions in (a).
(c) Evaluate A_{DM}, A_{CM}, and the $CMRR$.
(d) Determine R_{id} and R_{ic}.

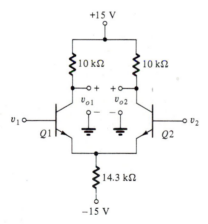

Prob. 10-76

10-77 The 14.3-kΩ resistance in Prob. 10-76 is replaced by a current mirror to establish the same bias currents as in Prob. 10-76. Transistor C is used for the mirror except that $V_A = 130$ V.
(a) Design the current mirror.
(b) What are the new values of A_{DM} and the $CMRR$?

10-78 The differential stage of Fig. 10-36 has a given A_{DM} and $CMRR$.
(a) For $v_1 = V_s$ and $v_2 = 0$, determine v_{o1} and v_{o2}.
(b) For $v_1 = 0$ and $v_2 = V_s$, determine v_{o1} and v_{o2}.

10-79 (*a*) For the IC differential stage shown, show that

$$|A_{DM}| = \left(\frac{V_T}{V_{An}} + \frac{V_T}{V_{Ap}}\right)^{-1}$$

where V_{An} and V_{Ap} are the Early voltages for the *npn* and *pnp* transistors, respectively.

(*b*) Given $V_{An} = 120$ V and $V_{Ap} = 50$ V, evaluate $|A_{DM}|$.

(*c*) Does changing the bias current alter the value of A_{DM}? Explain briefly.

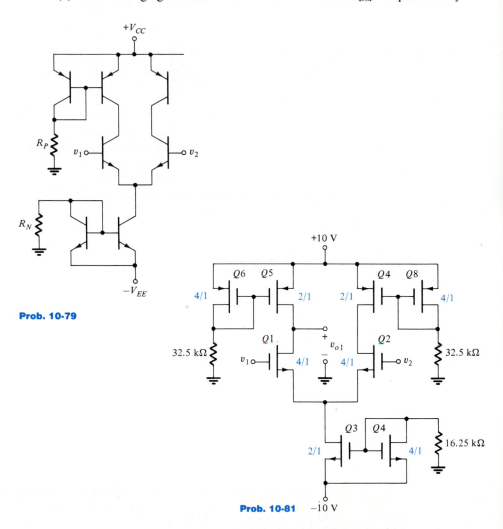

Prob. 10-79

Prob. 10-81

10-80 In the circuit in Prob. 10-79, evaluate the *CMRR* and A_{CM} given $V_{CC} = 15$ V, $V_{EE} = 15$ V, $R_P = 53.5$ kΩ, and $R_N = 28$ kΩ. The *npn* transistors have $\beta_F = \beta_o = 200$, and the *pnp* transistors have $\beta_F = \beta_o = 50$. The Early voltages are given in Prob. 10-79.

10-81 The circuit shown is a CMOS differential stage in which the aspect ratio W/L is indicated in color adjacent to the devices. The NMOS devices have $k = 25$ $\mu A/V^2$, $V_T = 1.5$ V, and $V_A = 1/\lambda = 50$ V; the PMOS devices have $k = 12.5$ $\mu A/V^2$, $V_T = -1.5$ V, and $V_A = 1/\lambda = 100$ V.
(a) Determine the bias drain currents in Q3, Q5, and Q7.
(b) Evaluate A_D and the *CMRR*.

10-82 The JFETs in a source-coupled pair have $g_m = 1$ m℧ and $r_d = 50$ kΩ. They are biased by a current source having an output resistance of 40 kΩ. The drain resistors $R_D = 30$ kΩ. Determine A_{DM}, A_{CM}, and the *CMRR*.

10-83 Verify Eq. (10-110).

10-84 In the circuit of Fig. 10-46, $v_1, v_2, \ldots v_n = 1$ V; $R_2 = 2R_1$, $R_3 = 2R_2, \ldots$ $R_n = 2R_{n-1}$ and $R' = R_1/2$.
(a) As $n \to \infty$, determine v_o.
(b) If $n = 4$, evaluate v_o.

10-85 The circuit in Fig. 10-46 is designed with $R_1 = R' = 1$ kΩ and $R_2 = 2R_1$, $R_3 = 2R_2 \ldots R_n = 2R_{n-1}$. The input voltages $v_1, v_2, \ldots v_n$ can be 0 or 10 volts.
(a) For $n = 4$, what is the smallest output voltage if at least one input is nonzero?
(b) For $n = 4$, what is the maximum output voltage?

10-86 (a) Determine the minimum output voltage, assuming at least one input is nonzero, for the conditions in Prob. 10-85. The maximum resistance allowable is 55 kΩ.
(b) What new value of n can be used if R_1 is reduced to 100 Ω?

10-87 We wish to use the circuit in Fig. 10-46 to provide the class average on a quiz. The class size is 25 and all grades are integers between 1 and 10. The maximum output voltage is 10 V, and the minimum resistance that can be used is 1 kΩ. The minimum value of input voltage that can be used is 250 μV.
(a) Design the circuit.
(b) Check your design with the following grade distribution.

Number of students	0	1	0	2	1	4	7	4	3	3
Quiz grade	1	2	3	4	5	6	7	8	9	10

10-88 The Op-Amp in Fig. 10-46 has a finite gain A_v but is otherwise ideal.
(a) Determine the transfer function i_L/v_s.
(b) If $R_1 = 10$ kΩ, what value of A_v must be used if the result in *a* is to be within percent of that given in Eq. (10-115).

10-89 Repeat Prob. 10-88 for the circuit in Fig. 10-48b. Use $R_2 = 10$ kΩ and compare the result with Eq. (10-116).

10-90 Repeat Prob. 10-88 for the integrator in Fig. 10-50 and described by Eq. (10-117).

CHAPTER 11

To the student: Biasing arrangements are not shown in many of the circuit diagrams for the problems in this chapter. You may assume that the devices are biased properly and those components used for biasing (and not shown) have a negligible effect on circuit performance.

The following transistors are used frequently in the problems that follow:

	Transistors				
	A	B	C	D	E
Type	npn	npn	npn	pnp	pnp
β_o	125	150	200	150	50
β_F	125	150	200	150	50
V_A, V	∞	100	∞	∞	50
f_T, MHz	300	400	400	100	10
C_μ, pF	0.5	0.3	0.3	0.5	0.5

11-1 An amplifier is excited by a signal $v_i = 0.1 \sin \omega_o t + 0.1 \sin 2\omega_o t$. With no frequency distortion, the output v_o is given in Fig. 11-1 as curve 1.

(a) With amplitude and phase distortion, $v_o = 1.0 \sin \omega_o t + 0.75 \sin (2\omega_o t - 30°)$. Plot one cycle of v_o and compare this with the undistorted waveform.

(b) Repeat (a) for $v_o = 1.0 \sin (\omega_o t - 15°) + 1.0 \sin (2\omega_o t - 30°)$. Comment on the result.

11-2 For the circuit shown, compute the upper half-power frequency, assuming that the low-frequency amplifier is a common-emitter stage.

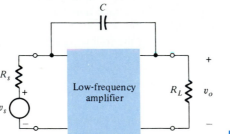

Prob. 11-2

11-3 Repeat Prob. 11-2, assuming that the low-frequency amplifier is an emitter follower.

11-4 In the circuit shown, determine the lower-half power frequency, assuming that the low-frequency amplifier is a source follower.

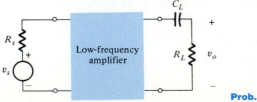

Prob. 11-4

11-5   Repeat Prob. 11-4 for use of a common-source stage with source resistance.

11-6   The input in Prob. 11-5 is a square wave.
    (*a*)   Determine the tilt on the output waveform.
    (*b*)   Repeat (*a*) for use of a simple common-source stage.

11-7   A transistor manufacturer provides the following data: at low frequencies $\beta_o = 160$; at $f = 50$ MHz, $|\beta(j\omega)| = 8$. Determine f_T and f_β.

11-8   At a bias current $I_C = 1$ mA, a semiconductor manufacturer indicates that a particular transistor has $\beta_o = 120$. At the same bias current, $|\beta(j\omega)| = 10$ at a frequency of 25 MHz. Determine C_π in the hybrid-π model at $I_C = 1$ mA, assuming $C_\mu = 1$ pF.

11-9   The transfer function of an amplifier is $A_o/(1 + s/\omega_o)$.
    (*a*)   Show that the response to a unit step of two such amplifiers in cascade (and noninteractive) is

$$v_o(t) = A_o^2 \left[1 - (1 + x)\,\epsilon^{-x} \right]$$

    where $x \equiv \omega_o t$.
    (*b*)   For $\omega_o t \ll 1$, show that the output varies quadratically with time.

11-10   Show that the common-base short-circuit current gain $\alpha(s)$ can be expressed as

$$\alpha(s) = \frac{\alpha_o}{1 + s/\omega_\alpha}$$

where $\alpha_o = \beta_o/(1 + \beta_o)$ and $\omega_\alpha \simeq \omega_\beta/(1 - \alpha_o)$.

11-11   A two-pole amplifier has the transfer function in Eq. (11-22) with $a_3 = 0$.
    (*a*)   Estimate the pole frequencies.
    (*b*)   With $n \equiv a_1^2/a_2$ defined as the approximate pole separation factor, show that for $n > 10$, the actual poles are separated by at least three octaves.

11-12   (*a*)   Determine the high-frequency output impedance $Z_o(s)$ of a common-emitter stage. Assume $r_o \to \infty$ but use the hybrid-π model.
    (*b*)   Repeat (*a*) for $r_o < \infty$.

11-13   (*a*)   Transistor A is used in a common-emitter circuit and biased at $I_{CQ} = 1$ mA. For $R_s = 300\ \Omega$ and $R_C = 1.2$ kΩ, determine the midband gain and upper half-power frequency ω_H.
    (*b*)   What is the input impedance at $s = j\omega_H$?
    (*Note:* This is the circuit in Prob. 10-35.)

11-14   A transistor having $g_m = 4$ m℧, $C_\pi = C_\mu = 1$ pF, and $\beta_o = 120$ is used in the CC configuration. Show that Z_o exhibits inductive behavior for $125\ \Omega < R_s < 30$ kΩ.

11-15   Transistor C, biased at $I_{CQ} = 0.5$ mA, is used in the CE configuration and driven by a voltage source having a 2-kΩ source resistance. The collector resistance is 6 kΩ. Determine the midband gain and upper half-power frequency.

11-16   Transistor D is used in the circuit shown and biased at $I_{CQ} = -2.5$ mA. Determine the midband gain and upper half-power frequency.

11-17   Transistor C, biased at $I_{CQ} = 1.0$ mA, is used in the CB configuration with $R_C = 5$ kΩ. The applied signal is $v_s(t) = 2.0 \sin \omega t$ mV and $R_s = 50\ \Omega$. Determine A_{VO} and the approximate value of f_H.

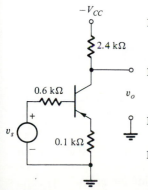

Prob. 11-16

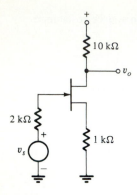

11-18 A JFET having $\mu = 50$ and $r_d = 10 \text{ k}\Omega$ is used in a source follower with $R_S = 1 \text{ k}\Omega$. The JFET capacitances are $C_{gs} = 5 \text{ pF}$, $C_{gd} = 2 \text{ pF}$, and $C_{ds} = 2 \text{ pF}$. Determine A_{VO} and the approximate value of f_H assuming $R_s = 5 \text{ k}\Omega$.

11-19 (a) Determine Z_o for the source follower in Prob. 11-18 as a function of R_s.
(b) Can Z_o appear inductive over some range of frequencies?

11-20 The JFET in Prob. 11-18 is used in the circuit shown. Determine A_{VO} and the approximate value of f_H.

11-21 Transistor A, biased at $I_{CQ} = -0.2 \text{ mA}$, is used in the CE configuration with emitter resistance. For $R_s = 2 \text{ k}\Omega$, $R_E = 0.1 \text{ k}\Omega$, and $R_C = 5 \text{ k}\Omega$, determine A_{VO} and f_H. (*Note:* This is the circuit in Prob. 10-43.)

11-22 A common-emitter stage uses the transistor whose parameters are given in Fig. 11-19. For $R_C = 1.5 \text{ k}\Omega$ and $R_s = 0.6 \text{ k}\Omega$, determine A_{VO} and f_H.

11-23 Verify Eqs. (11-47) and (11-48).

11-24 Verify Eqs. (11-53), (11-54), and (11-55).

11-25 Verify Eqs. (11-56) and (11-57).

11-26 (a) For the circuit shown, determine the a_1 and a_2 coefficients using the time constant method.
(b) For $R_1 = R_2 = R$, $C_1 = C_2 = C$, and $A_v = 2$, estimate the pole locations.
(c) Compare the results in (b) with the actual poles obtained from the roots of the quadratic equation.
(d) Comment on the validity of the dominant-pole approximation.

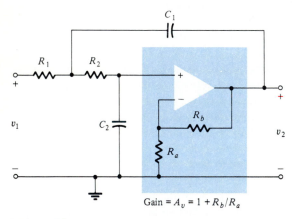

Gain = $A_v = 1 + R_b/R_a$

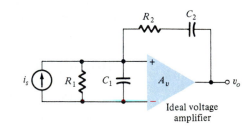

Ideal voltage amplifier

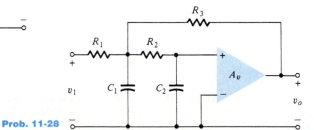

11-27 Repeat Prob. 11-26 for the circuit shown.

11-28 (a) Repeat Prob. 11-26, part (a) for the circuit shown. Assume that A_v is an ideal voltage amplifier.
(b) For $R_1 = R_2 = R_3 = R$, $C_1 = C_2$, $A_v = 2$, estimate the poles.
(c) Repeat Prob. 11-26, parts (c) and (d).

11-29   Each stage of a *CE-CE* configuration uses transistor *A*, and each is biased at $I_{CQ} = 1$ mA. The component values are $R_s = 0.6$ kΩ, $R_{C1} = R_{C2} = 1.2$ kΩ.
(*a*)   Determine A_{VO} and the approximate value of f_H.
(*b*)   Estimate the location of the closest nondominant pole.

11-30   (*a*)   The circuit shown uses the transistors described in Example 10-7. Assuming that each transistor has $f_T = 200$ MHz and $C_\mu = 1$ pF, determine A_{VO} and the approximate value of f_H.
(*b*)   Estimate the location of the nearest nondominant pole. (*Note:* Assume that the frequency response of the emitter follower is sufficiently high so that it has negligible effect on the value of f_H of the cascade.)

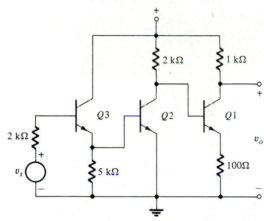

Prob. 11-30

11-31   The transistors in Example 10-7 have $f_T = 200$ MHz and $C_\mu = 1$ pF.
(*a*)   Estimate the value of f_H for the cascade.
(*b*)   Compare this result with that of the amplifier in Prob. 11-30.
(*c*)   Determine the approximate location of the nearest nondominant pole. (*Note:* Assume that the frequency response of the emitter follower has negligible effect on the frequency response of the cascade.)

11-32   Repeat Prob. 11-31 for the circuit shown.

11-33   The transistors in the circuit shown are identical and have $r_\pi = 1.5$ kΩ, $\beta_o = 150$, $C_\pi = 50$ pF, and $C_\mu = 1.0$ pF.

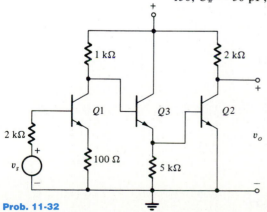

Prob. 11-32

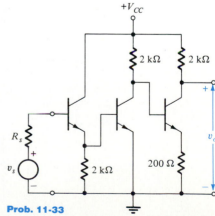

Prob. 11-33

(a) Determine A_{VO} and the approximate value of f_H, given $R_s = 20\text{ k}\Omega$. (*Note*: Assume that the frequency response of the common-collector stage is sufficiently high that it has no effect on the frequency response of the cascade.)

(b) Estimate the location of the nearest nondominant pole.

11-34 In the circuit shown, $Q1$ and $Q2$ have $\beta_F = 150$, $V_A = 120$ V, $f_T = 400$ MHz, and $C_\mu = 0.5$ pF at a bias current $I_{CQ} = 100\ \mu\text{A}$.

(a) For $R_s = 50\text{ k}\Omega$ and $R_C = 250\text{ k}\Omega$, determine the approximate value of f_H.

(b) Estimate the location of the nearest nondominant pole. (*Note*: A_{VO} was computed for this circuit in Prob. 10-66.)

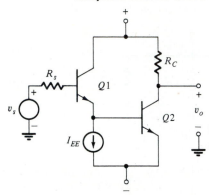

Prob. 11-34

11-35 The circuit in Prob. 11-34 uses transistor B with $Q1$ biased at 75 μA and $Q2$ biased at 250 μA. For $R_s = 500\text{ k}\Omega$ and $R_C = 500\text{ k}\Omega$, determine A_{VO} and the approximate value of f_H.

11-36 Transistor C is used in each stage of a cascode amplifier and each is biased at $I_{CQ} = 0.2$ mA.

(a) For $R_s = 1\text{ k}\Omega$ and $R_{C1} = R_{C2} = 5\text{ k}\Omega$, determine the approximate value of f_H.

(b) Compare the result in (a) with that of a common-emitter stage using transistor C, biased at $I_{CQ} = 0.2$ mA, and with $R_s = 1\text{ k}\Omega$ and $R_C = 5\text{ k}\Omega$.

11-37 (a) Transistors B and E are used in the IC cascode circuit shown and are biased at $|I_{CQ}| = 125\ \mu\text{A}$. Determine A_{vo} and the approximate value of f_H.

(b) Estimate the frequency of the nearest nondominant pole.

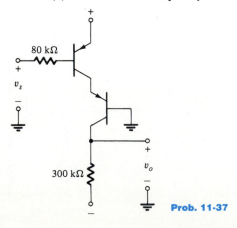

Prob. 11-37

11-38 A *CS-CS* cascade uses identical JFETs whose parameters are: $g_m = 1$ m$\mho$, $R_d = 40$ kΩ, $C_{gs} = 5$ pF, $C_{gd} = 1$ pF, and $C_{ds} = 1$ pF.

(a) With $R_s = 5$ kΩ, $R_{D1} = 40$ kΩ, and $R_{D2} = 10$ kΩ, determine A_{VO} and the approximate value of f_H.

(b) Estimate the frequency of the nearest nondominant pole.

11-39 The JFETs in the circuit shown have the parameters given in Prob. 11-38.

(a) Determine A_{VO} and the approximate value of f_H.

(b) Estimate the frequency of the nearest nondominant pole.

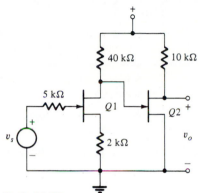

Prob. 11-39

11-40 The stages in the cascade of Prob. 11-39 are interchanged. Repeat Prob. 11-39.

11-41 The JFET in the circuit shown has $g_m = 2$ m$\mho$, $r_d = 30$ kΩ, $C_{gs} = 10$ pF, $C_{gd} = 5$ pF, and $C_{ds} = 5$ pF. The BJT parameters are: $r_\pi = 2.5$ kΩ, $\beta_o = 125$, $C_\pi = 100$ pF, and $C_\mu = 1.5$ pF.

(a) Determine A_{VO} and the approximate value of f_H.

(b) Estimate the frequency of the nearest nondominant pole.

11-42 Repeat Prob. 11-41 for the case where the BJT and FET stages are interchanged.

11-43 (a) The model of a differential amplifier with a difference signal v_d applied is shown. Given $\beta_o = 125$, $r_\pi = 25$ kΩ, $C_\pi = 5$ pF, $C_\mu = 1$ pF, and $r_o = 1$ MΩ, determine the approximate value of f_H.

(b) Estimate the frequency of the nearest nondominant pole.

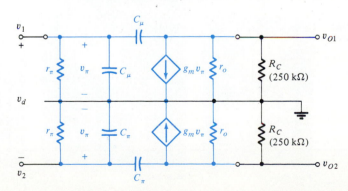

Prob. 11-43

11-44 The high-frequency model of the differential stage in Prob. 11-43 with a common-mode signal applied is shown.

(*a*) Determine the approximate value of f_H.

(*b*) Estimate the frequency of the nearest nondominant pole.

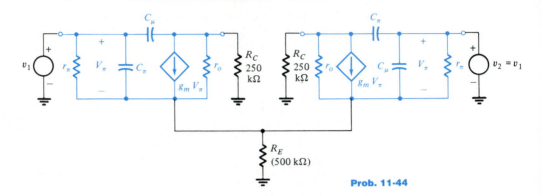

Prob. 11-44

11-45 The differential- and common-mode gains of a differential amplifier can be approximated as

$$A_{DM} = \frac{-2000}{1 + s/2\pi \times 10^6}; \qquad A_{CM} = \frac{-0.5}{1 + s/2\pi \times 10^8}$$

(*a*) Sketch the asymptotic Bode diagram of the *CMRR*.

(*b*) At what frequency is the *CMRR* one-half of its low-frequency value?

11-46 Verify Eqs. (11-73) and (11-74).

11-47 The Op-Amps in the circuit shown are identical and have $A_{vo} = 10^5$ and $f_h = 10$ Hz. In all other respects the Op-Amps are ideal.

(*a*) Determine the low-frequency gain and bandwidth of each stage.

(*b*) Use the results in (*a*) and write an equation for $A_{VH}(s)$ for the cascade.

(*c*) Plot the asymptotic Bode diagram of the function in (*b*) and estimate f_H.

(*d*) Use the dominant-pole approximation to determine f_H and compare with the result in (*c*). Comment on the comparison.

Prob. 11-47

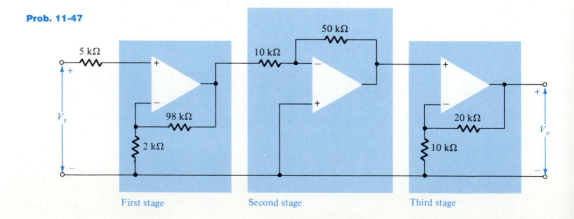

First stage   Second stage   Third stage

11-48 Manufacturing reliability and technological feasibility often dictate that the ratio of the largest resistance to the smallest resistance used in a circuit be equal to or less than 10.

(a) Using this restriction, find the maximum midband gain that can be realized by a three-stage Op-Amp cascade amplifier, assuming that the output is to be 180° out of phase with the input.

(b) Use this restriction and assume that the Op-Amp characteristics are $A_{vo} = 126$ dB and $f_h = 5$ Hz to find the bandwidth of the amplifier obtained in (a).

11-49 Repeat Prob. 11-48, assuming that the amplifier output and input signals are to be in phase.

11-50 Transistor C is used in the circuit shown.

(a) Determine the quiescent values of I_C and v_o.

(b) Assuming that C_E can be made arbitrarily large, determine C_B so that $f_L = 20$ Hz.

(c) Assuming that C_B can be made arbitrarily large, determine C_E for $f_L = 20$ Hz.

(d) Choose C_E and C_B for $f_L = 20$ Hz to minimize the total capacitance. Assume that the nondominant capacitive effect has a frequency of less than 2 Hz.

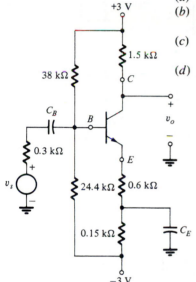

Prob. 11-50

11-51 Verify Eq. (11-81).

11-52 (a) For the circuit in Prob. 11-50, determine A_{VO} and the approximate value of f_H.

(b) A capacitor C is connected between B and C. Determine C so that f_H is reduced to 20 kHz.

(c) If $v_s(t) = V_m \sin (2\pi \times 10^3 t)$ is applied, what is the output voltage v_o if $V_m = 0.1$ V?

(d) How large can V_m be made before the stage displays distortion?

11-53 (a) Show that the gain of an FET stage with source bypass capacitance C_S is

$$A_{VL}(s) = \frac{A_{VO}}{1 + g_m R_S} \frac{1 + s/\omega_S}{1 + s/\omega_L}$$

where $A_{VO} = -g_m R_D$, $\omega_S = 1/R_S C_S$, and $\omega_L = (1 + g_m R_S)/R_S C_S$. Assume $R_S + R_D \ll r_d$.

(b) Given $g_m R_S \ll 1$ and $g_m = 3$ m℧, determine C_S so that a 60-Hz square wave will suffer no more than 10 percent tilt.

11-54 The transistor used has $\beta_o = 100$, $r_\pi = 1$ kΩ, and $r_o \to \infty$.
  (a) Determine the value of f_L.
  (b) Given $i(t) = 200$ Hz square wave, determine the percentage tilt in the output.
  (c) What is the lowest-frequency square wave that exhibits no more than 2 percent tilt?

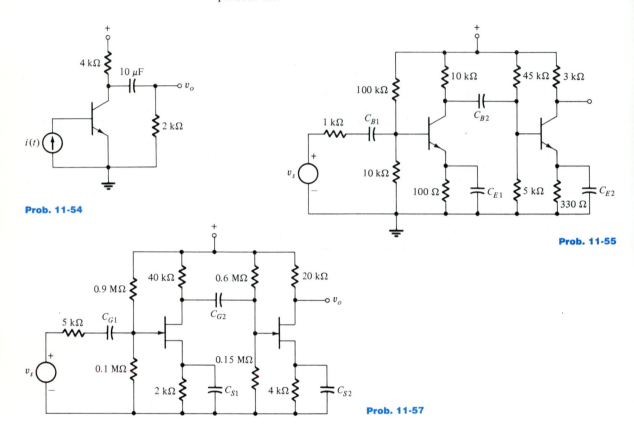

Prob. 11-54

Prob. 11-55

Prob. 11-57

11-55 The transistors in the circuit shown are identical and have $r_\pi = 4$ kΩ and $\beta_o = 200$.
  (a) Determine the value of f_L for each stage, assuming $C_{B1} = C_{B2} = 1$ μF and $C_{E1} = C_{E2} = 100$ μF.
  (b) What is the lower half-power frequency of the cascade?
  (c) Compare the value in (b) with the value of f_L obtained from the asymptotic Bode diagram for the cascade.

11-56 The circuit in Prob. 11-55 is to have an overall lower half-power frequency of 50 Hz. Select the values of C_{B1}, C_{B2}, C_{E1}, and C_{E2} that minimize the total capacitance used.

11-57 The JFETs in the circuit shown are identical and have $g_m = 2$ m$\mho$ and $r_d = 40$ kΩ.
  (a) For $C_{G1} = C_{G2} = 1$ μF and $C_{S1} = C_{S2} = 100$ μF, determine the lower half-power frequency for each stage.

(b) What is f_L for the cascade?

(c) Compare this value with that obtained from the asymptotic Bode diagram for the cascade.

11-58 The circuit in Prob. 11-57 is to have f_L = 50 Hz. Determine C_{G1}, C_{G2}, C_{E1}, and C_{E2} to minimize the total capacitance used.

CHAPTER 12

To the student: Biasing arrangements are not shown in many of the circuit diagrams for the problems in this chapter. You may assume that the devices are biased properly and those components used for biasing (and not indicated) have a negligible effect on circuit performance.

The following transistors are used frequently in the problems that follow:

	Transistors				
	A	B	C	D	E
Type	*npn*	*npn*	*npn*	*pnp*	*pnp*
β_o	125	150	200	150	50
β_F	125	150	200	150	50
V_A, V	∞	100	125	∞	50

The base-spreading resistance r_b = 0 except where otherwise indicated.

12-1 Which ideal amplifier types do each of the following approximate? Justify your answer.

(a) The emitter follower

(b) A common-source stage with source resistance

12-2 A transimpedance amplifier has Z_i = R_i = 50 Ω, Z_o = R_o = 50 Ω, and R_m = 10 kΩ. The characteristics of a certain transconductance amplifier are: Z_i = R_i = 50 kΩ, Z_o = R_o = 100 kΩ, and G_m = 0.1 $\mho$.

(a) Construct a current amplifier from the given circuits.

(b) What are the values of R_i, R_o, and A_i?

12-3 (a) Use the transimpedance and transconductance amplifiers in Prob. 12-2 to construct a voltage amplifier.

(b) What are the values of R_i, R_o, and A_V?

12-4 In the circuit shown, Q1 and Q2 are identical transistors having r_π = 1 kΩ and g_m = 0.1 $\mho$.

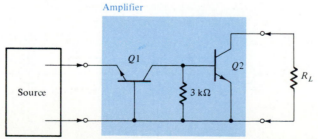

Prob. 12-4

    (*a*)   Which amplifier type does this circuit approximate?

    (*b*)   Determine the input and output resistances and the transfer ratio of the amplifier.

12-5   Repeat Prob. 12-4 for the circuit shown. The transistors are those given in Prob. 12-4.

12-6   (*a*)   For the circuit shown, find the signal component of the voltage v_i as a function of v_s and v_f. Assume that the inverting amplifier has infinite input resistance and a voltage gain of $A_v = 4000$. The feedback network is characterized by $\beta = V_f/V_o = \frac{1}{300}$. Circuit parameter values are $R_s = R_E = 2$ kΩ and $R_C = 6$ kΩ, and the transistor has $\beta_o = 200$ and $r_\pi = 4$ kΩ.

    (*b*)   Find $A_F = V_o/V_s$.

12-7   The block diagram represents a two-stage feedback system in which X_s is the signal to be amplified, X_1 is noise that is introduced with the signal, X_2 is a disturbance introduced within the amplifier (perhaps due to power-supply ripple), and X_3 is a disturbance introduced at the amplifier output.

    (*a*)   With $X_1 = X_2 = X_3 = 0$, find A_{OL}, T, and A_F.

    (*b*)   Determine the transfer ratios X_o/X_1, X_o/X_2, and X_o/X_3.

    (*c*)   Verify that

$$X_o = \frac{A_{OL}[X_s + X_1) + (X_2/A_1) + X_3/A_{OL}]}{1 + T}$$

    (*d*)   Let X_{os} be the output component due to X_s, X_{o1} be the output component due to X_2, and so on. Evaluate X_{os}/X_{o1}, X_{os}/X_{o2}, and X_{os}/X_{o3}.

    (*e*)   Repeat (*b*) with $\beta = 0$.

    (*f*)   Repeat (*d*) for $\beta = 0$ and compare the results. What is your conclusion?

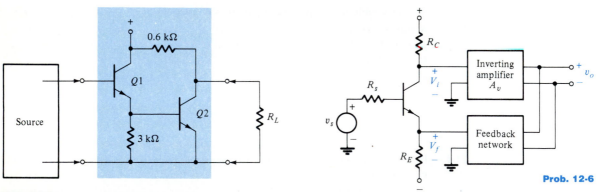

Prob. 12-5

Prob. 12-6

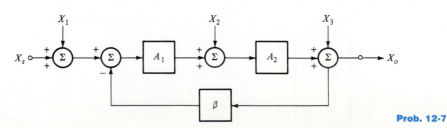

Prob. 12-7

12-8 A feedback amplifier is to be designed to have a closed-loop gain of 50 ± 0.1. The basic amplifier has a gain which can be controlled to within ± 10 percent. Determine the values of open-loop gain, the return ratio, and the reverse transmission β of the feedback network.

12-9 An amplifier without feedback provides an output signal of 15 V with 10 percent second-harmonic distortion when the input signal is 15 mV.
(a) If 1.5 percent of the output is fed back to the input in a negative series-shunt amplifier, what is the output voltage?
(b) If the fundamental output remains at 15 V but the second-harmonic distortion is reduced to 1 percent, what is the input voltage?

12-10 Use the approximate analysis to determine A_{OL}, β, T, and A_F for a source follower.

12-11 (a) Use the approximate analysis to determine A_{OL}, β, T, and A_F for a common-emitter stage with emitter resistance R_E.
(b) Compare the value of A_F with that in Table 10-3A and explain any differences.

12-12 (a) For the circuit shown, determine A_{OL}, T, β, and A_F. Use the approximate analysis.
(b) What is the feedback topology employed?

12-13 (a) Use Blackman's impedance formula to determine R_{IF} for the stage in Prob. 12-11.
(b) Repeat (a) for R_{OF}. Include r_o of the BJT.
(c) Compare these results with those given in Tables 10-3.

12-14 For the circuit in Prob. 12-12, obtain R_{IF} and R_{OF}.

12-15 The Op-Amp in the circuit shown has an open-loop gain A_v and output resistance R_o.
(a) Determine A_{OL}, β, T, and A_F for the stage.
(b) Determine R_{OF} of the circuit.
(c) What feedback topology is employed?

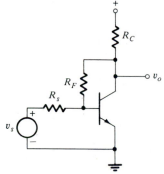

Prob. 12-12

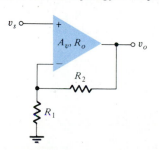

Prob. 12-15

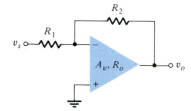

Prob. 12-18

12-16 Use the t parameters (Sec. 12-8) to obtain A_D, A_{OL}, T, and A_F of a source follower at low frequencies.

12-17 Repeat Prob. 12-16 for a common-emitter stage with emitter resistance R_E.

12-18 The Op-Amp in the circuit shown is characterized by A_v and R_o.
(a) Use the t parameters to obtain A_D, A_{OL}, T, and A_F.
(b) Use Blackman's impedance formula to obtain R_{IF} and R_{OF}.

12-19 In the source-follower shown, $v_R = V_{RM} \sin 2\pi \times 120t$ is the ripple voltage of the power supply and can be treated as a disturbance signal in the stage. For $R_S = 2$ kΩ, $g_m = 2$ m℧, $r_d = 20$ kΩ, and $R = 500$ kΩ, determine the maximum value of V_{RM}, assuming that the ripple component of the output is not to exceed 20 μV.

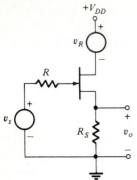

Prob. 12-19

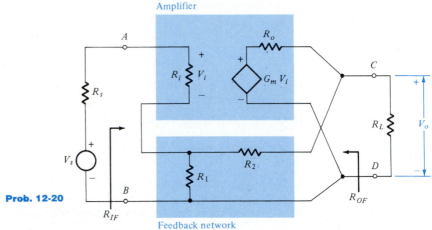

Prob. 12-20

12-20 (a) What topology is used in the circuit shown?
   (b) For $R_i = 1$ kΩ, $R_o = 5$ kΩ, $A_v = 10^3$, $R_2 = 50$ kΩ, $R_1 = 2$ kΩ, and $R_L = R_S = 0.6$ kΩ, determine A_{OL}, T, and A_F.
   (c) Given $A_v \to \infty$, find A_F.
   (d) Evaluate R_{IF} and R_{OF}.
   (e) What new value of A_v must be used if R_{OF} is to be 600 Ω?

12-21 (a) What topology is used?
   (b) Draw the circuit diagram of the amplifier without feedback.
   (c) For $R_i = 500$ Ω, $R_o = 20$ kΩ, $R_2 = 50$ kΩ, $R_1 = 1$ kΩ, $G_m = 100$ ℧, $R_s = 600$ Ω, and $R_L = 2$ kΩ, find A_{OL}, T, and A_F.
   (d) Determine R_{IF} and R_{OF}.
   (e) For $G_m \to \infty$, find A_F.

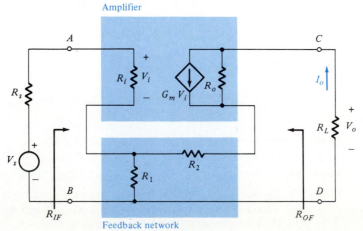

Prob. 12-21

12-22 (a) Repeat Prob. 12-21 for the circuit shown. The element values are: $R_i =$ 5 kΩ, $R_o = 0.5$ kΩ, $R_m = 100$ kΩ, $R_2 = 10$ kΩ, $R_1 = 1$ kΩ, $R_s = 50$ Ω, and $R_L = 2$ kΩ.

  (b) What new value of R_m is needed to make $R_{IF} = 50$ Ω?

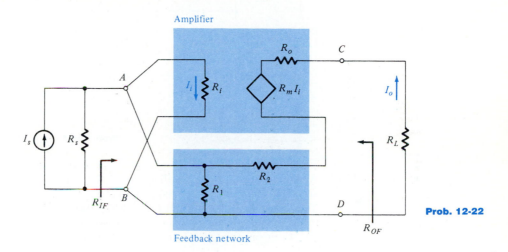

Prob. 12-22

12-23 Verify Eqs. (12-41) and (12-42).

12-24 The FETs in the circuit shown are identical and have $g_m = 2$ m℧ and $r_d = 20$ kΩ. The circuit parameters are $R_D = 12$ kΩ, $R_G = 500$ kΩ, $R_s = 50$ Ω, and $R_F = 5$ kΩ. Determine A_F and R_{OF}.

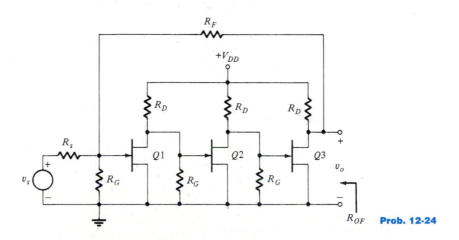

Prob. 12-24

12-25 Transistor A, biased at $I_{CQ} = 1.5$ mA, is used in the circuit shown.

  (a) Determine A_F and T.

  (b) Find R_{IF} and R_{OF}.

  (c) What value, if any, of R is needed to make $R_{OF} = 47$ Ω?

12-26 Verify Eq. (12-43).

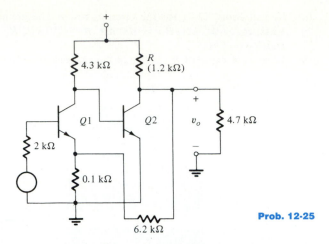

Prob. 12-25

12-27 A common-emitter stage with an emitter resistance R_E is designed with a transistor having $r_\pi = 2.5$ kΩ and $\beta_o = 125$.
    (a) Given $R_s = 2.5$ kΩ and $R_C = 3$ kΩ, find R_E so that $S_{\beta_o}^{A_F} = -\frac{1}{31}$.
    (b) Using the value of R_E in (a), determine A_F.
    (c) Compare the answer in (b) with A_F in Example 12-8.

12-28 The circuit in Prob. 12-12 uses the transistor described in Prob. 12-27.
    (a) Given $R_s = 2.5$ kΩ and $R_C = 3$ kΩ, find R_F so that $S_{\beta_o}^{A_F} = \frac{1}{31}$.
    (b) Evaluate A_F using the value of R_F in (a).

12-29 (a) For the circuit shown, determine T, A_{OL}, and A_F.
    (b) Evaluate R_{OF}. The MOSFETS have $g_m = 1$ m℧, $r_d = 20$ kΩ.

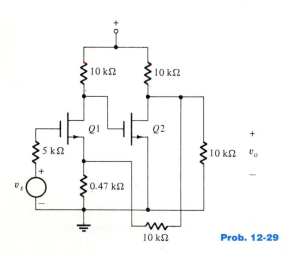

Prob. 12-29

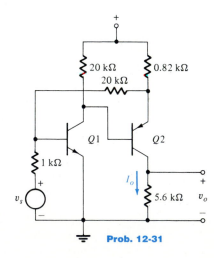

Prob. 12-31

12-30 Verify Eqs. (12-48) and (12-50).

12-31 (a) Repeat Prob. 12-29, part (a) for the circuit shown. Transistors C and D are used and biased at $I_{CQ} = 0.25$ mA and -0.5 mA, respectively.
    (b) Evaluate R_{IF}.

12-32 (a) What topology is used in the amplifier shown?

   (b) Assuming $T \gg 1$, show that $A_F = I_o/V_s \approx R_F/R_1R_2$ provided $R_F \gg (R_1 + R_2)$.

   (c) Given that $Q1$, $Q2$, and $Q3$ are each transistor C with $I_{CQ1} = 0.25$ mA, $I_{CQ2} = 1.0$ mA, and $I_{CQ3} = 0.5$ mA. The circuit elements are $R_{C1} = 5$ kΩ, $R_{C2} = 7.5$ kΩ, $R_{C3} = 10$ kΩ, $R_1 = 0.2$ kΩ, $R_2 = 0.33$ kΩ, and $R_s = 0.6$ kΩ. Evaluate A_F and T if $R_F = 20$ kΩ.

   (d) Find the voltage gain V_o/V_s.

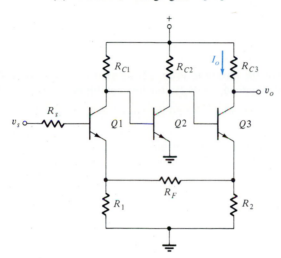

Prob. 12-32

12-33 Determine R_{IF} and R_{OF} for the circuit in Prob. 12-32.

12-34 Verify Eqs. (12-55) and (12-56).

12-35 Use the t parameters to evaluate A_{OL}, T, and A_F for the circuit in Example 12-6.

12-36 Repeat Prob. 12-35 for the circuit in Example 12-7.

12-37 Repeat Prob. 12-35 for the circuit in Example 12-9.

12-38 Repeat Prob. 12-35 for the circuit in Prob. 12-29.

12-39 (a) Verify Eq. (12-60).

   (b) Let $A_2A_3 = A$ in Eq. (12-60). Determine the sensitivity $S_A^{A_F}$ using Eq. (12-8).

   (c) Evaluate the result in (b) for $Af_2 = 1$.

12-40 For the multiloop circuit in Fig. 12-42, let $A_1 = a_1/(1 + \tau_1s)$, $A_2 = a_2/(1 + \tau_2s)$, and $A_3 = a_3/(1 + \tau_2s)$. The feedback transmissions f_1, f_2, and f_3 are real constants.

   (a) Obtain the transfer function $A_F(s)$.

   (b) Verify that each coefficient in $A_F(s)$ can be specified by adjusting the loop gain (A_F) of only one loop.

12-41 Repeat Prob. 12-40 for the leap-frog structure in Fig. 12-43.

12-42 The transistor in the circuit shown has parameters r_π, r_o, and β_o.

   (a) What relationship must exist if R_{IF} is to be r_π?

   (b) Evaluate V_o/V_s for the conditions in (a).

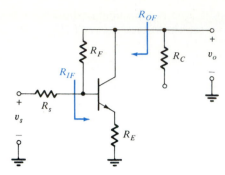

Prob. 12-42

12-43   (*a*)   For the circuit in Prob. 12-42, what relationship must exist if $R_{OF} = r_o$?
         (*b*)   Evaluate $A_F = V_o/V_s$ for this condition.

12-44   Using transistor C, design the circuit in Prob. 12-20.

CHAPTER 13   **13-1**   A single-stage amplifier without feedback can be represented by

$$A_{VH}(s) = \frac{A_{vo}}{1 + s/\omega_h}$$

at high frequencies and by

$$A_{VL}(s) = \frac{A_{vo}\, s/\omega_L}{1 + s/\omega_L}$$

at low frequencies where A_{vo} is the midband gain.

     A feedback network β is inserted between output and input of the basic amplifier.

   (*a*)   For $A_{vo} = 500$, $\beta = 0.01$, $f_h = 50$ kHz, and $f_L = 1$ kHz, determine the values of the upper and lower 3-dB frequencies of the feedback amplifier.

   (*b*)   Determine the gain-bandwidth products of the amplifier with and without feedback and compare the results.

13-2   Two identical amplifier blocks, each having $A = 200$ and $f_h = 100$ kHz, are available for designing an amplifier whose overall gain is to be 1000 and whose upper half-power frequency $f_H \geq 2$ MHz. Assume that the stages are noninteractive. What range of β values satisfies the design requirements?

13-3   The two feedback amplifiers shown are to be compared. Assume that $A_1 = a_1/(1 + s/\omega_1)$; $A_2 = a_2/(1 + s/\omega_2)$; and β_1, β_2, and β are frequency-independent.

   (*a*)   Assuming that $A_1\beta_1$, $A_2\beta_2$, and $A_1A_2\beta$ are each much greater than unity, determine β so that the closed-loop gains of the two amplifiers are equal at low frequencies.

   (*b*)   Compare $S_{A_1}^{A_F}$ at low frequencies for both amplifiers.

   (*c*)   If $A_1 = A_2$ and $\beta_1 = \beta_2$, which amplifier has the greater bandwidth?

   (*d*)   Which amplifier has the better overall performance? Explain.

13-4   The return ratio of a feedback amplifier is given by

$$T(s) = \frac{T_O}{(1 + s/\omega_1)(1 + s/\omega_2)(1 + s/\omega_3)}$$

(a)   For $T_O = 10^3$, $\omega_1 = 0.1$ Mrad/s, $\omega_2 = 1$ Mrad/s, and $\omega_3 = 10$ Mrad/s, determine whether the closed-loop amplifier is stable.

(b)   What are the gain and phase margins in (a)? Use the asymptotic Bode diagram.

13-5   Repeat Prob. 13-4 for $T_O = 2 \times 10^4$, $\omega_1 = 0.2$ Mrad/s, $\omega_2 = 40$ Mrad/s, and $\omega_3 = 200$ Mrad/s.

13-6   The low-frequency return ratio of a feedback amplifier is given by

$$T(s) = \frac{2 \times 10^{-3}\, s^3}{(1 + s)\,(1 + s/100)\,(1 + s/1000)}$$

(a)   Is the closed-loop amplifier stable?

(b)   What are the gain and phase margins?

13-7   The return ratio is given by

$$T(s) = \frac{T_O\,(1 + s/10^{6.5})}{(1 + s/10^5)\,(1 + s/10^6)\,(1 + s/10^7)^2}$$

(a)   Determine the largest value of T_O for which the amplifier is stable.

(b)   What value of T_O results in $GM = 10$ dB?

(c)   What value of T_O results in $\phi_M = 45°$? Use the asymptotic Bode diagram.

13-8   The amplifier in Prob. 13-4 is required to have $\phi_M \approx 90°$.

(a)   What new value of ω_1 is needed?

(b)   What is GM for the condition in (a)?

(c)   What is the approximate closed-loop bandwidth?

(d)   Using the analytic expression for $T(s)$, evaluate $\angle\, T(j\omega_G)$ and $T(j\omega_\phi)$. Use the values of ω_G and ω_ϕ obtained from the asymptotic Bode diagram.

13-9   The amplifier in Prob. 13-5 is to be compensated by using pole-zero cancellation to have a phase margin of 45°. The network used to provide pole-zero cancellation has a transfer function $(1 + s/z_1)/(1 + s/\omega_A)$.

(a)   Determine z_1 and ω_A.

(b)   What is the new gain margin?

13-10   Assume that after a pole-zero compensation network is added, $T(s)$ in Prob. 13-5 can be approximated by the *two-pole* function

$$T(s) = \frac{T_O}{(1 + s/\omega_A)\,(1 + s/\omega_2)}$$

(a)   Determine ω_A so that $\phi_M = 45°$.

(b)   Compare the value of ω_A obtained in (a) with the result in Prob. 13-9.

13-11   The return ratio of an amplifier after narrow-banding is

$$T(s) = \frac{4 \times 10^5}{[1 + (s/\omega_1)]\,[1 + (s/10^7)]\,[1 + (s/10^9)]}$$

(a)   Determine the value of ω_1 needed to obtain $\phi_M = 45°$.

(b)   An alternative method of compensation is to add a network in the feedback loop which makes

$$T(s) = \frac{4 \times 10^5\,(1 + s/z_1)}{[1 + (s/\omega_1)^2]\,[1 + (s/10^7)]\,[1 + (s/10^9)]}$$

For $z_1 > \omega_1$ and $z_1 \le 10^5$, determine ω_1 and z_1 to obtain $\phi_M = 45°$.

   (c) By comparing the results in (a) and (b), determine which compensation technique results in the larger closed-loop bandwidth. Explain.

13-12 (a) For $\omega_1 = 20$ rad/s, determine the gain margin and phase margin for $T(s)$ in Prob. 13-11, part (a) using the asymptotic Bode diagram.

   (b) From the analytic expression for $T(s)$, evaluate $\angle\ T(j\omega_G)$ and $T(j\omega_\phi)$ in decibels. Use the values of ω_G and ω_ϕ obtained from the asymptotic Bode diagram.

   (c) Compare the values obtained in (b) with the values obtained from the Bode diagram and comment on any differences.

13-13 (a) A two-pole amplifier has corner frequencies of $f_1 = 400$ kHz and $f_2 = 1.6$ MHz. What value of T (in decibels) is needed to give the fastest rise time without overshoot?

   (b) Compare the result in (a) with the rise time of an amplifier having the two corner frequencies given.

13-14 If $T = 30$ dB is applied to the amplifier in Prob. 13-13, what are the rise time and the overshoot (if any)?

13-15 (a) The two corner frequencies of an amplifier are $f_1 = 50$ kHz and $f_2 = 5$ MHz. Determine the maximum value of T_o for which the step-response overshoot is 5 percent.

   (b) At what time does the peak occur?

   (c) Calculate the magnitude of the first minimum and the time it occurs.

13-16 Derive Eqs. (13-28) to (13-31) for the step response of a two-pole amplifier. [*Hint:* For the overdamped case, assume $k^2 \gg 1$ and $(1 - 1/k^2)^{1/2}$ in a Taylor series.]

13-17 For a two-pole amplifier, calculate the phase margin corresponding to $k = 0.4$, 0.6, 0.707, 0.8, and 1.0.

13-18 In the circuit shown, transistor A^1 is used and biased at $I_{CQ} = 1.5$ mA. The parameter values used are: $R_s = 2.5$ kΩ, $R_C = 3$ kΩ, and $R_F = 20$ kΩ. Determine the open- and closed-loop dominant-pole frequencies.

---

[1]Data for transistors A to E is given at the beginning of the problems for Chap. 12.

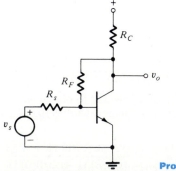

                Prob. 13-18

13-19   Using feedback analysis, verify that the open-loop and closed-loop gain-band-width products of a noninverting Op-Amp stage are equal.

13-20   Using feedback techniques, verify that for an inverting Op-Amp stage

$$\omega_H = \frac{A_{vo}\omega_h}{1 + |A_{FO}|}$$

where $A_{vo}\omega_h$ is the gain-bandwidth product of the Op-Amp and A_{FO} is the low-frequency gain of the inverting stage.

13-21   For the circuit shown, show that the closed-loop dominant pole is the open-loop dominant pole multiplied by $(1 + T_O)$. Use transistor C biased at $I_{CQ} = 0.5$ mA.

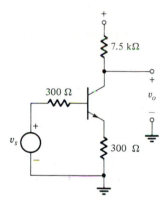

Prob. 13-21

13-22   The transistors used in Example 12-7 each have $f_t = 200$ MHz and $C_\mu = 1$ pF.
(a)   Determine the approximate value of the two dominant open-loop pole frequencies.
(b)   On the basis of (a), estimate the closed-loop pole frequencies.
(c)   Sketch the asymptotic Bode diagram on the basis of (a) and estimate the phase margin.

13-23   The JFETs in Example 12-9 each have $C_{gs} = 5$ pF, $C_{gd} = 2$ pF, and $C_{ds} = 1$ pF.
(a)   Estimate the approximate open-loop pole frequencies for the two poles nearest the origin.
(b)   On the basis of (a), sketch the asymptotic Bode diagram and determine the phase margin ϕ_M.
(c)   Assess the stability of the overall amplifier from the result in (b).

13-24   The values of FET capacitances in the circuit of Prob. 12-24 are given in Prob. 13-23. Repeat Prob. 13-23.

13-25   Repeat Prob. 13-22 for the circuit in Prob. 12-25.

13-26   Repeat Prob. 13-22 for the circuit in Prob. 12-29. The JFET capacitances are given in Prob. 13-23.

13-27   Repeat Prob. 13-22 for the circuit in Prob. 12-31.

13-28   Repeat Prob. 13-23 for the circuit in Prob. 12-32.

13-29 A capacitance C_C is connected between gate and drain of $Q2$ in the circuit of Probs. 12-24 and 13-23.
  (a) Determine C_C to obtain $k = 0.8$.
  (b) What is the phase margin for the situation in (a)?
  (c) Estimate the closed-loop bandwidth.

13-30 A capacitance C_C is connected between base and collector of $Q2$ in the circuit in Probs. 12-32 and 13-28.
  (a) Determine C_C so that $Q^2 = 0.1$.
  (b) Estimate the closed-loop bandwidth.

13-31 A compensating capacitor C_C is to be added to the circuit in Example 13-7 in order to make $Q^2 = 0.1$.
  (a) Determine C_C, assuming that it is connected between base and collector of the input transistor.
  (b) Repeat (a), assuming that C_C is connected between base and collector of the output BJT.

13-32 The circuit discussed in Probs. 12-32, 13-28, and 13-30 is to be compensated by connecting C_C between the base of $Q2$ and ground.
  (a) Determine C_C so that $Q^2 = 0.1$.
  (b) Estimate the closed-loop bandwidth and compare with the result in Prob. 13-30.
  (c) Compare the values of C_C obtained in (a) and in Prob. 13-30, part (a). Comment on the feasibility of each.

13-33 The circuit in Example 13-7 is required to drive a capacitive load $C_L = 20$ pF.
  (a) Determine the new values of ω_1 and ω_2.
  (b) Estimate the new pole frequencies (closed-loop) and the resultant closed-loop bandwidth.
  (c) What is the output rise time and overshoot, if any, to a step input voltage.

13-34 Consider the simple common-emitter stage shown to which a feedback capacitance $C_C \gg C_\mu$ is added.
  (a) Estimate the frequency of the dominant pole.
  (b) What is the new location of the zero in the transfer function?
  (c) Estimate the location of the nondominant pole.
  (d) Evaluate the pole and zero locations for the following parameters: $g_m = 1.5$ m℧, $r_\pi = 100$ kΩ, $C_\pi = 1.5$ pF, $C_\mu = 0.5$ pF, $C_C = 25$ pF, $R_s = 10$ kΩ, and $R_L = 10$ kΩ. Comment on the separation.
  (e) Compare the values in (d) to those obtained if $C_C = 0$.

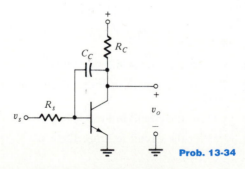

Prob. 13-34

13-35 The Op-Amp in the circuit shown is ideal; the transistor and circuit parameters have the values given in Prob. 13-34.

(a) Show that the dominant pole is given by

$$p_1 \approx - \frac{1}{R_s' R_C g_m C_C}$$

where $R_s' = R_s \| r_\pi$, $g_m R_L \gg 1$, and $C_C \gg C_\mu + C_\pi$.

(b) Verify that the zero occurs at

$$s = g_m/C_\mu$$

(c) Compare the values (a) and (b) with that obtained when $C_C = 0$.

(d) Compare the effectiveness of this circuit to the circuit in Prob. 11-34.

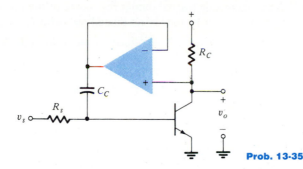

Prob. 13-35

13-36 (a) Show that dominant-pole compensation for which $Q^2 \leq 10/121$ results in at least a 1-decade separation in the closed-loop poles.

(b) If $T_O = 100$, what is the appropriate open-loop pole separation?

13-37 The *CC-CE* cascade uses identical transistors having $\beta_o = 250$, $V_A = 125$ V, $f_T = 400$ MHz, and $C_\mu = 0.5$ pF; $Q1$ is biased at $I_{CQ1} = 5$ μA and $I_{CQ2} = 250$ μA.

(a) At $\omega = 0$, determine R_i, R_o, and the transfer ratio V_o/I_s.

(b) If i_s is developed from a voltage source $v_s = i_s R_s$ and having source resistance R_s, what is the voltage gain V_o/V_s at $\omega = 0$?

(c) Determine the approximate location of the dominant pole of the cascade.

(d) What value of C_C must be used to make the dominant pole frequency 10 Hz?

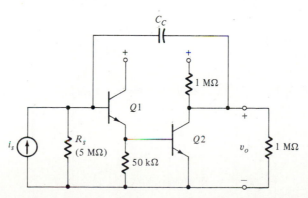

Prob. 13-37

13-38 (a) Analyze the circuit in Prob. 13-37 using feedback techniques and obtain $A_{OL}(s)$, $\beta(s)$, $T(s)$, and $A_F(s)$. Note that these are all functions of C_C.

(b) Evaluate T at $s = j\omega = 0$. Is this result reasonable?

(c) Given that i_s is a step current, sketch the output voltage $v_o(t)$.

13-39 The dominant pole $-\omega_1$ of the stage shown in Prob. 11-37 is one of the two poles in the open-loop gain $A_{OL}(s) = 10^5/(1 + s/\omega_1)(1 + s/2\pi \times 10^7)$ of an Op-Amp. Determine C_C to give a $45°$ phase margin for a unity closed-loop gain.

CHAPTER 14

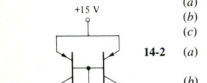

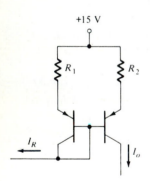

Prob. 14-1

14-1 The *pnp* transistors in the current mirror shown have $\beta_F = 50$, $\beta_o = 50$, and $V_A = 50$ V.

(a) Determine R so that $I_o = 100\ \mu A$.

(b) What is the small-signal output resistance of the mirror?

(c) Can this circuit be readily fabricated on a chip?

14-2 (a) In the circuit shown $I_R = 50\ \mu A$, what is the ratio R_1/R_2 needed for $I_o = 100\ \mu A$?

(b) What is the output resistance of the source?

14-3 (a) The *pnp* transistors in Fig. 14-2 have an Early voltage of $V_{AP} = 50$ V; the *npn* transistors have an Early voltage $V_{AN} = 120$ V. At a bias current I_{CQ}, show that the effective load resistance of the *npn* transistor is

$$R_L = \frac{V_{AN} V_{AP}}{I_{CQ}(V_{AN} + V_{AP})}$$

(b) Evaluate R_L for $I_{CQ} = 0.1$ mA.

(c) For $r_\pi \gg R_s$ and $I_{CQ} = 0.1$ mA, evaluate V_o/V_s for the stage.

(d) Is the result in (c) affected if I_{CQ} is varied?

14-4 (a) Verify Eq. (14-2). Assume that the base currents are negligible.

(b) What ratio of areas is needed for a 300-μA source if the reference current is 50 μA?

(c) How can such a current be achieved with a *pnp* mirror?

(d) For a reference current of 50 μA, what must be required for the source current to lie between 50 and 100 μA?

Prob. 14-2

14-5 A two-stage Op-Amp is shown. (See also Probs. 14-6 to 14-9, and 14-10.)

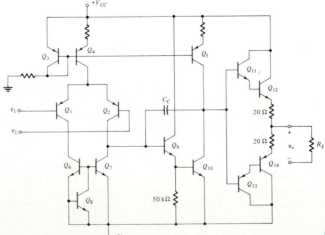

Prob. 14-5

(a)   Identify the function of each transistor.

(b)   Verify that, with no input signal (the bases of $Q1$ and $Q2$ are both at ground), $v_o = 0$. All *npn* transistors have $\beta_o = \beta_F = 200$, $V_A = 120$ V, $C_\mu = 0.5$ pF, and $f_T = 400$ MHz. All *pnp* transistors have $\beta_F = \beta_o = 50$, $V_A = 60$ V, $C_\mu = 0.5$ pF, and $f_T = 10$ MHz. Use $V_{CC} = 15$ V.

14-6   Design the current repeater so that $|I_{CQ2}| = I_{CQ10} = 12$ μA in the Op-Amp in Prob. 14-5.

14-7   (a)   Determine the low-frequency gain of the *CC-CE* cascade. Assume $I_{CQ2} = -12$ μA, $I_{CQ10} = 12$ μA, and $I_{CQ9} = 14$ μA.

(b)   What are the input and output resistances of this stage?

14-8   (a)   Evaluate the *CMRR*, A_{DM}, and R_o of the differential stage in Prob. 14-5.

(b)   What is the differential input resistance?

14-9   (a)   Determine the voltage gain and output resistance of the emitter follower. Assume that $Q10$ is biased at $I_C = 12$ μA. Assume that the input resistance to the emitter follower can be approximated by βR_E, where β is the current gain of the composite transistor and R_E is the effective resistance.

14-10   (a)   Compute the open-loop gain of the Op-Amp.

(b)   Determine the dominant pole (approximately) of both the differential and gain stages.

(c)   Determine C_C for a gain crossover frequency of 1 MH$_z$.

14-11   (a)   Determine the output resistance of the circuit in Fig. 14-14a.

(b)   Use the result in (a) to obtain the output resistance in Fig. 14-14b.

14-12   For the two-stage Op-Amp in Prob. 14-5,

(a)   Determine the open-loop gain, output resistance, and differential input resistance of the Op-Amp.

(b)   Compensate the Op-Amp to give a gain-crossover frequency of 1 MHz.

14-13   (a)   Plot the transfer characteristics I_{C1} versus v_D and I_{C2} versus v_D for the circuit shown.

(b)   Using the transistor in Prob. 14-5, and with $R_1 = 100$ Ω, determine the currents in the transistors. Use $I_{EE} = 1$ mA.

14-14   For the level-shift network shown, determine $V_2 - V_1$.

14-15   Show that $V_O = (V_Z + V_{BE})(1 + R_1/R_2)$ for the circuit shown.

14-16   (a)   For the current source shown, prove that

$$\frac{I_2}{I_1} = \frac{R_1}{R_2}\left[1 - \frac{V_T \ln (I_2/I_1)}{R_1 I_1}\right]$$

Neglect base currents.

(b)   If $0.1 < I_2/I_1 < 10$, what error is made if it is assumed that $I_2/I_1 = R_1/R_2$? Use $I_1 R_1 = 1$ V.

(c)   How is the answer to (b) affected if $I_1 R_1$ is increased?

---

[1]This circuit is analyzed in Probs. 14-5 to 14-9. Problem 14-10 involves the overall amplifier performance.

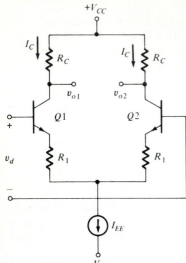

Prob. 14-13

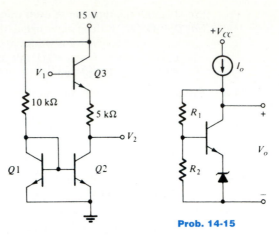

Prob. 14-14

Prob. 14-15

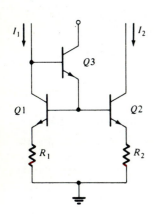

Prob. 14-16

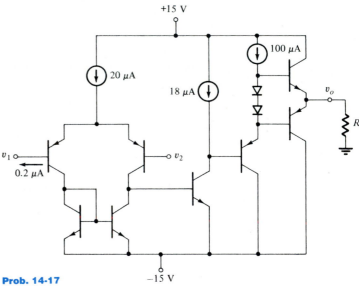

Prob. 14-17

14-17 In the simple Op-Amp shown, all *npn* devices have $\beta_F = 200$ and all *pnp* devices have $\beta_F = 50$. The current sources shown are realized by *pnp* current sources. Verify that $v_o = 0$ when $v_1 = v_2 = 0$.

14-18 The three current sources are fed from the same reference current $I_R = 300\ \mu A$. Design the circuit.

14-19 Assuming $\beta_F = 250$ for *npn* devices and $\beta_F = 50$ for *pnp* devices, determine the bias currents in the differential and gain stages of the 741-type Op-Amp in Fig. 14-19.

14-20 The circuit shown is a model of an Op-Amp stage in which the offset voltage and current are shown.

(*a*) Determine the signal component of v_o in terms of the difference signal $v_1 - v_2$.

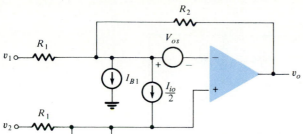

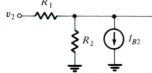

Prob. 14-20

(b) Determine the component of v_o caused by $I_{io}/2$.
(c) Repeat (b) for V'_{io}.
(d) For $v_1 = v_2$, determine the total offset voltage at the output.
(e) Evaluate the output offset voltage for $V_{io} = 6$ mV, $I_{io} = 0.2$ μA, $I_B = 0.5$ μA, $R_1 = 50$ kΩ, and $R_2 = 500$ kΩ.

14-21 (a) Use feedback analysis to determine the signal component of the output.
(b) What is the offset component of the output?

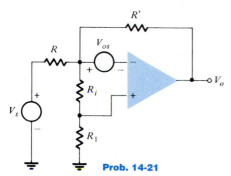

Prob. 14-21

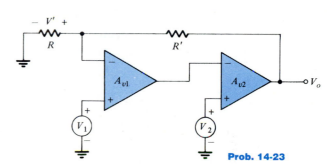

Prob. 14-23

14-22 Consider the Op-Amp in Fig. 14-22b with the model shown in Fig. 14-21. Assume $R_o = 0$ and $V_{io} = 0$. Prove that:
(a) The output voltage V_{o2} due to the bias current I_{B2} is

$$V_{o2} = \frac{-R'RR_iA_v}{(R_i + R_1)(R' + R) + RR' - A_vRR_i}I_{B2}$$

(b) The output voltage V_{o1} due to the bias current I_{B1} is

$$V_{o1} = \frac{R_iR_1(R + R')A_v}{(R + R')(R_1 + R_i) - A_vRR_i + RR'}I_{B1}$$

(c) Show that if $I_{B2}/I_{B1} \approx 1$, then $V_{o1} + V_{o2}$ is minimized by taking $R_1 = RR'/(R + R')$.

14-23 For the amplifier shown, V_1 and V_2 represent undesirable voltages. Show that, if $R_i = \infty$, $R_o = 0$, and $A_{v1} < 0$ and $A_{v2} < 0$, then

$$V_o = A_{v2}[A_{v1}(V' - V_1) - V_2] \quad \text{where} \quad V' = V_o\frac{R}{R + R'}$$

Show also that, if $A_{v2}A_{v1}R/(R + R') \gg 1$, then

$$V_o = \left(1 + \frac{R'}{R}\right)\left(V_1 + \frac{V_2}{A_{v1}}\right)$$

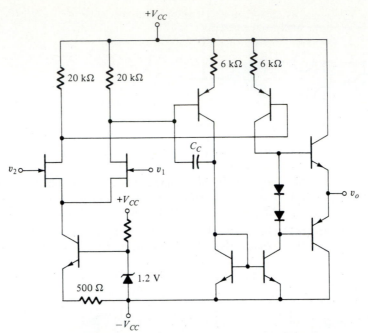

Prob. 14-24

14-24 The JFET differential stage shown is used in hybrid Op-Amps. That is, the JFET input stage is fabricated by using discrete components, and the BJTs and the remaining stages are monolithic. The JFET has $I_{DSS} = 3$ mA and $V_P = -3$ V. The BJT has $\beta_F = \beta_o = 200$ and $V_A = 100$ V. Determine the *CMRR* and A_{DM} for the stage. Use $\lambda = 0.01 \ V^{-1}$.

14-25 The BIMOS differential stage shown uses MOSFETs with $V_T = 0.5$ V and $kW/L = 400, \ \mu A/V^2$. The BJTs have $\beta_o = 200$ and $V_A = 100$ V. Determine A_{DM}. Use the half-circuit concept.

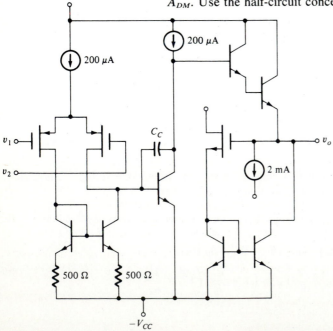

Prob. 14-25

14-26 The circuit shown is referred to as a current-differencing amplifier and is often used in operational transconductance amplifiers.

(a) Verify that i_o is proportional to $i_1 - i_2$.

(b) Determine the small-signal output resistance of the stage.

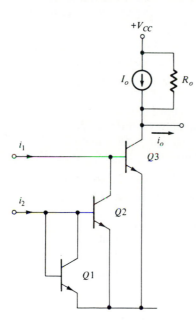

Prob. 14-26

14-27 An Op-Amp has an open-loop gain

$$A_{\text{OL}} = \frac{10^5}{(1 + s/\omega_1)(1 + s/\omega_2)}$$

(a) The amplifier is used as a unity gain buffer. Determine ω_1 for a 45° phase margin, given $\omega_2 = 10^7$ rad/s.

(b) What is the phase margin if the amplifier has a closed-loop gain of 5 and is compensated as in (b)?

(c) What value of ω_1 is needed if the amplifier is to have a 45° phase margin with $A_{FO} = 5$?

14-28 Repeat Prob. 14-27 for an inverting Op-Amp stage.

14-29 The two-pole approximation of A_{OL} is

$$A_{\text{OL}} = \frac{2 \times 10^5}{(1 + s/\omega_1)(1 + s/\omega_2)}$$

(a) For $\omega_2 = 10^{7.5}$ rad/s, repeat Prob. 14-27, part (a).

(b) What is the phase margin if $A_{FO} = 2$ and ω_1 is given as determined in (a)?

(c) What value of ω_1 is needed if $A_{FO} = 2$ and a phase margin of 60° is required?

14-30 Some Op-Amps are "partially" compensated. That is, internal compensation is used to provide a phase margin of 45° for $A_{FO} = 5$. For the Op-Amp in Prob. 14-29, $\omega_2 = 10^7$ rad/s and

$$\omega_1 = \frac{1}{R_{eq}(C_1 + C_C)}$$

where C_1 is the internal capacitor and C_C is the external compensating capacitor.
(a)   With $C_C = 0$ and $C_1 = 2$ pF, determine R_{eq}.
(b)   The amplifier is to be used as a unity gain buffer and to have $\phi_M \approx 90°$. Find C_C.

14-31   The amplifier in Prob. 14-30 is used as an inverting amplifier. Determine C_C, assuming $|A_{FO}| = 1$ and that a 60° phase margin is required.

14-32   The open-loop gain of the Op-Amp in Fig. 14-32a can be approximated as $A_{vo}/(1 + s/\omega_1)(1 + s/\omega_2)$.
(a)   Determine $T(s)$ for the circuit.
(b)   What relationship must exist if pole-zero cancellation is to be used?
(c)   Given $A_{vo} = 10^5$, $\omega_1 = 10^6$ rad/s, and $\omega_2 = 10^7$ rad/s, design the circuit to give $\phi_M = 45°$ with $A_{FO} = 1$.

14-33   Repeat Prob. 14-32 for the circuit in Fig. 14-32b.

14-34   An idealized model of a three-stage Op-Amp with unity closed-loop gain is shown.
(a)   What must ω_1 be if $\phi_M = 45°$?
(b)   A feedforward path is added and has a transfer function $\tau s/(1 + \tau s)$ as shown in (b). With ω_1 obtained in (a), what range of τ values will increase ϕ_M?

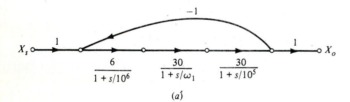

(a)

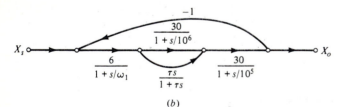

(b)   **Prob. 14-34**

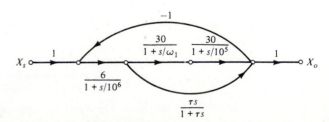

Prob. 14-35

14-35   Repeat Prob. 14-35 for the flow graph shown.

14-36   An Op-Amp has a slew rate of 0.5 V/μs.
  (*a*)   What is the maximum frequency of an output sinusoid of 5-V peak value before slew-rate distortion exists?
  (*b*)   Repeat (*a*) for a 15-V signal.

14-37   Repeat Prob. 14-36 for an Op-Amp with a slew rate of 10 V/μs.

14-38   Repeat Prob. 14-36 for an Op-Amp with a slew rate of 50 V/μs.

14-39   (*a*)   For the amplifier shown, determine $T(s)$.
  (*b*)   Can this arrangement be used to compensate the amplifier? Explain.

14-40   For the circuit in Fig. 14-39, express A_D and the *CMRR* in terms of R_1, R_2, R_3, and R_4.

14-41   In the circuit of Fig. 14-39, R_4 is changed to 110 kΩ.
  (*a*)   Evaluate the *CMRR*.
  (*b*)   Determine V_D for $V_1 = 5$ V and $V_2 = 5.001$ V.

14-42   (*a*)   Verify Eq. (14-37).
  (*b*)   Determine the *CMRR*.

14-43   The transistors in Fig. 14-42 have $\beta_o = 150$.
  (*a*)   For $R_L = 1$ kΩ and $V_{\text{ref}} = 5$ V, determine R_R to provide a gain of 100. Assume $V_{CC} = V_{EE} = 15$ V.
  (*b*)   If V_{ref} changes by 10 percent, what is the percentage change in the gain?

14-44   In Fig. 14-43, the differential amplifier has a gain of 12, the gain stage and the output amplifier each have a gain of 30, and the source followers have a gain of 0.9.
  (*a*)   Determine the gain V_o/V_{in}.
  (*b*)   The Op-Amp is used as a noninverting unity gain buffer. The source-follower frequency response can be neglected and the stages can be considered isolated. The two dominant poles of the amplifier are at $s = -10^6$ and -10^7 rad/s. The gain stage has $g_m = 1$ m℧, $r_o = 60$ kΩ, and $R_D = 60$ kΩ. The output resistance of source follower 2 is 1 kΩ. Determine C_C for a 45° phase margin.

14-45   In the circuit in Fig. 14-46, $R_s = 5$ kΩ, $R_L = 30$ kΩ, $g_m = 1$ m℧, $C_{gs} = 5$ pF, and $C_{gd} = 2$ pF.
  (*a*)   With $C_C = 0$, determine the pole frequency of the dominant pole.
  (*b*)   What is the frequency of the zero of the transfer function?
  (*c*)   Determine C_C to obtain a dominant-pole frequency of 1 kHz.
  (*d*)   Given C_C obtained in (*c*), what would be the frequency of the zero if the buffer were removed?

14-46   For the instrumentation amplifier shown, verify that

$$V_o = \left(1 + \frac{R_2}{R_1} + \frac{2R_2}{R}\right)(V_2 - V_1)$$

Note that the gain may be adjusted by varying R.

14-47   Given the signals v_1 and v_2, use two inverting amplifiers to obtain an output $v_o = k(v_2 - v_1)$, where k is a positive number.

14-48   An instrumentation amplifier is often used to amplify the output from a transducer bridge as shown. For example, in a strain gauge, the resistances R_1 are precision-fixed resistors. The resistance $R_2 + \Delta R$ is the transducer attached to the struc-

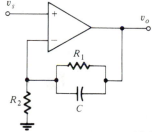

v_s

v_o

R_1

R_2

C

Prob. 14-39

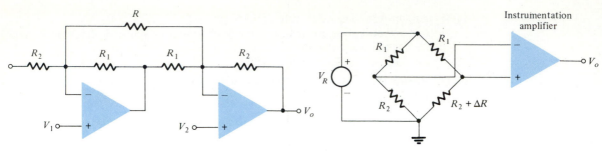

Prob. 14-46

Prob. 14-48

tural member under test. Because of the load on the structure, the resistance changes. The resistance R_2 in the remaining arm of the bridge is a dummy gauge (unloaded) and is used to balance out resistance variations caused by variations in temperature and other parameters.

(a) Assuming that the gain of the instrumentation amplifier is A_D, show that for $\delta = \Delta R/R_2 \ll 1$,

$$V_o = \frac{A_D V_R \delta}{4}$$

(b) Consider the instrumentation amplifier to have a *CMRR* and A_D. Express V_o in terms of A_D, *CMRR*, V_R, and δ.

(c) Let $R_1 = R_2$, $A_D = 10$, and $V_R = 12$ V. What must the *CMRR* be if the differential component of the output is to be 100 times the common-mode output component for $\delta = 10^{-4}$? Is this value realistic?

CHAPTER 15

15-1 Verify Eq. (15-4).

15-2 (a) For the network in Fig. 15-2a, use feedback techniques to show that the input impedance is

$$Z_i = R \frac{1 - 5\omega_N^2 - j\omega_N(6 - \omega_N^2)}{3 - \omega_N^2 - j4\omega_N}$$

(b) Evaluate Z_i at the frequency of oscillation.

15-3 Determine the required gain and frequency of oscillation of the circuit in Fig. 15-2a, including the loading effect of the JFET stage.

15-4 For the transistor phase-shift oscillator shown, the bias resistors R_1 and R_2 have negligible effect and C is sufficiently large that it acts as a perfect bypass.

(a) Determine $T(s)$.

(b) Show that the conditions for oscillation give

$$f_o = \frac{1}{2\pi RC} \sqrt{\frac{1}{6 + 4k}}$$

$$\beta_o \geq 4k + \frac{29}{k} + 23$$

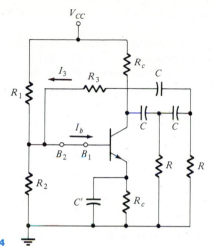

Prob. 15-4

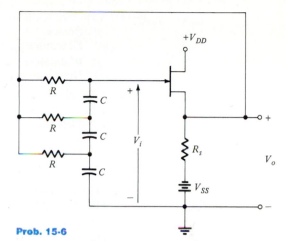

Prob. 15-6

where $k = R_c/R$.

(c)   Show that the minimum value of β_o required exceeds 44.5.

15-5   Design a phase-shift oscillator to operate at a frequency of 8 kHz. Use a MOS-FET with $\mu = 59$ and $r_d = 10$ kΩ. The phase-shift network is not to load down the amplifier.

(a)   Find the minimum value of the drain-circuit resistance R_D for which the circuit will oscillate.

(b)   Find the product RC.

(c)   Choose a reasonable value for R, and find C.

15-6   For the FET oscillator shown, find (a) V_i/V_D, (b) the frequency of oscillation, (c) the minimum gain of the source follower required for oscillations.

15-7   The circuit in Fig. 15-2a uses a JFET having parameters $g_m = 5$ m℧ and $r_o = 50$ kΩ. The capacitances C_{ds} and C_{gs} have negligible effect at the operating frequency. Determine R_D and C, for $R = 100$ kΩ, necessary to sustain oscillation at 10 kHz.

15-8   (a)   Verify Eq. (15-6).

(b)   Determine the frequency of oscillation and required gain for the Wien bridge oscillator in Fig. 10-7.

15-9   In the design of a particular variable-frequency Wien bridge oscillator, the values of R_1 and C_2 are adjustable according to the relations $0.1 \leq R_1/R_2 \leq 10$ and $0.1 \leq C_2/C_1 \leq 10$.

(a)   Find the minimum and maximum values of the oscillation frequency for $R_2 = 10$ kΩ and $C_1 = 0.1$ μF.

(b)   What must be the minimum gain-bandwidth product of the voltage amplifier if the amplifier frequency response is to have a negligible effect on oscillator performance?

15-10   (a)   Design the circuit in Fig. 15-3 to oscillate at 2 kHz. Choose $R_1 = R_2$ and $C_1 = C_2$ with the minimum resistance allowable equal to 1 kΩ.

(b)   If the technology used to fabricate the circuit allows the resistors and capacitors to be matched within 1 percent of nominal values, what is the range of oscillation frequency?

15-11 A Hartley oscillator is designed with $L_1 = 2$ mH and $L_2 = 20$ μH and a variable capacitance.

    (a) Determine the range of capacitance values for the case where the frequency of oscillation is varied between 950 and 2050 kHz.

    (b) Design an appropriate Op-Amp stage to realize Av.

15-12 A Colpitts oscillator is designed with $C_1 = 100$ pF and $C_2 = 7,500$ pF. The inductance is variable.

    (a) Determine the range of inductance values if the frequency of oscillation is to vary between 950 and 2050 kHz.

    (b) Design an appropriate Op-Amp stage to realize A_v.

15-13 An alternate realization of a Hartley oscillator is shown and in which $Z_1 = j\omega L_1$, $Z_2 = j\omega L_2$ and $Z_3 = -j/\omega C$. The voltage amplifier has infinite input resistance and an output resistance R_0. Determine the value of A_i needed to sustain oscillation and the oscillation frequency.

15-14 The circuit shown in Prob. 15-13 can be used as a Colpitts oscillator when $Z_1 = -j\omega C_1$, $Z_2 = -j\omega C_2$, and $Z_3 = j\omega L$. The current amplifier has an input resistance R_i, and an infinite output resistance. Determine the frequency of oscillation and the value of A_i needed to sustain oscillation.

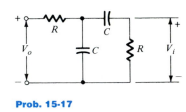

Prob. 15-13

Prob. 15-17

15-15 In this problem, the effect of practical elements on oscillator performance is examined. Consider the Colpitts circuit in Fig. 15-5 in which the capacitors are 500 pF and 0.05 μF and the 20-μH inductance contains series resistance $R_L = 4$ Ω. Determine the frequency of oscillation and the value of A_v needed to sustain oscillation.

15-16 The effect of the series resistance associated with practical inductors on the performance of the Hartley circuit in Fig. 15-5 is to be investigated. For $C = 100$ pF, $L_1 = 99$ μH in series with 10 Ω, and $L_2 = 1$ μH in series with 1 Ω, find the frequency of oscillation and the value of A_v needed to sustain oscillation.

15-17 (a) For the network shown prove that

$$\frac{V_j}{V_o} = \frac{1}{3 + j(\omega RC - 1/\omega RC)}$$

    (b) This network is used with an Op-Amp to form an oscillator. Show that the frequency of oscillation if $f = 2\pi RC$ and that the gain must exceed 3.

    (c) Draw the oscillator circuit

15-18　In the Wien bridge topology of Fig. 15-3, Z_1 consists of R, C, and L in series, and Z_2 is a resistor R_3. Find the frequency of oscillation and the minimum ratio R_1/R_2.

15-19　For the oscillator shown, find the frequency of oscillation and the minimum value of R.

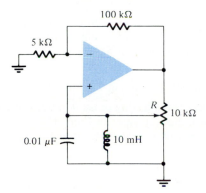

Prob. 15-19

15-20　(a)　Verify Eq. (15-14) for the reactance of a crystal.
　　(b)　Prove that the ratio of the parallel- to series-resonant frequencies is given approximately by $1 + \frac{1}{2}C/C'$.
　　(c)　If $C = 0.025$ pF and $C' = 1.25$ pF, by what percent is the parallel-resonant frequency greater than the series-resonant frequency?

15-21　(a)　The NOR gates in Fig. 15-13 can only sink or source 5 mA. For $V_{DD} = 5$ V and $V_T = 2.5$ V, determine the maximum pulse duration, given $C = 1000$ pF.
　　(b)　The catching diode has $V_\gamma = 0.5$ V and R_f if the diode is 20 Ω. What is the maximum value of v_x?

15-22　The NOR gates in Fig. 15-13 are fabricated in NMOS technology. The supply voltage is 5 V, and the logic levels are $V(0) = 0.2$ V and $V(1) = 5$ V. Consider $V_T = 2.5$ V.
　　(a)　Derive the equation for the pulse duration. Assume that catching diodes are used and have $V_y = 0.6$ V and $R_f = 20$ Ω.
　　(b)　Sketch the waveforms for v_{o1}, v_{o2}, and v_x.

15-23　If V_T in Prob. 15-22 can vary by 10 percent from unit to unit, what is the range of pulse durations that can be obtained?

15-24　The NOR gate in Fig. 15-13 has parameters that vary in manufacture.
　　(a)　Suppose that V_T of the MOSFET varies by 20 percent and V_{DD} by 5 percent, and that R and C can each vary by 20 percent.
　　(b)　What is the worst-case value of the pulse width if the nominal values of $R = 10$ kΩ and $C = 200$ pF?

15-25　The astable multivibrator in Fig. 15-15 is designed with $R = 50$ kΩ and $C = 0.01$ μF. The CMOS NOR gates have $V_T = 2.5$ V and internally connected catching diodes with $V_y = 0.5$ V. The circuit is supplied by $V_{DD} = 5$ V.
　　(a)　Determine the period of the output waveform.
　　(b)　What is the maximum instantaneous current that the CMOS gates must source or sink?

15-26 The astable circuit in Prob. 15-25 is now supplied by $V_{DD} = 6$ V.
  (a) Determine the period of the output square wave.
  (b) Sketch the waveforms for v_{o1}, v_{o2}, v_x, and v_C as functions of time. Carefully identify time constants.

15-27 The astable multivibrator in Prob. 15-25 is now supplied by $V_{DD} = 3.5$ V. Repeat Prob. 15-25.

15-28 The circuit in Fig. 15-17 is designed with $C = 0.01$ μF. The CMOS gates have $V_T = 2.5$ V and are supplied from a 5-V source. Diodes $D1$ and $D2$ are considered ideal. Determine R_1 and R_2 so that a 50-μs square-wave is obtained in which one pulse is 10 μs.

15-29 Repeat Prob. 15-28 for diodes having $V_\gamma = 0.5$ V.

15-30 An asymmetric square wave having pulse durations of 20 μs and period 20 ms is to be designed with the NOR gates in Prob. 15-28. Assuming that the largest resistance that can be used is 1 MΩ, determine R_1, R_2, and C. Is the value of C "practical"?

15-31 (a) Design the circuit in Prob. 15-30, with the condition that the resistance values must lie between 10 kΩ and 22 MΩ and the capacitance is to be minimized. Choose standard 5 percent resistance values and capacitance values.
  (b) What is the worst-case percentage error in the period?
  (c) What is the minimum value of the narrower pulse?

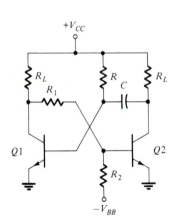

Prob. 15-32

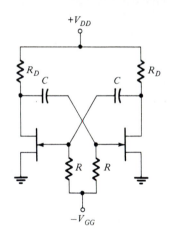

Prob. 15-33

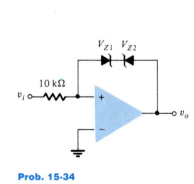

Prob. 15-34

15-32 The circuit shown is one form of discrete-component BJT monostable multivibrator. When conducting, the transistors are saturated.
  (a) Sketch the collector and base voltages V_{C1}, V_{C2}, V_{B1}, and V_{B2} from $t = 0^-$, just prior to the application of the trigger, to $t = T^+$, just after the circuit has returned to its stable state
  (b) Derive an expression for T, the pulse width.

15-33 A JFET astable circuit is shown.
  (a) Repeat Prob. 15-32, part (a).
  (b) Derive the relationships for the period and duration of the output pulses.

15-34 (a) For the comparator circuit shown plot the transfer characteristic, assuming that the Op-Amp gain is infinite and $V_{Z1} = V_{Z2} = 5$ V. Explain.

(b) Repeat part (a) if the large-signal gain is 10,000.

(c) Repeat part (a) if a voltage of 2 V is applied between the *negative terminal and ground*.

15-35 (a) Using two comparators and an AND gate, draw a system whose output is logic 1 if and only if the input lies in the window between V_{R1} and V_{R2}. Explain the operation.

(b) It is desired to determine the height of a pulse which may vary between 0 and 5 V with an uncertainty of 50 mV. Modify the system in (a) in order to obtain this pulse-height analyzer.

15-36 In the regenerative comparator in Fig. 15-22 it is desired that the threshold voltage V_1 equal the reference voltage V_R and that 0.1-V hysteresis be obtained. For $A_v = 100,000$, the loop gain is 2000, and $R_2 = 0.5$ kΩ, find V_R, V_Z, and R_1.

15-37 (a) The Schmitt trigger in Fig. 15-22 uses 6-V Zener diodes, with $V_D = 0.7$ V. Assuming that the threshold voltage V_1 is zero and the hysteresis is $V_H = 0.2$ V, calculate R_1/R_2 and V_R.

(b) This comparator converts a 4-kHz sine wave whose peak-to-peak value is 2 V into a square wave. Calculate the time duration of the negative and of the positive portions of the output waveform.

15-38 (a) In the Schmitt trigger in Fig. 15-22, $V_o = 4$ V, $V_1 = 2$ V, and $V_2 = 1.5$ V. Find R_1/R_2 and V_R.

(b) How must V_R be chosen so that V_2 is negative?

(c) How must V_R be chosen if $V_1 = -V_2$?

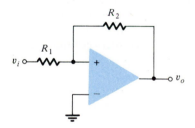

Prob. 15-39

15-39 (a) For the comparator shown find expressions for the threshold voltages V_1 and V_2 in terms of R_1, R_2, and the Op-Amp limited value (of magnitude V_o). Explain your calculations.

(b) Draw the transfer characteristic (similar to Fig. 15-22d).

(c) The peak output is 20 V and $R_2 = 5 R_1$. If a 8-V peak sinusoid is applied, draw the output waveform. Use the same time scale as for the input.

15-40 (a) Consider the square-wave generator of Fig. 15-25, where nonidentical avalanche diodes V_{Z1} and V_{Z2} are used. Assuming that the output is either $+V_{o1}$ or $-V_{o2}$, where $V_{o1} \equiv V_{Z1} + V_D$ and $V_{o2} \equiv V_{Z2} + V_d$, verify that the duration of the positive section is given by

$$T_1 = RC \ln \frac{1 + \beta V_{o2}/V_{o1}}{1 - \beta}$$

(b)   Verify that T_2 (the duration of the negative section) is given by the same equation with V_{o1} and V_{o2} interchanged.

(c)   If $V_{o1} > V_{o2}$, is T_1 greater or less than T_2? Explain.

15-41   The triangular waveform generator in Fig. 15-27 has a symmetry control voltage V_S added to the noninverting terminal of the integrator.

(a)   Verify that the sweep speed for the positive ramp is $(V_o + V_S)/RC$.

(b)   Find T_1, T_2, and f.

(c)   Verify that the duty cycle is given by Eq. (15-34).

15-42   Verify Eq. (15-37) for the pulse width T of the monostable multi in Fig. 15-29.

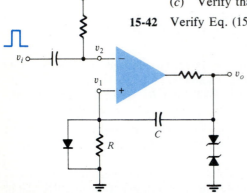

Prob. 15-43

15-43   (a)   Consider the pulse generator shown. In the quiescent state (before a trigger is applied) find v_2, v_o, and v_1.

(b)   At $t = 0$ a narrow, positive, triggering pulse v_t whose magnitude exceeds V_R is applied. At $t = 0 +$ find v_o and v_1. (Remember that the voltage across a capacitor cannot change instantaneously.) Now plot the waveforms v_o and v_1 as a function of time. Demonstrate that the circuit behaves as a monostable multivibrator with a pulse width T.

(c)   Find v_o and v_1 at $t = T +$ and continue the waveforms until the steady state is reached. What is the recovery-time constant? (Is the diode ON or OFF?)

(d)   Verify that T is given by

$$T = RC \ln \frac{2V_o}{V_R}$$

15-44   The circuit in Fig. 15-24 is designed with the following parameter values: $V_{CC} = 5$ V, $R_1 = 3.9$ kΩ, $R_2 = 2.6$ kΩ, and $R_E = 1$ kΩ. Transistors $Q1$ and $Q2$ are identical, with $\beta_F = 100$ in the active region and $\beta_{forced} = 50$ into saturation.

(a)   Determine the high and low output voltage levels.

(b)   Evaluate the hysteresis voltage.

15-45   For the circuit in Fig. 15-24, $R_1 = 7.5$ kΩ, $R_2 = 5.1$ kΩ, and $R_E = 2$ kΩ. The transistors are identical and conduct for $V_{BE} = 0.7$ V. Sketch the transfer characteristic V_o versus V_{in}, carefully indicating the threshold voltages and output levels. Identify the hysteresis voltage on your sketch.

15-46   Verify Eqs. (15-42) and (15-43).

15-47   (a)   Use the 555 timer to design a monostable circuit having a pulse duration of 20 μs. The parameters are $V_{CC} = 5$ V, $V(0) = 0$, and $R = 91$ kΩ.

(b)   With the values in (a), what is the pulse duration if $V(0) = 0.2$ V?

15-48 (a)   A square wave whose period is 100 μs is to be designed so that the positive and negative pulses are in a ratio 3:1. Given $C = 0.001$ μF, determine R_A and R_B.

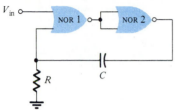

Prob. 15-49

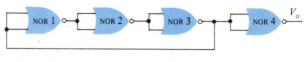

Prob. 15-50

15-49 (a)   For the CMOS monostable circuit shown, determine the pulse duration.
    (b)   Sketch v_{o1}, v_{o2}, v_x, and v_C as functions of time.

15-50 The circuit shown is used as an astable circuit which can be used to measure the propagation delay of a gate.
    (a)   Assuming that each gate has a propagation delay of 10 ns, sketch the voltages v_{o1}, v_{o2}, v_{o3}, and v_{o4} as functions of time. Each gate is supplied from a 5-V source and has a 2.5-V threshold.
    (b)   Can this technique be extended to have four gates in the feedback loop? Explain.
    (c)   Repeat (b) for five gates.

15-51 (a)   Verify Eq. (15-46) for the sweep-speed error e_1 of an exponential sweep.
    (b)   In the circuit of Fig. 15-37a a resistor R_1 is shunted across C_1. Show that e_s is now multiplied by $(R_1 + R_2)/R_2$.

15-52 For the Op-Amp sweep generator in Fig. 15-35 assume finite R_i, finite A_v, and nonzero R_o.
    (a)   Draw the amplifier model with R_i at the input and A_v in series with R_o across the output.
    (b)   Apply Miller's theorem to the impedance consisting of C_1 in series with R_o.
    (c)   Making reasonable order-of-magnitude approximations, show that the expression for the slope error is

$$e_x = \frac{V_s}{A_v} \frac{R_1 + R_i}{R_i}$$

15-53 In the Miller sweep shown in Fig. 15-35, $V = 30$ V, $R_i = 1$ MΩ, and the Op-Amp gain is $A_v = 10,000$. The output sweep amplitude is 10 V. The longest sweep is 1 s and the shortest is 1 μs.
    (a)   For $T_s = 1$ s, find C_1.
    (b)   Calculate the sweep-speed error e_s.
    (c)   Repeat parts (a) and (b) for $T_s = 1$ μs.
    (d)   How can a 1-μs sweep be obtained with a more reasonable value of C_1 than that found in (c), say, 100 times as large.
    (e)   What is the maximum value of T_s which may be measured with this system?

15-54 (a)   For the *bootstrap sweep* shown in part (a) of the figure, the capacitor C_1 may be taken as arbitrarily large. The drop across the ideal diode D may be neglected during conduction, and it may be assumed that any negative

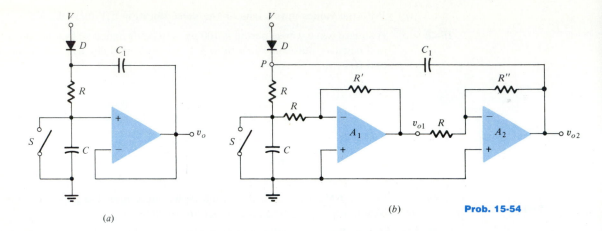

(a)

(b)   **Prob. 15-54**

voltage turns D OFF. The Op-Amp is ideal ($R_i = \infty$, $R_o = 0$, and $A = \infty$). With the switch S closed, what is the voltage across C_2 and R? With S open and C charged to v_c, what is the voltage across C_1 and R? Show that a precisely linear sweep is obtained and that $v_c = Vt/RC$.

(b)   A linear sweep with a pair of symmetrical outputs ($v_{o1} = -v_{o2}$) is to be obtained from the system shown in part (b) of the figure. Find the values of R'/R and R''/R.

15-55   (a)   Consider the chopper modulator of Fig. 15-41 with S_1 controlled by $+v_o$ instead of $-v_o$. For the modulating waveform v_m in Fig. 15-41b, sketch the first five pulses. Call this waveform v_+. Lined up in time with v_+ draw v_-, the output of the chopper, when S_t is controlled by $-v_o$ ($v_- = v$ in Fig. 15-41b).

(b)   Indicate how to combine v_+ and v_- with Op-Amps to obtain the AM signal shown in Fig. 15-40c.

CHAPTER 16   16-1   The gate control voltage applied in the circuit in Fig. 16-7b is shown. Sketch the output voltage for each waveform in Fig. 16-6.

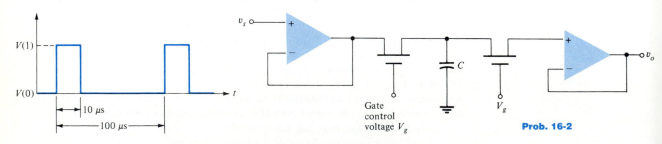

Prob. 16-1

16-2   Repeat Prob. 16-1 for the circuit shown.

16-3   For the input waveform shown, repeat Prob. 16-1.

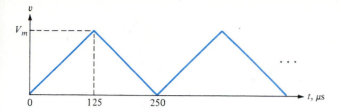

Prob. 16-3

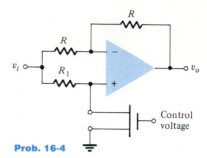

Prob. 16-4

16-4 (a) The switch S in the circuit shown is open for one half of the period and closed for the other half of the period. Determine the gain for each half-period.

(b) The circuit shown is used to generate the input signal to the sample-and-hold (S-H) circuit shown in Fig. 16-7b. The switch S is operated at an 8-kHz rate, and the gate control voltage is given in Prob. 16-1. Determine the output of this system, assuming that the input voltage is the 3.5-kHz sinusoid in Fig. 16-6.

16-5 (a) In the S-H module shown in Fig. 16-7b the negative input terminal of $A1$ is removed from its output and is connected to the output v_o of the second Op-Amp. Will the system function properly? Explain.

(b) A resistor R_2 is connected from the output in series with R_1 to ground. The inverting terminal of A_1 is now connected to the junction of R_2 and R_1. Show that this configuration operates as a noninverting S-H system with gain. What is the expression for the gain?

(c) Modify the connections so as to obtain an inverting S-H system with gain. Evaluate the gain.

16-6 (a) For the D/A converter of Fig. 16-13 the third most-significant bit (MSB) $N-3$ is 1 and all other bits are zero. Find the voltages at nodes $N-3$, $N-2$, $N-1$, and at the output V_o in terms of V_R and the resistors.

(b) For an 8-bit DAC with the least-significant bit (LSB) equal to 1 and all other bits equal to 0, find the voltage at all nodes, 0, 1, 2, . . . , and at the output.

16-7 In the inverted-ladder DAC shown the switches are connected directly to the Op-Amp input.

(a) Show that the current I drawn from V_R is a constant independent of the digital word. Explain why propagation-delay-time transients are eliminated with this system.

(b) What is the switch current and V_o if the MSB is 1 and all other bits are zero?

(c) Repeat (b), assuming that the next MSB is 1 and all other bits are zero.

(d) Calculate V_o for the LSB in the 4-bit D/A converter with all other bits zero.

16-8 The switches in the DAC shown in Fig. 16-11 are implemented as shown in Fig. 16-12b. The Op-Amp is supplied by a 10-V dc source. The desired analog output voltage is to be a maximum of 10 V and the largest resistance used is $\approx$32 kΩ (32×2^{10} Ω).

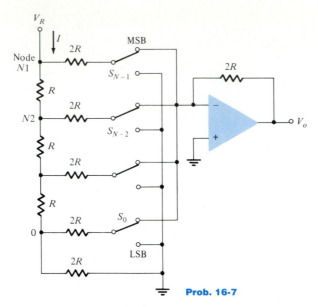

Prob. 16-7

    (*a*)   Specify the element values for an 8-bit DAC.
    (*b*)   What is the maximum current in the feedback resistance of the Op-Amp?
    (*c*)   What is the minimum voltage that can be resolved?

16-9   (*a*)   Draw the circuit diagram of a 6-bit inverted $R - 2R$ ladder DAC.
    (*b*)   For $V(1) = 5$ V, what is the maximum output voltage?
    (*c*)   What is the minimum voltage that can be resolved?

16-10   Describe how A/D and/or D/A building blocks can be used in a digital voltmeter. Sketch a schematic diagram in block form.

16-11   The integrator in Fig. 16-19 is designed with $R = 10$ kΩ, $C = 1000$ pF, and an Op-Amp whose open-loop gain and bandwidth are 106 dB and 5 Hz, respectively.
    (*a*)   Sketch the asymptotic Bode diagram and indicate the frequency range over which the circuit behaves as an integrator.
    (*b*)   Sketch the output voltage waveform to a step input voltage. Indicate the range of the times for which the output is the integral of the input.

16-12   Obtain the transfer function of the network shown. Verify that $v_o = (1/RC) \int v_s \, dt$ so that noninverting integration is performed.

16-13   Prove that the network shown is a noninverting integrator with $v_o = (2/RC) \int v_s(t) \, dt$.

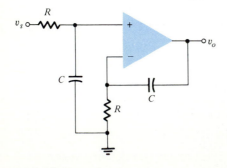

Prob. 16-13

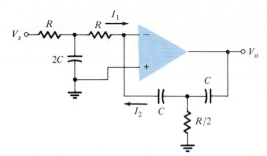

Prob. 16-14

16-14 Verify that the system shown, which uses only one Op-Amp, is a double integrator. In other words, prove that the transfer gain is

$$\frac{V_o}{V_s} = -\frac{1}{(RCs)^2}$$

(*Hint:* Evaluate I_1 and I_2 independently and set $I_1 = I_2$.) Why?

16-15 Verify Eq. (16-9).

16-16 A low-pass filter is to be designed to have a 3-dB bandwidth of 4 kHz and an attenuation of at least 30 dB at 6 kHz.
(*a*) What order of Butterworth filter is needed?
(*b*) Repeat (*a*) for a 0.5-dB Chebyshev filter.
(*c*) What is the band-pass function corresponding to (*a*) if the center frequency is 40 kHz?

16-17 A low-pass filter is to be designed to have a 3-dB bandwidth of 200 Hz and an attenuation of 50 dB at 400 Hz.
(*a*) Determine the order of the Butterworth filter required.
(*b*) Repeat (*a*) for a 1-dB Chebyshev filter.
(*c*) Convert the function in (*b*) to its high-pass equivalent having a bandwidth of 200 Hz.

16-18 Verify Eq. (16-22).

16-19 Design a fourth-order Butterworth low-pass filter whose bandwidth is 1 kHz. Select all capacitors equal to 1000 pF. If some resistance values can be chosen arbitrarily, which are these?

16-20 Verify Eq. (16-23).

16-21 Verify Eq. (16-24).

16-22 (*a*) An alternative to the design given in Eqs. (16-26) and (16-27) is to set $A_v = 2$ in all sections. Design a sixth-order, 0.5-dB Chebyshev low-pass filter having a cutoff frequency of 2 kHz. Select all component values. Note that some element values are arbitrary; indicate which these are but use reasonable values.
(*b*) What is the 3-dB frequency of the filter?

16-23 Verify Eq. (16-28).

16-24 Repeat Prob. 16-19 using the circuit in Fig. 16-30*b*.

16-25 Repeat Prob. 16-22 for a high-pass filter.

16-26 Design a band-pass circuit having a center frequency of 8 kHz and a bandwidth of 1.5 kHz. The attenuation one octave beyond the passband is to be 30 dB.
   (a) Design the circuit, assuming that no passband ripple is allowed. Use reasonable values for the elements that can be selected arbitrarily. Indicate which components these are.
   (b) What is the attenuation at 9 kHz?

16-27 Repeat Prob. 16-26, assuming that 0.5-dB passband ripple is permitted.

16-28 Verify Eq. (16-30).

16-29 Verify Eq. (16-31).

16-30 Use the biquad section in Fig. 16-34b to design the circuit described in Prob. 16-26.

16-31 Verify Eq. (16-32).

16-32 An ideal low-pass filter whose cutoff frequency is 5 kHz is cascaded with an ideal high-pass filter having f_c = 4.8 kHz.
   (a) Sketch the frequency response of the cascade.
   (b) Suppose that the cutoff frequency of each filter in the cascade can be controlled to within +1 percent. Sketch the frequency responses corresponding to the worst-case conditions. Comment of the effectiveness of this realization for practical filters.

16-33 Design a notch network having Q = 10 at f_o = 8 kHz. Choose C = 500 pF and select resistor values so that the spread-in values (the ratio of the largest to the smallest resistance used) does not exceed 10.

16-34 Verify Eq. (16-34).

16-35 (a) Determine the transfer function of the circuit in Fig. 16-39. Assume Y = 0.
   (b) Verify Eq. (16-33).

16-36 (a) Repeat Prob. 16-35, part (a) for $Y \neq 0$.
   (b) Show that $Y = 1/R_4$ results in a low-pass notch.
   (c) Show that $Y = sC$ results in a high-pass notch.

16-37 Show that the circuit in Fig. 16-42 has the response given by Eq. (16-36).

16-38 (a) The circuit shown is called a generalized-impedance converter (GIC). Show that if the Op-Amps are ideal, then

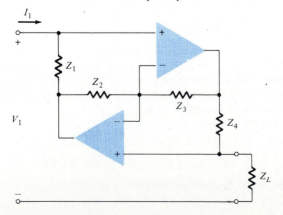

Prob. 16-38

$$Y_i = \frac{I_1}{V_1} = \frac{Y_1 Y_3}{Y_2 Y_4} Y_L$$

(b) The circuit is often used to simulate an inductance on a chip. Show that if $Y_4 = sC_4$ and all other components are resistive, Y_i is inductive.

(c) Assuming that any resistance can lie between 0.1 and 10 kΩ and $10 \le C \le 500$ pF, what is the range of inductance values possible?

16-39 Verify that the circuit shown simulates an inductor in series with a resistor. In other words, show that $V_i/I_i = R + Ls$.

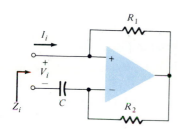

Prob. 16-39

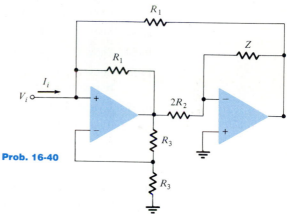

Prob. 16-40

16-40 The gyrator shown is used to simulate an inductance.
(a) Show that $V_i/I_i = R_1 R_2/Z$.
(b) Repeat Prob. 16-38, part (c).

16-41 An alternate form of the gyrator is shown. Show that V_i/I_i is inductive. Assume ideal Op-Amps.

Prob. 16-43

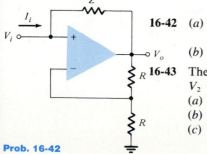

Prob. 16-41

Prob. 16-42

16-42 (a) Verify that the input impedance $V_i/I_i = -Z$. This circuit is called a negative-impedance converter (NIC).
(b) Determine V_o/V_i.

16-43 The NIC in the circuit shown has the following properties: $Z_1 = -Z_L$ and $V_2 = V_1$. The load used is $R_L \| C_L$ with the restriction that $R_L C_L = 1$.
(a) Determine V_2/V_i.
(b) Select R_L so that the circuit is a low-pass biquad section.
(c) Repeat (b) to obtain a band-pass section.

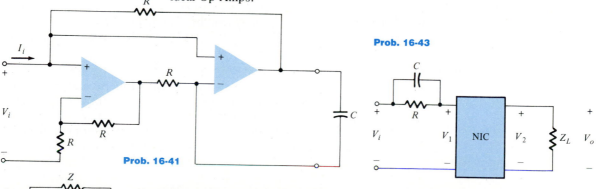

16-44 (a) Verify Eq. (16-37).
(b) Verify Table 16-5.

16-45 Verify Eqs. (16-38) and (16-39).

16-46 Design the filter in Prob. 16-19 using the biquad section in Fig. 16-44. Choose $R_2 = R_3 = R_4 = R_6 = 10$ kΩ.

16-47 Repeat Prob. 16-26 using the filter section in Fig. 16-44. Selected component values are given in Prob. 16-46.

16-48 Repeat Prob. 16-27 using the filter section in Fig. 16-44. Selected component values are given in Prob. 16-46.

16-49 Show a switched-capacitor equivalent of the circuit in Fig. 16-22.

16-50 Show the switched-capacitor equivalent of the circuit in Fig. 16-44.

16-51 Show that if the reference voltage V_R in Fig. 14-42 is time-varying, the operational transconductance amplifier functions as a multiplier.

16-52 Use one or more multipliers to generate a sinusoidal waveform of frequency $3f_o$ from a sinusoid of frequency f_o.

16-53 (a) Use the multiplier to generate a signal $v_o(t)$ that is proportional to v_i^3.
(b) Repeat (a) for v_o proportional to $v_i^{\frac{1}{3}}$.

Modulated carrier

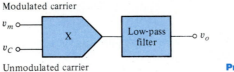

Unmodulated carrier

Prob. 16-54

16-54 The input signals to the multiplier circuit shown are an AM-modulated signal $v_m = V(t) \cos \omega_C t$ and the carrier signal $V_C \cos \omega_C t$.
(a) Show that v_o can be used to obtain the information $V(t)$.
(b) If $V(t) = V_m \cos \omega_s t$ where $\omega_s \ll \omega_C$, what should the cutoff frequency of the filter be?

16-55 Sketch a sinusoidal waveform of peak value V_m, which is the input to a half-wave rectifier. Directly below it draw the output waveform and indicate its positive and negative peak values, if the system is (a) that given in Fig. 16-60a; (b) the same system with the two diodes reversed; (c) the system obtained from Fig. 16-60a with the left-hand side of R grounded and v_i impressed on the non-inverting terminal; (d) the system in part (c) with the diodes reversed.

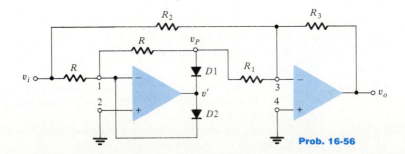

Prob. 16-56

16-56 (*a*) Verify that the circuit shown gives full-wave rectification provided that $R_2 = KR_1$. Find K.

(*b*) What is the peak value of the rectified output?

(*c*) Draw carefully the waveforms $v_i = 10 \sin \omega t$, v_r, and v_o if $R_3 = R_2$.

16-57 A waveform has a positive peak of magnitude V_1 and a negative peak of magnitude V_2. Draw a circuit using two peak detectors whose output is equal to the peak-to-peak value $V_1 - V_2$.

16-58 (*a*) The exponential amplifier in Fig. 16-54 is cascaded with the logarithmic amplifier in Fig. 16-53. If V_s is the input to the logarithmic amplifier and V_o' is the output of the exponential amplifier, how would you prove that $V_o' = V_s$?

(*b*) Assume that the resistors R_1, R_2, R_3, and R_4 in Fig. 16-53 are not identical with the corresponding resistors in Fig. 16-54. Designate the constants in Eq. (16-60) by $K_i \neq K_1$ and $K_2' \neq K_2$. For the cascaded arrangement in (*a*), prove that V_o' is proportional to a power n of V_s, where $n = K_1/K_1'$.

(*c*) Assume that R_3 in the exponential amplifier is adjustable, but that all other resistance values are as indicated in Figs. 16-53 and 16-54. Calculate R_3 so that $n = 3$. Repeat for $n = \frac{1}{3}$.

CHAPTER 17

17-1 A diode whose internal resistance is 20 Ω is to supply power to a 200 Ω load from a 110-V (rms) source of supply. Calculate (*a*) the peak load current; (*b*) the dc load current; (*c*) the ac load current; (*d*) the dc diode voltage; (*e*) the total input power to the circuit; (*f*) the percentage regulation from no load to the given load.

17-2 Verify Eqs. (17-15) and (17-16).

17-3 Show that the maximum dc output power $P_{dc} \equiv V_{dc}I_{dc}$ in a half-wave single-phase circuit occurs when the load resistance equals the diode resistance R_f.

17-4 The efficiency of rectification η_r is defined as the ratio of the dc output power $P_{dc} \equiv V_{dc}I_{dc}$ to the input power $P_i = (1/2\pi) \int_0^{2\pi} v_i i \, d\alpha$.

(*a*) Show that, for the half-wave-rectifier circuit

$$\eta_r = \frac{40.5}{1 + R_f/R_L} \text{ percent}$$

(*b*) Show that, for the full-wave rectifier, η_r has twice the value given in (*a*).

17-5 Prove that the regulation of both the half-wave and the full-wave rectifier is given by

$$\text{Percent regulation} = \frac{R_f}{R_L} \times 100 \text{ percent}$$

17-6 In the full-wave single-phase bridge, can the transformer and the load be interchanged? Explain carefully.

17-7 The bridge-rectifier system shown in Fig. 17-8 is used to construct an ac voltmeter. The forward resistance of the diodes is 50 Ω, the resistance R is 25 Ω, and the ammeter resistance is negligible. The signal voltage is given by $v_s = 200 \sin \omega t$.

(*a*) Sketch the waveform of the current i_L through the ammeter. Calculate maximum instantaneous value on your sketch.

(*b*) Write down an integral whose value will give the reading of the dc ammeter. Evaluate this expression and find I_{dc}.

(*c*) Sketch realistically the voltage waveform across diode $D1$. Indicate maximum instantaneous values on your sketch. Evaluate the average diode voltage.

(*d*) Write an integral whose value will give the reading of an rms voltmeter placed across $D1$. (This meter does not have a series blocking capacitor.) Find the value of this rms diode voltage.

17-8 A 5-mA dc meter whose resistance is 40 Ω is calibrated to read rms volts when used in a bridge circuit with semiconductor diodes. The effective resistance of each element may be considered to be zero in the forward direction and infinite in the inverse direction. The sinusoidal input voltage is applied in series with a 20-kΩ resistance. What is the full-scale reading of this meter?

17-9 (*a*) Consider the bridge voltage-doubler circuit in Fig. 17-9 with $R_L = \infty$. Show that in the steady state each capacitor charges up to peak transformer voltage V_m and, hence, that $v_o = 2V_m$. Assume ideal diodes.

(*b*) What is the peak inverse voltage across each diode?

17-10 The circuit shown is a half-wave voltage doubler. Analyze the operation of this circuit by sketching as a function of time the waveforms v_i, v_{c1}, v_{D1}, v_{D2}, and v_o. Assume that the capacitors are uncharged at $t = 0$. Calculate (*a*) the maximum possible voltage across each capacitor and (*b*) the peak inverse voltage of each diode. Compare this circuit with the bridge voltage doubler of Fig. 17-9. In this circuit the output voltage is negative with respect to ground. Show that if the connections to the cathode and anode of each diode are interchanged, the output voltage will be positive with respect to ground.

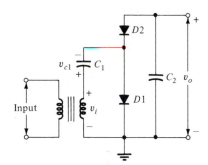

Prob. 17-10

17-11 The circuit in Prob. 17-10 can be extended from a doubler to a quadrupler by adding two diodes and two capacitors as shown. In the figure, parts (*a*) and (*b*) are alternative ways of drawing the same circuit.

(*a*) Analyze the operation of this circuit.

(*b*) Answer the same questions as asked in parts (*a*) and (*b*) of Prob. 17-10.

(*c*) Generalize the circuit of this and of Prob. 17-10 so as to obtain n-fold multiplication when n is any even number. In particular, sketch the circuit for sixfold multiplication.

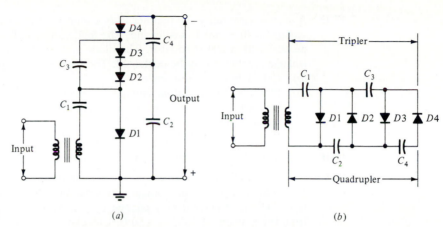

(*a*)   (*b*)

(*d*) Show that *n*-fold multiplication, with *n* odd, can also be obtained provided that the output is properly chosen.

17-12 (*a*) Consider the capacitor filter of Fig. 17-10. Show that, during the interval when the diode conducts, the diode current is given by $i = I_m \sin(\omega t + \psi)$, where

$$I_m \equiv V_m \sqrt{\frac{1}{R_L^2} + \omega^2 C^2} \qquad \text{and} \qquad \psi \equiv \arctan \omega C R_L$$

(*b*) Find the cutout angle ωt_1 in Fig. 17-12.

17-13 (*a*) Determine the transfer function V_2/V_1 for the rectifier circuit and the capacitor input filter shown.

(*b*) For $R = 25 \ \Omega$, $R_L = 200 \ \Omega$, $C = 200 \ \mu\text{F}$, and $L = 20 \ \text{H}$, determine the output voltage, assuming that the input can be represented by the first two terms of its Fourier series as

$$v_1(t) = \frac{110\sqrt{2}}{\pi}\left(1 - \frac{4}{3}\cos 754t + \cdots\right)$$

(*c*) The ripple factor is defined as the ratio of the ac component (rms) of the output voltage to the average value of the output voltage. Evaluate the ripple factor for the conditions in (*b*).

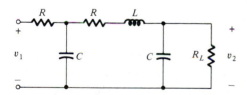

Prob. 17-13

Prob. 17-14

17-14 Repeat Prob. 17-13 for the rectifier circuit and the choke (inductance) input filter shown. The element values are $R_s = 25 \ \Omega$, $R = 50 \ \Omega$, $R_L = 500 \ \Omega$, $C = 100 \ \mu\text{F}$, and $L = 10 \ \text{H}$. The input voltage can be represented as

$$v_1(t) = \frac{220\sqrt{2}}{\pi}\left(1 - \frac{4}{3}\cos 100\pi t + \cdots\right)$$

17-15 A single-phase full-wave rectifier uses a semiconductor diode. The transformer voltage is 40 V rms to center tap. The load consists of a 25-μF capacitance in parallel with a 600-Ω resistor. The diode and the transformer resistances and leakage reactance may be neglected. The frequency of operation is 60 Hz.
(a) Calculate the cutout angle.
(b) Plot to scale the output voltage and the diode current. Determine the cut-in point graphically from this plot, and find the peak diode current corresponding to this point.
(c) Repeat (a) and (b), using a 75-μF instead of a 25-μF capacitance.

17-16 Repeat Prob. 17-15, assuming that the frequency of operation is 50 Hz.

17-17 Typical stabilization coefficients for a monolithic regulator are given in Sec. 17-6. The unregulated dc voltage varies by ±0.5 V due to line voltage fluctuations. The load current may change by ±2 A. The peak temperature change from the ambient of 30°C is ±50°C. Calculate the total maximum excursion in output voltage from that at 30°C.

17-18 In Fig. 17-17, $A_v = 10^5$, $R_1 = R_2$, $V_R = 6$ V, and the input offset voltage drift of the Op-Amp is 10 μV/°C.
(a) What is the approximate output voltage?
(b) What is S_T due to the input offset voltage drift of the Op-Amp?
(c) What is S_T caused by the base-emitter temperature drift of $Q1$? Assume $S_T = 0$ in (b).

17-19 The output voltage V_{reg} of the monolithic regulator of Fig. 17-16 may be adjusted to a higher value of V_o by the circuits shown. Find expressions for V_o in terms of V_{reg} and I_Q, defined in Fig. 17-16. What is the advantage of (b) over (a)?

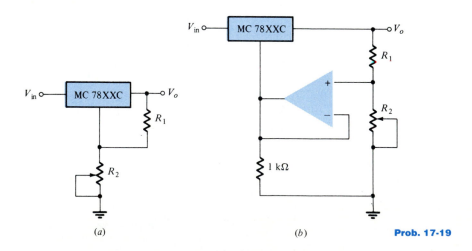

(a)   (b)   **Prob. 17-19**

Prob. 17-20

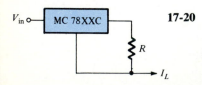

17-20 The three-terminal fixed-voltage regulator is converted into a current regulator by the circuit shown. If the output voltage of the regulator is 5 V, if $R = 5$ Ω, and if $I_Q = 10$ mA, what is the output current I_L? Note that I_L is independent of the load. How can I_L be made independent of I_Q? [*Hint:* See circuit (b) in Prob. 17-19.]

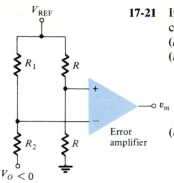

V_{REF}

R_1

R

$+$

v_m

R_2

R

Error
amplifier

$V_O < 0$

Prob. 17-21

17-21 If the output voltage V_O of a switching regulator is negative, the level-shifting circuit shown is used at the input of the error amplifier in Fig. 17-22.
(a) What is the effective feedback voltage?
(b) Verify that

$$V_O = \frac{1}{2} V_{ref}\left(1 - \frac{R_2}{R_1}\right)$$

Note that for $V_O < 0$, $R_2/R_1 > 1$.
(c) Show that for given values of V_O and V_{ref}, the ratio R_2/R_1 must be chosen to be

$$\frac{R_2}{R_1} = 1 - \frac{2V_O}{V_{ref}}$$

which indicates that for a negative V_O, $R_2/R_1 > 1$.

17-22 (a) Nonlinear distortion results in the generation of frequencies in the output that are not present in the input. Assuming that the dynamic curve can be represented by Eq. (17-35), and if the input signal is given by

$$i_b = I_1 \cos \omega_1 t + I_2 \cos \omega_2 t$$

show that the output will contain a dc term and sinusoidal terms of frequency ω_1, ω_2, $2\omega_1$, $2\omega_2$, $\omega_1 + \omega_2$, and $\omega_1 - \omega_2$.
(b) Generalize the results of (a) by showing that if the dynamic curve must be represented by higher-order terms in i_b, the output will contain intermodulation frequencies, given by the sum and difference of integral multiples of ω_1 and ω_2; for example, $2\omega_1 \pm 2\omega_2$, $2\omega_1 \pm \omega_2$, $3\omega_1 \pm \omega_2$, and so on.

17-23 A transistor supplies 2 W to a 4-kΩ load. The zero-signal dc collector current is 35 mA, and the dc collector current with signal is 39 mA. Determine the percent second-harmonic distortion.

17-24 The input excitation of an amplifier is $i_b = I_{bm} \sin \omega t$. Prove that the output current can be represented by a Fourier series which contains only odd sine components and even cosine components.

17-25 (a) Consider an ideal transistor with no distortion even if the transistor is driven from cutoff to the edge of saturation, where $v_C = V_{min}$. Verify that the conversion efficiency is given by

$$\eta = \frac{25(V_{CC} - V_{min})}{V_{CC}} \times 100 \text{ percent}$$

(b) What is the maximum possible efficiency and under what circumstances is this maximum value obtained?

17-26 For an ideal class B push-pull amplifier show that the collector dissipation P_C is zero at no signal ($V_m = 0$), rises as V_m increases, and passes through a maximum given by Eq. (17-58) at $V_m = 2V_{CC}/\pi$.

17-27 Mirror symmetry requires that the bottom portion of a waveform, when shifted $180°$ along the time axis, will be the mirror image of the top portion. The condition for mirror symmetry is represented mathematically by the equation

$$i(\omega t) = -i(\omega t + \pi)$$

(a) Verify that a class B push-pull system possesses mirror symmetry by using Eq. (17-64).

(b) Without recourse to a Fourier series, prove that a class B push-pull system has mirror symmetry.

17-28 For the ideal class B push-pull amplifier in Fig. 17-29, $V_{CC} = 15$ V and $R_L = 8$ Ω. The input is sinusoidal. Determine (a) the maximum output signal power; (b) the collector dissipation in each transistor at the power output; (c) the conversion efficiency. (d) What is the maximum dissipation of each transistor, and what is the efficiency under this condition?

17-29 The ideal class B push-pull amplifer in Fig. 17-29 is operating at the sinusoidal amplitude for which the dissipation is a maximum. Verify that the conversion efficiency is 50 percent.

17-30 In the circuit shown the base-emitter voltage may be assumed to remain constant at the cut-in value V_γ for all values of forward bias. The biasing voltage is idealized by two batteries of voltage kV_γ, where $0 < k \leqslant 1$. Assume that $v_i = V_s \sin \omega t$.

(a) For $V_\gamma = 0.6$ V and $V_s = 1$ V, sketch the output v_o as a function of time for $k = 0, 0.5,$ and 1. Calculate the cut-in angle for each value of k.

(b) What happens to the distortion as V_s is increased?

(c) What happens if k exceeds unity?

(d) If a resistor R is added between the two emitters, what happens if $k > 1$?

(e) Is the push-pull operation class A, B, AB, or C in (a) and (d)?

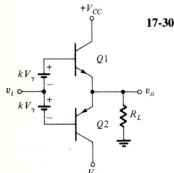

Prob. 17-31

Appendix E
ANSWERS TO SELECTED PROBLEMS

CHAPTER 1

1-2   62.5 V.

1-4   43 mV.

1-6   (*a*) 41.4 kV/cm.

1-8   10.0 cm^2/V·s.

1-10   (*a*) $p = n = 1.45 \times 10^{10}$ cm^{-3}; (*b*) the sample is intrinsic.

1-12   (*a*) $p = 6.58 \times 10^{17}$ cm^{-3}, $n = 3.20 \times 10^2$ cm^{-3}; (*b*) $n = 2.08 \times 10^{17}$ cm^{-3}, $p = 1.01 \times 10^3$ cm^{-3}.

1-14 0.833×10^{-7}.

1-16   (*c*) -173 mV.

1-20   196 mV.

1-22   594 mV.

CHAPTER 2

2-3   (*a*) -150 mV; (*b*) 54.6; (*c*) 220 μA, 1.63 mA, 12.0 mA.

2-5   (*a*) 6.82 mA; (*b*) 1.42 nA; (*c*) 46.8 mA.

2-7   100 V.

2-9   (*a*) 51 mA, 0.9 V; (*b*) 18.9 Ω.

2-11   (*a*) 9.38 V, 2.98 mW; (*b*) 2.2 mA; (*c*) 9.7 mA.

2-13   410 mW.

2-15   2.84 mA.

2-17   (*a*) $V_\gamma = 0.9$ V, $R_f = 3.33$ Ω; (*b*) $I_D = 49.4$ mA, $V_D = 1.06$ V.

2-19   (*a*) 150 Ω; (*b*) 50 Ω.

2-31   (*a*) $\Delta I_R = [r_d/(r_d + R)] \Delta I$; (*b*) 5.65 mA.

2-33   (*a*) .00965 sin ωt; (*b*) $6.70 + .00965$ sin ωt.

2-35 (b) $\Delta V_o = \frac{1}{2}\Delta V_i$.

2-39 (a) 4 Ω; (b) 6 W.

2-41 (a) -0.229 pF; (b) 0.276 pF.

2-43 (a) 8.37 V; (b) 3.87 V; (c) 8.37 V; (d) 4.07 V.

CHAPTER 3

3-1 (c) 9.5 mV.

3-5 (b) 0.10 nA, -0.10 nA.

3-8 (a) $R_B = 188$ kΩ, $R_C = 0.47$ kΩ; (b) 250.

3-10 20.8 mA, 4.33 V.

3-12 (a) 1.75 mA, 0.732 V.

3-13 (a) -1.99 mA, -4.09 V; (b) -0.926 mA, -7.19 V.

3-14 (a) 2.05 mA, 4.42 V; (b) 2.57 mA, 3.03 V.

3-15 (a) $R_C = 1.52$ kΩ, $R_F = 85.1$ kΩ; (b) 3.65 mA, 7.05 V.

3-17 (a) $V_{o1} = 10.1$ V, $V_{o2} = -5.04$ V; (b) 1.52 kΩ; (c) 1.51 kΩ.

3-21 6.69 V, 0.33 mA.

3-22 (a) 500 μA, 7 V; (b) 527 μA, 6.73 V.

3-24 (a) 14.3 V; (b) 87.8 V.

3-35 (a) 20 mΩ, 7.5 kΩ; (b) 100 Ω; (c) 38.

3-37 $R_i = r_\pi + r_b + (\beta_o + 1)R_E$.

3-44 (b) $R_{in} = 457$ kΩ, $R_o = 31.5$ Ω; (c) 0.99.

CHAPTER 4

4-4 2.11 V.

4-6 (a) 550 Ω; (b) 18.6 kΩ.

4-8 $R_D = 5$ kΩ, $R_S = 0.4$ kΩ.

4-10 6.32 V, 2.92 mA.

4-12 $R_1 = 10.6$ MΩ, $R_2 = 100$ kΩ, $R_D = 3.8$ kΩ.

4-14 (a) 2 mA, 4 V; (b) 2.15 kΩ.

4-16 (a) 20 μA, 3 V; (b) 3 V.

4-18 $V_{DS1} = 4$ V, $V_{DS2} = 2$ V.

4-20 $V_{DS1} \simeq 2.5$ V, $V_{DS2} \simeq 3.5$ V.

4-23 20 μA, 5.8 V.

4-25 80 μA, 5.2 V.

4-31 (b) 26.3 kΩ.

4-33 (a) $R_S \parallel (R_D + r_d)/(1 + \mu)$; (b) 461 Ω; (c) 425 Ω.

4-41 (b) $V_o = \mu_1 r_{d2} V_i/[r_{d1}(1 - \mu_2) + r_{d2}(1 + \mu_1)]$.

CHAPTER 5

5-4 37.8 Ω/square.

5-5 (a) 2.5 mm; (b) 1 μm.

5-7 307 pF.

CHAPTER 6

6-2 11011110, 100101110, 1111110000.

6-3 177, 550, 2052.

6-4 DE, 12E, 6FO.

6-6 $1.2 \leq R \leq 5.56$ kΩ.

6-8 (a) $R_{OFF} \geq 120$ kΩ; (b) $R_{ON} \leq 208$ Ω.

6-13 (b) $Y = \overline{ABC}$.

6-15 (b) $Y = AB\overline{C}$.

6-24 22.

6-25 (a) 52.6 kΩ.

6-45 (a) $Y = (\overline{A + B})$.

6-54 (a) 4.39; (b) 7.1 mW.

6-56 (a) 1.7 V; (b) 27.

6-62 0.3 V, $i_b = 51.7$ μA, $i_s = 131.3$ μA.

6-64 (b) 50; (c) 9.63 mW.

6-67 (c) $V(0) = 0.2$ V, $V(1) = 2.746$ V; (d) 88.

6-69 (a) 0; (b) 43.5 mA.

CHAPTER 7

7-17 (a) 145; (b) 145.

7-18 (a) 80; (b) 60.

7-22 (b) 9.

7-25 8, $X_0 = X_5 = \overline{D}$, $X_1 = X_4 = D$, $X_2 = X_6 = 1$, $X_3 = X_7 = 0$.

7-26 (b) $X_7 = 1$, $X_3 = X_5 = X_6 = D$, all other $X = 0$.

7-28 (b) Three transistors; $Q0$ and $Q1$ each have two emitters, $Q2$ has three emitters.

7-33 $Y_1 = W_6 + W_7 + W_2\overline{W_4}\overline{W_5} + W_3\overline{W_4}\overline{W_5}$.

7-36 (a) 10; (b) 7.

7-39 (a) $Y_5 = \overline{DCBA} + \overline{DCB}\overline{A} + \overline{DC}B\overline{A} + \overline{D}CB\overline{A} + D\overline{CB}\overline{A} + D\overline{C}B\overline{A} + DCBA$.

7-41 (a) 6; (b) 4.

7-42 (a) 7; (b) 3; (c) 6, 4; (d) 1024.

CHAPTER 8

8-5 0.3 V, 2.7 V, 533 mV.

8-20 (b) One pulse, $N = 5$.

8-21 (*b*) 2.

8-29 (*a*) $N = 4$.

8-36 1.98 s.

CHAPTER 9

9-12 (*a*) 1835; (*b*) 3606; (*c*) 101000101000.

9-16 (*a*) 0.397 mil²; (*b*) 0.492.

CHAPTER 10

10-1 (*a*) 18.3 kΩ; (*b*) 42.3°C.

10-2 (*a*) 0.85 mA; (*b*) $\beta_{max} = 1145$, $\beta_{min} = 91$.

10-3 (*a*) 14.1 kΩ; (*b*) 0.192%.

10-4 $(I_C)_{min} = 0.459$ mA, $(I_C)_{max} = 0.462$ mA.

10-5 (*a*) $I_{C1} = I_{C2} = 0.248$ mA; (*b*) 21.0 kΩ.

10-6 (*a*) 781 kΩ; (*b*) -1.95 V.

10-7 1.12 kΩ.

10-8 (*a*) 0.0541 mA; (*b*) 0.554%.

10-11 1.31 kΩ.

10-12 1.31 kΩ.

10-15 (*a*) 10.6 kΩ; (*b*) 0.07%.

10-16 (*a*) 10.6 kΩ; (*b*) 0.08%.

10-17 (*a*) $[\beta_F^2/(\beta_F^2 + 4\beta_F + 2)][(V_{CC} - 2V_{BE})/R]$; (*b*) 44.15 kΩ; (*c*) 309°C.

10-18 (*a*) $I_C = 1.21$ mA, $I_B = 6.06$ mA, $V_{CE} = 9.38$ V; (*b*) $I_C = 1.07$ mA, $V_{CE} = 9.59$ V.

10-19 (*a*) $I_C = 1.51$ mA, $I_B = 0.012$ mA, $V_{CE} = 6.83$ V; (*b*) $I_C = 1.57$ mA, $I_B = 0.00629$ mA, $V_{CE} = 6.51$ V.

10-21 (*a*) $I_{CQ} = 5.92$ mA, $V_{CEQ} = 3.13$ V; (*b*) 9.08 kΩ; (*c*) 6.36 V $\leq V_{CE} \leq$ 8.40 V.

10-23 54.0 kΩ.

10-24 $I_{C1} = I_{C2} = 0.271$ mA, $I_{C3} = 0.0287$ mA.

10-25 (*a*) 21.3 kΩ; (*b*) 152 μA; (*c*) 20 μA.

10-26 7 kΩ.

10-27 4 kΩ.

10-28 100 μA.

10-29 (*a*) 1.85 V; (*b*) 15.95 V; (*c*) -2 V.

10-30 $I_{DQ} = 2.098$ mA, $V_{DSQ} = 7.55$ V, $V_{GSQ} = -0.491$ V.

10-31 $2.083 \leq I_{DQ} \leq 2.113$ mA, $7.18 \leq V_{DSQ} \leq 7.93$ V.

10-32 $R_1 = 1.41$ MΩ, $R_2 = 108$ kΩ, $R_D = 2.17$ kΩ, $R_S = 1.43$ kΩ.

10-35 $A_V = -43.78$, $R_i = 3.425$ kΩ.

10-36   (a) 6 kΩ; (b) -111.4; (c) -80.

10-40   (a) 1.42 kΩ; (b) 0.924.

10-41   (a) 3.96 kΩ; (b) $A_V = 0.988$, $R_o = 52$ Ω, $R'_o = 51$ Ω.

10-43   (a) $A_V = -20.7$, $R_i = 15.7$ kΩ; (b) $-21.2 \leq A_V \leq -18.7$.

10-44   (a) $A_V = -20.9$, $R_i = 33.9$ kΩ, $-21.4 \leq A_V \leq -19.2$; (b) $R_o = \infty$, $R'_o = 5$ kΩ.

10-53   (a) -8.80; (b) $R_o = 92.84$ kΩ, $R'_o = 13.65$ kΩ.

10-54   (a) 0.550; (b) $R_o = 789$ Ω, $R'_o = 441$ Ω.

10-57   (a) -6.58; (b) -6.97.

10-58   (a) 36.2 kΩ; (b) $R_o = 976$ Ω, $R'_o = 950$ Ω.

10-61 $A_{V1} = -40.3$, $A_{V2} = -34.7$, $A_V = 1398$.

10-62   (a) $-38{,}000$; (b) $-48{,}500$; (c) $-44{,}200$.

10-65   (a) 25.3 kΩ; (b) 126.8 V.

10-67   (a) $A_{V1} = -49.75$, $A_{V2} = -0.985$; (b) 49.0.

10-71   (a) $A_{V1} = -9.88$, $A_{V2} = -8$, $A_{V3} = 0.816$; (b) $A_V = 64.5$; (c) $R_o = 976$ Ω, $R'_o = 817$ Ω.

10-74 $R_C = 125$ kΩ, $R_E = 1.25$ MΩ.

10-75   (a) $-2500 \sin (2\pi \times 10^3)t + 0.75 \sin 120\pi t$ mV; (b) $2500 \sin (2\pi \times 10^3)t + 0.75 \sin 120\pi t$ mV.

10-78   (a) $v_{o1} = A_{DM}(V_s/2)(1 + 1/CMRR)$, $v_{o2} = -A_{DM}(V_s/2)(1 - 1/CMRR)$; (b) $v_{o1} = A_{DM}(V_s/2)(-1 + 1/CMRR)$, $v_{o2} = A_{DM}(V_s/2)(1 + 1/CMRR)$.

10-82 $A_{DM} = -18.75$, $A_{CM} = -0.361$, $CMRR = 34.3$ dB.

10-84   (a) -2 V; (b) -1.875 V.

10-85   (a) -1.25 V; (b) -18.75 V.

CHAPTER 11

11-2 $1/[(R_s \| r\pi)C(1 + g_m R'_L) + R'_L C]$, where $R'_L = R_L \| R_C$.

11-5 $1/(R'_o + R_L)C_L$, where $R'_o = R_D \| [r_d + (1 + \mu)R_S]$.

11-7 $f_\beta = 2.5$ MHz, $f_T = 400$ MHz.

11-13 -43.8, 77.0 Mrad/s.

11-15 -97.1, 2.17 MHz.

11-17 -333, 90.4 MHz.

11-18   0.82, 10.1 MHz.

11-20 -7.04, 3.77 MHz.

11-26   (a) $a_1 = (R_1 + R_2)C_2 + (1 - A)R_1 C_1$, $a_2 = R_1 R_2 C_1 C_2$; (b) $-1/RC$, $-1/RC$; (c) $s = (1/2RC)(1 \pm j\sqrt{3})$.

11-30   (a) 2860, 3.38 MHz; (b) 18.0 MHz.

11-32   (a) 2870, 3.60 MHz; (c) 311 MHz.

11-35 -1340, 32.1 kHz.

11-38   (*a*) 320, 342 kHz; (*b*) 5.31 MHz.

11-42   (*a*) 417, 439 kHz; (*b*) 2.21 MHz.

11-45   (*b*) 2 MHz.

11-47   (*a*) $A_{V1} = 50$, $A_{V2} = -5$, $A_{V3} = 3$, $f_{H1} = 20$ kHz, $f_{H2} = 167$ kHz, $f_{H3} = 333$ kHz; (*c*) 20 kHz; (*d*) 16.95 kHz.

11-49   (*a*) 1331; (*b*) 455 kHz (Bode diagram), 303 kHz (dominant pole).

11-50   (*a*) 2 mA, 0 V; (*b*) 0.594 μF; (*c*) 66 μF; (*d*) $C_B = 0.6$ μF, $C_E = 6.6$ μF.

11-54   (*a*) 2.65 Hz; (*b*) 4.16%; (*c*) 416 Hz.

CHAPTER 12   **12-4**   (*a*) Current amplifier; (*b*) $R_i = 10$ Ω, $R_o = \infty$, $A_i = 7.5$

12-6   (*a*) $v_i = -200(v_s - v^f)$; (*b*) ≈ 300.

12-9   (*a*) 938 mV; (*b*) 150 mV.

12-19   1.02 mV.

12-21   (*a*) Series-series; (*c*) $A_{OL} = -6.57$, $T = 6.57 \times 10^3$, $A_F = -10^{-3}$ $\mho$; (*d*) $R_{IF} = 1.38$ MΩ, $R_{OF} = 480$ MΩ; (*e*) -10^{-3} $\mho$.

12-24 -93.6, 219 Ω.

12-29   (*a*) $T = 0.925$, $A_{OL} = 20.6$, $A_F = 10.7$; (*b*) 2.70 kΩ.

12-31   (*a*) $T = 7.12$, $A_{OL} = 181$, $A_F = 22.3$; (*b*) 5.03 Ω.

12-32   (*a*) Series-series; (*c*) $A_F = 0.295$ $\mho$, $T = 18.2$; (*d*) 2830.

CHAPTER 13   **13-1**   (*a*) 300 kHz, 167 Hz; (*b*) without feedback 24.5 MHz, with feedback 25 MHz.

13-4   (*a*) Unstable; (*b*) -20 dB, $-45°$.

13-7   (*a*) 71.6 dB; (*b*) 61.6 dB; (*c*) 30 dB.

13-10   (*a*) 2 krad/s; (*b*) the results differ by one-half octave.

13-11   (*a*) 25 rad/s; (*b*) $\omega_1 = 1.58$ krad/s, $z_1 = 10^5$ rad/s.

13-14   Approximately 45%.

13-17   43.1°, 59.2°, 65.5°, 69.9°, 76.3°.

13-22   (*a*) 0.858 Mrad/s, 3.16 Mrad/s; (*b*) $(-2.01 \times 10^6)(1 \pm j3.62)$ rad/s; (*c*) 25.3°.

13-27   (*a*) 1.56 Mrad/s, 37.0 Mrad/s; (*b*) $(-19.3 \times 10^6)(1 \pm j0.504)$ rad/s; (*c*) $\approx 55°$.

13-37   (*a*) $R_i = 5.43$ MΩ, $R_o = 333$ kΩ, $v_o/i_s = -3200$ mΩ; (*c*) $1/2\pi(12.66 + 3930C_C)$ MHz; (*d*) 4.05 pF.

CHAPTER 14   **14-1**   (*a*) 143 kΩ; (*b*) 500 kΩ; (*c*) No.

14-4   (*b*) 5:1.

14-7   (*a*) -1620; (*b*) $R_i = 9.3$ MΩ, $R_o = 2.7$ MΩ.

14-8   (*a*) $A_{DM} = -1600$, $CMRR = 75.5$ dB, $R_o = 3.33$ MΩ.

14-10   (*a*) 99 dB; (*c*) 13 pF.

14-16   (*b*) 5.98%.

14-25 -1200.

14-27   (*a*) 100 rad/s; (*b*) 66.5°; (*c*) 500 rad/s.

14-30   (*a*) 632 MΩ; (*b*) 98 pF.

14-34   (*a*) 18.5 rad/s; (*b*) $\tau > 10^{-5}$ s.

14-48   (*c*) 126 dB.

CHAPTER 15

15-5   (*a*) 6.13 kΩ; (*b*) 13.0 μs; (*c*) 30 kΩ, 430 pF.

15-7   6.56 kΩ, 65 pF.

15-9   (*a*) 100 rad/s, 10^4 rad/s; (*b*) 1.2×10^5 rad/s.

15-11   (*a*) 3.0 pF, 13.9 pF; (*b*) $A_V = 100$.

15-13 $\omega_o = 1/\sqrt{(L_1 + L_2)C}$, $A_V = L_2/L_1$.

15-16 $\omega_o = 9.14$ Mrad/s, $A_i = 119$.

15-18 R/R_3.

15-19   15.9 kHz, $R_{\min} = 476$ Ω.

15-24   0.70 μs, 2.44 μs.

15-26   (*a*) 539 ns.

15-28   1.44 kΩ, 5.77 kΩ.

15-36   5 V, 4.3 V, 99 kΩ.

15-38   (*a*) 15, 3.73 V; (*b*) $V_R < R_2 V_o/R_1$.

15-41   (*b*) $T_1 = (2R_2RC/R_1)[V_o/(V_o + V_S)]$, $f = (R_1/4R_2RC)[1 - (V_S/V_o)^2]$.

15-44   (*a*) 1.98 V, 5 V; (*b*) 1.6 V.

15-53   (*a*) 9 μF; (*b*) 1.1×10^{-3}%; (*c*) 9 pF, 1.1×10^{-3}%.

15-54   (*b*) 2, 1.

CHAPTER 16

16-4   (*a*) $A_V = -1$ with switch closed, $A_V = +1$ with switch open.

16-6   (*a*) $V_{N-1} = \frac{1}{2}V_{N-2} = \frac{1}{4}V_{N-3} = V_R/12$.

16-7   (*b*) $V_R/2R$, $-V_R$; (*c*) $V_R/4R$, $-V_R/2$; (*d*) $-V_R/8$.

16-8   (*a*) $R' = 128.5$ Ω. The binary-weighted resistors are 2^{15} Ω, 2^{14} Ω, 2^{13} Ω, 2^{12} Ω, 2^{11} Ω, 2^{10} Ω, 2^9 Ω, and 2^8 Ω corresponding to the LSB, . . . , MSB; (*b*) 77.8 mA; (*c*) 39.2 mV.

16-16   (*a*) 9; (*b*) 6.

16-17   (*a*) 9; (*b*) 6.

16-38   (*b*) $L = R_2C_4/R_1R_2R_L$; (*c*) 1 fH $\leq L \leq$ 5 μH.

16-43 (b) $C = C_L$; (c) $R = R_L$.

16-55 (a) $R'V_m/R$, 0; (b) 0, $-R'V_m/R$; (c) $(R' + R)V_m/R$, $-V_m$; (d) V_m, $-(R' + R)V_m/R$.

16-58 (c) 1.55 kΩ, -8.

CHAPTER 17

17-7 (a) 1.6 A; (b) 1.02 A; (c) 80 V, 0; (d) 56.6 V.

17-11 (b) V_m, $2V_m$, $2V_m$, and $2V_m$ are the maximum voltages across C_1, C_2, C_3, and C_4, respectively, where V_m is the peak value of applied voltage. The peak inverse voltage for each diode is $2V_m$.

17-14 (b) $86.1 + 4.85 \times 10^{-3} \cos(100\pi t - 269°)$ V; (b) 3.98×10^{-5}.

17-15 (a) 100°; (b) 44°, 0.452 A; (c) 93.4°, 62°, 0.835 A.

17-17 112 mV.

17-18 (a) 12 V; (b) 20 μV/°C; (c) -0.05 μV/°C.

17-19 (a) $V_O = I_Q R_2 + V_{\text{reg}} (1 + R_2/R_1)$.

17-23 12.65.

17-28 (a) 28.1 W; (b) 3.84 W; (c) 78.6%; (d) 5.70 W, 50%.

17-30 (a) 37°, 17.5°, 0.

INDEX